To the Student

As you begin, you may feel anxious _____, definitions, procedures, and equations. You may wonder if you can learn it all in time. Don't worry—your concerns are normal. This textbook was written with you in mind. If you attend class, work hard, and read and study this text, you will build the knowledge and skills you need to be successful. Here's how you can use the text to your benefit.

Read Carefully

When you get busy, it's easy to skip reading and go right to the problems. Don't . . . the text has a large number of examples and clear explanations to help you break down the mathematics into easy-to-understand steps. Reading will provide you with a clearer understanding, beyond simple memorization. Read before class (not after) so you can ask questions about anything you didn't understand. You'll be amazed at how much more you'll get out of class if you do this.

Use the Features

I use many different methods in the classroom to communicate. Those methods, when incorporated into the text, are called "features." The features serve many purposes, from providing timely review of material you learned before (just when you need it) to providing organized review sessions to help you prepare for quizzes and tests. Take advantage of the features and you will master the material.

To make this easier, we've provided a brief guide to getting the most from this text. Refer to "Prepare for Class," "Practice," and "Review" on the following three pages. Spend fifteen minutes reviewing the guide and familiarizing yourself with the features by flipping to the page numbers provided. Then, as you read, use them. This is the best way to make the most of your text.

Please do not hesitate to contact us, through Pearson Education, with any questions, comments, or suggestions for improving this text. I look forward to hearing from you, and good luck with all of your studies.

Best Wishes!
Michael Sullivan

Prepare for Class "Read the Book"

Feature	Description	Benefit	Page
Every Chapter Opener begins with . . .			
Chapter-Opening Topic & Project	Each chapter begins with a discussion of a topic of current interest and ends with a related project.	The Project lets you apply what you learned to solve a problem related to the topic.	402
Internet-Based Projects	The projects allow for the integration of spreadsheet technology that you will need to be a productive member of the workforce.	The projects give you an opportunity to collaborate and use mathematics to deal with issues of current interest.	503
Every Section begins with . . .			
Learning Objectives ✓ 2	Each section begins with a list of objectives. Objectives also appear in the text where the objective is covered.	These focus your studying by emphasizing what's most important and where to find it.	423
Sections contain . . .			
PREPARING FOR THIS SECTION	Most sections begin with a list of key concepts to review with page numbers.	Ever forget what you've learned? This feature highlights previously learned material to be used in this section. Review it, and you'll always be prepared to move forward.	423
Now Work the **'Are You Prepared?' Problems**	Problems that assess whether you have the prerequisite knowledge for the upcoming section.	Not sure you need the Preparing for This Section review? Work the 'Are You Prepared?' problems. If you get one wrong, you'll know exactly what you need to review and where to review it!	423, 434
Now Work PROBLEMS	These follow most examples and direct you to a related exercise.	We learn best by doing. You'll solidify your understanding of examples if you try a similar problem right away, to be sure you understand what you've just read.	430, 435
WARNING	Warnings are provided in the text.	These point out common mistakes and help you to avoid them.	456
Exploration and **Seeing the Concept**	These graphing utility activities foreshadow a concept or solidify a concept just presented.	You will obtain a deeper and more intuitive understanding of theorems and definitions.	418, 443
In Words	These provide alternative descriptions of select definitions and theorems.	Does math ever look foreign to you? This feature translates math into plain English.	440
Calculus	These appear next to information essential for the study of calculus.	Pay attention–if you spend extra time now, you'll do better later!	205, 407, 431
SHOWCASE EXAMPLES	These examples provide "how-to" instruction by offering a guided, step-by-step approach to solving a problem.	With each step presented on the left and the mathematics displayed on the right, you can immediately see how each step is employed.	334
Model It! Examples and Problems	These examples and problems require you to build a mathematical model from either a verbal description or data. The homework Model It! problems are marked by purple headings.	It is rare for a problem to come in the form "Solve the following equation." Rather, the equation must be developed based on an explanation of the problem. These problems require you to develop models that will allow you to describe the problem mathematically and suggest a solution to the problem.	447, 475

Practice "Work the Problems"

Feature	Description	Benefit	Page
'Are You Prepared?' Problems	These assess your retention of the prerequisite material you'll need. Answers are given at the end of the section exercises. This feature is related to the Preparing for This Section feature.	Do you always remember what you've learned? Working these problems is the best way to find out. If you get one wrong, you'll know exactly what you need to review and where to review it!	434, 440
Concepts and Vocabulary	These short-answer questions, mainly Fill-in-the-Blank, Multiple-Choice and True/False items, assess your understanding of key definitions and concepts in the current section.	It is difficult to learn math without knowing the language of mathematics. These problems test your understanding of the formulas and vocabulary.	434
Skill Building	Correlated with section examples, these problems provide straightforward practice.	It's important to dig in and develop your skills. These problems provide you with ample opportunity to do so.	434–436
Mixed Practice	These problems offer comprehensive assessment of the skills learned in the section by asking problems that relate to more than one concept or objective. These problems may also require you to utilize skills learned in previous sections.	Learning mathematics is a building process. Many concepts are interrelated. These problems help you see how mathematics builds on itself and also see how the concepts tie together.	436–437
Applications and Extensions	These problems allow you to apply your skills to real-world problems. They also allow you to extend concepts learned in the section.	You will see that the material learned within the section has many uses in everyday life.	437–439
Explaining Concepts: Discussion and Writing	"Discussion and Writing" problems are colored red. They support class discussion, verbalization of mathematical ideas, and writing and research projects.	To verbalize an idea, or to describe it clearly in writing, shows real understanding. These problems nurture that understanding. Many are challenging, but you'll get out what you put in.	439
NEW! Retain Your Knowledge	These problems allow you to practice content learned earlier in the course.	Remembering how to solve all the different kinds of problems that you encounter throughout the course is difficult. This practice helps you remember.	439
Now Work PROBLEMS	Many examples refer you to a related homework problem. These related problems are marked by a pencil and orange numbers.	If you get stuck while working problems, look for the closest Now Work problem, and refer to the related example to see if it helps.	432, 435, 436
Review Exercises	Every chapter concludes with a comprehensive list of exercises to pratice. Use the list of objectives to determine the objective and examples that correspond to the problems.	Work these problems to ensure that you understand all the skills and concepts of the chapter. Think of it as a comprehensive review of the chapter.	499–501

Review "Study for Quizzes and Tests"

Feature	Description	Benefit	Page
The Chapter Review at the end of each chapter contains . . .			
Things to Know	A detailed list of important theorems, formulas, and definitions from the chapter.	Review these and you'll know the most important material in the chapter!	497–498
You Should Be Able to . . .	Contains a complete list of objectives by section, examples that illustrate the objective, and practice exercises that test your understanding of the objective.	Do the recommended exercises and you'll have mastered the key material. If you get something wrong, go back and work through the example listed and try again.	498–499
Review Exercises	These provide comprehensive review and practice of key skills, matched to the Learning Objectives for each section.	Practice makes perfect. These problems combine exercises from all sections, giving you a comprehensive review in one place.	499–501
Chapter Test	About 15–20 problems that can be taken as a Chapter Test. Be sure to take the Chapter Test under test conditions—no notes!	Be prepared. Take the sample practice test under test conditions. This will get you ready for your instructor's test. If you get a problem wrong, you can watch the Chapter Test Prep Video.	502
Cumulative Review	These problem sets appear at the end of each chapter, beginning with Chapter 2. They combine problems from previous chapters, providing an ongoing cumulative review. When you use them in conjunction with the Retain Your Knowledge problems, you will be ready for the final exam.	These problem sets are really important. Completing them will ensure that you are not forgetting anything as you go. This will go a long way toward keeping you primed for the final exam.	502–503
Chapter Projects	The Chapter Projects apply to what you've learned in the chapter. Additional projects are available on the Instructor's Resource Center (IRC).	The Chapter Projects give you an opportunity to apply what you've learned in the chapter to the opening topic. If your instructor allows, these make excellent opportunities to work in a group, which is often the best way of learning math.	503–504
Internet-Based Projects	In selected chapters, a Web-based project is given.	These projects give you an opportunity to collaborate and use mathematics to deal with issues of current interest by using the Internet to research and collect data.	503

Achieve Your Potential

The author, Michael Sullivan, has developed specific content in MyMathLab® to ensure you have many resources to help you achieve success in mathematics - and beyond! The MyMathLab features described here will help you:

- **Review** math skills and concepts you may have forgotten
- **Retain** new concepts as you move through your math course
- **Develop** skills that will help with your transition to college

Adaptive Study Plan

The Study Plan will help you study more efficiently and effectively.

Your performance and activity are assessed continually in real time, providing a personalized experience based on your individual needs.

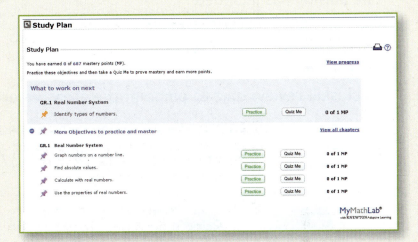

Skills for Success

The Skills for Success Modules support your continued success in college. These modules provide tutorials and guidance on a variety of topics, including transitioning to college, online learning, time management, and more.

Additional content is provided to help with the development of professional skills such as resume writing and interview preparation.

Getting Ready

Are you frustrated when you know you learned a math concept in the past, but you can't quite remember the skill when it's time to use it? Don't worry!

The author has included Getting Ready material so you can brush up on forgotten material efficiently by taking a quick skill review quiz to pinpoint the areas where you need help.

Then, a personalized homework assignment provides additional practice on those forgotten concepts, right when you need it.

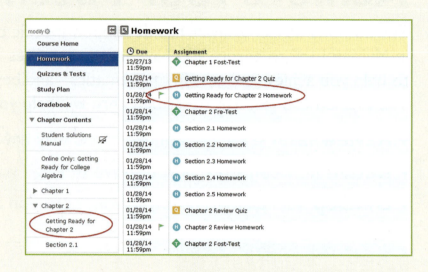

Retain Your Knowledge

As you work through your math course, these MyMathLab® exercises support ongoing review to help you maintain essential skills.

The ability to recall important math concepts as you continually acquire new mathematical skills will help you be successful in this math course and in your future math courses.

Algebra & Trigonometry

Tenth Edition

Michael Sullivan
Chicago State University

PEARSON

Boston   Columbus   Indianapolis   New York   San Francisco   Hoboken
Amsterdam   Cape Town   Dubai   London   Madrid   Milan   Munich   Paris   Montreal   Toronto
Delhi   Mexico City   São Paulo   Sydney   Hong Kong   Seoul   Singapore   Taipei   Tokyo

Editor in Chief: *Anne Kelly*
Acquisitions Editor: *Dawn Murrin*
Assistant Editor: *Joseph Colella*
Program Team Lead: *Marianne Stepanian*
Program Manager: *Chere Bemelmans*
Project Team Lead: *Peter Silvia*
Project Manager: *Peggy McMahon*
Associate Media Producer: *Marielle Guiney*
Senior Project Manager, MyMathLab: *Kristina Evans*
QA Manager, Assessment Content: *Marty Wright*
Senior Marketing Manager: *Michelle Cook*
Marketing Manager: *Peggy Sue Lucas*

Marketing Assistant: *Justine Goulart*
Senior Author Support/Technology Specialist: *Joe Vetere*
Procurement Manager: *Vincent Scelta*
Procurement Specialist: *Carol Melville*
Text Design: *Tamara Newnam*
Production Coordination,
Composition, Illustrations: *Cenveo® Publisher Services*
Associate Director of Design,
USHE EMSS/HSC/EDU: *Andrea Nix*
Project Manager, Rights and Permissions: *Diahanne Lucas Dowridge*
Art Director: *Heather Scott*
Cover Design and Cover Illustration: *Tamara Newnam*

Acknowledgments of third-party content appear on page C1, which constitutes an extension of this copyright page.

Unless otherwise indicated herein, any third-party trademarks that may appear in this work are the property of their respective owners, and any references to third-party trademarks, logos or other trade dress are for demonstrative or descriptive purposes only. Such references are not intended to imply any sponsorship, endorsement, authorization, or promotion of Pearson's products by the owners of such marks, or any relationship between the owner and Pearson Education, Inc. or its affiliates, authors, licensees or distributors.

MICROSOFT® AND WINDOWS® ARE REGISTERED TRADEMARKS OF THE MICROSOFT CORPORATION IN THE U.S.A. AND OTHER COUNTRIES. SCREEN SHOTS AND ICONS REPRINTED WITH PERMISSION FROM THE MICROSOFT CORPORATION. THIS BOOK IS NOT SPONSORED OR ENDORSED BY OR AFFILIATED WITH THE MICROSOFT CORPORATION.

MICROSOFT AND /OR ITS RESPECTIVE SUPPLIERS MAKE NO REPRESENTATIONS ABOUT THE SUITABILITY OF THE INFORMATION CONTAINED IN THE DOCUMENTS AND RELATED GRAPHICS PUBLISHED AS PART OF THE SERVICES FOR ANY PURPOSE. ALL SUCH DOCUMENTS AND RELATED GRAPHICS ARE PROVIDED "AS IS" WITHOUT WARRANTY OF ANY KIND. MICROSOFT AND /OR ITS RESPECTIVE SUPPLIERS HEREBY DISCLAIM ALL WARRANTIES AND CONDITIONS WITH REGARD TO THIS INFORMATION, INCLUDING ALL WARRANTIES AND CONDITIONS OF MERCHANTABILITY, WHETHER EXPRESS, IMPLIED OR STATUTORY, FITNESS FOR A PARTICULAR PURPOSE, TITLE AND NON-INFRINGEMENT. IN NO EVENT SHALL MICROSOFT AND /OR ITS RESPECTIVE SUPPLIERS BE LIABLE FOR ANY SPECIAL, INDIRECT OR CONSEQUENTIAL DAMAGES OR ANY DAMAGES WHATSOEVER RESULTING FROM LOSS OF USE, DATA OR PROFITS, WHETHER IN AN ACTION OF CONTRACT, NEGLIGENCE OR OTHER TORTIOUS ACTION, ARISING OUT OF OR IN CONNECTION WITH THE USE OR PERFORMANCE OF INFORMATION AVAILABLE FROM THE SERVICES. THE DOCUMENTS AND RELATED GRAPHICS CONTAINED HEREIN COULD INCLUDE TECHNICAL INACCURACIES OR TYPOGRAPHICAL ERRORS. CHANGES ARE PERIODICALLY ADDED TO THE INFORMATION HEREIN. MICROSOFT AND/OR ITS RESPECTIVE SUPPLIERS MAY MAKE IMPROVEMENTS AND /OR CHANGES IN THE PRODUCT (S) AND /OR THE PROGRAM (S) DESCRIBED HEREIN AT ANY TIME. PARTIAL SCREEN SHOTS MAY BE VIEWED IN FULL WITHIN THE SOFTWARE VERSION SPECIFIED.

The student edition of this text has been cataloged as follows:
Library of Congress Cataloging-in-Publication Data
Sullivan, Michael, 1942-
  Algebra & trigonometry / Michael Sullivan, Chicago State University. -- Tenth edition.
     pages cm.
  ISBN 978-0-321-99859-0
  1. Algebra--Textbooks. 2. Algebra--Study and teaching (Higher) 3. Trigonometry--Textbooks.
  4. Trigonometry--Study and teaching (Higher) I. Title. II. Title: Algebra and trigonometry.
  QA154.3.S73 2016
  512' .13--dc23

                                                                    2014021731

PEARSON, ALWAYS LEARNING, and MYMATHLAB are exclusive trademarks in the U.S. and/or other countries owned by Pearson Education, Inc. or its affiliates.

7 18

www.pearsonhighered.com

ISBN-10: **0-321-99859-6**
ISBN-13: **978-0-321-99859-0**

Contents

7 **Trigonometric Functions** **505**

8 Analytic Trigonometry 606

For the family

Katy (Murphy) and Pat	Shannon, Patrick, Ryan
Mike and Yola	Michael, Kevin, Marissa
Dan and Sheila	Maeve, Sean, Nolan
Colleen (O'Hara) and Bill	Kaleigh, Billy, Timmy

Three Distinct Series

Students have different goals, learning styles, and levels of preparation. Instructors have different teaching philosophies, styles, and techniques. Rather than write one series to fit all, the Sullivans have written three distinct series. All share the same goal—to develop a high level of mathematical understanding and an appreciation for the way mathematics can describe the world around us. The manner of reaching that goal, however, differs from series to series.

Contemporary Series, Tenth Edition

The Contemporary Series is the most traditional in approach yet modern in its treatment of precalculus mathematics. Graphing utility coverage is optional and can be included or excluded at the discretion of the instructor: *College Algebra, Algebra & Trigonometry, Trigonometry: A Unit Circle Approach, Precalculus.*

Enhanced with Graphing Utilities Series, Sixth Edition

This series provides a thorough integration of graphing utilities into topics, allowing students to explore mathematical concepts and encounter ideas usually studied in later courses. Using technology, the approach to solving certain problems differs from the Contemporary Series, while the emphasis on understanding concepts and building strong skills does not: *College Algebra, Algebra & Trigonometry, Precalculus.*

Concepts through Functions Series, Third Edition

This series differs from the others, utilizing a functions approach that serves as the organizing principle tying concepts together. Functions are introduced early in various formats. This approach supports the Rule of Four, which states that functions are represented symbolically, numerically, graphically, and verbally. Each chapter introduces a new type of function and then develops all concepts pertaining to that particular function. The solutions of equations and inequalities, instead of being developed as stand-alone topics, are developed in the context of the underlying functions. Graphing utility coverage is optional and can be included or excluded at the discretion of the instructor: *College Algebra; Precalculus, with a Unit Circle Approach to Trigonometry; Precalculus, with a Right Triangle Approach to Trigonometry.*

The Contemporary Series

College Algebra, Tenth Edition

This text provides a contemporary approach to college algebra, with three chapters of review material preceding the chapters on functions. Graphing calculator usage is provided, but is optional. After completing this book, a student will be adequately prepared for trigonometry, finite mathematics, and business calculus.

Algebra & Trigonometry, Tenth Edition

This text contains all the material in *College Algebra*, but also develops the trigonometric functions using a right triangle approach and showing how it relates to the unit circle approach. Graphing techniques are emphasized, including a thorough discussion of polar coordinates, parametric equations, and conics using polar coordinates. Graphing calculator usage is provided, but is optional. After completing this book, a student will be adequately prepared for finite mathematics, business calculus, and engineering calculus.

Precalculus, Tenth Edition

This text contains one review chapter before covering the traditional precalculus topic of functions and their graphs, polynomial and rational functions, and exponential and logarithmic functions. The trigonometric functions are introduced using a unit circle approach and showing how it relates to the right triangle approach. Graphing techniques are emphasized, including a thorough discussion of polar coordinates, parametric equations, and conics using polar coordinates. Graphing calculator usage is provided, but is optional. The final chapter provides an introduction to calculus, with a discussion of the limit, the derivative, and the integral of a function. After completing this book, a student will be adequately prepared for finite mathematics, business calculus, and engineering calculus.

Trigonometry: a Unit Circle Approach, Tenth Edition

This text, designed for stand-alone courses in trigonometry, develops the trigonometric functions using a unit circle approach and showing how it relates to the right triangle approach. Graphing techniques are emphasized, including a thorough discussion of polar coordinates, parametric equations, and conics using polar coordinates. Graphing calculator usage is provided, but is optional. After completing this book, a student will be adequately prepared for finite mathematics, business calculus, and engineering calculus.

Preface to the Instructor

As a professor of mathematics at an urban public university for 35 years, I understand the varied needs of algebra and trigonometry students. Students range from being underprepared, with little mathematical background and a fear of mathematics, to being highly prepared and motivated. For some, this is their final course in mathematics. For others, it is preparation for future mathematics courses. I have written this text with both groups in mind.

A tremendous benefit of authoring a successful series is the broad-based feedback I receive from teachers and students who have used previous editions. I am sincerely grateful for their support. Virtually every change to this edition is the result of their thoughtful comments and suggestions. I hope that I have been able to take their ideas and, building upon a successful foundation of the ninth edition, make this series an even better learning and teaching tool for students and teachers.

Features in the Tenth Edition

A descriptive list of the many special features of *Algebra & Trigonometry* can be found on the endpapers in the front of this text.

This list places the features in their proper context, as building blocks of an overall learning system that has been carefully crafted over the years to help students get the most out of the time they put into studying. Please take the time to review this and to discuss it with your students at the beginning of your course. My experience has been that when students utilize these features, they are more successful in the course.

New to the Tenth Edition

- **Retain Your Knowledge** This new category of problems in the exercise set are based on the article "To Retain New Learning, Do the Math" published in the *Edurati Review*. In this article, Kevin Washburn suggests that "the more students are required to recall new content or skills, the better their memory will be." It is frustrating when students cannot recall skills learned earlier in the course. To alleviate this recall problem, we have created "Retain Your Knowledge" problems. These are problems considered to be "final exam material" that students can use to maintain their skills. All the answers to these problems appear in the back of the text, and all are programmed in MyMathLab.

- **Guided Lecture Notes** Ideal for online, emporium/redesign courses, inverted classrooms, or traditional lecture classrooms. These lecture notes help students take thorough, organized, and understandable notes as they watch the Author in Action videos. They ask students to complete definitions, procedures, and examples based on the content of the videos and text. In addition, experience suggests that students learn by doing and understanding the why/how of the concept or

property. Therefore, many sections will have an exploration activity to motivate student learning. These explorations introduce the topic and/or connect it to either a real-world application or a previous section. For example, when the vertical-line test is discussed in Section 3.2, after the theorem statement, the notes ask the students to explain why the vertical-line test works by using the definition of a function. This challenge helps students process the information at a higher level of understanding.

- **Illustrations** Many of the figures now have captions to help connect the illustrations to the explanations in the body of the text.

- **TI Screen Shots** In this edition we have replaced all the screen shots from the ninth edition with screen shots using TI-84Plus C. These updated screen shots help students visualize concepts clearly and help make stronger connections between equations, data, and graphs in full color.

- **Chapter Projects,** which apply the concepts of each chapter to a real-world situation, have been enhanced to give students an up-to-the-minute experience. Many projects are new and Internet-based, requiring the student to research information online in order to solve problems.

- **Exercise Sets** All the exercises in the text have been reviewed and analyzed for this edition, some have been removed, and new ones have been added. All time-sensitive problems have been updated to the most recent information available. The problem sets remain classified according to purpose.

  The *'Are You Prepared?'* problems have been improved to better serve their purpose as a just-in-time review of concepts that the student will need to apply in the upcoming section.

  The *Concepts and Vocabulary* problems have been expanded and now include multiple-choice exercises. Together with the fill-in-the-blank and True/False problems, these exercises have been written to serve as reading quizzes.

  Skill Building problems develop the student's computational skills with a large selection of exercises that are directly related to the objectives of the section. *Mixed Practice* problems offer a comprehensive assessment of skills that relate to more than one objective. Often these require skills learned earlier in the course.

  Applications and Extensions problems have been updated. Further, many new application-type exercises have been added, especially ones involving information and data drawn from sources the student will recognize, to improve relevance and timeliness.

  The *Explaining Concepts: Discussion and Writing* exercises have been improved and expanded to provide more opportunity for classroom discussion and group projects.

New to this edition, ***Retain Your Knowledge*** exercises consist of a collection of four problems in each exercise set that are based on material learned earlier in the course. They serve to keep information that has already been learned "fresh" in the mind of the student. Answers to all these problems appear in the Student Edition.

The ***Review Exercises*** in the Chapter Review have been streamlined, but they remain tied to the clearly expressed objectives of the chapter. Answers to all these problems appear in the Student Edition.

- **Annotated Instructor's Edition** As a guide, the author's suggestions for homework assignments are indicated by a blue underscore below the problem number. These problems are assignable in the MyMathLab as part of a "Ready-to-Go" course.

Content Changes in the Tenth Edition

- **Section 3.1** The objective Find the Difference Quotient of a Function has been added.
- **Section 5.1** The subsection Behavior of the Graph of a Polynomial Function Near a Zero has been removed.
- **Section 5.3** A subsection has been added that discusses the role of multiplicity of the zeros of the denominator of a rational function as it relates to the graph near a vertical asymptote.
- **Section 5.5** The objective Use Descartes' Rule of Signs has been included.
- **Section 5.5** The theorem Bounds on the Zeros of a Polynomial Function is now based on the traditional method of using synthetic division.

Using the Tenth Edition Effectively with Your Syllabus

To meet the varied needs of diverse syllabi, this text contains more content than is likely to be covered in an *Algebra & Trigonometry* course. As the chart illustrates, this text has been organized with flexibility of use in mind. Within a given chapter, certain sections are optional (see the details that follow the figure below) and can be omitted without loss of continuity.

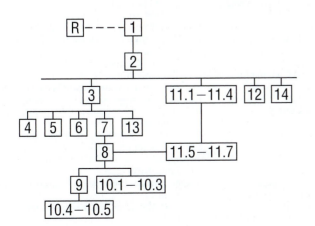

Chapter R   Review
This chapter consists of review material. It may be used as the first part of the course or later as a just-in-time review when the content is required. Specific references to this chapter occur throughout the text to assist in the review process.

Chapter 1   Equations and Inequalities
Primarily a review of Intermediate Algebra topics, this material is a prerequisite for later topics. The coverage of complex numbers and quadratic equations with a negative discriminant is optional and may be postponed or skipped entirely without loss of continuity.

Chapter 2   Graphs
This chapter lays the foundation for functions. Section 2.5 is optional.

Chapter 3   Functions and Their Graphs
Perhaps the most important chapter. Section 3.6 is optional.

Chapter 4   Linear and Quadratic Functions
Topic selection depends on your syllabus. Sections 4.2 and 4.4 may be omitted without loss of continuity.

Chapter 5   Polynomial and Rational Functions
Topic selection depends on your syllabus.

Chapter 6   Exponential and Logarithmic Functions
Sections 6.1–6.6 follow in sequence. Sections 6.7, 6.8, and 6.9 are optional.

Chapter 7   Trigonometric Functions
Section 7.8 may be omitted in a brief course.

Chapter 8   Analytic Trigonometry
Sections 8.2, 8.6, and 8.8 may be omitted in a brief course.

Chapter 9   Applications of Trigonometric Functions
Sections 9.4 and 9.5 may be omitted in a brief course.

Chapter 10   Polar Coordinates; Vectors
Sections 10.1–10.3 and Sections 10.4–10.5 are independent and may be covered separately.

Chapter 11   Analytic Geometry
Sections 11.1–11.4 follow in sequence. Sections 11.5, 11.6, and 11.7 are independent of each other, but each requires Sections 11.1–11.4.

Chapter 12   Systems of Equations and Inequalities
Sections 12.2–12.7 may be covered in any order, but each requires Section 12.1. Section 12.8 requires Section 12.7.

Chapter 13   Sequences; Induction; The Binomial Theorem
There are three independent parts: Sections 13.1–13.3; Section 13.4; and Section 13.5.

Chapter 14   Counting and Probability
The sections follow in sequence.

Acknowledgments

Textbooks are written by authors, but evolve from an idea to final form through the efforts of many people. It was Don Dellen who first suggested this text and series to me. Don is remembered for his extensive contributions to publishing and mathematics.

Thanks are due to the following people for their assistance and encouragement to the preparation of this edition:

- From Pearson Education: Anne Kelly for her substantial contributions, ideas, and enthusiasm; Dawn Murrin, for her unmatched talent at getting the details right; Joseph Colella for always getting the reviews and pages to me on time; Peggy McMahon for directing the always difficult production process; Rose Kernan for handling liaison between the compositor and author; Peggy Lucas for her genuine interest in marketing this text; Chris Hoag for her continued support and genuine interest; Greg Tobin for his leadership and commitment to excellence; and the Pearson Math and Science Sales team, for their continued confidence and personal support of our texts.

- Accuracy checkers: C. Brad Davis, who read the entire manuscript and accuracy checked answers. His attention to detail is amazing; Timothy Britt, for creating the Solutions Manuals and accuracy checking answers.

Finally, I offer my grateful thanks to the dedicated users and reviewers of my texts, whose collective insights form the backbone of each textbook revision.

James Africh, College of DuPage
Steve Agronsky, Cal Poly State University
Gererdo Aladro, Florida International University
Grant Alexander, Joliet Junior College
Dave Anderson, South Suburban College
Richard Andrews, Florida A&M University
Joby Milo Anthony, University of Central Florida
James E. Arnold, University of Wisconsin-Milwaukee
Adel Arshaghi, Center for Educational Merit
Carolyn Autray, University of West Georgia
Agnes Azzolino, Middlesex County College
Wilson P. Banks, Illinois State University
Sudeshna Basu, Howard University
Dale R. Bedgood, East Texas State University
Beth Beno, South Suburban College
Carolyn Bernath, Tallahassee Community College
Rebecca Berthiaume, Edison State College
William H. Beyer, University of Akron
Annette Blackwelder, Florida State University
Richelle Blair, Lakeland Community College
Kevin Bodden, Lewis and Clark College
Jeffrey Boerner, University of Wisconsin-Stout
Barry Booten, Florida Atlantic University
Larry Bouldin, Roane State Community College
Bob Bradshaw, Ohlone College
Trudy Bratten, Grossmont College
Tim Bremer, Broome Community College
Tim Britt, Jackson State Community College
Michael Brook, University of Delaware
Joanne Brunner, Joliet Junior College
Warren Burch, Brevard Community College
Mary Butler, Lincoln Public Schools
Melanie Butler, West Virginia University
Jim Butterbach, Joliet Junior College
William J. Cable, University of Wisconsin-Stevens Point
Lois Calamia, Brookdale Community College
Jim Campbell, Lincoln Public Schools
Roger Carlsen, Moraine Valley Community College
Elena Catoiu, Joliet Junior College
Mathews Chakkanakuzhi, Palomar College
Tim Chappell, Penn Valley Community College
John Collado, South Suburban College
Alicia Collins, Mesa Community College
Nelson Collins, Joliet Junior College
Rebecca Connell, Troy University
Jim Cooper, Joliet Junior College
Denise Corbett, East Carolina University

Carlos C. Corona, San Antonio College
Theodore C. Coskey, South Seattle Community College
Rebecca Connell, Troy University
Donna Costello, Plano Senior High School
Paul Crittenden, University of Nebraska at Lincoln
John Davenport, East Texas State University
Faye Dang, Joliet Junior College
Antonio David, Del Mar College
Stephanie Deacon, Liberty University
Duane E. Deal, Ball State University
Jerry DeGroot, Purdue North Central
Timothy Deis, University of Wisconsin-Platteville
Joanna DelMonaco, Middlesex Community College
Vivian Dennis, Eastfield College
Deborah Dillon, R. L. Turner High School
Guesna Dohrman, Tallahassee Community College
Cheryl Doolittle, Iowa State University
Karen R. Dougan, University of Florida
Jerrett Dumouchel, Florida Community College at Jacksonville
Louise Dyson, Clark College
Paul D. East, Lexington Community College
Don Edmondson, University of Texas-Austin
Erica Egizio, Joliet Junior College
Jason Eltrevoog, Joliet Junior College
Christopher Ennis, University of Minnesota
Kathy Eppler, Salt Lake Community College
Ralph Esparza, Jr., Richland College
Garret J. Etgen, University of Houston
Scott Fallstrom, Shoreline Community College
Pete Falzone, Pensacola Junior College
Arash Farahmand, Skyline College
W.A. Ferguson, University of Illinois-Urbana/Champaign
Iris B. Fetta, Clemson University
Mason Flake, student at Edison Community College
Timothy W. Flood, Pittsburg State University
Robert Frank, Westmoreland County Community College
Merle Friel, Humboldt State University
Richard A. Fritz, Moraine Valley Community College
Dewey Furness, Ricks College
Mary Jule Gabiou, North Idaho College
Randy Gallaher, Lewis and Clark College
Tina Garn, University of Arizona
Dawit Getachew, Chicago State University
Wayne Gibson, Rancho Santiago College

Loran W. Gierhart, University of Texas at San Antonio and Palo Alto College
Robert Gill, University of Minnesota Duluth
Nina Girard, University of Pittsburgh at Johnstown
Sudhir Kumar Goel, Valdosta State University
Adrienne Goldstein, Miami Dade College, Kendall Campus
Joan Goliday, Sante Fe Community College
Lourdes Gonzalez, Miami Dade College, Kendall Campus
Frederic Gooding, Goucher College
Donald Goral, Northern Virginia Community College
Sue Graupner, Lincoln Public Schools
Mary Beth Grayson, Liberty University
Jennifer L. Grimsley, University of Charleston
Ken Gurganus, University of North Carolina
James E. Hall, University of Wisconsin-Madison
Judy Hall, West Virginia University
Edward R. Hancock, DeVry Institute of Technology
Julia Hassett, DeVry Institute, Dupage
Christopher Hay-Jahans, University of South Dakota
Michah Heibel, Lincoln Public Schools
LaRae Helliwell, San Jose City College
Celeste Hernandez, Richland College
Gloria P. Hernandez, Louisiana State University at Eunice
Brother Herron, Brother Rice High School
Robert Hoburg, Western Connecticut State University
Lynda Hollingsworth, Northwest Missouri State University
Deltrye Holt, Augusta State University
Charla Holzbog, Denison High School
Lee Hruby, Naperville North High School
Miles Hubbard, St. Cloud State University
Kim Hughes, California State College-San Bernardino
Stanislav, Jabuka, University of Nevada, Reno
Ron Jamison, Brigham Young University
Richard A. Jensen, Manatee Community College
Glenn Johnson, Middlesex Community College
Sandra G. Johnson, St. Cloud State University
Tuesday Johnson, New Mexico State University
Susitha Karunaratne, Purdue University North Central
Moana H. Karsteter, Tallahassee Community College
Donna Katula, Joliet Junior College

Arthur Kaufman, College of Staten Island
Thomas Kearns, North Kentucky University
Jack Keating, Massasoit Community College
Shelia Kellenbarger, Lincoln Public Schools
Rachael Kenney, North Carolina State University
John B. Klassen, North Idaho College
Debra Kopcso, Louisiana State University
Lynne Kowski, Raritan Valley Community College
Yelena Kravchuk, University of Alabama at Birmingham
Ray S. Kuan, Skyline College
Keith Kuchar, Manatee Community College
Tor Kwembe, Chicago State University
Linda J. Kyle, Tarrant Country Jr. College
H.E. Lacey, Texas A & M University
Harriet Lamm, Coastal Bend College
James Lapp, Fort Lewis College
Matt Larson, Lincoln Public Schools
Christopher Lattin, Oakton Community College
Julia Ledet, Lousiana State University
Adele LeGere, Oakton Community College
Kevin Leith, University of Houston
JoAnn Lewin, Edison College
Jeff Lewis, Johnson County Community College
Janice C. Lyon, Tallahassee Community College
Jean McArthur, Joliet Junior College
Virginia McCarthy, Iowa State University
Karla McCavit, Albion College
Michael McClendon, University of Central Oklahoma
Tom McCollow, DeVry Institute of Technology
Marilyn McCollum, North Carolina State University
Jill McGowan, Howard University
Will McGowant, Howard University
Angela McNulty, Joliet Junior College
Laurence Maher, North Texas State University
Jay A. Malmstrom, Oklahoma City Community College
Rebecca Mann, Apollo High School
Lynn Marecek, Santa Ana College
Sherry Martina, Naperville North High School
Alec Matheson, Lamar University
Nancy Matthews, University of Oklahoma
James Maxwell, Oklahoma State University-Stillwater
Marsha May, Midwestern State University
James McLaughlin, West Chester University
Judy Meckley, Joliet Junior College
David Meel, Bowling Green State University
Carolyn Meitler, Concordia University
Samia Metwali, Erie Community College
Rich Meyers, Joliet Junior College
Eldon Miller, University of Mississippi
James Miller, West Virginia University
Michael Miller, Iowa State University
Kathleen Miranda, SUNY at Old Westbury
Chris Mirbaha, The Community College of Baltimore County
Val Mohanakumar, Hillsborough Community College
Thomas Monaghan, Naperville North High School
Miguel Montanez, Miami Dade College, Wolfson Campus
Maria Montoya, Our Lady of the Lake University
Susan Moosai, Florida Atlantic University

Craig Morse, Naperville North High School
Samad Mortabit, Metropolitan State University
Pat Mower, Washburn University
Tammy Muhs, University of Central Florida
A. Muhundan, Manatee Community College
Jane Murphy, Middlesex Community College
Richard Nadel, Florida International University
Gabriel Nagy, Kansas State University
Bill Naegele, South Suburban College
Karla Neal, Lousiana State University
Lawrence E. Newman, Holyoke Community College
Dwight Newsome, Pasco-Hernando Community College
Denise Nunley, Maricopa Community Colleges
James Nymann, University of Texas-El Paso
Mark Omodt, Anoka-Ramsey Community College
Seth F. Oppenheimer, Mississippi State University
Leticia Oropesa, University of Miami
Linda Padilla, Joliet Junior College
Sanja Pantic, University of Illinois at Chicago
E. James Peake, Iowa State University
Kelly Pearson, Murray State University
Dashamir Petrela, Florida Atlantic University
Philip Pina, Florida Atlantic University
Charlotte Pisors, Baylor University
Michael Prophet, University of Northern Iowa
Laura Pyzdrowski, West Virginia University
Carrie Quesnell, Weber State University
Neal C. Raber, University of Akron
Thomas Radin, San Joaquin Delta College
Aibeng Serene Radulovic, Florida Atlantic University
Ken A. Rager, Metropolitan State College
Kenneth D. Reeves, San Antonio College
Elsi Reinhardt, Truckee Meadows Community College
Jose Remesar, Miami Dade College, Wolfson Campus
Jane Ringwald, Iowa State University
Douglas F. Robertson, University of Minnesota, MPLS
Stephen Rodi, Austin Community College
William Rogge, Lincoln Northeast High School
Howard L. Rolf, Baylor University
Mike Rosenthal, Florida International University
Phoebe Rouse, Lousiana State University
Edward Rozema, University of Tennessee at Chattanooga
Dennis C. Runde, Manatee Community College
Alan Saleski, Loyola University of Chicago
Susan Sandmeyer, Jamestown Community College
Brenda Santistevan, Salt Lake Community College
Linda Schmidt, Greenville Technical College
Ingrid Scott, Montgomery College
A.K. Shamma, University of West Florida
Zachery Sharon, University of Texas at San Antonio
Martin Sherry, Lower Columbia College
Carmen Shershin, Florida International University
Tatrana Shubin, San Jose State University
Anita Sikes, Delgado Community College
Timothy Sipka, Alma College

Charlotte Smedberg, University of Tampa
Lori Smellegar, Manatee Community College
Gayle Smith, Loyola Blakefield
Cindy Soderstrom, Salt Lake Community College
Leslie Soltis, Mercyhurst College
John Spellman, Southwest Texas State University
Karen Spike, University of North Carolina
Rajalakshmi Sriram, Okaloosa-Walton Community College
Katrina Staley, North Carolina Agricultural and Technical State University
Becky Stamper, Western Kentucky University
Judy Staver, Florida Community College-South
Robin Steinberg, Pima Community College
Neil Stephens, Hinsdale South High School
Sonya Stephens, Florida A&M Univeristy
Patrick Stevens, Joliet Junior College
John Sumner, University of Tampa
Matthew TenHuisen, University of North Carolina, Wilmington
Christopher Terry, Augusta State University
Diane Tesar, South Suburban College
Tommy Thompson, Brookhaven College
Martha K. Tietze, Shawnee Mission Northwest High School
Richard J. Tondra, Iowa State University
Florentina Tone, University of West Florida
Suzanne Topp, Salt Lake Community College
Marilyn Toscano, University of Wisconsin, Superior
Marvel Townsend, University of Florida
Jim Trudnowski, Carroll College
Robert Tuskey, Joliet Junior College
Mihaela Vajiac, Chapman University-Orange
Julia Varbalow, Thomas Nelson Community College-Leesville
Richard G. Vinson, University of South Alabama
Jorge Viola-Prioli, Florida Atlantic University
Mary Voxman, University of Idaho
Jennifer Walsh, Daytona Beach Community College
Donna Wandke, Naperville North High School
Timothy L.Warkentin, Cloud County Community College
Melissa J. Watts, Virginia State University
Hayat Weiss, Middlesex Community College
Kathryn Wetzel, Amarillo College
Darlene Whitkenack, Northern Illinois University
Suzanne Williams, Central Piedmont Community College
Larissa Williamson, University of Florida
Christine Wilson, West Virginia University
Brad Wind, Florida International University
Anna Wiodarczyk, Florida International University
Mary Wolyniak, Broome Community College
Canton Woods, Auburn University
Tamara S. Worner, Wayne State College
Terri Wright, New Hampshire Community Technical College, Manchester
Aletheia Zambesi, University of West Florida
George Zazi, Chicago State University
Steve Zuro, Joliet Junior College

Michael Sullivan
Chicago State University

Resources for Success

MyMathLab® Online Course (access code required)

MyMathLab delivers **proven results** in helping individual students succeed. It provides **engaging experiences** that personalize, stimulate, and measure learning for each student. And it comes from an **experienced partner** with educational expertise and an eye on the future. MyMathLab helps prepare students and gets them thinking more conceptually and visually through the following features:

Adaptive Study Plan

The Study Plan makes studying more efficient and effective for every student. Performance and activity are assessed continually in real time. The data and analytics are used to provide personalized content—reinforcing concepts that target each student's strengths and weaknesses.

Getting Ready

Students refresh prerequisite topics through assignable skill review quizzes and personalized homework integrated in MyMathLab.

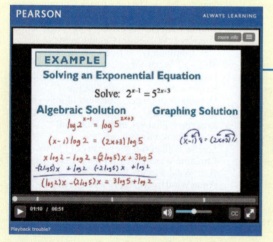

Video Assessment

Video assessment is tied to key Author in Action videos to check students' conceptual understanding of important math concepts.

Enhanced Graphing Functionality

New functionality within the graphing utility allows graphing of 3-point quadratic functions, 4-point cubic graphs, and transformations in exercises.

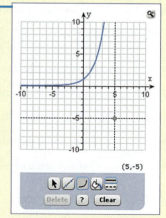

Skills for Success Modules are integrated within the MyMathLab course to help students succeed in collegiate courses and prepare for future professions.

Retain Your Knowledge These new exercises support ongoing review at the course level and help students maintain essential skills.

Instructor Resources

Additional resources can be downloaded from **www.pearsonhighered.com** or hardcopy resources can be ordered from your sales representative.

Ready to Go MyMathLab® Course

Now it is even easier to get started with MyMathLab. The Ready to Go MyMathLab course option includes author-chosen preassigned homework, integrated review, and more.

TestGen®

TestGen® (www.pearsoned.com/testgen) enables instructors to build, edit, print, and administer tests using a computerized bank of questions developed to cover all the objectives of the text.

PowerPoint® Lecture Slides

Fully editable slides correlated with the text.

Annotated Instructor's Edition

Shorter answers are on the page beside the exercises. Longer answers are in the back of the text.

Instructor Solutions Manual

Includes fully worked solutions to all exercises in the text.

Mini Lecture Notes

Includes additional examples and helpful teaching tips, by section.

Online Chapter Projects

Additional projects that give students an opportunity to apply what they learned in the chapter.

Student Resources

Additional resources to enhance student success:

Lecture Video

Author in Action videos are actual classroom lectures with fully worked out examples presented by Michael Sullivan, III. All video is assignable within MyMathlab.

Chapter Test Prep Videos

Students can watch instructors work through step-by-step solutions to all chapter test exercises from the text. These are available in MyMathlab and on YouTube.

Student Solutions Manual

Provides detailed worked-out solutions to odd-numbered exercises.

Guided Lecture Notes

These lecture notes assist students in taking thorough, organized, and understandable notes while watching Author in Action videos. Students actively participate in learning the how/why of important concepts through explorations and activities. The Guided Lecture Notes are available as PDF's and customizable Word files in MyMathLab. They can also be packaged with the text and the MyMathLab access code.

Algebra Review

Four chapters of Intermediate Algebra review. Perfect for a slower-paced course or for individual review.

Applications Index

Review

R

A Look Ahead •••

Chapter R, as the title states, contains review material. Your instructor may choose to cover all or part of it as a regular chapter at the beginning of your course or later as a just-in-time review when the content is required. Regardless, when information in this chapter is needed, a specific reference to this chapter will be made so you can review.

Outline

R.1 Real Numbers

PREPARING FOR THIS TEXT *Before getting started, read "To the Student" at the front of this text.*

OBJECTIVES **1** Work with Sets (p. 2)
2 Classify Numbers (p. 4)
3 Evaluate Numerical Expressions (p. 8)
4 Work with Properties of Real Numbers (p. 9)

1 Work with Sets

A **set** is a well-defined collection of distinct objects. The objects of a set are called its **elements**. By **well-defined**, we mean that there is a rule that enables us to determine whether a given object is an element of the set. If a set has no elements, it is called the **empty set**, or **null set**, and is denoted by the symbol $\varnothing$.

For example, the set of **digits** consists of the collection of numbers 0, 1, 2, 3, 4, 5, 6, 7, 8, and 9. If we use the symbol D to denote the set of digits, then we can write

$$D = \{0, 1, 2, 3, 4, 5, 6, 7, 8, 9\}$$

In this notation, the braces { } are used to enclose the objects, or **elements**, in the set. This method of denoting a set is called the **roster method**. A second way to denote a set is to use **set-builder notation**, where the set D of digits is written as

$$D = \{\quad x \quad | \quad x \text{ is a digit}\}$$

Read as "D is the set of all x such that x is a digit."

EXAMPLE 1 **Using Set-builder Notation and the Roster Method**

(a) $E = \{x | x \text{ is an even digit}\} = \{0, 2, 4, 6, 8\}$
(b) $O = \{x | x \text{ is an odd digit}\} = \{1, 3, 5, 7, 9\}$

Because the elements of a set are distinct, we never repeat elements. For example, we would never write $\{1, 2, 3, 2\}$; the correct listing is $\{1, 2, 3\}$. Because a set is a collection, the order in which the elements are listed is immaterial. $\{1, 2, 3\}$, $\{1, 3, 2\}$, $\{2, 1, 3\}$, and so on, all represent the same set.

If every element of a set A is also an element of a set B, then A is a **subset** of B, which is denoted $A \subseteq B$. If two sets A and B have the same elements, then A **equals** B, which is denoted $A = B$.

For example, $\{1, 2, 3\} \subseteq \{1, 2, 3, 4, 5\}$ and $\{1, 2, 3\} = \{2, 3, 1\}$.

DEFINITION

If A and B are sets, the **intersection** of A with B, denoted $A \cap B$, is the set consisting of elements that belong to both A and B. The **union** of A with B, denoted $A \cup B$, is the set consisting of elements that belong to either A or B, or both.

EXAMPLE 2 **Finding the Intersection and Union of Sets**

Let $A = \{1, 3, 5, 8\}$, $B = \{3, 5, 7\}$, and $C = \{2, 4, 6, 8\}$. Find:

(a) $A \cap B$ (b) $A \cup B$ (c) $B \cap (A \cup C)$

Solution   (a) $A \cap B = \{1, 3, 5, 8\} \cap \{3, 5, 7\} = \{3, 5\}$
(b) $A \cup B = \{1, 3, 5, 8\} \cup \{3, 5, 7\} = \{1, 3, 5, 7, 8\}$
(c) $B \cap (A \cup C) = \{3, 5, 7\} \cap [\{1, 3, 5, 8\} \cup \{2, 4, 6, 8\}]$
$= \{3, 5, 7\} \cap \{1, 2, 3, 4, 5, 6, 8\} = \{3, 5\}$ ●

Now Work PROBLEM 15

Usually, in working with sets, we designate a **universal set** U, the set consisting of all the elements that we wish to consider. Once a universal set has been designated, we can consider elements of the universal set not found in a given set.

DEFINITION

If A is a set, the **complement** of A, denoted $\overline{A}$, is the set consisting of all the elements in the universal set that are not in A.*

EXAMPLE 3   **Finding the Complement of a Set**

If the universal set is $U = \{1, 2, 3, 4, 5, 6, 7, 8, 9\}$ and if $A = \{1, 3, 5, 7, 9\}$, then $\overline{A} = \{2, 4, 6, 8\}$. ●

It follows from the definition of complement that $A \cup \overline{A} = U$ and $A \cap \overline{A} = \varnothing$. Do you see why?

Now Work PROBLEM 19

It is often helpful to draw pictures of sets. Such pictures, called **Venn diagrams**, represent sets as circles enclosed in a rectangle, which represents the universal set. Such diagrams often help us to visualize various relationships among sets. See Figure 1.

If we know that $A \subseteq B$, we might use the Venn diagram in Figure 2(a). If we know that A and B have no elements in common—that is, if $A \cap B = \varnothing$—we might use the Venn diagram in Figure 2(b). The sets A and B in Figure 2(b) are said to be **disjoint**.

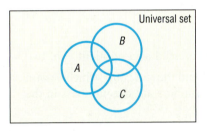

Figure 1 Venn diagram

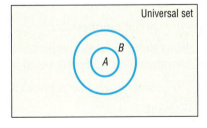   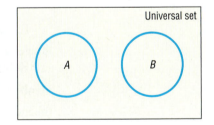

Figure 2   (a) $A \subseteq B$ (b) $A \cap B = \varnothing$
                      subset   disjoint sets

Figures 3(a), 3(b), and 3(c) use Venn diagrams to illustrate the definitions of intersection, union, and complement, respectively.

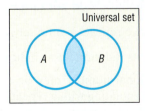   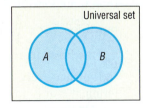   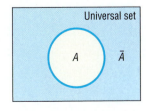

Figure 3   (a) $A \cap B$ (b) $A \cup B$ (c) $\overline{A}$
                 intersection   union   complement

*Some books use the notation A' for the complement of A.

2 Classify Numbers

It is helpful to classify the various kinds of numbers that we deal with as sets. The **counting numbers**, or **natural numbers**, are the numbers in the set $\{1, 2, 3, 4, \ldots\}$. (The three dots, called an **ellipsis**, indicate that the pattern continues indefinitely.) As their name implies, these numbers are often used to count things. For example, there are 26 letters in our alphabet; there are 100 cents in a dollar. The **whole numbers** are the numbers in the set $\{0, 1, 2, 3, \ldots\}$ —that is, the counting numbers together with 0. The set of counting numbers is a subset of the set of whole numbers.

DEFINITION

> The **integers** are the set of numbers $\{\ldots, -3, -2, -1, 0, 1, 2, 3, \ldots\}$.

These numbers are useful in many situations. For example, if your checking account has \$10 in it and you write a check for \$15, you can represent the current balance as −\$5.

Each time we expand a number system, such as from the whole numbers to the integers, we do so in order to be able to handle new, and usually more complicated, problems. The integers enable us to solve problems requiring both positive and negative counting numbers, such as profit/loss, height above/below sea level, temperature above/below 0°F, and so on.

But integers alone are not sufficient for *all* problems. For example, they do not answer the question "What part of a dollar is 38 cents?" To answer such a question, we enlarge our number system to include *rational numbers*. For example, $\dfrac{38}{100}$ answers the question "What part of a dollar is 38 cents?"

DEFINITION

> A **rational number** is a number that can be expressed as a quotient $\dfrac{a}{b}$ of two integers. The integer a is called the **numerator**, and the integer b, which cannot be 0, is called the **denominator**. The rational numbers are the numbers in the set $\left\{ x \,\middle|\, x = \dfrac{a}{b}, \text{ where } a, b \text{ are integers and } b \neq 0 \right\}$.

Examples of rational numbers are $\dfrac{3}{4}, \dfrac{5}{2}, \dfrac{0}{4}, -\dfrac{2}{3}$, and $\dfrac{100}{3}$. Since $\dfrac{a}{1} = a$ for any integer a, it follows that the set of integers is a subset of the set of rational numbers.

Rational numbers may be represented as **decimals**. For example, the rational numbers $\dfrac{3}{4}, \dfrac{5}{2}, -\dfrac{2}{3}$, and $\dfrac{7}{66}$ may be represented as decimals by merely carrying out the indicated division:

$$\frac{3}{4} = 0.75 \qquad \frac{5}{2} = 2.5 \qquad -\frac{2}{3} = -0.666\ldots = -0.\overline{6} \qquad \frac{7}{66} = 0.1060606\ldots = 0.1\overline{06}$$

Notice that the decimal representations of $\dfrac{3}{4}$ and $\dfrac{5}{2}$ terminate, or end. The decimal representations of $-\dfrac{2}{3}$ and $\dfrac{7}{66}$ do not terminate, but they do exhibit a pattern of repetition. For $-\dfrac{2}{3}$, the 6 repeats indefinitely, as indicated by the bar over the 6; for $\dfrac{7}{66}$, the block 06 repeats indefinitely, as indicated by the bar over the 06. It can be shown that every rational number may be represented by a decimal that either terminates or is nonterminating with a repeating block of digits, and vice versa.

On the other hand, some decimals do not fit into either of these categories. Such decimals represent **irrational numbers**. Every irrational number may be represented by a decimal that neither repeats nor terminates. In other words, irrational numbers cannot be written in the form $\dfrac{a}{b}$, where a, b are integers and $b \neq 0$.

Irrational numbers occur naturally. For example, consider the isosceles right triangle whose legs are each of length 1. See Figure 4. The length of the hypotenuse is $\sqrt{2}$, an irrational number.

Also, the number that equals the ratio of the circumference C to the diameter d of any circle, denoted by the symbol π (the Greek letter pi), is an irrational number. See Figure 5.

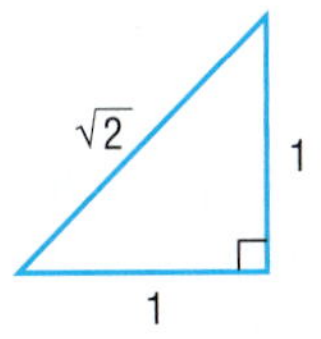

Figure 4

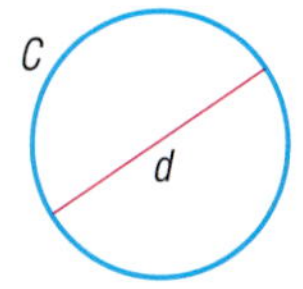

Figure 5 $\pi = \dfrac{C}{d}$

DEFINITION

> The set of **real numbers** is the union of the set of rational numbers with the set of irrational numbers.

Figure 6 shows the relationship of various types of numbers.*

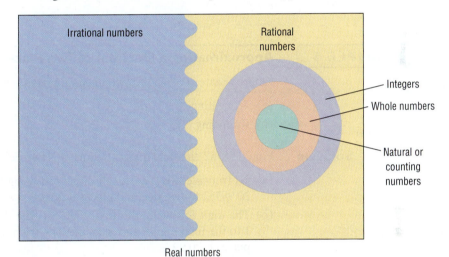

Real numbers

Figure 6

EXAMPLE 4

Classifying the Numbers in a Set

List the numbers in the set

$$\left\{ -3, \frac{4}{3}, 0.12, \sqrt{2}, \pi, 10, 2.151515\ldots \text{(where the block 15 repeats)} \right\}$$

that are

(a) Natural numbers   (b) Integers   (c) Rational numbers
(d) Irrational numbers   (e) Real numbers

Solution

(a) 10 is the only natural number.

(b) -3 and 10 are integers.

(c) $-3, 10, \dfrac{4}{3}, 0.12,$ and $2.151515\ldots$ are rational numbers.

(d) $\sqrt{2}$ and π are irrational numbers.

(e) All the numbers listed are real numbers. ●

 Now Work PROBLEM 25

*The set of real numbers is a subset of the set of complex numbers. We discuss complex numbers in Chapter 1, Section 1.3.

Approximations

Every decimal may be represented by a real number (either rational or irrational), and every real number may be represented by a decimal.

In practice, the decimal representation of an irrational number is given as an approximation. For example, using the symbol $\approx$ (read as "approximately equal to"), we can write

$$\sqrt{2} \approx 1.4142 \qquad \pi \approx 3.1416$$

In approximating decimals, we either *round off* or *truncate* to a given number of decimal places.* The number of places establishes the location of the *final digit* in the decimal approximation.

> **Truncation:** Drop all of the digits that follow the specified final digit in the decimal.
>
> **Rounding:** Identify the specified final digit in the decimal. If the next digit is 5 or more, add 1 to the final digit; if the next digit is 4 or less, leave the final digit as it is. Then truncate following the final digit.

EXAMPLE 5 **Approximating a Decimal to Two Places**

Approximate 20.98752 to two decimal places by

(a) Truncating
(b) Rounding

Solution For 20.98752, the final digit is 8, since it is two decimal places from the decimal point.

(a) To truncate, we remove all digits following the final digit 8. The truncation of 20.98752 to two decimal places is 20.98.

(b) The digit following the final digit 8 is the digit 7. Since 7 is 5 or more, we add 1 to the final digit 8 and truncate. The rounded form of 20.98752 to two decimal places is 20.99.

●

EXAMPLE 6 **Approximating a Decimal to Two and Four Places**

Number	Rounded to Two Decimal Places	Rounded to Four Decimal Places	Truncated to Two Decimal Places	Truncated to Four Decimal Places
(a) 3.14159	3.14	3.1416	3.14	3.1415
(b) 0.056128	0.06	0.0561	0.05	0.0561
(c) 893.46125	893.46	893.4613	893.46	893.4612

●

 Now Work PROBLEM 29

Calculators

Calculators are incapable of displaying decimals that contain a large number of digits. For example, some calculators are capable of displaying only eight digits. When a number requires more than eight digits, the calculator either truncates or rounds.

---

* Sometimes we say "correct to a given number of decimal places" instead of "truncate."

To see how your calculator handles decimals, divide 2 by 3. How many digits do you see? Is the last digit a 6 or a 7? If it is a 6, your calculator truncates; if it is a 7, your calculator rounds.

There are different kinds of calculators. An **arithmetic** calculator can only add, subtract, multiply, and divide numbers; therefore, this type is not adequate for this course. **Scientific** calculators have all the capabilities of arithmetic calculators and also contain **function keys** labeled ln, log, sin, cos, tan, x^y, inv, and so on. As you proceed through this text, you will discover how to use many of the function keys. **Graphing** calculators have all the capabilities of scientific calculators and contain a screen on which graphs can be displayed.

For those who have access to a graphing calculator, we have included comments, examples, and exercises marked with a 🖩, indicating that a graphing calculator is required. We have also included an appendix that explains some of the capabilities of a graphing calculator. The 🖩 comments, examples, and exercises may be omitted without loss of continuity, if so desired.

Operations

In algebra, we use letters such as $x, y, a, b,$ and c to represent numbers. The symbols used in algebra for the operations of addition, subtraction, multiplication, and division are $+, -, \cdot,$ and $/$. The words used to describe the results of these operations are **sum**, **difference**, **product**, and **quotient**. Table 1 summarizes these ideas.

Table 1

Operation	Symbol	Words
Addition	$a + b$	Sum: a plus b
Subtraction	$a - b$	Difference: a minus b
Multiplication	$a \cdot b, (a) \cdot b, a \cdot (b), (a) \cdot (b),$ $ab, (a)b, a(b), (a)(b)$	Product: a times b
Division	a/b or $\dfrac{a}{b}$	Quotient: a divided by b

In algebra, we generally avoid using the multiplication sign $\times$ and the division sign $\div$ so familiar in arithmetic. Notice also that when two expressions are placed next to each other without an operation symbol, as in ab, or in parentheses, as in $(a)(b)$, it is understood that the expressions, called **factors**, are to be multiplied.

We also prefer not to use mixed numbers in algebra. When mixed numbers are used, addition is understood; for example, $2\frac{3}{4}$ means $2 + \frac{3}{4}$. In algebra, use of a mixed number may be confusing because the absence of an operation symbol between two terms is generally taken to mean multiplication. The expression $2\frac{3}{4}$ is therefore written instead as 2.75 or as $\frac{11}{4}$.

The symbol $=$, called an **equal sign** and read as "equals" or "is," is used to express the idea that the number or expression on the left of the equal sign is equivalent to the number or expression on the right.

EXAMPLE 7

Writing Statements Using Symbols

(a) The sum of 2 and 7 equals 9. In symbols, this statement is written as $2 + 7 = 9$.

(b) The product of 3 and 5 is 15. In symbols, this statement is written as $3 \cdot 5 = 15$.

●

Now Work PROBLEM 41

3 Evaluate Numerical Expressions

Consider the expression $2 + 3 \cdot 6$. It is not clear whether we should add 2 and 3 to get 5, and then multiply by 6 to get 30; or first multiply 3 and 6 to get 18, and then add 2 to get 20. To avoid this ambiguity, we have the following agreement.

In Words
Multiply first, then add.

We agree that whenever the two operations of addition and multiplication separate three numbers, the multiplication operation will always be performed first, followed by the addition operation.

For $2 + 3 \cdot 6$, then, we have

$$2 + 3 \cdot 6 = 2 + 18 = 20$$

EXAMPLE 8

Finding the Value of an Expression

Evaluate each expression.

(a) $3 + 4 \cdot 5$ (b) $8 \cdot 2 + 1$ (c) $2 + 2 \cdot 2$

Solution

(a) $3 + 4 \cdot 5 = 3 + 20 = 23$ (b) $8 \cdot 2 + 1 = 16 + 1 = 17$
       ↑   ↑
   Multiply first   Multiply first

(c) $2 + 2 \cdot 2 = 2 + 4 = 6$

Now Work PROBLEM 53

When we want to indicate adding 3 and 4 and then multiplying the result by 5, we use parentheses and write $(3 + 4) \cdot 5$. Whenever parentheses appear in an expression, it means "perform the operations within the parentheses first!"

EXAMPLE 9

Finding the Value of an Expression

(a) $(5 + 3) \cdot 4 = 8 \cdot 4 = 32$
(b) $(4 + 5) \cdot (8 - 2) = 9 \cdot 6 = 54$

When we divide two expressions, as in

$$\frac{2 + 3}{4 + 8}$$

it is understood that the division bar acts like parentheses; that is,

$$\frac{2 + 3}{4 + 8} = \frac{(2 + 3)}{(4 + 8)}$$

Rules for the Order of Operations

1. Begin with the innermost parentheses and work outward. Remember that in dividing two expressions, we treat the numerator and denominator as if they were enclosed in parentheses.
2. Perform multiplications and divisions, working from left to right.
3. Perform additions and subtractions, working from left to right.

| EXAMPLE 10 | **Finding the Value of an Expression** |

Evaluate each expression.

(a) $8 \cdot 2 + 3$ (b) $5 \cdot (3 + 4) + 2$

(c) $\dfrac{2 + 5}{2 + 4 \cdot 7}$ (d) $2 + [4 + 2 \cdot (10 + 6)]$

Solution   (a) $8 \cdot 2 + 3 = 16 + 3 = 19$

Multiply first

(b) $5 \cdot (3 + 4) + 2 = 5 \cdot 7 + 2 = 35 + 2 = 37$

Parentheses first   Multiply before adding

(c) $\dfrac{2 + 5}{2 + 4 \cdot 7} = \dfrac{2 + 5}{2 + 28} = \dfrac{7}{30}$

(d) $2 + [4 + 2 \cdot (10 + 6)] = 2 + [4 + 2 \cdot (16)]$
$$= 2 + [4 + 32] = 2 + [36] = 38$$

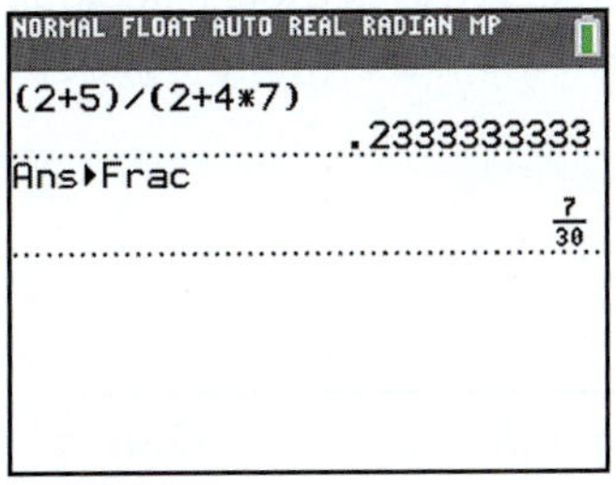

```
NORMAL FLOAT AUTO REAL RADIAN MP
(2+5)/(2+4*7)
                       .2333333333
Ans▶Frac
                                7
                               30
```

Figure 7

Be careful if you use a calculator. For Example 10(c), you need to use parentheses. See Figure 7.* If you don't, the calculator will compute the expression

$$2 + \frac{5}{2} + 4 \cdot 7 = 2 + 2.5 + 28 = 32.5$$

giving a wrong answer.

Now Work PROBLEMS **59** AND **67**

4 Work with Properties of Real Numbers

The equal sign is used to mean that one expression is equivalent to another. Four important properties of equality are listed next. In this list, a, b, and c represent real numbers.

1. The **reflexive property** states that a number equals itself; that is, $a = a$.
2. The **symmetric property** states that if $a = b$, then $b = a$.
3. The **transitive property** states that if $a = b$ and $b = c$, then $a = c$.
4. The **principle of substitution** states that if $a = b$, then we may substitute b for a in any expression containing a.

Now, let's consider some other properties of real numbers.

| EXAMPLE 11 | **Commutative Properties** |

(a) $3 + 5 = 8$ (b) $2 \cdot 3 = 6$
$\quad 5 + 3 = 8$ $\quad 3 \cdot 2 = 6$
$\quad 3 + 5 = 5 + 3$ $\quad 2 \cdot 3 = 3 \cdot 2$

This example illustrates the **commutative property** of real numbers, which states that the order in which addition or multiplication takes place does not affect the final result.

* Notice that we converted the decimal to its fraction form. Another option, when using a TI-84 Plus C, is to use the fraction template under the MATH button to enter the expression as it appears in Example 10(c). Consult your manual to see how to enter such expressions on your calculator.

Commutative Properties

$$a + b = b + a \tag{1a}$$
$$a \cdot b = b \cdot a \tag{1b}$$

Here, and in the properties listed next and on pages 11–13, a, b, and c represent real numbers.

EXAMPLE 12 | **Associative Properties**

(a) $2 + (3 + 4) = 2 + 7 = 9$ (b) $2 \cdot (3 \cdot 4) = 2 \cdot 12 = 24$
$(2 + 3) + 4 = 5 + 4 = 9$ $(2 \cdot 3) \cdot 4 = 6 \cdot 4 = 24$
$2 + (3 + 4) = (2 + 3) + 4$ $2 \cdot (3 \cdot 4) = (2 \cdot 3) \cdot 4$ ●

The way we add or multiply three real numbers does not affect the final result. Expressions such as $2 + 3 + 4$ and $3 \cdot 4 \cdot 5$ present no ambiguity, even though addition and multiplication are performed on one pair of numbers at a time. This property is called the **associative property**.

Associative Properties

$$a + (b + c) = (a + b) + c = a + b + c \tag{2a}$$
$$a \cdot (b \cdot c) = (a \cdot b) \cdot c = a \cdot b \cdot c \tag{2b}$$

Distributive Property

$$a \cdot (b + c) = a \cdot b + a \cdot c \tag{3a}$$
$$(a + b) \cdot c = a \cdot c + b \cdot c \tag{3b}$$

The **distributive property** may be used in two different ways.

EXAMPLE 13 | **Distributive Property**

(a) $2 \cdot (x + 3) = 2 \cdot x + 2 \cdot 3 = 2x + 6$ Use to remove parentheses.
(b) $3x + 5x = (3 + 5)x = 8x$ Use to combine two expressions.
(c) $(x + 2)(x + 3) = x(x + 3) + 2(x + 3) = (x^2 + 3x) + (2x + 6)$
$= x^2 + (3x + 2x) + 6 = x^2 + 5x + 6$ ●

Now Work PROBLEM 89

The real numbers 0 and 1 have unique properties called the *identity properties*.

EXAMPLE 14 | **Identity Properties**

(a) $4 + 0 = 0 + 4 = 4$ (b) $3 \cdot 1 = 1 \cdot 3 = 3$ ●

Identity Properties

$$0 + a = a + 0 = a \qquad \text{(4a)}$$

$$a \cdot 1 = 1 \cdot a = a \qquad \text{(4b)}$$

We call 0 the **additive identity** and 1 the **multiplicative identity**.

For each real number a, there is a real number $-a$, called the **additive inverse** of a, having the following property:

Additive Inverse Property

$$a + (-a) = -a + a = 0 \qquad \text{(5a)}$$

EXAMPLE 15

Finding an Additive Inverse

(a) The additive inverse of 6 is -6, because $6 + (-6) = 0$.

(b) The additive inverse of -8 is $-(-8) = 8$, because $-8 + 8 = 0$. ●

The additive inverse of a, that is, $-a$, is often called the *negative* of a or the *opposite* of a. The use of such terms can be dangerous, because they suggest that the additive inverse is a negative number, which may not be the case. For example, the additive inverse of -3, or $-(-3)$, equals 3, a positive number.

For each *nonzero* real number a, there is a real number $\dfrac{1}{a}$, called the **multiplicative inverse** of a, having the following property:

Multiplicative Inverse Property

$$a \cdot \frac{1}{a} = \frac{1}{a} \cdot a = 1 \qquad \text{if } a \neq 0 \qquad \text{(5b)}$$

The multiplicative inverse $\dfrac{1}{a}$ of a nonzero real number a is also referred to as the **reciprocal** of a.

EXAMPLE 16

Finding a Reciprocal

(a) The reciprocal of 6 is $\dfrac{1}{6}$, because $6 \cdot \dfrac{1}{6} = 1$.

(b) The reciprocal of -3 is $\dfrac{1}{-3}$, because $-3 \cdot \dfrac{1}{-3} = 1$.

(c) The reciprocal of $\dfrac{2}{3}$ is $\dfrac{3}{2}$, because $\dfrac{2}{3} \cdot \dfrac{3}{2} = 1$. ●

With these properties for adding and multiplying real numbers, we can define the operations of subtraction and division as follows:

DEFINITION

The **difference** $a - b$, also read "a less b" or "a minus b," is defined as

$$a - b = a + (-b) \qquad \text{(6)}$$

To subtract b from a, add the opposite of b to a.

DEFINITION

If b is a nonzero real number, the **quotient** $\dfrac{a}{b}$, also read as "a divided by b" or "the ratio of a to b," is defined as

$$\frac{a}{b} = a \cdot \frac{1}{b} \qquad \text{if } b \neq 0 \tag{7}$$

EXAMPLE 17 **Working with Differences and Quotients**

(a) $8 - 5 = 8 + (-5) = 3$

(b) $4 - 9 = 4 + (-9) = -5$

(c) $\dfrac{5}{8} = 5 \cdot \dfrac{1}{8}$ •

For any number a, the product of a times 0 is always 0; that is,

In Words
The result of multiplying by zero is zero.

Multiplication by Zero

$$a \cdot 0 = 0 \tag{8}$$

For a nonzero number a,

Division Properties

$$\frac{0}{a} = 0 \qquad \frac{a}{a} = 1 \quad \text{if } a \neq 0 \tag{9}$$

NOTE Division by 0 is not defined. One reason is to avoid the following difficulty: $\dfrac{2}{0} = x$ means to find x such that $0 \cdot x = 2$. But $0 \cdot x$ equals 0 for all x, so there is no unique number x such that $\dfrac{2}{0} = x$. ■

Rules of Signs

$$a(-b) = -(ab) \qquad (-a)b = -(ab) \qquad (-a)(-b) = ab$$
$$-(-a) = a \qquad \frac{a}{-b} = \frac{-a}{b} = -\frac{a}{b} \qquad \frac{-a}{-b} = \frac{a}{b} \tag{10}$$

EXAMPLE 18 **Applying the Rules of Signs**

(a) $2(-3) = -(2 \cdot 3) = -6$ (b) $(-3)(-5) = 3 \cdot 5 = 15$

(c) $\dfrac{3}{-2} = \dfrac{-3}{2} = -\dfrac{3}{2}$ (d) $\dfrac{-4}{-9} = \dfrac{4}{9}$

(e) $\dfrac{x}{-2} = \dfrac{1}{-2} \cdot x = -\dfrac{1}{2}x$ •

Cancellation Properties

$$ac = bc \quad \text{implies} \quad a = b \quad \text{if } c \neq 0$$
$$\frac{ac}{bc} = \frac{a}{b} \qquad\qquad \text{if } b \neq 0, c \neq 0 \qquad \textbf{(11)}$$

EXAMPLE 19 | **Using the Cancellation Properties**

(a) If $2x = 6$, then

$$2x = 6$$
$$2x = 2 \cdot 3 \qquad \text{Factor 6.}$$
$$x = 3 \qquad \text{Cancel the 2's.}$$

NOTE We follow the common practice of using slash marks to indicate cancellations. ∎

(b) $\dfrac{18}{12} = \dfrac{3 \cdot \not6}{2 \cdot \not6} = \dfrac{3}{2}$

Cancel the 6's.

In Words
If a product equals 0, then one or both of the factors is 0.

Zero-Product Property

$$\text{If } ab = 0, \text{ then } a = 0, \text{ or } b = 0, \text{ or both.} \qquad \textbf{(12)}$$

EXAMPLE 20 | **Using the Zero-Product Property**

If $2x = 0$, then either $2 = 0$ or $x = 0$. Since $2 \neq 0$, it follows that $x = 0$.

Arithmetic of Quotients

$$\frac{a}{b} + \frac{c}{d} = \frac{ad}{bd} + \frac{bc}{bd} = \frac{ad + bc}{bd} \qquad \text{if } b \neq 0, d \neq 0 \qquad \textbf{(13)}$$

$$\frac{a}{b} \cdot \frac{c}{d} = \frac{ac}{bd} \qquad \text{if } b \neq 0, d \neq 0 \qquad \textbf{(14)}$$

$$\frac{\dfrac{a}{b}}{\dfrac{c}{d}} = \frac{a}{b} \cdot \frac{d}{c} = \frac{ad}{bc} \qquad \text{if } b \neq 0, c \neq 0, d \neq 0 \qquad \textbf{(15)}$$

EXAMPLE 21 | **Adding, Subtracting, Multiplying, and Dividing Quotients**

(a) $\dfrac{2}{3} + \dfrac{5}{2} = \dfrac{2 \cdot 2}{3 \cdot 2} + \dfrac{3 \cdot 5}{3 \cdot 2} = \dfrac{2 \cdot 2 + 3 \cdot 5}{3 \cdot 2} = \dfrac{4 + 15}{6} = \dfrac{19}{6}$

↑ By equation (13)

(b) $\dfrac{3}{5} - \dfrac{2}{3} = \dfrac{3}{5} + \left(-\dfrac{2}{3}\right) = \dfrac{3}{5} + \dfrac{-2}{3}$

↑ By equation (6)   ↑ By equation (10)

$= \dfrac{3 \cdot 3 + 5 \cdot (-2)}{5 \cdot 3} = \dfrac{9 + (-10)}{15} = \dfrac{-1}{15} = -\dfrac{1}{15}$

↑ By equation (13)

NOTE Slanting the cancellation marks in different directions for different factors, as shown here, is a good practice to follow, since it will help in checking for errors. ∎

(c) $\dfrac{8}{3}\cdot\dfrac{15}{4}=\dfrac{8\cdot15}{3\cdot4}=\dfrac{2\cdot\cancel{4}\cdot3\cdot5}{3\cdot\cancel{4}\cdot1}=\dfrac{2\cdot5}{1}=10$

$\qquad\qquad\qquad\qquad$ By equation (14) $\qquad$ By equation (11)

(d) $\dfrac{\frac{3}{5}}{\frac{7}{9}}=\dfrac{3}{5}\cdot\dfrac{9}{7}=\dfrac{3\cdot9}{5\cdot7}=\dfrac{27}{35}$

$\qquad\qquad\qquad$ By equation (14)

By equation (15)

NOTE In writing quotients, we shall follow the usual convention and write the quotient in lowest terms. That is, we write it so that any common factors of the numerator and the denominator have been removed using the cancellation properties, equation (11). As examples,

$$\frac{90}{24}=\frac{15\cdot\cancel{6}}{4\cdot\cancel{6}}=\frac{15}{4}$$

$$\frac{24x^2}{18x}=\frac{4\cdot\cancel{6}\cdot x\cdot\cancel{x}}{3\cdot\cancel{6}\cdot\cancel{x}}=\frac{4x}{3}\qquad x\neq0$$ ∎

Now Work PROBLEMS **69, 73,** AND **83**

Sometimes it is easier to add two fractions using *least common multiples* (LCM). The LCM of two numbers is the smallest number that each has as a common multiple.

EXAMPLE 22 **Finding the Least Common Multiple of Two Numbers**

Find the least common multiple of 15 and 12.

Solution To find the LCM of 15 and 12, we look at multiples of 15 and 12.

$$15,\ 30,\ 45,\ 60,\ 75,\ 90,\ 105,\ 120,\ldots$$
$$12,\ 24,\ 36,\ 48,\ 60,\ 72,\ 84,\ 96,\ 108,\ 120,\ldots$$

The *common* multiples are in blue. The *least* common multiple is 60. ●

EXAMPLE 23 **Using the Least Common Multiple to Add Two Fractions**

Find: $\dfrac{8}{15}+\dfrac{5}{12}$

Solution We use the LCM of the denominators of the fractions and rewrite each fraction using the LCM as a common denominator. The LCM of the denominators (12 and 15) is 60. Rewrite each fraction using 60 as the denominator.

$$\frac{8}{15}+\frac{5}{12}=\frac{8}{15}\cdot\frac{4}{4}+\frac{5}{12}\cdot\frac{5}{5}$$
$$=\frac{32}{60}+\frac{25}{60}$$
$$=\frac{32+25}{60}$$
$$=\frac{57}{60}$$
$$=\frac{19}{20}$$ ●

Now Work PROBLEM **77**

Historical Feature

The real number system has a history that stretches back at least to the ancient Babylonians (1800 BC). It is remarkable how much the ancient Babylonian attitudes resemble our own. As we stated in the text, the fundamental difficulty with irrational numbers is that they cannot be written as quotients of integers or, equivalently, as repeating or terminating decimals. The Babylonians wrote their numbers in a system based on 60 in the same way that we write ours based on 10. They would carry as many places for π as the accuracy of the problem demanded, just as we now use

$$\pi \approx 3\frac{1}{7} \quad \text{or} \quad \pi \approx 3.1416 \quad \text{or} \quad \pi \approx 3.14159$$

$$\text{or} \quad \pi \approx 3.14159265358979$$

depending on how accurate we need to be.

Things were very different for the Greeks, whose number system allowed only rational numbers. When it was discovered that $\sqrt{2}$ was not a rational number, this was regarded as a fundamental flaw in the number concept. So serious was the matter that the Pythagorean Brotherhood (an early mathematical society) is said to have drowned one of its members for revealing this terrible secret. Greek mathematicians then turned away from the number concept, expressing facts about whole numbers in terms of line segments.

In astronomy, however, Babylonian methods, including the Babylonian number system, continued to be used. Simon Stevin (1548–1620), probably using the Babylonian system as a model, invented the decimal system, complete with rules of calculation, in 1585. [Others, for example, al-Kashi of Samarkand (d. 1429), had made some progress in the same direction.] The decimal system so effectively conceals the difficulties that the need for more logical precision began to be felt only in the early 1800s. Around 1880, Georg Cantor (1845–1918) and Richard Dedekind (1831–1916) gave precise definitions of real numbers. Cantor's definition, although more abstract and precise, has its roots in the decimal (and hence Babylonian) numerical system.

Sets and set theory were a spin-off of the research that went into clarifying the foundations of the real number system. Set theory has developed into a large discipline of its own, and many mathematicians regard it as the foundation upon which modern mathematics is built. Cantor's discoveries that infinite sets can also be counted and that there are different sizes of infinite sets are among the most astounding results of modern mathematics.

R.1 Assess Your Understanding

Concepts and Vocabulary

1. The numbers in the set $\left\{ x \,\middle|\, x = \dfrac{a}{b}, \text{ where } a, b \text{ are integers and } b \neq 0 \right\}$ are called _____ numbers.

2. The value of the expression $4 + 5 \cdot 6 - 3$ is ___.

3. The fact that $2x + 3x = (2 + 3)x$ is a consequence of the _____ Property.

4. "The product of 5 and $x + 3$ equals 6" may be written as _____.

5. The intersection of sets A and B is denoted by which of the following?
   (a) $A \cap B$ (b) $A \cup B$ (c) $A \subseteq B$ (d) $A \varnothing B$

6. Choose the correct name for the set of numbers $\{ 0, 1, 2, 3, \dots \}$.
   (a) Counting numbers   (b) Whole numbers
   (c) Integers   (d) Irrational numbers

7. *True or False* Rational numbers have decimals that either terminate or are nonterminating with a repeating block of digits.

8. *True or False* The Zero-Product Property states that the product of any number and zero equals zero.

9. *True or False* The least common multiple of 12 and 18 is 6.

10. *True or False* No real number is both rational and irrational.

Skill Building

In Problems 11–22, use $U = $ universal set $ = \{0, 1, 2, 3, 4, 5, 6, 7, 8, 9\}$, $A = \{1, 3, 4, 5, 9\}$, $B = \{2, 4, 6, 7, 8\}$, and $C = \{1, 3, 4, 6\}$ to find each set.

11. $A \cup B$	**12.** $A \cup C$	**13.** $A \cap B$	**14.** $A \cap C$
15. $(A \cup B) \cap C$	**16.** $(A \cap B) \cup C$	**17.** $\overline{A}$	**18.** $\overline{C}$
19. $\overline{A \cap B}$	**20.** $\overline{B \cup C}$	**21.** $\overline{A} \cup \overline{B}$	**22.** $\overline{B} \cap \overline{C}$

In Problems 23–28, list the numbers in each set that are (a) Natural numbers, (b) Integers, (c) Rational numbers, (d) Irrational numbers, (e) Real numbers.

23. $A = \left\{ -6, \dfrac{1}{2}, -1.333\ldots \text{(the 3's repeat)}, \pi, 2, 5 \right\}$

24. $B = \left\{ -\dfrac{5}{3}, 2.060606\ldots \text{(the block 06 repeats)}, 1.25, 0, 1, \sqrt{5} \right\}$

25. $C = \left\{ 0, 1, \dfrac{1}{2}, \dfrac{1}{3}, \dfrac{1}{4} \right\}$

26. $D = \{ -1, -1.1, -1.2, -1.3 \}$

27. $E = \left\{ \sqrt{2}, \pi, \sqrt{2} + 1, \pi + \dfrac{1}{2} \right\}$

28. $F = \left\{ -\sqrt{2}, \pi + \sqrt{2}, \dfrac{1}{2} + 10.3 \right\}$

In Problems 29–40, approximate each number (a) rounded and (b) truncated to three decimal places.

29. 18.9526   **30.** 25.86134   **31.** 28.65319   **32.** 99.05249   **33.** 0.06291   **34.** 0.05388

35. 9.9985   **36.** 1.0006   **37.** $\dfrac{3}{7}$ **38.** $\dfrac{5}{9}$ **39.** $\dfrac{521}{15}$ **40.** $\dfrac{81}{5}$

In Problems 41–50, write each statement using symbols.

41. The sum of 3 and 2 equals 5.

42. The product of 5 and 2 equals 10.

43. The sum of x and 2 is the product of 3 and 4.

44. The sum of 3 and y is the sum of 2 and 2.

45. The product of 3 and y is the sum of 1 and 2.

46. The product of 2 and x is the product of 4 and 6.

47. The difference x less 2 equals 6.

48. The difference 2 less y equals 6.

49. The quotient x divided by 2 is 6.

50. The quotient 2 divided by x is 6.

In Problems 51–88, evaluate each expression.

51. $9 - 4 + 2$

52. $6 - 4 + 3$

53. $-6 + 4 \cdot 3$

54. $8 - 4 \cdot 2$

55. $4 + 5 - 8$

56. $8 - 3 - 4$

57. $4 + \dfrac{1}{3}$

58. $2 - \dfrac{1}{2}$

59. $6 - [3 \cdot 5 + 2 \cdot (3 - 2)]$

60. $2 \cdot [8 - 3(4 + 2)] - 3$

61. $2 \cdot (3 - 5) + 8 \cdot 2 - 1$

62. $1 - (4 \cdot 3 - 2 + 2)$

63. $10 - [6 - 2 \cdot 2 + (8 - 3)] \cdot 2$

64. $2 - 5 \cdot 4 - [6 \cdot (3 - 4)]$

65. $(5 - 3)\dfrac{1}{2}$

66. $(5 + 4)\dfrac{1}{3}$

67. $\dfrac{4 + 8}{5 - 3}$

68. $\dfrac{2 - 4}{5 - 3}$

69. $\dfrac{3}{5} \cdot \dfrac{10}{21}$

70. $\dfrac{5}{9} \cdot \dfrac{3}{10}$

71. $\dfrac{6}{25} \cdot \dfrac{10}{27}$

72. $\dfrac{21}{25} \cdot \dfrac{100}{3}$

73. $\dfrac{3}{4} + \dfrac{2}{5}$

74. $\dfrac{4}{3} + \dfrac{1}{2}$

75. $\dfrac{5}{6} + \dfrac{9}{5}$

76. $\dfrac{8}{9} + \dfrac{15}{2}$

77. $\dfrac{5}{18} + \dfrac{1}{12}$

78. $\dfrac{2}{15} + \dfrac{8}{9}$

79. $\dfrac{1}{30} - \dfrac{7}{18}$

80. $\dfrac{3}{14} - \dfrac{2}{21}$

81. $\dfrac{3}{20} - \dfrac{2}{15}$

82. $\dfrac{6}{35} - \dfrac{3}{14}$

83. $\dfrac{\frac{5}{18}}{\frac{11}{27}}$

84. $\dfrac{\frac{5}{21}}{\frac{2}{35}}$

85. $\dfrac{1}{2} \cdot \dfrac{3}{5} + \dfrac{7}{10}$

86. $\dfrac{2}{3} + \dfrac{4}{5} \cdot \dfrac{1}{6}$

87. $2 \cdot \dfrac{3}{4} + \dfrac{3}{8}$

88. $3 \cdot \dfrac{5}{6} - \dfrac{1}{2}$

In Problems 89–100, use the Distributive Property to remove the parentheses.

89. $6(x + 4)$

90. $4(2x - 1)$

91. $x(x - 4)$

92. $4x(x + 3)$

93. $2\left(\dfrac{3}{4}x - \dfrac{1}{2}\right)$

94. $3\left(\dfrac{2}{3}x + \dfrac{1}{6}\right)$

95. $(x + 2)(x + 4)$

96. $(x + 5)(x + 1)$

97. $(x - 2)(x + 1)$

98. $(x - 4)(x + 1)$

99. $(x - 8)(x - 2)$

100. $(x - 4)(x - 2)$

Explaining Concepts: Discussion and Writing

101. Explain to a friend how the Distributive Property is used to justify the fact that $2x + 3x = 5x$.

102. Explain to a friend why $2 + 3 \cdot 4 = 14$, whereas $(2 + 3) \cdot 4 = 20$.

103. Explain why $2(3 \cdot 4)$ is not equal to $(2 \cdot 3) \cdot (2 \cdot 4)$.

104. Explain why $\dfrac{4 + 3}{2 + 5}$ is not equal to $\dfrac{4}{2} + \dfrac{3}{5}$.

105. Is subtraction commutative? Support your conclusion with an example.

106. Is subtraction associative? Support your conclusion with an example.

107. Is division commutative? Support your conclusion with an example.

108. Is division associative? Support your conclusion with an example.

109. If $2 = x$, why does $x = 2$?

110. If $x = 5$, why does $x^2 + x = 30$?

111. Are there any real numbers that are both rational and irrational? Are there any real numbers that are neither? Explain your reasoning.

112. Explain why the sum of a rational number and an irrational number must be irrational.

113. A rational number is defined as the quotient of two integers. When written as a decimal, the decimal will either repeat or terminate. By looking at the denominator of the rational number, there is a way to tell in advance whether its decimal representation will repeat or terminate. Make a list of rational numbers and their decimals. See if you can discover the pattern. Confirm your conclusion by consulting books on number theory at the library. Write a brief essay on your findings.

114. The current time is 12 noon CST. What time (CST) will it be 12,997 hours from now?

115. Both $\dfrac{a}{0}(a \neq 0)$ and $\dfrac{0}{0}$ are undefined, but for different reasons. Write a paragraph or two explaining the different reasons.

R.2 Algebra Essentials

OBJECTIVES **1** Graph Inequalities (p. 18)
2 Find Distance on the Real Number Line (p. 19)
3 Evaluate Algebraic Expressions (p. 20)
4 Determine the Domain of a Variable (p. 21)
5 Use the Laws of Exponents (p. 21)
6 Evaluate Square Roots (p. 23)
7 Use a Calculator to Evaluate Exponents (p. 24)
8 Use Scientific Notation (p. 24)

The Real Number Line

Real numbers can be represented by points on a line called the **real number line**. There is a one-to-one correspondence between real numbers and points on a line. That is, every real number corresponds to a point on the line, and each point on the line has a unique real number associated with it.

Pick a point on a line somewhere in the center, and label it O. This point, called the **origin**, corresponds to the real number 0. See Figure 8. The point 1 unit to the right of O corresponds to the number 1. The distance between 0 and 1 determines the **scale** of the number line. For example, the point associated with the number 2 is twice as far from O as 1. Notice that an arrowhead on the right end of the line indicates the direction in which the numbers increase. Points to the left of the origin correspond to the real numbers $-1, -2$, and so on. Figure 8 also shows the points associated with the rational numbers $-\dfrac{1}{2}$ and $\dfrac{1}{2}$ and with the irrational numbers $\sqrt{2}$ and π.

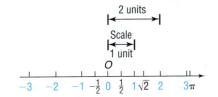

Figure 8 Real number line

DEFINITION

> The real number associated with a point P is called the **coordinate** of P, and the line whose points have been assigned coordinates is called the **real number line**.

➤ **Now Work** PROBLEM 13

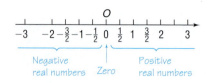

Figure 9

The real number line consists of three classes of real numbers, as shown in Figure 9.

1. The **negative real numbers** are the coordinates of points to the left of the origin O.

2. The real number **zero** is the coordinate of the origin O.

3. The **positive real numbers** are the coordinates of points to the right of the origin O.

Multiplication Properties of Positive and Negative Numbers

1. The product of two positive numbers is a positive number.

2. The product of two negative numbers is a positive number.

3. The product of a positive number and a negative number is a negative number.

1 Graph Inequalities

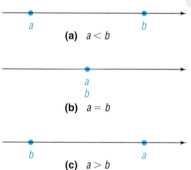

Figure 10

An important property of the real number line follows from the fact that, given two numbers (points) a and b, either a is to the left of b, or a is at the same location as b, or a is to the right of b. See Figure 10.

If a is to the left of b, then "a is less than b," which is written $a < b$. If a is to the right of b, then "a is greater than b," which is written $a > b$. If a is at the same location as b, then $a = b$. If a is either less than or equal to b, then $a \leq b$. Similarly, $a \geq b$ means that a is either greater than or equal to b. Collectively, the symbols $<, >, \leq$, and $\geq$ are called **inequality symbols**.

Note that $a < b$ and $b > a$ mean the same thing. It does not matter whether we write $2 < 3$ or $3 > 2$.

Furthermore, if $a < b$ or if $b > a$, then the difference $b - a$ is positive. Do you see why?

EXAMPLE 1   **Using Inequality Symbols**

(a) $3 < 7$ (b) $-8 > -16$ (c) $-6 < 0$
(d) $-8 < -4$ (e) $4 > -1$ (f) $8 > 0$ ●

In Example 1(a), we conclude that $3 < 7$ either because 3 is to the left of 7 on the real number line or because the difference, $7 - 3 = 4$, is a positive real number.

Similarly, we conclude in Example 1(b) that $-8 > -16$ either because -8 lies to the right of -16 on the real number line or because the difference, $-8 - (-16) = -8 + 16 = 8$, is a positive real number.

Look again at Example 1. Note that the inequality symbol always points in the direction of the smaller number.

An **inequality** is a statement in which two expressions are related by an inequality symbol. The expressions are referred to as the **sides** of the inequality. Inequalities of the form $a < b$ or $b > a$ are called **strict inequalities**, whereas inequalities of the form $a \leq b$ or $b \geq a$ are called **nonstrict inequalities**.

Based on the discussion so far, we conclude that

$a > 0$ is equivalent to a is positive

$a < 0$ is equivalent to a is negative

We sometimes read $a > 0$ by saying that "a is positive." If $a \geq 0$, then either $a > 0$ or $a = 0$, and we may read this as "a is nonnegative."

✏️ **Now Work** PROBLEMS 17 AND 27

| EXAMPLE 2 | **Graphing Inequalities** |

(a) On the real number line, graph all numbers x for which $x > 4$.

(b) On the real number line, graph all numbers x for which $x \leq 5$.

Solution

(a) See Figure 11. Notice that we use a left parenthesis to indicate that the number 4 is *not* part of the graph.

(b) See Figure 12. Notice that we use a right bracket to indicate that the number 5 *is* part of the graph. ●

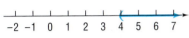

Figure 11 $x > 4$

Figure 12 $x \leq 5$

✏️ **Now Work** PROBLEM 33

2 Find Distance on the Real Number Line

The *absolute value* of a number a is the distance from 0 to a on the number line. For example, -4 is 4 units from 0, and 3 is 3 units from 0. See Figure 13. That is, the absolute value of -4 is 4, and the absolute value of 3 is 3.

A more formal definition of absolute value is given next.

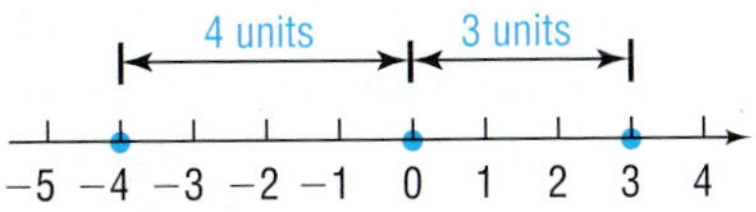

Figure 13

DEFINITION

The **absolute value** of a real number a, denoted by the symbol $|a|$, is defined by the rules

$$|a| = a \ \text{ if } a \geq 0 \qquad \text{and} \qquad |a| = -a \ \text{ if } a < 0$$

For example, because $-4 < 0$, the second rule must be used to get $|-4| = -(-4) = 4$.

| EXAMPLE 3 | **Computing Absolute Value** |

(a) $|8| = 8$ (b) $|0| = 0$ (c) $|-15| = -(-15) = 15$ ●

Look again at Figure 13. The distance from -4 to 3 is 7 units. This distance is the difference $3 - (-4)$, obtained by subtracting the smaller coordinate from the larger. However, since $|3 - (-4)| = |7| = 7$ and $|-4 - 3| = |-7| = 7$, we can use absolute value to calculate the distance between two points without being concerned about which is smaller.

DEFINITION

If P and Q are two points on a real number line with coordinates a and b, respectively, the **distance between P and Q**, denoted by $d(P, Q)$, is

$$d(P, Q) = |b - a|$$

Since $|b - a| = |a - b|$, it follows that $d(P, Q) = d(Q, P)$.

EXAMPLE 4

Finding Distance on a Number Line

Let P, Q, and R be points on a real number line with coordinates -5, 7, and -3, respectively. Find the distance

(a) between P and Q (b) between Q and R

Solution   See Figure 14.

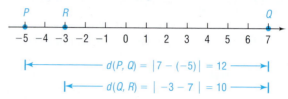

Figure 14

(a) $d(P, Q) = |7 - (-5)| = |12| = 12$
(b) $d(Q, R) = |-3 - 7| = |-10| = 10$

Now Work PROBLEM 39

3 Evaluate Algebraic Expressions

Remember, in algebra we use letters such as $x, y, a, b,$ and c to represent numbers. If the letter used is to represent *any* number from a given set of numbers, it is called a **variable**. A **constant** is either a fixed number, such as 5 or $\sqrt{3}$, or a letter that represents a fixed (possibly unspecified) number.

Constants and variables are combined using the operations of addition, subtraction, multiplication, and division to form *algebraic expressions*. Examples of algebraic expressions include

$$x + 3 \qquad \frac{3}{1 - t} \qquad 7x - 2y$$

To evaluate an algebraic expression, substitute a numerical value for each variable.

EXAMPLE 5

Evaluating an Algebraic Expression

Evaluate each expression if $x = 3$ and $y = -1$.

(a) $x + 3y$ (b) $5xy$ (c) $\dfrac{3y}{2 - 2x}$ (d) $|-4x + y|$

Solution   (a) Substitute 3 for x and -1 for y in the expression $x + 3y$.

$$x + 3y = \underset{\underset{x = 3, y = -1}{\uparrow}}{3 + 3(-1)} = 3 + (-3) = 0$$

(b) If $x = 3$ and $y = -1$, then

$$5xy = 5(3)(-1) = -15$$

(c) If $x = 3$ and $y = -1$, then

$$\frac{3y}{2 - 2x} = \frac{3(-1)}{2 - 2(3)} = \frac{-3}{2 - 6} = \frac{-3}{-4} = \frac{3}{4}$$

(d) If $x = 3$ and $y = -1$, then

$$|-4x + y| = |-4(3) + (-1)| = |-12 + (-1)| = |-13| = 13$$

Now Work PROBLEMS 41 AND 49

4 Determine the Domain of a Variable

In working with expressions or formulas involving variables, the variables may be allowed to take on values from only a certain set of numbers. For example, in the formula for the area A of a circle of radius r, $A = \pi r^2$, the variable r is necessarily restricted to the positive real numbers. In the expression $\dfrac{1}{x}$, the variable x cannot take on the value 0, since division by 0 is not defined.

DEFINITION

> The set of values that a variable may assume is called the **domain of the variable**.

EXAMPLE 6   **Finding the Domain of a Variable**

The domain of the variable x in the expression

$$\frac{5}{x - 2}$$

is $\{x \mid x \neq 2\}$ since, if $x = 2$, the denominator becomes 0, which is not defined. ●

EXAMPLE 7   **Circumference of a Circle**

In the formula for the circumference C of a circle of radius r,

$$C = 2\pi r$$

the domain of the variable r, representing the radius of the circle, is the set of positive real numbers, $\{r \mid r > 0\}$. The domain of the variable C, representing the circumference of the circle, is also the set of positive real numbers, $\{C \mid C > 0\}$. ●

In describing the domain of a variable, we may use either set notation or words, whichever is more convenient.

Now Work PROBLEM 59

5 Use the Laws of Exponents

Integer exponents provide a shorthand notation for representing repeated multiplications of a real number. For example,

$$3^4 = 3 \cdot 3 \cdot 3 \cdot 3 = 81$$

Additionally, many formulas have exponents. For example,

- The formula for the horsepower rating H of an engine is

$$H = \frac{D^2 N}{2.5}$$

where D is the diameter of a cylinder and N is the number of cylinders.

- A formula for the resistance R of blood flowing in a blood vessel is

$$R = C \frac{L}{r^4}$$

where L is the length of the blood vessel, r is the radius, and C is a positive constant.

In Problems 97–108, find the value of each expression if x = 2 and y = −1.

97. $2xy^{-1}$

98. $-3x^{-1}y$

99. $x^2 + y^2$

100. x^2y^2

101. $(xy)^2$

102. $(x + y)^2$

103. $\sqrt{x^2}$

104. $(\sqrt{x})^2$

105. $\sqrt{x^2 + y^2}$

106. $\sqrt{x^2} + \sqrt{y^2}$

107. x^y

108. y^x

109. Find the value of the expression $2x^3 - 3x^2 + 5x - 4$ if $x = 2$. What is the value if $x = 1$?

110. Find the value of the expression $4x^3 + 3x^2 - x + 2$ if $x = 1$. What is the value if $x = 2$?

111. What is the value of $\dfrac{(666)^4}{(222)^4}$?

112. What is the value of $(0.1)^3(20)^3$?

In Problems 113–120, use a calculator to evaluate each expression. Round your answer to three decimal places.

113. $(8.2)^6$

114. $(3.7)^5$

115. $(6.1)^{-3}$

116. $(2.2)^{-5}$

117. $(-2.8)^6$

118. $-(2.8)^6$

119. $(-8.11)^{-4}$

120. $-(8.11)^{-4}$

In Problems 121–128, write each number in scientific notation.

121. 454.2

122. 32.14

123. 0.013

124. 0.00421

125. 32,155

126. 21,210

127. 0.000423

128. 0.0514

In Problems 129–136, write each number as a decimal.

129. 6.15×10^4

130. 9.7×10^3

131. 1.214×10^{-3}

132. 9.88×10^{-4}

133. 1.1×10^8

134. 4.112×10^2

135. 8.1×10^{-2}

136. 6.453×10^{-1}

Applications and Extensions

In Problems 137–146, express each statement as an equation involving the indicated variables.

137. Area of a Rectangle The area A of a rectangle is the product of its length l and its width w.

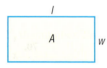

138. Perimeter of a Rectangle The perimeter P of a rectangle is twice the sum of its length l and its width w.

139. Circumference of a Circle The circumference C of a circle is the product of π and its diameter d.

140. Area of a Triangle The area A of a triangle is one-half the product of its base b and its height h.

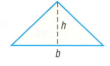

141. Area of an Equilateral Triangle The area A of an equilateral triangle is $\dfrac{\sqrt{3}}{4}$ times the square of the length x of one side.

142. Perimeter of an Equilateral Triangle The perimeter P of an equilateral triangle is 3 times the length x of one side.

143. Volume of a Sphere The volume V of a sphere is $\dfrac{4}{3}$ times π times the cube of the radius r.

144. Surface Area of a Sphere The surface area S of a sphere is 4 times π times the square of the radius r.

145. **Volume of a Cube** The volume V of a cube is the cube of the length x of a side.

146. **Surface Area of a Cube** The surface area S of a cube is 6 times the square of the length x of a side.

147. **Manufacturing Cost** The weekly production cost C of manufacturing x watches is given by the formula $C = 4000 + 2x$, where the variable C is in dollars.
    (a) What is the cost of producing 1000 watches?
    (b) What is the cost of producing 2000 watches?

148. **Balancing a Checkbook** At the beginning of the month, Mike had a balance of $210 in his checking account. During the next month, he deposited $80, wrote a check for $120, made another deposit of $25, and wrote two checks: one for $60 and the other for $32. He was also assessed a monthly service charge of $5. What was his balance at the end of the month?

In Problems 149 and 150, write an inequality using an absolute value to describe each statement.

149. x is at least 6 units from 4.

150. x is more than 5 units from 2.

151. **U.S. Voltage** In the United States, normal household voltage is 110 volts. It is acceptable for the actual voltage x to differ from normal by at most 5 volts. A formula that describes this is

$$|x - 110| \le 5$$

    (a) Show that a voltage of 108 volts is acceptable.
    (b) Show that a voltage of 104 volts is not acceptable.

152. **Foreign Voltage** In other countries, normal household voltage is 220 volts. It is acceptable for the actual voltage x to differ from normal by at most 8 volts. A formula that describes this is

$$|x - 220| \le 8$$

    (a) Show that a voltage of 214 volts is acceptable.
    (b) Show that a voltage of 209 volts is not acceptable.

153. **Making Precision Ball Bearings** The FireBall Company manufactures ball bearings for precision equipment. One

of its products is a ball bearing with a stated radius of 3 centimeters (cm). Only ball bearings with a radius within 0.01 cm of this stated radius are acceptable. If x is the radius of a ball bearing, a formula describing this situation is

$$|x - 3| \le 0.01$$

    (a) Is a ball bearing of radius $x = 2.999$ acceptable?
    (b) Is a ball bearing of radius $x = 2.89$ acceptable?

154. **Body Temperature** Normal human body temperature is 98.6°F. A temperature x that differs from normal by at least 1.5°F is considered unhealthy. A formula that describes this is

$$|x - 98.6| \ge 1.5$$

    (a) Show that a temperature of 97°F is unhealthy.
    (b) Show that a temperature of 100°F is not unhealthy.

155. **Distance from Earth to Its Moon** The distance from Earth to the Moon is about 4×10^8 meters.* Express this distance as a whole number.

156. **Height of Mt. Everest** The height of Mt. Everest is 8848 meters.* Express this height in scientific notation.

157. **Wavelength of Visible Light** The wavelength of visible light is about 5×10^{-7} meter.* Express this wavelength as a decimal.

158. **Diameter of an Atom** The diameter of an atom is about 1×10^{-10} meter.* Express this diameter as a decimal.

159. **Diameter of Copper Wire** The smallest commercial copper wire is about 0.0005 inch in diameter.† Express this diameter using scientific notation.

160. **Smallest Motor** The smallest motor ever made is less than 0.05 centimeter wide.† Express this width using scientific notation.

161. **Astronomy** One light-year is defined by astronomers to be the distance that a beam of light will travel in 1 year (365 days). If the speed of light is 186,000 miles per second, how many miles are in a light-year? Express your answer in scientific notation.

162. **Astronomy** How long does it take a beam of light to reach Earth from the Sun when the Sun is 93,000,000 miles from Earth? Express your answer in seconds, using scientific notation.

163. Does $\frac{1}{3}$ equal 0.333? If not, which is larger? By how much?

164. Does $\frac{2}{3}$ equal 0.666? If not, which is larger? By how much?

Explaining Concepts: Discussion and Writing

165. Is there a positive real number "closest" to 0?

166. **Number game** I'm thinking of a number! It lies between 1 and 10; its square is rational and lies between 1 and 10. The number is larger than π. Correct to two decimal places (that is, truncated to two decimal places), name the number. Now think of your own number, describe it, and challenge a fellow student to name it.

167. Write a brief paragraph that illustrates the similarities and differences between "less than" $(<)$ and "less than or equal to" $(\le)$.

168. Give a reason why the statement $5 < 8$ is true.

---

* *Powers of Ten*, Philip and Phylis Morrison.
† *2011 Information Please Almanac.*

R.3 Geometry Essentials

OBJECTIVES 1 Use the Pythagorean Theorem and Its Converse (p. 30)
2 Know Geometry Formulas (p. 31)
3 Understand Congruent Triangles and Similar Triangles (p. 32)

1 Use the Pythagorean Theorem and Its Converse

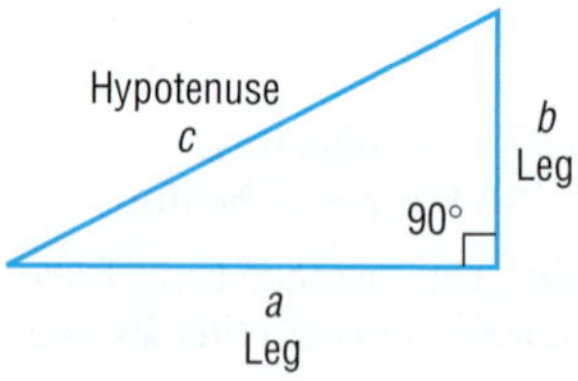

Figure 16 A right triangle

The *Pythagorean Theorem* is a statement about *right triangles*. A **right triangle** is one that contains a **right angle**—that is, an angle of 90°. The side of the triangle opposite the 90° angle is called the **hypotenuse**; the remaining two sides are called **legs**. In Figure 16 we have used c to represent the length of the hypotenuse and a and b to represent the lengths of the legs. Notice the use of the symbol ⌐ to show the 90° angle. We now state the Pythagorean Theorem.

PYTHAGOREAN THEOREM

> In a right triangle, the square of the length of the hypotenuse is equal to the sum of the squares of the lengths of the legs. That is, in the right triangle shown in Figure 16,
>
> $$c^2 = a^2 + b^2 \qquad (1)$$

A proof of the Pythagorean Theorem is given at the end of this section.

EXAMPLE 1 **Finding the Hypotenuse of a Right Triangle**

In a right triangle, one leg has length 4 and the other has length 3. What is the length of the hypotenuse?

Solution Since the triangle is a right triangle, we use the Pythagorean Theorem with $a = 4$ and $b = 3$ to find the length c of the hypotenuse. From equation (1),

$$c^2 = a^2 + b^2$$
$$c^2 = 4^2 + 3^2 = 16 + 9 = 25$$
$$c = \sqrt{25} = 5$$

Now Work PROBLEM 15

The converse of the Pythagorean Theorem is also true.

CONVERSE OF THE PYTHAGOREAN THEOREM

> In a triangle, if the square of the length of one side equals the sum of the squares of the lengths of the other two sides, the triangle is a right triangle. The 90° angle is opposite the longest side.

A proof is given at the end of this section.

EXAMPLE 2 **Verifying That a Triangle Is a Right Triangle**

Show that a triangle whose sides are of lengths 5, 12, and 13 is a right triangle. Identify the hypotenuse.

Solution Square the lengths of the sides.

$$5^2 = 25 \qquad 12^2 = 144 \qquad 13^2 = 169$$

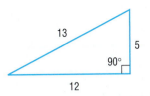

Figure 17

Notice that the sum of the first two squares (25 and 144) equals the third square (169). That is, because $5^2 + 12^2 = 13^2$, the triangle is a right triangle. The longest side, 13, is the hypotenuse. See Figure 17.

Now Work PROBLEM 23

EXAMPLE 3

Applying the Pythagorean Theorem

The tallest building in the world is Burj Khalifa in Dubai, United Arab Emirates, at 2717 feet and 163 floors. The observation deck is 1483 feet above ground level. How far can a person standing on the observation deck see (with the aid of a telescope)? Use 3960 miles for the radius of Earth.

Source: Council on Tall Buildings and Urban Habitat

Solution   From the center of Earth, draw two radii: one through Burj Khalifa and the other to the farthest point a person can see from the observation deck. See Figure 18. Apply the Pythagorean Theorem to the right triangle.

Since 1 mile = 5280 feet,   1483 feet = $\dfrac{1483}{5280}$ mile. Then

$$d^2 + (3960)^2 = \left(3960 + \frac{1483}{5280}\right)^2$$

$$d^2 = \left(3960 + \frac{1483}{5280}\right)^2 - (3960)^2 \approx 2224.58$$

$$d \approx 47.17$$

A person can see more than 47 miles from the observation deck.

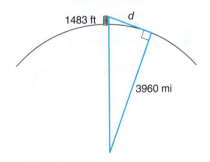

Figure 18

Now Work PROBLEM 55

2 Know Geometry Formulas

Certain formulas from geometry are useful in solving algebra problems.
For a rectangle of length l and width w,

$$\text{Area} = lw \qquad \text{Perimeter} = 2l + 2w$$

For a triangle with base b and altitude h,

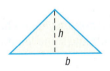

$$\text{Area} = \frac{1}{2}bh$$

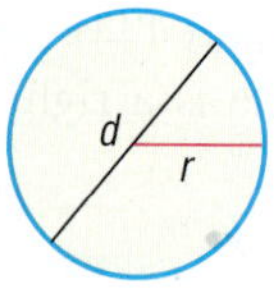

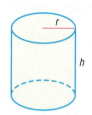

For a circle of radius r (diameter $d = 2r$),

$$\text{Area} = \pi r^2 \qquad \text{Circumference} = 2\pi r = \pi d$$

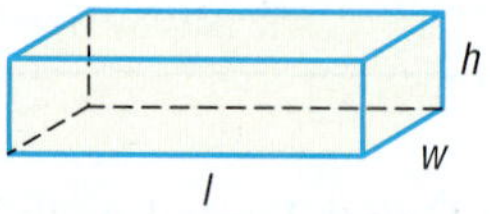

For a closed rectangular box of length l, width w, and height h,

$$\text{Volume} = lwh \qquad \text{Surface area} = 2lh + 2wh + 2lw$$

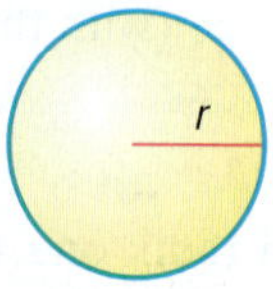

For a sphere of radius r,

$$\text{Volume} = \frac{4}{3}\pi r^3 \qquad \text{Surface area} = 4\pi r^2$$

For a closed right circular cylinder of height h and radius r,

$$\text{Volume} = \pi r^2 h \qquad \text{Surface area} = 2\pi r^2 + 2\pi rh$$

---Now Work PROBLEM 31

EXAMPLE 4 | **Using Geometry Formulas**

A Christmas tree ornament is in the shape of a semicircle on top of a triangle. How many square centimeters (cm^2) of copper is required to make the ornament if the height of the triangle is 6 cm and the base is 4 cm?

Solution See Figure 19. The amount of copper required equals the shaded area. This area is the sum of the areas of the triangle and the semicircle. The triangle has height $h = 6$ and base $b = 4$. The semicircle has diameter $d = 4$, so its radius is $r = 2$.

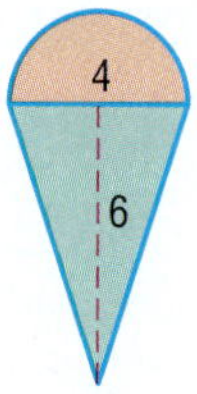

$$\text{Area} = \text{Area of triangle} + \text{Area of semicircle}$$

$$= \frac{1}{2}bh + \frac{1}{2}\pi r^2 = \frac{1}{2}(4)(6) + \frac{1}{2}\pi \cdot 2^2 \quad b = 4; h = 6; r = 2$$

$$= 12 + 2\pi \approx 18.28 \ \text{cm}^2$$

Figure 19

About 18.28 cm^2 of copper is required. ●

---Now Work PROBLEM 49

3 Understand Congruent Triangles and Similar Triangles

Throughout the text we will make reference to triangles. We begin with a discussion of *congruent* triangles. According to dictionary.com, the word **congruent** means "coinciding exactly when superimposed." For example, two angles are congruent if they have the same measure, and two line segments are congruent if they have the same length.

In Words

Two triangles are congruent if they have the same size and shape.

DEFINITION

Two triangles are **congruent** if each pair of corresponding angles have the same measure and each pair of corresponding sides are the same length.

In Figure 20, corresponding angles are equal and the corresponding sides are equal in length: $a = d$, $b = e$, and $c = f$. As a result, these triangles are congruent.

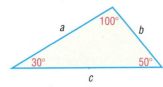

Figure 20   Congruent triangles

Actually, it is not necessary to verify that all three angles and all three sides are the same measure to determine whether two triangles are congruent.

Determining Congruent Triangles

1. **Angle–Side–Angle Case**   Two triangles are congruent if two of the angles are equal and the lengths of the corresponding sides between the two angles are equal.

    For example, in Figure 21(a), the two triangles are congruent because two angles and the included side are equal.

2. **Side–Side–Side Case**   Two triangles are congruent if the lengths of the corresponding sides of the triangles are equal.

    For example, in Figure 21(b), the two triangles are congruent because the three corresponding sides are all equal.

3. **Side–Angle–Side Case**   Two triangles are congruent if the lengths of two corresponding sides are equal and the angles between the two sides are the same.

    For example, in Figure 21(c), the two triangles are congruent because two sides and the included angle are equal.

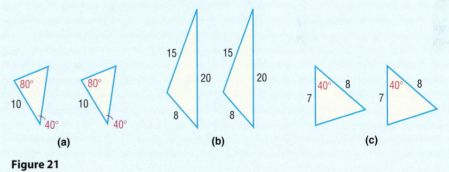

Figure 21

We contrast congruent triangles with *similar* triangles.

DEFINITION   Two triangles are **similar** if the corresponding angles are equal and the lengths of the corresponding sides are proportional.

In Words

Two triangles are similar if they have the same shape, but (possibly) different sizes.

For example, the triangles in Figure 22 (on the next page) are similar because the corresponding angles are equal. In addition, the lengths of the corresponding sides are proportional because each side in the triangle on the right is twice as long as each corresponding side in the triangle on the left. That is, the ratio of the corresponding sides is a constant: $\dfrac{d}{a} = \dfrac{e}{b} = \dfrac{f}{c} = 2$.

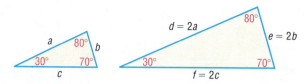

Figure 22 Similar triangles

It is not necessary to verify that all three angles are equal and all three sides are proportional to determine whether two triangles are congruent.

Determining Similar Triangles

1. **Angle–Angle Case** Two triangles are similar if two of the corresponding angles are equal.

   For example, in Figure 23(a), the two triangles are similar because two angles are equal.

2. **Side–Side–Side Case** Two triangles are similar if the lengths of all three sides of each triangle are proportional.

   For example, in Figure 23(b), the two triangles are similar because

   $$\frac{10}{30} = \frac{5}{15} = \frac{6}{18} = \frac{1}{3}$$

3. **Side–Angle–Side Case** Two triangles are similar if two corresponding sides are proportional and the angles between the two sides are equal.

   For example, in Figure 23(c), the two triangles are similar because $\frac{4}{6} = \frac{12}{18} = \frac{2}{3}$ and the angles between the sides are equal.

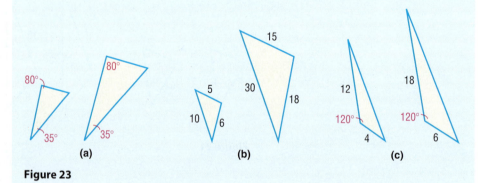

Figure 23

EXAMPLE 5

Using Similar Triangles

Given that the triangles in Figure 24 are similar, find the missing length x and the angles A, B, and C.

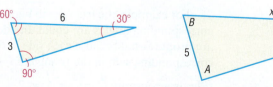

Figure 24

Solution   Because the triangles are similar, corresponding angles are equal. So $A = 90°$, $B = 60°$, and $C = 30°$. Also, the corresponding sides are proportional. That is, $\frac{3}{5} = \frac{6}{x}$. We solve this equation for x.

$$\frac{3}{5} = \frac{6}{x}$$

$$5x \cdot \frac{3}{5} = 5x \cdot \frac{6}{x} \quad \text{Multiply both sides by 5x.}$$

$$3x = 30 \quad \text{Simplify.}$$

$$x = 10 \quad \text{Divide both sides by 3.}$$

The missing length is 10 units.

Now Work PROBLEM **43**

Proof of the Pythagorean Theorem Begin with a square, each side of length $a + b$. In this square, form four right triangles, each having legs equal in length to a and b. See Figure 25. All these triangles are congruent (two sides and their included angle are equal). As a result, the hypotenuse of each is the same, say c, and the pink shading in Figure 25 indicates a square with an area equal to c^2.

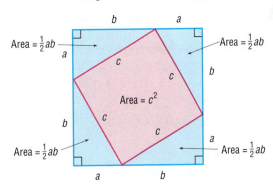

Figure 25

The area of the original square with sides $a + b$ equals the sum of the areas of the four triangles (each of area $\frac{1}{2}ab$) plus the area of the square with side c. That is,

$$(a + b)^2 = \frac{1}{2}ab + \frac{1}{2}ab + \frac{1}{2}ab + \frac{1}{2}ab + c^2$$

$$a^2 + 2ab + b^2 = 2ab + c^2$$

$$a^2 + b^2 = c^2$$

The proof is complete.   ■

Proof of the Converse of the Pythagorean Theorem Begin with two triangles: one a right triangle with legs a and b and the other a triangle with sides a, b, and c for which $c^2 = a^2 + b^2$. See Figure 26. By the Pythagorean Theorem, the length x of the third side of the first triangle is

$$x^2 = a^2 + b^2$$

But $c^2 = a^2 + b^2$. Then,

$$x^2 = c^2$$

$$x = c$$

The two triangles have the same sides and are therefore congruent. This means corresponding angles are equal, so the angle opposite side c of the second triangle equals 90°.

The proof is complete.   ■

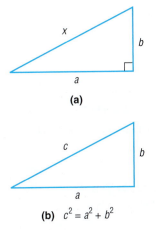

(a)

(b) $c^2 = a^2 + b^2$

Figure 26

R.3 Assess Your Understanding

Concepts and Vocabulary

1. A(n) _____ triangle is one that contains an angle of 90 degrees. The longest side is called the _____.

2. For a triangle with base b and altitude h, a formula for the area A is _____.

3. The formula for the circumference C of a circle of radius r is _____.

4. Two triangles are _____ if corresponding angles are equal and the lengths of the corresponding sides are proportional.

5. Which of the following is not a case for determining congruent triangles?
   (a) Angle–Side–Angle   (b) Side–Angle–Side
   (c) Angle–Angle–Angle   (d) Side-Side-Side

6. Choose the formula for the volume of a sphere of radius r.
   (a) $\frac{4}{3}\pi r^2$ (b) $\frac{4}{3}\pi r^3$ (c) $4\pi r^3$ (d) $4\pi r^2$

7. *True or False* In a right triangle, the square of the length of the longest side equals the sum of the squares of the lengths of the other two sides.

8. *True or False* The triangle with sides of lengths 6, 8, and 10 is a right triangle.

9. *True or False* The surface area of a sphere of radius r is $\frac{4}{3}\pi r^2$.

10. *True or False* The triangles shown are congruent.

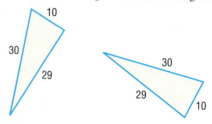

11. *True or False* The triangles shown are similar.

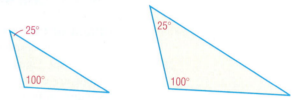

12. *True or False* The triangles shown are similar.

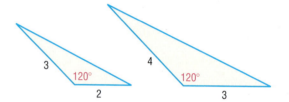

Skill Building

In Problems 13–18, the lengths of the legs of a right triangle are given. Find the hypotenuse.

13. $a = 5$, $b = 12$ **14.** $a = 6$, $b = 8$ **15.** $a = 10$, $b = 24$

16. $a = 4$, $b = 3$ **17.** $a = 7$, $b = 24$ **18.** $a = 14$, $b = 48$

In Problems 19–26, the lengths of the sides of a triangle are given. Determine which are right triangles. For those that are, identify the hypotenuse.

19. $3, 4, 5$ **20.** $6, 8, 10$ **21.** $4, 5, 6$ **22.** $2, 2, 3$

23. $7, 24, 25$ **24.** $10, 24, 26$ **25.** $6, 4, 3$ **26.** $5, 4, 7$

27. Find the area A of a rectangle with length 4 inches and width 2 inches.

28. Find the area A of a rectangle with length 9 centimeters and width 4 centimeters.

29. Find the area A of a triangle with height 4 inches and base 2 inches.

30. Find the area A of a triangle with height 9 centimeters and base 4 centimeters.

31. Find the area A and circumference C of a circle of radius 5 meters.

32. Find the area A and circumference C of a circle of radius 2 feet.

33. Find the volume V and surface area S of a closed rectangular box with length 8 feet, width 4 feet, and height 7 feet.

34. Find the volume V and surface area S of a closed rectangular box with length 9 inches, width 4 inches, and height 8 inches.

35. Find the volume V and surface area S of a sphere of radius 4 centimeters.

36. Find the volume V and surface area S of a sphere of radius 3 feet.

37. Find the volume V and surface area S of a closed right circular cylinder with radius 9 inches and height 8 inches.

38. Find the volume V and surface area S of a closed right circular cylinder with radius 8 inches and height 9 inches.

In Problems 39–42, find the area of the shaded region.

39.
2

2

40.
2

2

41.
2

2

42.
2

2

In Problems 43–46, the triangles in each pair are similar. Find the missing length x and the missing angles A, B, and C.

43.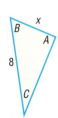
60° 2 90° 4 30°

B x A 8 C

44.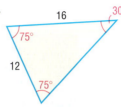
16 30° 75° 12 75°

x B A 6 C

45.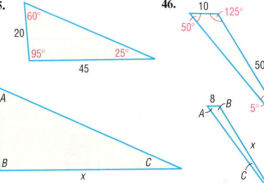
60° 20 95° 25° 45

A 30 B x C

46.
10 125° 50° 50

8 B A 5° x C

Applications and Extensions

47. How many feet has a wheel with a diameter of 16 inches traveled after four revolutions?

48. How many revolutions will a circular disk with a diameter of 4 feet have completed after it has rolled 20 feet?

49. In the figure shown, *ABCD* is a square, with each side of length 6 feet. The width of the border (shaded portion) between the outer square *EFGH* and *ABCD* is 2 feet. Find the area of the border.

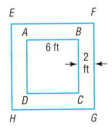

E F
A B
6 ft
2 ft
D C
H G

50. Refer to the figure. Square *ABCD* has an area of 100 square feet; square *BEFG* has an area of 16 square feet. What is the area of the triangle *CGF*?

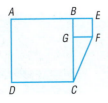

A B E
G F
D C

51. Architecture A **Norman window** consists of a rectangle surmounted by a semicircle. Find the area of the Norman window shown in the illustration. How much wood frame is needed to enclose the window?

6'

4'

52. Construction A circular swimming pool that is 20 feet in diameter is enclosed by a wooden deck that is 3 feet wide. What is the area of the deck? How much fence is required to enclose the deck?

3'

20'

COMMENT Vertical subtraction will be used when we divide polynomials. ∎

Vertical Subtraction: We line up like terms, change the sign of each coefficient of the second polynomial, and add.

$$
\begin{array}{ll}
\quad\;\, x^4 \quad\; x^3 \quad\; x^2 \quad\;\; x^1 \quad x^0 & \qquad x^4 \quad\; x^3 \quad\; x^2 \quad\;\; x^1 \quad x^0\\
\quad\; 3x^4 - 4x^3 + 6x^2 \qquad\quad -1 = & \quad 3x^4 - 4x^3 + \;\; 6x^2 \qquad\quad -1\\
- \;[\;2x^4 \qquad\quad -8x^2 - 6x + 5\;] = + & -2x^4 \qquad\quad +\;\; 8x^2 + 6x - 5\\
\hline
& \quad\;\; x^4 - 4x^3 + 14x^2 + 6x - 6
\end{array}
$$
●

Which of these methods to use for adding and subtracting polynomials is up to you. To save space, we shall most often use the horizontal format.

✏ **Now Work** PROBLEM 31

4 Multiply Polynomials

Two monomials may be multiplied using the Laws of Exponents and the Commutative and Associative Properties. For example,

$$(2x^3)\cdot(5x^4) = (2\cdot 5)\cdot(x^3\cdot x^4) = 10x^{3+4} = 10x^7$$

Products of polynomials are found by repeated use of the Distributive Property and the Laws of Exponents. Again, you have a choice of horizontal or vertical format.

EXAMPLE 6 **Multiplying Polynomials**

Find the product: $(2x + 5)(x^2 - x + 2)$

Solution *Horizontal Multiplication:*

$$(2x + 5)(x^2 - x + 2) = 2x(x^2 - x + 2) + 5(x^2 - x + 2)$$
↑
Distributive Property

$$= (2x\cdot x^2 - 2x\cdot x + 2x\cdot 2) + (5\cdot x^2 - 5\cdot x + 5\cdot 2)$$
↑
Distributive Property

$$= (2x^3 - 2x^2 + 4x) + (5x^2 - 5x + 10)$$
↑
Law of Exponents

$$= 2x^3 + 3x^2 - x + 10$$
↑
Combine like terms.

Vertical Multiplication: The idea here is very much like multiplying a two-digit number by a three-digit number.

$$
\begin{array}{r}
x^2 - \;\; x + \;\; 2\\
2x + \;\; 5\\
\hline
2x^3 - 2x^2 + 4x \qquad\qquad\\
(+)\quad\quad 5x^2 - 5x + 10\\
\hline
2x^3 + 3x^2 - \;\; x + 10
\end{array}
$$

This line is $2x(x^2 - x + 2)$.
This line is $5(x^2 - x + 2)$.
Sum of the above two lines

●

✏ **Now Work** PROBLEM 47

5 Know Formulas for Special Products

Certain products, which we call **special products**, occur frequently in algebra. We can calculate them easily using the **FOIL** (*First, Outer, Inner, Last*) method of multiplying two binomials.

$$(ax + b)(cx + d) = ax(cx + d) + b(cx + d)$$

$$= \underbrace{ax \cdot cx}_{\text{First}} + \underbrace{ax \cdot d}_{\text{Outer}} + \underbrace{b \cdot cx}_{\text{Inner}} + \underbrace{b \cdot d}_{\text{Last}}$$

$$= acx^2 + adx + bcx + bd$$

$$= acx^2 + (ad + bc)x + bd$$

EXAMPLE 7

Using FOIL

(a) $(x - 3)(x + 3) = x^2 + 3x - 3x - 9 = x^2 - 9$
  $ $ F O I L

(b) $(x + 2)^2 = (x + 2)(x + 2) = x^2 + 2x + 2x + 4 = x^2 + 4x + 4$

(c) $(x - 3)^2 = (x - 3)(x - 3) = x^2 - 3x - 3x + 9 = x^2 - 6x + 9$

(d) $(x + 3)(x + 1) = x^2 + x + 3x + 3 = x^2 + 4x + 3$

(e) $(2x + 1)(3x + 4) = 6x^2 + 8x + 3x + 4 = 6x^2 + 11x + 4$ ●

Now Work PROBLEMS 49 AND 57

Some products have been given special names because of their form. The following special products are based on Examples 7(a), (b), and (c).

Difference of Two Squares

$$(x - a)(x + a) = x^2 - a^2 \tag{2}$$

Squares of Binomials, or Perfect Squares

$$(x + a)^2 = x^2 + 2ax + a^2 \tag{3a}$$

$$(x - a)^2 = x^2 - 2ax + a^2 \tag{3b}$$

EXAMPLE 8

Using Special Product Formulas

(a) $(x - 5)(x + 5) = x^2 - 5^2 = x^2 - 25$ *Difference of two squares*

(b) $(x + 7)^2 = x^2 + 2 \cdot 7 \cdot x + 7^2 = x^2 + 14x + 49$ *Square of a binomial*

(c) $(2x + 1)^2 = (2x)^2 + 2 \cdot 1 \cdot 2x + 1^2 = 4x^2 + 4x + 1$ *Notice that we used 2x in place of x in formula (3a).*

(d) $(3x - 4)^2 = (3x)^2 - 2 \cdot 4 \cdot 3x + 4^2 = 9x^2 - 24x + 16$ *Replace x by 3x in formula (3b).* ●

Now Work PROBLEMS 67, 69, AND 71

Let's look at some more examples that lead to general formulas.

EXAMPLE 9   **Cubing a Binomial**

(a) $(x + 2)^3 = (x + 2)(x + 2)^2 = (x + 2)(x^2 + 4x + 4)$ Formula (3a)

$$= (x^3 + 4x^2 + 4x) + (2x^2 + 8x + 8)$$

$$= x^3 + 6x^2 + 12x + 8$$

(b) $(x - 1)^3 = (x - 1)(x - 1)^2 = (x - 1)(x^2 - 2x + 1)$ Formula (3b)

$$= (x^3 - 2x^2 + x) - (x^2 - 2x + 1)$$

$$= x^3 - 3x^2 + 3x - 1$$

Cubes of Binomials, or Perfect Cubes

$$(x + a)^3 = x^3 + 3ax^2 + 3a^2x + a^3 \tag{4a}$$

$$(x - a)^3 = x^3 - 3ax^2 + 3a^2x - a^3 \tag{4b}$$

Now Work PROBLEM 87

EXAMPLE 10   **Forming the Difference of Two Cubes**

$$(x - 1)(x^2 + x + 1) = x(x^2 + x + 1) - 1(x^2 + x + 1)$$

$$= x^3 + x^2 + x - x^2 - x - 1$$

$$= x^3 - 1$$

EXAMPLE 11   **Forming the Sum of Two Cubes**

$$(x + 2)(x^2 - 2x + 4) = x(x^2 - 2x + 4) + 2(x^2 - 2x + 4)$$

$$= x^3 - 2x^2 + 4x + 2x^2 - 4x + 8$$

$$= x^3 + 8$$

Examples 10 and 11 lead to two more special products.

Difference of Two Cubes

$$(x - a)(x^2 + ax + a^2) = x^3 - a^3 \tag{5}$$

Sum of Two Cubes

$$(x + a)(x^2 - ax + a^2) = x^3 + a^3 \tag{6}$$

---

6 Divide Polynomials Using Long Division

The procedure for dividing two polynomials is similar to the procedure for dividing two integers.

EXAMPLE 12 | **Dividing Two Integers**

Divide 842 by 15.

Solution

$$
\begin{array}{r}
56 \quad \leftarrow \text{Quotient} \\
\text{Divisor} \rightarrow 15\overline{)842} \quad \leftarrow \text{Dividend} \\
75 \quad \leftarrow 5\cdot 15 \text{ (subtract)} \\
\hline
92 \\
90 \quad \leftarrow 6\cdot 15 \text{ (subtract)} \\
\hline
2 \quad \leftarrow \text{Remainder}
\end{array}
$$

So, $\dfrac{842}{15} = 56 + \dfrac{2}{15}$.

In the long-division process detailed in Example 12, the number 15 is called the **divisor**, the number 842 is called the **dividend**, the number 56 is called the **quotient**, and the number 2 is called the **remainder**.

To check the answer obtained in a division problem, multiply the quotient by the divisor and add the remainder. The answer should be the dividend.

$$
(\text{Quotient})(\text{Divisor}) + \text{Remainder} = \text{Dividend}
$$

For example, we can check the results obtained in Example 12 as follows:

$$
(56)(15) + 2 = 840 + 2 = 842
$$

To divide two polynomials, we first must write each polynomial in standard form. The process then follows a pattern similar to that of Example 12. The next example illustrates the procedure.

EXAMPLE 13 | **Dividing Two Polynomials**

Find the quotient and the remainder when

$$3x^3 + 4x^2 + x + 7 \quad \text{is divided by} \quad x^2 + 1$$

Solution

Each polynomial is in standard form. The dividend is $3x^3 + 4x^2 + x + 7$, and the divisor is $x^2 + 1$.

NOTE Remember, a polynomial is in standard form when its terms are written in descending powers of x. ■

STEP 1: Divide the leading term of the dividend, $3x^3$, by the leading term of the divisor, x^2. Enter the result, $3x$, over the term $3x^3$, as follows:

$$
\begin{array}{r}
3x \qquad\qquad\quad \\
x^2 + 1\overline{)3x^3 + 4x^2 + x + 7}
\end{array}
$$

STEP 2: Multiply $3x$ by $x^2 + 1$, and enter the result below the dividend.

$$
\begin{array}{r}
3x \qquad\qquad\quad \\
x^2 + 1\overline{)3x^3 + 4x^2 + \ x + 7} \\
3x^3 \qquad\quad + 3x \qquad\quad \leftarrow 3x\cdot(x^2 + 1) = 3x^3 + 3x
\end{array}
$$

Align the $3x$ term under the x to make the next step easier.

STEP 3: Subtract and bring down the remaining terms.

$$
\begin{array}{r}
3x \qquad\qquad\quad \\
x^2 + 1\overline{)3x^3 + 4x^2 + \ x + 7} \\
\underline{3x^3 \qquad\quad + 3x} \qquad\quad \leftarrow \text{Subtract (change the signs and add).} \\
4x^2 - 2x + 7 \quad \leftarrow \text{Bring down the } 4x^2 \text{ and the } 7.
\end{array}
$$

STEP 4: Repeat Steps 1–3 using $4x^2 - 2x + 7$ as the dividend.

$$
\begin{array}{r}
3x\ \ + 4 \\
x^2 + 1\overline{)3x^3 + 4x^2 +\ \ x + 7} \\
\underline{3x^3\qquad\quad + 3x} \\
4x^2 - 2x + 7 \\
\underline{4x^2\qquad\ \ + 4} \\
-2x + 3
\end{array}
$$

← Divide $4x^2$ by x^2 to get 4.
← Multiply $x^2 + 1$ by 4; subtract.

COMMENT If the degree of the divisor is greater than the degree of the dividend, then the process ends. ∎

Since x^2 does not divide $-2x$ evenly (that is, the result is not a monomial), the process ends. The quotient is $3x + 4$, and the remainder is $-2x + 3$.

✓ **Check:** (Quotient)(Divisor) + Remainder

$$
\begin{aligned}
&= (3x + 4)(x^2 + 1) + (-2x + 3) \\
&= 3x^3 + 3x + 4x^2 + 4 + (-2x + 3) \\
&= 3x^3 + 4x^2 + x + 7 = \text{Dividend}
\end{aligned}
$$

Then

$$
\frac{3x^3 + 4x^2 + x + 7}{x^2 + 1} = 3x + 4 + \frac{-2x + 3}{x^2 + 1}
$$

The next example combines the steps involved in long division.

EXAMPLE 14

Dividing Two Polynomials

Find the quotient and the remainder when

$$
x^4 - 3x^3 + 2x - 5 \quad \text{is divided by} \quad x^2 - x + 1
$$

Solution In setting up this division problem, it is necessary to leave a space for the missing x^2 term in the dividend.

$$
\begin{array}{r}
x^2 - 2x\ - 3 \\
x^2 - x + 1\overline{)x^4 - 3x^3\qquad\ \ + 2x - 5} \\
\underline{x^4 -\ \ x^3 +\ \ x^2} \\
-2x^3 -\ \ x^2 + 2x - 5 \\
\underline{-2x^3 + 2x^2 - 2x} \\
-3x^2 + 4x - 5 \\
\underline{-3x^2 + 3x - 3} \\
x - 2
\end{array}
$$

← Quotient
← Dividend
Subtract →
Subtract →
Subtract →
← Remainder

✓ **Check:** (Quotient)(Divisor) + Remainder

$$
\begin{aligned}
&= (x^2 - 2x - 3)(x^2 - x + 1) + x - 2 \\
&= x^4 - x^3 + x^2 - 2x^3 + 2x^2 - 2x - 3x^2 + 3x - 3 + x - 2 \\
&= x^4 - 3x^3 + 2x - 5 = \text{Dividend}
\end{aligned}
$$

As a result,

$$
\frac{x^4 - 3x^3 + 2x - 5}{x^2 - x + 1} = x^2 - 2x - 3 + \frac{x - 2}{x^2 - x + 1}
$$

The process of dividing two polynomials leads to the following result:

THEOREM   Let Q be a polynomial of positive degree, and let P be a polynomial whose degree is greater than or equal to the degree of Q. The remainder after dividing P by Q is either the zero polynomial or a polynomial whose degree is less than the degree of the divisor Q.

Now Work PROBLEM 95

7 Work with Polynomials in Two Variables

A **monomial in two variables** x and y has the form $ax^n y^m$, where a is a constant, x and y are variables, and n and m are nonnegative integers. The **degree** of a monomial is the sum of the powers of the variables.

For example,

$$2xy^3, \quad x^2 y^2, \quad \text{and} \quad x^3 y$$

are all monomials that have degree 4.

A **polynomial in two variables** x and y is the sum of one or more monomials in two variables. The **degree of a polynomial** in two variables is the highest degree of all the monomials with nonzero coefficients.

EXAMPLE 15   **Examples of Polynomials in Two Variables**

$$3x^2 + 2x^3 y + 5 \qquad \pi x^3 - y^2 \qquad x^4 + 4x^3 y - xy^3 + y^4$$

Two variables,   Two variables,   Two variables,
degree is 4.   degree is 3.   degree is 4.

Multiplying polynomials in two variables is handled in the same way as multiplying polynomials in one variable.

EXAMPLE 16   **Using a Special Product Formula**

To multiply $(2x - y)^2$, use the Squares of Binomials formula (3b) with $2x$ instead of x and with y instead of a.

$$(2x - y)^2 = (2x)^2 - 2 \cdot y \cdot 2x + y^2$$
$$= 4x^2 - 4xy + y^2$$

Now Work PROBLEM 81

R.4 Assess Your Understanding

Concepts and Vocabulary

1. The polynomial $3x^4 - 2x^3 + 13x^2 - 5$ is of degree __. The leading coefficient is __.

2. $(x^2 - 4)(x^2 + 4) =$ _____.

3. $(x - 2)(x^2 + 2x + 4) =$ _____.

4. The monomials that make up a polynomial are called which of the following?
(a) terms   (b) variables   (c) factors   (d) coefficients

5. Choose the degree of the monomial $3x^4 y^2$.
(a) 3   (b) 8   (c) 6   (d) 2

6. *True or False* $4x^{-2}$ is a monomial of degree -2.

7. *True or False* The degree of the product of two nonzero polynomials equals the sum of their degrees.

8. *True or False* $(x + a)(x^2 + ax + a) = x^3 + a^3$.

Skill Building

In Problems 9–18, tell whether the expression is a monomial. If it is, name the variable(s) and the coefficient, and give the degree of the monomial. If it is not a monomial, state why not.

9. $2x^3$

10. $-4x^2$

11. $\dfrac{8}{x}$

12. $-2x^{-3}$

13. $-2xy^2$

14. $5x^2y^3$

15. $\dfrac{8x}{y}$

16. $-\dfrac{2x^2}{y^3}$

17. $x^2 + y^2$

18. $3x^2 + 4$

In Problems 19–28, tell whether the expression is a polynomial. If it is, give its degree. If it is not, state why not.

19. $3x^2 - 5$

20. $1 - 4x$

21. 5

22. $-\pi$

23. $3x^2 - \dfrac{5}{x}$

24. $\dfrac{3}{x} + 2$

25. $2y^3 - \sqrt{2}$

26. $10z^2 + z$

27. $\dfrac{x^2 + 5}{x^3 - 1}$

28. $\dfrac{3x^3 + 2x - 1}{x^2 + x + 1}$

In Problems 29–48, add, subtract, or multiply, as indicated. Express your answer as a single polynomial in standard form.

29. $(x^2 + 4x + 5) + (3x - 3)$

30. $(x^3 + 3x^2 + 2) + (x^2 - 4x + 4)$

31. $(x^3 - 2x^2 + 5x + 10) - (2x^2 - 4x + 3)$

32. $(x^2 - 3x - 4) - (x^3 - 3x^2 + x + 5)$

33. $(6x^5 + x^3 + x) + (5x^4 - x^3 + 3x^2)$

34. $(10x^5 - 8x^2) + (3x^3 - 2x^2 + 6)$

35. $(x^2 - 3x + 1) + 2(3x^2 + x - 4)$

36. $-2(x^2 + x + 1) + (-5x^2 - x + 2)$

37. $6(x^3 + x^2 - 3) - 4(2x^3 - 3x^2)$

38. $8(4x^3 - 3x^2 - 1) - 6(4x^3 + 8x - 2)$

39. $(x^2 - x + 2) + (2x^2 - 3x + 5) - (x^2 + 1)$

40. $(x^2 + 1) - (4x^2 + 5) + (x^2 + x - 2)$

41. $9(y^2 - 3y + 4) - 6(1 - y^2)$

42. $8(1 - y^3) + 4(1 + y + y^2 + y^3)$

43. $x(x^2 + x - 4)$

44. $4x^2(x^3 - x + 2)$

45. $-2x^2(4x^3 + 5)$

46. $5x^3(3x - 4)$

47. $(x + 1)(x^2 + 2x - 4)$

48. $(2x - 3)(x^2 + x + 1)$

In Problems 49–66, multiply the polynomials using the FOIL method. Express your answer as a single polynomial in standard form.

49. $(x + 2)(x + 4)$

50. $(x + 3)(x + 5)$

51. $(2x + 5)(x + 2)$

52. $(3x + 1)(2x + 1)$

53. $(x - 4)(x + 2)$

54. $(x + 4)(x - 2)$

55. $(x - 3)(x - 2)$

56. $(x - 5)(x - 1)$

57. $(2x + 3)(x - 2)$

58. $(2x - 4)(3x + 1)$

59. $(-2x + 3)(x - 4)$

60. $(-3x - 1)(x + 1)$

61. $(-x - 2)(-2x - 4)$

62. $(-2x - 3)(3 - x)$

63. $(x - 2y)(x + y)$

64. $(2x + 3y)(x - y)$

65. $(-2x - 3y)(3x + 2y)$

66. $(x - 3y)(-2x + y)$

In Problems 67–90, multiply the polynomials using the special product formulas. Express your answer as a single polynomial in standard form.

67. $(x - 7)(x + 7)$

68. $(x - 1)(x + 1)$

69. $(2x + 3)(2x - 3)$

70. $(3x + 2)(3x - 2)$

71. $(x + 4)^2$

72. $(x + 5)^2$

73. $(x - 4)^2$

74. $(x - 5)^2$

75. $(3x + 4)(3x - 4)$

76. $(5x - 3)(5x + 3)$

77. $(2x - 3)^2$

78. $(3x - 4)^2$

79. $(x + y)(x - y)$ **80.** $(x + 3y)(x - 3y)$ **81.** $(3x + y)(3x - y)$ **82.** $(3x + 4y)(3x - 4y)$

83. $(x + y)^2$ **84.** $(x - y)^2$ **85.** $(x - 2y)^2$ **86.** $(2x + 3y)^2$

87. $(x - 2)^3$ **88.** $(x + 1)^3$ **89.** $(2x + 1)^3$ **90.** $(3x - 2)^3$

In Problems 91–106, find the quotient and the remainder. Check your work by verifying that

$$(Quotient)(Divisor) + Remainder = Dividend$$

91. $4x^3 - 3x^2 + x + 1$ divided by $x + 2$

92. $3x^3 - x^2 + x - 2$ divided by $x + 2$

93. $4x^3 - 3x^2 + x + 1$ divided by x^2

94. $3x^3 - x^2 + x - 2$ divided by x^2

95. $5x^4 - 3x^2 + x + 1$ divided by $x^2 + 2$

96. $5x^4 - x^2 + x - 2$ divided by $x^2 + 2$

97. $4x^5 - 3x^2 + x + 1$ divided by $2x^3 - 1$

98. $3x^5 - x^2 + x - 2$ divided by $3x^3 - 1$

99. $2x^4 - 3x^3 + x + 1$ divided by $2x^2 + x + 1$

100. $3x^4 - x^3 + x - 2$ divided by $3x^2 + x + 1$

101. $-4x^3 + x^2 - 4$ divided by $x - 1$

102. $-3x^4 - 2x - 1$ divided by $x - 1$

103. $1 - x^2 + x^4$ divided by $x^2 + x + 1$

104. $1 - x^2 + x^4$ divided by $x^2 - x + 1$

105. $x^3 - a^3$ divided by $x - a$

106. $x^5 - a^5$ divided by $x - a$

Explaining Concepts: Discussion and Writing

107. Explain why the degree of the product of two nonzero polynomials equals the sum of their degrees.

108. Explain why the degree of the sum of two polynomials of different degrees equals the larger of their degrees.

109. Give a careful statement about the degree of the sum of two polynomials of the same degree.

110. Do you prefer adding two polynomials using the horizontal method or the vertical method? Write a brief position paper defending your choice.

111. Do you prefer to memorize the rule for the square of a binomial $(x + a)^2$ or to use FOIL to obtain the product? Write a brief position paper defending your choice.

R.5 Factoring Polynomials

OBJECTIVES   **1** Factor the Difference of Two Squares and the Sum and Difference of Two Cubes (p. 50)

       2 Factor Perfect Squares (p. 51)

       3 Factor a Second-Degree Polynomial: $x^2 + Bx + C$ (p. 52)

       4 Factor by Grouping (p. 53)

       5 Factor a Second-Degree Polynomial: $Ax^2 + Bx + C$, $A \neq 1$ (p. 54)

       6 Complete the Square (p. 56)

Consider the following product:

$$(2x + 3)(x - 4) = 2x^2 - 5x - 12$$

The two polynomials on the left side are called **factors** of the polynomial on the right side. Expressing a given polynomial as a product of other polynomials—that is, finding the factors of a polynomial—is called **factoring**.

We shall restrict our discussion here to factoring polynomials in one variable into products of polynomials in one variable, where all coefficients are integers. We call this **factoring over the integers**.

Any polynomial can be written as the product of 1 times itself or as -1 times its additive inverse. If a polynomial cannot be written as the product of two other polynomials (excluding 1 and -1), then the polynomial is **prime**. When a polynomial has been written as a product consisting only of prime factors, it is **factored completely**. Examples of prime polynomials (over the integers) are

$$2, \quad 3, \quad 5, \quad x, \quad x+1, \quad x-1, \quad 3x+4, \quad x^2+4$$

The first factor to look for in a factoring problem is a common monomial factor present in each term of the polynomial. If one is present, use the Distributive Property to factor it out. Continue factoring out monomial factors until none are left.

COMMENT Over the real numbers, $3x+4$ factors into $3\left(x+\frac{4}{3}\right)$. It is the noninteger $\frac{4}{3}$ that causes $3x+4$ to be prime over the integers. In most instances, we will be factoring over the integers. ∎

EXAMPLE 1 | **Identifying Common Monomial Factors**

Polynomial	Common Monomial Factor	Remaining Factor	Factored Form
$2x+4$	2	$x+2$	$2x+4 = 2(x+2)$
$3x-6$	3	$x-2$	$3x-6 = 3(x-2)$
$2x^2-4x+8$	2	x^2-2x+4	$2x^2-4x+8 = 2(x^2-2x+4)$
$8x-12$	4	$2x-3$	$8x-12 = 4(2x-3)$
x^2+x	x	$x+1$	$x^2+x = x(x+1)$
x^3-3x^2	x^2	$x-3$	$x^3-3x^2 = x^2(x-3)$
$6x^2+9x$	$3x$	$2x+3$	$6x^2+9x = 3x(2x+3)$

Notice that once all common monomial factors have been removed from a polynomial, the remaining factor is either a prime polynomial of degree 1 or a polynomial of degree 2 or higher. (Do you see why?)

Now Work PROBLEM 9

---

1 Factor the Difference of Two Squares and the Sum and Difference of Two Cubes

When you factor a polynomial, first check for common monomial factors. Then see whether you can use one of the special formulas discussed in the previous section.

Difference of Two Squares	$x^2 - a^2 = (x-a)(x+a)$
Perfect Squares	$x^2 + 2ax + a^2 = (x+a)^2$
	$x^2 - 2ax + a^2 = (x-a)^2$
Sum of Two Cubes	$x^3 + a^3 = (x+a)(x^2-ax+a^2)$
Difference of Two Cubes	$x^3 - a^3 = (x-a)(x^2+ax+a^2)$

EXAMPLE 2 | **Factoring the Difference of Two Squares**

Factor completely: $x^2 - 4$

Solution Note that $x^2 - 4$ is the difference of two squares, x^2 and 2^2.

$$x^2 - 4 = (x-2)(x+2)$$

| EXAMPLE 3 | **Factoring the Difference of Two Cubes** |

Factor completely: $x^3 - 1$

Solution   Because $x^3 - 1$ is the difference of two cubes, x^3 and 1^3,

$$x^3 - 1 = (x - 1)(x^2 + x + 1)$$

| EXAMPLE 4 | **Factoring the Sum of Two Cubes** |

Factor completely: $x^3 + 8$

Solution   Because $x^3 + 8$ is the sum of two cubes, x^3 and 2^3,

$$x^3 + 8 = (x + 2)(x^2 - 2x + 4)$$

| EXAMPLE 5 | **Factoring the Difference of Two Squares** |

Factor completely: $x^4 - 16$

Solution   Because $x^4 - 16$ is the difference of two squares, $x^4 = (x^2)^2$ and $16 = 4^2$,

$$x^4 - 16 = (x^2 - 4)(x^2 + 4)$$

But $x^2 - 4$ is also the difference of two squares. Then,

$$x^4 - 16 = (x^2 - 4)(x^2 + 4) = (x - 2)(x + 2)(x^2 + 4)$$

Now Work PROBLEMS **19 AND 37**

2 Factor Perfect Squares

When the first term and third term of a trinomial are both positive and are perfect squares, such as x^2, $9x^2$, 1, and 4, check to see whether the trinomial is a perfect square.

| EXAMPLE 6 | **Factoring a Perfect Square** |

Factor completely: $x^2 + 6x + 9$

Solution   The first term, x^2, and the third term, $9 = 3^2$, are perfect squares. Because the middle term, $6x$, is twice the product of x and 3, we have a perfect square.

$$x^2 + 6x + 9 = (x + 3)^2$$

| EXAMPLE 7 | **Factoring a Perfect Square** |

Factor completely: $9x^2 - 6x + 1$

Solution   The first term, $9x^2 = (3x)^2$, and the third term, $1 = 1^2$, are perfect squares. Because the middle term, $-6x$, is -2 times the product of $3x$ and 1, we have a perfect square.

$$9x^2 - 6x + 1 = (3x - 1)^2$$

| EXAMPLE 8 | **Factoring a Perfect Square** |

Factor completely: $25x^2 + 30x + 9$

Solution   The first term, $25x^2 = (5x)^2$, and the third term, $9 = 3^2$, are perfect squares. Because the middle term, $30x$, is twice the product of $5x$ and 3, we have a perfect square.

$$25x^2 + 30x + 9 = (5x + 3)^2$$

Now Work PROBLEMS **29 AND 103**

If a trinomial is not a perfect square, it may be possible to factor it using the technique discussed next.

3 Factor a Second-Degree Polynomial: $x^2 + Bx + C$

The idea behind factoring a second-degree polynomial like $x^2 + Bx + C$ is to see whether it can be made equal to the product of two (possibly equal) first-degree polynomials.

For example, consider

$$(x + 3)(x + 4) = x^2 + 7x + 12$$

The factors of $x^2 + 7x + 12$ are $x + 3$ and $x + 4$. Notice the following:

$$x^2 + 7x + 12 = (x + 3)(x + 4)$$

12 is the product of 3 and 4

7 is the sum of 3 and 4

In general, if $x^2 + Bx + C = (x + a)(x + b) = x^2 + (a + b)x + ab$, then $ab = C$ and $a + b = B$.

To factor a second-degree polynomial $x^2 + Bx + C$, find integers whose product is C and whose sum is B. That is, if there are numbers a, b, where $ab = C$ and $a + b = B$, then

$$x^2 + Bx + C = (x + a)(x + b)$$

EXAMPLE 9 | **Factoring a Trinomial**

Factor completely: $x^2 + 7x + 10$

Solution   First determine all pairs of integers whose product is 10, and then compute their sums.

Integers whose product is 10	1, 10	$-1, -10$	2, 5	$-2, -5$
Sum	11	-11	7	-7

The integers 2 and 5 have a product of 10 and add up to 7, the coefficient of the middle term. As a result,

$$x^2 + 7x + 10 = (x + 2)(x + 5)$$

●

EXAMPLE 10 | **Factoring a Trinomial**

Factor completely: $x^2 - 6x + 8$

Solution   First determine all pairs of integers whose product is 8, and then compute each sum.

Integers whose product is 8	1, 8	$-1, -8$	2, 4	$-2, -4$
Sum	9	-9	6	-6

Since -6 is the coefficient of the middle term,

$$x^2 - 6x + 8 = (x - 2)(x - 4)$$

●

EXAMPLE 11

Factoring a Trinomial

Factor completely: $x^2 - x - 12$

Solution

First determine all pairs of integers whose product is -12, and then compute each sum.

Integers whose product is -12	$1, -12$	$-1, 12$	$2, -6$	$-2, 6$	$3, -4$	$-3, 4$
Sum	-11	11	-4	4	-1	1

Since -1 is the coefficient of the middle term,

$$x^2 - x - 12 = (x + 3)(x - 4)$$

●

EXAMPLE 12

Factoring a Trinomial

Factor completely: $x^2 + 4x - 12$

Solution

The integers -2 and 6 have a product of -12 and have the sum 4. So,

$$x^2 + 4x - 12 = (x - 2)(x + 6)$$

●

To avoid errors in factoring, always check your answer by multiplying it out to see whether the result equals the original expression.

When none of the possibilities works, the polynomial is prime.

EXAMPLE 13

Identifying a Prime Polynomial

Show that $x^2 + 9$ is prime.

Solution

First list the pairs of integers whose product is 9, and then compute their sums.

Integers whose product is 9	$1, 9$	$-1, -9$	$3, 3$	$-3, -3$
Sum	10	-10	6	-6

Since the coefficient of the middle term in $x^2 + 9 = x^2 + 0x + 9$ is 0 and none of the sums equals 0, we conclude that $x^2 + 9$ is prime.

●

Example 13 demonstrates a more general result:

THEOREM

Any polynomial of the form $x^2 + a^2$, a real, is prime.

── Now Work PROBLEMS 43 AND 87

4 Factor by Grouping

Sometimes a common factor does not occur in every term of the polynomial but does occur in each of several groups of terms that together make up the polynomial. When this happens, the common factor can be factored out of each group by means of the Distributive Property. This technique is called **factoring by grouping**.

EXAMPLE 14

Factoring by Grouping

Factor completely by grouping: $(x^2 + 2)x + (x^2 + 2) \cdot 3$

Solution

Notice the common factor $x^2 + 2$. Applying the Distributive Property yields

$$(x^2 + 2)x + (x^2 + 2) \cdot 3 = (x^2 + 2)(x + 3)$$

Since $x^2 + 2$ and $x + 3$ are prime, the factorization is complete.

●

The next example shows a factoring problem that occurs in calculus.

EXAMPLE 15 **Factoring by Grouping**

Factor completely by grouping: $3(x-1)^2(x+2)^4 + 4(x-1)^3(x+2)^3$

Solution Here, $(x-1)^2(x+2)^3$ is a common factor of both $3(x-1)^2(x+2)^4$ and $4(x-1)^3(x+2)^3$. As a result,

$$3(x-1)^2(x+2)^4 + 4(x-1)^3(x+2)^3 = (x-1)^2(x+2)^3[3(x+2) + 4(x-1)]$$
$$= (x-1)^2(x+2)^3[3x+6+4x-4]$$
$$= (x-1)^2(x+2)^3(7x+2)$$ ●

EXAMPLE 16 **Factoring by Grouping**

Factor completely by grouping: $x^3 - 4x^2 + 2x - 8$

Solution To see whether factoring by grouping will work, group the first two terms and the last two terms. Then look for a common factor in each group. In this example, factor x^2 from $x^3 - 4x^2$ and 2 from $2x - 8$. The remaining factor in each case is the same, $x - 4$. This means that factoring by grouping will work, as follows:

$$x^3 - 4x^2 + 2x - 8 = (x^3 - 4x^2) + (2x - 8)$$
$$= x^2(x-4) + 2(x-4)$$
$$= (x-4)(x^2+2)$$

Since $x^2 + 2$ and $x - 4$ are prime, the factorization is complete. ●

 Now Work PROBLEMS 55 AND 131

5 Factor a Second-Degree Polynomial: $Ax^2 + Bx + C, A \neq 1$

To factor a second-degree polynomial $Ax^2 + Bx + C$, when $A \neq 1$ and A, B, and C have no common factors, follow these steps:

> **Steps for Factoring $Ax^2 + Bx + C$, When $A \neq 1$ and A, B, and C Have No Common Factors**
>
> **Step 1:** Find the value of AC.
> **Step 2:** Find a pair of integers whose product is AC and that add up to B. That is, find a and b such that $ab = AC$ and $a + b = B$.
> **Step 3:** Write $Ax^2 + Bx + C = Ax^2 + ax + bx + C$.
> **Step 4:** Factor this last expression by grouping.

EXAMPLE 17 **Factoring a Trinomial**

Factor completely: $2x^2 + 5x + 3$

Solution Comparing $2x^2 + 5x + 3$ to $Ax^2 + Bx + C$, we find that $A = 2$, $B = 5$, and $C = 3$.

Step 1: The value of AC is $2 \cdot 3 = 6$.

Step 2: Determine the pairs of integers whose product is $AC = 6$ and compute their sums.

Integers whose product is 6	1, 6	−1, −6	2, 3	−2, −3
Sum	7	−7	5	−5

STEP 3: The integers whose product is 6 that add up to $B = 5$ are 2 and 3.

$$2x^2 + 5x + 3 = 2x^2 + 2x + 3x + 3$$

STEP 4: Factor by grouping.

$$2x^2 + 2x + 3x + 3 = (2x^2 + 2x) + (3x + 3)$$
$$= 2x(x + 1) + 3(x + 1)$$
$$= (x + 1)(2x + 3)$$

As a result,

$$2x^2 + 5x + 3 = (x + 1)(2x + 3)$$

EXAMPLE 18

Factoring a Trinomial

Factor completely: $2x^2 - x - 6$

Solution

Comparing $2x^2 - x - 6$ to $Ax^2 + Bx + C$, we find that $A = 2, B = -1,$ and $C = -6$.

STEP 1: The value of AC is $2 \cdot (-6) = -12$.

STEP 2: Determine the pairs of integers whose product is $AC = -12$ and compute their sums.

Integers whose product is -12	$1, -12$	$-1, 12$	$2, -6$	$-2, 6$	$3, -4$	$-3, 4$
Sum	-11	11	-4	4	-1	1

STEP 3: The integers whose product is -12 that add up to $B = -1$ are -4 and 3.

$$2x^2 - x - 6 = 2x^2 - 4x + 3x - 6$$

STEP 4: Factor by grouping.

$$2x^2 - 4x + 3x - 6 = (2x^2 - 4x) + (3x - 6)$$
$$= 2x(x - 2) + 3(x - 2)$$
$$= (x - 2)(2x + 3)$$

As a result,

$$2x^2 - x - 6 = (x - 2)(2x + 3)$$

 Now Work PROBLEM 61

SUMMARY

Type of Polynomial	Method	Example
Any polynomial	Look for common monomial factors. (Always do this first!)	$6x^2 + 9x = 3x(2x + 3)$
Binomials of degree 2 or higher	Check for a special product: Difference of two squares, $x^2 - a^2$ Difference of two cubes, $x^3 - a^3$ Sum of two cubes, $x^3 + a^3$	$x^2 - 16 = (x - 4)(x + 4)$ $x^3 - 64 = (x - 4)(x^2 + 4x + 16)$ $x^3 + 27 = (x + 3)(x^2 - 3x + 9)$
Trinomials of degree 2	Check for a perfect square, $(x \pm a)^2$ Factoring $x^2 + Bx + C$ (p. 52) Factoring $Ax^2 + Bx + C$ (p. 54)	$x^2 + 8x + 16 = (x + 4)^2$ $x^2 - 10x + 25 = (x - 5)^2$ $x^2 - x - 2 = (x - 2)(x + 1)$ $6x^2 + x - 1 = (2x + 1)(3x - 1)$
Four or more terms	Grouping	$2x^3 - 3x^2 + 4x - 6 = (2x - 3)(x^2 + 2)$

6 Complete the Square

The idea behind completing the square in one variable is to "adjust" an expression of the form $x^2 + bx$ to make it a perfect square. Perfect squares are trinomials of the form

$$x^2 + 2ax + a^2 = (x + a)^2 \text{ or } x^2 - 2ax + a^2 = (x - a)^2$$

For example, $x^2 + 6x + 9$ is a perfect square because $x^2 + 6x + 9 = (x + 3)^2$. And $p^2 - 12p + 36$ is a perfect square because $p^2 - 12p + 36 = (p - 6)^2$.

So how do we "adjust" $x^2 + bx$ to make it a perfect square? We do it by adding a number. For example, to make $x^2 + 6x$ a perfect square, add 9. But how do we know to add 9? If we divide the coefficient of the first-degree term, 6, by 2, and then square the result, we obtain 9. This approach works in general.

WARNING To use $\left(\dfrac{1}{2}b\right)^2$ to complete the square, the coefficient of the x^2 term must be 1. ∎

Completing the Square of $x^2 + bx$

Identify the coefficient of the first-degree term. Multiply this coefficient by $\dfrac{1}{2}$ and then square the result. That is, determine the value of b in $x^2 + bx$ and compute $\left(\dfrac{1}{2}b\right)^2$.

EXAMPLE 19

Completing the Square

Determine the number that must be added to each expression to complete the square. Then factor the expression.

Start	Add	Result	Factored Form
$y^2 + 8y$	$\left(\dfrac{1}{2} \cdot 8\right)^2 = 16$	$y^2 + 8y + 16$	$(y + 4)^2$
$x^2 + 12x$	$\left(\dfrac{1}{2} \cdot 12\right)^2 = 36$	$x^2 + 12x + 36$	$(x + 6)^2$
$a^2 - 20a$	$\left(\dfrac{1}{2} \cdot (-20)\right)^2 = 100$	$a^2 - 20a + 100$	$(a - 10)^2$
$p^2 - 5p$	$\left(\dfrac{1}{2} \cdot (-5)\right)^2 = \dfrac{25}{4}$	$p^2 - 5p + \dfrac{25}{4}$	$\left(p - \dfrac{5}{2}\right)^2$

●

Notice that the factored form of a perfect square is either

$$x^2 + bx + \left(\frac{b}{2}\right)^2 = \left(x + \frac{b}{2}\right)^2 \text{ or } x^2 - bx + \left(\frac{b}{2}\right)^2 = \left(x - \frac{b}{2}\right)^2$$

Now Work PROBLEM 73

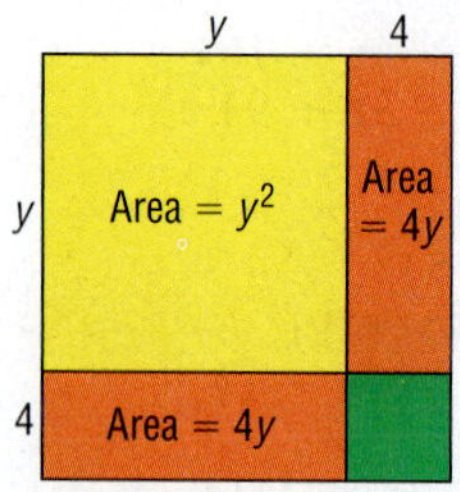

Are you wondering why we refer to making an expression a perfect square as "completing the square"? Look at the square in Figure 27. Its area is $(y + 4)^2$. The yellow area is y^2 and each orange area is $4y$ (for a total area of $8y$). The sum of these areas is $y^2 + 8y$. To complete the square, we need to add the area of the green region: $4 \cdot 4 = 16$. As a result, $y^2 + 8y + 16 = (y + 4)^2$.

	y	4
y	Area = y^2	Area = $4y$
4	Area = $4y$	

Figure 27

R.5 Assess Your Understanding

Concepts and Vocabulary

1. If factored completely, $3x^3 - 12x =$ _____.

2. If a polynomial cannot be written as the product of two other polynomials (excluding 1 and -1), then the polynomial is said to be _____.

3. For $x^2 + Bx + C = (x + a)(x + b)$, which of the following must be true?
   (a) $ab = B$ and $a + b = C$
   (b) $a + b = C$ and $a - b = B$
   (c) $ab = C$ and $a + b = B$
   (d) $ab = B$ and $a - b = C$

4. Choose the best description of $x^2 - 64$.
   (a) Prime   (b) Difference of two squares
   (c) Difference of two cubes   (d) Perfect Square

5. Choose the complete factorization of $4x^2 - 8x - 60$.
   (a) $2(x + 3)(x - 5)$ (b) $4(x^2 - 2x - 15)$
   (c) $(2x + 6)(2x - 10)$ (d) $4(x + 3)(x - 5)$

6. To complete the square of $x^2 + bx$, use which of the following?
   (a) $(2b)^2$ (b) $2b^2$ (c) $\left(\dfrac{1}{2}b\right)^2$ (d) $\dfrac{1}{2}b^2$

7. *True or False* The polynomial $x^2 + 4$ is prime.

8. *True or False* $3x^3 - 2x^2 - 6x + 4 = (3x - 2)(x^2 + 2)$.

Skill Building

In Problems 9–18, factor each polynomial by removing the common monomial factor.

9. $3x + 6$

10. $7x - 14$

11. $ax^2 + a$

12. $ax - a$

13. $x^3 + x^2 + x$

14. $x^3 - x^2 + x$

15. $2x^2 - 2x$

16. $3x^2 - 3x$

17. $3x^2y - 6xy^2 + 12xy$

18. $60x^2y - 48xy^2 + 72x^3y$

In Problems 19–26, factor the difference of two squares.

19. $x^2 - 1$

20. $x^2 - 4$

21. $4x^2 - 1$

22. $9x^2 - 1$

23. $x^2 - 16$

24. $x^2 - 25$

25. $25x^2 - 4$

26. $36x^2 - 9$

In Problems 27–36, factor the perfect squares.

27. $x^2 + 2x + 1$

28. $x^2 - 4x + 4$

29. $x^2 + 4x + 4$

30. $x^2 - 2x + 1$

31. $x^2 - 10x + 25$

32. $x^2 + 10x + 25$

33. $4x^2 + 4x + 1$

34. $9x^2 + 6x + 1$

35. $16x^2 + 8x + 1$

36. $25x^2 + 10x + 1$

In Problems 37–42, factor the sum or difference of two cubes.

37. $x^3 - 27$

38. $x^3 + 125$

39. $x^3 + 27$

40. $27 - 8x^3$

41. $8x^3 + 27$

42. $64 - 27x^3$

In Problems 43–54, factor each polynomial.

43. $x^2 + 5x + 6$

44. $x^2 + 6x + 8$

45. $x^2 + 7x + 6$

46. $x^2 + 9x + 8$

47. $x^2 + 7x + 10$

48. $x^2 + 11x + 10$

49. $x^2 - 10x + 16$

50. $x^2 - 17x + 16$

51. $x^2 - 7x - 8$

52. $x^2 - 2x - 8$

53. $x^2 + 7x - 8$

54. $x^2 + 2x - 8$

In Problems 55–60, factor by grouping.

55. $2x^2 + 4x + 3x + 6$

56. $3x^2 - 3x + 2x - 2$

57. $2x^2 - 4x + x - 2$

58. $3x^2 + 6x - x - 2$

59. $6x^2 + 9x + 4x + 6$

60. $9x^2 - 6x + 3x - 2$

In Problems 61–72, factor each polynomial.

61. $3x^2 + 4x + 1$

62. $2x^2 + 3x + 1$

63. $2z^2 + 5z + 3$

64. $6z^2 + 5z + 1$

65. $3x^2 + 2x - 8$

66. $3x^2 + 10x + 8$

67. $3x^2 - 2x - 8$

68. $3x^2 - 10x + 8$

69. $3x^2 + 14x + 8$

70. $3x^2 - 14x + 8$

71. $3x^2 + 10x - 8$

72. $3x^2 - 10x - 8$

In Problems 73–78, determine what number should be added to complete the square of each expression. Then factor each expression.

73. $x^2 + 10x$

74. $p^2 + 14p$

75. $y^2 - 6y$

76. $x^2 - 4x$

77. $x^2 - \dfrac{1}{2}x$

78. $x^2 + \dfrac{1}{3}x$

Mixed Practice

In Problems 79–126, factor each polynomial completely. If the polynomial cannot be factored, say it is prime.

79. $x^2 - 36$

80. $x^2 - 9$

81. $2 - 8x^2$

82. $3 - 27x^2$

83. $x^2 + 11x + 10$

84. $x^2 + 5x + 4$

85. $x^2 - 10x + 21$

86. $x^2 - 6x + 8$

87. $4x^2 - 8x + 32$

88. $3x^2 - 12x + 15$

89. $x^2 + 4x + 16$

90. $x^2 + 12x + 36$

91. $15 + 2x - x^2$

92. $14 + 6x - x^2$

93. $3x^2 - 12x - 36$

94. $x^3 + 8x^2 - 20x$

95. $y^4 + 11y^3 + 30y^2$

96. $3y^3 - 18y^2 - 48y$

97. $4x^2 + 12x + 9$

98. $9x^2 - 12x + 4$

99. $6x^2 + 8x + 2$

100. $8x^2 + 6x - 2$

101. $x^4 - 81$

102. $x^4 - 1$

103. $x^6 - 2x^3 + 1$

104. $x^6 + 2x^3 + 1$

105. $x^7 - x^5$

106. $x^8 - x^5$

107. $16x^2 + 24x + 9$

108. $9x^2 - 24x + 16$

109. $5 + 16x - 16x^2$

110. $5 + 11x - 16x^2$

111. $4y^2 - 16y + 15$

112. $9y^2 + 9y - 4$

113. $1 - 8x^2 - 9x^4$

114. $4 - 14x^2 - 8x^4$

115. $x(x + 3) - 6(x + 3)$

116. $5(3x - 7) + x(3x - 7)$

117. $(x + 2)^2 - 5(x + 2)$

118. $(x - 1)^2 - 2(x - 1)$

119. $(3x - 2)^3 - 27$

120. $(5x + 1)^3 - 1$

121. $3(x^2 + 10x + 25) - 4(x + 5)$

122. $7(x^2 - 6x + 9) + 5(x - 3)$

123. $x^3 + 2x^2 - x - 2$

124. $x^3 - 3x^2 - x + 3$

125. $x^4 - x^3 + x - 1$

126. $x^4 + x^3 + x + 1$

Applications and Extensions

In Problems 127–136, expressions that occur in calculus are given. Factor each expression completely.

127. $2(3x + 4)^2 + (2x + 3) \cdot 2(3x + 4) \cdot 3$

128. $5(2x + 1)^2 + (5x - 6) \cdot 2(2x + 1) \cdot 2$

129. $2x(2x + 5) + x^2 \cdot 2$

130. $3x^2(8x - 3) + x^3 \cdot 8$

131. $2(x + 3)(x - 2)^3 + (x + 3)^2 \cdot 3(x - 2)^2$

132. $4(x + 5)^3(x - 1)^2 + (x + 5)^4 \cdot 2(x - 1)$

133. $(4x - 3)^2 + x \cdot 2(4x - 3) \cdot 4$

134. $3x^2(3x + 4)^2 + x^3 \cdot 2(3x + 4) \cdot 3$

135. $2(3x - 5) \cdot 3(2x + 1)^3 + (3x - 5)^2 \cdot 3(2x + 1)^2 \cdot 2$

136. $3(4x + 5)^2 \cdot 4(5x + 1)^2 + (4x + 5)^3 \cdot 2(5x + 1) \cdot 5$

137. Show that $x^2 + 4$ is prime.

138. Show that $x^2 + x + 1$ is prime.

Explaining Concepts: Discussion and Writing

139. Make up a polynomial that factors into a perfect square.

140. Explain to a fellow student what you look for first when presented with a factoring problem. What do you do next?

R.6 Synthetic Division

> **OBJECTIVE 1** Divide Polynomials Using Synthetic Division (p. 58)

1 Divide Polynomials Using Synthetic Division

To find the quotient as well as the remainder when a polynomial of degree 1 or higher is divided by $x - c$, a shortened version of long division, called **synthetic division**, makes the task simpler.

To see how synthetic division works, first consider long division for dividing the polynomial $2x^3 - x^2 + 3$ by $x - 3$.

$$
\begin{array}{r}
2x^2 + 5x + 15 \quad \leftarrow \text{Quotient} \\
x - 3 \overline{)\, 2x^3 - x^2 \qquad\quad + 3} \\
\underline{2x^3 - 6x^2} \\
5x^2 \\
\underline{5x^2 - 15x} \\
15x + 3 \\
\underline{15x - 45} \\
48 \quad \leftarrow \text{Remainder}
\end{array}
$$

✔**Check:** $(\text{Divisor}) \cdot (\text{Quotient}) + \text{Remainder}$

$$
\begin{aligned}
&= (x - 3)(2x^2 + 5x + 15) + 48 \\
&= 2x^3 + 5x^2 + 15x - 6x^2 - 15x - 45 + 48 \\
&= 2x^3 - x^2 + 3
\end{aligned}
$$

The process of synthetic division arises from rewriting the long division in a more compact form, using simpler notation. For example, in the long division above, the terms in blue are not really necessary because they are identical to the terms directly above them. With these terms removed, we have

$$
\begin{array}{r}
2x^2 + 5x + 15 \\
x - 3 \overline{)\, 2x^3 - x^2 \qquad\quad + 3} \\
- 6x^2 \\
5x^2 \\
- 15x \\
15x \\
- 45 \\
48
\end{array}
$$

Most of the x's that appear in this process can also be removed, provided that we are careful about positioning each coefficient. In this regard, we will need to use 0 as the coefficient of x in the dividend, because that power of x is missing. Now we have

$$
\begin{array}{r}
2x^2 + 5x + 15 \\
x - 3 \overline{)\, 2 \quad -1 \quad\ 0 \quad\ 3} \\
- 6 \\
5 \\
- 15 \\
15 \\
- 45 \\
48
\end{array}
$$

We can make this display more compact by moving the lines up until the numbers in blue align horizontally.

$$
\begin{array}{r}
2x^2 + 5x + 15 \qquad \text{Row 1} \\
x - 3 \overline{)\, 2 \quad -1 \quad\ 0 \quad\ 3} \qquad \text{Row 2} \\
\underline{-6 \ -15 \ -45} \qquad \text{Row 3} \\
\bigcirc \quad\ 5 \quad 15 \quad 48 \qquad \text{Row 4}
\end{array}
$$

Because the leading coefficient of the divisor is always 1, the leading coefficient of the dividend will also be the leading coefficient of the quotient. So we place the leading coefficient of the quotient, 2, in the circled position. Now, the first three numbers in row 4 are precisely the coefficients of the quotient, and the last number in row 4 is the remainder. Since row 1 is not really needed, we can compress the process to three rows, where the bottom row contains both the coefficients of the quotient and the remainder.

$$
\begin{array}{r|rrrr}
x-3) & 2 & -1 & 0 & 3 \\
& & -6 & -15 & -45 \\
\hline
& 2 & 5 & 15 & 48
\end{array}
\quad
\begin{array}{l}
\text{Row 1} \\
\text{Row 2 (subtract)} \\
\text{Row 3}
\end{array}
$$

Recall that the entries in row 3 are obtained by subtracting the entries in row 2 from those in row 1. Rather than subtracting the entries in row 2, we can change the sign of each entry and add. With this modification, our display will look like this:

$$
\begin{array}{r|rrrr}
x-3) & 2 & -1 & 0 & 3 \\
& & 6 & 15 & 45 \\
\hline
& 2 & 5 & 15 & 48
\end{array}
\quad
\begin{array}{l}
\text{Row 1} \\
\text{Row 2 (add)} \\
\text{Row 3}
\end{array}
$$

Notice that the entries in row 2 are three times the prior entries in row 3. Our last modification to the display replaces the $x - 3$ by 3. The entries in row 3 give the quotient and the remainder, as shown next.

$$
\begin{array}{r|rrrr}
3) & 2 & -1 & 0 & 3 \\
& & 6 & 15 & 45 \\
\hline
& 2 & 5 & 15 & 48
\end{array}
\quad
\begin{array}{l}
\text{Row 1} \\
\text{Row 2 (add)} \\
\text{Row 3}
\end{array}
$$

$$\underbrace{2x^2 + 5x + 15}_{\text{Quotient}} \quad \underbrace{48}_{\text{Remainder}}$$

Let's go through an example step by step.

EXAMPLE I   **Using Synthetic Division to Find the Quotient and Remainder**

Use synthetic division to find the quotient and remainder when

$$x^3 - 4x^2 - 5 \quad \text{is divided by} \quad x - 3$$

Solution   **STEP 1:** Write the dividend in descending powers of x. Then copy the coefficients, remembering to insert a 0 for any missing powers of x.

$$1 \quad -4 \quad 0 \quad -5 \quad \text{Row 1}$$

STEP 2: Insert the usual division symbol. In synthetic division, the divisor is of the form $x - c$, and c is the number placed to the left of the division symbol. Here, since the divisor is $x - 3$, insert 3 to the left of the division symbol.

$$
\begin{array}{r|rrrr}
3) & 1 & -4 & 0 & -5
\end{array}
\quad \text{Row 1}
$$

STEP 3: Bring the 1 down two rows, and enter it in row 3.

$$
\begin{array}{r|rrrr}
3) & 1 & -4 & 0 & -5 \\
& \downarrow & & & \\
\hline
& 1 & & &
\end{array}
\quad
\begin{array}{l}
\text{Row 1} \\
\text{Row 2} \\
\text{Row 3}
\end{array}
$$

STEP 4: Multiply the latest entry in row 3 by 3, and place the result in row 2, one column over to the right.

$$
\begin{array}{r|rrrr}
3) & 1 & -4 & 0 & -5 \\
& & 3 & & \\
\hline
& 1 & & &
\end{array}
\quad
\begin{array}{l}
\text{Row 1} \\
\text{Row 2} \\
\text{Row 3}
\end{array}
$$

STEP 5: Add the entry in row 2 to the entry above it in row 1, and enter the sum in row 3.

$$
\begin{array}{r|rrrr}
3) & 1 & -4 & 0 & -5 \\
& & 3 & & \\
\hline
& 1 & -1 & &
\end{array}
\quad
\begin{array}{l}
\text{Row 1} \\
\text{Row 2} \\
\text{Row 3}
\end{array}
$$

STEP 6: Repeat Steps 4 and 5 until no more entries are available in row 1.

$$3)\overline{1 \quad -4 \quad 0 \quad -5} \quad \text{Row 1}$$
$$\underline{ 3 \quad -3 \quad -9} \quad \text{Row 2}$$
$$1 \quad -1 \quad -3 \quad -14 \quad \text{Row 3}$$

STEP 7: The final entry in row 3, the -14, is the remainder; the other entries in row 3, the 1, -1, and -3, are the coefficients (in descending order) of a polynomial whose degree is 1 less than that of the dividend. This is the quotient. That is,

$$\text{Quotient} = x^2 - x - 3 \qquad \text{Remainder} = -14$$

✓**Check:** $(\text{Divisor})(\text{Quotient}) + \text{Remainder}$

$$= (x - 3)(x^2 - x - 3) + (-14)$$
$$= (x^3 - x^2 - 3x - 3x^2 + 3x + 9) + (-14)$$
$$= x^3 - 4x^2 - 5 = \text{Dividend} \qquad \bullet$$

Let's do an example in which all seven steps are combined.

EXAMPLE 2   **Using Synthetic Division to Verify a Factor**

Use synthetic division to show that $x + 3$ is a factor of

$$2x^5 + 5x^4 - 2x^3 + 2x^2 - 2x + 3$$

Solution   The divisor is $x + 3 = x - (-3)$, so place -3 to the left of the division symbol. Then the row 3 entries will be multiplied by -3, entered in row 2, and added to row 1.

$$-3)\overline{2 \quad 5 \quad -2 \quad 2 \quad -2 \quad 3} \quad \text{Row 1}$$
$$\underline{ -6 \quad 3 \quad -3 \quad 3 \quad -3} \quad \text{Row 2}$$
$$2 \quad -1 \quad 1 \quad -1 \quad 1 \quad 0 \quad \text{Row 3}$$

Because the remainder is 0, we have

$$(\text{Divisor})(\text{Quotient}) + \text{Remainder}$$

$$= (x + 3)(2x^4 - x^3 + x^2 - x + 1) = 2x^5 + 5x^4 - 2x^3 + 2x^2 - 2x + 3$$

As we see, $x + 3$ is a factor of $2x^5 + 5x^4 - 2x^3 + 2x^2 - 2x + 3$. $\qquad \bullet$

As Example 2 illustrates, the remainder after division gives information about whether the divisor is, or is not, a factor. We shall have more to say about this in Chapter 5.

✏ **Now Work** PROBLEMS **9** AND **19**

R.6 Assess Your Understanding

Concepts and Vocabulary

1. To check division, the expression that is being divided, the dividend, should equal the product of the _____ and the _____ plus the _____.

2. To divide $2x^3 - 5x + 1$ by $x + 3$ using synthetic division, the first step is to write ___)‾‾‾‾‾‾‾‾‾.

3. Choose the division problem that cannot be done using synthetic division.
   (a) $2x^3 - 4x^2 + 6x - 8$ is divided by $x - 8$
   (b) $x^4 - 3$ is divided by $x + 1$
   (c) $x^5 + 3x^2 - 9x + 2$ is divided by $x + 10$
   (d) $x^4 - 5x^3 + 3x^2 - 9x + 13$ is divided by $x^2 + 5$

4. Choose the correct conclusion based on the following synthetic division:
   $$-5)\overline{2 \quad 3 \quad -38 \quad -15}$$
   $$\underline{ -10 \quad 35 \quad 15}$$
   $$2 \quad -7 \quad -3 \quad 0$$
   (a) $x + 5$ is a factor of $2x^3 + 3x^2 - 38x - 15$
   (b) $x - 5$ is a factor of $2x^3 + 3x^2 - 38x - 15$
   (c) $x + 5$ is not a factor of $2x^3 + 3x^2 - 38x - 15$
   (d) $x - 5$ is not a factor of $2x^3 + 3x^2 - 38x - 15$

5. *True or False* In using synthetic division, the divisor is always a polynomial of degree 1, whose leading coefficient is 1.

6. *True or False*
   $$-2)\overline{5 \quad 3 \quad 2 \quad 1}$$
   $$\underline{ -10 \quad 14 \quad -32}$$
   $$5 \quad -7 \quad 16 \quad -31$$
   means $\dfrac{5x^3 + 3x^2 + 2x + 1}{x + 2} = 5x^2 - 7x + 16 + \dfrac{-31}{x + 2}$.

Skill Building

In Problems 7–18, use synthetic division to find the quotient and remainder when:

7. $x^3 - x^2 + 2x + 4$ is divided by $x - 2$

8. $x^3 + 2x^2 - 3x + 1$ is divided by $x + 1$

9. $3x^3 + 2x^2 - x + 3$ is divided by $x - 3$

10. $-4x^3 + 2x^2 - x + 1$ is divided by $x + 2$

11. $x^5 - 4x^3 + x$ is divided by $x + 3$

12. $x^4 + x^2 + 2$ is divided by $x - 2$

13. $4x^6 - 3x^4 + x^2 + 5$ is divided by $x - 1$

14. $x^5 + 5x^3 - 10$ is divided by $x + 1$

15. $0.1x^3 + 0.2x$ is divided by $x + 1.1$

16. $0.1x^2 - 0.2$ is divided by $x + 2.1$

17. $x^5 - 1$ is divided by $x - 1$

18. $x^5 + 1$ is divided by $x + 1$

In Problems 19–28, use synthetic division to determine whether $x - c$ is a factor of the given polynomial.

19. $4x^3 - 3x^2 - 8x + 4$; $x - 2$

20. $-4x^3 + 5x^2 + 8$; $x + 3$

21. $3x^4 - 6x^3 - 5x + 10$; $x - 2$

22. $4x^4 - 15x^2 - 4$; $x - 2$

23. $3x^6 + 82x^3 + 27$; $x + 3$

24. $2x^6 - 18x^4 + x^2 - 9$; $x + 3$

25. $4x^6 - 64x^4 + x^2 - 15$; $x + 4$

26. $x^6 - 16x^4 + x^2 - 16$; $x + 4$

27. $2x^4 - x^3 + 2x - 1$; $x - \dfrac{1}{2}$

28. $3x^4 + x^3 - 3x + 1$; $x + \dfrac{1}{3}$

Applications and Extensions

29. Find the sum of $a, b, c,$ and d if

$$\frac{x^3 - 2x^2 + 3x + 5}{x + 2} = ax^2 + bx + c + \frac{d}{x + 2}$$

Explaining Concepts: Discussion and Writing

30. When dividing a polynomial by $x - c$, do you prefer to use long division or synthetic division? Does the value of c make a difference to you in choosing? Give reasons.

R.7 Rational Expressions

OBJECTIVES **1** Reduce a Rational Expression to Lowest Terms (p. 62)
  2 Multiply and Divide Rational Expressions (p. 63)
  3 Add and Subtract Rational Expressions (p. 64)
  4 Use the Least Common Multiple Method (p. 66)
  5 Simplify Complex Rational Expressions (p. 68)

1 Reduce a Rational Expression to Lowest Terms

If we form the quotient of two polynomials, the result is called a **rational expression**. Some examples of rational expressions are

(a) $\dfrac{x^3 + 1}{x}$ (b) $\dfrac{3x^2 + x - 2}{x^2 + 5}$ (c) $\dfrac{x}{x^2 - 1}$ (d) $\dfrac{xy^2}{(x - y)^2}$

Expressions (a), (b), and (c) are rational expressions in one variable, x, whereas (d) is a rational expression in two variables, x and y.

Rational expressions are described in the same manner as rational numbers. In expression (a), the polynomial $x^3 + 1$ is the **numerator**, and x is the **denominator**. When the numerator and denominator of a rational expression contain no common factors (except 1 and -1), we say that the rational expression is **reduced to lowest terms**, or **simplified**.

The polynomial in the denominator of a rational expression cannot be equal to 0 because division by 0 is not defined. For example, for the expression $\dfrac{x^3 + 1}{x}$, x cannot take on the value 0. The domain of the variable x is $\{x | x \neq 0\}$.

A rational expression is reduced to lowest terms by factoring the numerator and the denominator completely and canceling any common factors using the Cancellation Property:

$$\frac{a\cancel{c}}{b\cancel{c}} = \frac{a}{b} \qquad \text{if } b \neq 0, c \neq 0 \tag{1}$$

EXAMPLE 1

Reducing a Rational Expression to Lowest Terms

Reduce to lowest terms: $\dfrac{x^2 + 4x + 4}{x^2 + 3x + 2}$

Solution

Begin by factoring the numerator and the denominator.

$$x^2 + 4x + 4 = (x + 2)(x + 2)$$
$$x^2 + 3x + 2 = (x + 2)(x + 1)$$

WARNING Apply the Cancellation Property only to rational expressions written in factored form. Be sure to cancel only common factors, not common terms! ∎

Since a common factor, $x + 2$, appears, the original expression is not in lowest terms. To reduce it to lowest terms, use the Cancellation Property:

$$\frac{x^2 + 4x + 4}{x^2 + 3x + 2} = \frac{\cancel{(x + 2)}(x + 2)}{\cancel{(x + 2)}(x + 1)} = \frac{x + 2}{x + 1} \qquad x \neq -2, -1$$

●

EXAMPLE 2

Reducing Rational Expressions to Lowest Terms

Reduce each rational expression to lowest terms.

(a) $\dfrac{x^3 - 8}{x^3 - 2x^2}$
(b) $\dfrac{8 - 2x}{x^2 - x - 12}$

Solution

(a) $\dfrac{x^3 - 8}{x^3 - 2x^2} = \dfrac{\cancel{(x - 2)}(x^2 + 2x + 4)}{x^2\cancel{(x - 2)}} = \dfrac{x^2 + 2x + 4}{x^2} \qquad x \neq 0, 2$

(b) $\dfrac{8 - 2x}{x^2 - x - 12} = \dfrac{2(4 - x)}{(x - 4)(x + 3)} = \dfrac{2(-1)\cancel{(x - 4)}}{\cancel{(x - 4)}(x + 3)} = \dfrac{-2}{x + 3} \qquad x \neq -3, 4$

●

 Now Work PROBLEM 7

---

2 Multiply and Divide Rational Expressions

The rules for multiplying and dividing rational expressions are the same as the rules for multiplying and dividing rational numbers. If $\dfrac{a}{b}$ and $\dfrac{c}{d}$, $b \neq 0, d \neq 0$, are two rational expressions, then

$$\frac{a}{b} \cdot \frac{c}{d} = \frac{ac}{bd} \qquad \text{if } b \neq 0, d \neq 0 \tag{2}$$

$$\frac{\dfrac{a}{b}}{\dfrac{c}{d}} = \frac{a}{b} \cdot \frac{d}{c} = \frac{ad}{bc} \qquad \text{if } b \neq 0, c \neq 0, d \neq 0 \tag{3}$$

In using equations (2) and (3) with rational expressions, be sure first to factor each polynomial completely so that common factors can be canceled. Leave your answer in factored form.

EXAMPLE 3

Multiplying and Dividing Rational Expressions

Perform the indicated operation and simplify the result. Leave your answer in factored form.

(a) $\dfrac{x^2 - 2x + 1}{x^3 + x} \cdot \dfrac{4x^2 + 4}{x^2 + x - 2}$

(b) $\dfrac{\dfrac{x + 3}{x^2 - 4}}{\dfrac{x^2 - x - 12}{x^3 - 8}}$

Solution

(a) $\dfrac{x^2 - 2x + 1}{x^3 + x} \cdot \dfrac{4x^2 + 4}{x^2 + x - 2} = \dfrac{(x - 1)^2}{x(x^2 + 1)} \cdot \dfrac{4(x^2 + 1)}{(x + 2)(x - 1)}$

$= \dfrac{(x - 1)^2 (4)(x^2 + 1)}{x(x^2 + 1)(x + 2)(x - 1)}$

$= \dfrac{4(x - 1)}{x(x + 2)} \qquad x \neq -2, 0, 1$

(b) $\dfrac{\dfrac{x + 3}{x^2 - 4}}{\dfrac{x^2 - x - 12}{x^3 - 8}} = \dfrac{x + 3}{x^2 - 4} \cdot \dfrac{x^3 - 8}{x^2 - x - 12}$

$= \dfrac{x + 3}{(x - 2)(x + 2)} \cdot \dfrac{(x - 2)(x^2 + 2x + 4)}{(x - 4)(x + 3)}$

$= \dfrac{(x + 3)(x - 2)(x^2 + 2x + 4)}{(x - 2)(x + 2)(x - 4)(x + 3)}$

$= \dfrac{x^2 + 2x + 4}{(x + 2)(x - 4)} \qquad x \neq -3, -2, 2, 4$

✏️ **Now Work** PROBLEMS 19 AND 27

3 Add and Subtract Rational Expressions

The rules for adding and subtracting rational expressions are the same as the rules for adding and subtracting rational numbers. If the denominators of two rational expressions to be added (or subtracted) are equal, then add (or subtract) the numerators and keep the common denominator.

In Words
To add (or subtract) two rational expressions with the same denominator, keep the common denominator and add (or subtract) the numerators.

If $\dfrac{a}{b}$ and $\dfrac{c}{b}$ are two rational expressions, then

$$\dfrac{a}{b} + \dfrac{c}{b} = \dfrac{a + c}{b} \qquad \dfrac{a}{b} - \dfrac{c}{b} = \dfrac{a - c}{b} \qquad \text{if } b \neq 0 \tag{4}$$

EXAMPLE 4

Adding and Subtracting Rational Expressions with Equal Denominators

Perform the indicated operation and simplify the result. Leave your answer in factored form.

(a) $\dfrac{2x^2 - 4}{2x + 5} + \dfrac{x + 3}{2x + 5} \qquad x \neq -\dfrac{5}{2}$

(b) $\dfrac{x}{x - 3} - \dfrac{3x + 2}{x - 3} \qquad x \neq 3$

Solution

(a) $\dfrac{2x^2 - 4}{2x + 5} + \dfrac{x + 3}{2x + 5} = \dfrac{(2x^2 - 4) + (x + 3)}{2x + 5}$

$= \dfrac{2x^2 + x - 1}{2x + 5} = \dfrac{(2x - 1)(x + 1)}{2x + 5}$

(b) $\dfrac{x}{x-3} - \dfrac{3x+2}{x-3} = \dfrac{x-(3x+2)}{x-3} = \dfrac{x-3x-2}{x-3}$

$$= \dfrac{-2x-2}{x-3} = \dfrac{-2(x+1)}{x-3}$$

●

EXAMPLE 5

Adding Rational Expressions Whose Denominators Are Additive Inverses of Each Other

Perform the indicated operation and simplify the result. Leave your answer in factored form.

$$\dfrac{2x}{x-3} + \dfrac{5}{3-x} \qquad x \neq 3$$

Solution

Notice that the denominators of the two rational expressions are different. However, the denominator of the second expression is the additive inverse of the denominator of the first. That is,

$$3 - x = -x + 3 = -1 \cdot (x - 3) = -(x - 3)$$

Then

$$\dfrac{2x}{x-3} + \dfrac{5}{3-x} = \dfrac{2x}{x-3} + \dfrac{5}{-(x-3)} = \dfrac{2x}{x-3} + \dfrac{-5}{x-3}$$

$$\underset{\substack{\uparrow \\ 3-x = -(x-3)}}{} \qquad \underset{\substack{\uparrow \\ \frac{a}{-b} = \frac{-a}{b}}}{}$$

$$= \dfrac{2x + (-5)}{x-3} = \dfrac{2x-5}{x-3}$$

●

✏️ **Now Work** PROBLEMS **39** AND **45**

If the denominators of two rational expressions to be added or subtracted are not equal, we can use the general formulas for adding and subtracting rational expressions.

$$\dfrac{a}{b} + \dfrac{c}{d} = \dfrac{a \cdot d}{b \cdot d} + \dfrac{b \cdot c}{b \cdot d} = \dfrac{ad + bc}{bd} \qquad \text{if } b \neq 0, d \neq 0 \qquad \textbf{(5a)}$$

$$\dfrac{a}{b} - \dfrac{c}{d} = \dfrac{a \cdot d}{b \cdot d} - \dfrac{b \cdot c}{b \cdot d} = \dfrac{ad - bc}{bd} \qquad \text{if } b \neq 0, d \neq 0 \qquad \textbf{(5b)}$$

EXAMPLE 6

Adding and Subtracting Rational Expressions with Unequal Denominators

Perform the indicated operation and simplify the result. Leave your answer in factored form.

(a) $\dfrac{x-3}{x+4} + \dfrac{x}{x-2} \qquad x \neq -4, 2$ (b) $\dfrac{x^2}{x^2-4} - \dfrac{1}{x} \qquad x \neq -2, 0, 2$

Solution

(a) $\dfrac{x-3}{x+4} + \dfrac{x}{x-2} = \dfrac{x-3}{x+4} \cdot \dfrac{x-2}{x-2} + \dfrac{x+4}{x+4} \cdot \dfrac{x}{x-2}$

$$\underset{(5a)}{\uparrow}$$

$$= \dfrac{(x-3)(x-2) + (x+4)(x)}{(x+4)(x-2)}$$

$$= \dfrac{x^2 - 5x + 6 + x^2 + 4x}{(x+4)(x-2)} = \dfrac{2x^2 - x + 6}{(x+4)(x-2)}$$

(b) $\dfrac{x^2}{x^2-4}-\dfrac{1}{x}=\dfrac{x^2}{x^2-4}\cdot\dfrac{x}{x}-\dfrac{x^2-4}{x^2-4}\cdot\dfrac{1}{x}=\dfrac{x^2(x)-(x^2-4)(1)}{(x^2-4)(x)}$

(5b)

$$=\dfrac{x^3-x^2+4}{(x-2)(x+2)(x)}$$

●

Now Work PROBLEM 49

4 Use the Least Common Multiple Method

If the denominators of two rational expressions to be added (or subtracted) have common factors, we usually do not use the general rules given by equations (5a) and (5b). Just as with fractions, we apply the **least common multiple (LCM) method**. The LCM method uses the polynomial of least degree that has each denominator polynomial as a factor.

The LCM Method for Adding or Subtracting Rational Expressions

The Least Common Multiple (LCM) Method requires four steps:

STEP 1:   Factor completely the polynomial in the denominator of each rational expression.

STEP 2:   The LCM of the denominators is the product of each of these factors raised to a power equal to the greatest number of times that the factor occurs in the polynomials.

STEP 3:   Write each rational expression using the LCM as the common denominator.

STEP 4:   Add or subtract the rational expressions using equation (4).

We begin with an example that requires only Steps 1 and 2.

EXAMPLE 7

Finding the Least Common Multiple

Find the least common multiple of the following pair of polynomials:

$$x(x-1)^2(x+1)\quad\text{and}\quad 4(x-1)(x+1)^3$$

Solution   STEP 1:   The polynomials are already factored completely as

$$x(x-1)^2(x+1)\quad\text{and}\quad 4(x-1)(x+1)^3$$

STEP 2:   Start by writing the factors of the left-hand polynomial. (Or you could start with the one on the right.)

$$x(x-1)^2(x+1)$$

Now look at the right-hand polynomial. Its first factor, 4, does not appear in our list, so we insert it.

$$4x(x-1)^2(x+1)$$

The next factor, $x-1$, is already in our list, so no change is necessary. The final factor is $(x+1)^3$. Since our list has $x+1$ to the first power only, we replace $x+1$ in the list by $(x+1)^3$. The LCM is

$$4x(x-1)^2(x+1)^3$$

Notice that the LCM is, in fact, the polynomial of least degree that contains $x(x-1)^2(x+1)$ and $4(x-1)(x+1)^3$ as factors. ●

⟶ **Now Work** PROBLEM 55

EXAMPLE 8

Using the Least Common Multiple to Add Rational Expressions

Perform the indicated operation and simplify the result. Leave your answer in factored form.

$$\frac{x}{x^2 + 3x + 2} + \frac{2x - 3}{x^2 - 1} \qquad x \neq -2, -1, 1$$

Solution

STEP 1: Factor completely the polynomials in the denominators.

$$x^2 + 3x + 2 = (x + 2)(x + 1)$$

$$x^2 - 1 = (x - 1)(x + 1)$$

STEP 2: The LCM is $(x + 2)(x + 1)(x - 1)$. Do you see why?

STEP 3: Write each rational expression using the LCM as the denominator.

$$\frac{x}{x^2 + 3x + 2} = \frac{x}{(x + 2)(x + 1)} = \frac{x}{(x + 2)(x + 1)} \cdot \frac{x - 1}{x - 1} = \frac{x(x - 1)}{(x + 2)(x + 1)(x - 1)}$$

⬆ Multiply numerator and denominator by $x - 1$ to get the LCM in the denominator.

$$\frac{2x - 3}{x^2 - 1} = \frac{2x - 3}{(x - 1)(x + 1)} = \frac{2x - 3}{(x - 1)(x + 1)} \cdot \frac{x + 2}{x + 2} = \frac{(2x - 3)(x + 2)}{(x - 1)(x + 1)(x + 2)}$$

⬆ Multiply numerator and denominator by $x + 2$ to get the LCM in the denominator.

STEP 4: Now add by using equation (4).

$$\frac{x}{x^2 + 3x + 2} + \frac{2x - 3}{x^2 - 1} = \frac{x(x - 1)}{(x + 2)(x + 1)(x - 1)} + \frac{(2x - 3)(x + 2)}{(x + 2)(x + 1)(x - 1)}$$

$$= \frac{(x^2 - x) + (2x^2 + x - 6)}{(x + 2)(x + 1)(x - 1)}$$

$$= \frac{3x^2 - 6}{(x + 2)(x + 1)(x - 1)} = \frac{3(x^2 - 2)}{(x + 2)(x + 1)(x - 1)}$$

●

EXAMPLE 9

Using the Least Common Multiple to Subtract Rational Expressions

Perform the indicated operation and simplify the result. Leave your answer in factored form.

$$\frac{3}{x^2 + x} - \frac{x + 4}{x^2 + 2x + 1} \qquad x \neq -1, 0$$

Solution

STEP 1: Factor completely the polynomials in the denominators.

$$x^2 + x = x(x + 1)$$

$$x^2 + 2x + 1 = (x + 1)^2$$

STEP 2: The LCM is $x(x + 1)^2$.

STEP 3: Write each rational expression using the LCM as the denominator.

$$\frac{3}{x^2 + x} = \frac{3}{x(x + 1)} = \frac{3}{x(x + 1)} \cdot \frac{x + 1}{x + 1} = \frac{3(x + 1)}{x(x + 1)^2}$$

$$\frac{x + 4}{x^2 + 2x + 1} = \frac{x + 4}{(x + 1)^2} = \frac{x + 4}{(x + 1)^2} \cdot \frac{x}{x} = \frac{x(x + 4)}{x(x + 1)^2}$$

STEP 4: Subtract, using equation (4).

$$\frac{3}{x^2 + x} - \frac{x + 4}{x^2 + 2x + 1} = \frac{3(x + 1)}{x(x + 1)^2} - \frac{x(x + 4)}{x(x + 1)^2}$$

$$= \frac{3(x + 1) - x(x + 4)}{x(x + 1)^2}$$

$$= \frac{3x + 3 - x^2 - 4x}{x(x + 1)^2}$$

$$= \frac{-x^2 - x + 3}{x(x + 1)^2}$$

 **Now Work** PROBLEM 65

5 Simplify Complex Rational Expressions

When sums and/or differences of rational expressions appear as the numerator and/or denominator of a quotient, the quotient is called a **complex rational expression**.* For example,

$$\frac{1 + \dfrac{1}{x}}{1 - \dfrac{1}{x}} \quad \text{and} \quad \frac{\dfrac{x^2}{x^2 - 4} - 3}{\dfrac{x - 3}{x + 2} - 1}$$

are complex rational expressions. To **simplify** a complex rational expression means to write it as a rational expression reduced to lowest terms. This can be accomplished in either of two ways.

Simplifying a Complex Rational Expression

OPTION 1: Treat the numerator and denominator of the complex rational expression separately, performing whatever operations are indicated and simplifying the results. Follow this by simplifying the resulting rational expression.

OPTION 2: Find the LCM of the denominators of all rational expressions that appear in the complex rational expression. Multiply the numerator and denominator of the complex rational expression by the LCM and simplify the result.

We use both options in the next example. By carefully studying each option, you can discover situations in which one may be easier to use than the other.

EXAMPLE 10   **Simplifying a Complex Rational Expression**

Simplify: $\dfrac{\dfrac{1}{2} + \dfrac{3}{x}}{\dfrac{x + 3}{4}}$ $x \neq -3, 0$

* Some texts use the term **complex fraction**.

Solution   *Option 1:* First, we perform the indicated operation in the numerator, and then we divide.

$$\frac{\dfrac{1}{2}+\dfrac{3}{x}}{\dfrac{x+3}{4}}=\frac{\dfrac{1\cdot x+2\cdot 3}{2\cdot x}}{\dfrac{x+3}{4}}=\frac{\dfrac{x+6}{2x}}{\dfrac{x+3}{4}}=\frac{x+6}{2x}\cdot\frac{4}{x+3}$$

↑ Rule for adding quotients   ↑ Rule for dividing quotients

$$=\frac{(x+6)\cdot 4}{2\cdot x\cdot(x+3)}=\frac{2\cdot 2\cdot(x+6)}{2\cdot x\cdot(x+3)}=\frac{2(x+6)}{x(x+3)}$$

↑ Rule for multiplying quotients

Option 2: The rational expressions that appear in the complex rational expression are

$$\frac{1}{2},\ \frac{3}{x},\ \frac{x+3}{4}$$

The LCM of their denominators is $4x$. We multiply the numerator and denominator of the complex rational expression by $4x$ and then simplify.

$$\frac{\dfrac{1}{2}+\dfrac{3}{x}}{\dfrac{x+3}{4}}=\frac{4x\cdot\left(\dfrac{1}{2}+\dfrac{3}{x}\right)}{4x\cdot\left(\dfrac{x+3}{4}\right)}=\frac{4x\cdot\dfrac{1}{2}+4x\cdot\dfrac{3}{x}}{\dfrac{4x\cdot(x+3)}{4}}$$

↑ Multiply the numerator and denominator by $4x$.   ↑ Use the Distributive Property in the numerator.

$$=\frac{2\cdot 2x\cdot\dfrac{1}{2}+4x\cdot\dfrac{3}{x}}{\dfrac{4x\cdot(x+3)}{4}}=\frac{2x+12}{x(x+3)}=\frac{2(x+6)}{x(x+3)}$$

↑ Simplify   ↑ Factor   ●

EXAMPLE 11   **Simplifying a Complex Rational Expression**

Simplify: $\dfrac{\dfrac{x^2}{x-4}+2}{\dfrac{2x-2}{x}-1}\qquad x\neq 0,2,4$

Solution   We will use Option 1.

$$\frac{\dfrac{x^2}{x-4}+2}{\dfrac{2x-2}{x}-1}=\frac{\dfrac{x^2}{x-4}+\dfrac{2(x-4)}{x-4}}{\dfrac{2x-2}{x}-\dfrac{x}{x}}=\frac{\dfrac{x^2+2x-8}{x-4}}{\dfrac{2x-2-x}{x}}$$

$$=\frac{\dfrac{(x+4)(x-2)}{x-4}}{\dfrac{x-2}{x}}=\frac{(x+4)\,(x-2)}{x-4}\cdot\frac{x}{x-2}$$

$$=\frac{(x+4)\cdot x}{x-4}$$ ●

🖉 **Now Work** PROBLEM 75

Application

| EXAMPLE 12 | **Solving an Application in Electricity** |

An electrical circuit contains two resistors connected in parallel, as shown in Figure 28. If these two resistors provide resistance of R_1 and R_2 ohms, respectively, their combined resistance R is given by the formula

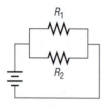

R_1

R_2

$$R = \dfrac{1}{\dfrac{1}{R_1} + \dfrac{1}{R_2}}$$

Express R as a rational expression; that is, simplify the right-hand side of this formula. Evaluate the rational expression if $R_1 = 6$ ohms and $R_2 = 10$ ohms.

Figure 28

Solution   We will use Option 2. If we consider 1 as the fraction $\dfrac{1}{1}$, the rational expressions in the complex rational expression are

$$\frac{1}{1}, \quad \frac{1}{R_1}, \quad \frac{1}{R_2}$$

The LCM of the denominators is $R_1 R_2$. We multiply the numerator and denominator of the complex rational expression by $R_1 R_2$ and simplify.

$$\frac{1}{\dfrac{1}{R_1} + \dfrac{1}{R_2}} = \frac{1 \cdot R_1 R_2}{\left(\dfrac{1}{R_1} + \dfrac{1}{R_2}\right) \cdot R_1 R_2} = \frac{R_1 R_2}{\dfrac{1}{R_1} \cdot R_1 R_2 + \dfrac{1}{R_2} \cdot R_1 R_2} = \frac{R_1 R_2}{R_2 + R_1}$$

So,

$$R = \frac{R_1 R_2}{R_2 + R_1}$$

If $R_1 = 6$ and $R_2 = 10$, then

$$R = \frac{6 \cdot 10}{10 + 6} = \frac{60}{16} = \frac{15}{4} \quad \text{ohms}$$

R.7 Assess Your Understanding

Concepts and Vocabulary

1. When the numerator and denominator of a rational expression contain no common factors (except 1 and -1), the rational expression is in _____ _____.

2. LCM is an abbreviation for _____ _____ _____.

3. Choose the statement that is not true. Assume $b \neq 0, c \neq 0$, and $d \neq 0$ as necessary.

   (a) $\dfrac{ac}{bc} = \dfrac{a}{b}$ (b) $\dfrac{a}{b} + \dfrac{c}{b} = \dfrac{a + c}{b}$

   (c) $\dfrac{a}{b} - \dfrac{c}{d} = \dfrac{ad - bc}{bd}$ (d) $\dfrac{\dfrac{a}{b}}{\dfrac{c}{d}} = \dfrac{ac}{bd}$

4. Choose the rational expression that simplifies to -1.

   (a) $\dfrac{a - b}{b - a}$ (b) $\dfrac{a - b}{a - b}$

   (c) $\dfrac{a + b}{a - b}$ (d) $\dfrac{b - a}{b + a}$

5. **True or False** The rational expression $\dfrac{2x^3 - 4x}{x - 2}$ is reduced to lowest terms.

6. **True or False** The LCM of $2x^3 + 6x^2$ and $6x^4 + 4x^3$ is $4x^3(x + 1)$.

Skill Building

In Problems 7–18, reduce each rational expression to lowest terms.

 7. $\dfrac{3x + 9}{x^2 - 9}$

8. $\dfrac{4x^2 + 8x}{12x + 24}$

9. $\dfrac{x^2 - 2x}{3x - 6}$

10. $\dfrac{15x^2 + 24x}{3x^2}$

11. $\dfrac{24x^2}{12x^2 - 6x}$

12. $\dfrac{x^2 + 4x + 4}{x^2 - 4}$

13. $\dfrac{y^2 - 25}{2y^2 - 8y - 10}$

14. $\dfrac{3y^2 - y - 2}{3y^2 + 5y + 2}$

15. $\dfrac{x^2 + 4x - 5}{x^2 - 2x + 1}$

16. $\dfrac{x - x^2}{x^2 + x - 2}$

17. $\dfrac{x^2 + 5x - 14}{2 - x}$

18. $\dfrac{2x^2 + 5x - 3}{1 - 2x}$

In Problems 19–36, perform the indicated operation and simplify the result. Leave your answer in factored form.

19. $\dfrac{3x+6}{5x^2} \cdot \dfrac{x}{x^2-4}$

20. $\dfrac{3}{2x} \cdot \dfrac{x^2}{6x+10}$

21. $\dfrac{4x^2}{x^2-16} \cdot \dfrac{x^3-64}{2x}$

22. $\dfrac{12}{x^2+x} \cdot \dfrac{x^3+1}{4x-2}$

23. $\dfrac{4x-8}{-3x} \cdot \dfrac{12}{12-6x}$

24. $\dfrac{6x-27}{5x} \cdot \dfrac{2}{4x-18}$

25. $\dfrac{x^2-3x-10}{x^2+2x-35} \cdot \dfrac{x^2+4x-21}{x^2+9x+14}$

26. $\dfrac{x^2+x-6}{x^2+4x-5} \cdot \dfrac{x^2-25}{x^2+2x-15}$

27. $\dfrac{\dfrac{6x}{x^2-4}}{\dfrac{3x-9}{2x+4}}$

28. $\dfrac{\dfrac{12x}{5x+20}}{\dfrac{4x^2}{x^2-16}}$

29. $\dfrac{\dfrac{8x}{x^2-1}}{\dfrac{10x}{x+1}}$

30. $\dfrac{\dfrac{x-2}{4x}}{\dfrac{x^2-4x+4}{12x}}$

31. $\dfrac{\dfrac{4-x}{4+x}}{\dfrac{4x}{x^2-16}}$

32. $\dfrac{\dfrac{3+x}{3-x}}{\dfrac{x^2-9}{9x^3}}$

33. $\dfrac{\dfrac{x^2+7x+12}{x^2-7x+12}}{\dfrac{x^2+x-12}{x^2-x-12}}$

34. $\dfrac{\dfrac{x^2+7x+6}{x^2+x-6}}{\dfrac{x^2+5x-6}{x^2+5x+6}}$

35. $\dfrac{\dfrac{2x^2-x-28}{3x^2-x-2}}{\dfrac{4x^2+16x+7}{3x^2+11x+6}}$

36. $\dfrac{\dfrac{9x^2+3x-2}{12x^2+5x-2}}{\dfrac{9x^2-6x+1}{8x^2-10x-3}}$

In Problems 37–54, perform the indicated operation and simplify the result. Leave your answer in factored form.

37. $\dfrac{x}{2}+\dfrac{5}{2}$

38. $\dfrac{3}{x}-\dfrac{6}{x}$

39. $\dfrac{x^2}{2x-3}-\dfrac{4}{2x-3}$

40. $\dfrac{3x^2}{2x-1}-\dfrac{9}{2x-1}$

41. $\dfrac{x+1}{x-3}+\dfrac{2x-3}{x-3}$

42. $\dfrac{2x-5}{3x+2}+\dfrac{x+4}{3x+2}$

43. $\dfrac{3x+5}{2x-1}-\dfrac{2x-4}{2x-1}$

44. $\dfrac{5x-4}{3x+4}-\dfrac{x+1}{3x+4}$

45. $\dfrac{4}{x-2}+\dfrac{x}{2-x}$

46. $\dfrac{6}{x-1}-\dfrac{x}{1-x}$

47. $\dfrac{4}{x-1}-\dfrac{2}{x+2}$

48. $\dfrac{2}{x+5}-\dfrac{5}{x-5}$

49. $\dfrac{x}{x+1}+\dfrac{2x-3}{x-1}$

50. $\dfrac{3x}{x-4}+\dfrac{2x}{x+3}$

51. $\dfrac{x-3}{x+2}-\dfrac{x+4}{x-2}$

52. $\dfrac{2x-3}{x-1}-\dfrac{2x+1}{x+1}$

53. $\dfrac{x}{x^2-4}+\dfrac{1}{x}$

54. $\dfrac{x-1}{x^3}+\dfrac{x}{x^2+1}$

In Problems 55–62, find the LCM of the given polynomials.

55. $x^2-4, \;\; x^2-x-2$

56. $x^2-x-12, \;\; x^2-8x+16$

57. $x^3-x, \;\; x^2-x$

58. $3x^2-27, \;\; 2x^2-x-15$

59. $4x^3-4x^2+x, \;\; 2x^3-x^2, \;\; x^3$

60. $x-3, \;\; x^2+3x, \;\; x^3-9x$

61. $x^3-x, \;\; x^3-2x^2+x, \;\; x^3-1$

62. $x^2+4x+4, \;\; x^3+2x^2, \;\; (x+2)^3$

In Problems 63–74, perform the indicated operations and simplify the result. Leave your answer in factored form.

63. $\dfrac{x}{x^2-7x+6}-\dfrac{x}{x^2-2x-24}$

64. $\dfrac{x}{x-3}-\dfrac{x+1}{x^2+5x-24}$

65. $\dfrac{4x}{x^2-4}-\dfrac{2}{x^2+x-6}$

66. $\dfrac{3x}{x-1}-\dfrac{x-4}{x^2-2x+1}$

67. $\dfrac{3}{(x-1)^2(x+1)}+\dfrac{2}{(x-1)(x+1)^2}$

68. $\dfrac{2}{(x+2)^2(x-1)}-\dfrac{6}{(x+2)(x-1)^2}$

69. $\dfrac{x+4}{x^2-x-2}-\dfrac{2x+3}{x^2+2x-8}$

70. $\dfrac{2x-3}{x^2+8x+7}-\dfrac{x-2}{(x+1)^2}$

71. $\dfrac{1}{x}-\dfrac{2}{x^2+x}+\dfrac{3}{x^3-x^2}$

72. $\dfrac{x}{(x-1)^2}+\dfrac{2}{x}-\dfrac{x+1}{x^3-x^2}$

73. $\dfrac{1}{h}\left(\dfrac{1}{x+h}-\dfrac{1}{x}\right)$

74. $\dfrac{1}{h}\left[\dfrac{1}{(x+h)^2}-\dfrac{1}{x^2}\right]$

In Problems 75–86, perform the indicated operations and simplify the result. Leave your answer in factored form.

75. $\dfrac{1 + \dfrac{1}{x}}{1 - \dfrac{1}{x}}$

76. $\dfrac{4 + \dfrac{1}{x^2}}{3 - \dfrac{1}{x^2}}$

77. $\dfrac{2 - \dfrac{x+1}{x}}{3 + \dfrac{x-1}{x+1}}$

78. $\dfrac{1 - \dfrac{x}{x+1}}{2 - \dfrac{x-1}{x}}$

79. $\dfrac{\dfrac{x+4}{x-2} - \dfrac{x-3}{x+1}}{x+1}$

80. $\dfrac{\dfrac{x-2}{x+1} - \dfrac{x}{x-2}}{x+3}$

81. $\dfrac{\dfrac{x-2}{x+2} + \dfrac{x-1}{x+1}}{\dfrac{x}{x+1} - \dfrac{2x-3}{x}}$

82. $\dfrac{\dfrac{2x+5}{x} - \dfrac{x}{x-3}}{\dfrac{x^2}{x-3} - \dfrac{(x+1)^2}{x+3}}$

83. $1 - \dfrac{1}{1 - \dfrac{1}{x}}$

84. $1 - \dfrac{1}{1 - \dfrac{1}{1-x}}$

85. $\dfrac{2(x-1)^{-1} + 3}{3(x-1)^{-1} + 2}$

86. $\dfrac{4(x+2)^{-1} - 3}{3(x+2)^{-1} - 1}$

In Problems 87–94, expressions that occur in calculus are given. Reduce each expression to lowest terms.

87. $\dfrac{(2x+3)\cdot 3 - (3x-5)\cdot 2}{(3x-5)^2}$

88. $\dfrac{(4x+1)\cdot 5 - (5x-2)\cdot 4}{(5x-2)^2}$

89. $\dfrac{x\cdot 2x - (x^2+1)\cdot 1}{(x^2+1)^2}$

90. $\dfrac{x\cdot 2x - (x^2-4)\cdot 1}{(x^2-4)^2}$

91. $\dfrac{(3x+1)\cdot 2x - x^2\cdot 3}{(3x+1)^2}$

92. $\dfrac{(2x-5)\cdot 3x^2 - x^3\cdot 2}{(2x-5)^2}$

93. $\dfrac{(x^2+1)\cdot 3 - (3x+4)\cdot 2x}{(x^2+1)^2}$

94. $\dfrac{(x^2+9)\cdot 2 - (2x-5)\cdot 2x}{(x^2+9)^2}$

Applications and Extensions

95. The Lensmaker's Equation The focal length f of a lens with index of refraction n is

$$\frac{1}{f} = (n-1)\left[\frac{1}{R_1} + \frac{1}{R_2}\right]$$

where R_1 and R_2 are the radii of curvature of the front and back surfaces of the lens. Express f as a rational expression. Evaluate the rational expression for $n = 1.5$, $R_1 = 0.1$ meter, and $R_2 = 0.2$ meter.

96. Electrical Circuits An electrical circuit contains three resistors connected in parallel. If these three resistors provide resistance of R_1, R_2, and R_3 ohms, respectively, their combined resistance R is given by the formula

$$\frac{1}{R} = \frac{1}{R_1} + \frac{1}{R_2} + \frac{1}{R_3}$$

Express R as a rational expression. Evaluate R for $R_1 = 5$ ohms, $R_2 = 4$ ohms, and $R_3 = 10$ ohms.

Explaining Concepts: Discussion and Writing

97. The following expressions are called **continued fractions**:

$$1 + \frac{1}{x}, \quad 1 + \frac{1}{1 + \dfrac{1}{x}}, \quad 1 + \frac{1}{1 + \dfrac{1}{1 + \dfrac{1}{x}}}, \quad 1 + \frac{1}{1 + \dfrac{1}{1 + \dfrac{1}{1 + \dfrac{1}{x}}}}, \cdots$$

Each simplifies to an expression of the form

$$\frac{ax + b}{bx + c}$$

Trace the successive values of a, b, and c as you "continue" the fraction. Can you discover the patterns that these values follow? Go to the library and research Fibonacci numbers. Write a report on your findings.

98. Explain to a fellow student when you would use the LCM method to add two rational expressions. Give two examples of adding two rational expressions: one in which you use the LCM and the other in which you do not.

99. Which of the two options given in the text for simplifying complex rational expressions do you prefer? Write a brief paragraph stating the reasons for your choice.

R.8 *nth* Roots; Rational Exponents

PREPARING FOR THIS SECTION *Before getting started, review the following:*

- Exponents, Square Roots (Section R.2, pp. 21–24)

Now Work the **'Are You Prepared?'** problems on page 78.

OBJECTIVES **1** Work with *nth* Roots (p. 73)
  2 Simplify Radicals (p. 74)
  3 Rationalize Denominators (p. 75)
  4 Simplify Expressions with Rational Exponents (p. 76)

1 Work with *nth* Roots

DEFINITION

The **principal *nth* root of a real number** a, $n \geq 2$ an integer, symbolized by $\sqrt[n]{a}$, is defined as follows:

$$\sqrt[n]{a} = b \quad \text{means} \quad a = b^n$$

where $a \geq 0$ and $b \geq 0$ if n is even and a, b are any real numbers if n is odd.

In Words
The symbol $\sqrt[n]{a}$ means "give me the number that, when raised to the power n, equals a."

Notice that if a is negative and n is even, then $\sqrt[n]{a}$ is not defined. When it is defined, the principal *nth* root of a number is unique.

The symbol $\sqrt[n]{a}$ for the principal *nth* root of a is called a **radical**; the integer n is called the **index**, and a is called the **radicand**. If the index of a radical is 2, we call $\sqrt[2]{a}$ the **square root** of a and omit the index 2 by simply writing $\sqrt{a}$. If the index is 3, we call $\sqrt[3]{a}$ the **cube root** of a.

EXAMPLE 1 **Simplifying Principal *nth* Roots**

(a) $\sqrt[3]{8} = \sqrt[3]{2^3} = 2$

(b) $\sqrt[3]{-64} = \sqrt[3]{(-4)^3} = -4$

(c) $\sqrt[4]{\dfrac{1}{16}} = \sqrt[4]{\left(\dfrac{1}{2}\right)^4} = \dfrac{1}{2}$

(d) $\sqrt[6]{(-2)^6} = |-2| = 2$

These are examples of **perfect roots**, since each simplifies to a rational number. Notice the absolute value in Example 1(d). If n is even, then the principal *nth* root must be nonnegative.

In general, if $n \geq 2$ is an integer and a is a real number, we have

$$\sqrt[n]{a^n} = a \qquad \text{if } n \geq 3 \text{ is odd} \tag{1a}$$

$$\sqrt[n]{a^n} = |a| \qquad \text{if } n \geq 2 \text{ is even} \tag{1b}$$

Now Work PROBLEM 11

Radicals provide a way of representing many irrational real numbers. For example, it can be shown that there is no rational number whose square is 2. Using radicals, we can say that $\sqrt{2}$ is the positive number whose square is 2.

EXAMPLE 2

Using a Calculator to Approximate Roots

Use a calculator to approximate $\sqrt[5]{16}$.

Solution Figure 29 shows the result using a TI-84 Plus C graphing calculator. ●

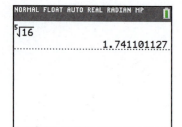

Figure 29

━━━━ **Now Work** PROBLEM 113

2 Simplify Radicals

Let $n \geq 2$ and $m \geq 2$ denote integers, and let a and b represent real numbers. Assuming that all radicals are defined, we have the following properties:

Properties of Radicals

$$\sqrt[n]{ab} = \sqrt[n]{a}\,\sqrt[n]{b} \qquad \textbf{(2a)}$$

$$\sqrt[n]{\frac{a}{b}} = \frac{\sqrt[n]{a}}{\sqrt[n]{b}} \qquad b \neq 0 \qquad \textbf{(2b)}$$

$$\sqrt[n]{a^m} = (\sqrt[n]{a})^m \qquad \textbf{(2c)}$$

When used in reference to radicals, the direction to "simplify" will mean to remove from the radicals any perfect roots that occur as factors.

EXAMPLE 3

Simplifying Radicals

(a) $\sqrt{32} = \sqrt{16 \cdot 2} = \sqrt{16} \cdot \sqrt{2} = 4\sqrt{2}$

    ↑ Factor out 16,   ↑ (2a)
    a perfect square.

(b) $\sqrt[3]{16} = \sqrt[3]{8 \cdot 2} = \sqrt[3]{8} \cdot \sqrt[3]{2} = \sqrt[3]{2^3} \cdot \sqrt[3]{2} = 2\sqrt[3]{2}$

    ↑ Factor out 8,   ↑ (2a)
    a perfect cube.

(c) $\sqrt[3]{-16x^4} = \sqrt[3]{-8 \cdot 2 \cdot x^3 \cdot x} = \sqrt[3]{(-8x^3)(2x)}$

    ↑ Factor perfect   ↑ Group perfect
    cubes inside radical.   cubes.

$\qquad = \sqrt[3]{(-2x)^3 \cdot 2x} = \sqrt[3]{(-2x)^3} \cdot \sqrt[3]{2x} = -2x\sqrt[3]{2x}$

                    ↑ (2a)

(d) $\sqrt[4]{\dfrac{16x^5}{81}} = \sqrt[4]{\dfrac{2^4 x^4 x}{3^4}} = \sqrt[4]{\left(\dfrac{2x}{3}\right)^4 \cdot x} = \sqrt[4]{\left(\dfrac{2x}{3}\right)^4} \cdot \sqrt[4]{x} = \left|\dfrac{2x}{3}\right| \sqrt[4]{x}$ ●

━━━━ **Now Work** PROBLEMS 15 AND 21

Two or more radicals can be combined, provided that they have the same index and the same radicand. Such radicals are called **like radicals**.

EXAMPLE 4

Combining Like Radicals

(a) $-8\sqrt{12} + \sqrt{3} = -8\sqrt{4 \cdot 3} + \sqrt{3}$

$\qquad\qquad\qquad = -8 \cdot \sqrt{4}\,\sqrt{3} + \sqrt{3}$

$\qquad\qquad\qquad = -16\sqrt{3} + \sqrt{3} = -15\sqrt{3}$

(b) $\sqrt[3]{8x^4} + \sqrt[3]{-x} + 4\sqrt[3]{27x} = \sqrt[3]{2^3x^3x} + \sqrt[3]{-1 \cdot x} + 4\sqrt[3]{3^3x}$

$\qquad = \sqrt[3]{(2x)^3} \cdot \sqrt[3]{x} + \sqrt[3]{-1} \cdot \sqrt[3]{x} + 4\sqrt[3]{3^3} \cdot \sqrt[3]{x}$

$\qquad = 2x\sqrt[3]{x} - 1 \cdot \sqrt[3]{x} + 12\sqrt[3]{x}$

$\qquad = (2x + 11)\sqrt[3]{x}$

●

Now Work PROBLEM 39

3 Rationalize Denominators

When radicals occur in quotients, it is customary to rewrite the quotient so that the new denominator contains no radicals. This process is referred to as **rationalizing the denominator**.

The idea is to multiply by an appropriate expression so that the new denominator contains no radicals. For example:

If a Denominator Contains the Factor	Multiply by	To Obtain a Denominator Free of Radicals
$\sqrt{3}$	$\sqrt{3}$	$(\sqrt{3})^2 = 3$
$\sqrt{3} + 1$	$\sqrt{3} - 1$	$(\sqrt{3})^2 - 1^2 = 3 - 1 = 2$
$\sqrt{2} - 3$	$\sqrt{2} + 3$	$(\sqrt{2})^2 - 3^2 = 2 - 9 = -7$
$\sqrt{5} - \sqrt{3}$	$\sqrt{5} + \sqrt{3}$	$(\sqrt{5})^2 - (\sqrt{3})^2 = 5 - 3 = 2$
$\sqrt[3]{4}$	$\sqrt[3]{2}$	$\sqrt[3]{4} \cdot \sqrt[3]{2} = \sqrt[3]{8} = 2$

In rationalizing the denominator of a quotient, be sure to multiply both the numerator and the denominator by the expression.

EXAMPLE 5

Rationalizing Denominators

Rationalize the denominator of each expression:

(a) $\dfrac{1}{\sqrt{3}}$ (b) $\dfrac{5}{4\sqrt{2}}$ (c) $\dfrac{\sqrt{2}}{\sqrt{3} - 3\sqrt{2}}$

Solution (a) The denominator contains the factor $\sqrt{3}$, so we multiply the numerator and denominator by $\sqrt{3}$ to obtain

$$\frac{1}{\sqrt{3}} = \frac{1}{\sqrt{3}} \cdot \frac{\sqrt{3}}{\sqrt{3}} = \frac{\sqrt{3}}{(\sqrt{3})^2} = \frac{\sqrt{3}}{3}$$

(b) The denominator contains the factor $\sqrt{2}$, so we multiply the numerator and denominator by $\sqrt{2}$ to obtain

$$\frac{5}{4\sqrt{2}} = \frac{5}{4\sqrt{2}} \cdot \frac{\sqrt{2}}{\sqrt{2}} = \frac{5\sqrt{2}}{4(\sqrt{2})^2} = \frac{5\sqrt{2}}{4 \cdot 2} = \frac{5\sqrt{2}}{8}$$

(c) The denominator contains the factor $\sqrt{3} - 3\sqrt{2}$, so we multiply the numerator and denominator by $\sqrt{3} + 3\sqrt{2}$ to obtain

$$\frac{\sqrt{2}}{\sqrt{3} - 3\sqrt{2}} = \frac{\sqrt{2}}{\sqrt{3} - 3\sqrt{2}} \cdot \frac{\sqrt{3} + 3\sqrt{2}}{\sqrt{3} + 3\sqrt{2}} = \frac{\sqrt{2}(\sqrt{3} + 3\sqrt{2})}{(\sqrt{3})^2 - (3\sqrt{2})^2}$$

$$= \frac{\sqrt{2}\sqrt{3} + 3(\sqrt{2})^2}{3 - 18} = \frac{\sqrt{6} + 6}{-15} = -\frac{6 + \sqrt{6}}{15}$$

●

Now Work PROBLEM 53

4 Simplify Expressions with Rational Exponents

Radicals are used to define rational exponents.

DEFINITION

If a is a real number and $n \geq 2$ is an integer, then

$$a^{1/n} = \sqrt[n]{a} \tag{3}$$

provided that $\sqrt[n]{a}$ exists.

Note that if n is even and $a < 0$, then $\sqrt[n]{a}$ and $a^{1/n}$ do not exist.

EXAMPLE 6 **Writing Expressions Containing Fractional Exponents as Radicals**

(a) $4^{1/2} = \sqrt{4} = 2$ (b) $8^{1/2} = \sqrt{8} = 2\sqrt{2}$

(c) $(-27)^{1/3} = \sqrt[3]{-27} = -3$ (d) $16^{1/3} = \sqrt[3]{16} = 2\sqrt[3]{2}$ •

DEFINITION

If a is a real number and m and n are integers containing no common factors, with $n \geq 2$, then

$$a^{m/n} = \sqrt[n]{a^m} = (\sqrt[n]{a})^m \tag{4}$$

provided that $\sqrt[n]{a}$ exists.

We have two comments about equation (4):

1. The exponent $\dfrac{m}{n}$ must be in lowest terms, and $n \geq 2$ must be positive.
2. In simplifying the rational expression $a^{m/n}$, either $\sqrt[n]{a^m}$ or $(\sqrt[n]{a})^m$ may be used, the choice depending on which is easier to simplify. Generally, taking the root first, as in $(\sqrt[n]{a})^m$, is easier.

EXAMPLE 7 **Using Equation (4)**

(a) $4^{3/2} = (\sqrt{4})^3 = 2^3 = 8$ (b) $(-8)^{4/3} = (\sqrt[3]{-8})^4 = (-2)^4 = 16$

(c) $(32)^{-2/5} = (\sqrt[5]{32})^{-2} = 2^{-2} = \dfrac{1}{4}$ (d) $256^{6/4} = 25^{3/2} = (\sqrt{25})^3 = 5^3 = 125$ •

Now Work PROBLEM 63

It can be shown that the Laws of Exponents hold for rational exponents. The next example illustrates using the Laws of Exponents to simplify.

EXAMPLE 8 **Simplifying Expressions Containing Rational Exponents**

Simplify each expression. Express your answer so that only positive exponents occur. Assume that the variables are positive.

(a) $(x^{2/3}y)(x^{-2}y)^{1/2}$ (b) $\left(\dfrac{2x^{1/3}}{y^{2/3}}\right)^{-3}$ (c) $\left(\dfrac{9x^2y^{1/3}}{x^{1/3}y}\right)^{1/2}$

Solution   (a) $(x^{2/3}y)(x^{-2}y)^{1/2} = (x^{2/3}y)\left[(x^{-2})^{1/2}y^{1/2}\right]$

$$= x^{2/3}yx^{-1}y^{1/2}$$

$$= (x^{2/3}\cdot x^{-1})(y\cdot y^{1/2})$$

$$= x^{-1/3}y^{3/2}$$

$$= \frac{y^{3/2}}{x^{1/3}}$$

(b) $\left(\dfrac{2x^{1/3}}{y^{2/3}}\right)^{-3} = \left(\dfrac{y^{2/3}}{2x^{1/3}}\right)^3 = \dfrac{(y^{2/3})^3}{(2x^{1/3})^3} = \dfrac{y^2}{2^3(x^{1/3})^3} = \dfrac{y^2}{8x}$

(c) $\left(\dfrac{9x^2y^{1/3}}{x^{1/3}y}\right)^{1/2} = \left(\dfrac{9x^{2-(1/3)}}{y^{1-(1/3)}}\right)^{1/2} = \left(\dfrac{9x^{5/3}}{y^{2/3}}\right)^{1/2} = \dfrac{9^{1/2}(x^{5/3})^{1/2}}{(y^{2/3})^{1/2}} = \dfrac{3x^{5/6}}{y^{1/3}}$

 Now Work PROBLEM 83

 The next two examples illustrate some algebra that you will need to know for certain calculus problems.

EXAMPLE 9   **Writing an Expression as a Single Quotient**

Write the following expression as a single quotient in which only positive exponents appear.

$$(x^2+1)^{1/2} + x\cdot\frac{1}{2}(x^2+1)^{-1/2}\cdot 2x$$

Solution $(x^2+1)^{1/2} + x\cdot\dfrac{1}{2}(x^2+1)^{-1/2}\cdot 2x = (x^2+1)^{1/2} + \dfrac{x^2}{(x^2+1)^{1/2}}$

$$= \frac{(x^2+1)^{1/2}(x^2+1)^{1/2} + x^2}{(x^2+1)^{1/2}}$$

$$= \frac{(x^2+1) + x^2}{(x^2+1)^{1/2}}$$

$$= \frac{2x^2+1}{(x^2+1)^{1/2}}$$

 Now Work PROBLEM 89

EXAMPLE 10   **Factoring an Expression Containing Rational Exponents**

Factor and simplify: $\dfrac{4}{3}x^{1/3}(2x+1) + 2x^{4/3}$

Solution   Begin by writing $2x^{4/3}$ as a fraction with 3 as the denominator.

$$\frac{4}{3}x^{1/3}(2x+1) + 2x^{4/3} = \frac{4x^{1/3}(2x+1)}{3} + \frac{6x^{4/3}}{3} \underset{\underset{\text{Add the two fractions}}{\uparrow}}{=} \frac{4x^{1/3}(2x+1) + 6x^{4/3}}{3}$$

$$= \underset{\underset{2 \text{ and } x^{1/3} \text{ are common factors}}{\uparrow}}{\frac{2x^{1/3}[2(2x+1)+3x]}{3}} = \underset{\underset{\text{Simplify}}{\uparrow}}{\frac{2x^{1/3}(7x+2)}{3}}$$

 Now Work PROBLEM 101

Historical Note

The radical sign, $\sqrt{}$, was first used in print by Christoff Rudolff in 1525. It is thought to be the manuscript form of the letter *r* (for the Latin word *radix = root*), although this has not been quite conclusively confirmed. It took a long time for $\sqrt{}$ to become the standard symbol for a square root and much longer to standardize $\sqrt[3]{}$, $\sqrt[4]{}$, $\sqrt[5]{}$, and so on. The indexes of the root were placed in every conceivable position, with

$$\sqrt[3]{8}, \quad \sqrt{③8}, \quad \text{and} \quad \underset{3}{\sqrt{}}8$$

all being variants for $\sqrt[3]{8}$. The notation $\sqrt{}\sqrt{}16$ was popular for $\sqrt[4]{16}$. By the 1700s, the index had settled where we now put it.

The bar on top of the present radical symbol, as follows,

$$\sqrt{a^2 + 2ab + b^2}$$

is the last survivor of the **vinculum**, a bar placed atop an expression to indicate what we would now indicate with parentheses. For example,

$$\overline{ab + c} = a(b + c)$$

R.8 Assess Your Understanding

'Are You Prepared?' *Answers are given at the end of these exercises. If you get a wrong answer, read the pages in red.*

1. $(-3)^2 = $ ___; $-3^2 = $ ___ (pp. 21–24)

2. $\sqrt{16} = $ ___; $\sqrt{(-4)^2} = $ ___ (pp. 21–24)

Concepts and Vocabulary

3. In the symbol $\sqrt[n]{a}$, the integer n is called the _____.

4. We call $\sqrt[3]{a}$ the _____ _____ of a.

5. Let $n \geq 2$ and $m \geq 2$ be integers, and let a and b be real numbers. Which of the following is not a property of radicals? Assume all radicals are defined.

(a) $\sqrt[n]{\dfrac{a}{b}} = \dfrac{\sqrt[n]{a}}{\sqrt[n]{b}}$ (b) $\sqrt[n]{a + b} = \sqrt[n]{a} + \sqrt[n]{b}$

(c) $\sqrt[n]{ab} = \sqrt[n]{a}\sqrt[n]{b}$ (d) $\sqrt[n]{a^m} = (\sqrt[n]{a})^m$

6. If a is a real number and $n \geq 2$ is an integer, then which of the following expressions is equivalent to $\sqrt[n]{a}$, provided that it exists?

(a) a^{-n} (b) a^n (c) $\dfrac{1}{a^n}$ (d) $a^{1/n}$

7. Which of the following phrases best defines like radicals?
   (a) Radical expressions that have the same index
   (b) Radical expressions that have the same radicand
   (c) Radical expressions that have the same index and the same radicand
   (d) Radical expressions that have the same variable

8. To rationalize the denominator of the expression $\dfrac{\sqrt{2}}{1 - \sqrt{3}}$, multiply both the numerator and the denominator by which of the following?

(a) $\sqrt{3}$ (b) $\sqrt{2}$ (c) $1 + \sqrt{3}$ (d) $1 - \sqrt{3}$

9. *True or False* $\sqrt[5]{-32} = -2$

10. *True or False* $\sqrt[4]{(-3)^4} = -3$

Skill Building

In Problems 11–48, simplify each expression. Assume that all variables are positive when they appear.

11. $\sqrt[3]{27}$

12. $\sqrt[4]{16}$

13. $\sqrt[3]{-8}$

14. $\sqrt[3]{-1}$

15. $\sqrt{8}$

16. $\sqrt[3]{54}$

17. $\sqrt[3]{-8x^4}$

18. $\sqrt[4]{48x^5}$

19. $\sqrt[4]{x^{12}y^8}$

20. $\sqrt[5]{x^{10}y^5}$

21. $\sqrt[4]{\dfrac{x^9y^7}{xy^3}}$

22. $\sqrt[3]{\dfrac{3xy^2}{81x^4y^2}}$

23. $\sqrt{36x}$

24. $\sqrt{9x^5}$

25. $\sqrt[4]{162x^9\,y^{12}}$

26. $\sqrt[3]{-40x^{14}\,y^{10}}$

27. $\sqrt{3x^2}\sqrt{12x}$

28. $\sqrt{5x}\sqrt{20x^3}$

29. $(\sqrt{5}\sqrt[3]{9})^2$

30. $(\sqrt[3]{3}\sqrt{10})^4$

31. $(3\sqrt{6})(2\sqrt{2})$

32. $(5\sqrt{8})(-3\sqrt{3})$

33. $3\sqrt{2} + 4\sqrt{2}$

34. $6\sqrt{5} - 4\sqrt{5}$

35. $-\sqrt{18} + 2\sqrt{8}$

36. $2\sqrt{12} - 3\sqrt{27}$

37. $(\sqrt{3} + 3)(\sqrt{3} - 1)$

38. $(\sqrt{5} - 2)(\sqrt{5} + 3)$

39. $5\sqrt[3]{2} - 2\sqrt[3]{54}$

40. $9\sqrt[3]{24} - \sqrt[3]{81}$

41. $(\sqrt{x} - 1)^2$

42. $(\sqrt{x} + \sqrt{5})^2$

43. $\sqrt[3]{16x^4} - \sqrt[3]{2x}$

44. $\sqrt[4]{32x} + \sqrt[4]{2x^5}$

45. $\sqrt{8x^3} - 3\sqrt{50x}$

46. $3x\sqrt{9y} + 4\sqrt{25y}$

47. $\sqrt[3]{16x^4y} - 3x\sqrt[3]{2xy} + 5\sqrt[3]{-2xy^4}$

48. $8xy - \sqrt{25x^2y^2} + \sqrt[3]{8x^3y^3}$

In Problems 49–62, rationalize the denominator of each expression. Assume that all variables are positive when they appear.

49. $\dfrac{1}{\sqrt{2}}$

50. $\dfrac{2}{\sqrt{3}}$

51. $\dfrac{-\sqrt{3}}{\sqrt{5}}$

52. $\dfrac{-\sqrt{3}}{\sqrt{8}}$

53. $\dfrac{\sqrt{3}}{5-\sqrt{2}}$

54. $\dfrac{\sqrt{2}}{\sqrt{7}+2}$

55. $\dfrac{2-\sqrt{5}}{2+3\sqrt{5}}$

56. $\dfrac{\sqrt{3}-1}{2\sqrt{3}+3}$

57. $\dfrac{5}{\sqrt{2}-1}$

58. $\dfrac{-3}{\sqrt{5}+4}$

59. $\dfrac{5}{\sqrt[3]{2}}$

60. $\dfrac{-2}{\sqrt[3]{9}}$

61. $\dfrac{\sqrt{x+h}-\sqrt{x}}{\sqrt{x+h}+\sqrt{x}}$

62. $\dfrac{\sqrt{x+h}+\sqrt{x-h}}{\sqrt{x+h}-\sqrt{x-h}}$

In Problems 63–78, simplify each expression.

63. $8^{2/3}$

64. $4^{3/2}$

65. $(-27)^{1/3}$

66. $16^{3/4}$

67. $16^{3/2}$

68. $25^{3/2}$

69. $9^{-3/2}$

70. $16^{-3/2}$

71. $\left(\dfrac{9}{8}\right)^{3/2}$

72. $\left(\dfrac{27}{8}\right)^{2/3}$

73. $\left(\dfrac{8}{9}\right)^{-3/2}$

74. $\left(\dfrac{8}{27}\right)^{-2/3}$

75. $(-1000)^{-1/3}$

76. $-25^{-1/2}$

77. $\left(-\dfrac{64}{125}\right)^{-2/3}$

78. $-81^{-3/4}$

In Problems 79–86, simplify each expression. Express your answer so that only positive exponents occur. Assume that the variables are positive.

79. $x^{3/4}x^{1/3}x^{-1/2}$

80. $x^{2/3}x^{1/2}x^{-1/4}$

81. $(x^3y^6)^{1/3}$

82. $(x^4y^8)^{3/4}$

83. $\dfrac{(x^2y)^{1/3}(xy^2)^{2/3}}{x^{2/3}y^{2/3}}$

84. $\dfrac{(xy)^{1/4}(x^2y^2)^{1/2}}{(x^2y)^{3/4}}$

85. $\dfrac{(16x^2y^{-1/3})^{3/4}}{(xy^2)^{1/4}}$

86. $\dfrac{(4x^{-1}y^{1/3})^{3/2}}{(xy)^{3/2}}$

Applications and Extensions

In Problems 87–100, expressions that occur in calculus are given. Write each expression as a single quotient in which only positive exponents and/or radicals appear.

87. $\dfrac{x}{(1+x)^{1/2}}+2(1+x)^{1/2}$ $x>-1$

88. $\dfrac{1+x}{2x^{1/2}}+x^{1/2}$ $x>0$

89. $2x(x^2+1)^{1/2}+x^2\cdot\dfrac{1}{2}(x^2+1)^{-1/2}\cdot 2x$

90. $(x+1)^{1/3}+x\cdot\dfrac{1}{3}(x+1)^{-2/3}$ $x\neq -1$

91. $\sqrt{4x+3}\cdot\dfrac{1}{2\sqrt{x-5}}+\sqrt{x-5}\cdot\dfrac{1}{5\sqrt{4x+3}}$ $x>5$

92. $\dfrac{\sqrt[3]{8x+1}}{3\sqrt[3]{(x-2)^2}}+\dfrac{\sqrt[3]{x-2}}{24\sqrt[3]{(8x+1)^2}}$ $x\neq 2, x\neq -\dfrac{1}{8}$

93. $\dfrac{\sqrt{1+x}-x\cdot\dfrac{1}{2\sqrt{1+x}}}{1+x}$ $x>-1$

94. $\dfrac{\sqrt{x^2+1}-x\cdot\dfrac{2x}{2\sqrt{x^2+1}}}{x^2+1}$

95. $\dfrac{(x+4)^{1/2}-2x(x+4)^{-1/2}}{x+4}$ $x>-4$

96. $\dfrac{(9-x^2)^{1/2}+x^2(9-x^2)^{-1/2}}{9-x^2}$ $-3<x<3$

97. $\dfrac{\dfrac{x^2}{(x^2-1)^{1/2}}-(x^2-1)^{1/2}}{x^2}$ $x<-1$ or $x>1$

98. $\dfrac{(x^2+4)^{1/2}-x^2(x^2+4)^{-1/2}}{x^2+4}$

99. $\dfrac{\dfrac{1+x^2}{2\sqrt{x}}-2x\sqrt{x}}{(1+x^2)^2}$ $x>0$

100. $\dfrac{2x(1-x^2)^{1/3}+\dfrac{2}{3}x^3(1-x^2)^{-2/3}}{(1-x^2)^{2/3}}$ $x\neq -1, x\neq 1$

1.1 Linear Equations

PREPARING FOR THIS SECTION *Before getting started, review the following:*

- Properties of Real Numbers (Section R.1, pp. 9–13)
- Domain of a Variable (Section R.2, p. 21)

✎ **Now Work** the *'Are You Prepared?'* problems on page 90.

OBJECTIVES **1** Solve a Linear Equation (p. 84)
          2 Solve Equations That Lead to Linear Equations (p. 86)
          3 Solve Problems That Can Be Modeled by Linear Equations (p. 87)

An **equation in one variable** is a statement in which two expressions, at least one containing the variable, are set equal. The expressions are called the **sides** of the equation. Since an equation is a statement, it may be true or false, depending on the value of the variable. Unless otherwise restricted, the admissible values of the variable are those in the domain of the variable. These admissible values of the variable, if any, that result in a true statement are called **solutions**, or **roots**, of the equation. To **solve an equation** means to find all the solutions of the equation.

For example, the following are all equations in one variable, x:

$$x + 5 = 9 \qquad x^2 + 5x = 2x - 2 \qquad \frac{x^2 - 4}{x + 1} = 0 \qquad \sqrt{x^2 + 9} = 5$$

The first of these statements, $x + 5 = 9$, is true when $x = 4$ and false for any other choice of x. That is, 4 is a solution of the equation $x + 5 = 9$. We also say that 4 **satisfies** the equation $x + 5 = 9$, because, when 4 is substituted for x, a true statement results.

Sometimes an equation will have more than one solution. For example, the equation

$$\frac{x^2 - 4}{x + 1} = 0$$

has $x = -2$ and $x = 2$ as solutions.

Usually, we will write the solutions of an equation in set notation. This set is called the **solution set** of the equation. For example, the solution set of the equation $x^2 - 9 = 0$ is $\{-3, 3\}$.

Some equations have no real solution. For example, $x^2 + 9 = 5$ has no real solution, because there is no real number whose square, when added to 9, equals 5.

An equation that is satisfied for every value of the variable for which both sides are defined is called an **identity**. For example, the equation

$$3x + 5 = x + 3 + 2x + 2$$

is an identity, because this statement is true for any real number x.

> One method for solving an equation is to replace the original equation by a succession of **equivalent equations**, equations having the same solution set, until an equation with an obvious solution is obtained.

For example, consider the following succession of equivalent equations:

$$2x + 3 = 13$$
$$2x = 10$$
$$x = 5$$

We conclude that the solution set of the original equation is $\{5\}$.

How do we obtain equivalent equations? In general, there are five ways.

Procedures That Result in Equivalent Equations

1. Interchange the two sides of the equation:

   Replace $\quad 3 = x \quad$ by $\quad x = 3$

2. Simplify the sides of the equation by combining like terms, eliminating parentheses, and so on:

   Replace $\quad (x + 2) + 6 = 2x + (x + 1)$

   by $\qquad\qquad x + 8 = 3x + 1$

3. Add or subtract the same expression on both sides of the equation:

   Replace $\qquad 3x - 5 = 4$

   by $\quad (3x - 5) + 5 = 4 + 5$

4. Multiply or divide both sides of the equation by the same nonzero expression:

   Replace $\qquad \dfrac{3x}{x - 1} = \dfrac{6}{x - 1} \qquad x \neq 1$

   by $\quad \dfrac{3x}{x - 1} \cdot (x - 1) = \dfrac{6}{x - 1} \cdot (x - 1)$

WARNING Squaring both sides of an equation does not necessarily lead to an equivalent equation. For example, $x = 3$ has one solution, but $x^2 = 9$ has two solutions, $x = -3$ and $x = 3$. ■

5. If one side of the equation is 0 and the other side can be factored, then we may use the Zero-Product Property* and set each factor equal to 0:

   Replace $\qquad x(x - 3) = 0$

   by $\quad x = 0 \quad$ or $\quad x - 3 = 0$

Whenever it is possible to solve an equation in your head, do so. For example,

The solution of $2x = 8$ is $x = 4$.
The solution of $3x - 15 = 0$ is $x = 5$.

━ **Now Work** PROBLEM 11

EXAMPLE 1

Solving an Equation

Solve the equation: $3x - 5 = 4$

Solution

Replace the original equation by a succession of equivalent equations.

$$3x - 5 = 4$$
$$(3x - 5) + 5 = 4 + 5 \qquad \text{Add 5 to both sides.}$$
$$3x = 9 \qquad \text{Simplify.}$$
$$\frac{3x}{3} = \frac{9}{3} \qquad \text{Divide both sides by 3.}$$
$$x = 3 \qquad \text{Simplify.}$$

The last equation, $x = 3$, has the single solution 3. All these equations are equivalent, so 3 is the only solution of the original equation, $3x - 5 = 4$.

*The Zero-Product Property says that if $ab = 0$, then $a = 0$ or $b = 0$ or both equal 0.

✓**Check:** Check the solution by substituting 3 for x in the original equation.

$$3x - 5 = 4$$
$$3(3) - 5 \overset{?}{=} 4$$
$$9 - 5 \overset{?}{=} 4$$
$$4 = 4$$

The solution checks. The solution set is $\{3\}$. ●

✏— **Now Work** PROBLEM 25

Steps for Solving Equations

STEP 1: List any restrictions on the domain of the variable.

STEP 2: Simplify the equation by replacing the original equation by a succession of equivalent equations using the procedures listed earlier.

STEP 3: If the result of Step 2 is a product of factors equal to 0, use the Zero-Product Property and set each factor equal to 0 (procedure 5).

STEP 4: Check your solution(s).

---

1 Solve a Linear Equation

Linear equations are equations such as

$$3x + 12 = 0 \qquad -2x + 5 = 0 \qquad \frac{1}{2}x - \sqrt{3} = 0$$

DEFINITION

A **linear equation in one variable** is an equation equivalent in form to

$$ax + b = 0$$

where a and b are real numbers and $a \neq 0$.

Sometimes a linear equation is called a **first-degree equation**, because the left side is a polynomial in x of degree 1.

It is relatively easy to solve a linear equation. The idea is to *isolate* the variable:

$$ax + b = 0 \qquad a \neq 0$$
$$ax = -b \qquad \text{Subtract } b \text{ from both sides.}$$
$$x = \frac{-b}{a} \qquad \text{Divide both sides by } a, a \neq 0.$$

The linear equation $ax + b = 0, a \neq 0$, has the single solution given by the formula $x = -\dfrac{b}{a}$.

EXAMPLE 2   **Solving a Linear Equation**

Solve the equation: $\dfrac{1}{2}(x + 5) - 4 = \dfrac{1}{3}(2x - 1)$

Solution   To clear the equation of fractions, multiply both sides by 6, the least common multiple (LCM) of the denominators of the fractions $\frac{1}{2}$ and $\frac{1}{3}$.

$$\frac{1}{2}(x + 5) - 4 = \frac{1}{3}(2x - 1)$$

$$6\left[\frac{1}{2}(x + 5) - 4\right] = 6\left[\frac{1}{3}(2x - 1)\right] \quad \text{Multiply both sides by 6, the LCM of 2 and 3.}$$

$$3(x + 5) - 6 \cdot 4 = 2(2x - 1) \quad \text{Use the Distributive Property on the left and the Associative Property on the right.}$$

$$3x + 15 - 24 = 4x - 2 \quad \text{Use the Distributive Property.}$$

$$3x - 9 = 4x - 2 \quad \text{Combine like terms.}$$

$$3x - 9 + 9 = 4x - 2 + 9 \quad \text{Add 9 to each side.}$$

$$3x = 4x + 7 \quad \text{Simplify.}$$

$$3x - 4x = 4x + 7 - 4x \quad \text{Subtract 4x from each side.}$$

$$-x = 7 \quad \text{Simplify.}$$

$$x = -7 \quad \text{Multiply both sides by } -1.$$

✔**Check:** Substitute -7 for x in the expressions on the left and right sides of the original equation, and simplify. If the two expressions are equal, the solution checks.

$$\frac{1}{2}(x + 5) - 4 = \frac{1}{2}(-7 + 5) - 4 = \frac{1}{2}(-2) - 4 = -1 - 4 = -5$$

$$\frac{1}{3}(2x - 1) = \frac{1}{3}[2(-7) - 1] = \frac{1}{3}(-14 - 1) = \frac{1}{3}(-15) = -5$$

Since the two expressions are equal, the solution checks. The solution set is $\{-7\}$. ●

    Now Work PROBLEM 35

EXAMPLE 3   **Solving a Linear Equation Using a Calculator**

Solve the equation: $2.78x + \dfrac{2}{17.931} = 54.06$

Round the answer to two decimal places.

Solution   To avoid rounding errors, solve for x before using a calculator.

$$2.78x + \frac{2}{17.931} = 54.06$$

$$2.78x = 54.06 - \frac{2}{17.931} \quad \text{Subtract } \frac{2}{17.931} \text{ from each side.}$$

$$x = \frac{54.06 - \dfrac{2}{17.931}}{2.78} \quad \text{Divide each side by 2.78.}$$

Now use your calculator. The solution, rounded to two decimal places, is 19.41.

✔**Check:** Store the calculator solution 19.40592134 in memory, and proceed to evaluate $2.78x + \dfrac{2}{17.931}$.

$$(2.78)(19.40592134) + \frac{2}{17.931} = 54.06$$

                                                          ●

    Now Work PROBLEM 67

2 Solve Equations That Lead to Linear Equations

EXAMPLE 4

Solving an Equation That Leads to a Linear Equation

Solve the equation: $(2y + 1)(y - 1) = (y + 5)(2y - 5)$

Solution

$$(2y + 1)(y - 1) = (y + 5)(2y - 5)$$

$2y^2 - y - 1 = 2y^2 + 5y - 25$ Multiply and combine like terms.

$-y - 1 = 5y - 25$ Subtract $2y^2$ from each side.

$-y = 5y - 24$ Add 1 to each side.

$-6y = -24$ Subtract 5y from each side.

$y = 4$ Divide both sides by -6.

✓**Check:** $(2y + 1)(y - 1) = [2(4) + 1](4 - 1) = (8 + 1)(3) = (9)(3) = 27$

$(y + 5)(2y - 5) = (4 + 5)[2(4) - 5] = (9)(8 - 5) = (9)(3) = 27$

The two expressions are equal, so the solution checks. The solution set is $\{4\}$.   ●

EXAMPLE 5

Solving an Equation That Leads to a Linear Equation

Solve the equation: $\dfrac{3}{x - 2} = \dfrac{1}{x - 1} + \dfrac{7}{(x - 1)(x - 2)}$

Solution

First, notice that the domain of the variable is $\{x \mid x \neq 1, x \neq 2\}$. Clear the equation of fractions by multiplying both sides by the least common multiple of the denominators of the three fractions, $(x - 1)(x - 2)$.

$$\dfrac{3}{x - 2} = \dfrac{1}{x - 1} + \dfrac{7}{(x - 1)(x - 2)}$$

$(x - 1)(x - 2) \cdot \dfrac{3}{x - 2} = (x - 1)(x - 2)\left[\dfrac{1}{x - 1} + \dfrac{7}{(x - 1)(x - 2)}\right]$ Multiply both sides by $(x - 1)(x - 2)$; cancel on the left.

$3x - 3 = (x - 1)(x - 2)\dfrac{1}{x - 1} + (x - 1)(x - 2)\dfrac{7}{(x - 1)(x - 2)}$ Use the Distributive Property on each side; cancel on the right.

$3x - 3 = (x - 2) + 7$ Simplify.

$3x - 3 = x + 5$ Combine like terms.

$2x = 8$ Add 3 to each side; subtract x from each side.

$x = 4$ Divide by 2.

✓**Check:** $\dfrac{3}{x - 2} = \dfrac{3}{4 - 2} = \dfrac{3}{2}$

$\dfrac{1}{x - 1} + \dfrac{7}{(x - 1)(x - 2)} = \dfrac{1}{4 - 1} + \dfrac{7}{(4 - 1)(4 - 2)} = \dfrac{1}{3} + \dfrac{7}{3 \cdot 2} = \dfrac{2}{6} + \dfrac{7}{6} = \dfrac{9}{6} = \dfrac{3}{2}$

The solution checks. The solution set is $\{4\}$.   ●

━━ **Now Work** PROBLEM 61

EXAMPLE 6

An Equation with No Solution

Solve the equation: $\dfrac{3x}{x - 1} + 2 = \dfrac{3}{x - 1}$

Solution   First, note that the domain of the variable is $\{x \mid x \neq 1\}$. Since the two quotients in the equation have the same denominator, $x - 1$, simplify by multiplying both sides by $x - 1$. The resulting equation is equivalent to the original equation, since we are multiplying by $x - 1$, which is not 0. (Remember, $x \neq 1$.)

$$\frac{3x}{x - 1} + 2 = \frac{3}{x - 1}$$

$$\left(\frac{3x}{x - 1} + 2\right) \cdot (x - 1) = \frac{3}{x - 1} \cdot (x - 1) \qquad \text{Multiply both sides by } x - 1; \text{ cancel on the right.}$$

$$\frac{3x}{x - 1} \cdot (x - 1) + 2 \cdot (x - 1) = 3 \qquad \text{Use the Distributive Property on the left side; cancel on the left.}$$

$$3x + 2x - 2 = 3 \qquad \text{Simplify.}$$

$$5x - 2 = 3 \qquad \text{Combine like terms.}$$

$$5x = 5 \qquad \text{Add 2 to each side.}$$

$$x = 1 \qquad \text{Divide both sides by 5.}$$

NOTE Example 6 illustrates the appearance of an *extraneous solution*. This will be discussed more in Section 1.4. ∎

The solution appears to be 1. But recall that $x = 1$ is not in the domain of the variable, so this value must be discarded. The equation has no solution. The solution set is ∅. ●

═══ **Now Work** PROBLEM **51**

EXAMPLE 7   **Converting to Fahrenheit from Celsius**

In the United States we measure temperature in both degrees Fahrenheit (°F) and degrees Celsius (°C), which are related by the formula $C = \frac{5}{9}(F - 32)$. What are the Fahrenheit temperatures corresponding to Celsius temperatures of 0°, 10°, 20°, and 30°C?

Solution   We could solve four equations for F by replacing C each time by 0, 10, 20, and 30. Instead, it is much easier and faster first to solve the equation $C = \frac{5}{9}(F - 32)$ for F and then to substitute in the values of C.

$$C = \frac{5}{9}(F - 32)$$

$$9C = 5(F - 32) \qquad \text{Multiply both sides by 9.}$$

$$9C = 5F - 160 \qquad \text{Use the Distributive Property.}$$

$$5F - 160 = 9C \qquad \text{Interchange sides.}$$

$$5F = 9C + 160 \qquad \text{Add 160 to each side.}$$

$$F = \frac{9}{5}C + 32 \qquad \text{Divide both sides by 5.}$$

Now do the required arithmetic.

$$0°C: \quad F = \frac{9}{5}(0) + 32 = 32°F$$

$$10°C: \quad F = \frac{9}{5}(10) + 32 = 50°F$$

$$20°C: \quad F = \frac{9}{5}(20) + 32 = 68°F$$

NOTE The icon is a Model It! icon. It indicates that the discussion or problem involves modeling. ∎

$$30°C: \quad F = \frac{9}{5}(30) + 32 = 86°F$$

●

3 Solve Problems That Can Be Modeled by Linear Equations

Although each situation has its unique features, we can provide an outline of the steps to follow when solving applied problems.

Steps for Solving Applied Problems

STEP 1: Read the problem carefully, perhaps two or three times. Pay particular attention to the question being asked in order to identify what you are looking for. Identify any relevant formulas you may need ($d = rt$, $A = \pi r^2$, etc.). If you can, determine realistic possibilities for the answer.

STEP 2: Assign a letter (variable) to represent what you are looking for, and, if necessary, express any remaining unknown quantities in terms of this variable.

STEP 3: Make a list of all the known facts, and translate them into mathematical expressions. These may take the form of an equation (or, later, an inequality) involving the variable. The equation (or inequality) is called the **model**. If possible, draw an appropriately labeled diagram to assist you. Sometimes a table or chart helps.

STEP 4: Solve the equation for the variable, and then answer the question, usually using a complete sentence.

STEP 5: Check the answer with the facts in the problem. If it agrees, congratulations! If it does not agree, try again.

NOTE It is a good practice to choose a variable that reminds you of the unknown. For example, use t for time. ∎

EXAMPLE 8

Investments

A total of $18,000 is invested, some in stocks and some in bonds. If the amount invested in bonds is half that invested in stocks, how much is invested in each category?

Step-by-Step Solution

Step 1: Determine what you are looking for.

We are being asked to find the amount of two investments. These amounts must total $18,000. (Do you see why?)

Step 2: Assign a variable to represent what you are looking for. If necessary, express any remaining unknown quantities in terms of this variable.

If x equals the amount invested in stocks, then the rest of the money, $18,000 - x$, is the amount invested in bonds.

Step 3: Translate the English into mathematical statements. It may be helpful to draw a figure that represents the situation. Sometimes a table can be used to organize the information. Use the information to build your model.

Set up a table:

Amount in Stocks	Amount in Bonds	Reason
x	$18,000 - x$	Total invested is $18,000.

We also know that:

Total amount invested in bonds	is	one-half that in stocks
$18,000 - x$	$=$	$\frac{1}{2}(x)$

Step 4: Solve the equation and answer the original question.

$$18,000 - x = \frac{1}{2}x$$

$$18,000 = x + \frac{1}{2}x \qquad \text{Add } x \text{ to both sides.}$$

$$18,000 = \frac{3}{2}x \qquad \text{Simplify.}$$

$$\left(\frac{2}{3}\right)18,000 = \left(\frac{2}{3}\right)\left(\frac{3}{2}x\right) \qquad \text{Multiply both sides by } \frac{2}{3}.$$

$$12,000 = x \qquad \text{Simplify.}$$

So $12,000 is invested in stocks, and $18,000 - $12,000 = $6000 is invested in bonds.

Step 5: Check your answer with the facts presented in the problem.

The total invested is $12,000 + $6000 = $18,000, and the amount in bonds, $6000, is half that in stocks, $12,000. ●

Now Work PROBLEM **83**

| EXAMPLE 9 | **Determining an Hourly Wage** |

Shannon grossed \$725 one week by working 52 hours. Her employer pays time-and-a-half for all hours worked in excess of 40 hours. With this information, can you determine Shannon's regular hourly wage?

Solution

STEP 1: We are looking for an hourly wage. Our answer will be expressed in dollars per hour.

STEP 2: Let x represent the regular hourly wage, measured in dollars per hour. Then $1.5x$ is the overtime hourly wage.

STEP 3: Set up a table:

	Hours Worked	**Hourly Wage**	**Salary**
Regular	40	x	$40x$
Overtime	12	$1.5x$	$12(1.5x) = 18x$

The sum of regular salary plus overtime salary will equal \$725. From the table, $40x + 18x = 725$.

STEP 4:
$$40x + 18x = 725$$
$$58x = 725$$
$$x = 12.50$$

Shannon's regular hourly wage is \$12.50 per hour.

STEP 5: Forty hours yields a salary of $40(12.50) = \$500$, and 12 hours of overtime yields a salary of $12(1.5)(12.50) = \$225$, for a total of \$725.

 Now Work PROBLEM 85

SUMMARY

Steps for Solving a Linear Equation

To solve a linear equation, follow these steps:

STEP 1: List any restrictions on the variable.

STEP 2: If necessary, clear the equation of fractions by multiplying both sides by the least common multiple (LCM) of the denominators of all the fractions.

STEP 3: Remove all parentheses and simplify.

STEP 4: Collect all terms containing the variable on one side and all remaining terms on the other side.

STEP 5: Simplify and solve.

STEP 6: Check your solution(s).

Historical Feature

Solving equations is among the oldest of mathematical activities, and efforts to systematize this activity determined much of the shape of modern mathematics.

Consider the following problem and its solution using only words: Solve the problem of how many apples Jim has, given that

"Bob's five apples and Jim's apples together make twelve apples" by thinking,

"Jim's apples are all twelve apples less Bob's five apples" and then concluding,

"Jim has seven apples."

The mental steps translated into algebra are

$$5 + x = 12$$
$$x = 12 - 5$$
$$= 7$$

The solution of this problem using only words is the earliest form of algebra. Such problems were solved exactly this way in Babylonia in 1800 BC. We know almost nothing of mathematical work before this date, although most authorities believe the sophistication of the earliest known texts indicates that there must have been a long period of previous development. The method of writing out equations in words persisted for thousands of years, and although it now seems extremely cumbersome, it was used very effectively by many generations of mathematicians. The Arabs developed a good deal of the theory of cubic equations while writing out all the equations in words. About AD 1500, the tendency to abbreviate words in the written equations began to lead in the direction of modern notation; for example, the Latin word *et* (meaning "and") developed into the plus sign, $+$. Although the occasional use of letters to represent variables dates back to AD 1200, the practice did not become common until about AD 1600. Development thereafter was rapid, and by 1635 algebraic notation did not differ essentially from what we use now.

1.1 Assess Your Understanding

'Are You Prepared?' *Answers are given at the end of these exercises. If you get a wrong answer, read the pages listed in* red.

1. The fact that $2(x + 3) = 2x + 6$ is attributable to the _____ Property. (pp. 9–13)

2. The fact that $3x = 0$ implies that $x = 0$ is a result of the _____ Property. (pp. 9–13)

3. The domain of the variable in the expression $\dfrac{x}{x-4}$ is _____ . (p. 21)

Concepts and Vocabulary

4. *True or False* Multiplying both sides of an equation by any number results in an equivalent equation.

5. An equation that is satisfied for every value of the variable for which both sides are defined is called a(n) _____ .

6. An equation of the form $ax + b = 0$ is called a(n) _____ equation or a(n) ___ _____ equation.

7. *True or False* The solution of the equation $3x - 8 = 0$ is $\dfrac{3}{8}$.

8. *True or False* Some equations have no solution.

9. An admissible value for the variable that makes the equation a true statement is called a(n) _____ of the equation.
   (a) identity   (b) solution   (c) degree   (d) model

10. A chemist mixes 10 liters of a 20% solution with x liters of a 35% solution. Which of the following expressions represents the total number of liters in the mixture?
    (a) x (b) $20 - x$ (c) $\dfrac{35}{x}$ (d) $10 + x$

Skill Building

In Problems 11–18, mentally solve each equation.

11. $7x = 21$

12. $6x = -24$

13. $3x + 15 = 0$

14. $6x + 18 = 0$

15. $2x - 3 = 0$

16. $3x + 4 = 0$

17. $\dfrac{1}{3}x = \dfrac{5}{12}$

18. $\dfrac{2}{3}x = \dfrac{9}{2}$

In Problems 19–66, solve each equation.

19. $3x + 4 = x$

20. $2x + 9 = 5x$

21. $2t - 6 = 3 - t$

22. $5y + 6 = -18 - y$

23. $6 - x = 2x + 9$

24. $3 - 2x = 2 - x$

25. $3 + 2n = 4n + 7$

26. $6 - 2m = 3m + 1$

27. $2(3 + 2x) = 3(x - 4)$

28. $3(2 - x) = 2x - 1$

29. $8x - (3x + 2) = 3x - 10$

30. $7 - (2x - 1) = 10$

31. $\dfrac{3}{2}x + 2 = \dfrac{1}{2} - \dfrac{1}{2}x$

32. $\dfrac{1}{3}x = 2 - \dfrac{2}{3}x$

33. $\dfrac{1}{2}x - 5 = \dfrac{3}{4}x$

34. $1 - \dfrac{1}{2}x = 6$

35. $\dfrac{2}{3}p = \dfrac{1}{2}p + \dfrac{1}{3}$

36. $\dfrac{1}{2} - \dfrac{1}{3}p = \dfrac{4}{3}$

37. $0.9t = 0.4 + 0.1t$

38. $0.9t = 1 + t$

39. $\dfrac{x+1}{3} + \dfrac{x+2}{7} = 2$

40. $\dfrac{2x+1}{3} + 16 = 3x$

41. $\dfrac{2}{y} + \dfrac{4}{y} = 3$

42. $\dfrac{4}{y} - 5 = \dfrac{5}{2y}$

43. $\dfrac{1}{2} + \dfrac{2}{x} = \dfrac{3}{4}$

44. $\dfrac{3}{x} - \dfrac{1}{3} = \dfrac{1}{6}$

45. $(x + 7)(x - 1) = (x + 1)^2$

46. $(x + 2)(x - 3) = (x + 3)^2$

47. $x(2x - 3) = (2x + 1)(x - 4)$

48. $x(1 + 2x) = (2x - 1)(x - 2)$

49. $z(z^2 + 1) = 3 + z^3$

50. $w(4 - w^2) = 8 - w^3$

51. $\dfrac{x}{x-2} + 3 = \dfrac{2}{x-2}$

52. $\dfrac{2x}{x+3} = \dfrac{-6}{x+3} - 2$

53. $\dfrac{2x}{x^2-4} = \dfrac{4}{x^2-4} - \dfrac{3}{x+2}$

54. $\dfrac{x}{x^2-9} + \dfrac{4}{x+3} = \dfrac{3}{x^2-9}$

55. $\dfrac{x}{x+2} = \dfrac{3}{2}$

56. $\dfrac{3x}{x-1} = 2$

57. $\dfrac{5}{2x-3} = \dfrac{3}{x+5}$

58. $\dfrac{-4}{x+4} = \dfrac{-3}{x+6}$

59. $\dfrac{6t+7}{4t-1} = \dfrac{3t+8}{2t-4}$

60. $\dfrac{8w+5}{10w-7} = \dfrac{4w-3}{5w+7}$

61. $\dfrac{4}{x-2} = \dfrac{-3}{x+5} + \dfrac{7}{(x+5)(x-2)}$

62. $\dfrac{-4}{2x+3} + \dfrac{1}{x-1} = \dfrac{1}{(2x+3)(x-1)}$

63. $\dfrac{2}{y+3} + \dfrac{3}{y-4} = \dfrac{5}{y+6}$

64. $\dfrac{5}{5z-11} + \dfrac{4}{2z-3} = \dfrac{-3}{5-z}$

65. $\dfrac{x}{x^2-1} - \dfrac{x+3}{x^2-x} = \dfrac{-3}{x^2+x}$

66. $\dfrac{x+1}{x^2+2x} - \dfrac{x+4}{x^2+x} = \dfrac{-3}{x^2+3x+2}$

In Problems 67–70, use a calculator to solve each equation. Round the solution to two decimal places.

67. $3.2x + \dfrac{21.3}{65.871} = 19.23$

68. $6.2x - \dfrac{19.1}{83.72} = 0.195$

69. $14.72 - 21.58x = \dfrac{18}{2.11}x + 2.4$

70. $18.63x - \dfrac{21.2}{2.6} = \dfrac{14}{2.32}x - 20$

Applications and Extensions

In Problems 71–74, solve each equation. The letters a, b, and c are constants.

71. $ax - b = c, \quad a \neq 0$

72. $1 - ax = b, \quad a \neq 0$

73. $\dfrac{x}{a} + \dfrac{x}{b} = c, \quad a \neq 0, b \neq 0, a \neq -b$

74. $\dfrac{a}{x} + \dfrac{b}{x} = c, \quad c \neq 0$

75. Find the number a for which $x = 4$ is a solution of the equation
$$x + 2a = 16 + ax - 6a$$

76. Find the number b for which $x = 2$ is a solution of the equation
$$x + 2b = x - 4 + 2bx$$

Problems 77–82 list some formulas that occur in applications. Solve each formula for the indicated variable.

77. **Electricity** $\dfrac{1}{R} = \dfrac{1}{R_1} + \dfrac{1}{R_2}$ for R

78. **Finance** $A = P(1 + rt)$ for r

79. **Mechanics** $F = \dfrac{mv^2}{R}$ for R

80. **Chemistry** $PV = nRT$ for T

81. **Mathematics** $S = \dfrac{a}{1-r}$ for r

82. **Mechanics** $v = -gt + v_0$ for t

83. **Finance** A total of $20,000 is to be invested, some in bonds and some in certificates of deposit (CDs). If the amount invested in bonds is to exceed that in CDs by $3000, how much will be invested in each type of investment?

84. **Finance** A total of $10,000 is to be divided between Sean and George, with George to receive $3000 less than Sean. How much will each receive?

85. **Computing Hourly Wages** Sandra, who is paid time-and-a-half for hours worked in excess of 40 hours, had gross weekly wages of $546 for 48 hours worked. What is her regular hourly rate?

86. **Computing Hourly Wages** Leigh is paid time-and-a-half for hours worked in excess of 40 hours and double-time for hours worked on Sunday. If Leigh had gross weekly wages of $627 for working 50 hours, 4 of which were on Sunday, what is her regular hourly rate?

87. **Computing Grades** Going into the final exam, which will count as two tests, Brooke has test scores of 80, 83, 71, 61, and 95. What score does Brooke need on the final in order to have an average score of 80?

88. **Computing Grades** Going into the final exam, which will count as two-thirds of the final grade, Mike has test scores of 86, 80, 84, and 90. What minimum score does Mike need on the final in order to earn a B, which requires an average score of 80? What does he need to earn an A, which requires an average of 90?

89. **Business: Discount Pricing** A builder of tract homes reduced the price of a model by 15%. If the new price is $425,000, what was its original price? How much can be saved by purchasing the model?

90. **Business: Discount Pricing** A car dealer, at a year-end clearance, reduces the list price of last year's models by 15%. If a certain four-door model has a discounted price of $8000, what was its list price? How much can be saved by purchasing last year's model?

91. **Personal Finance: Concession Markup** A movie theater marks up the candy it sells by 275%. If a box of candy sells for $3.00 at the theater, how much did the theater pay for the box?

When the left side factors into two linear equations with the same solution, the quadratic equation is said to have a **repeated solution**. This solution is also called a **root of multiplicity 2**, or a **double root**.

EXAMPLE 2 | **Solving a Quadratic Equation by Factoring**

Solve the equation: $9x^2 - 6x + 1 = 0$

Solution

This equation is already in standard form, and the left side can be factored.

$$9x^2 - 6x + 1 = 0$$
$$(3x - 1)(3x - 1) = 0$$

so

$$x = \frac{1}{3} \quad \text{or} \quad x = \frac{1}{3}$$

This equation has only the repeated solution $\frac{1}{3}$. The solution set is $\left\{\frac{1}{3}\right\}$.

●

Now Work PROBLEMS **13** AND **23**

The Square Root Method

Suppose that we wish to solve the quadratic equation

$$x^2 = p \tag{2}$$

where $p \geq 0$ is a nonnegative number. Proceeding as in the earlier examples,

$$x^2 - p = 0 \qquad \text{Put in standard form.}$$
$$(x - \sqrt{p})(x + \sqrt{p}) = 0 \qquad \text{Factor (over the real numbers).}$$
$$x = \sqrt{p} \quad \text{or} \quad x = -\sqrt{p} \qquad \text{Solve.}$$

We have the following result:

> If $x^2 = p$ and $p \geq 0$, then $x = \sqrt{p}$ or $x = -\sqrt{p}$. $\tag{3}$

When statement (3) is used, it is called the **Square Root Method**. In statement (3), note that if $p > 0$ the equation $x^2 = p$ has two solutions, $x = \sqrt{p}$ and $x = -\sqrt{p}$. We usually abbreviate these solutions as $x = \pm\sqrt{p}$, which is read as "x equals plus or minus the square root of p."

For example, the two solutions of the equation

$$x^2 = 4$$

are

$$x = \pm\sqrt{4} \qquad \text{Use the Square Root Method.}$$

and, since $\sqrt{4} = 2$, we have

$$x = \pm 2$$

The solution set is $\{-2, 2\}$.

EXAMPLE 3 | **Solving a Quadratic Equation Using the Square Root Method**

Solve each equation.

(a) $x^2 = 5$ (b) $(x - 2)^2 = 16$

Solution

(a) Use the Square Root Method to get

$$x^2 = 5$$
$$x = \pm\sqrt{5} \qquad \text{Use the Square Root Method.}$$
$$x = \sqrt{5} \quad \text{or} \quad x = -\sqrt{5}$$

The solution set is $\{-\sqrt{5}, \sqrt{5}\}$.

(b) Use the Square Root Method to get

$$(x - 2)^2 = 16$$

$$x - 2 = \pm\sqrt{16} \qquad\qquad \text{Use the Square Root Method.}$$

$$x - 2 = \pm 4 \qquad\qquad\qquad \sqrt{16} = 4$$

$$x - 2 = 4 \quad \text{or} \quad x - 2 = -4$$

$$x = 6 \quad \text{or} \qquad x = -2$$

The solution set is $\{-2, 6\}$.

🖉 **Now Work** PROBLEM 33

2 Solve a Quadratic Equation by Completing the Square

EXAMPLE 4   **Solving a Quadratic Equation by Completing the Square**

Solve by completing the square: $x^2 + 5x + 4 = 0$

Solution   Always begin this procedure by rearranging the equation so that the constant is on the right side.

$$x^2 + 5x + 4 = 0$$

$$x^2 + 5x = -4$$

NOTE If the coefficient of the square term is not 1, divide both sides by the coefficient of the square term before attempting to complete the square. ∎

Since the coefficient of x^2 is 1, complete the square on the left side by adding $\left(\dfrac{1}{2} \cdot 5\right)^2 = \dfrac{25}{4}$. Of course, in an equation, whatever is added to the left side must also be added to the right side. So add $\dfrac{25}{4}$ to *both* sides.

$$x^2 + 5x + \frac{25}{4} = -4 + \frac{25}{4} \qquad \text{Add } \frac{25}{4} \text{ to both sides.}$$

$$\left(x + \frac{5}{2}\right)^2 = \frac{9}{4} \qquad\qquad \text{Factor the left side.}$$

$$x + \frac{5}{2} = \pm\sqrt{\frac{9}{4}} \qquad\qquad \text{Use the Square Root Method.}$$

$$x + \frac{5}{2} = \pm\frac{3}{2} \qquad\qquad \sqrt{\frac{9}{4}} = \frac{\sqrt{9}}{\sqrt{4}} = \frac{3}{2}$$

$$x = -\frac{5}{2} \pm \frac{3}{2}$$

$$x = -\frac{5}{2} + \frac{3}{2} = -1 \quad \text{or} \quad x = -\frac{5}{2} - \frac{3}{2} = -4$$

The solution set is $\{-4, -1\}$.

🖉 THE SOLUTION OF THE EQUATION IN EXAMPLE **4** CAN ALSO BE OBTAINED BY FACTORING. REWORK EXAMPLE **4** USING FACTORING.

The next example illustrates an equation that cannot be solved by factoring.

EXAMPLE 5   **Solving a Quadratic Equation by Completing the Square**

Solve by completing the square: $2x^2 - 8x - 5 = 0$

Solution   First, rewrite the equation so that the constant is on the right side.

$$2x^2 - 8x - 5 = 0$$

$$2x^2 - 8x = 5 \quad \text{Add 5 to both sides.}$$

Next, divide both sides by 2 so that the coefficient of x^2 is 1. (This enables us to complete the square at the next step.)

$$x^2 - 4x = \frac{5}{2}$$

Finally, complete the square by adding 4 to both sides.

$$x^2 - 4x + 4 = \frac{5}{2} + 4 \qquad \text{Add 4 to both sides.}$$

$$(x - 2)^2 = \frac{13}{2} \qquad \text{Factor on the left; simplify on the right.}$$

$$x - 2 = \pm\sqrt{\frac{13}{2}} \qquad \text{Use the Square Root Method.}$$

$$x - 2 = \pm\frac{\sqrt{26}}{2} \qquad \sqrt{\frac{13}{2}} = \frac{\sqrt{13}}{\sqrt{2}} = \frac{\sqrt{13}}{\sqrt{2}} \cdot \frac{\sqrt{2}}{\sqrt{2}} = \frac{\sqrt{26}}{2}$$

$$x = 2 \pm \frac{\sqrt{26}}{2}$$

NOTE If we wanted an approximation, say rounded to two decimal places, of these solutions, we would use a calculator to get $\{-0.55, 4.55\}$. ∎

The solution set is $\left\{ 2 - \dfrac{\sqrt{26}}{2}, 2 + \dfrac{\sqrt{26}}{2} \right\}$.   ●

Now Work PROBLEM 37

3 Solve a Quadratic Equation Using the Quadratic Formula

We can use the method of completing the square to obtain a general formula for solving any quadratic equation

$$ax^2 + bx + c = 0 \qquad a \neq 0$$

NOTE There is no loss in generality to assume that $a > 0$, since if $a < 0$ we can multiply by -1 to obtain an equivalent equation with a positive leading coefficient. ∎

As in Examples 4 and 5, rearrange the terms as

$$ax^2 + bx = -c \quad a > 0$$

Since $a > 0$, divide both sides by a to get

$$x^2 + \frac{b}{a}x = -\frac{c}{a}$$

Now the coefficient of x^2 is 1. To complete the square on the left side, add the square of $\frac{1}{2}$ of the coefficient of x; that is, add

$$\left(\frac{1}{2} \cdot \frac{b}{a}\right)^2 = \frac{b^2}{4a^2}$$

to both sides. Then

$$x^2 + \frac{b}{a}x + \frac{b^2}{4a^2} = \frac{b^2}{4a^2} - \frac{c}{a}$$

$$\left(x + \frac{b}{2a}\right)^2 = \frac{b^2 - 4ac}{4a^2} \qquad \frac{b^2}{4a^2} - \frac{c}{a} = \frac{b^2}{4a^2} - \frac{4ac}{4a^2} = \frac{b^2 - 4ac}{4a^2} \qquad \textbf{(4)}$$

Provided that $b^2 - 4ac \geq 0$, we can now use the Square Root Method to get

$$x + \frac{b}{2a} = \pm\sqrt{\frac{b^2 - 4ac}{4a^2}}$$

$$x + \frac{b}{2a} = \frac{\pm\sqrt{b^2 - 4ac}}{2a} \qquad \begin{array}{l} \text{The square root of a quotient equals} \\ \text{the quotient of the square roots.} \\ \text{Also, } \sqrt{4a^2} = 2a \text{ since } a > 0. \end{array}$$

$$x = -\frac{b}{2a} \pm \frac{\sqrt{b^2 - 4ac}}{2a} \quad \text{Add } -\frac{b}{2a}\text{ to both sides.}$$

$$= \frac{-b \pm \sqrt{b^2 - 4ac}}{2a} \quad \text{Combine the quotients on the right.}$$

What if $b^2 - 4ac$ is negative? Then equation (4) states that the left expression (a real number squared) equals the right expression (a negative number). Since this is impossible for real numbers, we conclude that if $b^2 - 4ac < 0$, the quadratic equation has no *real* solution. (We discuss quadratic equations for which the quantity $b^2 - 4ac < 0$ in detail in the next section.)

THEOREM

Quadratic Formula

Consider the quadratic equation

$$ax^2 + bx + c = 0 \quad a \neq 0$$

If $b^2 - 4ac < 0$, this equation has no real solution.
If $b^2 - 4ac \geq 0$, the real solution(s) of this equation is (are) given by the **quadratic formula**:

$$x = \frac{-b \pm \sqrt{b^2 - 4ac}}{2a} \tag{5}$$

The quantity $b^2 - 4ac$ is called the **discriminant** of the quadratic equation, because its value tells us whether the equation has real solutions. In fact, it also tells us how many solutions to expect.

Discriminant of a Quadratic Equation

For a quadratic equation $ax^2 + bx + c = 0, a \neq 0$:

1. If $b^2 - 4ac > 0$, there are two unequal real solutions.
2. If $b^2 - 4ac = 0$, there is a repeated solution, a double root.
3. If $b^2 - 4ac < 0$, there is no real solution.

When asked to find the real solutions, of a quadratic equation, always evaluate the discriminant first to see if there are any real solutions.

EXAMPLE 6

Solving a Quadratic Equation Using the Quadratic Formula

Use the quadratic formula to find the real solutions, if any, of the equation

$$3x^2 - 5x + 1 = 0$$

Solution

The equation is in standard form, so compare it to $ax^2 + bx + c = 0$ to find a, b, and c.

$$3x^2 - 5x + 1 = 0$$
$$ax^2 + bx + c = 0 \quad a = 3, b = -5, c = 1$$

With $a = 3, b = -5$, and $c = 1$, evaluate the discriminant $b^2 - 4ac$.

$$b^2 - 4ac = (-5)^2 - 4(3)(1) = 25 - 12 = 13$$

Since $b^2 - 4ac > 0$, there are two real solutions, which can be found using the quadratic formula.

$$x = \frac{-b \pm \sqrt{b^2 - 4ac}}{2a} = \frac{-(-5) \pm \sqrt{13}}{2(3)} = \frac{5 \pm \sqrt{13}}{6}$$

The solution set is $\left\{ \dfrac{5 - \sqrt{13}}{6}, \dfrac{5 + \sqrt{13}}{6} \right\}$.

EXAMPLE 7

Solving a Quadratic Equation Using the Quadratic Formula

Use the quadratic formula to find the real solutions, if any, of the equation

$$\frac{25}{2}x^2 - 30x + 18 = 0$$

Solution

The equation is given in standard form. However, to simplify the arithmetic, clear the fraction.

$$\frac{25}{2}x^2 - 30x + 18 = 0$$

$25x^2 - 60x + 36 = 0$ *Clear fraction; multiply by 2.*

$ax^2 + bx + c = 0$ *Compare to standard form.*

With $a = 25, b = -60$, and $c = 36$, evaluate the discriminant.

$$b^2 - 4ac = (-60)^2 - 4(25)(36) = 3600 - 3600 = 0$$

The equation has a repeated solution, which is found by using the quadratic formula.

$$x = \frac{-b \pm \sqrt{b^2 - 4ac}}{2a} = \frac{60 \pm \sqrt{0}}{50} = \frac{60}{50} = \frac{6}{5}$$

The solution set is $\left\{ \dfrac{6}{5} \right\}$.

EXAMPLE 8

Solving a Quadratic Equation Using the Quadratic Formula

Use the quadratic formula to find the real solutions, if any, of the equation

$$3x^2 + 2 = 4x$$

Solution

The equation, as given, is not in standard form.

$$3x^2 + 2 = 4x$$

$3x^2 - 4x + 2 = 0$ *Put in standard form.*

$ax^2 + bx + c = 0$ *Compare to standard form.*

With $a = 3, b = -4$, and $c = 2$, the discriminant is

$$b^2 - 4ac = (-4)^2 - 4(3)(2) = 16 - 24 = -8$$

Since $b^2 - 4ac < 0$, the equation has no real solution.

Now Work PROBLEMS **47** AND **57**

| EXAMPLE 9 | **Solving a Quadratic Equation Using the Quadratic Formula** |

Find the real solutions, if any, of the equation: $9 + \dfrac{3}{x} - \dfrac{2}{x^2} = 0, \quad x \neq 0$

Solution   In its present form, the equation

$$9 + \frac{3}{x} - \frac{2}{x^2} = 0$$

is not a quadratic equation. However, it can be transformed into one by multiplying each side by x^2. The result is

$$9x^2 + 3x - 2 = 0$$

Although we multiplied each side by x^2, we know that $x^2 \neq 0$ (do you see why?), so this quadratic equation is equivalent to the original equation.

Using $a = 9$, $b = 3$, and $c = -2$, the discriminant is

$$b^2 - 4ac = 3^2 - 4(9)(-2) = 9 + 72 = 81$$

Since $b^2 - 4ac > 0$, the new equation has two real solutions.

$$x = \frac{-b \pm \sqrt{b^2 - 4ac}}{2a} = \frac{-3 \pm \sqrt{81}}{2(9)} = \frac{-3 \pm 9}{18}$$

$$x = \frac{-3 + 9}{18} = \frac{6}{18} = \frac{1}{3} \quad \text{or} \quad x = \frac{-3 - 9}{18} = \frac{-12}{18} = -\frac{2}{3}$$

The solution set is $\left\{ -\dfrac{2}{3}, \dfrac{1}{3} \right\}$. ●

SUMMARY

Procedure for Solving a Quadratic Equation
_____

To solve a quadratic equation, first put it in standard form:

$$ax^2 + bx + c = 0$$

Then:

STEP 1: Identify a, b, and c.

STEP 2: Evaluate the discriminant, $b^2 - 4ac$.

STEP 3: (a) If the discriminant is negative, the equation has no real solution.

(b) If the discriminant is zero, the equation has one real solution, a double root.

(c) If the discriminant is positive, the equation has two distinct real solutions.

If you can easily spot factors, use the factoring method to solve the equation. Otherwise, use the quadratic formula or the method of completing the square.

 4 Solve Problems That Can Be Modeled by Quadratic Equations

Many applied problems require the solution of a quadratic equation. Let's look at one that you will probably see again in a slightly different form if you study calculus.

 | EXAMPLE 10 | **Constructing a Box** |

From each corner of a square piece of sheet metal, remove a square of side 9 centimeters. Turn up the edges to form an open box. If the box is to hold 144 cubic centimeters (cm³), what should be the dimensions of the piece of sheet metal?

Solution   Use Figure 1 as a guide. We have labeled by x the length of a side of the square piece of sheet metal. The box will be of height 9 centimeters, and its square base will measure $x - 18$ on each side. The volume V (Length × Width × Height) of the box is therefore

$$V = (x - 18)(x - 18) \cdot 9 = 9(x - 18)^2$$

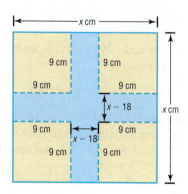

Figure 1

Since the volume of the box is to be 144 cm³, we have

$$9(x - 18)^2 = 144 \qquad \textcolor{blue}{V = 144}$$
$$(x - 18)^2 = 16 \qquad \textcolor{blue}{\text{Divide each side by 9.}}$$
$$x - 18 = \pm 4 \qquad \textcolor{blue}{\text{Use the Square Root Method.}}$$
$$x = 18 \pm 4$$
$$x = 22 \quad \text{or} \quad x = 14$$

✓ Check: If we take a piece of sheet metal 22 centimeters by 22 centimeters, cut out a 9-centimeter square from each corner, and fold up the edges, we get a box whose dimensions are 9 by 4 by 4, with volume $9 \times 4 \times 4 = 144$ cm³, as required.

Discard the solution $x = 14$ (do you see why?) and conclude that the sheet metal should be 22 centimeters by 22 centimeters.   ●

🖉 **Now Work** PROBLEM 99

Historical Feature

Problems using quadratic equations are found in the oldest known mathematical literature. Babylonians and Egyptians were solving such problems before 1800 BC. Euclid solved quadratic equations geometrically in his *Data* (300 BC), and the Hindus and Arabs gave rules for solving any quadratic equation with real roots. Because negative numbers were not freely used before AD 1500, there were several different types of quadratic equations, each with its own rule. Thomas Harriot (1560–1621) introduced the method of factoring to obtain solutions, and François Viète (1540–1603) introduced a method that is essentially completing the square.

Until modern times it was usual to neglect the negative roots (if there were any), and equations involving square roots of negative quantities were regarded as unsolvable until the 1500s.

Historical Problems

1. **One of al-Khwărízmí solutions**   Solve $x^2 + 12x = 85$ by drawing the square shown. The area of the four white rectangles and the yellow square is $x^2 + 12x$. We then set this expression equal to 85 to get the equation $x^2 + 12x = 85$. If we add the four blue squares, we will have a larger square of known area. Complete the solution.

2. **Viète's method**   Solve $x^2 + 12x - 85 = 0$ by letting $x = u + z$. Then

$$(u + z)^2 + 12(u + z) - 85 = 0$$
$$u^2 + (2z + 12)u + (z^2 + 12z - 85) = 0$$

Now select z so that $2z + 12 = 0$ and finish the solution.

3. **Another method to get the quadratic formula**   Look at equation (4) on page 96. Rewrite the right side as $\left(\dfrac{\sqrt{b^2 - 4ac}}{2a}\right)^2$ and then subtract it from each side. The right side is now 0 and the left side is a difference of two squares. If you factor this difference of two squares, you will easily be able to get the quadratic formula, and moreover, the quadratic expression is factored, which is sometimes useful.

1.2 Assess Your Understanding

'Are You Prepared?' *Answers are given at the end of these exercises. If you get a wrong answer, read the pages listed in* red.

1. Factor: $x^2 - 5x - 6$ (pp. 49–55)

2. Factor: $2x^2 - x - 3$ (pp. 49–55)

3. The solution set of the equation $(x - 3)(3x + 5) = 0$ is _____ . (p. 13)

4. *True or False* $\sqrt{x^2} = |x|$. (pp. 23–24)

5. Complete the square of $x^2 + 5x$. Factor the new expression. (p. 56)

Concepts and Vocabulary

6. The quantity $b^2 - 4ac$ is called the _____ of a quadratic equation. If it is _____, the equation has no real solution.

7. *True or False* Quadratic equations always have two real solutions.

8. *True or False* If the discriminant of a quadratic equation is positive, then the equation has two solutions that are negatives of one another.

9. A quadratic equation is sometimes called a _____ equation.
(a) first-degree   (b) second-degree
(c) third-degree   (d) fourth-degree

10. Which of the following quadratic equations is in standard form?
(a) $x^2 - 7x = 5$ (b) $9 = x^2$
(c) $(x + 5)(x - 4) = 0$ (d) $0 = 5x^2 - 6x - 1$

Skill Building

In Problems 11–30, solve each equation by factoring.

11. $x^2 - 9x = 0$

12. $x^2 + 4x = 0$

13. $x^2 - 25 = 0$

14. $x^2 - 9 = 0$

15. $z^2 + z - 6 = 0$

16. $v^2 + 7v + 6 = 0$

17. $2x^2 - 5x - 3 = 0$

18. $3x^2 + 5x + 2 = 0$

19. $3t^2 - 48 = 0$

20. $2y^2 - 50 = 0$

21. $x(x - 8) + 12 = 0$

22. $x(x + 4) = 12$

23. $4x^2 + 9 = 12x$

24. $25x^2 + 16 = 40x$

25. $6(p^2 - 1) = 5p$

26. $2(2u^2 - 4u) + 3 = 0$

27. $6x - 5 = \dfrac{6}{x}$

28. $x + \dfrac{12}{x} = 7$

29. $\dfrac{4(x - 2)}{x - 3} + \dfrac{3}{x} = \dfrac{-3}{x(x - 3)}$

30. $\dfrac{5}{x + 4} = 4 + \dfrac{3}{x - 2}$

In Problems 31–36, solve each equation by the Square Root Method.

31. $x^2 = 25$

32. $x^2 = 36$

33. $(x - 1)^2 = 4$

34. $(x + 2)^2 = 1$

35. $(2y + 3)^2 = 9$

36. $(3z - 2)^2 = 4$

In Problems 37–42, solve each equation by completing the square.

37. $x^2 + 4x = 21$

38. $x^2 - 6x = 13$

39. $x^2 - \dfrac{1}{2}x - \dfrac{3}{16} = 0$

40. $x^2 + \dfrac{2}{3}x - \dfrac{1}{3} = 0$

41. $3x^2 + x - \dfrac{1}{2} = 0$

42. $2x^2 - 3x - 1 = 0$

In Problems 43–66, find the real solutions, if any, of each equation. Use the quadratic formula.

43. $x^2 - 4x + 2 = 0$

44. $x^2 + 4x + 2 = 0$

45. $x^2 - 4x - 1 = 0$

46. $x^2 + 6x + 1 = 0$

47. $2x^2 - 5x + 3 = 0$

48. $2x^2 + 5x + 3 = 0$

49. $4y^2 - y + 2 = 0$

50. $4t^2 + t + 1 = 0$

51. $4x^2 = 1 - 2x$

52. $2x^2 = 1 - 2x$

53. $4x^2 = 9x$

54. $5x = 4x^2$

55. $9t^2 - 6t + 1 = 0$

56. $4u^2 - 6u + 9 = 0$

57. $\dfrac{3}{4}x^2 - \dfrac{1}{4}x - \dfrac{1}{2} = 0$

58. $\frac{2}{3}x^2 - x - 3 = 0$ –

59. $\frac{5}{3}x^2 - x = \frac{1}{3}$

60. $\frac{3}{5}x^2 - x = \frac{1}{5}$

61. $2x(x + 2) = 3$

62. $3x(x + 2) = 1$

63. $4 - \frac{1}{x} - \frac{2}{x^2} = 0$

64. $4 + \frac{1}{x} - \frac{1}{x^2} = 0$

65. $\frac{3x}{x - 2} + \frac{1}{x} = 4$

66. $\frac{2x}{x - 3} + \frac{1}{x} = 4$

In Problems 67–72, find the real solutions, if any, of each equation. Use the quadratic formula and a calculator. Express any solutions rounded to two decimal places.

67. $x^2 - 4.1x + 2.2 = 0$

68. $x^2 + 3.9x + 1.8 = 0$

69. $x^2 + \sqrt{3}x - 3 = 0$

70. $x^2 + \sqrt{2}x - 2 = 0$

71. $\pi x^2 - x - \pi = 0$

72. $\pi x^2 + \pi x - 2 = 0$

In Problems 73–78, use the discriminant to determine whether each quadratic equation has two unequal real solutions, a repeated real solution (a double root), or no real solution, without solving the equation.

73. $2x^2 - 6x + 7 = 0$

74. $x^2 + 4x + 7 = 0$

75. $9x^2 - 30x + 25 = 0$

76. $25x^2 - 20x + 4 = 0$

77. $3x^2 + 5x - 8 = 0$

78. $2x^2 - 3x - 7 = 0$

Mixed Practice

In Problems 79–92, find the real solutions, if any, of each equation. Use any method.

79. $x^2 - 5 = 0$

80. $x^2 - 6 = 0$

81. $16x^2 - 8x + 1 = 0$

82. $9x^2 - 12x + 4 = 0$

83. $10x^2 - 19x - 15 = 0$

84. $6x^2 + 7x - 20 = 0$

85. $2 + z = 6z^2$

86. $2 = y + 6y^2$

87. $x^2 + \sqrt{2}x = \frac{1}{2}$

88. $\frac{1}{2}x^2 = \sqrt{2}x + 1$

89. $x^2 + x = 4$

90. $x^2 + x = 1$

91. $\frac{x}{x - 2} + \frac{2}{x + 1} = \frac{7x + 1}{x^2 - x - 2}$

92. $\frac{3x}{x + 2} + \frac{1}{x - 1} = \frac{4 - 7x}{x^2 + x - 2}$

Applications and Extensions

93. Pythagorean Theorem How many right triangles have a hypotenuse that measures $2x + 3$ meters and legs that measure $2x - 5$ meters and $x + 7$ meters? What are their dimensions?

94. Pythagorean Theorem How many right triangles have a hypotenuse that measures $4x + 5$ inches and legs that measure $3x + 13$ inches and x inches? What are the dimensions of the triangle(s)?

95. Dimensions of a Window The area of the opening of a rectangular window is to be 143 square feet. If the length is to be 2 feet more than the width, what are the dimensions?

96. Dimensions of a Window The area of a rectangular window is to be 306 square centimeters. If the length exceeds the width by 1 centimeter, what are the dimensions?

97. Geometry Find the dimensions of a rectangle whose perimeter is 26 meters and whose area is 40 square meters.

98. Watering a Field An adjustable water sprinkler that sprays water in a circular pattern is placed at the center of a square field whose area is 1250 square feet (see the figure). What is

the shortest radius setting that can be used if the field is to be completely enclosed within the circle?

99. Constructing a Box An open box is to be constructed from a square piece of sheet metal by removing a square of side 1 foot from each corner and turning up the edges. If the box is to hold 4 cubic feet, what should be the dimensions of the sheet metal?

100. Constructing a Box Rework Problem 99 if the piece of sheet metal is a rectangle whose length is twice its width.

101. Physics A ball is thrown vertically upward from the top of a building 96 feet tall with an initial velocity of

80 feet per second. The distance s (in feet) of the ball from the ground after t seconds is $s = 96 + 80t - 16t^2$.
(a) After how many seconds does the ball strike the ground?
(b) After how many seconds will the ball pass the top of the building on its way down?

102. **Physics** An object is propelled vertically upward with an initial velocity of 20 meters per second. The distance s (in meters) of the object from the ground after t seconds is $s = -4.9t^2 + 20t$.
(a) When will the object be 15 meters above the ground?
(b) When will it strike the ground?
(c) Will the object reach a height of 100 meters?

103. **Reducing the Size of a Candy Bar** A jumbo chocolate bar with a rectangular shape measures 12 centimeters in length, 7 centimeters in width, and 3 centimeters in thickness. Due to escalating costs of cocoa, management decides to reduce the volume of the bar by 10%. To accomplish this reduction, management decides that the new bar should have the same 3-centimeter thickness, but the length and width of each should be reduced an equal number of centimeters. What should be the dimensions of the new candy bar?

104. **Reducing the Size of a Candy Bar** Rework Problem 103 if the reduction is to be 20%.

105. **Constructing a Border around a Pool** A circular pool measures 10 feet across. One cubic yard of concrete is to be used to create a circular border of uniform width around the pool. If the border is to have a depth of 3 inches, how wide will the border be? (1 cubic yard = 27 cubic feet) See the illustration.

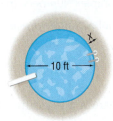

106. **Constructing a Border around a Pool** Rework Problem 105 if the depth of the border is 4 inches.

107. **Constructing a Border around a Garden** A landscaper, who just completed a rectangular flower garden measuring 6 feet by 10 feet, orders 1 cubic yard of premixed cement, all of which is to be used to create a border of uniform width around the garden. If the border is to have a depth of 3 inches, how wide will the border be? (1 cubic yard = 27 cubic feet)

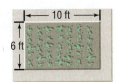

108. **Dimensions of a Patio** A contractor orders 8 cubic yards of premixed cement, all of which is to be used to pour a patio that will be 4 inches thick. If the length of the patio is specified to be twice the width, what will be the patio dimensions? (1 cubic yard = 27 cubic feet)

109. **Comparing Tablets** The screen size of a tablet is determined by the length of the diagonal of the rectangular screen. The 9.7-inch iPad Air™ comes in a 4:3 format, which means that the ratio of the length to the width of the rectangular screen is 4:3. What is the area of the iPad's screen? What is the area of a 10-inch Google Nexus™ if its screen is in a 16:10 format? Which screen is larger? (**Hint:** If x is the length of a 4:3 format screen, then $\frac{3}{4}x$ is the width.)

iPad mini 4:3   Google Nexus 16:10

110. **Comparing Tablets** Refer to Problem 109. Find the screen area of a 7.9-inch iPad mini with Retina™ in a 4:3 format, and compare it with an 8-inch Dell Venue Pro™ if its screen is in a 16:9 format. Which screen is larger?

111. **Field Design** A football field is sloped from the center toward the sides for drainage. The height h, in feet, of the field, x feet from the side, is given by $h = -0.00025x^2 + 0.04x$. Find the height of the field a distance of 35 feet from the side. Round to the nearest tenth of a foot.

112. **College Value** The difference, d, in median earnings, in $1000s, between high school graduates and college graduates can be approximated by $d = -0.002x^2 + 0.319x + 7.512$, where x is the number of years after 1965. Based on this model, estimate to the nearest year when the difference in median earnings was $15,000. (*Source: Current Population Survey*)

113. **Student Working** A study found that a student's GPA, g, is related to the number of hours worked each week, h, by the equation $g = -0.0006h^2 + 0.015h + 3.04$. Estimate the number of hours worked each week for a student with a GPA of 2.97. Round to the nearest whole hour.

114. **Fraternity Purchase** A fraternity wants to buy a new LED Smart TV that costs $1470. If 7 members of the fraternity are not able to contribute, the share for the remaining members increases by $5. How many members are in the fraternity?

115. The sum of the consecutive integers $1, 2, 3, \ldots, n$ is given by the formula $\frac{1}{2}n(n + 1)$. How many consecutive integers, starting with 1, must be added to get a sum of 703?

116. **Geometry** If a polygon of n sides has $\frac{1}{2}n(n - 3)$ diagonals, how many sides will a polygon with 65 diagonals have? Is there a polygon with 80 diagonals?

117. Show that the sum of the roots of a quadratic equation is $-\frac{b}{a}$.

118. Show that the product of the roots of a quadratic equation is $\frac{c}{a}$.

119. Find k such that the equation $kx^2 + x + k = 0$ has a repeated real solution.

120. Find k such that the equation $x^2 - kx + 4 = 0$ has a repeated real solution.

121. Show that the real solutions of the equation $ax^2 + bx + c = 0$ are the negatives of the real solutions of the equation $ax^2 - bx + c = 0$. Assume that $b^2 - 4ac \geq 0$.

122. Show that the real solutions of the equation $ax^2 + bx + c = 0$ are the reciprocals of the real solutions of the equation $cx^2 + bx + a = 0$. Assume that $b^2 - 4ac \geq 0$.

Explaining Concepts: Discussion and Writing

123. Which of the following pairs of equations are equivalent? Explain.
(a) $x^2 = 9$; $x = 3$ (b) $x = \sqrt{9}$; $x = 3$
(c) $(x - 1)(x - 2) = (x - 1)^2$; $x - 2 = x - 1$

124. Describe three ways that you might solve a quadratic equation. State your preferred method; explain why you chose it.

125. Explain the benefits of evaluating the discriminant of a quadratic equation before attempting to solve it.

126. Create three quadratic equations: one having two distinct real solutions, one having no real solution, and one having exactly one real solution.

127. The word *quadratic* seems to imply four (*quad*), yet a quadratic equation is an equation that involves a polynomial of degree 2. Investigate the origin of the term *quadratic* as it is used in the expression *quadratic equation*. Write a brief essay on your findings.

'Are You Prepared?' Answers

1. $(x - 6)(x + 1)$ **2.** $(2x - 3)(x + 1)$ **3.** $\left\{-\dfrac{5}{3}, 3\right\}$ **4.** True   **5.** $x^2 + 5x + \dfrac{25}{4} = \left(x + \dfrac{5}{2}\right)^2$

1.3 Complex Numbers; Quadratic Equations in the Complex Number System*

PREPARING FOR THIS SECTION *Before getting started, review the following:*

- Classification of Numbers (Section R.1, pp. 4–5)
- Rationalizing Denominators (Section R.8, p. 75)

Now Work the **'Are You Prepared?'** problems on page 111.

OBJECTIVES **1** Add, Subtract, Multiply, and Divide Complex Numbers (p. 105)

2 Solve Quadratic Equations in the Complex Number System (p. 109)

Complex Numbers

One property of a real number is that its square is nonnegative. For example, there is no real number x for which

$$x^2 = -1$$

To remedy this situation, we introduce a new number called the *imaginary unit.*

DEFINITION

The **imaginary unit**, which we denote by **i**, is the number whose square is -1. That is,

$$i^2 = -1$$

This should not surprise you. If our universe were to consist only of integers, there would be no number x for which $2x = 1$. This was remedied by introducing numbers such as $\dfrac{1}{2}$ and $\dfrac{2}{3}$, the *rational numbers*. If our universe were to consist only of rational numbers, there would be no x whose square equals 2. That is, there would be no number x for which $x^2 = 2$. To remedy this, we introduced numbers such as $\sqrt{2}$ and $\sqrt[3]{5}$, the *irrational numbers*. Recall that the *real numbers* consist of the rational numbers and the irrational numbers. Now, if our universe were to consist only of real numbers, then there would be no number x whose square is -1. To remedy this, we introduce the number i, whose square is -1.

*This section may be omitted without any loss of continuity.

In the progression outlined, each time we encountered a situation that was unsuitable, a new number system was introduced to remedy the situation. The number system that results from introducing the number i is called the **complex number system**.

DEFINITION

> **Complex numbers** are numbers of the form $a + bi$, where a and b are real numbers. The real number a is called the **real part** of the number $a + bi$; the real number b is called the **imaginary part** of $a + bi$; and i is the imaginary unit, so $i^2 = -1$.

For example, the complex number $-5 + 6i$ has the real part -5 and the imaginary part 6.

When a complex number is written in the form $a + bi$, where a and b are real numbers, it is in **standard form**. However, if the imaginary part of a complex number is negative, such as in the complex number $3 + (-2)i$, we agree to write it instead in the form $3 - 2i$.

Also, the complex number $a + 0i$ is usually written merely as a. This serves to remind us that the real numbers are a subset of the complex numbers. Similarly, the complex number $0 + bi$ is usually written as bi. Sometimes the complex number bi is called a **pure imaginary number**.

1 Add, Subtract, Multiply, and Divide Complex Numbers

Equality, addition, subtraction, and multiplication of complex numbers are defined so as to preserve the familiar rules of algebra for real numbers. Two complex numbers are equal if and only if their real parts are equal and their imaginary parts are equal.

Equality of Complex Numbers

$$a + bi = c + di \quad \text{if and only if} \quad a = c \text{ and } b = d \qquad \textbf{(1)}$$

Two complex numbers are added by forming the complex number whose real part is the sum of the real parts and whose imaginary part is the sum of the imaginary parts.

Sum of Complex Numbers

$$(a + bi) + (c + di) = (a + c) + (b + d)i \qquad \textbf{(2)}$$

To subtract two complex numbers, use this rule:

Difference of Complex Numbers

$$(a + bi) - (c + di) = (a - c) + (b - d)i \qquad \textbf{(3)}$$

EXAMPLE 1 | **Adding and Subtracting Complex Numbers**

(a) $(3 + 5i) + (-2 + 3i) = [3 + (-2)] + (5 + 3)i = 1 + 8i$

(b) $(6 + 4i) - (3 + 6i) = (6 - 3) + (4 - 6)i = 3 + (-2)i = 3 - 2i$ ●

Now Work PROBLEM 15

Products of complex numbers are calculated as illustrated in Example 2.

EXAMPLE 2

Multiplying Complex Numbers

$$(5 + 3i) \cdot (2 + 7i) = 5 \cdot (2 + 7i) + 3i(2 + 7i) = 10 + 35i + 6i + 21i^2$$

Distributive Property   Distributive Property

$$= 10 + 41i + 21(-1)$$

$i^2 = -1$

$$= -11 + 41i$$ ●

Based on the procedure of Example 2, the **product** of two complex numbers is defined as follows:

Product of Complex Numbers

$$(a + bi) \cdot (c + di) = (ac - bd) + (ad + bc)i \qquad (4)$$

Do not bother to memorize formula (4). Instead, whenever it is necessary to multiply two complex numbers, follow the usual rules for multiplying two binomials, as in Example 2, remembering that $i^2 = -1$. For example,

$$(2i)(2i) = 4i^2 = 4(-1) = -4$$

$$(2 + i)(1 - i) = 2 - 2i + i - i^2 = 3 - i$$

✏️ **Now Work** PROBLEM 21

Algebraic properties for addition and multiplication, such as the commutative, associative, and distributive properties, hold for complex numbers. The property that every nonzero complex number has a multiplicative inverse, or reciprocal, requires a closer look.

DEFINITION

If $z = a + bi$ is a complex number, then its **conjugate**, denoted by $\bar{z}$, is defined as

NOTE *The conjugate of a complex number can be found by changing the sign of the imaginary part.* ■

$$\bar{z} = \overline{a + bi} = a - bi$$

For example, $\overline{2 + 3i} = 2 - 3i$ and $\overline{-6 - 2i} = -6 + 2i$.

EXAMPLE 3

Multiplying a Complex Number by Its Conjugate

Find the product of the complex number $z = 3 + 4i$ and its conjugate $\bar{z}$.

Solution

Since $\bar{z} = 3 - 4i$, we have

$$z\bar{z} = (3 + 4i)(3 - 4i) = 9 - 12i + 12i - 16i^2 = 9 + 16 = 25$$ ●

The result obtained in Example 3 has an important generalization.

THEOREM

The product of a complex number and its conjugate is a nonnegative real number. That is, if $z = a + bi$, then

$$z\bar{z} = a^2 + b^2 \qquad (5)$$

Proof If $z = a + bi$, then

$$z\bar{z} = (a + bi)(a - bi) = a^2 - (bi)^2 = a^2 - b^2i^2 = a^2 + b^2 \qquad \blacksquare$$

To express the reciprocal of a nonzero complex number z in standard form, multiply the numerator and denominator of $\dfrac{1}{z}$ by $\bar{z}$. That is, if $z = a + bi$ is a nonzero complex number, then

$$\frac{1}{a + bi} = \frac{1}{z} = \frac{1}{z} \cdot \frac{\bar{z}}{\bar{z}} = \frac{\bar{z}}{z\bar{z}} = \frac{a - bi}{a^2 + b^2}$$

$$\uparrow$$
$$\text{Use (5).}$$

$$= \frac{a}{a^2 + b^2} - \frac{b}{a^2 + b^2}i$$

EXAMPLE 4   **Writing the Reciprocal of a Complex Number in Standard Form**

Write $\dfrac{1}{3 + 4i}$ in standard form $a + bi$; that is, find the reciprocal of $3 + 4i$.

Solution   The idea is to multiply the numerator and denominator by the conjugate of $3 + 4i$, that is, by the complex number $3 - 4i$. The result is

$$\frac{1}{3 + 4i} = \frac{1}{3 + 4i} \cdot \frac{3 - 4i}{3 - 4i} = \frac{3 - 4i}{9 + 16} = \frac{3}{25} - \frac{4}{25}i \qquad \bullet$$

To express the quotient of two complex numbers in standard form, multiply the numerator and denominator of the quotient by the conjugate of the denominator.

EXAMPLE 5   **Writing the Quotient of Two Complex Numbers in Standard Form**

Write each of the following in standard form.

(a) $\dfrac{1 + 4i}{5 - 12i}$ (b) $\dfrac{2 - 3i}{4 - 3i}$

Solution   (a) $\dfrac{1 + 4i}{5 - 12i} = \dfrac{1 + 4i}{5 - 12i} \cdot \dfrac{5 + 12i}{5 + 12i} = \dfrac{5 + 12i + 20i + 48i^2}{25 + 144}$

$$= \frac{-43 + 32i}{169} = -\frac{43}{169} + \frac{32}{169}i$$

(b) $\dfrac{2 - 3i}{4 - 3i} = \dfrac{2 - 3i}{4 - 3i} \cdot \dfrac{4 + 3i}{4 + 3i} = \dfrac{8 + 6i - 12i - 9i^2}{16 + 9} = \dfrac{17 - 6i}{25} = \dfrac{17}{25} - \dfrac{6}{25}i \qquad \bullet$

Now Work PROBLEM 29

EXAMPLE 6   **Writing Other Expressions in Standard Form**

If $z = 2 - 3i$ and $w = 5 + 2i$, write each of the following expressions in standard form.

(a) $\dfrac{z}{w}$ (b) $\overline{z + w}$ (c) $z + \bar{z}$

Solution   (a) $\dfrac{z}{w} = \dfrac{z \cdot \overline{w}}{w \cdot \overline{w}} = \dfrac{(2 - 3i)(5 - 2i)}{(5 + 2i)(5 - 2i)} = \dfrac{10 - 4i - 15i + 6i^2}{25 + 4}$

$= \dfrac{4 - 19i}{29} = \dfrac{4}{29} - \dfrac{19}{29}i$

(b) $\overline{z + w} = \overline{(2 - 3i) + (5 + 2i)} = \overline{7 - i} = 7 + i$

(c) $z + \overline{z} = (2 - 3i) + (2 + 3i) = 4$ ●

The conjugate of a complex number has certain general properties that will be useful later.

For a real number $a = a + 0i$, the conjugate is $\overline{a} = \overline{a + 0i} = a - 0i = a$.

THEOREM   The conjugate of a real number is the real number itself.

Other properties that are direct consequences of the definition of the conjugate are given next. In each statement, z and w represent complex numbers.

THEOREM   The conjugate of the conjugate of a complex number is the complex number itself.

$$(\overline{\overline{z}}) = z \tag{6}$$

The conjugate of the sum of two complex numbers equals the sum of their conjugates.

$$\overline{z + w} = \overline{z} + \overline{w} \tag{7}$$

The conjugate of the product of two complex numbers equals the product of their conjugates.

$$\overline{z \cdot w} = \overline{z} \cdot \overline{w} \tag{8}$$

The proofs of equations (6), (7), and (8) are left as exercises.

Powers of *i*

The **powers of *i*** follow a pattern that is useful to know.

$i^1 = i$ $\qquad\qquad\qquad\qquad$ $i^5 = i^4 \cdot i = 1 \cdot i = i$

$i^2 = -1$ $\qquad\qquad\qquad\qquad$ $i^6 = i^4 \cdot i^2 = -1$

$i^3 = i^2 \cdot i = -1 \cdot i = -i$ $\qquad$ $i^7 = i^4 \cdot i^3 = -i$

$i^4 = i^2 \cdot i^2 = (-1)(-1) = 1$ $\qquad$ $i^8 = i^4 \cdot i^4 = 1$

And so on. The powers of i repeat with every fourth power.

EXAMPLE 7   **Evaluating Powers of *i***

(a) $i^{27} = i^{24} \cdot i^3 = (i^4)^6 \cdot i^3 = 1^6 \cdot i^3 = -i$

(b) $i^{101} = i^{100} \cdot i^1 = (i^4)^{25} \cdot i = 1^{25} \cdot i = i$ ●

EXAMPLE 8	**Writing the Power of a Complex Number in Standard Form**

Write $(2 + i)^3$ in standard form.

Solution   Use the special product formula for $(x + a)^3$.

$$(x + a)^3 = x^3 + 3ax^2 + 3a^2x + a^3$$

NOTE Another way to find $(2 + i)^3$ is to multiply out $(2 + i)^2 (2 + i)$. ∎

Using this special product formula,

$$(2 + i)^3 = 2^3 + 3 \cdot i \cdot 2^2 + 3 \cdot i^2 \cdot 2 + i^3$$
$$= 8 + 12i + 6(-1) + (-i)$$
$$= 2 + 11i$$

➤ **Now Work** PROBLEM **43**

2 Solve Quadratic Equations in the Complex Number System

Quadratic equations with a negative discriminant have no real number solution. However, if we extend our number system to allow complex numbers, quadratic equations will always have a solution. Since the solution to a quadratic equation involves the square root of the discriminant, we begin with a discussion of square roots of negative numbers.

DEFINITION
> If N is a positive real number, we define the **principal square root of** $-N$, denoted by $\sqrt{-N}$, as
>
> $$\boxed{\sqrt{-N} = \sqrt{N}\,i}$$
>
> where i is the imaginary unit and $i^2 = -1$.

WARNING In writing $\sqrt{-N} = \sqrt{N}\,i$, be sure to place i outside the $\sqrt{}$ symbol. ∎

EXAMPLE 9	**Evaluating the Square Root of a Negative Number**

(a) $\sqrt{-1} = \sqrt{1}\,i = i$
(b) $\sqrt{-4} = \sqrt{4}\,i = 2i$
(c) $\sqrt{-8} = \sqrt{8}\,i = 2\sqrt{2}\,i$

EXAMPLE 10	**Solving Equations**

Solve each equation in the complex number system.

(a) $x^2 = 4$ (b) $x^2 = -9$

Solution   (a) $x^2 = 4$
$$x = \pm\sqrt{4} = \pm 2$$

The equation has two solutions, -2 and 2. The solution set is $\{-2, 2\}$.

(b) $x^2 = -9$
$$x = \pm\sqrt{-9} = \pm\sqrt{9}\,i = \pm 3i$$

The equation has two solutions, $-3i$ and $3i$. The solution set is $\{-3i, 3i\}$.

➤ **Now Work** PROBLEMS **51** AND **55**

WARNING When working with square roots of negative numbers, do not set the square root of a product equal to the product of the square roots (which can be done with positive real numbers). To see why, look at this calculation: We know that $\sqrt{100} = 10$. However, it is also true that $100 = (-25)(-4)$, so

$$10 = \sqrt{100} = \sqrt{(-25)(-4)} = \sqrt{-25}\sqrt{-4} = \left(\sqrt{25}i\right)\left(\sqrt{4}i\right) = (5i)(2i) = 10i^2 = -10$$

↑
Here is the error. ∎

Because we have defined the square root of a negative number, we can now restate the quadratic formula without restriction.

THEOREM

Quadratic Formula

In the complex number system, the solutions of the quadratic equation $ax^2 + bx + c = 0$, where a, b, and c are real numbers and $a \neq 0$, are given by the formula

$$x = \frac{-b \pm \sqrt{b^2 - 4ac}}{2a} \qquad \textbf{(9)}$$

EXAMPLE 11

Solving a Quadratic Equation in the Complex Number System

Solve the equation $x^2 - 4x + 8 = 0$ in the complex number system.

Solution   Here $a = 1, b = -4, c = 8,$ and $b^2 - 4ac = (-4)^2 - 4(1)(8) = -16.$ Using equation (9), we find that

$$x = \frac{-(-4) \pm \sqrt{-16}}{2(1)} = \frac{4 \pm \sqrt{16}i}{2} = \frac{4 \pm 4i}{2} = \frac{2(2 \pm 2i)}{2} = 2 \pm 2i$$

The equation has two solutions: $2 - 2i$ and $2 + 2i$.
The solution set is $\{2 - 2i, 2 + 2i\}$.

✓**Check:** $2 + 2i$: $(2 + 2i)^2 - 4(2 + 2i) + 8 = 4 + 8i + 4i^2 - 8 - 8i + 8$
$$= 4 + 4i^2$$
$$= 4 - 4 = 0$$

  $2 - 2i$: $(2 - 2i)^2 - 4(2 - 2i) + 8 = 4 - 8i + 4i^2 - 8 + 8i + 8$
$$= 4 - 4 = 0 \qquad \bullet$$

Now Work PROBLEM 61

The discriminant $b^2 - 4ac$ of a quadratic equation still serves as a way to determine the character of the solutions.

Character of the Solutions of a Quadratic Equation

In the complex number system, consider a quadratic equation $ax^2 + bx + c = 0$ with real coefficients.

1. If $b^2 - 4ac > 0$, the equation has two unequal real solutions.
2. If $b^2 - 4ac = 0$, the equation has a repeated real solution, a double root.
3. If $b^2 - 4ac < 0$, the equation has two complex solutions that are not real. The solutions are conjugates of each other.

The third conclusion in the display is a consequence of the fact that if $b^2 - 4ac = -N < 0$, then by the quadratic formula, the solutions are

$$x = \frac{-b + \sqrt{b^2 - 4ac}}{2a} = \frac{-b + \sqrt{-N}}{2a} = \frac{-b + \sqrt{N}i}{2a} = \frac{-b}{2a} + \frac{\sqrt{N}}{2a}i$$

and

$$x = \frac{-b - \sqrt{b^2 - 4ac}}{2a} = \frac{-b - \sqrt{-N}}{2a} = \frac{-b - \sqrt{N}i}{2a} = \frac{-b}{2a} - \frac{\sqrt{N}}{2a}i$$

which are conjugates of each other.

EXAMPLE 12

Determining the Character of the Solutions of a Quadratic Equation

Without solving, determine the character of the solutions of each equation.

(a) $3x^2 + 4x + 5 = 0$ (b) $2x^2 + 4x + 1 = 0$ (c) $9x^2 - 6x + 1 = 0$

Solution
(a) Here $a = 3, b = 4$, and $c = 5$, so $b^2 - 4ac = 16 - 4(3)(5) = -44$. The solutions are two complex numbers that are not real and are conjugates of each other.

(b) Here $a = 2, b = 4$, and $c = 1$, so $b^2 - 4ac = 16 - 8 = 8$. The solutions are two unequal real numbers.

(c) Here $a = 9, b = -6$, and $c = 1$, so $b^2 - 4ac = 36 - 4(9)(1) = 0$. The solution is a repeated real number—that is, a double root. ●

Now Work PROBLEM 75

1.3 Assess Your Understanding

'Are You Prepared?' *Answers are given at the end of these exercises. If you get a wrong answer, read the pages listed in* red.

1. Name the integers and the rational numbers in the set $\left\{-3, 0, \sqrt{2}, \dfrac{6}{5}, \pi\right\}$. (pp. 4–5)

2. *True or False* Rational numbers and irrational numbers are in the set of real numbers. (pp. 4–5)

3. Rationalize the denominator of $\dfrac{3}{2 + \sqrt{3}}$. (p. 75)

Concepts and Vocabulary

4. In the complex number $5 + 2i$, the number 5 is called the _____ part; the number 2 is called the _____ part; the number i is called the _____ _____.

5. *True or False* The conjugate of $2 + 5i$ is $-2 - 5i$.

6. *True or False* All real numbers are complex numbers.

7. *True or False* If $2 - 3i$ is a solution of a quadratic equation with real coefficients, then $-2 + 3i$ is also a solution.

8. Which of the following is the principal square root of -4?
(a) $-2i$ (b) $2i$ (c) -2 (d) 2

9. Which operation involving complex numbers requires the use of a conjugate?
(a) division   (b) multiplication
(c) subtraction   (d) addition

10. Powers of i repeat every _____ power.
(a) second   (b) third   (c) fourth   (d) fifth

Skill Building

In Problems 11–48, perform the indicated operation, and write each expression in the standard form $a + bi$.

11. $(2 - 3i) + (6 + 8i)$

12. $(4 + 5i) + (-8 + 2i)$

13. $(-3 + 2i) - (4 - 4i)$

14. $(3 - 4i) - (-3 - 4i)$

15. $(2 - 5i) - (8 + 6i)$

16. $(-8 + 4i) - (2 - 2i)$

17. $3(2 - 6i)$

18. $-4(2 + 8i)$

19. $2i(2 - 3i)$

20. $3i(-3 + 4i)$

21. $(3 - 4i)(2 + i)$

22. $(5 + 3i)(2 - i)$

23. $(-6 + i)(-6 - i)$

24. $(-3 + i)(3 + i)$

25. $\dfrac{10}{3 - 4i}$

26. $\dfrac{13}{5 - 12i}$

27. $\dfrac{2 + i}{i}$

28. $\dfrac{2 - i}{-2i}$

29. $\dfrac{6 - i}{1 + i}$

30. $\dfrac{2 + 3i}{1 - i}$

31. $\left(\dfrac{1}{2} + \dfrac{\sqrt{3}}{2}i\right)^2$

32. $\left(\dfrac{\sqrt{3}}{2} - \dfrac{1}{2}i\right)^2$

33. $(1 + i)^2$

34. $(1 - i)^2$

35. i^{23} **36.** i^{14} **37.** i^{-15} **38.** i^{-23} **39.** $i^6 - 5$

40. $4 + i^3$ **41.** $6i^3 - 4i^5$ **42.** $4i^3 - 2i^2 + 1$ **43.** $(1 + i)^3$ **44.** $(3i)^4 + 1$

45. $i^7(1 + i^2)$ **46.** $2i^4(1 + i^2)$ **47.** $i^6 + i^4 + i^2 + 1$ **48.** $i^7 + i^5 + i^3 + i$

In Problems 49–54, perform the indicated operations, and express your answer in the form $a + bi$.

49. $\sqrt{-4}$ **50.** $\sqrt{-9}$ **51.** $\sqrt{-25}$

52. $\sqrt{-64}$ **53.** $\sqrt{(3 + 4i)(4i - 3)}$ **54.** $\sqrt{(4 + 3i)(3i - 4)}$

In Problems 55–74, solve each equation in the complex number system.

55. $x^2 + 4 = 0$ **56.** $x^2 - 4 = 0$ **57.** $x^2 - 16 = 0$ **58.** $x^2 + 25 = 0$

59. $x^2 - 6x + 13 = 0$ **60.** $x^2 + 4x + 8 = 0$ **61.** $x^2 - 6x + 10 = 0$ **62.** $x^2 - 2x + 5 = 0$

63. $8x^2 - 4x + 1 = 0$ **64.** $10x^2 + 6x + 1 = 0$ **65.** $5x^2 + 1 = 2x$ **66.** $13x^2 + 1 = 6x$

67. $x^2 + x + 1 = 0$ **68.** $x^2 - x + 1 = 0$ **69.** $x^3 - 8 = 0$ **70.** $x^3 + 27 = 0$

71. $x^4 = 16$ **72.** $x^4 = 1$ **73.** $x^4 + 13x^2 + 36 = 0$ **74.** $x^4 + 3x^2 - 4 = 0$

In Problems 75–80, without solving, determine the character of the solutions of each equation in the complex number system.

75. $3x^2 - 3x + 4 = 0$ **76.** $2x^2 - 4x + 1 = 0$ **77.** $2x^2 + 3x = 4$

78. $x^2 + 6 = 2x$ **79.** $9x^2 - 12x + 4 = 0$ **80.** $4x^2 + 12x + 9 = 0$

81. $2 + 3i$ is a solution of a quadratic equation with real coefficients. Find the other solution.

82. $4 - i$ is a solution of a quadratic equation with real coefficients. Find the other solution.

In Problems 83–86, $z = 3 - 4i$ and $w = 8 + 3i$. Write each expression in the standard form $a + bi$.

83. $z + \bar{z}$ **84.** $w - \bar{w}$ **85.** $z\bar{z}$ **86.** $\overline{z - w}$

Applications and Extensions

87. Electrical Circuits The impedance Z, in ohms, of a circuit element is defined as the ratio of the phasor voltage V, in volts, across the element to the phasor current I, in amperes, through the elements. That is, $Z = \dfrac{V}{I}$. If the voltage across a circuit element is $18 + i$ volts and the current through the element is $3 - 4i$ amperes, determine the impedance.

88. Parallel Circuits In an ac circuit with two parallel pathways, the total impedance Z, in ohms, satisfies the formula $\dfrac{1}{Z} = \dfrac{1}{Z_1} + \dfrac{1}{Z_2}$, where Z_1 is the impedance of the first pathway

and Z_2 is the impedance of the second pathway. Determine the total impedance if the impedances of the two pathways are $Z_1 = 2 + i$ ohms and $Z_2 = 4 - 3i$ ohms.

89. Use $z = a + bi$ to show that $z + \bar{z} = 2a$ and $z - \bar{z} = 2bi$.

90. Use $z = a + bi$ to show that $\bar{\bar{z}} = z$.

91. Use $z = a + bi$ and $w = c + di$ to show that $\overline{z + w} = \bar{z} + \bar{w}$.

92. Use $z = a + bi$ and $w = c + di$ to show that $\overline{z \cdot w} = \bar{z} \cdot \bar{w}$.

Explaining Concepts: Discussion and Writing

93. Explain to a friend how you would add two complex numbers and how you would multiply two complex numbers. Explain any differences between the two explanations.

94. Write a brief paragraph that compares the method used to rationalize the denominator of a radical expression and the method used to write the quotient of two complex numbers in standard form.

95. Use an Internet search engine to investigate the origins of complex numbers. Write a paragraph describing what you find, and present it to the class.

96. Explain how the method of multiplying two complex numbers is related to multiplying two binomials.

97. What Went Wrong? A student multiplied $\sqrt{-9}$ and $\sqrt{-9}$ as follows:

$$\sqrt{-9} \cdot \sqrt{-9} = \sqrt{(-9)(-9)}$$
$$= \sqrt{81}$$
$$= 9$$

The instructor marked the problem incorrect. Why?

'Are You Prepared?' Answers

1. Integers: $\{-3, 0\}$; rational numbers: $\left\{-3, 0, \dfrac{6}{5}\right\}$ **2.** True   **3.** $3(2 - \sqrt{3})$

1.4 Radical Equations; Equations Quadratic in Form; Factorable Equations

PREPARING FOR THIS SECTION *Before getting started, review the following:*

- Square Roots (Section R.2, pp. 23–24)
- Factoring Polynomials (Section R.5, pp. 49–55)
- nth Roots; Rational Exponents (Section R.8, pp. 73–77)

Now Work the **'Are You Prepared?'** problems on page 117.

OBJECTIVES 1 Solve Radical Equations (p. 113)
2 Solve Equations Quadratic in Form (p. 114)
3 Solve Equations by Factoring (p. 116)

1 Solve Radical Equations

When the variable in an equation occurs in a square root, cube root, and so on—that is, when it occurs in a radical—the equation is called a **radical equation**. Sometimes a suitable operation will change a radical equation to one that is linear or quadratic. A commonly used procedure is to isolate the most complicated radical on one side of the equation and then eliminate it by raising each side to a power equal to the index of the radical. Care must be taken, however, because apparent solutions that are not, in fact, solutions of the original equation may result. These are called **extraneous solutions**. Therefore, we need to check all answers when working with radical equations, and we check them in the *original* equation.

EXAMPLE 1

Solving a Radical Equation

Find the real solutions of the equation: $\sqrt[3]{2x - 4} - 2 = 0$

Solution

The equation contains a radical whose index is 3. Isolate it on the left side.

$$\sqrt[3]{2x - 4} - 2 = 0$$
$$\sqrt[3]{2x - 4} = 2 \quad \text{Add 2 to both sides.}$$

Now raise each side to the third power (the index of the radical is 3) and solve.

$$\left(\sqrt[3]{2x - 4}\right)^3 = 2^3 \quad \text{Raise each side to the power 3.}$$
$$2x - 4 = 8 \quad \text{Simplify.}$$
$$2x = 12 \quad \text{Add 4 to both sides.}$$
$$x = 6 \quad \text{Divide both sides by 2.}$$

✓**Check:** $\sqrt[3]{2(6) - 4} - 2 = \sqrt[3]{12 - 4} - 2 = \sqrt[3]{8} - 2 = 2 - 2 = 0$

The solution set is $\{6\}$.

 Now Work PROBLEM 9

EXAMPLE 2

Solving a Radical Equation

Find the real solutions of the equation: $\sqrt{x - 1} = x - 7$

Solution

Square both sides since the index of a square root is 2.

$$\sqrt{x - 1} = x - 7$$
$$\left(\sqrt{x - 1}\right)^2 = (x - 7)^2 \quad \text{Square both sides.}$$
$$x - 1 = x^2 - 14x + 49 \quad \text{Remove parentheses.}$$
$$x^2 - 15x + 50 = 0 \quad \text{Put in standard form.}$$

$$(x - 10)(x - 5) = 0 \qquad \text{\color{teal}Factor.}$$
$$x = 10 \quad \text{or} \quad x = 5 \qquad \text{\color{teal}Apply the Zero-Product Property and solve.}$$

✓**Check:**

$$x = 10: \quad \sqrt{x - 1} = \sqrt{10 - 1} = \sqrt{9} = 3 \text{ and } x - 7 = 10 - 7 = 3$$
$$x = 5: \quad \sqrt{x - 1} = \sqrt{5 - 1} = \sqrt{4} = 2 \text{ and } x - 7 = 5 - 7 = -2$$

The solution $x = 5$ does not check, so it is extraneous; the only solution of the equation is $x = 10$. The solution set is {10}. ●

━━━ **Now Work** PROBLEM **19**

Sometimes it is necessary to raise each side to a power more than once in order to solve a radical equation.

EXAMPLE 3

Solving a Radical Equation

Find the real solutions of the equation: $\sqrt{2x + 3} - \sqrt{x + 2} = 2$

Solution

First, isolate the more complicated radical expression (in this case, $\sqrt{2x + 3}$) on the left side.

$$\sqrt{2x + 3} = \sqrt{x + 2} + 2$$

Now square both sides (the index of the radical on the left is 2).

$$\left(\sqrt{2x + 3}\right)^2 = \left(\sqrt{x + 2} + 2\right)^2 \qquad \text{\color{teal}Square both sides.}$$
$$2x + 3 = \left(\sqrt{x + 2}\right)^2 + 4\sqrt{x + 2} + 4 \qquad \text{\color{teal}Multiply out.}$$
$$2x + 3 = x + 2 + 4\sqrt{x + 2} + 4 \qquad \text{\color{teal}Simplify.}$$
$$2x + 3 = x + 6 + 4\sqrt{x + 2} \qquad \text{\color{teal}Combine like terms.}$$

Because the equation still contains a radical, isolate the remaining radical on the right side and again square both sides.

$$x - 3 = 4\sqrt{x + 2} \qquad \text{\color{teal}Isolate the radical on the right side.}$$
$$(x - 3)^2 = 16(x + 2) \qquad \text{\color{teal}Square both sides.}$$
$$x^2 - 6x + 9 = 16x + 32 \qquad \text{\color{teal}Multiply out.}$$
$$x^2 - 22x - 23 = 0 \qquad \text{\color{teal}Put in standard form.}$$
$$(x - 23)(x + 1) = 0 \qquad \text{\color{teal}Factor.}$$
$$x = 23 \quad \text{or} \quad x = -1 \qquad \text{\color{teal}Apply the Zero-Product Property and solve.}$$

The original equation appears to have the solution set $\{-1, 23\}$. However, we have not yet checked.

✓**Check:**

$$x = 23: \quad \sqrt{2x + 3} - \sqrt{x + 2} = \sqrt{2(23) + 3} - \sqrt{23 + 2} = \sqrt{49} - \sqrt{25} = 7 - 5 = 2$$
$$x = -1: \quad \sqrt{2x + 3} - \sqrt{x + 2} = \sqrt{2(-1) + 3} - \sqrt{-1 + 2} = \sqrt{1} - \sqrt{1} = 1 - 1 = 0$$

The equation has only one solution, 23; the solution -1 is extraneous. The solution set is {23}. ●

━━━ **Now Work** PROBLEM **31**

2 Solve Equations Quadratic in Form

The equation $x^4 + x^2 - 12 = 0$ is not quadratic in x, but it is quadratic in x^2. That is, if we let $u = x^2$, we get $u^2 + u - 12 = 0$, a quadratic equation. This equation can be solved for u, and in turn, by using $u = x^2$, we can find the solutions x of the original equation.

In general, if an appropriate substitution u transforms an equation into one of the form

$$au^2 + bu + c = 0 \qquad a \neq 0$$

then the original equation is called an **equation of the quadratic type** or an **equation quadratic in form**.

The difficulty of solving such an equation lies in the determination that the equation is, in fact, quadratic in form. After you are told an equation is quadratic in form, it is easy enough to see it, but some practice is needed to enable you to recognize such equations on your own.

EXAMPLE 4

Solving an Equation Quadratic in Form

Find the real solutions of the equation: $(x + 2)^2 + 11(x + 2) - 12 = 0$

Solution For this equation, let $u = x + 2$. Then $u^2 = (x + 2)^2$, and the original equation,

$$(x + 2)^2 + 11(x + 2) - 12 = 0$$

becomes

$$u^2 + 11u - 12 = 0 \qquad \text{Let } u = x + 2. \text{ Then } u^2 = (x + 2)^2.$$
$$(u + 12)(u - 1) = 0 \qquad \text{Factor.}$$
$$u = -12 \quad \text{or} \quad u = 1 \qquad \text{Solve.}$$

But we want to solve for x. Because $u = x + 2$, we have

$$x + 2 = -12 \quad \text{or} \quad x + 2 = 1$$
$$x = -14 \qquad\qquad x = -1$$

✓**Check:** $x = -14$: $\quad (-14 + 2)^2 + 11(-14 + 2) - 12$
$$= (-12)^2 + 11(-12) - 12 = 144 - 132 - 12 = 0$$
$\qquad\qquad x = -1$: $\quad (-1 + 2)^2 + 11(-1 + 2) - 12 = 1 + 11 - 12 = 0$

The original equation has the solution set $\{-14, -1\}$. ●

EXAMPLE 5

Solving an Equation Quadratic in Form

Find the real solutions of the equation: $(x^2 - 1)^2 + (x^2 - 1) - 12 = 0$

Solution For the equation $(x^2 - 1)^2 + (x^2 - 1) - 12 = 0$, let $u = x^2 - 1$ so that $u^2 = (x^2 - 1)^2$. Then the original equation,

$$(x^2 - 1)^2 + (x^2 - 1) - 12 = 0$$

becomes

$$u^2 + u - 12 = 0 \qquad \text{Let } u = x^2 - 1. \text{ Then } u^2 = (x^2 - 1)^2.$$
$$(u + 4)(u - 3) = 0 \qquad \text{Factor.}$$
$$u = -4 \quad \text{or} \quad u = 3 \qquad \text{Solve.}$$

But remember that we want to solve for x. Because $u = x^2 - 1$, we have

$$x^2 - 1 = -4 \quad \text{or} \quad x^2 - 1 = 3$$
$$x^2 = -3 \qquad\qquad x^2 = 4$$

The first of these has no real solution; the second has the solution set $\{-2, 2\}$.

✓**Check:** $x = -2$: $\quad ((-2)^2 - 1)^2 + ((-2)^2 - 1) - 12 = 9 + 3 - 12 = 0$
$\qquad\qquad x = 2$: $\quad ((2)^2 - 1)^2 + ((2)^2 - 1) - 12 = 9 + 3 - 12 = 0$

The original equation has the solution set $\{-2, 2\}$. ●

EXAMPLE 6

Solving an Equation Quadratic in Form

Find the real solutions of the equation: $x + 2\sqrt{x} - 3 = 0$

Solution
For the equation $x + 2\sqrt{x} - 3 = 0$, let $u = \sqrt{x}$. Then $u^2 = x$, and the original equation,

$$x + 2\sqrt{x} - 3 = 0$$

becomes

$$u^2 + 2u - 3 = 0 \quad \text{Let } u = \sqrt{x}. \text{ Then } u^2 = x.$$
$$(u + 3)(u - 1) = 0 \quad \text{Factor.}$$
$$u = -3 \quad \text{or} \quad u = 1 \quad \text{Solve.}$$

Since $u = \sqrt{x}$, we have $\sqrt{x} = -3$ or $\sqrt{x} = 1$. The first of these, $\sqrt{x} = -3$, has no real solution, since the principal square root of a real number is never negative. The second, $\sqrt{x} = 1$, has the solution $x = 1$.

✓**Check:** $1 + 2\sqrt{1} - 3 = 1 + 2 - 3 = 0$

The original equation has the solution set {1}.   ●

ANOTHER METHOD FOR SOLVING EXAMPLE **6** WOULD BE TO TREAT IT AS A RADICAL EQUATION. SOLVE IT THIS WAY FOR PRACTICE.

The idea should now be clear. If an equation contains an expression and that same expression squared, make a substitution for the expression. You may get a quadratic equation.

Now Work PROBLEM **53**

3 Solve Equations by Factoring

We have already solved certain quadratic equations using factoring. Let's look at examples of other kinds of equations that can be solved by factoring.

EXAMPLE 7

Solving an Equation by Factoring

Solve the equation: $x^4 = 4x^2$

Solution
Begin by collecting all terms on one side. This results in 0 on one side and an expression to be factored on the other.

$$x^4 = 4x^2$$
$$x^4 - 4x^2 = 0$$
$$x^2(x^2 - 4) = 0 \quad \text{Factor.}$$
$$x^2 = 0 \quad \text{or} \quad x^2 - 4 = 0 \quad \text{Apply the Zero-Product Property.}$$
$$x = 0 \quad \text{or} \quad x^2 = 4$$
$$x = 0 \quad \text{or} \quad x = -2 \quad \text{or} \quad x = 2$$

✓**Check:** $x = -2$: $(-2)^4 = 16$ and $4(-2)^2 = 16$ -2 is a solution.

$x = 0$: $0^4 = 0$ and $4 \cdot 0^2 = 0$ 0 is a solution.

$x = 2$: $2^4 = 16$ and $4 \cdot 2^2 = 16$ 2 is a solution.

The solution set is $\{-2, 0, 2\}$.   ●

EXAMPLE 8   **Solving an Equation by Factoring**

Solve the equation: $x^3 - x^2 - 4x + 4 = 0$

Solution   Do you recall the method of factoring by grouping? (If not, review pp. 53–54.) Group the terms of $x^3 - x^2 - 4x + 4 = 0$ as follows:

$$(x^3 - x^2) - (4x - 4) = 0$$

Factor out x^2 from the first grouping and 4 from the second.

$$x^2(x - 1) - 4(x - 1) = 0$$

This reveals the common factor $(x - 1)$, so we have

$$(x^2 - 4)(x - 1) = 0$$
$$(x - 2)(x + 2)(x - 1) = 0 \qquad \text{Factor again.}$$
$$x - 2 = 0 \quad \text{or} \quad x + 2 = 0 \quad \text{or} \quad x - 1 = 0 \quad \text{Set each factor equal to 0.}$$
$$x = 2 \qquad\qquad x = -2 \qquad\qquad x = 1 \quad \text{Solve.}$$

✓**Check:**

$x = -2$: $(-2)^3 - (-2)^2 - 4(-2) + 4 = -8 - 4 + 8 + 4 = 0$ -2 is a solution.
$x = 1$: $1^3 - 1^2 - 4(1) + 4 = 1 - 1 - 4 + 4 = 0$ 1 is a solution.
$x = 2$: $2^3 - 2^2 - 4(2) + 4 = 8 - 4 - 8 + 4 = 0$ 2 is a solution.

The solution set is $\{-2, 1, 2\}$.

 Now Work PROBLEM 81

1.4 Assess Your Understanding

'Are You Prepared?' *Answers are given at the end of these exercises. If you get a wrong answer, read the pages listed in* red.

1. *True or False* The principal square root of any nonnegative real number is always nonnegative. (pp. 23–24)

2. $\sqrt[3]{-8} =$ ___ (pp. 73–77)

3. Factor $6x^3 - 2x^2$ (pp. 49–55)

Concepts and Vocabulary

4. *True or False* Factoring can only be used to solve quadratic equations or equations that are quadratic in form.

5. If u is an expression that involves x, then the equation $au^2 + bu + c = 0, a \neq 0$, is called an equation _____ _____.

6. *True or False* Radical equations sometimes have extraneous solutions.

7. An apparent solution that does not satisfy the original equation is called a(n) _____ solution.
   (a) extraneous   (b) imaginary
   (c) radical   (d) conditional

8. Which equation is likely to require squaring each side more than once?
   (a) $\sqrt{x + 2} = \sqrt{3x - 5}$ (b) $x^4 - 3x^2 = 10$
   (c) $\sqrt{x + 1} + \sqrt{x - 4} = 8$ (d) $\sqrt{3x + 1} = 5$

Skill Building

In Problems 9–42, find the real solutions of each equation.

9. $\sqrt{2t - 1} = 1$

10. $\sqrt{3t + 4} = 2$

11. $\sqrt{3t + 4} = -6$

12. $\sqrt{5t + 3} = -2$

13. $\sqrt[3]{1 - 2x} - 3 = 0$

14. $\sqrt[3]{1 - 2x} - 1 = 0$

15. $\sqrt[5]{x^2 + 2x} = -1$

16. $\sqrt[4]{x^2 + 16} = \sqrt{5}$

17. $x = 8\sqrt{x}$

18. $x = 3\sqrt{x}$

19. $\sqrt{15 - 2x} = x$

20. $\sqrt{12 - x} = x$

21. $x = 2\sqrt{x - 1}$

22. $x = 2\sqrt{-x - 1}$

23. $\sqrt{x^2 - x - 4} = x + 2$

24. $\sqrt{3 - x + x^2} = x - 2$

25. $3 + \sqrt{3x + 1} = x$

26. $2 + \sqrt{12 - 2x} = x$

27. $\sqrt{3(x + 10)} - 4 = x$

28. $\sqrt{1 - x} - 3 = x + 2$

29. $\sqrt{2x + 3} - \sqrt{x + 1} = 1$

30. $\sqrt{3x + 7} + \sqrt{x + 2} = 1$

31. $\sqrt{3x + 1} - \sqrt{x - 1} = 2$

32. $\sqrt{3x - 5} - \sqrt{x + 7} = 2$

33. $\sqrt{3 - 2\sqrt{x}} = \sqrt{x}$　　　　　　**34.** $\sqrt{10 + 3\sqrt{x}} = \sqrt{x}$　　　　　　**35.** $(3x + 1)^{1/2} = 4$

36. $(3x - 5)^{1/2} = 2$　　　　　　**37.** $(5x - 2)^{1/3} = 2$　　　　　　**38.** $(2x + 1)^{1/3} = -1$

39. $(x^2 + 9)^{1/2} = 5$　　　　**40.** $(x^2 - 16)^{1/2} = 9$　　　　**41.** $x^{3/2} - 3x^{1/2} = 0$　　　　**42.** $x^{3/4} - 9x^{1/4} = 0$

In Problems 43–74, find the real solutions of each equation.

43. $x^4 - 5x^2 + 4 = 0$　　　　　　**44.** $x^4 - 10x^2 + 25 = 0$　　　　　　**45.** $3x^4 - 2x^2 - 1 = 0$

46. $2x^4 - 5x^2 - 12 = 0$　　　　　　**47.** $x^6 + 7x^3 - 8 = 0$　　　　　　**48.** $x^6 - 7x^3 - 8 = 0$

49. $(x + 2)^2 + 7(x + 2) + 12 = 0$　　　　　　**50.** $(2x + 5)^2 - (2x + 5) - 6 = 0$　　　　　　**51.** $(3x + 4)^2 - 6(3x + 4) + 9 = 0$

52. $(2 - x)^2 + (2 - x) - 20 = 0$　　　　✎ **53.** $2(s + 1)^2 - 5(s + 1) = 3$　　　　　　**54.** $3(1 - y)^2 + 5(1 - y) + 2 = 0$

55. $x - 4x\sqrt{x} = 0$　　　　　　**56.** $x + 8\sqrt{x} = 0$　　　　　　**57.** $x + \sqrt{x} = 20$

58. $x + \sqrt{x} = 6$　　　　　　**59.** $t^{1/2} - 2t^{1/4} + 1 = 0$　　　　　　**60.** $z^{1/2} - 4z^{1/4} + 4 = 0$

61. $4x^{1/2} - 9x^{1/4} + 4 = 0$　　　　　　**62.** $x^{1/2} - 3x^{1/4} + 2 = 0$　　　　　　**63.** $\sqrt[4]{5x^2 - 6} = x$

64. $\sqrt[4]{4 - 5x^2} = x$　　　　　　**65.** $x^2 + 3x + \sqrt{x^2 + 3x} = 6$　　　　　　**66.** $x^2 - 3x - \sqrt{x^2 - 3x} = 2$

67. $\dfrac{1}{(x + 1)^2} = \dfrac{1}{x + 1} + 2$　　　　　　**68.** $\dfrac{1}{(x - 1)^2} + \dfrac{1}{x - 1} = 12$　　　　　　**69.** $3x^{-2} - 7x^{-1} - 6 = 0$

70. $2x^{-2} - 3x^{-1} - 4 = 0$　　　　　　**71.** $2x^{2/3} - 5x^{1/3} - 3 = 0$　　　　　　**72.** $3x^{4/3} + 5x^{2/3} - 2 = 0$

73. $\left(\dfrac{v}{v + 1}\right)^2 + \dfrac{2v}{v + 1} = 8$　　　　　　**74.** $\left(\dfrac{y}{y - 1}\right)^2 = 6\left(\dfrac{y}{y - 1}\right) + 7$

In Problems 75–90, find the real solutions of each equation by factoring.

75. $x^3 - 9x = 0$　　　　　　**76.** $x^4 - x^2 = 0$　　　　　　**77.** $4x^3 = 3x^2$　　　　　　**78.** $x^5 = 4x^3$

79. $x^3 + x^2 - 20x = 0$　　　　　　**80.** $x^3 + 6x^2 - 7x = 0$　　　　✎ **81.** $x^3 + x^2 - x - 1 = 0$

82. $x^3 + 4x^2 - x - 4 = 0$　　　　　　**83.** $x^3 - 3x^2 - 4x + 12 = 0$　　　　　　**84.** $x^3 - 3x^2 - x + 3 = 0$

85. $2x^3 + 4 = x^2 + 8x$　　　　　　**86.** $3x^3 + 4x^2 = 27x + 36$　　　　　　**87.** $5x^3 + 45x = 2x^2 + 18$

88. $3x^3 + 12x = 5x^2 + 20$　　　　　　**89.** $x(x^2 - 3x)^{1/3} + 2(x^2 - 3x)^{4/3} = 0$　　　　　　**90.** $3x(x^2 + 2x)^{1/2} - 2(x^2 + 2x)^{3/2} = 0$

In Problems 91–96, find the real solutions of each equation. Use a calculator to express any solutions rounded to two decimal places.

91. $x - 4x^{1/2} + 2 = 0$　　　　　　**92.** $x^{2/3} + 4x^{1/3} + 2 = 0$　　　　　　**93.** $x^4 + \sqrt{3}x^2 - 3 = 0$

94. $x^4 + \sqrt{2}x^2 - 2 = 0$　　　　　　**95.** $\pi(1 + t)^2 = \pi + 1 + t$　　　　　　**96.** $\pi(1 + r)^2 = 2 + \pi(1 + r)$

Mixed Practice

97. If $k = \dfrac{x + 3}{x - 3}$ and $k^2 - k = 12$, find x.

98. If $k = \dfrac{x + 3}{x - 4}$ and $k^2 - 3k = 28$, find x.

Applications

99. Physics: Using Sound to Measure Distance The distance to the surface of
the water in a well can sometimes be found by dropping an object into the
well and measuring the time elapsed until a sound is heard. If t_1 is the time
(measured in seconds) that it takes for the object to strike the water, then t_1
will obey the equation $s = 16t_1^2$, where s is the distance (measured in feet). It
follows that $t_1 = \dfrac{\sqrt{s}}{4}$. Suppose that t_2 is the time that it takes for the sound
of the impact to reach your ears. Because sound waves are known to travel
at a speed of approximately 1100 feet per second, the time t_2 to travel the
distance s will be $t_2 = \dfrac{s}{1100}$. See the illustration.

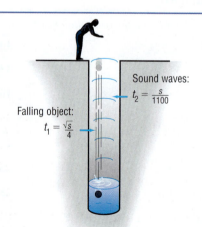

Sound waves:
$t_2 = \dfrac{s}{1100}$

Falling object:
$t_1 = \dfrac{\sqrt{s}}{4}$

Now $t_1 + t_2$ is the total time that elapses from the moment that the object is dropped to the moment that a sound is heard. We have the equation

$$\text{Total time elapsed} = \frac{\sqrt{s}}{4} + \frac{s}{1100}$$

Find the distance to the water's surface if the total time elapsed from dropping a rock to hearing it hit water is 4 seconds.

100. Crushing Load A civil engineer relates the thickness T, in inches, and height H, in feet, of a square wooden pillar to its crushing load L, in tons, using the model $T = \sqrt[4]{\dfrac{LH^2}{25}}$. If a square wooden pillar is 4 inches thick and 10 feet high, what is its crushing load?

101. Foucault's Pendulum The period of a pendulum is the time it takes the pendulum to make one full swing back and forth. The period T, in seconds, is given by the formula $T = 2\pi\sqrt{\dfrac{l}{32}}$, where l is the length, in feet, of the pendulum. In 1851, Jean Bernard Leon Foucault demonstrated the axial rotation of Earth using a large pendulum that he hung in the Panthéon in Paris. The period of Foucault's pendulum was approximately 16.5 seconds. What was its length?

Explaining Concepts: Discussion and Writing

102. Make up a radical equation that has no solution.

103. Make up a radical equation that has an extraneous solution.

104. Discuss the step in the solving process for radical equations that leads to the possibility of extraneous solutions. Why is there no such possibility for linear and quadratic equations?

105. What Went Wrong? On an exam, Jane solved the equation $\sqrt{2x+3} - x = 0$ and wrote that the solution set was $\{-1, 3\}$. Jane received 3 out of 5 points for the problem. Jane asks you why she received 3 out of 5 points. Provide an explanation.

'Are You Prepared?' Answers

1. True   **2.** -2 **3.** $2x^2(3x - 1)$

1.5 Solving Inequalities

PREPARING FOR THIS SECTION *Before getting started, review the following:*

* Algebra Essentials (Section R.2, pp. 17–26)

Now Work the **'Are You Prepared?'** problems on page 127.

OBJECTIVES **1** Use Interval Notation (p. 120)
                2 Use Properties of Inequalities (p. 121)
                3 Solve Inequalities (p. 123)
                4 Solve Combined Inequalities (p. 124)

Suppose that a and b are two real numbers and $a < b$. The notation $a < x < b$ means that x is a number *between* a and b. The expression $a < x < b$ is equivalent to the two inequalities $a < x$ and $x < b$. Similarly, the expression $a \le x \le b$ is equivalent to the two inequalities $a \le x$ and $x \le b$. The remaining two possibilities, $a \le x < b$ and $a < x \le b$, are defined similarly.

Although it is acceptable to write $3 \ge x \ge 2$, it is preferable to reverse the inequality symbols and write instead $2 \le x \le 3$ so that the values go from smaller to larger, reading from left to right.

A statement such as $2 \le x \le 1$ is false because there is no number x for which $2 \le x$ and $x \le 1$. Finally, never mix inequality symbols, as in $2 \le x \ge 3$.

1 Use Interval Notation

Let a and b represent two real numbers with $a < b$.

DEFINITION

> An **open interval**, denoted by **(a, b)**, consists of all real numbers x for which $a < x < b$.
>
> A **closed interval**, denoted by **[a, b]**, consists of all real numbers x for which $a \leq x \leq b$.
>
> The **half-open**, or **half-closed**, **intervals** are **(a, b]**, consisting of all real numbers x for which $a < x \leq b$, and **[a, b)**, consisting of all real numbers x for which $a \leq x < b$.

In Words

The notation [a, b] represents all real numbers between a and b, inclusive. The notation (a, b) represents all real numbers between a and b, not including either a or b.

In each of these definitions, a is called the **left endpoint** and b the **right endpoint** of the interval.

The symbol ∞ (read as "infinity") is not a real number, but notation used to indicate unboundedness in the positive direction. The symbol $-\infty$ (read as "negative infinity") also is not a real number, but notation used to indicate unboundedness in the negative direction. The symbols ∞ and $-\infty$ are used to define five other kinds of intervals:

[a, ∞)	Consists of all real numbers x for which $x \geq a$
(a, ∞)	Consists of all real numbers x for which $x > a$
($-\infty$, a]	Consists of all real numbers x for which $x \leq a$
($-\infty$, a)	Consists of all real numbers x for which $x < a$
($-\infty$, ∞)	Consists of all real numbers

Note that ∞ and $-\infty$ are never included as endpoints, since neither is a real number.

Table 1 summarizes interval notation, corresponding inequality notation, and their graphs.

Table 1

Interval	Inequality	Graph
The open interval (a, b)	$a < x < b$	
The closed interval [a, b]	$a \leq x \leq b$	
The half-open interval [a, b)	$a \leq x < b$	
The half-open interval (a, b]	$a < x \leq b$	
The interval [a, ∞)	$x \geq a$	
The interval (a, ∞)	$x > a$	
The interval ($-\infty$, a]	$x \leq a$	
The interval ($-\infty$, a)	$x < a$	
The interval ($-\infty$, ∞)	All real numbers	

EXAMPLE 1   **Writing Inequalities Using Interval Notation**

Write each inequality using interval notation.

(a) $1 \leq x \leq 3$ (b) $-4 < x < 0$ (c) $x > 5$ (d) $x \leq 1$

Solution   (a) $1 \leq x \leq 3$ describes all real numbers x between 1 and 3, inclusive. In interval notation, we write $[1, 3]$.

(b) In interval notation, $-4 < x < 0$ is written $(-4, 0)$.

(c) In interval notation, $x > 5$ is written $(5, \infty)$.

(d) In interval notation, $x \le 1$ is written $(-\infty, 1]$.

EXAMPLE 2

Writing Intervals Using Inequality Notation

Write each interval as an inequality involving x.

(a) $[1, 4)$ (b) $(2, \infty)$ (c) $[2, 3]$ (d) $(-\infty, -3]$

Solution

(a) $[1, 4)$ consists of all real numbers x for which $1 \le x < 4$.

(b) $(2, \infty)$ consists of all real numbers x for which $x > 2$.

(c) $[2, 3]$ consists of all real numbers x for which $2 \le x \le 3$.

(d) $(-\infty, -3]$ consists of all real numbers x for which $x \le -3$.

 Now Work PROBLEMS **13, 25, AND 33**

2 Use Properties of Inequalities

The product of two positive real numbers is positive, the product of two negative real numbers is positive, and the product of 0 and 0 is 0. For any real number a, the value of a^2 is 0 or positive; that is, a^2 is nonnegative. This is called the **nonnegative property**.

> **In Words**
>
> The square of a real number is never negative.

Nonnegative Property

For any real number a,

$$a^2 \ge 0 \tag{1}$$

When the same number is added to both sides of an inequality, an equivalent inequality is obtained. For example, since $3 < 5$, then $3 + 4 < 5 + 4$ or $7 < 9$. This is called the **addition property** of inequalities.

> **In Words**
>
> The addition property states that the sense, or direction, of an inequality remains unchanged if the same number is added to each side.

Addition Property of Inequalities

For real numbers a, b, and c,

$$\text{If } a < b, \text{ then } a + c < b + c. \tag{2a}$$
$$\text{If } a > b, \text{ then } a + c > b + c. \tag{2b}$$

Figure 2 illustrates the addition property (2a). In Figure 2(a), we see that a lies to the left of b. If c is positive, then $a + c$ and $b + c$ lie c units to the right of a and c units to the right of b, respectively. Consequently, $a + c$ must lie to the left of $b + c$; that is, $a + c < b + c$. Figure 2(b) illustrates the situation if c is negative.

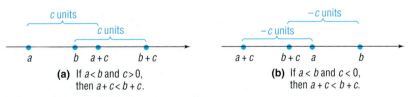

(a) If $a < b$ and $c > 0$, then $a + c < b + c$.

(b) If $a < b$ and $c < 0$, then $a + c < b + c$.

Figure 2 Addition property of inequalities

 DRAW AN ILLUSTRATION SIMILAR TO FIGURE **2** THAT ILLUSTRATES THE ADDITION PROPERTY (**2b**).

EXAMPLE 3

Addition Property of Inequalities

(a) If $x < -5$, then $x + 5 < -5 + 5$ or $x + 5 < 0$.
(b) If $x > 2$, then $x + (-2) > 2 + (-2)$ or $x - 2 > 0$.   ●

 Now Work PROBLEM 41

EXAMPLE 4

Multiplying an Inequality by a Positive Number

Express as an inequality the result of multiplying each side of the inequality $3 < 7$ by 2.

Solution   Begin with

$$3 < 7$$

Multiplying each side by 2 yields the numbers 6 and 14, so we have

$$6 < 14$$ ●

EXAMPLE 5

Multiplying an Inequality by a Negative Number

Express as an inequality the result of multiplying each side of the inequality $9 > 2$ by -4.

Solution   Begin with

$$9 > 2$$

Multiplying each side by -4 yields the numbers -36 and -8, so we have

$$-36 < -8$$ ●

In Words

Multiplying by a negative number reverses the inequality.

Note that the effect of multiplying both sides of $9 > 2$ by the negative number -4 is that the direction of the inequality symbol is reversed.

Examples 4 and 5 illustrate the following general **multiplication properties** for inequalities:

In Words

The multiplication properties state that the sense, or direction, of an inequality *remains the same* if each side is multiplied by a *positive* real number, whereas the direction is *reversed* if each side is multiplied by a *negative* real number.

Multiplication Properties for Inequalities

For real numbers $a, b,$ and c,

If $a < b$ and if $c > 0$, then $ac < bc$.
If $a < b$ and if $c < 0$, then $ac > bc$.   **(3a)**

If $a > b$ and if $c > 0$, then $ac > bc$.
If $a > b$ and if $c < 0$, then $ac < bc$.   **(3b)**

EXAMPLE 6

Multiplication Property of Inequalities

(a) If $2x < 6$, then $\dfrac{1}{2}(2x) < \dfrac{1}{2}(6)$ or $x < 3$.

(b) If $\dfrac{x}{-3} > 12$, then $-3\left(\dfrac{x}{-3}\right) < -3(12)$ or $x < -36$.

(c) If $-4x < -8$, then $\dfrac{-4x}{-4} > \dfrac{-8}{-4}$ or $x > 2$.

(d) If $-x > 8$, then $(-1)(-x) < (-1)(8)$ or $x < -8$.   ●

 Now Work PROBLEM 47

Reciprocal Property for Inequalities

$$\text{If } a > 0, \text{ then } \frac{1}{a} > 0. \qquad \text{If } \frac{1}{a} > 0, \text{ then } a > 0. \tag{4a}$$

$$\text{If } a < 0, \text{ then } \frac{1}{a} < 0. \qquad \text{If } \frac{1}{a} < 0, \text{ then } a < 0. \tag{4b}$$

3 Solve Inequalities

An **inequality in one variable** is a statement involving two expressions, at least one containing the variable, separated by one of the inequality symbols: $<, \leq, >$, or $\geq$. To **solve an inequality** means to find all values of the variable for which the statement is true. These values are called **solutions** of the inequality.

For example, the following are all inequalities involving one variable x:

$$x + 5 < 8 \qquad 2x - 3 \geq 4 \qquad x^2 - 1 \leq 3 \qquad \frac{x + 1}{x - 2} > 0$$

As with equations, one method for solving an inequality is to replace it by a series of equivalent inequalities until an inequality with an obvious solution, such as $x < 3$, is obtained. Equivalent inequalities are obtained by applying some of the same properties that are used to find equivalent equations. The addition property and the multiplication properties for inequalities form the basis for the following procedures.

Procedures That Leave the Inequality Symbol Unchanged

1. Simplify both sides of the inequality by combining like terms and eliminating parentheses:

$$\text{Replace} \quad x + 2 + 6 > 2x + 5(x + 1)$$
$$\text{by} \qquad x + 8 > 7x + 5$$

2. Add or subtract the same expression on both sides of the inequality:

$$\text{Replace} \qquad 3x - 5 < 4$$
$$\text{by} \quad (3x - 5) + 5 < 4 + 5$$

3. Multiply or divide both sides of the inequality by the same *positive* expression:

$$\text{Replace} \quad 4x > 16 \quad \text{by} \quad \frac{4x}{4} > \frac{16}{4}$$

Procedures That Reverse the Sense or Direction of the Inequality Symbol

1. Interchange the two sides of the inequality:

$$\text{Replace} \quad 3 < x \quad \text{by} \quad x > 3$$

2. Multiply or divide both sides of the inequality by the same *negative* expression:

$$\text{Replace} \quad -2x > 6 \quad \text{by} \quad \frac{-2x}{-2} < \frac{6}{-2}$$

As the examples that follow illustrate, we solve inequalities using many of the same steps that we would use to solve equations. In writing the solution of an inequality, either set notation or interval notation may be used, whichever is more convenient.

EXAMPLE 7

Solving an Inequality

Solve the inequality $3 - 2x < 5$, and graph the solution set

Solution

$$3 - 2x < 5$$

$$3 - 2x - 3 < 5 - 3 \qquad \text{Subtract 3 from both sides.}$$

$$-2x < 2 \qquad \text{Simplify.}$$

$$\frac{-2x}{-2} > \frac{2}{-2} \qquad \text{Divide both sides by } -2. \text{ (The sense of the inequality symbol is reversed.)}$$

$$x > -1 \qquad \text{Simplify.}$$

Figure 3 $x > -1$

The solution set is $\{x \mid x > -1\}$ or, using interval notation, all numbers in the interval $(-1, \infty)$. See Figure 3 for the graph. ●

EXAMPLE 8

Solving an Inequality

Solve the inequality $4x + 7 \geq 2x - 3$, and graph the solution set.

Solution

$$4x + 7 \geq 2x - 3$$

$$4x + 7 - 7 \geq 2x - 3 - 7 \qquad \text{Subtract 7 from both sides.}$$

$$4x \geq 2x - 10 \qquad \text{Simplify.}$$

$$4x - 2x \geq 2x - 10 - 2x \qquad \text{Subtract } 2x \text{ from both sides.}$$

$$2x \geq -10 \qquad \text{Simplify.}$$

$$\frac{2x}{2} \geq \frac{-10}{2} \qquad \text{Divide both sides by 2. (The direction of the inequality symbol is unchanged.)}$$

$$x \geq -5 \qquad \text{Simplify.}$$

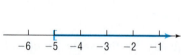

Figure 4 $x \geq -5$

The solution set is $\{x \mid x \geq -5\}$ or, using interval notation, all numbers in the interval $[-5, \infty)$. See Figure 4 for the graph. ●

═══ **Now Work** PROBLEM 55

4 Solve Combined Inequalities

EXAMPLE 9

Solving a Combined Inequality

Solve the inequality $-5 < 3x - 2 < 1$, and graph the solution set.

Solution

Recall that the inequality

$$-5 < 3x - 2 < 1$$

is equivalent to the two inequalities

$$-5 < 3x - 2 \quad \text{and} \quad 3x - 2 < 1$$

Solve each of these inequalities separately.

$$-5 < 3x - 2$$
$$-5 + 2 < 3x - 2 + 2 \qquad \text{Add 2 to both sides.}$$
$$-3 < 3x \qquad \text{Simplify.}$$
$$\frac{-3}{3} < \frac{3x}{3} \qquad \text{Divide both sides by 3.}$$
$$-1 < x \qquad \text{Simplify.}$$

$$3x - 2 < 1$$
$$3x - 2 + 2 < 1 + 2$$
$$3x < 3$$
$$\frac{3x}{3} < \frac{3}{3}$$
$$x < 1$$

The solution set of the original pair of inequalities consists of all x for which

$$-1 < x \quad \text{and} \quad x < 1$$

Figure 5 $-1 < x < 1$

This may be written more compactly as $\{x \,|\, -1 < x < 1\}$. In interval notation, the solution is $(-1, 1)$. See Figure 5 for the graph. ●

Observe in the preceding process that solving each of the two inequalities required exactly the same steps. A shortcut to solving the original inequality algebraically is to deal with the two inequalities at the same time, as follows:

$$-5 < \quad 3x - 2 \quad < 1$$
$$-5 + 2 < 3x - 2 + 2 < 1 + 2 \qquad \text{Add 2 to each part.}$$
$$-3 < \quad 3x \quad < 3 \qquad \text{Simplify.}$$
$$\frac{-3}{3} < \quad \frac{3x}{3} \quad < \frac{3}{3} \qquad \text{Divide each part by 3.}$$
$$-1 < \quad x \quad < 1 \qquad \text{Simplify.}$$

EXAMPLE 10   **Solving a Combined Inequality**

Solve the inequality $-1 \le \dfrac{3 - 5x}{2} \le 9$, and graph the solution set.

Solution

$$-1 \le \quad \frac{3 - 5x}{2} \quad \le 9$$

$$2(-1) \le 2\left(\frac{3 - 5x}{2}\right) \le 2(9) \qquad \text{Multiply each part by 2 to remove the denominator.}$$

$$-2 \le \quad 3 - 5x \quad \le 18 \qquad \text{Simplify.}$$

$$-2 - 3 \le 3 - 5x - 3 \le 18 - 3 \qquad \text{Subtract 3 from each part to isolate the term containing } x.$$

$$-5 \le \quad -5x \quad \le 15 \qquad \text{Simplify.}$$

$$\frac{-5}{-5} \ge \quad \frac{-5x}{-5} \quad \ge \frac{15}{-5} \qquad \text{Divide each part by } -5 \text{ (reverse the direction of each inequality symbol).}$$

$$1 \ge \quad x \quad \ge -3 \qquad \text{Simplify.}$$

$$-3 \le \quad x \quad \le 1 \qquad \text{Reverse the order so that the numbers get larger as you read from left to right.}$$

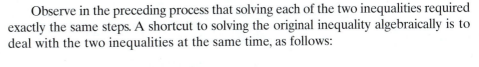

Figure 6 $-3 \le x \le 1$

The solution set is $\{x \,|\, -3 \le x \le 1\}$, that is, all x in the interval $[-3, 1]$. Figure 6 illustrates the graph. ●

━━━ **Now Work** PROBLEM 75

EXAMPLE 11

Using the Reciprocal Property to Solve an Inequality

Solve the inequality $(4x - 1)^{-1} > 0$, and graph the solution set.

Solution

Recall that $(4x - 1)^{-1} = \dfrac{1}{4x - 1}$. The Reciprocal Property states that when $\dfrac{1}{a} > 0$, then $a > 0$.

$$(4x - 1)^{-1} > 0$$

$$\frac{1}{4x - 1} > 0$$

$$4x - 1 > 0 \quad \text{Reciprocal Property}$$

$$4x > 1 \quad \text{Add 1 to both sides.}$$

$$x > \frac{1}{4} \quad \text{Divide both sides by 4.}$$

Figure 7 $x > \dfrac{1}{4}$

The solution set is $\left\{ x \mid x > \dfrac{1}{4} \right\}$, that is, all x in the interval $\left(\dfrac{1}{4}, \infty \right)$. Figure 7 illustrates the graph.

Now Work PROBLEM 85

EXAMPLE 12

Creating Equivalent Inequalities

If $-1 < x < 4$, find a and b so that $a < 2x + 1 < b$.

Solution

The idea here is to change the middle part of the combined inequality from x to $2x + 1$, using properties of inequalities.

$$-1 < \quad x \quad < 4$$
$$-2 < \quad 2x \quad < 8 \quad \text{Multiply each part by 2.}$$
$$-1 < 2x + 1 < 9 \quad \text{Add 1 to each part.}$$

Now we see that $a = -1$ and $b = 9$.

Now Work PROBLEM 95

Application

EXAMPLE 13

Physics: Ohm's Law

In electricity, Ohm's law states that $E = IR$, where E is the voltage (in volts), I is the current (in amperes), and R is the resistance (in ohms). An air-conditioning unit is rated at a resistance of 10 ohms. If the voltage varies from 110 to 120 volts, inclusive, what corresponding range of current will the air conditioner draw?

Solution

The voltage lies between 110 and 120, inclusive, so

$$110 \leq \quad E \quad \leq 120$$
$$110 \leq \quad IR \quad \leq 120 \quad \text{Ohm's law, } E = IR$$
$$110 \leq I(10) \leq 120 \quad R = 10$$
$$\frac{110}{10} \leq \frac{I(10)}{10} \leq \frac{120}{10} \quad \text{Divide each part by 10.}$$
$$11 \leq \quad I \quad \leq 12 \quad \text{Simplify.}$$

The air conditioner will draw between 11 and 12 amperes of current, inclusive.

1.5 Assess Your Understanding

'Are You Prepared?' *Answers are given at the end of these exercises. If you get a wrong answer, read the pages listed in* red.

1. Graph the inequality: $x \geq -2$. (pp. 17–26)

2. *True or False* $-5 > -3$ (pp. 17–26)

Concepts and Vocabulary

3. A(n) _____ _____, denoted $[a, b]$, consists of all real numbers x for which $a \leq x \leq b$.

4. The _____ _____ state that the sense, or direction, of an inequality remains the same if each side is multiplied by a positive number, while the direction is reversed if each side is multiplied by a negative number.

In Problems 5–8, assume that $a < b$ and $c < 0$.

5. *True or False* $a + c < b + c$

6. *True or False* $a - c < b - c$

7. *True or False* $ac > bc$

8. *True or False* $\dfrac{a}{c} < \dfrac{b}{c}$

9. *True or False* The square of any real number is always nonnegative.

10. *True or False* A half-closed interval must have an endpoint of either $-\infty$ or ∞.

11. Which of the following will change the direction, or sense, of an inequality?
 (a) Dividing each side by a positive number
 (b) Interchanging sides
 (c) Adding a negative number to each side
 (d) Subtracting a positive number from each side

12. Which pair of inequalities is equivalent to $0 < x \leq 3$?
 (a) $x > 0$ and $x \geq 3$ (b) $x < 0$ and $x \geq 3$
 (c) $x > 0$ and $x \leq 3$ (d) $x < 0$ and $x \leq 3$

Skill Building

In Problems 13–18, express the graph shown in blue using interval notation. Also express each as an inequality involving x.

13.

14.

15.

16.

17.

18.

In Problems 19–24, an inequality is given. Write the inequality obtained by:
 (a) *Adding 3 to each side of the given inequality.*
 (b) *Subtracting 5 from each side of the given inequality.*
 (c) *Multiplying each side of the given inequality by 3.*
 (d) *Multiplying each side of the given inequality by -2.*

19. $3 < 5$ **20.** $2 > 1$ **21.** $4 > -3$ **22.** $-3 > -5$ **23.** $2x + 1 < 2$ **24.** $1 - 2x > 5$

In Problems 25–32, write each inequality using interval notation, and graph each inequality on the real number line.

25. $0 \leq x \leq 4$ **26.** $-1 < x < 5$ **27.** $4 \leq x < 6$ **28.** $-2 < x < 0$

29. $x \geq 4$ **30.** $x \leq 5$ **31.** $x < -4$ **32.** $x > 1$

In Problems 33–40, write each interval as an inequality involving x, and graph each inequality on the real number line.

33. $[2, 5]$ **34.** $(1, 2)$ **35.** $(-3, -2)$ **36.** $[0, 1)$

37. $[4, \infty)$ **38.** $(-\infty, 2]$ **39.** $(-\infty, -3)$ **40.** $(-8, \infty)$

In Problems 41–54, fill in the blank with the correct inequality symbol.

41. If $x < 5$, then $x - 5$ ___ 0.   **42.** If $x < -4$, then $x + 4$ ___ 0.

43. If $x > -4$, then $x + 4$ ___ 0.   **44.** If $x > 6$, then $x - 6$ ___ 0.

45. If $x \geq -4$, then $3x$ ___ -12.   **46.** If $x \leq 3$, then $2x$ ___ 6.

47. If $x > 6$, then $-2x$ ___ -12.   **48.** If $x > -2$, then $-4x$ ___ 8.

49. If $x \geq 5$, then $-4x$ ___ -20.   **50.** If $x \leq -4$, then $-3x$ ___ 12.

51. If $2x > 6$, then x ___ 3.   **52.** If $3x \leq 12$, then x ___ 4.

53. If $-\dfrac{1}{2}x \leq 3$, then x ___ -6.   **54.** If $-\dfrac{1}{4}x > 1$, then x ___ -4.

In Problems 55–92, solve each inequality. Express your answer using set notation or interval notation. Graph the solution set.

55. $x + 1 < 5$

56. $x - 6 < 1$

57. $1 - 2x \le 3$

58. $2 - 3x \le 5$

59. $3x - 7 > 2$

60. $2x + 5 > 1$

61. $3x - 1 \ge 3 + x$

62. $2x - 2 \ge 3 + x$

63. $-2(x + 3) < 8$

64. $-3(1 - x) < 12$

65. $4 - 3(1 - x) \le 3$

66. $8 - 4(2 - x) \le -2x$

67. $\frac{1}{2}(x - 4) > x + 8$

68. $3x + 4 > \frac{1}{3}(x - 2)$

69. $\frac{x}{2} \ge 1 - \frac{x}{4}$

70. $\frac{x}{3} \ge 2 + \frac{x}{6}$

71. $0 \le 2x - 6 \le 4$

72. $4 \le 2x + 2 \le 10$

73. $-5 \le 4 - 3x \le 2$

74. $-3 \le 3 - 2x \le 9$

75. $-3 < \frac{2x - 1}{4} < 0$

76. $0 < \frac{3x + 2}{2} < 4$

77. $1 < 1 - \frac{1}{2}x < 4$

78. $0 < 1 - \frac{1}{3}x < 1$

79. $(x + 2)(x - 3) > (x - 1)(x + 1)$

80. $(x - 1)(x + 1) > (x - 3)(x + 4)$

81. $x(4x + 3) \le (2x + 1)^2$

82. $x(9x - 5) \le (3x - 1)^2$

83. $\frac{1}{2} \le \frac{x + 1}{3} < \frac{3}{4}$

84. $\frac{1}{3} < \frac{x + 1}{2} \le \frac{2}{3}$

85. $(4x + 2)^{-1} < 0$

86. $(2x - 1)^{-1} > 0$

87. $(2 - 7x)^{-1} \ge 5$

88. $2(3x + 5)^{-1} \le -3$

89. $0 < \frac{2}{x} < \frac{3}{5}$

90. $0 < \frac{4}{x} < \frac{2}{3}$

91. $0 < (2x - 4)^{-1} < \frac{1}{2}$

92. $0 < (3x + 6)^{-1} < \frac{1}{3}$

Applications and Extensions

In Problems 93–102, find a and b.

93. If $-1 < x < 1$, then $a < x + 4 < b$.

94. If $-3 < x < 2$, then $a < x - 6 < b$.

95. If $2 < x < 3$, then $a < -4x < b$.

96. If $-4 < x < 0$, then $a < \frac{1}{2}x < b$.

97. If $0 < x < 4$, then $a < 2x + 3 < b$.

98. If $-3 < x < 3$, then $a < 1 - 2x < b$.

99. If $-3 < x < 0$, then $a < \frac{1}{x + 4} < b$.

100. If $2 < x < 4$, then $a < \frac{1}{x - 6} < b$.

101. If $6 < 3x < 12$, then $a < x^2 < b$.

102. If $0 < 2x < 6$, then $a < x^2 < b$.

103. What is the domain of the variable in the expression $\sqrt{3x + 6}$?

104. What is the domain of the variable in the expression $\sqrt{8 + 2x}$?

105. A young adult may be defined as someone older than 21 but less than 30 years of age. Express this statement using inequalities.

106. Middle-aged may be defined as being 40 or more and less than 60. Express this statement using inequalities.

107. Life Expectancy The Social Security Administration determined that an average 30-year-old male in 2014 could expect to live at least 51.9 more years and that an average 30-year-old female in 2014 could expect to live at least 55.6 more years.
(a) To what age could an average 30-year-old male expect to live? Express your answer as an inequality.
(b) To what age could an average 30-year-old female expect to live? Express your answer as an inequality.
(c) Who can expect to live longer, a male or a female? By how many years?

Source: Social Security Administration, 2014

108. General Chemistry For a certain ideal gas, the volume V (in cubic centimeters) equals 20 times the temperature T (in degrees Celsius). If the temperature varies from 80° to 120°C, inclusive, what is the corresponding range of the volume of the gas?

109. Real Estate A real estate agent agrees to sell an apartment complex according to the following commission schedule: $45,000 plus 25% of the selling price in excess of $900,000. Assuming that the complex will sell at some price between

$900,000 and $1,100,000, inclusive, over what range does the agent's commission vary? How does the commission vary as a percent of selling price?

110. Sales Commission A used car salesperson is paid a commission of $25 plus 40% of the selling price in excess of owner's cost. The owner claims that used cars typically sell for at least owner's cost plus $200 and at most owner's cost plus $3000. For each sale made, over what range can the salesperson expect the commission to vary?

111. Federal Tax Withholding The percentage method of withholding for federal income tax (2014) states that a single person whose weekly wages, after subtracting withholding allowances, are over $753, but not over $1762, shall have $97.75 plus 25% of the excess over $753 withheld. Over what range does the amount withheld vary if the weekly wages vary from $900 to $1100, inclusive?

Source: Employer's Tax Guide. Internal Revenue Service, 2014.

112. Exercising Sue wants to lose weight. For healthy weight loss, the American College of Sports Medicine (ACSM) recommends 200 to 300 minutes of exercise per week. For the first six days of the week, Sue exercised 40, 45, 0, 50, 25, and 35 minutes. How long should Sue exercise on the seventh day in order to stay within the ACSM guidelines?

113. Electricity Rates Commonwealth Edison Company's charge for electricity in January 2014 was 8.21¢ per kilowatt-hour. In addition, each monthly bill contains a customer charge of $15.37. If last year's bills ranged from a low of $72.84 to a high of $237.04, over what range did usage vary (in kilowatt-hours)?

Source: Commonwealth Edison Co., 2014.

114. Water Bills The Village of Oak Lawn charges homeowners $57.07 per quarter-year plus $5.81 per 1000 gallons for water usage in excess of 10,000 gallons. In 2014 one homeowner's quarterly bill ranged from a high of $150.03 to a low of $97.74. Over what range did water usage vary?

Source: Village of Oak Lawn, Illinois, January 2014.

115. Markup of a New Car The markup over dealer's cost of a new car ranges from 12% to 18%. If the sticker price is $18,000, over what range will the dealer's cost vary?

116. IQ Tests A standard intelligence test has an average score of 100. According to statistical theory, of the people who take the test, the 2.5% with the highest scores will have scores of more than 1.96σ above the average, where σ (sigma, a number called the **standard deviation**) depends on the nature of the test. If $\sigma = 12$ for this test and there is (in principle) no upper limit to the score possible on the test, write the interval of possible test scores of the people in the top 2.5%.

117. Computing Grades In your Economics 101 class, you have scores of 68, 82, 87, and 89 on the first four of five tests. To get a grade of B, the average of the first five test scores must be greater than or equal to 80 and less than 90.
  (a) Solve an inequality to find the least score you can get on the last test and still earn a B.
  (b) What score do you need if the fifth test counts double?

What do I need to get a B?

118. "Light" Foods For food products to be labeled "light," the U.S. Food and Drug Administration requires that the altered product must either contain at least one-third fewer calories than the regular product or contain at least one-half less fat than the regular product. If a serving of Miracle Whip® Light contains 20 calories and 1.5 grams of fat, then what must be true about either the number of calories or the grams of fat in a serving of regular Miracle Whip®?

119. Arithmetic Mean If $a < b$, show that $a < \dfrac{a + b}{2} < b$. The number $\dfrac{a + b}{2}$ is called the **arithmetic mean** of a and b.

120. Refer to Problem 119. Show that the arithmetic mean of a and b is equidistant from a and b.

121. Geometric Mean If $0 < a < b$, show that $a < \sqrt{ab} < b$. The number $\sqrt{ab}$ is called the **geometric mean** of a and b.

122. Refer to Problems 119 and 121. Show that the geometric mean of a and b is less than the arithmetic mean of a and b.

123. Harmonic Mean For $0 < a < b$, let h be defined by
$$\frac{1}{h} = \frac{1}{2}\left(\frac{1}{a} + \frac{1}{b}\right)$$
Show that $a < h < b$. The number h is called the **harmonic mean** of a and b.

124. Refer to Problems 119, 121, and 123. Show that the harmonic mean of a and b equals the geometric mean squared, divided by the arithmetic mean.

125. Another Reciprocal Property Prove that if $0 < a < b$, then $0 < \dfrac{1}{b} < \dfrac{1}{a}$.

Explaining Concepts: Discussion and Writing

126. Make up an inequality that has no solution. Make up an inequality that has exactly one solution.

127. The inequality $x^2 + 1 < -5$ has no real solution. Explain why.

128. Do you prefer to use inequality notation or interval notation to express the solution to an inequality? Give your reasons. Are there particular circumstances when you prefer one to the other? Cite examples.

129. How would you explain to a fellow student the underlying reason for the multiplication properties for inequalities (page 122)? That is, the sense or direction of an inequality remains the same if each side is multiplied by a positive real number, whereas the direction is reversed if each side is multiplied by a negative real number.

'Are You Prepared?' Answers

1.

          |————|————————→
        −4   −2   0

2. False

Solving an Inequality Involving Absolute Value

Solve the inequality $|x| > 3$, and graph the solution set.

Solution

Figure 12 $|x| > 3$

We are looking for all points whose coordinate x is a distance greater than 3 units from the origin. Figure 12 illustrates the situation. Any number x less than -3 or greater than 3 satisfies the condition $|x| > 3$. The solution set consists of all numbers x for which $x < -3$ or $x > 3$, that is, all x in $(-\infty, -3) \cup (3, \infty)$.*

THEOREM

If a is a positive number and u is an algebraic expression, then

$$|u| > a \quad \text{is equivalent to} \quad u < -a \quad \text{or} \quad u > a \quad (4)$$

$$|u| \geq a \quad \text{is equivalent to} \quad u \leq -a \quad \text{or} \quad u \geq a \quad (5)$$

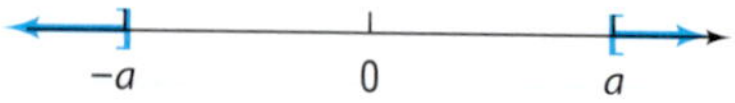

Figure 13 $|u| \geq a, a > 0$

See Figure 13 for an illustration of statement (5).

EXAMPLE 6

Solving an Inequality Involving Absolute Value

Solve the inequality $|2x - 5| > 3$, and graph the solution set.

Solution

$$|2x - 5| > 3 \qquad \text{This follows the form of statement (4); the expression } u = 2x - 5 \text{ is inside the absolute value bars.}$$

$$2x - 5 < -3 \qquad \text{or} \qquad 2x - 5 > 3 \qquad \text{Apply statement (4).}$$
$$2x - 5 + 5 < -3 + 5 \quad \text{or} \quad 2x - 5 + 5 > 3 + 5 \qquad \text{Add 5 to each part.}$$
$$2x < 2 \qquad \text{or} \qquad 2x > 8 \qquad \text{Simplify.}$$
$$\frac{2x}{2} < \frac{2}{2} \qquad \text{or} \qquad \frac{2x}{2} > \frac{8}{2} \qquad \text{Divide each part by 2.}$$
$$x < 1 \qquad \text{or} \qquad x > 4 \qquad \text{Simplify.}$$

Figure 14 $|2x - 5| > 3$

The solution set is $\{x | x < 1 \text{ or } x > 4\}$, that is, all x in $(-\infty, 1) \cup (4, \infty)$. See Figure 14 for the graph of the solution set.

WARNING A common error to be avoided is to attempt to write the solution $x < 1$ or $x > 4$ as the combined inequality $1 > x > 4$, which is incorrect, since there are no numbers x for which $1 > x$ and $x > 4$. Another common error is to "mix" the symbols and write $1 < x > 4$, which makes no sense. ■

Now Work PROBLEM 45

*Recall that the symbol $\cup$ stands for the union of two sets. Refer to page 2 if necessary.

1.6 Assess Your Understanding

'Are You Prepared?' *Answers are given at the end of these exercises. If you get a wrong answer, read the pages listed in* red.

1. $|-2| =$ _____ (p. 19)

2. True or False $|x| \geq 0$ for any real number x. (p. 19)

Concepts and Vocabulary

3. The solution set of the equation $|x| = 5$ is $\{$ _____ $\}$.

4. The solution set of the inequality $|x| < 5$ is $\{x|$ _____ $\}$.

5. True or False The equation $|x| = -2$ has no solution.

6. True or False The inequality $|x| \geq -2$ has the set of real numbers as its solution set.

7. Which of the following pairs of inequalities is equivalent to $|x| > 4$?
(a) $x > -4$ and $x < 4$ (b) $x < -4$ and $x < 4$
(c) $x > -4$ or $x > 4$ (d) $x < -4$ or $x > 4$

8. Which of the following has no solution?
(a) $|x| < -5$ (b) $|x| \leq 0$ (c) $|x| > 0$ (d) $|x| \geq 0$

Skill Building

In Problems 9–36, find the real solutions, if any, of each equation.

9. $|2x| = 6$

10. $|3x| = 12$

11. $|2x + 3| = 5$

12. $|3x - 1| = 2$

13. $|1 - 4t| + 8 = 13$

14. $|1 - 2z| + 6 = 9$

15. $|-2x| = |8|$

16. $|-x| = |1|$

17. $|-2|x = 4$

18. $|3|x = 9$

19. $\frac{2}{3}|x| = 9$

20. $\frac{3}{4}|x| = 9$

21. $\left|\frac{x}{3} + \frac{2}{5}\right| = 2$

22. $\left|\frac{x}{2} - \frac{1}{3}\right| = 1$

23. $|u - 2| = -\frac{1}{2}$

24. $|2 - v| = -1$

25. $4 - |2x| = 3$

26. $5 - \left|\frac{1}{2}x\right| = 3$

27. $|x^2 - 9| = 0$

28. $|x^2 - 16| = 0$

29. $|x^2 - 2x| = 3$

30. $|x^2 + x| = 12$

31. $|x^2 + x - 1| = 1$

32. $|x^2 + 3x - 2| = 2$

33. $\left|\frac{3x - 2}{2x - 3}\right| = 2$

34. $\left|\frac{2x + 1}{3x + 4}\right| = 1$

35. $|x^2 + 3x| = |x^2 - 2x|$

36. $|x^2 - 2x| = |x^2 + 6x|$

In Problems 37–64, solve each inequality. Express your answer using set notation or interval notation. Graph the solution set.

37. $|2x| < 8$

38. $|3x| < 15$

39. $|3x| > 12$

40. $|2x| > 6$

41. $|x - 2| + 2 < 3$

42. $|x + 4| + 3 < 5$

43. $|3t - 2| \le 4$

44. $|2u + 5| \le 7$

45. $|2x - 3| \ge 2$

46. $|3x + 4| \ge 2$

47. $|1 - 4x| - 7 < -2$

48. $|1 - 2x| - 4 < -1$

49. $|1 - 2x| > 3$

50. $|2 - 3x| > 1$

51. $|-4x| + |-5| \le 1$

52. $|-x| - |4| \le 2$

53. $|-2x| > |-3|$

54. $|-x - 2| \ge 1$

55. $-|2x - 1| \ge -3$

56. $-|1 - 2x| \ge -3$

57. $|2x| < -1$

58. $|3x| \ge 0$

59. $|5x| \ge -1$

60. $|6x| < -2$

61. $\left|\frac{2x + 3}{3} - \frac{1}{2}\right| < 1$

62. $3 - |x + 1| < \frac{1}{2}$

63. $5 + |x - 1| > \frac{1}{2}$

64. $\left|\frac{2x - 3}{2} + \frac{1}{3}\right| > 1$

Applications and Extensions

65. Body Temperature "Normal" human body temperature is 98.6°F. If a temperature x that differs from normal by at least 1.5° is considered unhealthy, write the condition for an unhealthy temperature x as an inequality involving an absolute value, and solve for x.

66. Household Voltage In the United States, normal household voltage is 110 volts. However, it is not uncommon for actual voltage to differ from normal voltage by at most 5 volts. Express this situation as an inequality involving an absolute value. Use x as the actual voltage and solve for x.

67. Reading Books A HuffPost/YouGov poll conducted September 27–28, 2013, found that Americans read an average of 13.6 books per year. Suppose HuffPost/YouGov is 99% confident that the result from this poll is off by fewer than 1.8 books from the actual average x. Express this

situation as an inequality involving absolute value, and solve the inequality for x to determine the interval in which the actual average is likely to fall. [**Note:** In statistics, this interval is called a 99% **confidence interval**.]

68. Speed of Sound According to data from the Hill Aerospace Museum (Hill Air Force Base, Utah), the speed of sound varies depending on altitude, barometric pressure, and temperature. For example, at 20,000 feet, 13.75 inches of mercury, and $-12.3°$F, the speed of sound is about 707 miles per hour, but the speed can vary from this result by as much as 55 miles per hour as conditions change.

(a) Express this situation as an inequality involving an absolute value.

(b) Using x for the speed of sound, solve for x to find an interval for the speed of sound.

69. Express the fact that x differs from 3 by less than $\frac{1}{2}$ as an inequality involving an absolute value. Solve for x.

70. Express the fact that x differs from -4 by less than 1 as an inequality involving an absolute value. Solve for x.

71. Express the fact that x differs from -3 by more than 2 as an inequality involving an absolute value. Solve for x.

72. Express the fact that x differs from 2 by more than 3 as an inequality involving an absolute value. Solve for x.

In Problems 73–78, find a and b.

73. If $|x - 1| < 3$, then $a < x + 4 < b$.

74. If $|x + 2| < 5$, then $a < x - 2 < b$.

75. If $|x + 4| \leq 2$, then $a \leq 2x - 3 \leq b$.

76. If $|x - 3| \leq 1$, then $a \leq 3x + 1 \leq b$.

77. If $|x - 2| \leq 7$, then $a \leq \dfrac{1}{x - 10} \leq b$.

78. If $|x + 1| \leq 3$, then $a \leq \dfrac{1}{x + 5} \leq b$.

79. Show that: if $a > 0, b > 0$, and $\sqrt{a} < \sqrt{b}$, then $a < b$.

    [**Hint:** $b - a = (\sqrt{b} - \sqrt{a})(\sqrt{b} + \sqrt{a})$.]

80. Show that $a \leq |a|$.

81. Prove the *triangle inequality* $|a + b| \leq |a| + |b|$.

    [**Hint:** Expand $|a + b|^2 = (a + b)^2$, and use the result of Problem 80.]

82. Prove that $|a - b| \geq |a| - |b|$.

    [**Hint:** Apply the triangle inequality from Problem 81 to $|a| = |(a - b) + b|$.]

83. If $a > 0$, show that the solution set of the inequality

$$x^2 < a$$

consists of all numbers x for which

$$-\sqrt{a} < x < \sqrt{a}$$

84. If $a > 0$, show that the solution set of the inequality

$$x^2 > a$$

consists of all numbers x for which

$$x < -\sqrt{a} \quad \text{or} \quad x > \sqrt{a}$$

In Problems 85–92, use the results found in Problems 83 and 84 to solve each inequality.

85. $x^2 < 1$

86. $x^2 < 4$

87. $x^2 \geq 9$

88. $x^2 \geq 1$

89. $x^2 \leq 16$

90. $x^2 \leq 9$

91. $x^2 > 4$

92. $x^2 > 16$

93. Solve $|3x - |2x + 1|| = 4$.

94. Solve $|x + |3x - 2|| = 2$.

Explaining Concepts: Discussion and Writing

95. The equation $|x| = -2$ has no solution. Explain why.

96. The inequality $|x| > -0.5$ has all real numbers as solutions. Explain why.

97. The inequality $|x| > 0$ has as solution set $\{x | x \neq 0\}$. Explain why.

'Are You Prepared?' Answers

  1. 2   **2.** True

1.7 Problem Solving: Interest, Mixture, Uniform Motion, Constant Rate Job Applications

OBJECTIVES   **1** Translate Verbal Descriptions into Mathematical Expressions (p. 135)

          2 Solve Interest Problems (p. 136)

          3 Solve Mixture Problems (p. 137)

          4 Solve Uniform Motion Problems (p. 138)

          5 Solve Constant Rate Job Problems (p. 140)

Applied (word) problems do not come in the form "Solve the equation . . . ." Instead, they supply information using words, a verbal description of the real problem. So, to solve applied problems, we must be able to translate the verbal description into the language of mathematics. This can be done by using variables to represent unknown quantities and then finding relationships (such as equations) that involve these variables. The process of doing all this is called **mathematical modeling**. An equation or inequality that describes a relationship among the variables is called a **model**.

Any solution to the mathematical problem must be checked against the mathematical problem, the verbal description, and the real problem. See Figure 15 for an illustration of the **modeling process**.

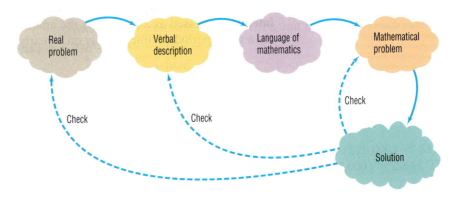

Figure 15

1 Translate Verbal Descriptions into Mathematical Expressions

EXAMPLE 1 | **Translating Verbal Descriptions into Mathematical Expressions**

(a) For uniform motion, the average speed of an object equals the distance traveled divided by the time required.

Translation: If r is the speed, d the distance, and t the time, then $r = \dfrac{d}{t}$.

(b) Let x denote a number.

The number 5 times as large as x is $5x$.

The number 3 less than x is $x - 3$.

The number that exceeds x by 4 is $x + 4$.

The number that, when added to x, gives 5 is $5 - x$.   ●

Now Work   PROBLEM 9

Always check the units used to measure the variables of an applied problem. In Example (1a), if r is measured in miles per hour, then the distance d must be expressed in miles, and the time t must be expressed in hours. It is a good practice to check units to be sure that they are consistent and make sense.

The steps to follow for solving applied problems, given earlier, are repeated next.

Steps for Solving Applied Problems

STEP 1: Read the problem carefully, perhaps two or three times. Pay particular attention to the question being asked in order to identify what you are looking for. Identify any relevant formulas you may need ($d = rt$, $A = \pi r^2$, etc.). If you can, determine realistic possibilities for the answer.

STEP 2: Assign a letter (variable) to represent what you are looking for, and if necessary, express any remaining unknown quantities in terms of this variable.

STEP 3: Make a list of all the known facts, and translate them into mathematical expressions. These may take the form of an equation or an inequality involving the variable. If possible, draw an appropriately labeled diagram to assist you. Sometimes, creating a table or chart helps.

STEP 4: Solve for the variable, and then answer the question.

STEP 5: Check the answer with the facts in the problem. If it agrees, congratulations! If it does not agree, try again.

Now solve the equation:

$$5c + 10(100 - c) = 700$$
$$5c + 1000 - 10c = 700$$
$$-5c = -300$$
$$c = 60$$

The manager should blend 60 pounds of B grade Colombian coffee with $100 - 60 = 40$ pounds of A grade Arabica coffee to get the desired blend.

✓**Check:** The 60 pounds of B grade coffee would sell for $(\$5)(60) = \$300$, and the 40 pounds of A grade coffee would sell for $(\$10)(40) = \$400$; the total revenue, \$700, equals the revenue obtained from selling the blend, as desired.

●

Now Work PROBLEM 23

4 Solve Uniform Motion Problems

Objects that move at a constant speed are said to be in **uniform motion**. When the average speed of an object is known, it can be interpreted as that object's constant speed. For example, a bicyclist traveling at an average speed of 25 miles per hour can be considered in uniform motion with a constant speed of 25 miles per hour.

Uniform Motion Formula

If an object moves at an average speed (rate) r, the distance d covered in time t is given by the formula

$$d = rt \tag{2}$$

That is, Distance = Rate · Time.

EXAMPLE 5

Physics: Uniform Motion

Tanya, who is a long-distance runner, runs at an average speed of 8 miles per hour (mi/h). Two hours after Tanya leaves your house, you leave in your Honda and follow the same route. If your average speed is 40 mi/h, how long will it be before you catch up to Tanya? How far will each of you be from your home?

Solution

Refer to Figure 17. We use t to represent the time (in hours) that it takes the Honda to catch up to Tanya. When this occurs, the total time elapsed for Tanya is $t + 2$ hours because she left 2 hours earlier.

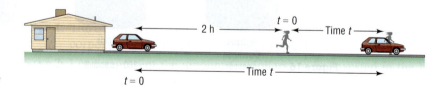

Figure 17

Set up the following table:

	Rate mi/h	Time h	Distance mi
Tanya	8	$t + 2$	$8(t + 2)$
Honda	40	t	$40t$

The distance traveled is the same for both, which leads to the equation

$$8(t + 2) = 40t$$

$$8t + 16 = 40t$$

$$32t = 16$$

$$t = \frac{1}{2} \text{ hour}$$

It will take the Honda $\frac{1}{2}$ hour to catch up to Tanya. Each will have gone 20 miles.

✓ **Check:** In 2.5 hours, Tanya travels a distance of $(2.5)(8) = 20$ miles. In $\frac{1}{2}$ hour, the Honda travels a distance of $\left(\frac{1}{2}\right)(40) = 20$ miles. ●

EXAMPLE 6

Physics: Uniform Motion

A motorboat heads upstream a distance of 24 miles on a river whose current is running at 3 miles per hour (mi/h). The trip up and back takes 6 hours. Assuming that the motorboat maintained a constant speed relative to the water, what was its speed?

Solution

See Figure 18. Use r to represent the constant speed of the motorboat relative to the water. Then the true speed going upstream is $r - 3$ mi/h, and the true speed going downstream is $r + 3$ mi/h. Since Distance = Rate · Time, then Time = $\dfrac{\text{Distance}}{\text{Rate}}$. Set up a table.

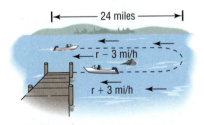

	Rate mi/h	Distance mi	Time = $\dfrac{\text{Distance}}{\text{Rate}}$ h
Upstream	$r - 3$	24	$\dfrac{24}{r - 3}$
Downstream	$r + 3$	24	$\dfrac{24}{r + 3}$

Figure 18

The total time up and back is 6 hours, which gives the equation

$$\frac{24}{r - 3} + \frac{24}{r + 3} = 6$$

$$\frac{24(r + 3) + 24(r - 3)}{(r - 3)(r + 3)} = 6 \qquad \text{Add the quotients on the left.}$$

$$\frac{48r}{r^2 - 9} = 6 \qquad \text{Simplify.}$$

$$48r = 6(r^2 - 9) \qquad \text{Multiply both sides by } r^2 - 9.$$

$$6r^2 - 48r - 54 = 0 \qquad \text{Place in standard form.}$$

$$r^2 - 8r - 9 = 0 \qquad \text{Divide by 6.}$$

$$(r - 9)(r + 1) = 0 \qquad \text{Factor.}$$

$$r = 9 \quad \text{or} \quad r = -1 \qquad \text{Apply the Zero-Product Property and solve.}$$

Discard the solution $r = -1$ mi/h and conclude that the speed of the motorboat relative to the water is 9 mi/h. ●

Now Work PROBLEM 29

5 Solve Constant Rate Job Problems

Here we look at jobs that are performed at a **constant rate**. The assumption is that if a job can be done in t units of time, then $\frac{1}{t}$ of the job is done in 1 unit of time. In other words, if a job takes 4 hours, then $\frac{1}{4}$ of the job is done in 1 hour.

EXAMPLE 7

Working Together to Do a Job

At 10 AM Danny is asked by his father to weed the garden. From past experience, Danny knows that this will take him 4 hours, working alone. His older brother Mike, when it is his turn to do this job, requires 6 hours. Since Mike wants to go golfing with Danny and has a reservation for 1 PM, he agrees to help Danny. Assuming no gain or loss of efficiency, when will they finish if they work together? Can they make the golf date?

Solution

Set up Table 2. In 1 hour, Danny does $\frac{1}{4}$ of the job, and in 1 hour, Mike does $\frac{1}{6}$ of the job. Let t be the time (in hours) that it takes them to do the job together. In 1 hour, then, $\frac{1}{t}$ of the job is completed. Reason as follows:

$$\left(\begin{array}{c}\text{Part done by Danny}\\\text{in 1 hour}\end{array}\right) + \left(\begin{array}{c}\text{Part done by Mike}\\\text{in 1 hour}\end{array}\right) = \left(\begin{array}{c}\text{Part done together}\\\text{in 1 hour}\end{array}\right)$$

From Table 2,

Table 2

	Hours to Do Job	Part of Job Done in 1 Hour
Danny	4	$\frac{1}{4}$
Mike	6	$\frac{1}{6}$
Together	t	$\frac{1}{t}$

$$\frac{1}{4} + \frac{1}{6} = \frac{1}{t} \qquad \textit{The model.}$$

$$\frac{3}{12} + \frac{2}{12} = \frac{1}{t} \qquad \textit{LCD} = 12 \text{ on the left.}$$

$$\frac{5}{12} = \frac{1}{t} \qquad \textit{Simplify.}$$

$$5t = 12 \qquad \textit{Multiply both sides by 12t.}$$

$$t = \frac{12}{5} \qquad \textit{Divide each side by 5.}$$

Working together, Mike and Danny can do the job in $\frac{12}{5}$ hours, or 2 hours, 24 minutes. They should make the golf date, since they will finish at 12:24 PM.   ●

Now Work PROBLEM 35

1.7 Assess Your Understanding

Concepts and Vocabulary

1. The process of using variables to represent unknown quantities and then finding relationships that involve these variables is referred to as _____ _____.

2. The money paid for the use of money is _____.

3. Objects that move at a constant speed are said to be in _____ _____.

4. *True or False* The amount charged for the use of principal for a given period of time is called the rate of interest.

5. *True or False* If an object moves at an average speed r, the distance d covered in time t is given by the formula $d = rt$.

6. Suppose that you want to mix two coffees in order to obtain 100 pounds of a blend. If x represents the number of pounds of coffee A, write an algebraic expression that represents the number of pounds of coffee B.

    (a) $100 - x$ (b) $x - 100$ (c) $100x$ (d) $100 + x$

7. Which of the following is the simple interest formula?

    (a) $I = \frac{rt}{P}$ (b) $I = Prt$ (c) $I = \frac{P}{rt}$ (d) $I = P + rt$

8. If it takes 5 hours to complete a job, what fraction of the job is done in 1 hour?

    (a) $\frac{4}{5}$ (b) $\frac{5}{4}$ (c) $\frac{1}{5}$ (d) $\frac{1}{4}$

Applications and Extensions

In Problems 9–18, translate each sentence into a mathematical equation. Be sure to identify the meaning of all symbols.

9. **Geometry** The area of a circle is the product of the number π and the square of the radius.

10. **Geometry** The circumference of a circle is the product of the number π and twice the radius.

11. **Geometry** The area of a square is the square of the length of a side.

12. **Geometry** The perimeter of a square is four times the length of a side.

13. **Physics** Force equals the product of mass and acceleration.

14. **Physics** Pressure is force per unit area.

15. **Physics** Work equals force times distance.

16. **Physics** Kinetic energy is one-half the product of the mass and the square of the velocity.

17. **Business** The total variable cost of manufacturing x dishwashers is $150 per dishwasher times the number of dishwashers manufactured.

18. **Business** The total revenue derived from selling x dishwashers is $250 per dishwasher times the number of dishwashers sold.

19. **Financial Planning** Betsy, a recent retiree, requires $6000 per year in extra income. She has $50,000 to invest and can invest in B-rated bonds paying 15% per year or in a certificate of deposit (CD) paying 7% per year. How much money should Betsy invest in each to realize exactly $6000 in interest per year?

20. **Financial Planning** After 2 years, Betsy (see Problem 19) finds that she will now require $7000 per year. Assuming that the remaining information is the same, how should the money be reinvested?

21. **Banking** A bank loaned out $12,000, part of it at the rate of 8% per year and the rest at the rate of 18% per year. If the interest received totaled $1000, how much was loaned at 8%?

22. **Banking** Wendy, a loan officer at a bank, has $1,000,000 to lend and is required to obtain an average return of 18% per year. If she can lend at the rate of 19% or at the rate of 16%, how much can she lend at the 16% rate and still meet her requirement?

23. **Blending Teas** The manager of a store that specializes in selling tea decides to experiment with a new blend. She will mix some Earl Grey tea that sells for $5 per pound with some Orange Pekoe tea that sells for $3 per pound to get 100 pounds of the new blend. The selling price of the new blend is to be $4.50 per pound, and there is to be no difference in revenue between selling the new blend and selling the other types. How many pounds of the Earl Grey tea and of the Orange Pekoe tea are required?

24. **Business: Blending Coffee** A coffee manufacturer wants to market a new blend of coffee that sells for $3.90 per pound by mixing two coffees that sell for $2.75 and $5 per pound, respectively. What amounts of each coffee should be blended to obtain the desired mixture?

[**Hint:** Assume that the total weight of the desired blend is 100 pounds.]

25. **Business: Mixing Nuts** A nut store normally sells cashews for $9.00 per pound and almonds for $3.50 per pound. But at the end of the month the almonds had not sold well, so, in order to sell 60 pounds of almonds, the manager decided to mix the 60 pounds of almonds with some cashews and sell the mixture for $7.50 per pound. How many pounds of cashews should be mixed with the almonds to ensure no change in the revenue?

26. **Business: Mixing Candy** A candy store sells boxes of candy containing caramels and cremes. Each box sells for $12.50 and holds 30 pieces of candy (all pieces are the same size). If the caramels cost $0.25 to produce and the cremes cost $0.45 to produce, how many of each should be in a box to yield a profit of $3?

27. **Physics: Uniform Motion** A motorboat can maintain a constant speed of 16 miles per hour relative to the water. The boat travels upstream to a certain point in 20 minutes; the return trip takes 15 minutes. What is the speed of the current? See the figure.

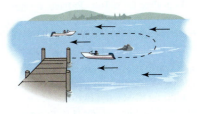

28. **Physics: Uniform Motion** A motorboat heads upstream on a river that has a current of 3 miles per hour. The trip upstream takes 5 hours, and the return trip takes 2.5 hours. What is the speed of the motorboat? (Assume that the boat maintains a constant speed relative to the water.)

29. **Physics: Uniform Motion** A motorboat maintained a constant speed of 15 miles per hour relative to the water in going 10 miles upstream and then returning. The total time for the trip was 1.5 hours. Use this information to find the speed of the current.

30. **Physics: Uniform Motion** Two cars enter the Florida Turnpike at Commercial Boulevard at 8:00 AM, each heading for Wildwood. One car's average speed is 10 miles per hour more than the other's. The faster car arrives at Wildwood at 11:00 AM, $\frac{1}{2}$ hour before the other car. What was the average speed of each car? How far did each travel?

31. **Moving Walkways** The speed of a moving walkway is typically about 2.5 feet per second. Walking on such a moving walkway, it takes Karen a total of 40 seconds to travel 50 feet with the movement of the walkway and then back again against the movement of the walkway. What is Karen's normal walking speed?

Source: Answers.com

32. High-Speed Walkways Toronto's Pearson International Airport has a high-speed version of a moving walkway. If Liam walks while riding this moving walkway, he can travel 280 meters in 60 seconds less time than if he stands still on the moving walkway. If Liam walks at a normal rate of 1.5 meters per second, what is the speed of the walkway?

Source: Answers.com

33. Tennis A regulation doubles tennis court has an area of 2808 square feet. If it is 6 feet longer than twice its width, determine the dimensions of the court.

Source: United States Tennis Association

34. Laser Printers It takes an HP LaserJet M451dw laser printer 16 minutes longer to complete an 840-page print job by itself than it takes an HP LaserJet CP4025dn to complete the same job by itself. Together the two printers can complete the job in 15 minutes. How long does it take each printer to complete the print job alone? What is the speed of each printer?

Source: Hewlett-Packard

35. Working Together on a Job Trent can deliver his newspapers in 30 minutes. It takes Lois 20 minutes to do the same route. How long would it take them to deliver the newspapers if they worked together?

36. Working Together on a Job Patrice, by himself, can paint four rooms in 10 hours. If he hires April to help, they can do the same job together in 6 hours. If he lets April work alone, how long will it take her to paint four rooms?

37. Enclosing a Garden A gardener has 46 feet of fencing to be used to enclose a rectangular garden that has a border 2 feet wide surrounding it. See the figure.
(a) If the length of the garden is to be twice its width, what will be the dimensions of the garden?
(b) What is the area of the garden?
(c) If the length and width of the garden are to be the same, what will be the dimensions of the garden?
(d) What will be the area of the square garden?

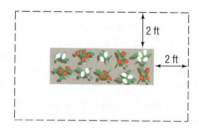

38. Construction A pond is enclosed by a wooden deck that is 3 feet wide. The fence surrounding the deck is 100 feet long.
(a) If the pond is square, what are its dimensions?
(b) If the pond is rectangular and the length of the pond is to be three times its width, what are its dimensions?
(c) If the pond is circular, what is its diameter?
(d) Which pond has the larger area?

39. Football A tight end can run the 100-yard dash in 12 seconds. A defensive back can do it in 10 seconds. The tight end catches a pass at his own 20-yard line with the defensive back at the 15-yard line. (See the figure-top, right.) If no other players are nearby, at what yard line will the defensive back catch up to the tight end?

[**Hint:** At time $t = 0$, the defensive back is 5 yards behind the tight end.]

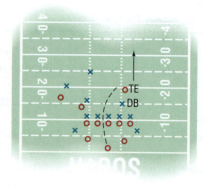

40. Computing Business Expense Therese, an outside salesperson, uses her car for both business and pleasure. Last year, she traveled 30,000 miles, using 900 gallons of gasoline. Her car gets 40 miles per gallon on the highway and 25 in the city. She can deduct all highway travel, but no city travel, on her taxes. How many miles should Therese deduct as a business expense?

41. Mixing Water and Antifreeze How much water should be added to 1 gallon of pure antifreeze to obtain a solution that is 60% antifreeze?

42. Mixing Water and Antifreeze The cooling system of a certain foreign-made car has a capacity of 15 liters. If the system is filled with a mixture that is 40% antifreeze, how much of this mixture should be drained and replaced by pure antifreeze so that the system is filled with a solution that is 60% antifreeze?

43. Chemistry: Salt Solutions How much water must be evaporated from 32 ounces of a 4% salt solution to make a 6% salt solution?

44. Chemistry: Salt Solutions How much water must be evaporated from 240 gallons of a 3% salt solution to produce a 5% salt solution?

45. Purity of Gold The purity of gold is measured in karats, with pure gold being 24 karats. Other purities of gold are expressed as proportional parts of pure gold. Thus, 18-karat gold is $\frac{18}{24}$, or 75% pure gold; 12-karat gold is $\frac{12}{24}$, or 50% pure gold; and so on. How much 12-karat gold should be mixed with pure gold to obtain 60 grams of 16-karat gold?

46. Chemistry: Sugar Molecules A sugar molecule has twice as many atoms of hydrogen as it does oxygen and one more atom of carbon than of oxygen. If a sugar molecule has a total of 45 atoms, how many are oxygen? How many are hydrogen?

47. Running a Race Mike can run the mile in 6 minutes, and Dan can run the mile in 9 minutes. If Mike gives Dan a head start of 1 minute, how far from the start will Mike pass Dan? How long does it take? See the figure.

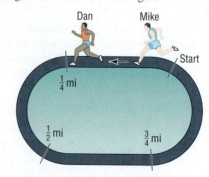

48. Range of an Airplane An air rescue plane averages 300 miles per hour in still air. It carries enough fuel for 5 hours of flying time. If, upon takeoff, it encounters a head wind of 30 mi/h, how far can it fly and return safely? (Assume that the wind remains constant.)

49. Emptying Oil Tankers An oil tanker can be emptied by the main pump in 4 hours. An auxiliary pump can empty the tanker in 9 hours. If the main pump is started at 9 AM, when should the auxiliary pump be started so that the tanker is emptied by noon?

50. Cement Mix A 20-pound bag of Economy brand cement mix contains 25% cement and 75% sand. How much pure cement must be added to produce a cement mix that is 40% cement?

51. Filling a Tub A bathroom tub will fill in 15 minutes with both faucets open and the stopper in place. With both faucets closed and the stopper removed, the tub will empty in 20 minutes. How long will it take for the tub to fill if both faucets are open and the stopper is removed?

52. Using Two Pumps A 5-horsepower (hp) pump can empty a pool in 5 hours. A smaller, 2-hp pump empties the same pool in 8 hours. The pumps are used together to begin emptying this pool. After two hours, the 2-hp pump breaks down. How long will it take the larger pump to finish emptying the pool?

53. A Biathlon Suppose that you have entered an 87-mile biathlon that consists of a run and a bicycle race. During your run, your average speed is 6 miles per hour, and during your bicycle race, your average speed is 25 miles per hour. You finish the race in 5 hours. What is the distance of the run? What is the distance of the bicycle race?

54. Cyclists Two cyclists leave a city at the same time, one going east and the other going west. The westbound cyclist bikes 5 mph faster than the eastbound cyclist. After 6 hours they are 246 miles apart. How fast is each cyclist riding?

55. Comparing Olympic Heroes In the 2012 Olympics, Usain Bolt of Jamaica won the gold medal in the 100-meter race with a time of 9.69 seconds. In the 1896 Olympics, Thomas Burke of the United States won the gold medal in the 100-meter race in 12.0 seconds. If they ran in the same race, repeating their respective times, by how many meters would Bolt beat Burke?

56. Constructing a Coffee Can A 39-ounce can of Hills Bros.® coffee requires 188.5 square inches of aluminum. If its height is 7 inches, what is its radius? [**Hint:** The surface area S of a right cylinder is $S = 2\pi r^2 + 2\pi rh$, where r is the radius and h is the height.]

Explaining Concepts: Discussion and Writing

57. Critical Thinking You are the manager of a clothing store and have just purchased 100 dress shirts for $20.00 each. After 1 month of selling the shirts at the regular price, you plan to have a sale giving 40% off the original selling price. However, you still want to make a profit of $4 on each shirt at the sale price. What should you price the shirts at initially to ensure this? If, instead of 40% off at the sale, you give 50% off, by how much is your profit reduced?

58. Critical Thinking Make up a word problem that requires solving a linear equation as part of its solution. Exchange problems with a friend. Write a critique of your friend's problem.

59. Critical Thinking Without solving, explain what is wrong with the following mixture problem: How many liters of 25% ethanol should be added to 20 liters of 48% ethanol to obtain a solution of 58% ethanol? Now go through an algebraic solution. What happens?

60. Computing Average Speed In going from Chicago to Atlanta, a car averages 45 miles per hour, and in going from Atlanta to Miami, it averages 55 miles per hour. If Atlanta is halfway between Chicago and Miami, what is the average speed from Chicago to Miami? Discuss an intuitive solution. Write a paragraph defending your intuitive solution. Then solve the problem algebraically. Is your intuitive solution the same as the algebraic one? If not, find the flaw.

61. Speed of a Plane On a recent flight from Phoenix to Kansas City, a distance of 919 nautical miles, the plane arrived 20 minutes early. On leaving the aircraft, I asked the captain, "What was our tail wind?" He replied, "I don't know, but our ground speed was 550 knots." Has enough information been provided for you to find the tail wind? If possible, find the tail wind. (1 knot = 1 nautical mile per hour)

Chapter Review

Things to Know

Quadratic formula (pp. 97 and 110)

If $ax^2 + bx + c = 0, a \neq 0$, then $x = \dfrac{-b \pm \sqrt{b^2 - 4ac}}{2a}$.

If $b^2 - 4ac < 0$, there are no real solutions.

Discriminant (pp. 97 and 110)

If $b^2 - 4ac > 0$, there are two unequal real solutions.

If $b^2 - 4ac = 0$, there is one repeated real solution, a double root.

If $b^2 - 4ac < 0$, there are no real solutions, but there are two distinct complex solutions that are not real; the complex solutions are conjugates of each other.

Interval notation (p. 120)

$[a, b]$	$\{x \mid a \leq x \leq b\}$	$[a, \infty)$	$\{x \mid x \geq a\}$
$[a, b)$	$\{x \mid a \leq x < b\}$	(a, ∞)	$\{x \mid x > a\}$
$(a, b]$	$\{x \mid a < x \leq b\}$	$(-\infty, a]$	$\{x \mid x \leq a\}$
(a, b)	$\{x \mid a < x < b\}$	$(-\infty, a)$	$\{x \mid x < a\}$
		$(-\infty, \infty)$	All real numbers

Properties of inequalities

Addition property (p. 121)

If $a < b$, then $a + c < b + c$.

If $a > b$, then $a + c > b + c$.

Multiplication properties (p. 122)

(a) If $a < b$ and if $c > 0$, then $ac < bc$.   (b) If $a > b$ and if $c > 0$, then $ac > bc$.

If $a < b$ and if $c < 0$, then $ac > bc$.   If $a > b$ and if $c < 0$, then $ac < bc$.

Reciprocal properties (p. 123)

If $a > 0$, then $\dfrac{1}{a} > 0$.   If $a < 0$, then $\dfrac{1}{a} < 0$.

If $\dfrac{1}{a} > 0$, then $a > 0$.   If $\dfrac{1}{a} < 0$, then $a < 0$.

Absolute value

If $|u| = a, a > 0$, then $u = -a$ or $u = a$. (p. 130)

If $|u| \leq a, a > 0$, then $-a \leq u \leq a$. (p. 131)

If $|u| \geq a, a > 0$, then $u \leq -a$ or $u \geq a$. (p. 132)

Objectives

Section		You should be able to . . .	Examples	Review Exercises
1.1	1	Solve a linear equation (p. 84)	1–3	1–3, 7, 8
	2	Solve equations that lead to linear equations (p. 86)	4–6	4
	3	Solve problems that can be modeled by linear equations (p. 87)	8, 9	45, 57
1.2	1	Solve a quadratic equation by factoring (p. 93)	1, 2	6, 9, 21, 22
	2	Solve a quadratic equation by completing the square (p. 95)	4, 5	5, 6, 9, 10, 13, 21, 22
	3	Solve a quadratic equation using the quadratic formula (p. 96)	6–9	5, 6, 9, 10, 13, 21, 22
	4	Solve problems that can be modeled by quadratic equations (p. 99)	10	50, 54, 56
1.3	1	Add, subtract, multiply, and divide complex numbers (p. 105)	1–8	35–39
	2	Solve quadratic equations in the complex number system (p. 109)	9–12	40–43
1.4	1	Solve radical equations (p. 113)	1–3	11, 12, 15–19, 23
	2	Solve equations quadratic in form (p. 114)	4–6	14, 20
	3	Solve equations by factoring (p. 116)	7, 8	26, 27
1.5	1	Use interval notation (p. 120)	1, 2	28–34
	2	Use properties of inequalities (p. 121)	3–6	28–34, 47
	3	Solve inequalities (p. 123)	7, 8	28, 47
	4	Solve combined inequalities (p. 124)	9, 10	29, 30
1.6	1	Solve equations involving absolute value (p. 130)	1	24, 25
	2	Solve inequalities involving absolute value (p. 130)	2–6	31–34
1.7	1	Translate verbal descriptions into mathematical expressions (p. 135)	1	44
	2	Solve interest problems (p. 136)	2, 3	45
	3	Solve mixture problems (p. 137)	4	53, 55
	4	Solve uniform motion problems (p. 138)	5, 6	46, 48, 49, 59, 60
	5	Solve constant rate job problems (p. 140)	7	51, 52, 58

Review Exercises

In Problems 1–27, find the real solutions, if any, of each equation. (Where they appear, a, b, m, and n are positive constants.)

1. $2 - \dfrac{x}{3} = 8$

2. $-2(5 - 3x) + 8 = 4 + 5x$

3. $\dfrac{3x}{4} - \dfrac{x}{3} = \dfrac{1}{12}$

4. $\dfrac{x}{x - 1} = \dfrac{6}{5} \quad x \neq 1$

5. $x(1 - x) = 6$

6. $x(1 + x) = 6$

7. $\dfrac{1}{2}\left(x - \dfrac{1}{3}\right) = \dfrac{3}{4} - \dfrac{x}{6}$

8. $\dfrac{1 - 3x}{4} = \dfrac{x + 6}{3} + \dfrac{1}{2}$

9. $(x - 1)(2x + 3) = 3$

10. $2x + 3 = 4x^2$

11. $\sqrt[3]{x^2 - 1} = 2$

12. $\sqrt{1 + x^3} = 3$

13. $x(x + 1) + 2 = 0$

14. $x^4 - 5x^2 + 4 = 0$

15. $\sqrt{2x - 3} + x = 3$

16. $\sqrt[4]{2x + 3} = 2$

17. $\sqrt{x + 1} + \sqrt{x - 1} = \sqrt{2x + 1}$

18. $\sqrt{2x - 1} - \sqrt{x - 5} = 3$

19. $2x^{1/2} - 3 = 0$

20. $x^{-6} - 7x^{-3} - 8 = 0$

21. $x^2 + m^2 = 2mx + (nx)^2 \quad n \neq 1$

22. $10a^2x^2 - 2abx - 36b^2 = 0$

23. $\sqrt{x^2 + 3x + 7} - \sqrt{x^2 - 3x + 9} + 2 = 0$

24. $|2x + 3| = 7$

25. $|2 - 3x| + 2 = 9$

26. $2x^3 = 3x^2$

27. $2x^3 + 5x^2 - 8x - 20 = 0$

In Problems 28–34, solve each inequality. Express your answer using set notation or interval notation. Graph the solution set.

28. $\dfrac{2x - 3}{5} + 2 \leq \dfrac{x}{2}$

29. $-9 \leq \dfrac{2x + 3}{-4} \leq 7$

30. $2 < \dfrac{3 - 3x}{12} < 6$

31. $|3x + 4| < \dfrac{1}{2}$

32. $|2x - 5| \geq 9$

33. $2 + |2 - 3x| \leq 4$

34. $1 - |2 - 3x| < -4$

In Problems 35–39, use the complex number system and write each expression in the standard form a + bi.

35. $(6 + 3i) - (2 - 4i)$

36. $4(3 - i) + 3(-5 + 2i)$

37. $\dfrac{3}{3 + i}$

38. i^{50}

39. $(2 + 3i)^3$

In Problems 40–43, solve each equation in the complex number system.

40. $x^2 + x + 1 = 0$

41. $2x^2 + x - 2 = 0$

42. $x^2 + 3 = x$

43. $x(1 - x) = 6$

---

44. Translate the following statement into a mathematical expression: The total cost C of manufacturing x bicycles in one day is $50,000 plus $95 times the number of bicycles manufactured.

45. **Financial Planning** Steve, a recent retiree, requires $5000 per year in extra income. He has $70,000 to invest and can invest in A-rated bonds paying 8% per year or in a certificate of deposit (CD) paying 5% per year. How much money should be invested in each to realize exactly $5000 in interest per year?

46. **Lightning and Thunder** A flash of lightning is seen, and the resulting thunderclap is heard 3 seconds later. If the speed of sound averages 1100 feet per second, how far away is the storm?

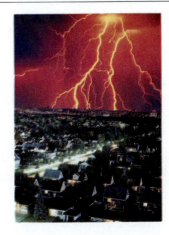

47. Physics: Intensity of Light The intensity I (in candlepower) of a certain light source obeys the equation $I = \dfrac{900}{x^2}$, where x is the distance (in meters) from the light. Over what range of distances can an object be placed from this light source so that the range of intensity of light is from 1600 to 3600 candlepower, inclusive?

48. Extent of Search and Rescue A search plane has a cruising speed of 250 miles per hour and carries enough fuel for at most 5 hours of flying. If there is a wind that averages 30 miles per hour and the direction of the search is with the wind one way and against it the other, how far can the search plane travel before it has to turn back?

49. Rescue at Sea A life raft, set adrift from a sinking ship 150 miles offshore, travels directly toward a Coast Guard station at the rate of 5 miles per hour. At the time that the raft is set adrift, a rescue helicopter is dispatched from the Coast Guard station. If the helicopter's average speed is 90 miles per hour, how long will it take the helicopter to reach the life raft?

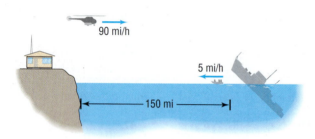

90 mi/h

5 mi/h

← 150 mi →

50. Physics An object is thrown down from the top of a building 1280 feet tall with an initial velocity of 32 feet per second. The distance s (in feet) of the object from the ground after t seconds is $s = 1280 - 32t - 16t^2$.
(a) When will the object strike the ground?
(b) What is the height of the object after 4 seconds?

51. Working Together to Get a Job Done Clarissa and Shawna, working together, can paint the exterior of a house in 6 days. Clarissa by herself can complete this job in 5 days less than Shawna. How long will it take Clarissa to complete the job by herself?

52. Emptying a Tank Two pumps of different sizes, working together, can empty a fuel tank in 5 hours. The larger pump can empty this tank in 4 hours less than the smaller one. If the larger pump is out of order, how long will it take the smaller one to do the job alone?

53. Chemistry: Salt Solutions How much water should be added to 64 ounces of a 10% salt solution to make a 2% salt solution?

54. Geometry The diagonal of a rectangle measures 10 inches. If the length is 2 inches more than the width, find the dimensions of the rectangle.

55. Chemistry: Mixing Acids A laboratory has 60 cubic centimeters (cm^3) of a solution that is 40% HCl acid. How many cubic centimeters of a 15% solution of HCl acid should be mixed with the 60 cm^3 of 40% acid to obtain a solution of 25% HCl? How much of the 25% solution is there?

56. Framing a Painting An artist has 50 inches of oak trim to frame a painting. The frame is to have a border 3 inches wide surrounding the painting.
(a) If the painting is square, what are its dimensions? What are the dimensions of the frame?
(b) If the painting is rectangular with a length twice its width, what are the dimensions of the painting? What are the dimensions of the frame?

57. Finance An inheritance of $900,000 is to be divided among Scott, Alice, and Tricia in the following manner: Alice is to receive $\dfrac{3}{4}$ of what Scott gets, while Tricia gets $\dfrac{1}{2}$ of what Scott gets. How much does each receive?

58. Utilizing Copying Machines A new copying machine can do a certain job in 1 hour less than an older copier. Together they can do this job in 72 minutes. How long would it take the older copier by itself to do the job?

59. Evening Up a Race In a 100-meter race, Todd crosses the finish line 5 meters ahead of Scott. To even things up, Todd suggests to Scott that they race again, this time with Todd lining up 5 meters behind the start.
(a) Assuming that Todd and Scott run at the same pace as before, does the second race end in a tie?
(b) If not, who wins?
(c) By how many meters does he win?
(d) How far back should Todd start so that the race ends in a tie?

After running the race a second time, Scott, to even things up, suggests to Todd that he (Scott) line up 5 meters in front of the start.
(e) Assuming again that they run at the same pace as in the first race, does the third race result in a tie?
(f) If not, who wins?
(g) By how many meters?
(h) How far ahead should Scott start so that the race ends in a tie?

60. Physics: Uniform Motion A man is walking at an average speed of 4 miles per hour alongside a railroad track. A freight train, going in the same direction at an average speed of 30 miles per hour, requires 5 seconds to pass the man. How long is the freight train? Give your answer in feet.

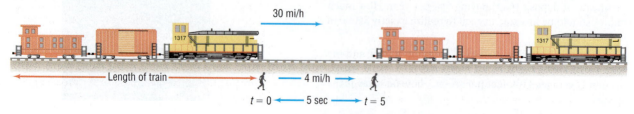

30 mi/h

← Length of train →   ← 4 mi/h →
$t = 0$ ← 5 sec → $t = 5$

Chapter Test

CHAPTER Test Prep VIDEOS | The Chapter Test Prep Videos are step-by-step solutions available in **MyMathLab**, or on this text's **You Tube** Channel. Flip back to the Resources for Success page for a link to this text's YouTube channel.

In Problems 1–7, solve each equation.

1. $\dfrac{2x}{3} - \dfrac{x}{2} = \dfrac{5}{12}$

2. $x(x - 1) = 6$

3. $x^4 - 3x^2 - 4 = 0$

4. $\sqrt{2x - 5} + 2 = 4$

5. $|2x - 3| + 7 = 10$

6. $3x^3 + 2x^2 - 12x - 8 = 0$

7. $3x^2 - x + 1 = 0$

In Problems 8–10, solve each inequality. Express your answer using interval notation. Graph the solution set.

8. $-3 \le \dfrac{3x - 4}{2} \le 6$

9. $|3x + 4| < 8$

10. $2 + |2x - 5| \ge 9$

11. Write $\dfrac{-2}{3 - i}$ in the standard form $a + bi$.

12. Solve the equation $4x^2 - 4x + 5 = 0$ in the complex number system.

13. **Blending Coffee** A coffee house has 20 pounds of a coffee that sells for \$4 per pound. How many pounds of a coffee that sells for \$8 per pound should be mixed with the 20 pounds of \$4-per-pound coffee to obtain a blend that will sell for \$5 per pound? How much of the \$5-per-pound coffee is there to sell?

Chapter Projects

Internet-based Project

I. Financing a Purchase At some point in your life, you are likely to need to borrow money to finance a purchase. For example, most of us will finance the purchase of a car or a home. What is the mathematics behind financing a purchase? When you borrow money from a bank, the bank uses a rather complex equation (or formula) to determine how much you need to pay each month to repay the loan. There are a number of variables that determine the monthly payment. These variables include the amount borrowed, the interest rate, and the length of the loan. The interest rate is based on current economic conditions, the length of the loan, the type of item being purchased, and your credit history.

The following formula gives the monthly payment P required to pay off a loan amount L at an annual interest rate r, expressed as a decimal, but usually given as a percent. The time t, measured in months, is the length of the loan. For example, a 30-year loan requires $12 \times 30 = 360$ monthly payments.

$$P = L\left[\frac{\dfrac{r}{12}}{1 - \left(1 + \dfrac{r}{12}\right)^{-t}}\right]$$

P = monthly payment
L = loan amount
r = annual rate of interest expressed as a decimal
t = length of loan, in months

1. Interest rates change daily. Many websites post current interest rates on loans. Go to *www.bankrate.com* (or some other website that posts lenders' interest rates) and find the current best interest rate on a 60-month new-car purchase loan. Use this rate to determine the monthly payment on a \$30,000 automobile loan.

2. Determine the total amount paid for the loan by multiplying the loan payment by the term of the loan.

3. Determine the total amount of interest paid by subtracting the loan amount from the total amount paid from question 2.

4. More often than not, we decide how much of a payment we can afford and use that information to determine the loan amount. Suppose you can afford a monthly payment of \$500. Use the interest rate from question 1 to determine the maximum amount you can borrow. If you have \$5000 to put down on the car, what is the maximum value of a car you can purchase?

5. Repeat questions 1 through 4 using a 72-month new-car purchase loan, a 60-month used-car purchase loan, and a 72-month used-car purchase loan.

6. We can use the power of a spreadsheet, such as Excel, to create a loan amortization schedule. A loan amortization schedule is a list of the monthly payments, a breakdown of interest and principal, along with a current loan balance. Create a loan amortization schedule for each of the four loan scenarios discussed on the previous page, using the following as a guide. You may want to use an Internet search engine to research specific keystrokes for creating an amortization schedule in a spreadsheet. We supply a sample spreadsheet with formulas included as a guide. Use the spreadsheet to verify your results from questions 1 through 5.

Loan Information		Payment Number	Payment Amount	Interest	Principal	Balance	Total Interest Paid
Loan Amount	$30,000.00	1	=PMT($B$3/12,$B$5,−$B$2,0)	=B2*$B$3/12	=E2−F2	=B2−G2	=B2*$B$3/12
Annual Interest Rate	0.045	2	=PMT($B$3/12,$B$5,−$B$2,0)	=H2*$B$3/12	=E3−F3	=H2−G3	=I2+F3
Length of Loan (years)	5	3	=PMT($B$3/12,$B$5,−$B$2,0)	=H3*$B$3/12	=E4−F4	=H3−G4	=I3+F4
Number of Payments	=B4*12	.	.	.	.	.	.
		.	.	.	.	.	.
		.		.	.	.	.

7. Go to an online automobile website such as *www.cars.com*, *www.edmunds.com*, or *www.autobytel.com*. Research the types of vehicles you can afford for a monthly payment of $500. Decide on a vehicle you would purchase based on your analysis in questions 1–6. Be sure to justify your decision, and include the impact the term of the loan has on your decision. You might consider other factors in your decision, such as expected maintenance costs and insurance costs.

Citation: Excel ©2013 Microsoft Corporation. Used with permission from Microsoft.

The following project is also available on the Instructor's Resource Center (IRC):

II. Project at Motorola   How Many Cellular Phones Can I Make?   An industrial engineer uses a model involving equations to be sure production levels meet customer demand.

Graphs

2

How to Value a House

Two things to consider in valuing a home are, first, how does it compare to similar homes that have sold recently? Is the asking price fair? And second, what value do you place on the advertised features and amenities? Yes, other people might value them highly, but do you?

Zestimate home valuation, RealEstateABC.com, and Reply.com are among the many algorithmic (generated by a computer model) starting points in figuring out the value of a home. They show you how the home is priced relative to other homes in the area, but you need to add in all the things that only someone who has seen the house knows. You can do that using My Estimator, and then you create your own estimate and see how it stacks up against the asking price.

Looking at "Comps"

Knowing whether an asking price is fair will be important when you're ready to make an offer on a house. It will be even more important when your mortgage lender hires an appraiser to determine whether the house is worth the loan you want.

Check with your agent, Zillow.com, propertyshark.com, or other websites to see recent sales of homes in the area that are similar, or comparable, to what you're looking for. Print them out and keep these "comps" in a three-ring binder; you'll be referring to them quite a bit.

Note that "recent sales" usually means within the last six months. A sales price from a year ago may bear little or no relation to what is going on in your area right now. In fact, some lenders will not accept comps older than three months.

Market activity also determines how easy or difficult it is to find accurate comps. In a "hot" or busy market, with sales happening all the time, you're likely to have lots of comps to choose from. In a less active market, finding reasonable comps becomes harder. And if the home you're looking at has special design features, finding a comparable property is harder still. It's also necessary to know what's going on in a given sub-segment. Maybe large, high-end homes are selling like hotcakes, but owners of smaller houses are staying put, or vice versa.

Source: http://allmyhome.blogspot.com/2008/07/how-to-value-house.html

 —See the Internet-based Chapter Project—

••• A Look Back

Chapter R and Chapter 1 review skills from intermediate algebra.

A Look Ahead •••

Here we connect algebra and geometry using the rectangular coordinate system. In the 1600s, algebra had developed to the point that René Descartes (1596–1650) and Pierre de Fermat (1601–1665) were able to use rectangular coordinates to translate geometry problems into algebra problems, and vice versa. This enabled both geometers and algebraists to gain new insights into their subjects, which had been thought to be separate but now were seen as connected.

Outline

149

value of the difference of the x-coordinates, $|x_2 - x_1|$. The vertical distance from P_3 to P_2 is the absolute value of the difference of the y-coordinates, $|y_2 - y_1|$. See Figure 6(b). The distance $d(P_1, P_2)$ is the length of the hypotenuse of the right triangle, so, by the Pythagorean Theorem, it follows that

$$[d(P_1, P_2)]^2 = |x_2 - x_1|^2 + |y_2 - y_1|^2$$
$$= (x_2 - x_1)^2 + (y_2 - y_1)^2$$
$$d(P_1, P_2) = \sqrt{(x_2 - x_1)^2 + (y_2 - y_1)^2}$$

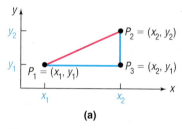 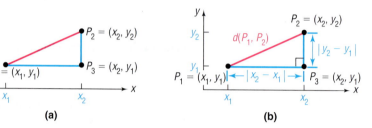

(a)   (b)

Figure 6

Now, if the line joining P_1 and P_2 is horizontal, then the y-coordinate of P_1 equals the y-coordinate of P_2; that is, $y_1 = y_2$. Refer to Figure 7(a). In this case, the distance formula (1) still works, because for $y_1 = y_2$, it reduces to

$$d(P_1, P_2) = \sqrt{(x_2 - x_1)^2 + 0^2} = \sqrt{(x_2 - x_1)^2} = |x_2 - x_1|$$

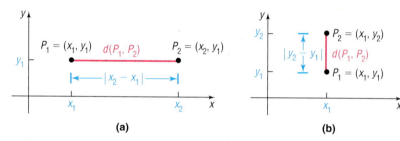 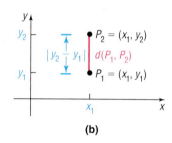

(a)   (b)

Figure 7

A similar argument holds if the line joining P_1 and P_2 is vertical. See Figure 7(b). ∎

EXAMPLE 2

Using the Distance Formula

Find the distance d between the points $(-4, 5)$ and $(3, 2)$.

Solution

Using the distance formula, equation (1), reveals that the distance d is

$$d = \sqrt{[3 - (-4)]^2 + (2 - 5)^2} = \sqrt{7^2 + (-3)^2}$$
$$= \sqrt{49 + 9} = \sqrt{58} \approx 7.62$$

●

✏️ **Now Work** PROBLEMS 19 AND 23

The distance between two points $P_1 = (x_1, y_1)$ and $P_2 = (x_2, y_2)$ is never a negative number. Also, the distance between two points is 0 only when the points are identical—that is, when $x_1 = x_2$ and $y_1 = y_2$. And, because $(x_2 - x_1)^2 = (x_1 - x_2)^2$ and $(y_2 - y_1)^2 = (y_1 - y_2)^2$, it makes no difference whether the distance is computed from P_1 to P_2 or from P_2 to P_1; that is, $d(P_1, P_2) = d(P_2, P_1)$.

The introduction to this chapter mentioned that rectangular coordinates enable us to translate geometry problems into algebra problems, and vice versa. The next example shows how algebra (the distance formula) can be used to solve geometry problems.

| EXAMPLE 3 | **Using Algebra to Solve Geometry Problems** |

Consider the three points $A = (-2, 1)$, $B = (2, 3)$, and $C = (3, 1)$.

(a) Plot each point and form the triangle ABC.
(b) Find the length of each side of the triangle.
(c) Show that the triangle is a right triangle.
(d) Find the area of the triangle.

Solution

(a) Figure 8 shows the points A, B, C and the triangle ABC.

(b) To find the length of each side of the triangle, use the distance formula, equation (1).

$$d(A, B) = \sqrt{[2 - (-2)]^2 + (3 - 1)^2} = \sqrt{16 + 4} = \sqrt{20} = 2\sqrt{5}$$
$$d(B, C) = \sqrt{(3 - 2)^2 + (1 - 3)^2} = \sqrt{1 + 4} = \sqrt{5}$$
$$d(A, C) = \sqrt{[3 - (-2)]^2 + (1 - 1)^2} = \sqrt{25 + 0} = 5$$

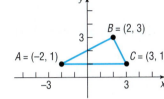

Figure 8

(c) If the sum of the squares of the lengths of two of the sides equals the square of the length of the third side, then the triangle is a right triangle. Looking at Figure 8, it seems reasonable to conjecture that the angle at vertex B might be a right angle. We shall check to see whether

$$[d(A, B)]^2 + [d(B, C)]^2 = [d(A, C)]^2$$

Using the results in part (b) yields

$$[d(A, B)]^2 + [d(B, C)]^2 = (2\sqrt{5})^2 + (\sqrt{5})^2$$
$$= 20 + 5 = 25 = [d(A, C)]^2$$

It follows from the converse of the Pythagorean Theorem that triangle ABC is a right triangle.

(d) Because the right angle is at vertex B, the sides AB and BC form the base and height of the triangle. Its area is

$$\text{Area} = \frac{1}{2}(\text{Base})(\text{Height}) = \frac{1}{2}(2\sqrt{5})(\sqrt{5}) = 5 \text{ square units}$$

●

Now Work PROBLEM 31

2 Use the Midpoint Formula

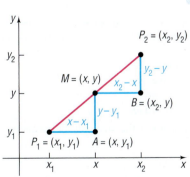

Figure 9

We now derive a formula for the coordinates of the **midpoint of a line segment**. Let $P_1 = (x_1, y_1)$ and $P_2 = (x_2, y_2)$ be the endpoints of a line segment, and let $M = (x, y)$ be the point on the line segment that is the same distance from P_1 as it is from P_2. See Figure 9. The triangles P_1AM and MBP_2 are congruent. [Do you see why? $d(P_1, M) = d(M, P_2)$ is given; also, $\angle AP_1M = \angle BMP_2$* and $\angle P_1MA = \angle MP_2B$. Thus, we have angle–side–angle.] Because triangles P_1AM and MBP_2 are congruent, corresponding sides are equal in length. That is,

$$x - x_1 = x_2 - x \quad \text{and} \quad y - y_1 = y_2 - y$$
$$2x = x_1 + x_2 \qquad\qquad 2y = y_1 + y_2$$
$$x = \frac{x_1 + x_2}{2} \qquad\qquad y = \frac{y_1 + y_2}{2}$$

*A postulate from geometry states that the transversal $\overline{P_1P_2}$ forms congruent corresponding angles with the parallel line segments $\overline{P_1A}$ and $\overline{MB}$.

THEOREM **Midpoint Formula**

The midpoint $M = (x, y)$ of the line segment from $P_1 = (x_1, y_1)$ to $P_2 = (x_2, y_2)$ is

In Words
To find the midpoint of a line segment, average the x-coordinates of the endpoints, and average the y-coordinates of the endpoints.

$$M = (x, y) = \left(\frac{x_1 + x_2}{2}, \frac{y_1 + y_2}{2} \right) \qquad \textbf{(2)}$$

EXAMPLE 4 **Finding the Midpoint of a Line Segment**

Find the midpoint of the line segment from $P_1 = (-5, 5)$ to $P_2 = (3, 1)$. Plot the points P_1 and P_2 and their midpoint.

Solution Apply the midpoint formula (2) using $x_1 = -5$, $y_1 = 5$, $x_2 = 3$, and $y_2 = 1$. Then the coordinates (x, y) of the midpoint M are

$$x = \frac{x_1 + x_2}{2} = \frac{-5 + 3}{2} = -1 \quad \text{and} \quad y = \frac{y_1 + y_2}{2} = \frac{5 + 1}{2} = 3$$

That is, $M = (-1, 3)$. See Figure 10.

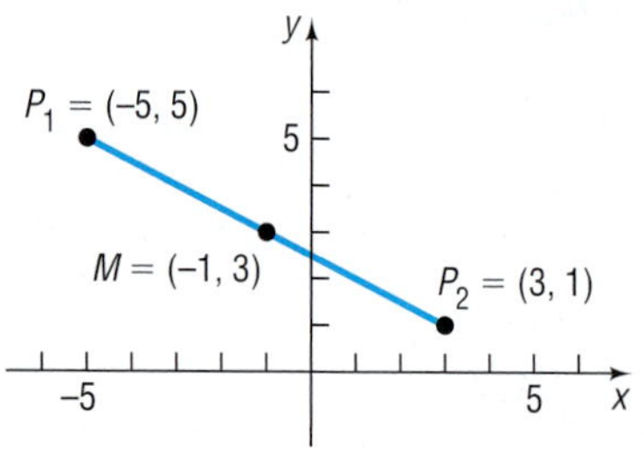

Figure 10

➤ **Now Work** PROBLEM 37

2.1 Assess Your Understanding

'Are You Prepared?' *Answers are given at the end of these exercises. If you get a wrong answer, read the pages listed in red.*

1. On the real number line, the origin is assigned the number _____. (p. 17)

2. If -3 and 5 are the coordinates of two points on the real number line, the distance between these points is _____. (pp. 19–20)

3. If 3 and 4 are the legs of a right triangle, the hypotenuse is _____. (p. 30)

4. Use the converse of the Pythagorean Theorem to show that a triangle whose sides are of lengths 11, 60, and 61 is a right triangle. (pp. 30–31)

5. The area A of a triangle whose base is b and whose altitude is h is $A =$ _____. (p. 31)

6. *True or False* Two triangles are congruent if two angles and the included side of one equals two angles and the included side of the other. (pp. 32–33).

Concepts and Vocabulary

7. If (x, y) are the coordinates of a point P in the xy-plane, then x is called the _____ of P, and y is the _____ of P.

8. The coordinate axes divide the xy-plane into four sections called _____.

9. If three distinct points P, Q, and R all lie on a line, and if $d(P, Q) = d(Q, R)$, then Q is called the _____ of the line segment from P to R.

10. *True or False* The distance between two points is sometimes a negative number.

11. *True or False* The point $(-1, 4)$ lies in quadrant IV of the Cartesian plane.

12. *True or False* The midpoint of a line segment is found by averaging the x-coordinates and averaging the y-coordinates of the endpoints.

13. Which of the following statements is true for a point (x, y) that lies in quadrant III?
    (a) Both x and y are positive.
    (b) Both x and y are negative.
    (c) x is positive, and y is negative.
    (d) x is negative, and y is positive.

14. Choose the formula that gives the distance between two points (x_1, y_1) and (x_2, y_2).
    (a) $\sqrt{(x_2 - x_1)^2 + (y_2 - y_1)^2}$
    (b) $\sqrt{(x_2 + x_1)^2 - (y_2 + y_1)^2}$
    (c) $\sqrt{(x_2 - x_1)^2 - (y_2 - y_1)^2}$
    (d) $\sqrt{(x_2 + x_1)^2 + (y_2 + y_1)^2}$

Skill Building

In Problems 15 and 16, plot each point in the xy-plane. Tell in which quadrant or on what coordinate axis each point lies.

15. (a) $A = (-3, 2)$ (d) $D = (6, 5)$
    (b) $B = (6, 0)$ (e) $E = (0, -3)$
    (c) $C = (-2, -2)$ (f) $F = (6, -3)$

16. (a) $A = (1, 4)$ (d) $D = (4, 1)$
    (b) $B = (-3, -4)$ (e) $E = (0, 1)$
    (c) $C = (-3, 4)$ (f) $F = (-3, 0)$

17. Plot the points $(2, 0), (2, -3), (2, 4), (2, 1)$, and $(2, -1)$. Describe the set of all points of the form $(2, y)$, where y is a real number.

18. Plot the points $(0, 3), (1, 3), (-2, 3), (5, 3)$, and $(-4, 3)$. Describe the set of all points of the form $(x, 3)$, where x is a real number.

In Problems 19–30, find the distance $d(P_1, P_2)$ between the points P_1 and P_2.

19.

20.
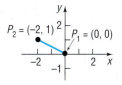

21.
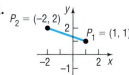

22.
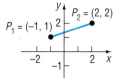

23. $P_1 = (3, -4);$ $P_2 = (5, 4)$

24. $P_1 = (-1, 0);$ $P_2 = (2, 4)$

25. $P_1 = (-3, 2);$ $P_2 = (6, 0)$

26. $P_1 = (2, -3);$ $P_2 = (4, 2)$

27. $P_1 = (4, -3);$ $P_2 = (6, 4)$

28. $P_1 = (-4, -3);$ $P_2 = (6, 2)$

29. $P_1 = (a, b);$ $P_2 = (0, 0)$

30. $P_1 = (a, a);$ $P_2 = (0, 0)$

In Problems 31–36, plot each point and form the triangle ABC. Show that the triangle is a right triangle. Find its area.

31. $A = (-2, 5);$ $B = (1, 3);$ $C = (-1, 0)$

32. $A = (-2, 5);$ $B = (12, 3);$ $C = (10, -11)$

33. $A = (-5, 3);$ $B = (6, 0);$ $C = (5, 5)$

34. $A = (-6, 3);$ $B = (3, -5);$ $C = (-1, 5)$

35. $A = (4, -3);$ $B = (0, -3);$ $C = (4, 2)$

36. $A = (4, -3);$ $B = (4, 1);$ $C = (2, 1)$

In Problems 37–44, find the midpoint of the line segment joining the points P_1 and P_2.

37. $P_1 = (3, -4);$ $P_2 = (5, 4)$

38. $P_1 = (-2, 0);$ $P_2 = (2, 4)$

39. $P_1 = (-3, 2);$ $P_2 = (6, 0)$

40. $P_1 = (2, -3);$ $P_2 = (4, 2)$

41. $P_1 = (4, -3);$ $P_2 = (6, 1)$

42. $P_1 = (-4, -3);$ $P_2 = (2, 2)$

43. $P_1 = (a, b);$ $P_2 = (0, 0)$

44. $P_1 = (a, a);$ $P_2 = (0, 0)$

Applications and Extensions

45. If the point $(2, 5)$ is shifted 3 units to the right and 2 units down, what are its new coordinates?

46. If the point $(-1, 6)$ is shifted 2 units to the left and 4 units up, what are its new coordinates?

47. Find all points having an x-coordinate of 3 whose distance from the point $(-2, -1)$ is 13.
    (a) By using the Pythagorean Theorem.
    (b) By using the distance formula.

48. Find all points having a y-coordinate of -6 whose distance from the point $(1, 2)$ is 17.
    (a) By using the Pythagorean Theorem.
    (b) By using the distance formula.

49. Find all points on the x-axis that are 6 units from the point $(4, -3)$.

50. Find all points on the y-axis that are 6 units from the point $(4, -3)$.

51. Suppose that $A = (2, 5)$ are the coordinates of a point in the xy-plane.

(a) Find the coordinates of the point if A is shifted 3 units to the left and 4 units down.
(b) Find the coordinates of the point if A is shifted 2 units to the left and 8 units up.

52. Plot the points $A = (-1, 8)$ and $M = (2, 3)$ in the xy-plane. If M is the midpoint of a line segment AB, find the coordinates of B.

53. The midpoint of the line segment from P_1 to P_2 is $(-1, 4)$. If $P_1 = (-3, 6)$, what is P_2?

54. The midpoint of the line segment from P_1 to P_2 is $(5, -4)$. If $P_2 = (7, -2)$, what is P_1?

55. **Geometry** The **medians** of a triangle are the line segments from each vertex to the midpoint of the opposite side (see the figure). Find the lengths of the medians of the triangle with vertices at $A = (0, 0), B = (6, 0)$, and $C = (4, 4)$.

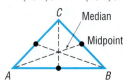

56. Geometry An **equilateral triangle** is one in which all three sides are of equal length. If two vertices of an equilateral triangle are $(0, 4)$ and $(0, 0)$, find the third vertex. How many of these triangles are possible?

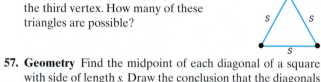

57. Geometry Find the midpoint of each diagonal of a square with side of length s. Draw the conclusion that the diagonals of a square intersect at their midpoints.
[**Hint:** Use $(0, 0)$, $(0, s)$, $(s, 0)$, and (s, s) as the vertices of the square.]

58. Geometry Verify that the points $(0, 0)$, $(a, 0)$, and $\left(\dfrac{a}{2}, \dfrac{\sqrt{3}a}{2} \right)$ are the vertices of an equilateral triangle. Then show that the midpoints of the three sides are the vertices of a second equilateral triangle (refer to Problem 56).

*In Problems 59–62, find the length of each side of the triangle determined by the three points P_1, P_2, and P_3. State whether the triangle is an isosceles triangle, a right triangle, neither of these, or both. (An **isosceles triangle** is one in which at least two of the sides are of equal length.)*

59. $P_1 = (2, 1); \quad P_2 = (-4, 1); \quad P_3 = (-4, -3)$

60. $P_1 = (-1, 4); \quad P_2 = (6, 2); \quad P_3 = (4, -5)$

61. $P_1 = (-2, -1); \quad P_2 = (0, 7); \quad P_3 = (3, 2)$

62. $P_1 = (7, 2); \quad P_2 = (-4, 0); \quad P_3 = (4, 6)$

63. Baseball A major league baseball "diamond" is actually a square 90 feet on a side (see the figure). What is the distance directly from home plate to second base (the diagonal of the square)?

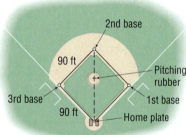

64. Little League Baseball The layout of a Little League playing field is a square 60 feet on a side. How far is it directly from home plate to second base (the diagonal of the square)?
Source: Little League Baseball, Official Regulations and Playing Rules, 2014.

65. Baseball Refer to Problem 63. Overlay a rectangular coordinate system on a major league baseball diamond so that the origin is at home plate, the positive x-axis lies in the direction from home plate to first base, and the positive y-axis lies in the direction from home plate to third base.
(a) What are the coordinates of first base, second base, and third base? Use feet as the unit of measurement.
(b) If the right fielder is located at $(310, 15)$, how far is it from the right fielder to second base?
(c) If the center fielder is located at $(300, 300)$, how far is it from the center fielder to third base?

66. Little League Baseball Refer to Problem 64. Overlay a rectangular coordinate system on a Little League baseball diamond so that the origin is at home plate, the positive x-axis lies in the direction from home plate to first base, and the positive y-axis lies in the direction from home plate to third base.
(a) What are the coordinates of first base, second base, and third base? Use feet as the unit of measurement.
(b) If the right fielder is located at $(180, 20)$, how far is it from the right fielder to second base?
(c) If the center fielder is located at $(220, 220)$, how far is it from the center fielder to third base?

67. Distance between Moving Objects A Ford Focus and a Freightliner truck leave an intersection at the same time. The Focus heads east at an average speed of 30 miles per hour, while the truck heads south at an average speed of 40 miles per hour. Find an expression for their distance apart d (in miles) at the end of t hours.

68. Distance of a Moving Object from a Fixed Point A hot-air balloon, headed due east at an average speed of 15 miles per hour and at a constant altitude of 100 feet, passes over an intersection (see the figure). Find an expression for the distance d (measured in feet) from the balloon to the intersection t seconds later.

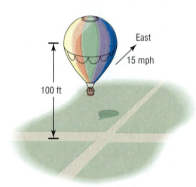

69. Drafting Error When a draftsman draws three lines that are to intersect at one point, the lines may not intersect as intended and subsequently will form an **error triangle**. If this error triangle is long and thin, one estimate for the location of the desired point is the midpoint of the shortest side. The figure shows one such error triangle.

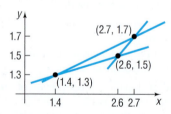

(a) Find an estimate for the desired intersection point.
(b) Find the length of the median for the midpoint found in part (a). See Problem 55.

70. Net Sales The figure on page 157 illustrates how net sales of Wal-Mart Stores, Inc., grew from 2007 through 2013. Use the midpoint formula to estimate the net sales of Wal-Mart Stores, Inc., in 2010. How does your result compare to the reported value of $405 billion?
Source: Wal-Mart Stores, Inc., 2013 Annual Report

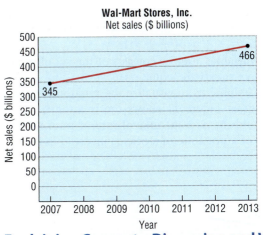

Wal-Mart Stores, Inc.
Net sales ($ billions)

71. Poverty Threshold Poverty thresholds are determined by the U.S. Census Bureau. A poverty threshold represents the minimum annual household income for a family not to be considered poor. In 2003, the poverty threshold for a family of four with two children under the age of 18 years was $18,660. In 2013, the poverty threshold for a family of four with two children under the age of 18 years was $23,624. Assuming that poverty thresholds increase in a straight-line fashion, use the midpoint formula to estimate the poverty threshold for a family of four with two children under the age of 18 in 2008. How does your result compare to the actual poverty threshold in 2008 of $21,834?

Source: U.S. Census Bureau

Explaining Concepts: Discussion and Writing

72. Write a paragraph that describes a Cartesian plane. Then write a second paragraph that describes how to plot points in the Cartesian plane. Your paragraphs should include the terms "coordinate axes," "ordered pair," "coordinates," "plot," "*x*-coordinate," and "*y*-coordinate."

Retain Your Knowledge

Problems 73–76 are based on material learned earlier in the course. The purpose of these problems is to keep the material fresh in your mind so that you are better prepared for the final exam.

73. Determine the domain of the variable x in the expression
$$\frac{3x + 1}{2x - 5}.$$

74. Find the real solution(s), if any, of the equation
$$3x^2 - 7x - 20 = 0.$$

75. Multiply $(7 + 3i)(1 - 2i)$. Write the answer in the form $a + bi$.

76. Solve the inequality $5(x - 3) + 2x \geq 6(2x - 3) - 7$. Express the solution using interval notation. Graph the solution set.

'Are You Prepared?' Answers

1. 0   **2.** 8   **3.** 5   **4.** $11^2 + 60^2 = 121 + 3600 = 3721 = 61^2$ **5.** $A = \frac{1}{2}bh$ **6.** True

2.2 Graphs of Equations in Two Variables; Intercepts; Symmetry

PREPARING FOR THIS SECTION *Before getting started, review the following:*

- Solving Linear Equations (Section 1.1, pp. 82–85)
- Solve a Quadratic Equation by Factoring (Section 1.2, pp. 93–94)

Now Work the 'Are You Prepared?' problems on page 164.

OBJECTIVES **1** Graph Equations by Plotting Points (p. 157)
                    2 Find Intercepts from a Graph (p. 159)
                    3 Find Intercepts from an Equation (p. 160)
                    4 Test an Equation for Symmetry with Respect to the *x*-Axis, the *y*-Axis, and the Origin (p. 160)
                    5 Know How to Graph Key Equations (p. 163)

1 Graph Equations by Plotting Points

An **equation in two variables**, say x and y, is a statement in which two expressions involving x and y are equal. The expressions are called the **sides** of the equation. Since an equation is a statement, it may be true or false, depending on the value of the variables. Any values of x and y that result in a true statement are said to **satisfy** the equation.

For example, the following are all equations in two variables x and y:

$$x^2 + y^2 = 5 \qquad 2x - y = 6 \qquad y = 2x + 5 \qquad x^2 = y$$

The first of these, $x^2 + y^2 = 5$, is satisfied for $x = 1, y = 2$, since $1^2 + 2^2 = 5$. Other choices of x and y, such as $x = -1, y = -2$, also satisfy this equation. It is not satisfied for $x = 2$ and $y = 3$, since $2^2 + 3^2 = 4 + 9 = 13 \neq 5$.

The **graph of an equation in two variables** x and y consists of the set of points in the xy-plane whose coordinates (x, y) satisfy the equation.

Graphs play an important role in helping us to visualize the relationships that exist between two variables or quantities. Figure 11 shows the relation between the level of risk in a stock portfolio and the average annual rate of return. From the graph, we can see that when 30% of a portfolio of stocks is invested in foreign companies, risk is minimized.

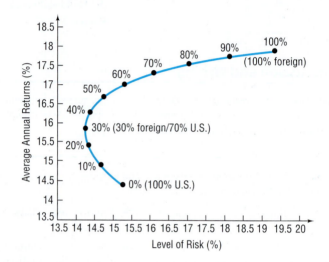

Figure 11
Source: T. Rowe Price

EXAMPLE 1

Determining Whether a Point Is on the Graph of an Equation

Determine if the following points are on the graph of the equation $2x - y = 6$.

(a) $(2, 3)$ (b) $(2, -2)$

Solution

(a) For the point $(2, 3)$, check to see whether $x = 2, y = 3$ satisfies the equation $2x - y = 6$.

$$2x - y = 2(2) - 3 = 4 - 3 = 1 \neq 6$$

The equation is not satisfied, so the point $(2, 3)$ is not on the graph of $2x - y = 6$.

(b) For the point $(2, -2)$,

$$2x - y = 2(2) - (-2) = 4 + 2 = 6$$

The equation is satisfied, so the point $(2, -2)$ is on the graph of $2x - y = 6$.

 **Now Work** PROBLEM 13

EXAMPLE 2

Graphing an Equation by Plotting Points

Graph the equation: $y = 2x + 5$

Solution

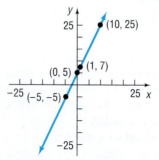

Figure 12 $y = 2x + 5$

The graph consists of all points (x, y) that satisfy the equation. To locate some of these points (and get an idea of the pattern of the graph), assign some numbers to x, and find corresponding values for y.

If	Then	Point on Graph
$x = 0$	$y = 2(0) + 5 = 5$	$(0, 5)$
$x = 1$	$y = 2(1) + 5 = 7$	$(1, 7)$
$x = -5$	$y = 2(-5) + 5 = -5$	$(-5, -5)$
$x = 10$	$y = 2(10) + 5 = 25$	$(10, 25)$

By plotting these points and then connecting them, we obtain the graph of the equation (a *line*), as shown in Figure 12.

| EXAMPLE 3 | **Graphing an Equation by Plotting Points** |

Graph the equation: $y = x^2$

Solution Table 1 provides several points on the graph. Plotting these points and connecting them with a smooth curve gives the graph (a *parabola*) shown in Figure 13.

Table 1

x	$y = x^2$	(x, y)
-4	16	$(-4, 16)$
-3	9	$(-3, 9)$
-2	4	$(-2, 4)$
-1	1	$(-1, 1)$
0	0	$(0, 0)$
1	1	$(1, 1)$
2	4	$(2, 4)$
3	9	$(3, 9)$
4	16	$(4, 16)$

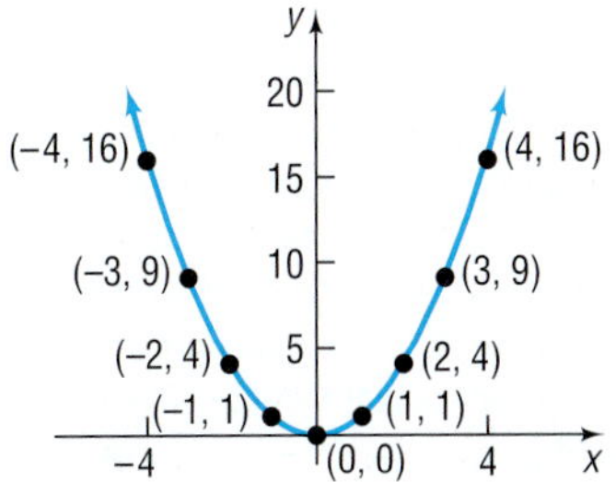

Figure 13 $y = x^2$

The graphs of the equations shown in Figures 12 and 13 do not show all points. For example, in Figure 12, the point $(20, 45)$ is a part of the graph of $y = 2x + 5$, but it is not shown. Since the graph of $y = 2x + 5$ can be extended out indefinitely, we use arrows to indicate that the pattern shown continues. It is important when illustrating a graph to present enough of the graph so that any viewer of the illustration will "see" the rest of it as an obvious continuation of what is actually there. This is referred to as a **complete graph**.

⬛ COMMENT Another way to obtain the graph of an equation is to use a graphing utility. Read Section 2, *Using a Graphing Utility to Graph Equations*, in the Appendix. ∎

One way to obtain the complete graph of an equation is to plot enough points on the graph for a pattern to become evident. Then these points are connected with a smooth curve following the suggested pattern. But how many points are sufficient? Sometimes knowledge about the equation tells us. For example, we will learn in the next section that if an equation is of the form $y = mx + b$, then its graph is a line. In this case, only two points are needed to obtain the graph.

One purpose of this book is to investigate the properties of equations in order to decide whether a graph is complete. Sometimes we shall graph equations by plotting points. Shortly, we shall investigate various techniques that will enable us to graph an equation without plotting so many points.

Two techniques that sometimes reduce the number of points required to graph an equation involve finding *intercepts* and checking for *symmetry*.

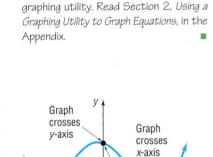

Figure 14

2 Find Intercepts from a Graph

The points, if any, at which a graph crosses or touches the coordinate axes are called the **intercepts**. See Figure 14. The x-coordinate of a point at which the graph crosses or touches the x-axis is an **x-intercept**, and the y-coordinate of a point at which the graph crosses or touches the y-axis is a **y-intercept**.

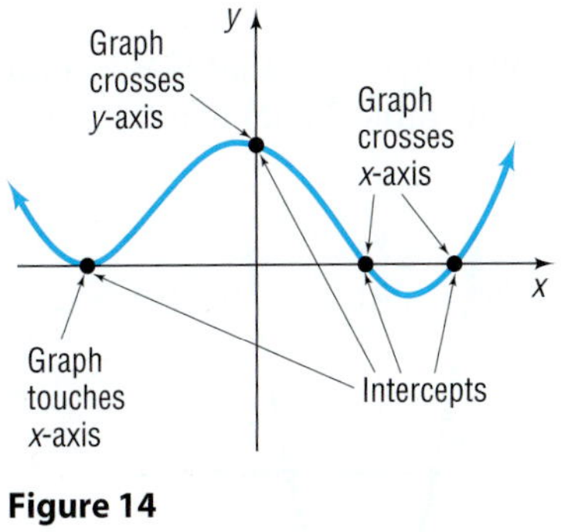

| EXAMPLE 4 | **Finding Intercepts from a Graph** |

Find the intercepts of the graph in Figure 15. What are its x-intercepts? What are its y-intercepts?

Solution The intercepts of the graph are the points

$$(-3, 0), \quad (0, 3), \quad \left(\frac{3}{2}, 0\right), \quad \left(0, -\frac{4}{3}\right), \quad (0, -3.5), \quad (4.5, 0)$$

The x-intercepts are $-3, \frac{3}{2}$, and 4.5; the y-intercepts are $-3.5, -\frac{4}{3}$, and 3.

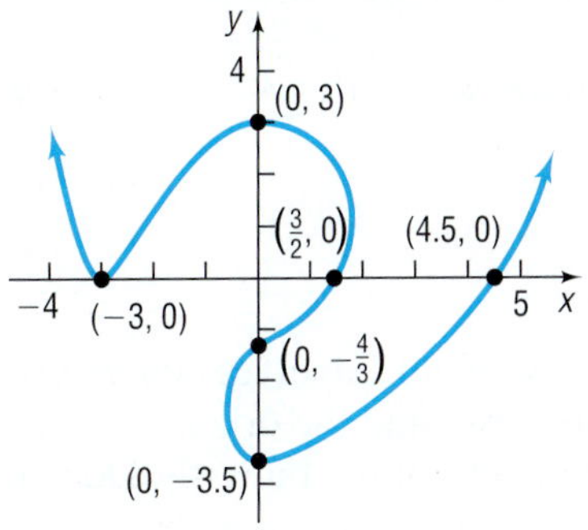

Figure 15

In Example 4, note the following usage: If the type of intercept (x- versus y-) is not specified, then report the intercept as an ordered pair. However, if the type of intercept is specified, then report the coordinate of the specified intercept. For *x*-intercepts, report the *x*-coordinate of the intercept; for *y*-intercepts, report the *y*-coordinate of the intercept.

Now Work PROBLEM 41(a)

3 Find Intercepts from an Equation

The intercepts of a graph can be found from its equation by using the fact that points on the *x*-axis have *y*-coordinates equal to 0, and points on the *y*-axis have *x*-coordinates equal to 0.

COMMENT For many equations, finding intercepts may not be so easy. In such cases, a graphing utility can be used. Read the first part of Section 3, *Using a Graphing Utility to Locate Intercepts and Check for Symmetry*, in the Appendix, to find out how to locate intercepts using a graphing utility. ∎

Procedure for Finding Intercepts

1. To find the *x*-intercept(s), if any, of the graph of an equation, let $y = 0$ in the equation and solve for x, where x is a real number.
2. To find the *y*-intercept(s), if any, of the graph of an equation, let $x = 0$ in the equation and solve for y, where y is a real number.

EXAMPLE 5

Finding Intercepts from an Equation

Find the *x*-intercept(s) and the *y*-intercept(s) of the graph of $y = x^2 - 4$. Then graph $y = x^2 - 4$ by plotting points.

Solution

To find the *x*-intercept(s), let $y = 0$ and obtain the equation

$$x^2 - 4 = 0 \quad \text{{\color{magenta}} } y = x^2 - 4 \text{ with } y = 0$$
$$(x + 2)(x - 2) = 0 \quad \text{Factor.}$$
$$x + 2 = 0 \quad \text{or} \quad x - 2 = 0 \quad \text{Zero-Product Property}$$
$$x = -2 \quad \text{or} \quad x = 2 \quad \text{Solve.}$$

The equation has two solutions, -2 and 2. The *x*-intercepts are -2 and 2.

To find the *y*-intercept(s), let $x = 0$ in the equation.

$$y = x^2 - 4$$
$$= 0^2 - 4 = -4$$

The *y*-intercept is -4.

Since $x^2 \geq 0$ for all x, we deduce from the equation $y = x^2 - 4$ that $y \geq -4$ for all x. This information, the intercepts, and the points from Table 2 enable us to graph $y = x^2 - 4$. See Figure 16.

Table 2

x	y = x² − 4	(x, y)
−3	5	(−3, 5)
−1	−3	(−1, −3)
1	−3	(1, −3)
3	5	(3, 5)

Figure 16 $y = x^2 - 4$

Now Work PROBLEM 23

4 Test an Equation for Symmetry with Respect to the *x*-Axis, the *y*-Axis, and the Origin

Another helpful tool for graphing equations by hand involves *symmetry*, particularly symmetry with respect to the *x*-axis, the *y*-axis, and the origin.

Symmetry often occurs in nature. Consider the picture of the butterfly. Do you see the symmetry?

DEFINITION

A graph is said to be **symmetric with respect to the x-axis** if, for every point (x, y) on the graph, the point $(x, -y)$ is also on the graph.

Figure 17 illustrates the definition. Note that when a graph is symmetric with respect to the x-axis, the part of the graph above the x-axis is a reflection (or mirror image) of the part below it, and vice versa.

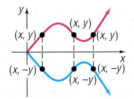

Figure 17
Symmetry with respect to the x-axis

EXAMPLE 6

Points Symmetric with Respect to the x-Axis

If a graph is symmetric with respect to the x-axis, and the point $(3, 2)$ is on the graph, then the point $(3, -2)$ is also on the graph.   ●

DEFINITION

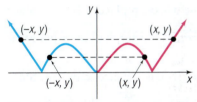

Figure 18
Symmetry with respect to the y-axis

A graph is said to be **symmetric with respect to the y-axis** if, for every point (x, y) on the graph, the point $(-x, y)$ is also on the graph.

Figure 18 illustrates the definition. When a graph is symmetric with respect to the y-axis, the part of the graph to the right of the y-axis is a reflection of the part to the left of it, and vice versa.

EXAMPLE 7

Points Symmetric with Respect to the y-Axis

If a graph is symmetric with respect to the y-axis and the point $(5, 8)$ is on the graph, then the point $(-5, 8)$ is also on the graph.   ●

DEFINITION

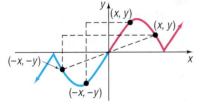

Figure 19
Symmetry with respect to the origin

A graph is said to be **symmetric with respect to the origin** if, for every point (x, y) on the graph, the point $(-x, -y)$ is also on the graph.

Figure 19 illustrates the definition. Symmetry with respect to the origin may be viewed in three ways:

1. As a reflection about the y-axis, followed by a reflection about the x-axis
2. As a projection along a line through the origin so that the distances from the origin are equal
3. As half of a complete revolution about the origin

EXAMPLE 8

Points Symmetric with Respect to the Origin

If a graph is symmetric with respect to the origin, and the point $(4, 2)$ is on the graph, then the point $(-4, -2)$ is also on the graph.   ●

Now Work PROBLEMS **31** AND **41(b)**

When the graph of an equation is symmetric with respect to a coordinate axis or the origin, the number of points that you need to plot in order to see the pattern is reduced. For example, if the graph of an equation is symmetric with respect to the *y*-axis, then once points to the right of the *y*-axis are plotted, an equal number of points on the graph can be obtained by reflecting them about the *y*-axis. Because of this, before we graph an equation, we should first determine whether it has any symmetry. The following tests are used for this purpose.

Tests for Symmetry

To test the graph of an equation for symmetry with respect to the

x-Axis   Replace *y* by −*y* in the equation and simplify. If an equivalent equation results, the graph of the equation is symmetric with respect to the *x*-axis.

y-Axis   Replace *x* by −*x* in the equation and simplify. If an equivalent equation results, the graph of the equation is symmetric with respect to the *y*-axis.

Origin   Replace *x* by −*x* and *y* by −*y* in the equation and simplify. If an equivalent equation results, the graph of the equation is symmetric with respect to the origin.

EXAMPLE 9

Testing an Equation for Symmetry

Test $y = \dfrac{4x^2}{x^2 + 1}$ for symmetry.

Solution

x-Axis: To test for symmetry with respect to the *x*-axis, replace *y* by −*y*. Since $-y = \dfrac{4x^2}{x^2 + 1}$ is not equivalent to $y = \dfrac{4x^2}{x^2 + 1}$, the graph of the equation is not symmetric with respect to the *x*-axis.

y-Axis: To test for symmetry with respect to the *y*-axis, replace *x* by −*x*. Since $y = \dfrac{4(-x)^2}{(-x)^2 + 1} = \dfrac{4x^2}{x^2 + 1}$ is equivalent to $y = \dfrac{4x^2}{x^2 + 1}$, the graph of the equation is symmetric with respect to the *y*-axis.

Origin: To test for symmetry with respect to the origin, replace *x* by −*x* and *y* by −*y*.

$$-y = \frac{4(-x)^2}{(-x)^2 + 1} \qquad \text{Replace } x \text{ by } -x \text{ and } y \text{ by } -y.$$

$$-y = \frac{4x^2}{x^2 + 1} \qquad \text{Simplify.}$$

$$y = -\frac{4x^2}{x^2 + 1} \qquad \text{Multiply both sides by } -1.$$

Since the result is not equivalent to the original equation, the graph of the equation $y = \dfrac{4x^2}{x^2 + 1}$ is not symmetric with respect to the origin. ●

Seeing the Concept

Figure 20 shows the graph of $y = \dfrac{4x^2}{x^2 + 1}$ using a graphing utility. Do you see the symmetry with respect to the *y*-axis?

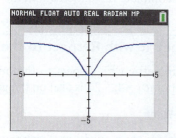

Figure 20 $y = \dfrac{4x^2}{x^2 + 1}$

➤ **Now Work** PROBLEM 61

5 Know How to Graph Key Equations

The next three examples use intercepts, symmetry, and point plotting to obtain the graphs of key equations. It is important to know the graphs of these key equations because we use them later. The first of these is $y = x^3$.

EXAMPLE 10

Graphing the Equation $y = x^3$ by Finding Intercepts, Checking for Symmetry, and Plotting Points

Graph the equation $y = x^3$ by plotting points. Find any intercepts and check for symmetry first.

Solution

First, find the intercepts. When $x = 0$, then $y = 0$; and when $y = 0$, then $x = 0$. The origin $(0, 0)$ is the only intercept. Now test for symmetry.

x-Axis: Replace y by $-y$. Since $-y = x^3$ is not equivalent to $y = x^3$, the graph is not symmetric with respect to the *x*-axis.

y-Axis: Replace x by $-x$. Since $y = (-x)^3 = -x^3$ is not equivalent to $y = x^3$, the graph is not symmetric with respect to the *y*-axis.

Origin: Replace x by $-x$ and y by $-y$. Since $-y = (-x)^3 = -x^3$ is equivalent to $y = x^3$ (multiply both sides by -1), the graph is symmetric with respect to the origin.

To graph $y = x^3$, use the equation to obtain several points on the graph. Because of the symmetry, we need to locate only points on the graph for which $x \geq 0$. See Table 3. Since $(1, 1)$ is on the graph, and the graph is symmetric with respect to the origin, the point $(-1, -1)$ is also on the graph. Plot the points from Table 3 and use the symmetry. Figure 21 shows the graph.

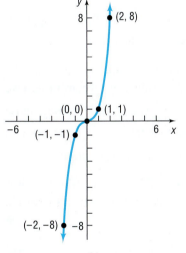

Figure 21 $y = x^3$

Table 3

x	$y = x^3$	(x, y)
0	0	$(0, 0)$
1	1	$(1, 1)$
2	8	$(2, 8)$
3	27	$(3, 27)$

EXAMPLE 11

Graphing the Equation $x = y^2$

(a) Graph the equation $x = y^2$. Find any intercepts and check for symmetry first.

(b) Graph $x = y^2, y \geq 0$.

Solution

(a) The lone intercept is $(0, 0)$. The graph is symmetric with respect to the *x*-axis. (Do you see why? Replace y by $-y$.) Figure 22 shows the graph.

(b) If we restrict y so that $y \geq 0$, the equation $x = y^2, y \geq 0$, may be written equivalently as $y = \sqrt{x}$. The portion of the graph of $x = y^2$ in quadrant I is therefore the graph of $y = \sqrt{x}$. See Figure 23.

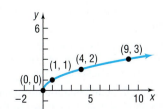

Figure 22 $x = y^2$

Figure 23 $y = \sqrt{x}$

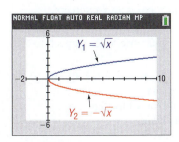

Figure 24

COMMENT To see the graph of the equation $x = y^2$ on a graphing calculator, you will need to graph two equations: $Y_1 = \sqrt{x}$ and $Y_2 = -\sqrt{x}$. We discuss why in Chapter 3. See Figure 24. ∎

EXAMPLE 12

Graphing the Equation $y = \dfrac{1}{x}$

Graph the equation $y = \dfrac{1}{x}$. First, find any intercepts and check for symmetry.

Solution Check for intercepts first. If we let $x = 0$, we obtain 0 in the denominator, which makes y undefined. We conclude that there is no y-intercept. If we let $y = 0$, we get the equation $\dfrac{1}{x} = 0$, which has no solution. We conclude that there is no x-intercept.

The graph of $y = \dfrac{1}{x}$ does not cross or touch the coordinate axes.

Next check for symmetry:

x-Axis: Replacing y by $-y$ yields $-y = \dfrac{1}{x}$, which is not equivalent to $y = \dfrac{1}{x}$.

y-Axis: Replacing x by $-x$ yields $y = \dfrac{1}{-x} = -\dfrac{1}{x}$, which is not equivalent to $y = \dfrac{1}{x}$.

Origin: Replacing x by $-x$ and y by $-y$ yields $-y = -\dfrac{1}{x}$, which is equivalent to $y = \dfrac{1}{x}$. The graph is symmetric with respect to the origin.

Now set up Table 4, listing several points on the graph. Because of the symmetry with respect to the origin, we use only positive values of x. From Table 4 we infer that if x is a large and positive number, then $y = \dfrac{1}{x}$ is a positive number close to 0. We also infer that if x is a positive number close to 0, then $y = \dfrac{1}{x}$ is a large and positive number. Armed with this information, we can graph the equation.

Figure 25 illustrates some of these points and the graph of $y = \dfrac{1}{x}$. Observe how the absence of intercepts and the existence of symmetry with respect to the origin were utilized. ●

Table 4

x	$y = \dfrac{1}{x}$	(x, y)
$\dfrac{1}{10}$	10	$\left(\dfrac{1}{10}, 10\right)$
$\dfrac{1}{3}$	3	$\left(\dfrac{1}{3}, 3\right)$
$\dfrac{1}{2}$	2	$\left(\dfrac{1}{2}, 2\right)$
1	1	$(1, 1)$
2	$\dfrac{1}{2}$	$\left(2, \dfrac{1}{2}\right)$
3	$\dfrac{1}{3}$	$\left(3, \dfrac{1}{3}\right)$
10	$\dfrac{1}{10}$	$\left(10, \dfrac{1}{10}\right)$

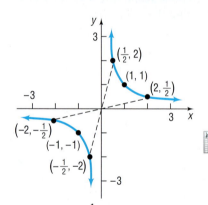

Figure 25 $y = \dfrac{1}{x}$

COMMENT Refer to Example 2 in the Appendix, Section 3, for the graph of $y = \dfrac{1}{x}$ found using a graphing utility. ■

2.2 Assess Your Understanding

'Are You Prepared?' *Answers are given at the end of these exercises. If you get a wrong answer, read the pages listed in* red.

1. Solve the equation $2(x + 3) - 1 = -7$. (pp. 82–85)

2. Solve the equation $x^2 - 9 = 0$. (pp. 93–94)

Concepts and Vocabulary

3. The points, if any, at which a graph crosses or touches the coordinate axes are called _____.

4. The x-intercepts of the graph of an equation are those x-values for which _____.

5. If for every point (x, y) on the graph of an equation the point $(-x, y)$ is also on the graph, then the graph is symmetric with respect to the _____.

6. If the graph of an equation is symmetric with respect to the y-axis and -4 is an x-intercept of this graph, then ___ is also an x-intercept.

7. If the graph of an equation is symmetric with respect to the origin and $(3, -4)$ is a point on the graph, then _____ is also a point on the graph.

8. **True or False** To find the y-intercepts of the graph of an equation, let $x = 0$ and solve for y.

9. **True or False** The y-coordinate of a point at which the graph crosses or touches the x-axis is an x-intercept.

10. **True or False** If a graph is symmetric with respect to the x-axis, then it cannot be symmetric with respect to the y-axis.

11. Given that the intercepts of a graph are $(-4, 0)$ and $(0, 5)$, choose the statement that is true.
    (a) The y-intercept is -4, and the x-intercept is 5.
    (b) The y-intercepts are -4 and 5.
    (c) The x-intercepts are -4 and 5.
    (d) The x-intercept is -4, and the y-intercept is 5.

12. To test whether the graph of an equation is symmetric with respect to the origin, replace _____ in the equation and simplify. If an equivalent equation results, then the graph is symmetric with respect to the origin.

(a) x by $-x$
(b) y by $-y$
(c) x by $-x$ and y by $-y$
(d) x by $-y$ and y by $-x$

Skill Building

In Problems 13–18, determine which of the given points are on the graph of the equation.

13. Equation: $y = x^4 - \sqrt{x}$
Points: $(0,0)$; $(1,1)$; $(2,4)$

14. Equation: $y = x^3 - 2\sqrt{x}$
Points: $(0,0)$; $(1,1)$; $(1,-1)$

15. Equation: $y^2 = x^2 + 9$
Points: $(0,3)$; $(3,0)$; $(-3,0)$

16. Equation: $y^3 = x + 1$
Points: $(1,2)$; $(0,1)$; $(-1,0)$

17. Equation: $x^2 + y^2 = 4$
Points: $(0,2)$; $(-2,2)$; $\left(\sqrt{2}, \sqrt{2}\right)$

18. Equation: $x^2 + 4y^2 = 4$
Points: $(0,1)$; $(2,0)$; $\left(2, \frac{1}{2}\right)$

In Problems 19–30, find the intercepts and graph each equation by plotting points. Be sure to label the intercepts.

19. $y = x + 2$

20. $y = x - 6$

21. $y = 2x + 8$

22. $y = 3x - 9$

23. $y = x^2 - 1$

24. $y = x^2 - 9$

25. $y = -x^2 + 4$

26. $y = -x^2 + 1$

27. $2x + 3y = 6$

28. $5x + 2y = 10$

29. $9x^2 + 4y = 36$

30. $4x^2 + y = 4$

In Problems 31–40, plot each point. Then plot the point that is symmetric to it with respect to (a) the x-axis; (b) the y-axis; (c) the origin.

31. $(3,4)$

32. $(5,3)$

33. $(-2,1)$

34. $(4,-2)$

35. $(5,-2)$

36. $(-1,-1)$

37. $(-3,-4)$

38. $(4,0)$

39. $(0,-3)$

40. $(-3,0)$

In Problems 41–52, the graph of an equation is given. (a) Find the intercepts. (b) Indicate whether the graph is symmetric with respect to the x-axis, the y-axis, or the origin.

41.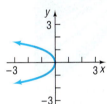

42.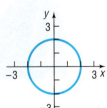

43.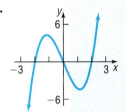

44.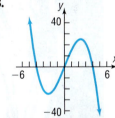

45.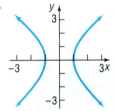

46.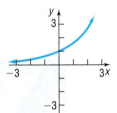

47.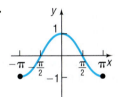

48.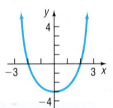

49.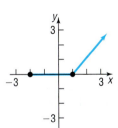

50.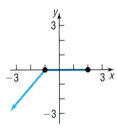

51.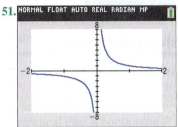

52.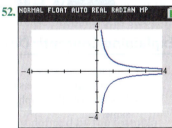

In Problems 53–56, draw a complete graph so that it has the type of symmetry indicated.

53. y-axis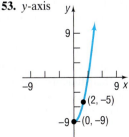

54. x-axis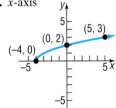

55. Origin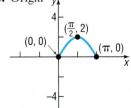

56. y-axis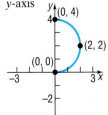

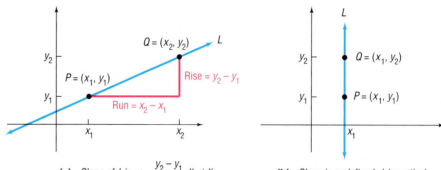

Figure 27

(a) Slope of L is $m = \dfrac{y_2 - y_1}{x_2 - x_1}$, $x_1 \neq x_2$

(b) Slope is undefined; L is vertical

In Words

The symbol Δ is the Greek uppercase letter delta. In mathematics, Δ is read "change in," so $\dfrac{\Delta y}{\Delta x}$ is read "change in y divided by change in x."

As Figure 27(a) illustrates, the slope m of a nonvertical line may be viewed as

$$m = \frac{y_2 - y_1}{x_2 - x_1} = \frac{\text{Rise}}{\text{Run}} \quad \text{or as} \quad m = \frac{y_2 - y_1}{x_2 - x_1} = \frac{\text{Change in } y}{\text{Change in } x} = \frac{\Delta y}{\Delta x}$$

That is, the slope m of a nonvertical line measures the amount y changes when x changes from x_1 to x_2. The expression $\dfrac{\Delta y}{\Delta x}$ is called the **average rate of change** of y with respect to x.

Two comments about computing the slope of a nonvertical line may prove helpful:

1. Any two distinct points on the line can be used to compute the slope of the line. (See Figure 28 for justification.) Since any two distinct points can be used to compute the slope of a line, the average rate of change of a line is always the same number.

Figure 28
Triangles ABC and PQR are similar (equal angles), so ratios of corresponding sides are proportional. Then

Slope using P and Q $= \dfrac{y_2 - y_1}{x_2 - x_1} =$

$\dfrac{d(B, C)}{d(A, C)} =$ Slope using A and B

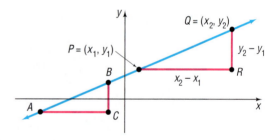

2. The slope of a line may be computed from $P = (x_1, y_1)$ to $Q = (x_2, y_2)$ or from Q to P because

$$\frac{y_2 - y_1}{x_2 - x_1} = \frac{y_1 - y_2}{x_1 - x_2}$$

EXAMPLE 1

Finding and Interpreting the Slope of a Line Given Two Points

The slope m of the line containing the points $(1, 2)$ and $(5, -3)$ may be computed as

$$m = \frac{-3 - 2}{5 - 1} = \frac{-5}{4} = -\frac{5}{4} \quad \text{or as} \quad m = \frac{2 - (-3)}{1 - 5} = \frac{5}{-4} = -\frac{5}{4}$$

For every 4-unit change in x, y will change by -5 units. That is, if x increases by 4 units, then y will decrease by 5 units. The average rate of change of y with respect to x is $-\dfrac{5}{4}$.

Now Work PROBLEMS 13 AND 19

EXAMPLE 2

Finding the Slopes of Various Lines Containing the Same Point (2, 3)

Compute the slopes of the lines L_1, L_2, L_3, and L_4 containing the following pairs of points. Graph all four lines on the same set of coordinate axes.

$$L_1: \quad P = (2,3) \qquad Q_1 = (-1,-2)$$
$$L_2: \quad P = (2,3) \qquad Q_2 = (3,-1)$$
$$L_3: \quad P = (2,3) \qquad Q_3 = (5,3)$$
$$L_4: \quad P = (2,3) \qquad Q_4 = (2,5)$$

Solution

Let m_1, m_2, m_3, and m_4 denote the slopes of the lines L_1, L_2, L_3, and L_4, respectively. Then

$$m_1 = \frac{-2-3}{-1-2} = \frac{-5}{-3} = \frac{5}{3} \qquad \text{A rise of 5 divided by a run of 3}$$

$$m_2 = \frac{-1-3}{3-2} = \frac{-4}{1} = -4$$

$$m_3 = \frac{3-3}{5-2} = \frac{0}{3} = 0$$

m_4 is undefined because $x_1 = x_2 = 2$

The graphs of these lines are given in Figure 29.

Figure 29 illustrates the following facts:

1. When the slope of a line is positive, the line slants upward from left to right (L_1).
2. When the slope of a line is negative, the line slants downward from left to right (L_2).
3. When the slope is 0, the line is horizontal (L_3).
4. When the slope is undefined, the line is vertical (L_4).

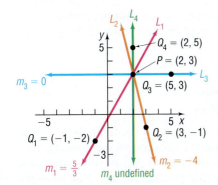

Figure 29

Seeing the Concept

On the same screen, graph the following equations:

$Y_1 = 0$	Slope of line is 0.
$Y_2 = \frac{1}{4}x$	Slope of line is $\frac{1}{4}$.
$Y_3 = \frac{1}{2}x$	Slope of line is $\frac{1}{2}$.
$Y_4 = x$	Slope of line is 1.
$Y_5 = 2x$	Slope of line is 2.
$Y_6 = 6x$	Slope of line is 6.

See Figure 30.

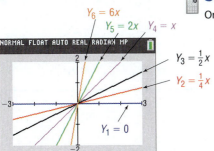

Figure 30

Seeing the Concept

On the same screen, graph the following equations:

$Y_1 = 0$	Slope of line is 0.
$Y_2 = -\frac{1}{4}x$	Slope of line is $-\frac{1}{4}$.
$Y_3 = -\frac{1}{2}x$	Slope of line is $-\frac{1}{2}$.
$Y_4 = -x$	Slope of line is -1.
$Y_5 = -2x$	Slope of line is -2.
$Y_6 = -6x$	Slope of line is -6.

See Figure 31.

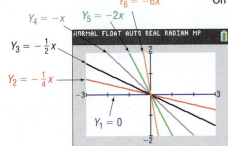

Figure 31

Figures 30 and 31 on page 169 illustrate that the closer the line is to the vertical position, the greater the magnitude of the slope.

2 Graph Lines Given a Point and the Slope

EXAMPLE 3

Graphing a Line Given a Point and a Slope

Draw a graph of the line that contains the point $(3, 2)$ and has a slope of:

(a) $\dfrac{3}{4}$

(b) $-\dfrac{4}{5}$

Solution

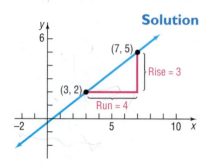

Figure 32

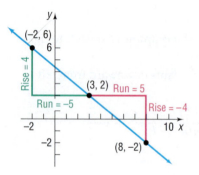

Figure 33

(a) Slope $= \dfrac{\text{Rise}}{\text{Run}}$. The slope $\dfrac{3}{4}$ means that for every horizontal movement (run) of 4 units to the right, there will be a vertical movement (rise) of 3 units. Start at the given point $(3, 2)$ and move 4 units to the right and 3 units up, arriving at the point $(7, 5)$. Drawing the line through this point and the point $(3, 2)$ gives the graph. See Figure 32.

(b) The fact that the slope is

$$-\frac{4}{5} = \frac{-4}{5} = \frac{\text{Rise}}{\text{Run}}$$

means that for every horizontal movement of 5 units to the right, there will be a corresponding vertical movement of -4 units (a downward movement). Start at the given point $(3, 2)$ and move 5 units to the right and then 4 units down, arriving at the point $(8, -2)$. Drawing the line through these points gives the graph. See Figure 33.

Alternatively, consider that

$$-\frac{4}{5} = \frac{4}{-5} = \frac{\text{Rise}}{\text{Run}}$$

so for every horizontal movement of -5 units (a movement to the left), there will be a corresponding vertical movement of 4 units (upward). This approach leads to the point $(-2, 6)$, which is also on the graph of the line in Figure 33. ●

Now Work PROBLEM 25 (graph the line)

3 Find the Equation of a Vertical Line

EXAMPLE 4

Graphing a Line

Graph the equation: $x = 3$

Solution

To graph $x = 3$, we find all points (x, y) in the plane for which $x = 3$. No matter what y-coordinate is used, the corresponding x-coordinate always equals 3. Consequently, the graph of the equation $x = 3$ is a vertical line with x-intercept 3 and undefined slope. See Figure 34.

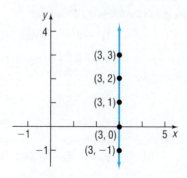

Figure 34 $x = 3$

Example 4 suggests the following result:

THEOREM

Equation of a Vertical Line

A vertical line is given by an equation of the form

$$x = a$$

where a is the x-intercept.

COMMENT To graph an equation using a graphing utility, we need to express the equation in the form $y = \{$ expression in $x\}$. But $x = 3$ cannot be put in this form. To overcome this, most graphing utilities have special commands for drawing vertical lines. DRAW, LINE, PLOT, and VERT are among the more common ones. Consult your manual to determine the correct methodology for your graphing utility. ∎

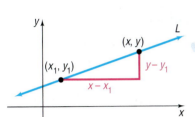

Figure 35

4 Use the Point–Slope Form of a Line; Identify Horizontal Lines

Let L be a nonvertical line with slope m that contains the point (x_1, y_1). See Figure 35. For any other point (x, y) on L, we have

$$m = \frac{y - y_1}{x - x_1} \quad \text{or} \quad y - y_1 = m(x - x_1)$$

THEOREM

Point–Slope Form of an Equation of a Line

An equation of a nonvertical line with slope m that contains the point (x_1, y_1) is

$$y - y_1 = m(x - x_1) \tag{2}$$

EXAMPLE 5

Using the Point–Slope Form of a Line

An equation of the line with slope 4 that contains the point $(1, 2)$ can be found by using the point–slope form with $m = 4$, $x_1 = 1$, and $y_1 = 2$.

$$y - y_1 = m(x - x_1)$$
$$y - 2 = 4(x - 1) \qquad m = 4, x_1 = 1, y_1 = 2$$
$$y = 4x - 2 \qquad \text{Solve for y.}$$

See Figure 36 for the graph.

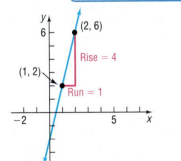

Figure 36 $y = 4x - 2$

Now Work PROBLEM 25 (find the point-slope form)

EXAMPLE 6

Finding the Equation of a Horizontal Line

Find an equation of the horizontal line containing the point $(3, 2)$.

Solution

Because all the y-values are equal on a horizontal line, the slope of a horizontal line is 0. To get an equation, use the point–slope form with $m = 0$, $x_1 = 3$, and $y_1 = 2$.

$$y - y_1 = m(x - x_1)$$
$$y - 2 = 0 \cdot (x - 3) \qquad m = 0, x_1 = 3, \text{and } y_1 = 2$$
$$y - 2 = 0$$
$$y = 2$$

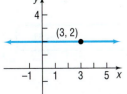

Figure 37 $y = 2$

See Figure 37 for the graph.

Example 6 suggests the following result:

THEOREM

Equation of a Horizontal Line

A horizontal line is given by an equation of the form

$$y = b$$

where b is the y-intercept.

5 Find the Equation of a Line Given Two Points

EXAMPLE 7

Finding an Equation of a Line Given Two Points

Find an equation of the line containing the points $(2, 3)$ and $(-4, 5)$. Graph the line.

Solution   First compute the slope of the line.

$$m = \frac{5 - 3}{-4 - 2} = \frac{2}{-6} = -\frac{1}{3} \qquad m = \frac{y_2 - y_1}{x_2 - x_1}$$

Use the point $(2, 3)$ and the slope $m = -\dfrac{1}{3}$ to get the point–slope form of the equation of the line.

$$y - 3 = -\frac{1}{3}(x - 2) \qquad y - y_1 = m(x - x_1)$$

See Figure 38 for the graph.

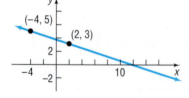

Figure 38 $y - 3 = -\dfrac{1}{3}(x - 2)$

In the solution to Example 7, we could have used the other point, $(-4, 5)$, instead of the point $(2, 3)$. The equation that results, although it looks different, is equivalent to the equation that we obtained in the example. (Try it for yourself.)

Now Work PROBLEM 39

6 Write the Equation of a Line in Slope–Intercept Form

Another useful equation of a line is obtained when the slope m and y-intercept b are known. In this event, both the slope m of the line and a point $(0, b)$ on the line are known; then use the point–slope form, equation (2), to obtain the following equation:

$$y - b = m(x - 0) \quad \text{or} \quad y = mx + b$$

THEOREM

Slope–Intercept Form of an Equation of a Line

An equation of a line with slope m and y-intercept b is

$$y = mx + b \tag{3}$$

Now Work PROBLEMS 47 AND 53 (express answer in slope–intercept form)

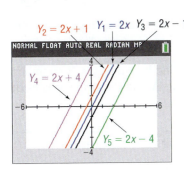

Figure 39 $y = mx + 2$

$Y_5 = -3x + 2$ $Y_4 = 3x + 2$ $Y_2 = x + 2$
$Y_3 = -x + 2$ $Y_1 = 2$

$Y_2 = 2x + 1$ $Y_1 = 2x$ $Y_3 = 2x - 1$
$Y_4 = 2x + 4$
$Y_5 = 2x - 4$

Figure 40 $y = 2x + b$

Seeing the Concept

To see the role that the slope m plays, graph the following lines on the same screen.

$$Y_1 = 2 \qquad m = 0$$
$$Y_2 = x + 2 \qquad m = 1$$
$$Y_3 = -x + 2 \qquad m = -1$$
$$Y_4 = 3x + 2 \qquad m = 3$$
$$Y_5 = -3x + 2 \qquad m = -3$$

See Figure 39. What do you conclude about the lines $y = mx + 2$?

Seeing the Concept

To see the role of the y-intercept b, graph the following lines on the same screen.

$$Y_1 = 2x \qquad b = 0$$
$$Y_2 = 2x + 1 \qquad b = 1$$
$$Y_3 = 2x - 1 \qquad b = -1$$
$$Y_4 = 2x + 4 \qquad b = 4$$
$$Y_5 = 2x - 4 \qquad b = -4$$

See Figure 40. What do you conclude about the lines $y = 2x + b$?

7 Identify the Slope and y-Intercept of a Line from Its Equation

When the equation of a line is written in slope–intercept form, it is easy to find the slope m and y-intercept b of the line. For example, suppose that the equation of a line is

$$y = -2x + 7$$

Compare this equation to $y = mx + b$.

$$y = -2x + 7$$
$$y = \quad mx \ + b$$

The slope of this line is -2 and its y-intercept is 7.

Now Work PROBLEM **73**

EXAMPLE 8

Finding the Slope and y-Intercept

Find the slope m and y-intercept b of the equation $2x + 4y = 8$. Graph the equation.

Solution

To obtain the slope and y-intercept, write the equation in slope–intercept form by solving for y.

$$2x + 4y = 8$$
$$4y = -2x + 8$$
$$y = -\frac{1}{2}x + 2 \qquad y = mx + b$$

The coefficient of x, $-\dfrac{1}{2}$, is the slope, and the constant, 2, is the y-intercept. Graph the line with y-intercept 2 and with slope $-\dfrac{1}{2}$. Starting at the point $(0, 2)$, go to the right 2 units and then down 1 unit to the point $(2, 1)$. Draw the line through these points. See Figure 41.

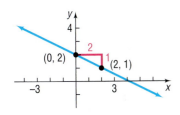

Figure 41 $y = -\dfrac{1}{2}x + 2$

Now Work PROBLEM **79**

8 Graph Lines Written in General Form Using Intercepts

Refer to Example 8. The form of the equation of the line $2x + 4y = 8$ is called the *general form*.

DEFINITION

The equation of a line is in **general form*** when it is written as

$$Ax + By = C \qquad (4)$$

where A, B, and C are real numbers and A and B are not both 0.

If $B = 0$ in equation (4), then $A \neq 0$ and the graph of the equation is a vertical line: $x = \dfrac{C}{A}$. If $B \neq 0$ in equation (4), then we can solve the equation for y and write the equation in slope–intercept form as we did in Example 8.

Another approach to graphing equation (4) is to find its intercepts. Remember, the intercepts of the graph of an equation are the points where the graph crosses or touches a coordinate axis.

EXAMPLE 9

Graphing an Equation in General Form Using Its Intercepts

Graph the equation $2x + 4y = 8$ by finding its intercepts.

Solution

To obtain the x-intercept, let $y = 0$ in the equation and solve for x.

$$2x + 4y = 8$$
$$2x + 4(0) = 8 \qquad \text{Let } y = 0.$$
$$2x = 8$$
$$x = 4 \qquad \text{Divide both sides by 2.}$$

The x-intercept is 4, and the point $(4, 0)$ is on the graph of the equation. To obtain the y-intercept, let $x = 0$ in the equation and solve for y.

$$2x + 4y = 8$$
$$2(0) + 4y = 8 \qquad \text{Let } x = 0.$$
$$4y = 8$$
$$y = 2 \qquad \text{Divide both sides by 4.}$$

The y-intercept is 2, and the point $(0, 2)$ is on the graph of the equation. Plot the points $(4, 0)$ and $(0, 2)$ and draw the line through the points. See Figure 42.

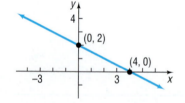

Figure 42 $2x + 4y = 8$

Now Work PROBLEM 93

Every line has an equation that is equivalent to an equation written in general form. For example, a vertical line whose equation is

$$x = a$$

can be written in the general form

$$1 \cdot x + 0 \cdot y = a \qquad A = 1, B = 0, C = a$$

A horizontal line whose equation is

$$y = b$$

can be written in the general form

$$0 \cdot x + 1 \cdot y = b \qquad A = 0, B = 1, C = b$$

*Some texts use the term **standard form**.

Lines that are neither vertical nor horizontal have general equations of the form

$$Ax + By = C \qquad A \neq 0 \text{ and } B \neq 0$$

Because the equation of every line can be written in general form, any equation equivalent to equation (4) is called a **linear equation**.

9 Find Equations of Parallel Lines

When two lines (in the plane) do not intersect (that is, they have no points in common), they are **parallel**. Look at Figure 43. There we have drawn two parallel lines and have constructed two right triangles by drawing sides parallel to the coordinate axes. The right triangles are similar. (Do you see why? Two angles are equal.) Because the triangles are similar, the ratios of corresponding sides are equal.

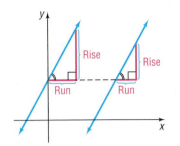

Figure 43 Parallel lines

> **THEOREM Criteria for Parallel Lines**
>
> Two nonvertical lines are parallel if and only if their slopes are equal and they have different y-intercepts.

The use of the phrase "if and only if" in the preceding theorem means that actually two statements are being made, one the converse of the other.

> If two nonvertical lines are parallel, then their slopes are equal and they have different y-intercepts.
>
> If two nonvertical lines have equal slopes and they have different y-intercepts, then they are parallel.

EXAMPLE 10

Showing That Two Lines Are Parallel

Show that the lines given by the following equations are parallel.

$$L_1: \ 2x + 3y = 6 \qquad L_2: \ 4x + 6y = 0$$

Solution

To determine whether these lines have equal slopes and different y-intercepts, write each equation in slope–intercept form.

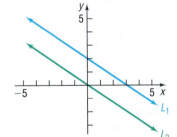

Figure 44

$$
\begin{array}{ll}
L_1: \ 2x + 3y = 6 & L_2: \ 4x + 6y = 0 \\
\quad\quad 3y = -2x + 6 & \quad\quad 6y = -4x \\
\quad\quad\quad y = -\dfrac{2}{3}x + 2 & \quad\quad\quad y = -\dfrac{2}{3}x
\end{array}
$$

$$\text{Slope} = -\frac{2}{3}; \quad y\text{-intercept} = 2 \qquad \text{Slope} = -\frac{2}{3}; \quad y\text{-intercept} = 0$$

Because these lines have the same slope, $-\dfrac{2}{3}$, but different y-intercepts, the lines are parallel. See Figure 44.

●

EXAMPLE 11

Finding a Line That Is Parallel to a Given Line

Find an equation for the line that contains the point $(2, -3)$ and is parallel to the line $2x + y = 6$.

Solution

Since the two lines are to be parallel, the slope of the line being sought equals the slope of the line $2x + y = 6$. Begin by writing the equation of the line $2x + y = 6$ in slope–intercept form.

$$
\begin{aligned}
2x + y &= 6 \\
y &= -2x + 6
\end{aligned}
$$

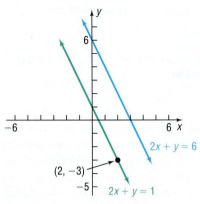

Figure 45

The slope is -2. Since the line being sought also has slope -2 and contains the point $(2, -3)$, use the point–slope form to obtain its equation.

$$y - y_1 = m(x - x_1) \quad \text{Point–slope form}$$
$$y - (-3) = -2(x - 2) \quad m = -2, x_1 = 2, y_1 = -3$$
$$y + 3 = -2x + 4 \quad \text{Simplify.}$$
$$y = -2x + 1 \quad \text{Slope–intercept form}$$
$$2x + y = 1 \quad \text{General form}$$

This line is parallel to the line $2x + y = 6$ and contains the point $(2, -3)$. See Figure 45.

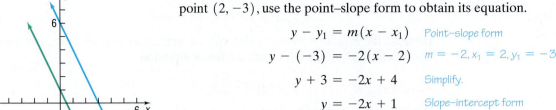
Now Work PROBLEM 61

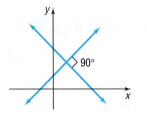

Figure 46 Perpendicular lines

10 Find Equations of Perpendicular Lines

When two lines intersect at a right angle (90°), they are **perpendicular**. See Figure 46.

The following result gives a condition, in terms of their slopes, for two lines to be perpendicular.

THEOREM

Criterion for Perpendicular Lines

Two nonvertical lines are perpendicular if and only if the product of their slopes is -1.

Here we shall prove the "only if" part of the statement:

If two nonvertical lines are perpendicular, then the product of their slopes is -1.

In Problem 130 you are asked to prove the "if" part of the theorem:

If two nonvertical lines have slopes whose product is -1, then the lines are perpendicular.

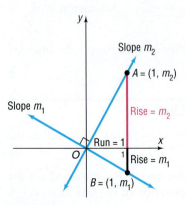

Figure 47

Proof Let m_1 and m_2 denote the slopes of the two lines. There is no loss in generality (that is, neither the angle nor the slopes are affected) if we situate the lines so that they meet at the origin. See Figure 47. The point $A = (1, m_2)$ is on the line having slope m_2, and the point $B = (1, m_1)$ is on the line having slope m_1. (Do you see why this must be true?)

Suppose that the lines are perpendicular. Then triangle OAB is a right triangle. As a result of the Pythagorean Theorem, it follows that

$$[d(O, A)]^2 + [d(O, B)]^2 = [d(A, B)]^2 \qquad (5)$$

Using the distance formula, the squares of these distances are

$$[d(O, A)]^2 = (1 - 0)^2 + (m_2 - 0)^2 = 1 + m_2^2$$
$$[d(O, B)]^2 = (1 - 0)^2 + (m_1 - 0)^2 = 1 + m_1^2$$
$$[d(A, B)]^2 = (1 - 1)^2 + (m_2 - m_1)^2 = m_2^2 - 2m_1m_2 + m_1^2$$

Using these facts in equation (5), we get

$$\left(1 + m_2^2\right) + \left(1 + m_1^2\right) = m_2^2 - 2m_1m_2 + m_1^2$$

which, upon simplification, can be written as

$$m_1m_2 = -1$$

If the lines are perpendicular, the product of their slopes is -1. ∎

You may find it easier to remember the condition for two nonvertical lines to be perpendicular by observing that the equality $m_1m_2 = -1$ means that m_1 and m_2 are negative reciprocals of each other; that is, $m_1 = -\dfrac{1}{m_2}$ and $m_2 = -\dfrac{1}{m_1}$.

EXAMPLE 12

Finding the Slope of a Line Perpendicular to Another Line

If a line has slope $\dfrac{3}{2}$, any line having slope $-\dfrac{2}{3}$ is perpendicular to it. ●

EXAMPLE 13

Finding the Equation of a Line Perpendicular to a Given Line

Find an equation of the line that contains the point $(1, -2)$ and is perpendicular to the line $x + 3y = 6$. Graph the two lines.

Solution

First write the equation of the given line in slope–intercept form to find its slope.

$$x + 3y = 6$$
$$3y = -x + 6 \qquad \textit{Proceed to solve for y.}$$
$$y = -\frac{1}{3}x + 2 \qquad \textit{Place in the form } y = mx + b.$$

The given line has slope $-\dfrac{1}{3}$. Any line perpendicular to this line will have slope 3.

Because the point $(1, -2)$ is on this line with slope 3, use the point–slope form of the equation of a line.

$$y - y_1 = m(x - x_1) \qquad \textit{Point–slope form}$$
$$y - (-2) = 3(x - 1) \qquad \textit{m = 3, } x_1 = 1, y_1 = -2$$
$$y + 2 = 3(x - 1)$$

To obtain other forms of the equation, proceed as follows:

$$y + 2 = 3x - 3 \qquad \textit{Simplify.}$$
$$y = 3x - 5 \qquad \textit{Slope-intercept form}$$
$$3x - y = 5 \qquad \textit{General form}$$

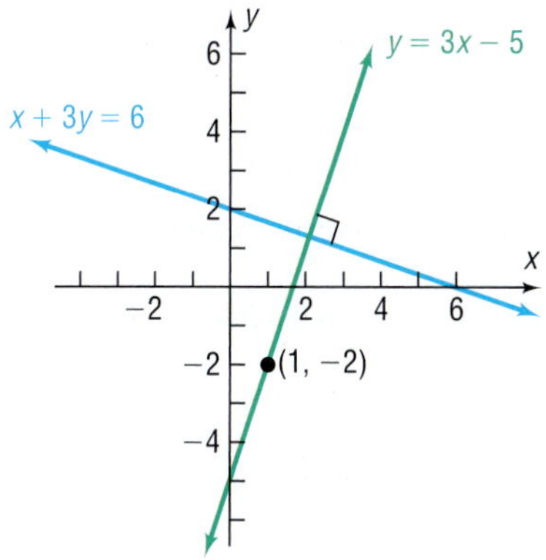

Figure 48 shows the graphs. ●

Figure 48

Now Work PROBLEM 67

WARNING Be sure to use a square screen when you use a graphing calculator to graph perpendicular lines. Otherwise, the angle between the two lines will appear distorted. A discussion of square screens is given in Section 5 of the Appendix. ∎

2.3 Assess Your Understanding

Concepts and Vocabulary

1. The slope of a vertical line is _____; the slope of a horizontal line is ____.

2. For the line $2x + 3y = 6$, the x-intercept is ____ and the y-intercept is ____.

3. **True or False** The equation $3x + 4y = 6$ is written in general form.

4. **True or False** The slope of the line $2y = 3x + 5$ is 3.

5. **True or False** The point $(1, 2)$ is on the line $2x + y = 4$.

6. Two nonvertical lines have slopes m_1 and m_2, respectively. The lines are parallel if _____ and the _____ are unequal; the lines are perpendicular if _____.

7. The lines $y = 2x + 3$ and $y = ax + 5$ are parallel if $a = $ ____.

8. The lines $y = 2x - 1$ and $y = ax + 2$ are perpendicular if $a = $ ____.

9. **True or False** Perpendicular lines have slopes that are reciprocals of one another.

10. Choose the formula for finding the slope m of a nonvertical line that contains the two distinct points (x_1, y_1) and (x_2, y_2).
    (a) $m = \dfrac{y_2 - x_2}{y_1 - x_1} \quad x_1 \neq y_1$
    (b) $m = \dfrac{y_2 - x_1}{x_2 - y_1} \quad y_1 \neq x_2$
    (c) $m = \dfrac{x_2 - x_1}{y_2 - y_1} \quad y_1 \neq y_2$
    (d) $m = \dfrac{y_2 - y_1}{x_2 - x_1} \quad x_1 \neq x_2$

11. If a line slants downward from left to right, then which of the following describes its slope?
    (a) positive   (b) zero
    (c) negative   (d) undefined

12. Choose the correct statement about the graph of the line $y = -3$.
    (a) The graph is vertical with x-intercept -3.
    (b) The graph is horizontal with y-intercept -3.
    (c) The graph is vertical with y-intercept -3.
    (d) The graph is horizontal with x-intercept -3.

Skill Building

In Problems 13–16, (a) find the slope of the line and (b) interpret the slope.

13.

14.

15.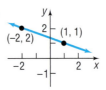

16.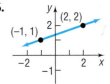

In Problems 17–24, plot each pair of points and determine the slope of the line containing them. Graph the line.

17. $(2, 3)$; $(4, 0)$

18. $(4, 2)$; $(3, 4)$

19. $(-2, 3)$; $(2, 1)$

20. $(-1, 1)$; $(2, 3)$

21. $(-3, -1)$; $(2, -1)$

22. $(4, 2)$; $(-5, 2)$

23. $(-1, 2)$; $(-1, -2)$

24. $(2, 0)$; $(2, 2)$

In Problems 25–32, graph the line that contains the point P and has slope m. In Problems 25–30, find the point-slope form of the equation of the line. In Problems 31 and 32, find an equation of the line.

25. $P = (1, 2)$; $m = 3$

26. $P = (2, 1)$; $m = 4$

27. $P = (2, 4)$; $m = -\dfrac{3}{4}$

28. $P = (1, 3)$; $m = -\dfrac{2}{5}$

29. $P = (-1, 3)$; $m = 0$

30. $P = (2, -4)$; $m = 0$

31. $P = (0, 3)$; slope undefined

32. $P = (-2, 0)$; slope undefined

In Problems 33–38, the slope and a point on a line are given. Use this information to locate three additional points on the line. Answers may vary. [**Hint:** *It is not necessary to find the equation of the line. See Example 3.*]

33. Slope 4; point $(1, 2)$

34. Slope 2; point $(-2, 3)$

35. Slope $-\dfrac{3}{2}$; point $(2, -4)$

36. Slope $\dfrac{4}{3}$; point $(-3, 2)$

37. Slope -2; point $(-2, -3)$

38. Slope -1; point $(4, 1)$

In Problems 39–46, find an equation of the line L.

39.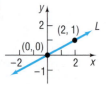

40.

41.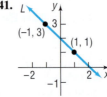

42.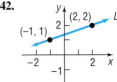

43.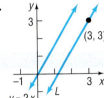

L is parallel to $y = 2x$

44.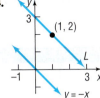

L is parallel to $y = -x$

45.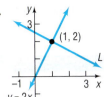

L is perpendicular to $y = 2x$

46.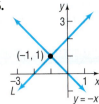

L is perpendicular to $y = -x$

In Problems 47–72, find an equation for the line with the given properties. Express your answer using either the general form or the slope–intercept form of the equation of a line, whichever you prefer.

47. Slope = 3; containing the point $(-2, 3)$

48. Slope = 2; containing the point $(4, -3)$

49. Slope = $-\dfrac{2}{3}$; containing the point $(1, -1)$

50. Slope = $\dfrac{1}{2}$; containing the point $(3, 1)$

51. Containing the points $(1, 3)$ and $(-1, 2)$

52. Containing the points $(-3, 4)$ and $(2, 5)$

53. Slope = -3; y-intercept = 3

54. Slope = -2; y-intercept = -2

55. x-intercept = 2; y-intercept = -1

56. x-intercept = -4; y-intercept = 4

57. Slope undefined; containing the point $(2, 4)$

58. Slope undefined; containing the point $(3, 8)$

59. Horizontal; containing the point $(-3, 2)$

60. Vertical; containing the point $(4, -5)$

61. Parallel to the line $y = 2x$; containing the point $(-1, 2)$

62. Parallel to the line $y = -3x$; containing the point $(-1, 2)$

63. Parallel to the line $2x - y = -2$; containing the point $(0, 0)$

64. Parallel to the line $x - 2y = -5$; containing the point $(0, 0)$

65. Parallel to the line $x = 5$; containing the point $(4, 2)$

66. Parallel to the line $y = 5$; containing the point $(4, 2)$

67. Perpendicular to the line $y = \dfrac{1}{2}x + 4$; containing the point $(1, -2)$

68. Perpendicular to the line $y = 2x - 3$; containing the point $(1, -2)$

69. Perpendicular to the line $2x + y = 2$; containing the point $(-3, 0)$

70. Perpendicular to the line $x - 2y = -5$; containing the point $(0, 4)$

71. Perpendicular to the line $x = 8$; containing the point $(3, 4)$

72. Perpendicular to the line $y = 8$; containing the point $(3, 4)$

In Problems 73–92, find the slope and y-intercept of each line. Graph the line.

73. $y = 2x + 3$

74. $y = -3x + 4$

75. $\dfrac{1}{2}y = x - 1$

76. $\dfrac{1}{3}x + y = 2$

77. $y = \dfrac{1}{2}x + 2$

78. $y = 2x + \dfrac{1}{2}$

79. $x + 2y = 4$

80. $-x + 3y = 6$

81. $2x - 3y = 6$

82. $3x + 2y = 6$

83. $x + y = 1$

84. $x - y = 2$

85. $x = -4$

86. $y = -1$

87. $y = 5$

88. $x = 2$

89. $y - x = 0$

90. $x + y = 0$

91. $2y - 3x = 0$

92. $3x + 2y = 0$

In Problems 93–102, (a) find the intercepts of the graph of each equation and (b) graph the equation.

93. $2x + 3y = 6$

94. $3x - 2y = 6$

95. $-4x + 5y = 40$

96. $6x - 4y = 24$

97. $7x + 2y = 21$

98. $5x + 3y = 18$

99. $\dfrac{1}{2}x + \dfrac{1}{3}y = 1$

100. $x - \dfrac{2}{3}y = 4$

101. $0.2x - 0.5y = 1$

102. $-0.3x + 0.4y = 1.2$

103. Find an equation of the x-axis.

104. Find an equation of the y-axis.

In Problems 105–108, the equations of two lines are given. Determine whether the lines are parallel, perpendicular, or neither.

105. $y = 2x - 3$
$y = 2x + 4$

106. $y = \dfrac{1}{2}x - 3$
$y = -2x + 4$

107. $y = 4x + 5$
$y = -4x + 2$

108. $y = -2x + 3$
$y = -\dfrac{1}{2}x + 2$

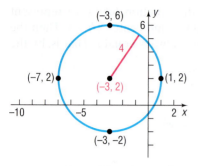

Figure 51 $(x + 3)^2 + (y - 2)^2 = 16$

$$(x + 3)^2 + (y - 2)^2 = 16$$
$$(x - (-3))^2 + (y - 2)^2 = 4^2$$
$$(x - h)^2 + (y - k)^2 = r^2$$

We see that $h = -3$, $k = 2$, and $r = 4$. The circle has center $(-3, 2)$ and a radius of 4 units. To graph this circle, first plot the center $(-3, 2)$. Since the radius is 4, we can locate four points on the circle by plotting points 4 units to the left, to the right, up, and down from the center. These four points can then be used as guides to obtain the graph. See Figure 51.

Now Work PROBLEMS 25(a) AND (b)

EXAMPLE 3

Finding the Intercepts of a Circle

For the circle $(x + 3)^2 + (y - 2)^2 = 16$, find the intercepts, if any, of its graph.

Solution

This is the equation discussed and graphed in Example 2. To find the x-intercepts, if any, let $y = 0$. Then

$$(x + 3)^2 + (y - 2)^2 = 16$$
$$(x + 3)^2 + (0 - 2)^2 = 16 \qquad y = 0$$
$$(x + 3)^2 + 4 = 16 \qquad \text{Simplify.}$$
$$(x + 3)^2 = 12 \qquad \text{Simplify.}$$
$$x + 3 = \pm\sqrt{12} \qquad \text{Apply the Square Root Method.}$$
$$x = -3 \pm 2\sqrt{3} \qquad \text{Solve for } x.$$

> **In Words**
> The symbol $\pm$ is read "plus or minus." It means to add and subtract the quantity following the $\pm$ symbol. For example, 5 ± 2 means "$5 - 2 = 3$ or $5 + 2 = 7$."

The x-intercepts are $-3 - 2\sqrt{3} \approx -6.46$ and $-3 + 2\sqrt{3} \approx 0.46$.

To find the y-intercepts, if any, let $x = 0$. Then

$$(x + 3)^2 + (y - 2)^2 = 16$$
$$(0 + 3)^2 + (y - 2)^2 = 16$$
$$9 + (y - 2)^2 = 16$$
$$(y - 2)^2 = 7$$
$$y - 2 = \pm\sqrt{7}$$
$$y = 2 \pm \sqrt{7}$$

The y-intercepts are $2 - \sqrt{7} \approx -0.65$ and $2 + \sqrt{7} \approx 4.65$.

Look back at Figure 51 to verify the approximate locations of the intercepts.

Now Work PROBLEM 25(c)

3 Work with the General Form of the Equation of a Circle

If we eliminate the parentheses from the standard form of the equation of the circle given in Example 2, we get

$$(x + 3)^2 + (y - 2)^2 = 16$$
$$x^2 + 6x + 9 + y^2 - 4y + 4 = 16$$

which simplifies to

$$x^2 + y^2 + 6x - 4y - 3 = 0 \qquad (2)$$

It can be shown that any equation of the form

$$x^2 + y^2 + ax + by + c = 0$$

has a graph that is a circle, is a point, or has no graph at all. For example, the graph of the equation $x^2 + y^2 = 0$ is the single point $(0, 0)$. The equation $x^2 + y^2 + 5 = 0$, or $x^2 + y^2 = -5$, has no graph, because sums of squares of real numbers are never negative.

DEFINITION

When its graph is a circle, the equation

$$x^2 + y^2 + ax + by + c = 0$$

is the **general form of the equation of a circle.**

―――― **Now Work** PROBLEM 15

If an equation of a circle is in general form, we use the method of completing the square to put the equation in standard form so that we can identify its center and radius.

EXAMPLE 4

Graphing a Circle Whose Equation Is in General Form

Graph the equation: $x^2 + y^2 + 4x - 6y + 12 = 0$

Solution

Group the terms involving x, group the terms involving y, and put the constant on the right side of the equation. The result is

$$(x^2 + 4x) + (y^2 - 6y) = -12$$

Next, complete the square of each expression in parentheses. Remember that any number added on the left side of the equation must also be added on the right.

$$(x^2 + 4x + 4) + (y^2 - 6y + 9) = -12 + 4 + 9$$

$$\left(\frac{4}{2}\right)^2 = 4 \qquad \left(\frac{-6}{2}\right)^2 = 9$$

$$(x + 2)^2 + (y - 3)^2 = 1 \quad \text{Factor.}$$

This equation is the standard form of the equation of a circle with radius 1 and center $(-2, 3)$. To graph the equation, use the center $(-2, 3)$ and the radius 1. See Figure 52.

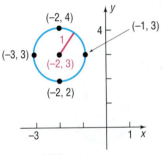

Figure 52 $(x + 2)^2 + (y - 3)^2 = 1$

―――― **Now Work** PROBLEM 29

 EXAMPLE 5

Using a Graphing Utility to Graph a Circle

Graph the equation: $x^2 + y^2 = 4$

Solution

This is the equation of a circle with center at the origin and radius 2. To graph this equation, solve for y.

$$x^2 + y^2 = 4$$
$$y^2 = 4 - x^2 \qquad \text{Subtract } x^2 \text{ from each side.}$$
$$y = \pm\sqrt{4 - x^2} \quad \text{Apply the Square Root Method to solve for } y.$$

There are two equations to graph: first graph $Y_1 = \sqrt{4 - x^2}$ and then graph $Y_2 = -\sqrt{4 - x^2}$ on the same square screen. (Your circle will appear oval if you do not use a square screen.*) See Figure 53.

$Y_1 = \sqrt{4-x^2}$

$Y_2 = -\sqrt{4-x^2}$

Figure 53 $x^2 + y^2 = 4$

Overview

The discussion in Sections 2.3 and 2.4 about lines and circles dealt with two main types of problems that can be generalized as follows:

1. Given an equation, classify it and graph it.

2. Given a graph, or information about a graph, find its equation.

This text deals with both types of problems. We shall study various equations, classify them, and graph them. The second type of problem is usually more difficult to solve than the first.

*The square screen ratio for the TI-84 Plus C calculator is 8:5.

2.4 Assess Your Understanding

Are You Prepared? *Answers are given at the end of these exercises. If you get a wrong answer, read the pages listed in* red.

1. To complete the square of $x^2 + 10x$, you would _____ (*add/subtract*) the number _____. (p. 56)

2. Use the Square Root Method to solve the equation $(x - 2)^2 = 9$. (pp. 94–95)

Concepts and Vocabulary

3. *True or False* Every equation of the form
$$x^2 + y^2 + ax + by + c = 0$$
has a circle as its graph.

4. For a circle, the _____ is the distance from the center to any point on the circle.

5. *True or False* The radius of the circle $x^2 + y^2 = 9$ is 3.

6. *True or False* The center of the circle
$$(x + 3)^2 + (y - 2)^2 = 13$$
is $(3, -2)$.

7. Choose the equation of a circle with radius 6 and center $(3, -5)$.
  (a) $(x - 3)^2 + (y + 5)^2 = 6$
  (b) $(x + 3)^2 + (y - 5)^2 = 36$
  (c) $(x + 3)^2 + (y - 5)^2 = 6$
  (d) $(x - 3)^2 + (y + 5)^2 = 36$

8. The equation of a circle can be changed from general form to standard from by doing which of the following?
  (a) completing the squares
  (b) solving for x
  (c) solving for y
  (d) squaring both sides

Skill Building

In Problems 9–12, find the center and radius of each circle. Write the standard form of the equation.

9.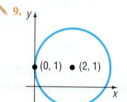

10.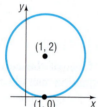

11.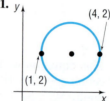

12.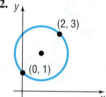

In Problems 13–22, write the standard form of the equation and the general form of the equation of each circle of radius r and center (h, k). Graph each circle.

13. $r = 2$; $(h, k) = (0, 0)$
14. $r = 3$; $(h, k) = (0, 0)$
15. $r = 2$; $(h, k) = (0, 2)$
16. $r = 3$; $(h, k) = (1, 0)$

17. $r = 5$; $(h, k) = (4, -3)$
18. $r = 4$; $(h, k) = (2, -3)$
19. $r = 4$; $(h, k) = (-2, 1)$
20. $r = 7$; $(h, k) = (-5, -2)$

21. $r = \dfrac{1}{2}$; $(h, k) = \left(\dfrac{1}{2}, 0\right)$

22. $r = \dfrac{1}{2}$; $(h, k) = \left(0, -\dfrac{1}{2}\right)$

In Problems 23–36, (a) find the center (h, k) and radius r of each circle; (b) graph each circle; (c) find the intercepts, if any.

23. $x^2 + y^2 = 4$
24. $x^2 + (y - 1)^2 = 1$
25. $2(x - 3)^2 + 2y^2 = 8$

26. $3(x + 1)^2 + 3(y - 1)^2 = 6$
27. $x^2 + y^2 - 2x - 4y - 4 = 0$
28. $x^2 + y^2 + 4x + 2y - 20 = 0$

29. $x^2 + y^2 + 4x - 4y - 1 = 0$
30. $x^2 + y^2 - 6x + 2y + 9 = 0$
31. $x^2 + y^2 - x + 2y + 1 = 0$

32. $x^2 + y^2 + x + y - \dfrac{1}{2} = 0$
33. $2x^2 + 2y^2 - 12x + 8y - 24 = 0$
34. $2x^2 + 2y^2 + 8x + 7 = 0$

35. $2x^2 + 8x + 2y^2 = 0$
36. $3x^2 + 3y^2 - 12y = 0$

In Problems 37–44, find the standard form of the equation of each circle.

37. Center at the origin and containing the point $(-2, 3)$

38. Center $(1, 0)$ and containing the point $(-3, 2)$

39. Center $(2, 3)$ and tangent to the x-axis

40. Center $(-3, 1)$ and tangent to the y-axis

41. With endpoints of a diameter at $(1, 4)$ and $(-3, 2)$

42. With endpoints of a diameter at $(4, 3)$ and $(0, 1)$

43. Center $(-1, 3)$ and tangent to the line $y = 2$

44. Center $(4, -2)$ and tangent to the line $x = 1$

In Problems 45–48, match each graph with the correct equation.

(a) $(x - 3)^2 + (y + 3)^2 = 9$ (b) $(x + 1)^2 + (y - 2)^2 = 4$ (c) $(x - 1)^2 + (y + 2)^2 = 4$ (d) $(x + 3)^2 + (y - 3)^2 = 9$

45.
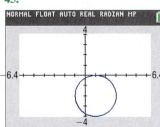

46.
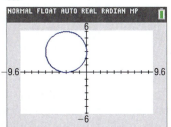

47.
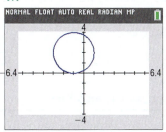

48.
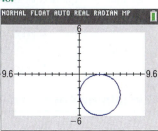

Applications and Extensions

49. Find the area of the square in the figure.

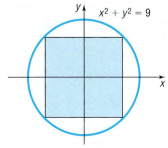

$x^2 + y^2 = 9$

50. Find the area of the blue shaded region in the figure, assuming the quadrilateral inside the circle is a square.

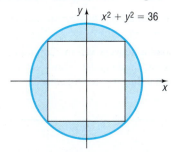

$x^2 + y^2 = 36$

51. Ferris Wheel The original Ferris wheel was built in 1893 by Pittsburgh, Pennsylvania bridge builder George W. Ferris. The Ferris wheel was originally built for the 1893 World's Fair in Chicago, but it was also later reconstructed for the 1904 World's Fair in St. Louis. It had a maximum height of 264 feet and a wheel diameter of 250 feet. Find an equation for the wheel if the center of the wheel is on the y-axis.

Source: guinnessworldrecords.com

52. Ferris Wheel Opening in 2014 in Las Vegas, The High Roller observation wheel has a maximum height of 550 feet and a diameter of 520 feet, with one full rotation taking approximately 30 minutes. Find an equation for the wheel if the center of the wheel is on the y-axis.

Source: Las Vegas Review Journal

53. Weather Satellites Earth is represented on a map of a portion of the solar system so that its surface is the circle with equation $x^2 + y^2 + 2x + 4y - 4091 = 0$. A weather satellite circles 0.6 unit above Earth with the center of its circular orbit at the center of Earth. Find the equation for the orbit of the satellite on this map.

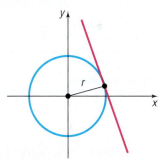

54. The **tangent line** to a circle may be defined as the line that intersects the circle in a single point, called the **point of tangency**. See the figure.

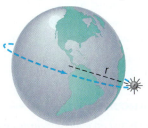

If the equation of the circle is $x^2 + y^2 = r^2$ and the equation of the tangent line is $y = mx + b$, show that:

(a) $r^2(1 + m^2) = b^2$

   [**Hint:** The quadratic equation $x^2 + (mx + b)^2 = r^2$ has exactly one solution.]

(b) The point of tangency is $\left(\dfrac{-r^2 m}{b}, \dfrac{r^2}{b} \right)$.

(c) The tangent line is perpendicular to the line containing the center of the circle and the point of tangency.

55. The Greek Method The Greek method for finding the equation of the tangent line to a circle uses the fact that at any point on a circle, the lines containing the center and the tangent line are perpendicular (see Problem 54). Use this method to find an equation of the tangent line to the circle $x^2 + y^2 = 9$ at the point $(1, 2\sqrt{2})$.

56. Use the Greek method described in Problem 55 to find an equation of the tangent line to the circle $x^2 + y^2 - 4x + 6y + 4 = 0$ at the point $(3, 2\sqrt{2} - 3)$.

57. Refer to Problem 54. The line $x - 2y + 4 = 0$ is tangent to a circle at $(0, 2)$. The line $y = 2x - 7$ is tangent to the same circle at $(3, -1)$. Find the center of the circle.

58. Find an equation of the line containing the centers of the two circles

$$x^2 + y^2 - 4x + 6y + 4 = 0$$

and

$$x^2 + y^2 + 6x + 4y + 9 = 0$$

59. If a circle of radius 2 is made to roll along the x-axis, what is an equation for the path of the center of the circle?

60. If the circumference of a circle is 6π, what is its radius?

Explaining Concepts: Discussion and Writing

61. Which of the following equations might have the graph shown? (More than one answer is possible.)
(a) $(x - 2)^2 + (y + 3)^2 = 13$
(b) $(x - 2)^2 + (y - 2)^2 = 8$
(c) $(x - 2)^2 + (y - 3)^2 = 13$
(d) $(x + 2)^2 + (y - 2)^2 = 8$
(e) $x^2 + y^2 - 4x - 9y = 0$
(f) $x^2 + y^2 + 4x - 2y = 0$
(g) $x^2 + y^2 - 9x - 4y = 0$
(h) $x^2 + y^2 - 4x - 4y = 4$

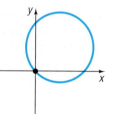

62. Which of the following equations might have the graph shown? (More than one answer is possible.)
(a) $(x - 2)^2 + y^2 = 3$
(b) $(x + 2)^2 + y^2 = 3$
(c) $x^2 + (y - 2)^2 = 3$
(d) $(x + 2)^2 + y^2 = 4$
(e) $x^2 + y^2 + 10x + 16 = 0$
(f) $x^2 + y^2 + 10x - 2y = 1$
(g) $x^2 + y^2 + 9x + 10 = 0$
(h) $x^2 + y^2 - 9x - 10 = 0$

63. Explain how the center and radius of a circle can be used to graph the circle.

64. **What Went Wrong?** A student stated that the center and radius of the graph whose equation is $(x + 3)^2 + (y - 2)^2 = 16$ are $(3, -2)$ and 4, respectively. Why is this incorrect?

Retain Your Knowledge

Problems 65–68 are based on material learned earlier in the course. The purpose of these problems is to keep the material fresh in your mind so that you are better prepared for the final exam.

65. Find the area and circumference of a circle of radius 13 cm.

66. Multiply $(3x - 2)(x^2 - 2x + 3)$. Express the answer as a polynomial in standard form.

67. Solve the equation: $\sqrt{2x^2 + 3x - 1} = x + 1$

68. Aaron can load a delivery van in 22 minutes. Elizabeth can load the same van in 28 minutes. How long would it take them to load the van if they worked together?

'Are You Prepared?' Answers

1. add; 25   **2.** $\{-1, 5\}$

2.5 Variation

OBJECTIVES 1 Construct a Model Using Direct Variation (p. 189)
              2 Construct a Model Using Inverse Variation (p. 189)
              3 Construct a Model Using Joint Variation or Combined Variation (p. 190)

 When a mathematical model is developed for a real-world problem, it often involves relationships between quantities that are expressed in terms of proportionality:

Force is proportional to acceleration.

When an ideal gas is held at a constant temperature, pressure and volume are inversely proportional.

The force of attraction between two heavenly bodies is inversely proportional to the square of the distance between them.

Revenue is directly proportional to sales.

Each of the preceding statements illustrates the idea of **variation**, or how one quantity varies in relation to another quantity. Quantities may vary *directly*, *inversely*, or *jointly*.

1 Construct a Model Using Direct Variation

DEFINITION

Let x and y denote two quantities. Then y **varies directly** with x, or y is **directly proportional to** x, if there is a nonzero number k such that

$$y = kx$$

The number k is called the **constant of proportionality**.

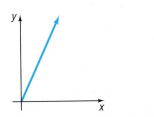

Figure 54 $y = kx; k > 0, x \geq 0$

The graph in Figure 54 illustrates the relationship between y and x if y varies directly with x and $k > 0, x \geq 0$. Note that the constant of proportionality is, in fact, the slope of the line.

If two quantities vary directly, then knowing the value of each quantity in one instance enables us to write a formula that is true in all cases.

EXAMPLE 1 **Mortgage Payments**

The monthly payment p on a mortgage varies directly with the amount borrowed B. If the monthly payment on a 30-year mortgage is \$6.65 for every \$1000 borrowed, find a formula that relates the monthly payment p to the amount borrowed B for a mortgage with these terms. Then find the monthly payment p when the amount borrowed B is \$120,000.

Solution Because p varies directly with B, we know that

$$p = kB$$

for some constant k. Because $p = 6.65$ when $B = 1000$, it follows that

$$6.65 = k(1000)$$

$$k = 0.00665 \quad \text{Solve for } k.$$

Since $p = kB$,

$$p = 0.00665B$$

In particular, when $B = \$120,000$,

$$p = 0.00665(\$120,000) = \$798$$

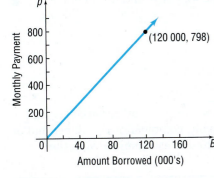

Figure 55

Figure 55 illustrates the relationship between the monthly payment p and the amount borrowed B.

━━ **Now Work** PROBLEMS **5 AND 23**

2 Construct a Model Using Inverse Variation

DEFINITION

Let x and y denote two quantities. Then y **varies inversely** with x, or y is **inversely proportional to** x, if there is a nonzero constant k such that

$$y = \frac{k}{x}$$

Figure 56 $y = \dfrac{k}{x}; k > 0, x > 0$

The graph in Figure 56 illustrates the relationship between y and x if y varies inversely with x and $k > 0, x > 0$.

EXAMPLE 2

Maximum Weight That Can Be Supported by a Piece of Pine

See Figure 57. The maximum weight W that can be safely supported by a 2-inch by 4-inch piece of pine varies inversely with its length l. Experiments indicate that the maximum weight that a 10-foot-long 2-by-4 piece of pine can support is 500 pounds. Write a general formula relating the maximum weight W (in pounds) to length l (in feet). Find the maximum weight W that can be safely supported by a length of 25 feet.

Solution Because W varies inversely with l, we know that

$$W = \frac{k}{l}$$

for some constant k. Because $W = 500$ when $l = 10$, we have

$$500 = \frac{k}{10}$$

$$k = 5000$$

Since $W = \frac{k}{l}$,

$$W = \frac{5000}{l}$$

In particular, the maximum weight W that can be safely supported by a piece of pine 25 feet in length is

$$W = \frac{5000}{25} = 200 \text{ pounds}$$

Figure 58 illustrates the relationship between the weight W and the length l.

Figure 57

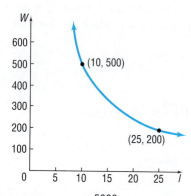

Figure 58 $W = \dfrac{5000}{l}$

(10, 500)

(25, 200)

Now Work PROBLEM 33

3 Construct a Model Using Joint Variation or Combined Variation

When a variable quantity Q is proportional to two or more other variables, we say that Q **varies jointly** with these quantities. Finally, combinations of direct and/or inverse variation may occur. This is usually referred to as **combined variation**.

EXAMPLE 3

Loss of Heat through a Wall

The loss of heat through a wall varies jointly with the area of the wall and the difference between the inside and outside temperatures and varies inversely with the thickness of the wall. Write an equation that relates these quantities.

Solution Begin by assigning symbols to represent the quantities:

$$L = \text{Heat loss} \qquad T = \text{Temperature difference}$$
$$A = \text{Area of wall} \qquad d = \text{Thickness of wall}$$

Then

$$L = k\frac{AT}{d}$$

where k is the constant of proportionality.

In direct or inverse variation, the quantities that vary may be raised to powers. For example, in the early seventeenth century, Johannes Kepler (1571–1630) discovered that the square of the period of revolution T of a planet around the Sun varies directly with the cube of its mean distance a from the Sun. That is, $T^2 = ka^3$, where k is the constant of proportionality.

EXAMPLE 4   **Force of the Wind on a Window**

The force F of the wind on a flat surface positioned at a right angle to the direction of the wind varies jointly with the area A of the surface and the square of the speed v of the wind. A wind of 30 miles per hour blowing on a window measuring 4 feet by 5 feet has a force of 150 pounds. See Figure 59. What force does a wind of 50 miles per hour exert on a window measuring 3 feet by 4 feet?

Solution   Since F varies jointly with A and v^2, we have

$$F = kAv^2$$

where k is the constant of proportionality. We are told that $F = 150$ when $A = 4 \cdot 5 = 20$ and $v = 30$. Then

$$150 = k(20)(900) \quad \text{\textcolor{blue}{$F = kAv^2, F = 150, A = 20, v = 30$}}$$

$$k = \frac{1}{120}$$

Since $F = kAv^2$,

$$F = \frac{1}{120}Av^2$$

For a wind of 50 miles per hour blowing on a window whose area is $A = 3 \cdot 4 = 12$ square feet, the force F is

$$F = \frac{1}{120}(12)(2500) = 250 \text{ pounds}$$

Figure 59

 Now Work **PROBLEM 41**

2.5 Assess Your Understanding

Concepts and Vocabulary

1. If x and y are two quantities, then y is directly proportional to x if there is a nonzero number k such that _____.

2. *True or False* If y varies directly with x, then $y = \dfrac{k}{x}$, where k is a constant.

3. Which equation represents a joint variation model?
  (a) $y = 5x$ (b) $y = 5xzw$
  (c) $y = \dfrac{5}{x}$ (d) $y = \dfrac{5xz}{w}$

4. Choose the best description for the model $y = \dfrac{kx}{z}$, if k is a nonzero constant.
  (a) y varies jointly with x and z.
  (b) y is inversely proportional to x and z.
  (c) y varies directly with x and inversely with z.
  (d) y is directly proportion to z and inversely proportional to x.

Skill Building

In Problems 5–16, write a general formula to describe each variation.

5. y varies directly with x; $y = 2$ when $x = 10$

6. v varies directly with t; $v = 16$ when $t = 2$

7. A varies directly with x^2; $A = 4\pi$ when $x = 2$

8. V varies directly with x^3; $V = 36\pi$ when $x = 3$

9. F varies inversely with d^2; $F = 10$ when $d = 5$

10. y varies inversely with $\sqrt{x}$; $y = 4$ when $x = 9$

11. z varies directly with the sum of the squares of x and y; $z = 5$ when $x = 3$ and $y = 4$

12. T varies jointly with the cube root of x and the square of d; $T = 18$ when $x = 8$ and $d = 3$

13. M varies directly with the square of d and inversely with the square root of x; $M = 24$ when $x = 9$ and $d = 4$

14. z varies directly with the sum of the cube of x and the square of y; $z = 1$ when $x = 2$ and $y = 3$

15. The square of T varies directly with the cube of a and inversely with the square of d; $T = 2$ when $a = 2$ and $d = 4$

16. The cube of z varies directly with the sum of the squares of x and y; $z = 2$ when $x = 9$ and $y = 4$

Applications and Extensions

In Problems 17–22, write an equation that relates the quantities.

17. Geometry The volume V of a sphere varies directly with the cube of its radius r. The constant of proportionality is $\dfrac{4\pi}{3}$.

18. Geometry The square of the length of the hypotenuse c of a right triangle varies jointly with the sum of the squares of the lengths of its legs a and b. The constant of proportionality is 1.

19. Geometry The area A of a triangle varies jointly with the product of the lengths of the base b and the height h. The constant of proportionality is $\dfrac{1}{2}$.

20. Geometry The perimeter p of a rectangle varies jointly with the sum of the lengths of its sides l and w. The constant of proportionality is 2.

21. Physics: Newton's Law The force F (in newtons) of attraction between two bodies varies jointly with their masses m and M (in kilograms) and inversely with the square of the distance d (in meters) between them. The constant of proportionality is $G = 6.67 \times 10^{-11}$.

22. Physics: Simple Pendulum The **period** of a pendulum is the time required for one oscillation; the pendulum is usually referred to as **simple** when the angle made to the vertical is less than $5°$. The period T of a simple pendulum (in seconds) varies directly with the square root of its length l (in feet). The constant of proportionality is $\dfrac{2\pi}{\sqrt{32}}$.

23. Mortgage Payments The monthly payment p on a mortgage varies directly with the amount borrowed B. If the monthly payment on a 30-year mortgage is $6.49 for every $1000 borrowed, find a linear equation that relates the monthly payment p to the amount borrowed B for a mortgage with these terms. Then find the monthly payment p when the amount borrowed B is $145,000.

24. Mortgage Payments The monthly payment p on a mortgage varies directly with the amount borrowed B. If the monthly payment on a 15-year mortgage is $8.99 for every $1000 borrowed, find a linear equation that relates the monthly payment p to the amount borrowed B for a mortgage with these terms. Then find the monthly payment p when the amount borrowed B is $175,000.

25. Physics: Falling Objects The distance s that an object falls is directly proportional to the square of the time t of the fall. If an object falls 16 feet in 1 second, how far will it fall in 3 seconds? How long will it take an object to fall 64 feet?

26. Physics: Falling Objects The velocity v of a falling object is directly proportional to the time t of the fall. If, after 2 seconds, the velocity of the object is 64 feet per second, what will its velocity be after 3 seconds?

27. Physics: Stretching a Spring The elongation E of a spring balance varies directly with the applied weight W (see the figure). If $E = 3$ when $W = 20$, find E when $W = 15$.

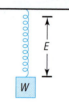

28. Physics: Vibrating String The rate of vibration of a string under constant tension varies inversely with the length of the string. If a string is 48 inches long and vibrates 256 times per second, what is the length of a string that vibrates 576 times per second?

29. Revenue Equation At the corner Shell station, the revenue R varies directly with the number g of gallons of gasoline sold. If the revenue is $47.40 when the number of gallons sold is 12, find a linear equation that relates revenue R to the number g of gallons of gasoline sold. Then find the revenue R when the number of gallons of gasoline sold is 10.5.

30. Cost Equation The cost C of roasted almonds varies directly with the number A of pounds of almonds purchased. If the cost is $23.75 when the number of pounds of roasted almonds purchased is 5, find a linear equation that relates the cost C to the number A of pounds of almonds purchased. Then find the cost C when the number of pounds of almonds purchased is 3.5.

31. Demand Suppose that the demand D for candy at the movie theater is inversely related to the price p.
(a) When the price of candy is $2.75 per bag, the theater sells 156 bags of candy. Express the demand for candy in terms of its price.
(b) Determine the number of bags of candy that will be sold if the price is raised to $3 a bag.

32. Driving to School The time t that it takes to get to school varies inversely with your average speed s.
(a) Suppose that it takes you 40 minutes to get to school when your average speed is 30 miles per hour. Express the driving time to school in terms of average speed.
(b) Suppose that your average speed to school is 40 miles per hour. How long will it take you to get to school?

33. Pressure The volume of a gas V held at a constant temperature in a closed container varies inversely with its pressure P. If the volume of a gas is 600 cubic centimeters (cm^3) when the pressure is 150 millimeters of mercury (mm Hg), find the volume when the pressure is 200 mm Hg.

34. **Resistance** The current i in a circuit is inversely proportional to its resistance Z measured in ohms. Suppose that when the current in a circuit is 30 amperes, the resistance is 8 ohms. Find the current in the same circuit when the resistance is 10 ohms.

35. **Weight** The weight of an object above the surface of Earth varies inversely with the square of the distance from the center of Earth. If Maria weighs 125 pounds when she is on the surface of Earth (3960 miles from the center), determine Maria's weight when she is at the top of Mount McKinley (3.8 miles from the surface of Earth).

36. **Weight of a Body** The weight of a body above the surface of Earth varies inversely with the square of the distance from the center of Earth. If a certain body weighs 55 pounds when it is 3960 miles from the center of Earth, how much will it weigh when it is 3965 miles from the center?

37. **Geometry** The volume V of a right circular cylinder varies jointly with the square of its radius r and its height h. The constant of proportionality is π. See the figure. Write an equation for V.

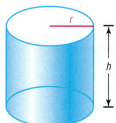

38. **Geometry** The volume V of a right circular cone varies jointly with the square of its radius r and its height h. The constant of proportionality is $\dfrac{\pi}{3}$. See the figure. Write an equation for V.

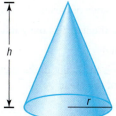

39. **Intensity of Light** The intensity I of light (measured in foot-candles) varies inversely with the square of the distance from the bulb. Suppose that the intensity of a 100-watt light bulb at a distance of 2 meters is 0.075 foot-candle. Determine the intensity of the bulb at a distance of 5 meters.

40. **Force of the Wind on a Window** The force exerted by the wind on a plane surface varies jointly with the area of the surface and the square of the velocity of the wind. If the force on an area of 20 square feet is 11 pounds when the wind velocity is 22 miles per hour, find the force on a surface area of 47.125 square feet when the wind velocity is 36.5 miles per hour.

41. **Horsepower** The horsepower (hp) that a shaft can safely transmit varies jointly with its speed (in revolutions per minute, rpm) and the cube of its diameter. If a shaft of a certain material 2 inches in diameter can transmit 36 hp at 75 rpm, what diameter must the shaft have in order to transmit 45 hp at 125 rpm?

42. **Chemistry: Gas Laws** The volume V of an ideal gas varies directly with the temperature T and inversely with the pressure P. Write an equation relating V, T, and P using k as the constant of proportionality. If a cylinder contains oxygen at a temperature of 300 K and a pressure of 15 atmospheres in a volume of 100 liters, what is the constant of proportionality k? If a piston is lowered into the cylinder, decreasing the volume occupied by the gas to 80 liters and raising the temperature to 310 K, what is the gas pressure?

43. **Physics: Kinetic Energy** The kinetic energy K of a moving object varies jointly with its mass m and the square of its velocity v. If an object weighing 25 kilograms and moving with a velocity of 10 meters per second has a kinetic energy of 1250 joules, find its kinetic energy when the velocity is 15 meters per second.

44. **Electrical Resistance of a Wire** The electrical resistance of a wire varies directly with the length of the wire and inversely with the square of the diameter of the wire. If a wire 432 feet long and 4 millimeters in diameter has a resistance of 1.24 ohms, find the length of a wire of the same material whose resistance is 1.44 ohms and whose diameter is 3 millimeters.

45. **Measuring the Stress of Materials** The stress in the material of a pipe subject to internal pressure varies jointly with the internal pressure and the internal diameter of the pipe and inversely with the thickness of the pipe. The stress is 100 pounds per square inch when the diameter is 5 inches, the thickness is 0.75 inch, and the internal pressure is 25 pounds per square inch. Find the stress when the internal pressure is 40 pounds per square inch if the diameter is 8 inches and the thickness is 0.50 inch.

46. **Safe Load for a Beam** The maximum safe load for a horizontal rectangular beam varies jointly with the width of the beam and the square of the thickness of the beam and inversely with its length. If an 8-foot beam will support up to 750 pounds when the beam is 4 inches wide and 2 inches thick, what is the maximum safe load in a similar beam 10 feet long, 6 inches wide, and 2 inches thick?

Explaining Concepts: Discussion and Writing

47. In the early seventeenth century, Johannes Kepler discovered that the square of the period T of the revolution of a planet around the Sun varies directly with the cube of its mean distance a from the Sun. Go to the library and research this law and Kepler's other two laws. Write a brief paper about these laws and Kepler's place in history.

48. Using a situation that has not been discussed in the text, write a real-world problem that you think involves two variables that vary directly. Exchange your problem with another student's to solve and critique.

49. Using a situation that has not been discussed in the text, write a real-world problem that you think involves two variables that vary inversely. Exchange your problem with another student's to solve and critique.

50. Using a situation that has not been discussed in the text, write a real-world problem that you think involves three variables that vary jointly. Exchange your problem with another student's to solve and critique.

Retain Your Knowledge

Problems 51–54 are based on material learned earlier in the course. The purpose of these problems is to keep the material fresh in your mind so that you are better prepared for the final exam.

51. Factor $3x^3 + 25x^2 - 12x - 100$ completely.

52. Add $\dfrac{5}{x+3} + \dfrac{x-2}{x^2 + 7x + 12}$ and simplify the result.

53. Simplify: $\left(\dfrac{4}{25}\right)^{3/2}$

54. Rationalize the denominator of $\dfrac{3}{\sqrt{7}-2}$.

Chapter Review

Things to Know

Formulas

Distance formula (p. 151)	$d = \sqrt{(x_2 - x_1)^2 + (y_2 - y_1)^2}$
Midpoint formula (p. 154)	$(x, y) = \left(\dfrac{x_1 + x_2}{2}, \dfrac{y_1 + y_2}{2}\right)$
Slope (p. 167)	$m = \dfrac{y_2 - y_1}{x_2 - x_1}$ if $x_1 \neq x_2$; undefined if $x_1 = x_2$
Parallel lines (p. 175)	Equal slopes ($m_1 = m_2$) and different y-intercepts ($b_1 \neq b_2$)
Perpendicular lines (p. 176)	Product of slopes is -1 ($m_1 \cdot m_2 = -1$)
Direct variation (p. 189)	$y = kx$
Inverse variation (p. 189)	$y = \dfrac{k}{x}$

Equations of Lines and Circles

Vertical line (p. 171)	$x = a$; a is the x-intercept
Horizontal line (p. 172)	$y = b$; b is the y-intercept
Point–slope form of the equation of a line (p. 171)	$y - y_1 = m(x - x_1)$; m is the slope of the line, (x_1, y_1) is a point on the line
Slope–intercept form of the equation of a line (p. 172)	$y = mx + b$; m is the slope of the line, b is the y-intercept
General form of the equation of a line (p. 174)	$Ax + By = C$; A, B not both 0
Standard form of the equation of a circle (p. 183)	$(x - h)^2 + (y - k)^2 = r^2$; r is the radius of the circle, (h, k) is the center of the circle
Equation of the unit circle (p. 183)	$x^2 + y^2 = 1$
General form of the equation of a circle (p. 185)	$x^2 + y^2 + ax + by + c = 0$, with restrictions on $a, b,$ and c

Objectives

Section		You should be able to …	Examples	Review Exercises
2.1	1	Use the distance formula (p. 151)	1–3	1(a)–3(a), 29, 30(a), 31
	2	Use the midpoint formula (p. 153)	4	1(b)–3(b), 31
2.2	1	Graph equations by plotting points (p. 157)	1–3	4
	2	Find intercepts from a graph (p. 159)	4	5
	3	Find intercepts from an equation (p. 160)	5	6–10
	4	Test an equation for symmetry with respect to the x-axis, the y-axis, and the origin (p. 160)	6–9	6–10
	5	Know how to graph key equations (p. 163)	10–12	26, 27
2.3	1	Calculate and interpret the slope of a line (p. 167)	1, 2	1(c)–3(c), 1(d)–3(d), 32
	2	Graph lines given a point and the slope (p. 170)	3	28
	3	Find the equation of a vertical line (p. 170)	4	17

Section		You should be able to . . .	Examples	Review Exercises
	4	Use the point–slope form of a line; identify horizontal lines (p. 171)	5, 6	16
	5	Find the equation of a line given two points (p. 172)	7	18, 19
	6	Write the equation of a line in slope–intercept form (p. 172)	8	16, 18–21
	7	Identify the slope and y-intercept of a line from its equation (p. 173)	8	22, 23
	8	Graph lines written in general form using intercepts (p. 174)	9	24, 25
	9	Find equations of parallel lines (p. 175)	10, 11	20
	10	Find equations of perpendicular lines (p. 176)	12, 13	21, 30(b)
2.4	1	Write the standard form of the equation of a circle (p. 182)	1	11, 12, 31
	2	Graph a circle (p. 183)	2, 3, 5	13–15
	3	Work with the general form of the equation of a circle (p. 184)	4	14, 15
2.5	1	Construct a model using direct variation (p. 189)	1	33
	2	Construct a model using inverse variation (p. 189)	2	34
	3	Construct a model using joint or combined variation (p. 190)	3, 4	35

Review Exercises

In Problems 1–3, find the following for each pair of points:
 (a) *The distance between the points*
 (b) *The midpoint of the line segment connecting the points*
 (c) *The slope of the line containing the points*
 (d) *Interpret the slope found in part (c)*

1. $(0, 0)$; $(4, 2)$

2. $(1, -1)$; $(-2, 3)$

3. $(4, -4)$; $(4, 8)$

4. Graph $y = x^2 + 4$ by plotting points.

5. List the intercepts of the graph below.

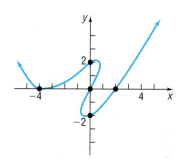

In Problems 6–10, list the intercepts and test for symmetry with respect to the x-axis, the y-axis, and the origin.

6. $2x = 3y^2$

7. $x^2 + 4y^2 = 16$

8. $y = x^4 + 2x^2 + 1$

9. $y = x^3 - x$

10. $x^2 + x + y^2 + 2y = 0$

In Problems 11 and 12, find the standard form of the equation of the circle whose center and radius are given.

11. $(h, k) = (-2, 3)$; $r = 4$

12. $(h, k) = (-1, -2)$; $r = 1$

In Problems 13–15, find the center and radius of each circle. Graph each circle. Find the intercepts, if any, of each circle.

13. $x^2 + (y - 1)^2 = 4$

14. $x^2 + y^2 - 2x + 4y - 4 = 0$

15. $3x^2 + 3y^2 - 6x + 12y = 0$

In Problems 16–21, find an equation of the line having the given characteristics. Express your answer using either the general form or the slope–intercept form of the equation of a line, whichever you prefer.

16. Slope $= -2$; containing the point $(3, -1)$

17. Vertical; containing the point $(-3, 4)$

18. y-intercept $= -2$; containing the point $(5, -3)$

19. Containing the points $(3, -4)$ and $(2, 1)$

20. Parallel to the line $2x - 3y = -4$; containing the point $(-5, 3)$

21. Perpendicular to the line $x + y = 2$; containing the point $(4, -3)$

In Problems 22 and 23, find the slope and y-intercept of each line. Graph the line, labeling any intercepts.

22. $4x - 5y = -20$

23. $\dfrac{1}{2}x - \dfrac{1}{3}y = -\dfrac{1}{6}$

In Problems 24 and 25, find the intercepts and graph each line.

24. $2x - 3y = 12$

25. $\dfrac{1}{2}x + \dfrac{1}{3}y = 2$

26. Sketch a graph of $y = x^3$.

27. Sketch a graph of $y = \sqrt{x}$.

28. Graph the line with slope $\dfrac{2}{3}$ containing the point $(1, 2)$.

29. Show that the points $A = (3, 4), B = (1, 1)$, and $C = (-2, 3)$ are the vertices of an isosceles triangle.

30. Show that the points $A = (-2, 0), B = (-4, 4)$, and $C = (8, 5)$ are the vertices of a right triangle in two ways:
    (a) By using the converse of the Pythagorean Theorem
    (b) By using the slopes of the lines joining the vertices

31. The endpoints of the diameter of a circle are $(-3, 2)$ and $(5, -6)$. Find the center and radius of the circle. Write the standard equation of this circle.

32. Show that the points $A = (2, 5), B = (6, 1)$, and $C = (8, -1)$ lie on a line by using slopes.

33. **Mortgage Payments** The monthly payment p on a mortgage varies directly with the amount borrowed B. If the monthly payment on a 30-year mortgage is $854.00 when $130,000 is borrowed, find an equation that relates the monthly payment p to the amount borrowed B for a mortgage with these terms. Then find the monthly payment p when the amount borrowed B is $165,000.

34. **Weight of a Body** The weight of a body varies inversely with the square of its distance from the center of Earth. Assuming that the radius of Earth is 3960 miles, how much would a man weigh at an altitude of 1 mile above Earth's surface if he weighs 200 pounds on Earth's surface?

35. **Heat Loss** The amount of heat transferred per hour through a glass window varies jointly with the surface area of the window and the difference in temperature between the areas separated by the glass. A window with a surface area of 7.5 square feet loses 135 Btu per hour when the temperature difference is 40°F. How much heat is lost per hour for a similar window with a surface are of 12 square feet when the temperature difference is 35°F?

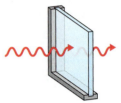

Chapter Test

CHAPTER **Test Prep** VIDEOS  The Chapter Test Prep Videos are step-by-step solutions available in **MyMathLab**, or on this text's You**Tube** Channel. Flip back to the Resources for Success page for a link to this text's YouTube channel.

In Problems 1–3, use $P_1 = (-1, 3)$ and $P_2 = (5, -1)$.

1. Find the distance from P_1 to P_2.

2. Find the midpoint of the line segment joining P_1 and P_2.

3. (a) Find the slope of the line containing P_1 and P_2.
   (b) Interpret this slope.

4. Graph $y = x^2 - 9$ by plotting points.

5. Sketch the graph of $y^2 = x$.

6. List the intercepts and test for symmetry: $x^2 + y = 9$

7. Write the slope–intercept form of the line with slope -2 containing the point $(3, -4)$. Graph the line.

8. Write the general form of the circle with center $(4, -3)$ and radius 5.

9. Find the center and radius of the circle $x^2 + y^2 + 4x - 2y - 4 = 0$. Graph this circle.

10. For the line $2x + 3y = 6$, find a line parallel to it containing the point $(1, -1)$. Also find a line perpendicular to it containing the point $(0, 3)$.

11. **Resistance Due to a Conductor** The resistance (in ohms) of a circular conductor varies directly with the length of the conductor and inversely with the square of the radius of the conductor. If 50 feet of wire with a radius of 6×10^{-3} inch has a resistance of 10 ohms, what would be the resistance of 100 feet of the same wire if the radius were increased to 7×10^{-3} inch?

Cumulative Review

In Problems 1–8, find the real solution(s), if any, of each equation.

1. $3x - 5 = 0$

2. $x^2 - x - 12 = 0$

3. $2x^2 - 5x - 3 = 0$

4. $x^2 - 2x - 2 = 0$

5. $x^2 + 2x + 5 = 0$

6. $\sqrt{2x + 1} = 3$

7. $|x - 2| = 1$

8. $\sqrt{x^2 + 4x} = 2$

In Problems 9 and 10, solve each equation in the complex number system.

9. $x^2 = -9$

10. $x^2 - 2x + 5 = 0$

In Problems 11–14, solve each inequality. Graph the solution set.

11. $2x - 3 \leq 7$

12. $-1 < x + 4 < 5$

13. $|x - 2| \leq 1$

14. $|2 + x| > 3$

15. Find the distance between the points $P = (-1, 3)$ and $Q = (4, -2)$. Find the midpoint of the line segment from P to Q.

16. Which of the following points are on the graph of $y = x^3 - 3x + 1$?
(a) $(-2, -1)$ (b) $(2, 3)$ (c) $(3, 1)$

17. Sketch the graph of $y = x^3$.

18. Find the equation of the line containing the points $(-1, 4)$ and $(2, -2)$. Express your answer in slope–intercept form.

19. Find the equation of the line perpendicular to the line $y = 2x + 1$ and containing the point $(3, 5)$. Express your answer in slope–intercept form and graph the line.

20. Graph the equation $x^2 + y^2 - 4x + 8y - 5 = 0$.

Chapter Project

Internet-based Project

Determining the Selling Price of a Home Determining how much to pay for a home is one of the more difficult decisions that must be made when purchasing a home. There are many factors that play a role in a home's value. Location, size, number of bedrooms, number of bathrooms, lot size, and building materials are just a few. Fortunately, the website Zillow.com has developed its own formula for predicting the selling price of a home. This information is a great tool for predicting the actual sale price. For example, the data below show the "zestimate"—the selling price of a home as predicted by the folks at Zillow—and the actual selling price of the home, for homes in Oak Park, Illinois.

Zestimate ($ thousands)	Sale Price ($ thousands)
291.5	268
320	305
371.5	375
303.5	283
351.5	350
314	275
332.5	356
295	300
313	285
368	385

The graph below, called a scatter diagram, shows the points (291.5, 268), (320, 305), ..., (368, 385) in a Cartesian plane. From the graph, it appears that the data follow a linear relation.

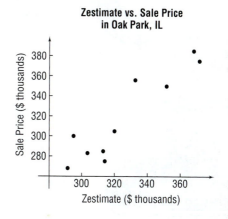

Zestimate vs. Sale Price in Oak Park, IL

1. Imagine drawing a line through the data that appears to fit the data well. Do you believe the slope of the line would be positive, negative, or close to zero? Why?

2. Pick two points from the scatter diagram. Treat the zestimate as the value of x, and treat the sale price as the corresponding value of y. Find the equation of the line through the two points you selected.

3. Interpret the slope of the line.

4. Use your equation to predict the selling price of a home whose zestimate is $335,000.

5. Do you believe it would be a good idea to use the equation you found in part 2 if the zestimate were $950,000? Why or why not?

6. Choose a location in which you would like to live. Go to www.zillow.com and randomly select at least ten homes that have recently sold.
(a) Draw a scatter diagram of your data.
(b) Select two points from the scatter diagram and find the equation of the line through the points.
(c) Interpret the slope.
(d) Find a home from the Zillow website that interests you under the "Make Me Move" option for which a zestimate is available. Use your equation to predict the sale price based on the zestimate.

3

Functions and Their Graphs

Choosing a Wireless Data Plan

Most consumers choose a cellular provider first and then select an appropriate data plan from that provider. The choice as to the type of plan selected depends on your use of the device. For example, is online gaming important? Do you want to stream audio or video? The mathematics learned in this chapter can help you decide what plan is best suited to your particular needs.

—See the Internet-based Chapter Project—

Outline

••• A Look Back

So far, our discussion has focused on techniques for graphing equations containing two variables.

A Look Ahead •••

In this chapter, we look at a special type of equation involving two variables called a *function*. This chapter deals with what a function is, how to graph functions, properties of functions, and how functions are used in applications. The word *function* apparently was introduced by René Descartes in 1637. For him, a function was simply any positive integral power of a variable *x*. Gottfried Wilhelm Leibniz (1646–1716), who always emphasized the geometric side of mathematics, used the word *function* to denote any quantity associated with a curve, such as the coordinates of a point on the curve. Leonhard Euler (1707–1783) employed the word to mean any equation or formula involving variables and constants. His idea of a function is similar to the one most often seen in courses that precede calculus. Later, the use of functions in investigating heat flow equations led to a very broad definition that originated with Lejeune Dirichlet (1805–1859), which describes a function as a correspondence between two sets. That is the definition used in this text.

3.1 Functions

PREPARING FOR THIS SECTION *Before getting started, review the following:*

- Intervals (Section 1.5, pp. 120–121)
- Solving Inequalities (Section 1.5, pp. 123–126)
- Evaluating Algebraic Expressions, Domain of a Variable (Chapter R, Section R.2, pp. 20–23)

- Rationalizing Denominators (Chapter R, Section R.8, p. 75)

✎ **Now Work** the **'Are You Prepared?'** problems on page 210.

OBJECTIVES 1 Determine Whether a Relation Represents a Function (p. 199)
2 Find the Value of a Function (p. 202)
3 Find the Difference Quotient of a Function (p. 205)
4 Find the Domain of a Function Defined by an Equation (p. 206)
5 Form the Sum, Difference, Product, and Quotient of Two Functions (p. 208)

1 Determine Whether a Relation Represents a Function

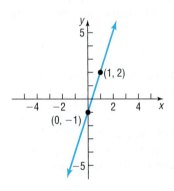

Figure 1 $y = 3x - 1$

Often there are situations where the value of one variable is somehow linked to the value of another variable. For example, an individual's level of education is linked to annual income. Engine size is linked to gas mileage. When the value of one variable is related to the value of a second variable, we have a *relation*. A **relation** is a correspondence between two sets. If x and y are two elements, one from each of these sets, and if a relation exists between x and y, then we say that x **corresponds** to y or that y **depends on** x, and we write $x \rightarrow y$.

There are a number of ways to express relations between two sets. For example, the equation $y = 3x - 1$ shows a relation between x and y. It says that if we take some number x, multiply it by 3, and then subtract 1, we obtain the corresponding value of y. In this sense, x serves as the **input** to the relation, and y is the **output** of the relation. This relation, expressed as a graph, is shown in Figure 1.

The set of all inputs for a relation is called the **domain** of the relation, and the set of all outputs is called the **range**.

In addition to being expressed as equations and graphs, relations can be expressed through a technique called *mapping*. A **map** illustrates a relation as a set of inputs with an arrow drawn from each element in the set of inputs to the corresponding element in the set of outputs. **Ordered pairs** can be used to represent $x \rightarrow y$ as (x, y).

EXAMPLE 1 **Maps and Ordered Pairs as Relations**

Figure 2 shows a relation between states and the number of representatives each state has in the House of Representatives. (***Source:*** www.house.gov). The relation might be named "number of representatives."

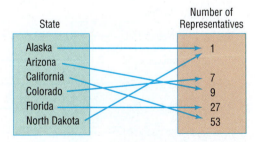

Figure 2 Number of representatives

In this relation, Alaska corresponds to 1, Arizona corresponds to 9, and so on. Using ordered pairs, this relation would be expressed as

{ (Alaska, 1), (Arizona, 9), (California, 53), (Colorado, 7), (Florida, 27), (North Dakota, 1) }

Person   Phone Number

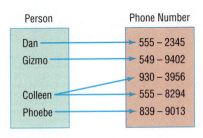

Figure 3 Phone numbers

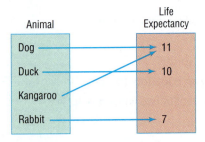

Figure 4 Animal life expectancy

DEFINITION

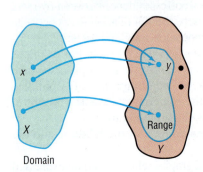

Figure 5

The domain of the relation is {Alaska, Arizona, California, Colorado, Florida, North Dakota}, and the range is {1, 7, 9, 27, 53}. Note that the output "1" is listed only once in the range. ●

One of the most important concepts in algebra is the *function*. A function is a special type of relation. To understand the idea behind a function, let's revisit the relation presented in Example 1. If we were to ask, "How many representatives does Alaska have?" you would respond "1." In fact, each input *state* corresponds to a single output *number of representatives*.

Let's consider a second relation, one that involves a correspondence between four people and their phone numbers. See Figure 3. Notice that Colleen has two telephone numbers. There is no single answer to the question "What is Colleen's phone number?"

Let's look at one more relation. Figure 4 is a relation that shows a correspondence between type of *animal* and *life expectancy*. If asked to determine the life expectancy of a dog, we would all respond, "11 years." If asked to determine the life expectancy of a rabbit, we would all respond, "7 years."

Notice that the relations presented in Figures 2 and 4 have something in common. What is it? In both of these relations, each input corresponds to exactly one output. This leads to the definition of a *function*.

> Let X and Y be two nonempty sets.* A **function** from X into Y is a relation that associates with each element of X exactly one element of Y.

The set X is called the **domain** of the function. For each element x in X, the corresponding element y in Y is called the **value** of the function at x, or the **image** of x. The set of all images of the elements in the domain is called the **range** of the function. See Figure 5.

Since there may be some elements in Y that are not the image of some x in X, it follows that the range of a function may be a subset of Y, as shown in Figure 5.

Not all relations between two sets are functions. The next example shows how to determine whether a relation is a function.

EXAMPLE 2

Determining Whether a Relation Is a Function

For each relation in Figures 6, 7, and 8, state the domain and range. Then determine whether the relation is a function.

(a) See Figure 6. For this relation, the input is the number of calories in a fast-food sandwich, and the output is the fat content (in grams).

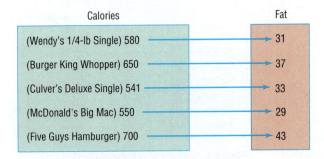

Figure 6 Fat content
Source: Each company's Website

---

*The sets X and Y will usually be sets of real numbers, in which case a (real) function results. The two sets can also be sets of complex numbers, and then we have defined a complex function. In the broad definition (proposed by Lejeune Dirichlet), X and Y can be any two sets.

(b) See Figure 7. For this relation, the inputs are gasoline stations in Harris County, Texas, and the outputs are the price per gallon of unleaded regular in March 2014.

(c) See Figure 8. For this relation, the inputs are the weight (in carats) of pear-cut diamonds and the outputs are the price (in dollars).

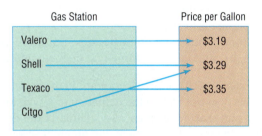

Figure 7 Unleaded price per gallon

Figure 8 Diamond price

Source: Used with permission of Diamonds.com

Solution

(a) The domain of the relation is {541, 550, 580, 650, 700}, and the range of the relation is {29, 31, 33, 37, 43}. The relation in Figure 6 is a function because each element in the domain corresponds to exactly one element in the range.

(b) The domain of the relation is {Citgo, Shell, Texaco, Valero}. The range of the relation is {$3.19, $3.29, $3.35}. The relation in Figure 7 is a function because each element in the domain corresponds to exactly one element in the range. Notice that it is okay for more than one element in the domain to correspond to the same element in the range (Shell and Citgo both sell gas for $3.29 a gallon).

(c) The domain of the relation is {0.70, 0.71, 0.75, 0.78} and the range is {$1529, $1575, $1765, $1798, $1952}. The relation in Figure 8 is not a function because not every element in the domain corresponds to exactly one element in the range. If a 0.71-carat diamond is chosen from the domain, a single price cannot be assigned to it. ●

 Now Work PROBLEM 19

The idea behind a function is its predictability. If the input is known, we can use the function to determine the output. With "nonfunctions," we don't have this predictability. Look back at Figure 6. If asked, "How many grams of fat are in a 580-calorie sandwich?" we could use the correspondence to answer, "31." Now consider Figure 8. If asked, "What is the price of a 0.71-carat diamond?" we could not give a single response because two outputs result from the single input "0.71." For this reason, the relation in Figure 8 is not a function.

We may also think of a function as a set of ordered pairs (x, y) in which no ordered pairs have the same first element and different second elements. The set of all first elements x is the domain of the function, and the set of all second elements y is its range. Each element x in the domain corresponds to exactly one element y in the range.

In Words

For a function, no input has more than one output. The domain of a function is the set of all inputs; the range is the set of all outputs.

EXAMPLE 3

Determining Whether a Relation Is a Function

For each relation, state the domain and range. Then determine whether the relation is a function.

(a) { (1, 4), (2, 5), (3, 6), (4, 7) }

(b) { (1, 4), (2, 4), (3, 5), (6, 10) }

(c) { (−3, 9), (−2, 4), (0, 0), (1, 1), (−3, 8) }

Solution   (a) The domain of this relation is {1, 2, 3, 4}, and its range is {4, 5, 6, 7}. This relation is a function because there are no ordered pairs with the same first element and different second elements.

(b) The domain of this relation is {1, 2, 3, 6}, and its range is {4, 5, 10}. This relation is a function because there are no ordered pairs with the same first element and different second elements.

(c) The domain of this relation is {−3, −2, 0, 1}, and its range is {0, 1, 4, 8, 9}. This relation is not a function because there are two ordered pairs, $(-3, 9)$ and $(-3, 8)$, that have the same first element and different second elements.   ●

In Example 3(b), notice that 1 and 2 in the domain both have the same image in the range. This does not violate the definition of a function; two different first elements can have the same second element. A violation of the definition occurs when two ordered pairs have the same first element and different second elements, as in Example 3(c).

➤ **Now Work** PROBLEM 23

Up to now we have shown how to identify when a relation is a function for relations defined by mappings (Example 2) and ordered pairs (Example 3). But relations can also be expressed as equations. The circumstances under which equations are functions are discussed next.

To determine whether an equation, where y depends on x, is a function, it is often easiest to solve the equation for y. If any value of x in the domain corresponds to more than one y, the equation does not define a function; otherwise, it does define a function.

EXAMPLE 4   **Determining Whether an Equation Is a Function**

Determine whether the equation $y = 2x - 5$ defines y as a function of x.

Solution   The equation tells us to take an input x, multiply it by 2, and then subtract 5. For any input x, these operations yield only one output y, so the equation is a function. For example, if $x = 1$, then $y = 2(1) - 5 = -3$. If $x = 3$, then $y = 2(3) - 5 = 1$. The graph of the equation $y = 2x - 5$ is a line with slope 2 and y-intercept −5. The function is called a *linear function*.   ●

EXAMPLE 5   **Determining Whether an Equation Is a Function**

Determine whether the equation $x^2 + y^2 = 1$ defines y as a function of x.

Solution   To determine whether the equation $x^2 + y^2 = 1$, which defines the unit circle, is a function, solve the equation for y.

$$x^2 + y^2 = 1$$
$$y^2 = 1 - x^2$$
$$y = \pm\sqrt{1 - x^2}$$

For values of x for which $-1 < x < 1$, two values of y result. For example, if $x = 0$, then $y = \pm 1$, so two different outputs result from the same input. This means that the equation $x^2 + y^2 = 1$ does not define a function.   ●

➤ **Now Work** PROBLEM 37

2 Find the Value of a Function

Functions are often denoted by letters such as f, F, g, G, and others. If f is a function, then for each number x in its domain, the corresponding image in the range is designated by the symbol $f(x)$, read as "f of x" or as "f at x." We refer to $f(x)$ as the **value of f at the number x**; $f(x)$ is the number that results when x is given and the function f is applied; $f(x)$ is the output corresponding to x or $f(x)$ is the image of x; $f(x)$

does *not* mean "*f* times *x*." For example, the function given in Example 4 may be written as $y = f(x) = 2x - 5$. Then $f(1) = -3$ and $f(3) = 1$.

Figure 9 illustrates some other functions. Notice that in every function, for each *x* in the domain, there is one value in the range.

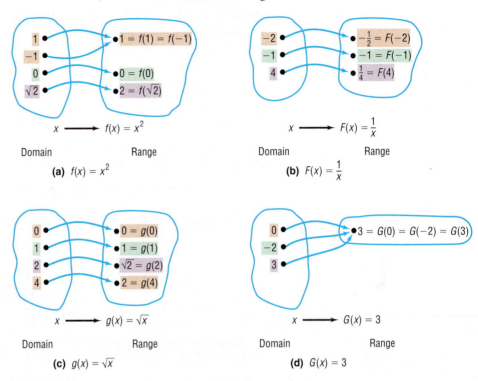

(a) $f(x) = x^2$

(b) $F(x) = \dfrac{1}{x}$

(c) $g(x) = \sqrt{x}$

(d) $G(x) = 3$

Figure 9

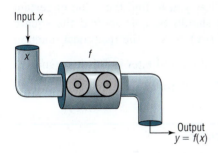

Input *x*

Output
$y = f(x)$

Figure 10 Input/output machine

Sometimes it is helpful to think of a function *f* as a machine that receives as input a number from the domain, manipulates it, and outputs a value. See Figure 10.

The restrictions on this input/output machine are as follows:

1. It accepts only numbers from the domain of the function.

2. For each input, there is exactly one output (which may be repeated for different inputs).

For a function $y = f(x)$, the variable *x* is called the **independent variable**, because it can be assigned any of the permissible numbers from the domain. The variable *y* is called the **dependent variable**, because its value depends on *x*.

Any symbols can be used to represent the independent and dependent variables. For example, if *f* is the *cube function,* then *f* can be given by $f(x) = x^3$ or $f(t) = t^3$ or $f(z) = z^3$. All three functions are the same. Each says to cube the independent variable to get the output. In practice, the symbols used for the independent and dependent variables are based on common usage, such as using *C* for cost in business.

The independent variable is also called the **argument** of the function. Thinking of the independent variable as an argument can sometimes make it easier to find the value of a function. For example, if *f* is the function defined by $f(x) = x^3$, then *f* tells us to cube the argument. Thus $f(2)$ means to cube 2, $f(a)$ means to cube the number *a*, and $f(x + h)$ means to cube the quantity $x + h$.

EXAMPLE 6

Finding Values of a Function

For the function *f* defined by $f(x) = 2x^2 - 3x$, evaluate

(a) $f(3)$ (b) $f(x) + f(3)$ (c) $3f(x)$ (d) $f(-x)$

(e) $-f(x)$ (f) $f(3x)$ (g) $f(x + 3)$

Solution   (a) Substitute 3 for x in the equation for f, $f(x) = 2x^2 - 3x$, to get

$$f(3) = 2(3)^2 - 3(3) = 18 - 9 = 9$$

The image of 3 is 9.

(b) $f(x) + f(3) = (2x^2 - 3x) + (9) = 2x^2 - 3x + 9$

(c) Multiply the equation for f by 3.

$$3f(x) = 3(2x^2 - 3x) = 6x^2 - 9x$$

(d) Substitute $-x$ for x in the equation for f and simplify.

$$f(-x) = 2(-x)^2 - 3(-x) = 2x^2 + 3x \quad \textcolor{blue}{\textit{Notice the use of parentheses here.}}$$

(e) $-f(x) = -(2x^2 - 3x) = -2x^2 + 3x$

(f) Substitute $3x$ for x in the equation for f and simplify.

$$f(3x) = 2(3x)^2 - 3(3x) = 2(9x^2) - 9x = 18x^2 - 9x$$

(g) Substitute $x + 3$ for x in the equation for f and simplify.

$$f(x + 3) = 2(x + 3)^2 - 3(x + 3)$$
$$= 2(x^2 + 6x + 9) - 3x - 9$$
$$= 2x^2 + 12x + 18 - 3x - 9$$
$$= 2x^2 + 9x + 9 \qquad \bullet$$

Notice in this example that $f(x + 3) \neq f(x) + f(3)$, $f(-x) \neq -f(x)$, and $3f(x) \neq f(3x)$.

 Now Work   PROBLEM 43

Most calculators have special keys that allow you to find the value of certain commonly used functions. For example, you should be able to find the square function $f(x) = x^2$, the square root function $f(x) = \sqrt{x}$, the reciprocal function $f(x) = \dfrac{1}{x} = x^{-1}$, and many others that will be discussed later in this text (such as $\ln x$ and $\log x$). Verify the results of Example 7, which follows, on your calculator.

EXAMPLE 7   **Finding Values of a Function on a Calculator**

(a) $f(x) = x^2$ $f(1.234) = 1.234^2 = 1.522756$

(b) $F(x) = \dfrac{1}{x}$ $F(1.234) = \dfrac{1}{1.234} \approx 0.8103727715$

(c) $g(x) = \sqrt{x}$ $g(1.234) = \sqrt{1.234} \approx 1.110855526 \qquad \bullet$

COMMENT Graphing calculators can be used to evaluate any function. Figure 11 shows the result obtained in Example 6(a) on a TI-84 Plus C graphing calculator with the function to be evaluated, $f(x) = 2x^2 - 3x$, in Y_1.

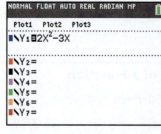
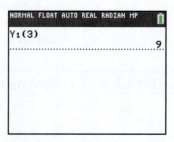

Figure 11 Evaluating $f(x) = 2x^2 - 3x$ for $x = 3$

Implicit Form of a Function

COMMENT The explicit form of a function is the form required by a graphing calculator. ∎

In general, when a function f is defined by an equation in x and y, we say that the function f is given **implicitly**. If it is possible to solve the equation for y in terms of x, then we write $y = f(x)$ and say that the function is given **explicitly**. For example,

Implicit Form	**Explicit Form**
$3x + y = 5$	$y = f(x) = -3x + 5$
$x^2 - y = 6$	$y = f(x) = x^2 - 6$
$xy = 4$	$y = f(x) = \dfrac{4}{x}$

SUMMARY

Important Facts about Functions

(a) For each x in the domain of a function f, there is exactly one image $f(x)$ in the range; however, an element in the range can result from more than one x in the domain.

(b) f is the symbol that we use to denote the function. It is symbolic of the equation (rule) that we use to get from an x in the domain to $f(x)$ in the range.

(c) If $y = f(x)$, then x is called the independent variable or argument of f, and y is called the dependent variable or the value of f at x.

3 Find the Difference Quotient of a Function

An important concept in calculus involves looking at a certain quotient. For a given function $y = f(x)$, the inputs x and $x + h$, $h \neq 0$, result in the images $f(x)$ and $f(x + h)$. The quotient of their differences

$$\frac{f(x + h) - f(x)}{(x + h) - x} = \frac{f(x + h) - f(x)}{h}$$

with $h \neq 0$, is called the *difference quotient of f* at x.

DEFINITION

The **difference quotient** of a function f at x is given by

$$\frac{f(x + h) - f(x)}{h} \qquad h \neq 0 \tag{1}$$

The difference quotient is used in calculus to define the derivative, which leads to applications such as the velocity of an object and optimization of resources.

When finding a difference quotient, it is necessary to simplify the expression in order to cancel the h in the denominator, as illustrated in the following example.

EXAMPLE 8

Finding the Difference Quotient of a Function

Find the difference quotient of each function.

(a) $f(x) = 2x^2 - 3x$

(b) $f(x) = \dfrac{4}{x}$

(c) $f(x) = \sqrt{x}$

Solution   (a) $\dfrac{f(x + h) - f(x)}{h} = \dfrac{[2(x + h)^2 - 3(x + h)] - [2x^2 - 3x]}{h}$

$$\uparrow$$
$$f(x + h) = 2(x + h)^2 - 3(x + h)$$

$$= \dfrac{2(x^2 + 2xh + h^2) - 3x - 3h - 2x^2 + 3x}{h} \qquad \text{Simplify.}$$

$$= \dfrac{2x^2 + 4xh + 2h^2 - 3h - 2x^2}{h} \qquad \text{Distribute and combine like terms.}$$

$$= \dfrac{4xh + 2h^2 - 3h}{h} \qquad \text{Combine like terms.}$$

$$= \dfrac{h(4x + 2h - 3)}{h} \qquad \text{Factor out } h.$$

$$= 4x + 2h - 3 \qquad \text{Divide out the } h\text{'s.}$$

(b) $\dfrac{f(x + h) - f(x)}{h} = \dfrac{\dfrac{4}{x + h} - \dfrac{4}{x}}{h} \qquad f(x + h) = \dfrac{4}{x + h}$

$$= \dfrac{\dfrac{4x - 4(x + h)}{x(x + h)}}{h} \qquad \text{Subtract.}$$

$$= \dfrac{4x - 4x - 4h}{x(x + h)h} \qquad \text{Divide and distribute.}$$

$$= \dfrac{-4h}{x(x + h)h} \qquad \text{Simplify.}$$

$$= -\dfrac{4}{x(x + h)} \qquad \text{Divide out the factor } h.$$

(c) $\dfrac{f(x + h) - f(x)}{h} = \dfrac{\sqrt{x + h} - \sqrt{x}}{h} \qquad f(x + h) = \sqrt{x + h}$

$$= \dfrac{\sqrt{x + h} - \sqrt{x}}{h} \cdot \dfrac{\sqrt{x + h} + \sqrt{x}}{\sqrt{x + h} + \sqrt{x}} \qquad \text{Rationalize the numerator.}$$

$$= \dfrac{(\sqrt{x + h})^2 - (\sqrt{x})^2}{h(\sqrt{x + h} + \sqrt{x})} \qquad (A - B)(A + B) = A^2 - B^2$$

$$= \dfrac{h}{h(\sqrt{x + h} + \sqrt{x})} \qquad (\sqrt{x + h})^2 - (\sqrt{x})^2 = x + h - x = h$$

$$= \dfrac{1}{\sqrt{x + h} + \sqrt{x}} \qquad \text{Divide out the factor } h.$$

●

✏️ **Now Work** PROBLEM **79**

4 Find the Domain of a Function Defined by an Equation

Often the domain of a function f is not specified; instead, only the equation defining the function is given. In such cases, we agree that the **domain of f** is the largest set of real numbers for which the value $f(x)$ is a real number. The domain of a function f is the same as the domain of the variable x in the expression $f(x)$.

EXAMPLE 9

Finding the Domain of a Function

Find the domain of each of the following functions.

(a) $f(x) = x^2 + 5x$

(b) $g(x) = \dfrac{3x}{x^2 - 4}$

(c) $h(t) = \sqrt{4 - 3t}$

(d) $F(x) = \dfrac{\sqrt{3x + 12}}{x - 5}$

Solution (a) The function says to square a number and then add five times the number. Since these operations can be performed on any real number, the domain of f is the set of all real numbers.

(b) The function g says to divide $3x$ by $x^2 - 4$. Since division by 0 is not defined, the denominator $x^2 - 4$ can never be 0, so x can never equal -2 or 2. The domain of the function g is $\{x \mid x \neq -2, x \neq 2\}$.

(c) The function h says to take the square root of $4 - 3t$. But only nonnegative numbers have real square roots, so the expression under the square root (the radicand) must be nonnegative (greater than or equal to zero). This requires that

$$4 - 3t \geq 0$$
$$-3t \geq -4$$
$$t \leq \frac{4}{3}$$

The domain of h is $\left\{t \,\middle|\, t \leq \dfrac{4}{3}\right\}$, or the interval $\left(-\infty, \dfrac{4}{3}\right]$.

(d) The function F says to take the square root of $3x + 12$ and divide this result by $x - 5$. This requires that $3x + 12 \geq 0$, so $x \geq -4$, and also that $x - 5 \neq 0$, so $x \neq 5$. Combining these two restrictions, the domain of F is

$$\{x \mid x \geq -4, \quad x \neq 5\}.$$

The following steps may prove helpful for finding the domain of a function that is defined by an equation and whose domain is a subset of the real numbers.

Finding the Domain of a Function Defined by an Equation

1. Start with the domain as the set of real numbers.
2. If the equation has a denominator, exclude any numbers that give a zero denominator.
3. If the equation has a radical of even index, exclude any numbers that cause the expression inside the radical (the radicand) to be negative.

Now Work PROBLEM 55

If x is in the domain of a function f, we shall say that f **is defined at** x, or $f(x)$ **exists**. If x is not in the domain of f, we say that f **is not defined at** x, or $f(x)$ **does not exist**. For example, if $f(x) = \dfrac{x}{x^2 - 1}$, then $f(0)$ exists, but $f(1)$ and $f(-1)$ do not exist. (Do you see why?)

We have not said much about finding the range of a function. We will say more about finding the range when we look at the graph of a function in the next section. When a function is defined by an equation, it can be difficult to find the range. Therefore, we shall usually be content to find just the domain of a function when the function is defined by an equation. We shall express the domain of a function using inequalities, interval notation, set notation, or words, whichever is most convenient.

When we use functions in applications, the domain may be restricted by physical or geometric considerations. For example, the domain of the function f defined by $f(x) = x^2$ is the set of all real numbers. However, if f is used to obtain the area of a square when the length x of a side is known, then we must restrict the domain of f to the positive real numbers, since the length of a side can never be 0 or negative.

EXAMPLE 10	**Finding the Domain in an Application**

Express the area of a circle as a function of its radius. Find the domain.

Solution

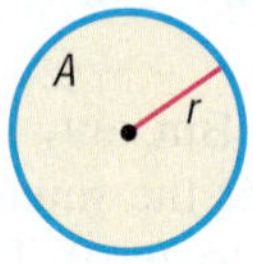

Figure 12 Circle of radius r

See Figure 12. The formula for the area A of a circle of radius r is $A = \pi r^2$. Using r to represent the independent variable and A to represent the dependent variable, the function expressing this relationship is

$$A(r) = \pi r^2$$

In this setting, the domain is $\{r \,|\, r > 0\}$. (Do you see why?) •

Observe, in the solution to Example 10, that the symbol A is used in two ways: It is used to name the function, and it is used to symbolize the dependent variable. This double use is common in applications and should not cause any difficulty.

Now Work PROBLEM 97

5 Form the Sum, Difference, Product, and Quotient of Two Functions

Next we introduce some operations on functions. Functions, like numbers, can be added, subtracted, multiplied, and divided. For example, if $f(x) = x^2 + 9$ and $g(x) = 3x + 5$, then

$$f(x) + g(x) = (x^2 + 9) + (3x + 5) = x^2 + 3x + 14$$

The new function $y = x^2 + 3x + 14$ is called the *sum function* $f + g$. Similarly,

$$f(x) \cdot g(x) = (x^2 + 9)(3x + 5) = 3x^3 + 5x^2 + 27x + 45$$

The new function $y = 3x^3 + 5x^2 + 27x + 45$ is called the *product function* $f \cdot g$.

The general definitions are given next.

DEFINITION

If f and g are functions:
The **sum $f + g$** is the function defined by

$$(f + g)(x) = f(x) + g(x)$$

In Words
Remember, the symbol ∩ stands for intersection. It means you should find the elements that are common to two sets.

The domain of $f + g$ consists of the numbers x that are in the domains of both f and g. That is, domain of $f + g = $ domain of $f \cap$ domain of g.

DEFINITION

The **difference $f - g$** is the function defined by

$$(f - g)(x) = f(x) - g(x)$$

The domain of $f - g$ consists of the numbers x that are in the domains of both f and g. That is, domain of $f - g = $ domain of $f \cap$ domain of g.

DEFINITION

The **product $f \cdot g$** is the function defined by

$$(f \cdot g)(x) = f(x) \cdot g(x)$$

The domain of $f \cdot g$ consists of the numbers x that are in the domains of both f and g. That is, domain of $f \cdot g = $ domain of $f \cap$ domain of g.

DEFINITION

The **quotient** $\dfrac{f}{g}$ is the function defined by

$$\left(\frac{f}{g}\right)(x) = \frac{f(x)}{g(x)} \qquad g(x) \neq 0$$

The domain of $\dfrac{f}{g}$ consists of the numbers x for which $g(x) \neq 0$ and that are in the domains of both f and g. That is,

$$\text{domain of } \frac{f}{g} = \{x \,|\, g(x) \neq 0\} \cap \text{domain of } f \cap \text{domain of } g$$

EXAMPLE 11

Operations on Functions

Let f and g be two functions defined as

$$f(x) = \frac{1}{x + 2} \quad \text{and} \quad g(x) = \frac{x}{x - 1}$$

Find the following functions, and determine the domain in each case.

(a) $(f + g)(x)$ (b) $(f - g)(x)$ (c) $(f \cdot g)(x)$ (d) $\left(\dfrac{f}{g}\right)(x)$

Solution

The domain of f is $\{x \,|\, x \neq -2\}$ and the domain of g is $\{x \,|\, x \neq 1\}$.

(a) $(f + g)(x) = f(x) + g(x) = \dfrac{1}{x + 2} + \dfrac{x}{x - 1}$

$$= \frac{x - 1}{(x + 2)(x - 1)} + \frac{x(x + 2)}{(x + 2)(x - 1)} = \frac{x^2 + 3x - 1}{(x + 2)(x - 1)}$$

The domain of $f + g$ consists of those numbers x that are in the domains of both f and g. Therefore, the domain of $f + g$ is $\{x \,|\, x \neq -2, x \neq 1\}$.

(b) $(f - g)(x) = f(x) - g(x) = \dfrac{1}{x + 2} - \dfrac{x}{x - 1}$

$$= \frac{x - 1}{(x + 2)(x - 1)} - \frac{x(x + 2)}{(x + 2)(x - 1)} = \frac{-(x^2 + x + 1)}{(x + 2)(x - 1)}$$

The domain of $f - g$ consists of those numbers x that are in the domains of both f and g. Therefore, the domain of $f - g$ is $\{x \,|\, x \neq -2, x \neq 1\}$.

(c) $(f \cdot g)(x) = f(x) \cdot g(x) = \dfrac{1}{x + 2} \cdot \dfrac{x}{x - 1} = \dfrac{x}{(x + 2)(x - 1)}$

The domain of $f \cdot g$ consists of those numbers x that are in the domains of both f and g. Therefore, the domain of $f \cdot g$ is $\{x \,|\, x \neq -2, x \neq 1\}$.

(d) $\left(\dfrac{f}{g}\right)(x) = \dfrac{f(x)}{g(x)} = \dfrac{\dfrac{1}{x + 2}}{\dfrac{x}{x - 1}} = \dfrac{1}{x + 2} \cdot \dfrac{x - 1}{x} = \dfrac{x - 1}{x(x + 2)}$

The domain of $\dfrac{f}{g}$ consists of the numbers x for which $g(x) \neq 0$ and that are in the domains of both f and g. Since $g(x) = 0$ when $x = 0$, we exclude 0 as well as -2 and 1 from the domain. The domain of $\dfrac{f}{g}$ is $\{x \,|\, x \neq -2, x \neq 0, x \neq 1\}$.

Now Work PROBLEM **67**

In calculus, it is sometimes helpful to view a complicated function as the sum, difference, product, or quotient of simpler functions. For example,

$$F(x) = x^2 + \sqrt{x} \text{ is the sum of } f(x) = x^2 \text{ and } g(x) = \sqrt{x}.$$

$$H(x) = \frac{x^2 - 1}{x^2 + 1} \text{ is the quotient of } f(x) = x^2 - 1 \text{ and } g(x) = x^2 + 1.$$

SUMMARY

Function	A relation between two sets of real numbers so that each number x in the first set, the domain, has corresponding to it exactly one number y in the second set.
	A set of ordered pairs (x, y) or $(x, f(x))$ in which no first element is paired with two different second elements.
	The range is the set of y-values of the function that are the images of the x-values in the domain.
	A function f may be defined implicitly by an equation involving x and y or explicitly by writing $y = f(x)$.
Unspecified domain	If a function f is defined by an equation and no domain is specified, then the domain will be taken to be the largest set of real numbers for which the equation defines a real number.
Function notation	$y = f(x)$
	f is a symbol for the function.
	x is the independent variable, or argument.
	y is the dependent variable.
	$f(x)$ is the value of the function at x, or the image of x.

3.1 Assess Your Understanding

'Are You Prepared?' *Answers are given at the end of these exercises. If you get a wrong answer, read the pages listed in* red.

1. The inequality $-1 < x < 3$ can be written in interval notation as _____. (pp. 120–121)

2. If $x = -2$, the value of the expression $3x^2 - 5x + \dfrac{1}{x}$ is _____. (pp. 20–23)

3. The domain of the variable in the expression $\dfrac{x-3}{x+4}$ is _____. (pp. 20–23)

4. Solve the inequality: $3 - 2x > 5$. Graph the solution set. (pp. 123–126)

5. To rationalize the denominator of $\dfrac{3}{\sqrt{5}-2}$, multiply the numerator and denominator by _____. (p. 75)

6. A quotient is considered rationalized if its denominator contains no _____. (p. 75)

Concepts and Vocabulary

7. If f is a function defined by the equation $y = f(x)$, then x is called the _____ variable, and y is the _____ variable.

8. If the domain of f is all real numbers in the interval $[0, 7]$, and the domain of g is all real numbers in the interval $[-2, 5]$, then the domain of $f + g$ is all real numbers in the interval _____.

9. The domain of $\dfrac{f}{g}$ consists of numbers x for which $g(x)$ ___ 0 that are in the domains of both ___ and ___.

10. If $f(x) = x + 1$ and $g(x) = x^3$, then _____ $= x^3 - (x + 1)$.

11. *True or False* Every relation is a function.

12. *True or False* The domain of $(f \cdot g)(x)$ consists of the numbers x that are in the domains of both f and g.

13. *True or False* If no domain is specified for a function f, then the domain of f is taken to be the set of real numbers.

14. *True or False* The domain of the function $f(x) = \dfrac{x^2 - 4}{x}$ is $\{x \mid x \neq \pm 2\}$.

15. The set of all images of the elements in the domain of a function is called the _____.
    (a) range (b) domain (c) solution set (d) function

16. The independent variable is sometimes referred to as the _____ of the function.
    (a) range (b) value (c) argument (d) definition

17. The expression $\dfrac{f(x+h) - f(x)}{h}$ is called the _____ of f.
    (a) radicand   (b) image
    (c) correspondence (d) difference quotient

18. When written as $y = f(x)$, a function is said to be defined _____.
    (a) explicitly   (b) consistently
    (c) implicitly   (d) rationally

Skill Building

In Problems 19–30, state the domain and range for each relation. Then determine whether each relation represents a function.

19.

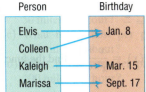

20.

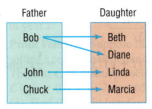

21.

22.

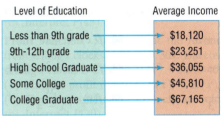

23. $\{(2,6),(-3,6),(4,9),(2,10)\}$ **24.** $\{(-2,5),(-1,3),(3,7),(4,12)\}$ **25.** $\{(1,3),(2,3),(3,3),(4,3)\}$

26. $\{(0,-2),(1,3),(2,3),(3,7)\}$ **27.** $\{(-2,4),(-2,6),(0,3),(3,7)\}$ **28.** $\{(-4,4),(-3,3),(-2,2),(-1,1),(-4,0)\}$

29. $\{(-2,4),(-1,1),(0,0),(1,1)\}$ **30.** $\{(-2,16),(-1,4),(0,3),(1,4)\}$

In Problems 31–42, determine whether the equation defines y as a function of x.

31. $y = 2x^2 - 3x + 4$ **32.** $y = x^3$ **33.** $y = \dfrac{1}{x}$ **34.** $y = |x|$

35. $y^2 = 4 - x^2$ **36.** $y = \pm\sqrt{1 - 2x}$ **37.** $x = y^2$ **38.** $x + y^2 = 1$

39. $y = \sqrt[3]{x}$ **40.** $y = \dfrac{3x - 1}{x + 2}$ **41.** $2x^2 + 3y^2 = 1$ **42.** $x^2 - 4y^2 = 1$

In Problems 43–50, find the following for each function:

 (a) $f(0)$ (b) $f(1)$ (c) $f(-1)$ (d) $f(-x)$ (e) $-f(x)$ (f) $f(x + 1)$ (g) $f(2x)$ (h) $f(x + h)$

43. $f(x) = 3x^2 + 2x - 4$ **44.** $f(x) = -2x^2 + x - 1$ **45.** $f(x) = \dfrac{x}{x^2 + 1}$ **46.** $f(x) = \dfrac{x^2 - 1}{x + 4}$

47. $f(x) = |x| + 4$ **48.** $f(x) = \sqrt{x^2 + x}$ **49.** $f(x) = \dfrac{2x + 1}{3x - 5}$ **50.** $f(x) = 1 - \dfrac{1}{(x + 2)^2}$

In Problems 51–66, find the domain of each function.

51. $f(x) = -5x + 4$ **52.** $f(x) = x^2 + 2$ **53.** $f(x) = \dfrac{x}{x^2 + 1}$ **54.** $f(x) = \dfrac{x^2}{x^2 + 1}$

55. $g(x) = \dfrac{x}{x^2 - 16}$ **56.** $h(x) = \dfrac{2x}{x^2 - 4}$ **57.** $F(x) = \dfrac{x - 2}{x^3 + x}$ **58.** $G(x) = \dfrac{x + 4}{x^3 - 4x}$

59. $h(x) = \sqrt{3x - 12}$ **60.** $G(x) = \sqrt{1 - x}$ **61.** $p(x) = \sqrt{\dfrac{2}{x - 1}}$ **62.** $f(x) = \dfrac{4}{\sqrt{x - 9}}$

63. $f(x) = \dfrac{x}{\sqrt{x - 4}}$ **64.** $f(x) = \dfrac{-x}{\sqrt{-x - 2}}$ **65.** $P(t) = \dfrac{\sqrt{t - 4}}{3t - 21}$ **66.** $h(z) = \dfrac{\sqrt{z + 3}}{z - 2}$

In Problems 67–76, for the given functions f and g, find the following. For parts (a)–(d), also find the domain.

 (a) $(f + g)(x)$ (b) $(f - g)(x)$ (c) $(f \cdot g)(x)$ (d) $\left(\dfrac{f}{g}\right)(x)$

 (e) $(f + g)(3)$ (f) $(f - g)(4)$ (g) $(f \cdot g)(2)$ (h) $\left(\dfrac{f}{g}\right)(1)$

67. $f(x) = 3x + 4;\ \ g(x) = 2x - 3$ **68.** $f(x) = 2x + 1;\ \ g(x) = 3x - 2$

69. $f(x) = x - 1;\ \ g(x) = 2x^2$ **70.** $f(x) = 2x^2 + 3;\ \ g(x) = 4x^3 + 1$

3.2 The Graph of a Function

PREPARING FOR THIS SECTION *Before getting started, review the following:*

- Graphs of Equations (Section 2.2, pp. 157–159)
- Intercepts (Section 2.2, pp. 159–160)

Now Work the **'Are You Prepared?'** problems on page 218.

OBJECTIVES 1 Identify the Graph of a Function (p. 214)
2 Obtain Information from or about the Graph of a Function (p. 215)

In applications, a graph often demonstrates more clearly the relationship between two variables than, say, an equation or table. For example, Table 1 shows the average price of gasoline in the United States for the years 1985–2014 (adjusted for inflation, based on 2014 dollars). If we plot these data and then connect the points, we obtain Figure 13.

Table 1

Year	Price	Year	Price	Year	Price
1985	2.55	1995	1.72	2005	2.74
1986	1.90	1996	1.80	2006	3.01
1987	1.89	1997	1.76	2007	3.19
1988	1.81	1998	1.49	2008	3.56
1989	1.87	1999	1.61	2009	2.58
1990	2.03	2000	2.03	2010	3.00
1991	1.91	2001	1.90	2011	3.69
1992	1.82	2002	1.76	2012	3.72
1993	1.74	2003	1.99	2013	3.54
1994	1.71	2004	2.31	2014	3.43

Source: U.S. Energy Information Administration

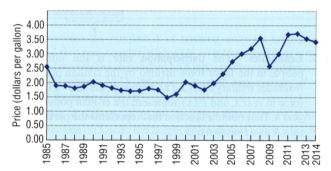

Figure 13 Average retail price of gasoline (2014 dollars)
Source: U.S. Energy Information Administration

We can see from the graph that the price of gasoline (adjusted for inflation) stayed roughly the same from 1986 to 1991 and rose rapidly from 2002 to 2008. The graph also shows that the lowest price occurred in 1998. To learn information such as this from an equation requires that some calculations be made.

Look again at Figure 13. The graph shows that for each date on the horizontal axis, there is only one price on the vertical axis. The graph represents a function, although the exact rule for getting from date to price is not given.

When a function is defined by an equation in x and y, the **graph of the function** is the graph of the equation; that is, it is the set of points (x, y) in the xy-plane that satisfy the equation.

1 Identify the Graph of a Function

Not every collection of points in the xy-plane represents the graph of a function. Remember, for a function, each number x in the domain has exactly one image y in the range. This means that the graph of a function cannot contain two points with the same x-coordinate and different y-coordinates. Therefore, the graph of a function must satisfy the following **vertical-line test**.

In Words
If any vertical line intersects a graph at more than one point, the graph is not the graph of a function.

THEOREM **Vertical-Line Test**

A set of points in the xy-plane is the graph of a function if and only if every vertical line intersects the graph in at most one point.

EXAMPLE 1

Identifying the Graph of a Function

Which of the graphs in Figure 14 are graphs of functions?

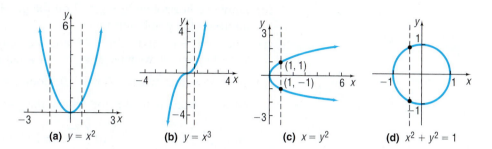

Figure 14

(a) $y = x^2$ **(b)** $y = x^3$ **(c)** $x = y^2$ **(d)** $x^2 + y^2 = 1$

Solution   The graphs in Figures 14(a) and 14(b) are graphs of functions, because every vertical line intersects each graph in at most one point. The graphs in Figures 14(c) and 14(d) are not graphs of functions, because there is a vertical line that intersects each graph in more than one point. Notice in Figure 14(c) that the input 1 corresponds to two outputs, -1 and 1. This is why the graph does not represent a function. ●

 **Now Work** PROBLEM 17

2 Obtain Information from or about the Graph of a Function

If (x, y) is a point on the graph of a function f, then y is the value of f at x; that is, $y = f(x)$. Also if $y = f(x)$, then (x, y) is a point on the graph of f. For example, if $(-2, 7)$ is on the graph of f, then $f(-2) = 7$, and if $f(5) = 8$, then the point $(5, 8)$ is on the graph of $y = f(x)$. The next example illustrates how to obtain information about a function if its graph is given.

EXAMPLE 2

Obtaining Information from the Graph of a Function

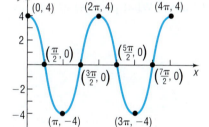

Figure 15

Let f be the function whose graph is given in Figure 15. (The graph of f might represent the distance y that the bob of a pendulum is from its *at-rest* position at time x. Negative values of y mean that the pendulum is to the left of the at-rest position, and positive values of y mean that the pendulum is to the right of the at-rest position.)

(a) What are $f(0)$, $f\left(\dfrac{3\pi}{2}\right)$, and $f(3\pi)$?

(b) What is the domain of f?

(c) What is the range of f?

(d) List the intercepts. (Recall that these are the points, if any, where the graph crosses or touches the coordinate axes.)

(e) How many times does the line $y = 2$ intersect the graph?

(f) For what values of x does $f(x) = -4$?

(g) For what values of x is $f(x) > 0$?

Solution   (a) Since $(0, 4)$ is on the graph of f, the y-coordinate 4 is the value of f at the x-coordinate 0; that is, $f(0) = 4$. In a similar way, when $x = \dfrac{3\pi}{2}$, then $y = 0$, so $f\left(\dfrac{3\pi}{2}\right) = 0$. When $x = 3\pi$, then $y = -4$, so $f(3\pi) = -4$.

(b) To determine the domain of f, notice that the points on the graph of f have x-coordinates between 0 and 4π, inclusive; and for each number x between 0 and 4π, there is a point $(x, f(x))$ on the graph. The domain of f is $\{x \mid 0 \le x \le 4\pi\}$ or the interval $[0, 4\pi]$.

(c) The points on the graph all have y-coordinates between -4 and 4, inclusive; and for each such number y, there is at least one number x in the domain. The range of f is $\{y \mid -4 \le y \le 4\}$ or the interval $[-4, 4]$.

(d) The intercepts are the points

$$(0, 4), \left(\frac{\pi}{2}, 0\right), \left(\frac{3\pi}{2}, 0\right), \left(\frac{5\pi}{2}, 0\right), \quad \text{and} \quad \left(\frac{7\pi}{2}, 0\right)$$

(e) Draw the horizontal line $y = 2$ on the graph in Figure 15. Notice that the line intersects the graph four times.

(f) Since $(\pi, -4)$ and $(3\pi, -4)$ are the only points on the graph for which $y = f(x) = -4$, we have $f(x) = -4$ when $x = \pi$ and $x = 3\pi$.

(g) To determine where $f(x) > 0$, look at Figure 15 and determine the x-values from 0 to 4π for which the y-coordinate is positive. This occurs on $\left[0, \frac{\pi}{2}\right) \cup \left(\frac{3\pi}{2}, \frac{5\pi}{2}\right) \cup \left(\frac{7\pi}{2}, 4\pi\right]$. Using inequality notation, $f(x) > 0$ for $0 \le x < \frac{\pi}{2}$ or $\frac{3\pi}{2} < x < \frac{5\pi}{2}$ or $\frac{7\pi}{2} < x \le 4\pi$.   ●

When the graph of a function is given, its domain may be viewed as the shadow created by the graph on the x-axis by vertical beams of light. Its range can be viewed as the shadow created by the graph on the y-axis by horizontal beams of light. Try this technique with the graph given in Figure 15.

✏ **Now Work** PROBLEMS **11** AND **15**

EXAMPLE 3

Obtaining Information about the Graph of a Function

Consider the function: $f(x) = \dfrac{x + 1}{x + 2}$

(a) Find the domain of f.

(b) Is the point $\left(1, \dfrac{1}{2}\right)$ on the graph of f?

(c) If $x = 2$, what is $f(x)$? What point is on the graph of f?

(d) If $f(x) = 2$, what is x? What point is on the graph of f?

(e) What are the x-intercepts of the graph of f (if any)? What point(s) are on the graph of f?

Solution

(a) The domain of f is $\{x \mid x \ne -2\}$.

(b) When $x = 1$, then

$$f(1) = \frac{1 + 1}{1 + 2} = \frac{2}{3} \qquad f(x) = \frac{x + 1}{x + 2}$$

The point $\left(1, \dfrac{2}{3}\right)$ is on the graph of f; the point $\left(1, \dfrac{1}{2}\right)$ is not.

(c) If $x = 2$, then

$$f(2) = \frac{2 + 1}{2 + 2} = \frac{3}{4}$$

The point $\left(2, \dfrac{3}{4}\right)$ is on the graph of f.

(d) If $f(x) = 2$, then

$$\frac{x + 1}{x + 2} = 2 \qquad\qquad f(x) = 2$$
$$x + 1 = 2(x + 2) \qquad \text{Multiply both sides by } x + 2.$$
$$x + 1 = 2x + 4 \qquad \text{Distribute.}$$
$$x = -3 \qquad\qquad \text{Solve for } x.$$

If $f(x) = 2$, then $x = -3$. The point $(-3, 2)$ is on the graph of f.

(e) The x-intercepts of the graph of f are the real solutions of the equation $f(x) = 0$ that are in the domain of f.

$$\frac{x + 1}{x + 2} = 0$$

$x + 1 = 0$ *Multiply both sides by x + 2.*

$x = -1$ *Subtract 1 from both sides.*

The only real solution of the equation $f(x) = \dfrac{x + 1}{x + 2} = 0$ is $x = -1$, so -1 is the only x-intercept. Since $f(-1) = 0$, the point $(-1, 0)$ is on the graph of f. ●

Now Work PROBLEM 27

EXAMPLE 4

Average Cost Function

The average cost $\overline{C}$ per computer of manufacturing x computers per day is given by the function

$$\overline{C}(x) = 0.56x^2 - 34.39x + 1212.57 + \frac{20,000}{x}$$

Determine the average cost of manufacturing:

(a) 30 computers in a day
(b) 40 computers in a day
(c) 50 computers in a day
(d) Graph the function $\overline{C} = \overline{C}(x), 0 < x \leq 80$.
(e) Create a TABLE with TblStart $= 1$ and ΔTbl $= 1$. Which value of x minimizes the average cost?

Solution

(a) The average cost per computer of manufacturing $x = 30$ computers is

$$\overline{C}(30) = 0.56(30)^2 - 34.39(30) + 1212.57 + \frac{20,000}{30} = \$1351.54$$

(b) The average cost per computer of manufacturing $x = 40$ computers is

$$\overline{C}(40) = 0.56(40)^2 - 34.39(40) + 1212.57 + \frac{20,000}{40} = \$1232.97$$

(c) The average cost per computer of manufacturing $x = 50$ computers is

$$\overline{C}(50) = 0.56(50)^2 - 34.39(50) + 1212.57 + \frac{20,000}{50} = \$1293.07$$

(d) See Figure 16 for the graph of $\overline{C} = \overline{C}(x)$.
(e) With the function $\overline{C} = \overline{C}(x)$ in Y_1, we create Table 2. We scroll down until we find a value of x for which Y_1 is smallest. Table 3 shows that manufacturing $x = 41$ computers minimizes the average cost at \$1231.74 per computer.

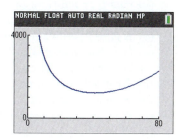

Figure 16 $\overline{C}(x) = 0.56x^2 - 34.39x + 1212.57 + \dfrac{20,000}{x}$ **Table 2**   **Table 3**   ●

Now Work PROBLEM 35

SUMMARY

Graph of a Function The collection of points (x, y) that satisfies the equation $y = f(x)$.

Vertical-Line Test A collection of points is the graph of a function if and only if every vertical line intersects the graph in at most one point.

3.2 Assess Your Understanding

'Are You Prepared?' *Answers are given at the end of these exercises. If you get a wrong answer, read the pages listed in* red.

1. The intercepts of the equation $x^2 + 4y^2 = 16$ are _____. (pp. 159–160)

2. *True or False* The point $(-2, -6)$ is on the graph of the equation $x = 2y - 2$. (pp. 157–159)

Concepts and Vocabulary

3. A set of points in the xy-plane is the graph of a function if and only if every _____ line intersects the graph in at most one point.

4. If the point $(5, -3)$ is a point on the graph of f, then $f(\underline{}) = \underline{}$.

5. Find a so that the point $(-1, 2)$ is on the graph of $f(x) = ax^2 + 4$.

6. *True or False* Every graph represents a function.

7. *True or False* The graph of a function $y = f(x)$ always crosses the y-axis.

8. *True or False* The y-intercept of the graph of the function $y = f(x)$, whose domain is all real numbers, is $f(0)$.

9. If a function is defined by an equation in x and y, then the set of points (x, y) in the xy-plane that satisfy the equation is called _____.
   (a) the domain of the function   (b) the range of the function
   (c) the graph of the function   (d) the relation of the function

10. The graph of a function $y = f(x)$ can have more than one of which type of intercept?
    (a) x-intercept   (b) y-intercept   (c) both   (d) neither

Skill Building

11. Use the given graph of the function f to answer parts (a)–(n).

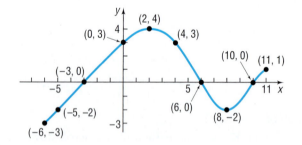

(a) Find $f(0)$ and $f(-6)$.
(b) Find $f(6)$ and $f(11)$.
(c) Is $f(3)$ positive or negative?
(d) Is $f(-4)$ positive or negative?
(e) For what values of x is $f(x) = 0$?
(f) For what values of x is $f(x) > 0$?
(g) What is the domain of f?
(h) What is the range of f?
(i) What are the x-intercepts?
(j) What is the y-intercept?

(k) How often does the line $y = \dfrac{1}{2}$ intersect the graph?

(l) How often does the line $x = 5$ intersect the graph?
(m) For what values of x does $f(x) = 3$?
(n) For what values of x does $f(x) = -2$?

12. Use the given graph of the function f to answer parts (a)–(n).

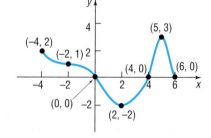

(a) Find $f(0)$ and $f(6)$.
(b) Find $f(2)$ and $f(-2)$.
(c) Is $f(3)$ positive or negative?
(d) Is $f(-1)$ positive or negative?
(e) For what values of x is $f(x) = 0$?
(f) For what values of x is $f(x) < 0$?
(g) What is the domain of f?
(h) What is the range of f?
(i) What are the x-intercepts?
(j) What is the y-intercept?
(k) How often does the line $y = -1$ intersect the graph?
(l) How often does the line $x = 1$ intersect the graph?
(m) For what value of x does $f(x) = 3$?
(n) For what value of x does $f(x) = -2$?

In Problems 13–24, determine whether the graph is that of a function by using the vertical-line test. If it is, use the graph to find:
   (a) The domain and range   *(b) The intercepts, if any*   *(c) Any symmetry with respect to the x-axis, the y-axis, or the origin*

13.
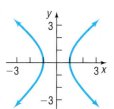

14.
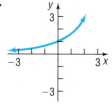

15.
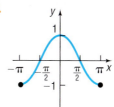

16.
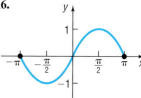

17.
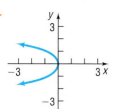

18.
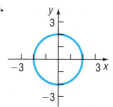

19.
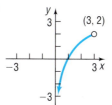

20.
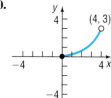

21.
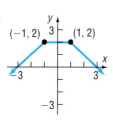

22.
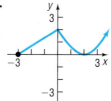

23.
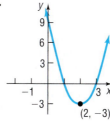

24.
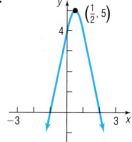

In Problems 25–30, answer the questions about the given function.

25. $f(x) = 2x^2 - x - 1$
  (a) Is the point $(-1, 2)$ on the graph of f?
  (b) If $x = -2$, what is $f(x)$? What point is on the graph of f?
  (c) If $f(x) = -1$, what is x? What point(s) are on the graph of f?
  (d) What is the domain of f?
  (e) List the x-intercepts, if any, of the graph of f.
  (f) List the y-intercept, if there is one, of the graph of f.

26. $f(x) = -3x^2 + 5x$
  (a) Is the point $(-1, 2)$ on the graph of f?
  (b) If $x = -2$, what is $f(x)$? What point is on the graph of f?
  (c) If $f(x) = -2$, what is x? What point(s) are on the graph of f?
  (d) What is the domain of f?
  (e) List the x-intercepts, if any, of the graph of f.
  (f) List the y-intercept, if there is one, of the graph of f.

27. $f(x) = \dfrac{x + 2}{x - 6}$
  (a) Is the point $(3, 14)$ on the graph of f?
  (b) If $x = 4$, what is $f(x)$? What point is on the graph of f?
  (c) If $f(x) = 2$, what is x? What point(s) are on the graph of f?
  (d) What is the domain of f?
  (e) List the x-intercepts, if any, of the graph of f.
  (f) List the y-intercept, if there is one, of the graph of f.

28. $f(x) = \dfrac{x^2 + 2}{x + 4}$
  (a) Is the point $\left(1, \dfrac{3}{5}\right)$ on the graph of f?

  (b) If $x = 0$, what is $f(x)$? What point is on the graph of f?
  (c) If $f(x) = \dfrac{1}{2}$, what is x? What point(s) are on the graph of f?
  (d) What is the domain of f?
  (e) List the x-intercepts, if any, of the graph of f.
  (f) List the y-intercept, if there is one, of the graph of f.

29. $f(x) = \dfrac{2x^2}{x^4 + 1}$
  (a) Is the point $(-1, 1)$ on the graph of f?
  (b) If $x = 2$, what is $f(x)$? What point is on the graph of f?
  (c) If $f(x) = 1$, what is x? What point(s) are on the graph of f?
  (d) What is the domain of f?
  (e) List the x-intercepts, if any, of the graph of f.
  (f) List the y-intercept, if there is one, of the graph of f.

30. $f(x) = \dfrac{2x}{x - 2}$
  (a) Is the point $\left(\dfrac{1}{2}, -\dfrac{2}{3}\right)$ on the graph of f?
  (b) If $x = 4$, what is $f(x)$? What point is on the graph of f?
  (c) If $f(x) = 1$, what is x? What point(s) are on the graph of f?
  (d) What is the domain of f?
  (e) List the x-intercepts, if any, of the graph of f.
  (f) List the y-intercept, if there is one, of the graph of f.

A function f is even if and only if, whenever the point (x, y) is on the graph of f, the point $(-x, y)$ is also on the graph. Using function notation, we define an even function as follows:

DEFINITION

A function f is **even** if, for every number x in its domain, the number $-x$ is also in the domain and

$$f(-x) = f(x)$$

A function f is odd if and only if, whenever the point (x, y) is on the graph of f, the point $(-x, -y)$ is also on the graph. Using function notation, we define an odd function as follows:

DEFINITION

A function f is **odd** if, for every number x in its domain, the number $-x$ is also in the domain and

$$f(-x) = -f(x)$$

Refer to page 162, where the tests for symmetry are listed. The following results are then evident.

THEOREM

A function is even if and only if its graph is symmetric with respect to the y-axis. A function is odd if and only if its graph is symmetric with respect to the origin.

EXAMPLE 1   **Determining Even and Odd Functions from the Graph**

Determine whether each graph given in Figure 17 is the graph of an even function, an odd function, or a function that is neither even nor odd.

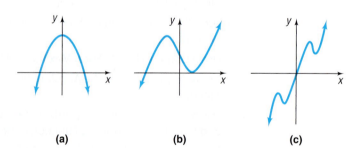

Figure 17   (a)   (b)   (c)

Solution   (a) The graph in Figure 17(a) is that of an even function, because the graph is symmetric with respect to the y-axis.

(b) The function whose graph is given in Figure 17(b) is neither even nor odd, because the graph is neither symmetric with respect to the y-axis nor symmetric with respect to the origin.

(c) The function whose graph is given in Figure 17(c) is odd, because its graph is symmetric with respect to the origin.

━━━ **Now Work** PROBLEMS 25(a), (b), AND (d)

2 Identify Even and Odd Functions from an Equation

EXAMPLE 2

Identifying Even and Odd Functions Algebraically

Determine whether each of the following functions is even, odd, or neither. Then determine whether the graph is symmetric with respect to the y-axis, with respect to the origin, or neither.

(a) $f(x) = x^2 - 5$　　　　　(b) $g(x) = x^3 - 1$
(c) $h(x) = 5x^3 - x$　　　　　(d) $F(x) = |x|$

Solution

(a) To determine whether f is even, odd, or neither, replace x by $-x$ in $f(x) = x^2 - 5$.

$$f(-x) = (-x)^2 - 5 = x^2 - 5 = f(x)$$

Since $f(-x) = f(x)$, the function is even, and the graph of f is symmetric with respect to the y-axis.

(b) Replace x by $-x$ in $g(x) = x^3 - 1$.

$$g(-x) = (-x)^3 - 1 = -x^3 - 1$$

Since $g(-x) \neq g(x)$ and $g(-x) \neq -g(x) = -(x^3 - 1) = -x^3 + 1$, the function is neither even nor odd. The graph of g is not symmetric with respect to the y-axis, nor is it symmetric with respect to the origin.

(c) Replace x by $-x$ in $h(x) = 5x^3 - x$.

$$h(-x) = 5(-x)^3 - (-x) = -5x^3 + x = -(5x^3 - x) = -h(x)$$

Since $h(-x) = -h(x)$, h is an odd function, and the graph of h is symmetric with respect to the origin.

(d) Replace x by $-x$ in $F(x) = |x|$.

$$F(-x) = |-x| = |-1| \cdot |x| = |x| = F(x)$$

Since $F(-x) = F(x)$, F is an even function, and the graph of F is symmetric with respect to the y-axis. ●

 Now Work PROBLEM 37

3 Use a Graph to Determine Where a Function Is Increasing, Decreasing, or Constant

 Consider the graph given in Figure 18. If you look from left to right along the graph of the function, you will notice that parts of the graph are going up, parts are going down, and parts are horizontal. In such cases, the function is described as *increasing*, *decreasing*, or *constant*, respectively.

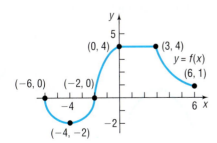

Figure 18

EXAMPLE 3

Determining Where a Function Is Increasing, Decreasing, or Constant from Its Graph

Determine the values of x for which the function in Figure 18 is increasing. Where is it decreasing? Where is it constant?

The absolute maximum and absolute minimum of a function f are sometimes called the **extreme values** of f on I.

The absolute maximum or absolute minimum of a function f may not exist. Let's look at some examples.

EXAMPLE 5

Finding the Absolute Maximum and the Absolute Minimum from the Graph of a Function

For each graph of a function $y = f(x)$ in Figure 23, find the absolute maximum and the absolute minimum, if they exist. Also, find any local maxima or local minima.

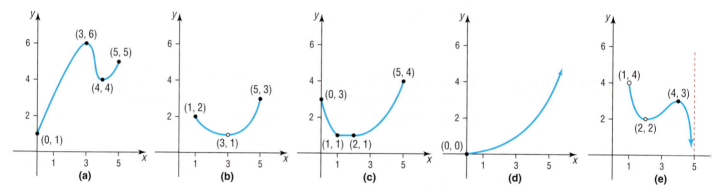

Figure 23

Solution

(a) The function f whose graph is given in Figure 23(a) has the closed interval $[0, 5]$ as its domain. The largest value of f is $f(3) = 6$, the absolute maximum. The smallest value of f is $f(0) = 1$, the absolute minimum. The function has a local maximum of 6 at $x = 3$ and a local minimum of 4 at $x = 4$.

WARNING A function may have an absolute maximum or an absolute minimum at an endpoint but not a local maximum or a local minimum. Why? Local maxima and local minima are found over some open interval I, and this interval cannot be created around an endpoint. ∎

(b) The function f whose graph is given in Figure 23(b) has the domain $\{x \mid 1 \leq x \leq 5, x \neq 3\}$. Note that we exclude 3 from the domain because of the "hole" at $(3, 1)$. The largest value of f on its domain is $f(5) = 3$, the absolute maximum. There is no absolute minimum. Do you see why? As you trace the graph, getting closer to the point $(3, 1)$, there is no single smallest value. [As soon as you claim a smallest value, we can trace closer to $(3, 1)$ and get a smaller value!] The function has no local maxima or minima.

(c) The function f whose graph is given in Figure 23(c) has the interval $[0, 5]$ as its domain. The absolute maximum of f is $f(5) = 4$. The absolute minimum is 1. Notice that the absolute minimum 1 occurs at any number in the interval $[1, 2]$. The function has a local minimum value of 1 at every x in the interval $[1, 2]$, but it has no local maximum value.

(d) The function f given in Figure 23(d) has the interval $[0, \infty)$ as its domain. The function has no absolute maximum; the absolute minimum is $f(0) = 0$. The function has no local maximum or local minimum.

(e) The function f in Figure 23(e) has the domain $\{x \mid 1 < x < 5, x \neq 2\}$. The function has no absolute maximum and no absolute minimum. Do you see why? The function has a local maximum value of 3 at $x = 4$, but no local minimum value. ●

In calculus, there is a theorem with conditions that guarantee a function will have an absolute maximum and an absolute minimum.

THEOREM

Extreme Value Theorem

If f is a continuous function* whose domain is a closed interval $[a, b]$, then f has an absolute maximum and an absolute minimum on $[a, b]$.

*Although a precise definition requires calculus, we'll agree for now that a continuous function is one whose graph has no gaps or holes and can be traced without lifting the pencil from the paper.

The absolute maximum (minimum) can be found by selecting the largest (smallest) value of f from the following list:

1. The values of f at any local maxima or local minima of f in $[a, b]$.
2. The value of f at each endpoint of $[a, b]$ — that is, $f(a)$ and $f(b)$.

For example, the graph of the function f given in Figure 23(a) is continuous on the closed interval $[0, 5]$. The Extreme Value Theorem guarantees that f has extreme values on $[0, 5]$. To find them, we list

1. The value of f at the local extrema: $f(3) = 6, f(4) = 4$
2. The value of f at the endpoints: $f(0) = 1, f(5) = 5$

The largest of these, 6, is the absolute maximum; the smallest of these, 1, is the absolute minimum.

 Now Work PROBLEM 49

 6 Use a Graphing Utility to Approximate Local Maxima and Local Minima and to Determine Where a Function Is Increasing or Decreasing

 To locate the exact value at which a function f has a local maximum or a local minimum usually requires calculus. However, a graphing utility may be used to approximate these values using the MAXIMUM and MINIMUM features.

| EXAMPLE 6 | **Using a Graphing Utility to Approximate Local Maxima and Minima and to Determine Where a Function Is Increasing or Decreasing** |

(a) Use a graphing utility to graph $f(x) = 6x^3 - 12x + 5$ for $-2 < x < 2$. Approximate where f has a local maximum and where f has a local minimum.
(b) Determine where f is increasing and where it is decreasing.

Solution (a) Graphing utilities have a feature that finds the maximum or minimum point of a graph within a given interval. Graph the function f for $-2 < x < 2$. The MAXIMUM and MINIMUM commands require us to first determine the open interval I. The graphing utility will then approximate the maximum or minimum value in the interval. Using MAXIMUM, we find that the local maximum value is 11.53 and that it occurs at $x = -0.82$, rounded to two decimal places. See Figure 24(a). Using MINIMUM, we find that the local minimum value is -1.53 and that it occurs at $x = 0.82$, rounded to two decimal places. See Figure 24(b).

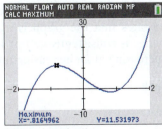 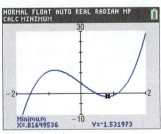

Figure 24   **(a)** Local maximum   **(b)** Local minimum

(b) Looking at Figures 24(a) and (b), we see that the graph of f is increasing from $x = -2$ to $x = -0.82$ and from $x = 0.82$ to $x = 2$, so f is increasing on the intervals $(-2, -0.82)$ and $(0.82, 2)$, or for $-2 < x < -0.82$ and $0.82 < x < 2$. The graph is decreasing from $x = -0.82$ to $x = 0.82$, so f is decreasing on the interval $(-0.82, 0.82)$, or for $-0.82 < x < 0.82$. ●

 Now Work PROBLEM 57

7 Find the Average Rate of Change of a Function

In Section 2.3, we said that the slope of a line can be interpreted as the average rate of change. To find the average rate of change of a function between any two points on its graph, calculate the slope of the line containing the two points.

DEFINITION

If a and $b, a \neq b$, are in the domain of a function $y = f(x)$, the **average rate of change of f** from a to b is defined as

$$\text{Average rate of change} = \frac{\Delta y}{\Delta x} = \frac{f(b) - f(a)}{b - a} \qquad a \neq b \qquad \textbf{(1)}$$

> **In Words**
> The symbol Δ is the Greek capital letter delta and is read "change in."

The symbol Δy in equation (1) is the "change in y," and Δx is the "change in x." The average rate of change of f is the change in y divided by the change in x.

EXAMPLE 7

Finding the Average Rate of Change

Find the average rate of change of $f(x) = 3x^2$:

(a) From 1 to 3   (b) From 1 to 5   (c) From 1 to 7

Solution

(a) The average rate of change of $f(x) = 3x^2$ from 1 to 3 is

$$\frac{\Delta y}{\Delta x} = \frac{f(3) - f(1)}{3 - 1} = \frac{27 - 3}{3 - 1} = \frac{24}{2} = 12$$

(b) The average rate of change of $f(x) = 3x^2$ from 1 to 5 is

$$\frac{\Delta y}{\Delta x} = \frac{f(5) - f(1)}{5 - 1} = \frac{75 - 3}{5 - 1} = \frac{72}{4} = 18$$

(c) The average rate of change of $f(x) = 3x^2$ from 1 to 7 is

$$\frac{\Delta y}{\Delta x} = \frac{f(7) - f(1)}{7 - 1} = \frac{147 - 3}{7 - 1} = \frac{144}{6} = 24$$

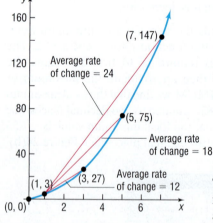

Figure 25 $f(x) = 3x^2$

See Figure 25 for a graph of $f(x) = 3x^2$. The function f is increasing for $x > 0$. The fact that the average rate of change is positive for any $x_1, x_2, x_1 \neq x_2$, in the interval $(1, 7)$ indicates that the graph is increasing on $1 < x < 7$. Further, the average rate of change is consistently getting larger for $1 < x < 7$, which indicates that the graph is increasing at an increasing rate.

Now Work PROBLEM 65

The Secant Line

The average rate of change of a function has an important geometric interpretation. Look at the graph of $y = f(x)$ in Figure 26. Two points are labeled on the graph: $(a, f(a))$ and $(b, f(b))$. The line containing these two points is called the **secant line**; its slope is

$$m_{\text{sec}} = \frac{f(b) - f(a)}{b - a}$$

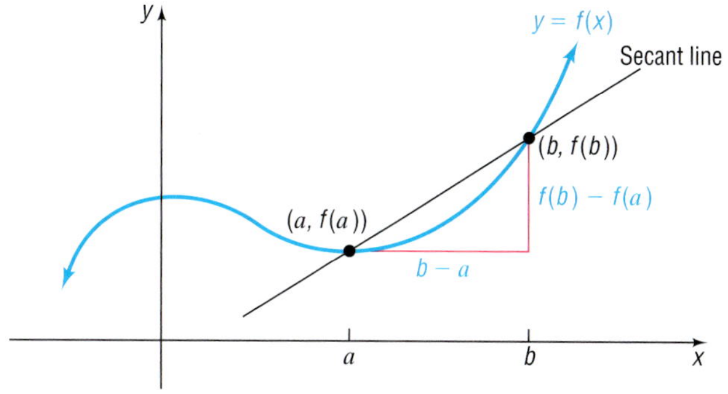

Figure 26 Secant line

THEOREM

Slope of the Secant Line

The average rate of change of a function from a to b equals the slope of the secant line containing the two points $(a, f(a))$ and $(b, f(b))$ on its graph.

EXAMPLE 8   **Finding the Equation of a Secant Line**

Suppose that $g(x) = 3x^2 - 2x + 3$.

(a) Find the average rate of change of g from -2 to 1.

(b) Find an equation of the secant line containing $(-2, g(-2))$ and $(1, g(1))$.

(c) Using a graphing utility, draw the graph of g and the secant line obtained in part (b) on the same screen.

Solution   (a) The average rate of change of $g(x) = 3x^2 - 2x + 3$ from -2 to 1 is

$$\text{Average rate of change} = \frac{g(1) - g(-2)}{1 - (-2)}$$

$$= \frac{4 - 19}{3} \qquad \begin{aligned} g(1) &= 3(1)^2 - 2(1) + 3 = 4 \\ g(-2) &= 3(-2)^2 - 2(-2) + 3 = 19 \end{aligned}$$

$$= -\frac{15}{3} = -5$$

(b) The slope of the secant line containing $(-2, g(-2)) = (-2, 19)$ and $(1, g(1)) = (1, 4)$ is $m_{\text{sec}} = -5$. Use the point–slope form to find an equation of the secant line.

$$y - y_1 = m_{\text{sec}}(x - x_1) \qquad \textit{Point–slope form of the secant line}$$

$$y - 19 = -5(x - (-2)) \qquad x_1 = -2, y_1 = g(-2) = 19, m_{\text{sec}} = -5$$

$$y - 19 = -5x - 10 \qquad \textit{Distribute.}$$

$$y = -5x + 9 \qquad \textit{Slope–intercept form of the secant line}$$

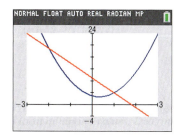

Figure 27 Graph of g and the secant line

(c) Figure 27 shows the graph of g along with the secant line $y = -5x + 9$.

Now Work   PROBLEM 71

65. Find the average rate of change of $f(x) = -2x^2 + 4$:
  (a) From 0 to 2
  (b) From 1 to 3
  (c) From 1 to 4

66. Find the average rate of change of $f(x) = -x^3 + 1$:
  (a) From 0 to 2
  (b) From 1 to 3
  (c) From -1 to 1

67. Find the average rate of change of $g(x) = x^3 - 2x + 1$:
  (a) From -3 to -2
  (b) From -1 to 1
  (c) From 1 to 3

68. Find the average rate of change of $h(x) = x^2 - 2x + 3$:
  (a) From -1 to 1
  (b) From 0 to 2
  (c) From 2 to 5

69. $f(x) = 5x - 2$
  (a) Find the average rate of change from 1 to 3.
  (b) Find an equation of the secant line containing $(1, f(1))$ and $(3, f(3))$.

70. $f(x) = -4x + 1$
  (a) Find the average rate of change from 2 to 5.
  (b) Find an equation of the secant line containing $(2, f(2))$ and $(5, f(5))$.

71. $g(x) = x^2 - 2$
  (a) Find the average rate of change from -2 to 1.
  (b) Find an equation of the secant line containing $(-2, g(-2))$ and $(1, g(1))$.

72. $g(x) = x^2 + 1$
  (a) Find the average rate of change from -1 to 2.
  (b) Find an equation of the secant line containing $(-1, g(-1))$ and $(2, g(2))$.

73. $h(x) = x^2 - 2x$
  (a) Find the average rate of change from 2 to 4.
  (b) Find an equation of the secant line containing $(2, h(2))$ and $(4, h(4))$.

74. $h(x) = -2x^2 + x$
  (a) Find the average rate of change from 0 to 3.
  (b) Find an equation of the secant line containing $(0, h(0))$ and $(3, h(3))$.

Mixed Practice

75. $g(x) = x^3 - 27x$
  (a) Determine whether g is even, odd, or neither.
  (b) There is a local minimum value of -54 at 3. Determine the local maximum value.

76. $f(x) = -x^3 + 12x$
  (a) Determine whether f is even, odd, or neither.
  (b) There is a local maximum value of 16 at 2. Determine the local minimum value.

77. $F(x) = -x^4 + 8x^2 + 8$
  (a) Determine whether F is even, odd, or neither.
  (b) There is a local maximum value of 24 at $x = 2$. Determine a second local maximum value.
  (c) Suppose the area under the graph of F between $x = 0$ and $x = 3$ that is bounded from below by the x-axis is 47.4 square units. Using the result from part (a), determine the area under the graph of F between $x = -3$ and $x = 0$ that is bounded from below by the x-axis.

78. $G(x) = -x^4 + 32x^2 + 144$
  (a) Determine whether G is even, odd, or neither.
  (b) There is a local maximum value of 400 at $x = 4$. Determine a second local maximum value.
  (c) Suppose the area under the graph of G between $x = 0$ and $x = 6$ that is bounded from below by the x-axis is 1612.8 square units. Using the result from part (a), determine the area under the graph of G between $x = -6$ and $x = 0$ that is bounded from below by the x-axis.

Applications and Extensions

79. Minimum Average Cost The average cost per hour in dollars, $\overline{C}$, of producing x riding lawn mowers can be modeled by the function

$$\overline{C}(x) = 0.3x^2 + 21x - 251 + \frac{2500}{x}$$

  (a) Use a graphing utility to graph $\overline{C} = \overline{C}(x)$.
  (b) Determine the number of riding lawn mowers to produce in order to minimize average cost.
  (c) What is the minimum average cost?

80. Medicine Concentration The concentration C of a medication in the bloodstream t hours after being administered is modeled by the function

$$C(t) = -0.002x^4 + 0.039t^3 - 0.285t^2 + 0.766t + 0.085$$

  (a) After how many hours will the concentration be highest?
  (b) A woman nursing a child must wait until the concentration is below 0.5 before she can feed him. After taking the medication, how long must she wait before feeding her child?

81. Data Plan Cost The monthly cost C, in dollars, for wireless data plans with x gigabytes of data included is shown in the table below. Since each input value for x corresponds to exactly one output value for C, the plan cost is a function of the number of data gigabytes. Thus $C(x)$ represents the monthly cost for a wireless data plan with x gigabytes included.

GB	Cost ($)	GB	Cost ($)
4	70	20	150
6	80	30	225
10	100	40	300
15	130	50	375

  (a) Plot the points (4, 70), (6, 80), (10, 100), and so on in a Cartesian plane.
  (b) Draw a line segment from the point (10, 100) to (30, 225). What does the slope of this line segment represent?

(c) Find the average rate of change of the monthly cost from 4 to 10 gigabytes.

(d) Find the average rate of change of the monthly cost from 10 to 30 gigabytes.

(e) Find the average rate of change of the monthly cost from 30 to 50 gigabytes.

(f) What is happening to the average rate of change as the gigabytes of data increase?

82. **National Debt** The size of the total debt owed by the United States federal government continues to grow. In fact, according to the Department of the Treasury, the debt per person living in the United States is approximately $53,000 (or over $140,000 per U.S. household). The following data represent the U.S. debt for the years 2001–2013. Since the debt D depends on the year y, and each input corresponds to exactly one output, the debt is a function of the year. So $D(y)$ represents the debt for each year y.

Year	Debt (billions of dollars)	Year	Debt (billions of dollars)
2001	5807	2008	10,025
2002	6228	2009	11,910
2003	6783	2010	13,562
2004	7379	2011	14,790
2005	7933	2012	16,066
2006	8507	2013	16,738
2007	9008		

Source: www.treasurydirect.gov

(a) Plot the points (2001, 5807), (2002, 6228), and so on in a Cartesian plane.

(b) Draw a line segment from the point (2001, 5807) to (2006, 8507). What does the slope of this line segment represent?

(c) Find the average rate of change of the debt from 2002 to 2004.

(d) Find the average rate of change of the debt from 2006 to 2008.

(e) Find the average rate of change of the debt from 2010 to 2012.

(f) What appears to be happening to the average rate of change as time passes?

83. **E. coli Growth** A strain of *E. coli* Beu 397-recA441 is placed into a nutrient broth at 30° Celsius and allowed to grow. The data shown in the table are collected. The population is measured in grams and the time in hours. Since population P depends on time t, and each input corresponds to exactly one output, we can say that population is a function of time. Thus $P(t)$ represents the population at time t.

Time (hours), t	Population (grams), P
0	0.09
2.5	0.18
3.5	0.26
4.5	0.35
6	0.50

(a) Find the average rate of change of the population from 0 to 2.5 hours.

(b) Find the average rate of change of the population from 4.5 to 6 hours.

(c) What is happening to the average rate of change as time passes?

84. **e-Filing Tax Returns** The Internal Revenue Service Restructuring and Reform Act (RRA) was signed into law by President Bill Clinton in 1998. A major objective of the RRA was to promote electronic filing of tax returns. The data in the table that follows show the percentage of individual income tax returns filed electronically for filing years 2004–2012. Since the percentage P of returns filed electronically depends on the filing year y, and each input corresponds to exactly one output, the percentage of returns filed electronically is a function of the filing year; so $P(y)$ represents the percentage of returns filed electronically for filing year y.

(a) Find the average rate of change of the percentage of e-filed returns from 2004 to 2006.

(b) Find the average rate of change of the percentage of e-filed returns from 2007 to 2009.

(c) Find the average rate of change of the percentage of e-filed returns from 2010 to 2012.

(d) What is happening to the average rate of change as time passes?

Year	Percentage of returns e-filed
2004	46.5
2005	51.1
2006	53.8
2007	57.1
2008	58.5
2009	67.2
2010	69.8
2011	77.2
2012	82.7

Source: Internal Revenue Service

85. For the function $f(x) = x^2$, compute the average rate of change:

(a) From 0 to 1

(b) From 0 to 0.5

(c) From 0 to 0.1

(d) From 0 to 0.01

(e) From 0 to 0.001

(f) Use a graphing utility to graph each of the secant lines along with f.

(g) What do you think is happening to the secant lines?

(h) What is happening to the slopes of the secant lines? Is there some number that they are getting closer to? What is that number?

Table 4

x	$y = f(x) = \sqrt[3]{x}$	(x, y)
0	0	(0, 0)
$\frac{1}{8}$	$\frac{1}{2}$	$\left(\frac{1}{8}, \frac{1}{2}\right)$
1	1	(1, 1)
2	$\sqrt[3]{2} \approx 1.26$	$\left(2, \sqrt[3]{2}\right)$
8	2	(8, 2)

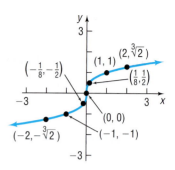

Figure 29 Cube Root Function

From the results of Example 1 and Figure 29, we have the following properties of the cube root function.

> **Properties of $f(x) = \sqrt[3]{x}$**
>
> 1. The domain and the range are the set of all real numbers.
> 2. The x-intercept of the graph of $f(x) = \sqrt[3]{x}$ is 0. The y-intercept of the graph of $f(x) = \sqrt[3]{x}$ is also 0.
> 3. The function is odd. The graph is symmetric with respect to the origin.
> 4. The function is increasing on the interval $(-\infty, \infty)$.
> 5. The function does not have any local minima or any local maxima.

EXAMPLE 2

Graphing the Absolute Value Function

(a) Determine whether $f(x) = |x|$ is even, odd, or neither. State whether the graph of f is symmetric with respect to the y-axis, symmetric with respect to the origin, or neither.
(b) Determine the intercepts, if any, of the graph of $f(x) = |x|$.
(c) Graph $f(x) = |x|$.

Solution

(a) Because

$$f(-x) = |-x|$$
$$= |x| = f(x)$$

the function is even. The graph of f is symmetric with respect to the y-axis.
(b) The y-intercept is $f(0) = |0| = 0$. The x-intercept is found by solving the equation $f(x) = 0$, or $|x| = 0$. The x-intercept is 0.
(c) Use the function to form Table 5 and obtain some points on the graph. Because of the symmetry with respect to the y-axis, we only need to find points (x, y) for which $x \geq 0$. Figure 30 shows the graph of $f(x) = |x|$.

Table 5

x	$y = f(x) = |x|$	(x, y)
0	0	(0, 0)
1	1	(1, 1)
2	2	(2, 2)
3	3	(3, 3)

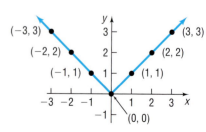

Figure 30 Absolute Value Function

From the results of Example 2 and Figure 30, we have the following properties of the absolute value function.

Properties of f(x) = |x|

1. The domain is the set of all real numbers. The range of f is $\{y \mid y \geq 0\}$.
2. The x-intercept of the graph of $f(x) = |x|$ is 0. The y-intercept of the graph of $f(x) = |x|$ is also 0.
3. The function is even. The graph is symmetric with respect to the y-axis.
4. The function is decreasing on the interval $(-\infty, 0)$. It is increasing on the interval $(0, \infty)$.
5. The function has an absolute minimum of 0 at $x = 0$.

Seeing the Concept

Graph $y = |x|$ on a square screen and compare what you see with Figure 30. Note that some graphing calculators use abs(x) for absolute value.

Below is a list of the key functions that we have discussed. In going through this list, pay special attention to the properties of each function, particularly to the shape of each graph. Knowing these graphs, along with key points on each graph, will lay the foundation for further graphing techniques.

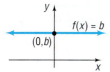

Figure 31 Constant Function

Constant Function

$$f(x) = b \qquad b \text{ is a real number}$$

See Figure 31.

The domain of a **constant function** is the set of all real numbers; its range is the set consisting of a single number b. Its graph is a horizontal line whose y-intercept is b. The constant function is an even function.

Figure 32 Identity Function

Identity Function

$$f(x) = x$$

See Figure 32.

The domain and the range of the **identity function** are the set of all real numbers. Its graph is a line whose slope is 1 and whose y-intercept is 0. The line consists of all points for which the x-coordinate equals the y-coordinate. The identity function is an odd function that is increasing over its domain. Note that the graph bisects quadrants I and III.

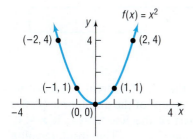

Figure 33 Square Function

Square Function

$$f(x) = x^2$$

See Figure 33.

The domain of the **square function** is the set of all real numbers; its range is the set of nonnegative real numbers. The graph of this function is a parabola whose intercept is at $(0, 0)$. The square function is an even function that is decreasing on the interval $(-\infty, 0)$ and increasing on the interval $(0, \infty)$.

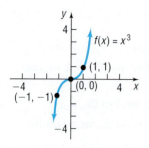

Figure 34 Cube Function

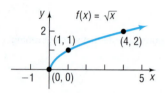

Figure 35 Square Root Function

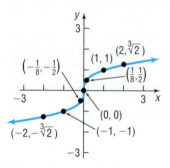

Figure 36 Cube Root Function

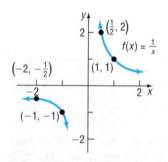

Figure 37 Reciprocal Function

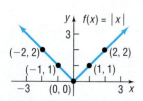

Figure 38 Absolute Value Function

Cube Function

$$f(x) = x^3$$

See Figure 34.

The domain and the range of the **cube function** are the set of all real numbers. The intercept of the graph is at $(0, 0)$. The cube function is odd and is increasing on the interval $(-\infty, \infty)$.

Square Root Function

$$f(x) = \sqrt{x}$$

See Figure 35.

The domain and the range of the **square root function** are the set of nonnegative real numbers. The intercept of the graph is at $(0, 0)$. The square root function is neither even nor odd and is increasing on the interval $(0, \infty)$.

Cube Root Function

$$f(x) = \sqrt[3]{x}$$

See Figure 36.

The domain and the range of the **cube root function** are the set of all real numbers. The intercept of the graph is at $(0, 0)$. The cube root function is an odd function that is increasing on the interval $(-\infty, \infty)$.

Reciprocal Function

$$f(x) = \frac{1}{x}$$

Refer to Example 12, page 164, for a discussion of the equation $y = \frac{1}{x}$. See Figure 37.

The domain and the range of the **reciprocal function** are the set of all nonzero real numbers. The graph has no intercepts. The reciprocal function is decreasing on the intervals $(-\infty, 0)$ and $(0, \infty)$ and is an odd function.

Absolute Value Function

$$f(x) = |x|$$

See Figure 38.

The domain of the **absolute value function** is the set of all real numbers; its range is the set of nonnegative real numbers. The intercept of the graph is at $(0, 0)$. If $x \geq 0$, then $f(x) = x$, and the graph of f is part of the line $y = x$; if $x < 0$, then $f(x) = -x$, and the graph of f is part of the line $y = -x$. The absolute value function is an even function; it is decreasing on the interval $(-\infty, 0)$ and increasing on the interval $(0, \infty)$.

The notation $\text{int}(x)$ stands for the largest integer less than or equal to x. For example,

$$\text{int}(1) = 1, \quad \text{int}(2.5) = 2, \quad \text{int}\left(\frac{1}{2}\right) = 0, \quad \text{int}\left(-\frac{3}{4}\right) = -1, \quad \text{int}(\pi) = 3$$

This type of correspondence occurs frequently enough in mathematics that we give it a name.

DEFINITION **Greatest Integer Function**

$$f(x) = \text{int}(x)^* = \text{greatest integer less than or equal to } x$$

Table 6

	$y = f(x)$	
x	$= \text{int}(x)$	(x, y)
-1	-1	$(-1, -1)$
$-\dfrac{1}{2}$	-1	$\left(-\dfrac{1}{2}, -1\right)$
$-\dfrac{1}{4}$	-1	$\left(-\dfrac{1}{4}, -1\right)$
0	0	$(0, 0)$
$\dfrac{1}{4}$	0	$\left(\dfrac{1}{4}, 0\right)$
$\dfrac{1}{2}$	0	$\left(\dfrac{1}{2}, 0\right)$
$\dfrac{3}{4}$	0	$\left(\dfrac{3}{4}, 0\right)$

We obtain the graph of $f(x) = \text{int}(x)$ by plotting several points. See Table 6. For values of x, $-1 \le x < 0$, the value of $f(x) = \text{int}(x)$ is -1; for values of x, $0 \le x < 1$, the value of f is 0. See Figure 39 for the graph.

The domain of the **greatest integer function** is the set of all real numbers; its range is the set of integers. The y-intercept of the graph is 0. The x-intercepts lie in the interval $[0, 1)$. The greatest integer function is neither even nor odd. It is constant on every interval of the form $[k, k + 1)$, for k an integer. In Figure 39, a solid dot is used to indicate, for example, that at $x = 1$ the value of f is $f(1) = 1$; an open circle is used to illustrate that the function does not assume the value of 0 at $x = 1$.

Although a precise definition requires the idea of a limit (discussed in calculus), in a rough sense, a function is said to be *continuous* if its graph has no gaps or holes and can be drawn without lifting a pencil from the paper on which the graph is drawn. We contrast this with a *discontinuous* function. A function is discontinuous if its graph has gaps or holes and so cannot be drawn without lifting a pencil from the paper.

From the graph of the greatest integer function, we can see why it is also called a **step function**. At $x = 0, x = \pm 1, x = \pm 2$, and so on, this function is discontinuous because, at integer values, the graph suddenly "steps" from one value to another without taking on any of the intermediate values. For example, to the immediate left of $x = 3$, the y-coordinates of the points on the graph are 2, and at $x = 3$ and to the immediate right of $x = 3$, the y-coordinates of the points on the graph are 3. Consequently, the graph has gaps in it.

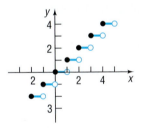

Figure 39 Greatest Integer Function

COMMENT When graphing a function using a graphing utility, typically you can choose either **connected mode**, in which points plotted on the screen are connected, making the graph appear without any breaks, or **dot mode**, in which only the points plotted appear. When graphing the greatest integer function with a graphing utility, it may be necessary to be in **dot mode**. This is to prevent the utility from "connecting the dots" when $f(x)$ changes from one integer value to the next. However, some utilities will display the gaps even when in "connected" mode. See Figure 40. ∎

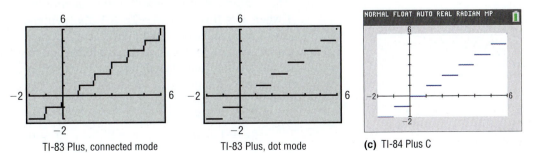

Figure 40
$f(x) = \text{int}(x)$

TI-83 Plus, connected mode   TI-83 Plus, dot mode   **(c)** TI-84 Plus C

*Some texts use the notation $f(x) = [x]$ instead of $\text{int}(x)$.

The functions discussed so far are basic. Whenever you encounter one of them, you should see a mental picture of its graph. For example, if you encounter the function $f(x) = x^2$, you should see in your mind's eye a picture like Figure 33.

> **Now Work** PROBLEMS 11 THROUGH 18

2 Graph Piecewise-defined Functions

Sometimes a function is defined using different equations on different parts of its domain. For example, the absolute value function $f(x) = |x|$ is actually defined by two equations: $f(x) = x$ if $x \geq 0$ and $f(x) = -x$ if $x < 0$. For convenience, these equations are generally combined into one expression as

$$f(x) = |x| = \begin{cases} x & \text{if } x \geq 0 \\ -x & \text{if } x < 0 \end{cases}$$

When a function is defined by different equations on different parts of its domain, it is called a **piecewise-defined** function.

EXAMPLE 3

Analyzing a Piecewise-defined Function

The function f is defined as

$$f(x) = \begin{cases} -2x + 1 & \text{if } -3 \leq x < 1 \\ 2 & \text{if } x = 1 \\ x^2 & \text{if } x > 1 \end{cases}$$

(a) Find $f(-2)$, $f(1)$, and $f(2)$. (b) Determine the domain of f.
(c) Locate any intercepts. (d) Graph f.
(e) Use the graph to find the range of f. (f) Is f continuous on its domain?

Solution

(a) To find $f(-2)$, observe that when $x = -2$, the equation for f is given by $f(x) = -2x + 1$. So

$$f(-2) = -2(-2) + 1 = 5$$

When $x = 1$, the equation for f is $f(x) = 2$. So,

$$f(1) = 2$$

When $x = 2$, the equation for f is $f(x) = x^2$. So

$$f(2) = 2^2 = 4$$

(b) To find the domain of f, look at its definition. Since f is defined for all x greater than or equal to -3, the domain of f is $\{x \mid x \geq -3\}$, or the interval $[-3, \infty)$.

(c) The y-intercept of the graph of the function is $f(0)$. Because the equation for f when $x = 0$ is $f(x) = -2x + 1$, the y-intercept is $f(0) = -2(0) + 1 = 1$. The x-intercepts of the graph of a function f are the real solutions to the equation $f(x) = 0$. To find the x-intercepts of f, solve $f(x) = 0$ for each "piece" of the function, and then determine what values of x, if any, satisfy the condition that defines the piece.

$$f(x) = 0 \qquad\qquad f(x) = 0 \qquad\qquad f(x) = 0$$
$$-2x + 1 = 0 \quad {\scriptstyle -3 \leq x < 1} \qquad 2 = 0 \quad {\scriptstyle x = 1} \qquad x^2 = 0 \quad {\scriptstyle x > 1}$$
$$-2x = -1 \qquad\qquad \text{No solution} \qquad x = 0$$
$$x = \frac{1}{2}$$

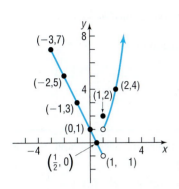

Figure 41

The first potential x-intercept, $x = \dfrac{1}{2}$, satisfies the condition $-3 \le x < 1$, so $x = \dfrac{1}{2}$ is an x-intercept. The second potential x-intercept, $x = 0$, does not satisfy the condition $x > 1$, so $x = 0$ is not an x-intercept. The only x-intercept is $\dfrac{1}{2}$. The intercepts are $(0, 1)$ and $\left(\dfrac{1}{2}, 0\right)$.

(d) To graph f, graph each "piece." First graph the line $y = -2x + 1$ and keep only the part for which $-3 \le x < 1$. Then plot the point $(1, 2)$ because, when $x = 1, f(x) = 2$. Finally, graph the parabola $y = x^2$ and keep only the part for which $x > 1$. See Figure 41.

(e) From the graph, we conclude that the range of f is $\{y \mid y > -1\}$, or the interval $(-1, \infty)$.

(f) The function f is not continuous because there is a "jump" in the graph at $x = 1$. ●

 Now Work PROBLEM **31**

EXAMPLE 4

Cost of Electricity

In the spring of 2014, Duke Energy Progress supplied electricity to residences in South Carolina for a monthly customer charge of $6.50 plus 9.971¢ per kilowatt-hour (kWh) for the first 800 kWh supplied in the month and 8.971¢ per kWh for all usage over 800 kWh in the month.

(a) What is the charge for using 300 kWh in a month?
(b) What is the charge for using 1500 kWh in a month?
(c) If C is the monthly charge for x kWh, develop a model relating the monthly charge and kilowatt-hours used. That is, express C as a function of x.

Source: Duke Energy Progress, 2014

Solution

(a) For 300 kWh, the charge is $6.50 plus (9.971¢ = $0.09971) per kWh. That is,

$$\text{Charge} = \$6.50 + \$0.09971(300) = \$36.41$$

(b) For 1500 kWh, the charge is $6.50 plus 9.971¢ per kWh for the first 800 kWh plus 8.971¢ per kWh for the 700 in excess of 800. That is,

$$\text{Charge} = \$6.50 + \$0.09971(800) + \$0.08971(700) = \$149.07$$

(c) Let x represent the number of kilowatt-hours used. If $0 \le x \le 800$, then the monthly charge C (in dollars) can be found by multiplying x times $0.09971 and adding the monthly customer charge of $6.50. So if $0 \le x \le 800$, then

$$C(x) = 0.09971x + 6.50$$

For $x > 800$, the charge is $0.09971(800) + 6.50 + 0.08971(x - 800)$, since $(x - 800)$ equals the usage in excess of 800 kWh, which costs $0.08971 per kWh. That is, if $x > 800$, then

$$C(x) = 0.09971(800) + 6.50 + 0.08971(x - 800)$$
$$= 79.768 + 6.50 + 0.08971x - 71.768$$
$$= 0.08971x + 14.50$$

The rule for computing C follows two equations:

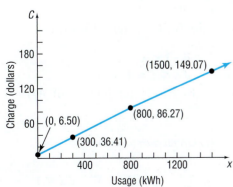

Figure 42

$$C(x) = \begin{cases} 0.09971x + 6.50 & \text{if } 0 \le x \le 800 \\ 0.08971x + 14.50 & \text{if } x > 800 \end{cases} \quad \textit{The Model}$$

See Figure 42 for the graph. Note that the two "pieces" are linear, but they have different slopes (rates), and meet at the point $(800, 86.27)$. ●

3.4 Assess Your Understanding

'Are You Prepared?' *Answers are given at the end of these exercises. If you get a wrong answer, read the pages listed in* red.

1. Sketch the graph of $y = \sqrt{x}$. (p. 163)

2. Sketch the graph of $y = \dfrac{1}{x}$. (p. 164)

3. List the intercepts of the equation $y = x^3 - 8$. (pp. 159–160)

Concepts and Vocabulary

4. The function $f(x) = x^2$ is decreasing on the interval _____.

5. When functions are defined by more than one equation, they are called _____ functions.

6. *True or False* The cube function is odd and is increasing on the interval $(-\infty, \infty)$.

7. *True or False* The cube root function is odd and is decreasing on the interval $(-\infty, \infty)$.

8. *True or False* The domain and the range of the reciprocal function are the set of all real numbers.

9. Which of the following functions has a graph that is symmetric about the y-axis?

(a) $y = \sqrt{x}$ (b) $y = |x|$ (c) $y = x^3$ (d) $y = \dfrac{1}{x}$

10. Consider the following function.
$$f(x) = \begin{cases} 3x - 2 & \text{if} & x < 2 \\ x^2 + 5 & \text{if} & 2 \le x < 10 \\ 3 & \text{if} & x \ge 10 \end{cases}$$
Which "piece(s)" should be used to find the y-intercept?
(a) $3x - 2$ (b) $x^2 + 5$ (c) 3 (d) all three

Skill Building

In Problems 11–18, match each graph to its function.

A. *Constant function* B. *Identity function* C. *Square function* D. *Cube function*
E. *Square root function* F. *Reciprocal function* G. *Absolute value function* H. *Cube root function*

11. **12.** **13.** **14.**

15. **16.** **17.** **18.**

In Problems 19–26, sketch the graph of each function. Be sure to label three points on the graph.

19. $f(x) = x$ **20.** $f(x) = x^2$ **21.** $f(x) = x^3$ **22.** $f(x) = \sqrt{x}$

23. $f(x) = \dfrac{1}{x}$ **24.** $f(x) = |x|$ **25.** $f(x) = \sqrt[3]{x}$ **26.** $f(x) = 3$

27. If $f(x) = \begin{cases} x^2 & \text{if } x < 0 \\ 2 & \text{if } x = 0 \\ 2x + 1 & \text{if } x > 0 \end{cases}$

find: (a) $f(-2)$ (b) $f(0)$ (c) $f(2)$

28. If $f(x) = \begin{cases} -3x & \text{if } x < -1 \\ 0 & \text{if } x = -1 \\ 2x^2 + 1 & \text{if } x > -1 \end{cases}$

find: (a) $f(-2)$ (b) $f(-1)$ (c) $f(0)$

29. If $f(x) = \begin{cases} 2x - 4 & \text{if } -1 \le x \le 2 \\ x^3 - 2 & \text{if } 2 < x \le 3 \end{cases}$

find: (a) $f(0)$ (b) $f(1)$ (c) $f(2)$ (d) $f(3)$

30. If $f(x) = \begin{cases} x^3 & \text{if } -2 \le x < 1 \\ 3x + 2 & \text{if } 1 \le x \le 4 \end{cases}$

find: (a) $f(-1)$ (b) $f(0)$ (c) $f(1)$ (d) $f(3)$

In Problems 31–42:

(a) *Find the domain of each function.*   (b) *Locate any intercepts.*   (c) *Graph each function.*
(d) *Based on the graph, find the range.*   (e) *Is f continuous on its domain?*

31. $f(x) = \begin{cases} 2x & \text{if } x \ne 0 \\ 1 & \text{if } x = 0 \end{cases}$

32. $f(x) = \begin{cases} 3x & \text{if } x \ne 0 \\ 4 & \text{if } x = 0 \end{cases}$

33. $f(x) = \begin{cases} -2x + 3 & \text{if } x < 1 \\ 3x - 2 & \text{if } x \ge 1 \end{cases}$

34. $f(x) = \begin{cases} x + 3 & \text{if } x < -2 \\ -2x - 3 & \text{if } x \geq -2 \end{cases}$

35. $f(x) = \begin{cases} x + 3 & \text{if } -2 \leq x < 1 \\ 5 & \text{if } x = 1 \\ -x + 2 & \text{if } x > 1 \end{cases}$

36. $f(x) = \begin{cases} 2x + 5 & \text{if } -3 \leq x < 0 \\ -3 & \text{if } x = 0 \\ -5x & \text{if } x > 0 \end{cases}$

37. $f(x) = \begin{cases} 1 + x & \text{if } x < 0 \\ x^2 & \text{if } x \geq 0 \end{cases}$

38. $f(x) = \begin{cases} \dfrac{1}{x} & \text{if } x < 0 \\ \sqrt[3]{x} & \text{if } x \geq 0 \end{cases}$

39. $f(x) = \begin{cases} |x| & \text{if } -2 \leq x < 0 \\ x^3 & \text{if } x > 0 \end{cases}$

40. $f(x) = \begin{cases} 2 - x & \text{if } -3 \leq x < 1 \\ \sqrt{x} & \text{if } x > 1 \end{cases}$

41. $f(x) = 2 \text{ int}(x)$

42. $f(x) = \text{int}(2x)$

In Problems 43–46, the graph of a piecewise-defined function is given. Write a definition for each function.

43.
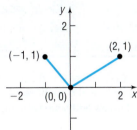

44.
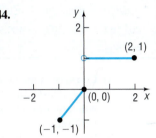

45.
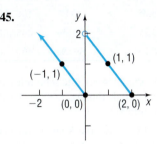

46.
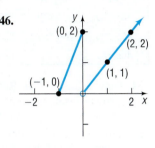

47. If $f(x) = \text{int}(2x)$, find
  (a) $f(1.2)$ (b) $f(1.6)$ (c) $f(-1.8)$

48. If $f(x) = \text{int}\left(\dfrac{x}{2}\right)$, find
  (a) $f(1.2)$ (b) $f(1.6)$ (c) $f(-1.8)$

Applications and Extensions

49. Tablet Service Sprint offers a monthly tablet plan for $34.99. It includes 3 gigabytes of data and charges $15 per gigabyte for additional gigabytes. The following function is used to compute the monthly cost for a subscriber.

$$C(x) = \begin{cases} 34.99 & \text{if } 0 \leq x \leq 3 \\ 15x - 10.01 & \text{if } x > 3 \end{cases}$$

Compute the monthly cost for each of the following gigabytes of use.
(a) 2   (b) 5   (c) 13
Source: Sprint, April 2014

50. Parking at O'Hare International Airport The short-term (no more than 24 hours) parking fee F (in dollars) for parking x hours on a weekday at O'Hare International Airport's main parking garage can be modeled by the function

$$F(x) = \begin{cases} 3 & \text{if } 0 < x \leq 3 \\ 5 \text{ int}(x + 1) + 1 & \text{if } 3 < x < 9 \\ 50 & \text{if } 9 \leq x \leq 24 \end{cases}$$

Determine the fee for parking in the short-term parking garage for
(a) 2 hours   (b) 7 hours
(c) 15 hours   (d) 8 hours and 24 minutes
Source: O'Hare International Airport

51. Cost of Natural Gas In March 2014, Laclede Gas had the rate schedule (on, right) for natural gas usage in single-family residences.
(a) What is the charge for using 20 therms in a month?
(b) What is the charge for using 150 therms in a month?

(c) Develop a function that models the monthly charge C for x therms of gas.
(d) Graph the function found in part (c).

Monthly service charge	$19.50
Delivery charge	
First 30 therms	$0.91686/therm
Over 30 therms	$0
Natural gas cost	
First 30 therms –	$0.3313/therm
Over 30 therms	$0.5757/therm

Source: Laclede Gas

52. Cost of Natural Gas In April 2014, Nicor Gas had the following rate schedule for natural gas usage in small businesses.

Monthly customer charge	$72.60
Distribution charge	
1st 150 therms	$0.1201/therm
Next 4850 therms	$0.0549/therm
Over 5000 therms	$0.0482/therm
Gas supply charge	$0.68/therm

(a) What is the charge for using 1000 therms in a month?
(b) What is the charge for using 6000 therms in a month?
(c) Develop a function that models the monthly charge C for x therms of gas.
(d) Graph the function found in part (c).
Source: Nicor Gas, 2014

53. **Federal Income Tax** Two 2014 Tax Rate Schedules are given in the accompanying table. If x equals taxable income and y equals the tax due, construct a function $y = f(x)$ for Schedule X.

2014 Tax Rate Schedules										
Schedule X—Single					**Schedule Y-1—Married Filing Jointly or Qualified Widow(er)**					
If Taxable Income is Over	**But Not Over**	**The Tax is This Amount**		**Plus This %**	**Of the Excess Over**	**If Taxable Income is Over**	**But Not Over**	**The Tax is This Amount**		**Plus This %**	**Of the Excess Over**
$0	$9,075	$0	+	10%	$0	$0	$18,150	$0	+	10%	$0
9,075	36,900	907.50	+	15%	9,075	18,150	73,800	1,815	+	15%	18,150
36,900	89,350	5,081.25	+	25%	36,900	73,800	148,850	10,162.50	+	25%	73,800
89,350	186,350	18,193.75	+	28%	89,350	148,850	226,850	28,925.00	+	28%	148,850
186,350	405,100	45,353.75	+	33%	186,350	226,850	405,100	50,765.00	+	33%	226,850
405,100	406,750	117,541.25	+	35%	405,100	405,100	457,600	109,587.50	+	35%	405,100
406,750	—	118,188.75	+	39.6%	406,750	457,600	—	127,962.50	+	39.6%	457,600

54. **Federal Income Tax** Refer to the 2014 tax rate schedules. If x equals taxable income and y equals the tax due, construct a function $y = f(x)$ for Schedule Y-1.

55. **Cost of Transporting Goods** A trucking company transports goods between Chicago and New York, a distance of 960 miles. The company's policy is to charge, for each pound, $0.50 per mile for the first 100 miles, $0.40 per mile for the next 300 miles, $0.25 per mile for the next 400 miles, and no charge for the remaining 160 miles.
    (a) Graph the relationship between the cost of transportation in dollars and mileage over the entire 960-mile route.
    (b) Find the cost as a function of mileage for hauls between 100 and 400 miles from Chicago.
    (c) Find the cost as a function of mileage for hauls between 400 and 800 miles from Chicago.

56. **Car Rental Costs** An economy car rented in Florida from Enterprise® on a weekly basis costs $185 per week. Extra days cost $37 per day until the day rate exceeds the weekly rate, in which case the weekly rate applies. Also, any part of a day used counts as a full day. Find the cost C of renting an economy car as a function of the number x of days used, where $7 \leq x \leq 14$. Graph this function.

57. **Mortgage Fees** Fannie Mae charges a loan-level price adjustment (LLPA) on all mortgages, which represents a fee homebuyers seeking a loan must pay. The rate paid depends on the credit score of the borrower, the amount borrowed, and the loan-to-value (LTV) ratio. The LTV ratio is the ratio of amount borrowed to appraised value of the home. For example, a homebuyer who wishes to borrow $250,000 with a credit score of 730 and an LTV ratio of 80% will pay 0.5% (0.005) of $250,000, or $1250. The table shows the LLPA for various credit scores and an LTV ratio of 80%.

Credit Score	Loan-Level Price Adjustment Rate
≤659	3.00%
660–679	2.50%
680–699	1.75%
700–719	1%
720–739	0.5%
≥740	0.25%

Source: Fannie Mae.

   (a) Construct a function $C = C(s)$, where C is the loan-level price adjustment (LLPA) and s is the credit score of an individual who wishes to borrow $300,000 with an 80% LTV ratio.
   (b) What is the LLPA on a $300,000 loan with an 80% LTV ratio for a borrower whose credit score is 725?
   (c) What is the LLPA on a $300,000 loan with an 80% LTV ratio for a borrower whose credit score is 670?

58. **Minimum Payments for Credit Cards** Holders of credit cards issued by banks, department stores, oil companies, and so on receive bills each month that state minimum amounts that must be paid by a certain due date. The minimum due depends on the total amount owed. One such credit card company uses the following rules: For a bill of less than $10, the entire amount is due. For a bill of at least $10 but less than $500, the minimum due is $10. A minimum of $30 is due on a bill of at least $500 but less than $1000, a minimum of $50 is due on a bill of at least $1000 but less than $1500, and a minimum of $70 is due on bills of $1500 or more. Find the function f that describes the minimum payment due on a bill of x dollars. Graph f.

59. **Wind Chill** The wind chill factor represents the air temperature at a standard wind speed that would produce the same heat loss as the given temperature and wind speed. One formula for computing the equivalent temperature is

$$W = \begin{cases} t & 0 \leq v < 1.79 \\ 33 - \dfrac{(10.45 + 10\sqrt{v} - v)(33 - t)}{22.04} & 1.79 \leq v \leq 20 \\ 33 - 1.5958(33 - t) & v > 20 \end{cases}$$

where v represents the wind speed (in meters per second) and t represents the air temperature (°C). Compute the wind chill for the following:
   (a) An air temperature of 10°C and a wind speed of 1 meter per second (m/sec)
   (b) An air temperature of 10°C and a wind speed of 5 m/sec
   (c) An air temperature of 10°C and a wind speed of 15 m/sec
   (d) An air temperature of 10°C and a wind speed of 25 m/sec
   (e) Explain the physical meaning of the equation corresponding to $0 \leq v < 1.79$.
   (f) Explain the physical meaning of the equation corresponding to $v > 20$.

60. Wind Chill Redo Problem 59(a)–(d) for an air temperature of $-10°C$.

61. First-class Mail In 2014 the U.S. Postal Service charged $0.98 postage for first-class mail retail flats (such as an 8.5" by 11" envelope) weighing up to 1 ounce, plus $0.21 for each

additional ounce up to 13 ounces. First-class rates do not apply to flats weighing more than 13 ounces. Develop a model that relates C, the first-class postage charged, for a flat weighing x ounces. Graph the function.

Source: United States Postal Service

Explaining Concepts: Discussion and Writing

 In Problems 62–69, use a graphing utility.

62. Exploration Graph $y = x^2$. Then on the same screen graph $y = x^2 + 2$, followed by $y = x^2 + 4$, followed by $y = x^2 - 2$. What pattern do you observe? Can you predict the graph of $y = x^2 - 4$? Of $y = x^2 + 5$?

63. Exploration Graph $y = x^2$. Then on the same screen graph $y = (x - 2)^2$, followed by $y = (x - 4)^2$, followed by $y = (x + 2)^2$. What pattern do you observe? Can you predict the graph of $y = (x + 4)^2$? Of $y = (x - 5)^2$?

64. Exploration Graph $y = |x|$. Then on the same screen graph $y = 2|x|$, followed by $y = 4|x|$, followed by $y = \frac{1}{2}|x|$. What pattern do you observe? Can you predict the graph of $y = \frac{1}{4}|x|$? Of $y = 5|x|$?

65. Exploration Graph $y = x^2$. Then on the same screen graph $y = -x^2$. Now try $y = |x|$ and $y = -|x|$. What do you conclude?

66. Exploration Graph $y = \sqrt{x}$. Then on the same screen graph $y = \sqrt{-x}$. Now try $y = 2x + 1$ and $y = 2(-x) + 1$. What do you conclude?

67. Exploration Graph $y = x^3$. Then on the same screen graph $y = (x - 1)^3 + 2$. Could you have predicted the result?

68. Exploration Graph $y = x^2$, $y = x^4$, and $y = x^6$ on the same screen. What do you notice is the same about each graph? What do you notice is different?

69. Exploration Graph $y = x^3$, $y = x^5$, and $y = x^7$ on the same screen. What do you notice is the same about each graph? What do you notice is different?

70. Consider the equation

$$y = \begin{cases} 1 & \text{if } x \text{ is rational} \\ 0 & \text{if } x \text{ is irrational} \end{cases}$$

Is this a function? What is its domain? What is its range? What is its y-intercept, if any? What are its x-intercepts, if any? Is it even, odd, or neither? How would you describe its graph?

71. Define some functions that pass through $(0, 0)$ and $(1, 1)$ and are increasing for $x \geq 0$. Begin your list with $y = \sqrt{x}, y = x$, and $y = x^2$. Can you propose a general result about such functions?

Retain Your Knowledge

Problems 72–75 are based on material learned earlier in the course. The purpose of these problems is to keep the material fresh in your mind so that you are better prepared for the final exam.

72. Simplify: $(3 + 2i)(4 - 5i)$

73. Find the center and radius of the circle $x^2 + y^2 = 6y + 16$.

74. Solve: $4x - 5(2x - 1) = 4 - 7(x + 1)$

75. Ethan has $60,000 to invest. He puts part of the money in a CD that earns 3% simple interest per year and the rest in a mutual fund that earns 8% simple interest per year. How much did he invest in each if his earned interest the first year was $3700?

'Are You Prepared?' Answers

1.

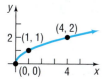

2.

3. $(0, -8), (2, 0)$

3.5 Graphing Techniques: Transformations

OBJECTIVES 1 Graph Functions Using Vertical and Horizontal Shifts (p. 248)

2 Graph Functions Using Compressions and Stretches (p. 251)

3 Graph Functions Using Reflections about the x-Axis and the y-Axis (p. 253)

At this stage, if you were asked to graph any of the functions defined by $y = x$, $y = x^2, y = x^3$, $y = \sqrt{x}, y = \sqrt[3]{x}$, $y = \frac{1}{x}$, or $y = |x|$, your response should be, "Yes, I recognize these functions and know the general shapes of their graphs." (If this is not your answer, review the previous section, Figures 32 through 38.)

Sometimes we are asked to graph a function that is "almost" like one that we already know how to graph. In this section, we develop techniques for graphing such functions. Collectively, these techniques are referred to as **transformations**.

1 Graph Functions Using Vertical and Horizontal Shifts

EXAMPLE 1 **Vertical Shift Up**

Use the graph of $f(x) = x^2$ to obtain the graph of $g(x) = x^2 + 3$. Find the domain and range of g.

Solution Begin by obtaining some points on the graphs of f and g. For example, when $x = 0$, then $y = f(0) = 0$ and $y = g(0) = 3$. When $x = 1$, then $y = f(1) = 1$ and $y = g(1) = 4$. Table 7 lists these and a few other points on each graph. Notice that each y-coordinate of a point on the graph of g is 3 units larger than the y-coordinate of the corresponding point on the graph of f. We conclude that the graph of g is identical to that of f, except that it is shifted vertically up 3 units. See Figure 43.

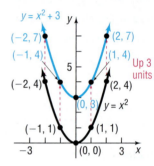

Table 7

x	$y = f(x)$ $= x^2$	$y = g(x)$ $= x^2 + 3$
-2	4	7
-1	1	4
0	0	3
1	1	4
2	4	7

Figure 43

The domain of g is all real numbers, or $(-\infty, \infty)$. The range of g is $[3, \infty)$. ●

EXAMPLE 2 **Vertical Shift Down**

Use the graph of $f(x) = x^2$ to obtain the graph of $g(x) = x^2 - 4$. Find the domain and range of g.

Solution Table 8 lists some points on the graphs of f and g. Notice that each y-coordinate of g is 4 units less than the corresponding y-coordinate of f.

To obtain the graph of g from the graph of f, subtract 4 from each y-coordinate on the graph of f. The graph of g is identical to that of f, except that it is shifted down 4 units. See Figure 44.

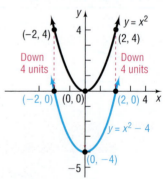

Table 8

x	$y = f(x)$ $= x^2$	$y = g(x)$ $= x^2 - 4$
-2	4	0
-1	1	-3
0	0	-4
1	1	-3
2	4	0

Figure 44

The domain of g is all real numbers, or $(-\infty, \infty)$. The range of g is $[-4, \infty)$. ●

Note that a vertical shift affects only the range of a function, not the domain. For example, the range of $f(x) = x^2$ is $[0, \infty)$. In Example 1 the range of g is $[3, \infty)$, whereas in Example 2 the range of g is $[-4, \infty)$. The domain for all three functions is all real numbers.

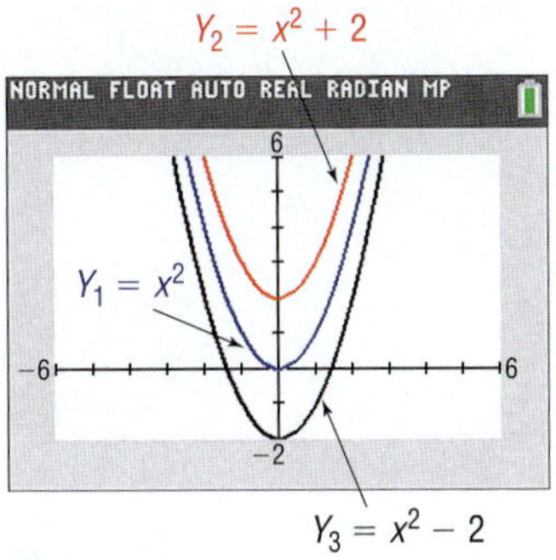

$Y_2 = x^2 + 2$

NORMAL FLOAT AUTO REAL RADIAN MP

$Y_1 = x^2$

$Y_3 = x^2 - 2$

Figure 45

Exploration On the same screen, graph each of the following functions:

$$Y_1 = x^2$$
$$Y_2 = x^2 + 2$$
$$Y_3 = x^2 - 2$$

Figure 45 illustrates the graphs. You should have observed a general pattern. With $Y_1 = x^2$ on the screen, the graph of $Y_2 = x^2 + 2$ is identical to that of $Y_1 = x^2$, except that it is shifted vertically up 2 units. The graph of $Y_3 = x^2 - 2$ is identical to that of $Y_1 = x^2$, except that it is shifted vertically down 2 units.

We are led to the following conclusions:

> **In Words**
> For $y = f(x) + k, k > 0$, add k to each y-coordinate on the graph of $y = f(x)$ to shift the graph up k units.
> For $y = f(x) - k, k > 0$, subtract k from each y-coordinate to shift the graph down k units.

If a positive real number k is added to the output of a function $y = f(x)$, the graph of the new function $y = f(x) + k$ is the graph of f **shifted vertically up** k units.

If a positive real number k is subtracted from the output of a function $y = f(x)$, the graph of the new function $y = f(x) - k$ is the graph of f **shifted vertically down** k units.

Now Work PROBLEM 39

EXAMPLE 3

Horizontal Shift to the Right

Use the graph of $f(x) = \sqrt{x}$ to obtain the graph of $g(x) = \sqrt{x - 2}$. Find the domain and range of g.

Solution

The function $g(x) = \sqrt{x - 2}$ is basically a square root function. Table 9 lists some points on the graphs of f and g. Note that when $f(x) = 0$, then $x = 0$, and when $g(x) = 0$, then $x = 2$. Also, when $f(x) = 2$, then $x = 4$, and when $g(x) = 2$, then $x = 6$. Notice that the x-coordinates on the graph of g are 2 units larger than the corresponding x-coordinates on the graph of f for any given y-coordinate. We conclude that the graph of g is identical to that of f, except that it is shifted horizontally 2 units to the right. See Figure 46.

Table 9

	$y = f(x)$ $= \sqrt{x}$		$y = g(x)$ $= \sqrt{x - 2}$
x		x	
0	0	2	0
1	1	3	1
4	2	6	2
9	3	11	3

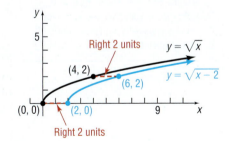

Figure 46

The domain of g is $[2, \infty)$ and the range is $[0, \infty)$.

EXAMPLE 4

Horizontal Shift to the Left

Use the graph of $f(x) = \sqrt{x}$ to obtain the graph of $g(x) = \sqrt{x + 4}$. Find the domain and range of g.

Solution

The function $g(x) = \sqrt{x + 4}$ is basically a square root function. Its graph is the same as that of f, except that it is shifted horizontally 4 units to the left. See Figure 47 on page 250.

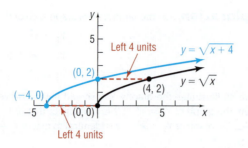

Figure 47

The domain of g is $[-4, \infty)$ and the range is $[0, \infty)$.

━━━━➤ **Now Work** PROBLEM 43

Note that a horizontal shift affects only the domain of a function, not the range. For example, the domain of $f(x) = \sqrt{x}$ is $[0, \infty)$. In Example 3 the domain of g is $[2, \infty)$, whereas in Example 4 the domain of g is $[-4, \infty)$. The range for all three functions is $[0, \infty)$.

Exploration On the same screen, graph each of the following functions:

$$Y_1 = x^2$$
$$Y_2 = (x - 3)^2$$
$$Y_3 = (x + 2)^2$$

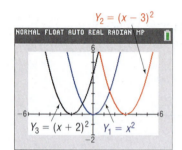

Figure 48

Figure 48 illustrates the graphs.

You should have observed the following pattern. With the graph of $Y_1 = x^2$ on the screen, the graph of $Y_2 = (x - 3)^2$ is identical to that of $Y_1 = x^2$, except that it is shifted horizontally to the right 3 units. The graph of $Y_3 = (x + 2)^2$ is identical to that of $Y_1 = x^2$, except that it is shifted horizontally to the left 2 units.

We are led to the following conclusions:

> If the argument x of a function f is replaced by $x - h$, $h > 0$, the graph of the new function $y = f(x - h)$ is the graph of f **shifted horizontally right** h units.

In Words

For $y = f(x - h)$, $h > 0$, add h to each x-coordinate on the graph of $y = f(x)$ to shift the graph right h units. For $y = f(x + h)$, $h > 0$, subtract h from each x-coordinate on the graph of $y = f(x)$ to shift the graph left h units.

> If the argument x of a function f is replaced by $x + h$, $h > 0$, the graph of the new function $y = f(x + h)$ is the graph of f **shifted horizontally left** h units.

Observe the distinction between vertical and horizontal shifts. The graph of $f(x) = x^3 + 2$ is obtained by shifting the graph of $y = x^3$ *up* 2 units, because we evaluate the cube function first and then add 2. The graph of $g(x) = (x + 2)^3$ is obtained by shifting the graph of $y = x^3$ *left* 2 units, because we add 2 to x before we evaluate the cube function.

Vertical and horizontal shifts are sometimes combined.

EXAMPLE 5

Combining Vertical and Horizontal Shifts

Graph the function $f(x) = |x + 3| - 5$. Find the domain and range of f.

Solution

We graph f in steps. First, note that the rule for f is basically an absolute value function, so begin with the graph of $y = |x|$ as shown in Figure 49(a). Next, to get the graph of $y = |x + 3|$, shift the graph of $y = |x|$ horizontally 3 units to the left. See Figure 49(b). Finally, to get the graph of $y = |x + 3| - 5$, shift the graph of $y = |x + 3|$ vertically down 5 units. See Figure 49(c). Note the points plotted on each graph. Using key points can be helpful in keeping track of the transformation that has taken place.

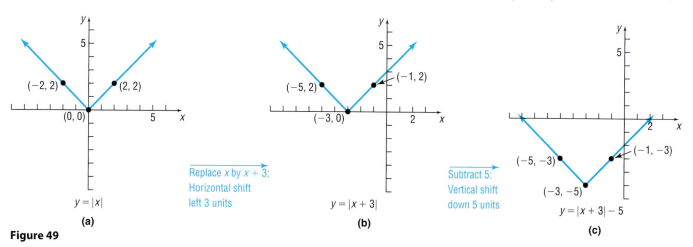

Figure 49

The domain of f is all real numbers, or $(-\infty, \infty)$. The range of f is $[-5, \infty)$. ●

✓**Check:** Graph $Y_1 = f(x) = |x + 3| - 5$ and compare the graph to Figure 49(c).

In Example 5, if the vertical shift had been done first, followed by the horizontal shift, the final graph would have been the same. Try it for yourself.

Now Work PROBLEMS 45 AND 69

2 Graph Functions Using Compressions and Stretches

EXAMPLE 6 | **Vertical Stretch**

Use the graph of $f(x) = \sqrt{x}$ to obtain the graph of $g(x) = 2\sqrt{x}$.

Solution To see the relationship between the graphs of f and g, we form Table 10, listing points on each graph. For each x, the y-coordinate of a point on the graph of g is 2 times as large as the corresponding y-coordinate on the graph of f. The graph of $f(x) = \sqrt{x}$ is vertically stretched by a factor of 2 to obtain the graph of $g(x) = 2\sqrt{x}$. For example, $(1, 1)$ is on the graph of f, but $(1, 2)$ is on the graph of g. See Figure 50.

Table 10

x	$y = f(x)$ $= \sqrt{x}$	$y = g(x)$ $= 2\sqrt{x}$
0	0	0
1	1	2
4	2	4
9	3	6

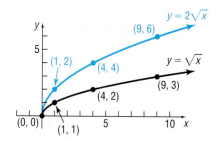

Figure 50

●

EXAMPLE 7 | **Vertical Compression**

Use the graph of $f(x) = |x|$ to obtain the graph of $g(x) = \dfrac{1}{2}|x|$.

Solution For each x, the y-coordinate of a point on the graph of g is $\dfrac{1}{2}$ as large as the corresponding y-coordinate on the graph of f. The graph of $f(x) = |x|$ is vertically compressed by a factor of $\dfrac{1}{2}$ to obtain the graph of $g(x) = \dfrac{1}{2}|x|$. For example, $(2, 2)$ is on the graph of f, but $(2, 1)$ is on the graph of g. See Table 11 and Figure 51 on page 252.

Table 11

x	$y = f(x)$ $= \|x\|$	$y = g(x)$ $= \frac{1}{2}\|x\|$
−2	2	1
−1	1	$\frac{1}{2}$
0	0	0
1	1	$\frac{1}{2}$
2	2	1

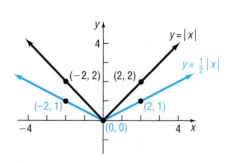

Figure 51

When the right side of a function $y = f(x)$ is multiplied by a positive number a, the graph of the new function $y = af(x)$ is obtained by multiplying each y-coordinate on the graph of $y = f(x)$ by a. The new graph is a **vertically compressed** (if $0 < a < 1$) or a **vertically stretched** (if $a > 1$) version of the graph of $y = f(x)$.

Now Work PROBLEM 47

What happens if the argument x of a function $y = f(x)$ is multiplied by a positive number a, creating a new function $y = f(ax)$? To find the answer, look at the following Exploration.

Exploration On the same screen, graph each of the following functions:

$$Y_1 = f(x) = \sqrt{x} \qquad Y_2 = f(2x) = \sqrt{2x} \qquad Y_3 = f\left(\frac{1}{2}x\right) = \sqrt{\frac{1}{2}x} = \sqrt{\frac{x}{2}}$$

Create a table of values to explore the relation between the x- and y-coordinates of each function.

Result You should have obtained the graphs in Figure 52. Look at Table 12(a). Note that (1, 1), (4, 2), and (9, 3) are points on the graph of $Y_1 = \sqrt{x}$. Also, (0.5, 1), (2, 2), and (4.5, 3) are points on the graph of $Y_2 = \sqrt{2x}$. For a given y-coordinate, the x-coordinate on the graph of Y_2 is $\frac{1}{2}$ of the x-coordinate on Y_1.

Table 12

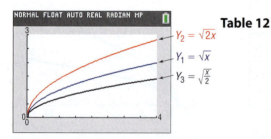

NORMAL FLOAT AUTO REAL RADIAN MP
PRESS ENTER TO EDIT

X	Y1	Y2
0	0	0
.5	.70711	1
1	1	1.4142
2	1.4142	2
4	2	2.8284
4.5	2.1213	3
8	2.8284	4
9	3	4.2426
16	4	5.6569
12.5	3.5355	5
25	5	7.0711

$Y_2 = \sqrt{2X}$

(a)

NORMAL FLOAT AUTO REAL RADIAN MP
PRESS ENTER TO EDIT

X	Y1	Y3
0	0	0
1	1	.70711
2	1.4142	1
4	2	1.4142
8	2.8284	2
9	3	2.1213
16	4	2.8284
18	4.2426	3
25	5	3.5355
32	5.6569	4
50	7.0711	5

$Y_3 = \sqrt{X/2}$

(b)

Figure 52

We conclude that the graph of $Y_2 = \sqrt{2x}$ is obtained by multiplying the x-coordinate of each point on the graph of $Y_1 = \sqrt{x}$ by $\frac{1}{2}$. The graph of $Y_2 = \sqrt{2x}$ is the graph of $Y_1 = \sqrt{x}$ compressed horizontally.

Look at Table 12(b). Notice that (1, 1), (4, 2), and (9, 3) are points on the graph of $Y_1 = \sqrt{x}$. Also notice that (2, 1), (8, 2), and (18, 3) are points on the graph of $Y_3 = \sqrt{\frac{x}{2}}$. For a given y-coordinate, the x-coordinate on the graph of Y_3 is 2 times the x-coordinate on Y_1. We conclude that the graph of $Y_3 = \sqrt{\frac{x}{2}}$ is obtained by multiplying the x-coordinate of each point on the graph of $Y_1 = \sqrt{x}$ by 2. The graph of $Y_3 = \sqrt{\frac{x}{2}}$ is the graph of $Y_1 = \sqrt{x}$ stretched horizontally.

Based on the Exploration, we have the following result:

If the argument x of a function $y = f(x)$ is multiplied by a positive number a, then the graph of the new function $y = f(ax)$ is obtained by multiplying each x-coordinate of $y = f(x)$ by $\frac{1}{a}$. A **horizontal compression** results if $a > 1$, and a **horizontal stretch** results if $0 < a < 1$.

Let's look at an example.

EXAMPLE 8

Graphing Using Stretches and Compressions

The graph of $y = f(x)$ is given in Figure 53. Use this graph to find the graphs of

(a) $y = 2f(x)$ (b) $y = f(3x)$

Solution (a) The graph of $y = 2f(x)$ is obtained by multiplying each y-coordinate of $y = f(x)$ by 2. See Figure 54.

(b) The graph of $y = f(3x)$ is obtained from the graph of $y = f(x)$ by multiplying each x-coordinate of $y = f(x)$ by $\frac{1}{3}$. See Figure 55.

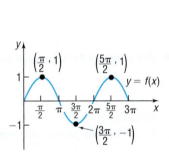

Figure 53 $y = f(x)$

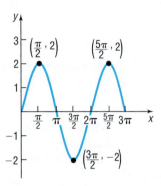

Figure 54 $y = 2f(x)$

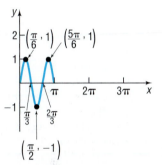

Figure 55 $y = f(3x)$ •

Now Work PROBLEMS 63(e) AND (g)

3 Graph Functions Using Reflections about the x-Axis and the y-Axis

EXAMPLE 9

Reflection about the x-Axis

Graph the function $f(x) = -x^2$. Find the domain and range of f.

Solution Begin with the graph of $y = x^2$, as shown in black in Figure 56. For each point (x, y) on the graph of $y = x^2$, the point $(x, -y)$ is on the graph of $y = -x^2$, as indicated in Table 13. Draw the graph of $y = -x^2$ by reflecting the graph of $y = x^2$ about the x-axis. See Figure 56.

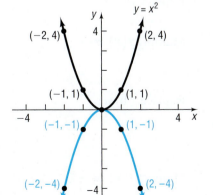

Figure 56

Table 13

x	$y = x^2$	$y = -x^2$
-2	4	-4
-1	1	-1
0	0	0
1	1	-1
2	4	-4

The domain of f is all real numbers, or $(-\infty, \infty)$. The range of f is $(-\infty, 0]$. •

When the right side of the function $y = f(x)$ is multiplied by -1, the graph of the new function $y = -f(x)$ is the **reflection about the x-axis** of the graph of the function $y = f(x)$.

Now Work PROBLEM 49

EXAMPLE 10

Reflection about the y-Axis

Graph the function $f(x) = \sqrt{-x}$. Find the domain and range of f.

Solution

To get the graph of $f(x) = \sqrt{-x}$, begin with the graph of $y = \sqrt{x}$, as shown in Figure 57. For each point (x, y) on the graph of $y = \sqrt{x}$, the point $(-x, y)$ is on the graph of $y = \sqrt{-x}$. Obtain the graph of $y = \sqrt{-x}$ by reflecting the graph of $y = \sqrt{x}$ about the y-axis. See Figure 57.

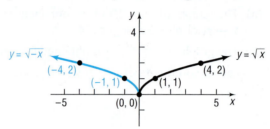

Figure 57

The domain of f is $(-\infty, 0]$. The range of f is the set of all nonnegative real numbers, or $[0, \infty)$.

In Words

For $y = -f(x)$, multiply each y-coordinate on the graph of $y = f(x)$ by -1.
For $y = f(-x)$, multiply each x-coordinate by -1.

When the graph of the function $y = f(x)$ is known, the graph of the new function $y = f(-x)$ is the **reflection about the y-axis** of the graph of the function $y = f(x)$.

SUMMARY OF GRAPHING TECHNIQUES

To Graph:	Draw the Graph of f and:	Functional Change to $f(x)$
Vertical shifts		
$y = f(x) + k$, $k > 0$	Raise the graph of f by k units.	Add k to $f(x)$.
$y = f(x) - k$, $k > 0$	Lower the graph of f by k units.	Subtract k from $f(x)$.
Horizontal shifts		
$y = f(x + h)$, $h > 0$	Shift the graph of f to the left h units.	Replace x by $x + h$.
$y = f(x - h)$, $h > 0$	Shift the graph of f to the right h units.	Replace x by $x - h$.
Compressing or stretching		
$y = af(x)$, $a > 0$	Multiply each y-coordinate of $y = f(x)$ by a. Stretch the graph of f vertically if $a > 1$. Compress the graph of f vertically if $0 < a < 1$.	Multiply $f(x)$ by a.
$y = f(ax)$, $a > 0$	Multiply each x-coordinate of $y = f(x)$ by $\frac{1}{a}$. Stretch the graph of f horizontally if $0 < a < 1$. Compress the graph of f horizontally if $a > 1$.	Replace x by ax.
Reflection about the x-axis		
$y = -f(x)$	Reflect the graph of f about the x-axis.	Multiply $f(x)$ by -1.
Reflection about the y-axis		
$y = f(-x)$	Reflect the graph of f about the y-axis.	Replace x by $-x$.

EXAMPLE 11

Determining the Function Obtained from a Series of Transformations

Find the function that is finally graphed after the following three transformations are applied to the graph of $y = |x|$.

1. Shift left 2 units
2. Shift up 3 units
3. Reflect about the y-axis

Solution

1. Shift left 2 units: Replace x by $x + 2$. $y = |x + 2|$
2. Shift up 3 units: Add 3. $y = |x + 2| + 3$
3. Reflect about the y-axis: Replace x by $-x$. $y = |-x + 2| + 3$ ●

Now Work PROBLEM 27

EXAMPLE 12

Combining Graphing Procedures

Graph the function $f(x) = \dfrac{3}{x - 2} + 1$. Find the domain and range of f.

Solution

It is helpful to write f as $f(x) = 3\left(\dfrac{1}{x - 2}\right) + 1$. Now use the following steps to obtain the graph of f.

STEP 1: $y = \dfrac{1}{x}$ Reciprocal function

STEP 2: $y = 3 \cdot \left(\dfrac{1}{x}\right) = \dfrac{3}{x}$ Multiply by 3; vertical stretch by a factor of 3.

STEP 3: $y = \dfrac{3}{x - 2}$ Replace x by $x - 2$; horizontal shift to the right 2 units.

STEP 4: $y = \dfrac{3}{x - 2} + 1$ Add 1; vertical shift up 1 unit.

See Figure 58.

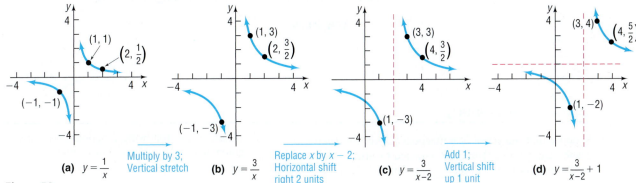

Figure 58

(a) $y = \dfrac{1}{x}$ Multiply by 3; Vertical stretch   (b) $y = \dfrac{3}{x}$ Replace x by $x - 2$; Horizontal shift right 2 units   (c) $y = \dfrac{3}{x-2}$ Add 1; Vertical shift up 1 unit   (d) $y = \dfrac{3}{x-2} + 1$

The domain of $y = \dfrac{1}{x}$ is $\{x | x \neq 0\}$ and its range is $\{y | y \neq 0\}$. Because we shifted right 2 units and up 1 unit to obtain f, the domain of f is $\{x | x \neq 2\}$ and its range is $\{y | y \neq 1\}$.   ●

HINT: Although the order in which transformations are performed can be altered, consider using the following order for consistency:
1. Reflections
2. Compressions and stretches
3. Shifts

Other orderings of the steps shown in Example 12 would also result in the graph of f. For example, try this one:

STEP 1: $y = \dfrac{1}{x}$ *Reciprocal function*

STEP 2: $y = \dfrac{1}{x - 2}$ *Replace x by x − 2; horizontal shift to the right 2 units.*

STEP 3: $y = \dfrac{3}{x - 2}$ *Multiply by 3; vertical stretch by a factor of 3.*

STEP 4: $y = \dfrac{3}{x - 2} + 1$ *Add 1; vertical shift up 1 unit.*

EXAMPLE 13

Combining Graphing Procedures

Graph the function $f(x) = \sqrt{1 - x} + 2$. Find the domain and range of f.

Solution   Because horizontal shifts require the form $x - h$, begin by rewriting $f(x)$ as
$$f(x) = \sqrt{1 - x} + 2 = \sqrt{-(x - 1)} + 2.$$ Now use the following steps.

STEP 1: $y = \sqrt{x}$ *Square root function*

STEP 2: $y = \sqrt{-x}$ *Replace x by −x; reflect about the y-axis.*

STEP 3: $y = \sqrt{-(x - 1)} = \sqrt{1 - x}$ *Replace x by x − 1; horizontal shift to the right 1 unit.*

STEP 4: $y = \sqrt{1 - x} + 2$ *Add 2; vertical shift up 2 units.*

See Figure 59.

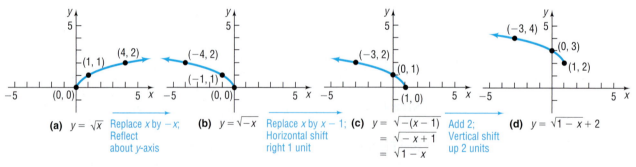

(a) $y = \sqrt{x}$ Replace x by −x; Reflect about y-axis   **(b)** $y = \sqrt{-x}$ Replace x by x − 1; Horizontal shift right 1 unit;   **(c)** $y = \sqrt{-(x - 1)}$ $= \sqrt{-x + 1}$ $= \sqrt{1 - x}$ Add 2; Vertical shift up 2 units   **(d)** $y = \sqrt{1 - x} + 2$

Figure 59

The domain of f is $(-\infty, 1]$ and the range is $[2, \infty)$.   ●

Now Work PROBLEM 55

3.5 Assess Your Understanding

Concepts and Vocabulary

1. Suppose that the graph of a function f is known. Then the graph of $y = f(x - 2)$ may be obtained by a(n) _____ shift of the graph of f to the _____ a distance of 2 units.

2. Suppose that the graph of a function f is known. Then the graph of $y = f(-x)$ may be obtained by a reflection about the _____-axis of the graph of the function $y = f(x)$.

3. *True or False* The graph of $y = \dfrac{1}{3}g(x)$ is the graph of $y = g(x)$ vertically stretched by a factor of 3.

4. *True or False* The graph of $y = -f(x)$ is the reflection about the x-axis of the graph of $y = f(x)$.

5. Which of the following functions has a graph that is the graph of $y = \sqrt{x}$ shifted down 3 units?
   (a) $y = \sqrt{x + 3}$ (b) $y = \sqrt{x - 3}$
   (c) $y = \sqrt{x} + 3$ (d) $y = \sqrt{x} - 3$

6. Which of the following functions has a graph that is the graph of $y = f(x)$ compressed horizontally by a factor of 4?
   (a) $y = f(4x)$ (b) $y = f\left(\dfrac{1}{4}x\right)$
   (c) $y = 4f(x)$ (d) $y = \dfrac{1}{4}f(x)$

Skill Building

In Problems 7–18, match each graph to one of the following functions:

A. $y = x^2 + 2$ B. $y = -x^2 + 2$ C. $y = |x| + 2$ D. $y = -|x| + 2$

E. $y = (x - 2)^2$ F. $y = -(x + 2)^2$ G. $y = |x - 2|$ H. $y = -|x + 2|$

I. $y = 2x^2$ J. $y = -2x^2$ K. $y = 2|x|$ L. $y = -2|x|$

7.
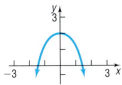

8.
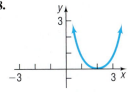

9.
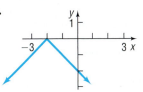

10.
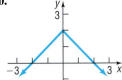

11.
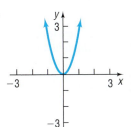

12.
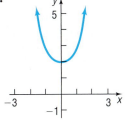

13.
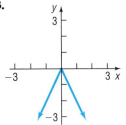

14.
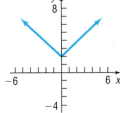

15.
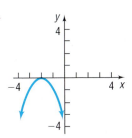

16.
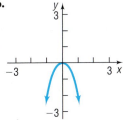

17.
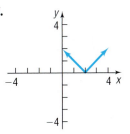

18.
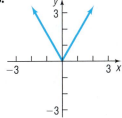

In Problems 19–26, write the function whose graph is the graph of $y = x^3$, but is:

19. Shifted to the right 4 units

20. Shifted to the left 4 units

21. Shifted up 4 units

22. Shifted down 4 units

23. Reflected about the *y*-axis

24. Reflected about the *x*-axis

25. Vertically stretched by a factor of 4

26. Horizontally stretched by a factor of 4

In Problems 27–30, find the function that is finally graphed after each of the following transformations is applied to the graph of $y = \sqrt{x}$ in the order stated.

27. (1) Shift up 2 units
(2) Reflect about the *x*-axis
(3) Reflect about the *y*-axis

28. (1) Reflect about the *x*-axis
(2) Shift right 3 units
(3) Shift down 2 units

29. (1) Reflect about the *x*-axis
(2) Shift up 2 units
(3) Shift left 3 units

30. (1) Shift up 2 units
(2) Reflect about the *y*-axis
(3) Shift left 3 units

31. If $(3, 6)$ is a point on the graph of $y = f(x)$, which of the following points must be on the graph of $y = -f(x)$?
(a) $(6, 3)$
(b) $(6, -3)$
(c) $(3, -6)$
(d) $(-3, 6)$

32. If $(3, 6)$ is a point on the graph of $y = f(x)$, which of the following points must be on the graph of $y = f(-x)$?
(a) $(6, 3)$
(b) $(6, -3)$
(c) $(3, -6)$
(d) $(-3, 6)$

33. If $(1, 3)$ is a point on the graph of $y = f(x)$, which of the following points must be on the graph of $y = 2f(x)$?
(a) $\left(1, \dfrac{3}{2}\right)$
(b) $(2, 3)$
(c) $(1, 6)$
(d) $\left(\dfrac{1}{2}, 3\right)$

34. If $(4, 2)$ is a point on the graph of $y = f(x)$, which of the following points must be on the graph of $y = f(2x)$?
(a) $(4, 1)$
(b) $(8, 2)$
(c) $(2, 2)$
(d) $(4, 4)$

35. Suppose that the x-intercepts of the graph of $y = f(x)$ are -5 and 3.
 (a) What are the x-intercepts of the graph of $y = f(x+2)$?
 (b) What are the x-intercepts of the graph of $y = f(x-2)$?
 (c) What are the x-intercepts of the graph of $y = 4f(x)$?
 (d) What are the x-intercepts of the graph of $y = f(-x)$?

36. Suppose that the x-intercepts of the graph of $y = f(x)$ are -8 and 1.
 (a) What are the x-intercepts of the graph of $y = f(x+4)$?
 (b) What are the x-intercepts of the graph of $y = f(x-3)$?
 (c) What are the x-intercepts of the graph of $y = 2f(x)$?
 (d) What are the x-intercepts of the graph of $y = f(-x)$?

37. Suppose that the function $y = f(x)$ is increasing on the interval $(-1, 5)$.
 (a) Over what interval is the graph of $y = f(x+2)$ increasing?
 (b) Over what interval is the graph of $y = f(x-5)$ increasing?
 (c) What can be said about the graph of $y = -f(x)$?
 (d) What can be said about the graph of $y = f(-x)$?

38. Suppose that the function $y = f(x)$ is decreasing on the interval $(-2, 7)$.
 (a) Over what interval is the graph of $y = f(x+2)$ decreasing?
 (b) Over what interval is the graph of $y = f(x-5)$ decreasing?
 (c) What can be said about the graph of $y = -f(x)$?
 (d) What can be said about the graph of $y = f(-x)$?

In Problems 39–62, graph each function using the techniques of shifting, compressing, stretching, and/or reflecting. Start with the graph of the basic function (for example, $y = x^2$) and show all stages. Be sure to show at least three key points. Find the domain and the range of each function.

39. $f(x) = x^2 - 1$

40. $f(x) = x^2 + 4$

41. $g(x) = x^3 + 1$

42. $g(x) = x^3 - 1$

43. $h(x) = \sqrt{x+2}$

44. $h(x) = \sqrt{x+1}$

45. $f(x) = (x-1)^3 + 2$

46. $f(x) = (x+2)^3 - 3$

47. $g(x) = 4\sqrt{x}$

48. $g(x) = \dfrac{1}{2}\sqrt{x}$

49. $f(x) = -\sqrt[3]{x}$

50. $f(x) = -\sqrt{x}$

51. $f(x) = 2(x+1)^2 - 3$

52. $f(x) = 3(x-2)^2 + 1$

53. $g(x) = 2\sqrt{x-2} + 1$

54. $g(x) = 3|x+1| - 3$

55. $h(x) = \sqrt{-x} - 2$

56. $h(x) = \dfrac{4}{x} + 2$

57. $f(x) = -(x+1)^3 - 1$

58. $f(x) = -4\sqrt{x-1}$

59. $g(x) = 2|1 - x|$

60. $g(x) = 4\sqrt{2-x}$

61. $h(x) = \dfrac{1}{2x}$

62. $h(x) = \sqrt[3]{x-1} + 3$

In Problems 63–66, the graph of a function f is illustrated. Use the graph of f as the first step toward graphing each of the following functions:

 (a) $F(x) = f(x) + 3$
 (b) $G(x) = f(x+2)$
 (c) $P(x) = -f(x)$
 (d) $H(x) = f(x+1) - 2$
 (e) $Q(x) = \dfrac{1}{2}f(x)$
 (f) $g(x) = f(-x)$
 (g) $h(x) = f(2x)$

63.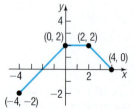

64.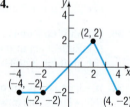

65.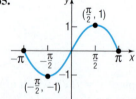

66.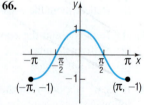

Mixed Practice

In Problems 67–74, complete the square of each quadratic expression. Then graph each function using the technique of shifting. (If necessary, refer to Chapter R, Section R.5 to review completing the square.)

67. $f(x) = x^2 + 2x$

68. $f(x) = x^2 - 6x$

69. $f(x) = x^2 - 8x + 1$

70. $f(x) = x^2 + 4x + 2$

71. $f(x) = 2x^2 - 12x + 19$

72. $f(x) = 3x^2 + 6x + 1$

73. $f(x) = -3x^2 - 12x - 17$

74. $f(x) = -2x^2 - 12x - 13$

Applications and Extensions

75. The graph of a function f is illustrated in the figure.
   (a) Draw the graph of $y = |f(x)|$.
   (b) Draw the graph of $y = f(|x|)$.

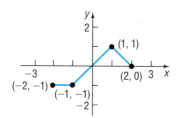

76. The graph of a function f is illustrated in the figure.
   (a) Draw the graph of $y = |f(x)|$.
   (b) Draw the graph of $y = f(|x|)$.

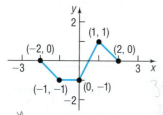

77. Suppose $(1, 3)$ is a point on the graph of $y = f(x)$.
   (a) What point is on the graph of $y = f(x + 3) - 5$?
   (b) What point is on the graph of $y = -2f(x - 2) + 1$?
   (c) What point is on the graph of $y = f(2x + 3)$?

78. Suppose $(-3, 5)$ is a point on the graph of $y = g(x)$.
   (a) What point is on the graph of $y = g(x + 1) - 3$?
   (b) What point is on the graph of $y = -3g(x - 4) + 3$?
   (c) What point is on the graph of $y = g(3x + 9)$?

79. Graph the following functions using transformations.
   (a) $f(x) = \text{int}(-x)$ (b) $g(x) = -\text{int}(x)$

80. Graph the following functions using transformations
   (a) $f(x) = \text{int}(x - 1)$ (b) $g(x) = \text{int}(1 - x)$

81. (a) Graph $f(x) = |x - 3| - 3$ using transformations.
   (b) Find the area of the region that is bounded by f and the x-axis and lies below the x-axis.

82. (a) Graph $f(x) = -2|x - 4| + 4$ using transformations.
   (b) Find the area of the region that is bounded by f and the x-axis and lies above the x-axis.

83. Thermostat Control Energy conservation experts estimate that homeowners can save 5% to 10% on winter heating bills by programming their thermostats 5 to 10 degrees lower while sleeping. In the graph (top, right), the temperature T (in degrees Fahrenheit) of a home is given as a function of time t (in hours after midnight) over a 24-hour period.
   (a) At what temperature is the thermostat set during daytime hours? At what temperature is the thermostat set overnight?
   (b) The homeowner reprograms the thermostat to $y = T(t) - 2$. Explain how this affects the temperature in the house. Graph this new function.
   (c) The homeowner reprograms the thermostat to $y = T(t + 1)$. Explain how this affects the temperature in the house. Graph this new function.

Source: Roger Albright, *547 Ways to Be Fuel Smart,* 2000

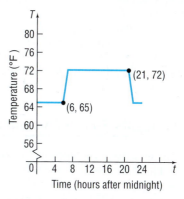

84. Digital Music Revenues The total projected worldwide digital music revenues R, in millions of dollars, for the years 2012 through 2017 can be estimated by the function

$$R(x) = 28.6x^2 + 300x + 4843$$

where x is the number of years after 2012.
   (a) Find $R(0)$, $R(3)$, and $R(5)$ and explain what each value represents.
   (b) Find $r(x) = R(x - 2)$.
   (c) Find $r(2)$, $r(5)$, and $r(7)$ and explain what each value represents.
   (d) In the model $r = r(x)$, what does x represent?
   (e) Would there be an advantage in using the model r when estimating the projected revenues for a given year instead of the model R?

Source: IFPI Digital Music Report 2013

85. Temperature Measurements The relationship between the Celsius (°C) and Fahrenheit (°F) scales for measuring temperature is given by the equation

$$F = \frac{9}{5}C + 32$$

The relationship between the Celsius (°C) and Kelvin (K) scales is $K = C + 273$. Graph the equation $F = \frac{9}{5}C + 32$ using degrees Fahrenheit on the y-axis and degrees Celsius on the x-axis. Use the techniques introduced in this section to obtain the graph showing the relationship between Kelvin and Fahrenheit temperatures.

86. Period of a Pendulum The period T (in seconds) of a simple pendulum is a function of its length l (in feet) defined by the equation

$$T = 2\pi\sqrt{\frac{l}{g}}$$

where $g \approx 32.2$ feet per second per second is the acceleration due to gravity. (*Continued on page 260.*)

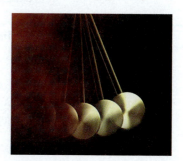

(a) Use a graphing utility to graph the function $T = T(l)$.

(b) Now graph the functions $T = T(l + 1)$, $T = T(l + 2)$, and $T = T(l + 3)$.

(c) Discuss how adding to the length l changes the period T.

(d) Now graph the functions $T = T(2l)$, $T = T(3l)$, and $T = T(4l)$.

(e) Discuss how multiplying the length l by factors of 2, 3, and 4 changes the period T.

87. The equation $y = (x - c)^2$ defines a *family of parabolas*, one parabola for each value of c. On one set of coordinate axes, graph the members of the family for $c = 0, c = 3$, and $c = -2$.

88. Repeat Problem 87 for the family of parabolas $y = x^2 + c$.

Explaining Concepts: Discussion and Writing

89. Suppose that the graph of a function f is known. Explain how the graph of $y = 4f(x)$ differs from the graph of $y = f(4x)$.

90. Suppose that the graph of a function f is known. Explain how the graph of $y = f(x) - 2$ differs from the graph of $y = f(x - 2)$.

91. The area under the curve $y = \sqrt{x}$ bounded from below by the x-axis and on the right by $x = 4$ is $\dfrac{16}{3}$ square units. Using the

ideas presented in this section, what do you think is the area under the curve of $y = \sqrt{-x}$ bounded from below by the x-axis and on the left by $x = -4$? Justify your answer.

92. Explain how the range of the function $f(x) = x^2$ compares to the range of $g(x) = f(x) + k$.

93. Explain how the domain of $g(x) = \sqrt{x}$ compares to the domain of $g(x - k)$, where $k \geq 0$.

Retain Your Knowledge

Problems 94–97 are based on material learned earlier in the course. The purpose of these problems is to keep the material fresh in your mind so that you are better prepared for the final exam.

94. Determine the slope and y-intercept of the graph of $3x - 5y = 30$.

95. Simplify $\dfrac{(x^{-2}y^3)^4}{(x^2y^{-5})^{-2}}$

96. The amount of water used when taking a shower varies directly with the number of minutes the shower is run. If a 4-minute shower uses 7 gallons of water, how much water is used in a 9-minute shower?

97. List the intercepts and test for symmetry: $y^2 = x + 4$

3.6 Mathematical Models: Building Functions

OBJECTIVE 1 Build and Analyze Functions (p. 260)

1 Build and Analyze Functions

Real-world problems often result in mathematical models that involve functions. These functions need to be constructed or built based on the information given. In building functions, we must be able to translate the verbal description into the language of mathematics. This is done by assigning symbols to represent the independent and dependent variables and then by finding the function or rule that relates these variables.

EXAMPLE 1 **Finding the Distance from the Origin to a Point on a Graph**

Let $P = (x, y)$ be a point on the graph of $y = x^2 - 1$.

(a) Express the distance d from P to the origin O as a function of x.

(b) What is d if $x = 0$?

(c) What is d if $x = 1$?

(d) What is d if $x = \dfrac{\sqrt{2}}{2}$?

(e) Use a graphing utility to graph the function $d = d(x), x \geq 0$. Rounding to two decimal places, find the value(s) of x at which d has a local minimum. [This gives the point(s) on the graph of $y = x^2 - 1$ closest to the origin.]

Solution (a) Figure 60 illustrates the graph of $y = x^2 - 1$. The distance d from P to O is

$$d = \sqrt{(x-0)^2 + (y-0)^2} = \sqrt{x^2 + y^2}$$

Since P is a point on the graph of $y = x^2 - 1$, substitute $x^2 - 1$ for y. Then

$$d(x) = \sqrt{x^2 + (x^2-1)^2} = \sqrt{x^4 - x^2 + 1}$$

The distance d is expressed as a function of x.

(b) If $x = 0$, the distance d is

$$d(0) = \sqrt{0^4 - 0^2 + 1} = \sqrt{1} = 1$$

(c) If $x = 1$, the distance d is

$$d(1) = \sqrt{1^4 - 1^2 + 1} = 1$$

(d) If $x = \dfrac{\sqrt{2}}{2}$, the distance d is

$$d\left(\frac{\sqrt{2}}{2}\right) = \sqrt{\left(\frac{\sqrt{2}}{2}\right)^4 - \left(\frac{\sqrt{2}}{2}\right)^2 + 1} = \sqrt{\frac{1}{4} - \frac{1}{2} + 1} = \frac{\sqrt{3}}{2}$$

(e) Figure 61 shows the graph of $Y_1 = \sqrt{x^4 - x^2 + 1}$. Using the MINIMUM feature on a graphing utility, we find that when $x \approx 0.71$ the value of d is smallest. The local minimum is $d \approx 0.87$ rounded to two decimal places. Since $d(x)$ is even, it follows by symmetry that when $x \approx -0.71$, the value of d is also a local minimum. Since $(\pm 0.71)^2 - 1 \approx -0.50$, the points $(-0.71, -0.50)$ and $(0.71, -0.50)$ on the graph of $y = x^2 - 1$ are closest to the origin. ●

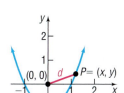

Figure 60 $y = x^2 - 1$

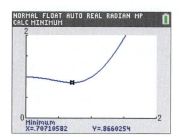

Figure 61 $d(x) = \sqrt{x^4 - x^2 + 1}$

Now Work PROBLEM 1

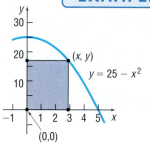

Figure 62

EXAMPLE 2

Area of a Rectangle

A rectangle has one corner in quadrant I on the graph of $y = 25 - x^2$, another at the origin, a third on the positive y-axis, and the fourth on the positive x-axis. See Figure 62.

(a) Express the area A of the rectangle as a function of x.
(b) What is the domain of A?
(c) Graph $A = A(x)$.
(d) For what value of x is the area largest?

Solution (a) The area A of the rectangle is $A = xy$, where $y = 25 - x^2$. Substituting this expression for y, we obtain $A(x) = x(25 - x^2) = 25x - x^3$.

(b) Since (x, y) is in quadrant I, we have $x > 0$. Also, $y = 25 - x^2 > 0$, which implies that $x^2 < 25$, so $-5 < x < 5$. Combining these restrictions, we have the domain of A as $\{x | 0 < x < 5\}$, or $(0, 5)$ using interval notation.

(c) See Figure 63 on page 262 for the graph of $A = A(x)$.

(d) Using MAXIMUM, we find that the maximum area is 48.11 square units at $x = 2.89$ units, each rounded to two decimal places. See Figure 64.

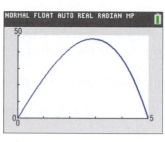

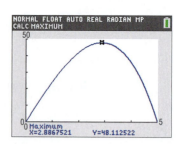

Figure 63 $A(x) = 25x - x^3$ **Figure 64**

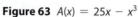

Now Work PROBLEM 7

EXAMPLE 3

Close Call?

Suppose two planes flying at the same altitude are headed toward each other. One plane is flying due south at a groundspeed of 400 miles per hour and is 600 miles from the potential intersection point of the planes. The other plane is flying due west with a groundspeed of 250 miles per hour and is 400 miles from the potential intersection point of the planes. See Figure 65.

(a) Build a model that expresses the distance d between the planes as a function of time t.

(b) Use a graphing utility to graph $d = d(t)$. How close do the planes come to each other? At what time are the planes closest?

Solution

(a) Refer to Figure 65. The distance d between the two planes is the hypotenuse of a right triangle. At any time t, the length of the north/south leg of the triangle is $600 - 400t$. At any time t, the length of the east/west leg of the triangle is $400 - 250t$. Use the Pythagorean Theorem to find that the square of the distance between the two planes is

$$d^2 = (600 - 400t)^2 + (400 - 250t)^2$$

Therefore, the distance between the two planes as a function of time is given by the model

$$d(t) = \sqrt{(600 - 400t)^2 + (400 - 250t)^2}$$

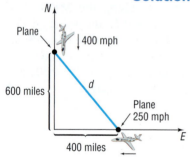

Figure 65

(b) Figure 66(a) shows the graph of $d = d(t)$. Using MINIMUM, the minimum distance between the planes is 21.20 miles, and the time at which the planes are closest is after 1.53 hours, each rounded to two decimal places. See Figure 66(b).

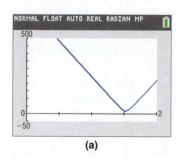

Figure 66   (a)   (b)

Now Work PROBLEM 19

3.6 Assess Your Understanding

Applications and Extensions

1. Let $P = (x, y)$ be a point on the graph of $y = x^2 - 8$.
   (a) Express the distance d from P to the origin as a function of x.
   (b) What is d if $x = 0$?
   (c) What is d if $x = 1$?
    (d) Use a graphing utility to graph $d = d(x)$.
   (e) For what values of x is d smallest?

2. Let $P = (x, y)$ be a point on the graph of $y = x^2 - 8$.
   (a) Express the distance d from P to the point $(0, -1)$ as a function of x.
   (b) What is d if $x = 0$?
   (c) What is d if $x = -1$?
    (d) Use a graphing utility to graph $d = d(x)$.
   (e) For what values of x is d smallest?

3. Let $P = (x, y)$ be a point on the graph of $y = \sqrt{x}$.
   (a) Express the distance d from P to the point $(1, 0)$ as a function of x.
    (b) Use a graphing utility to graph $d = d(x)$.
   (c) For what values of x is d smallest?

4. Let $P = (x, y)$ be a point on the graph of $y = \dfrac{1}{x}$.
   (a) Express the distance d from P to the origin as a function of x.
    (b) Use a graphing utility to graph $d = d(x)$.
   (c) For what values of x is d smallest?

5. A right triangle has one vertex on the graph of $y = x^3, x > 0$, at (x, y), another at the origin, and the third on the positive y-axis at $(0, y)$, as shown in the figure. Express the area A of the triangle as a function of x.

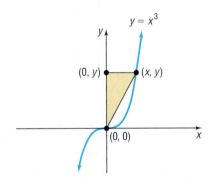

6. A right triangle has one vertex on the graph of $y = 9 - x^2, x > 0$, at (x, y), another at the origin, and the third on the positive x-axis at $(x, 0)$. Express the area A of the triangle as a function of x.

7. A rectangle has one corner in quadrant I on the graph of $y = 16 - x^2$, another at the origin, a third on the positive y-axis, and the fourth on the positive x-axis. See the figure (top, right).
   (a) Express the area A of the rectangle as a function of x.
   (b) What is the domain of A?
    (c) Graph $A = A(x)$. For what value of x is A largest?

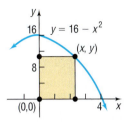

8. A rectangle is inscribed in a semicircle of radius 2. See the figure. Let $P = (x, y)$ be the point in quadrant I that is a vertex of the rectangle and is on the circle.

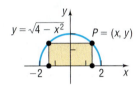

   (a) Express the area A of the rectangle as a function of x.
   (b) Express the perimeter p of the rectangle as a function of x.
    (c) Graph $A = A(x)$. For what value of x is A largest?
   (d) Graph $p = p(x)$. For what value of x is p largest?

9. A rectangle is inscribed in a circle of radius 2. See the figure. Let $P = (x, y)$ be the point in quadrant I that is a vertex of the rectangle and is on the circle.

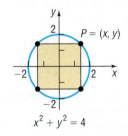

   (a) Express the area A of the rectangle as a function of x.
   (b) Express the perimeter p of the rectangle as a function of x.
    (c) Graph $A = A(x)$. For what value of x is A largest?
   (d) Graph $p = p(x)$. For what value of x is p largest?

10. A circle of radius r is inscribed in a square. See the figure.

   (a) Express the area A of the square as a function of the radius r of the circle.
   (b) Express the perimeter p of the square as a function of r.

11. Geometry A wire 10 meters long is to be cut into two pieces. One piece will be shaped as a square, and the other piece will be shaped as a circle. See the figure.

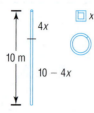

(a) Express the total area A enclosed by the pieces of wire as a function of the length x of a side of the square.
(b) What is the domain of A ?
(c) Graph $A = A(x)$. For what value of x is A smallest?

12. Geometry A wire 10 meters long is to be cut into two pieces. One piece will be shaped as an equilateral triangle, and the other piece will be shaped as a circle.
(a) Express the total area A enclosed by the pieces of wire as a function of the length x of a side of the equilateral triangle.
(b) What is the domain of A ?
(c) Graph $A = A(x)$. For what value of x is A smallest?

13. Geometry A wire of length x is bent into the shape of a circle.
(a) Express the circumference C of the circle as a function of x.
(b) Express the area A of the circle as a function of x.

14. Geometry A wire of length x is bent into the shape of a square.
(a) Express the perimeter p of the square as a function of x.
(b) Express the area A of the square as a function of x.

15. Geometry A semicircle of radius r is inscribed in a rectangle so that the diameter of the semicircle is the length of the rectangle. See the figure.

(a) Express the area A of the rectangle as a function of the radius r of the semicircle.
(b) Express the perimeter p of the rectangle as a function of r.

16. Geometry An equilateral triangle is inscribed in a circle of radius r. See the figure. Express the circumference C of the circle as a function of the length x of a side of the triangle.

[**Hint:** First show that $r^2 = \dfrac{x^2}{3}$.]

17. Geometry An equilateral triangle is inscribed in a circle of radius r. See the figure in Problem 16. Express the area A within the circle, but outside the triangle, as a function of the length x of a side of the triangle.

18. Uniform Motion Two cars leave an intersection at the same time. One is headed south at a constant speed of 30 miles per hour, and the other is headed west at a constant speed of

40 miles per hour (see the figure). Build a model that expresses the distance d between the cars as a function of the time t.

[**Hint:** At $t = 0$, the cars leave the intersection.]

19. Uniform Motion Two cars are approaching an intersection. One is 2 miles south of the intersection and is moving at a constant speed of 30 miles per hour. At the same time, the other car is 3 miles east of the intersection and is moving at a constant speed of 40 miles per hour.
(a) Build a model that expresses the distance d between the cars as a function of time t.
[**Hint:** At $t = 0$, the cars are 2 miles south and 3 miles east of the intersection, respectively.]
(b) Use a graphing utility to graph $d = d(t)$. For what value of t is d smallest?

20. Inscribing a Cylinder in a Sphere Inscribe a right circular cylinder of height h and radius r in a sphere of fixed radius R. See the illustration. Express the volume V of the cylinder as a function of h.

[**Hint:** $V = \pi r^2 h$. Note also the right triangle.]

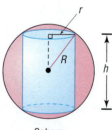

Sphere

21. Inscribing a Cylinder in a Cone Inscribe a right circular cylinder of height h and radius r in a cone of fixed radius R and fixed height H. See the illustration. Express the volume V of the cylinder as a function of r.

[**Hint:** $V = \pi r^2 h$. Note also the similar triangles.]

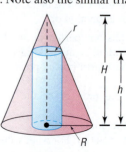

Cone

22. Installing Cable TV MetroMedia Cable is asked to provide service to a customer whose house is located 2 miles from the road along which the cable is buried. The nearest connection box for the cable is located 5 miles down the road. See the figure.

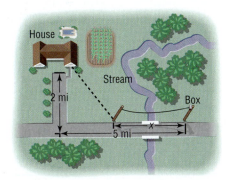

(a) If the installation cost is $500 per mile along the road and $700 per mile off the road, build a model that expresses the total cost C of installation as a function of the distance x (in miles) from the connection box to the point where the cable installation turns off the road. Find the domain of $C = C(x)$.
(b) Compute the cost if $x = 1$ mile.
(c) Compute the cost if $x = 3$ miles.
(d) Graph the function $C = C(x)$. Use TRACE to see how the cost C varies as x changes from 0 to 5.
(e) What value of x results in the least cost?

23. Time Required to Go from an Island to a Town An island is 2 miles from the nearest point P on a straight shoreline. A town is 12 miles down the shore from P. See the illustration.

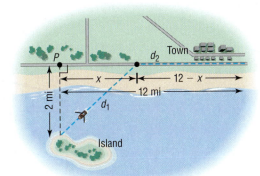

(a) If a person can row a boat at an average speed of 3 miles per hour and the same person can walk 5 miles per hour, build a model that expresses the time T that it takes to go from the island to town as a function of the distance x from P to where the person lands the boat.
(b) What is the domain of T?
(c) How long will it take to travel from the island to town if the person lands the boat 4 miles from P?
(d) How long will it take if the person lands the boat 8 miles from P?

24. Filling a Conical Tank Water is poured into a container in the shape of a right circular cone with radius 4 feet and height 16 feet. See the figure. Express the volume V of the water in the cone as a function of the height h of the water. [**Hint:** The volume V of a cone of radius r and height h is $V = \dfrac{1}{3}\pi r^2 h$.]

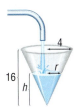

25. Constructing an Open Box An open box with a square base is to be made from a square piece of cardboard 24 inches on a side by cutting out a square from each corner and turning up the sides. See the figure.

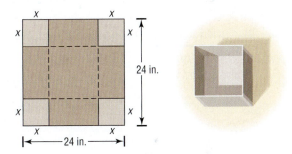

(a) Express the volume V of the box as a function of the length x of the side of the square cut from each corner.
(b) What is the volume if a 3-inch square is cut out?
(c) What is the volume if a 10-inch square is cut out?
(d) Graph $V = V(x)$. For what value of x is V largest?

26. Constructing an Open Box An open box with a square base is required to have a volume of 10 cubic feet.
(a) Express the amount A of material used to make such a box as a function of the length x of a side of the square base.
(b) How much material is required for a base 1 foot by 1 foot?
(c) How much material is required for a base 2 feet by 2 feet?
(d) Use a graphing utility to graph $A = A(x)$. For what value of x is A smallest?

Retain Your Knowledge

Problems 27–30 are based on material learned earlier in the course. The purpose of these problems is to keep the material fresh in your mind so that you are better prepared for the final exam.

27. Solve: $|2x - 3| - 5 = -2$

28. A 16-foot long Ford Fusion wants to pass a 50-foot truck traveling at 55 mi/h. How fast must the car travel to completely pass the truck in 5 seconds?

29. Find the slope of the line containing the points $(3, -2)$ and $(1, 6)$.

30. Find the missing length x for the given pair of similar triangles.

Chapter Review

Library of Functions

Constant function (p. 239)

$f(x) = b$

The graph is a horizontal line with y-intercept b.

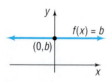

Identity function (p. 239)

$f(x) = x$

The graph is a line with slope 1 and y-intercept 0.

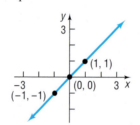

Square function (p. 239)

$f(x) = x^2$

The graph is a parabola with intercept at $(0, 0)$.

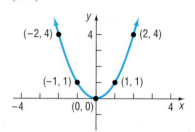

Cube function (p. 240)

$f(x) = x^3$

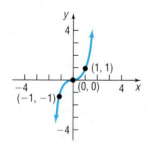

Square root function (p. 240)

$f(x) = \sqrt{x}$

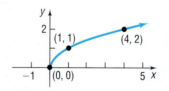

Cube root function (p. 240)

$f(x) = \sqrt[3]{x}$

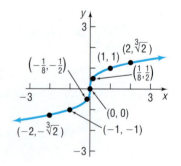

Reciprocal function (p. 240)

$f(x) = \dfrac{1}{x}$

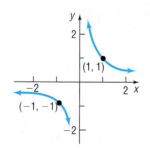

Absolute value function (p. 240)

$f(x) = |x|$

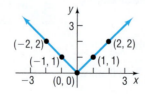

Greatest integer function (p. 241)

$f(x) = \text{int}(x)$

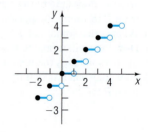

Things to Know

Function (pp. 199–202)

A relation between two sets so that each element x in the first set, the domain, has corresponding to it exactly one element y in the second set, the range. The range is the set of y-values of the function for the x-values in the domain.

A function can also be characterized as a set of ordered pairs (x, y) in which no first element is paired with two different second elements.

Function notation (pp. 202–205)

$y = f(x)$

f is a symbol for the function.

x is the argument, or independent variable.

y is the dependent variable.

$f(x)$ is the value of the function at x, or the image of x.

A function f may be defined implicitly by an equation involving x and y or explicitly by writing $y = f(x)$.

Difference quotient of f (p. 205)

$\dfrac{f(x + h) - f(x)}{h} \quad h \neq 0$

Domain (pp. 206–208)

If unspecified, the domain of a function f defined by an equation is the largest set of real numbers for which $f(x)$ is a real number.

Vertical-line test (p. 214)

A set of points in the xy-plane is the graph of a function if and only if every vertical line intersects the graph in at most one point.

Even function f (p. 224)

$f(-x) = f(x)$ for every x in the domain ($-x$ must also be in the domain).

Odd function f (p. 224)

$f(-x) = -f(x)$ for every x in the domain ($-x$ must also be in the domain).

Increasing function (p. 226)

A function f is increasing on an open interval I if, for any choice of x_1 and x_2 in I, with $x_1 < x_2$, we have $f(x_1) < f(x_2)$.

Decreasing function (p. 226)

A function f is decreasing on an open interval I if, for any choice of x_1 and x_2 in I, with $x_1 < x_2$, we have $f(x_1) > f(x_2)$.

Constant function (p. 226)

A function f is constant on an open interval I if, for all choices of x in I, the values of $f(x)$ are equal.

Local maximum (p. 227)

A function f, defined on some interval I, has a local maximum at c if there is an open interval in I containing c such that, for all x in this open interval, $f(x) \leq f(c)$.

Local minimum (p. 227)

A function f, defined on some interval I, has a local minimum at c if there is an open interval in I containing c such that, for all x in this open interval, $f(x) \geq f(c)$.

Absolute maximum and Absolute minimum (p. 227)

Let f denote a function defined on some interval I.

If there is a number u in I for which $f(x) \leq f(u)$ for all x in I, then f has an absolute maximum at u, and the number $f(u)$ is the absolute maximum of f on I.

If there is a number v in I for which $f(x) \geq f(v)$, for all x in I, then f has an absolute minimum at v, and the number $f(v)$ is the absolute minimum of f on I.

Average rate of change of a function (p. 230)

The average rate of change of f from a to b is

$$\frac{\Delta y}{\Delta x} = \frac{f(b) - f(a)}{b - a} \quad a \neq b$$

Objectives

Section		You should be able to ...	Examples	Review Exercises
3.1	1	Determine whether a relation represents a function (p. 199)	1–5	1, 2
	2	Find the value of a function (p. 202)	6, 7	3–5, 39
	3	Find the difference quotient of a function (p. 205)	8	15
	4	Find the domain of a function defined by an equation (p. 206)	9, 10	6–11
	5	Form the sum, difference, product, and quotient of two functions (p. 208)	11	12–14
3.2	1	Identify the graph of a function (p. 214)	1	27, 28
	2	Obtain information from or about the graph of a function (p. 215)	2–4	16(a)–(e), 17(a), 17(e), 17(g)

Section	You should be able to ...	Examples	Review Exercises
3.3	**1** Determine even and odd functions from a graph (p. 223)	1	17(f)
	2 Identify even and odd functions from an equation (p. 225)	2	18–21
	3 Use a graph to determine where a function is increasing, decreasing, or constant (p. 225)	3	17(b)
	4 Use a graph to locate local maxima and local minima (p. 226)	4	17(c)
	5 Use a graph to locate the absolute maximum and the absolute minimum (p. 227)	5	17(d)
	6 Use a graphing utility to approximate local maxima and local minima and to determine where a function is increasing or decreasing (p. 229)	6	22, 23, 40(d), 41(b)
	7 Find the average rate of change of a function (p. 230)	7, 8	24–26
3.4	**1** Graph the functions listed in the library of functions (p. 237)	1, 2	29, 30
	2 Graph piecewise-defined functions (p. 242)	3, 4	37, 38
3.5	**1** Graph functions using vertical and horizontal shifts (p. 248)	1–5, 11–13	16(f), 31, 33–36
	2 Graph functions using compressions and stretches (p. 251)	6–8, 12	16(g), 32, 36
	3 Graph functions using reflections about the x-axis and the y-axis (p. 253)	9, 10, 11, 13	16(h), 32, 34, 36
3.6	**1** Build and analyze functions (p. 260)	1–3	40, 41

Review Exercises

In Problems 1 and 2, determine whether each relation represents a function. For each function, state the domain and range.

1. $\{(-1,0), (2,3), (4,0)\}$

2. $\{(4,-1), (2,1), (4,2)\}$

In Problems 3–5, find the following for each function:

(a) $f(2)$ (b) $f(-2)$ (c) $f(-x)$ (d) $-f(x)$ (e) $f(x-2)$ (f) $f(2x)$

3. $f(x) = \dfrac{3x}{x^2 - 1}$

4. $f(x) = \sqrt{x^2 - 4}$

5. $f(x) = \dfrac{x^2 - 4}{x^2}$

In Problems 6–11, find the domain of each function.

6. $f(x) = \dfrac{x}{x^2 - 9}$

7. $f(x) = \sqrt{2 - x}$

8. $g(x) = \dfrac{|x|}{x}$

9. $f(x) = \dfrac{x}{x^2 + 2x - 3}$

10. $f(x) = \dfrac{\sqrt{x + 1}}{x^2 - 4}$

11. $g(x) = \dfrac{x}{\sqrt{x + 8}}$

In Problems 12–14, find $f + g$, $f - g$, $f \cdot g$, and $\dfrac{f}{g}$ for each pair of functions. State the domain of each of these functions.

12. $f(x) = 2 - x;\ g(x) = 3x + 1$

13. $f(x) = 3x^2 + x + 1;\ g(x) = 3x$

14. $f(x) = \dfrac{x + 1}{x - 1};\ g(x) = \dfrac{1}{x}$

15. Find the difference quotient of $f(x) = -2x^2 + x + 1$; that is, find $\dfrac{f(x + h) - f(x)}{h}, h \neq 0$.

16. Consider the graph of the function f on the right.
(a) Find the domain and the range of f.
(b) List the intercepts.
(c) Find $f(-2)$.
(d) For what value of x does $f(x) = -3$?
(e) Solve $f(x) > 0$.
(f) Graph $y = f(x - 3)$.
(g) Graph $y = f\left(\dfrac{1}{2}x\right)$.
(h) Graph $y = -f(x)$.

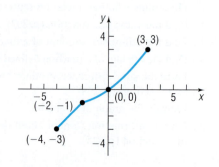

17. Use the graph of the function f shown to find:
  (a) The domain and the range of f.
  (b) The intervals on which f is increasing, decreasing, or constant.
  (c) The local minimum values and local maximum values.
  (d) The absolute maximum and absolute minimum.
  (e) Whether the graph is symmetric with respect to the x-axis, the y-axis, or the origin.
  (f) Whether the function is even, odd, or neither.
  (g) The intercepts, if any.

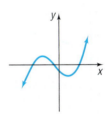

In Problems 18–21, determine (algebraically) whether the given function is even, odd, or neither.

18. $f(x) = x^3 - 4x$

19. $g(x) = \dfrac{4 + x^2}{1 + x^4}$

20. $G(x) = 1 - x + x^3$

21. $f(x) = \dfrac{x}{1 + x^2}$

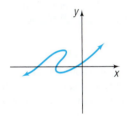 In Problems 22 and 23, use a graphing utility to graph each function over the indicated interval. Approximate any local maximum values and local minimum values. Determine where the function is increasing and where it is decreasing.

22. $f(x) = 2x^3 - 5x + 1 \quad (-3, 3)$

23. $f(x) = 2x^4 - 5x^3 + 2x + 1 \quad (-2, 3)$

24. Find the average rate of change of $f(x) = 8x^2 - x$:
  (a) From 1 to 2   (b) From 0 to 1   (c) From 2 to 4

In Problems 25 and 26, find the average rate of change from 2 to 3 for each function f. Be sure to simplify.

25. $f(x) = 2 - 5x$

26. $f(x) = 3x - 4x^2$

In Problems 27 and 28, is the graph shown the graph of a function?

27.

28.

In Problems 29 and 30, graph each function. Be sure to label at least three points.

29. $f(x) = |x|$

30. $f(x) = \sqrt{x}$

In Problems 31–36, graph each function using the techniques of shifting, compressing or stretching, and reflections. Identify any intercepts of the graph. State the domain and, based on the graph, find the range.

31. $F(x) = |x| - 4$

32. $g(x) = -2|x|$

33. $h(x) = \sqrt{x - 1}$

34. $f(x) = \sqrt{1 - x}$

35. $h(x) = (x - 1)^2 + 2$

36. $g(x) = -2(x + 2)^3 - 8$

In Problems 37 and 38:
  (a) Find the domain of each function.
  (b) Locate any intercepts.
  (c) Graph each function.
  (d) Based on the graph, find the range.
  (e) Is f continuous on its domain?

37. $f(x) = \begin{cases} 3x & \text{if } -2 < x \le 1 \\ x + 1 & \text{if } x > 1 \end{cases}$

38. $f(x) = \begin{cases} x & \text{if } -4 \le x < 0 \\ 1 & \text{if } x = 0 \\ 3x & \text{if } x > 0 \end{cases}$

39. A function f is defined by
$$f(x) = \frac{Ax + 5}{6x - 2}$$
  If $f(1) = 4$, find A.

40. Constructing a Closed Box A closed box with a square base is required to have a volume of 10 cubic feet.
  (a) Build a model that expresses the amount A of material used to make such a box as a function of the length x of a side of the square base.
  (b) How much material is required for a base 1 foot by 1 foot?
  (c) How much material is required for a base 2 feet by 2 feet?
   (d) Graph $A = A(x)$. For what value of x is A smallest?

41. Area of a Rectangle A rectangle has one vertex in quadrant I on the graph of $y = 10 - x^2$, another at the origin, one on the positive x-axis, and one on the positive y-axis.
  (a) Express the area A of the rectangle as a function of x.
  (b) Find the largest area A that can be enclosed by the rectangle.

Chapter Test

The Chapter Test Prep Videos are step-by-step solutions available in **MyMathLab**, or on this text's **You Tube** Channel. Flip back to the Resources for Success page for a link to this text's YouTube channel.

1. Determine whether each relation represents a function. For each function, state the domain and the range.
(a) $\{(2,5),(4,6),(6,7),(8,8)\}$
(b) $\{(1,3),(4,-2),(-3,5),(1,7)\}$
(c)

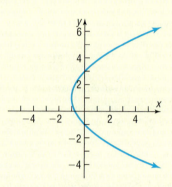

(d)

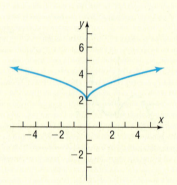

In Problems 2–4, find the domain of each function and evaluate each function at $x = -1$.

2. $f(x) = \sqrt{4 - 5x}$

3. $g(x) = \dfrac{x+2}{|x+2|}$

4. $h(x) = \dfrac{x-4}{x^2 + 5x - 36}$

5. Consider the graph of the function f below.

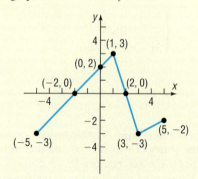

(a) Find the domain and the range of f.
(b) List the intercepts.
(c) Find $f(1)$.
(d) For what value(s) of x does $f(x) = -3$?
(e) Solve $f(x) < 0$.

6. Use a graphing utility to graph the function $f(x) = -x^4 + 2x^3 + 4x^2 - 2$ on the interval $(-5,5)$. Then approximate any local maximum values and local minimum values rounded to two decimal places. Determine where the function is increasing and where it is decreasing.

7. Consider the function $g(x) = \begin{cases} 2x+1 & \text{if } x < -1 \\ x-4 & \text{if } x \geq -1 \end{cases}$

(a) Graph the function.
(b) List the intercepts.
(c) Find $g(-5)$.
(d) Find $g(2)$.

8. For the function $f(x) = 3x^2 - 2x + 4$, find the average rate of change of f from 3 to 4.

9. For the functions $f(x) = 2x^2 + 1$ and $g(x) = 3x - 2$, find the following and simplify.
(a) $(f - g)(x)$
(b) $(f \cdot g)(x)$
(c) $f(x + h) - f(x)$

10. Graph each function using the techniques of shifting, compressing or stretching, and reflecting. Start with the graph of the basic function and show all stages.
(a) $h(x) = -2(x+1)^3 + 3$
(b) $g(x) = |x+4| + 2$

11. The variable interest rate on a student loan changes each July 1 based on the bank prime loan rate. For the years 1992–2007, this rate can be approximated by the model

$$r(x) = -0.115x^2 + 1.183x + 5.623,$$

where x is the number of years since 1992 and r is the interest rate as a percent.

(a) Use a graphing utility to estimate the highest rate during this time period. During which year was the interest rate the highest?
(b) Use the model to estimate the rate in 2010. Does this value seem reasonable?

Source: U.S. Federal Reserve

12. A community skating rink is in the shape of a rectangle with semicircles attached at the ends. The length of the rectangle is 20 feet less than twice the width. The thickness of the ice is 0.75 inch.
(a) Build a model that expresses the ice volume, V, as a function of the width, x.
(b) How much ice is in the rink if the width is 90 feet?

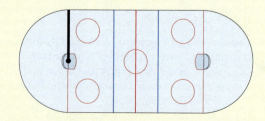

Cumulative Review

In Problems 1–6, find the real solutions of each equation.

1. $3x - 8 = 10$ **2.** $3x^2 - x = 0$

3. $x^2 - 8x - 9 = 0$ **4.** $6x^2 - 5x + 1 = 0$

5. $|2x + 3| = 4$ **6.** $\sqrt{2x + 3} = 2$

In Problems 7–9, solve each inequality. Graph the solution set.

7. $2 - 3x > 6$ **8.** $|2x - 5| < 3$ **9.** $|4x + 1| \geq 7$

10. (a) Find the distance from $P_1 = (-2, -3)$ to $P_2 = (3, -5)$.
    (b) What is the midpoint of the line segment from P_1 to P_2?
    (c) What is the slope of the line containing the points P_1 and P_2?

In Problems 11–14, graph each equation.

11. $3x - 2y = 12$ **12.** $x = y^2$

13. $x^2 + (y - 3)^2 = 16$ **14.** $y = \sqrt{x}$

15. For the equation $3x^2 - 4y = 12$, find the intercepts and check for symmetry.

16. Find the slope–intercept form of the equation of the line containing the points $(-2, 4)$ and $(6, 8)$.

In Problems 17–19, graph each function.

17. $f(x) = (x + 2)^2 - 3$

18. $f(x) = \dfrac{1}{x}$

19. $f(x) = \begin{cases} 2 - x & \text{if } x \leq 2 \\ |x| & \text{if } x > 2 \end{cases}$

Chapter Projects

🛜 **Internet-based Project**

I. Choosing a Wireless Data Plan Collect information from your family, friends, or consumer agencies such as Consumer Reports. Then decide on a cellular provider, choosing the company that you feel offers the best service. Once you have selected a service provider, research the various types of individual plans offered by the company by visiting the provider's website. Many providers offer family plans that include unlimited talk and text. The monthly cost is primarily determined by the amount of data used and the number of devices.

1. Suppose you expect to use 10 gigabytes of data for a single smartphone. What would be the monthly cost of each plan you are considering?

2. Suppose you expect to use 30 gigabytes of data and want a personal hotspot, but you still have only a single smartphone. What would be the monthly cost of each plan you are considering?

3. Suppose you expect to use 20 gigabytes of data with three smartphones sharing the data. What would be the monthly cost of each plan you are considering?

4. Suppose you expect to use 20 gigabytes of data with a single smartphone and a personal hotspot. What would be the monthly cost of each plan you are considering?

5. Build a model that describes the monthly cost C, in dollars, as a function of the number of data gigabytes used, g, assuming a single smartphone and a personal hotspot for each plan you are considering.

6. Graph each function from Problem 5.

7. Based on your particular usage, which plan is best for you?

8. Now, develop an Excel spreadsheet to analyze the various plans you are considering. Suppose you want a family plan with unlimited talk and text that offers 10 gigabytes of shared data and costs $100 per month. Additional gigabytes of data cost $15 per gigabyte, extra phones can be added to the plan for $15 each per month, and each hotspot costs $20 per month. Because wireless

data plans have a cost structure based on piecewise-defined functions, we need an "if/then" statement within Excel to analyze the cost of the plan. Use the accompanying Excel spreadsheet as a guide in developing your spreadsheet. Enter into your spreadsheet a variety of possible amounts of data and various numbers of additional phones and hotspots.

	A	B	C	D	
1					
2	Monthly fee	$100			
3	Allotted data per month (GB)	10			
4	Data used (GB)	12			
5	Cost per additional GB of data	$15			
6					
7	Monthly cost of hotspot	$20			
8	Number of hotspots	1			
9	Monthly cost of additional phone	$15			
10	Number of additional phones	2			
11					
12	Cost of data	=IF(B4<B3,B2,B2+B5*(B4-B3))			
13	Cost of additional devices/hotspots	=B8*B7+B10*B9			
14					
15	Total Cost	=B12+B13			
16					

9. Write a paragraph supporting the choice in plans that best meets your needs.

10. How are "if/then" loops similar to a piecewise-defined function?

Citation: Excel © 2013 Microsoft Corporation. Used with permission from Microsoft.

The following projects are available on the Instructor's Resource Center (IRC).

II. Project at Motorola: *Wireless Internet Service* Use functions and their graphs to analyze the total cost of various wireless Internet service plans.

III. Cost of Cable When government regulations and customer preference influence the path of a new cable line, the Pythagorean Theorem can be used to assess the cost of installation.

IV. Oil Spill Functions are used to analyze the size and spread of an oil spill from a leaking tanker.

Linear and Quadratic Functions

4

The Beta of a Stock

Investing in the stock market can be rewarding and fun, but how does one go about selecting which stocks to purchase? Financial investment firms hire thousands of analysts who track individual stocks (equities) and assess the value of the underlying company. One measure the analysts consider is the *beta* of the stock. **Beta** measures the risk of an individual company's equity relative to that of a market basket of stocks, such as the Standard & Poor's 500. But how is beta computed?

 —See the Internet-based Chapter Project I—

••• A Look Back

Up to now, our discussion has focused on graphs of equations and functions. We learned how to graph equations using the point-plotting method, intercepts, and the tests for symmetry. In addition, we learned what a function is and how to identify whether a relation represents a function. We also discussed properties of functions, such as domain/range, increasing/decreasing, even/odd, and average rate of change.

A Look Ahead •••

Going forward, we will look at classes of functions. This chapter focuses on linear and quadratic functions, their properties, and their applications.

Outline

273

4.1 Properties of Linear Functions and Linear Models

PREPARING FOR THIS SECTION *Before getting started, review the following:*

- Lines (Section 2.3, pp. 167–175)
- Graphs of Equations in Two Variables; Intercepts; Symmetry (Section 2.2, pp. 157–164)
- Linear Equations (Section 1.1, pp. 82–87)

- Functions (Section 3.1, pp. 199–208)
- The Graph of a Function (Section 3.2, pp. 214–217)
- Properties of Functions (Section 3.3, pp. 223–231)

Now Work the *'Are You Prepared?'* problems on page 280.

OBJECTIVES
1 Graph Linear Functions (p. 274)
2 Use Average Rate of Change to Identify Linear Functions (p. 274)
3 Determine Whether a Linear Function Is Increasing, Decreasing, or Constant (p. 277)
4 Build Linear Models from Verbal Descriptions (p. 278)

1 Graph Linear Functions

In Section 2.3 we discussed lines. In particular, for nonvertical lines we developed the slope–intercept form of the equation of a line $y = mx + b$. When the slope–intercept form of a line is written using function notation, the result is a *linear function*.

DEFINITION

A **linear function** is a function of the form

$$f(x) = mx + b$$

The graph of a linear function is a line with slope m and y-intercept b. Its domain is the set of all real numbers.

Functions that are not linear are said to be **nonlinear**.

EXAMPLE 1 **Graphing a Linear Function**

Graph the linear function $f(x) = -3x + 7$. What are the domain and the range of f?

Solution This is a linear function with slope $m = -3$ and y-intercept $b = 7$. To graph this function, plot the point $(0, 7)$, the y-intercept, and use the slope to find an additional point by moving right 1 unit and down 3 units. See Figure 1. The domain and the range of f are each the set of all real numbers. ●

Alternatively, an additional point could have been found by evaluating the function at some $x \neq 0$. For $x = 1$, $f(1) = -3(1) + 7 = 4$ and the point $(1, 4)$ lies on the graph.

Now Work PROBLEMS 13(a) AND (b)

Figure 1 $f(x) = -3x + 7$

2 Use Average Rate of Change to Identify Linear Functions

Look at Table 1, which shows certain values of the independent variable x and corresponding values of the dependent variable y for the function $f(x) = -3x + 7$. Notice that as the value of the independent variable, x, increases by 1, the value of the dependent variable y decreases by 3. That is, the average rate of change of y with respect to x is a constant, -3.

Table 1

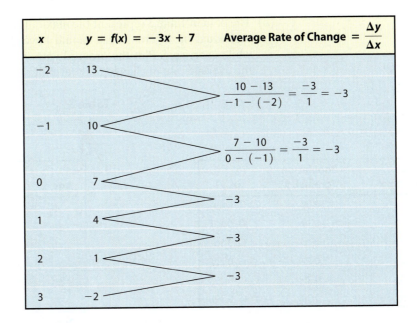

x	$y = f(x) = -3x + 7$	Average Rate of Change $= \dfrac{\Delta y}{\Delta x}$
-2	13	
		$\dfrac{10 - 13}{-1 - (-2)} = \dfrac{-3}{1} = -3$
-1	10	
		$\dfrac{7 - 10}{0 - (-1)} = \dfrac{-3}{1} = -3$
0	7	
		-3
1	4	
		-3
2	1	
		-3
3	-2	

It is not a coincidence that the average rate of change of the linear function $f(x) = -3x + 7$ is the slope of the linear function. That is, $\dfrac{\Delta y}{\Delta x} = m = -3$. The following theorem states this fact.

THEOREM

> **Average Rate of Change of a Linear Function**
>
> Linear functions have a constant average rate of change. That is, the average rate of change of a linear function $f(x) = mx + b$ is
>
> $$\frac{\Delta y}{\Delta x} = m$$

Proof The average rate of change of $f(x) = mx + b$ from x_1 to x_2, $x_1 \neq x_2$, is

$$\frac{\Delta y}{\Delta x} = \frac{f(x_2) - f(x_1)}{x_2 - x_1} = \frac{(mx_2 + b) - (mx_1 + b)}{x_2 - x_1}$$

$$= \frac{mx_2 - mx_1}{x_2 - x_1} = \frac{m(x_2 - x_1)}{x_2 - x_1} = m \qquad \blacksquare$$

Based on the theorem just proved, the average rate of change of the function $g(x) = -\dfrac{2}{5}x + 5$ is $-\dfrac{2}{5}$.

━━➤ **Now Work** PROBLEM **13(c)**

As it turns out, only linear functions have a constant average rate of change. Because of this, the average rate of change can be used to determine whether a function is linear. This is especially useful if the function is defined by a data set.

EXAMPLE 2

Using the Average Rate of Change to Identify Linear Functions

(a) A strain of *E. coli* known as Beu 397-recA441 is placed into a Petri dish at 30° Celsius and allowed to grow. The data shown in Table 2 on the next page are collected. The population is measured in grams and the time in hours. Plot the ordered pairs (x, y) in the Cartesian plane, and use the average rate of change to determine whether the function is linear.

(b) The data in Table 3 represent the maximum number of heartbeats that a healthy individual of different ages should have during a 15-second interval of time while exercising. Plot the ordered pairs (x, y) in the Cartesian plane, and use the average rate of change to determine whether the function is linear.

Table 2

Time (hours), x	Population (grams), y	(x, y)
0	0.09	(0, 0.09)
1	0.12	(1, 0.12)
2	0.16	(2, 0.16)
3	0.22	(3, 0.22)
4	0.29	(4, 0.29)
5	0.39	(5, 0.39)

Table 3

Age, x	Maximum Number of Heartbeats, y	(x, y)
20	50	(20, 50)
30	47.5	(30, 47.5)
40	45	(40, 45)
50	42.5	(50, 42.5)
60	40	(60, 40)
70	37.5	(70, 37.5)

Source: American Heart Association

Solution Compute the average rate of change of each function. If the average rate of change is constant, the function is linear. If the average rate of change is not constant, the function is nonlinear.

(a) Figure 2 shows the points listed in Table 2 plotted in the Cartesian plane. Note that it is impossible to draw a straight line that contains all the points. Table 4 displays the average rate of change of the population.

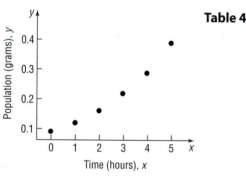

Figure 2

Table 4

Time (hours), x	Population (grams), y	Average Rate of Change $= \dfrac{\Delta y}{\Delta x}$
0	0.09	
		$\dfrac{0.12 - 0.09}{1 - 0} = 0.03$
1	0.12	
		0.04
2	0.16	
		0.06
3	0.22	
		0.07
4	0.29	
		0.10
5	0.39	

Because the average rate of change is not constant, the function is not linear. In fact, because the average rate of change is increasing as the value of the independent variable increases, the function is increasing at an increasing rate. So not only is the population increasing over time, but it is also growing more rapidly as time passes.

(b) Figure 3 shows the points listed in Table 3 plotted in the Cartesian plane. Note that the data in Figure 3 lie on a straight line. Table 5 displays the average rate of change of the maximum number of heartbeats. The average rate of change of the heartbeat data is constant, -0.25 beat per year, so the function is linear.

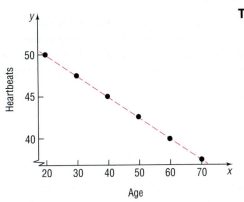

Figure 3

Table 5	Age, x	Maximum Number of Heartbeats, y	Average Rate of Change $= \dfrac{\Delta y}{\Delta x}$
	20	50	
			$\dfrac{47.5 - 50}{30 - 20} = -0.25$
	30	47.5	
			-0.25
	40	45	
			-0.25
	50	42.5	
			-0.25
	60	40	
			-0.25
	70	37.5	

Now Work PROBLEM 21

3 Determine Whether a Linear Function Is Increasing, Decreasing, or Constant

Look back at the Seeing the Concept on page 169. When the slope m of a linear function is positive ($m > 0$), the line slants upward from left to right. When the slope m of a linear function is negative ($m < 0$), the line slants downward from left to right. When the slope m of a linear function is zero ($m = 0$), the line is horizontal.

THEOREM

Increasing, Decreasing, and Constant Linear Functions

A linear function $f(x) = mx + b$ is increasing over its domain if its slope, m, is positive. It is decreasing over its domain if its slope, m, is negative. It is constant over its domain if its slope, m, is zero.

EXAMPLE 3

Determining Whether a Linear Function Is Increasing, Decreasing, or Constant

Determine whether the following linear functions are increasing, decreasing, or constant.

(a) $f(x) = 5x - 2$ (b) $g(x) = -2x + 8$

(c) $s(t) = \dfrac{3}{4}t - 4$ (d) $h(z) = 7$

Solution

(a) For the linear function $f(x) = 5x - 2$, the slope is 5, which is positive. The function f is increasing on the interval $(-\infty, \infty)$.

(b) For the linear function $g(x) = -2x + 8$, the slope is -2, which is negative. The function g is decreasing on the interval $(-\infty, \infty)$.

(c) For the linear function $s(t) = \dfrac{3}{4}t - 4$, the slope is $\dfrac{3}{4}$, which is positive. The function s is increasing on the interval $(-\infty, \infty)$.

(d) The linear function h can be written as $h(z) = 0z + 7$. Because the slope is 0, the function h is constant on the interval $(-\infty, \infty)$.

Now Work PROBLEM 13 (d)

4 Build Linear Models from Verbal Descriptions

When the average rate of change of a function is constant, a linear function can model the relation between the two variables. For example, if a recycling company pays $0.52 per pound for aluminum cans, then the relation between the price paid p and the pounds recycled x can be modeled as the linear function $p(x) = 0.52x$, with slope $m = \dfrac{0.52 \, \text{dollar}}{1 \, \text{pound}}$.

Modeling with a Linear Function

If the average rate of change of a function is a constant m, a linear function f can be used to model the relation between the two variables as follows:

$$f(x) = mx + b$$

where b is the value of f at 0; that is, $b = f(0)$.

EXAMPLE 4

Straight-line Depreciation

Book value is the value of an asset that a company uses to create its balance sheet. Some companies depreciate assets using straight-line depreciation so that the value of the asset declines by a fixed amount each year. The amount of the decline depends on the useful life that the company assigns to the asset. Suppose a company just purchased a fleet of new cars for its sales force at a cost of $31,500 per car. The company chooses to depreciate each vehicle using the straight-line method over 7 years. This means that each car will depreciate by $\dfrac{\$31,500}{7} = \$4500$ per year.

(a) Write a linear function that expresses the book value V of each car as a function of its age, x, in years.
(b) Graph the linear function.
(c) What is the book value of each car after 3 years?
(d) Interpret the slope.
(e) When will the book value of each car be $9000?
   [**Hint:** Solve the equation $V(x) = 9000$.]

Solution

(a) If we let $V(x)$ represent the value of each car after x years, then $V(0)$ represents the original value of each car, so $V(0) = \$31,500$. The y-intercept of the linear function is $31,500. Because each car depreciates by $4500 per year, the slope of the linear function is -4500. The linear function that represents the book value V of each car after x years is

$$V(x) = -4500x + 31,500$$

(b) Figure 4 shows the graph of V.

(c) The book value of each car after 3 years is

$$V(3) = -4500(3) + 31,500$$
$$= \$18,000$$

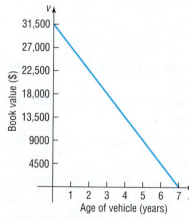

Figure 4 $V(x) = -4500x + 31,500$

(d) Since the slope of $V(x) = -4500x + 31,500$ is -4500, the average rate of change of the book value is $-\$4500/\text{year}$. So for each additional year that passes, the book value of the car decreases by $4500.

(e) To find when the book value will be $9000, solve the equation

$$V(x) = 9000$$

$$-4500x + 31{,}500 = 9000$$

$$-4500x = -22{,}500 \qquad \text{Subtract 31,500 from each side.}$$

$$x = \frac{-22{,}500}{-4500} = 5 \qquad \text{Divide by } -4500.$$

The car will have a book value of $9000 when it is 5 years old.

Now Work PROBLEM 45

EXAMPLE 5

Supply and Demand

The **quantity supplied** of a good is the amount of a product that a company is willing to make available for sale at a given price. The **quantity demanded** of a good is the amount of a product that consumers are willing to purchase at a given price. Suppose that the quantity supplied, S, and the quantity demanded, D, of cellular telephones each month are given by the following functions:

$$S(p) = 60p - 900$$

$$D(p) = -15p + 2850$$

where p is the price (in dollars) of the telephone.

(a) The **equilibrium price** of a product is defined as the price at which quantity supplied equals quantity demanded. That is, the equilibrium price is the price at which $S(p) = D(p)$. Find the equilibrium price of cellular telephones. What is the **equilibrium quantity**, the amount demanded (or supplied) at the equilibrium price?

(b) Determine the prices for which quantity supplied is greater than quantity demanded. That is, solve the inequality $S(p) > D(p)$.

(c) Graph $S = S(p)$ and $D = D(p)$, and label the **equilibrium point**, the point of intersection of S and D.

Solution

(a) To find the equilibrium price, solve the equation $S(p) = D(p)$.

$$60p - 900 = -15p + 2850 \qquad \begin{array}{l} S(p) = 60p - 900; \\ D(p) = -15p + 2850 \end{array}$$

$$60p = -15p + 3750 \qquad \text{Add 900 to each side.}$$

$$75p = 3750 \qquad \text{Add 15p to each side.}$$

$$p = 50 \qquad \text{Divide each side by 75.}$$

The equilibrium price is $50 per cellular phone. To find the equilibrium quantity, evaluate either $S(p)$ or $D(p)$ at $p = 50$.

$$S(50) = 60(50) - 900 = 2100$$

The equilibrium quantity is 2100 cellular phones. At a price of $50 per phone, the company will produce and sell 2100 phones each month and have no shortages or excess inventory.

(b) The inequality $S(p) > D(p)$ is

$$60p - 900 > -15p + 2850 \qquad S(p) > D(p)$$

$$60p > -15p + 3750 \qquad \text{Add 900 to each side.}$$

$$75p > 3750 \qquad \text{Add 15p to each side.}$$

$$p > 50 \qquad \text{Divide each side by 75.}$$

If the company charges more than $50 per phone, quantity supplied will exceed quantity demanded. In this case the company will have excess phones in inventory.

(c) Figure 5 shows the graphs of $S = S(p)$ and $D = D(p)$ with the equilibrium point labeled.

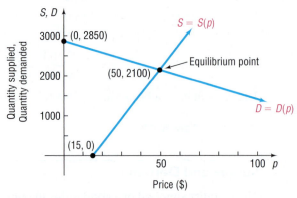

Figure 5 Supply and demand functions

Now Work PROBLEM 39

4.1 Assess Your Understanding

'Are You Prepared?' *Answers are given at the end of these exercises. If you get a wrong answer, read the pages listed in red.*

1. Graph $y = x^2 - 1$. (pp. 157–164)

2. Find the slope of the line joining the points $(2, 5)$ and $(-1, 3)$. (pp. 167–175)

3. Find the average rate of change of $f(x) = 3x^2 - 2$, from 2 to 4. (pp. 223–231)

4. Solve: $60x - 900 = -15x + 2850$. (pp. 82–87)

5. If $f(x) = x^2 - 4$, find $f(-2)$. (pp. 199–208)

6. *True or False* The graph of the function $f(x) = x^2$ is increasing on the interval $(0, \infty)$. (pp. 214–217)

Concepts and Vocabulary

7. For the graph of the linear function $f(x) = mx + b$, m is the _____ and b is the _____.

8. If the slope m of the graph of a linear function is _____, the function is increasing over its domain.

9. *True or False* The slope of a nonvertical line is the average rate of change of the linear function.

10. *True or False* The average rate of change of $f(x) = 2x + 8$ is 8.

11. What is the only type of function that has a constant average rate of change?
    (a) linear function   (b) quadratic function
    (c) step function   (d) absolute value function

12. A car has 12,500 miles on its odometer. Say the car is driven an average of 40 miles per day. Choose the model that expresses the number of miles N that will be on its odometer after x days.
    (a) $N(x) = -40x + 12{,}500$ (b) $N(x) = 40x - 12{,}500$
    (c) $N(x) = 12{,}500x + 40$ (d) $N(x) = 40x + 12{,}500$

Skill Building

In Problems 13–20, a linear function is given.
  (a) *Determine the slope and y-intercept of each function.*
  (b) *Use the slope and y-intercept to graph the linear function.*
  (c) *Determine the average rate of change of each function.*
  (d) *Determine whether the linear function is increasing, decreasing, or constant.*

13. $f(x) = 2x + 3$

14. $g(x) = 5x - 4$

15. $h(x) = -3x + 4$

16. $p(x) = -x + 6$

17. $f(x) = \frac{1}{4}x - 3$

18. $h(x) = -\frac{2}{3}x + 4$

19. $F(x) = 4$

20. $G(x) = -2$

In Problems 21–28, determine whether the given function is linear or nonlinear. If it is linear, determine the slope.

21.

x	y = f(x)
−2	4
−1	1
0	−2
1	−5
2	−8

22.

x	y = f(x)
−2	1/4
−1	1/2
0	1
1	2
2	4

23.

x	y = f(x)
−2	−8
−1	−3
0	0
1	1
2	0

24.

x	y = f(x)
−2	−4
−1	0
0	4
1	8
2	12

25.

x	y = f(x)
−2	−26
−1	−4
0	2
1	−2
2	−10

26.

x	y = f(x)
−2	−4
−1	−3.5
0	−3
1	−2.5
2	−2

27.

x	y = f(x)
−2	8
−1	8
0	8
1	8
2	8

28.

x	y = f(x)
−2	0
−1	1
0	4
1	9
2	16

Applications and Extensions

29. Suppose that $f(x) = 4x - 1$ and $g(x) = -2x + 5$.
(a) Solve $f(x) = 0$.   (b) Solve $f(x) > 0$.
(c) Solve $f(x) = g(x)$.   (d) Solve $f(x) \leq g(x)$.
(e) Graph $y = f(x)$ and $y = g(x)$ and label the point that represents the solution to the equation $f(x) = g(x)$.

30. Suppose that $f(x) = 3x + 5$ and $g(x) = -2x + 15$.
(a) Solve $f(x) = 0$.   (b) Solve $f(x) < 0$.
(c) Solve $f(x) = g(x)$.   (d) Solve $f(x) \geq g(x)$.
(e) Graph $y = f(x)$ and $y = g(x)$ and label the point that represents the solution to the equation $f(x) = g(x)$.

31. In parts (a)–(f), use the following figure.

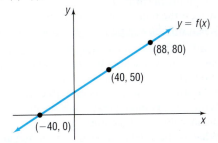

(a) Solve $f(x) = 50$.   (b) Solve $f(x) = 80$.
(c) Solve $f(x) = 0$.   (d) Solve $f(x) > 50$.
(e) Solve $f(x) \leq 80$.   (f) Solve $0 < f(x) < 80$.

32. In parts (a)–(f), use the following figure.

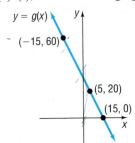

(a) Solve $g(x) = 20$.   (b) Solve $g(x) = 60$.
(c) Solve $g(x) = 0$.   (d) Solve $g(x) > 20$.
(e) Solve $g(x) \leq 60$.   (f) Solve $0 < g(x) < 60$.

33. In parts (a) and (b), use the following figure.

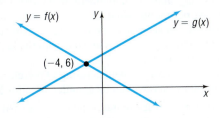

(a) Solve the equation: $f(x) = g(x)$.
(b) Solve the inequality: $f(x) > g(x)$.

34. In parts (a) and (b), use the following figure.

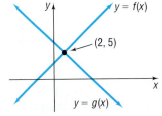

(a) Solve the equation: $f(x) = g(x)$.
(b) Solve the inequality: $f(x) \leq g(x)$.

35. In parts (a) and (b), use the following figure.

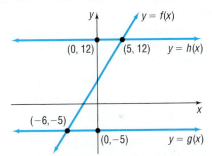

(a) Solve the equation: $f(x) = g(x)$.
(b) Solve the inequality: $g(x) \leq f(x) < h(x)$.

36. In parts (a) and (b), use the following figure.

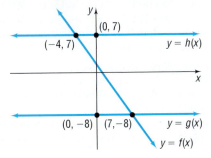

(a) Solve the equation: $f(x) = g(x)$.
(b) Solve the inequality: $g(x) < f(x) \leq h(x)$.

37. Truck Rentals The cost C, in dollars, of a one-day truck rental is modeled by the function $C(x) = 0.35x + 45$, where x is the number of miles driven.
(a) What is the cost if you drive $x = 40$ miles?
(b) If the cost of renting the truck is $108, how many miles did you drive?
(c) Suppose that you want the cost to be no more than $150. What is the maximum number of miles that you can drive?
(d) What is the implied domain of C?

38. Phone Charges The monthly cost C, in dollars, for calls from the United States to Germany on a certain phone plan is modeled by the function $C(x) = 0.26x + 5$, where x is the number of minutes used.
  (a) What is the cost if you talk on the phone for $x = 50$ minutes?
  (b) Suppose that your monthly bill is $21.64. How many minutes did you use the phone?
  (c) Suppose that you budget yourself $50 per month for the phone. What is the maximum number of minutes that you can talk?
  (d) What is the implied domain of C if there are 30 days in the month?

39. Supply and Demand Suppose that the quantity supplied S and the quantity demanded D of T-shirts at a concert are given by the following functions:
$$S(p) = -600 + 50p$$
$$D(p) = 1200 - 25p$$
where p is the price of a T-shirt.
  (a) Find the equilibrium price for T-shirts at this concert. What is the equilibrium quantity?
  (b) Determine the prices for which quantity demanded is greater than quantity supplied.
  (c) What do you think will eventually happen to the price of T-shirts if quantity demanded is greater than quantity supplied?

40. Supply and Demand Suppose that the quantity supplied S and the quantity demanded D of hot dogs at a baseball game are given by the following functions:
$$S(p) = -2000 + 3000p$$
$$D(p) = 10,000 - 1000p$$
where p is the price of a hot dog.
  (a) Find the equilibrium price for hot dogs at the baseball game. What is the equilibrium quantity?
  (b) Determine the prices for which quantity demanded is less than quantity supplied.
  (c) What do you think will eventually happen to the price of hot dogs if quantity demanded is less than quantity supplied?

41. Taxes The function $T(x) = 0.15(x - 9075) + 907.50$ represents the tax bill T of a single person whose adjusted gross income is x dollars for income between $9075 and $36,900, inclusive, in 2014.
  Source: Internal Revenue Service
  (a) What is the domain of this linear function?
  (b) What is a single filer's tax bill if adjusted gross income is $20,000?
  (c) Which variable is independent and which is dependent?
  (d) Graph the linear function over the domain specified in part (a).
  (e) What is a single filer's adjusted gross income if the tax bill is $3671.25?

42. Luxury Tax In 2011, major league baseball signed a labor agreement with the players. In this agreement, any team whose payroll exceeded $189 million in 2014 had to pay a luxury tax of 50%. The linear function $T(p) = 0.50(p - 189)$ describes the luxury tax T for a team whose payroll was p (in millions of dollars).
  Source: Major League Baseball
  (a) What is the implied domain of this linear function?
  (b) What was the luxury tax for the New York Yankees, whose 2014 payroll was $203.4 million?

  (c) Graph the linear function.
  (d) What is the payroll of a team that pays a luxury tax of $15.7 million?

*The point at which a company's profits equal zero is called the company's **break-even point.** For Problems 43 and 44, let R represent a company's revenue, let C represent the company's costs, and let x represent the number of units produced and sold each day.*
  (a) *Find the firm's break-even point; that is, find x so that $R = C$.*
  (b) *Find the values of x such that $R(x) > C(x)$. This represents the number of units that the company must sell to earn a profit.*

43. $R(x) = 8x$
  $C(x) = 4.5x + 17,500$

44. $R(x) = 12x$
  $C(x) = 10x + 15,000$

45. Straight-line Depreciation Suppose that a company has just purchased a new computer for $3000. The company chooses to depreciate the computer using the straight-line method over 3 years.
  (a) Write a linear model that expresses the book value V of the computer as a function of its age x.
  (b) What is the implied domain of the function found in part (a)?
  (c) Graph the linear function.
  (d) What is the book value of the computer after 2 years?
  (e) When will the computer have a book value of $2000?

46. Straight-line Depreciation Suppose that a company has just purchased a new machine for its manufacturing facility for $120,000. The company chooses to depreciate the machine using the straight-line method over 10 years.
  (a) Write a linear model that expresses the book value V of the machine as a function of its age x.
  (b) What is the implied domain of the function found in part (a)?
  (c) Graph the linear function.
  (d) What is the book value of the machine after 4 years?
  (e) When will the machine have a book value of $72,000?

47. Cost Function The simplest cost function is the linear cost function, $C(x) = mx + b$, where the y-intercept b represents the fixed costs of operating a business and the slope m represents the cost of each item produced. Suppose that a small bicycle manufacturer has daily fixed costs of $1800, and each bicycle costs $90 to manufacture.
  (a) Write a linear model that expresses the cost C of manufacturing x bicycles in a day.
  (b) Graph the model.
  (c) What is the cost of manufacturing 14 bicycles in a day?
  (d) How many bicycles could be manufactured for $3780?

48. Cost Function Refer to Problem 47. Suppose that the landlord of the building increases the bicycle manufacturer's rent by $100 per month.
  (a) Assuming that the manufacturer is open for business 20 days per month, what are the new daily fixed costs?
  (b) Write a linear model that expresses the cost C of manufacturing x bicycles in a day with the higher rent.
  (c) Graph the model.
  (d) What is the cost of manufacturing 14 bicycles in a day?
  (e) How many bicycles can be manufactured for $3780?

49. Truck Rentals A truck rental company rents a truck for one day by charging $31.95 plus $0.89 per mile.
  (a) Write a linear model that relates the cost C, in dollars, of renting the truck to the number x of miles driven.

(b) What is the cost of renting the truck if the truck is driven 110 miles? 230 miles?

50. **International Call Plan** A cell phone company offers an international plan by charging $30 for the first 80 minutes, plus $0.50 for each minute over 80.

(a) Write a linear model that relates the cost C, in dollars, of talking x minutes, assuming $x \geq 80$.

(b) What is the cost of talking 105 minutes? 120 minutes?

Mixed Practice

51. **Building a Linear Model from Data** How many songs can an iPod hold? The following data represent the memory m and the number of songs, n.

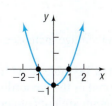

Memory, m (gigabytes)	Number of Songs, n
8	1750
16	3500
32	7000
64	14,000

(a) Plot the ordered pairs (m, n) in a Cartesian plane.
(b) Show that the number of songs n is a linear function of the memory m.
(c) Determine the linear function that describes the relation between m and n.
(d) What is the implied domain of the linear function?
(e) Graph the linear function in the Cartesian plane drawn in part (a).
(f) Interpret the slope.

52. **Building a Linear Model from Data** The following data represent the various combinations of soda and hot dogs that Yolanda can buy at a baseball game with $60.

Soda, s	Hot Dogs, h
20	0
15	3
10	6
5	9

(a) Plot the ordered pairs (s, h) in a Cartesian plane.
(b) Show that the number of hot dogs purchased h is a linear function of the number of sodas purchased s.
(c) Determine the linear function that describes the relation between s and h.
(d) What is the implied domain of the linear function?
(e) Graph the linear function in the Cartesian plane drawn in part (a).
(f) Interpret the slope.
(g) Interpret the values of the intercepts.

Explaining Concepts: Discussion and Writing

53. Which of the following functions might have the graph shown? (More than one answer is possible.)
(a) $f(x) = 2x - 7$
(b) $g(x) = -3x + 4$
(c) $H(x) = 5$
(d) $F(x) = 3x + 4$
(e) $G(x) = \dfrac{1}{2}x + 2$

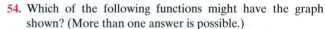

54. Which of the following functions might have the graph shown? (More than one answer is possible.)
(a) $f(x) = 3x + 1$
(b) $g(x) = -2x + 3$
(c) $H(x) = 3$
(d) $F(x) = -4x - 1$
(e) $G(x) = -\dfrac{2}{3}x + 3$

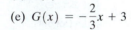

55. Under what circumstances is a linear function $f(x) = mx + b$ odd? Can a linear function ever be even?

56. Explain how the graph of $f(x) = mx + b$ can be used to solve $mx + b > 0$.

Retain Your Knowledge

Problems 57–60 are based on material learned earlier in the course. The purpose of these problems is to keep the material fresh in your mind so that you are better prepared for the final exam.

57. Graph $x^2 - 4x + y^2 + 10y - 7 = 0$.

58. If $f(x) = \dfrac{2x + B}{x - 3}$ and $f(5) = 8$, what is the value of B?

59. Find the average rate of change of $f(x) = 3x^2 - 5x$ from 1 to 3.

60. Graph $g(x) = \begin{cases} x^2 & x \leq 0 \\ \sqrt{x} + 1 & x > 0 \end{cases}$

'Are You Prepared?' Answers

1.

2. $\dfrac{2}{3}$

3. 18

4. $\{50\}$

5. 0

6. True

4.2 Building Linear Models from Data

PREPARING FOR THIS SECTION *Before getting started, review the following:*

- Rectangular Coordinates (Section 2.1, pp. 150–151)
- Functions (Section 3.1, pp. 199–208)
- Lines (Section 2.3, pp. 167–175)

Now Work the **'Are You Prepared?'** problems on page 287.

OBJECTIVES 1 Draw and Interpret Scatter Diagrams (p. 284)
2 Distinguish between Linear and Nonlinear Relations (p. 285)
3 Use a Graphing Utility to Find the Line of Best Fit (p. 286)

1 Draw and Interpret Scatter Diagrams

In Section 4.1, we built linear models from verbal descriptions. Linear models can also be constructed by fitting a linear function to data. The first step is to plot the ordered pairs using rectangular coordinates. The resulting graph is a **scatter diagram**.

EXAMPLE I **Drawing and Interpreting a Scatter Diagram**

In baseball, the on-base percentage for a team represents the percentage of time that the players safely reach base. The data given in Table 6 represent the number of runs scored y and the on-base percentage x for teams in the National League during the 2013 baseball season.

Table 6

Team	On-Base Percentage, x	Runs Scored, y	(x, y)
Arizona	32.3	685	(32.3, 685)
Atlanta	32.1	688	(32.1, 688)
Chicago Cubs	30.0	602	(30.0, 602)
Cincinnati	32.7	698	(32.7, 698)
Colorado	32.3	706	(32.3, 706)
LA Dodgers	32.6	649	(32.6, 649)
Miami	29.3	513	(29.3, 513)
Milwaukee	31.1	640	(31.1, 640)
NY Mets	30.6	619	(30.6, 619)
Philadelphia	30.6	610	(30.6, 610)
Pittsburgh	31.3	634	(31.3, 634)
San Diego	30.8	618	(30.8, 618)
San Francisco	32.0	629	(32.0, 629)
St. Louis	33.2	783	(33.2, 783)
Washington	31.3	656	(31.3, 656)

Source: espn.go.com

(a) Draw a scatter diagram of the data, treating on-base percentage as the independent variable.

(b) Use a graphing utility to draw a scatter diagram.

(c) Describe what happens to runs scored as the on-base percentage increases.

Solution (a) To draw a scatter diagram, plot the ordered pairs listed in Table 6, with the on-base percentage as the x-coordinate and the runs scored as the y-coordinate. See Figure 6(a). Notice that the points in the scatter diagram are not connected.

(b) Figure 6(b) shows a scatter diagram using a TI-84 Plus C graphing calculator.

(c) The scatter diagrams show that as the on-base percentage increases, the number of runs scored also increases.

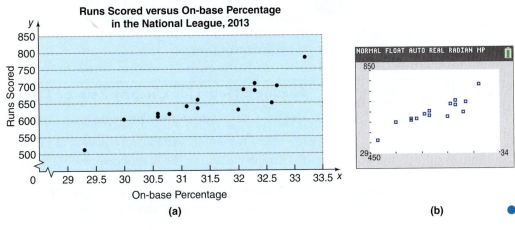

Figure 6

(a)   (b)

Now Work PROBLEM 11(a)

2 Distinguish between Linear and Nonlinear Relations

Notice that the points in Figure 6 do not follow a perfect linear relation (as they do in Figure 3 in Section 4.1). However, the data do exhibit a linear pattern. There are numerous possible explanations why the data are not perfectly linear, but one easy explanation is the fact that other variables besides on-base percentage (such as number of home runs hit) play a role in determining runs scored.

Scatter diagrams are used to help us to see the type of relation that exists between two variables. In this text, we will discuss a variety of different relations that may exist between two variables. For now, we concentrate on distinguishing between linear and nonlinear relations. See Figure 7.

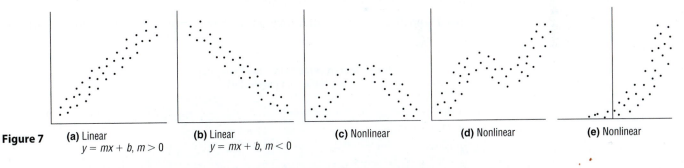

Figure 7

(a) Linear
$y = mx + b, m > 0$

(b) Linear
$y = mx + b, m < 0$

(c) Nonlinear

(d) Nonlinear

(e) Nonlinear

EXAMPLE 2

Distinguishing between Linear and Nonlinear Relations

Determine whether the relation between the two variables in each scatter diagram in Figure 8 is linear or nonlinear.

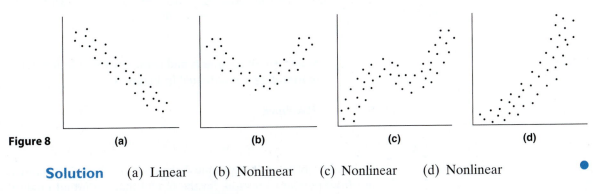

Figure 8   (a)   (b)   (c)   (d)

Solution   (a) Linear   (b) Nonlinear   (c) Nonlinear   (d) Nonlinear

Now Work PROBLEM 5

This section considers data whose scatter diagrams suggest that a linear relation exists between the two variables.

Suppose that the scatter diagram of a set of data appears to indicate a linear relationship, as in Figure 7(a) or (b). We might want to model the data by finding an equation of a line that relates the two variables. One way to obtain a model for such data is to draw a line through two points on the scatter diagram and determine the equation of the line.

EXAMPLE 3

Finding a Model for Linearly Related Data

Use the data in Table 6 from Example 1.

(a) Select two points and find an equation of the line containing the points.
(b) Graph the line on the scatter diagram obtained in Example 1(a).

Solution

(a) Select two points, say $(30.6, 610)$ and $(32.1, 688)$. The slope of the line joining the points $(30.6, 610)$ and $(32.1, 688)$ is

$$m = \frac{688 - 610}{32.1 - 30.6} = \frac{78}{1.5} = 52$$

The equation of the line that has slope 52 and passes through $(30.6, 610)$ is found using the point–slope form with $m = 52$, $x_1 = 30.6$, and $y_1 = 610$.

$$y - y_1 = m(x - x_1) \qquad \text{Point–slope form of a line}$$

$$y - 610 = 52(x - 30.6) \qquad x_1 = 30.6,\ y_1 = 610,\ m = 52$$

$$y - 610 = 52x - 1591.2$$

$$y = 52x - 981.2 \qquad \text{The model}$$

(b) Figure 9 shows the scatter diagram with the graph of the line found in part (a).

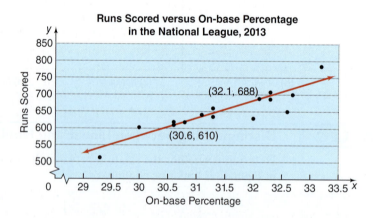

Figure 9

 Select two other points and complete the solution. Graph the line on the scatter diagram obtained in Figure 6.

 Now Work PROBLEMS 11(b) AND (c)

3 Use a Graphing Utility to Find the Line of Best Fit

The model obtained in Example 3 depends on the selection of points, which will vary from person to person. So the model that we found might be different from the model you found. Although the model in Example 3 appears to fit the data well, there may be a model that "fits them better." Do you think your model fits the data better? Is there a *line of best fit*? As it turns out, there is a method for finding a model that best fits linearly related data (called the **line of best fit**).*

EXAMPLE 4

Finding a Model for Linearly Related Data

Use the data in Table 6 from Example 1.

(a) Use a graphing utility to find the line of best fit that models the relation between on-base percentage and runs scored.

(b) Graph the line of best fit on the scatter diagram obtained in Example 1(b).

(c) Interpret the slope.

(d) Use the line of best fit to predict the number of runs a team will score if their on-base percentage is 33.1.

Solution

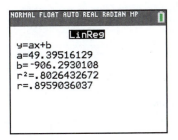

Figure 10

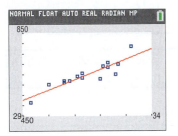

Figure 11

(a) Graphing utilities contain built-in programs that find the line of best fit for a collection of points in a scatter diagram. Executing the LINear REGression program provides the results shown in Figure 10. This output shows the equation $y = ax + b$, where a is the slope of the line and b is the y-intercept. The line of best fit that relates on-base percentage to runs scored may be expressed as the line

$$y = 49.40x - 906.29 \quad \text{The model}$$

(b) Figure 11 shows the graph of the line of best fit, along with the scatter diagram.

(c) The slope of the line of best fit is 49.40, which means that, for every 1 percent increase in the on-base percentage, runs scored increase by 49.40, on average.

(d) Letting $x = 33.1$ in the equation of the line of best fit, we obtain $y = 49.40(33.1) - 906.29 \approx 729$ runs. ●

Now Work PROBLEMS **11(d)** AND **(e)**

Does the line of best fit appear to be a good fit? In other words, does it appear to accurately describe the relation between on-base percentage and runs scored?

And just how "good" is this line of best fit? Look again at Figure 10. The last line of output is $r = 0.896$. This number, called the **correlation coefficient**, r, $-1 \le r \le 1$, is a measure of the strength of the linear relation that exists between two variables. The closer $|r|$ is to 1, the more nearly perfect the linear relationship is. If r is close to 0, there is little or no linear relationship between the variables. A negative value of r, $r < 0$, indicates that as x increases, y decreases; a positive value of r, $r > 0$, indicates that as x increases, y does also. The data given in Table 6, which have a correlation coefficient of 0.896, are indicative of a linear relationship with positive slope.

4.2 Assess Your Understanding

'Are You Prepared?' *Answers are given at the end of these exercises. If you get a wrong answer, read the pages listed in red.*

1. Plot the points $(1, 5)$, $(2, 6)$, $(3, 9)$, $(1, 12)$ in the Cartesian plane. Is the relation $\{(1, 5), (2, 6), (3, 9), (1, 12)\}$ a function? Why? (pp. 150 and 199–208)

2. Find an equation of the line containing the points $(1, 4)$ and $(3, 8)$. (pp. 167–175)

Concepts and Vocabulary

3. A _____ is used to help us to see what type of relation, if any, may exist between two variables.

4. *True or False* The correlation coefficient is a measure of the strength of a linear relation between two variables and must lie between -1 and 1, inclusive.

*We shall not discuss the underlying mathematics of lines of best fit in this text.

Skill Building

In Problems 5–10, examine the scatter diagram and determine whether the type of relation is linear or nonlinear.

5.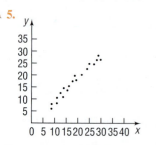

6.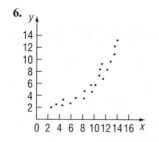

7.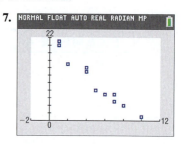

8.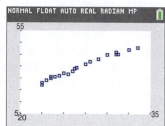

9.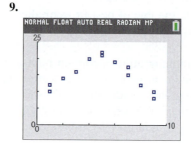

10.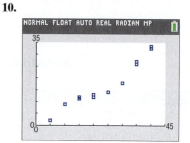

In Problems 11–16:
(a) Draw a scatter diagram.
(b) Select two points from the scatter diagram, and find the equation of the line containing the points selected.
(c) Graph the line found in part (b) on the scatter diagram.
(d) Use a graphing utility to find the line of best fit.
(e) Use a graphing utility to draw the scatter diagram and graph the line of best fit on it.

11.

x	3	4	5	6	7	8	9
y	4	6	7	10	12	14	16

12.

x	3	5	7	9	11	13
y	0	2	3	6	9	11

13.

x	−2	−1	0	1	2
y	−4	0	1	4	5

14.

x	−2	−1	0	1	2
y	7	6	3	2	0

15.

x	−20	−17	−15	−14	−10
y	100	120	118	130	140

16.

x	−30	−27	−25	−20	−14
y	10	12	13	13	18

Applications and Extensions

17. Candy The following data represent the weight (in grams) of various candy bars and the corresponding number of calories.

Candy Bar	Weight, x	Calories, y
Hershey's Milk Chocolate®	44.28	230
Nestle's Crunch®	44.84	230
Butterfinger®	61.30	270
Baby Ruth®	66.45	280
Almond Joy®	47.33	220
Twix® (with caramel)	58.00	280
Snickers®	61.12	280
Heath®	39.52	210

Source: Megan Pocius, student at Joliet Junior College

(a) Draw a scatter diagram of the data, treating weight as the independent variable.
(b) What type of relation appears to exist between the weight of a candy bar and the number of calories?
(c) Select two points and find a linear model that contains the points.

(d) Graph the line on the scatter diagram drawn in part (a).
(e) Use the linear model to predict the number of calories in a candy bar that weighs 62.3 grams.
(f) Interpret the slope of the line found in part (c).

18. Raisins The following data represent the weight (in grams) of a box of raisins and the number of raisins in the box.

Weight (grams), w	Number of Raisins, N
42.3	87
42.7	91
42.8	93
42.4	87
42.6	89
42.4	90
42.3	82
42.5	86
42.7	86
42.5	86

Source: Jennifer Maxwell, student at Joliet Junior College

(a) Draw a scatter diagram of the data, treating weight as the independent variable.
(b) What type of relation appears to exist between the weight of a box of raisins and the number of raisins?
(c) Select two points and find a linear model that contains the points.
(d) Graph the line on the scatter diagram drawn in part (b).
(e) Use the linear model to predict the number of raisins in a box that weighs 42.5 grams.
(f) Interpret the slope of the line found in part (c).

19. Video Games and Grade-Point Average Professor Grant Alexander wanted to find a linear model that relates the number of hours a student plays video games each week, h, to the cumulative grade-point average, G, of the student. He obtained a random sample of 10 full-time students at his college and asked each student to disclose the number of hours spent playing video games and the student's cumulative grade-point average.

Hours of Video Games per Week, h	Grade-point Average, G
0	3.49
0	3.05
2	3.24
3	2.82
3	3.19
5	2.78
8	2.31
8	2.54
10	2.03
12	2.51

(a) Explain why the number of hours spent playing video games is the independent variable and cumulative grade-point average is the dependent variable.
(b) Use a graphing utility to draw a scatter diagram.
(c) Use a graphing utility to find the line of best fit that models the relation between number of hours of video game playing each week and grade-point average. Express the model using function notation.
(d) Interpret the slope.
(e) Predict the grade-point average of a student who plays video games for 8 hours each week.
(f) How many hours of video game playing do you think a student plays whose grade-point average is 2.40?

20. Height versus Head Circumference A pediatrician wanted to find a linear model that relates a child's height, H, to head circumference, C. She randomly selects nine children from her practice, measures their height and head circumference, and obtains the data shown. Let H represent the independent variable and C the dependent variable.
(a) Use a graphing utility to draw a scatter diagram.
(b) Use a graphing utility to find the line of best fit that models the relation between height and head circumference. Express the model using function notation.
(c) Interpret the slope.

(d) Predict the head circumference of a child who is 26 inches tall.
(e) What is the height of a child whose head circumference is 17.4 inches?

Height, H (inches)	Head Circumference, C (inches)
25.25	16.4
25.75	16.9
25	16.9
27.75	17.6
26.5	17.3
27	17.5
26.75	17.3
26.75	17.5
27.5	17.5

Source: Denise Slucki, student at Joliet Junior College

21. Flight Time and Ticket Price The following data represent nonstop flight time (in minutes) and one-way ticket price (in dollars) for flying from Chicago to various cities on Southwest Airlines.

City	Time (minutes), t	Price (dollars), P
Atlanta, GA	110	140
Los Angeles, CA	260	199
Nashville, TN	80	108
Oklahoma City, OK	125	130
Omaha, NE	85	89
Ft. Myers, FL	170	175
Phoenix, AZ	225	191
Houston, TX	155	147
Seattle, WA	265	196
St. Louis, MO	65	93

Source: Southwest.com, for midmorning flights in May 2014

(a) Use a graphing utility to draw a scatter diagram.
(b) Use a graphing utility to find the line of best fit that models the relation between flight time and airfare. Express the model using function notation.
(c) Interpret the slope.
(d) Predict the airfare for a flight from Chicago to Kansas City, Missouri, if the flight time is 85 minutes. Round to the nearest dollar.
(e) Predict the flight time from Chicago to Baltimore, Maryland if the airfare is $120. Round to the nearest minute.

Explaining Concepts: Discussion and Writing

22. **Maternal Age versus Down Syndrome** A biologist would like to know how the age of the mother affects the incidence of Down syndrome. The data to the right represent the age of the mother and the incidence of Down syndrome per 1000 pregnancies. Draw a scatter diagram treating age of the mother as the independent variable. Would it make sense to find the line of best fit for these data? Why or why not?

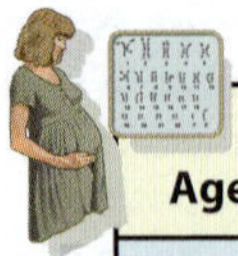

23. Find the line of best fit for the ordered pairs $(1, 5)$ and $(3, 8)$. What is the correlation coefficient for these data? Why is this result reasonable?

24. What does a correlation coefficient of 0 imply?

25. Explain why it does not make sense to interpret the y-intercept in Problem 17.

26. Refer to Problem 19. Solve $G(h) = 0$. Provide an interpretation of this result. Find $G(0)$. Provide an interpretation of this result.

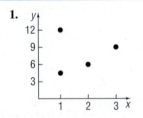

Age of Mother, x	Incidence of Down Syndrome, y
33	2.4
34	3.1
35	4
36	5
37	6.7
38	8.3
39	10
40	13.3
41	16.7
42	22.2
43	28.6
44	33.3
45	50

Source: Hook, E.B., *Journal of the American Medical Association*, 249, 2034-2038, 1983.

Retain Your Knowledge

Problems 27–30 are based on material learned earlier in the course. The purpose of these problems is to keep the material fresh in your mind so that you are better prepared for the final exam.

27. Find an equation for the line containing the points $(-1, 5)$ and $(3, -3)$. Express your answer using either the general form or the slope-intercept form of the equation of a line, whichever you prefer.

28. Find the domain of $f(x) = \dfrac{x - 1}{x^2 - 25}$.

29. For $f(x) = 5x - 8$ and $g(x) = x^2 - 3x + 4$, find $(g - f)(x)$.

30. Write the function whose graph is the graph of $y = x^2$, but shifted to the left 3 units and shifted down 4 units.

'Are You Prepared?' Answers

1.

(graph with points plotted, y-axis marked 3, 6, 9, 12; x-axis marked 1, 2, 3)

No, because the input, 1, corresponds to two different outputs.

2. $y = 2x + 2$

4.3 Quadratic Functions and Their Properties

PREPARING FOR THIS SECTION *Before getting started, review the following:*

- Intercepts (Section 2.2, pp. 159–160)
- Graphing Techniques: Transformations (Section 3.5, pp. 247–256)
- Completing the Square (Section R.5, p. 56)
- Quadratic Equations (Section 1.2, pp. 92–99)

Now Work the **'Are You Prepared?'** problems on page 299.

OBJECTIVES
1 Graph a Quadratic Function Using Transformations (p. 292)
2 Identify the Vertex and Axis of Symmetry of a Quadratic Function (p. 294)
3 Graph a Quadratic Function Using Its Vertex, Axis, and Intercepts (p. 294)
4 Find a Quadratic Function Given Its Vertex and One Other Point (p. 297)
5 Find the Maximum or Minimum Value of a Quadratic Function (p. 298)

Quadratic Functions

Here are some examples of quadratic functions.

$$F(x) = 3x^2 - 5x + 1 \quad g(x) = -6x^2 + 1 \quad H(x) = \frac{1}{2}x^2 + \frac{2}{3}x$$

DEFINITION

A **quadratic function** is a function of the form

$$f(x) = ax^2 + bx + c$$

where a, b, and c are real numbers and $a \neq 0$. The domain of a quadratic function is the set of all real numbers.

In Words

A quadratic function is a function defined by a second-degree polynomial in one variable.

Many applications require a knowledge of quadratic functions. For example, suppose that Texas Instruments collects the data shown in Table 7, which relate the number of calculators sold to the price p (in dollars) per calculator. Since the price of a product determines the quantity that will be purchased, we treat price as the independent variable. The relationship between the number x of calculators sold and the price p per calculator is given by the linear equation

$$x = 21{,}000 - 150p$$

Table 7

Price p per Calculator, (in dollars)	Number of Calculators, x
60	12,000
65	11,250
70	10,500
75	9,750
80	9,000
85	8,250
90	7,500

Then the revenue R derived from selling x calculators at the price p per calculator is equal to the unit selling price p of the calculator times the number x of units actually sold. That is,

$$R = xp$$
$$R(p) = (21{,}000 - 150p)p \quad x = 21{,}000 - 150p$$
$$= -150p^2 + 21{,}000p$$

So the revenue R is a quadratic function of the price p. Figure 12 illustrates the graph of this revenue function, whose domain is $0 \leq p \leq 140$, since both x and p must be nonnegative.

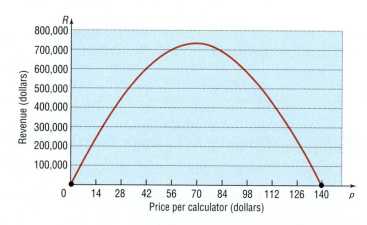

Figure 12
$R(p) = -150p^2 + 21{,}000p$

Figure 13
Path of a cannonball

A second situation in which a quadratic function appears involves the motion of a projectile. Based on Newton's Second Law of Motion (force equals mass times acceleration, $F = ma$), it can be shown that, ignoring air resistance, the path of a projectile propelled upward at an inclination to the horizontal is the graph of a quadratic function. See Figure 13 for an illustration.

1 Graph a Quadratic Function Using Transformations

We know how to graph the square function $f(x) = x^2$. Figure 14 shows the graph of three functions of the form $f(x) = ax^2, a > 0$, for $a = 1$, $a = \dfrac{1}{2}$, and $a = 3$. Note that the larger the value of a, the "narrower" the graph is, and the smaller the value of a, the "wider" the graph is.

Figure 15 shows the graphs of $f(x) = ax^2$ for $a < 0$. Notice that these graphs are reflections about the x-axis of the graphs in Figure 14. Based on the results of these two figures, general conclusions can be drawn about the graph of $f(x) = ax^2$. First, as $|a|$ increases, the graph is stretched vertically (becomes "taller"), and as $|a|$ gets closer to zero, the graph is compressed vertically (becomes "shorter"). Second, if a is positive, the graph opens "up," and if a is negative, the graph opens "down."

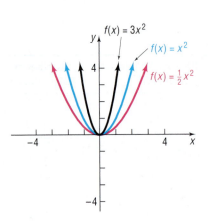

Figure 14

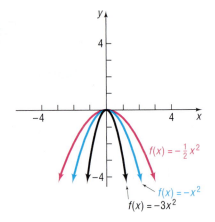

Figure 15

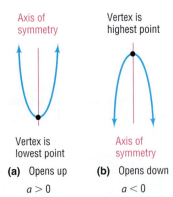

Axis of symmetry

Vertex is highest point

Vertex is lowest point

Axis of symmetry

(a) Opens up
$a > 0$

(b) Opens down
$a < 0$

Figure 16
Graphs of a quadratic function,
$f(x) = ax^2 + bx + c, a \neq 0$

The graphs in Figures 14 and 15 are typical of the graphs of all quadratic functions, which are called **parabolas.*** Refer to Figure 16, where two parabolas are pictured. The one on the left **opens up** and has a lowest point; the one on the right **opens down** and has a highest point. The lowest or highest point of a parabola is called the **vertex.** The vertical line passing through the vertex in each parabola in Figure 16 is called the **axis of symmetry** (usually abbreviated to **axis**) of the parabola. Because the parabola is symmetric about its axis, the axis of symmetry of a parabola can be used to find additional points on the parabola.

The parabolas shown in Figure 16 are the graphs of a quadratic function $f(x) = ax^2 + bx + c, a \neq 0$. Notice that the coordinate axes are not included in the figure. Depending on the values of a, b, and c, the axes could be placed anywhere. The important fact is that the shape of the graph of a quadratic function will look like one of the parabolas in Figure 16.

In the following example, techniques from Section 3.5 are used to graph a quadratic function $f(x) = ax^2 + bx + c, a \neq 0$. The method of completing the square is used to write the function f in the form $f(x) = a(x - h)^2 + k$.

EXAMPLE 1

Graphing a Quadratic Function Using Transformations

Graph the function $f(x) = 2x^2 + 8x + 5$. Find the vertex and axis of symmetry.

---

* Parabolas will be studied using a geometric definition later in this text.

Solution   Begin by completing the square on the right side.

$$f(x) = 2x^2 + 8x + 5$$

$$= 2(x^2 + 4x) + 5 \qquad \text{Factor out the 2 from } 2x^2 + 8x.$$

$$= 2(x^2 + 4x + 4) + 5 - 8 \qquad \text{Complete the square of } x^2 + 4x \text{ by}$$
$$\qquad\qquad\qquad\qquad\qquad \text{adding 4. Notice that the factor of 2}$$
$$= 2(x + 2)^2 - 3 \qquad\qquad \text{requires that 8 be added and subtracted.}$$

The graph of f can be obtained from the graph of $y = x^2$ in three stages, as shown in Figure 17. Now compare this graph to the graph in Figure 16(a). The graph of $f(x) = 2x^2 + 8x + 5$ is a parabola that opens up and has its vertex (lowest point) at $(-2, -3)$. Its axis of symmetry is the line $x = -2$.

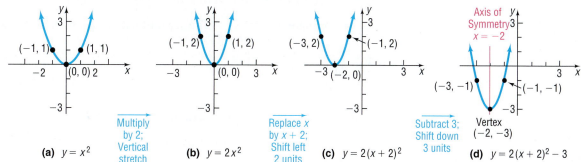

Figure 17   (a) $y = x^2$ — Multiply by 2; Vertical stretch   (b) $y = 2x^2$ — Replace x by $x + 2$; Shift left 2 units   (c) $y = 2(x + 2)^2$ — Subtract 3; Shift down 3 units   (d) $y = 2(x + 2)^2 - 3$ — Vertex $(-2, -3)$

Now Work PROBLEM 25

The method used in Example 1 can be used to graph any quadratic function $f(x) = ax^2 + bx + c, a \neq 0$, as follows:

$$f(x) = ax^2 + bx + c$$

$$= a\left(x^2 + \frac{b}{a}x\right) + c \qquad \text{Factor out } a \text{ from } ax^2 + bx.$$

$$= a\left(x^2 + \frac{b}{a}x + \frac{b^2}{4a^2}\right) + c - a\left(\frac{b^2}{4a^2}\right) \qquad \text{Complete the square by adding } \frac{b^2}{4a^2}.$$
$$\qquad\qquad\qquad\qquad\qquad\qquad\qquad\qquad \text{Look closely at this step!}$$

$$= a\left(x + \frac{b}{2a}\right)^2 + c - \frac{b^2}{4a} \qquad \text{Factor; simplify.}$$

$$= a\left(x + \frac{b}{2a}\right)^2 + \frac{4ac - b^2}{4a} \qquad c - \frac{b^2}{4a} = c \cdot \frac{4a}{4a} - \frac{b^2}{4a} = \frac{4ac - b^2}{4a}$$

These results lead to the following conclusion:

If $h = -\dfrac{b}{2a}$ and $k = \dfrac{4ac - b^2}{4a}$, then

$$f(x) = ax^2 + bx + c = a(x - h)^2 + k \qquad\qquad (1)$$

The graph of $f(x) = a(x - h)^2 + k$ is the parabola $y = ax^2$ shifted horizontally h units (replace x by $x - h$) and vertically k units (add k). As a result, the vertex is at (h, k), and the graph opens up if $a > 0$ and down if $a < 0$. The axis of symmetry is the vertical line $x = h$.

For example, compare equation (1) with the solution given in Example 1.

$$f(x) = 2(x + 2)^2 - 3$$
$$= 2(x - (-2))^2 + (-3)$$
$$= a(x - h)^2 + k$$

Because $a = 2$, the graph opens up. Also, because $h = -2$ and $k = -3$, its vertex is at $(-2, -3)$.

2 Identify the Vertex and Axis of Symmetry of a Quadratic Function

We do not need to complete the square to obtain the vertex. In almost every case, it is easier to obtain the vertex of a quadratic function f by remembering that its x-coordinate is $h = -\dfrac{b}{2a}$. The y-coordinate k can then be found by evaluating f at $-\dfrac{b}{2a}$. That is, $k = f\left(-\dfrac{b}{2a}\right)$.

Properties of the Graph of a Quadratic Function

$$f(x) = ax^2 + bx + c \qquad a \neq 0$$

$$\text{Vertex} = \left(-\frac{b}{2a}, f\left(-\frac{b}{2a}\right)\right) \quad \text{Axis of symmetry: the vertical line } x = -\frac{b}{2a} \quad \text{(2)}$$

Parabola opens up if $a > 0$; the vertex is a minimum point.
Parabola opens down if $a < 0$; the vertex is a maximum point.

> **EXAMPLE 2**

Locating the Vertex without Graphing

Without graphing, locate the vertex and axis of symmetry of the parabola defined by $f(x) = -3x^2 + 6x + 1$. Does it open up or down?

Solution

For this quadratic function, $a = -3$, $b = 6$, and $c = 1$. The x-coordinate of the vertex is

$$h = -\frac{b}{2a} = -\frac{6}{2(-3)} = 1$$

The y-coordinate of the vertex is

$$k = f\left(-\frac{b}{2a}\right) = f(1) = -3(1)^2 + 6(1) + 1 = 4$$

The vertex is located at the point $(1, 4)$. The axis of symmetry is the line $x = 1$. Because $a = -3 < 0$, the parabola opens down. ●

3 Graph a Quadratic Function Using Its Vertex, Axis, and Intercepts

The location of the vertex and intercepts, along with knowledge of whether the graph opens up or down, usually provides enough information to graph $f(x) = ax^2 + bx + c, a \neq 0$.

The y-intercept is the value of f at $x = 0$; that is, the y-intercept is $f(0) = c$.

The x-intercepts, if there are any, are found by solving the quadratic equation

$$ax^2 + bx + c = 0$$

This equation has two, one, or no real solutions, depending on whether the discriminant $b^2 - 4ac$ is positive, 0, or negative. Depending on the value of the discriminant, the graph of f has x-intercepts, as follows:

> **The x-Intercepts of a Quadratic Function**
>
> 1. If the discriminant $b^2 - 4ac > 0$, the graph of $f(x) = ax^2 + bx + c$ has two distinct x-intercepts so it crosses the x-axis in two places.
> 2. If the discriminant $b^2 - 4ac = 0$, the graph of $f(x) = ax^2 + bx + c$ has one x-intercept so it touches the x-axis at its vertex.
> 3. If the discriminant $b^2 - 4ac < 0$, the graph of $f(x) = ax^2 + bx + c$ has no x-intercepts so it does not cross or touch the x-axis.

Figure 18 illustrates these possibilities for parabolas that open up.

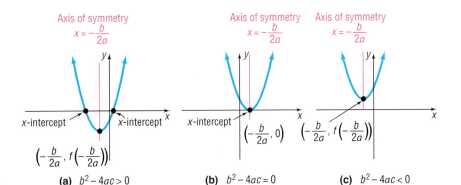

Figure 18
$f(x) = ax^2 + bx + c, a > 0$

(a) $b^2 - 4ac > 0$
Two x-intercepts

(b) $b^2 - 4ac = 0$
One x-intercept

(c) $b^2 - 4ac < 0$
No x-intercepts

EXAMPLE 3

Graphing a Quadratic Function Using Its Vertex, Axis, and Intercepts

(a) Use the information from Example 2 and the locations of the intercepts to graph $f(x) = -3x^2 + 6x + 1$.

(b) Determine the domain and the range of f.

(c) Determine where f is increasing and where it is decreasing.

Solution

(a) In Example 2, we found the vertex to be at $(1, 4)$ and the axis of symmetry to be $x = 1$. The y-intercept is found by letting $x = 0$. The y-intercept is $f(0) = 1$. The x-intercepts are found by solving the equation $f(x) = 0$. This results in the equation

$$-3x^2 + 6x + 1 = 0 \quad a = -3, b = 6, c = 1$$

The discriminant $b^2 - 4ac = (6)^2 - 4(-3)(1) = 36 + 12 = 48 > 0$, so the equation has two real solutions and the graph has two x-intercepts. Use the quadratic formula to find that

$$x = \frac{-b + \sqrt{b^2 - 4ac}}{2a} = \frac{-6 + \sqrt{48}}{-6} = \frac{-6 + 4\sqrt{3}}{-6} \approx -0.15$$

and

$$x = \frac{-b - \sqrt{b^2 - 4ac}}{2a} = \frac{-6 - \sqrt{48}}{-6} = \frac{-6 - 4\sqrt{3}}{-6} \approx 2.15$$

The x-intercepts are approximately -0.15 and 2.15.

The graph is illustrated in Figure 19. Notice how we used the y-intercept and the axis of symmetry, $x = 1$, to obtain the additional point $(2, 1)$ on the graph.

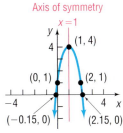

Figure 19 $f(x) = -3x^2 + 6x + 1$

5 Find the Maximum or Minimum Value of a Quadratic Function

The graph of a quadratic function

$$f(x) = ax^2 + bx + c \qquad a \neq 0$$

is a parabola with vertex at $\left(-\dfrac{b}{2a}, f\left(-\dfrac{b}{2a}\right)\right)$. This vertex is the highest point on the graph if $a < 0$ and the lowest point on the graph if $a > 0$. If the vertex is the highest point $(a < 0)$, then $f\left(-\dfrac{b}{2a}\right)$ is the **maximum value** of f. If the vertex is the lowest point $(a > 0)$, then $f\left(-\dfrac{b}{2a}\right)$ is the **minimum value** of f.

EXAMPLE 7

Finding the Maximum or Minimum Value of a Quadratic Function

Determine whether the quadratic function

$$f(x) = x^2 - 4x - 5$$

has a maximum or a minimum value. Then find the maximum or minimum value.

Solution Compare $f(x) = x^2 - 4x - 5$ to $f(x) = ax^2 + bx + c$. Then $a = 1$, $b = -4$, and $c = -5$. Because $a > 0$, the graph of f opens up, which means the vertex is a minimum point. The minimum value occurs at

$$x = -\frac{b}{2a} = -\frac{-4}{2(1)} = \frac{4}{2} = 2$$
$$\uparrow$$
$$a = 1, b = -4$$

The minimum value is

$$f\left(-\frac{b}{2a}\right) = f(2) = 2^2 - 4(2) - 5 = 4 - 8 - 5 = -9$$

$\bullet$

Now Work PROBLEM 57

SUMMARY

Steps for Graphing a Quadratic Function $f(x) = ax^2 + bx + c$, $a \neq 0$

Option 1

STEP 1: Complete the square in x to write the quadratic function in the form $f(x) = a(x - h)^2 + k$.

STEP 2: Graph the function in stages using transformations.

Option 2

STEP 1: Determine whether the parabola opens up $(a > 0)$ or down $(a < 0)$.

STEP 2: Determine the vertex $\left(-\dfrac{b}{2a}, f\left(-\dfrac{b}{2a}\right)\right)$.

STEP 3: Determine the axis of symmetry, $x = -\dfrac{b}{2a}$.

STEP 4: Determine the y-intercept, $f(0)$, and the x-intercepts, if any.

    (a) If $b^2 - 4ac > 0$, the graph of the quadratic function has two x-intercepts, which are found by solving the equation $ax^2 + bx + c = 0$.

    (b) If $b^2 - 4ac = 0$, the vertex is the x-intercept.

    (c) If $b^2 - 4ac < 0$, there are no x-intercepts.

STEP 5: Determine an additional point by using the y-intercept and the axis of symmetry.

STEP 6: Plot the points and draw the graph.

4.3 Assess Your Understanding

'Are You Prepared?' *Answers are given at the end of these exercises. If you get a wrong answer, read the pages listed in red.*

1. List the intercepts of the equation $y = x^2 - 9$. (pp. 159–160)
2. Find the real solutions of the equation $2x^2 + 7x - 4 = 0$. (pp. 92–99)
3. To complete the square of $x^2 - 5x$, you add the number _____. (p. 56)
4. To graph $y = (x - 4)^2$, you shift the graph of $y = x^2$ to the _____ a distance of ____ units. (pp. 247–256)

Concepts and Vocabulary

5. The graph of a quadratic function is called a(n) _____.
6. The vertical line passing through the vertex of a parabola is called the _____ .
7. The x-coordinate of the vertex of $f(x) = ax^2 + bx + c$, $a \neq 0$, is _____.
8. **True or False** The graph of $f(x) = 2x^2 + 3x - 4$ opens up.
9. **True or False** The y-coordinate of the vertex of $f(x) = -x^2 + 4x + 5$ is $f(2)$.
10. **True or False** If the discriminant $b^2 - 4ac = 0$, the graph of $f(x) = ax^2 + bx + c, a \neq 0$, will touch the x-axis at its vertex.

11. If $b^2 - 4ac > 0$, which of the following conclusions can be made about the graph of $f(x) = ax^2 + bx + c$, $a \neq 0$?
    (a) The graph has two distinct x-intercepts.
    (b) The graph has no x-intercepts.
    (c) The graph has three distinct x-intercepts.
    (d) The graph has one x-intercept.
12. If the graph of $f(x) = ax^2 + bx + c$, $a \neq 0$, has a maximum value at its vertex, which of the following conditions must be true?
    (a) $-\dfrac{b}{2a} > 0$ (b) $-\dfrac{b}{2a} < 0$
    (c) $a > 0$ (d) $a < 0$

Skill Building

In Problems 13–20, match each graph to one the following functions.

13. $f(x) = x^2 - 1$
14. $f(x) = -x^2 - 1$
15. $f(x) = x^2 - 2x + 1$
16. $f(x) = x^2 + 2x + 1$
17. $f(x) = x^2 - 2x + 2$
18. $f(x) = x^2 + 2x$
19. $f(x) = x^2 - 2x$
20. $f(x) = x^2 + 2x + 2$

A.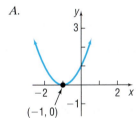
B.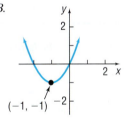
C.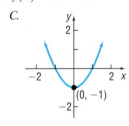
D.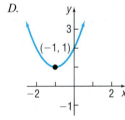

E.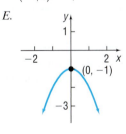
F.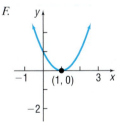
G.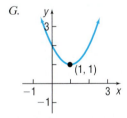
H.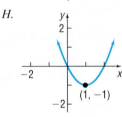

In Problems 21–32, graph the function f by starting with the graph of $y = x^2$ and using transformations (shifting, compressing, stretching, and/or reflecting).
[Hint: If necessary, write f in the form $f(x) = a(x - h)^2 + k$.]

21. $f(x) = \dfrac{1}{4}x^2$
22. $f(x) = 2x^2 + 4$
23. $f(x) = (x + 2)^2 - 2$
24. $f(x) = (x - 3)^2 - 10$
25. $f(x) = x^2 + 4x + 2$
26. $f(x) = x^2 - 6x - 1$
27. $f(x) = 2x^2 - 4x + 1$
28. $f(x) = 3x^2 + 6x$
29. $f(x) = -x^2 - 2x$
30. $f(x) = -2x^2 + 6x + 2$
31. $f(x) = \dfrac{1}{2}x^2 + x - 1$
32. $f(x) = \dfrac{2}{3}x^2 + \dfrac{4}{3}x - 1$

In Problems 33–48, (a) graph each quadratic function by determining whether its graph opens up or down and by finding its vertex, axis of symmetry, y-intercept, and x-intercepts, if any. (b) Determine the domain and the range of the function. (c) Determine where the function is increasing and where it is decreasing.

33. $f(x) = x^2 + 2x$
34. $f(x) = x^2 - 4x$
35. $f(x) = -x^2 - 6x$
36. $f(x) = -x^2 + 4x$
37. $f(x) = x^2 + 2x - 8$
38. $f(x) = x^2 - 2x - 3$
39. $f(x) = x^2 + 2x + 1$
40. $f(x) = x^2 + 6x + 9$

41. $f(x) = 2x^2 - x + 2$ **42.** $f(x) = 4x^2 - 2x + 1$ **43.** $f(x) = -2x^2 + 2x - 3$ **44.** $f(x) = -3x^2 + 3x - 2$
45. $f(x) = 3x^2 + 6x + 2$ **46.** $f(x) = 2x^2 + 5x + 3$ **47.** $f(x) = -4x^2 - 6x + 2$ **48.** $f(x) = 3x^2 - 8x + 2$

In Problems 49–54, determine the quadratic function whose graph is given.

49.

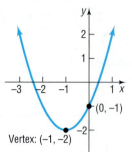

50.

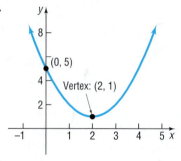

51.

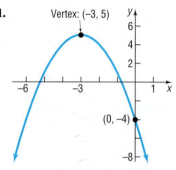

52.

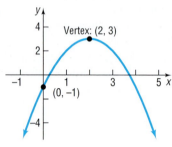

53.

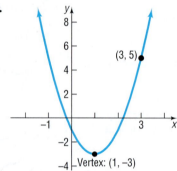

54.

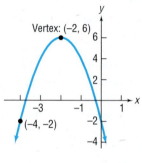

In Problems 55–62, determine, without graphing, whether the given quadratic function has a maximum value or a minimum value, and then find the value.

55. $f(x) = 2x^2 + 12x$ **56.** $f(x) = -2x^2 + 12x$ **57.** $f(x) = 2x^2 + 12x - 3$ **58.** $f(x) = 4x^2 - 8x + 3$
59. $f(x) = -x^2 + 10x - 4$ **60.** $f(x) = -2x^2 + 8x + 3$ **61.** $f(x) = -3x^2 + 12x + 1$ **62.** $f(x) = 4x^2 - 4x$

┌─ Mixed Practice

In Problems 63–70, (a) graph each function. (b) Determine the domain and the range of the function. (c) Determine where the function is increasing and where it is decreasing.

63. $f(x) = x^2 - 2x - 15$ **64.** $g(x) = x^2 - 2x - 8$ **65.** $h(x) = -\frac{2}{5}x + 4$ **66.** $f(x) = \frac{3}{2}x - 2$

67. $g(x) = -2(x - 3)^2 + 2$ **68.** $h(x) = -3(x + 1)^2 + 4$ **69.** $f(x) = 2x^2 + x + 1$ **70.** $F(x) = -4x^2 + 20x - 25$

Applications and Extensions

71. The graph of the function $f(x) = ax^2 + bx + c$ has vertex at $(0, 2)$ and passes through the point $(1, 8)$. Find a, b, and c.

72. The graph of the function $f(x) = ax^2 + bx + c$ has vertex at $(1, 4)$ and passes through the point $(-1, -8)$. Find a, b, and c.

In Problems 73–78, for the given functions f and g:
   (a) Graph f and g on the same Cartesian plane.   *(b) Solve $f(x) = g(x)$.*
   (c) Use the result of part (b) to label the points of intersection of the graphs of f and g.
   (d) Shade the region for which $f(x) > g(x)$ —that is, the region below f and above g.

73. $f(x) = 2x - 1$; $g(x) = x^2 - 4$ **74.** $f(x) = -2x - 1$; $g(x) = x^2 - 9$
75. $f(x) = -x^2 + 4$; $g(x) = -2x + 1$ **76.** $f(x) = -x^2 + 9$; $g(x) = 2x + 1$
77. $f(x) = -x^2 + 5x$; $g(x) = x^2 + 3x - 4$ **78.** $f(x) = -x^2 + 7x - 6$; $g(x) = x^2 + x - 6$

Answer Problems 79 and 80 using the following: A quadratic function of the form $f(x) = ax^2 + bx + c$ with $b^2 - 4ac > 0$ may also be written in the form $f(x) = a(x - r_1)(x - r_2)$, where r_1 and r_2 are the x-intercepts of the graph of the quadratic function.

79. (a) Find a quadratic function whose x-intercepts are -3 and 1 with $a = 1; a = 2; a = -2; a = 5$.
   (b) How does the value of a affect the intercepts?
   (c) How does the value of a affect the axis of symmetry?
   (d) How does the value of a affect the vertex?
   (e) Compare the x-coordinate of the vertex with the midpoint of the x-intercepts. What might you conclude?

80. (a) Find a quadratic function whose x-intercepts are -5 and 3 with $a = 1; a = 2; a = -2; a = 5$.
   (b) How does the value of a affect the intercepts?
   (c) How does the value of a affect the axis of symmetry?
   (d) How does the value of a affect the vertex?
   (e) Compare the x-coordinate of the vertex with the midpoint of the x-intercepts. What might you conclude?

81. Suppose that $f(x) = x^2 + 4x - 21$.
    (a) What is the vertex of f?
    (b) What are the x-intercepts of the graph of f?
    (c) Solve $f(x) = -21$ for x. What points are on the graph of f?
    (d) Use the information obtained in parts (a)–(c) to graph $f(x) = x^2 + 4x - 21$.

82. Suppose that $f(x) = x^2 + 2x - 8$.
    (a) What is the vertex of f?
    (b) What are the x-intercepts of the graph of f?
    (c) Solve $f(x) = -8$ for x. What points are on the graph of f?
    (d) Use the information obtained in parts (a)–(c) to graph $f(x) = x^2 + 2x - 8$.

83. Find the point on the line $y = x$ that is closest to the point $(3, 1)$.

    [**Hint:** Express the distance d from the point to the line as a function of x, and then find the minimum value of $[d(x)]^2$.

84. Find the point on the line $y = x + 1$ that is closest to the point $(4, 1)$.

85. **Maximizing Revenue** Suppose that the manufacturer of a gas clothes dryer has found that when the unit price is p dollars, the revenue R (in dollars) is

    $$R(p) = -4p^2 + 4000p$$

    What unit price should be established for the dryer to maximize revenue? What is the maximum revenue?

86. **Maximizing Revenue** The John Deere company has found that the revenue, in dollars, from sales of riding mowers is a function of the unit price p, in dollars, that it charges. If the revenue R is

    $$R(p) = -\frac{1}{2}p^2 + 1900p$$

    what unit price p should be charged to maximize revenue? What is the maximum revenue?

87. **Minimizing Marginal Cost** The **marginal cost** of a product can be thought of as the cost of producing one additional unit of output. For example, if the marginal cost of producing the 50th product is $6.20, it cost $6.20 to increase production from 49 to 50 units of output. Suppose the marginal cost C (in dollars) to produce x thousand digital music players is given by the function

    $$C(x) = x^2 - 140x + 7400$$

    (a) How many players should be produced to minimize the marginal cost?
    (b) What is the minimum marginal cost?

88. **Minimizing Marginal Cost** (See Problem 87.) The marginal cost C (in dollars) of manufacturing x cell phones (in thousands) is given by

    $$C(x) = 5x^2 - 200x + 4000$$

    (a) How many cell phones should be manufactured to minimize the marginal cost?
    (b) What is the minimum marginal cost?

89. **Business** The monthly revenue R achieved by selling x wristwatches is figured to be $R(x) = 75x - 0.2x^2$. The monthly cost C of selling x wristwatches is

    $$C(x) = 32x + 1750$$

    (a) How many wristwatches must the firm sell to maximize revenue? What is the maximum revenue?
    (b) Profit is given as $P(x) = R(x) - C(x)$. What is the profit function?
    (c) How many wristwatches must the firm sell to maximize profit? What is the maximum profit?
    (d) Provide a reasonable explanation as to why the answers found in parts (a) and (c) differ. Explain why a quadratic function is a reasonable model for revenue.

90. **Business** The daily revenue R achieved by selling x boxes of candy is figured to be $R(x) = 9.5x - 0.04x^2$. The daily cost C of selling x boxes of candy is $C(x) = 1.25x + 250$.
    (a) How many boxes of candy must the firm sell to maximize revenue? What is the maximum revenue?
    (b) Profit is given as $P(x) = R(x) - C(x)$. What is the profit function?
    (c) How many boxes of candy must the firm sell to maximize profit? What is the maximum profit?
    (d) Provide a reasonable explanation as to why the answers found in parts (a) and (c) differ. Explain why a quadratic function is a reasonable model for revenue.

91. **Stopping Distance** An accepted relationship between stopping distance, d (in feet), and the speed of a car, v (in mph), is $d = 1.1v + 0.06v^2$ on dry, level concrete.
    (a) How many feet will it take a car traveling 45 mph to stop on dry, level concrete?
    (b) If an accident occurs 200 feet ahead of you, what is the maximum speed you can be traveling to avoid being involved?
    (c) What might the term $1.1v$ represent?

92. **Birth Rate of Unmarried Women** In the United States, the birth rate B of unmarried women (births per 1000 unmarried women) for women whose age is a is modeled by the function $B(a) = -0.30a^2 + 16.26a - 158.90$
    (a) What is the age of unmarried women with the highest birth rate?
    (b) What is the highest birth rate of unmarried women?
    (c) Evaluate and interpret $B(40)$.

    Source: National Vital Statistics System, 2013

93. Let $f(x) = ax^2 + bx + c$, where $a, b,$ and c are odd integers. If x is an integer, show that $f(x)$ must be an odd integer.

    [**Hint:** x is either an even integer or an odd integer.]

Explaining Concepts: Discussion and Writing

94. Make up a quadratic function that opens down and has only one x-intercept. Compare yours with others in the class. What are the similarities? What are the differences?

95. On one set of coordinate axes, graph the family of parabolas $f(x) = x^2 + 2x + c$ for $c = -3$, $c = 0$, and $c = 1$. Describe the characteristics of a member of this family.

96. On one set of coordinate axes, graph the family of parabolas $f(x) = x^2 + bx + 1$ for $b = -4$, $b = 0$, and $b = 4$. Describe the general characteristics of this family.

97. State the circumstances that cause the graph of a quadratic function $f(x) = ax^2 + bx + c$ to have no x-intercepts.

98. Why does the graph of a quadratic function open up if $a > 0$ and down if $a < 0$?

99. Can a quadratic function have a range of $(-\infty, \infty)$? Justify your answer.

100. What are the possibilities for the number of times the graphs of two different quadratic functions intersect?

Retain Your Knowledge

Problems 101–104 are based on material learned earlier in the course. The purpose of these problems is to keep the material fresh in your mind so that you are better prepared for the final exam.

101. Determine whether $x^2 + 4y^2 = 16$ is symmetric respect to the x-axis, the y-axis, and/or the origin.

102. Solve the inequality $27 - x \geq 5x + 3$. Write the solution in both set notation and interval notation.

103. Find the center and radius of the circle $x^2 + y^2 - 10x + 4y + 20 = 0$.

104. Write the function whose graph is the graph of $y = \sqrt{x}$, but reflected about the y-axis.

'Are You Prepared?' Answers

1. $(0, -9), (-3, 0), (3, 0)$ **2.** $\left\{-4, \dfrac{1}{2}\right\}$ **3.** $\dfrac{25}{4}$ **4.** right; 4

4.4 Build Quadratic Models from Verbal Descriptions and from Data

PREPARING FOR THIS SECTION *Before getting started, review the following:*

- Problem Solving (Section 1.7, pp. 134–140)
- Building Linear Models from Data (Section 4.2, pp. 284–287)

 Now Work the **'Are You Prepared?'** problems on page 307.

OBJECTIVES **1** Build Quadratic Models from Verbal Descriptions (p. 302)
 2 Build Quadratic Models from Data (p. 306)

In this section we will first discuss models in the form of a quadratic function when a verbal description of the problem is given. We end the section by fitting a quadratic function to data, which is another form of modeling.

When a mathematical model is in the form of a quadratic function, the properties of the graph of the function can provide important information about the model. In particular, we can use the quadratic function to determine the maximum or minimum value of the function. The fact that the graph of a quadratic function has a maximum or minimum value enables us to answer questions involving **optimization**—that is, finding the maximum or minimum values in models.

1 Build Quadratic Models from Verbal Descriptions

In economics, revenue R, in dollars, is defined as the amount of money received from the sale of an item and is equal to the unit selling price p, in dollars, of the item times the number x of units actually sold. That is,

$$R = xp$$

The Law of Demand states that p and x are related: As one increases, the other decreases. The equation that relates p and x is called the **demand equation**. When the demand equation is linear, the revenue model is a quadratic function.

EXAMPLE I **Maximizing Revenue**

The marketing department at Texas Instruments has found that when certain calculators are sold at a price of p dollars per unit, the number x of calculators sold is given by the demand equation

$$x = 21{,}000 - 150p$$

(a) Find a model that expresses the revenue R as a function of the price p.

(b) What is the domain of R?

(c) What unit price should be used to maximize revenue?

(d) If this price is charged, what is the maximum revenue?

(e) How many units are sold at this price?

(f) Graph R.

(g) What price should Texas Instruments charge to collect at least $675,000 in revenue?

Solution

(a) The revenue is $R = xp$, where $x = 21{,}000 - 150p$.

$$R = xp = (21{,}000 - 150p)p = -150p^2 + 21{,}000p \quad \text{The model}$$

(b) Because x represents the number of calculators sold, we have $x \geq 0$, so $21{,}000 - 150p \geq 0$. Solving this linear inequality gives $p \leq 140$. In addition, Texas Instruments will charge only a positive price for the calculator, so $p > 0$. Combining these inequalities gives the domain of R, which is $\{p \mid 0 < p \leq 140\}$.

(c) The function R is a quadratic function with $a = -150$, $b = 21{,}000$, and $c = 0$. Because $a < 0$, the vertex is the highest point on the parabola. The revenue R is a maximum when the price p is

$$p = -\frac{b}{2a} \underset{\underset{\textcolor{blue}{a = -150, b = 21{,}000}}{\uparrow}}{=} -\frac{21{,}000}{2(-150)} = \$70.00$$

(d) The maximum revenue R is

$$R(70) = -150(70)^2 + 21{,}000(70) = \$735{,}000$$

(e) The number of calculators sold is given by the demand equation $x = 21{,}000 - 150p$. At a price of $p = \$70$,

$$x = 21{,}000 - 150(70) = 10{,}500$$

calculators are sold.

(f) To graph R, plot the intercept $(140, 0)$ and the vertex $(70, 735\,000)$. See Figure 23 for the graph.

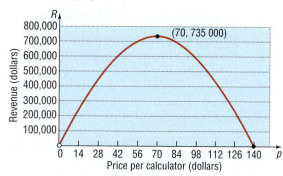

Figure 23

(g) Graph $R = 675{,}000$ and $R(p) = -150p^2 + 21{,}000p$ on the same Cartesian plane. See Figure 24. We find where the graphs intersect by solving

$$675{,}000 = -150p^2 + 21{,}000p$$

$$150p^2 - 21{,}000p + 675{,}000 = 0 \quad \textcolor{blue}{\text{Add } 150p^2 - 21{,}000p \text{ to both sides.}}$$

$$p^2 - 140p + 4500 = 0 \quad \textcolor{blue}{\text{Divide both sides by 150.}}$$

$$(p - 50)(p - 90) = 0 \quad \textcolor{blue}{\text{Factor.}}$$

$$p = 50 \text{ or } p = 90 \quad \textcolor{blue}{\text{Use the Zero-Product Property.}}$$

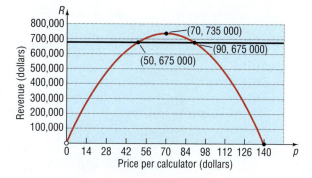

Figure 24

The graphs intersect at $(50, 675\,000)$ and $(90, 675\,000)$. Based on the graph in Figure 24, Texas Instruments should charge between \$50 and \$90 to earn at least \$675,000 in revenue. ●

Now Work PROBLEM 3

EXAMPLE 2

Maximizing the Area Enclosed by a Fence

A farmer has 2000 yards of fence to enclose a rectangular field. What are the dimensions of the rectangle that encloses the most area?

Solution Figure 25 illustrates the situation. The available fence represents the perimeter of the rectangle. If x is the length and w is the width, then

$$2x + 2w = 2000 \tag{1}$$

The area A of the rectangle is

$$A = xw$$

To express A in terms of a single variable, solve equation (1) for w and substitute the result in $A = xw$. Then A involves only the variable x. [You could also solve equation (1) for x and express A in terms of w alone. Try it!]

$$2x + 2w = 2000$$
$$2w = 2000 - 2x$$
$$w = \frac{2000 - 2x}{2} = 1000 - x$$

Then the area A is

$$A = xw = x(1000 - x) = -x^2 + 1000x$$

Now, A is a quadratic function of x.

$$A(x) = -x^2 + 1000x \quad a = -1, b = 1000, c = 0$$

Figure 26 shows the graph of $A(x) = -x^2 + 1000x$. Because $a < 0$, the vertex is a maximum point on the graph of A. The maximum value occurs at

$$x = -\frac{b}{2a} = -\frac{1000}{2(-1)} = 500$$

The maximum value of A is

$$A\left(-\frac{b}{2a}\right) = A(500) = -500^2 + 1000(500) = -250,000 + 500,000 = 250,000$$

The largest rectangle that can be enclosed by 2000 yards of fence has an area of 250,000 square yards. Its dimensions are 500 yards by 500 yards. ●

Now Work PROBLEM 7

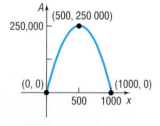

Figure 25

Figure 26 $A(x) = -x^2 + 1000x$

EXAMPLE 3

Analyzing the Motion of a Projectile

A projectile is fired from a cliff 500 feet above the water at an inclination of 45° to the horizontal, with a muzzle velocity of 400 feet per second. From physics, the height h of the projectile above the water can be modeled by

$$h(x) = \frac{-32x^2}{(400)^2} + x + 500$$

where x is the horizontal distance of the projectile from the base of the cliff. See Figure 27.

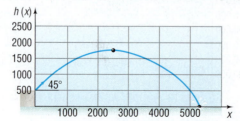

Figure 27

(a) Find the maximum height of the projectile.

(b) How far from the base of the cliff will the projectile strike the water?

Solution   (a) The height of the projectile is given by a quadratic function.

$$h(x) = \frac{-32x^2}{(400)^2} + x + 500 = \frac{-1}{5000}x^2 + x + 500$$

We are looking for the maximum value of h. Because $a < 0$, the maximum value occurs at the vertex, whose x-coordinate is

$$x = -\frac{b}{2a} = -\frac{1}{2\left(\dfrac{-1}{5000}\right)} = \frac{5000}{2} = 2500$$

The maximum height of the projectile is

$$h(2500) = \frac{-1}{5000}(2500)^2 + 2500 + 500 = -1250 + 2500 + 500 = 1750 \text{ ft}$$

(b) The projectile will strike the water when the height is zero. To find the distance x traveled, solve the equation

$$h(x) = \frac{-1}{5000}x^2 + x + 500 = 0$$

The discriminant of this quadratic equation is

$$b^2 - 4ac = 1^2 - 4\left(\frac{-1}{5000}\right)(500) = 1.4$$

Then

$$x = \frac{-b \pm \sqrt{b^2 - 4ac}}{2a} = \frac{-1 \pm \sqrt{1.4}}{2\left(\dfrac{-1}{5000}\right)} \approx \begin{cases} -458 \\ 5458 \end{cases}$$

Discard the negative solution. The projectile will strike the water at a distance of about 5458 feet from the base of the cliff.   ●

 **Now Work** PROBLEM 11

Seeing the Concept

Graph

$$h(x) = \frac{-1}{5000}x^2 + x + 500$$
$$0 \le x \le 5500$$

Use MAXIMUM to find the maximum height of the projectile, and use ROOT or ZERO to find the distance from the base of the cliff to where it strikes the water. Compare your results with those obtained in Example 3.

EXAMPLE 4   **The Golden Gate Bridge**

The Golden Gate Bridge, a suspension bridge, spans the entrance to San Francisco Bay. Its 746-foot-tall towers are 4200 feet apart. The bridge is suspended from two huge cables more than 3 feet in diameter; the 90-foot-wide roadway is 220 feet above the water. The cables are parabolic in shape* and touch the road surface at the center of the bridge. Find the height of the cable above the road at a distance of 1000 feet from the center.

Solution   See Figure 28 on page 306. Begin by choosing the placement of the coordinate axes so that the x-axis coincides with the road surface and the origin coincides with the center of the bridge. As a result, the 746-foot towers will be vertical (height $746 - 220 = 526$ feet above the road) and located 2100 feet from the center. Also, the cable, which has the shape of a parabola, will extend from the towers, open up, and have its vertex at $(0,0)$. This choice of placement of the axes enables the equation of the parabola to have the form $y = ax^2, a > 0$. Note that the points $(-2100, 526)$ and $(2100, 526)$ are on the graph.

*A cable suspended from two towers is in the shape of a **catenary**, but when a horizontal roadway is suspended from the cable, the cable takes the shape of a parabola.

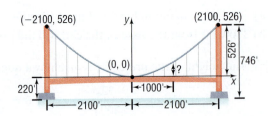

Figure 28

Use these facts to find the value of a in $y = ax^2$.

$$y = ax^2$$

$$526 = a(2100)^2 \qquad {\color{blue} x = 2100, y = 526}$$

$$a = \frac{526}{(2100)^2}$$

The equation of the parabola is

$$y = \frac{526}{(2100)^2}x^2$$

When $x = 1000$, the height of the cable is

$$y = \frac{526}{(2100)^2}(1000)^2 \approx 119.3 \text{ feet}$$

The cable is 119.3 feet above the road at a distance of 1000 feet from the center of the bridge. ●

 Now Work PROBLEM 13

 2 Build Quadratic Models from Data

In Section 4.2, we found the line of best fit for data that appeared to be linearly related. It was noted that data may also follow a nonlinear relation. Figures 29(a) and (b) show scatter diagrams of data that follow a quadratic relation.

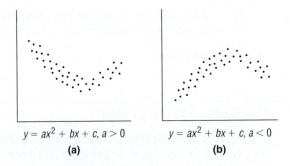

$y = ax^2 + bx + c, a > 0$ $y = ax^2 + bx + c, a < 0$

Figure 29   (a)   (b)

EXAMPLE 5

Fitting a Quadratic Function to Data

The data in Table 8 on page 307 represent the percentage D of the population that is divorced for various ages x in 2012.

(a) Draw a scatter diagram of the data, treating age as the independent variable. Comment on the type of relation that may exist between age and percentage of the population divorced.

(b) Use a graphing utility to find the quadratic function of best fit that models the relation between age and percentage of the population divorced.

(c) Use the model found in part (b) to approximate the age at which the percentage of the population divorced is greatest.

(d) Use the model found in part (b) to approximate the highest percentage of the population that is divorced.

(e) Use a graphing utility to draw the quadratic function of best fit on the scatter diagram.

Table 8

Age, x	Percentage Divorced, D
22	0.9
27	3.6
32	7.4
37	10.4
42	12.7
50	15.7
60	16.2
70	13.1
80	6.5

Source: United States Statistical Abstract, 2012

Solution

(a) Figure 30 shows the scatter diagram, from which it appears the data follow a quadratic relation, with $a < 0$.

(b) Execute the QUADratic REGression program to obtain the results shown in Figure 31. The output shows the equation $y = ax^2 + bx + c$. The quadratic function of best fit that models the relation between age and percentage divorced is

$$D(x) = -0.0143x^2 + 1.5861x - 28.1886 \quad \text{The model}$$

where x represents age and D represents the percentage divorced.

(c) Based on the quadratic function of best fit, the age with the greatest percentage divorced is

$$-\frac{b}{2a} = -\frac{1.5861}{2(-0.0143)} \approx 55 \text{ years}$$

(d) Evaluate the function $D(x)$ at $x = 55$.

$$D(55) = -0.0143(55)^2 + 1.5861(55) - 28.1886 \approx 15.8 \text{ percent}$$

According to the model, 55-year-olds have the highest percentage divorced at 15.8 percent.

(e) Figure 32 shows the graph of the quadratic function found in part (b) drawn on the scatter diagram. ●

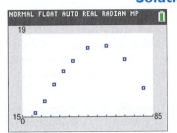

Figure 30

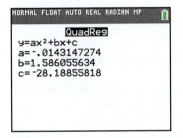

Figure 31

Look again at Figure 31. Notice that the output given by the graphing calculator does not include r, the correlation coefficient. Recall that the correlation coefficient is a measure of the strength of a linear relation that exists between two variables. The graphing calculator does not provide an indication of how well the function fits the data in terms of r, since a quadratic function cannot be expressed as a linear function.

━ **Now Work** PROBLEM 25

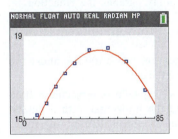

Figure 32

4.4 Assess Your Understanding

'Are You Prepared?' *Answers are given at the end of these exercises. If you get a wrong answer, read the pages listed in* red.

1. Translate the following sentence into a mathematical equation: The total revenue R from selling x hot dogs is $3 times the number of hot dogs sold. (pp. 134–140)

2. Use a graphing utility to find the line of best fit for the following data: (pp. 284–287)

x	3	5	5	6	7	8
y	10	13	12	15	16	19

Applications and Extensions

3. Maximizing Revenue The price p (in dollars) and the quantity x sold of a certain product obey the demand equation

$$p = -\frac{1}{6}x + 100$$

(a) Find a model that expresses the revenue R as a function of x. (Remember, $R = xp$.)
(b) What is the domain of R?
(c) What is the revenue if 200 units are sold?
(d) What quantity x maximizes revenue? What is the maximum revenue?
(e) What price should the company charge to maximize revenue?

4. Maximizing Revenue The price p (in dollars) and the quantity x sold of a certain product obey the demand equation

$$p = -\frac{1}{3}x + 100$$

(a) Find a model that expresses the revenue R as a function of x.
(b) What is the domain of R?
(c) What is the revenue if 100 units are sold?
(d) What quantity x maximizes revenue? What is the maximum revenue?
(e) What price should the company charge to maximize revenue?

5. Maximizing Revenue The price p (in dollars) and the quantity x sold of a certain product obey the demand equation
$$x = -5p + 100 \qquad 0 < p \le 20$$
(a) Express the revenue R as a function of x.
(b) What is the revenue if 15 units are sold?
(c) What quantity x maximizes revenue? What is the maximum revenue?
(d) What price should the company charge to maximize revenue?
(e) What price should the company charge to earn at least $480 in revenue?

6. Maximizing Revenue The price p (in dollars) and the quantity x sold of a certain product obey the demand equation
$$x = -20p + 500 \qquad 0 < p \le 25$$
(a) Express the revenue R as a function of x.
(b) What is the revenue if 20 units are sold?
(c) What quantity x maximizes revenue? What is the maximum revenue?
(d) What price should the company charge to maximize revenue?
(e) What price should the company charge to earn at least $3000 in revenue?

7. Enclosing a Rectangular Field David has 400 yards of fencing and wishes to enclose a rectangular area.
(a) Express the area A of the rectangle as a function of the width w of the rectangle.
(b) For what value of w is the area largest?
(c) What is the maximum area?

8. Enclosing a Rectangular Field Beth has 3000 feet of fencing available to enclose a rectangular field.
(a) Express the area A of the rectangle as a function of x, where x is the length of the rectangle.
(b) For what value of x is the area largest?
(c) What is the maximum area?

9. Enclosing the Most Area with a Fence A farmer with 4000 meters of fencing wants to enclose a rectangular plot that borders on a river. If the farmer does not fence the side along the river, what is the largest area that can be enclosed? (See the figure.)

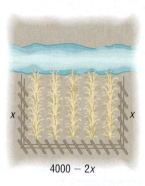

$4000 - 2x$

10. Enclosing the Most Area with a Fence A farmer with 2000 meters of fencing wants to enclose a rectangular plot that borders on a straight highway. If the farmer does not fence the side along the highway, what is the largest area that can be enclosed?

11. Analyzing the Motion of a Projectile A projectile is fired from a cliff 200 feet above the water at an inclination of 45° to the horizontal, with a muzzle velocity of 50 feet per second. The height h of the projectile above the water is modeled by

$$h(x) = \frac{-32x^2}{(50)^2} + x + 200$$

where x is the horizontal distance of the projectile from the face of the cliff.
(a) At what horizontal distance from the face of the cliff is the height of the projectile a maximum?
(b) Find the maximum height of the projectile.
(c) At what horizontal distance from the face of the cliff will the projectile strike the water?
(d) Using a graphing utility, graph the function h, $0 \le x \le 200$.
(e) Use a graphing utility to verify the solutions found in parts (b) and (c).
(f) When the height of the projectile is 100 feet above the water, how far is it from the cliff?

12. Analyzing the Motion of a Projectile A projectile is fired at an inclination of 45° to the horizontal, with a muzzle velocity of 100 feet per second. The height h of the projectile is modeled by

$$h(x) = \frac{-32x^2}{(100)^2} + x$$

where x is the horizontal distance of the projectile from the firing point.
(a) At what horizontal distance from the firing point is the height of the projectile a maximum?
(b) Find the maximum height of the projectile.
(c) At what horizontal distance from the firing point will the projectile strike the ground?
(d) Using a graphing utility, graph the function h, $0 \le x \le 350$.

(e) Use a graphing utility to verify the results obtained in parts (b) and (c).

(f) When the height of the projectile is 50 feet above the ground, how far has it traveled horizontally?

13. Suspension Bridge A suspension bridge with weight uniformly distributed along its length has twin towers that extend 75 meters above the road surface and are 400 meters apart. The cables are parabolic in shape and are suspended from the tops of the towers. The cables touch the road surface at the center of the bridge. Find the height of the cables at a point 100 meters from the center. (Assume that the road is level.)

14. Architecture A parabolic arch has a span of 120 feet and a maximum height of 25 feet. Choose suitable rectangular coordinate axes and find the equation of the parabola. Then calculate the height of the arch at points 10 feet, 20 feet, and 40 feet from the center.

15. Constructing Rain Gutters A rain gutter is to be made of aluminum sheets that are 12 inches wide by turning up the edges 90°. See the illustration.

(a) What depth will provide maximum cross-sectional area and hence allow the most water to flow?

(b) What depths will allow at least 16 square inches of water to flow?

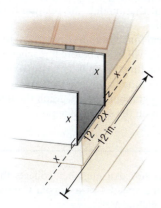

16. Norman Windows A **Norman window** has the shape of a rectangle surmounted by a semicircle of diameter equal to the width of the rectangle. See the figure. If the perimeter of the window is 20 feet, what dimensions will admit the most light (maximize the area)?

[**Hint:** Circumference of a circle $= 2\pi r$; area of a circle $= \pi r^2$, where r is the radius of the circle.]

17. Constructing a Stadium A track-and-field playing area is in the shape of a rectangle with semicircles at each end. See the figure (top, right). The inside perimeter of the track is to be 1500 meters. What should the dimensions of the rectangle be so that the area of the rectangle is a maximum?

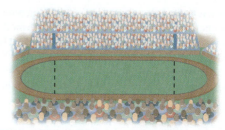

18. Architecture A special window has the shape of a rectangle surmounted by an equilateral triangle. See the figure. If the perimeter of the window is 16 feet, what dimensions will admit the most light?

[**Hint:** Area of an equilateral triangle $= \left(\dfrac{\sqrt{3}}{4}\right)x^2$, where x is the length of a side of the triangle.]

19. Chemical Reactions A self-catalytic chemical reaction results in the formation of a compound that causes the formation ratio to increase. If the reaction rate V is modeled by

$$V(x) = kx(a - x), \qquad 0 \le x \le a$$

where k is a positive constant, a is the initial amount of the compound, and x is the variable amount of the compound, for what value of x is the reaction rate a maximum?

20. Calculus: Simpson's Rule The figure shows the graph of $y = ax^2 + bx + c$. Suppose that the points $(-h, y_0)$, $(0, y_1)$, and (h, y_2) are on the graph. It can be shown that the area enclosed by the parabola, the x-axis, and the lines $x = -h$ and $x = h$ is

$$\text{Area} = \frac{h}{3}(2ah^2 + 6c)$$

Show that this area may also be given by

$$\text{Area} = \frac{h}{3}(y_0 + 4y_1 + y_2)$$

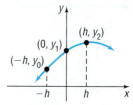

21. Use the result obtained in Problem 20 to find the area enclosed by $f(x) = -5x^2 + 8$, the x-axis, and the lines $x = -1$ and $x = 1$.

22. Use the result obtained in Problem 20 to find the area enclosed by $f(x) = 2x^2 + 8$, the x-axis, and the lines $x = -2$ and $x = 2$.

23. Use the result obtained in Problem 20 to find the area enclosed by $f(x) = x^2 + 3x + 5$, the x-axis, and the lines $x = -4$ and $x = 4$.

24. Use the result obtained in Problem 20 to find the area enclosed by $f(x) = -x^2 + x + 4$, the x-axis, and the lines $x = -1$ and $x = 1$.

25. Life Cycle Hypothesis An individual's income varies with his or her age. The following table shows the median income *I* of males of different age groups within the United States for 2012. For each age group, let the class midpoint represent the independent variable, *x*. For the class "65 years and older," we will assume that the class midpoint is 69.5.

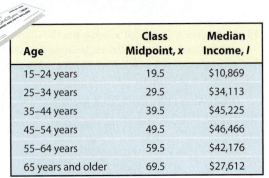

Age	Class Midpoint, x	Median Income, I
15–24 years	19.5	$10,869
25–34 years	29.5	$34,113
35–44 years	39.5	$45,225
45–54 years	49.5	$46,466
55–64 years	59.5	$42,176
65 years and older	69.5	$27,612

Source: U.S. Census Bureau

(a) Use a graphing utility to draw a scatter diagram of the data. Comment on the type of relation that may exist between the two variables.
(b) Use a graphing utility to find the quadratic function of best fit that models the relation between age and median income.
(c) Use the function found in part (b) to determine the age at which an individual can expect to earn the most income.
(d) Use the function found in part (b) to predict the peak income earned.
(e) With a graphing utility, graph the quadratic function of best fit on the scatter diagram.

26. Height of a Ball A shot-putter throws a ball at an inclination of 45° to the horizontal. The following data represent the height of the ball *h*, in feet, at the instant that it has traveled *x* feet horizontally.

Distance, x	Height, h
20	25
40	40
60	55
80	65
100	71
120	77
140	77
160	75
180	71
200	64

(a) Use a graphing utility to draw a scatter diagram of the data. Comment on the type of relation that may exist between the two variables.
(b) Use a graphing utility to find the quadratic function of best fit that models the relation between distance and height.
(c) Use the function found in part (b) to determine how far the ball will travel before it reaches its maximum height.
(d) Use the function found in part (b) to find the maximum height of the ball.
(e) With a graphing utility, graph the quadratic function of best fit on the scatter diagram.

Mixed Practice

27. Which Model? The following data represent the square footage and rents (dollars per month) for apartments in the La Jolla area of San Diego, California.

Square Footage, x	Rent per Month, R
520	$1525
621	$1750
718	$1785
753	$1850
850	$1900
968	$2130
1020	$2180

Source: apartments.com, 2014

(a) Using a graphing utility, draw a scatter diagram of the data treating square footage as the independent variable. What type of relation appears to exist between square footage and rent?
(b) Based on your response to part (a), find either a linear or a quadratic model that describes the relation between square footage and rent.
(c) Use your model to predict the rent for an apartment in San Diego that is 875 square feet.

28. Which Model? An engineer collects the following data showing the speed *s* of a Toyota Camry and its average miles per gallon, *M*.

Speed, s	Miles per Gallon, M
30	18
35	20
40	23
40	25
45	25
50	28
55	30
60	29
65	26
65	25
70	25

(a) Using a graphing utility, draw a scatter diagram of the data, treating speed as the independent variable. What type of relation appears to exist between speed and miles per gallon?

(b) Based on your response to part (a), find either a linear model or a quadratic model that describes the relation between speed and miles per gallon.

(c) Use your model to predict the miles per gallon for a Camry that is traveling 63 miles per hour.

29. Which Model? The following data represent the birth rate (births per 1000 population) for women whose age is a, in 2012.

Age, a	Birth Rate, B
16	14.1
19	51.4
22	83.1
27	106.5
32	97.3
37	48.3
42	10.4

Source: National Vital Statistics System, 2013

(a) Using a graphing utility, draw a scatter diagram of the data, treating age as the independent variable. What type of relation appears to exist between age and birth rate?

(b) Based on your response to part (a), find either a linear or a quadratic model that describes the relation between age and birth rate.

(c) Use your model to predict the birth rate for 35-year-old women.

30. Which Model? A cricket makes a chirping noise by sliding its wings together rapidly. Perhaps you have noticed that the number of chirps seems to increase with the temperature. The following data list the temperature (in degrees Fahrenheit) and the number of chirps per second for the striped ground cricket.

Temperature (°F), x	Chirps per Second, C
88.6	20.0
93.3	19.8
80.6	17.1
69.7	14.7
69.4	15.4
79.6	15.0
80.6	16.0
76.3	14.4
75.2	15.5

Source: Pierce, George W. *The Songs of Insects.* Cambridge, MA Harvard University Press, 1949, pp. 12 – 21

(a) Using a graphing utility, draw a scatter diagram of the data, treating temperature as the independent variable. What type of relation appears to exist between temperature and chirps per second?

(b) Based on your response to part (a), find either a linear or a quadratic model that best describes the relation between temperature and chirps per second.

(c) Use your model to predict the chirps per second if the temperature is 80°F.

Explaining Concepts: Discussion and Writing

31. Refer to Example 1 in this section. Notice that if the price charged for the calculators is $0 or $140, then the revenue is $0. It is easy to explain why revenue would be $0 if the price charged were $0, but how can revenue be $0 if the price charged is $140?

Retain Your Knowledge

Problems 32–35 are based on material learned earlier in the course. The purpose of these problems is to keep the material fresh in your mind so that you are better prepared for the final exam.

32. Express as a complex number: $\sqrt{-225}$

33. Find the distance between the points $P_1 = (4, -7)$ and $P_2 = (-1, 5)$.

34. Find the equation of the circle with center $(-6, 0)$ and radius $r = \sqrt{7}$.

35. Solve: $5x^2 + 8x - 3 = 0$

'Are You Prepared?' Answers

1. $R = 3x$ **2.** $y = 1.7826x + 4.0652$

4.5 Inequalities Involving Quadratic Functions

PREPARING FOR THIS SECTION *Before getting started, review the following:*

• Solve Inequalities (Section 1.5, pp. 123–126)   • Use Interval Notation (Section 1.5, pp. 120–121)

Now Work the **'Are You Prepared?'** problems on page 314.

OBJECTIVE 1 Solve Inequalities Involving a Quadratic Function (p. 312)

1 Solve Inequalities Involving a Quadratic Function

In this section we solve inequalities that involve quadratic functions. We will accomplish this by using their graphs. For example, to solve the inequality

$$ax^2 + bx + c > 0 \qquad a \neq 0$$

graph the function $f(x) = ax^2 + bx + c$ and, from the graph, determine where it is above the x-axis—that is, where $f(x) > 0$. To solve the inequality $ax^2 + bx + c < 0, a \neq 0$, graph the function $f(x) = ax^2 + bx + c$ and determine where the graph is below the x-axis. If the inequality is not strict, include the x-intercepts, if any, in the solution.

EXAMPLE 1

Solving an Inequality

Solve the inequality $x^2 - 4x - 12 \leq 0$ and graph the solution set.

Solution

Graph the function $f(x) = x^2 - 4x - 12$.

y-intercept: $\qquad\qquad\qquad\quad f(0) = -12$ *Evaluate f at 0.*

x-intercepts (if any): $\quad x^2 - 4x - 12 = 0$ *Solve f(x) = 0.*

$\qquad\qquad\qquad (x - 6)(x + 2) = 0$ *Factor.*

$\qquad\qquad\quad x - 6 = 0 \quad\text{or}\quad x + 2 = 0$ *Apply the Zero-Product Property.*

$\qquad\qquad\qquad x = 6 \quad\text{or}\qquad x = -2$

The y-intercept is -12; the x-intercepts are -2 and 6.

The vertex is at $x = -\dfrac{b}{2a} = -\dfrac{-4}{2} = 2$. Because $f(2) = -16$, the vertex is at $(2, -16)$.

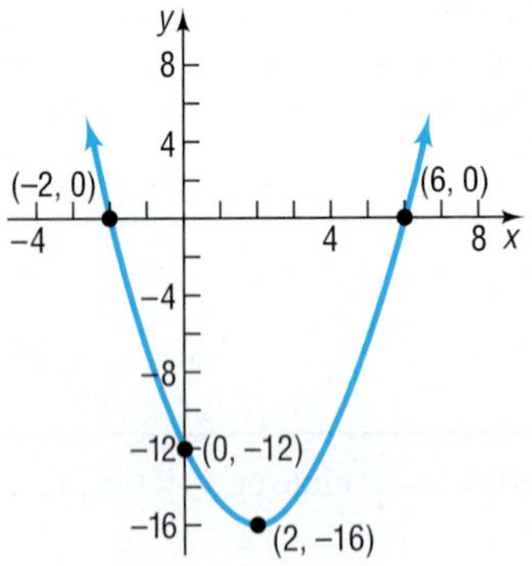

See Figure 33 for the graph.

The graph is below the x-axis for $-2 < x < 6$. Because the original inequality is not strict, include the x-intercepts. The solution set is $\{x | -2 \leq x \leq 6\}$ or, using interval notation, $[-2, 6]$. See Figure 34 for the graph of the solution set. ●

Figure 33 $f(x) = x^2 - 4x - 12$

Figure 34

Now Work PROBLEM 9

EXAMPLE 2

Solving an Inequality

Solve the inequality $2x^2 < x + 10$ and graph the solution set.

Solution

Option 1 Rearrange the inequality so that 0 is on the right side.

$$2x^2 < x + 10$$

$$2x^2 - x - 10 < 0 \qquad \text{Subtract x + 10 from both sides.}$$

This inequality is equivalent to the original inequality.

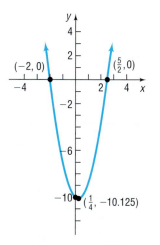

Figure 35 $f(x) = 2x^2 - x - 10$

Next graph the function $f(x) = 2x^2 - x - 10$ to find where $f(x) < 0$.

y-intercept: $\qquad$ $f(0) = -10$ $\qquad$ Evaluate f at 0.

x-intercepts (if any): $\qquad$ $2x^2 - x - 10 = 0$ $\qquad$ Solve $f(x) = 0$.

$$(2x - 5)(x + 2) = 0 \qquad \text{Factor.}$$

$$2x - 5 = 0 \quad \text{or} \quad x + 2 = 0 \qquad \text{Apply the Zero-Product Property.}$$

$$x = \frac{5}{2} \quad \text{or} \quad x = -2$$

The y-intercept is -10; the x-intercepts are -2 and $\dfrac{5}{2}$.

The vertex is at $x = -\dfrac{b}{2a} = -\dfrac{-1}{4} = \dfrac{1}{4}$. Because $f\left(\dfrac{1}{4}\right) = -10.125$, the vertex is $\left(\dfrac{1}{4}, -10.125\right)$. See Figure 35 for the graph.

The graph is below the x-axis ($f(x) < 0$) between $x = -2$ and $x = \dfrac{5}{2}$. Because the inequality is strict, the solution set is $\left\{x \,\middle|\, -2 < x < \dfrac{5}{2}\right\}$ or, using interval notation, $\left(-2, \dfrac{5}{2}\right)$.

Option 2 If $f(x) = 2x^2$ and $g(x) = x + 10$, then the inequality to be solved is $f(x) < g(x)$. Graph the functions $f(x) = 2x^2$ and $g(x) = x + 10$. See Figure 36. The graphs intersect where $f(x) = g(x)$. Then

$$2x^2 = x + 10 \qquad f(x) = g(x)$$

$$2x^2 - x - 10 = 0$$

$$(2x - 5)(x + 2) = 0 \qquad \text{Factor.}$$

$$2x - 5 = 0 \quad \text{or} \quad x + 2 = 0 \qquad \text{Apply the Zero-Product Property.}$$

$$x = \frac{5}{2} \quad \text{or} \qquad x = -2$$

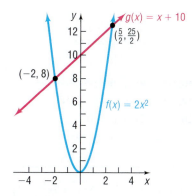

Figure 36

The graphs intersect at the points $(-2, 8)$ and $\left(\dfrac{5}{2}, \dfrac{25}{2}\right)$. To solve $f(x) < g(x)$, find where the graph of f is below the graph of g. This happens between the points of intersection. Because the inequality is strict, the solution set is $\left\{x \,\middle|\, -2 < x < \dfrac{5}{2}\right\}$ or, using interval notation, $\left(-2, \dfrac{5}{2}\right)$.

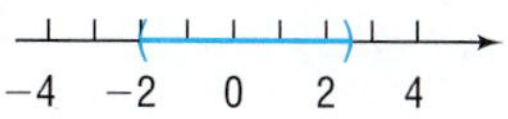

See Figure 37 for the graph of the solution set. $\qquad\qquad\qquad\qquad$ ●

Figure 37

Now Work PROBLEMS 5 AND 13

EXAMPLE 3 | **Solving an Inequality**

Solve the inequality $x^2 + x + 1 > 0$ and graph the solution set.

Solution Graph the function $f(x) = x^2 + x + 1$. The y-intercept is 1; there are no x-intercepts (Do you see why? Check the discriminant). The vertex is at $x = -\dfrac{b}{2a} = -\dfrac{1}{2}$. Since $f\left(-\dfrac{1}{2}\right) = \dfrac{3}{4}$, the vertex is at $\left(-\dfrac{1}{2}, \dfrac{3}{4}\right)$. The points $(1, 3)$ and $(-1, 1)$ are also on the graph. See Figure 38.

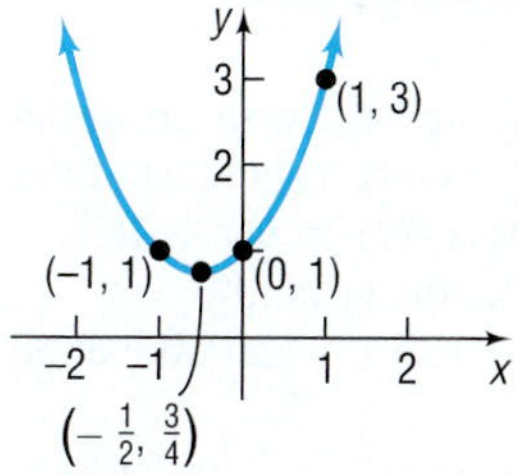

The graph of f lies above the x-axis for all x. The solution set is the set of all real numbers, or $(-\infty, \infty)$. See Figure 39. $\qquad\qquad\qquad$ ●

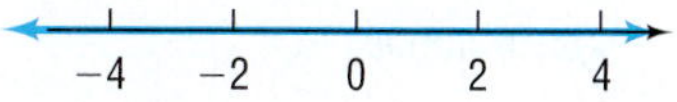

Figure 38 $f(x) = x^2 + x + 1$

Figure 39

Now Work PROBLEM 17

4.5 Assess Your Understanding

'Are You Prepared?' *Answers are given at the end of these exercises. If you get a wrong answer, read the pages listed in red.*

1. Solve the inequality $-3x - 2 < 7$ (pp. 123–126)

2. Write $(-2, 7]$ using inequality notation. (pp. 120–121)

Skill Building

In Problems 3–6, use the figure to solve each inequality.

3.
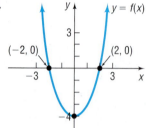

4.
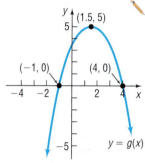

5.
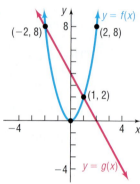

6.
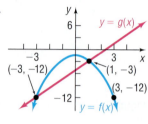

(a) $f(x) > 0$
(b) $f(x) \le 0$

(a) $g(x) < 0$
(b) $g(x) \ge 0$

(a) $g(x) \ge f(x)$
(b) $f(x) > g(x)$

(a) $f(x) < g(x)$
(b) $f(x) \ge g(x)$

In Problems 7–22, solve each inequality.

7. $x^2 - 3x - 10 < 0$

8. $x^2 + 3x - 10 > 0$

9. $x^2 - 4x > 0$

10. $x^2 + 8x > 0$

11. $x^2 - 9 < 0$

12. $x^2 - 1 < 0$

13. $x^2 + x > 12$

14. $x^2 + 7x < -12$

15. $2x^2 < 5x + 3$

16. $6x^2 < 6 + 5x$

17. $x^2 - x + 1 \le 0$

18. $x^2 + 2x + 4 > 0$

19. $4x^2 + 9 < 6x$

20. $25x^2 + 16 < 40x$

21. $6(x^2 - 1) > 5x$

22. $2(2x^2 - 3x) > -9$

Mixed Practice

23. What is the domain of the function $f(x) = \sqrt{x^2 - 16}$?

24. What is the domain of the function $f(x) = \sqrt{x - 3x^2}$?

In Problems 25–32, use the given functions f and g.

(a) *Solve* $f(x) = 0$.
(b) *Solve* $g(x) = 0$.
(c) *Solve* $f(x) = g(x)$.
(d) *Solve* $f(x) > 0$.
(e) *Solve* $g(x) \le 0$.
(f) *Solve* $f(x) > g(x)$.
(g) *Solve* $f(x) \ge 1$.

25. $f(x) = x^2 - 1$
$g(x) = 3x + 3$

26. $f(x) = -x^2 + 3$
$g(x) = -3x + 3$

27. $f(x) = -x^2 + 1$
$g(x) = 4x + 1$

28. $f(x) = -x^2 + 4$
$g(x) = -x - 2$

29. $f(x) = x^2 - 4$
$g(x) = -x^2 + 4$

30. $f(x) = x^2 - 2x + 1$
$g(x) = -x^2 + 1$

31. $f(x) = x^2 - x - 2$
$g(x) = x^2 + x - 2$

32. $f(x) = -x^2 - x + 1$
$g(x) = -x^2 + x + 6$

Applications and Extensions

33. Physics A ball is thrown vertically upward with an initial velocity of 80 feet per second. The distance *s* (in feet) of the ball from the ground after *t* seconds is $s(t) = 80t - 16t^2$.

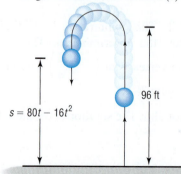

$s = 80t - 16t^2$

96 ft

(a) At what time *t* will the ball strike the ground?
(b) For what time *t* is the ball more than 96 feet above the ground?

34. Physics A ball is thrown vertically upward with an initial velocity of 96 feet per second. The distance *s* (in feet) of the ball from the ground after *t* seconds is $s(t) = 96t - 16t^2$.
(a) At what time *t* will the ball strike the ground?
(b) For what time *t* is the ball more than 128 feet above the ground?

35. Revenue Suppose that the manufacturer of a gas clothes dryer has found that when the unit price is *p* dollars, the revenue *R* (in dollars) is

$$R(p) = -4p^2 + 4000p$$

(a) At what prices p is revenue zero?

(b) For what range of prices will revenue exceed $800,000?

36. Revenue The John Deere company has found that the revenue from sales of heavy-duty tractors is a function of the unit price p, in dollars, that it charges. The revenue R, in dollars, is given by

$$R(p) = -\frac{1}{2}p^2 + 1900p$$

(a) At what prices p is revenue zero?

(b) For what range of prices will revenue exceed $1,200,000?

37. Artillery A projectile fired from the point $(0, 0)$ at an angle to the positive x-axis has a trajectory given by

$$y = cx - (1 + c^2)\left(\frac{g}{2}\right)\left(\frac{x}{v}\right)^2$$

where

x = horizontal distance in meters

y = height in meters

v = initial muzzle velocity in meters per second (m/sec)

g = acceleration due to gravity = 9.81 meters per second squared (m/sec^2)

$c > 0$ is a constant determined by the angle of elevation.

A howitzer fires an artillery round with a muzzle velocity of 897 m/sec.

(a) If the round must clear a hill 200 meters high at a distance of 2000 meters in front of the howitzer, what c values are permitted in the trajectory equation?

(b) If the goal in part (a) is to hit a target on the ground 75 kilometers away, is it possible to do so? If so, for what values of c? If not, what is the maximum distance the round will travel?

Source: www.answers.com

38. Runaway Car Using Hooke's Law, we can show that the *work* done in compressing a spring a distance of x feet from its at-rest position is $W = \frac{1}{2}kx^2$, where k is a stiffness constant depending on the spring. It can also be shown that the work done by a body in motion before it comes to rest is given by $\tilde{W} = \frac{w}{2g}v^2$, where w = weight of the object (in lbs), g = acceleration due to gravity (32.2 ft/sec^2), and v = object's velocity (in ft/sec). A parking garage has a spring shock absorber at the end of a ramp to stop runaway cars. The spring has a stiffness constant $k = 9450$ lb/ft and must be able to stop a 4000-lb car traveling at 25 mph. What is the least compression required of the spring? Express your answer using feet to the nearest tenth.

[**Hint:** Solve $W > \tilde{W}, x \geq 0$].

Source: www.sciforums.com

Explaining Concepts: Discussion and Writing

39. Show that the inequality $(x - 4)^2 \leq 0$ has exactly one solution.

40. Show that the inequality $(x - 2)^2 > 0$ has one real number that is not a solution.

41. Explain why the inequality $x^2 + x + 1 > 0$ has all real numbers as the solution set.

42. Explain why the inequality $x^2 - x + 1 < 0$ has the empty set as the solution set.

43. Explain the circumstances under which the x-intercepts of the graph of a quadratic function are included in the solution set of a quadratic inequality.

Retain Your Knowledge

Problems 44–47 are based on material learned earlier in the course. The purpose of these problems is to keep the material fresh in your mind so that you are better prepared for the final exam.

44. Determine the domain of $f(x) = \sqrt{10 - 2x}$.

45. Determine algebraically whether $f(x) = \dfrac{-x}{x^2 + 9}$ is even, odd, or neither.

46. Consider the linear function $f(x) = \dfrac{2}{3}x - 6$.

(a) Find the intercepts of the graph of f.

(b) Graph f.

47. Multiply $(6 - 5i)(4 - i)$. Write the answer in the form $a + bi$.

'Are You Prepared?' Answers

1. $\{x | x > -3\}$ or $(-3, \infty)$

2. $-2 < x \leq 7$

Chapter Review

Things to Know

Linear function (p. 274)

$f(x) = mx + b$ Average rate of change = m

The graph is a line with slope m and y-intercept b.

Quadratic function (pp. 291–295)

$f(x) = ax^2 + bx + c, a \neq 0$

The graph is a parabola that opens up if $a > 0$ and opens down if $a < 0$.

Vertex: $\left(-\dfrac{b}{2a}, f\left(-\dfrac{b}{2a}\right)\right)$

Axis of symmetry: $x = -\dfrac{b}{2a}$

y-intercept: $f(0) = c$

x-intercept(s): If any, found by finding the real solutions of the equation $ax^2 + bx + c = 0$

Objectives

Section		You should be able to...	Examples	Review Exercises
4.1	1	Graph linear functions (p. 274)	1	1(a)–3(a), 1(c)–3(c)
	2	Use average rate of change to identify linear functions (p. 274)	2	1(b)–3(b), 4, 5
	3	Determine whether a linear function is increasing, decreasing, or constant (p. 277)	3	1(d)–3(d)
	4	Build linear models from verbal descriptions (p. 278)	4, 5	21
4.2	1	Draw and interpret scatter diagrams (p. 284)	1	29(a), 30(a)
	2	Distinguish between linear and nonlinear relations (p. 285)	2	29(b), 30(a)
	3	Use a graphing utility to find the line of best fit (p. 286)	4	29(c)
4.3	1	Graph a quadratic function using transformations (p. 292)	1	6–8
	2	Identify the vertex and axis of symmetry of a quadratic function (p. 294)	2	9–13
	3	Graph a quadratic function using its vertex, axis, and intercepts (p. 294)	3–5	9–13
	4	Find a quadratic function given its vertex and one other point (p. 297)	6	19, 20
	5	Find the maximum or minimum value of a quadratic function (p. 298)	7	14–16, 22–27
4.4	1	Build quadratic models from verbal descriptions (p. 302)	1–4	22–28
	2	Build quadratic models from data (p. 306)	5	30
4.5	3	Solve inequalities involving a quadratic function (p. 312)	1–3	17, 18

Review Exercises

In Problems 1–3:

   (a) Determine the slope and y-intercept of each linear function.
   (b) Find the average rate of change of each function.
   (c) Graph each function. Label the intercepts.
   (d) Determine whether the function is increasing, decreasing, or constant.

1. $f(x) = 2x - 5$ **2.** $h(x) = \dfrac{4}{5}x - 6$ **3.** $G(x) = 4$

In Problems 4 and 5, determine whether the function is linear or nonlinear. If the function is linear, state its slope.

4.

x	y = f(x)
−1	−2
0	3
1	8
2	13
3	18

5.

x	y = g(x)
−1	−3
0	4
1	7
2	6
3	1

In Problems 6–8, graph each quadratic function using transformations (shifting, compressing, stretching, and/or reflecting).

6. $f(x) = (x + 1)^2 - 4$ **7.** $f(x) = -(x - 4)^2$ **8.** $f(x) = -3(x + 2)^2 + 1$

In Problems 9–13, (a) graph each quadratic function by determining whether its graph opens up or down and by finding its vertex, axis of symmetry, y-intercept, and x-intercepts, if any. (b) Determine the domain and the range of the function. (c) Determine where the function is increasing and where it is decreasing.

9. $f(x) = (x - 2)^2 + 2$

10. $f(x) = \frac{1}{4}x^2 - 16$

11. $f(x) = -4x^2 + 4x$

12. $f(x) = \frac{9}{2}x^2 + 3x + 1$

13. $f(x) = 3x^2 + 4x - 1$

In Problems 14–16, determine whether the given quadratic function has a maximum value or a minimum value, and then find the value.

14. $f(x) = 3x^2 - 6x + 4$

15. $f(x) = -x^2 + 8x - 4$

16. $f(x) = -3x^2 + 12x + 4$

In Problems 17 and 18, solve each quadratic inequality.

17. $x^2 + 6x - 16 < 0$

18. $3x^2 \geq 14x + 5$

In Problems 19 and 20, find the quadratic function for which:

19. Vertex is $(2, -4)$; y-intercept is -16

20. Vertex is $(-1, 2)$; contains the point $(1, 6)$

21. Sales Commissions Bill was just offered a sales position for a computer company. His salary would be $25,000 per year plus 1% of his total annual sales.
  (a) Find a linear function that relates Bill's annual salary, S, to his total annual sales, x.
  (b) If Bill's total annual sales were $1,000,000, what would be Bill's salary?
  (c) What would Bill have to sell to earn $100,000?
  (d) Determine the sales required of Bill for his salary to exceed $150,000.

22. Demand Equation The price p (in dollars) and the quantity x sold of a certain product obey the demand equation

$$p = -\frac{1}{10}x + 150 \qquad 0 \leq x \leq 1500$$

  (a) Express the revenue R as a function of x.
  (b) What is the revenue if 100 units are sold?
  (c) What quantity x maximizes revenue? What is the maximum revenue?
  (d) What price should the company charge to maximize revenue?

23. Landscaping A landscape engineer has 200 feet of border to enclose a rectangular pond. What dimensions will result in the largest pond?

24. Enclosing the Most Area with a Fence A farmer with 10,000 meters of fencing wants to enclose a rectangular field and then divide it into two plots with a fence parallel to one of the sides. See the figure. What is the largest area that can be enclosed?

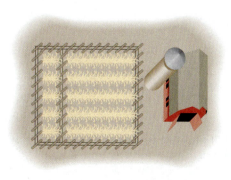

25. Architecture A special window in the shape of a rectangle with semicircles at each end is to be constructed so that the outside perimeter is 100 feet. See the illustration. Find the dimensions of the rectangle that maximizes its area.

26. Minimizing Marginal Cost Callaway Golf Company has determined that the marginal cost C of manufacturing x Big Bertha golf clubs may be expressed by the quadratic function

$$C(x) = 4.9x^2 - 617.4x + 19,600$$

  (a) How many clubs should be manufactured to minimize the marginal cost?
  (b) At this level of production, what is the marginal cost?

27. Maximizing Area A rectangle has one vertex on the line $y = 10 - x$, $x > 0$, another at the origin, one on the positive x-axis, and one on the positive y-axis. Express the area A of the rectangle as a function of x. Find the largest area A that can be enclosed by the rectangle.

28. Parabolic Arch Bridge A horizontal bridge is in the shape of a parabolic arch. Given the information shown in the figure, what is the height h of the arch 2 feet from shore?

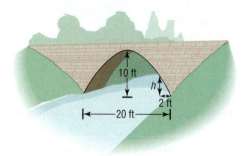

29. Bone Length Research performed at NASA, led by Dr. Emily R. Morey-Holton, measured the lengths of the right humerus and right tibia in 11 rats that were sent to space on Spacelab Life Sciences 2. The data on page 318 were collected.
  (a) Draw a scatter diagram of the data, treating length of the right humerus as the independent variable.
  (b) Based on the scatter diagram, do you think that there is a linear relation between the length of the right humerus and the length of the right tibia?

(c) Use a graphing utility to find the line of best fit relating length of the right humerus and length of the right tibia.

(d) Predict the length of the right tibia on a rat whose right humerus is 26.5 millimeters (mm).

Right Humerus (mm), x	Right Tibia (mm), y
24.80	36.05
24.59	35.57
24.59	35.57
24.29	34.58
23.81	34.20
24.87	34.73
25.90	37.38
26.11	37.96
26.63	37.46
26.31	37.75
26.84	38.50

Source: NASA Life Sciences Data Archive

30. Advertising  A small manufacturing firm collected the following data on advertising expenditures A (in thousands of dollars) and total revenue R (in thousands of dollars).

(a) Draw a scatter diagram of the data. Comment on the type of relation that may exist between the two variables.

Advertising Expenditures ($1000s)	Total Revenue ($1000s)
20	6101
22	6222
25	6350
25	6378
27	6453
28	6423
29	6360
31	6231

(b) The quadratic function of best fit to these data is

$$R(A) = -7.76A^2 + 411.88A + 942.72$$

Use this function to determine the optimal level of advertising.

(c) Use the function to predict the total revenue when the optimal level of advertising is spent.

(d) Use a graphing utility to verify that the function given in part (b) is the quadratic function of best fit.

(e) Use a graphing utility to draw a scatter diagram of the data, and then graph the quadratic function of best fit on the scatter diagram.

Chapter Test

CHAPTER
Test Prep
VIDEOS

The Chapter Test Prep Videos are step-by-step solutions available in *MyMathLab*, or on this text's **YouTube** Channel. Flip back to the Resources for Success page for a link to this text's YouTube channel.

1. Consider the linear function $f(x) = -4x + 3$:
   (a) Find the slope and y-intercept.
   (b) What is the average rate of change of f?
   (c) Determine whether f is increasing, decreasing, or constant.
   (d) Graph f.

In Problems 2 and 3, find the intercepts, if any, of each quadratic function.

2. $f(x) = 3x^2 - 2x - 8$

3. $G(x) = -2x^2 + 4x + 1$

4. Given that $f(x) = x^2 + 3x$ and $g(x) = 5x + 3$, solve $f(x) = g(x)$. Graph each function and label the points of intersection.

5. Graph $f(x) = (x - 3)^2 - 2$ using transformations.

6. Consider the quadratic function $f(x) = 3x^2 - 12x + 4$:
   (a) Determine whether the graph opens up or down.
   (b) Determine the vertex.
   (c) Determine the axis of symmetry.
   (d) Determine the intercepts.
   (e) Use the information from parts (a)–(d) to graph f.

7. Determine whether $f(x) = -2x^2 + 12x + 3$ has a maximum or a minimum. Then find the maximum or minimum value.

8. Solve $x^2 - 10x + 24 \geq 0$.

9. **RV Rental** The weekly rental cost of a 20-foot recreational vehicle is $129.50 plus $0.15 per mile.
   (a) Find a linear function that expresses the cost C as a function of miles driven m.
   (b) What is the rental cost if 860 miles are driven?
   (c) How many miles were driven if the rental cost is $213.80?

Cumulative Review

1. Find the distance between the points $P = (-1, 3)$ and $Q = (4, -2)$. Find the midpoint of the line segment P to Q.

2. Which of the following points are on the graph of $y = x^3 - 3x + 1$?
   (a) $(-2, -1)$ (b) $(2, 3)$ (c) $(3, 1)$

3. Solve the inequality $5x + 3 \geq 0$ and graph the solution set.

4. Find the equation of the line containing the points $(-1, 4)$ and $(2, -2)$. Express your answer in slope–intercept form and graph the line.

5. Find the equation of the line perpendicular to the line $y = 2x + 1$ and containing the point $(3, 5)$. Express your answer in slope–intercept form and graph both lines.

6. Graph the equation $x^2 + y^2 - 4x + 8y - 5 = 0$.

7. Does the following relation represent a function?
   $\{(-3, 8), (1, 3), (2, 5), (3, 8)\}$.

8. For the function f defined by $f(x) = x^2 - 4x + 1$, find:
   (a) $f(2)$ (b) $f(x) + f(2)$
   (c) $f(-x)$ (d) $-f(x)$
   (e) $f(x + 2)$ (f) $\dfrac{f(x + h) - f(x)}{h}$, $h \neq 0$

9. Find the domain of $h(z) = \dfrac{3z - 1}{6z - 7}$.

10. Is the following graph the graph of a function?

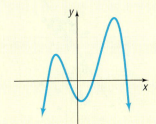

11. Consider the function $f(x) = \dfrac{x}{x + 4}$.
    (a) Is the point $\left(1, \dfrac{1}{4}\right)$ on the graph of f?
    (b) If $x = -2$, what is $f(x)$? What point is on the graph of f?
    (c) If $f(x) = 2$, what is x? What point is on the graph of f?

12. Is the function $f(x) = \dfrac{x^2}{2x + 1}$ even, odd, or neither?

13. Approximate the local maximum values and local minimum values of $f(x) = x^3 - 5x + 1$ on $(-4, 4)$. Determine where the function is increasing and where it is decreasing.

14. If $f(x) = 3x + 5$ and $g(x) = 2x + 1$:
    (a) Solve $f(x) = g(x)$.
    (b) Solve $f(x) > g(x)$.

15. Consider the graph below of the function f.
    (a) Find the domain and the range of f.
    (b) Find the intercepts.
    (c) Is the graph of f symmetric with respect to the x-axis, the y-axis, or the origin?
    (d) Find $f(2)$.
    (e) For what value(s) of x is $f(x) = 3$?
    (f) Solve $f(x) < 0$.
    (g) Graph $y = f(x) + 2$.
    (h) Graph $y = f(-x)$.
    (i) Graph $y = 2f(x)$.
    (j) Is f even, odd, or neither?
    (k) Find the interval(s) on which f is increasing.

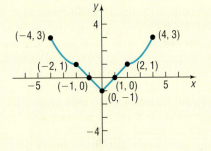

Chapter Projects

📡 **Internet-based Project**

I. The Beta of a Stock You want to invest in the stock market but are not sure which stock to purchase. Information is the key to making an informed investment decision. One piece of information that many stock analysts use is the beta of the stock. Go to Wikipedia (*http://en.wikipedia.org/wiki/Beta_%28finance%29*) and research what beta measures and what it represents.

1. **Approximating the beta of a stock.** Choose a well-known company such as Google or Coca-Cola. Go to a website such as Yahoo! Finance (*http://finance.yahoo.com/*) and find the weekly closing price of the company's stock for the past year. Then find the closing price of the Standard & Poor's 500 (S&P500) for the same time period.

   To get the historical prices in Yahoo! Finance, select Historical Prices from the left menu. Choose the appropriate time period. Select Weekly and Get Prices. Finally, select Download to Spreadsheet. Repeat this for the S&P500, and copy the data into the same spreadsheet. Finally, rearrange the data in chronological order. Be sure to expand the selection to sort all the data. Now, using the adjusted close price, compute the percentage change in price for each week, using the formula $\% \text{ change} = \dfrac{P_1 - P_o}{P_o}$. For example, if week 1 price is in cell D1 and week 2 price is in cell D2, then $\% \text{ change} = \dfrac{D2 - D1}{D1}$. Repeat this for the S&P500 data.

2. **Using Excel to draw a scatter diagram.** Treat the percentage change in the S&P500 as the independent variable and the percentage change in the stock you chose as the dependent variable. The easiest way to draw a scatter diagram in Excel is to place the two columns of data next to each other (for example, have the percentage change in the S&P500 in column F and the percentage change in the stock you chose in column G). Then highlight the data and select the Scatter Diagram icon under Insert. Comment on the type of relation that appears to exist between the two variables.

3. **Finding beta.** To find beta requires that we find the line of best fit using least-squares regression. The easiest approach is to click inside the scatter diagram. Select the Chart Elements icon (+). Check the box for Trendline, select the arrow to the right, and choose More Options. Select Linear and check the box for Display Equation on chart. The line of best fit appears on the scatter diagram. See below.

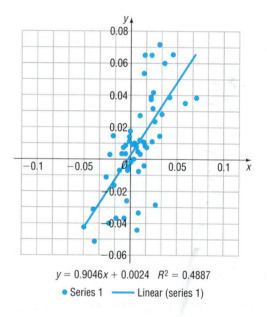

$y = 0.9046x + 0.0024 \quad R^2 = 0.4887$

● Series 1 —— Linear (series 1)

The line of best fit for this data is $y = 0.9046x + 0.0024$. You may click on Chart Title or either axis title and insert the appropriate names. The beta is the slope of the line of best fit, 0.9046. We interpret this by saying, "If the S&P500 increases by 1%, then this stock will increase by 0.9%, on average." Find the beta of your stock and provide an interpretation. NOTE: Another way to use Excel to find the line of best fit requires using the Data Analysis Tool Pack under add-ins.

The following projects are available on the Instructor's Resource Center (IRC):

II. Cannons A battery commander uses the weight of a missile, its initial velocity, and the position of its gun to determine where the missile will travel.

III. First and Second Differences Finite differences provide a numerical method that is used to estimate the graph of an unknown function.

IV. CBL Experiment Computer simulation is used to study the physical properties of a bouncing ball.

Polynomial and Rational Functions

5

Day Length

Day length is the length of time each day from the moment the upper limb of the sun's disk appears above the horizon during sunrise to the moment when the upper limb disappears below the horizon during sunset. The length of a day depends on the day of the year as well as the latitude of the location. Latitude gives the location of a point on Earth north or south of the equator. In the Internet Project at the end of this chapter, we use information from the chapter to investigate the relation between day length and latitude for a specific day of the year.

 —See the Internet-based Chapter Project I—

••• A Look Back

In Chapter 3, we began our discussion of functions. We defined domain, range, and independent and dependent variables, found the value of a function, and graphed functions. We continued our study of functions by listing the properties that a function might have, such as being even or odd, and created a library of functions, naming key functions and listing their properties, including their graphs.

In Chapter 4, we discussed linear functions and quadratic functions, which belong to the class of *polynomial functions*.

A Look Ahead •••

In this chapter, we look at two general classes of functions, polynomial functions and rational functions, and examine their properties. Polynomial functions are arguably the simplest expressions in algebra. For this reason, they are often used to approximate other, more complicated functions. Rational functions are ratios of polynomial functions.

Outline

5.1 Polynomial Functions and Models

PREPARING FOR THIS SECTION *Before getting started, review the following:*

- Polynomials (Chapter R, Section R.4, pp. 39–47)
- Using a Graphing Utility to Approximate Local Maxima and Local Minima (Section 3.3, p. 229)
- Intercepts of a Function (Section 3.2, pp. 215–217)

- Graphing Techniques: Transformations (Section 3.5, pp. 247–256)
- Intercepts (Section 2.2, pp. 159–160)

Now Work the 'Are You Prepared?' problems on page 338.

OBJECTIVES 1 Identify Polynomial Functions and Their Degree (p. 322)
2 Graph Polynomial Functions Using Transformations (p. 326)
3 Identify the Real Zeros of a Polynomial Function and Their Multiplicity (p. 327)
4 Analyze the Graph of a Polynomial Function (p. 334)
5 Build Cubic Models from Data (p. 336)

1 Identify Polynomial Functions and Their Degree

In Chapter 4, we studied the linear function $f(x) = mx + b$, which can be written as

$$f(x) = a_1 x + a_0$$

and the quadratic function $f(x) = ax^2 + bx + c, a \neq 0$, which can be written as

$$f(x) = a_2 x^2 + a_1 x + a_0 \qquad a_2 \neq 0$$

Each of these functions is an example of a *polynomial function*.

DEFINITION

> A **polynomial function** in one variable is a function of the form
>
> $$f(x) = a_n x^n + a_{n-1} x^{n-1} + \cdots + a_1 x + a_0 \qquad (1)$$
>
> where $a_n, a_{n-1}, \ldots, a_1, a_0$ are constants, called the **coefficients** of the polynomial, $n \geq 0$ is an integer, and x is the variable. If $a_n \neq 0$, it is called the **leading coefficient**, and n is the **degree** of the polynomial.
> The domain of a polynomial function is the set of all real numbers.

In Words

A polynomial function is a sum of monomials.

The monomials that make up a polynomial are called its **terms**. If $a_n \neq 0$, $a_n x^n$ is called the **leading term**; a_0 is called the **constant term**. If all of the coefficients are 0, the polynomial is called the **zero polynomial**, which has no degree.

Polynomials are usually written in **standard form**, beginning with the nonzero term of highest degree and continuing with terms in descending order according to degree. If a power of x is missing, it is because its coefficient is zero.

Polynomial functions are among the simplest expressions in algebra. They are easy to evaluate: only addition and repeated multiplication are required. Because of this, they are often used to approximate other, more complicated functions. In this section, we investigate properties of this important class of functions.

EXAMPLE 1 Identifying Polynomial Functions

Determine which of the following are polynomial functions. For those that are, state the degree; for those that are not, tell why not. Write each polynomial in standard form, and then identify the leading term and the constant term.

(a) $p(x) = 5x^3 - \dfrac{1}{4}x^2 - 9$ (b) $f(x) = x + 2 - 3x^4$ (c) $g(x) = \sqrt{x}$

(d) $h(x) = \dfrac{x^2 - 2}{x^3 - 1}$ (e) $G(x) = 8$ (f) $H(x) = -2x^3(x - 1)^2$

Solution
(a) p is a polynomial function of degree 3, and it is already in standard form. The leading term is $5x^3$, and the constant term is -9.

(b) f is a polynomial function of degree 4. Its standard form is $f(x) = -3x^4 + x + 2$. The leading term is $-3x^4$, and the constant term is 2.

(c) g is not a polynomial function because $g(x) = \sqrt{x} = x^{\frac{1}{2}}$, so the variable x is raised to the $\dfrac{1}{2}$ power, which is not a nonnegative integer.

(d) h is not a polynomial function. It is the ratio of two distinct polynomials, and the polynomial in the denominator is of positive degree.

(e) G is a nonzero constant polynomial function so it is of degree 0. The polynomial is in standard form. The leading term and constant term are both 8.

(f) $H(x) = -2x^3(x-1)^2 = -2x^3(x^2 - 2x + 1) = -2x^5 + 4x^4 - 2x^3$. So, H is a polynomial function of degree 5. Because $H(x) = -2x^5 + 4x^4 - 2x^3$, the leading term is $-2x^5$. Since no constant term is shown, the constant term is 0. Do you see a way to find the degree of H without multiplying it out? ●

Now Work PROBLEMS **17 AND 21**

We have already discussed in detail polynomial functions of degrees 0, 1, and 2. See Table 1 for a summary of the properties of the graphs of these polynomial functions.

Table 1

Degree	Form	Name	Graph
No degree	$f(x) = 0$	Zero function	The x-axis
0	$f(x) = a_0, \quad a_0 \neq 0$	Constant function	Horizontal line with y-intercept a_0
1	$f(x) = a_1 x + a_0, \quad a_1 \neq 0$	Linear function	Nonvertical, nonhorizontal line with slope a_1 and y-intercept a_0
2	$f(x) = a_2 x^2 + a_1 x + a_0, \quad a_2 \neq 0$	Quadratic function	Parabola: graph opens up if $a_2 > 0$; graph opens down if $a_2 < 0$

One objective of this section is to analyze the graph of a polynomial function. If you take a course in calculus, you will learn that the graph of every polynomial function is both smooth and continuous. By **smooth**, we mean that the graph contains no sharp corners or cusps; by **continuous**, we mean that the graph has no gaps or holes and can be drawn without lifting your pencil from the paper. See Figures 1(a) and (b).

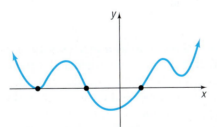

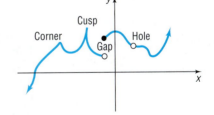

Figure 1

(a) Graph of a polynomial function: smooth, continuous

(b) Cannot be the graph of a polynomial function

Power Functions

We begin the analysis of the graph of a polynomial function by discussing *power functions*, a special kind of polynomial function.

DEFINITION

In Words
A power function is defined by a single monomial.

A **power function of degree n** is a monomial function of the form

$$f(x) = ax^n \tag{2}$$

where a is a real number, $a \neq 0$, and $n > 0$ is an integer.

Examples of power functions are

$$f(x) = 3x \qquad f(x) = -5x^2 \qquad f(x) = 8x^3 \qquad f(x) = -5x^4$$

degree 1   degree 2   degree 3   degree 4

The graph of a power function of degree 1, $f(x) = ax$, is a straight line, with slope a, that passes through the origin. The graph of a power function of degree 2, $f(x) = ax^2$, is a parabola, with vertex at the origin, that opens up if $a > 0$ and opens down if $a < 0$.

If we know how to graph a power function of the form $f(x) = x^n$, a compression or stretch and, perhaps, a reflection about the x-axis will enable us to obtain the graph of $g(x) = ax^n$. Consequently, we shall concentrate on graphing power functions of the form $f(x) = x^n$.

We begin with power functions of even degree of the form $f(x) = x^n$, $n \geq 2$ and n even. The domain of f is the set of all real numbers, and the range is the set of nonnegative real numbers. Such a power function is an even function. (Do you see why?). Its graph is symmetric with respect to the y-axis. Its graph always contains the origin and the points $(-1, 1)$ and $(1, 1)$.

If $n = 2$, the graph is the familiar parabola $y = x^2$ that opens up, with vertex at the origin. If $n \geq 4$, the graph of $f(x) = x^n$, n even, will be closer to the x-axis than the parabola $y = x^2$ if $-1 < x < 1$, $x \neq 0$, and farther from the x-axis than the parabola $y = x^2$ if $x < -1$ or if $x > 1$. Figure 2(a) illustrates this conclusion. Figure 2(b) shows the graphs of $y = x^4$ and $y = x^8$ for comparison.

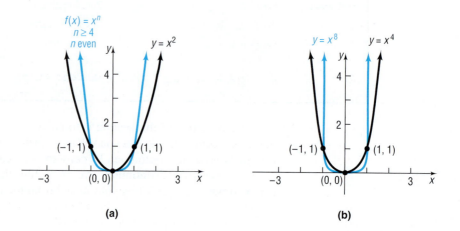

Figure 2   (a)   (b)

Figure 2 shows that as n increases, the graph of $f(x) = x^n$, $n \geq 2$ and n even, tends to flatten out near the origin and is steeper when x is far from 0. For large n, it may appear that the graph coincides with the x-axis near the origin, but it does not; the graph actually touches the x-axis only at the origin (see Table 2). Also, for large n, it may appear that for $x < -1$ or for $x > 1$ the graph is vertical, but it is not; it is only increasing very rapidly in these intervals. If the graphs were enlarged many times, these distinctions would be clear.

Table 2

	$x = 0.1$	$x = 0.3$	$x = 0.5$
$f(x) = x^8$	10^{-8}	0.0000656	0.0039063
$f(x) = x^{20}$	10^{-20}	$3.487 \cdot 10^{-11}$	0.000001
$f(x) = x^{40}$	10^{-40}	$1.216 \cdot 10^{-21}$	$9.095 \cdot 10^{-13}$

Seeing the Concept

Graph $Y_1 = x^4$, $Y_2 = x^8$, and $Y_3 = x^{12}$ using the viewing rectangle $-2 \leq x \leq 2$, $-4 \leq y \leq 16$. Then graph each again using the viewing rectangle $-1 \leq x \leq 1$, $0 \leq y \leq 1$. See Figure 3. TRACE along one of the graphs to confirm that for x close to 0 the graph is above the x-axis and that for $x > 0$ the graph is increasing.

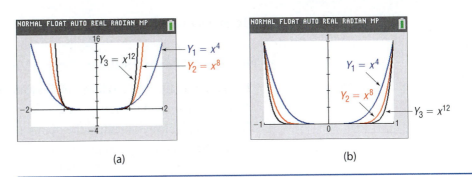

Figure 3

(a)   (b)

Properties of Power Functions, $f(x) = x^n$, n Is a Positive Even Integer

1. f is an even function, so its graph is symmetric with respect to the y-axis.
2. The domain is the set of all real numbers. The range is the set of nonnegative real numbers.
3. The graph always contains the points $(-1, 1)$, $(0, 0)$, and $(1, 1)$.
4. As the exponent n increases in magnitude, the graph is steeper when $x < -1$ or $x > 1$; but for x near the origin, the graph tends to flatten out and lie closer to the x-axis.

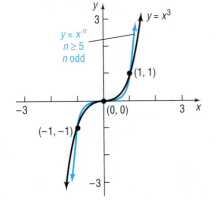

Figure 4

Now we consider power functions of odd degree of the form $f(x) = x^n$, $n \geq 3$ and n odd. The domain and the range of f are the set of real numbers. Such a power function is an odd function. (Do you see why?). Its graph is symmetric with respect to the origin. Its graph always contains the origin and the points $(-1, -1)$ and $(1, 1)$.

The graph of $f(x) = x^n$ when $n = 3$ has been shown several times and is repeated in Figure 4. If $n \geq 5$, the graph of $f(x) = x^n$, n odd, will be closer to the x-axis than that of $y = x^3$ if $-1 < x < 1$ and farther from the x-axis than that of $y = x^3$ if $x < -1$ or if $x > 1$. Figure 4 illustrates this conclusion. Figure 5 shows the graphs of $y = x^5$ and $y = x^9$ for further comparison.

It appears that each graph coincides with the x-axis near the origin, but it does not; each graph actually crosses the x-axis at the origin. Also, it appears that as x increases the graphs become vertical, but they do not; each graph is just increasing very rapidly.

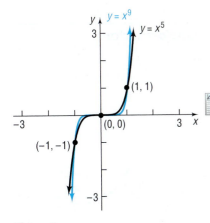

Figure 5

Seeing the Concept

Graph $Y_1 = x^3$, $Y_2 = x^7$, and $Y_3 = x^{11}$ using the viewing rectangle $-2 \leq x \leq 2$, $-16 \leq y \leq 16$. Then graph each again using the viewing rectangle $-1 \leq x \leq 1$, $-1 \leq y \leq 1$. See Figure 6. TRACE along one of the graphs to confirm that the graph is increasing and crosses the x-axis at the origin.

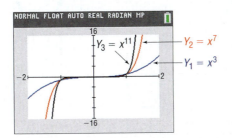 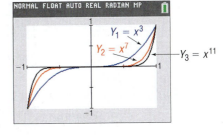

Figure 6   (a)   (b)

To summarize:

> **Properties of Power Functions, $f(x) = x^n$, n Is a Positive Odd Integer**
>
> 1. f is an odd function, so its graph is symmetric with respect to the origin.
> 2. The domain and the range are the set of all real numbers.
> 3. The graph always contains the points $(-1, -1)$, $(0, 0)$, and $(1, 1)$.
> 4. As the exponent n increases in magnitude, the graph is steeper when $x < -1$ or $x > 1$; but for x near the origin, the graph tends to flatten out and lie closer to the x-axis.

2 Graph Polynomial Functions Using Transformations

The methods of shifting, compression, stretching, and reflection (studied in Section 3.5), when used with the facts just presented, will enable us to graph polynomial functions that are transformations of power functions.

EXAMPLE 2 **Graphing a Polynomial Function Using Transformations**

Graph: $f(x) = 1 - x^5$

Solution It is helpful to rewrite f as $f(x) = -x^5 + 1$. Figure 7 shows the required stages.

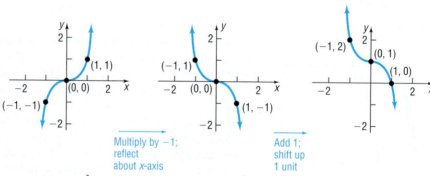

Figure 7 (a) $y = x^5$ (b) $y = -x^5$ (c) $y = -x^5 + 1 = 1 - x^5$ ●

Multiply by -1; reflect about x-axis

Add 1; shift up 1 unit

EXAMPLE 3 **Graphing a Polynomial Function Using Transformations**

Graph: $f(x) = \dfrac{1}{2}(x - 1)^4$

Solution Figure 8 shows the required stages.

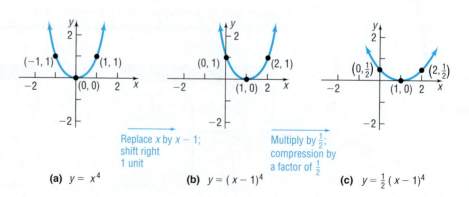

Figure 8 (a) $y = x^4$ (b) $y = (x - 1)^4$ (c) $y = \frac{1}{2}(x - 1)^4$ ●

Replace x by $x - 1$; shift right 1 unit

Multiply by $\frac{1}{2}$; compression by a factor of $\frac{1}{2}$

- **Now Work** PROBLEMS 29 AND 35

3 Identify the Real Zeros of a Polynomial Function and Their Multiplicity

Figure 9 shows the graph of a polynomial function with four x-intercepts. Notice that at the x-intercepts, the graph must either cross the x-axis or touch the x-axis. Consequently, between consecutive x-intercepts the graph is either above the x-axis or below the x-axis.

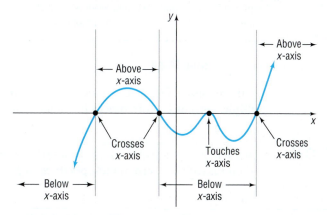

Figure 9 Graph of a polynomial function

If a polynomial function f is factored completely, it is easy to locate the x-intercepts of the graph by solving the equation $f(x) = 0$ and using the Zero-Product Property. For example, if $f(x) = (x - 1)^2(x + 3)$, then the solutions of the equation

$$f(x) = (x - 1)^2(x + 3) = 0$$

are identified as 1 and -3. That is, $f(1) = 0$ and $f(-3) = 0$.

DEFINITION

> If f is a function and r is a real number for which $f(r) = 0$, then r is called a **real zero** of f.

As a consequence of this definition, the following statements are equivalent.

> 1. r is a real zero of a polynomial function f.
> 2. r is an x-intercept of the graph of f.
> 3. $x - r$ is a factor of f.
> 4. r is a solution to the equation $f(x) = 0$.

So the real zeros of a polynomial function are the x-intercepts of its graph, and they are found by solving the equation $f(x) = 0$.

EXAMPLE 4

Finding a Polynomial Function from Its Zeros

(a) Find a polynomial function of degree 3 whose zeros are $-3, 2$, and 5.

(b) Use a graphing utility to graph the polynomial found in part (a) to verify your result.

Solution

(a) If r is a real zero of a polynomial function f, then $x - r$ is a factor of f. This means that $x - (-3) = x + 3, x - 2$, and $x - 5$ are factors of f. As a result, any polynomial function of the form

$$f(x) = a(x + 3)(x - 2)(x - 5)$$

where a is a nonzero real number, qualifies. The value of a causes a stretch, compression, or reflection, but it does not affect the x-intercepts of the graph. Do you know why?

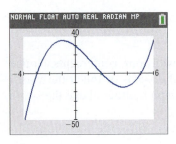

Figure 10 $f(x) = x^3 - 4x^2 - 11x + 30$

 (b) We choose to graph f with $a = 1$. Then

$$f(x) = (x + 3)(x - 2)(x - 5) = x^3 - 4x^2 - 11x + 30$$

Figure 10 shows the graph of f. Notice that the x-intercepts are $-3, 2,$ and 5. ●

Seeing the Concept

Graph the function found in Example 4 for $a = 2$ and $a = -1$. Does the value of a affect the zeros of f? How does the value of a affect the graph of f?

 Now Work PROBLEM 43

If the same factor $x - r$ occurs more than once, r is called a **repeated**, or **multiple**, **zero of** f. More precisely, we have the following definition.

DEFINITION

> If $(x - r)^m$ is a factor of a polynomial f and $(x - r)^{m+1}$ is not a factor of f, then r is called a **zero of multiplicity m of** f.*

EXAMPLE 5

Identifying Zeros and Their Multiplicities

For the polynomial

$$f(x) = 5x^2(x + 2)\left(x - \frac{1}{2}\right)^4$$

> **In Words**
> The multiplicity of a zero is the number of times its corresponding factor occurs.

- -2 is a zero of multiplicity 1 because the exponent on the factor $x + 2$ is 1.
- 0 is a zero of multiplicity 2 because the exponent on the factor x is 2.
- $\frac{1}{2}$ is a zero of multiplicity 4 because the exponent on the factor $x - \frac{1}{2}$ is 4. ●

 Now Work PROBLEM 57(a)

Suppose that it is possible to completely factor a polynomial function and, as a result, locate all the x-intercepts of its graph (the real zeros of the function). These x-intercepts then divide the x-axis into open intervals and, on each such interval, the graph of the polynomial will be either above or below the x-axis over the entire interval. Let's look at an example.

EXAMPLE 6

Graphing a Polynomial Using Its x-Intercepts

Consider the following polynomial: $f(x) = (x + 1)^2(x - 2)$

(a) Find the x- and y-intercepts of the graph of f.
(b) Use the x-intercepts to find the intervals on which the graph of f is above the x-axis and the intervals on which the graph of f is below the x-axis.
(c) Locate other points on the graph, and connect all the points plotted with a smooth, continuous curve.

Solution   (a) The y-intercept is $f(0) = (0 + 1)^2(0 - 2) = -2$.
The x-intercepts satisfy the equation

$$f(x) = (x + 1)^2(x - 2) = 0$$

from which we find

$$(x + 1)^2 = 0 \quad \text{or} \quad x - 2 = 0$$
$$x = -1 \quad \text{or} \quad x = 2$$

The x-intercepts are -1 and 2.

*Some books use the terms **multiple root** and **root of multiplicity** m.

(b) The two x-intercepts divide the x-axis into three intervals:

$$(-\infty, -1) \qquad (-1, 2) \qquad (2, \infty)$$

Since the graph of f crosses or touches the x-axis only at $x = -1$ and $x = 2$, it follows that the graph of f is either above the x-axis $[f(x) > 0]$ or below the x-axis $[f(x) < 0]$ on each of these three intervals. To see where the graph lies, we need only pick a number in each interval, evaluate f there, and see whether the value is positive (above the x-axis) or negative (below the x-axis). See Table 3.

(c) In constructing Table 3, we obtained three additional points on the graph: $(-2, -4)$, $(1, -4)$ and $(3, 16)$. Figure 11 illustrates these points, the intercepts, and a smooth, continuous curve (the graph of f) connecting them.

Table 3

Interval	$(-\infty, -1)$	$(-1, 2)$	$(2, \infty)$
Number chosen	-2	1	3
Value of f	$f(-2) = -4$	$f(1) = -4$	$f(3) = 16$
Location of graph	Below x-axis	Below x-axis	Above x-axis
Point on graph	$(-2, -4)$	$(1, -4)$	$(3, 16)$

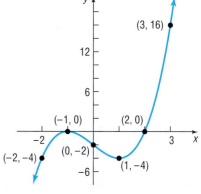

Figure 11 $f(x) = (x + 1)^2(x - 2)$

Look again at Table 3. Since the graph of $f(x) = (x + 1)^2(x - 2)$ is below the x-axis on both sides of -1, the graph of f *touches* the x-axis at $x = -1$, a *zero of multiplicity 2*. Since the graph of f is below the x-axis for $x < 2$ and above the x-axis for $x > 2$, the graph of f *crosses* the x-axis at $x = 2$, a *zero of multiplicity 1*.

This suggests the following results:

> **If r Is a Zero of Even Multiplicity**
>
> Numerically: The sign of $f(x)$ does not change from one side to the other side of r.
> Graphically: The graph of f **touches** the x-axis at r.

> **If r Is a Zero of Odd Multiplicity**
>
> Numerically: The sign of $f(x)$ changes from one side to the other side of r.
> Graphically: The graph of f **crosses** the x-axis at r.

 **Now Work** PROBLEM 57(b)

Turning Points

 Look again at Figure 11 above. We cannot be sure just how low the graph actually goes between $x = -1$ and $x = 2$. But we do know that somewhere in the interval $(-1, 2)$ the graph of f must change direction (from decreasing to increasing). The points at which a graph changes direction are called **turning points**.* Each turning point yields either a **local maximum** or a **local minimum** (see Section 3.3). The following result from calculus tells us the maximum number of turning points that the graph of a polynomial function can have.

---

*Graphing utilities can be used to approximate turning points. For most polynomials, calculus is needed to find the exact turning points.

THEOREM

> **Turning Points**
>
> If f is a polynomial function of degree n, then the graph of f has at most $n - 1$ turning points.
> If the graph of a polynomial function f has $n - 1$ turning points, then the degree of f is at least n.

Based on the first part of the theorem, a polynomial function of degree 5 will have at most $5 - 1 = 4$ turning points. Based on the second part of the theorem, if the graph of a polynomial function has three turning points, then the degree of the function must be at least 4.

Exploration

A graphing utility can be used to locate the turning points of a graph. Graph $Y_1 = (x + 1)^2(x - 2)$. Use MINIMUM to find the location of the turning point for $0 < x < 2$. See Figure 12.

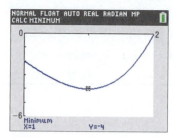

Figure 12 $Y_1 = (x + 1)^2(x - 2)$

 Now Work PROBLEM 57(c)

EXAMPLE 7

Identifying the Graph of a Polynomial Function

Which of the graphs in Figure 13 could be the graph of a polynomial function? For those that could, list the real zeros and state the least degree the polynomial can have. For those that could not, say why not.

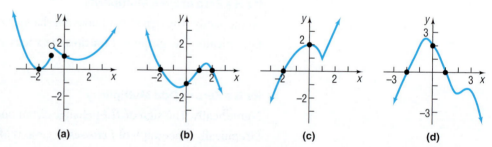

Figure 13   (a)   (b)   (c)   (d)

Solution

(a) The graph in Figure 13(a) cannot be the graph of a polynomial function because of the gap that occurs at $x = -1$. Remember, the graph of a polynomial function is continuous—no gaps or holes. (See Figure 1.)

(b) The graph in Figure 13(b) could be the graph of a polynomial function because the graph is smooth and continuous. It has three real zeros: $-2, 1$, and 2. Since the graph has two turning points, the degree of the polynomial function must be at least 3.

(c) The graph in Figure 13(c) cannot be the graph of a polynomial function because of the cusp at $x = 1$. Remember, the graph of a polynomial function is smooth.

(d) The graph in Figure 13(d) could be the graph of a polynomial function. It has two real zeros: -2 and 1. Since the graph has three turning points, the degree of the polynomial function is at least 4. ●

 Now Work PROBLEM 69

End Behavior

One last remark about Figure 11. For very large values of x, either positive or negative, the graph of $f(x) = (x + 1)^2(x - 2)$ looks like the graph of $y = x^3$. To see why, we write f in the form

$$f(x) = (x + 1)^2(x - 2) = x^3 - 3x - 2 = x^3\left(1 - \frac{3}{x^2} - \frac{2}{x^3}\right)$$

Now, for large values of x, either positive or negative, the terms $\frac{3}{x^2}$ and $\frac{2}{x^3}$ are close to 0, so for large values of x,

$$f(x) = x^3 - 3x - 2 = x^3\left(1 - \frac{3}{x^2} - \frac{2}{x^3}\right) \approx x^3$$

The behavior of the graph of a function for large values of x, either positive or negative, is referred to as its **end behavior**.

THEOREM

In Words
The end behavior of a polynomial function resembles that of its leading term.

> **End Behavior**
>
> For large values of x, either positive or negative, the graph of the polynomial function
>
> $$f(x) = a_n x^n + a_{n-1}x^{n-1} + \cdots + a_1 x + a_0 \quad a_n \neq 0$$
>
> resembles the graph of the power function
>
> $$y = a_n x^n$$

For example, if $f(x) = -2x^3 + 5x^2 + x - 4$, then the graph of f will behave like the graph of $y = -2x^3$ for very large values of x, either positive or negative. We can see that the graphs of f and $y = -2x^3$ "behave" the same by considering Table 4 and Figure 14.

Table 4

x	$f(x)$	$y = -2x^3$
10	$-1{,}494$	$-2{,}000$
100	$-1{,}949{,}904$	$-2{,}000{,}000$
500	$-248{,}749{,}504$	$-250{,}000{,}000$
1,000	$-1{,}994{,}999{,}004$	$-2{,}000{,}000{,}000$

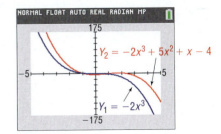

Figure 14

NOTE Infinity (∞) and negative infinity ($-\infty$) are not numbers. Rather, they are symbols that represent unboundedness. ∎

Notice that as x becomes a larger and larger positive number, the values of f become larger and larger negative numbers. When this happens, we say that f is **unbounded in the negative direction**. Rather than using words to describe the behavior of the graph of the function, we explain its behavior using notation. We can symbolize "the value of f becomes a larger and larger negative number as x becomes a larger and larger positive number" by writing $f(x) \to -\infty$ as $x \to \infty$ (read "the values of f approach negative infinity as x approaches infinity").

In calculus, **limits** are used to convey these ideas. There we use the symbolism $\lim\limits_{x \to \infty} f(x) = -\infty$, read "the limit of $f(x)$ as x approaches infinity equals negative infinity," to mean that $f(x) \to -\infty$ as $x \to \infty$.

When we say that the value of a limit equals infinity (or negative infinity), we mean that the values of the function are unbounded in the positive (or negative) direction and call the limit an **infinite limit**. When we discuss limits as x becomes unbounded in the negative direction or unbounded in the positive direction, we are discussing **limits at infinity**.

Look back at Figures 2 and 4. Based on the preceding theorem and the previous discussion on power functions, the end behavior of a polynomial function can be of only four types. See Figure 15.

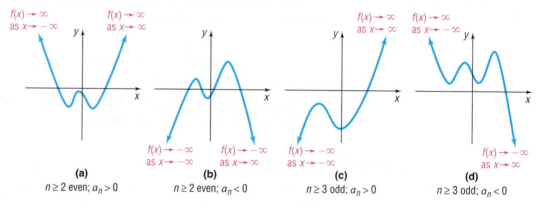

(a)
$n \geq 2$ even; $a_n > 0$

(b)
$n \geq 2$ even; $a_n < 0$

(c)
$n \geq 3$ odd; $a_n > 0$

(d)
$n \geq 3$ odd; $a_n < 0$

Figure 15 End behavior of $f(x) = a_n x^n + a_{n-1} x^{n-1} + \cdots + a_1 x + a_0$

For example, if $f(x) = -2x^4 + x^3 + 4x^2 - 7x + 1$, the graph of f will resemble the graph of the power function $y = -2x^4$ for large $|x|$. The graph of f will behave like Figure 15(b) for large $|x|$.

➤ **Now Work** PROBLEM 57(d)

EXAMPLE 8

Identifying the Graph of a Polynomial Function

Which of the graphs in Figure 16 could be the graph of

$$f(x) = x^4 + ax^3 + bx^2 - 5x - 6$$

where $a > 0, b > 0$?

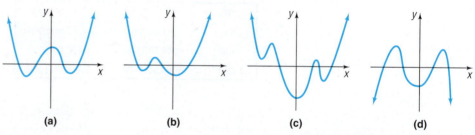

(a) **(b)** **(c)** **(d)**

Figure 16

Solution   The y-intercept of f is $f(0) = -6$. We can eliminate the graph in Figure 16(a), whose y-intercept is positive.

We are not able to solve $f(x) = 0$ to find the x-intercepts of f, so we move on to investigate the turning points of each graph. Since f is of degree 4, the graph of f has at most 3 turning points. We eliminate the graph in Figure 16(c) because that graph has 5 turning points.

Now we look at end behavior. For large values of x, the graph of f will behave like the graph of $y = x^4$. This eliminates the graph in Figure 16(d), whose end behavior is like the graph of $y = -x^4$.

Only the graph in Figure 16(b) could be the graph of

$$f(x) = x^4 + ax^3 + bx^2 - 5x - 6$$

where $a > 0, b > 0$.

EXAMPLE 9

Writing a Polynomial Function from Its Graph

Write a polynomial function whose graph is shown in Figure 17 (use the smallest degree possible).

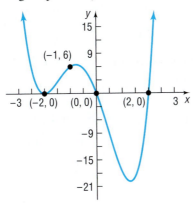

Figure 17

Solution   The x-intercepts are -2, 0, and 2. Therefore, the polynomial must have the factors $(x + 2)$, x, and $(x - 2)$, respectively. There are three turning points, so the degree of the polynomial must be at least 4. The graph touches the x-axis at $x = -2$, so -2 must have an even multiplicity. The graph crosses the x-axis at $x = 0$ and $x = 2$, so 0 and 2 must have odd multiplicities. Using the smallest degree possible (1 for odd multiplicity and 2 for even multiplicity), we can write

$$f(x) = ax(x + 2)^2(x - 2)$$

All that remains is to find the leading coefficient, a. From Figure 17, the point $(-1, 6)$ must lie on the graph.

$$6 = a(-1)(-1 + 2)^2(-1 - 2) \quad {\color{teal}f(-1) = 6}$$
$$6 = 3a$$
$$2 = a$$

The polynomial function $f(x) = 2x(x + 2)^2(x - 2)$ would have the graph in Figure 17.

Check: Graph $Y_1 = 2x(x + 2)^2(x - 2)$ using a graphing utility to verify this result.

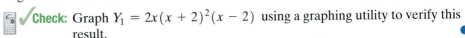 **Now Work** PROBLEMS **73** AND **77**

SUMMARY

Graph of a Polynomial Function $f(x) = a_n x^n + a_{n-1} x^{n-1} + \cdots + a_1 x + a_0 \quad a_n \neq 0$

Degree of the polynomial function f: n

y-intercept: $f(0) = a_0$.

Graph is smooth and continuous.

Maximum number of turning points: $n - 1$

At a zero of even multiplicity: The graph of f touches the x-axis.

At a zero of odd multiplicity: The graph of f crosses the x-axis.

Between zeros, the graph of f is either above or below the x-axis.

End behavior: For large $|x|$, the graph of f behaves like the graph of $y = a_n x^n$.

4 Analyze the Graph of a Polynomial Function

EXAMPLE 10

How to Analyze the Graph of a Polynomial Function

Analyze the factored form of the polynomial function $f(x) = (2x + 1)(x - 3)^2$.

Step-by-Step Solution

Step 1: Determine the end behavior of the graph of the function.

Expand the polynomial:

$$f(x) = (2x + 1)(x - 3)^2 = (2x + 1)(x^2 - 6x + 9)$$
$$= 2x^3 - 12x^2 + 18x + x^2 - 6x + 9 \quad \text{Multiply.}$$
$$= 2x^3 - 11x^2 + 12x + 9 \quad \text{Combine like terms.}$$

The polynomial function f is of degree 3. The graph of f behaves like $y = 2x^3$ for large values of $|x|$.

Step 2: Find the x- and y-intercepts of the graph of the function.

The y-intercept is $f(0) = 9$. To find the x-intercepts, solve $f(x) = 0$.

$$f(x) = 0$$
$$(2x + 1)(x - 3)^2 = 0$$
$$2x + 1 = 0 \quad \text{or} \quad (x - 3)^2 = 0$$
$$x = -\frac{1}{2} \quad \text{or} \quad x = 3$$

The x-intercepts are $-\frac{1}{2}$ and 3.

Step 3: Determine the zeros of the function and their multiplicity. Use this information to determine whether the graph crosses or touches the x-axis at each x-intercept.

The zeros of f are $-\frac{1}{2}$ and 3. The zero $-\frac{1}{2}$ is a zero of multiplicity 1, so the graph of f crosses the x-axis at $x = -\frac{1}{2}$. The zero 3 is a zero of multiplicity 2, so the graph of f touches the x-axis at $x = 3$.

Step 4: Determine the maximum number of turning points on the graph of the function.

Because the polynomial function is of degree 3 (Step 1), the graph of the function will have at most $3 - 1 = 2$ turning points.

Step 5: Put all the information from Steps 1 through 4 together to obtain the graph of f. To help establish the y-axis scale, find additional points on the graph on each side of any x-intercept.

Figure 18(a) illustrates the information obtained from Steps 1 through 4. We evaluate f at $-1, 1,$ and 4 to help establish the scale on the y-axis.

We find that $f(-1) = -16, f(1) = 12,$ and $f(4) = 9$, so we plot the points $(-1, -16), (1, 12),$ and $(4, 9)$. The graph of f is given in Figure 18(b).

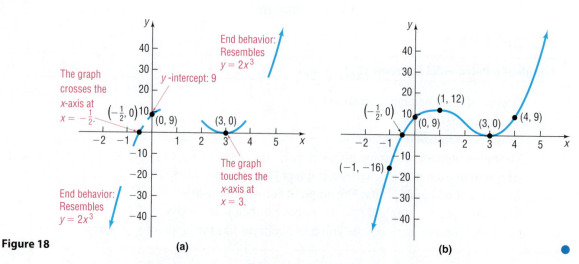

Figure 18 (a) (b)

SUMMARY

Analyzing the Graph of a Polynomial Function

STEP 1: Determine the end behavior of the graph of the function.

STEP 2: Find the x- and y-intercepts of the graph of the function.

STEP 3: Determine the zeros of the function and their multiplicity. Use this information to determine whether the graph crosses or touches the x-axis at each x-intercept.

STEP 4: Determine the maximum number of turning points on the graph of the function.

STEP 5: Use the information in Steps 1 through 4 to draw a complete graph of the function. To help establish the y-axis scale, find additional points on the graph on each side of any x-intercept.

 **Now Work** PROBLEM 81

For polynomial functions that have noninteger coefficients and for polynomials that are not easily factored, we use a graphing utility early in the analysis. This is because the amount of information that can be obtained from algebraic analysis is limited.

EXAMPLE 11

How to Use a Graphing Utility to Analyze the Graph of a Polynomial Function

Analyze the graph of the polynomial function

$$f(x) = x^3 + 2.48x^2 - 4.3155x + 2.484406$$

Step-by-Step Solution

Step 1: Determine the end behavior of the graph of the function.

The polynomial function f is of degree 3. The graph of f behaves like $y = x^3$ for large values of $|x|$.

Step 2: Graph the function using a graphing utility.

See Figure 19 for the graph of f.

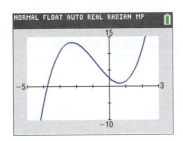

Figure 19 $f(y_1) = x^3 + 2.48x^2 - 4.3155x + 2.484406$

Step 3: Use a graphing utility to approximate the x- and y-intercepts of the graph.

The y-intercept is $f(0) = 2.484406$.

In Example 10, the polynomial function was factored, so it was easy to find the x-intercepts algebraically. However, it is not readily apparent how to factor f in this example. Therefore, we use a graphing utility's ZERO (or ROOT or SOLVE) feature and find the lone x-intercept to be -3.79, rounded to two decimal places.

Step 4: Use a graphing utility to create a TABLE to find points on the graph around each x-intercept.

Table 5 below shows values of x on each side of the x-intercept. The points $(-4, -4.57)$ and $(-2, 13.04)$ are on the graph.

Table 5

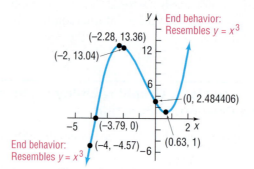

Y₁⊟X³+2.48X²-4.3155X+2.4844

Step 5: Approximate the turning points of the graph.

From the graph of f shown in Figure 19, we can see that f has two turning points. Using MAXIMUM reveals one turning point is at $(-2.28, 13.36)$, rounded to two decimal places. Using MINIMUM shows that the other turning point is at $(0.63, 1)$, rounded to two decimal places.

Step 6: Use the information in Steps 1 through 5 to draw a complete graph of the function by hand.

Figure 20 shows a graph of f drawn by hand using the information in Steps 1 through 5.

Figure 20

End behavior: Resembles $y = x^3$
(−2.28, 13.36)
(−2, 13.04)
12
6
(0, 2.484406)
−5
(−3.79, 0)
2 x
End behavior: Resembles $y = x^3$
(−4, −4.57)
−6
(0.63, 1)

Step 7: Find the domain and the range of the function.

The domain and the range of f are the set of all real numbers.

Step 8: Use the graph to determine where the function is increasing and where it is decreasing.

Based on the graph, f is increasing on the intervals $(-\infty, -2.28)$ and $(0.63, \infty)$. Also, f is decreasing on the interval $(-2.28, 0.63)$. ●

SUMMARY

Using a Graphing Utility to Analyze the Graph of a Polynomial Function

STEP 1: Determine the end behavior of the graph of the function.
STEP 2: Graph the function using a graphing utility.
STEP 3: Use a graphing utility to approximate the x- and y-intercepts of the graph.
STEP 4: Use a graphing utility to create a TABLE to find points on the graph around each x-intercept.
STEP 5: Approximate the turning points of the graph.
STEP 6: Use the information in Steps 1 through 5 to draw a complete graph of the function by hand.
STEP 7: Find the domain and the range of the function.
STEP 8: Use the graph to determine where the function is increasing and where it is decreasing.

 Now Work PROBLEM 99

5 Build Cubic Models from Data

In Section 4.2 we found the line of best fit from data, and in Section 4.4 we found the quadratic function of best fit. It is also possible to find polynomial functions of best fit. However, most statisticians do not recommend finding polynomials of best fit of degree higher than 3.

Data that follow a cubic relation should look like Figure 21(a) or (b).

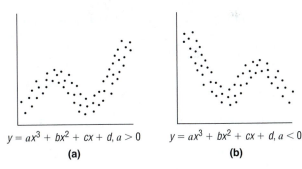

$y = ax^3 + bx^2 + cx + d, a > 0$

(a)

$y = ax^3 + bx^2 + cx + d, a < 0$

(b)

Figure 21 Cubic relation

EXAMPLE 12

A Cubic Function of Best Fit

The data in Table 6 represent the weekly cost C (in thousands of dollars) of printing x thousand textbooks.

(a) Draw a scatter diagram of the data using x as the independent variable and C as the dependent variable. Comment on the type of relation that may exist between the two variables x and C.

(b) Using a graphing utility, find the cubic function of best fit $C = C(x)$ that models the relation between number of texts and cost.

(c) Graph the cubic function of best fit on your scatter diagram.

(d) Use the function found in part (b) to predict the cost of printing 22 thousand texts per week.

Solution

(a) Figure 22 shows the scatter diagram. A cubic relation may exist between the two variables.

(b) Upon executing the CUBIC REGression program, we obtain the results shown in Figure 23. The output that the utility provides shows us the equation $y = ax^3 + bx^2 + cx + d$. The cubic function of best fit to the data is $C(x) = 0.0155x^3 - 0.5951x^2 + 9.1502x + 98.4327$.

(c) Figure 24 shows the graph of the cubic function of best fit on the scatter diagram. The function fits the data reasonably well.

Table 6

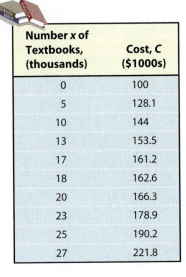

Number x of Textbooks, (thousands)	Cost, C ($1000s)
0	100
5	128.1
10	144
13	153.5
17	161.2
18	162.6
20	166.3
23	178.9
25	190.2
27	221.8

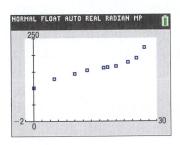

Figure 22

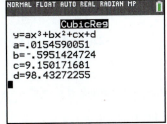

Figure 23

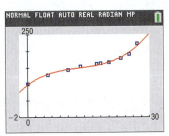

Figure 24

(d) Evaluate the function $C(x)$ at $x = 22$.

$$C(22) = 0.0155(22)^3 - 0.5951(22)^2 + 9.1502(22) + 98.4327 \approx 176.8$$

The model predicts that the cost of printing 22 thousand textbooks in a week will be 176.8 thousand dollars—that is, $176,800.

5.1 Assess Your Understanding

'Are You Prepared?' *Answers are given at the end of these exercises. If you get a wrong answer, read the pages listed in red.*

1. The intercepts of the equation $9x^2 + 4y = 36$ are _____. (pp. 159–160)

2. Is the expression $4x^3 - 3.6x^2 - \sqrt{2}$ a polynomial? If so, what is its degree? (pp. 39–47)

3. To graph $y = x^2 - 4$, you would shift the graph of $y = x^2$ _____ a distance of ____ units. (pp. 247–256)

 4. Use a graphing utility to approximate (rounded to two decimal places) the local maximum value and local minimum value of $f(x) = x^3 - 2x^2 - 4x + 5$, for $-3 < x < 3$. (p. 229)

5. *True or False* The x-intercepts of the graph of a function $y = f(x)$ are the real solutions of the equation $f(x) = 0$. (pp. 215–217)

6. If $g(5) = 0$, what point is on the graph of g? What is the corresponding x-intercept of the graph of g? (pp. 215–217)

Concepts and Vocabulary

7. The graph of every polynomial function is both _____ and _____.

8. If r is a real zero of even multiplicity of a polynomial function f, then the graph of f _____ (crosses/touches) the x-axis at r.

9. The graphs of power functions of the form $f(x) = x^n$, where n is an even integer, always contain the points _____, _____, and _____.

10. If r is a solution to the equation $f(x) = 0$, name three additional statements that can be made about f and r, assuming f is a polynomial function.

11. The points at which a graph changes direction (from increasing to decreasing or decreasing to increasing) are called _____.

12. The graph of the function $f(x) = 3x^4 - x^3 + 5x^2 - 2x - 7$ will behave like the graph of _____ for large values of $|x|$.

13. If $f(x) = -2x^5 + x^3 - 5x^2 + 7$, then $\lim\limits_{x \to -\infty} f(x) =$ ____ and $\lim\limits_{x \to \infty} f(x) =$ _____.

14. Explain what the notation $\lim\limits_{x \to \infty} f(x) = -\infty$ means.

15. The _____ of a zero is the number of times its corresponding factor occurs.
(a) degree (b) multiplicity (c) turning point (d) limit

16. The graph of $y = 5x^6 - 3x^4 + 2x - 9$ has at most how many turning points?
(a) -9 (b) 14 (c) 6 (d) 5

Skill Building

In Problems 17–28, determine which functions are polynomial functions. For those that are, state the degree. For those that are not, tell why not. Write each polynomial in standard form. Then identify the leading term and the constant term.

17. $f(x) = 4x + x^3$

18. $f(x) = 5x^2 + 4x^4$

19. $g(x) = \dfrac{1 - x^2}{2}$

20. $h(x) = 3 - \dfrac{1}{2}x$

21. $f(x) = 1 - \dfrac{1}{x}$

22. $f(x) = x(x - 1)$

23. $g(x) = x^{3/2} - x^2 + 2$

24. $h(x) = \sqrt{x}(\sqrt{x} - 1)$

25. $F(x) = 5x^4 - \pi x^3 + \dfrac{1}{2}$

26. $F(x) = \dfrac{x^2 - 5}{x^3}$

27. $G(x) = 2(x - 1)^2(x^2 + 1)$

28. $G(x) = -3x^2(x + 2)^3$

In Problems 29–42, use transformations of the graph of $y = x^4$ or $y = x^5$ to graph each function.

29. $f(x) = (x + 1)^4$

30. $f(x) = (x - 2)^5$

31. $f(x) = x^5 - 3$

32. $f(x) = x^4 + 2$

33. $f(x) = \dfrac{1}{2}x^4$

34. $f(x) = 3x^5$

35. $f(x) = -x^5$

36. $f(x) = -x^4$

37. $f(x) = (x - 1)^5 + 2$

38. $f(x) = (x + 2)^4 - 3$

39. $f(x) = 2(x + 1)^4 + 1$

40. $f(x) = \dfrac{1}{2}(x - 1)^5 - 2$

41. $f(x) = 4 - (x - 2)^5$

42. $f(x) = 3 - (x + 2)^4$

In Problems 43–50, form a polynomial function whose real zeros and degree are given. Answers will vary depending on the choice of a leading coefficient.

43. Zeros: $-1, 1, 3$; degree 3

44. Zeros: $-2, 2, 3$; degree 3

45. Zeros: $-3, 0, 4$; degree 3

46. Zeros: $-4, 0, 2$; degree 3

47. Zeros: $-4, -1, 2, 3$; degree 4

48. Zeros: $-3, -1, 2, 5$; degree 4

49. Zeros: -1, multiplicity 1; 3, multiplicity 2; degree 3

50. Zeros: -2, multiplicity 2; 4, multiplicity 1; degree 3

In Problems 51–56, find the polynomial function with the given zeros whose graph passes through the given point.

51. Zeros: $-3, 1, 4$
Point: $(6, 180)$

52. Zeros: $-2, 0, 2$
Point: $(-4, 16)$

53. Zeros: $-1, 0, 2, 4$
Point: $\left(\frac{1}{2}, 63 \right)$

54. Zeros: $-5, -1, 2, 6$
Point: $\left(\frac{5}{2}, 15 \right)$

55. Zeros: -1 (multiplicity 2),
1 (multiplicity 2)
Point: $(-2, 45)$

56. Zeros: -1 (multiplicity 2),
0, 3 (multiplicity 2)
Point: $(1, -48)$

In Problems 57–68, for each polynomial function:
(a) List each real zero and its multiplicity.
(b) Determine whether the graph crosses or touches the x-axis at each x-intercept.
(c) Determine the maximum number of turning points on the graph.
(d) Determine the end behavior; that is, find the power function that the graph of f resembles for large values of $|x|$.

57. $f(x) = 3(x - 7)(x + 3)^2$

58. $f(x) = 4(x + 4)(x + 3)^3$

59. $f(x) = 4(x^2 + 1)(x - 2)^3$

60. $f(x) = 2(x - 3)(x^2 + 4)^3$

61. $f(x) = -2\left(x + \frac{1}{2}\right)^2 (x + 4)^3$

62. $f(x) = \left(x - \frac{1}{3}\right)^2 (x - 1)^3$

63. $f(x) = (x - 5)^3 (x + 4)^2$

64. $f(x) = (x + \sqrt{3})^2 (x - 2)^4$

65. $f(x) = 3(x^2 + 8)(x^2 + 9)^2$

66. $f(x) = -2(x^2 + 3)^3$

67. $f(x) = -2x^2(x^2 - 2)$

68. $f(x) = 4x(x^2 - 3)$

In Problems 69–72, identify which of the graphs could be the graph of a polynomial function. For those that could, list the real zeros and state the least degree the polynomial can have. For those that could not, say why not.

69.
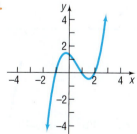

70.
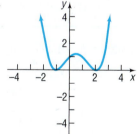

71.
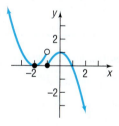

72.
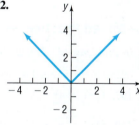

In Problems 73–76, construct a polynomial function that might have the given graph. (More than one answer may be possible.)

73.
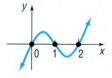

74.

75.
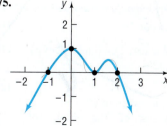

76.
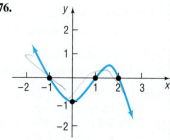

In Problems 77–80, write a polynomial function whose graph is shown (use the smallest degree possible).

77.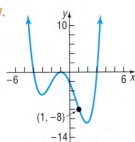

78.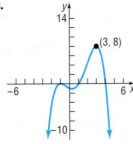

79.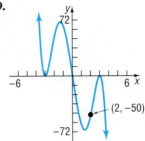

80.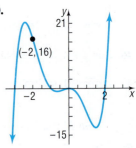

In Problems 81–98, analyze each polynomial function by following Steps 1 through 5 on page 335.

81. $f(x) = x^2(x - 3)$

82. $f(x) = x(x + 2)^2$

83. $f(x) = (x + 4)^2(1 - x)$

84. $f(x) = (x - 1)(x + 3)^2$

85. $f(x) = -2(x + 2)(x - 2)^3$

86. $f(x) = -\frac{1}{2}(x + 4)(x - 1)^3$

87. $f(x) = (x + 1)(x - 2)(x + 4)$

88. $f(x) = (x - 1)(x + 4)(x - 3)$

89. $f(x) = x^2(x - 2)(x + 2)$

90. $f(x) = x^2(x - 3)(x + 4)$

91. $f(x) = (x + 1)^2(x - 2)^2$

92. $f(x) = (x - 4)^2(x + 2)^2$

93. $f(x) = x^2(x + 3)(x + 1)$

94. $f(x) = x^2(x - 3)(x - 1)$

95. $f(x) = 5x(x^2 - 4)(x + 3)$

96. $f(x) = (x - 2)^2(x + 2)(x + 4)$

97. $f(x) = x^2(x - 2)(x^2 + 3)$

98. $f(x) = x^2(x^2 + 1)(x + 4)$

 In Problems 99–106, analyze each polynomial function f by following Steps 1 through 8 on page 336.

99. $f(x) = x^3 + 0.2x^2 - 1.5876x - 0.31752$

100. $f(x) = x^3 - 0.8x^2 - 4.6656x + 3.73248$

101. $f(x) = x^3 + 2.56x^2 - 3.31x + 0.89$

102. $f(x) = x^3 - 2.91x^2 - 7.668x - 3.8151$

103. $f(x) = x^4 - 2.5x^2 + 0.5625$

104. $f(x) = x^4 - 18.5x^2 + 50.2619$

105. $f(x) = 2x^4 - \pi x^3 + \sqrt{5}x - 4$

106. $f(x) = -1.2x^4 + 0.5x^2 - \sqrt{3}x + 2$

Mixed Practice

In Problems 107–114, analyze each polynomial function by following Steps 1 through 5 on page 335.

[**Hint:** You will need to first factor the polynomial].

107. $f(x) = 4x - x^3$

108. $f(x) = x - x^3$

109. $f(x) = x^3 + x^2 - 12x$

110. $f(x) = x^3 + 2x^2 - 8x$

111. $f(x) = 2x^4 + 12x^3 - 8x^2 - 48x$

112. $f(x) = 4x^3 + 10x^2 - 4x - 10$

113. $f(x) = -x^5 - x^4 + x^3 + x^2$

114. $f(x) = -x^5 + 5x^4 + 4x^3 - 20x^2$

In Problems 115–118, construct a polynomial function f with the given characteristics.

115. Zeros: $-3, 1, 4$; degree 3; y-intercept: 36

116. Zeros: $-4, -1, 2$; degree 3; y-intercept: 16

117. Zeros: -5(multiplicity 2); 2 (multiplicity 1); 4 (multiplicity 1); degree 4; contains the point $(3, 128)$

118. Zeros: -4 (multiplicity 1); 0 (multiplicity 3); 2 (multiplicity 1); degree 5; contains the point $(-2, 64)$

119. $G(x) = (x + 3)^2(x - 2)$
(a) Identify the x-intercepts of the graph of G.
(b) What are the x-intercepts of the graph of $y = G(x + 3)$?

120. $h(x) = (x + 2)(x - 4)^3$
(a) Identify the x-intercepts of the graph of h.
(b) What are the x-intercepts of the graph of $y = h(x - 2)$?

Applications and Extensions

121. Hurricanes In 2012, Hurricane Sandy struck the East Coast of the United States, killing 147 people and causing an estimated $75 billion in damage. With a gale diameter of about 1000 miles, it was the largest ever to form over the Atlantic Basin. The accompanying data represent the number of major hurricane strikes in the Atlantic Basin (category 3, 4, or 5) each decade from 1921 to 2010.

Decade, x	Major Hurricanes Striking Atlantic Basin, H
1921–1930, 1	17
1931–1940, 2	16
1941–1950, 3	29
1951–1960, 4	33
1961–1970, 5	27
1971–1980, 6	16
1981–1990, 7	16
1991–2000, 8	27
2001–2010, 9	33

Source: National Oceanic & Atmospheric Administration

(a) Draw a scatter diagram of the data. Comment on the type of relation that may exist between the two variables.

(b) Use a graphing utility to find the cubic function of best fit that models the relation between decade and number of major hurricanes.

(c) Use the model found in part (b) to predict the number of major hurricanes that struck the Atlantic Basin between 1961 and 1970.

(d) With a graphing utility, draw a scatter diagram of the data and then graph the cubic function of best fit on the scatter diagram.

(e) Concern has risen about the increase in the number and intensity of hurricanes, but some scientists believe this is just a natural fluctuation that could last another decade or two. Use your model to predict the number of major hurricanes that will strike the Atlantic Basin between 2011 and 2020. Is your result reasonable?

122. Poverty Rates The following data represent the percentage of families in the United States whose income is below the poverty level.

(a) With a graphing utility, draw a scatter diagram of the data. Comment on the type of relation that appears to exist between the two variables.

(b) Decide on a function of best fit to these data (linear, quadratic, or cubic), and use this function to predict the percentage of U.S. families that were below the poverty level in 2012 ($t = 23$). Compare your prediction to the actual value of 15.9.

(c) Draw the function of best fit on the scatter diagram drawn in part (a).

Year, t	Percent below Poverty Level, p	Year, t	Percent below Poverty Level, p
1990, 1	13.5	2001, 12	11.7
1991, 2	14.2	2002, 13	12.1
1992, 3	14.8	2003, 14	12.5
1993, 4	15.1	2004, 15	12.7
1994, 5	14.5	2005, 16	12.6
1995, 6	13.8	2006, 17	12.3
1996, 7	13.7	2007, 18	12.5
1997, 8	13.3	2008, 19	13.2
1998, 9	12.7	2009, 20	14.3
1999, 10	11.9	2010, 21	15.3
2000, 11	11.3	2011, 22	15.9

Source: U.S. Census Bureau

123. Temperature The following data represent the temperature T (°Fahrenheit) in Kansas City, Missouri, x hours after midnight on April 15, 2014.

Hours after Midnight, x	Temperature (°F), T
3	36.1
6	32.0
9	39.0
12	46.2
15	52.0
18	55.0
21	52.0
24	48.9

Source: The Weather Underground

(a) Draw a scatter diagram of the data. Comment on the type of relation that may exist between the two variables.

(b) Find the average rate of change in temperature from 9 AM to 12 noon.

(c) What is the average rate of change in temperature from 3 PM to 6 PM ?

(d) Decide on a function of best fit to these data (linear, quadratic, or cubic) and use this function to predict the temperature at 5 PM.

(e) With a graphing utility, draw a scatter diagram of the data and then graph the function of best fit on the scatter diagram.

(f) Interpret the y-intercept.

124. Future Value of Money Suppose that you make deposits of $500 at the beginning of every year into an Individual Retirement Account (IRA) earning interest r (expressed as a decimal). At the beginning of the first year, the value of the account will be $500; at the beginning of the second year, the value of the account, will be

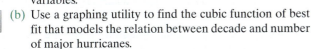

$$\underbrace{\$500 + \$500r}_{\text{Value of 1st deposit}} + \underbrace{\$500}_{\text{Value of 2nd deposit}} = \$500(1 + r) + \$500 = 500r + 1000$$

(a) Verify that the value of the account at the beginning of the third year is $T(r) = 500r^2 + 1500r + 1500$.

(b) The account value at the beginning of the fourth year is $F(r) = 500r^3 + 2000r^2 + 3000r + 2000$. If the annual rate of interest is $5\% = 0.05$, what will be the value of the account at the beginning of the fourth year?

 125. A Geometric Series In calculus, you will learn that certain functions can be approximated by polynomial functions. We will explore one such function now.

(a) Using a graphing utility, create a table of values with
$$Y_1 = f(x) = \frac{1}{1-x} \text{ and } Y_2 = g_2(x) = 1 + x + x^2 + x^3$$
for $-1 < x < 1$ with ΔTbl $= 0.1$.

(b) Using a graphing utility, create a table of values with
$$Y_1 = f(x) = \frac{1}{1-x} \text{ and}$$
$$Y_2 = g_3(x) = 1 + x + x^2 + x^3 + x^4$$
for $-1 < x < 1$ with ΔTbl $= 0.1$.

(c) Using a graphing utility, create a table of values with
$$Y_1 = f(x) = \frac{1}{1-x} \text{ and}$$
$$Y_2 = g_4(x) = 1 + x + x^2 + x^3 + x^4 + x^5$$
for $-1 < x < 1$ with ΔTbl $= 0.1$.

(d) What do you notice about the values of the function as more terms are added to the polynomial? Are there some values of x for which the approximations are better?

Explaining Concepts: Discussion and Writing

126. Can the graph of a polynomial function have no y-intercept? Can it have no x-intercepts? Explain.

127. Write a few paragraphs that provide a general strategy for graphing a polynomial function. Be sure to mention the following: degree, intercepts, end behavior, and turning points.

128. Make up a polynomial that has the following characteristics: crosses the x-axis at -1 and 4, touches the x-axis at 0 and 2, and is above the x-axis between 0 and 2. Give your polynomial to a fellow classmate and ask for a written critique.

129. Make up two polynomials, not of the same degree, with the following characteristics: crosses the x-axis at -2, touches the x-axis at 1, and is above the x-axis between -2 and 1. Give your polynomials to a fellow classmate and ask for a written critique.

130. The graph of a polynomial function is always smooth and continuous. Name a function studied earlier that is smooth but not continuous. Name one that is continuous but not smooth.

131. Which of the following statements are true regarding the graph of the cubic polynomial $f(x) = x^3 + bx^2 + cx + d$? (Give reasons for your conclusions.)
(a) It intersects the y-axis in one and only one point.
(b) It intersects the x-axis in at most three points.
(c) It intersects the x-axis at least once.
(d) For $|x|$ very large, it behaves like the graph of $y = x^3$.
(e) It is symmetric with respect to the origin.
(f) It passes through the origin.

132. The illustration shows the graph of a polynomial function.

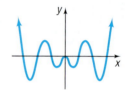

(a) Is the degree of the polynomial even or odd?
(b) Is the leading coefficient positive or negative?
(c) Is the function even, odd, or neither?
(d) Why is x^2 necessarily a factor of the polynomial?
(e) What is the minimum degree of the polynomial?
(f) Formulate five different polynomials whose graphs could look like the one shown. Compare yours to those of other students. What similarities do you see? What differences?

133. Design a polynomial function with the following characteristics: degree 6; four distinct real zeros, one of multiplicity 3; y-intercept 3; behaves like $y = -5x^6$ for large values of $|x|$. Is this polynomial unique? Compare your polynomial with those of other students. What terms will be the same as everyone else's? Add some more characteristics, such as symmetry or naming the real zeros. How does this modify the polynomial?

Retain Your Knowledge

Problems 134–137 are based on material learned earlier in the course. The purpose of these problems is to keep the material fresh in your mind so that you are better prepared for the final exam.

134. Find the equation of the line that contains the point $(2, -3)$ and is perpendicular to the line $5x - 2y = 6$.

135. Find the domain of the function $h(x) = \dfrac{x-3}{x+5}$.

136. Use the quadratic formula to find the zeros of the function $f(x) = 4x^2 + 8x - 3$.

137. Solve: $|5x - 3| = 7$.

'Are You Prepared?' Answers

1. $(-2, 0), (2, 0), (0, 9)$ **2.** Yes; 3   **3.** Down; 4   **4.** Local maximum value 6.48 at $x = -0.67$; local minimum value -3 at $x = 2$
5. True   **6.** $(5, 0); 5$

5.2 Properties of Rational Functions

PREPARING FOR THIS SECTION   *Before getting started, review the following:*

- Rational Expressions (Chapter R, Section R.7, pp. 62–69)
- Polynomial Division (Chapter R, Section R.4, pp. 44–47)

- Graph of $f(x) = \dfrac{1}{x}$ (Section 2.2, Example 12, p. 164)
- Graphing Techniques: Transformations (Section 3.5, pp. 247–256)

Now Work the '**Are You Prepared?**' problems on page 350.

OBJECTIVES   1 Find the Domain of a Rational Function (p. 343)
                    2 Find the Vertical Asymptotes of a Rational Function (p. 346)
                    3 Find the Horizontal or Oblique Asymptote of a Rational Function (p. 348)

Ratios of integers are called *rational numbers*. Similarly, ratios of polynomial functions are called *rational functions*. Examples of rational functions are

$$R(x) = \frac{x^2 - 4}{x^2 + x + 1} \qquad F(x) = \frac{x^3}{x^2 - 4} \qquad G(x) = \frac{3x^2}{x^4 - 1}$$

DEFINITION

A **rational function** is a function of the form

$$R(x) = \frac{p(x)}{q(x)}$$

where p and q are polynomial functions and q is not the zero polynomial. The domain of a rational function is the set of all real numbers except those for which the denominator q is 0.

1 Find the Domain of a Rational Function

EXAMPLE 1

Finding the Domain of a Rational Function

(a) The domain of $R(x) = \dfrac{2x^2 - 4}{x + 5}$ is the set of all real numbers x except -5; that is, the domain is $\{x \mid x \neq -5\}$.

(b) The domain of $R(x) = \dfrac{1}{x^2 - 4} = \dfrac{1}{(x + 2)(x - 2)}$ is the set of all real numbers x except -2 and 2; that is, the domain is $\{x \mid x \neq -2, x \neq 2\}$.

(c) The domain of $R(x) = \dfrac{x^3}{x^2 + 1}$ is the set of all real numbers.

(d) The domain of $R(x) = \dfrac{x^2 - 1}{x - 1}$ is the set of all real numbers x except 1; that is, the domain is $\{x \mid x \neq 1\}$. ●

Although $\dfrac{x^2 - 1}{x - 1}$ reduces to $x + 1$, it is important to observe that the functions

$$R(x) = \frac{x^2 - 1}{x - 1} \quad \text{and} \quad f(x) = x + 1$$

are not equal, since the domain of R is $\{x \mid x \neq 1\}$ and the domain of f is the set of all real numbers.

Now Work   PROBLEM 17

If $R(x) = \dfrac{p(x)}{q(x)}$ is a rational function, and if p and q have no common factors, then the rational function R is said to be in **lowest terms.** For a rational function $R(x) = \dfrac{p(x)}{q(x)}$ in lowest terms, the real zeros, if any, of the numerator in the domain of R are the x-intercepts of the graph of R and so will play a major role in the graph of R. The real zeros of the denominator of R [that is, the numbers x, if any, for which $q(x) = 0$], although not in the domain of R, also play a major role in the graph of R.

We have already discussed the properties of the rational function $y = \dfrac{1}{x}$. (Refer to Example 12, page 164). The next rational function that we take up is $H(x) = \dfrac{1}{x^2}$.

EXAMPLE 2

Graphing $y = \dfrac{1}{x^2}$

Analyze the graph of $H(x) = \dfrac{1}{x^2}$.

Solution

The domain of $H(x) = \dfrac{1}{x^2}$ is the set of all real numbers x except 0. The graph has no y-intercept, because x can never equal 0. The graph has no x-intercept because the equation $H(x) = 0$ has no solution. Therefore, the graph of H will not cross or touch either of the coordinate axes. Because

$$H(-x) = \frac{1}{(-x)^2} = \frac{1}{x^2} = H(x)$$

H is an even function, so its graph is symmetric with respect to the y-axis.

Table 7 shows the behavior of $H(x) = \dfrac{1}{x^2}$ for selected positive numbers x. (We use symmetry to obtain the graph of H when $x < 0$.) From the first three rows of Table 7, we see that as the values of x approach (get closer to) 0, the values of $H(x)$ become larger and larger positive numbers, so H is unbounded in the positive direction. In calculus we use limit notation, $\lim\limits_{x \to 0} H(x) = \infty$, which is read "the limit of $H(x)$ as x approaches zero equals infinity," to mean that $H(x) \to \infty$ as $x \to 0$.

Look at the last four rows of Table 7. As $x \to \infty$, the values of $H(x)$ approach 0 (the end behavior of the graph). In calculus, this is expressed by writing $\lim\limits_{x \to \infty} H(x) = 0$. Figure 25 shows the graph. Notice the use of red dashed lines to convey the ideas discussed above.

Table 7

x	$H(x) = \dfrac{1}{x^2}$
$\dfrac{1}{2}$	4
$\dfrac{1}{100}$	10,000
$\dfrac{1}{10,000}$	100,000,000
1	1
2	$\dfrac{1}{4}$
100	$\dfrac{1}{10,000}$
10,000	$\dfrac{1}{100,000,000}$

Figure 25 $H(x) = \dfrac{1}{x^2}$

EXAMPLE 3

Using Transformations to Graph a Rational Function

Graph the rational function: $R(x) = \dfrac{1}{(x-2)^2} + 1$

Solution   The domain of R is the set of all real numbers except $x = 2$. To graph R, start with the graph of $y = \dfrac{1}{x^2}$. See Figure 26 for the steps.

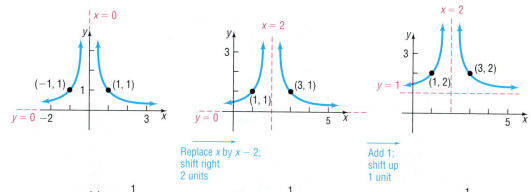

Figure 26

(a) $y = \dfrac{1}{x^2}$

Replace x by $x - 2$; shift right 2 units

(b) $y = \dfrac{1}{(x-2)^2}$

Add 1; shift up 1 unit

(c) $y = \dfrac{1}{(x-2)^2} + 1$

Now Work PROBLEMS 35(a) AND 35(b)

Asymptotes

Let's investigate the roles of the vertical line $x = 2$ and the horizontal line $y = 1$ in Figure 26(c).

First, we look at the end behavior of $R(x) = \dfrac{1}{(x-2)^2} + 1$. Table 8(a) shows the values of R at $x = 10, 100, 1000$, and $10,000$. Note that as x becomes unbounded in the positive direction, the values of R approach 1, so $\lim\limits_{x \to \infty} R(x) = 1$. From Table 8(b) we see that as x becomes unbounded in the negative direction, the values of R also approach 1, so $\lim\limits_{x \to -\infty} R(x) = 1$.

Even though $x = 2$ is not in the domain of R, the behavior of the graph of R near $x = 2$ is important. Table 8(c) shows the values of R at $x = 1.5, 1.9, 1.99, 1.999$, and 1.9999. We see that as x approaches 2 for $x < 2$, denoted $x \to 2^-$, the values of R are increasing without bound, so $\lim\limits_{x \to 2^-} R(x) = \infty$. From Table 8(d), we see that as x approaches 2 for $x > 2$, denoted $x \to 2^+$, the values of R are also increasing without bound, so $\lim\limits_{x \to 2^+} R(x) = \infty$.

Table 8

x	R(x)
10	1.0156
100	1.0001
1000	1.000001
10,000	1.00000001

(a)

x	R(x)
−10	1.0069
−100	1.0001
−1000	1.000001
−10,000	1.00000001

(b)

x	R(x)
1.5	5
1.9	101
1.99	10,001
1.999	1,000,001
1.9999	100,000,001

(c)

x	R(x)
2.5	5
2.1	101
2.01	10,001
2.001	1,000,001
2.0001	100,000,001

(d)

The vertical line $x = 2$ and the horizontal line $y = 1$ are called *asymptotes* of the graph of R.

DEFINITION

Let R denote a function.

If, as $x \to -\infty$ or as $x \to \infty$, the values of $R(x)$ approach some fixed number L, then the line $y = L$ is a **horizontal asymptote** of the graph of R. [Refer to Figures 27(a) and (b) on page 346.]

If, as x approaches some number c, the values $|R(x)| \to \infty$ [that is, $R(x) \to -\infty$ or $R(x) \to \infty$], then the line $x = c$ is a **vertical asymptote** of the graph of R. [Refer to Figures 27(c) and (d).]

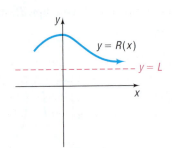

(a) End behavior:
As $x \to \infty$, the values of $R(x)$ approach L [$\lim\limits_{x\to\infty} R(x) = L$]. That is, the points on the graph of R are getting closer to the line $y = L$; $y = L$ is a horizontal asymptote.

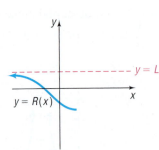

(b) End behavior:
As $x \to -\infty$, the values of $R(x)$ approach L [$\lim\limits_{x\to-\infty} R(x) = L$]. That is, the points on the graph of R are getting closer to the line $y = L$; $y = L$ is a horizontal asymptote.

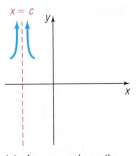

(c) As x approaches c, the values of $|R(x)| \to \infty$ [$\lim\limits_{x\to c^-} R(x) = \infty$; $\lim\limits_{x\to c^+} R(x) = \infty$]. That is, the points on the graph of R are getting closer to the line $x = c$; $x = c$ is a vertical asymptote.

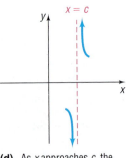

(d) As x approaches c, the values of $|R(x)| \to \infty$ [$\lim\limits_{x\to c^-} R(x) = -\infty$; $\lim\limits_{x\to c^+} R(x) = \infty$]. That is, the points on the graph of R are getting closer to the line $x = c$; $x = c$ is a vertical asymptote.

Figure 27

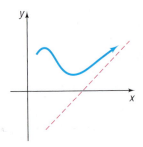

Figure 28 Oblique asymptote

A horizontal asymptote, when it occurs, describes the **end behavior** of the graph as $x \to \infty$ or as $x \to -\infty$. **The graph of a function may intersect a horizontal asymptote**.

A vertical asymptote, when it occurs, describes the behavior of the graph when x is close to some number c. **The graph of a rational function will never intersect a vertical asymptote**.

There is a third possibility. If, as $x \to -\infty$ or as $x \to \infty$, the value of a rational function $R(x)$ approaches a linear expression $ax + b$, $a \neq 0$, then the line $y = ax + b, a \neq 0$, is an **oblique (or slant) asymptote** of R. Figure 28 shows an oblique asymptote. An oblique asymptote, when it occurs, describes the end behavior of the graph. **The graph of a function may intersect an oblique asymptote**.

 Now Work PROBLEMS **27** AND **35(c)**

2 Find the Vertical Asymptotes of a Rational Function

The vertical asymptotes of a rational function $R(x) = \dfrac{p(x)}{q(x)}$, in lowest terms, are located at the real zeros of the denominator $q(x)$. Suppose that r is a real zero of q, so $x - r$ is a factor of q. As x approaches r, symbolized as $x \to r$, the values of $x - r$ approach 0, causing the ratio to become unbounded; that is, $|R(x)| \to \infty$. Based on the definition, we conclude that the line $x = r$ is a vertical asymptote.

THEOREM

WARNING If a rational function is not in lowest terms, an application of this theorem may result in an incorrect listing of vertical asymptotes. ∎

Locating Vertical Asymptotes

A rational function $R(x) = \dfrac{p(x)}{q(x)}$, *in lowest terms*, will have a vertical asymptote $x = r$ if r is a real zero of the *denominator* q. That is, if $x - r$ is a factor of the denominator q of a rational function $R(x) = \dfrac{p(x)}{q(x)}$, in lowest terms, R will have the vertical asymptote $x = r$.

EXAMPLE 4   **Finding Vertical Asymptotes**

Find the vertical asymptotes, if any, of the graph of each rational function.

(a) $F(x) = \dfrac{x + 3}{x - 1}$

(b) $R(x) = \dfrac{x}{x^2 - 4}$

(c) $H(x) = \dfrac{x^2}{x^2 + 1}$

(d) $G(x) = \dfrac{x^2 - 9}{x^2 + 4x - 21}$

Solution

(a) F is in lowest terms, and the only zero of the denominator is 1. The line $x = 1$ is the vertical asymptote of the graph of F.

(b) R is in lowest terms, and the zeros of the denominator $x^2 - 4$ are -2 and 2. The lines $x = -2$ and $x = 2$ are the vertical asymptotes of the graph of R.

(c) H is in lowest terms, and the denominator has no real zeros because the equation $x^2 + 1 = 0$ has no real solutions. The graph of H has no vertical asymptotes.

(d) Factor the numerator and denominator of $G(x)$ to determine whether it is in lowest terms.

$$G(x) = \frac{x^2 - 9}{x^2 + 4x - 21} = \frac{(x + 3)(x - 3)}{(x + 7)(x - 3)} = \frac{x + 3}{x + 7} \qquad x \neq 3$$

The only zero of the denominator of $G(x)$ in lowest terms is -7. The line $x = -7$ is the only vertical asymptote of the graph of G. ●

As Example 4 points out, rational functions can have no vertical asymptotes, one vertical asymptote, or more than one vertical asymptote.

Multiplicity and Vertical Asymptotes

Recall from Figure 15 in Section 5.1 that the end behavior of a polynomial function is always one of four types. For polynomials of odd degree, the ends of the graph go in opposite directions (one up and one down), whereas for polynomials of even degree, the ends go in the same direction (both up or both down).

For a rational function in lowest terms, the multiplicities of the zeros in the denominator can be used in a similar fashion to determine the behavior of the graph around each vertical asymptote. Consider the following four functions, each with a single vertical asymptote, $x = 2$.

$$R_1(x) = \frac{1}{x - 2} \qquad R_2(x) = -\frac{1}{x - 2} \qquad R_3(x) = \frac{1}{(x - 2)^2} \qquad R_4(x) = -\frac{1}{(x - 2)^2}$$

Figure 29 shows the graphs of each function. The graphs of R_1 and R_2 are transformations of the graph of $y = \frac{1}{x}$, and the graphs of R_3 and R_4 are transformations of the graph of $y = \frac{1}{x^2}$.

Based on Figure 29, we can make the following conclusions:

- If the multiplicity of the zero that gives rise to a vertical asymptote is odd, the graph approaches ∞ on one side of the vertical asymptote and approaches $-\infty$ on the other side.

- If the multiplicity of the zero that gives rise to the vertical asymptote is even, the graph approaches either ∞ or $-\infty$ on both sides of the vertical asymptote.

These results are true in general and will be helpful when graphing rational functions in the next section.

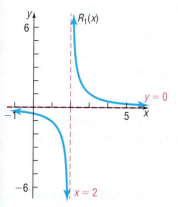

(a) Odd multiplicity
$$\lim_{x \to 2^-} R_1(x) = -\infty$$
$$\lim_{x \to 2^+} R_1(x) = \infty$$

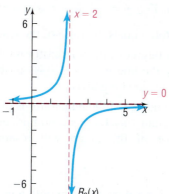

(b) Odd multiplicity
$$\lim_{x \to 2^-} R_2(x) = \infty$$
$$\lim_{x \to 2^+} R_2(x) = -\infty$$

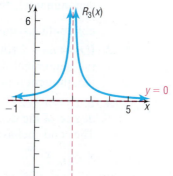

(c) Even multiplicity
$$\lim_{x \to 2^-} R_3(x) = \infty$$
$$\lim_{x \to 2^+} R_3(x) = \infty$$

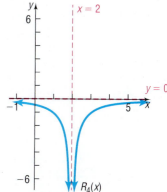

(d) Even multiplicity
$$\lim_{x \to 2^-} R_4(x) = -\infty$$
$$\lim_{x \to 2^+} R_4(x) = -\infty$$

Figure 29

3 Find the Horizontal or Oblique Asymptote of a Rational Function

To find horizontal or oblique asymptotes, we need to know how the value of the function behaves as $x \rightarrow -\infty$ or as $x \rightarrow \infty$. That is, we need to determine the end behavior of the function. This can be done by examining the degrees of the numerator and denominator, and the respective power functions that each resembles. For example, consider the rational function

$$R(x) = \frac{3x - 2}{5x^2 - 7x + 1}$$

The degree of the numerator, 1, is less than the degree of the denominator, 2. When $|x|$ is very large, the numerator of R can be approximated by the power function $y = 3x$, and the denominator can be approximated by the power function $y = 5x^2$. This means

$$R(x) = \frac{3x - 2}{5x^2 - 7x + 1} \approx \frac{3x}{5x^2} = \frac{3}{5x} \rightarrow 0$$

For $|x|$ very large   As $x \rightarrow -\infty$ or $x \rightarrow \infty$

which shows that the line $y = 0$ is a horizontal asymptote. This result is true for all rational functions that are **proper** (that is, the degree of the numerator is less than the degree of the denominator). If a rational function is **improper** (that is, if the degree of the numerator is greater than or equal to the degree of the denominator), there could be a horizontal asymptote, an oblique asymptote, or neither. The following summary details how to find horizontal or oblique asymptotes.

Finding a Horizontal or Oblique Asymptote of a Rational Function

Consider the rational function

$$R(x) = \frac{p(x)}{q(x)} = \frac{a_n x^n + a_{n-1}x^{n-1} + \cdots + a_1 x + a_0}{b_m x^m + b_{m-1}x^{m-1} + \cdots + b_1 x + b_0}$$

in which the degree of the numerator is n and the degree of the denominator is m.

1. If $n < m$ (the degree of the numerator is less than the degree of the denominator), the line $y = 0$ is a horizontal asymptote.

2. If $n = m$ (the degree of the numerator equals the degree of the denominator), the line $y = \dfrac{a_n}{b_m}$ is a horizontal asymptote. (That is, the horizontal asymptote equals the ratio of the leading coefficients.)

3. If $n = m + 1$ (the degree of the numerator is one more than the degree of the denominator), the line $y = ax + b$ is an oblique asymptote, which is the quotient found using long division.

4. If $n \geq m + 2$ (the degree of the numerator is two or more greater than the degree of the denominator), there are no horizontal or oblique asymptotes. The end behavior of the graph will resemble the power function $y = \dfrac{a_n}{b_m}x^{n-m}$.

Note: A rational function will never have both a horizontal asymptote and an oblique asymptote. A rational function may have neither a horizontal nor an oblique asymptote.

We illustrate each of the possibilities in Examples 5 through 8.

EXAMPLE 5

Finding a Horizontal Asymptote

Find the horizontal asymptote, if one exists, of the graph of

$$R(x) = \frac{4x^3 - 5x + 2}{7x^5 + 2x^4 - 3x}$$

Solution   Since the degree of the numerator, 3, is less than the degree of the denominator, 5, the rational function R is proper. The line $y = 0$ is a horizontal asymptote of the graph of R.   ●

EXAMPLE 6

Finding a Horizontal or Oblique Asymptote

Find the horizontal or oblique asymptote, if one exists, of the graph of

$$H(x) = \frac{3x^4 - x^2}{x^3 - x^2 + 1}$$

Solution   Since the degree of the numerator, 4, is exactly one greater than the degree of the denominator, 3, the rational function H has an oblique asymptote. Find the asymptote by using long division.

$$
\begin{array}{r}
3x + 3 \\
x^3 - x^2 + 1 \overline{)3x^4 \qquad\quad - x^2 \qquad\quad} \\
\underline{3x^4 - 3x^3 \qquad\quad + 3x} \\
3x^3 - x^2 - 3x \\
\underline{3x^3 - 3x^2 \qquad + 3} \\
2x^2 - 3x - 3
\end{array}
$$

As a result,

$$H(x) = \frac{3x^4 - x^2}{x^3 - x^2 + 1} = 3x + 3 + \frac{2x^2 - 3x - 3}{x^3 - x^2 + 1}$$

As $x \to -\infty$ or as $x \to \infty$,

$$\frac{2x^2 - 3x - 3}{x^3 - x^2 + 1} \approx \frac{2x^2}{x^3} = \frac{2}{x} \to 0$$

As $x \to -\infty$ or as $x \to \infty$, we have $H(x) \to 3x + 3$. The graph of the rational function H has an oblique asymptote $y = 3x + 3$. Put another way, as $x \to \pm\infty$, the graph of H will behave like the graph of $y = 3x + 3$.   ●

EXAMPLE 7

Finding a Horizontal or Oblique Asymptote

Find the horizontal or oblique asymptote, if one exists, of the graph of

$$R(x) = \frac{8x^2 - x + 2}{4x^2 - 1}$$

Solution   Since the degree of the numerator, 2, equals the degree of the denominator, 2, the rational function R has a horizontal asymptote equal to the ratio of the leading coefficients.

$$y = \frac{a_n}{b_m} = \frac{8}{4} = 2$$

To see why the horizontal asymptote equals the ratio of the leading coefficients, investigate the behavior of R as $x \to -\infty$ or as $x \to \infty$. When $|x|$ is very large, the numerator of R can be approximated by the power function $y = 8x^2$, and the denominator can be approximated by the power function $y = 4x^2$. This means that as $x \to -\infty$ or as $x \to \infty$,

$$R(x) = \frac{8x^2 - x + 2}{4x^2 - 1} \approx \frac{8x^2}{4x^2} = \frac{8}{4} = 2$$

The graph of the rational function R has a horizontal asymptote $y = 2$. The graph of R will behave like $y = 2$ as $x \to \pm\infty$. ●

EXAMPLE 8   **Finding a Horizontal or Oblique Asymptote**

Find the horizontal or oblique asymptote, if one exists, of the graph of

$$G(x) = \frac{2x^5 - x^3 + 2}{x^3 - 1}$$

Solution   Since the degree of the numerator, 5, is greater than the degree of the denominator, 3, by more than one, the rational function G has no horizontal or oblique asymptote. The end behavior of the graph will resemble the power function $y = 2x^{5-3} = 2x^2$.

To see why this is the case, investigate the behavior of G as $x \to -\infty$ or as $x \to \infty$. When $|x|$ is very large, the numerator of G can be approximated by the power function $y = 2x^5$, and the denominator can be approximated by the power function $y = x^3$. This means as $x \to -\infty$ or as $x \to \infty$,

$$G(x) = \frac{2x^5 - x^3 + 2}{x^3 - 1} \approx \frac{2x^5}{x^3} = 2x^{5-3} = 2x^2$$

Since this is not linear, the graph of G has no horizontal or oblique asymptote. The graph of G will behave like $y = 2x^2$ as $x \to \pm\infty$. ●

 Now Work   PROBLEMS **45, 47, AND 49**

5.2 Assess Your Understanding

'Are You Prepared?' *Answers are given at the end of these exercises. If you get a wrong answer, read the pages listed in* red.

1. *True or False* The quotient of two polynomial expressions is a rational expression. (pp. 62–69)

2. What are the quotient and remainder when $3x^4 - x^2$ is divided by $x^3 - x^2 + 1$. (pp. 44–47)

3. Graph $y = \dfrac{1}{x}$. (p. 164)

4. Graph $y = 2(x + 1)^2 - 3$ using transformations. (pp. 247–256)

Concepts and Vocabulary

5. *True or False* The domain of every rational function is the set of all real numbers.

6. If, as $x \to -\infty$ or as $x \to \infty$, the values of $R(x)$ approach some fixed number L, then the line $y = L$ is a _____ _____ of the graph of R.

7. If, as x approaches some number c, the values of $|R(x)| \to \infty$, then the line $x = c$ is a _____ _____ of the graph of R.

8. For a rational function R, if the degree of the numerator is less than the degree of the denominator, then R is _____.

9. *True or False* The graph of a rational function may intersect a horizontal asymptote.

10. *True or False* The graph of a rational function may intersect a vertical asymptote.

11. If a rational function is proper, then _____ is a horizontal asymptote.

12. *True or False* If the degree of the numerator of a rational function equals the degree of the denominator, then the ratio of the leading coefficients gives rise to the horizontal asymptote.

13. If $R(x) = \dfrac{p(x)}{q(x)}$ is a rational function and if p and q have no common factors, then R is _____.

    (a) improper   (b) proper
    (c) undefined   (d) in lowest terms

14. Which type of asymptote, when it occurs, describes the behavior of a graph when x is close to some number?

    (a) vertical   (b) horizontal   (c) oblique   (d) all of these

Skill Building

In Problems 15–26, find the domain of each rational function.

15. $R(x) = \dfrac{4x}{x - 3}$

16. $R(x) = \dfrac{5x^2}{3 + x}$

 17. $H(x) = \dfrac{-4x^2}{(x - 2)(x + 4)}$

18. $G(x) = \dfrac{6}{(x + 3)(4 - x)}$

19. $F(x) = \dfrac{3x(x - 1)}{2x^2 - 5x - 3}$

20. $Q(x) = \dfrac{-x(1 - x)}{3x^2 + 5x - 2}$

21. $R(x) = \dfrac{x}{x^3 - 8}$

22. $R(x) = \dfrac{x}{x^4 - 1}$

23. $H(x) = \dfrac{3x^2 + x}{x^2 + 4}$

24. $G(x) = \dfrac{x - 3}{x^4 + 1}$

25. $R(x) = \dfrac{3(x^2 - x - 6)}{4(x^2 - 9)}$

26. $F(x) = \dfrac{-2(x^2 - 4)}{3(x^2 + 4x + 4)}$

In Problems 27–32, use the graph shown to find
(a) The domain and range of each function   *(b) The intercepts, if any*   *(c) Horizontal asymptotes, if any*
(d) Vertical asymptotes, if any   *(e) Oblique asymptotes, if any*

27.
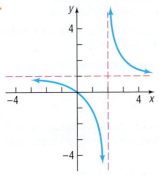

28.
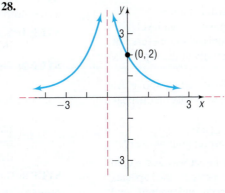

29.
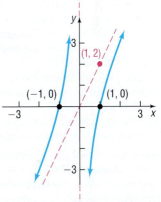

30.
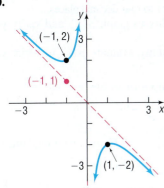

31.
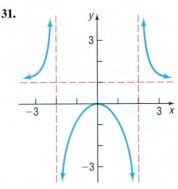

32.
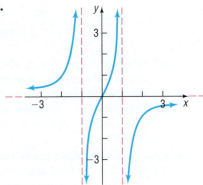

In Problems 33–44, (a) graph the rational function using transformations, (b) use the final graph to find the domain and range, and (c) use the final graph to list any vertical, horizontal, or oblique asymptotes.

33. $F(x) = 2 + \dfrac{1}{x}$

34. $Q(x) = 3 + \dfrac{1}{x^2}$

35. $R(x) = \dfrac{1}{(x - 1)^2}$

36. $R(x) = \dfrac{3}{x}$

37. $H(x) = \dfrac{-2}{x + 1}$

38. $G(x) = \dfrac{2}{(x + 2)^2}$

39. $R(x) = \dfrac{-1}{x^2 + 4x + 4}$

40. $R(x) = \dfrac{1}{x - 1} + 1$

41. $G(x) = 1 + \dfrac{2}{(x - 3)^2}$

42. $F(x) = 2 - \dfrac{1}{x + 1}$

43. $R(x) = \dfrac{x^2 - 4}{x^2}$

44. $R(x) = \dfrac{x - 4}{x}$

In Problems 45–56, find the vertical, horizontal, and oblique asymptotes, if any, of each rational function.

45. $R(x) = \dfrac{3x}{x + 4}$

46. $R(x) = \dfrac{3x + 5}{x - 6}$

47. $H(x) = \dfrac{x^3 - 8}{x^2 - 5x + 6}$

48. $G(x) = \dfrac{x^3 + 1}{x^2 - 5x - 14}$

49. $T(x) = \dfrac{x^3}{x^4 - 1}$

50. $P(x) = \dfrac{4x^2}{x^3 - 1}$

51. $Q(x) = \dfrac{2x^2 - 5x - 12}{3x^2 - 11x - 4}$

52. $F(x) = \dfrac{x^2 + 6x + 5}{2x^2 + 7x + 5}$

53. $R(x) = \dfrac{6x^2 + 7x - 5}{3x + 5}$

54. $R(x) = \dfrac{8x^2 + 26x - 7}{4x - 1}$

55. $G(x) = \dfrac{x^4 - 1}{x^2 - x}$

56. $F(x) = \dfrac{x^4 - 16}{x^2 - 2x}$

Applications and Extensions

57. Gravity In physics, it is established that the acceleration due to gravity, g (in meters/sec^2), at a height h meters above sea level is given by

$$g(h) = \frac{3.99 \times 10^{14}}{(6.374 \times 10^6 + h)^2}$$

where 6.374×10^6 is the radius of Earth in meters.
(a) What is the acceleration due to gravity at sea level?
(b) The Willis Tower in Chicago, Illinois, is 443 meters tall. What is the acceleration due to gravity at the top of the Willis Tower?
(c) The peak of Mount Everest is 8848 meters above sea level. What is the acceleration due to gravity on the peak of Mount Everest?
(d) Find the horizontal asymptote of $g(h)$.
(e) Solve $g(h) = 0$. How do you interpret your answer?

58. Population Model A rare species of insect was discovered in the Amazon Rain Forest. To protect the species, environmentalists declared the insect endangered and transplanted the insect into a protected area. The population P of the insect t months after being transplanted is

$$P(t) = \frac{50(1 + 0.5t)}{2 + 0.01t}$$

(a) How many insects were discovered? In other words, what was the population when $t = 0$?
(b) What will the population be after 5 years?
(c) Determine the horizontal asymptote of $P(t)$. What is the largest population that the protected area can sustain?

59. Resistance in Parallel Circuits From Ohm's Law for circuits, it follows that the total resistance R_{tot} of two components hooked in parallel is given by the equation

$$R_{tot} = \frac{R_1 R_2}{R_1 + R_2}$$

where R_1 and R_2 are the individual resistances.
(a) Let $R_1 = 10$ ohms, and graph R_{tot} as a function of R_2.
(b) Find and interpret any asymptotes of the graph obtained in part (a).
(c) If $R_2 = 2\sqrt{R_1}$, what value of R_1 will yield an R_{tot} of 17 ohms?

60. Newton's Method In calculus you will learn that if

$$p(x) = a_n x^n + a_{n-1} x^{n-1} + \cdots + a_1 x + a_0$$

is a polynomial function, then the *derivative* of $p(x)$ is

$$p'(x) = n a_n x^{n-1} + (n - 1) a_{n-1} x^{n-2} + \cdots + 2a_2 x + a_1$$

Newton's Method is an efficient method for approximating the x-intercepts (or real zeros) of a function, such as $p(x)$. The following steps outline Newton's Method.

STEP 1: Select an initial value x_0 that is somewhat close to the x-intercept being sought.

STEP 2: Find values for x using the relation

$$x_{n+1} = x_n - \frac{p(x_n)}{p'(x_n)} \quad n = 1, 2, \ldots$$

until you get two consecutive values x_n and x_{n+1} that agree to whatever decimal place accuracy you desire.

STEP 3: The approximate zero will be x_{n+1}.

Consider the polynomial $p(x) = x^3 - 7x - 40$.
(a) Evaluate $p(5)$ and $p(-3)$.
(b) What might we conclude about a zero of p? Explain.
(c) Use Newton's Method to approximate an x-intercept, r, $-3 < r < 5$, of $p(x)$ to four decimal places.
(d) Use a graphing utility to graph $p(x)$ and verify your answer in part (c).
(e) Using a graphing utility, evaluate $p(r)$ to verify your result.

61. Exploration The standard form of the rational function $R(x) = \dfrac{mx + b}{cx + d}$, where $c \neq 0$,

is $R(x) = a\left(\dfrac{1}{x - h}\right) + k$. To write a rational function in standard form requires long division.

(a) Write the rational function $R(x) = \dfrac{2x + 3}{x - 1}$ in standard form by writing R in the form

$$\text{Quotient} + \frac{\text{remainder}}{\text{divisor}}$$

(b) Graph R using transformations.
(c) Determine the vertical asymptote and the horizontal asymptote of R.

62. Exploration Repeat Problem 61 for the rational function $R(x) = \dfrac{-6x + 16}{2x - 7}$.

Explaining Concepts: Discussion and Writing

63. If the graph of a rational function R has the vertical asymptote $x = 4$, the factor $x - 4$ must be present in the denominator of R. Explain why.

64. If the graph of a rational function R has the horizontal asymptote $y = 2$, the degree of the numerator of R equals the degree of the denominator of R. Explain why.

65. The graph of a rational function cannot have both a horizontal and an oblique asymptote? Explain why.

66. Make up a rational function that has $y = 2x + 1$ as an oblique asymptote. Explain the methodology that you used.

Retain Your Knowledge

Problems 67–70 are based on material learned earlier in the course. The purpose of these problems is to keep the material fresh in your mind so that you are better prepared for the final exam.

67. Find the equation of a vertical line passing through the point $(5, -3)$.

68. Solve: $\dfrac{2}{5}(3x - 7) + 1 = \dfrac{x}{4} - 2$

69. Determine whether the graph of the equation $2x^3 - xy^2 = 4$ is symmetric with respect to the x-axis, the y-axis, the origin, or none of these.

70. What are the points of intersection of the graphs of the functions $f(x) = -3x + 2$ and $g(x) = x^2 - 2x - 4$?

'Are You Prepared?' Answers

1. True

2. Quotient: $3x + 3$; remainder: $2x^2 - 3x - 3$

3.

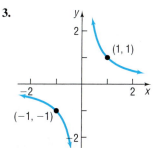

4.

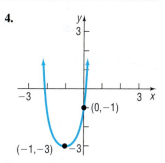

5.3 The Graph of a Rational Function

PREPARING FOR THIS SECTION *Before getting started, review the following:*

- Intercepts Section 2.2, pp. 159–160

Now Work the 'Are You Prepared?' problem on page 365.

OBJECTIVES **1** Analyze the Graph of a Rational Function (p. 353)
2 Solve Applied Problems Involving Rational Functions (p. 364)

1 Analyze the Graph of a Rational Function

We commented earlier that calculus provides the tools required to graph a polynomial function accurately. The same holds true for rational functions. However, we can gather together quite a bit of information about their graphs to get an idea of the general shape and position of the graph.

EXAMPLE 1

How to Analyze the Graph of a Rational Function

Analyze the graph of the rational function: $R(x) = \dfrac{x - 1}{x^2 - 4}$

Step-by-Step Solution

Step 1: Factor the numerator and denominator of R. Find the domain of the rational function.

$$R(x) = \frac{x - 1}{x^2 - 4} = \frac{x - 1}{(x + 2)(x - 2)}$$

The domain of R is $\{x \,|\, x \neq -2, x \neq 2\}$.

Step 2: Write R in lowest terms.

Because there are no common factors between the numerator and denominator, R is in lowest terms.

Step 3: Find and plot the intercepts of the graph. Use multiplicity to determine the behavior of the graph of R at each x-intercept.

Since 0 is in the domain of R, the y-intercept is $R(0) = \dfrac{1}{4}$. Plot the point $\left(0, \dfrac{1}{4}\right)$. The x-intercepts are found by determining the real zeros of the numerator of R written in lowest terms. By solving $x - 1 = 0$, we find that the only real zero of the numerator is 1, so the only x-intercept of the graph of R is 1. Plot the point $(1,0)$. The multiplicity of 1 is odd, so the graph will cross the x-axis at $x = 1$.

Step 4: Find the vertical asymptotes. Graph each vertical asymptote using a dashed line. Determine the behavior of the graph on either side of each vertical asymptote.

The vertical asymptotes are the zeros of the denominator with the rational function in lowest terms. With R written in lowest terms, we find that the graph of R has two vertical asymptotes: the lines $x = -2$ and $x = 2$. See Figure 30(a). The multiplicities of the zeros that give rise to the vertical asymptotes are both odd. Therefore, the graph will approach ∞ on one side of each vertical asymptote, and will approach $-\infty$ on the other side.

Step 5: Find the horizontal or oblique asymptote, if one exists. Find points, if any, at which the graph of R intersects this asymptote. Graph the asymptote using a dashed line. Plot any points at which the graph of R intersects the asymptote.

Because the degree of the numerator is less than the degree of the denominator, R is proper and the line $y = 0$ (the x-axis) is a horizontal asymptote of the graph. To determine whether the graph of R intersects the horizontal asymptote, solve the equation $R(x) = 0$:

$$\frac{x - 1}{x^2 - 4} = 0$$
$$x - 1 = 0$$
$$x = 1$$

The only solution is $x = 1$, so the graph of R intersects the horizontal asymptote at $(1, 0)$.

Step 6: Use the zeros of the numerator and denominator of R to divide the x-axis into intervals. Determine where the graph of R is above or below the x-axis by choosing a number in each interval and evaluating R there. Plot the points found.

The zero of the numerator, 1, and the zeros of the denominator, -2 and 2, divide the x-axis into four intervals:

$$(-\infty, -2) \qquad (-2, 1) \qquad (1, 2) \qquad (2, \infty)$$

Now construct Table 9.

Table 9

Interval	$(-\infty, -2)$	$(-2, 1)$	$(1, 2)$	$(2, \infty)$
Number chosen	-3	-1	$\dfrac{3}{2}$	3
Value of R	$R(-3) = -0.8$	$R(-1) = \dfrac{2}{3}$	$R\left(\dfrac{3}{2}\right) = -\dfrac{2}{7}$	$R(3) = 0.4$
Location of graph	Below x-axis	Above x-axis	Below x-axis	Above x-axis
Point on graph	$(-3, -0.8)$	$\left(-1, \dfrac{2}{3}\right)$	$\left(\dfrac{3}{2}, -\dfrac{2}{7}\right)$	$(3, 0.4)$

Figure 30(a) shows the asymptotes, the points from Table 9, the y-intercept, and the x-intercept.

Step 7: Use the results obtained in Steps 1 through 6 to graph R.

- The graph crosses the x-axis at $x = 1$, changing from being above the x-axis for $x < 1$ to below it for $x > 1$. Indicate this on the graph. See Figure 30(b).
- Since $y = 0$ (the x-axis) is a horizontal asymptote and the graph lies below the x-axis for $x < -2$, we can sketch a portion of the graph by placing a small arrow to the far left and under the x-axis.

- Since the line $x = -2$ is a vertical asymptote and the graph lies below the x-axis for $x < -2$, we place an arrow well below the x-axis and approaching the line $x = -2$ from the left $\left(\lim\limits_{x \to -2^-} R(x) = -\infty \right)$.

- Since the graph approaches $-\infty$ on one side of $x = -2$, and -2 is a zero of odd multiplicity, the graph will approach ∞ on the other side of $x = -2$. That is, $\lim\limits_{x \to -2^+} R(x) = \infty$. Similar analysis leads to $\lim\limits_{x \to 2^-} R(x) = -\infty$ and $\lim\limits_{x \to 2^+} R(x) = \infty$. Finally, $\lim\limits_{x \to -\infty} R(x) = 0$ and $\lim\limits_{x \to \infty} R(x) = 0$.

  Figure 30(b) illustrates these conclusions and Figure 30(c) shows the graph of R.

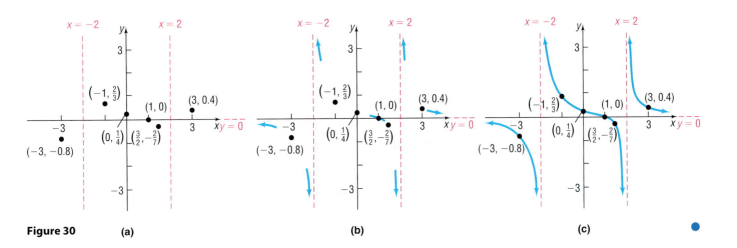

Figure 30   (a)   (b)   (c)

Exploration

Graph the rational function: $R(x) = \dfrac{x - 1}{x^2 - 4}$

Result The analysis just completed in Example 1 helps us to set the viewing rectangle to obtain a complete graph. Figure 31(a) shows the graph of $R(x) = \dfrac{x - 1}{x^2 - 4}$ in connected mode, and Figure 31(b) shows it in dot mode. Notice in Figure 31(a) that the graph has vertical lines at $x = -2$ and $x = 2$. This is due to the fact that when a graphing utility is in connected mode, some will connect the dots between consecutive pixels, and vertical lines may occur. We know that the graph of R does not cross the lines $x = -2$ and $x = 2$, since R is not defined at $x = -2$ or $x = 2$. So, when graphing rational functions, use dot mode if extraneous vertical lines are present in connected mode. Newer graphing utilities may not have extraneous vertical lines in connected mode. See Figure 31(c).

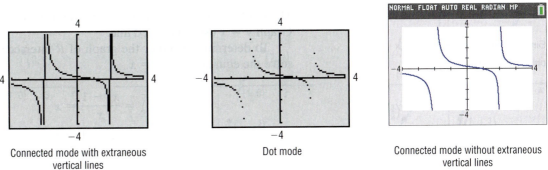

Connected mode with extraneous   Dot mode   Connected mode without extraneous
vertical lines   vertical lines

Figure 31   (a)   (b)   (c)

To find out whether the graph of R intersects the horizontal asymptote $y = 2$, solve the equation $R(x) = 2$.

$$R(x) = \frac{2x - 1}{x + 2} = 2$$
$$2x - 1 = 2(x + 2)$$
$$2x - 1 = 2x + 4$$
$$-1 = 4 \qquad \text{Impossible}$$

The graph does not intersect the line $y = 2$.

STEP 6: Look at the factored expression for R in Step 1. The real zeros of the numerator and denominator, $-2, \frac{1}{2}$, and 2, divide the x-axis into four intervals:

$$(-\infty, -2) \qquad \left(-2, \frac{1}{2}\right) \qquad \left(\frac{1}{2}, 2\right) \qquad (2, \infty)$$

Construct Table 13. Plot the points in Table 13.

Table 13

	-2	$1/2$	2	x
Interval	$(-\infty, -2)$	$\left(-2, \frac{1}{2}\right)$	$\left(\frac{1}{2}, 2\right)$	$(2, \infty)$
Number chosen	-3	-1	1	3
Value of R	$R(-3) = 7$	$R(-1) = -3$	$R(1) = \frac{1}{3}$	$R(3) = 1$
Location of graph	Above x-axis	Below x-axis	Above x-axis	Above x-axis
Point on graph	$(-3, 7)$	$(-1, -3)$	$\left(1, \frac{1}{3}\right)$	$(3, 1)$

STEP 7: From Table 13 we know that the graph of R is above the x-axis for $x < -2$. From Step 5 we know that the graph of R does not intersect the asymptote $y = 2$. Therefore, the graph of R will approach $y = 2$ from above as $x \to -\infty$ and will approach the vertical asymptote $x = -2$ at the top from the left.

Since the graph of R is below the x-axis for $-2 < x < \frac{1}{2}$, the graph will approach $x = -2$ at the bottom from the right. Finally, since the graph of R is above the x-axis for $x > \frac{1}{2}$ and does not intersect the horizontal asymptote $y = 2$, the graph of R will approach $y = 2$ from below as $x \to \infty$. The graph crosses the x-axis at $x = \frac{1}{2}$, changing from being below the x-axis to being above. See Figure 36(a).

See Figure 36(b) for the complete graph. Since R is not defined at 2, there is a hole at the point $\left(2, \frac{3}{4}\right)$.

NOTE The coordinates of the hole were obtained by evaluating R in lowest terms at $x = 2$. R in lowest terms is $\frac{2x - 1}{x + 2}$, which, at $x = 2$, is $\frac{2(2) - 1}{2 + 2} = \frac{3}{4}$. ∎

Exploration

Graph $R(x) = \frac{2x^2 - 5x + 2}{x^2 - 4}$. Do you see the hole at $\left(2, \frac{3}{4}\right)$? TRACE along the graph. Did you obtain an ERROR at $x = 2$? Are you convinced that an algebraic analysis of a rational function is required in order to accurately interpret the graph obtained with a graphing utility?

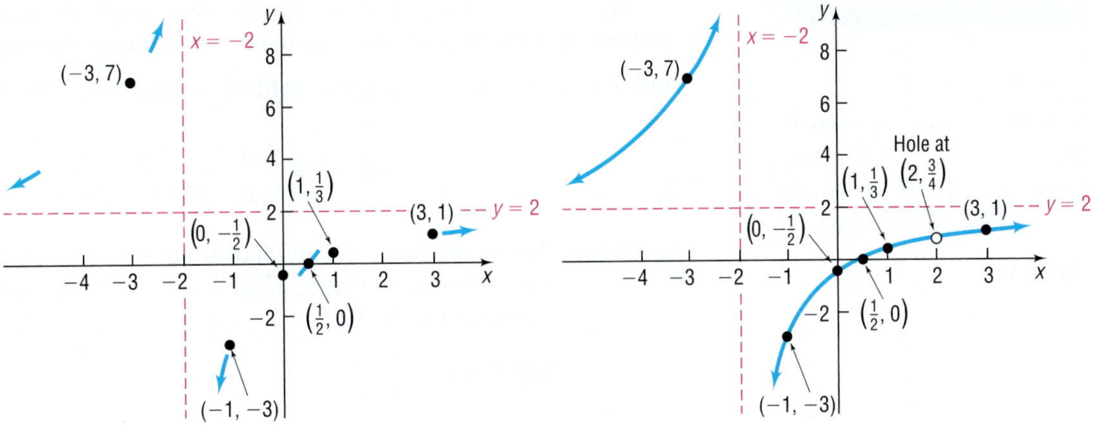

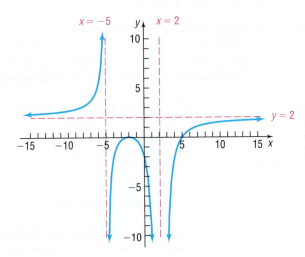

Figure 36

(a)   (b)

As Example 5 shows, **the zeros of the denominator of a rational function give rise to either vertical asymptotes or holes on the graph.**

 **Now Work** PROBLEM 33

EXAMPLE 6 | **Constructing a Rational Function from Its Graph**

Find a rational function that might have the graph shown in Figure 37.

Figure 37

Solution   The numerator of a rational function $R(x) = \dfrac{p(x)}{q(x)}$ in lowest terms determines the x-intercepts of its graph. The graph shown in Figure 37 has x-intercepts -2 (even multiplicity; graph touches the x-axis) and 5 (odd multiplicity; graph crosses the x-axis). So one possibility for the numerator is $p(x) = (x + 2)^2(x - 5)$.

The denominator of a rational function in lowest terms determines the vertical asymptotes of its graph. The vertical asymptotes of the graph are $x = -5$ and $x = 2$. Since $R(x)$ approaches ∞ to the left of $x = -5$ and $R(x)$ approaches $-\infty$ to the right of $x = -5$, we know that $(x + 5)$ is a factor of odd multiplicity in $q(x)$. Also, $R(x)$ approaches $-\infty$ on both sides of $x = 2$, so $(x - 2)$ is a factor of even multiplicity in $q(x)$. A possibility for the denominator is $q(x) = (x + 5)(x - 2)^2$.

So far we have $R(x) = \dfrac{(x + 2)^2(x - 5)}{(x + 5)(x - 2)^2}$.

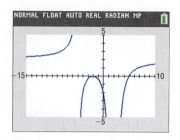

Figure 38

However, the horizontal asymptote of the graph given in Figure 37 is $y = 2$, so we know that the degree of the numerator must equal the degree of the denominator and that the quotient of leading coefficients must be $\frac{2}{1}$. This leads to

$$R(x) = \frac{2(x + 2)^2(x - 5)}{(x + 5)(x - 2)^2}$$

 Check: Figure 38 shows the graph of R on a graphing utility. Since Figure 38 looks similar to Figure 37, we have found a rational function R for the graph in Figure 37.

Now Work PROBLEM 51

2 Solve Applied Problems Involving Rational Functions

EXAMPLE 7

Finding the Least Cost of a Can

Reynolds Metal Company manufactures aluminum cans in the shape of a cylinder with a capacity of 500 cubic centimeters $\left(\frac{1}{2} \text{ liter}\right)$. The top and bottom of the can are made of a special aluminum alloy that costs 0.05¢ per square centimeter. The sides of the can are made of material that costs 0.02¢ per square centimeter.

(a) Express the cost of material for the can as a function of the radius r of the can.
(b) Use a graphing utility to graph the function $C = C(r)$.
(c) What value of r will result in the least cost?
(d) What is this least cost?

Solution

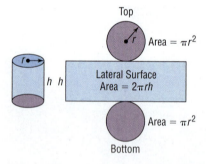

Figure 39

(a) Figure 39 illustrates the components of a can in the shape of a right circular cylinder. Notice that the material required to produce a cylindrical can of height h and radius r consists of a rectangle of area $2\pi rh$ and two circles, each of area πr^2. The total cost C (in cents) of manufacturing the can is therefore

$$C = \text{Cost of the top and bottom} + \text{Cost of the side}$$
$$= \underbrace{2(\pi r^2)}_{\substack{\text{Total area} \\ \text{of top and} \\ \text{bottom}}} \cdot \underbrace{(0.05)}_{\substack{\text{Cost/unit} \\ \text{area}}} + \underbrace{(2\pi rh)}_{\substack{\text{Total} \\ \text{area of} \\ \text{side}}} \cdot \underbrace{(0.02)}_{\substack{\text{Cost/unit} \\ \text{area}}}$$
$$= 0.10\pi r^2 + 0.04\pi rh$$

There is an additional restriction that the height h and radius r must be chosen so that the volume V of the can is 500 cubic centimeters. Since $V = \pi r^2 h$, we have

$$500 = \pi r^2 h \quad \text{so} \quad h = \frac{500}{\pi r^2}$$

Substituting this expression for h, we find that the cost C, in cents, as a function of the radius r is

$$C(r) = 0.10\pi r^2 + 0.04\pi r \cdot \frac{500}{\pi r^2} = 0.10\pi r^2 + \frac{20}{r} = \frac{0.10\pi r^3 + 20}{r}$$

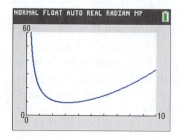

Figure 40

(b) See Figure 40 for the graph of $C = C(r)$.
(c) Using the MINIMUM command, the cost is least for a radius of about 3.17 centimeters.
(d) The least cost is $C(3.17) \approx 9.47$¢.

 Now Work PROBLEM 61

5.3 Assess Your Understanding

'Are You Prepared?' *The answer is given at the end of these exercises. If you get a wrong answer, read the pages listed in* red.

1. Find the intercepts of the graph of the equation $y = \dfrac{x^2 - 1}{x^2 - 4}$. (pp. 159–160)

Concepts and Vocabulary

2. *True or False* Every rational function has at least one asymptote.

3. Which type of asymptote will never intersect the graph of a rational function?
(a) horizontal (b) oblique (c) vertical (d) all of these

4. Identify the y-intercept of the graph of
$$R(x) = \frac{6(x - 1)}{(x + 1)(x + 2)}.$$
(a) -3 (b) -2 (c) -1 (d) 1

5. $R(x) = \dfrac{x(x - 2)^2}{x - 2}$
(a) Find the domain of R.
(b) Find the x-intercepts of R.

6. *True or False* The graph of a rational function sometimes has a hole.

Skill Building

In Problems 7–50, follow Steps 1 through 7 on page 356 to analyze the graph of each function.

7. $R(x) = \dfrac{x + 1}{x(x + 4)}$

8. $R(x) = \dfrac{x}{(x - 1)(x + 2)}$

9. $R(x) = \dfrac{3x + 3}{2x + 4}$

10. $R(x) = \dfrac{2x + 4}{x - 1}$

11. $R(x) = \dfrac{3}{x^2 - 4}$

12. $R(x) = \dfrac{6}{x^2 - x - 6}$

13. $P(x) = \dfrac{x^4 + x^2 + 1}{x^2 - 1}$

14. $Q(x) = \dfrac{x^4 - 1}{x^2 - 4}$

15. $H(x) = \dfrac{x^3 - 1}{x^2 - 9}$

16. $G(x) = \dfrac{x^3 + 1}{x^2 + 2x}$

17. $R(x) = \dfrac{x^2}{x^2 + x - 6}$

18. $R(x) = \dfrac{x^2 + x - 12}{x^2 - 4}$

19. $G(x) = \dfrac{x}{x^2 - 4}$

20. $G(x) = \dfrac{3x}{x^2 - 1}$

21. $R(x) = \dfrac{3}{(x - 1)(x^2 - 4)}$

22. $R(x) = \dfrac{-4}{(x + 1)(x^2 - 9)}$

23. $H(x) = \dfrac{x^2 - 1}{x^4 - 16}$

24. $H(x) = \dfrac{x^2 + 4}{x^4 - 1}$

25. $F(x) = \dfrac{x^2 - 3x - 4}{x + 2}$

26. $F(x) = \dfrac{x^2 + 3x + 2}{x - 1}$

27. $R(x) = \dfrac{x^2 + x - 12}{x - 4}$

28. $R(x) = \dfrac{x^2 - x - 12}{x + 5}$

29. $F(x) = \dfrac{x^2 + x - 12}{x + 2}$

30. $G(x) = \dfrac{x^2 - x - 12}{x + 1}$

31. $R(x) = \dfrac{x(x - 1)^2}{(x + 3)^3}$

32. $R(x) = \dfrac{(x - 1)(x + 2)(x - 3)}{x(x - 4)^2}$

33. $R(x) = \dfrac{x^2 + x - 12}{x^2 - x - 6}$

34. $R(x) = \dfrac{x^2 + 3x - 10}{x^2 + 8x + 15}$

35. $R(x) = \dfrac{6x^2 - 7x - 3}{2x^2 - 7x + 6}$

36. $R(x) = \dfrac{8x^2 + 26x + 15}{2x^2 - x - 15}$

37. $R(x) = \dfrac{x^2 + 5x + 6}{x + 3}$

38. $R(x) = \dfrac{x^2 + x - 30}{x + 6}$

39. $H(x) = \dfrac{3x - 6}{4 - x^2}$

40. $H(x) = \dfrac{2 - 2x}{x^2 - 1}$

41. $F(x) = \dfrac{x^2 - 5x + 4}{x^2 - 2x + 1}$

42. $F(x) = \dfrac{x^2 - 2x - 15}{x^2 + 6x + 9}$

43. $G(x) = \dfrac{x}{(x + 2)^2}$

44. $G(x) = \dfrac{2 - x}{(x - 1)^2}$

45. $f(x) = x + \dfrac{1}{x}$

46. $f(x) = 2x + \dfrac{9}{x}$

47. $f(x) = x^2 + \dfrac{1}{x}$

48. $f(x) = 2x^2 + \dfrac{16}{x}$

49. $f(x) = x + \dfrac{1}{x^3}$

50. $f(x) = 2x + \dfrac{9}{x^3}$

In Problems 51–54, find a rational function that might have the given graph. (More than one answer might be possible.)

51.

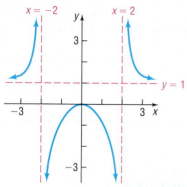

52.

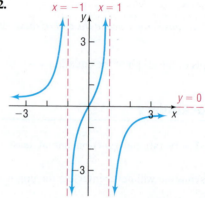

53.

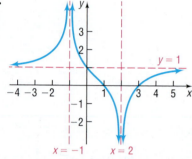

54.

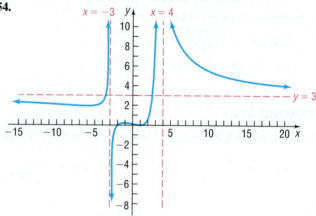

Applications and Extensions

55. Drug Concentration The concentration C of a certain drug in a patient's bloodstream t hours after injection is given by

$$C(t) = \frac{t}{2t^2 + 1}$$

(a) Find the horizontal asymptote of $C(t)$. What happens to the concentration of the drug as t increases?

(b) Using your graphing utility, graph $C = C(t)$.

(c) Determine the time at which the concentration is highest.

56. Drug Concentration The concentration C of a certain drug in a patient's bloodstream t minutes after injection is given by

$$C(t) = \frac{50t}{t^2 + 25}$$

(a) Find the horizontal asymptote of $C(t)$. What happens to the concentration of the drug as t increases?

(b) Using your graphing utility, graph $C = C(t)$.

(c) Determine the time at which the concentration is highest.

57. Minimum Cost A rectangular area adjacent to a river is to be fenced in; no fence is needed on the river side. The enclosed area is to be 1000 square feet. Fencing for the side parallel to the river is $5 per linear foot, and fencing for the other two sides is $8 per linear foot; the four corner posts are $25 apiece. Let x be the length of one of the sides perpendicular to the river.

(a) Write a function $C(x)$ that describes the cost of the project.

(b) What is the domain of C?

(c) Use a graphing utility to graph $C = C(x)$.

(d) Find the dimensions of the cheapest enclosure.

Source: http://dl.uncw.edu/digilib/mathematics/algebra/mat111hb/pandr/rational/rational.html

58. Doppler Effect The Doppler effect (named after Christian Doppler) is the change in the pitch (frequency) of the sound from a source (s) as heard by an observer (o) when one or both are in motion. If we assume both the source and the observer are moving in the same direction, the relationship is

$$f' = f_a\!\left(\frac{v - v_o}{v - v_s}\right)$$

where
$f' =$ perceived pitch by the observer
$f_a =$ actual pitch of the source
$v =$ speed of sound in air (assume 772.4 mph)
$v_o =$ speed of the observer
$v_s =$ speed of the source

Suppose that you are traveling down the road at 45 mph and you hear an ambulance (with siren) coming toward you from the rear. The actual pitch of the siren is 600 hertz (Hz).

(a) Write a function $f'(v_s)$ that describes this scenario.

(b) If $f' = 620$ Hz, find the speed of the ambulance.

(c) Use a graphing utility to graph the function.

(d) Verify your answer from part (b).

Source: www.acs.psu.edu/drussell/

59. Minimizing Surface Area United Parcel Service has contracted you to design a closed box with a square base that has a volume of 10,000 cubic inches. See the illustration.
  (a) Express the surface area S of the box as a function of x.
   (b) Using a graphing utility, graph the function found in part (a).
  (c) What is the minimum amount of cardboard that can be used to construct the box?
  (d) What are the dimensions of the box that minimize the surface area?
  (e) Why might UPS be interested in designing a box that minimizes the surface area?

60. Minimizing Surface Area United Parcel Service has contracted you to design an open box with a square base that has a volume of 5000 cubic inches. See the illustration.
  (a) Express the surface area S of the box as a function of x.
   (b) Using a graphing utility, graph the function found in part (a).
  (c) What is the minimum amount of cardboard that can be used to construct the box?
  (d) What are the dimensions of the box that minimize the surface area?
  (e) Why might UPS be interested in designing a box that minimizes the surface area?

61. Cost of a Can A can in the shape of a right circular cylinder is required to have a volume of 500 cubic centimeters. The top

and bottom are made of material that costs 6¢ per square centimeter, while the sides are made of material that costs 4¢ per square centimeter.
  (a) Express the total cost C of the material as a function of the radius r of the cylinder. (Refer to Figure 39.)
   (b) Graph $C = C(r)$. For what value of r is the cost C a minimum?

62. Material Needed to Make a Drum A steel drum in the shape of a right circular cylinder is required to have a volume of 100 cubic feet.

  (a) Express the amount A of material required to make the drum as a function of the radius r of the cylinder.
  (b) How much material is required if the drum's radius is 3 feet?
  (c) How much material is required if the drum's radius is 4 feet?
  (d) How much material is required if the drum's radius is 5 feet?
   (e) Graph $A = A(r)$. For what value of r is A smallest?

Discussion and Writing

63. Graph each of the following functions:
$$y = \frac{x^2 - 1}{x - 1} \qquad y = \frac{x^3 - 1}{x - 1}$$
$$y = \frac{x^4 - 1}{x - 1} \qquad y = \frac{x^5 - 1}{x - 1}$$
Is $x = 1$ a vertical asymptote? Why not? What is happening for $x = 1$? What do you conjecture about $y = \dfrac{x^n - 1}{x - 1}$, $n \geq 1$ an integer, for $x = 1$?

64. Graph each of the following functions:
$$y = \frac{x^2}{x - 1} \quad y = \frac{x^4}{x - 1} \quad y = \frac{x^6}{x - 1} \quad y = \frac{x^8}{x - 1}$$
What similarities do you see? What differences?

65. Write a few paragraphs that provide a general strategy for graphing a rational function. Be sure to mention the following: proper, improper, intercepts, and asymptotes.

66. Create a rational function that has the following characteristics: crosses the x-axis at 2; touches the x-axis

at −1; one vertical asymptote at $x = -5$ and another at $x = 6$; and one horizontal asymptote, $y = 3$. Compare your function to a fellow classmate's. How do they differ? What are their similarities?

67. Create a rational function that has the following characteristics: crosses the x-axis at 3; touches the x-axis at −2; one vertical asymptote, $x = 1$; and one horizontal asymptote, $y = 2$. Give your rational function to a fellow classmate and ask for a written critique of your rational function.

68. Create a rational function with the following characteristics: three real zeros, one of multiplicity 2; y-intercept 1; vertical asymptotes, $x = -2$ and $x = 3$; oblique asymptote, $y = 2x + 1$. Is this rational function unique? Compare your function with those of other students. What will be the same as everyone else's? Add some more characteristics, such as symmetry or naming the real zeros. How does this modify the rational function?

69. Explain the circumstances under which the graph of a rational function will have a hole.

Retain Your Knowledge

Problems 70–73 are based on material learned earlier in the course. The purpose of these problems is to keep the material fresh in your mind so that you are better prepared for the final exam.

70. Subtract: $(4x^3 - 7x + 1) - (5x^2 - 9x + 3)$

71. Solve: $\dfrac{3x}{3x + 1} = \dfrac{x - 2}{x + 5}$

72. Find the maximum value of $f(x) = -\dfrac{2}{3}x^2 + 6x - 5$.

73. Approximate $\dfrac{\sqrt{5} - 3}{\sqrt{7} + 2}$. Round your answer to three decimal places.

'Are You Prepared?' Answers

1. $\left(0, \dfrac{1}{4}\right)$, $(1, 0)$, $(-1, 0)$

5.4 Polynomial and Rational Inequalities

PREPARING FOR THIS SECTION *Before getting started, review the following:*

- Solving Linear Inequalities (Section 1.5, pp. 123–125)
- Solving Quadratic Inequalities (Section 4.5, pp. 312–313)

Now Work the **'Are You Prepared?'** problems on page 372.

OBJECTIVES 1 Solve Polynomial Inequalities (p. 368)
2 Solve Rational Inequalities (p. 370)

1 Solve Polynomial Inequalities

In this section we solve inequalities that involve polynomials of degree 3 and higher, along with inequalities that involve rational functions. To help understand the algebraic procedure for solving such inequalities, we use the information obtained in the previous three sections about the graphs of polynomial and rational functions. The approach follows the same methodology that we used to solve inequalities involving quadratic functions.

EXAMPLE I **Solving a Polynomial Inequality Using Its Graph**

Solve $(x + 3)(x - 1)^2 > 0$ by graphing $f(x) = (x + 3)(x - 1)^2$.

Solution Graph $f(x) = (x + 3)(x - 1)^2$ and determine the intervals of x for which the graph is above the x-axis. These values of x result in $f(x)$ being positive. Using Steps 1 through 5 on page 335, we obtain the graph shown in Figure 41.

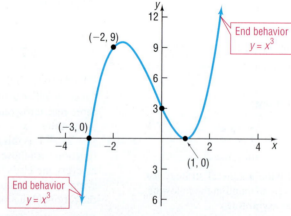

Figure 41 $f(x) = (x + 3)(x - 1)^2$

From the graph, we can see that $f(x) > 0$ for $-3 < x < 1$ or $x > 1$. The solution set is $\{x \mid -3 < x < 1 \text{ or } x > 1\}$ or, using interval notation, $(-3, 1) \cup (1, \infty)$. ●

Now Work PROBLEM 9

The results of Example 1 lead to the following approach to solving polynomial and rational inequalities algebraically. Suppose that the polynomial or rational inequality is in one of the forms

$$f(x) < 0 \qquad f(x) > 0 \qquad f(x) \le 0 \qquad f(x) \ge 0$$

Locate the zeros of f if f is a polynomial function, and locate the zeros of the numerator and the denominator if f is a rational function. If we use these zeros to divide the real number line into intervals, we know that on each interval, the graph of f is either above the x-axis $[f(x) > 0]$ or below the x-axis $[f(x) < 0]$. This will enable us to identify the solution of the inequality.

| EXAMPLE 2 | **How to Solve a Polynomial Inequality Algebraically** |

Solve the inequality $x^4 > x$ algebraically, and graph the solution set.

Step-by-Step Solution

Step 1: Write the inequality so that a polynomial expression f is on the left side and zero is on the right side.

Rearrange the inequality so that 0 is on the right side.

$$x^4 > x$$
$$x^4 - x > 0 \quad \text{Subtract } x \text{ from both sides of the inequality.}$$

This inequality is equivalent to the one we wish to solve.

Step 2: Determine the real zeros (x-intercepts of the graph) of f.

Find the real zeros of $f(x) = x^4 - x$ by solving $x^4 - x = 0$.

$$x^4 - x = 0$$
$$x(x^3 - 1) = 0 \quad \text{Factor out } x.$$
$$x(x - 1)(x^2 + x + 1) = 0 \quad \text{Factor the difference of two cubes.}$$
$$x = 0 \quad \text{or} \quad x - 1 = 0 \quad \text{or} \quad x^2 + x + 1 = 0 \quad \text{Set each factor equal to zero and solve.}$$
$$x = 0 \quad \text{or} \quad x = 1$$

The equation $x^2 + x + 1 = 0$ has no real solutions. Do you see why?

Step 3: Use the zeros found in Step 2 to divide the real number line into intervals.

Use the real zeros to separate the real number line into three intervals:

$$(-\infty, 0) \quad (0, 1) \quad (1, \infty)$$

Step 4: Select a number in each interval, evaluate f at the number, and determine whether $f(x)$ is positive or negative. If $f(x)$ is positive, all values of f in the interval are positive. If $f(x)$ is negative, all values of f in the interval are negative.

Select a test number in each interval found in Step 3 and evaluate $f(x) = x^4 - x$ at each number to determine whether $f(x)$ is positive or negative. See Table 14.

Table 14

Interval	$(-\infty, 0)$	$(0, 1)$	$(1, \infty)$
Number chosen	-1	$\frac{1}{2}$	2
Value of f	$f(-1) = 2$	$f\left(\frac{1}{2}\right) = -\frac{7}{16}$	$f(2) = 14$
Conclusion	Positive	Negative	Positive

Since we want to know where $f(x)$ is positive, conclude that $f(x) > 0$ for all numbers x for which $x < 0$ or $x > 1$. Because the original inequality is strict, numbers x that satisfy the equation $x^4 = x$ are not solutions. The solution set of the inequality $x^4 > x$ is $\{x \mid x < 0 \text{ or } x > 1\}$ or, using interval notation, $(-\infty, 0) \cup (1, \infty)$.

Figure 42 shows the graph of the solution set.

NOTE If the inequality is not strict (that is, if it is $\leq$ or $\geq$), include the solutions of $f(x) = 0$ in the solution set. ∎

$$\begin{array}{c} -2\ -1\ 0\ 1\ 2 \end{array} \quad x$$

Figure 42

The Role of Multiplicity in Solving Polynomial Inequalities

In Example 2, we used the number -1 and found that f is positive for all $x < 0$. Because the "cut point" of 0 is the result of a zero of odd multiplicity (x is a factor to the first power), we know that the sign of f will change on either side of 0, so

The remainder is $f(3) = 0$, so 3 is a zero and $x - 3$ is a factor of f. Use the bottom row of the synthetic division to continue the factoring of f.

$$f(x) = x^5 - 7x^4 + 19x^3 - 37x^2 + 60x - 36$$
$$= (x - 1)(x - 3)(x^3 - 3x^2 + 4x - 12)$$

The remaining zeros satisfy the new depressed equation

$$q_2(x) = x^3 - 3x^2 + 4x - 12 = 0$$

Notice that $q_2(x)$ can be factored by grouping. (Alternatively, Step 3 could be repeated to again check the potential rational zero 3. The potential rational zeros 1 or 2 would no longer be checked, because they have already been eliminated from further consideration.) Then

$$x^3 - 3x^2 + 4x - 12 = 0$$
$$x^2(x - 3) + 4(x - 3) = 0$$
$$(x^2 + 4)(x - 3) = 0$$
$$x^2 + 4 = 0 \text{ or } x - 3 = 0$$
$$x = 3$$

Since $x^2 + 4 = 0$ has no real solutions, the real zeros of f are 1 and 3, with 3 being a repeated zero of multiplicity 2. The factored form of f is

$$f(x) = x^5 - 7x^4 + 19x^3 - 37x^2 + 60x - 36$$
$$= (x - 1)(x - 3)^2(x^2 + 4)$$

● **Now Work** PROBLEM 45

5 Solve Polynomial Equations

EXAMPLE 7

Solving a Polynomial Equation

Find the real solutions of the equation: $x^5 - 7x^4 + 19x^3 - 37x^2 + 60x - 36 = 0$

Solution

The real solutions of this equation are the real zeros of the polynomial function

$$f(x) = x^5 - 7x^4 + 19x^3 - 37x^2 + 60x - 36$$

Using the result of Example 6, the real zeros of f are 1 and 3. The real solutions of the equation $x^5 - 7x^4 + 19x^3 - 37x^2 + 60x - 36 = 0$ are 1 and 3. ●

Now Work PROBLEM 57

In Example 6, the quadratic factor $x^2 + 4$ that appears in the factored form of f is called *irreducible*, because the polynomial $x^2 + 4$ cannot be factored over the real numbers. In general, a quadratic factor $ax^2 + bx + c$ is **irreducible** if it cannot be factored over the real numbers—that is, if it is prime over the real numbers.

Refer to Examples 5 and 6. The polynomial function of Example 5 has three real zeros, and its factored form contains three linear factors. The polynomial function of Example 6 has two distinct real zeros, and its factored form contains two distinct linear factors and one irreducible quadratic factor.

THEOREM

> Every polynomial function (with real coefficients) can be uniquely factored into a product of linear factors and/or irreducible quadratic factors.

We prove this result in Section 5.6, and in fact, we shall draw several additional conclusions about the zeros of a polynomial function. One conclusion is worth noting now. If a polynomial with real coefficients is of odd degree, it must contain at least one linear factor. (Do you see why? Consider the end behavior of polynomial functions of odd degree.) This means that it must have at least one real zero.

THEOREM

> A polynomial function (with real coefficients) of odd degree has at least one real zero.

6 Use the Theorem for Bounds on Zeros

The work involved in finding the zeros of a polynomial function can be reduced somewhat if upper and lower bounds to the zeros can be found. A number M is an **upper bound** to the zeros of a polynomial f if no zero of f is greater than M. The number m is a **lower bound** if no zero of f is less than m. Accordingly, if m is a lower bound and M is an upper bound to the zeros of a polynomial function f, then

$$m \leq \text{ any zero of } f \leq M$$

For polynomials with integer coefficients, knowing the values of a lower bound m and an upper bound M may enable you to eliminate some potential rational zeros—that is, any zeros outside of the interval $[m, M]$.

COMMENT The bounds on the zeros of a polynomial provide good choices for setting Xmin and Xmax of the viewing rectangle. With these choices, all the x-intercepts of the graph can be seen. ∎

THEOREM

> **Bounds on Zeros**
>
> Let f denote a polynomial function whose leading coefficient is positive.
>
> - If $M > 0$ is a real number and if the third row in the process of synthetic division of f by $x - M$ contains only numbers that are positive or zero, then M is an upper bound to the zeros of f.
> - If $m < 0$ is a real number and if the third row in the process of synthetic division of f by $x - m$ contains numbers that alternate positive (or 0) and negative (or 0), then m is a lower bound to the zeros of f.

NOTE When finding a lower bound, remember that a 0 can be treated as either positive or negative, but not both. For example, 3, 0, 5 would be considered to alternate sign, whereas 3, 0, −5 would not. ∎

Proof (Outline) We give only an outline of the proof of the first part of the theorem. Suppose that M is a positive real number, and the third row in the process of synthetic division of the polynomial f by $x - M$ contains only numbers that are positive or 0. Then there are a quotient q and a remainder R such that

$$f(x) = (x - M)q(x) + R$$

where the coefficients of $q(x)$ are positive or 0 and the remainder $R \geq 0$. Then, for any $x > M$, we must have $x - M > 0$, $q(x) > 0$, and $R \geq 0$, so that $f(x) > 0$. That is, there is no zero of f larger than M. The proof of the second part follows similar reasoning. ∎

In finding bounds, it is preferable to find the smallest upper bound and largest lower bound. This will require repeated synthetic division until a desired pattern is observed. For simplicity, we will consider only potential rational zeros that are integers. If a bound is not found using these values, continue checking positive and/or negative integers until you find both an upper and a lower bound.

EXAMPLE 8 **Finding Upper and Lower Bounds of Zeros**

For the polynomial function $f(x) = 2x^3 + 11x^2 - 7x - 6$, use the Bounds on Zeros Theorem to find integer upper and lower bounds to the zeros of f.

Solution From Example 4, the potential rational zeros of f are ± 1, ± 2, ± 3, ± 6, $\pm\dfrac{1}{2}$, $\pm\dfrac{3}{2}$.

To find an upper bound, start with the smallest positive integer that is a potential rational zero, which is 1. Continue checking 2, 3, and 6 (and then subsequent positive integers), if necessary, until an upper bound is found. To find a lower bound, start with the largest negative integer that is a potential rational zero, which is −1. Continue checking −2, −3, and −6 (and then subsequent negative integers), if necessary, until a lower bound is found. Table 16 summarizes the results of doing repeated synthetic divisions by showing only the third row of each division. For example, the first row of the table shows the result of dividing $f(x)$ by $x - 1$.

$$
\begin{array}{r|rrr}
1) & 2 & 11 & -7 & -6 \\
  &   & 2 & 13 & 6 \\
\hline
  & 2 & 13 & 6 & 0
\end{array}
$$

Table 16   Synthetic Division Summary

	r	Coefficients of $q(x)$			Remainder	
Upper bound	1	2	13	6	0	All nonnegative
	-1	2	9	-16	10	
	-2	2	7	-21	36	
	-3	2	5	-22	60	
	-6	2	-1	-1	0	
Lower bound	-7	2	-3	14	-104	Alternating Signs

NOTE Keep track of any zeros that are found when looking for bounds. ∎

For $r = 1$, the third row of synthetic division contains only numbers that are positive or 0, so we know there are no zeros greater than 1. Since the third row of synthetic division for $r = -7$ results in alternating positive (or 0) and negative (or 0) values, we know that -7 is a lower bound. There are no zeros less than -7. Notice that in looking for bounds, two zeros were discovered. These zeros are 1 and -6. ●

✏ **Now Work** PROBLEM 69

If the leading coefficient of f is negative, the upper and lower bounds can still be found by first multiplying the polynomial by -1. Since $-f(x) = (-1)f(x)$, the zeros of $-f(x)$ are the same as the zeros of $f(x)$.

7 Use the Intermediate Value Theorem

The next result, called the **Intermediate Value Theorem**, is based on the fact that the graph of a polynomial function is continuous; that is, it contains no "holes" or "gaps." Although the proof of this result requires advanced methods in calculus, it is easy to "see" why the result is true. Look at Figure 45.

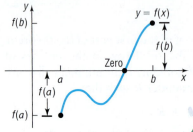

Figure 45 If $f(a) < 0$ and $f(b) > 0$, and if f is continuous, there is a zero between a and b.

THEOREM

Intermediate Value Theorem

Let f denote a polynomial function. If $a < b$ and if $f(a)$ and $f(b)$ are of opposite sign, there is at least one real zero of f between a and b.

EXAMPLE 9

Using the Intermediate Value Theorem to Locate a Real Zero

Show that $f(x) = x^5 - x^3 - 1$ has a zero between 1 and 2.

Solution   Evaluate f at 1 and at 2.

$$f(1) = -1 \quad \text{and} \quad f(2) = 23$$

Because $f(1) < 0$ and $f(2) > 0$, it follows from the Intermediate Value Theorem that the polynomial function f has at least one zero between 1 and 2. ●

✏ **Now Work** PROBLEM 79

Let's look at the polynomial f of Example 9 more closely. Based on Descartes' Rule of Signs, f has exactly one positive real zero. Based on the Rational Zeros Theorem, 1 is the only potential positive rational zero. Since $f(1) \neq 0$, the zero between 1 and 2 is irrational. The Intermediate Value Theorem can be used to approximate it.

Approximating the Real Zeros of a Polynomial Function

STEP 1: Find two consecutive integers a and $a + 1$ such that f has a zero between them.

STEP 2: Divide the interval $[a, a + 1]$ into 10 equal subintervals.

STEP 3: Evaluate f at each endpoint of the subintervals until the Intermediate Value Theorem applies; this interval then contains a zero.

STEP 4: Now divide the new interval into 10 equal subintervals and repeat Step 3.

STEP 5: Continue with Steps 3 and 4 until the desired accuracy is achieved.

EXAMPLE 10

Approximating a Real Zero of a Polynomial Function

Find the positive zero of $f(x) = x^5 - x^3 - 1$ correct to two decimal places.

Solution

From Example 9 we know that the positive zero is between 1 and 2. Divide the interval $[1, 2]$ into 10 equal subintervals: $[1, 1.1], [1.1, 1.2], [1.2, 1.3], [1.3, 1.4], [1.4, 1.5], [1.5, 1.6], [1.6, 1.7], [1.7, 1.8], [1.8, 1.9], [1.9, 2]$. Now find the value of f at each endpoint until the Intermediate Value Theorem applies.

$$f(x) = x^5 - x^3 - 1$$
$$f(1.0) = -1 \qquad\qquad f(1.2) = -0.23968$$
$$f(1.1) = -0.72049 \qquad f(1.3) = 0.51593$$

We can stop here and conclude that the zero is between 1.2 and 1.3. Now divide the interval $[1.2, 1.3]$ into 10 equal subintervals and proceed to evaluate f at each endpoint.

$$f(1.20) = -0.23968 \qquad f(1.23) \approx -0.0455613$$
$$f(1.21) \approx -0.1778185 \qquad f(1.24) \approx 0.025001$$
$$f(1.22) \approx -0.1131398$$

The zero lies between 1.23 and 1.24, and so, correct to two decimal places, the zero is 1.23. ●

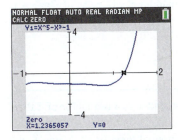

Figure 46

Exploration

We examine the polynomial function f given in Example 10. The Theorem on Bounds of Zeros tells us that every zero is between -1 and 2. If we graph f using $-1 \leq x \leq 2$ (see Figure 46), we see that f has exactly one x-intercept. Using ZERO or ROOT, we find this zero to be 1.24 rounded to two decimal places. Correct to two decimal places, the zero is 1.23.

Now Work PROBLEM 91

COMMENT The TABLE feature of a graphing calculator makes the computations in the solution to Example 10 a lot easier. ∎

There are many other numerical techniques for approximating the zeros of a polynomial. The one outlined in Example 10 (a variation of the *bisection method*) has the advantages that it will always work, it can be programmed rather easily on a computer, and each time it is used another decimal place of accuracy is achieved. See Problem 119 for the bisection method, which places the zero in a succession of intervals, with each new interval being half the length of the preceding one.

Historical Feature

ormulas for the solution of third- and fourth-degree polynomial equations exist, and although they are not very practical, they do have an interesting history.

In the 1500s in Italy, mathematical contests were a popular pastime, and people who possessed methods for solving problems kept them secret. (Solutions that were published were already common knowledge.) Niccolo of Brescia (1499–1557), commonly referred to as Tartaglia ("the stammerer"), had the secret for solving cubic (third-degree) equations, which gave him a decided advantage in the contests. Girolamo Cardano (1501–1576) found out that Tartaglia had the secret, and, being interested in cubics, he requested it from Tartaglia. The reluctant Tartaglia hesitated for some time, but finally, swearing Cardano to secrecy with midnight oaths by candlelight, told him the secret. Cardano then published the

solution in his book *Ars Magna* (1545), giving Tartaglia the credit but rather compromising the secrecy. Tartaglia exploded into bitter recriminations, and each wrote pamphlets that reflected on the other's mathematics, moral character, and ancestry.

The quartic (fourth-degree) equation was solved by Cardano's student Lodovico Ferrari, and this solution also was included, with credit and this time with permission, in the *Ars Magna*.

Attempts were made to solve the fifth-degree equation in similar ways, all of which failed. In the early 1800s, P. Ruffini, Niels Abel, and Evariste Galois all found ways to show that it is not possible to solve fifth-degree equations by formula, but the proofs required the introduction of new methods. Galois's methods eventually developed into a large part of modern algebra.

Historical Problems

Problems 1–8 develop the Tartaglia–Cardano solution of the cubic equation and show why it is not altogether practical.

1. Show that the general cubic equation $y^3 + by^2 + cy + d = 0$ can be transformed into an equation of the form $x^3 + px + q = 0$ by using the substitution $y = x - \dfrac{b}{3}$.

2. In the equation $x^3 + px + q = 0$, replace x by $H + K$. Let $3HK = -p$, and show that $H^3 + K^3 = -q$.

3. Based on Problem 2, we have the two equations

$$3HK = -p \quad \text{and} \quad H^3 + K^3 = -q$$

   Solve for K in $3HK = -p$ and substitute into $H^3 + K^3 = -q$. Then show that

$$H = \sqrt[3]{-\frac{q}{2} + \sqrt{\frac{q^2}{4} + \frac{p^3}{27}}}$$

   [**Hint:** Look for an equation that is quadratic in form.]

4. Use the solution for H from Problem 3 and the equation $H^3 + K^3 = -q$ to show that

$$K = \sqrt[3]{-\frac{q}{2} - \sqrt{\frac{q^2}{4} + \frac{p^3}{27}}}$$

5. Use the results from Problems 2 to 4 to show that the solution of $x^3 + px + q = 0$ is

$$x = \sqrt[3]{-\frac{q}{2} + \sqrt{\frac{q^2}{4} + \frac{p^3}{27}}} + \sqrt[3]{-\frac{q}{2} - \sqrt{\frac{q^2}{4} + \frac{p^3}{27}}}$$

6. Use the result of Problem 5 to solve the equation $x^3 - 6x - 9 = 0$.

7. Use a calculator and the result of Problem 5 to solve the equation $x^3 + 3x - 14 = 0$.

8. Use the methods of this section to solve the equation $x^3 + 3x - 14 = 0$.

5.5 Assess Your Understanding

'Are You Prepared?' *Answers are given at the end of these exercises. If you get a wrong answer, read the pages listed in* red.

1. Find $f(-1)$ if $f(x) = 2x^2 - x$. (pp. 202–204)
2. Factor the expression $6x^2 + x - 2$. (pp. 49–55)
3. Find the quotient and remainder if $3x^4 - 5x^3 + 7x - 4$ is divided by $x - 3$. (pp. 44–47 or 58–61)
4. Solve $x^2 + x - 3 = 0$. (pp. 92–99)

Concepts and Vocabulary

5. If $f(x) = q(x)g(x) + r(x)$, the function $r(x)$ is called the _____.
   (a) remainder (b) dividend (c) quotient (d) divisor

6. When a polynomial function f is divided by $x - c$, the remainder is _____.

7. Given $f(x) = 3x^4 - 2x^3 + 7x - 2$, how many sign changes are there in the coefficients of $f(-x)$?
   (a) 0 (b) 1 (c) 2 (d) 3

8. *True or False* Every polynomial function of degree 3 with real coefficients has exactly three real zeros.

9. If f is a polynomial function and $x - 4$ is a factor of f, then $f(4) =$ _____.

10. *True or False* If f is a polynomial function of degree 4 and if $f(2) = 5$, then

$$\frac{f(x)}{x - 2} = p(x) + \frac{5}{x - 2}$$

   where $p(x)$ is a polynomial of degree 3.

Skill Building

In Problems 11–20, use the Remainder Theorem to find the remainder when $f(x)$ is divided by $x - c$. Then use the Factor Theorem to determine whether $x - c$ is a factor of $f(x)$.

11. $f(x) = 4x^3 - 3x^2 - 8x + 4; x - 2$

12. $f(x) = -4x^3 + 5x^2 + 8; x + 3$

13. $f(x) = 3x^4 - 6x^3 - 5x + 10; x - 2$

14. $f(x) = 4x^4 - 15x^2 - 4; x - 2$

15. $f(x) = 3x^6 + 82x^3 + 27; x + 3$

16. $f(x) = 2x^6 - 18x^4 + x^2 - 9; x + 3$

17. $f(x) = 4x^6 - 64x^4 + x^2 - 15; x + 4$

18. $f(x) = x^6 - 16x^4 + x^2 - 16; x + 4$

19. $f(x) = 2x^4 - x^3 + 2x - 1; x - \frac{1}{2}$

20. $f(x) = 3x^4 + x^3 - 3x + 1; x + \frac{1}{3}$

In Problems 21–32, tell the maximum number of real zeros that each polynomial function may have. Then use Descartes' Rule of Signs to determine how many positive and how many negative zeros each polynomial function may have. Do not attempt to find the zeros.

21. $f(x) = -4x^7 + x^3 - x^2 + 2$

22. $f(x) = 5x^4 + 2x^2 - 6x - 5$

23. $f(x) = 2x^6 - 3x^2 - x + 1$

24. $f(x) = -3x^5 + 4x^4 + 2$

25. $f(x) = 3x^3 - 2x^2 + x + 2$

26. $f(x) = -x^3 - x^2 + x + 1$

27. $f(x) = -x^4 + x^2 - 1$

28. $f(x) = x^4 + 5x^3 - 2$

29. $f(x) = x^5 + x^4 + x^2 + x + 1$

30. $f(x) = x^5 - x^4 + x^3 - x^2 + x - 1$

31. $f(x) = x^6 - 1$

32. $f(x) = x^6 + 1$

In Problems 33–44, list the potential rational zeros of each polynomial function. Do not attempt to find the zeros.

33. $f(x) = 3x^4 - 3x^3 + x^2 - x + 1$

34. $f(x) = x^5 - x^4 + 2x^2 + 3$

35. $f(x) = x^5 - 6x^2 + 9x - 3$

36. $f(x) = 2x^5 - x^4 - x^2 + 1$

37. $f(x) = -4x^3 - x^2 + x + 2$

38. $f(x) = 6x^4 - x^2 + 2$

39. $f(x) = 6x^4 - x^2 + 9$

40. $f(x) = -4x^3 + x^2 + x + 6$

41. $f(x) = 2x^5 - x^3 + 2x^2 + 12$

42. $f(x) = 3x^5 - x^2 + 2x + 18$

43. $f(x) = 6x^4 + 2x^3 - x^2 + 20$

44. $f(x) = -6x^3 - x^2 + x + 10$

In Problems 45–56, use the Rational Zeros Theorem to find all the real zeros of each polynomial function. Use the zeros to factor f over the real numbers.

45. $f(x) = x^3 + 2x^2 - 5x - 6$

46. $f(x) = x^3 + 8x^2 + 11x - 20$

47. $f(x) = 2x^3 - x^2 + 2x - 1$

48. $f(x) = 2x^3 + x^2 + 2x + 1$

49. $f(x) = 2x^3 - 4x^2 - 10x + 20$

50. $f(x) = 3x^3 + 6x^2 - 15x - 30$

51. $f(x) = 2x^4 + x^3 - 7x^2 - 3x + 3$

52. $f(x) = 2x^4 - x^3 - 5x^2 + 2x + 2$

53. $f(x) = x^4 + x^3 - 3x^2 - x + 2$

54. $f(x) = x^4 - x^3 - 6x^2 + 4x + 8$

55. $f(x) = 4x^4 + 5x^3 + 9x^2 + 10x + 2$

56. $f(x) = 3x^4 + 4x^3 + 7x^2 + 8x + 2$

In Problems 57–68, solve each equation in the real number system.

57. $x^4 - x^3 + 2x^2 - 4x - 8 = 0$

58. $2x^3 + 3x^2 + 2x + 3 = 0$

59. $3x^3 + 4x^2 - 7x + 2 = 0$

60. $2x^3 - 3x^2 - 3x - 5 = 0$

61. $3x^3 - x^2 - 15x + 5 = 0$

62. $2x^3 - 11x^2 + 10x + 8 = 0$

63. $x^4 + 4x^3 + 2x^2 - x + 6 = 0$

64. $x^4 - 2x^3 + 10x^2 - 18x + 9 = 0$

65. $x^3 - \dfrac{2}{3}x^2 + \dfrac{8}{3}x + 1 = 0$

66. $x^3 + \dfrac{3}{2}x^2 + 3x - 2 = 0$

67. $2x^4 - 19x^3 + 57x^2 - 64x + 20 = 0$

68. $2x^4 + x^3 - 24x^2 + 20x + 16 = 0$

In Problems 69–78, find bounds on the real zeros of each polynomial function.

69. $f(x) = x^4 - 3x^2 - 4$

70. $f(x) = x^4 - 5x^2 - 36$

71. $f(x) = x^4 + x^3 - x - 1$

72. $f(x) = x^4 - x^3 + x - 1$

73. $f(x) = 3x^4 + 3x^3 - x^2 - 12x - 12$

74. $f(x) = 3x^4 - 3x^3 - 5x^2 + 27x - 36$

75. $f(x) = 4x^5 - x^4 + 2x^3 - 2x^2 + x - 1$

76. $f(x) = 4x^5 + x^4 + x^3 + x^2 - 2x - 2$

77. $f(x) = -x^4 + 3x^3 - 4x^2 - 2x + 9$

78. $f(x) = -4x^5 + 5x^3 + 9x^2 + 3x - 12$

In Problems 79–84, use the Intermediate Value Theorem to show that each polynomial function has a zero in the given interval.

79. $f(x) = 8x^4 - 2x^2 + 5x - 1;\ [0, 1]$

80. $f(x) = x^4 + 8x^3 - x^2 + 2;\ [-1, 0]$

81. $f(x) = 2x^3 + 6x^2 - 8x + 2;\ [-5, -4]$

82. $f(x) = 3x^3 - 10x + 9;\ [-3, -2]$

83. $f(x) = x^5 - x^4 + 7x^3 - 7x^2 - 18x + 18;\ [1.4, 1.5]$

84. $f(x) = x^5 - 3x^4 - 2x^3 + 6x^2 + x + 2;\ [1.7, 1.8]$

In Problems 85–88, each equation has a solution r in the interval indicated. Use the method of Example 10 to approximate this solution correct to two decimal places.

85. $8x^4 - 2x^2 + 5x - 1 = 0;\ 0 \le r \le 1$

86. $x^4 + 8x^3 - x^2 + 2 = 0;\ -1 \le r \le 0$

87. $2x^3 + 6x^2 - 8x + 2 = 0;\ -5 \le r \le -4$

88. $3x^3 - 10x + 9 = 0;\ -3 \le r \le -2$

In Problems 89–92, each polynomial function has exactly one positive zero. Use the method of Example 10 to approximate the zero correct to two decimal places.

89. $f(x) = x^3 + x^2 + x - 4$

90. $f(x) = 2x^4 + x^2 - 1$

91. $f(x) = 2x^4 - 3x^3 - 4x^2 - 8$

92. $f(x) = 3x^3 - 2x^2 - 20$

Mixed Practice

In Problems 93–104, graph each polynomial function.

93. $f(x) = x^3 + 2x^2 - 5x - 6$

94. $f(x) = x^3 + 8x^2 + 11x - 20$

95. $f(x) = 2x^3 - x^2 + 2x - 1$

96. $f(x) = 2x^3 + x^2 + 2x + 1$

97. $f(x) = x^4 + x^2 - 2$

98. $f(x) = x^4 - 3x^2 - 4$

99. $f(x) = 4x^4 + 7x^2 - 2$

100. $f(x) = 4x^4 + 15x^2 - 4$

101. $f(x) = x^4 + x^3 - 3x^2 - x + 2$

102. $f(x) = x^4 - x^3 - 6x^2 + 4x + 8$

103. $f(x) = 4x^5 - 8x^4 - x + 2$

104. $f(x) = 4x^5 + 12x^4 - x - 3$

105. Suppose that $f(x) = 3x^3 + 16x^2 + 3x - 10$. Find the zeros of $f(x + 3)$.

106. Suppose that $f(x) = 4x^3 - 11x^2 - 26x + 24$. Find the zeros of $f(x - 2)$.

Applications and Extensions

107. Find k such that $f(x) = x^3 - kx^2 + kx + 2$ has the factor $x - 2$.

108. Find k such that $f(x) = x^4 - kx^3 + kx^2 + 1$ has the factor $x + 2$.

109. What is the remainder when $f(x) = 2x^{20} - 8x^{10} + x - 2$ is divided by $x - 1$?

110. What is the remainder when $f(x) = -3x^{17} + x^9 - x^5 + 2x$ is divided by $x + 1$?

111. Use the Factor Theorem to prove that $x - c$ is a factor of $x^n - c^n$ for any positive integer n.

112. Use the Factor Theorem to prove that $x + c$ is a factor of $x^n + c^n$ if $n \geq 1$ is an odd integer.

113. One solution of the equation $x^3 - 8x^2 + 16x - 3 = 0$ is 3. Find the sum of the remaining solutions.

114. One solution of the equation $x^3 + 5x^2 + 5x - 2 = 0$ is -2. Find the sum of the remaining solutions.

115. Geometry What is the length of the edge of a cube if, after a slice 1 inch thick is cut from one side, the volume remaining is 294 cubic inches?

116. Geometry What is the length of the edge of a cube if its volume could be doubled by an increase of 6 centimeters in one edge, an increase of 12 centimeters in a second edge, and a decrease of 4 centimeters in the third edge?

117. Let $f(x)$ be a polynomial function whose coefficients are integers. Suppose that r is a real zero of f and that the leading coefficient of f is 1. Use the Rational Zeros Theorem to show that r is either an integer or an irrational number.

118. Prove the Rational Zeros Theorem.

[**Hint:** Let $\dfrac{p}{q}$, where p and q have no common factors except 1 and -1, be a zero of the polynomial function

$$f(x) = a_n x^n + a_{n-1} x^{n-1} + \cdots + a_1 x + a_0$$

whose coefficients are all integers. Show that

$$a_n p^n + a_{n-1} p^{n-1} q + \cdots + a_1 p q^{n-1} + a_0 q^n = 0$$

Now, because p is a factor of the first n terms of this equation, p must also be a factor of the term $a_0 q^n$. Since p is not a factor of q (why?), p must be a factor of a_0. Similarly, q must be a factor of a_n.]

119. Bisection Method for Approximating Zeros of a Function f We begin with two consecutive integers, a and $a + 1$, such that $f(a)$ and $f(a + 1)$ are of opposite sign. Evaluate f at the midpoint m_1 of a and $a + 1$. If $f(m_1) = 0$, then m_1 is the zero of f, and we are finished. Otherwise, $f(m_1)$ is of opposite sign to either $f(a)$ or $f(a + 1)$. Suppose that it is $f(a)$ and $f(m_1)$ that are of opposite sign. Now evaluate f at the midpoint m_2 of a and m_1. Repeat this process until the desired degree of accuracy is obtained. Note that each iteration places the zero in an interval whose length is half that of the previous interval. Use the bisection method to approximate the zero of $f(x) = 8x^4 - 2x^2 + 5x - 1$ in the interval $[0, 1]$ correct to three decimal places.

[**Hint:** The process ends when both endpoints agree to the desired number of decimal places.]

Discussion and Writing

120. Is $\dfrac{1}{3}$ a zero of $f(x) = 2x^3 + 3x^2 - 6x + 7$? Explain.

121. Is $\dfrac{1}{3}$ a zero of $f(x) = 4x^3 - 5x^2 - 3x + 1$? Explain.

122. Is $\dfrac{3}{5}$ a zero of $f(x) = 2x^6 - 5x^4 + x^3 - x + 1$? Explain.

123. Is $\dfrac{2}{3}$ a zero of $f(x) = x^7 + 6x^5 - x^4 + x + 2$? Explain.

Retain Your Knowledge

Problems 124–127 are based on material learned earlier in the course. The purpose of these problems is to keep the material fresh in your mind so that you are better prepared for the final exam.

124. Solve $2x - 5y = 3$ for y.

125. Express the inequality $3 \leq x < 8$ using interval notation.

126. Find the intercepts of the graph of the equation $3x + y^2 = 12$.

127. Use the figure to determine the interval(s) on which the function is increasing.

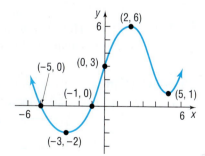

'Are You Prepared?' Answers

1. 3   **2.** $(3x + 2)(2x - 1)$ **3.** Quotient: $3x^3 + 4x^2 + 12x + 43$; Remainder: 125   **4.** $\dfrac{-1 - \sqrt{13}}{2}, \dfrac{-1 + \sqrt{13}}{2}$

5.6 Complex Zeros; Fundamental Theorem of Algebra

PREPARING FOR THIS SECTION *Before getting started, review the following:*

- Complex Numbers (Section 1.3, pp. 104–109)
- Complex Solutions of a Quadratic Equation (Section 1.3, pp. 109–111)

Now Work the **'Are You Prepared?'** problems on page 394.

OBJECTIVES 1 Use the Conjugate Pairs Theorem (p. 391)
          2 Find a Polynomial Function with Specified Zeros (p. 392)
          3 Find the Complex Zeros of a Polynomial Function (p. 393)

In Section 1.2, we found the real solutions of a quadratic equation. That is, we found the real zeros of a polynomial function of degree 2. Then, in Section 1.3 we found the complex solutions of a quadratic equation. That is, we found the complex zeros of a polynomial function of degree 2.

In Section 5.5, we found the real zeros of polynomial functions of degree 3 or higher. In this section we will find the *complex zeros* of polynomial functions of degree 3 or higher.

DEFINITION

A variable in the complex number system is referred to as a **complex variable.** A **complex polynomial function** f of degree n is a function of the form

$$f(x) = a_n x^n + a_{n-1} x^{n-1} + \cdots + a_1 x + a_0 \qquad (1)$$

where $a_n, a_{n-1}, \ldots, a_1, a_0$ are complex numbers, $a_n \neq 0$, n is a nonnegative integer, and x is a complex variable. As before, a_n is called the **leading coefficient** of f. A complex number r is called a **complex zero** of f if $f(r) = 0$.

In most of our work, the coefficients in (1) will be real numbers.

We have learned that some quadratic equations have no real solutions, but that in the complex number system every quadratic equation has a solution, either real or complex. The next result, proved by Karl Friedrich Gauss (1777–1855) when he was 22 years old,* gives an extension to complex polynomials. In fact, this result is so important and useful that it has become known as the **Fundamental Theorem of Algebra**.

FUNDAMENTAL THEOREM OF ALGEBRA

Every complex polynomial function f of degree $n \geq 1$ has at least one complex zero.

We shall not prove this result, as the proof is beyond the scope of this text. However, using the Fundamental Theorem of Algebra and the Factor Theorem, we can prove the following result:

THEOREM

Every complex polynomial function f of degree $n \geq 1$ can be factored into n linear factors (not necessarily distinct) of the form

$$f(x) = a_n(x - r_1)(x - r_2) \cdots \cdot (x - r_n) \qquad (2)$$

where $a_n, r_1, r_2, \ldots, r_n$ are complex numbers. That is, every complex polynomial function of degree $n \geq 1$ has exactly n complex zeros, some of which may repeat.

*In all, Gauss gave four different proofs of this theorem, the first one in 1799 being the subject of his doctoral dissertation.

Proof Let

$$f(x) = a_n x^n + a_{n-1} x^{n-1} + \cdots + a_1 x + a_0$$

By the Fundamental Theorem of Algebra, f has at least one zero, say r_1. Then, by the Factor Theorem, $x - r_1$ is a factor, and

$$f(x) = (x - r_1) q_1(x)$$

where $q_1(x)$ is a complex polynomial of degree $n - 1$ whose leading coefficient is a_n. Repeating this argument n times, we arrive at

$$f(x) = (x - r_1)(x - r_2) \cdot \cdots \cdot (x - r_n) q_n(x)$$

where $q_n(x)$ is a complex polynomial of degree $n - n = 0$ whose leading coefficient is also a_n. That is, $q_n(x) = a_n x^0 = a_n$, and so

$$f(x) = a_n (x - r_1)(x - r_2) \cdot \cdots \cdot (x - r_n)$$

We conclude that every complex polynomial function f of degree $n \geq 1$ has exactly n (not necessarily distinct) zeros. ∎

1 Use the Conjugate Pairs Theorem

The Fundamental Theorem of Algebra can be used to obtain valuable information about the complex zeros of polynomial functions whose coefficients are real numbers.

CONJUGATE PAIRS THEOREM

> Let f be a polynomial function whose coefficients are real numbers. If $r = a + bi$ is a zero of f, the complex conjugate $\bar{r} = a - bi$ is also a zero of f.

In other words, for polynomial functions whose coefficients are real numbers, the complex zeros occur in conjugate pairs. This result should not be all that surprising since the complex zeros of a quadratic function occurred in conjugate pairs.

Proof Let

$$f(x) = a_n x^n + a_{n-1} x^{n-1} + \cdots + a_1 x + a_0$$

where $a_n, a_{n-1}, \ldots, a_1, a_0$ are real numbers and $a_n \neq 0$. If $r = a + bi$ is a zero of f, then $f(r) = f(a + bi) = 0$, so

$$a_n r^n + a_{n-1} r^{n-1} + \cdots + a_1 r + a_0 = 0$$

Take the conjugate of both sides to get

$$\overline{a_n r^n + a_{n-1} r^{n-1} + \cdots + a_1 r + a_0} = \bar{0}$$

$$\overline{a_n r^n} + \overline{a_{n-1} r^{n-1}} + \cdots + \overline{a_1 r} + \overline{a_0} = \bar{0} \quad \text{\color{teal}\small The conjugate of a sum equals the sum of the conjugates (see Section 1.3).}$$

$$\overline{a_n}(\bar{r})^n + \overline{a_{n-1}}(\bar{r})^{n-1} + \cdots + \overline{a_1}\bar{r} + \overline{a_0} = \bar{0} \quad \text{\color{teal}\small The conjugate of a product equals the product of the conjugates.}$$

$$a_n(\bar{r})^n + a_{n-1}(\bar{r})^{n-1} + \cdots + a_1\bar{r} + a_0 = 0 \quad \text{\color{teal}\small The conjugate of a real number equals the real number.}$$

This last equation states that $f(\bar{r}) = 0$; that is, $\bar{r} = a - bi$ is a zero of f. ∎

The importance of this result should be clear. Once we know that, say, $3 + 4i$ is a zero of a polynomial function with real coefficients, then we know that $3 - 4i$ is also a zero. This result has an important corollary.

COROLLARY

> A polynomial function f of odd degree with real coefficients has at least one real zero.

Proof Because complex zeros occur as conjugate pairs in a polynomial function with real coefficients, there will always be an even number of zeros that are not real numbers. Consequently, since f is of odd degree, one of its zeros has to be a real number. ∎

For example, the polynomial function $f(x) = x^5 - 3x^4 + 4x^3 - 5$ has at least one zero that is a real number, since f is of degree 5 (odd) and has real coefficients.

EXAMPLE 1

Using the Conjugate Pairs Theorem

A polynomial function f of degree 5 whose coefficients are real numbers has the zeros $1, 5i$, and $1 + i$. Find the remaining two zeros.

Solution   Since f has coefficients that are real numbers, complex zeros appear as conjugate pairs. It follows that $-5i$, the conjugate of $5i$, and $1 - i$, the conjugate of $1 + i$, are the two remaining zeros. ●

 Now Work PROBLEM 7

2 Find a Polynomial Function with Specified Zeros

EXAMPLE 2

Finding a Polynomial Function Whose Zeros Are Given

Find a polynomial function f of degree 4 whose coefficients are real numbers that has the zeros $1, 1$, and $-4 + i$.

Solution   Since $-4 + i$ is a zero, by the Conjugate Pairs Theorem, $-4 - i$ must also be a zero of f. Because of the Factor Theorem, if $f(c) = 0$, then $x - c$ is a factor of $f(x)$. So f can now be written as

$$f(x) = a(x - 1)(x - 1)[x - (-4 + i)][x - (-4 - i)]$$

where a is any real number. Then

$$
\begin{aligned}
f(x) &= a(x - 1)(x - 1)[x - (-4 + i)][x - (-4 - i)]\\
&= a(x^2 - 2x + 1)[(x + 4) - i][(x + 4) + i]\\
&= a(x^2 - 2x + 1)[(x + 4)^2 - i^2]\\
&= a(x^2 - 2x + 1)[x^2 + 8x + 16 - (-1)]\\
&= a(x^2 - 2x + 1)(x^2 + 8x + 17)\\
&= a(x^4 + 8x^3 + 17x^2 - 2x^3 - 16x^2 - 34x + x^2 + 8x + 17)\\
&= a(x^4 + 6x^3 + 2x^2 - 26x + 17)
\end{aligned}
$$

●

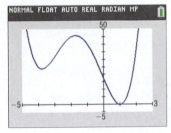

Figure 47
$f(x) = x^4 + 6x^3 + 2x^2 - 26x + 17$

Exploration

Graph the function f found in Example 2 for $a = 1$. Does the value of a affect the zeros of f? How does the value of a affect the graph of f? What information about f is sufficient to uniquely determine a?

Result A quick analysis of the polynomial function f tells us what to expect:

At most three turning points.

For large $|x|$, the graph will behave like $y = x^4$.

A repeated real zero at 1, so the graph will touch the x-axis at 1.

The only x-intercept is 1; the y-intercept is 17.

Figure 47 shows the complete graph. (Do you see why? The graph has exactly three turning points.) The value of a causes a stretch or compression; a reflection also occurs if $a < 0$. The zeros are not affected.

If any point other than an x-intercept on the graph of f is known, then a can be determined. For example, if $(2, 3)$ is on the graph, then $f(2) = 3 = a(37)$, so $a = 3/37$. Why won't an x-intercept work?

Now we can prove the theorem we conjectured in Section 5.5.

THEOREM   Every polynomial function with real coefficients can be uniquely factored over the real numbers into a product of linear factors and/or irreducible quadratic factors.

Proof Every complex polynomial function f of degree n has exactly n zeros and can be factored into a product of n linear factors. If its coefficients are real, those zeros that are complex numbers will always occur as conjugate pairs. As a result, if $r = a + bi$ is a complex zero, then so is $\bar{r} = a - bi$. Consequently, when the linear factors $x - r$ and $x - \bar{r}$ of f are multiplied, we have

$$(x - r)(x - \bar{r}) = x^2 - (r + \bar{r})x + r\bar{r} = x^2 - 2ax + a^2 + b^2$$

This second-degree polynomial has real coefficients and is irreducible (over the real numbers). So, the factors of f are either linear or irreducible quadratic factors.   ∎

Now Work PROBLEM 17

3 Find the Complex Zeros of a Polynomial Function

The steps for finding the complex zeros of a polynomial function are the same as those for finding the real zeros.

EXAMPLE 3

Finding the Complex Zeros of a Polynomial Function

Find the complex zeros of the polynomial function

$$f(x) = 3x^4 + 5x^3 + 25x^2 + 45x - 18$$

Write f in factored form.

Solution

STEP 1: The degree of f is 4, so f will have four complex zeros.

STEP 2: Since the coefficients of f are real numbers, Descartes' Rule of Signs can be used to obtain information about the real zeros. For this polynomial function, there is one positive real zero. There are three negative real zeros or one negative real zero, because

$$f(-x) = 3x^4 - 5x^3 + 25x^2 - 45x - 18$$

has three variations in sign.

STEP 3: Since the coefficients of f are integers, the Rational Zeros Theorem can be used to obtain information about the potential rational zeros of f. The potential rational zeros are

$$\pm \frac{1}{3}, \ \pm \frac{2}{3}, \ \pm 1, \ \pm 2, \ \pm 3, \ \pm 6, \ \pm 9, \ \pm 18$$

Table 17 summarizes some results of synthetic division.

Table 17

r	Coefficients of $q(x)$				Remainder	
1	3	8	33	78	60	1 is not a zero.
-1	3	2	23	22	-40	-1 is not a zero.
2	3	11	47	139	260	2 is not a zero.
-2	3	-1	27	-9	⓪	-2 is a zero.

Since $f(-2) = 0$, then -2 is a zero and $x + 2$ is a factor of f. The depressed equation is

$$3x^3 - x^2 + 27x - 9 = 0$$

REPEAT STEP 3: We factor the depressed equation by grouping.

$$3x^3 - x^2 + 27x - 9 = 0$$

$$x^2(3x - 1) + 9(3x - 1) = 0 \quad \text{Factor } x^2 \text{ from } 3x^3 - x^2 \text{ and 9 from } 27x - 9.$$

$$(x^2 + 9)(3x - 1) = 0 \quad \text{Factor out the common factor } 3x - 1.$$

$$x^2 + 9 = 0 \quad \text{or} \quad 3x - 1 = 0 \quad \text{Apply the Zero-Product Property.}$$

$$x^2 = -9 \quad \text{or} \quad x = \frac{1}{3}$$

$$x = -3i, \quad x = 3i \quad \text{or} \quad x = \frac{1}{3}$$

The four complex zeros of f are $-3i, 3i, -2,$ and $\dfrac{1}{3}$.

The factored form of f is

$$f(x) = 3x^4 + 5x^3 + 25x^2 + 45x - 18$$

$$= 3(x + 3i)(x - 3i)(x + 2)\left(x - \dfrac{1}{3}\right)$$

 **Now Work** PROBLEM 33

5.6 Assess Your Understanding

'Are You Prepared?' *Answers are given at the end of these exercises. If you get the wrong answer, read the pages listed in red.*

1. Find the sum and the product of the complex numbers $3 - 2i$ and $-3 + 5i$. (pp. 104–109)

2. In the complex number system, find the complex zeros of $f(x) = x^2 + 2x + 2$. (pp. 109–111)

Concepts and Vocabulary

3. Every polynomial function of odd degree with real coefficients will have at least _____ real zero(s).

4. If $3 + 4i$ is a zero of a polynomial function of degree 5 with real coefficients, then so is _____.

5. *True or False* A polynomial function of degree n with real coefficients has exactly n complex zeros. At most n of them are real zeros.

6. *True or False* A polynomial function of degree 4 with real coefficients could have $-3, 2 + i, 2 - i,$ and $-3 + 5i$ as its zeros.

Skill Building

In Problems 7–16, information is given about a polynomial function f whose coefficients are real numbers. Find the remaining zeros of f.

7. Degree 3; zeros: $3, 4 - i$

8. Degree 3; zeros: $4, 3 + i$

9. Degree 4; zeros: $i, 1 + i$

10. Degree 4; zeros: $1, 2, 2 + i$

11. Degree 5; zeros: $1, i, 2i$

12. Degree 5; zeros: $0, 1, 2, i$

13. Degree 4; zeros: $i, 2, -2$

14. Degree 4; zeros: $2 - i, -i$

15. Degree 6; zeros: $2, 2 + i, -3 - i, 0$

16. Degree 6; zeros: $i, 3 - 2i, -2 + i$

In Problems 17–22, find a polynomial function f with real coefficients having the given degree and zeros. Answers will vary depending on the choice of leading coefficient.

17. Degree 4; zeros: $3 + 2i; 4,$ multiplicity 2

18. Degree 4; zeros: $i, 1 + 2i$

19. Degree 5; zeros: $2; -i; 1 + i$

20. Degree 6; zeros: $i, 4 - i; 2 + i$

21. Degree 4; zeros: $3,$ multiplicity 2; $-i$

22. Degree 5; zeros: $1,$ multiplicity 3; $1 + i$

In Problems 23–30, use the given zero to find the remaining zeros of each polynomial function.

23. $f(x) = x^3 - 4x^2 + 4x - 16$; zero: $2i$

24. $g(x) = x^3 + 3x^2 + 25x + 75$; zero: $-5i$

25. $f(x) = 2x^4 + 5x^3 + 5x^2 + 20x - 12$; zero: $-2i$

26. $h(x) = 3x^4 + 5x^3 + 25x^2 + 45x - 18$; zero: $3i$

27. $h(x) = x^4 - 9x^3 + 21x^2 + 21x - 130$; zero: $3 - 2i$

28. $f(x) = x^4 - 7x^3 + 14x^2 - 38x - 60$; zero: $1 + 3i$

29. $h(x) = 3x^5 + 2x^4 + 15x^3 + 10x^2 - 528x - 352$; zero: $-4i$

30. $g(x) = 2x^5 - 3x^4 - 5x^3 - 15x^2 - 207x + 108$; zero: $3i$

In Problems 31–40, find the complex zeros of each polynomial function. Write f in factored form.

31. $f(x) = x^3 - 1$

32. $f(x) = x^4 - 1$

33. $f(x) = x^3 - 8x^2 + 25x - 26$

34. $f(x) = x^3 + 13x^2 + 57x + 85$

35. $f(x) = x^4 + 5x^2 + 4$

36. $f(x) = x^4 + 13x^2 + 36$

37. $f(x) = x^4 + 2x^3 + 22x^2 + 50x - 75$

38. $f(x) = x^4 + 3x^3 - 19x^2 + 27x - 252$

39. $f(x) = 3x^4 - x^3 - 9x^2 + 159x - 52$

40. $f(x) = 2x^4 + x^3 - 35x^2 - 113x + 65$

Mixed Practice

41. Given $f(x) = 2x^3 - 14x^2 + bx - 3$ with $f(2) = 0$, $g(x) = x^3 + cx^2 - 8x + 30$, with the zero $x = 3 - i$, and b and c real numbers, find $(f \cdot g)(1)$.[†]

42. Let f be the polynomial function of degree 4 with real coefficients, leading coefficient 1, and zeros $x = 3 + i$, 2, −2. Let g be the polynomial function of degree 4 with intercept $(0, -4)$ and zeros $x = i$, 2i. Find $(f + g)(1)$.[†]

43. The complex zeros of $f(x) = x^4 + 1$ For the function $f(x) = x^4 + 1$:
   (a) Factor f into the product of two irreducible quadratics. (**Hint:** complete the square by adding and subtracting $2x^2$.)
   (b) Find the zeros of f by finding the zeros of each irreducible quadratic.

[†]Courtesy of the Joliet Junior College Mathematics Department

Discussion and Writing

In Problems 44 and 45, explain why the facts given are contradictory.

44. f is a polynomial function of degree 3 whose coefficients are real numbers; its zeros are $2, i$, and $3 + i$.

45. f is a polynomial function of degree 3 whose coefficients are real numbers; its zeros are $4 + i, 4 - i$, and $2 + i$.

46. f is a polynomial function of degree 4 whose coefficients are real numbers; two of its zeros are -3 and $4 - i$. Explain why one of the remaining zeros must be a real number. Write down one of the missing zeros.

47. f is a polynomial function of degree 4 whose coefficients are real numbers; three of its zeros are $2, 1 + 2i$, and $1 - 2i$. Explain why the remaining zero must be a real number.

48. For the polynomial function $f(x) = x^2 + 2ix - 10$:
   (a) Verify that $3 - i$ is a zero of f.
   (b) Verify that $3 + i$ is not a zero of f.
   (c) Explain why these results do not contradict the Conjugate Pairs Theorem.

Retain Your Knowledge

Problems 49–52 are based on material learned earlier in the course. The purpose of these problems is to keep the material fresh in your mind so that you are better prepared for the final exam.

49. Draw a scatter diagram for the given data.

x	−1	1	2	5	8	10
y	−4	0	3	1	5	7

50. Solve: $\sqrt{3 - x} = 5$

51. Multiply: $(2x - 5)(3x^2 + x - 4)$

52. Find the area and circumference of a circle with a diameter of 6 feet.

'Are You Prepared?' Answers

1. Sum: $3i$; product: $1 + 21i$

2. $-1 - i, \ -1 + i$

Figure 3 provides a second illustration of the definition. Here x is the input to the function g, yielding $g(x)$. Then $g(x)$ is the input to the function f, yielding $f(g(x))$. Note that the "inside" function g in $f(g(x))$ is "processed" first.

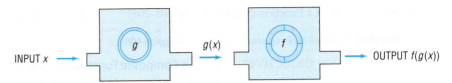

Figure 3

EXAMPLE 1

Evaluating a Composite Function

Suppose that $f(x) = 2x^2 - 3$ and $g(x) = 4x$. Find:

(a) $(f \circ g)(1)$ (b) $(g \circ f)(1)$ (c) $(f \circ f)(-2)$ (d) $(g \circ g)(-1)$

Solution   (a) $(f \circ g)(1) = f(g(1)) = f(4) = 2 \cdot 4^2 - 3 = 29$

$g(x) = 4x \quad f(x) = 2x^2 - 3$
$g(1) = 4$

(b) $(g \circ f)(1) = g(f(1)) = g(-1) = 4 \cdot (-1) = -4$

$f(x) = 2x^2 - 3 \quad g(x) = 4x$
$f(1) = -1$

(c) $(f \circ f)(-2) = f(f(-2)) = f(5) = 2 \cdot 5^2 - 3 = 47$

$f(-2) = 2(-2)^2 - 3 = 5$

(d) $(g \circ g)(-1) = g(g(-1)) = g(-4) = 4 \cdot (-4) = -16$

$g(-1) = -4$

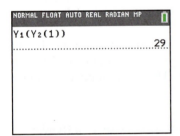

Figure 4

⌨ COMMENT *Graphing calculators can be used to evaluate composite functions.**
Let $Y_1 = f(x) = 2x^2 - 3$ and $Y_2 = g(x) = 4x$. Then, using a TI-84 Plus C graphing calculator, find $(f \circ g)(1)$ as shown in Figure 4. Note that this is the result obtained in Example 1(a).

✏ Now Work PROBLEM 13

2 Find the Domain of a Composite Function

EXAMPLE 2

Finding a Composite Function and Its Domain

Suppose that $f(x) = x^2 + 3x - 1$ and $g(x) = 2x + 3$.

Find: (a) $f \circ g$ (b) $g \circ f$

Then find the domain of each composite function.

Solution   The domain of f and the domain of g are the set of all real numbers.

(a) $(f \circ g)(x) = f(g(x)) = f(2x + 3) = (2x + 3)^2 + 3(2x + 3) - 1$

$f(x) = x^2 + 3x - 1$

$= 4x^2 + 12x + 9 + 6x + 9 - 1 = 4x^2 + 18x + 17$

Because the domains of both f and g are the set of all real numbers, the domain of $f \circ g$ is the set of all real numbers.

* Consult your owner's manual for the appropriate keystrokes.

(b) $(g \circ f)(x) = g(f(x)) = g(x^2 + 3x - 1) = 2\underset{\uparrow}{(x^2 + 3x - 1)} + 3$

$$g(x) = 2x + 3$$

$$= 2x^2 + 6x - 2 + 3 = 2x^2 + 6x + 1$$

Because the domains of both f and g are the set of all real numbers, the domain of $g \circ f$ is the set of all real numbers. ●

Example 2 illustrates that, in general, $f \circ g \neq g \circ f$. Sometimes $f \circ g$ does equal $g \circ f$, as we shall see in Example 5.

Look back at Figure 2 on page 403. In determining the domain of the composite function $(f \circ g)(x) = f(g(x))$, keep the following two thoughts in mind about the input x.

1. Any x not in the domain of g must be excluded.

2. Any x for which $g(x)$ is not in the domain of f must be excluded.

EXAMPLE 3 **Finding the Domain of $f \circ g$**

Find the domain of $f \circ g$ if $f(x) = \dfrac{1}{x + 2}$ and $g(x) = \dfrac{4}{x - 1}$.

Solution For $(f \circ g)(x) = f(g(x))$, first note that the domain of g is $\{x \mid x \neq 1\}$, so 1 is excluded from the domain of $f \circ g$. Next note that the domain of f is $\{x \mid x \neq -2\}$, which means that $g(x)$ cannot equal -2. Solve the equation $g(x) = -2$ to determine what additional value(s) of x to exclude.

$$\frac{4}{x - 1} = -2 \qquad\quad g(x) = -2$$

$$4 = -2(x - 1) \quad \text{Multiply both sides by } x - 1.$$

$$4 = -2x + 2 \quad\;\; \text{Apply the Distributive Property.}$$

$$2x = -2 \qquad\quad\; \text{Add } 2x \text{ to both sides. Subtract 4 from both sides.}$$

$$x = -1 \qquad\quad\;\; \text{Divide both sides by 2.}$$

Also exclude -1 from the domain of $f \circ g$.
The domain of $f \circ g$ is $\{x \mid x \neq -1, x \neq 1\}$.

✓**Check:** For $x = 1, g(x) = \dfrac{4}{x - 1}$ is not defined, so $(f \circ g)(x) = f(g(x))$ is not defined.

For $x = -1, g(-1) = -2$, and $(f \circ g)(-1) = f(g(-1)) = f(-2)$ is not defined. ●

EXAMPLE 4 **Finding a Composite Function and Its Domain**

Suppose that $f(x) = \dfrac{1}{x + 2}$ and $g(x) = \dfrac{4}{x - 1}$.

Find: (a) $f \circ g$ (b) $f \circ f$

Then find the domain of each composite function.

Solution The domain of f is $\{x \mid x \neq -2\}$ and the domain of g is $\{x \mid x \neq 1\}$.

(a) $(f \circ g)(x) = f(g(x)) = f\left(\dfrac{4}{x - 1}\right) = \dfrac{1}{\underset{\uparrow}{\dfrac{4}{x - 1}} + 2} = \dfrac{x - 1}{4 + 2(x - 1)} = \dfrac{x - 1}{2x + 2} = \dfrac{x - 1}{2(x + 1)}$

$$f(x) = \frac{1}{x + 2} \qquad\qquad \text{Multiply by } \frac{x - 1}{x - 1}.$$

In Example 3, the domain of $f \circ g$ was found to be $\{x \mid x \neq -1, x \neq 1\}$.

The domain of $f \circ g$ also can be found by first looking at the domain of g: $\{x \mid x \neq 1\}$. Exclude 1 from the domain of $f \circ g$ as a result. Then look at $f \circ g$ and note that x cannot equal -1, because $x = -1$ results in division by 0. So exclude -1 from the domain of $f \circ g$. Therefore, the domain of $f \circ g$ is $\{x \mid x \neq -1, x \neq 1\}$.

(b) $(f \circ f)(x) = f(f(x)) = f\left(\dfrac{1}{x+2}\right) = \dfrac{1}{\underset{\uparrow}{\dfrac{1}{x+2}} + 2} = \dfrac{x+2}{1+2(x+2)} = \dfrac{x+2}{2x+5}$

$$f(x) = \dfrac{1}{x+2} \qquad \text{Multiply by } \dfrac{x+2}{x+2}.$$

The domain of $f \circ f$ consists of all values of x in the domain of f, $\{x \mid x \neq -2\}$, for which

$$f(x) = \dfrac{1}{x+2} \neq -2 \qquad \dfrac{1}{x+2} = -2$$

$$1 = -2(x+2)$$
$$1 = -2x - 4$$
$$2x = -5$$
$$x = -\dfrac{5}{2}$$

or, equivalently,

$$x \neq -\dfrac{5}{2}$$

The domain of $f \circ f$ is $\left\{x \,\middle|\, x \neq -\dfrac{5}{2}, x \neq -2\right\}$.

The domain of $f \circ f$ also can be found by recognizing that -2 is not in the domain of f and so should be excluded from the domain of $f \circ f$. Then, looking at $f \circ f$, note that x cannot equal $-\dfrac{5}{2}$. Do you see why? Therefore, the domain of $f \circ f$ is $\left\{x \,\middle|\, x \neq -\dfrac{5}{2}, x \neq -2\right\}$. ●

══ **Now Work** PROBLEMS 27 AND 29

EXAMPLE 5

Showing That Two Composite Functions Are Equal

If $f(x) = 3x - 4$ and $g(x) = \dfrac{1}{3}(x+4)$, show that

$$(f \circ g)(x) = (g \circ f)(x) = x$$

for every x in the domain of $f \circ g$ and $g \circ f$.

Solution

$(f \circ g)(x) = f(g(x))$

$= f\left(\dfrac{x+4}{3}\right) \qquad g(x) = \dfrac{1}{3}(x+4) = \dfrac{x+4}{3}$

$= 3\left(\dfrac{x+4}{3}\right) - 4 \qquad f(x) = 3x - 4.$

$= x + 4 - 4 = x$

Seeing the Concept
Using a graphing calculator, let

$$Y_1 = f(x) = 3x - 4$$

$$Y_2 = g(x) = \frac{1}{3}(x + 4)$$

$$Y_3 = f \circ g, \ Y_4 = g \circ f$$

Using the viewing window $-3 \le x \le 3$, $-2 \le y \le 2$, graph only Y_3 and Y_4. What do you see? TRACE to verify that $Y_3 = Y_4$.

$$(g \circ f)(x) = g(f(x))$$
$$= g(3x - 4) \qquad f(x) = 3x - 4$$
$$= \frac{1}{3}\left[(3x - 4) + 4\right] \qquad g(x) = \frac{1}{3}(x + 4).$$
$$= \frac{1}{3}(3x) = x$$

We conclude that $(f \circ g)(x) = (g \circ f)(x) = x$.

In Section 6.2, we shall see that there is an important relationship between functions f and g for which $(f \circ g)(x) = (g \circ f)(x) = x$.

 Now Work PROBLEM 39

Calculus Application

 Some techniques in calculus require the ability to determine the components of a composite function. For example, the function $H(x) = \sqrt{x + 1}$ is the composition of the functions f and g, where $f(x) = \sqrt{x}$ and $g(x) = x + 1$, because $H(x) = (f \circ g)(x) = f(g(x)) = f(x + 1) = \sqrt{x + 1}$.

EXAMPLE 6	**Finding the Components of a Composite Function**

Find functions f and g such that $f \circ g = H$ if $H(x) = (x^2 + 1)^{50}$.

Solution The function H takes $x^2 + 1$ and raises it to the power 50. A natural way to decompose H is to raise the function $g(x) = x^2 + 1$ to the power 50. Let $f(x) = x^{50}$ and $g(x) = x^2 + 1$. Then

$$(f \circ g)(x) = f(g(x))$$
$$= f(x^2 + 1)$$
$$= (x^2 + 1)^{50} = H(x)$$

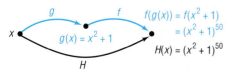

Figure 5

See Figure 5.

Other functions f and g may be found for which $f \circ g = H$ in Example 6. For instance, if $f(x) = x^2$ and $g(x) = (x^2 + 1)^{25}$, then

$$(f \circ g)(x) = f(g(x)) = f((x^2 + 1)^{25}) = [(x^2 + 1)^{25}]^2 = (x^2 + 1)^{50}$$

Although the functions f and g found as a solution to Example 6 are not unique, there is usually a "natural" selection for f and g that comes to mind first.

EXAMPLE 7	**Finding the Components of a Composite Function**

Find functions f and g such that $f \circ g = H$ if $H(x) = \dfrac{1}{x + 1}$.

Solution Here H is the reciprocal of $g(x) = x + 1$. Let $f(x) = \dfrac{1}{x}$ and $g(x) = x + 1$. Then

$$(f \circ g)(x) = f(g(x)) = f(x + 1) = \frac{1}{x + 1} = H(x)$$

 Now Work PROBLEM 47

6.1 Assess Your Understanding

'Are You Prepared?' *Answers are given at the end of these exercises. If you get a wrong answer, read the pages listed in* red.

1. Find $f(3)$ if $f(x) = -4x^2 + 5x$. (pp. 202–206)

2. Find $f(3x)$ if $f(x) = 4 - 2x^2$. (pp. 202–206)

3. Find the domain of the function $f(x) = \dfrac{x^2 - 1}{x^2 - 25}$. (pp. 206–208)

Concepts and Vocabulary

4. Given two functions f and g, the _____ _____, denoted $f \circ g$, is defined by $(f \circ g)(x) =$ _____.

5. *True or False* If $f(x) = x^2$ and $g(x) = \sqrt{x + 9}$, then $(f \circ g)(4) = 5$.

6. If $f(x) = \sqrt{x + 2}$ and $g(x) = \dfrac{3}{x}$, which of the following does $(f \circ g)(x)$ equal?

(a) $\dfrac{3}{\sqrt{x + 2}}$ (b) $\dfrac{3}{\sqrt{x}} + 2$ (c) $\sqrt{\dfrac{3}{x} + 2}$ (d) $\sqrt{\dfrac{3}{x + 2}}$

7. If $H = f \circ g$ and $H(x) = \sqrt{25 - 4x^2}$, which of the following cannot be the component functions f and g?

(a) $f(x) = \sqrt{25 - x^2};\ g(x) = 4x$

(b) $f(x) = \sqrt{x};\ g(x) = 25 - 4x^2$

(c) $f(x) = \sqrt{25 - x};\ g(x) = 4x^2$

(d) $f(x) = \sqrt{25 - 4x};\ g(x) = x^2$

8. *True or False* The domain of the composite function $(f \circ g)(x)$ is the same as the domain of $g(x)$.

Skill Building

In Problems 9 and 10, evaluate each expression using the values given in the table.

9.

x	−3	−2	−1	0	1	2	3
f(x)	−7	−5	−3	−1	3	5	7
g(x)	8	3	0	−1	0	3	8

(a) $(f \circ g)(1)$ (b) $(f \circ g)(-1)$
(c) $(g \circ f)(-1)$ (d) $(g \circ f)(0)$
(e) $(g \circ g)(-2)$ (f) $(f \circ f)(-1)$

10.

x	−3	−2	−1	0	1	2	3
f(x)	11	9	7	5	3	1	−1
g(x)	−8	−3	0	1	0	−3	−8

(a) $(f \circ g)(1)$ (b) $(f \circ g)(2)$
(c) $(g \circ f)(2)$ (d) $(g \circ f)(3)$
(e) $(g \circ g)(1)$ (f) $(f \circ f)(3)$

In Problems 11 and 12, evaluate each expression using the graphs of $y = f(x)$ and $y = g(x)$ shown in the figure.

11. (a) $(g \circ f)(-1)$ (b) $(g \circ f)(0)$
(c) $(f \circ g)(-1)$ (d) $(f \circ g)(4)$

12. (a) $(g \circ f)(1)$ (b) $(g \circ f)(5)$
(c) $(f \circ g)(0)$ (d) $(f \circ g)(2)$

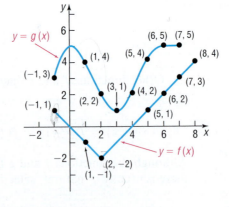

In Problems 13–22, for the given functions f and g, find:

(a) $(f \circ g)(4)$ (b) $(g \circ f)(2)$ (c) $(f \circ f)(1)$ (d) $(g \circ g)(0)$

13. $f(x) = 2x;\ g(x) = 3x^2 + 1$

14. $f(x) = 3x + 2;\ g(x) = 2x^2 - 1$

15. $f(x) = 4x^2 - 3;\ g(x) = 3 - \dfrac{1}{2}x^2$

16. $f(x) = 2x^2;\ g(x) = 1 - 3x^2$

17. $f(x) = \sqrt{x};\ g(x) = 2x$

18. $f(x) = \sqrt{x + 1};\ g(x) = 3x$

19. $f(x) = |x|;\ g(x) = \dfrac{1}{x^2 + 1}$

20. $f(x) = |x - 2|;\ g(x) = \dfrac{3}{x^2 + 2}$

21. $f(x) = \dfrac{3}{x + 1};\ g(x) = \sqrt[3]{x}$

22. $f(x) = x^{3/2};\ g(x) = \dfrac{2}{x + 1}$

In Problems 23–38, for the given functions f and g, find:
(a) $f \circ g$ (b) $g \circ f$ (c) $f \circ f$ (d) $g \circ g$
State the domain of each composite function.

23. $f(x) = 2x + 3$; $g(x) = 3x$

24. $f(x) = -x$; $g(x) = 2x - 4$

25. $f(x) = 3x + 1$; $g(x) = x^2$

26. $f(x) = x + 1$; $g(x) = x^2 + 4$

27. $f(x) = x^2$; $g(x) = x^2 + 4$

28. $f(x) = x^2 + 1$; $g(x) = 2x^2 + 3$

29. $f(x) = \dfrac{3}{x - 1}$; $g(x) = \dfrac{2}{x}$

30. $f(x) = \dfrac{1}{x + 3}$; $g(x) = -\dfrac{2}{x}$

31. $f(x) = \dfrac{x}{x - 1}$; $g(x) = -\dfrac{4}{x}$

32. $f(x) = \dfrac{x}{x + 3}$; $g(x) = \dfrac{2}{x}$

33. $f(x) = \sqrt{x}$; $g(x) = 2x + 3$

34. $f(x) = \sqrt{x - 2}$; $g(x) = 1 - 2x$

35. $f(x) = x^2 + 1$; $g(x) = \sqrt{x - 1}$

36. $f(x) = x^2 + 4$; $g(x) = \sqrt{x - 2}$

37. $f(x) = \dfrac{x - 5}{x + 1}$; $g(x) = \dfrac{x + 2}{x - 3}$

38. $f(x) = \dfrac{2x - 1}{x - 2}$; $g(x) = \dfrac{x + 4}{2x - 5}$

In Problems 39–46, show that $(f \circ g)(x) = (g \circ f)(x) = x$.

39. $f(x) = 2x$; $g(x) = \dfrac{1}{2}x$

40. $f(x) = 4x$; $g(x) = \dfrac{1}{4}x$

41. $f(x) = x^3$; $g(x) = \sqrt[3]{x}$

42. $f(x) = x + 5$; $g(x) = x - 5$

43. $f(x) = 2x - 6$; $g(x) = \dfrac{1}{2}(x + 6)$

44. $f(x) = 4 - 3x$; $g(x) = \dfrac{1}{3}(4 - x)$

45. $f(x) = ax + b$; $g(x) = \dfrac{1}{a}(x - b)$ $a \neq 0$

46. $f(x) = \dfrac{1}{x}$; $g(x) = \dfrac{1}{x}$

In Problems 47–52, find functions f and g so that $f \circ g = H$.

47. $H(x) = (2x + 3)^4$

48. $H(x) = (1 + x^2)^3$

49. $H(x) = \sqrt{x^2 + 1}$

50. $H(x) = \sqrt{1 - x^2}$

51. $H(x) = |2x + 1|$

52. $H(x) = |2x^2 + 3|$

Applications and Extensions

53. If $f(x) = 2x^3 - 3x^2 + 4x - 1$ and $g(x) = 2$, find $(f \circ g)(x)$ and $(g \circ f)(x)$.

54. If $f(x) = \dfrac{x + 1}{x - 1}$, find $(f \circ f)(x)$.

55. If $f(x) = 2x^2 + 5$ and $g(x) = 3x + a$, find a so that the graph of $f \circ g$ crosses the y-axis at 23.

56. If $f(x) = 3x^2 - 7$ and $g(x) = 2x + a$, find a so that the graph of $f \circ g$ crosses the y-axis at 68.

In Problems 57 and 58, use the functions f and g to find:
(a) $f \circ g$ (b) $g \circ f$
(c) *the domain of $f \circ g$ and of $g \circ f$*
(d) *the conditions for which $f \circ g = g \circ f$*

57. $f(x) = ax + b$ $g(x) = cx + d$

58. $f(x) = \dfrac{ax + b}{cx + d}$ $g(x) = mx$

59. Surface Area of a Balloon The surface area S (in square meters) of a hot-air balloon is given by
$$S(r) = 4\pi r^2$$
where r is the radius of the balloon (in meters). If the radius r is increasing with time t (in seconds) according to the formula $r(t) = \dfrac{2}{3}t^3, t \geq 0$, find the surface area S of the balloon as a function of the time t.

60. Volume of a Balloon The volume V (in cubic meters) of the hot-air balloon described in Problem 59 is given by $V(r) = \dfrac{4}{3}\pi r^3$. If the radius r is the same function of t as in Problem 59, find the volume V as a function of the time t.

61. Automobile Production The number N of cars produced at a certain factory in one day after t hours of operation is given by $N(t) = 100t - 5t^2, 0 \leq t \leq 10$. If the cost C (in dollars) of producing N cars is $C(N) = 15{,}000 + 8000N$, find the cost C as a function of the time t of operation of the factory.

62. Environmental Concerns The spread of oil leaking from a tanker is in the shape of a circle. If the radius r (in feet) of the spread after t hours is $r(t) = 200\sqrt{t}$, find the area A of the oil slick as a function of the time t.

63. Production Cost The price p, in dollars, of a certain product and the quantity x sold obey the demand equation
$$p = -\frac{1}{4}x + 100 \quad 0 \leq x \leq 400$$
Suppose that the cost C, in dollars, of producing x units is
$$C = \frac{\sqrt{x}}{25} + 600$$
Assuming that all items produced are sold, find the cost C as a function of the price p.
[**Hint:** Solve for x in the demand equation and then form the composite function.]

64. Cost of a Commodity The price p, in dollars, of a certain commodity and the quantity x sold obey the demand equation

$$p = -\frac{1}{5}x + 200 \quad 0 \le x \le 1000$$

Suppose that the cost C, in dollars, of producing x units is

$$C = \frac{\sqrt{x}}{10} + 400$$

Assuming that all items produced are sold, find the cost C as a function of the price p.

65. Volume of a Cylinder The volume V of a right circular cylinder of height h and radius r is $V = \pi r^2 h$. If the height is twice the radius, express the volume V as a function of r.

66. Volume of a Cone The volume V of a right circular cone is $V = \frac{1}{3}\pi r^2 h$. If the height is twice the radius, express the volume V as a function of r.

67. Foreign Exchange Traders often buy foreign currency in the hope of making money when the currency's value changes. For example, on April 28, 2014, one U.S. dollar could purchase 0.7235 euro, and one euro could purchase 141.119 yen. Let $f(x)$ represent the number of euros you can buy with x dollars, and let $g(x)$ represent the number of yen you can buy with x euros.
(a) Find a function that relates dollars to euros.
(b) Find a function that relates euros to yen.
(c) Use the results of parts (a) and (b) to find a function that relates dollars to yen. That is, find

$$(g \circ f)(x) = g(f(x)).$$

(d) What is $g(f(1000))$?

68. Temperature Conversion The function $C(F) = \frac{5}{9}(F - 32)$ converts a temperature in degrees Fahrenheit, F, to a temperature in degrees Celsius, C. The function

$K(C) = C + 273$, converts a temperature in degrees Celsius to a temperature in kelvins, K.
(a) Find a function that converts a temperature in degrees Fahrenheit to a temperature in kelvins.
(b) Determine 80 degrees Fahrenheit in kelvins.

69. Discounts The manufacturer of a computer is offering two discounts on last year's model computer. The first discount is a \$200 rebate and the second discount is 20% off the regular price, p.
(a) Write a function f that represents the sale price if only the rebate applies.
(b) Write a function g that represents the sale price if only the 20% discount applies.
(c) Find $f \circ g$ and $g \circ f$. What does each of these functions represent? Which combination of discounts represents a better deal for the consumer? Why?

70. Taxes Suppose that you work for \$15 per hour. Write a function that represents gross salary G as a function of hours worked h. Your employer is required to withhold taxes (federal income tax, Social Security, Medicare) from your paycheck. Suppose your employer withholds 20% of your income for taxes. Write a function that represents net salary N as a function of gross salary G. Find and interpret $N \circ G$.

71. Let $f(x) = ax + b$ and $g(x) = bx + a$, where a and b are integers. If $f(1) = 8$ and $f(g(20)) - g(f(20)) = -14$, find the product of a and b.*

72. If f and g are odd functions, show that the composite function $f \circ g$ is also odd.

73. If f is an odd function and g is an even function, show that the composite functions $f \circ g$ and $g \circ f$ are both even.

* Courtesy of the Joliet Junior College Mathematics Department

---

Retain Your Knowledge

Problems 74–77 are based on material learned earlier in the course. The purpose of these problems is to keep the material fresh in your mind so that you are better prepared for the final exam.

74. Given $f(x) = 3x + 8$ and $g(x) = x - 5$, find $(f + g)(x)$, $(f - g)(x)$, $(f \cdot g)(x)$, and $\left(\dfrac{f}{g}\right)(x)$. State the domain of each.

75. Find the real zeros of $f(x) = 2x - 5\sqrt{x} + 2$.

76. Use a graphing utility to graph $f(x) = -x^3 + 4x - 2$ over the interval $(-3, 3)$. Approximate any local maxima and local minima. Determine where the function is increasing and where it is decreasing.

77. Find the domain of $R(x) = \dfrac{x^2 + 6x + 5}{x - 3}$. Find any horizontal, vertical, or oblique asymptotes.

'Are You Prepared?' Answers

1. -21 **2.** $4 - 18x^2$ **3.** $\{x \mid x \ne -5, x \ne 5\}$

6.2 One-to-One Functions; Inverse Functions

PREPARING FOR THIS SECTION *Before getting started, review the following:*

- Functions (Section 3.1, pp. 199–208)
- Increasing/Decreasing Functions (Section 3.3, pp. 225–226)
- Rational Expressions (Chapter R, Section R.7, pp. 62–69)

Now Work the **'Are You Prepared?'** problems on page 419.

OBJECTIVES   **1** Determine Whether a Function Is One-to-One (p. 411)
   2 Determine the Inverse of a Function Defined by a Map or a Set of Ordered Pairs (p. 413)
   3 Obtain the Graph of the Inverse Function from the Graph of the Function (p. 416)
   4 Find the Inverse of a Function Defined by an Equation (p. 417)

1 Determine Whether a Function Is One-to-One

Section 3.1 presented four different ways to represent a function: (1) a map, (2) a set of ordered pairs, (3) a graph, and (4) an equation. For example, Figures 6 and 7 illustrate two different functions represented as mappings. The function in Figure 6 shows the correspondence between states and their populations (in millions). The function in Figure 7 shows a correspondence between animals and life expectancies (in years).

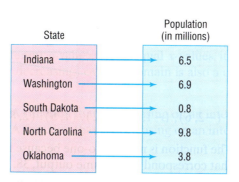

Figure 6

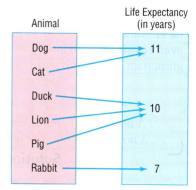

Figure 7

Suppose several people are asked to name a state that has a population of 0.8 million based on the function in Figure 6. Everyone will respond "South Dakota." Now, if the same people are asked to name an animal whose life expectancy is 11 years based on the function in Figure 7, some may respond "dog," while others may respond "cat." What is the difference between the functions in Figures 6 and 7? In Figure 6, no two elements in the domain correspond to the same element in the range. In Figure 7, this is not the case: Different elements in the domain correspond to the same element in the range. Functions such as the one in Figure 6 are given a special name.

DEFINITION

A function is **one-to-one** if any two different inputs in the domain correspond to two different outputs in the range. That is, if x_1 and x_2 are two different inputs of a function f, then f is one-to-one if $f(x_1) \neq f(x_2)$.

In Words
A function is not one-to-one if two different inputs correspond to the same output.

Put another way, a function f is one-to-one if no y in the range is the image of more than one x in the domain. A function is not one-to-one if any two (or more) different elements in the domain correspond to the same element in the range. So the function in Figure 7 is not one-to-one because two different elements in

EXAMPLE 9

Finding the Inverse Function

The function

$$f(x) = \frac{2x + 1}{x - 1} \qquad x \neq 1$$

is one-to-one. Find its inverse function and check the result.

Solution

STEP 1: Replace $f(x)$ with y and interchange the variables x and y in

$$y = \frac{2x + 1}{x - 1}$$

to obtain

$$x = \frac{2y + 1}{y - 1}$$

STEP 2: Solve for y.

$$x = \frac{2y + 1}{y - 1}$$

$x(y - 1) = 2y + 1$ Multiply both sides by $y - 1$.

$xy - x = 2y + 1$ Apply the Distributive Property.

$xy - 2y = x + 1$ Subtract $2y$ from both sides; add x to both sides.

$(x - 2)y = x + 1$ Factor.

$y = \dfrac{x + 1}{x - 2}$ Divide by $x - 2$.

The inverse function is

$$f^{-1}(x) = \frac{x + 1}{x - 2} \qquad x \neq 2 \qquad \text{Replace } y \text{ by } f^{-1}(x).$$

STEP 3: ✓ Check:

$$f^{-1}(f(x)) = f^{-1}\left(\frac{2x + 1}{x - 1}\right) = \frac{\dfrac{2x + 1}{x - 1} + 1}{\dfrac{2x + 1}{x - 1} - 2} = \frac{2x + 1 + x - 1}{2x + 1 - 2(x - 1)} = \frac{3x}{3} = x, \quad x \neq 1$$

$$f(f^{-1}(x)) = f\left(\frac{x + 1}{x - 2}\right) = \frac{2\left(\dfrac{x + 1}{x - 2}\right) + 1}{\dfrac{x + 1}{x - 2} - 1} = \frac{2(x + 1) + x - 2}{x + 1 - (x - 2)} = \frac{3x}{3} = x, \quad x \neq 2$$

Exploration

In Example 9, we found that if $f(x) = \dfrac{2x + 1}{x - 1}$, then $f^{-1}(x) = \dfrac{x + 1}{x - 2}$. Compare the vertical and horizontal asymptotes of f and f^{-1}.

Result The vertical asymptote of f is $x = 1$, and the horizontal asymptote is $y = 2$. The vertical asymptote of f^{-1} is $x = 2$, and the horizontal asymptote is $y = 1$.

══► **Now Work** PROBLEMS 53 AND 67

If a function is not one-to-one, it has no inverse function. Sometimes, though, an appropriate restriction on the domain of such a function will yield a new function that *is* one-to-one. Then the function defined on the restricted domain has an inverse function. Let's look at an example of this common practice.

EXAMPLE 10

Finding the Inverse of a Domain-restricted Function

Find the inverse of $y = f(x) = x^2$ if $x \geq 0$. Graph f and f^{-1}.

Solution

The function $y = x^2$ is not one-to-one. [Refer to Example 2(a).] However, restricting the domain of this function to $x \geq 0$, as indicated, results in a new function that is increasing and therefore is one-to-one. Consequently, the function defined by $y = f(x) = x^2, x \geq 0$, has an inverse function, f^{-1}.

Follow the steps given previously to find f^{-1}.

STEP 1: In the equation $y = x^2, x \geq 0$, interchange the variables x and y. The result is

$$x = y^2 \qquad y \geq 0$$

This equation defines the inverse function implicitly.

STEP 2: Solve for y to get the explicit form of the inverse. Because $y \geq 0$, only one solution for y is obtained: $y = \sqrt{x}$. So $f^{-1}(x) = \sqrt{x}$.

STEP 3: ✓ **Check:** $f^{-1}(f(x)) = f^{-1}(x^2) = \sqrt{x^2} = |x| = x$ because $x \geq 0$

$$f(f^{-1}(x)) = f(\sqrt{x}) = (\sqrt{x})^2 = x$$

Figure 17 illustrates the graphs of $f(x) = x^2, x \geq 0$, and $f^{-1}(x) = \sqrt{x}$. ●

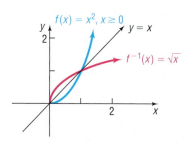

Figure 17

SUMMARY

1. If a function f is one-to-one, then it has an inverse function f^{-1}.
2. Domain of f = Range of f^{-1}; Range of f = Domain of f^{-1}.
3. To verify that f^{-1} is the inverse of f, show that $f^{-1}(f(x)) = x$ for every x in the domain of f and that $f(f^{-1}(x)) = x$ for every x in the domain of f^{-1}.
4. The graphs of f and f^{-1} are symmetric with respect to the line $y = x$.

6.2 Assess Your Understanding

'Are You Prepared?' *Answers are given at the end of these exercises. If you get a wrong answer, read the pages listed in* red.

1. Is the set of ordered pairs $\{(1,3),(2,3),(-1,2)\}$ a function? Why or why not? (pp. 199–208)
2. Where is the function $f(x) = x^2$ increasing? Where is it decreasing? (pp. 225–226)
3. What is the domain of $f(x) = \dfrac{x+5}{x^2 + 3x - 18}$? (pp. 199–208)
4. Simplify: $\dfrac{\frac{1}{x} + 1}{\frac{1}{x^2} - 1}$ (pp. 62–69)

Concepts and Vocabulary

5. If x_1 and x_2 are two different inputs of a function f, then f is one-to-one if _____.
6. If every horizontal line intersects the graph of a function f at no more than one point, then f is a(n) _____ function.
7. If f is a one-to-one function and $f(3) = 8$, then $f^{-1}(8) = $ _____.
8. If f^{-1} denotes the inverse of a function f, then the graphs of f and f^{-1} are symmetric with respect to the line _____.
9. If the domain of a one-to-one function f is $[4, \infty)$, then the range of its inverse function f^{-1} is _____.

10. **True or False** If f and g are inverse functions, then the domain of f is the same as the range of g.
11. If $(-2, 3)$ is a point on the graph of a one-to-one function f, which of the following points is on the graph of f^{-1}?
    (a) $(3, -2)$ (b) $(2, -3)$ (c) $(-3, 2)$ (d) $(-2, -3)$
12. Suppose f is a one-to-one function with a domain of $\{x | x \neq 3\}$ and a range of $\left\{y \middle| y \neq \dfrac{2}{3}\right\}$. Which of the following is the domain of f^{-1}?
    (a) $\{x | x \neq 3\}$ (b) All real numbers
    (c) $\left\{x \middle| x \neq \dfrac{2}{3}, x \neq 3\right\}$ (d) $\left\{x \middle| x \neq \dfrac{2}{3}\right\}$

Skill Building

In Problems 13–20, determine whether the function is one-to-one.

13.
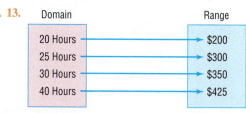

14.
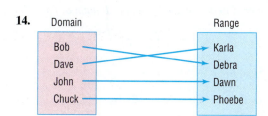

15.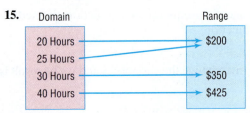

16.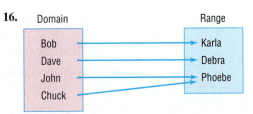

17. $\{(2, 6), (-3, 6), (4, 9), (1, 10)\}$

18. $\{(-2, 5), (-1, 3), (3, 7), (4, 12)\}$

19. $\{(0, 0), (1, 1), (2, 16), (3, 81)\}$

20. $\{(1, 2), (2, 8), (3, 18), (4, 32)\}$

In Problems 21–26, the graph of a function f is given. Use the horizontal-line test to determine whether f is one-to-one.

21.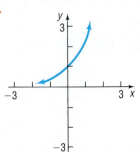

22.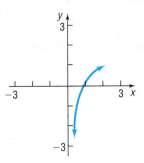

23.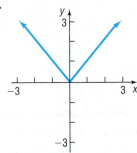

24.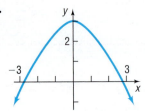

25.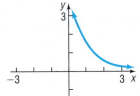

26.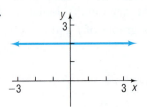

In Problems 27–34, find the inverse of each one-to-one function. State the domain and the range of each inverse function.

27.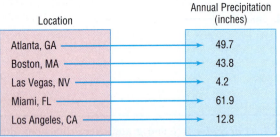

Source: currentresults.com

28.

Source: boxofficemojo.com

29.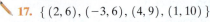

Source: tiaa-cref.org

30.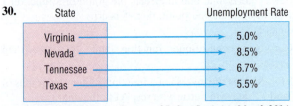

Source: United States Bureau of Labor Statistics, March 2014

31. $\{(-3, 5), (-2, 9), (-1, 2), (0, 11), (1, -5)\}$

32. $\{(-2, 2), (-1, 6), (0, 8), (1, -3), (2, 9)\}$

33. $\{(-2, 1), (-3, 2), (-10, 0), (1, 9), (2, 4)\}$

34. $\{(-2, -8), (-1, -1), (0, 0), (1, 1), (2, 8)\}$

In Problems 35–44, verify that the functions f and g are inverses of each other by showing that $f(g(x)) = x$ and $g(f(x)) = x$. Give any values of x that need to be excluded from the domain of f and the domain of g.

35. $f(x) = 3x + 4;\quad g(x) = \dfrac{1}{3}(x - 4)$

36. $f(x) = 3 - 2x;\quad g(x) = -\dfrac{1}{2}(x - 3)$

37. $f(x) = 4x - 8;\quad g(x) = \dfrac{x}{4} + 2$

38. $f(x) = 2x + 6;\quad g(x) = \dfrac{1}{2}x - 3$

39. $f(x) = x^3 - 8;\quad g(x) = \sqrt[3]{x + 8}$

40. $f(x) = (x - 2)^2, x \geq 2;\quad g(x) = \sqrt{x} + 2$

41. $f(x) = \dfrac{1}{x}$; $g(x) = \dfrac{1}{x}$

42. $f(x) = x$; $g(x) = x$

43. $f(x) = \dfrac{2x + 3}{x + 4}$; $g(x) = \dfrac{4x - 3}{2 - x}$

44. $f(x) = \dfrac{x - 5}{2x + 3}$; $g(x) = \dfrac{3x + 5}{1 - 2x}$

In Problems 45–50, the graph of a one-to-one function f is given. Draw the graph of the inverse function f^{-1}.

45.

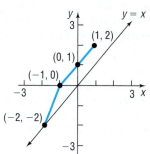

46.

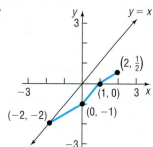

47.

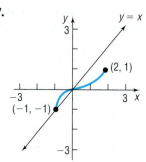

48.

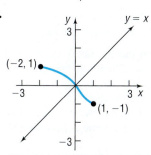

49.

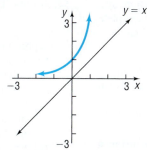

50.

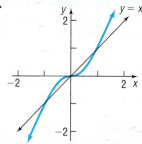

In Problems 51–62, the function f is one-to-one. (a) Find its inverse function f^{-1} and check your answer. (b) Find the domain and the range of f and f^{-1}. (c) Graph f, f^{-1}, and $y = x$ on the same coordinate axes.

51. $f(x) = 3x$

52. $f(x) = -4x$

53. $f(x) = 4x + 2$

54. $f(x) = 1 - 3x$

55. $f(x) = x^3 - 1$

56. $f(x) = x^3 + 1$

57. $f(x) = x^2 + 4$, $x \geq 0$

58. $f(x) = x^2 + 9$, $x \geq 0$

59. $f(x) = \dfrac{4}{x}$

60. $f(x) = -\dfrac{3}{x}$

61. $f(x) = \dfrac{1}{x - 2}$

62. $f(x) = \dfrac{4}{x + 2}$

In Problems 63–74, the function f is one-to-one. (a) Find its inverse function f^{-1} and check your answer. (b) Find the domain and the range of f and f^{-1}.

63. $f(x) = \dfrac{2}{3 + x}$

64. $f(x) = \dfrac{4}{2 - x}$

65. $f(x) = \dfrac{3x}{x + 2}$

66. $f(x) = -\dfrac{2x}{x - 1}$

67. $f(x) = \dfrac{2x}{3x - 1}$

68. $f(x) = -\dfrac{3x + 1}{x}$

69. $f(x) = \dfrac{3x + 4}{2x - 3}$

70. $f(x) = \dfrac{2x - 3}{x + 4}$

71. $f(x) = \dfrac{2x + 3}{x + 2}$

72. $f(x) = \dfrac{-3x - 4}{x - 2}$

73. $f(x) = \dfrac{x^2 - 4}{2x^2}$, $x > 0$

74. $f(x) = \dfrac{x^2 + 3}{3x^2}$ $x > 0$

Applications and Extensions

75. Use the graph of $y = f(x)$ given in Problem 45 to evaluate the following:
  (a) $f(-1)$ (b) $f(1)$ (c) $f^{-1}(1)$ (d) $f^{-1}(2)$

76. Use the graph of $y = f(x)$ given in Problem 46 to evaluate the following:
  (a) $f(2)$ (b) $f(1)$ (c) $f^{-1}(0)$ (d) $f^{-1}(-1)$

77. If $f(7) = 13$ and f is one-to-one, what is $f^{-1}(13)$?

78. If $g(-5) = 3$ and g is one-to-one, what is $g^{-1}(3)$?

79. The domain of a one-to-one function f is $[5, \infty)$, and its range is $[-2, \infty)$. State the domain and the range of f^{-1}.

80. The domain of a one-to-one function f is $[0, \infty)$, and its range is $[5, \infty)$. State the domain and the range of f^{-1}.

81. The domain of a one-to-one function g is $(-\infty, 0]$, and its range is $[0, \infty)$. State the domain and the range of g^{-1}.

82. The domain of a one-to-one function g is $[0, 15]$, and its range is $(0, 8)$. State the domain and the range of g^{-1}.

83. A function $y = f(x)$ is increasing on the interval $(0, 5)$. What conclusions can you draw about the graph of $y = f^{-1}(x)$?

84. A function $y = f(x)$ is decreasing on the interval $(0, 5)$. What conclusions can you draw about the graph of $y = f^{-1}(x)$?

85. Find the inverse of the linear function

$$f(x) = mx + b, \quad m \neq 0$$

86. Find the inverse of the function

$$f(x) = \sqrt{r^2 - x^2}, \quad 0 \leq x \leq r$$

87. A function f has an inverse function f^{-1}. If the graph of f lies in quadrant I, in which quadrant does the graph of f^{-1} lie?

88. A function f has an inverse function f^{-1}. If the graph of f lies in quadrant II, in which quadrant does the graph of f^{-1} lie?

89. The function $f(x) = |x|$ is not one-to-one. Find a suitable restriction on the domain of f so that the new function that results is one-to-one. Then find the inverse of the new function.

90. The function $f(x) = x^4$ is not one-to-one. Find a suitable restriction on the domain of f so that the new function that results is one-to-one. Then find the inverse of the new function.

In applications, the symbols used for the independent and dependent variables are often based on common usage. So, rather than using $y = f(x)$ to represent a function, an applied problem might use $C = C(q)$ to represent the cost C of manufacturing q units of a good. Because of this, the inverse notation f^{-1} used in a pure mathematics problem is not used when finding inverses of applied problems. Rather, the inverse of a function such as $C = C(q)$ will be $q = q(C)$. So $C = C(q)$ is a function that represents the cost C as a function of the number q of units manufactured, and $q = q(C)$ is a function that represents the number q as a function of the cost C. Problems 91–94 illustrate this idea.

91. Vehicle Stopping Distance Taking into account reaction time, the distance d (in feet) that a car requires to come to a complete stop while traveling r miles per hour is given by the function

$$d(r) = 6.97r - 90.39$$

(a) Express the speed r at which the car is traveling as a function of the distance d required to come to a complete stop.

(b) Verify that $r = r(d)$ is the inverse of $d = d(r)$ by showing that $r(d(r)) = r$ and $d(r(d)) = d$.

(c) Predict the speed that a car was traveling if the distance required to stop was 300 feet.

92. Height and Head Circumference The head circumference C of a child is related to the height H of the child (both in inches) through the function

$$H(C) = 2.15C - 10.53$$

(a) Express the head circumference C as a function of height H.

(b) Verify that $C = C(H)$ is the inverse of $H = H(C)$ by showing that $H(C(H)) = H$ and $C(H(C)) = C$.

(c) Predict the head circumference of a child who is 26 inches tall.

93. Ideal Body Weight One model for the ideal body weight W for men (in kilograms) as a function of height h (in inches) is given by the function

$$W(h) = 50 + 2.3(h - 60)$$

(a) What is the ideal weight of a 6-foot male?

(b) Express the height h as a function of weight W.

(c) Verify that $h = h(W)$ is the inverse of $W = W(h)$ by showing that $h(W(h)) = h$ and $W(h(W)) = W$.

(d) What is the height of a male who is at his ideal weight of 80 kilograms?

[**Note:** The ideal body weight W for women (in kilograms) as a function of height h (in inches) is given by $W(h) = 45.5 + 2.3(h - 60)$.]

94. Temperature Conversion The function $F(C) = \dfrac{9}{5}C + 32$ converts a temperature from C degrees Celsius to F degrees Fahrenheit.

(a) Express the temperature in degrees Celsius C as a function of the temperature in degrees Fahrenheit F.

(b) Verify that $C = C(F)$ is the inverse of $F = F(C)$ by showing that $C(F(C)) = C$ and $F(C(F)) = F$.

(c) What is the temperature in degrees Celsius if it is 70 degrees Fahrenheit?

95. Income Taxes The function

$$T(g) = 5081.25 + 0.25(g - 36,900)$$

represents the 2014 federal income tax T (in dollars) due for a "single" filer whose modified adjusted gross income is g dollars, where $36,900 \leq g \leq 89,350$.

(a) What is the domain of the function T?

(b) Given that the tax due T is an increasing linear function of modified adjusted gross income g, find the range of the function T.

(c) Find adjusted gross income g as a function of federal income tax T. What are the domain and the range of this function?

96. Income Taxes The function

$$T(g) = 1815 + 0.15(g - 18,150)$$

represents the 2014 federal income tax T (in dollars) due for a "married filing jointly" filer whose modified adjusted gross income is g dollars, where $18,150 \leq g \leq 73,800$.

(a) What is the domain of the function T?

(b) Given that the tax due T is an increasing linear function of modified adjusted gross income g, find the range of the function T.

(c) Find adjusted gross income g as a function of federal income tax T. What are the domain and the range of this function?

97. Gravity on Earth If a rock falls from a height of 100 meters on Earth, the height H (in meters) after t seconds is approximately

$$H(t) = 100 - 4.9t^2$$

(a) In general, quadratic functions are not one-to-one. However, the function H is one-to-one. Why?

(b) Find the inverse of H and verify your result.

(c) How long will it take a rock to fall 80 meters?

98. Period of a Pendulum The period T (in seconds) of a simple pendulum as a function of its length l (in feet) is given by

$$T(l) = 2\pi \sqrt{\frac{l}{32.2}}$$

(a) Express the length l as a function of the period T.

(b) How long is a pendulum whose period is 3 seconds?

99. Given

$$f(x) = \frac{ax + b}{cx + d}$$

find $f^{-1}(x)$. If $c \neq 0$, under what conditions on $a, b, c,$ and d is $f = f^{-1}$?

Explaining Concepts: Discussion and Writing

100. Can a one-to-one function and its inverse be equal? What must be true about the graph of *f* for this to happen? Give some examples to support your conclusion.

101. Draw the graph of a one-to-one function that contains the points $(-2, -3), (0, 0),$ and $(1, 5)$. Now draw the graph of its inverse. Compare your graph to those of other students. Discuss any similarities. What differences do you see?

102. Give an example of a function whose domain is the set of real numbers and that is neither increasing nor decreasing on its domain, but is one-to-one.

[**Hint:** Use a piecewise-defined function.]

103. Is every odd function one-to-one? Explain.

104. Suppose that $C(g)$ represents the cost C, in dollars, of manufacturing g cars. Explain what $C^{-1}(800,000)$ represents.

105. Explain why the horizontal-line test can be used to identify one-to-one functions from a graph.

106. Explain why a function must be one-to-one in order to have an inverse that is a function. Use the function $y = x^2$ to support your explanation.

Retain Your Knowledge

Problems 107–110 are based on material learned earlier in the course. The purpose of these problems is to keep the material fresh in your mind so that you are better prepared for the final exam.

107. If $f(x) = 3x^2 - 7x$, find $f(x + h) - f(x)$.

108. Use the techniques of shifting, compressing or stretching, and reflections to graph $f(x) = -|x + 2| + 3$.

109. Find the zeros of the quadratic function $f(x) = 3x^2 + 5x + 1$. What are the *x*-intercepts, if any, of the graph of the function?

110. Find the domain of $R(x) = \dfrac{6x^2 - 11x - 2}{2x^2 - x - 6}$. Find any horizontal, vertical, or oblique asymptotes.

'Are You Prepared?' Answers

1. Yes; for each input x there is one output y.

2. Increasing on $(0, \infty)$; decreasing on $(-\infty, 0)$

3. $\{x \mid x \neq -6, x \neq 3\}$

4. $\dfrac{x}{1 - x}, x \neq 0, x \neq -1$

6.3 Exponential Functions

PREPARING FOR THIS SECTION *Before getting started, review the following:*

- Exponents (Chapter R, Section R.2, pp. 21–24, and Section R.8, pp. 73–77)
- Graphing Techniques: Transformations (Section 3.5, pp. 247–256)
- Solving Equations (Section 1.1, pp. 82–87 and Section 1.2, pp. 92–99)

- Average Rate of Change (Section 3.3, pp. 230–231)
- Quadratic Functions (Section 4.3, pp. 290–298)
- Linear Functions (Section 4.1, pp. 274–277)
- Horizontal Asymptotes (Section 5.2, pp. 348–350)

Now Work the 'Are You Prepared?' problems on page 434.

OBJECTIVES **1** Evaluate Exponential Functions (p. 423)
2 Graph Exponential Functions (p. 427)
3 Define the Number *e* (p. 430)
4 Solve Exponential Equations (p. 432)

1 Evaluate Exponential Functions

Chapter R, Section R.8 gives a definition for raising a real number *a* to a rational power. That discussion provides meaning to expressions of the form

$$a^r$$

where the base *a* is a positive real number and the exponent *r* is a rational number.

 But what is the meaning of a^x, where the base *a* is a positive real number and the exponent *x* is an irrational number? Although a rigorous definition requires methods discussed in calculus, the basis for the definition is easy to follow: Select a rational number r that is formed by truncating (removing) all but a finite number of digits from the irrational number x. Then it is reasonable to expect that

$$a^x \approx a^r$$

For example, take the irrational number $\pi = 3.14159\ldots$. Then an approximation to a^π is

$$a^\pi \approx a^{3.14}$$

where the digits after the hundredths position have been removed from the value for π. A better approximation would be

$$a^\pi \approx a^{3.14159}$$

where the digits after the hundred-thousandths position have been removed. Continuing in this way, we can obtain approximations to a^π to any desired degree of accuracy.

Most calculators have an $\boxed{x^y}$ key or a caret key $\boxed{\wedge}$ for working with exponents. To evaluate expressions of the form a^x, enter the base a, then press the $\boxed{x^y}$ key (or the $\boxed{\wedge}$ key), enter the exponent x, and press $\boxed{=}$ (or $\boxed{\text{ENTER}}$).

EXAMPLE 1 **Using a Calculator to Evaluate Powers of 2**

Using a calculator, evaluate:

(a) $2^{1.4}$ (b) $2^{1.41}$ (c) $2^{1.414}$ (d) $2^{1.4142}$ (e) $2^{\sqrt{2}}$

Solution (a) $2^{1.4} \approx 2.639015822$ (b) $2^{1.41} \approx 2.657371628$
(c) $2^{1.414} \approx 2.66474965$ (d) $2^{1.4142} \approx 2.665119089$
(d) $2^{\sqrt{2}} \approx 2.665144143$ ●

 Now Work PROBLEM 15

It can be shown that the familiar laws for rational exponents hold for real exponents.

THEOREM

> **Laws of Exponents**
>
> If s, t, a, and b are real numbers with $a > 0$ and $b > 0$, then
>
> $$a^s \cdot a^t = a^{s+t} \qquad (a^s)^t = a^{st} \qquad (ab)^s = a^s \cdot b^s$$
>
> $$1^s = 1 \qquad a^{-s} = \frac{1}{a^s} = \left(\frac{1}{a}\right)^s \qquad a^0 = 1 \qquad \textbf{(1)}$$

Introduction to Exponential Growth

Suppose a function f has the following two properties:

1. The value of f doubles with every 1-unit increase in the independent variable x.
2. The value of f at $x = 0$ is 5, so $f(0) = 5$.

Table 1 shows values of the function f for $x = 0, 1, 2, 3$, and 4.

Let's find an equation $y = f(x)$ that describes this function f. The key fact is that the value of f doubles for every 1-unit increase in x.

Table 1

x	$f(x)$
0	5
1	10
2	20
3	40
4	80

$f(0) = 5$

$f(1) = 2f(0) = 2 \cdot 5 = 5 \cdot 2^1$ *Double the value of f at 0 to get the value at 1.*

$f(2) = 2f(1) = 2(5 \cdot 2) = 5 \cdot 2^2$ *Double the value of f at 1 to get the value at 2.*

$f(3) = 2f(2) = 2(5 \cdot 2^2) = 5 \cdot 2^3$

$f(4) = 2f(3) = 2(5 \cdot 2^3) = 5 \cdot 2^4$

The pattern leads to

$$f(x) = 2f(x-1) = 2(5 \cdot 2^{x-1}) = 5 \cdot 2^x$$

DEFINITION

An **exponential function** is a function of the form

$$f(x) = Ca^x$$

where a is a positive real number $(a > 0)$, $a \neq 1$, and $C \neq 0$ is a real number. The domain of f is the set of all real numbers. The base a is the **growth factor**, and, because $f(0) = Ca^0 = C$, C is called the **initial value.**

WARNING It is important to distinguish a power function, $g(x) = ax^n$, $n \geq 2$ an integer, from an exponential function, $f(x) = C \cdot a^x$, $a \neq 1$, $a > 0$. In a power function, the base is a variable and the exponent is a constant. In an exponential function, the base is a constant and the exponent is a variable. ∎

In the definition of an exponential function, the base $a = 1$ is excluded because this function is simply the constant function $f(x) = C \cdot 1^x = C$. Bases that are negative are also excluded; otherwise, many values of x would have to be excluded from the domain, such as $x = \dfrac{1}{2}$ and $x = \dfrac{3}{4}$. [Recall that $(-2)^{1/2} = \sqrt{-2}$, $(-3)^{3/4} = \sqrt[4]{(-3)^3} = \sqrt[4]{-27}$, and so on, are not defined in the set of real numbers.]

Transformations (vertical shifts, horizontal shifts, reflections, and so on) of a function of the form $f(x) = Ca^x$ also represent exponential functions. Some examples of exponential functions are

$$f(x) = 2^x \qquad F(x) = \left(\frac{1}{3}\right)^x + 5 \qquad G(x) = 2 \cdot 3^{x-3}$$

For each function, note that the base of the exponential expression is a constant and the exponent contains a variable.

In the function $f(x) = 5 \cdot 2^x$, notice that the ratio of consecutive outputs is constant for 1-unit increases in the input. This ratio equals the constant 2, the base of the exponential function. In other words,

$$\frac{f(1)}{f(0)} = \frac{5 \cdot 2^1}{5} = 2 \quad \frac{f(2)}{f(1)} = \frac{5 \cdot 2^2}{5 \cdot 2^1} = 2 \quad \frac{f(3)}{f(2)} = \frac{5 \cdot 2^3}{5 \cdot 2^2} = 2 \quad \text{and so on}$$

This leads to the following result.

THEOREM

For an exponential function $f(x) = Ca^x$, $a > 0$, $a \neq 1$, and $C \neq 0$, if x is any real number, then

$$\frac{f(x + 1)}{f(x)} = a \quad \text{or} \quad f(x + 1) = af(x)$$

In Words

For 1-unit changes in the input x of an exponential function $f(x) = C \cdot a^x$, the ratio of consecutive outputs is the constant a.

Proof

$$\frac{f(x + 1)}{f(x)} = \frac{Ca^{x+1}}{Ca^x} = a^{x+1-x} = a^1 = a \qquad ∎$$

EXAMPLE 2

Identifying Linear or Exponential Functions

Determine whether the given function is linear, exponential, or neither. For those that are linear, find a linear function that models the data. For those that are exponential, find an exponential function that models the data.

(a)

x	y
−1	5
0	2
1	−1
2	−4
3	−7

(b)

x	y
−1	32
0	16
1	8
2	4
3	2

(c)

x	y
−1	2
0	4
1	7
2	11
3	16

Solution

For each function, compute the average rate of change of y with respect to x and the ratio of consecutive outputs. If the average rate of change is constant, then the function is linear. If the ratio of consecutive outputs is constant, then the function is exponential.

Table 2

x	y	Average Rate of Change	Ratio of Consecutive Outputs
-1	5		
		$\dfrac{\Delta y}{\Delta x} = \dfrac{2-5}{0-(-1)} = -3$	$\dfrac{2}{5}$
0	2		
		$\dfrac{-1-2}{1-0} = -3$	$\dfrac{-1}{2} = -\dfrac{1}{2}$
1	-1		
		$\dfrac{-4-(-1)}{2-1} = -3$	$\dfrac{-4}{-1} = 4$
2	-4		
		$\dfrac{-7-(-4)}{3-2} = -3$	$\dfrac{-7}{-4} = \dfrac{7}{4}$
3	-7		

(a)

x	y	Average Rate of Change	Ratio of Consecutive Outputs
-1	32		
		$\dfrac{\Delta y}{\Delta x} = \dfrac{16-32}{0-(-1)} = -16$	$\dfrac{16}{32} = \dfrac{1}{2}$
0	16		
		-8	$\dfrac{8}{16} = \dfrac{1}{2}$
1	8		
		-4	$\dfrac{4}{8} = \dfrac{1}{2}$
2	4		
		-2	$\dfrac{2}{4} = \dfrac{1}{2}$
3	2		

(b)

x	y	Average Rate of Change	Ratio of Consecutive Outputs
-1	2		
		$\dfrac{\Delta y}{\Delta x} = \dfrac{4-2}{0-(-1)} = 2$	2
0	4		
		3	$\dfrac{7}{4}$
1	7		
		4	$\dfrac{11}{7}$
2	11		
		5	$\dfrac{16}{11}$
3	16		

(c)

(a) See Table 2(a). The average rate of change for every 1-unit increase in x is -3. Therefore, the function is a linear function. In a linear function the average rate of change is the slope m, so $m = -3$. The y-intercept b is the value of the function at $x = 0$, so $b = 2$. The linear function that models the data is $f(x) = mx + b = -3x + 2$.

(b) See Table 2(b). For this function, the average rate of change from -1 to 0 is -16, and the average rate of change from 0 to 1 is -8. Because the average rate of change is not constant, the function is not a linear function. The ratio of consecutive outputs for a 1-unit increase in the inputs is a constant, $\dfrac{1}{2}$. Because the ratio of consecutive outputs is constant, the function is an exponential function with growth factor $a = \dfrac{1}{2}$. The initial value C of the exponential function is $C = 16$, the value of the function at 0. Therefore, the exponential function that models the data is $g(x) = Ca^x = 16 \cdot \left(\dfrac{1}{2}\right)^x$.

(c) See Table 2(c). For this function, the average rate of change from -1 to 0 is 2, and the average rate of change from 0 to 1 is 3. Because the average rate of change is not constant, the function is not a linear function. The ratio of consecutive outputs from -1 to 0 is 2, and the ratio of consecutive outputs from 0 to 1 is $\dfrac{7}{4}$. Because the ratio of consecutive outputs is not a constant, the function is not an exponential function. ●

Now Work PROBLEM 27

2 Graph Exponential Functions

If we know how to graph an exponential function of the form $f(x) = a^x$, then we can use transformations (shifting, stretching, and so on) to obtain the graph of any exponential function.

First, let's graph the exponential function $f(x) = 2^x$.

EXAMPLE 3

Graphing an Exponential Function

Graph the exponential function: $f(x) = 2^x$

Solution   The domain of $f(x) = 2^x$ is the set of all real numbers. Begin by locating some points on the graph of $f(x) = 2^x$, as listed in Table 3.

Because $2^x > 0$ for all x, the range of f is $(0, \infty)$. Therefore, the graph has no x-intercepts, and in fact the graph will lie above the x-axis for all x. As Table 3 indicates, the y-intercept is 1. Table 3 also indicates that as $x \to -\infty$, the value of $f(x) = 2^x$ gets closer and closer to 0. Therefore, the x-axis ($y = 0$) is a horizontal asymptote to the graph as $x \to -\infty$. This provides the end behavior for x large and negative.

To determine the end behavior for x large and positive, look again at Table 3. As $x \to \infty, f(x) = 2^x$ grows very quickly, causing the graph of $f(x) = 2^x$ to rise very rapidly. It is apparent that f is an increasing function and so is one-to-one.

Using all this information, plot some of the points from Table 3 and connect them with a smooth, continuous curve, as shown in Figure 18.

Table 3

x	$f(x) = 2^x$
-10	$2^{-10} \approx 0.00098$
-3	$2^{-3} = \dfrac{1}{8}$
-2	$2^{-2} = \dfrac{1}{4}$
-1	$2^{-1} = \dfrac{1}{2}$
0	$2^0 = 1$
1	$2^1 = 2$
2	$2^2 = 4$
3	$2^3 = 8$
10	$2^{10} = 1024$

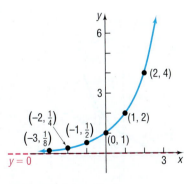

Figure 18 $f(x) = 2^x$ ●

Graphs that look like the one in Figure 18 occur very frequently in a variety of situations. For example, the graph in Figure 19 illustrates the number of Facebook subscribers by year from 2004 to 2013. One might conclude from this graph that the number of Facebook subscribers is growing *exponentially*.

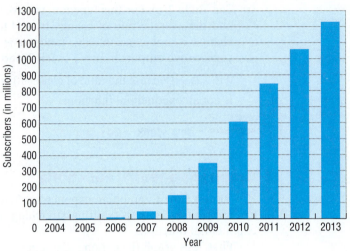

Source: Facebook Newsroom

Figure 19

Figure 20

Later in this chapter, more will be said about situations that lead to exponential growth. For now, let's continue to explore properties of exponential functions.

The graph of $f(x) = 2^x$ in Figure 18 is typical of all exponential functions of the form $f(x) = a^x$ with $a > 1$. Such functions are increasing functions and hence are one-to-one. Their graphs lie above the x-axis, pass through the point $(0, 1)$, and thereafter rise rapidly as $x \to \infty$. As $x \to -\infty$, the x-axis $(y = 0)$ is a horizontal asymptote. There are no vertical asymptotes. Finally, the graphs are smooth and continuous with no corners or gaps.

Figure 20 illustrates the graphs of two more exponential functions whose bases are larger than 1. Notice that the larger the base, the steeper the graph is when $x > 0$, and when $x < 0$, the larger the base, the closer the graph of the equation is to the x-axis.

Seeing the Concept

Graph $Y_1 = 2^x$ and compare what you see to Figure 18. Clear the screen, graph $Y_1 = 3^x$ and $Y_2 = 6^x$, and compare what you see to Figure 20. Clear the screen and graph $Y_1 = 10^x$ and $Y_2 = 100^x$.

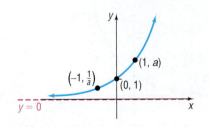

Figure 21 $f(x) = a^x, a > 1$

Properties of the Exponential Function $f(x) = a^x, a > 1$

1. The domain is the set of all real numbers, or $(-\infty, \infty)$ using interval notation; the range is the set of positive real numbers, or $(0, \infty)$ using interval notation.

2. There are no x-intercepts; the y-intercept is 1.

3. The x-axis $(y = 0)$ is a horizontal asymptote as $x \to -\infty$. $\left[\lim_{x \to -\infty} a^x = 0\right]$.

4. $f(x) = a^x, a > 1$, is an increasing function and is one-to-one.

5. The graph of f contains the points $\left(-1, \dfrac{1}{a}\right)$, $(0, 1)$, and $(1, a)$.

6. The graph of f is smooth and continuous, with no corners or gaps. See Figure 21.

Now consider $f(x) = a^x$ when $0 < a < 1$.

EXAMPLE 4

Graphing an Exponential Function

Graph the exponential function: $f(x) = \left(\dfrac{1}{2}\right)^x$

Solution

The domain of $f(x) = \left(\dfrac{1}{2}\right)^x$ consists of all real numbers. As before, locate some points on the graph as shown in Table 4. Because $\left(\dfrac{1}{2}\right)^x > 0$ for all x, the range of f is the interval $(0, \infty)$. The graph lies above the x-axis and has no x-intercepts. The y-intercept is 1. As $x \to -\infty$, $f(x) = \left(\dfrac{1}{2}\right)^x$ grows very quickly. As $x \to \infty$, the values of $f(x)$ approach 0. The x-axis $(y = 0)$ is a horizontal asymptote as $x \to \infty$. It is apparent that f is a decreasing function and so is one-to-one. Figure 22 illustrates the graph.

Table 4

x	$f(x) = \left(\dfrac{1}{2}\right)^x$
-10	$\left(\dfrac{1}{2}\right)^{-10} = 1024$
-3	$\left(\dfrac{1}{2}\right)^{-3} = 8$
-2	$\left(\dfrac{1}{2}\right)^{-2} = 4$
-1	$\left(\dfrac{1}{2}\right)^{-1} = 2$
0	$\left(\dfrac{1}{2}\right)^{0} = 1$
1	$\left(\dfrac{1}{2}\right)^{1} = \dfrac{1}{2}$
2	$\left(\dfrac{1}{2}\right)^{2} = \dfrac{1}{4}$
3	$\left(\dfrac{1}{2}\right)^{3} = \dfrac{1}{8}$
10	$\left(\dfrac{1}{2}\right)^{10} \approx 0.00098$

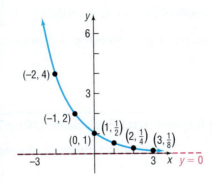

Figure 22 $f(x) = \left(\dfrac{1}{2}\right)^x$

The graph of $y = \left(\dfrac{1}{2}\right)^x$ also can be obtained from the graph of $y = 2^x$ using transformations. The graph of $y = \left(\dfrac{1}{2}\right)^x = 2^{-x}$ is a reflection about the y-axis of the graph of $y = 2^x$ (replace x by $-x$). See Figures 23(a) and (b).

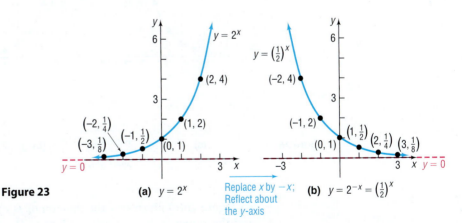

Figure 23 (a) $y = 2^x$ Replace x by $-x$; Reflect about the y-axis (b) $y = 2^{-x} = \left(\dfrac{1}{2}\right)^x$

Seeing the Concept

Using a graphing utility, simultaneously graph:

(a) $Y_1 = 3^x$, $Y_2 = \left(\dfrac{1}{3}\right)^x$

(b) $Y_1 = 6^x$, $Y_2 = \left(\dfrac{1}{6}\right)^x$

Conclude that the graph of $Y_2 = \left(\dfrac{1}{a}\right)^x$, for $a > 0$, is the reflection about the y-axis of the graph of $Y_1 = a^x$.

The graph of $f(x) = \left(\dfrac{1}{2}\right)^x$ in Figure 22 is typical of all exponential functions of the form $f(x) = a^x$ with $0 < a < 1$. Such functions are decreasing and one-to-one. Their graphs lie above the x-axis and pass through the point $(0, 1)$. The graphs rise rapidly as $x \to -\infty$. As $x \to \infty$, the x-axis $(y = 0)$ is a horizontal asymptote. There are no vertical asymptotes. Finally, the graphs are smooth and continuous, with no corners or gaps.

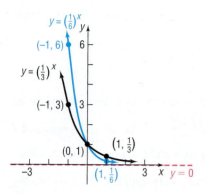

Figure 24

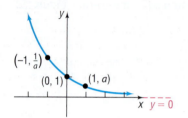

Figure 25 $f(x) = a^x, 0 < a < 1$

Figure 24 illustrates the graphs of two more exponential functions whose bases are between 0 and 1. Notice that the smaller base results in a graph that is steeper when $x < 0$. When $x > 0$, the graph of the equation with the smaller base is closer to the x-axis.

> **Properties of the Exponential Function $f(x) = a^x, 0 < a < 1$**
>
> 1. The domain is the set of all real numbers, or $(-\infty, \infty)$ using interval notation; the range is the set of positive real numbers, or $(0, \infty)$ using interval notation.
> 2. There are no x-intercepts; the y-intercept is 1.
> 3. The x-axis $(y = 0)$ is a horizontal asymptote as $x \rightarrow \infty$ $\left[\lim\limits_{x \to \infty} a^x = 0\right]$.
> 4. $f(x) = a^x, 0 < a < 1$, is a decreasing function and is one-to-one.
> 5. The graph of f contains the points $\left(-1, \dfrac{1}{a}\right)$, $(0, 1)$, and $(1, a)$.
> 6. The graph of f is smooth and continuous, with no corners or gaps. See Figure 25.

EXAMPLE 5

Graphing Exponential Functions Using Transformations

Graph $f(x) = 2^{-x} - 3$ and determine the domain, range, and horizontal asymptote of f.

Solution

Begin with the graph of $y = 2^x$. Figure 26 shows the stages.

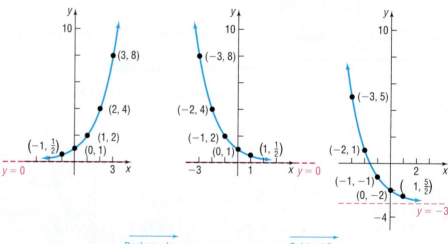

Figure 26

(a) $y = 2^x$ Replace x by $-x$; reflect about y-axis   (b) $y = 2^{-x}$ Subtract 3; shift down 3 units   (c) $y = 2^{-x} - 3$

As Figure 26(c) illustrates, the domain of $f(x) = 2^{-x} - 3$ is the interval $(-\infty, \infty)$ and the range is the interval $(-3, \infty)$. The horizontal asymptote of f is the line $y = -3$. ●

—— **Now Work** PROBLEM 43

3 Define the Number e

Many problems that occur in nature require the use of an exponential function whose base is a certain irrational number, symbolized by the letter e.

One way of arriving at this important number e is given next.

DEFINITION

The **number** e is defined as the number that the expression

$$\left(1 + \frac{1}{n}\right)^n \tag{2}$$

approaches as $n \rightarrow \infty$. In calculus, this is expressed, using limit notation, as

$$e = \lim_{n \to \infty}\left(1 + \frac{1}{n}\right)^n$$

Table 5 illustrates what happens to the defining expression (2) as n takes on increasingly large values. The last number in the right column in the table approximates e correct to nine decimal places. That is, $e = 2.718281828. \ldots$ Remember, the three dots indicate that the decimal places continue. Because these decimal places continue but do not repeat, e is an irrational number. The number e is often expressed as a decimal rounded to a specific number of places. For example, $e \approx 2.71828$ is rounded to five decimal places.

Table 5

n	$\dfrac{1}{n}$	$1 + \dfrac{1}{n}$	$\left(1 + \dfrac{1}{n}\right)^n$
1	1	2	2
2	0.5	1.5	2.25
5	0.2	1.2	2.48832
10	0.1	1.1	2.59374246
100	0.01	1.01	2.704813829
1,000	0.001	1.001	2.716923932
10,000	0.0001	1.0001	2.718145927
100,000	0.00001	1.00001	2.718268237
1,000,000	0.000001	1.000001	2.718280469
10,000,000,000	10^{-10}	$1 + 10^{-10}$	2.718281828

Table 6

x	e^x
-2	$e^{-2} \approx 0.14$
-1	$e^{-1} \approx 0.37$
0	$e^0 = 1$
1	$e^1 \approx 2.72$
2	$e^2 \approx 7.39$

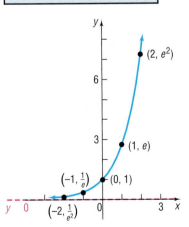

Figure 27 $y = e^x$

The exponential function $f(x) = e^x$, whose base is the number e, occurs with such frequency in applications that it is usually referred to as *the* exponential function. Indeed, most calculators have the key $\boxed{e^x}$ or $\boxed{\exp(x)}$, which may be used to evaluate the exponential function for a given value of x.*

Now use your calculator to approximate e^x for $x = -2, x = -1, x = 0, x = 1,$ and $x = 2$. See Table 6. The graph of the exponential function $f(x) = e^x$ is given in Figure 27. Since $2 < e < 3$, the graph of $y = e^x$ lies between the graphs of $y = 2^x$ and $y = 3^x$. Do you see why? (Refer to Figures 18 and 20.)

Seeing the Concept

Graph $Y_1 = e^x$ and compare what you see to Figure 27. Use eVALUEate or TABLE to verify the points on the graph shown in Figure 27. Now graph $Y_2 = 2^x$ and $Y_3 = 3^x$ on the same screen as $Y_1 = e^x$. Notice that the graph of $Y_1 = e^x$ lies between these two graphs.

EXAMPLE 6

Graphing Exponential Functions Using Transformations

Graph $f(x) = -e^{x-3}$ and determine the domain, range, and horizontal asymptote of f.

*If your calculator does not have one of these keys, refer to your owner's manual.

Solution   Begin with the graph of $y = e^x$. Figure 28 shows the stages.

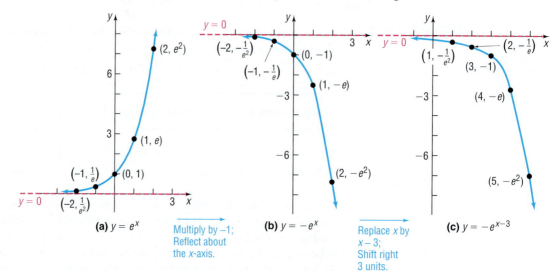

Figure 28

(a) $y = e^x$ Multiply by -1;
Reflect about
the x-axis.

(b) $y = -e^x$ Replace x by
$x - 3$;
Shift right
3 units.

(c) $y = -e^{x-3}$

As Figure 28(c) illustrates, the domain of $f(x) = -e^{x-3}$ is the interval $(-\infty, \infty)$, and the range is the interval $(-\infty, 0)$. The horizontal asymptote is the line $y = 0$.

Now Work   PROBLEM 55

4 Solve Exponential Equations

Equations that involve terms of the form a^x, where $a > 0$ and $a \ne 1$, are referred to as **exponential equations**. Such equations can sometimes be solved by appropriately applying the Laws of Exponents and property (3):

In Words
When two exponential expressions
with the same base are equal,
then their exponents are equal.

$$\text{If } \quad a^u = a^v, \quad \text{then} \quad u = v. \qquad (3)$$

Property (3) is a consequence of the fact that exponential functions are one-to-one. To use property (3), each side of the equality must be written with the same base.

EXAMPLE 7   Solving Exponential Equations

Solve each exponential equation.

(a) $3^{x+1} = 81$ (a) $4^{2x-1} = 8^{x+3}$

Solution   (a) Since $81 = 3^4$, write the equation as

$$3^{x+1} = 81 = 3^4$$

Now the expressions on both sides of the equation have the same base, 3. Set the exponents equal to each other to obtain

$$x + 1 = 4$$
$$x = 3$$

The solution set is $\{3\}$.

(b)
$$4^{2x-1} = 8^{x+3}$$
$$(2^2)^{(2x-1)} = (2^3)^{(x+3)} \qquad 4 = 2^2; 8 = 2^3$$
$$2^{2(2x-1)} = 2^{3(x+3)} \qquad (a^r)^s = a^{rs}$$
$$2(2x - 1) = 3(x + 3) \qquad \text{If } a^u = a^v, \text{ then } u = v.$$
$$4x - 2 = 3x + 9$$
$$x = 11$$

The solution set is $\{11\}$.

Now Work   PROBLEMS 65 AND 75

EXAMPLE 8

Solving an Exponential Equation

Solve: $e^{-x^2} = (e^x)^2 \cdot \dfrac{1}{e^3}$

Solution

Use the Laws of Exponents first to get a single expression with the base e on the right side.

$$(e^x)^2 \cdot \frac{1}{e^3} = e^{2x} \cdot e^{-3} = e^{2x-3}$$

As a result,

$$e^{-x^2} = e^{2x-3}$$
$$-x^2 = 2x - 3 \qquad \text{Apply property (3).}$$
$$x^2 + 2x - 3 = 0 \qquad \text{Place the quadratic equation in standard form.}$$
$$(x + 3)(x - 1) = 0 \qquad \text{Factor.}$$
$$x = -3 \quad \text{or} \quad x = 1 \qquad \text{Use the Zero-Product Property.}$$

The solution set is $\{-3, 1\}$.

●

Now Work PROBLEM 81

EXAMPLE 9

Exponential Probability

Between 9:00 PM and 10:00 PM, cars arrive at Burger King's drive-thru at the rate of 12 cars per hour (0.2 car per minute). The following formula from statistics can be used to determine the probability that a car will arrive within t minutes of 9:00 PM.

$$F(t) = 1 - e^{-0.2t}$$

(a) Determine the probability that a car will arrive within 5 minutes of 9 PM (that is, before 9:05 PM).

(b) Determine the probability that a car will arrive within 30 minutes of 9 PM (before 9:30 PM).

(c) Graph F using your graphing utility.

(d) What value does F approach as t increases without bound in the positive direction?

Solution

(a) The probability that a car will arrive within 5 minutes is found by evaluating $F(t)$ at $t = 5$.

$$F(5) = 1 - e^{-0.2(5)} \approx 0.63212$$
↑
Use a calculator.

There is a 63% probability that a car will arrive within 5 minutes.

(b) The probability that a car will arrive within 30 minutes is found by evaluating $F(t)$ at $t = 30$.

$$F(30) = 1 - e^{-0.2(30)} \approx 0.9975$$
↑
Use a calculator.

There is a 99.75% probability that a car will arrive within 30 minutes.

(c) See Figure 29 for the graph of F.

(d) As time passes, the probability that a car will arrive increases. The value that F approaches can be found by letting $t \to \infty$. Since $e^{-0.2t} = \dfrac{1}{e^{0.2t}}$, it follows that $e^{-0.2t} \to 0$ as $t \to \infty$. Therefore, F approaches 1 as t gets large. The algebraic analysis is confirmed by Figure 29.

●

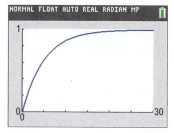

Figure 29 $F(t) = 1 - e^{-0.2t}$

Now Work PROBLEM 113

SUMMARY

Properties of the Exponential Function

$f(x) = a^x, \quad a > 1$	Domain: the interval $(-\infty, \infty)$; range: the interval $(0, \infty)$ x-intercepts: none; y-intercept: 1 Horizontal asymptote: x-axis $(y = 0)$ as $x \to -\infty$ Increasing; one-to-one; smooth; continuous See Figure 21 for a typical graph.
$f(x) = a^x, \quad 0 < a < 1$	Domain: the interval $(-\infty, \infty)$; range: the interval $(0, \infty)$ x-intercepts: none; y-intercept: 1 Horizontal asymptote: x-axis $(y = 0)$ as $x \to \infty$ Decreasing; one-to-one; smooth; continuous See Figure 25 for a typical graph.

If $a^u = a^v$, then $u = v$.

6.3 Assess Your Understanding

'Are You Prepared?' *Answers are given at the end of these exercises. If you get a wrong answer, read the pages listed in red.*

1. $4^3 =$ _____ ; $8^{2/3} =$ _____ ; $3^{-2} =$ _____ . (pp. 21–24 and pp. 73–77)

2. Solve: $x^2 + 3x = 4$ (pp. 92–99)

3. **True or False** To graph $y = (x - 2)^3$, shift the graph of $y = x^3$ to the left 2 units. (pp. 247–256)

4. Find the average rate of change of $f(x) = 3x - 5$ from $x = 0$ to $x = 4$. (pp. 230–231)

5. **True or False** The function $f(x) = \dfrac{2x}{x - 3}$ has $y = 2$ as a horizontal asymptote. (pp. 348–350)

Concepts and Vocabulary

6. A(n) _____ _____ is a function of the form $f(x) = Ca^x$, where $a > 0, a \neq 1$, and $C \neq 0$ are real numbers. The base a is the _____ _____ and C is the _____ _____ .

7. For an exponential function $f(x) = Ca^x$, $\dfrac{f(x + 1)}{f(x)} =$ ___ .

8. **True or False** The domain of the exponential function $f(x) = a^x$, where $a > 0$ and $a \neq 1$, is the set of all real numbers.

9. **True or False** The graph of the exponential function $f(x) = a^x$, where $a > 0$ and $a \neq 1$, has no x-intercept.

10. The graph of every exponential function $f(x) = a^x$, where $a > 0$ and $a \neq 1$, passes through three points: ___ , ___ , and ___ .

11. If $3^x = 3^4$, then $x =$ ___ .

12. **True or False** The graphs of $y = 3^x$ and $y = \left(\dfrac{1}{3}\right)^x$ are identical.

13. Which of the following exponential functions is an increasing function?
   (a) $f(x) = 0.5^x$
   (b) $f(x) = \left(\dfrac{5}{2}\right)^x$
   (c) $f(x) = \left(\dfrac{2}{3}\right)^x$
   (d) $f(x) = 0.9^x$

14. Which of the following is the range of the exponential function $f(x) = a^x, a > 0$ and $a \neq 1$?
   (a) $(-\infty, \infty)$
   (b) $(-\infty, 0)$
   (c) $(0, \infty)$
   (d) $(-\infty, 0) \cup (0, \infty)$

Skill Building

In Problems 15–26, approximate each number using a calculator. Express your answer rounded to three decimal places.

15. (a) $2^{3.14}$ (b) $2^{3.141}$ (c) $2^{3.1415}$ (d) 2^{π}

16. (a) $2^{2.7}$ (b) $2^{2.71}$ (c) $2^{2.718}$ (d) 2^{e}

17. (a) $3.1^{2.7}$ (b) $3.14^{2.71}$ (c) $3.141^{2.718}$ (d) π^{e}

18. (a) $2.7^{3.1}$ (b) $2.71^{3.14}$ (c) $2.718^{3.141}$ (d) e^{π}

19. $(1 + 0.04)^6$

20. $\left(1 + \dfrac{0.09}{12}\right)^{24}$

21. $8.4\left(\dfrac{1}{3}\right)^{2.9}$

22. $158\left(\dfrac{5}{6}\right)^{8.63}$

23. $e^{1.2}$

24. $e^{-1.3}$

25. $125e^{0.026(7)}$

26. $83.6e^{-0.157(9.5)}$

In Problems 27–34, determine whether the given function is linear, exponential, or neither. For those that are linear functions, find a linear function that models the data; for those that are exponential, find an exponential function that models the data.

27.

x	f(x)
−1	3
0	6
1	12
2	18
3	30

28.

x	g(x)
−1	2
0	5
1	8
2	11
3	14

29.

x	H(x)
−1	$\frac{1}{4}$
0	1
1	4
2	16
3	64

30.

x	F(x)
−1	$\frac{2}{3}$
0	1
1	$\frac{3}{2}$
2	$\frac{9}{4}$
3	$\frac{27}{8}$

31.

x	f(x)
−1	$\frac{3}{2}$
0	3
1	6
2	12
3	24

32.

x	g(x)
−1	6
0	1
1	0
2	3
3	10

33.

x	H(x)
−1	2
0	4
1	6
2	8
3	10

34.

x	F(x)
−1	$\frac{1}{2}$
0	$\frac{1}{4}$
1	$\frac{1}{8}$
2	$\frac{1}{16}$
3	$\frac{1}{32}$

In Problems 35–42, the graph of an exponential function is given. Match each graph to one of the following functions.

(A) $y = 3^x$ (B) $y = 3^{-x}$ (C) $y = -3^x$ (D) $y = -3^{-x}$

(E) $y = 3^x - 1$ (F) $y = 3^{x-1}$ (G) $y = 3^{1-x}$ (H) $y = 1 - 3^x$

35.
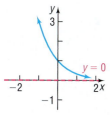

36.
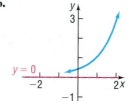

37.
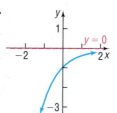

38.
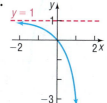

39.
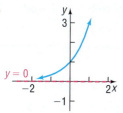

40.
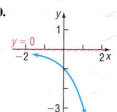

41.
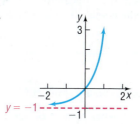

42.
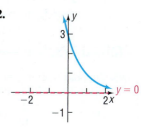

In Problems 43–54, use transformations to graph each function. Determine the domain, range, and horizontal asymptote of each function.

43. $f(x) = 2^x + 1$ **44.** $f(x) = 3^x - 2$ **45.** $f(x) = 3^{x-1}$ **46.** $f(x) = 2^{x+2}$

47. $f(x) = 3 \cdot \left(\frac{1}{2}\right)^x$ **48.** $f(x) = 4 \cdot \left(\frac{1}{3}\right)^x$ **49.** $f(x) = 3^{-x} - 2$ **50.** $f(x) = -3^x + 1$

51. $f(x) = 2 + 4^{x-1}$ **52.** $f(x) = 1 - 2^{x+3}$ **53.** $f(x) = 2 + 3^{x/2}$ **54.** $f(x) = 1 - 2^{-x/3}$

In Problems 55–62, begin with the graph of $y = e^x$ (Figure 27) and use transformations to graph each function. Determine the domain, range, and horizontal asymptote of each function.

55. $f(x) = e^{-x}$ **56.** $f(x) = -e^x$ **57.** $f(x) = e^{x+2}$ **58.** $f(x) = e^x - 1$

59. $f(x) = 5 - e^{-x}$ **60.** $f(x) = 9 - 3e^{-x}$ **61.** $f(x) = 2 - e^{-x/2}$ **62.** $f(x) = 7 - 3e^{2x}$

In Problems 63–82, solve each equation.

63. $7^x = 7^3$

64. $5^x = 5^{-6}$

65. $2^{-x} = 16$

66. $3^{-x} = 81$

67. $\left(\dfrac{1}{5}\right)^x = \dfrac{1}{25}$

68. $\left(\dfrac{1}{4}\right)^x = \dfrac{1}{64}$

69. $2^{2x-1} = 4$

70. $5^{x+3} = \dfrac{1}{5}$

71. $3^{x^3} = 9^x$

72. $4^{x^2} = 2^x$

73. $8^{-x+14} = 16^x$

74. $9^{-x+15} = 27^x$

75. $3^{x^2-7} = 27^{2x}$

76. $5^{x^2+8} = 125^{2x}$

77. $4^x \cdot 2^{x^2} = 16^2$

78. $9^{2x} \cdot 27^{x^2} = 3^{-1}$

79. $e^x = e^{3x+8}$

80. $e^{3x} = e^{2-x}$

81. $e^{x^2} = e^{3x} \cdot \dfrac{1}{e^2}$

82. $\left(e^4\right)^x \cdot e^{x^2} = e^{12}$

83. If $4^x = 7$, what does 4^{-2x} equal?

84. If $2^x = 3$, what does 4^{-x} equal?

85. If $3^{-x} = 2$, what does 3^{2x} equal?

86. If $5^{-x} = 3$, what does 5^{3x} equal?

87. If $9^x = 25$, what does 3^x equal?

88. If $2^{-3x} = \dfrac{1}{1000}$, what does 2^x equal?

In Problems 89–92, determine the exponential function whose graph is given.

89.

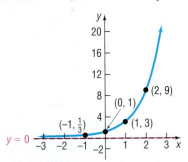

90.

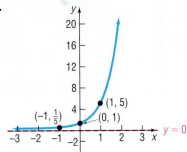

91.

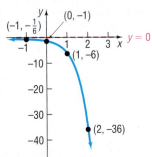

92.

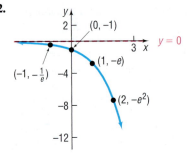

93. Find an exponential function with horizontal asymptote $y = 2$ whose graph contains the points $(0, 3)$ and $(1, 5)$.

94. Find an exponential function with horizontal asymptote $y = -3$ whose graph contains the points $(0, -2)$ and $(-2, 1)$.

Mixed Practice

95. Suppose that $f(x) = 2^x$.
 (a) What is $f(4)$? What point is on the graph of f?
 (b) If $f(x) = \dfrac{1}{16}$, what is x? What point is on the graph of f?

96. Suppose that $f(x) = 3^x$.
 (a) What is $f(4)$? What point is on the graph of f?
 (b) If $f(x) = \dfrac{1}{9}$, what is x? What point is on the graph of f?

97. Suppose that $g(x) = 4^x + 2$.
 (a) What is $g(-1)$? What point is on the graph of g?
 (b) If $g(x) = 66$, what is x? What point is on the graph of g?

98. Suppose that $g(x) = 5^x - 3$.
 (a) What is $g(-1)$? What point is on the graph of g?
 (b) If $g(x) = 122$, what is x? What point is on the graph of g?

99. Suppose that $H(x) = \left(\dfrac{1}{2}\right)^x - 4$.
 (a) What is $H(-6)$? What point is on the graph of H?
 (b) If $H(x) = 12$, what is x? What point is on the graph of H?
 (c) Find the zero of H.

100. Suppose that $F(x) = \left(\dfrac{1}{3}\right)^x - 3$.
 (a) What is $F(-5)$? What point is on the graph of F?
 (b) If $F(x) = 24$, what is x? What point is on the graph of F?
 (c) Find the zero of F.

In Problems 101–104, graph each function. Based on the graph, state the domain and the range, and find any intercepts.

101. $f(x) = \begin{cases} e^{-x} & \text{if } x < 0 \\ e^x & \text{if } x \geq 0 \end{cases}$

102. $f(x) = \begin{cases} e^x & \text{if } x < 0 \\ e^{-x} & \text{if } x \geq 0 \end{cases}$

103. $f(x) = \begin{cases} -e^x & \text{if } x < 0 \\ -e^{-x} & \text{if } x \geq 0 \end{cases}$

104. $f(x) = \begin{cases} -e^{-x} & \text{if } x < 0 \\ -e^x & \text{if } x \geq 0 \end{cases}$

Applications and Extensions

105. Optics If a single pane of glass obliterates 3% of the light passing through it, the percent p of light that passes through n successive panes is given approximately by the function

$$p(n) = 100(0.97)^n$$

(a) What percent of light will pass through 10 panes?
(b) What percent of light will pass through 25 panes?
(c) Explain the meaning of the base 0.97 in this problem.

106. Atmospheric Pressure The atmospheric pressure p on a balloon or airplane decreases with increasing height. This pressure, measured in millimeters of mercury, is related to the height h (in kilometers) above sea level by the function

$$p(h) = 760e^{-0.145h}$$

(a) Find the atmospheric pressure at a height of 2 km (over a mile).
(b) What is it at a height of 10 kilometers (over 30,000 feet)?

107. Depreciation The price p, in dollars, of a Honda Civic EX-L sedan that is x years old is modeled by

$$p(x) = 22,265(0.90)^x$$

(a) How much should a 3-year-old Civic EX-L sedan cost?
(b) How much should a 9-year-old Civic EX-L sedan cost?
(c) Explain the meaning of the base 0.90 in this problem.

108. Healing of Wounds The normal healing of wounds can be modeled by an exponential function. If A_0 represents the original area of the wound and if A equals the area of the wound, then the function

$$A(n) = A_0 e^{-0.35n}$$

describes the area of a wound after n days following an injury when no infection is present to retard the healing. Suppose that a wound initially had an area of 100 square millimeters.
(a) If healing is taking place, how large will the area of the wound be after 3 days?
(b) How large will it be after 10 days?

109. Advanced-Stage Pancreatic Cancer The percentage of patients P who have survived t years after initial diagnosis of advanced-stage pancreatic cancer is modeled by the function

$$P(t) = 100(0.3)^t$$

Source: Cancer Treatment Centers of America
(a) According to the model, what percent of patients survive 1 year after initial diagnosis?
(b) What percent of patients survive 2 years after initial diagnosis?
(c) Explain the meaning of the base 0.3 in the context of this problem.

110. Endangered Species In a protected environment, the population P of a certain endangered species recovers over time t (in years) according to the model

$$P(t) = 30(1.149)^t$$

(a) What is the size of the initial population of the species?
(b) According to the model, what will be the population of the species in 5 years?

(c) According to the model, what will be the population of the species in 10 years?
(d) According to the model, what will be the population of the species in 15 years?
(e) What is happening to the population every 5 years?

111. Drug Medication The function

$$D(h) = 5e^{-0.4h}$$

can be used to find the number of milligrams D of a certain drug that is in a patient's bloodstream h hours after the drug has been administered. How many milligrams will be present after 1 hour? After 6 hours?

112. Spreading of Rumors A model for the number N of people in a college community who have heard a certain rumor is

$$N = P(1 - e^{-0.15d})$$

where P is the total population of the community and d is the number of days that have elapsed since the rumor began. In a community of 1000 students, how many students will have heard the rumor after 3 days?

113. Exponential Probability Between 12:00 PM and 1:00 PM, cars arrive at Citibank's drive-thru at the rate of 6 cars per hour (0.1 car per minute). The following formula from probability can be used to determine the probability that a car will arrive within t minutes of 12:00 PM.

$$F(t) = 1 - e^{-0.1t}$$

(a) Determine the probability that a car will arrive within 10 minutes of 12:00 PM (that is, before 12:10 PM).
(b) Determine the probability that a car will arrive within 40 minutes of 12:00 PM (before 12:40 PM).
(c) What value does F approach as t becomes unbounded in the positive direction?
(d) Graph F using a graphing utility.
(e) Using INTERSECT, determine how many minutes are needed for the probability to reach 50%.

114. Exponential Probability Between 5:00 PM and 6:00 PM, cars arrive at Jiffy Lube at the rate of 9 cars per hour (0.15 car per minute). This formula from probability can be used to determine the probability that a car will arrive within t minutes of 5:00 PM:

$$F(t) = 1 - e^{-0.15t}$$

(a) Determine the probability that a car will arrive within 15 minutes of 5:00 PM (that is, before 5:15 PM).
(b) Determine the probability that a car will arrive within 30 minutes of 5:00 PM (before 5:30 PM).
(c) What value does F approach as t becomes unbounded in the positive direction?
(d) Graph F using a graphing utility.
(e) Using INTERSECT, determine how many minutes are needed for the probability to reach 60%.

115. Poisson Probability Between 5:00 PM and 6:00 PM, cars arrive at a McDonald's drive-thru at the rate of 20 cars per hour. The following formula from probability can be used to determine the probability that x cars will arrive between 5:00 PM and 6:00 PM.

$$P(x) = \frac{20^x e^{-20}}{x!}$$

where

$$x! = x \cdot (x-1) \cdot (x-2) \cdot \cdots \cdot 3 \cdot 2 \cdot 1$$

(a) Determine the probability that $x = 15$ cars will arrive between 5:00 PM and 6:00 PM.
(b) Determine the probability that $x = 20$ cars will arrive between 5:00 PM and 6:00 PM.

116. Poisson Probability People enter a line for the *Demon Roller Coaster* at the rate of 4 per minute. The following formula from probability can be used to determine the probability that x people will arrive within the next minute.

$$P(x) = \frac{4^x e^{-4}}{x!}$$

where

$$x! = x \cdot (x-1) \cdot (x-2) \cdot \cdots \cdot 3 \cdot 2 \cdot 1$$

(a) Determine the probability that $x = 5$ people will arrive within the next minute.
(b) Determine the probability that $x = 8$ people will arrive within the next minute.

117. Relative Humidity The relative humidity is the ratio (expressed as a percent) of the amount of water vapor in the air to the maximum amount that the air can hold at a specific temperature. The relative humidity, R, is found using the following formula:

$$R = 10^{\left(\frac{4221}{T+459.4} - \frac{4221}{D+459.4} + 2 \right)}$$

where T is the air temperature (in °F) and D is the dew point temperature (in °F).

(a) Determine the relative humidity if the air temperature is 50° Fahrenheit and the dew point temperature is 41° Fahrenheit.
(b) Determine the relative humidity if the air temperature is 68° Fahrenheit and the dew point temperature is 59° Fahrenheit.
(c) What is the relative humidity if the air temperature and the dew point temperature are the same?

118. Learning Curve Suppose that a student has 500 vocabulary words to learn. If the student learns 15 words after 5 minutes, the function

$$L(t) = 500(1 - e^{-0.0061t})$$

approximates the number of words L that the student will have learned after t minutes.

(a) How many words will the student have learned after 30 minutes?
(b) How many words will the student have learned after 60 minutes?

119. Current in an RL Circuit The equation governing the amount of current I (in amperes) after time t (in seconds) in a single RL circuit consisting of a resistance R (in ohms), an inductance L (in henrys), and an electromotive force E (in volts) is

$$I = \frac{E}{R}\left[1 - e^{-(R/L)t}\right]$$

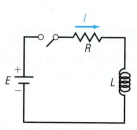

(a) If $E = 120$ volts, $R = 10$ ohms, and $L = 5$ henrys, how much current I_1 is flowing after 0.3 second? After 0.5 second? After 1 second?
(b) What is the maximum current?
(c) Graph this function $I = I_1(t)$, measuring I along the y-axis and t along the x-axis.
(d) If $E = 120$ volts, $R = 5$ ohms, and $L = 10$ henrys, how much current I_2 is flowing after 0.3 second? After 0.5 second? After 1 second?
(e) What is the maximum current?
(f) Graph the function $I = I_2(t)$ on the same coordinate axes as $I_1(t)$.

120. Current in an RC Circuit The equation governing the amount of current I (in amperes) after time t (in microseconds) in a single RC circuit consisting of a resistance R (in ohms), a capacitance C (in microfarads), and an electromotive force E (in volts) is

$$I = \frac{E}{R} e^{-t/(RC)}$$

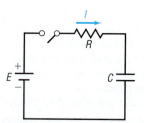

(a) If $E = 120$ volts, $R = 2000$ ohms, and $C = 1.0$ microfarad, how much current I_1 is flowing initially $(t = 0)$? After 1000 microseconds? After 3000 microseconds?
(b) What is the maximum current?
(c) Graph the function $I = I_1(t)$, measuring I along the y-axis and t along the x-axis.
(d) If $E = 120$ volts, $R = 1000$ ohms, and $C = 2.0$ microfarads, how much current I_2 is flowing initially? After 1000 microseconds? After 3000 microseconds?
(e) What is the maximum current?
(f) Graph the function $I = I_2(t)$ on the same coordinate axes as $I_1(t)$.

121. If f is an exponential function of the form $f(x) = Ca^x$ with growth factor 3, and if $f(6) = 12$, what is $f(7)$?

122. Another Formula for e Use a calculator to compute the values of

$$2 + \frac{1}{2!} + \frac{1}{3!} + \cdots + \frac{1}{n!}$$

for $n = 4, 6, 8,$ and 10. Compare each result with e.

[**Hint:** $1! = 1, 2! = 2 \cdot 1, 3! = 3 \cdot 2 \cdot 1,$
$n! = n(n-1) \cdot \cdots \cdot (3)(2)(1).$]

123. Another Formula for e Use a calculator to compute the various values of the expression. Compare the values to e.

$$2 + \cfrac{1}{1 + \cfrac{1}{2 + \cfrac{2}{3 + \cfrac{3}{4 + \cfrac{4}{\text{etc.}}}}}}$$

124. Difference Quotient If $f(x) = a^x$, show that

$$\frac{f(x+h) - f(x)}{h} = a^x \cdot \frac{a^h - 1}{h} \quad h \neq 0$$

125. If $f(x) = a^x$, show that $f(A + B) = f(A) \cdot f(B)$.

126. If $f(x) = a^x$, show that $f(-x) = \dfrac{1}{f(x)}$.

127. If $f(x) = a^x$, show that $f(\alpha x) = [f(x)]^\alpha$.

Problems 128 and 129 provide definitions for two other transcendental functions.

128. The **hyperbolic sine function,** designated by $\sinh x$, is defined as

$$\sinh x = \frac{1}{2}(e^x - e^{-x})$$

(a) Show that $f(x) = \sinh x$ is an odd function.
(b) Graph $f(x) = \sinh x$ using a graphing utility.

129. The **hyperbolic cosine function,** designated by $\cosh x$, is defined as

$$\cosh x = \frac{1}{2}(e^x + e^{-x})$$

(a) Show that $f(x) = \cosh x$ is an even function.
(b) Graph $f(x) = \cosh x$ using a graphing utility.
(c) Refer to Problem 128. Show that, for every x,

$$(\cosh x)^2 - (\sinh x)^2 = 1$$

130. Historical Problem Pierre de Fermat (1601–1665) conjectured that the function

$$f(x) = 2^{(2^x)} + 1$$

for $x = 1, 2, 3, \ldots,$ would always have a value equal to a prime number. But Leonhard Euler (1707–1783) showed that this formula fails for $x = 5$. Use a calculator to determine the prime numbers produced by f for $x = 1, 2, 3, 4$. Then show that $f(5) = 641 \times 6{,}700{,}417$, which is not prime.

Explaining Concepts: Discussion and Writing

131. The bacteria in a 4-liter container double every minute. After 60 minutes the container is full. How long did it take to fill half the container?

132. Explain in your own words what the number e is. Provide at least two applications that use this number.

133. Do you think that there is a power function that increases more rapidly than an exponential function whose base is greater than 1? Explain.

134. As the base a of an exponential function $f(x) = a^x$, where $a > 1$, increases, what happens to the behavior of its graph for $x > 0$? What happens to the behavior of its graph for $x < 0$?

135. The graphs of $y = a^{-x}$ and $y = \left(\dfrac{1}{a}\right)^x$ are identical. Why?

Retain Your Knowledge

Problems 136–139 are based on material learned earlier in the course. The purpose of these problems is to keep the material fresh in your mind so that you are better prepared for the final exam.

136. Solve the inequality: $x^3 + 5x^2 \le 4x + 20$.

137. Solve the inequality: $\dfrac{x+1}{x-2} \ge 1$.

138. Find the equation of the quadratic function f that has its vertex at $(3, 5)$ and contains the point $(2, 3)$.

139. Consider the quadratic function $f(x) = x^2 + 2x - 3$.
(a) Graph f by determining whether its graph opens up or down and by finding its vertex, axis of symmetry, y-intercept, and x-intercepts, if any.
(b) Determine the domain and range of f.
(c) Determine where f is increasing and where it is decreasing.

'Are You Prepared?' Answers

1. $64; 4; \dfrac{1}{9}$ **2.** $\{-4, 1\}$ **3.** False   **4.** 3   **5.** True

6.4 Logarithmic Functions

PREPARING FOR THIS SECTION *Before getting started, review the following:*

- Solving Inequalities (Section 1.5, pp. 119–126)
- Quadratic Inequalities (Section 4.5, pp. 312–313)

- Polynomial and Rational Inequalities (Section 5.4, pp. 370–372)
- Solve Linear Equations (Section 1.1, pp. 82–87)

Now Work the **'Are You Prepared?'** problems on page 448.

OBJECTIVES **1** Change Exponential Statements to Logarithmic Statements and Logarithmic Statements to Exponential Statements (p. 440)
      2 Evaluate Logarithmic Expressions (p. 441)
      3 Determine the Domain of a Logarithmic Function (p. 441)
      4 Graph Logarithmic Functions (p. 442)
      5 Solve Logarithmic Equations (p. 446)

Recall that a one-to-one function $y = f(x)$ has an inverse function that is defined implicitly by the equation $x = f(y)$. In particular, the exponential function $y = f(x) = a^x$, where $a > 0$ and $a \neq 1$, is one-to-one and hence has an inverse function that is defined implicitly by the equation

$$x = a^y \qquad a > 0 \qquad a \neq 1$$

This inverse function is so important that it is given a name, the *logarithmic function*.

DEFINITION

The **logarithmic function with base a**, where $a > 0$ and $a \neq 1$, is denoted by $y = \log_a x$ (read as " y is the logarithm with base a of x") and is defined by

$$y = \log_a x \quad \text{if and only if} \quad x = a^y$$

The domain of the logarithmic function $y = \log_a x$ is $x > 0$.

In Words
When you need to evaluate $\log_a x$, think to yourself "a raised to what power gives me x?"

As this definition illustrates, **a logarithm is a name for a certain exponent**. So $\log_a x$ represents the exponent to which a must be raised to obtain x.

EXAMPLE 1 **Relating Logarithms to Exponents**

(a) If $y = \log_3 x$, then $x = 3^y$. For example, the logarithmic statement $4 = \log_3 81$ is equivalent to the exponential statement $81 = 3^4$.

(b) If $y = \log_5 x$, then $x = 5^y$. For example, $-1 = \log_5\left(\dfrac{1}{5}\right)$ is equivalent to $\dfrac{1}{5} = 5^{-1}$.

1 Change Exponential Statements to Logarithmic Statements and Logarithmic Statements to Exponential Statements

The definition of a logarithm can be used to convert from exponential form to logarithmic form, and vice versa, as the following two examples illustrate.

EXAMPLE 2 **Changing Exponential Statements to Logarithmic Statements**

Change each exponential statement to an equivalent statement involving a logarithm.

(a) $1.2^3 = m$ (b) $e^b = 9$ (c) $a^4 = 24$

Solution Use the fact that $y = \log_a x$ and $x = a^y$, where $a > 0$ and $a \neq 1$, are equivalent.

(a) If $1.2^3 = m$, then $3 = \log_{1.2} m$.   (b) If $e^b = 9$, then $b = \log_e 9$.
(c) If $a^4 = 24$, then $4 = \log_a 24$.

Now Work PROBLEM 11

EXAMPLE 3

Changing Logarithmic Statements to Exponential Statements

Change each logarithmic statement to an equivalent statement involving an exponent.

(a) $\log_a 4 = 5$ (b) $\log_e b = -3$ (c) $\log_3 5 = c$

Solution

(a) If $\log_a 4 = 5$, then $a^5 = 4$.
(b) If $\log_e b = -3$, then $e^{-3} = b$.
(c) If $\log_3 5 = c$, then $3^c = 5$.

 Now Work PROBLEM 19

2 Evaluate Logarithmic Expressions

To find the exact value of a logarithm, write the logarithm in exponential notation using the fact that $y = \log_a x$ is equivalent to $a^y = x$, and use the fact that if $a^u = a^v$, then $u = v$.

EXAMPLE 4

Finding the Exact Value of a Logarithmic Expression

Find the exact value of:

(a) $\log_2 16$ (b) $\log_3 \dfrac{1}{27}$

Solution

(a) To evaluate $\log_2 16$, think "2 raised to what power yields 16?" Then,

$$y = \log_2 16$$
$$2^y = 16 \qquad \textit{Change to exponential form.}$$
$$2^y = 2^4 \qquad \textit{\textcolor{teal}{$16 = 2^4$}}$$
$$y = 4 \qquad \textit{Equate exponents.}$$

Therefore, $\log_2 16 = 4$.

(b) To evaluate $\log_3 \dfrac{1}{27}$, think "3 raised to what power yields $\dfrac{1}{27}$?" Then,

$$y = \log_3 \dfrac{1}{27}$$
$$3^y = \dfrac{1}{27} \qquad \textit{Change to exponential form.}$$
$$3^y = 3^{-3} \qquad \textit{\textcolor{teal}{$\dfrac{1}{27} = \dfrac{1}{3^3} = 3^{-3}$}}$$
$$y = -3 \qquad \textit{Equate exponents.}$$

Therefore, $\log_3 \dfrac{1}{27} = -3$.

Now Work PROBLEM 27

3 Determine the Domain of a Logarithmic Function

The logarithmic function $y = \log_a x$ has been defined as the inverse of the exponential function $y = a^x$. That is, if $f(x) = a^x$, then $f^{-1}(x) = \log_a x$. Based on the discussion in Section 6.2 on inverse functions, for a function f and its inverse f^{-1},

$$\text{Domain of } f^{-1} = \text{Range of } f \quad \text{and} \quad \text{Range of } f^{-1} = \text{Domain of } f$$

Consequently, it follows that

> Domain of the logarithmic function = Range of the exponential function = $(0, \infty)$
>
> Range of the logarithmic function = Domain of the exponential function = $(-\infty, \infty)$

The next box summarizes some properties of the logarithmic function.

> $y = \log_a x$ (defining equation: $x = a^y$)
>
> Domain: $0 < x < \infty$ Range: $-\infty < y < \infty$

The domain of a logarithmic function consists of the *positive* real numbers, so the argument of a logarithmic function must be greater than zero.

EXAMPLE 5   **Finding the Domain of a Logarithmic Function**

Find the domain of each logarithmic function.

(a) $F(x) = \log_2(x + 3)$ (b) $g(x) = \log_5\left(\dfrac{1 + x}{1 - x}\right)$ (c) $h(x) = \log_{1/2}|x|$

Solution   (a) The domain of F consists of all x for which $x + 3 > 0$, that is, $x > -3$. Using interval notation, the domain of F is $(-3, \infty)$.

(b) The domain of g is restricted to

$$\frac{1 + x}{1 - x} > 0$$

Solve this inequality to find that the domain of g consists of all x between -1 and 1, that is, $-1 < x < 1$, or, using interval notation, $(-1, 1)$.

(c) Since $|x| > 0$, provided that $x \neq 0$, the domain of h consists of all real numbers except zero, or, using interval notation, $(-\infty, 0) \cup (0, \infty)$. ●

Now Work PROBLEMS 41 AND 47

4 Graph Logarithmic Functions

Because exponential functions and logarithmic functions are inverses of each other, the graph of the logarithmic function $y = \log_a x$ is the reflection about the line $y = x$ of the graph of the exponential function $y = a^x$, as shown in Figure 30.

For example, to graph $y = \log_2 x$, graph $y = 2^x$ and reflect it about the line $y = x$. See Figure 31. To graph $y = \log_{1/3} x$, graph $y = \left(\dfrac{1}{3}\right)^x$ and reflect it about the line $y = x$. See Figure 32.

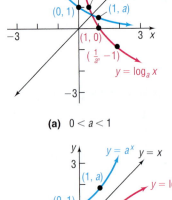

(a) $0 < a < 1$

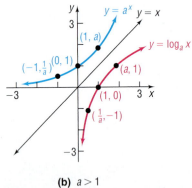

(b) $a > 1$

Figure 30

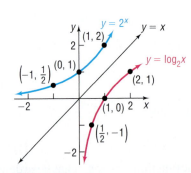

Figure 31

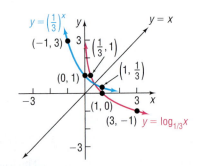

Figure 32

Now Work PROBLEM 61

The graphs of $y = \log_a x$ in Figures 30(a) and (b) lead to the following properties.

Properties of the Logarithmic Function $f(x) = \log_a x; a > 0, a \neq 1$

1. The domain is the set of positive real numbers, or $(0, \infty)$ using interval notation; the range is the set of all real numbers, or $(-\infty, \infty)$ using interval notation.

2. The x-intercept of the graph is 1. There is no y-intercept.

3. The y-axis $(x = 0)$ is a vertical asymptote of the graph.

4. A logarithmic function is decreasing if $0 < a < 1$ and is increasing if $a > 1$.

5. The graph of f contains the points $(1, 0)$, $(a, 1)$, and $\left(\dfrac{1}{a}, -1\right)$.

6. The graph is smooth and continuous, with no corners or gaps.

If the base of a logarithmic function is the number e, the result is the **natural logarithm function**. This function occurs so frequently in applications that it is given a special symbol, **ln** (from the Latin, *logarithmus naturalis*). That is,

$$y = \ln x \quad \text{if and only if} \quad x = e^y \qquad \textbf{(1)}$$

Because $y = \ln x$ and the exponential function $y = e^x$ are inverse functions, the graph of $y = \ln x$ can be obtained by reflecting the graph of $y = e^x$ about the line $y = x$. See Figure 33.

Using a calculator with an $\boxed{\ln}$ key, we can obtain other points on the graph of $f(x) = \ln x$. See Table 7.

Seeing the Concept

Graph $Y_1 = e^x$ and $Y_2 = \ln x$ on the same square screen. Use eVALUEate to verify the points on the graph given in Figure 33. Do you see the symmetry of the two graphs with respect to the line $y = x$?

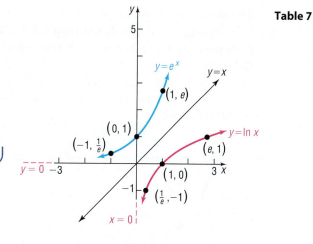

Figure 33

Table 7	x	$\ln x$
	$\dfrac{1}{2}$	-0.69
	2	0.69
	3	1.10

EXAMPLE 6 **Graphing a Logarithmic Function and Its Inverse**

(a) Find the domain of the logarithmic function $f(x) = -\ln(x - 2)$.
(b) Graph f.
(c) From the graph, determine the range and vertical asymptote of f.
(d) Find f^{-1}, the inverse of f.
(e) Find the domain and the range of f^{-1}.
(f) Graph f^{-1}.

Solution (a) The domain of f consists of all x for which $x - 2 > 0$, or equivalently, $x > 2$. The domain of f is $\{x \mid x > 2\}$, or $(2, \infty)$ in interval notation.

(b) To obtain the graph of $y = -\ln(x - 2)$, begin with the graph of $y = \ln x$ and use transformations. See Figure 34.

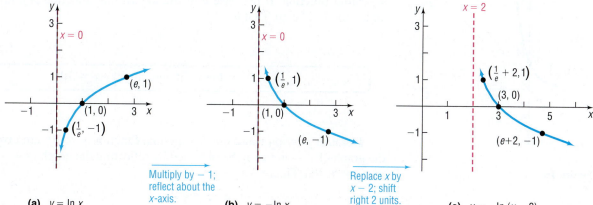

Figure 34   (a) $y = \ln x$ (b) $y = -\ln x$ (c) $y = -\ln(x - 2)$

(c) The range of $f(x) = -\ln(x - 2)$ is the set of all real numbers. The vertical asymptote is $x = 2$. [Do you see why? The original asymptote $(x = 0)$ is shifted to the right 2 units.]

(d) To find f^{-1}, begin with $y = -\ln(x - 2)$. The inverse function is defined implicitly by the equation

$$x = -\ln(y - 2)$$

Now solve for y.

$$\begin{aligned} -x &= \ln(y - 2) && \text{Isolate the logarithm.} \\ e^{-x} &= y - 2 && \text{Change to exponential form.} \\ y &= e^{-x} + 2 && \text{Solve for } y. \end{aligned}$$

The inverse of f is $f^{-1}(x) = e^{-x} + 2$.

(e) The domain of f^{-1} equals the range of f, which is the set of all real numbers, from part (c). The range of f^{-1} is the domain of f, which is $(2, \infty)$ in interval notation.

(f) To graph f^{-1}, use the graph of f in Figure 34(c) and reflect it about the line $y = x$. See Figure 35. We could also graph $f^{-1}(x) = e^{-x} + 2$ using transformations.

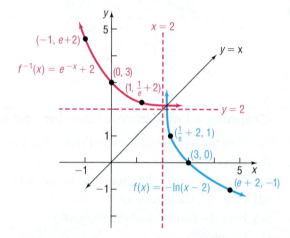

Figure 35

Now Work PROBLEM 73

If the base of a logarithmic function is the number 10, the result is the **common logarithm function**. If the base a of the logarithmic function is not indicated, it is understood to be 10. That is,

$$y = \log x \quad \text{if and only if} \quad x = 10^y$$

Because $y = \log x$ and the exponential function $y = 10^x$ are inverse functions, the graph of $y = \log x$ can be obtained by reflecting the graph of $y = 10^x$ about the line $y = x$. See Figure 36.

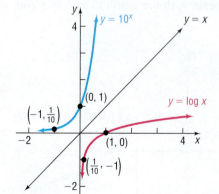

Figure 36

EXAMPLE 7 **Graphing a Logarithmic Function and Its Inverse**

(a) Find the domain of the logarithmic function $f(x) = 3 \log (x - 1)$.

(b) Graph f.

(c) From the graph, determine the range and vertical asymptote of f.

(d) Find f^{-1}, the inverse of f.

(e) Find the domain and the range of f^{-1}.

(f) Graph f^{-1}.

Solution (a) The domain of f consists of all x for which $x - 1 > 0$, or equivalently, $x > 1$. The domain of f is $\{x \mid x > 1\}$, or $(1, \infty)$ in interval notation.

(b) To obtain the graph of $y = 3 \log(x - 1)$, begin with the graph of $y = \log x$ and use transformations. See Figure 37.

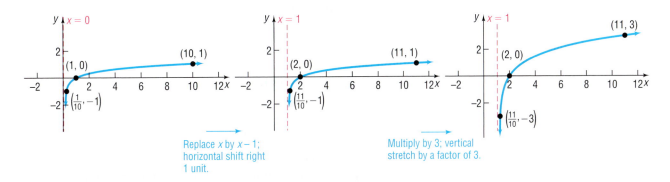

Figure 37

(a) $y = \log x$

Replace x by $x - 1$; horizontal shift right 1 unit.

(b) $y = \log (x - 1)$

Multiply by 3; vertical stretch by a factor of 3.

(c) $y = 3 \log (x - 1)$

(c) The range of $f(x) = 3 \log(x - 1)$ is the set of all real numbers. The vertical asymptote is $x = 1$.

(d) Begin with $y = 3 \log(x - 1)$. The inverse function is defined implicitly by the equation

$$x = 3 \log(y - 1)$$

Proceed to solve for y.

$$\frac{x}{3} = \log (y - 1) \quad \text{Isolate the logarithm.}$$

$$10^{x/3} = y - 1 \quad \text{Change to exponential form.}$$

$$y = 10^{x/3} + 1 \quad \text{Solve for } y.$$

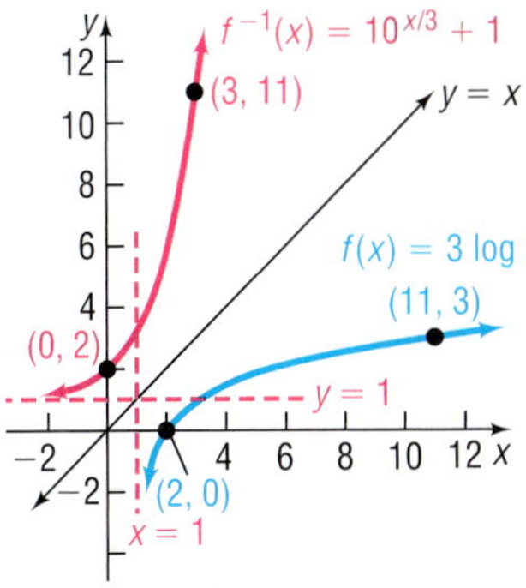

The inverse of f is $f^{-1}(x) = 10^{x/3} + 1$.

(e) The domain of f^{-1} is the range of f, which is the set of all real numbers, from part (c). The range of f^{-1} is the domain of f, which is $(1, \infty)$ in interval notation.

(f) To graph f^{-1}, use the graph of f in Figure 37(c) and reflect it about the line $y = x$. See Figure 38. We could also graph $f^{-1}(x) = 10^{x/3} + 1$ using transformations.

Figure 38

Now Work PROBLEM 81

SUMMARY

Properties of the Logarithmic Function

$f(x) = \log_a x, \quad a > 1$	Domain: the interval $(0, \infty)$; Range: the interval $(-\infty, \infty)$
$(y = \log_a x \text{ means } x = a^y)$	x-intercept: 1; y-intercept: none; vertical asymptote: $x = 0$ (y-axis); increasing; one-to-one
	See Figure 39(a) for a typical graph.
$f(x) = \log_a x, \quad 0 < a < 1$	Domain: the interval $(0, \infty)$; Range: the interval $(-\infty, \infty)$
$(y = \log_a x \text{ means } x = a^y)$	x-intercept: 1; y-intercept: none; vertical asymptote: $x = 0$ (y-axis); decreasing; one-to-one
	See Figure 39(b) for a typical graph.

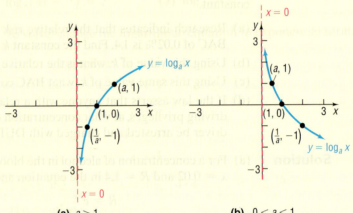

Figure 39   **(a)** $a > 1$ **(b)** $0 < a < 1$

6.4 Assess Your Understanding

'Are You Prepared?' *Answers are given at the end of these exercises. If you get a wrong answer, read the pages listed in* red.

1. Solve each inequality:
   (a) $3x - 7 \leq 8 - 2x$ (pp. 119–126)
   (b) $x^2 - x - 6 > 0$ (pp. 312–313)

2. Solve the inequality: $\dfrac{x-1}{x+4} > 0$ (pp. 370–372)

3. Solve: $2x + 3 = 9$ (pp. 82–87)

Concepts and Vocabulary

4. The domain of the logarithmic function $f(x) = \log_a x$ is _____.

5. The graph of every logarithmic function $f(x) = \log_a x$, where $a > 0$ and $a \neq 1$, passes through three points: _____, _____, and _____.

6. If the graph of a logarithmic function $f(x) = \log_a x$, where $a > 0$ and $a \neq 1$, is increasing, then its base must be larger than _____.

7. *True or False* If $y = \log_a x$, then $y = a^x$.

8. *True or False* The graph of $f(x) = \log_a x$, where $a > 0$ and $a \neq 1$, has an x-intercept equal to 1 and no y-intercept.

9. Select the answer that completes the statement: $y = \ln x$ if and only if _____.
   (a) $x = e^y$ (b) $y = e^x$ (c) $x = 10^y$ (d) $y = 10^x$

10. Choose the domain of $f(x) = \log_3(x + 2)$.
   (a) $(-\infty, \infty)$ (b) $(2, \infty)$ (c) $(-2, \infty)$ (d) $(0, \infty)$

Skill Building

In Problems 11–18, change each exponential statement to an equivalent statement involving a logarithm.

11. $9 = 3^2$

12. $16 = 4^2$

13. $a^2 = 1.6$

14. $a^3 = 2.1$

15. $2^x = 7.2$

16. $3^x = 4.6$

17. $e^x = 8$

18. $e^{2.2} = M$

In Problems 19–26, change each logarithmic statement to an equivalent statement involving an exponent.

19. $\log_2 8 = 3$

20. $\log_3\left(\dfrac{1}{9}\right) = -2$

21. $\log_a 3 = 6$

22. $\log_b 4 = 2$

23. $\log_3 2 = x$

24. $\log_2 6 = x$

25. $\ln 4 = x$

26. $\ln x = 4$

In Problems 27–38, find the exact value of each logarithm without using a calculator.

27. $\log_2 1$ **28.** $\log_8 8$ **29.** $\log_5 25$ **30.** $\log_3\left(\dfrac{1}{9}\right)$

31. $\log_{1/2} 16$ **32.** $\log_{1/3} 9$ **33.** $\log_{10} \sqrt{10}$ **34.** $\log_5 \sqrt[3]{25}$

35. $\log_{\sqrt{2}} 4$ **36.** $\log_{\sqrt{3}} 9$ **37.** $\ln \sqrt{e}$ **38.** $\ln e^3$

In Problems 39–50, find the domain of each function.

39. $f(x) = \ln(x - 3)$ **40.** $g(x) = \ln(x - 1)$ **41.** $F(x) = \log_2 x^2$

42. $H(x) = \log_5 x^3$ **43.** $f(x) = 3 - 2\log_4\left(\dfrac{x}{2} - 5\right)$ **44.** $g(x) = 8 + 5\ln(2x + 3)$

45. $f(x) = \ln\left(\dfrac{1}{x + 1}\right)$ **46.** $g(x) = \ln\left(\dfrac{1}{x - 5}\right)$ **47.** $g(x) = \log_5\left(\dfrac{x + 1}{x}\right)$

48. $h(x) = \log_3\left(\dfrac{x}{x - 1}\right)$ **49.** $f(x) = \sqrt{\ln x}$ **50.** $g(x) = \dfrac{1}{\ln x}$

In Problems 51–58, use a calculator to evaluate each expression. Round your answer to three decimal places.

51. $\ln \dfrac{5}{3}$ **52.** $\dfrac{\ln 5}{3}$ **53.** $\dfrac{\ln \dfrac{10}{3}}{0.04}$ **54.** $\dfrac{\ln \dfrac{2}{3}}{-0.1}$

55. $\dfrac{\ln 4 + \ln 2}{\log 4 + \log 2}$ **56.** $\dfrac{\log 15 + \log 20}{\ln 15 + \ln 20}$ **57.** $\dfrac{2\ln 5 + \log 50}{\log 4 - \ln 2}$ **58.** $\dfrac{3\log 80 - \ln 5}{\log 5 + \ln 20}$

59. Find a so that the graph of $f(x) = \log_a x$ contains the point $(2, 2)$.

60. Find a so that the graph of $f(x) = \log_a x$ contains the point $\left(\dfrac{1}{2}, -4\right)$.

In Problems 61–64, graph each function and its inverse on the same set of axes.

61. $f(x) = 3^x; f^{-1}(x) = \log_3 x$ **62.** $f(x) = 4^x; f^{-1}(x) = \log_4 x$

63. $f(x) = \left(\dfrac{1}{2}\right)^x; f^{-1}(x) = \log_{1/2} x$ **64.** $f(x) = \left(\dfrac{1}{3}\right)^x; f^{-1}(x) = \log_{1/3} x$

In Problems 65–72, the graph of a logarithmic function is given. Match each graph to one of the following functions:

(A) $y = \log_3 x$ (B) $y = \log_3(-x)$ (C) $y = -\log_3 x$ (D) $y = -\log_3(-x)$

(E) $y = \log_3 x - 1$ (F) $y = \log_3(x - 1)$ (G) $y = \log_3(1 - x)$ (H) $y = 1 - \log_3 x$

65. 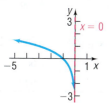   **66.** 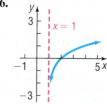   **67.** 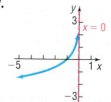   **68.**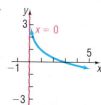

69. 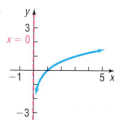   **70.** 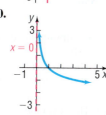   **71.** 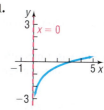   **72.**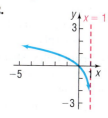

In Problems 73–88, use the given function f.

(a) Find the domain of f. (b) Graph f. (c) From the graph, determine the range and any asymptotes of f.

(d) Find f^{-1}, the inverse of f. (e) Find the domain and the range of f^{-1}. (f) Graph f^{-1}.

73. $f(x) = \ln(x + 4)$ **74.** $f(x) = \ln(x - 3)$ **75.** $f(x) = 2 + \ln x$ **76.** $f(x) = -\ln(-x)$

77. $f(x) = \ln(2x) - 3$ **78.** $f(x) = -2\ln(x + 1)$ **79.** $f(x) = \log(x - 4) + 2$ **80.** $f(x) = \dfrac{1}{2}\log x - 5$

81. $f(x) = \dfrac{1}{2}\log(2x)$ **82.** $f(x) = \log(-2x)$ **83.** $f(x) = 3 + \log_3(x + 2)$ **84.** $f(x) = 2 - \log_3(x + 1)$

85. $f(x) = e^{x+2} - 3$ **86.** $f(x) = 3e^x + 2$ **87.** $f(x) = 2^{x/3} + 4$ **88.** $f(x) = -3^{x+1}$

EXAMPLE 8   **Using the Change-of-Base Formula**

Approximate:

(a) $\log_5 89$ (b) $\log_{\sqrt{2}} \sqrt{5}$

Round answers to four decimal places.

Solution
(a) $\log_5 89 = \dfrac{\log 89}{\log 5} \approx \dfrac{1.949390007}{0.6989700043}$
≈ 2.7889

or

$\log_5 89 = \dfrac{\ln 89}{\ln 5} \approx \dfrac{4.48863637}{1.609437912}$
≈ 2.7889

(b) $\log_{\sqrt{2}} \sqrt{5} = \dfrac{\log \sqrt{5}}{\log \sqrt{2}} = \dfrac{\dfrac{1}{2}\log 5}{\dfrac{1}{2}\log 2}$

$= \dfrac{\log 5}{\log 2} \approx 2.3219$

or

$\log_{\sqrt{2}} \sqrt{5} = \dfrac{\ln \sqrt{5}}{\ln \sqrt{2}} = \dfrac{\dfrac{1}{2}\ln 5}{\dfrac{1}{2}\ln 2}$

$= \dfrac{\ln 5}{\ln 2} \approx 2.3219$

 Now Work PROBLEMS 23 AND 71

 COMMENT To graph logarithmic functions when the base is different from e or 10 requires the Change-of-Base Formula. For example, to graph $y = \log_2 x$, graph either $y = \dfrac{\ln x}{\ln 2}$ or $y = \dfrac{\log x}{\log 2}$. ■

Now Work PROBLEM 79

SUMMARY

Properties of Logarithms

In the list that follows, a, b, M, N, and r are real numbers. Also, $a > 0$, $a \neq 1$, $b > 0$, $b \neq 1$, $M > 0$, and $N > 0$.

Definition
$y = \log_a x$ means $x = a^y$

Properties of logarithms
$\log_a 1 = 0$ $\log_a a = 1$ $\log_a M^r = r \log_a M$

$a^{\log_a M} = M$ $\log_a a^r = r$ $a^r = e^{r \ln a}$

$\log_a(MN) = \log_a M + \log_a N$ $\log_a\left(\dfrac{M}{N}\right) = \log_a M - \log_a N$

If $M = N$, then $\log_a M = \log_a N$.   If $\log_a M = \log_a N$, then $M = N$.

Change-of-Base Formula
$\log_a M = \dfrac{\log_b M}{\log_b a}$

Historical Feature

John Napier
(1550–1617)

ogarithms were invented about 1590 by John Napier (1550–1617) and Joost Bürgi (1552–1632), working independently. Napier, whose work had the greater influence, was a Scottish lord, a secretive man whose neighbors were inclined to believe him to be in league with the devil. His approach to logarithms was very different from ours; it was based on the relationship between arithmetic and geometric sequences, discussed in a later chapter, and not on the inverse function relationship of logarithms to exponential functions (described in Section 6.4). Napier's tables, published in 1614, listed what would now be called *natural logarithms* of sines and were rather difficult to use. A London professor, Henry Briggs, became interested in the tables and visited Napier. In their conversations, they developed the idea of common logarithms, which were published in 1617. The importance of this tool for calculation was immediately recognized, and by 1650 common logarithms were being printed as far away as China. They remained an important calculation tool until the advent of the inexpensive handheld calculator about 1972, which has decreased their calculational—but not their theoretical—importance.

A side effect of the invention of logarithms was the popularization of the decimal system of notation for real numbers.

6.5 Assess Your Understanding

Concepts and Vocabulary

1. $\log_a 1 = $ _____

2. $a^{\log_a M} = $ _____

3. $\log_a a^r = $ _____

4. $\log_a (MN) = $ _____ $+$ _____

5. $\log_a \left(\dfrac{M}{N}\right) = $ _____ $-$ _____

6. $\log_a M^r = $ _____

7. If $\log_8 M = \dfrac{\log_5 7}{\log_5 8}$, then $M = $ _____ .

8. True or False $\ln(x + 3) - \ln(2x) = \dfrac{\ln(x + 3)}{\ln(2x)}$

9. True or False $\log_2(3x^4) = 4\log_2(3x)$

10. True or False $\log\left(\dfrac{2}{3}\right) = \dfrac{\log 2}{\log 3}$

11. Choose the expression equivalent to 2^x.
    (a) e^{2x} (b) $e^{x\ln 2}$ (c) $e^{\log_2 x}$ (d) $e^{2\ln x}$

12. Writing $\log_a x - \log_a y + 2\log_a z$ as a single logarithm results in which of the following?
    (a) $\log_a(x - y + 2z)$ (b) $\log_a\left(\dfrac{xz^2}{y}\right)$
    (c) $\log_a\left(\dfrac{2xz}{y}\right)$ (d) $\log_a\left(\dfrac{x}{yz^2}\right)$

Skill Building

In Problems 13–28, use properties of logarithms to find the exact value of each expression. Do not use a calculator.

13. $\log_3 3^{71}$

14. $\log_2 2^{-13}$

15. $\ln e^{-4}$

16. $\ln e^{\sqrt{2}}$

17. $2^{\log_2 7}$

18. $e^{\ln 8}$

19. $\log_8 2 + \log_8 4$

20. $\log_6 9 + \log_6 4$

21. $\log_6 18 - \log_6 3$

22. $\log_8 16 - \log_8 2$

23. $\log_2 6 \cdot \log_6 8$

24. $\log_3 8 \cdot \log_8 9$

25. $3^{\log_3 5 - \log_3 4}$

26. $5^{\log_5 6 + \log_5 7}$

27. $e^{\log_{e^2} 16}$

28. $e^{\log_{e^2} 9}$

In Problems 29–36, suppose that $\ln 2 = a$ and $\ln 3 = b$. Use properties of logarithms to write each logarithm in terms of a and b.

29. $\ln 6$

30. $\ln \dfrac{2}{3}$

31. $\ln 1.5$

32. $\ln 0.5$

33. $\ln 8$

34. $\ln 27$

35. $\ln \sqrt[5]{6}$

36. $\ln \sqrt[4]{\dfrac{2}{3}}$

In Problems 37–56, write each expression as a sum and/or difference of logarithms. Express powers as factors.

37. $\log_5 (25x)$

38. $\log_3 \dfrac{x}{9}$

39. $\log_2 z^3$

40. $\log_7 x^5$

41. $\ln(ex)$

42. $\ln \dfrac{e}{x}$

43. $\ln \dfrac{x}{e^x}$

44. $\ln(xe^x)$

45. $\log_a(u^2 v^3)$ $u > 0, v > 0$

46. $\log_2\left(\dfrac{a}{b^2}\right)$ $a > 0, b > 0$

47. $\ln(x^2\sqrt{1 - x})$ $0 < x < 1$

48. $\ln(x\sqrt{1 + x^2})$ $x > 0$

49. $\log_2\left(\dfrac{x^3}{x - 3}\right)$ $x > 3$

50. $\log_5\left(\dfrac{\sqrt[3]{x^2 + 1}}{x^2 - 1}\right)$ $x > 1$

51. $\log\left[\dfrac{x(x + 2)}{(x + 3)^2}\right]$ $x > 0$

52. $\log\left[\dfrac{x^3\sqrt{x + 1}}{(x - 2)^2}\right]$ $x > 2$

53. $\ln\left[\dfrac{x^2 - x - 2}{(x + 4)^2}\right]^{1/3}$ $x > 2$

54. $\ln\left[\dfrac{(x - 4)^2}{x^2 - 1}\right]^{2/3}$ $x > 4$

55. $\ln \dfrac{5x\sqrt{1 + 3x}}{(x - 4)^3}$ $x > 4$

56. $\ln\left[\dfrac{5x^2\sqrt[3]{1 - x}}{4(x + 1)^2}\right]$ $0 < x < 1$

In Problems 57–70, write each expression as a single logarithm.

57. $3\log_5 u + 4\log_5 v$

58. $2\log_3 u - \log_3 v$

59. $\log_3 \sqrt{x} - \log_3 x^3$

60. $\log_2\left(\dfrac{1}{x}\right) + \log_2\left(\dfrac{1}{x^2}\right)$

61. $\log_4(x^2 - 1) - 5\log_4(x + 1)$

62. $\log(x^2 + 3x + 2) - 2\log(x + 1)$

63. $\ln\left(\dfrac{x}{x - 1}\right) + \ln\left(\dfrac{x + 1}{x}\right) - \ln(x^2 - 1)$

64. $\log\left(\dfrac{x^2 + 2x - 3}{x^2 - 4}\right) - \log\left(\dfrac{x^2 + 7x + 6}{x + 2}\right)$

65. $8\log_2 \sqrt{3x - 2} - \log_2\left(\dfrac{4}{x}\right) + \log_2 4$

66. $21\log_3 \sqrt[3]{x} + \log_3(9x^2) - \log_3 9$

67. $2\log_a(5x^3) - \dfrac{1}{2}\log_a(2x + 3)$

68. $\dfrac{1}{3}\log(x^3 + 1) + \dfrac{1}{2}\log(x^2 + 1)$

69. $2\log_2(x + 1) - \log_2(x + 3) - \log_2(x - 1)$

70. $3\log_5(3x + 1) - 2\log_5(2x - 1) - \log_5 x$

In Problems 71–78, use the Change-of-Base Formula and a calculator to evaluate each logarithm. Round your answer to three decimal places.

71. $\log_3 21$

72. $\log_5 18$

73. $\log_{1/3} 71$

74. $\log_{1/2} 15$

75. $\log_{\sqrt{2}} 7$

76. $\log_{\sqrt{5}} 8$

77. $\log_\pi e$

78. $\log_\pi \sqrt{2}$

In Problems 79–84, graph each function using a graphing utility and the Change-of-Base Formula.

79. $y = \log_4 x$

80. $y = \log_5 x$

81. $y = \log_2(x + 2)$

82. $y = \log_4(x - 3)$

83. $y = \log_{x-1}(x + 1)$

84. $y = \log_{x+2}(x - 2)$

Mixed Practice

85. If $f(x) = \ln x$, $g(x) = e^x$, and $h(x) = x^2$, find:
   (a) $(f \circ g)(x)$. What is the domain of $f \circ g$?
   (b) $(g \circ f)(x)$. What is the domain of $g \circ f$?
   (c) $(f \circ g)(5)$
   (d) $(f \circ h)(x)$. What is the domain of $f \circ h$?
   (e) $(f \circ h)(e)$

86. If $f(x) = \log_2 x$, $g(x) = 2^x$, and $h(x) = 4x$, find:
   (a) $(f \circ g)(x)$. What is the domain of $f \circ g$?
   (b) $(g \circ f)(x)$. What is the domain of $g \circ f$?
   (c) $(f \circ g)(3)$
   (d) $(f \circ h)(x)$. What is the domain of $f \circ h$?
   (e) $(f \circ h)(8)$

Applications and Extensions

In Problems 87–96, express y as a function of x. The constant C is a positive number.

87. $\ln y = \ln x + \ln C$

88. $\ln y = \ln(x + C)$

89. $\ln y = \ln x + \ln(x + 1) + \ln C$

90. $\ln y = 2\ln x - \ln(x + 1) + \ln C$

91. $\ln y = 3x + \ln C$

92. $\ln y = -2x + \ln C$

93. $\ln(y - 3) = -4x + \ln C$

94. $\ln(y + 4) = 5x + \ln C$

95. $3\ln y = \dfrac{1}{2}\ln(2x + 1) - \dfrac{1}{3}\ln(x + 4) + \ln C$

96. $2\ln y = -\dfrac{1}{2}\ln x + \dfrac{1}{3}\ln(x^2 + 1) + \ln C$

97. Find the value of $\log_2 3 \cdot \log_3 4 \cdot \log_4 5 \cdot \log_5 6 \cdot \log_6 7 \cdot \log_7 8$.

98. Find the value of $\log_2 4 \cdot \log_4 6 \cdot \log_6 8$.

99. Find the value of $\log_2 3 \cdot \log_3 4 \cdot \cdots \cdot \log_n(n + 1) \cdot \log_{n+1} 2$.

100. Find the value of $\log_2 2 \cdot \log_2 4 \cdot \cdots \cdot \log_2 2^n$.

101. Show that $\log_a(x + \sqrt{x^2 - 1}) + \log_a(x - \sqrt{x^2 - 1}) = 0$.

102. Show that $\log_a(\sqrt{x} + \sqrt{x - 1}) + \log_a(\sqrt{x} - \sqrt{x - 1}) = 0$.

103. Show that $\ln(1 + e^{2x}) = 2x + \ln(1 + e^{-2x})$.

104. Difference Quotient If $f(x) = \log_a x$, show that $\dfrac{f(x + h) - f(x)}{h} = \log_a\left(1 + \dfrac{h}{x}\right)^{1/h}$, $h \neq 0$.

105. If $f(x) = \log_a x$, show that $-f(x) = \log_{1/a} x$.

106. If $f(x) = \log_a x$, show that $f(AB) = f(A) + f(B)$.

107. If $f(x) = \log_a x$, show that $f\left(\dfrac{1}{x}\right) = -f(x)$.

108. If $f(x) = \log_a x$, show that $f(x^\alpha) = \alpha f(x)$.

109. Show that $\log_a\left(\dfrac{M}{N}\right) = \log_a M - \log_a N$, where a, M, and N are positive real numbers and $a \neq 1$.

110. Show that $\log_a\left(\dfrac{1}{N}\right) = -\log_a N$, where a and N are positive real numbers and $a \neq 1$.

Explaining Concepts: Discussion and Writing

111. Graph $Y_1 = \log(x^2)$ and $Y_2 = 2\log(x)$ using a graphing utility. Are they equivalent? What might account for any differences in the two functions?

112. Write an example that illustrates why $(\log_a x)^r \neq r\log_a x$.

113. Write an example that illustrates why
$$\log_2(x + y) \neq \log_2 x + \log_2 y.$$

114. Does $3^{\log_3(-5)} = -5$? Why or why not?

Retain Your Knowledge

Problems 115–118 are based on material learned earlier in the course. The purpose of these problems is to keep the material fresh in your mind so that you are better prepared for the final exam.

115. Use a graphing utility to solve $x^3 - 3x^2 - 4x + 8 = 0$. Round answers to two decimal places.

116. Without solving, determine the character of the solution of the quadratic equation $4x^2 - 28x + 49 = 0$ in the complex number system.

117. Find the real zeros of
$$f(x) = 5x^5 + 44x^4 + 116x^3 + 95x^2 - 4x - 4$$

118. Graph $f(x) = \sqrt{2 - x}$ using the techniques of shifting, compressing or stretching, and reflecting. State the domain and the range of f.

6.6 Logarithmic and Exponential Equations

PREPARING FOR THIS SECTION *Before getting started, review the following:*

- Solving Equations Using a Graphing Utility (Appendix, Section 4, pp. A8–A10)
- Solving Quadratic Equations (Section 1.2, pp. 92–99)
- Solving Equations Quadratic in Form (Section 1.4, pp. 114–116)

 **Now Work** the *'Are You Prepared?'* problems on page 465.

> **OBJECTIVES**   **1** Solve Logarithmic Equations (p. 461)
>   **2** Solve Exponential Equations (p. 463)
>   **3** Solve Logarithmic and Exponential Equations Using a Graphing Utility (p. 464)

1 Solve Logarithmic Equations

In Section 6.4 we solved logarithmic equations by changing a logarithmic expression to an exponential expression. That is, we used the definition of a logarithm:

$$y = \log_a x \quad \text{is equivalent to} \quad x = a^y \quad a > 0 \quad a \neq 1$$

For example, to solve the equation $\log_2(1 - 2x) = 3$, write the logarithmic equation as an equivalent exponential equation $1 - 2x = 2^3$ and solve for x.

$$\log_2(1 - 2x) = 3$$
$$1 - 2x = 2^3 \qquad \textit{Change to exponential form.}$$
$$-2x = 7 \qquad \textit{Simplify.}$$
$$x = -\frac{7}{2} \qquad \textit{Solve.}$$

You should check this solution for yourself.

For most logarithmic equations, some manipulation of the equation (usually using properties of logarithms) is required to obtain a solution. Also, to avoid extraneous solutions with logarithmic equations, determine the domain of the variable first.

Let's begin with an example of a logarithmic equation that requires using the fact that a logarithmic function is a one-to-one function:

> If $\log_a M = \log_a N$, then $M = N$ $M, N,$ and a are positive and $a \neq 1$

EXAMPLE 1   **Solving a Logarithmic Equation**

Solve: $2 \log_5 x = \log_5 9$

Solution   The domain of the variable in this equation is $x > 0$. Note that each logarithm has the same base, 5. Then find the exact solution as follows:

$$2 \log_5 x = \log_5 9$$
$$\log_5 x^2 = \log_5 9 \qquad \textcolor{teal}{r \log_a M = \log_a M^r}$$
$$x^2 = 9 \qquad \textcolor{teal}{\text{If } \log_a M = \log_a N, \text{ then } M = N.}$$
$$x = 3 \quad \text{or} \quad x = -3$$

Recall that the domain of the variable is $x > 0$. Therefore, -3 is extraneous and must be discarded.

✔**Check:** $2 \log_5 3 \overset{2}{=} \log_5 9$

$$\log_5 3^2 \overset{2}{=} \log_5 9 \qquad {\color{teal} r\log_a M = \log_a M^r}$$

$$\log_5 9 = \log_5 9$$

The solution set is $\{3\}$.

Now Work PROBLEM 13

Often one or more properties of logarithms are needed to rewrite the equation as a single logarithm. In the next example, the log of a product property is used.

EXAMPLE 2 | **Solving a Logarithmic Equation**

Solve: $\log_5(x + 6) + \log_5(x + 2) = 1$

Solution The domain of the variable requires that $x + 6 > 0$ and $x + 2 > 0$, so $x > -6$ and $x > -2$. This means any solution must satisfy $x > -2$. To obtain an exact solution, first express the left side as a single logarithm. Then change the equation to an equivalent exponential equation.

$$\log_5(x + 6) + \log_5(x + 2) = 1$$

$$\log_5[(x + 6)(x + 2)] = 1 \qquad {\color{teal} \log_a M + \log_a N = \log_a(MN)}$$

$$(x + 6)(x + 2) = 5^1 = 5 \qquad {\color{teal} \text{Change to exponential form.}}$$

$$x^2 + 8x + 12 = 5 \qquad {\color{teal} \text{Multiply out.}}$$

$$x^2 + 8x + 7 = 0 \qquad {\color{teal} \text{Place the quadratic equation in standard form.}}$$

$$(x + 7)(x + 1) = 0 \qquad {\color{teal} \text{Factor.}}$$

$$x = -7 \quad \text{or} \quad x = -1 \qquad {\color{teal} \text{Zero-Product Property}}$$

WARNING A negative solution is not automatically extraneous. You must determine whether the potential solution causes the argument of any logarithmic expression in the equation to be negative or 0. ■

Only $x = -1$ satisfies the restriction that $x > -2$, so $x = -7$ is extraneous. The solution set is $\{-1\}$, which you should check.

Now Work PROBLEM 21

EXAMPLE 3 | **Solving a Logarithmic Equation**

Solve: $\ln x = \ln(x + 6) - \ln(x - 4)$

Solution The domain of the variable requires that $x > 0$, $x + 6 > 0$, and $x - 4 > 0$. As a result, the domain of the variable here is $x > 4$. Begin the solution using the log of a difference property.

$$\ln x = \ln(x + 6) - \ln(x - 4)$$

$$\ln x = \ln\left(\frac{x + 6}{x - 4}\right) \qquad {\color{teal} \ln M - \ln N = \ln\left(\frac{M}{N}\right)}$$

$$x = \frac{x + 6}{x - 4} \qquad {\color{teal} \text{If } \ln M = \ln N, \text{ then } M = N.}$$

$$x(x - 4) = x + 6 \qquad {\color{teal} \text{Multiply both sides by } x - 4.}$$

$$x^2 - 4x = x + 6 \qquad {\color{teal} \text{Multiply out.}}$$

$$x^2 - 5x - 6 = 0 \qquad {\color{teal} \text{Place the quadratic equation in standard form.}}$$

$$(x - 6)(x + 1) = 0 \qquad {\color{teal} \text{Factor.}}$$

$$x = 6 \quad \text{or} \quad x = -1 \qquad {\color{teal} \text{Zero-Product Property}}$$

Because the domain of the variable is $x > 4$, discard -1 as extraneous. The solution set is $\{6\}$, which you should check.

WARNING In using properties of logarithms to solve logarithmic equations, avoid using the property $\log_a x^r = r \log_a x$, when r is even. The reason can be seen in this example:

Solve: $\log_3 x^2 = 4$

Solution: The domain of the variable x is all real numbers except 0.

(a) $\log_3 x^2 = 4$
$x^2 = 3^4 = 81$
$x = -9 \text{ or } x = 9$

(b) $\log_3 x^2 = 4$
$2 \log_3 x = 4$ $x > 0$
$\log_3 x = 2$
$x = 9$

Both -9 and 9 are solutions of $\log_3 x^2 = 4$ (as you can verify). The solution in part (b) does not find the solution -9 because the domain of the variable was further restricted due to the application of the property $\log_a x^r = r \log_a x$. ∎

━━━ **Now Work** PROBLEM 31

2 Solve Exponential Equations

In Sections 6.3 and 6.4, we solved exponential equations algebraically by expressing each side of the equation using the same base. That is, we used the one-to-one property of the exponential function:

$$\text{If } a^u = a^v, \text{ then } u = v \qquad a > 0 \quad a \neq 1$$

For example, to solve the exponential equation $4^{2x+1} = 16$, notice that $16 = 4^2$ and apply the property above to obtain the equation $2x + 1 = 2$, from which we find $x = \dfrac{1}{2}$.

Not all exponential equations can be readily expressed so that each side of the equation has the same base. For such equations, algebraic techniques often can be used to obtain exact solutions.

EXAMPLE 4

Solving Exponential Equations

Solve: (a) $2^x = 5$ (b) $8 \cdot 3^x = 5$

Solution (a) Because 5 cannot be written as an integer power of 2 ($2^2 = 4$ and $2^3 = 8$), write the exponential equation as the equivalent logarithmic equation.

$$2^x = 5$$
$$x = \log_2 5 = \frac{\ln 5}{\ln 2}$$

↑ *Change-of-Base Formula (10), Section 6.5*

Alternatively, the equation $2^x = 5$ can be solved by taking the natural logarithm (or common logarithm) of each side.

$$2^x = 5$$
$$\ln 2^x = \ln 5 \qquad \text{If } M = N, \text{ then } \ln M = \ln N.$$
$$x \ln 2 = \ln 5 \qquad \ln M^r = r \ln M$$
$$x = \frac{\ln 5}{\ln 2} \qquad \text{Exact solution}$$
$$\approx 2.322 \quad \text{Approximate solution}$$

The solution set is $\left\{ \dfrac{\ln 5}{\ln 2} \right\}$.

(b) $8 \cdot 3^x = 5$
$$3^x = \frac{5}{8} \qquad \text{Solve for } 3^x.$$

$$x = \log_3\left(\frac{5}{8}\right) = \frac{\ln\left(\frac{5}{8}\right)}{\ln 3} \quad \text{\textcolor{blue}{Exact solution}}$$

$$\approx -0.428 \quad \text{\textcolor{blue}{Approximate solution}}$$

The solution set is $\left\{\dfrac{\ln\left(\dfrac{5}{8}\right)}{\ln 3}\right\}.$

Now Work PROBLEM 43

EXAMPLE 5

Solving an Exponential Equation

Solve: $5^{x-2} = 3^{3x+2}$

Solution

Because the bases are different, first apply property (7), Section 6.5 (take the natural logarithm of each side), and then use a property of logarithms. The result is an equation in x that can be solved.

$$5^{x-2} = 3^{3x+2}$$

$$\ln 5^{x-2} = \ln 3^{3x+2} \qquad \text{\textcolor{blue}{If } } M = N, \ln M = \ln N.$$

$$(x-2)\ln 5 = (3x+2)\ln 3 \qquad \text{\textcolor{blue}{$\ln M^r = r\ln M$}}$$

$$(\ln 5)x - 2\ln 5 = (3\ln 3)x + 2\ln 3 \qquad \text{\textcolor{blue}{Distribute.}}$$

$$(\ln 5)x - (3\ln 3)x = 2\ln 3 + 2\ln 5 \qquad \text{\textcolor{blue}{Place terms involving x on the left.}}$$

$$(\ln 5 - 3\ln 3)x = 2(\ln 3 + \ln 5) \qquad \text{\textcolor{blue}{Factor.}}$$

$$x = \frac{2(\ln 3 + \ln 5)}{\ln 5 - 3\ln 3} \qquad \text{\textcolor{blue}{Exact solution}}$$

$$\approx -3.212 \qquad \text{\textcolor{blue}{Approximate solution}}$$

NOTE: Because of the properties of logarithms, exact solutions involving logarithms often can be expressed in multiple ways. For example, the solution to $5^{x-2} = 3^{3x+2}$ from Example 5 can be expressed equivalently as $\dfrac{2\ln 15}{\ln 5 - \ln 27}$ or as $\dfrac{\ln 225}{\ln(5/27)}$, among others. Do you see why? ∎

The solution set is $\left\{\dfrac{2(\ln 3 + \ln 5)}{\ln 5 - 3\ln 3}\right\}.$

Now Work PROBLEM 53

EXAMPLE 6

Solving an Exponential Equation That Is Quadratic in Form

Solve: $4^x - 2^x - 12 = 0$

Solution

Note that $4^x = (2^2)^x = 2^{(2x)} = (2^x)^2$, so the equation is quadratic in form and can be written as

$$(2^x)^2 - 2^x - 12 = 0 \qquad \text{\textcolor{blue}{Let $u = 2^x$; then $u^2 - u - 12 = 0$.}}$$

Now factor as usual.

$$(2^x - 4)(2^x + 3) = 0 \qquad \text{\textcolor{blue}{$(u - 4)(u + 3) = 0$}}$$

$$2^x - 4 = 0 \quad \text{or} \quad 2^x + 3 = 0 \qquad \text{\textcolor{blue}{$u - 4 = 0$ \quad or \quad $u + 3 = 0$}}$$

$$2^x = 4 \qquad\qquad 2^x = -3 \qquad \text{\textcolor{blue}{$u = 2^x = 4$ \qquad $u = 2^x = -3$}}$$

The equation on the left has the solution $x = 2$, since $2^x = 4 = 2^2$; the equation on the right has no solution, since $2^x > 0$ for all x. The only solution is 2. The solution set is $\{2\}$.

Now Work PROBLEM 61

---

 3 Solve Logarithmic and Exponential Equations Using a Graphing Utility

 The algebraic techniques introduced in this section to obtain exact solutions apply only to certain types of logarithmic and exponential equations. Solutions for other types are generally studied in calculus, using numerical methods. For such types, we can use a graphing utility to approximate the solution.

EXAMPLE 7

Solving Equations Using a Graphing Utility

Solve: $x + e^x = 2$

Express the solution(s) rounded to two decimal places.

Solution

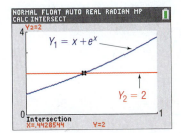

The solution is found by graphing $Y_1 = x + e^x$ and $Y_2 = 2$. Since Y_1 is an increasing function (do you know why?), there is only one point of intersection for Y_1 and Y_2. Figure 40 shows the graphs of Y_1 and Y_2. Using the INTERSECT command reveals that the solution is 0.44, rounded to two decimal places. ●

Now Work PROBLEM 71

Figure 40

6.6 Assess Your Understanding

'Are You Prepared?' *Answers are given at the end of these exercises. If you get a wrong answer, read the pages listed in* red.

1. Solve $x^2 - 7x - 30 = 0$. (pp. 92–99)

2. Solve $(x + 3)^2 - 4(x + 3) + 3 = 0$. (pp. 114–116)

3. Approximate the solution(s) to $x^3 = x^2 - 5$ using a graphing utility. (pp. A8–A10)

4. Approximate the solution(s) to $x^3 - 2x + 2 = 0$ using a graphing utility. (pp. A8–A10)

Skill Building

In Problems 5–40, solve each logarithmic equation. Express irrational solutions in exact form and as a decimal rounded to three decimal places.

5. $\log_4 x = 2$

6. $\log (x + 6) = 1$

7. $\log_2 (5x) = 4$

8. $\log_3 (3x - 1) = 2$

9. $\log_4 (x + 2) = \log_4 8$

10. $\log_5 (2x + 3) = \log_5 3$

11. $\frac{1}{2} \log_3 x = 2 \log_3 2$

12. $-2 \log_4 x = \log_4 9$

13. $3 \log_2 x = -\log_2 27$

14. $2 \log_5 x = 3 \log_5 4$

15. $3 \log_2 (x - 1) + \log_2 4 = 5$

16. $2 \log_3 (x + 4) - \log_3 9 = 2$

17. $\log x + \log (x + 15) = 2$

18. $\log x + \log (x - 21) = 2$

19. $\log (2x + 1) = 1 + \log (x - 2)$

20. $\log (2x) - \log (x - 3) = 1$

21. $\log_2 (x + 7) + \log_2 (x + 8) = 1$

22. $\log_6 (x + 4) + \log_6 (x + 3) = 1$

23. $\log_8 (x + 6) = 1 - \log_8 (x + 4)$

24. $\log_5 (x + 3) = 1 - \log_5 (x - 1)$

25. $\ln x + \ln (x + 2) = 4$

26. $\ln (x + 1) - \ln x = 2$

27. $\log_3 (x + 1) + \log_3 (x + 4) = 2$

28. $\log_2 (x + 1) + \log_2 (x + 7) = 3$

29. $\log_{1/3} (x^2 + x) - \log_{1/3} (x^2 - x) = -1$

30. $\log_4 (x^2 - 9) - \log_4 (x + 3) = 3$

31. $\log_a (x - 1) - \log_a (x + 6) = \log_a (x - 2) - \log_a (x + 3)$

32. $\log_a x + \log_a (x - 2) = \log_a (x + 4)$

33. $2 \log_5 (x - 3) - \log_5 8 = \log_5 2$

34. $\log_3 x - 2 \log_3 5 = \log_3 (x + 1) - 2 \log_3 10$

35. $2 \log_6 (x + 2) = 3 \log_6 2 + \log_6 4$

36. $3(\log_7 x - \log_7 2) = 2 \log_7 4$

37. $2 \log_{13} (x + 2) = \log_{13} (4x + 7)$

38. $\log (x - 1) = \frac{1}{3} \log 2$

39. $(\log_3 x)^2 - 5(\log_3 x) = 6$

40. $\ln x - 3 \sqrt{\ln x} + 2 = 0$

In Problems 41–68, solve each exponential equation. Express irrational solutions in exact form and as a decimal rounded to three decimal places.

41. $2^{x-5} = 8$

42. $5^{-x} = 25$

43. $2^x = 10$

44. $3^x = 14$

45. $8^{-x} = 1.2$

46. $2^{-x} = 1.5$

47. $5(2^{3x}) = 8$

48. $0.3(4^{0.2x}) = 0.2$

49. $3^{1-2x} = 4^x$

50. $2^{x+1} = 5^{1-2x}$

51. $\left(\dfrac{3}{5}\right)^x = 7^{1-x}$

52. $\left(\dfrac{4}{3}\right)^{1-x} = 5^x$

53. $1.2^x = (0.5)^{-x}$

54. $0.3^{1+x} = 1.7^{2x-1}$

55. $\pi^{1-x} = e^x$

56. $e^{x+3} = \pi^x$

57. $2^{2x} + 2^x - 12 = 0$ **58.** $3^{2x} + 3^x - 2 = 0$ **59.** $3^{2x} + 3^{x+1} - 4 = 0$ **60.** $2^{2x} + 2^{x+2} - 12 = 0$

61. $16^x + 4^{x+1} - 3 = 0$ **62.** $9^x - 3^{x+1} + 1 = 0$ **63.** $25^x - 8 \cdot 5^x = -16$ **64.** $36^x - 6 \cdot 6^x = -9$

65. $3 \cdot 4^x + 4 \cdot 2^x + 8 = 0$ **66.** $2 \cdot 49^x + 11 \cdot 7^x + 5 = 0$ **67.** $4^x - 10 \cdot 4^{-x} = 3$ **68.** $3^x - 14 \cdot 3^{-x} = 5$

In Problems 69–82, use a graphing utility to solve each equation. Express your answer rounded to two decimal places.

69. $\log_5(x + 1) - \log_4(x - 2) = 1$ **70.** $\log_2(x - 1) - \log_6(x + 2) = 2$

71. $e^x = -x$ **72.** $e^{2x} = x + 2$ **73.** $e^x = x^2$ **74.** $e^x = x^3$

75. $\ln x = -x$ **76.** $\ln(2x) = -x + 2$ **77.** $\ln x = x^3 - 1$ **78.** $\ln x = -x^2$

79. $e^x + \ln x = 4$ **80.** $e^x - \ln x = 4$ **81.** $e^{-x} = \ln x$ **82.** $e^{-x} = -\ln x$

Mixed Practice

In Problems 83–94, solve each equation. Express irrational solutions in exact form and as a decimal rounded to three decimal places.

83. $\log_2(x + 1) - \log_4 x = 1$
[**Hint:** Change $\log_4 x$ to base 2.]

84. $\log_2(3x + 2) - \log_4 x = 3$

85. $\log_{16} x + \log_4 x + \log_2 x = 7$

86. $\log_9 x + 3\log_3 x = 14$

87. $(\sqrt[3]{2})^{2-x} = 2^{x^2}$

88. $\log_2 x^{\log_2 x} = 4$

89. $\dfrac{e^x + e^{-x}}{2} = 1$
[**Hint:** Multiply each side by e^x.]

90. $\dfrac{e^x + e^{-x}}{2} = 3$

91. $\dfrac{e^x - e^{-x}}{2} = 2$

92. $\dfrac{e^x - e^{-x}}{2} = -2$

93. $\log_5 x + \log_3 x = 1$
[**Hint:** Use the Change-of-Base Formula.]

94. $\log_2 x + \log_6 x = 3$

---

95. $f(x) = \log_2(x + 3)$ and $g(x) = \log_2(3x + 1)$.
(a) Solve $f(x) = 3$. What point is on the graph of f?
(b) Solve $g(x) = 4$. What point is on the graph of g?
(c) Solve $f(x) = g(x)$. Do the graphs of f and g intersect? If so, where?
(d) Solve $(f + g)(x) = 7$.
(e) Solve $(f - g)(x) = 2$.

96. $f(x) = \log_3(x + 5)$ and $g(x) = \log_3(x - 1)$.
(a) Solve $f(x) = 2$. What point is on the graph of f?
(b) Solve $g(x) = 3$. What point is on the graph of g?
(c) Solve $f(x) = g(x)$. Do the graphs of f and g intersect? If so, where?
(d) Solve $(f + g)(x) = 3$.
(e) Solve $(f - g)(x) = 2$.

97. (a) If $f(x) = 3^{x+1}$ and $g(x) = 2^{x+2}$, graph f and g on the same Cartesian plane.
(b) Find the point(s) of intersection of the graphs of f and g by solving $f(x) = g(x)$. Round answers to three decimal places. Label any intersection points on the graph drawn in part (a).
(c) Based on the graph, solve $f(x) > g(x)$.

98. (a) If $f(x) = 5^{x-1}$ and $g(x) = 2^{x+1}$, graph f and g on the same Cartesian plane.
(b) Find the point(s) of intersection of the graphs of f and g by solving $f(x) = g(x)$. Label any intersection points on the graph drawn in part (a).
(c) Based on the graph, solve $f(x) > g(x)$.

99. (a) Graph $f(x) = 3^x$ and $g(x) = 10$ on the same Cartesian plane.
(b) Shade the region bounded by the y-axis, $f(x) = 3^x$, and $g(x) = 10$ on the graph drawn in part (a).
(c) Solve $f(x) = g(x)$ and label the point of intersection on the graph drawn in part (a).

100. (a) Graph $f(x) = 2^x$ and $g(x) = 12$ on the same Cartesian plane.
(b) Shade the region bounded by the y-axis, $f(x) = 2^x$, and $g(x) = 12$ on the graph drawn in part (a).
(c) Solve $f(x) = g(x)$ and label the point of intersection on the graph drawn in part (a).

101. (a) Graph $f(x) = 2^{x+1}$ and $g(x) = 2^{-x+2}$ on the same Cartesian plane.
(b) Shade the region bounded by the y-axis, $f(x) = 2^{x+1}$, and $g(x) = 2^{-x+2}$ on the graph drawn in part (a).
(c) Solve $f(x) = g(x)$ and label the point of intersection on the graph drawn in part (a).

102. (a) Graph $f(x) = 3^{-x+1}$ and $g(x) = 3^{x-2}$ on the same Cartesian plane.
(b) Shade the region bounded by the y-axis, $f(x) = 3^{-x+1}$, and $g(x) = 3^{x-2}$ on the graph drawn in part (a).
(c) Solve $f(x) = g(x)$ and label the point of intersection on the graph drawn in part (a).

103. (a) Graph $f(x) = 2^x - 4$.
(b) Find the zero of f.
(c) Based on the graph, solve $f(x) < 0$.

104. (a) Graph $g(x) = 3^x - 9$.
(b) Find the zero of g.
(c) Based on the graph, solve $g(x) > 0$.

Applications and Extensions

105. A Population Model The resident population of the United States in 2014 was 317 million people and was growing at a rate of 0.7% per year. Assuming that this growth rate continues, the model $P(t) = 317(1.007)^{t-2014}$ represents the population P (in millions of people) in year t.
   (a) According to this model, when will the population of the United States be 400 million people?
   (b) According to this model, when will the population of the United States be 435 million people?

Source: U.S. Census Bureau

106. A Population Model The population of the world in 2014 was 7.14 billion people and was growing at a rate of 1.1% per year. Assuming that this growth rate continues, the model $P(t) = 7.14(1.011)^{t-2014}$ represents the population P (in billions of people) in year t.
   (a) According to this model, when will the population of the world be 9 billion people?

(b) According to this model, when will the population of the world be 12.5 billion people?

Source: U.S. Census Bureau

107. Depreciation The value V of a Chevy Cruze that is t years old can be modeled by $V(t) = 18,700(0.84)^t$.
   (a) According to the model, when will the car be worth $9000?
   (b) According to the model, when will the car be worth $6000?
   (c) According to the model, when will the car be worth $2000?

Source: Kelley Blue Book

108. Depreciation The value V of a Honda Civic LX that is t years old can be modeled by $V(t) = 18,955(0.905)^t$.
   (a) According to the model, when will the car be worth $16,000?
   (b) According to the model, when will the car be worth $10,000?
   (c) According to the model, when will the car be worth $7500?

Source: Kelley Blue Book

Explaining Concepts: Discussion and Writing

109. Fill in the reason for each step in the following two solutions.

Solve: $\log_3(x-1)^2 = 2$

Solution A

$\log_3(x-1)^2 = 2$
$(x-1)^2 = 3^2 = 9$ _____
$(x-1) = \pm 3$ _____
$x - 1 = -3$ or $x - 1 = 3$ _____
$x = -2$ or $x = 4$ _____

Solution B

$\log_3(x-1)^2 = 2$
$2\log_3(x-1) = 2$ _____
$\log_3(x-1) = 1$ _____
$x - 1 = 3^1 = 3$ _____
$x = 4$ _____

Both solutions given in Solution A check. Explain what caused the solution $x = -2$ to be lost in Solution B.

Retain Your Knowledge

Problems 110–113 are based on material learned earlier in the course. The purpose of these problems is to keep the material fresh in your mind so that you are better prepared for the final exam.

110. Solve: $4x^3 + 3x^2 - 25x + 6 = 0$

111. Determine whether the function $\{(0, -4), (2, -2), (4, 0), (6, 2)\}$ is one-to-one.

112. For $f(x) = \dfrac{x}{x-2}$ and $g(x) = \dfrac{x+5}{x-3}$, find $f \circ g$. Then find the domain of $f \circ g$.

113. Find the domain of $f(x) = \sqrt{x+3} + \sqrt{x-1}$.

'Are You Prepared?' Answers

1. $\{-3, 10\}$ **2.** $\{-2, 0\}$ **3.** $\{-1.43\}$ **4.** $\{-1.77\}$

6.7 Financial Models

PREPARING FOR THIS SECTION *Before getting started, review the following:*

• Simple Interest (Section 1.7, p. 136)

Now Work the '**Are You Prepared?**' *problems on page 474.*

OBJECTIVES **1** Determine the Future Value of a Lump Sum of Money (p. 468)
  2 Calculate Effective Rates of Return (p. 471)
  3 Determine the Present Value of a Lump Sum of Money (p. 472)
  4 Determine the Rate of Interest or the Time Required to Double a Lump Sum of Money (p. 473)

1 Determine the Future Value of a Lump Sum of Money

Interest is money paid for the use of money. The total amount borrowed (whether by an individual from a bank in the form of a loan or by a bank from an individual in the form of a savings account) is called the **principal**. The **rate of interest**, expressed as a percent, is the amount charged for the use of the principal for a given period of time, usually on a yearly (that is, per annum) basis.

THEOREM

> **Simple Interest Formula**
>
> If a principal of P dollars is borrowed for a period of t years at a per annum interest rate r, expressed as a decimal, the interest I charged is
>
> $$I = Prt \qquad \qquad (1)$$

Interest charged according to formula (1) is called **simple interest**.

In problems involving interest, the term **payment period** is defined as follows.

Annually:	Once per year	**Monthly:**	12 times per year
Semiannually:	Twice per year	**Daily:**	365 times per year*
Quarterly:	Four times per year		

When the interest due at the end of a payment period is added to the principal so that the interest computed at the end of the next payment period is based on this new principal amount (old principal + interest), the interest is said to have been **compounded**. **Compound interest** is interest paid on the principal and on previously earned interest.

EXAMPLE 1   **Computing Compound Interest**

A credit union pays interest of 2% per annum compounded quarterly on a certain savings plan. If $1000 is deposited in such a plan and the interest is left to accumulate, how much is in the account after 1 year?

Solution   Use the simple interest formula, $I = Prt$. The principal P is $1000 and the rate of interest is 2% $= 0.02$. After the first quarter of a year, the time t is $\frac{1}{4}$ year, so the interest earned is

$$I = Prt = (\$1000)(0.02)\left(\frac{1}{4}\right) = \$5$$

* Most banks use a 360-day "year." Why do you think they do?

The new principal is $P + I = \$1000 + \$5 = \$1005$. At the end of the second quarter, the interest on this principal is

$$I = (\$1005)(0.02)\left(\frac{1}{4}\right) = \$5.03$$

At the end of the third quarter, the interest on the new principal of $\$1005 + \$5.03 = \$1010.03$ is

$$I = (\$1010.03)(0.02)\left(\frac{1}{4}\right) = \$5.05$$

Finally, after the fourth quarter, the interest is

$$I = (\$1015.08)(0.02)\left(\frac{1}{4}\right) = \$5.08$$

After 1 year the account contains $\$1015.08 + \$5.08 = \$1020.16$. ●

The pattern of the calculations performed in Example 1 leads to a general formula for compound interest. For this purpose, let P represent the principal to be invested at a per annum interest rate r that is compounded n times per year, so the time of each compounding period is $\frac{1}{n}$ years. (For computing purposes, r is expressed as a decimal.) The interest earned after each compounding period is given by formula (1).

$$\text{Interest} = \text{principal} \times \text{rate} \times \text{time} = P \cdot r \cdot \frac{1}{n} = P \cdot \left(\frac{r}{n}\right)$$

The amount A after one compounding period is

$$A = P + P \cdot \left(\frac{r}{n}\right) = P \cdot \left(1 + \frac{r}{n}\right)$$

After two compounding periods, the amount A, based on the new principal $P \cdot \left(1 + \frac{r}{n}\right)$, is

$$A = \underbrace{P \cdot \left(1 + \frac{r}{n}\right)}_{\text{New principal}} + \underbrace{P \cdot \left(1 + \frac{r}{n}\right)\left(\frac{r}{n}\right)}_{\text{Interest on new principal}} = P \cdot \left(1 + \frac{r}{n}\right) \cdot \left(1 + \frac{r}{n}\right) = P \cdot \left(1 + \frac{r}{n}\right)^2$$

Factor out $P \cdot \left(1 + \frac{r}{n}\right)$

After three compounding periods, the amount A is

$$A = P \cdot \left(1 + \frac{r}{n}\right)^2 + P \cdot \left(1 + \frac{r}{n}\right)^2\left(\frac{r}{n}\right) = P \cdot \left(1 + \frac{r}{n}\right)^2 \cdot \left(1 + \frac{r}{n}\right) = P \cdot \left(1 + \frac{r}{n}\right)^3$$

Continuing this way, after n compounding periods (1 year), the amount A is

$$A = P \cdot \left(1 + \frac{r}{n}\right)^n$$

Because t years will contain $n \cdot t$ compounding periods, the amount after t years is

$$A = P \cdot \left(1 + \frac{r}{n}\right)^{nt}$$

THEOREM

Compound Interest Formula

The amount A after t years due to a principal P invested at an annual interest rate r, expressed as a decimal, compounded n times per year is

$$A = P \cdot \left(1 + \frac{r}{n}\right)^{nt} \tag{2}$$

To observe the effects of compounding interest monthly on an initial deposit of $1,

graph $Y_1 = \left(1 + \dfrac{r}{12}\right)^{12x}$ with $r = 0.06$

and $r = 0.12$ for $0 \le x \le 30$. What is the future value of $1 in 30 years when the interest rate per annum is $r = 0.06$ (6%)? What is the future value of $1 in 30 years when the interest rate per annum is $r = 0.12$ (12%)? Does doubling the interest rate double the future value?

For example, to rework Example 1, use $P = \$1000$, $r = 0.02$, $n = 4$ (quarterly compounding), and $t = 1$ year to obtain

$$A = P \cdot \left(1 + \frac{r}{n}\right)^{nt} = 1000\left(1 + \frac{0.02}{4}\right)^{4\cdot 1} = \$1020.16$$

In equation (2), the amount A is typically referred to as the **future value** of the account, and P is called the **present value.**

✏️ **Now Work** PROBLEM 7

EXAMPLE 2

Comparing Investments Using Different Compounding Periods

Investing $1000 at an annual rate of 10% compounded annually, semiannually, quarterly, monthly, and daily will yield the following amounts after 1 year:

Annual compounding ($n = 1$): $A = P \cdot (1 + r)$

$$= (\$1000)(1 + 0.10) = \$1100.00$$

Semiannual compounding ($n = 2$): $A = P \cdot \left(1 + \dfrac{r}{2}\right)^2$

$$= (\$1000)(1 + 0.05)^2 = \$1102.50$$

Quarterly compounding ($n = 4$): $A = P \cdot \left(1 + \dfrac{r}{4}\right)^4$

$$= (\$1000)(1 + 0.025)^4 = \$1103.81$$

Monthly compounding ($n = 12$): $A = P \cdot \left(1 + \dfrac{r}{12}\right)^{12}$

$$= (\$1000)\left(1 + \frac{0.10}{12}\right)^{12} = \$1104.71$$

Daily compounding ($n = 365$): $A = P \cdot \left(1 + \dfrac{r}{365}\right)^{365}$

$$= (\$1000)\left(1 + \frac{0.10}{365}\right)^{365} = \$1105.16$$

From Example 2, note that the effect of compounding more frequently is that the amount after 1 year is higher: $1000 compounded 4 times a year at 10% results in $1103.81, $1000 compounded 12 times a year at 10% results in $1104.71, and $1000 compounded 365 times a year at 10% results in $1105.16. This leads to the following question: What would happen to the amount after 1 year if the number of times that the interest is compounded were increased without bound?

Let's find the answer. Suppose that P is the principal, r is the per annum interest rate, and n is the number of times that the interest is compounded each year. The amount A after 1 year is

$$A = P \cdot \left(1 + \frac{r}{n}\right)^n$$

Rewrite this expression as follows:

$$A = P \cdot \left(1 + \frac{r}{n}\right)^n = P \cdot \left(1 + \frac{1}{\frac{n}{r}}\right)^n = P \cdot \left[\left(1 + \frac{1}{\frac{n}{r}}\right)^{n/r}\right]^r = P \cdot \left[\left(1 + \frac{1}{h}\right)^h\right]^r \quad (3)$$

$$\uparrow$$
$$h = \frac{n}{r}$$

Now suppose that the number n of times that the interest is compounded per year gets larger and larger; that is, suppose that $n \to \infty$. Then $h = \dfrac{n}{r} \to \infty$, and the expression in brackets in equation (3) equals e. That is, $A \to Pe^r$.

Table 8 compares $\left(1 + \dfrac{r}{n}\right)^n$, for large values of n, to e^r for $r = 0.05$, $r = 0.10$, $r = 0.15$, and $r = 1$. The larger that n gets, the closer $\left(1 + \dfrac{r}{n}\right)^n$ gets to e^r. No matter how frequent the compounding, the amount after 1 year has the definite ceiling Pe^r.

Table 8

	$\left(1 + \frac{r}{n}\right)^n$			
	$n = 100$	$n = 1000$	$n = 10{,}000$	e^r
$r = 0.05$	1.0512580	1.0512698	1.051271	1.0512711
$r = 0.10$	1.1051157	1.1051654	1.1051704	1.1051709
$r = 0.15$	1.1617037	1.1618212	1.1618329	1.1618342
$r = 1$	2.7048138	2.7169239	2.7181459	2.7182818

When interest is compounded so that the amount after 1 year is Pe^r, the interest is said to be **compounded continuously**.

THEOREM

Continuous Compounding

The amount A after t years due to a principal P invested at an annual interest rate r compounded continuously is

$$A = Pe^{rt} \tag{4}$$

EXAMPLE 3

Using Continuous Compounding

The amount A that results from investing a principal P of $1000 at an annual rate r of 10% compounded continuously for a time t of 1 year is

$$A = \$1000 e^{0.10} = (\$1000)(1.10517) = \$1105.17$$

Now Work PROBLEM 13

2 Calculate Effective Rates of Return

Suppose that you have $1000 and a bank offers to pay you 3% annual interest on a savings account with interest compounded monthly. What annual interest rate must be earned for you to have the same amount at the end of the year as if the interest had been compounded annually (once per year)? To answer this question, first determine the value of the $1000 in the account that earns 3% compounded monthly.

$$A = \$1000 \left(1 + \frac{0.03}{12}\right)^{12} \quad \text{Use } A = P\left(1 + \frac{r}{n}\right)^n \text{ with } P = \$1000, r = 0.03, n = 12.$$

$$= \$1030.42$$

So the interest earned is $30.42. Using $I = Prt$ with $t = 1, I = \$30.42$, and $P = \$1000$, the annual simple interest rate is $0.03042 = 3.042\%$. This interest rate is known as the *effective rate of interest*.

The **effective rate of interest** is the annual simple interest rate that would yield the same amount as compounding n times per year, or continuously, after 1 year.

THEOREM

> **Effective Rate of Interest**
>
> The effective rate of interest r_e of an investment earning an annual interest rate r is given by
>
> Compounding n times per year: $r_e = \left(1 + \dfrac{r}{n}\right)^n - 1$
>
> Continuous compounding: $r_e = e^r - 1$

EXAMPLE 4

Computing the Effective Rate of Interest—Which Is the Best Deal?

Suppose you want to buy a 5-year certificate of deposit (CD). You visit three banks to determine their CD rates. American Express offers you 2.15% annual interest compounded monthly, and First Internet Bank offers you 2.20% compounded quarterly. Discover offers 2.12% compounded daily. Determine which bank is offering the best deal.

Solution

The bank that offers the best deal is the one with the highest effective interest rate.

American Express	**First Internet Bank**	**Discover**
$r_e = \left(1 + \dfrac{0.0215}{12}\right)^{12} - 1$	$r_e = \left(1 + \dfrac{0.022}{4}\right)^{4} - 1$	$r_e = \left(1 + \dfrac{0.0212}{365}\right)^{365} - 1$
$\approx 1.02171 - 1$	$\approx 1.02218 - 1$	$\approx 1.02143 - 1$
$= 0.02171$	$= 0.02218$	$= 0.02143$
$= 2.171\%$	$= 2.218\%$	$= 2.143\%$

The effective rate of interest is highest for First Internet Bank, so First Internet Bank is offering the best deal. ●

 Now Work PROBLEM 23

3 Determine the Present Value of a Lump Sum of Money

When people in finance speak of the "time value of money," they are usually referring to the *present value* of money. The **present value** of A dollars to be received at a future date is the principal that you would need to invest now so that it will grow to A dollars in the specified time period. The present value of money to be received at a future date is always less than the amount to be received, since the amount to be received will equal the present value (money invested now) *plus* the interest accrued over the time period.

The compound interest formula (2) is used to develop a formula for present value. If P is the present value of A dollars to be received after t years at a per annum interest rate r compounded n times per year, then, by formula (2),

$$A = P \cdot \left(1 + \frac{r}{n}\right)^{nt}$$

To solve for P, divide both sides by $\left(1 + \dfrac{r}{n}\right)^{nt}$. The result is

$$\frac{A}{\left(1 + \dfrac{r}{n}\right)^{nt}} = P \quad \text{or} \quad P = A \cdot \left(1 + \frac{r}{n}\right)^{-nt}$$

THEOREM

Present Value Formulas

The present value P of A dollars to be received after t years, assuming a per annum interest rate r compounded n times per year, is

$$P = A \cdot \left(1 + \frac{r}{n}\right)^{-nt} \qquad (5)$$

If the interest is compounded continuously, then

$$P = Ae^{-rt} \qquad (6)$$

To derive (6), solve formula (4) for P.

EXAMPLE 5

Computing the Value of a Zero-Coupon Bond

A zero-coupon (noninterest-bearing) bond can be redeemed in 10 years for $1000. How much should you be willing to pay for it now if you want a return of

(a) 8% compounded monthly?   (b) 7% compounded continuously?

Solution (a) To find the present value of $1000, use formula (5) with $A = \$1000$, $n = 12, r = 0.08$, and $t = 10$.

$$P = A \cdot \left(1 + \frac{r}{n}\right)^{-nt} = \$1000\left(1 + \frac{0.08}{12}\right)^{-12(10)} = \$450.52$$

For a return of 8% compounded monthly, pay $450.52 for the bond.

(b) Here use formula (6) with $A = \$1000$, $r = 0.07$, and $t = 10$.

$$P = Ae^{-rt} = \$1000e^{-(0.07)(10)} = \$496.59$$

For a return of 7% compounded continuously, pay $496.59 for the bond.   ●

 Now Work PROBLEM 15

---

4 Determine the Rate of Interest or the Time Required to Double a Lump Sum of Money

---

EXAMPLE 6

Rate of Interest Required to Double an Investment

What annual rate of interest compounded annually is needed in order to double an investment in 5 years?

Solution If P is the principal and P is to double, then the amount A will be $2P$. Use the compound interest formula with $n = 1$ and $t = 5$ to find r.

$$A = P \cdot \left(1 + \frac{r}{n}\right)^{nt}$$

$$2P = P \cdot (1 + r)^5 \qquad \textcolor{blue}{A = 2P, n = 1, t = 5}$$

$$2 = (1 + r)^5 \qquad \textcolor{blue}{\text{Divide both sides by } P.}$$

$$1 + r = \sqrt[5]{2} \qquad \textcolor{blue}{\text{Take the fifth root of each side.}}$$

$$r = \sqrt[5]{2} - 1 \approx 1.148698 - 1 = 0.148698$$

The annual rate of interest needed to double the principal in 5 years is 14.87%.   ●

Now Work PROBLEM 31

| EXAMPLE 7 | **Time Required to Double or Triple an Investment** |

(a) How long will it take for an investment to double in value if it earns 5% compounded continuously?

(b) How long will it take to triple at this rate?

Solution
(a) If P is the initial investment and P is to double, then the amount A will be $2P$. Use formula (4) for continuously compounded interest with $r = 0.05$.

$$A = Pe^{rt}$$
$$2P = Pe^{0.05t} \qquad \text{\textcolor{blue}{$A = 2P, r = 0.05$}}$$
$$2 = e^{0.05t} \qquad \text{\textcolor{blue}{Divide out the P's.}}$$
$$0.05t = \ln 2 \qquad \text{\textcolor{blue}{Rewrite as a logarithm.}}$$
$$t = \frac{\ln 2}{0.05} \approx 13.86 \quad \text{\textcolor{blue}{Solve for t.}}$$

It will take about 14 years to double the investment.

(b) To triple the investment, let $A = 3P$ in formula (4).

$$A = Pe^{rt}$$
$$3P = Pe^{0.05t} \qquad \text{\textcolor{blue}{$A = 3P, r = 0.05$}}$$
$$3 = e^{0.05t} \qquad \text{\textcolor{blue}{Divide out the P's.}}$$
$$0.05t = \ln 3 \qquad \text{\textcolor{blue}{Rewrite as a logarithm.}}$$
$$t = \frac{\ln 3}{0.05} \approx 21.97 \quad \text{\textcolor{blue}{Solve for t.}}$$

It will take about 22 years to triple the investment.

● **Now Work** PROBLEM 35

6.7 Assess Your Understanding

'Are You Prepared?' *Answers are given at the end of these exercises. If you get a wrong answer, read the pages listed in* red.

1. What is the interest due when $500 is borrowed for 6 months at a simple interest rate of 6% per annum? (p. 136)

2. If you borrow $5000 and, after 9 months, pay off the loan in the amount of $5500, what per annum rate of interest was charged? (p. 136)

Concepts and Vocabulary

3. The total amount borrowed (whether by an individual from a bank in the form of a loan or by a bank from an individual in the form of a savings account) is called the _____

4. If a principal of P dollars is borrowed for a period of t years at a per annum interest rate r, expressed as a decimal, the interest I charged is ____ = _____ . Interest charged according to this formula is called _____ _____ .

5. In problems involving interest, if the payment period of the interest is quarterly, then interest is paid _____ times per year.

6. The _____ ____ __ _____ is the annual simple interest rate that would yield the same amount as compounding n times per year, or continuously, after 1 year.

Skill Building

In Problems 7–14, find the amount that results from each investment.

7. $100 invested at 4% compounded quarterly after a period of 2 years

8. $50 invested at 6% compounded monthly after a period of 3 years

9. $500 invested at 8% compounded quarterly after a period of $2\frac{1}{2}$ years

10. $300 invested at 12% compounded monthly after a period of $1\frac{1}{2}$ years

11. $600 invested at 5% compounded daily after a period of 3 years

12. $700 invested at 6% compounded daily after a period of 2 years

13. $1000 invested at 11% compounded continuously after a period of 2 years

14. $400 invested at 7% compounded continuously after a period of 3 years

In Problems 15–22, find the principal needed now to get each amount; that is, find the present value.

15. To get $100 after 2 years at 6% compounded monthly

16. To get $75 after 3 years at 8% compounded quarterly

17. To get $1000 after $2\frac{1}{2}$ years at 6% compounded daily

18. To get $800 after $3\frac{1}{2}$ years at 7% compounded monthly

19. To get $600 after 2 years at 4% compounded quarterly

20. To get $300 after 4 years at 3% compounded daily

21. To get $80 after $3\frac{1}{4}$ years at 9% compounded continuously

22. To get $800 after $2\frac{1}{2}$ years at 8% compounded continuously

In Problems 23–26, find the effective rate of interest.

23. For 5% compounded quarterly

24. For 6% compounded monthly

25. For 5% compounded continuously

26. For 6% compounded continuously

In Problems 27–30, determine the rate that represents the better deal.

27. 6% compounded quarterly or $6\frac{1}{4}$% compounded annually

28. 9% compounded quarterly or $9\frac{1}{4}$% compounded annually

29. 9% compounded monthly or 8.8% compounded daily

30. 8% compounded semiannually or 7.9% compounded daily

31. What rate of interest compounded annually is required to double an investment in 3 years?

32. What rate of interest compounded annually is required to double an investment in 6 years?

33. What rate of interest compounded annually is required to triple an investment in 5 years?

34. What rate of interest compounded annually is required to triple an investment in 10 years?

35. (a) How long does it take for an investment to double in value if it is invested at 8% compounded monthly?
    (b) How long does it take if the interest is compounded continuously?

36. (a) How long does it take for an investment to triple in value if it is invested at 6% compounded monthly?
    (b) How long does it take if the interest is compounded continuously?

37. What rate of interest compounded quarterly will yield an effective interest rate of 7%?

38. What rate of interest compounded continuously will yield an effective interest rate of 6%?

Applications and Extensions

39. **Time Required to Reach a Goal** If Tanisha has $100 to invest at 4% per annum compounded monthly, how long will it be before she has $150? If the compounding is continuous, how long will it be?

40. **Time Required to Reach a Goal** If Angela has $100 to invest at 2.5% per annum compounded monthly, how long will it be before she has $175? If the compounding is continuous, how long will it be?

41. **Time Required to Reach a Goal** How many years will it take for an initial investment of $10,000 to grow to $25,000? Assume a rate of interest of 6% compounded continuously.

42. **Time Required to Reach a Goal** How many years will it take for an initial investment of $25,000 to grow to $80,000? Assume a rate of interest of 7% compounded continuously.

43. **Price Appreciation of Homes** What will a $90,000 condominium cost 5 years from now if the price appreciation for condos over that period averages 3% compounded annually?

44. **Credit Card Interest** A department store charges 1.25% per month on the unpaid balance for customers with charge accounts (interest is compounded monthly). A customer charges $200 and does not pay her bill for 6 months. What is the bill at that time?

45. **Saving for a Car** Jerome will be buying a used car for $15,000 in 3 years. How much money should he ask his parents for now so that, if he invests it at 5% compounded continuously, he will have enough to buy the car?

46. **Paying off a Loan** John requires $3000 in 6 months to pay off a loan that has no prepayment privileges. If he has the $3000 now, how much of it should he save in an account paying 3% compounded monthly so that in 6 months he will have exactly $3000?

47. **Return on a Stock** George contemplates the purchase of 100 shares of a stock selling for $15 per share. The stock pays no dividends. The history of the stock indicates that it should grow at an annual rate of 15% per year. How much should the 100 shares of stock be worth in 5 years?

48. **Return on an Investment** A business purchased for $650,000 in 2010 is sold in 2013 for $850,000. What is the annual rate of return for this investment?

49. **Comparing Savings Plans** Jim places $1000 in a bank account that pays 5.6% compounded continuously. After 1 year, will he have enough money to buy a computer system that costs $1060? If another bank will pay Jim 5.9% compounded monthly, is this a better deal?

50. **Savings Plans** On January 1, Kim places $1000 in a certificate of deposit that pays 6.8% compounded continuously and matures in 3 months. Then Kim places the $1000 and the interest in a passbook account that pays 5.25% compounded monthly. How much does Kim have in the passbook account on May 1?

(b) To find the growth rate k, note that the number of cells doubles in 3 hours, so

$$N(3) = 2N_0$$

But $N(3) = N_0 e^{k(3)}$, so

$$N_0 e^{k(3)} = 2N_0$$

$$e^{3k} = 2 \qquad \text{\color{teal}Divide both sides by } N_0.$$

$$3k = \ln 2 \qquad \text{\color{teal}Write the exponential equation as a logarithm.}$$

$$k = \frac{1}{3}\ln 2 \approx 0.23105$$

The function that models this growth process is therefore

$$N(t) = N_0\, e^{0.23105t}$$

(c) The time t needed for the size of the colony to triple requires that $N = 3N_0$. Substitute $3N_0$ for N to get

$$3N_0 = N_0\, e^{0.23105t}$$

$$3 = e^{0.23105t}$$

$$0.23105t = \ln 3$$

$$t = \frac{\ln 3}{0.23105} \approx 4.755 \text{ hours}$$

It will take about 4.755 hours, or 4 hours and 45 minutes, for the size of the colony to triple.

(d) If a population doubles in 3 hours, it will double a second time in 3 more hours, for a total time of 6 hours. ●

2 Find Equations of Populations That Obey the Law of Decay

Radioactive materials follow the law of uninhibited decay.

> **Uninhibited Radioactive Decay**
> The amount A of a radioactive material present at time t is given by
>
> $$A(t) = A_0 e^{kt} \qquad k < 0 \tag{3}$$
>
> where A_0 is the original amount of radioactive material and k is a negative number that represents the rate of decay.

All radioactive substances have a specific **half-life,** which is the time required for half of the radioactive substance to decay. **Carbon dating** uses the fact that all living organisms contain two kinds of carbon, carbon-12 (a stable carbon) and carbon-14 (a radioactive carbon with a half-life of 5730 years). While an organism is living, the ratio of carbon-12 to carbon-14 is constant. But when an organism dies, the original amount of carbon-12 present remains unchanged, whereas the amount of carbon-14 begins to decrease. This change in the amount of carbon-14 present relative to the amount of carbon-12 present makes it possible to calculate when the organism died.

EXAMPLE 3 **Estimating the Age of Ancient Tools**

Traces of burned wood along with ancient stone tools in an archeological dig in Chile were found to contain approximately 1.67% of the original amount of carbon-14. If the half-life of carbon-14 is 5730 years, approximately when was the tree cut and burned?

Solution Using formula (3), the amount A of carbon-14 present at time t is

$$A(t) = A_0 e^{kt}$$

where A_0 is the original amount of carbon-14 present and k is a negative number. We first seek the number k. To find it, we use the fact that after 5730 years, half of the original amount of carbon-14 remains, so $A(5730) = \frac{1}{2}A_0$. Then

$$\frac{1}{2}A_0 = A_0 e^{k(5730)}$$

$$\frac{1}{2} = e^{5730k} \qquad \qquad \text{Divide both sides of the equation by } A_0.$$

$$5730k = \ln\frac{1}{2} \qquad \qquad \text{Rewrite as a logarithm.}$$

$$k = \frac{1}{5730}\ln\frac{1}{2} \approx -0.000120968$$

Formula (3) therefore becomes

$$A(t) = A_0\, e^{-0.000120968t}$$

If the amount A of carbon-14 now present is 1.67% of the original amount, it follows that

$$0.0167A_0 = A_0\, e^{-0.000120968t}$$

$$0.0167 = e^{-0.000120968t} \qquad \qquad \text{Divide both sides of the equation by } A_0.$$

$$-0.000120968t = \ln 0.0167 \qquad \qquad \text{Rewrite as a logarithm.}$$

$$t = \frac{\ln 0.0167}{-0.000120968} \approx 33{,}830 \text{ years}$$

The tree was cut and burned about 33,830 years ago. Some archeologists use this conclusion to argue that humans lived in the Americas nearly 34,000 years ago, much earlier than is generally accepted. ●

➤ **Now Work** PROBLEM 3

3 Use Newton's Law of Cooling

Newton's Law of Cooling* states that the temperature of a heated object decreases exponentially over time toward the temperature of the surrounding medium.

> **Newton's Law of Cooling**
> The temperature u of a heated object at a given time t can be modeled by the following function:
>
> $$u(t) = T + (u_0 - T)e^{kt} \qquad k < 0 \qquad \qquad \textbf{(4)}$$
>
> where T is the constant temperature of the surrounding medium, u_0 is the initial temperature of the heated object, and k is a negative constant.

EXAMPLE 4 **Using Newton's Law of Cooling**

An object is heated to 100°C (degrees Celsius) and is then allowed to cool in a room whose air temperature is 30°C.

(a) If the temperature of the object is 80°C after 5 minutes, when will its temperature be 50°C?

(b) Determine the elapsed time before the temperature of the object is 35°C.

(c) What do you notice about the temperature as time passes?

*Named after Sir Isaac Newton (1643–1727), one of the cofounders of calculus.

Solution (a) Using formula (4) with $T = 30$ and $u_0 = 100$, the temperature $u(t)$ (in degrees Celsius) of the object at time t (in minutes) is

$$u(t) = 30 + (100 - 30)e^{kt} = 30 + 70e^{kt} \qquad \textbf{(5)}$$

where k is a negative constant. To find k, use the fact that $u = 80$ when $t = 5$. Then

$$u(t) = 30 + 70e^{kt}$$

$$80 = 30 + 70e^{k(5)} \qquad \color{blue}{u(5) = 80}$$

$$50 = 70e^{5k} \qquad \color{blue}{\text{Simplify.}}$$

$$e^{5k} = \frac{50}{70} \qquad \color{blue}{\text{Solve for } e^{5k}.}$$

$$5k = \ln\frac{5}{7} \qquad \color{blue}{\text{Rewrite as a logarithm.}}$$

$$k = \frac{1}{5}\ln\frac{5}{7} \approx -0.0673 \qquad \color{blue}{\text{Solve for } k.}$$

Formula (5) therefore becomes

$$u(t) = 30 + 70e^{-0.0673t} \qquad \textbf{(6)}$$

To find t when $u = 50°\text{C}$, solve the equation

$$50 = 30 + 70e^{-0.0673t}$$

$$20 = 70e^{-0.0673t} \qquad \color{blue}{\text{Simplify.}}$$

$$e^{-0.0673t} = \frac{20}{70}$$

$$-0.0673t = \ln\frac{2}{7} \qquad \color{blue}{\text{Rewrite as a logarithm.}}$$

$$t = \frac{\ln\dfrac{2}{7}}{-0.0673} \approx 18.6 \text{ minutes} \qquad \color{blue}{\text{Solve for } t.}$$

The temperature of the object will be $50°\text{C}$ after about 18.6 minutes, or 18 minutes, 36 seconds.

(b) Use equation (6) to find t when $u = 35°\text{C}$.

$$35 = 30 + 70e^{-0.0673t}$$

$$5 = 70e^{-0.0673t} \qquad \color{blue}{\text{Simplify.}}$$

$$e^{-0.0673t} = \frac{5}{70}$$

$$-0.0673t = \ln\frac{5}{70} \qquad \color{blue}{\text{Rewrite as a logarithm.}}$$

$$t = \frac{\ln\dfrac{5}{70}}{-0.0673} \approx 39.2 \text{ minutes} \qquad \color{blue}{\text{Solve for } t.}$$

The object will reach a temperature of $35°\text{C}$ after about 39.2 minutes.

(c) Look at equation (6). As t increases, the exponent $-0.0673t$ becomes unbounded in the negative direction. As a result, the value of $e^{-0.0673t}$ approaches zero, so the value of u, the temperature of the object, approaches $30°\text{C}$, the air temperature of the room. ●

Now Work PROBLEM 13

4 Use Logistic Models

The exponential growth model $A(t) = A_0 e^{kt}$, $k > 0$, assumes uninhibited growth, meaning that the value of the function grows without limit. Recall that cell division could be modeled using this function, assuming that no cells die and no by-products are produced. However, cell division eventually is limited by factors such as living space and food supply. The **logistic model**, given next, can describe situations where the growth or decay of the dependent variable is limited.

Logistic Model

In a logistic model, the population P after time t is given by the function

$$P(t) = \frac{c}{1 + ae^{-bt}} \tag{7}$$

where a, b, and c are constants with $a > 0$ and $c > 0$. The model is a growth model if $b > 0$; the model is a decay model if $b < 0$.

The number c is called the **carrying capacity** (for growth models) because the value $P(t)$ approaches c as t approaches infinity; that is, $\lim\limits_{t\to\infty} P(t) = c$. The number $|b|$ is the growth rate for $b > 0$ and the decay rate for $b < 0$. Figure 42(a) shows the graph of a typical logistic growth function, and Figure 42(b) shows the graph of a typical logistic decay function.

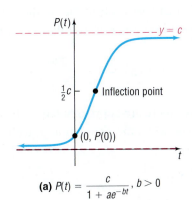

Figure 42

(a) $P(t) = \dfrac{c}{1 + ae^{-bt}}$, $b > 0$
Logistic growth

(b) $P(t) = \dfrac{c}{1 + ae^{-bt}}$, $b < 0$
Logistic decay

Based on the figures, the following properties of logistic functions emerge.

Properties of the Logistic Model, Equation (7)

1. The domain is the set of all real numbers. The range is the interval $(0, c)$, where c is the carrying capacity.

2. There are no x-intercepts; the y-intercept is $P(0)$.

3. There are two horizontal asymptotes: $y = 0$ and $y = c$.

4. $P(t)$ is an increasing function if $b > 0$ and a decreasing function if $b < 0$.

5. There is an **inflection point** where $P(t)$ equals $\dfrac{1}{2}$ of the carrying capacity.

   The inflection point is the point on the graph where the graph changes from being curved upward to being curved downward for growth functions, and the point where the graph changes from being curved downward to being curved upward for decay functions.

6. The graph is smooth and continuous, with no corners or gaps.

EXAMPLE 5

Fruit Fly Population

Fruit flies are placed in a half-pint milk bottle with a banana (for food) and yeast plants (for food and to provide a stimulus to lay eggs). Suppose that the fruit fly population after t days is given by

$$P(t) = \frac{230}{1 + 56.5e^{-0.37t}}$$

(a) State the carrying capacity and the growth rate.
(b) Determine the initial population.
(c) What is the population after 5 days?
(d) How long does it take for the population to reach 180?
(e) Use a graphing utility to determine how long it takes for the population to reach one-half of the carrying capacity.

Solution

(a) As $t \to \infty$, $e^{-0.37t} \to 0$ and $P(t) \to \dfrac{230}{1}$. The carrying capacity of the half-pint bottle is 230 fruit flies. The growth rate is $|b| = |0.37| = 37\%$ per day.

(b) To find the initial number of fruit flies in the half-pint bottle, evaluate $P(0)$.

$$P(0) = \frac{230}{1 + 56.5e^{-0.37(0)}} = \frac{230}{1 + 56.5} = 4$$

So, initially, there were 4 fruit flies in the half-pint bottle.

(c) After 5 days the number of fruit flies in the half-pint bottle is

$$P(5) = \frac{230}{1 + 56.5e^{-0.37(5)}} \approx 23 \text{ fruit flies}$$

After 5 days, there are approximately 23 fruit flies in the bottle.

(d) To determine when the population of fruit flies will be 180, solve the equation $P(t) = 180$.

$$\frac{230}{1 + 56.5e^{-0.37t}} = 180$$

$$230 = 180(1 + 56.5e^{-0.37t})$$

$1.2778 = 1 + 56.5e^{-0.37t}$ Divide both sides by 180.

$0.2778 = 56.5e^{-0.37t}$ Subtract 1 from both sides.

$0.0049 = e^{-0.37t}$ Divide both sides by 56.5.

$\ln(0.0049) = -0.37t$ Rewrite as a logarithmic expression.

$t \approx 14.4 \text{ days}$ Divide both sides by −0.37.

It will take approximately 14.4 days (14 days, 10 hours) for the population to reach 180 fruit flies.

(e) One-half of the carrying capacity is 115 fruit flies. Solve $P(t) = 115$ by graphing $Y_1 = \dfrac{230}{1 + 56.5e^{-0.37t}}$ and $Y_2 = 115$ and using INTERSECT. See Figure 43. The population will reach one-half of the carrying capacity in about 10.9 days (10 days, 22 hours).

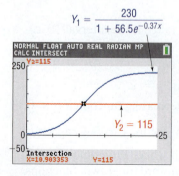

Figure 43

Look at Figure 43. Notice the point where the graph reaches 115 fruit flies (one-half of the carrying capacity): The graph changes from being curved upward to being curved downward. Using the language of calculus, we say the graph changes from increasing at an increasing rate to increasing at a decreasing rate. For any logistic growth function, when the population reaches one-half the carrying capacity, the population growth starts to slow down.

Now Work PROBLEM 23

Exploration

On the same viewing rectangle, graph

$$Y_1 = \frac{500}{1 + 24e^{-0.03t}} \quad \text{and} \quad Y_2 = \frac{500}{1 + 24e^{-0.08t}}$$

What effect does the growth rate $|b|$ have on the logistic growth function?

EXAMPLE 6

Wood Products

The EFISCEN wood product model classifies wood products according to their life-span. There are four classifications: short (1 year), medium short (4 years), medium long (16 years), and long (50 years). Based on data obtained from the European Forest Institute, the percentage of remaining wood products after t years for wood products with long life-spans (such as those used in the building industry) is given by

$$P(t) = \frac{100.3952}{1 + 0.0316e^{0.0581t}}$$

(a) What is the decay rate?

(b) What is the percentage of remaining wood products after 10 years?

(c) How long does it take for the percentage of remaining wood products to reach 50%?

(d) Explain why the numerator given in the model is reasonable.

Solution

(a) The decay rate is $|b| = |-0.0581| = 5.81\%$ per year.

(b) Evaluate $P(10)$.

$$P(10) = \frac{100.3952}{1 + 0.0316e^{0.0581(10)}} \approx 95.0$$

So 95% of long-life-span wood products remain after 10 years.

(c) Solve the equation $P(t) = 50$.

$$\frac{100.3952}{1 + 0.0316e^{0.0581t}} = 50$$

$$100.3952 = 50(1 + 0.0316e^{0.0581t})$$

$$2.0079 = 1 + 0.0316e^{0.0581t} \qquad \textcolor{teal}{\textit{Divide both sides by 50.}}$$

$$1.0079 = 0.0316e^{0.0581t} \qquad \textcolor{teal}{\textit{Subtract 1 from both sides.}}$$

$$31.8956 = e^{0.0581t} \qquad \textcolor{teal}{\textit{Divide both sides by 0.0316.}}$$

$$\ln(31.8956) = 0.0581t \qquad \textcolor{teal}{\textit{Rewrite as a logarithmic expression.}}$$

$$t \approx 59.6 \text{ years} \qquad \textcolor{teal}{\textit{Divide both sides by 0.0581.}}$$

It will take approximately 59.6 years for the percentage of long-life-span wood products remaining to reach 50%.

(d) The numerator of 100.3952 is reasonable because the maximum percentage of wood products remaining that is possible is 100%.

6.8 Assess Your Understanding

Applications and Extensions

1. **Growth of an Insect Population** The size P of a certain insect population at time t (in days) obeys the law of uninhibited growth $P(t) = 500e^{0.02t}$.
   (a) Determine the number of insects at $t = 0$ days.
   (b) What is the growth rate of the insect population?
   (c) What is the population after 10 days?
   (d) When will the insect population reach 800?
   (e) When will the insect population double?

2. **Growth of Bacteria** The number N of bacteria present in a culture at time t (in hours) obeys the law of uninhibited growth $N(t) = 1000e^{0.01t}$.
   (a) Determine the number of bacteria at $t = 0$ hours.
   (b) What is the growth rate of the bacteria?
   (c) What is the population after 4 hours?
   (d) When will the number of bacteria reach 1700?
   (e) When will the number of bacteria double?

3. **Radioactive Decay** Strontium-90 is a radioactive material that decays according to the function $A(t) = A_0 e^{-0.0244t}$, where A_0 is the initial amount present and A is the amount present at time t (in years). Assume that a scientist has a sample of 500 grams of strontium-90.
   (a) What is the decay rate of strontium-90?
   (b) How much strontium-90 is left after 10 years?
   (c) When will 400 grams of strontium-90 be left?
   (d) What is the half-life of strontium-90?

4. **Radioactive Decay** Iodine-131 is a radioactive material that decays according to the function $A(t) = A_0 e^{-0.087t}$, where A_0 is the initial amount present and A is the amount present at time t (in days). Assume that a scientist has a sample of 100 grams of iodine-131.
   (a) What is the decay rate of iodine-131?
   (b) How much iodine-131 is left after 9 days?
   (c) When will 70 grams of iodine-131 be left?
   (d) What is the half-life of iodine-131?

5. **Growth of a Colony of Mosquitoes** The population of a colony of mosquitoes obeys the law of uninhibited growth.
   (a) If N is the population of the colony and t is the time in days, express N as a function of t.
   (b) If there are 1000 mosquitoes initially and there are 1800 after 1 day, what is the size of the colony after 3 days?
   (c) How long is it until there are 10,000 mosquitoes?

6. **Bacterial Growth** A culture of bacteria obeys the law of uninhibited growth.
   (a) If N is the number of bacteria in the culture and t is the time in hours, express N as a function of t.
   (b) If 500 bacteria are present initially and there are 800 after 1 hour, how many will be present in the culture after 5 hours?
   (c) How long is it until there are 20,000 bacteria?

7. **Population Growth** The population of a southern city follows the exponential law.
   (a) If N is the population of the city and t is the time in years, express N as a function of t.
   (b) If the population doubled in size over an 18-month period and the current population is 10,000, what will the population be 2 years from now?

8. **Population Decline** The population of a midwestern city follows the exponential law.
   (a) If N is the population of the city and t is the time in years, express N as a function of t.
   (b) If the population decreased from 900,000 to 800,000 from 2008 to 2010, what will the population be in 2012?

9. **Radioactive Decay** The half-life of radium is 1690 years. If 10 grams is present now, how much will be present in 50 years?

10. **Radioactive Decay** The half-life of radioactive potassium is 1.3 billion years. If 10 grams is present now, how much will be present in 100 years? In 1000 years?

11. **Estimating the Age of a Tree** A piece of charcoal is found to contain 30% of the carbon-14 that it originally had. When did the tree die from which the charcoal came? Use 5730 years as the half-life of carbon-14.

12. **Estimating the Age of a Fossil** A fossilized leaf contains 70% of its normal amount of carbon-14. How old is the fossil?

13. **Cooling Time of a Pizza Pan** A pizza pan is removed at 5:00 PM from an oven whose temperature is fixed at 450°F into a room that is a constant 70°F. After 5 minutes, the temperature of the pan is 300°F.
   (a) At what time is the temperature of the pan 135°F?
   (b) Determine the time that needs to elapse before the temperature of the pan is 160°F.
   (c) What do you notice about the temperature as time passes?

14. **Newton's Law of Cooling** A thermometer reading 72°F is placed in a refrigerator where the temperature is a constant 38°F.
   (a) If the thermometer reads 60°F after 2 minutes, what will it read after 7 minutes?
   (b) How long will it take before the thermometer reads 39°F?
   (c) Determine the time that must elapse before the thermometer reads 45°F.
   (d) What do you notice about the temperature as time passes?

15. **Newton's Law of Heating** A thermometer reading 8°C is brought into a room with a constant temperature of 35°C. If the thermometer reads 15°C after 3 minutes, what will it read after being in the room for 5 minutes? For 10 minutes?
   [**Hint:** You need to construct a formula similar to equation (4).]

16. Warming Time of a Beer Stein A beer stein has a temperature of 28°F. It is placed in a room with a constant temperature of 70°F. After 10 minutes, the temperature of the stein has risen to 35°F. What will the temperature of the stein be after 30 minutes? How long will it take the stein to reach a temperature of 45°F? (See the hint given for Problem 15.)

17. Decomposition of Chlorine in a Pool Under certain water conditions, the free chlorine (hypochlorous acid, HOCl) in a swimming pool decomposes according to the law of uninhibited decay. After shocking his pool, Ben tested the water and found the amount of free chlorine to be 2.5 parts per million (ppm). Twenty-four hours later, Ben tested the water again and found the amount of free chlorine to be 2.2 ppm. What will be the reading after 3 days (that is, 72 hours)? When the chlorine level reaches 1.0 ppm, Ben must shock the pool again. How long can Ben go before he must shock the pool again?

18. Decomposition of Dinitrogen Pentoxide At 45°C, dinitrogen pentoxide (N_2O_5) decomposes into nitrous dioxide (NO_2) and oxygen (O_2) according to the law of uninhibited decay. An initial amount of 0.25 M N_2O_5 (M is a measure of concentration known as molarity) decomposes to 0.15 M N_2O_5 in 17 minutes. What concentration of N_2O_5 will remain after 30 minutes? How long will it take until only 0.01 M N_2O_5 remains?

19. Decomposition of Sucrose Reacting with water in an acidic solution at 35°C, sucrose ($C_{12}H_{22}O_{11}$) decomposes into glucose ($C_6H_{12}O_6$) and fructose ($C_6H_{12}O_6$)* according to the law of uninhibited decay. An initial concentration of 0.40 M of sucrose decomposes to 0.36 M sucrose in 30 minutes. What concentration of sucrose will remain after 2 hours? How long will it take until only 0.10 M sucrose remains?

20. Decomposition of Salt in Water Salt (NaCl) decomposes in water into sodium (Na^+) and chloride (Cl^-) ions according to the law of uninhibited decay. If the initial amount of salt is 25 kilograms and, after 10 hours, 15 kilograms of salt is left, how much salt is left after 1 day? How long does it take until $\frac{1}{2}$ kilogram of salt is left?

21. Radioactivity from Chernobyl After the release of radioactive material into the atmosphere from a nuclear power plant at Chernobyl (Ukraine) in 1986, the hay in Austria was contaminated by iodine 131 (half-life 8 days). If it is safe to feed the hay to cows when 10% of the iodine 131 remains, how long did the farmers need to wait to use this hay?

22. Word Users According to a survey by Olsten Staffing Services, the percentage of companies reporting usage of Microsoft Word t years since 1984 is given by

$$P(t) = \frac{99.744}{1 + 3.014e^{-0.799t}}$$

(a) What is the growth rate in the percentage of Microsoft Word users?
(b) Use a graphing utility to graph $P = P(t)$.
(c) What was the percentage of Microsoft Word users in 1990?
(d) During what year did the percentage of Microsoft Word users reach 90%?
(e) Explain why the numerator given in the model is reasonable. What does it imply?

23. Home Computers The logistic model

$$P(t) = \frac{95.4993}{1 + 0.0405e^{0.1968t}}$$

represents the percentage of households that do not own a personal computer t years since 1984.
(a) Evaluate and interpret $P(0)$.
(b) Use a graphing utility to graph $P = P(t)$.
(c) What percentage of households did not own a personal computer in 1995?
(d) In what year did the percentage of households that do not own a personal computer reach 10%?
Source: U.S. Department of Commerce

24. Farmers The logistic model

$$W(t) = \frac{14{,}656{,}248}{1 + 0.059e^{0.057t}}$$

represents the number of farm workers in the United States t years after 1910.
(a) Evaluate and interpret $W(0)$.
(b) Use a graphing utility to graph $W = W(t)$.
(c) How many farm workers were there in the United States in 2010?
(d) When did the number of farm workers in the United States reach 10,000,000?
(e) According to this model, what happens to the number of farm workers in the United States as t approaches ∞? Based on this result, do you think that it is reasonable to use this model to predict the number of farm workers in the United States in 2060? Why?
Source: U.S. Department of Agriculture

25. Birthdays The logistic model

$$P(n) = \frac{113.3198}{1 + 0.115e^{0.0912n}}$$

models the probability that, in a room of n people, no two people share the same birthday.
(a) Use a graphing utility to graph $P = P(n)$.
(b) In a room of $n = 15$ people, what is the probability that no two share the same birthday?
(c) How many people must be in a room before the probability that no two people share the same birthday falls below 10%?
(d) What happens to the probability as n increases? Explain what this result means.

26. Population of an Endangered Species Environmentalists often capture an endangered species and transport the species to a controlled environment where the species can produce offspring and regenerate its population. Suppose that six American bald eagles are captured, transported to Montana, and set free. Based on experience, the environmentalists expect the population to grow according to the model

$$P(t) = \frac{500}{1 + 83.33e^{-0.162t}}$$

where t is measured in years. (continued on the next page)

*Author's Note: Surprisingly, the chemical formulas for glucose and fructose are the same: This is not a typo.

(a) Determine the carrying capacity of the environment.
(b) What is the growth rate of the bald eagle?
(c) What is the population after 3 years?
(d) When will the population be 300 eagles?
(e) How long does it take for the population to reach one-half of the carrying capacity?

27. **The *Challenger* Disaster** After the *Challenger* disaster in 1986, a study was made of the 23 launches that preceded the fatal flight. A mathematical model was developed involving the relationship between the Fahrenheit temperature x around the O-rings and the number y of eroded or leaky primary O-rings. The model stated that

$$y = \frac{6}{1 + e^{-(5.085 - 0.1156x)}}$$

where the number 6 indicates the 6 primary O-rings on the spacecraft.
(a) What is the predicted number of eroded or leaky primary O-rings at a temperature of 100°F?
(b) What is the predicted number of eroded or leaky primary O-rings at a temperature of 60°F?
(c) What is the predicted number of eroded or leaky primary O-rings at a temperature of 30°F?
(d) Graph the equation using a graphing utility. At what temperature is the predicted number of eroded or leaky O-rings 1? 3? 5?

Source: Linda Tappin, "Analyzing Data Relating to the Challenger *Disaster," Mathematics Teacher, Vol. 87, No. 6, September 1994, pp. 423–426.*

Problems 28 and 29 use the following discussion: Uninhibited growth can be modeled by exponential functions other than $A(t) = A_0 e^{kt}$. For example, if an initial population P_0 requires n units of time to double, then the function $P(t) = P_0 \cdot 2^{t/n}$ models the size of the population at time t. Likewise, a population requiring n units of time to triple can be modeled by $P(t) = P_0 \cdot 3^{t/n}$.

28. **Growth of a Human Population** The population of a town is growing exponentially.
(a) If its population doubled in size over an 8-year period and the current population is 25,000, write an exponential function of the form $P(t) = P_0 \cdot 2^{t/n}$ that models the population.
(b) What will the population be in 3 years?
(c) When will the population reach 80,000?
(d) Express the model from part (a) in the form $A(t) = A_0 e^{kt}$.

29. **Growth of an Insect Population** An insect population grows exponentially.
(a) If the population triples in 20 days, and 50 insects are present initially, write an exponential function of the form $P(t) = P_0 \cdot 3^{t/n}$ that models the population.
(b) What will the population be in 47 days?
(c) When will the population reach 700?
(d) Express the model from part (a) in the form $A(t) = A_0 e^{kt}$.

Retain Your Knowledge

Problems 30–33 are based on material learned earlier in the course. The purpose of these problems is to keep the material fresh in your mind so that you are better prepared for the final exam.

30. Find the equation of the linear function f that passes through the points $(4, 1)$ and $(8, -5)$.

31. Determine whether the graphs of the linear functions $f(x) = 5x - 1$ and $g(x) = \frac{1}{5}x + 1$ are parallel, perpendicular, or neither.

32. Write the logarithmic expression $\ln\left(\frac{x^2\sqrt{y}}{z}\right)$ as the sum and/or difference of logarithms. Express powers as factors.

33. Rationalize the denominator of $\dfrac{10}{\sqrt[3]{25}}$.

6.9 Building Exponential, Logarithmic, and Logistic Models from Data

PREPARING FOR THIS SECTION *Before getting started, review the following:*

- Building Linear Models from Data (Section 4.2, pp. 284–287)
- Building Cubic Models from Data (Section 5.1, pp. 336–337)
- Building Quadratic Models from Data (Section 4.4, pp. 306–307)

OBJECTIVES 1 Build an Exponential Model from Data (p. 489)
     2 Build a Logarithmic Model from Data (p. 491)
     3 Build a Logistic Model from Data (p. 491)

In Section 4.2 we discussed how to find the linear function of best fit $(y = ax + b)$, in Section 4.4 we discussed how to find the quadratic function of best fit $(y = ax^2 + bx + c)$, and in Section 5.1 we discussed how to find the cubic function of best fit $(y = ax^3 + bx^2 + cx + d)$.

In this section we discuss how to use a graphing utility to find equations of best fit that describe the relation between two variables when the relation is thought to be exponential $(y = ab^x)$, logarithmic $(y = a + b \ln x)$, or logistic $\left(y = \dfrac{c}{1 + ae^{-bx}} \right)$. As before, we draw a scatter diagram of the data to help to determine the appropriate model to use.

Figure 44 shows scatter diagrams that will typically be observed for the three models. Below each scatter diagram are any restrictions on the values of the parameters.

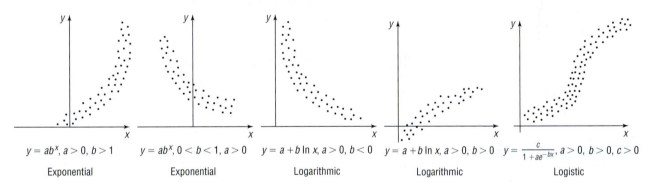

$y = ab^x, a > 0, b > 1$	$y = ab^x, 0 < b < 1, a > 0$	$y = a + b \ln x, a > 0, b < 0$	$y = a + b \ln x, a > 0, b > 0$	$y = \dfrac{c}{1 + ae^{-bx}}, a > 0, b > 0, c > 0$
Exponential	Exponential	Logarithmic	Logarithmic	Logistic

Figure 44

Most graphing utilities have REGression options that fit data to a specific type of curve. Once the data have been entered and a scatter diagram obtained, the type of curve that you want to fit to the data is selected. Then that REGression option is used to obtain the curve of *best fit* of the type selected.

The correlation coefficient r will appear only if the model can be written as a linear expression. As it turns out, r will appear for the linear, power, exponential, and logarithmic models, since these models can be written as a linear expression. Remember, the closer $|r|$ is to 1, the better the fit.

1 Build an Exponential Model from Data

We saw in Section 6.7 that the future value of money behaves exponentially, and we saw in Section 6.8 that growth and decay models also behave exponentially. The next example shows how data can lead to an exponential model.

Table 9

Year, x	Account Value, y
0	20,000
1	21,516
2	23,355
3	24,885
4	27,484
5	30,053
6	32,622

EXAMPLE 1   Fitting an Exponential Function to Data

Mariah deposited $20,000 in a well-diversified mutual fund 6 years ago. The data in Table 9 represent the value of the account at the beginning of each year for the last 7 years.

(a) Using a graphing utility, draw a scatter diagram with year as the independent variable.
(b) Using a graphing utility, build an exponential model from the data.
(c) Express the function found in part (b) in the form $A = A_0 e^{kt}$.
(d) Graph the exponential function found in part (b) or (c) on the scatter diagram.
(e) Using the solution to part (b) or (c), predict the value of the account after 10 years.
(f) Interpret the value of k found in part (c).

Solution

(a) Enter the data into the graphing utility and draw the scatter diagram as shown in Figure 45.

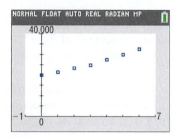

Figure 45

(b) A graphing utility fits the data in Table 9 to an exponential model of the form $y = ab^x$ using the EXPonential REGression option. Figure 46 shows that $y = ab^x = 19,820.43(1.085568)^x$. Notice that $|r| = 0.999$, which is close to 1, indicating a good fit.

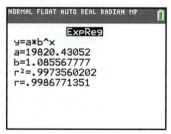

Figure 46

(c) To express $y = ab^x$ in the form $A = A_0 e^{kt}$, where $x = t$ and $y = A$, proceed as follows:

$$ab^x = A_0 e^{kt}$$

If $x = t = 0$, then $a = A_0$. This leads to

$$b^x = e^{kt}$$
$$b^x = (e^k)^t$$
$$b = e^k \qquad x = t$$

Because $y = ab^x = 19,820.43(1.085568)^x$, this means that $a = 19,820.43$ and $b = 1.085568$.

$$a = A_0 = 19,820.43 \quad \text{and} \quad b = e^k = 1.085568$$

To find k, rewrite $e^k = 1.085568$ as a logarithm to obtain

$$k = \ln(1.085568) \approx 0.08210$$

As a result, $A = A_0 e^{kt} = 19,820.43 e^{0.08210t}$.

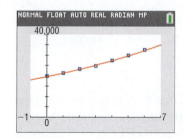

Figure 47

(d) See Figure 47 for the graph of the exponential function of best fit.
(e) Let $t = 10$ in the function found in part (c). The predicted value of the account after 10 years is

$$A = A_0 e^{kt} = 19,820.43 e^{0.08210(10)} \approx \$45,047$$

(f) The value of $k = 0.08210 = 8.210\%$ represents the annual growth rate of the account. It represents the rate of interest earned, assuming the account is growing continuously. ●

Now Work PROBLEM 1

2 Build a Logarithmic Model from Data

Some relations between variables follow a logarithmic model.

Table 10

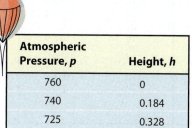

Atmospheric Pressure, p	Height, h
760	0
740	0.184
725	0.328
700	0.565
650	1.079
630	1.291
600	1.634
580	1.862
550	2.235

EXAMPLE 2   **Fitting a Logarithmic Function to Data**

Jodi, a meteorologist, is interested in finding a function that explains the relation between the height of a weather balloon (in kilometers) and the atmospheric pressure (measured in millimeters of mercury) on the balloon. She collects the data shown in Table 10.

(a) Using a graphing utility, draw a scatter diagram of the data with atmospheric pressure as the independent variable.

(b) It is known that the relation between atmospheric pressure and height follows a logarithmic model. Using a graphing utility, build a logarithmic model from the data.

(c) Draw the logarithmic function found in part (b) on the scatter diagram.

(d) Use the function found in part (b) to predict the height of the weather balloon if the atmospheric pressure is 560 millimeters of mercury.

Solution

(a) Enter the data into the graphing utility, and draw the scatter diagram. See Figure 48.

(b) A graphing utility fits the data in Table 10 to a logarithmic function of the form $y = a + b \ln x$ by using the LOGarithm REGression option. See Figure 49. The logarithmic model from the data is

$$h(p) = 45.7863 - 6.9025 \ln p$$

where h is the height of the weather balloon and p is the atmospheric pressure. Notice that $|r|$ is close to 1, indicating a good fit.

(c) Figure 50 shows the graph of $h(p) = 45.7863 - 6.9025 \ln p$ on the scatter diagram.

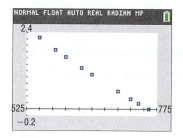

Figure 48

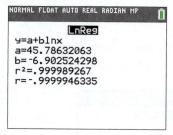

Figure 49

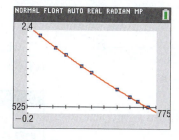

Figure 50

(d) Using the function found in part (b), Jodi predicts the height of the weather balloon when the atmospheric pressure is 560 to be

$$h(560) = 45.7863 - 6.9025 \ln 560$$

$$\approx 2.108 \text{ kilometers}$$

●

 Now Work PROBLEM 5

3 Build a Logistic Model from Data

Logistic growth models can be used to model situations for which the value of the dependent variable is limited. Many real-world situations conform to this scenario. For example, the population of the human race is limited by the availability of natural resources such as food and shelter. When the value of the dependent variable is limited, a logistic growth model is often appropriate.

EXAMPLE 3

Fitting a Logistic Function to Data

The data in Table 11 represent the amount of yeast biomass in a culture after t hours.

Table 11

Time (hours)	Yeast Biomass	Time (hours)	Yeast Biomass	Time (hours)	Yeast Biomass
0	9.6	7	257.3	14	640.8
1	18.3	8	350.7	15	651.1
2	29.0	9	441.0	16	655.9
3	47.2	10	513.3	17	659.6
4	71.1	11	559.7	18	661.8
5	119.1	12	594.8		
6	174.6	13	629.4		

Source: Tor Carlson (Über Geschwindigkeit and Grösse der Hefevermehrung in Würze, Biochemische Zeitschrift, Bd. 57, pp. 313–334, 1913)

(a) Using a graphing utility, draw a scatter diagram of the data with time as the independent variable.

(b) Using a graphing utility, build a logistic model from the data.

(c) Using a graphing utility, graph the function found in part (b) on the scatter diagram.

(d) What is the predicted carrying capacity of the culture?

(e) Use the function found in part (b) to predict the population of the culture at $t = 19$ hours.

Solution

(a) See Figure 51 for a scatter diagram of the data.

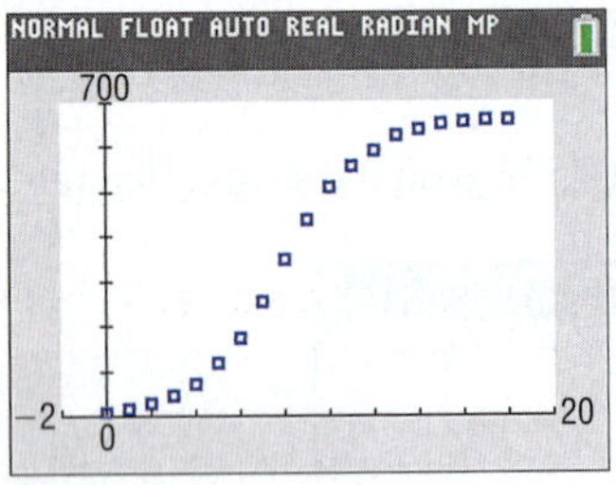

(b) A graphing utility fits the data in Table 11 to a logistic growth model of the form $y = \dfrac{c}{1 + ae^{-bx}}$ by using the LOGISTIC regression option. See Figure 52. The logistic model from the data is

$$y = \frac{663.0}{1 + 71.6e^{-0.5470x}}$$

where y is the amount of yeast biomass in the culture and x is the time.

(c) See Figure 53 for the graph of the logistic model.

Figure 51

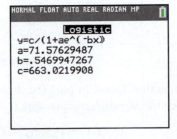

Figure 52

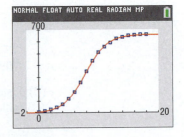

Figure 53

(d) Based on the logistic growth model found in part (b), the carrying capacity of the culture is 663.

(e) Using the logistic growth model found in part (b), the predicted amount of yeast biomass at $t = 19$ hours is

$$y = \frac{663.0}{1 + 71.6e^{-0.5470(19)}} \approx 661.5$$

━━━ **Now Work** PROBLEM 7

6.9 Assess Your Understanding

Applications and Extensions

1. Biology A strain of E-coli Beu 397-recA441 is placed into a nutrient broth at 30° Celsius and allowed to grow. The following data are collected. Theory states that the number of bacteria in the petri dish will initially grow according to the law of uninhibited growth. The population is measured using an optical device in which the amount of light that passes through the petri dish is measured.

Time (hours), x	Population, y
0	0.09
2.5	0.18
3.5	0.26
4.5	0.35
6	0.50

Source: Dr. Polly Lavery, Joliet Junior College

(a) Draw a scatter diagram treating time as the independent variable.
(b) Using a graphing utility, build an exponential model from the data.
(c) Express the function found in part (b) in the form $N(t) = N_0 e^{kt}$.
(d) Graph the exponential function found in part (b) or (c) on the scatter diagram.
(e) Use the exponential function from part (b) or (c) to predict the population at $x = 7$ hours.
(f) Use the exponential function from part (b) or (c) to predict when the population will reach 0.75.

2. Ethanol Production The data in the table below represent ethanol production (in billions of gallons) in the United States from 2000 to 2013.

Year	Ethanol Produced (billion gallons)	Year	Ethanol Produced (billion gallons)
2000 ($x = 0$)	1.6	2007 ($x = 7$)	6.5
2001 ($x = 1$)	1.8	2008 ($x = 8$)	9.3
2002 ($x = 2$)	2.1	2009 ($x = 9$)	10.9
2003 ($x = 3$)	2.8	2010 ($x = 10$)	13.3
2004 ($x = 4$)	3.4	2011 ($x = 11$)	13.9
2005 ($x = 5$)	3.9	2012 ($x = 12$)	13.2
2006 ($x = 6$)	4.9	2013 ($x = 13$)	13.3

Source: Renewable Fuels Association, 2014

(a) Using a graphing utility, draw a scatter diagram of the data using 0 for 2000, 1 for 2001, and so on, as the independent variable.
(b) Using a graphing utility, build an exponential model from the data.
(c) Express the function found in part (b) in the form $A(t) = A_0 e^{kt}$.

(d) Graph the exponential function found in part (b) or (c) on the scatter diagram.
(e) Use the model to predict the amount of ethanol that will be produced in 2015.
(f) Interpret the meaning of k in the function found in part (c).

3. Advanced-Stage Breast Cancer The data in the table below represent the percentage of patients who have survived after diagnosis of advanced-stage breast cancer at 6-month intervals of time.

Time after Diagnosis (years)	Percentage Surviving
0.5	95.7
1	83.6
1.5	74.0
2	58.6
2.5	47.4
3	41.9
3.5	33.6

Source: Cancer Treatment Centers of America

(a) Using a graphing utility, draw a scatter diagram of the data with time after diagnosis as the independent variable.
(b) Using a graphing utility, build an exponential model from the data.
(c) Express the function found in part (b) in the form $A(t) = A_0 e^{kt}$.
(d) Graph the exponential function found in part (b) or (c) on the scatter diagram.
(e) What percentage of patients diagnosed with advanced-stage cancer are expected to survive for 4 years after initial diagnosis?
(f) Interpret the meaning of k in the function found in part (c).

4. Chemistry A chemist has a 100-gram sample of a radioactive material. He records the amount of radioactive material every week for 7 weeks and obtains the following data: (continued on the next page)

Week	Weight (in grams)
0	100.0
1	88.3
2	75.9
3	69.4
4	59.1
5	51.8
6	45.5

(a) Using a graphing utility, draw a scatter diagram with week as the independent variable.
(b) Using a graphing utility, build an exponential model from the data.
(c) Express the function found in part (b) in the form $A(t) = A_0 e^{kt}$.
(d) Graph the exponential function found in part (b) or (c) on the scatter diagram.
(e) From the result found in part (b), determine the half-life of the radioactive material.
(f) How much radioactive material will be left after 50 weeks?
(g) When will there be 20 grams of radioactive material?

5. Milk Production The data in the table below represent the number of dairy farms (in thousands) and the amount of milk produced (in billions of pounds) in the United States for various years.

Year	Dairy Farms (thousands)	Milk Produced (billion pounds)
1980	334	128
1985	269	143
1990	193	148
1995	140	155
2000	105	167
2005	78	177
2010	63	193

Source: Statistical Abstract of the United States, 2012

(a) Using a graphing utility, draw a scatter diagram of the data with the number of dairy farms as the independent variable.
(b) Using a graphing utility, build a logarithmic model from the data.
(c) Graph the logarithmic function found in part (b) on the scatter diagram.
(d) In 2008, there were 67 thousand dairy farms in the United States. Use the function in part (b) to predict the amount of milk produced in 2008.
(e) The actual amount of milk produced in 2008 was 190 billion pounds. How does your prediction in part (d) compare to this?

6. Cable Rates The data (top, right) represent the average monthly rate charged for expanded basic cable television in the United States from 1995 to 2012. A market researcher believes that external factors, such as the growth of satellite television and internet programming, have affected the cost of basic cable. She is interested in building a model that will describe the average monthly cost of basic cable.
(a) Using a graphing utility, draw a scatter diagram of the data using 0 for 1995, 1 for 1996, and so on, as the independent variable and average monthly rate as the dependent variable.
(b) Using a graphing utility, build a logistic model from the data.
(c) Graph the logistic function found in part (b) on the scatter diagram.
(d) Based on the model found in part (b), what is the maximum possible average monthly rate for basic cable?
(e) Use the model to predict the average rate for basic cable in 2017.

Year	Average Monthly Rate (dollars)
1995 ($x = 0$)	22.35
1996 ($x = 1$)	24.28
1997 ($x = 2$)	26.31
1998 ($x = 3$)	27.88
1999 ($x = 4$)	28.94
2000 ($x = 5$)	31.22
2001 ($x = 6$)	33.75
2002 ($x = 7$)	36.47
2003 ($x = 8$)	38.95
2004 ($x = 9$)	41.04
2005 ($x = 10$)	43.04
2006 ($x = 11$)	45.26
2007 ($x = 12$)	47.27
2008 ($x = 13$)	49.65
2009 ($x = 14$)	52.37
2010 ($x = 15$)	54.44
2011 ($x = 16$)	57.46
2012 ($x = 17$)	61.63

Source: Federal Communications Commission, 2013

7. Population Model The following data represent the population of the United States. An ecologist is interested in building a model that describes the population of the United States.

Year	Population
1900	76,212,168
1910	92,228,496
1920	106,021,537
1930	123,202,624
1940	132,164,569
1950	151,325,798
1960	179,323,175
1970	203,302,031
1980	226,542,203
1990	248,709,873
2000	281,421,906
2010	308,745,538

Source: U.S. Census Bureau

(a) Using a graphing utility, draw a scatter diagram of the data using years since 1900 as the independent variable and population as the dependent variable.
(b) Using a graphing utility, build a logistic model from the data.
(c) Using a graphing utility, draw the function found in part (b) on the scatter diagram.
(d) Based on the function found in part (b), what is the carrying capacity of the United States?
(e) Use the function found in part (b) to predict the population of the United States in 2012.
(f) When will the United States population be 350,000,000?
(g) Compare actual U.S. Census figures to the predictions found in parts (e) and (f). Discuss any differences.

8. Population Model The data on the right represent the world population. An ecologist is interested in building a model that describes the world population.
  (a) Using a graphing utility, draw a scatter diagram of the data using years since 2000 as the independent variable and population as the dependent variable.
  (b) Using a graphing utility, build a logistic model from the data.
  (c) Using a graphing utility, draw the function found in part (b) on the scatter diagram.
  (d) Based on the function found in part (b), what is the carrying capacity of the world?
  (e) Use the function found in part (b) to predict the populationof the world in 2020.
  (f) When will world population be 10 billion?

Year	Population (billions)	Year	Population (billions)
2001	6.17	2008	6.71
2002	6.24	2009	6.79
2003	6.32	2010	6.86
2004	6.40	2011	6.94
2005	6.47	2012	7.02
2006	6.55	2013	7.10
2007	6.63		

Source: U.S. Census Bureau

9. Cell Phone Towers The following data represent the number of cell sites in service in the United States from 1985 to 2012 at the end of June each year.

Year	Cell Sites (thousands)	Year	Cell Sites (thousands)	Year	Cell Sites (thousands)
1985 ($x = 1$)	0.6	1995 ($x = 11$)	19.8	2004 ($x = 20$)	174.4
1986 ($x = 2$)	1.2	1996 ($x = 12$)	24.8	2005 ($x = 21$)	178.0
1987 ($x = 3$)	1.7	1997 ($x = 13$)	38.7	2006 ($x = 22$)	197.6
1988 ($x = 4$)	2.8	1998 ($x = 14$)	57.7	2007 ($x = 23$)	210.4
1989 ($x = 5$)	3.6	1999 ($x = 15$)	74.2	2008 ($x = 24$)	220.5
1990 ($x = 6$)	4.8	2000 ($x = 16$)	95.7	2009 ($x = 25$)	245.9
1991 ($x = 7$)	6.7	2001 ($x = 17$)	114.1	2010 ($x = 26$)	251.6
1992 ($x = 8$)	8.9	2002 ($x = 18$)	131.4	2011 ($x = 27$)	256.9
1993 ($x = 9$)	11.6	2003 ($x = 19$)	147.7	2012 ($x = 28$)	285.6
1994 ($x = 10$)	14.7				

Source: ©2013 CTIA-The Wireless Association®. All Rights Reserved.

  (a) Using a graphing utility, draw a scatter diagram of the data using 1 for 1985, 2 for 1986, and so on, as the independent variable and number of cell sites as the dependent variable.
  (b) Using a graphing utility, build a logistic model from the data.
  (c) Graph the logistic function found in part (b) on the scatter diagram.
  (d) What is the predicted carrying capacity for cell sites in the United States?
  (e) Use the model to predict the number of cell sites in the United States at the end of June 2017.

Mixed Practice

10. Age versus Total Cholesterol The following data represent the age and average total cholesterol for adult males at various ages.

Age	Total Cholesterol
27	189
40	205
50	215
60	210
70	210
80	194

  (a) Using a graphing utility, draw a scatter diagram of the data using age, x, as the independent variable and total cholesterol, y, as the dependent variable.
  (b) Based on the scatter diagram drawn in part (a), decide on a model (linear, quadratic, cubic, exponential, logarithmic, or logistic) that you think best describes the relation between age and total cholesterol. Be sure to justify your choice of model.
  (c) Using a graphing utility, find the model of best fit.
  (d) Using a graphing utility, draw the model of best fit on the scatter diagram drawn in part (a).
  (e) Use your model to predict the total cholesterol of a 35-year-old male.

11. Golfing The data below represent the expected percentage of putts that will be made by professional golfers on the PGA Tour depending on distance. For example, it is expected that 99.3% of 2-foot putts will be made.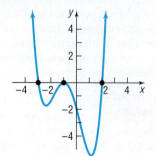

Distance (feet)	Expected Percentage	Distance (feet)	Expected Percentage
2	99.3	14	25.0
3	94.8	15	22.0
4	85.8	16	20.0
5	74.7	17	19.0
6	64.7	18	17.0
7	55.6	19	16.0
8	48.5	20	14.0
9	43.4	21	13.0
10	38.3	22	12.0
11	34.2	23	11.0
12	30.1	24	11.0
13	27.0	25	10.0

Source: TheSandTrap.com

(a) Using a graphing utility, draw a scatter diagram of the data with distance as the independent variable.
(b) Based on the scatter diagram drawn in part (a), decide on a model (linear, quadratic, cubic, exponential, logarithmic, or logistic) that you think best describes the relation between distance and expected percentage. Be sure to justify your choice of model.
(c) Using a graphing utility, find the model of best fit.
(d) Graph the function found in part (c) on the scatter diagram.
(e) Use the function found in part (c) to predict what percentage of 30-foot putts will be made.

12. Income versus Crime Rate The following data represent property crime rate against individuals (crimes per 1000 households) and their household income (in dollars) in the United States in 2009.

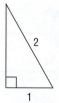

Income Level	Property Crime Rate
5000	201.1
11,250	157.0
20,000	141.6
30,000	134.1
42,500	139.7
62,500	120.0

Source: Statistical Abstract of the United States, 2012

(a) Using a graphing utility, draw a scatter diagram of the data using income, x, as the independent variable and crime rate, y, as the dependent variable.
(b) Based on the scatter diagram drawn in part (a), decide on a model (linear, quadratic, cubic, exponential, logarithmic, or logistic) that you think best describes the relation between income and crime rate. Be sure to justify your choice of model.
(c) Using a graphing utility, find the model of best fit.
(d) Using a graphing utility, draw the model of best fit on the scatter diagram you drew in part (a).
(e) Use your model to predict the crime rate of a household whose income is $55,000.

Retain Your Knowledge

Problems 13–16 are based on material learned earlier in the course. The purpose of these problems is to keep the material fresh in your mind so that you are better prepared for the final exam.

13. Construct a polynomial function that might have the graph shown. (More than one answer is possible.)

14. Rationalize the denominator of $\dfrac{3}{\sqrt{2}}$.

15. Use the Pythagorean Theorem to find the exact length of the unlabeled side in the given right triangle.

16. Graph the equation $(x - 3)^2 + y^2 = 25$.

Chapter Review

Things to Know

Composite function (p. 403)

$(f \circ g)(x) = f(g(x))$ The domain of $f \circ g$ is the set of all numbers x in the domain of g for which $g(x)$ is in the domain of f.

One-to-one function f (p. 411)

A function for which any two different inputs in the domain correspond to two different outputs in the range
For any choice of elements x_1, x_2 in the domain of f, if $x_1 \neq x_2$, then $f(x_1) \neq f(x_2)$.

Horizontal-line test (p. 412)

If every horizontal line intersects the graph of a function f in at most one point, f is one-to-one.

Inverse function f^{-1} of f (pp. 413–416)

Domain of f = range of f^{-1}; range of f = domain of f^{-1}
$f^{-1}(f(x)) = x$ for all x in the domain of f
$f(f^{-1}(x)) = x$ for all x in the domain of f^{-1}
The graphs of f and f^{-1} are symmetric with respect to the line $y = x$.

Properties of the exponential function (pp. 425, 428, 430)

$f(x) = Ca^x \quad a > 1, C > 0$

Domain: the interval $(-\infty, \infty)$
Range: the interval $(0, \infty)$
x-intercepts: none; y-intercept: C
Horizontal asymptote: x-axis ($y = 0$) as $x \to -\infty$
Increasing; one-to-one; smooth; continuous
See Figure 21 for a typical graph.

$f(x) = Ca^x \quad 0 < a < 1, C > 0$

Domain: the interval $(-\infty, \infty)$
Range: the interval $(0, \infty)$
x-intercepts: none; y-intercept: C
Horizontal asymptote: x-axis ($y = 0$) as $x \to \infty$
Decreasing; one-to-one; smooth; continuous
See Figure 25 for a typical graph.

Number e (p. 431)

Number approached by the expression $\left(1 + \dfrac{1}{n}\right)^n$ as $n \to \infty$; that is, $\lim\limits_{n \to \infty}\left(1 + \dfrac{1}{n}\right)^n = e$.

Property of exponents (p. 432)

If $a^u = a^v$, then $u = v$.

Properties of the logarithmic function (pp. 440–442)

$f(x) = \log_a x \quad a > 1$
($y = \log_a x$ means $x = a^y$)

Domain: the interval $(0, \infty)$
Range: the interval $(-\infty, \infty)$
x-intercept: 1; y-intercept: none
Vertical asymptote: $x = 0$ (y-axis)
Increasing; one-to-one; smooth; continuous
See Figure 39(a) for a typical graph.

$f(x) = \log_a x \quad 0 < a < 1$
($y = \log_a x$ means $x = a^y$)

Domain: the interval $(0, \infty)$
Range: the interval $(-\infty, \infty)$
x-intercept: 1; y-intercept: none
Vertical asymptote: $x = 0$ (y-axis)
Decreasing; one-to-one; smooth; continuous
See Figure 39(b) for a typical graph.

Natural logarithm (p. 443)

$y = \ln x$ means $x = e^y$.

Properties of logarithms (pp. 453–454, 456)

$\log_a 1 = 0 \qquad \log_a a = 1 \qquad a^{\log_a M} = M \qquad \log_a a^r = r \qquad a^r = e^{r \ln a}$

$\log_a(MN) = \log_a M + \log_a N \qquad \log_a\left(\dfrac{M}{N}\right) = \log_a M - \log_a N$
$\log_a M^r = r \log_a M$
If $M = N$, then $\log_a M = \log_a N$.
If $\log_a M = \log_a N$, then $M = N$.

Formulas

Change-of-Base Formula (p. 457) $\log_a M = \dfrac{\log_b M}{\log_b a}$

Compound Interest Formula (p. 469) $A = P \cdot \left(1 + \dfrac{r}{n}\right)^{nt}$

Continuous compounding (p. 471) $A = Pe^{rt}$

Effective rate of interest (p. 472)   Compounding n times per year: $r_e = \left(1 + \dfrac{r}{n}\right)^n - 1$

Continuous compounding: $r_e = e^r - 1$

Present Value Formulas (p. 473) $P = A \cdot \left(1 + \dfrac{r}{n}\right)^{-nt}$ or $P = Ae^{-rt}$

Uninhibited Growth and decay (pp. 478, 480) $A(t) = A_0 e^{kt}$

Newton's Law of Cooling (p. 481) $u(t) = T + (u_0 - T)e^{kt}$ $k < 0$

Logistic model (p. 483) $P(t) = \dfrac{c}{1 + ae^{-bt}}$

Objectives

Section		You should be able to . . .	Example(s)	Review Exercises
6.1	1	Form a composite function (p. 403)	1, 2, 4, 5	1–6
	2	Find the domain of a composite function (p. 404)	2–4	4–6
6.2	1	Determine whether a function is one-to-one (p. 411)	1, 2	7(a), 8
	2	Determine the inverse of a function defined by a map or a set of ordered pairs (p. 413)	3, 4	7(b)
	3	Obtain the graph of the inverse function from the graph of the function (p. 416)	7	8
	4	Find the inverse of a function defined by an equation (p. 417)	8, 9, 10	9–12
6.3	1	Evaluate exponential functions (p. 423)	1	13(a), (c), 46(a)
	2	Graph exponential functions (p. 427)	3–6	30–32
	3	Define the number e (p. 430)	p. 431	
	4	Solve exponential equations (p. 432)	7, 8	34, 35, 38, 40, 46(b)
6.4	1	Change exponential statements to logarithmic statements and logarithmic statements to exponential statements (p. 440)	2, 3	14, 15
	2	Evaluate logarithmic expressions (p. 441)	4	13(b), (d), 18, 45(b), 47(a), 48
	3	Determine the domain of a logarithmic function (p. 441)	5	16, 17, 33(a)
	4	Graph logarithmic functions (p. 442)	6, 7	29, 33(b), 45(a)
	5	Solve logarithmic equations (p. 446)	8, 9	36, 39, 45(c), 47(b)
6.5	1	Work with the properties of logarithms (p. 452)	1, 2	19, 20
	2	Write a logarithmic expression as a sum or difference of logarithms (p. 454)	3–5	21–24
	3	Write a logarithmic expression as a single logarithm (p. 455)	6	25–27
	4	Evaluate logarithms whose base is neither 10 nor e (p. 457)	7, 8	28
6.6	1	Solve logarithmic equations (p. 461)	1–3	36, 42
	2	Solve exponential equations (p. 463)	4–6	37, 41, 43, 44
	3	Solve logarithmic and exponential equations using a graphing utility (p. 464)	7	34–44
6.7	1	Determine the future value of a lump sum of money (p. 468)	1–3	49
	2	Calculate effective rates of return (p. 471)	4	49
	3	Determine the present value of a lump sum of money (p. 472)	5	50
	4	Determine the rate of interest or the time required to double a lump sum of money (p. 473)	6, 7	49

Section	You should be able to . . .	Example(s)	Review Exercises
6.8	**1** Find equations of populations that obey the law of uninhibited growth (p. 478)	1, 2	53
	2 Find equations of populations that obey the law of decay (p. 480)	3	51, 54
	3 Use Newton's Law of Cooling (p. 481)	4	52
	4 Use logistic models (p. 483)	5, 6	55
6.9	**1** Build an exponential model from data (p. 489)	1	56
	2 Build a logarithmic model from data (p. 491)	2	57
	3 Build a logistic model from data (p. 491)	3	58

Review Exercises

In Problems 1–3, for the given functions f and g, find:
(a) $(f \circ g)(2)$ *(b)* $(g \circ f)(-2)$ *(c)* $(f \circ f)(4)$ *(d)* $(g \circ g)(-1)$

1. $f(x) = 3x - 5$; $g(x) = 1 - 2x^2$ **2.** $f(x) = \sqrt{x + 2}$; $g(x) = 2x^2 + 1$ **3.** $f(x) = e^x$; $g(x) = 3x - 2$

In Problems 4–6, find $f \circ g$, $g \circ f$, $f \circ f$, and $g \circ g$ for each pair of functions. State the domain of each composite function.

4. $f(x) = 2 - x$; $g(x) = 3x + 1$ **5.** $f(x) = \sqrt{3x}$; $g(x) = 1 + x + x^2$ **6.** $f(x) = \dfrac{x + 1}{x - 1}$; $g(x) = \dfrac{1}{x}$

7. (a) Verify that the function below is one-to-one; (b) find its inverse.
$\{ (1, 2), (3, 5), (5, 8), (6, 10) \}$

8. State why the graph of the function is one-to-one. Then draw the graph of the inverse function f^{-1}.

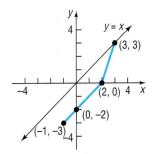

In Problems 9–12, each function is one-to-one. Find the inverse of each function and check your answer.

9. $f(x) = \dfrac{2x + 3}{5x - 2}$ **10.** $f(x) = \dfrac{1}{x - 1}$

11. $f(x) = \sqrt{x - 2}$ **12.** $f(x) = x^{1/3} + 1$

In Problem 13, $f(x) = 3^x$ and $g(x) = \log_3 x$.

13. Evaluate: (a) $f(4)$ (b) $g(9)$ (c) $f(-2)$ (d) $g\left(\dfrac{1}{27}\right)$

14. Change $5^2 = z$ to an equivalent statement involving a logarithm.
15. Change $\log_5 u = 13$ to an equivalent statement involving an exponent.

In Problems 16 and 17, find the domain of each logarithmic function.

16. $f(x) = \log(3x - 2)$ **17.** $H(x) = \log_2(x^2 - 3x + 2)$

In Problems 18–20, find the exact value of each expression. Do not use a calculator.

18. $\log_2\left(\dfrac{1}{8}\right)$ **19.** $\ln e^{\sqrt{2}}$ **20.** $2^{\log_2 0.4}$

In Problems 21–24, write each expression as the sum and/or difference of logarithms. Express powers as factors.

21. $\log_3\left(\dfrac{uv^2}{w}\right)$ $u > 0, v > 0, w > 0$ **22.** $\log_2(a^2 \sqrt{b})^4$ $a > 0, b > 0$

23. $\log(x^2 \sqrt{x^3 + 1})$ $x > 0$ **24.** $\ln\left(\dfrac{2x + 3}{x^2 - 3x + 2}\right)^2$ $x > 2$

In Problems 25–27, write each expression as a single logarithm.

25. $3 \log_4 x^2 + \dfrac{1}{2} \log_4 \sqrt{x}$

26. $\ln\left(\dfrac{x-1}{x}\right) + \ln\left(\dfrac{x}{x+1}\right) - \ln(x^2 - 1)$

27. $\dfrac{1}{2} \ln(x^2 + 1) - 4 \ln\dfrac{1}{2} - \dfrac{1}{2}\left[\ln(x-4) + \ln x\right]$

28. Use the Change-of-Base Formula and a calculator to evaluate $\log_4 19$. Round your answer to three decimal places.

29. Graph $y = \log_3 x$ using a graphing utility and the Change-of-Base Formula.

In Problems 30–33, use the given function f to:
    (a) Find the domain of f.   *(b) Graph f.* *(c) From the graph, determine the range and any asymptotes of f.*
    (d) Find f^{-1}, the inverse function of f. *(e) Find the domain and the range of f^{-1}.* *(f) Graph f^{-1}.*

30. $f(x) = 2^{x-3}$

31. $f(x) = 1 + 3^{-x}$

32. $f(x) = 3e^{x-2}$

33. $f(x) = \dfrac{1}{2}\ln(x+3)$

In Problems 34–44, solve each equation. Express any irrational solution in exact form and as a decimal rounded to three decimal places.

34. $8^{6+3x} = 4$

35. $3^{x^2+x} = \sqrt{3}$

36. $\log_x 64 = -3$

37. $5^x = 3^{x+2}$

38. $25^{2x} = 5^{x^2-12}$

39. $\log_3 \sqrt{x-2} = 2$

40. $8 = 4^{x^2} \cdot 2^{5x}$

41. $2^x \cdot 5 = 10^x$

42. $\log_6(x+3) + \log_6(x+4) = 1$

43. $e^{1-x} = 5$

44. $9^x + 4 \cdot 3^x - 3 = 0$

---

45. Suppose that $f(x) = \log_2(x-2) + 1$.
  (a) Graph f.
  (b) What is $f(6)$? What point is on the graph of f?
  (c) Solve $f(x) = 4$. What point is on the graph of f?
  (d) Based on the graph drawn in part (a), solve $f(x) > 0$.
  (e) Find $f^{-1}(x)$. Graph f^{-1} on the same Cartesian plane as f.

46. Amplifying Sound An amplifier's power output P (in watts) is related to its decibel voltage gain d by the formula
$$P = 25e^{0.1d}$$
  (a) Find the power output for a decibel voltage gain of 4 decibels.
  (b) For a power output of 50 watts, what is the decibel voltage gain?

47. Limiting Magnitude of a Telescope A telescope is limited in its usefulness by the brightness of the star that it is aimed at and by the diameter of its lens. One measure of a star's brightness is its *magnitude;* the dimmer the star, the larger its magnitude. A formula for the limiting magnitude L of a telescope—that is, the magnitude of the dimmest star that it can be used to view—is given by
$$L = 9 + 5.1 \log d$$
where d is the diameter (in inches) of the lens.
  (a) What is the limiting magnitude of a 3.5-inch telescope?
  (b) What diameter is required to view a star of magnitude 14?

48. Salvage Value The number of years n for a piece of machinery to depreciate to a known salvage value can be found using the formula
$$n = \dfrac{\log s - \log i}{\log(1-d)}$$
where s is the salvage value of the machinery, i is its initial value, and d is the annual rate of depreciation.

  (a) How many years will it take for a piece of machinery to decline in value from \$90,000 to \$10,000 if the annual rate of depreciation is 0.20 (20%)?
  (b) How many years will it take for a piece of machinery to lose half of its value if the annual rate of depreciation is 15%?

49. Funding a College Education A child's grandparents purchase a \$10,000 bond fund that matures in 18 years to be used for her college education. The bond fund pays 4% interest compounded semiannually. How much will the bond fund be worth at maturity? What is the effective rate of interest? How long will it take the bond to double in value under these terms?

50. Funding a College Education A child's grandparents wish to purchase a bond that matures in 18 years to be used for her college education. The bond pays 4% interest compounded semiannually. How much should they pay so that the bond will be worth \$85,000 at maturity?

51. Estimating the Date When a Prehistoric Man Died The bones of a prehistoric man found in the desert of New Mexico contain approximately 5% of the original amount of carbon-14. If the half-life of carbon-14 is 5730 years, approximately how long ago did the man die?

52. Temperature of a Skillet A skillet is removed from an oven where the temperature is 450°F and placed in a room where the temperature is 70°F. After 5 minutes, the temperature of the skillet is 400°F. How long will it be until its temperature is 150°F?

53. World Population The annual growth rate of the world's population in 2014 was $k = 1.1\% = 0.011$. The population of the world in 2014 was 7,137,577,750. Letting $t = 0$ represent 2014, use the uninhibited growth model to predict the world's population in the year 2024.

Source: U.S. Census Bureau

54. Radioactive Decay The half-life of radioactive cobalt is 5.27 years. If 100 grams of radioactive cobalt is present now, how much will be present in 20 years? In 40 years?

55. Logistic Growth The logistic growth model

$$P(t) = \frac{0.8}{1 + 1.67e^{-0.16t}}$$

represents the proportion of new cars with a global positioning system (GPS). Let $t = 0$ represent 2006, $t = 1$ represent 2007, and so on.

(a) What proportion of new cars in 2006 had a GPS?

(b) Determine the maximum proportion of new cars that have a GPS.

(c) Using a graphing utility, graph $P = P(t)$.

(d) When will 75% of new cars have a GPS?

56. Rising Tuition The following data represent the average in-state tuition and fees (in 2013 dollars) at public four-year colleges and universities in the United States from the academic year 1983–84 to the academic year 2013–14.

Academic Year	Tuition and Fees (2013 dollars)
1983–84 ($x = 0$)	2684
1988–89 ($x = 5$)	3111
1993–94 ($x = 10$)	4101
1998–99 ($x = 15$)	4648
2003–04 ($x = 20$)	5900
2008–09 ($x = 25$)	7008
2013–14 ($x = 30$)	8893

Source: The College Board

(a) Using a graphing utility, draw a scatter diagram with academic year as the independent variable.

(b) Using a graphing utility, build an exponential model from the data.

(c) Express the function found in part (b) in the form $A(t) = A_0 e^{kt}$.

(d) Graph the exponential function found in part (b) or (c) on the scatter diagram.

(e) Predict the academic year when the average tuition will reach $12,000.

57. Wind Chill Factor The data (top, right) represent the wind speed (mph) and the wind chill factor at an air temperature of 15°F.

(a) Using a graphing utility, draw a scatter diagram with wind speed as the independent variable.

(b) Using a graphing utility, build a logarithmic model from the data.

(c) Using a graphing utility, draw the logarithmic function found in part (b) on the scatter diagram.

(d) Use the function found in part (b) to predict the wind chill factor if the air temperature is 15°F and the wind speed is 23 mph.

Wind Speed (mph)	Wind Chill Factor (°F)
5	7
10	3
15	0
20	−2
25	−4
30	−5
35	−7

Source: U.S. National Weather Service

58. Spreading of a Disease Jack and Diane live in a small town of 50 people. Unfortunately, both Jack and Diane have a cold. Those who come in contact with someone who has this cold will themselves catch the cold. The following data represent the number of people in the small town who have caught the cold after t days.

Days, t	Number of People with Cold, C
0	2
1	4
2	8
3	14
4	22
5	30
6	37
7	42
8	44

(a) Using a graphing utility, draw a scatter diagram of the data. Comment on the type of relation that appears to exist between the number of days that have passed and the number of people with a cold.

(b) Using a graphing utility, build a logistic model from the data.

(c) Graph the function found in part (b) on the scatter diagram.

(d) According to the function found in part (b), what is the maximum number of people who will catch the cold? In reality, what is the maximum number of people who could catch the cold?

(e) Sometime between the second day and the third day, 10 people in the town had a cold. According to the model found in part (b), when did 10 people have a cold?

(f) How long will it take for 46 people to catch the cold?

Find the distance between Sioux Falls (43° 33′ north latitude) and Dallas (32° 46′ north latitude). See Figure 13(b). Assume that the radius of Earth is 3960 miles.

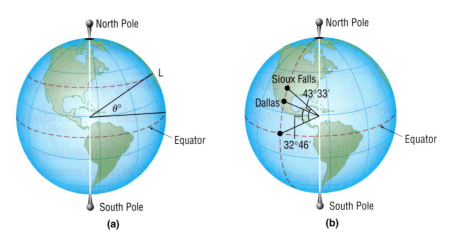

Figure 13

(a) (b)

Solution The measure of the central angle between the two cities is 43° 33′ − 32° 46′ = 10° 47′. Use equation (4), $s = r\theta$. But remember first to convert the angle of 10° 47′ to radians.

$$\theta = 10°47' \approx 10.7833° = 10.7833 \cdot \frac{\pi}{180} \text{ radian} \approx 0.188 \text{ radian}$$

$$47' = 47\left(\frac{1}{60}\right)°$$

Use $\theta = 0.188$ radian and $r = 3960$ miles in equation (4). The distance between the two cities is

$$s = r\theta = 3960 \cdot 0.188 \approx 744 \text{ miles}$$

When an angle is measured in degrees, the degree symbol will always be shown. However, when an angle is measured in radians, the usual practice is to omit the word *radians*. So if the measure of an angle is given as $\frac{\pi}{6}$, it is understood to mean $\frac{\pi}{6}$ radian.

NOTE If the measure of an angle is given as 5, it is understood to mean 5 radians; if the measure of an angle is given as 5°, it means 5 degrees. ∎

 Now Work PROBLEM 107

4 Find the Area of a Sector of a Circle

Consider a circle of radius r. Suppose that θ, measured in radians, is a central angle of this circle. See Figure 14. We seek a formula for the area A of the sector (shown in blue) formed by the angle θ.

Now consider a circle of radius r and two central angles θ and θ_1, both measured in radians. See Figure 15. From geometry, the ratio of the measures of the angles equals the ratio of the corresponding areas of the sectors formed by these angles. That is,

$$\frac{\theta}{\theta_1} = \frac{A}{A_1}$$

Suppose that $\theta_1 = 2\pi$ radians. Then $A_1 =$ area of the circle $= \pi r^2$. Solving for A, we find

$$A = A_1 \frac{\theta}{\theta_1} = \pi r^2 \frac{\theta}{2\pi} = \frac{1}{2} r^2 \theta$$

$$A_1 = \pi r^2$$

$$\theta_1 = 2\pi$$

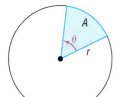

Figure 14 Sector of a Circle

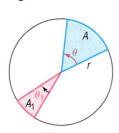

Figure 15 $\frac{\theta}{\theta_1} = \frac{A}{A_1}$

THEOREM

Area of a Sector

The area A of the sector of a circle of radius r formed by a central angle of θ radians is

$$A = \frac{1}{2}r^2\theta \qquad (8)$$

EXAMPLE 7

Finding the Area of a Sector of a Circle

Find the area of the sector of a circle of radius 2 feet formed by an angle of 30°. Round the answer to two decimal places.

Solution

Use equation (8) with $r = 2$ feet and $\theta = 30° = \dfrac{\pi}{6}$ radian. [Remember, in equation (8), θ must be in radians.]

$$A = \frac{1}{2}r^2\theta = \frac{1}{2}(2)^2\frac{\pi}{6} = \frac{\pi}{3} \approx 1.05$$

The area A of the sector is 1.05 square feet, rounded to two decimal places.

- **Now Work** PROBLEM 79

5 Find the Linear Speed of an Object Traveling in Circular Motion

In Chapter 1, Section 1.7, we defined the average speed of an object as the distance traveled divided by the elapsed time. For motion along a circle, we distinguish between *linear speed* and *angular speed*.

DEFINITION

Suppose that an object moves on a circle of radius r at a constant speed. If s is the distance traveled in time t on this circle, then the **linear speed** v of the object is defined as

$$v = \frac{s}{t} \qquad (9)$$

Figure 16 $v = \dfrac{s}{t}$

As this object travels on the circle, suppose that θ (measured in radians) is the central angle swept out in time t. See Figure 16.

DEFINITION

The **angular speed** ω (the Greek lowercase letter omega) of this object is the angle θ (measured in radians) swept out, divided by the elapsed time t; that is,

$$\omega = \frac{\theta}{t} \qquad (10)$$

Angular speed is the way the turning rate of an engine is described. For example, an engine idling at 900 rpm (revolutions per minute) is one that rotates at an angular speed of

$$900 \, \frac{\text{revolutions}}{\text{minute}} = 900 \, \frac{\text{revolutions}}{\text{minute}} \cdot 2\pi \, \frac{\text{radians}}{\text{revolution}} = 1800\pi \, \frac{\text{radians}}{\text{minute}}$$

There is an important relationship between linear speed and angular speed:

$$\text{linear speed} = v = \underset{\underset{(9)}{\uparrow}}{\frac{s}{t}} = \underset{\underset{s = r\theta}{\uparrow}}{\frac{r\theta}{t}} = r\left(\frac{\theta}{t}\right) = \underset{\underset{(10)}{\uparrow}}{r \cdot \omega}$$

$$v = r\omega \qquad\qquad (11)$$

where ω is measured in radians per unit time.

When using equation (11), remember that $v = \dfrac{s}{t}$ (the linear speed) has the dimensions of length per unit of time (such as feet per second or miles per hour), r (the radius of the circular motion) has the same length dimension as s, and ω (the angular speed) has the dimension of radians per unit of time. If the angular speed is given in terms of *revolutions* per unit of time (as is often the case), be sure to convert it to *radians* per unit of time, using the fact that 1 revolution $= 2\pi$ radians, before attempting to use equation (11).

EXAMPLE 8

Finding Linear Speed

A child is spinning a rock at the end of a 2-foot rope at the rate of 180 revolutions per minute (rpm). Find the linear speed of the rock when it is released.

Solution

Look at Figure 17. The rock is moving around a circle of radius $r = 2$ feet. The angular speed ω of the rock is

$$\omega = 180\ \frac{\text{revolutions}}{\text{minute}} = 180\ \frac{\text{revolutions}}{\text{minute}} \cdot 2\pi\ \frac{\text{radians}}{\text{revolution}} = 360\pi\ \frac{\text{radians}}{\text{minute}}$$

From equation (11), the linear speed v of the rock is

$$v = r\omega = 2\ \text{feet} \cdot 360\pi\ \frac{\text{radians}}{\text{minute}} = 720\pi\ \frac{\text{feet}}{\text{minute}} \approx 2262\ \frac{\text{feet}}{\text{minute}}$$

The linear speed of the rock when it is released is 2262 ft/min $\approx$ 25.7 mi/hr. ●

Figure 17

 Now Work PROBLEM **99**

Historical Feature

Trigonometry was developed by Greek astronomers, who regarded the sky as the inside of a sphere, so it was natural that triangles on a sphere were investigated early (by Menelaus of Alexandria about AD 100) and that triangles in the plane were studied much later. The first book containing a systematic treatment of plane and spherical trigonometry was written by the Persian astronomer Nasir Eddin (about AD 1250).

Regiomontanus (1436–1476) is the person most responsible for moving trigonometry from astronomy into mathematics. His work was improved by Copernicus (1473–1543) and Copernicus' student

Rhaeticus (1514–1576). Rhaeticus's book was the first to define the six trigonometric functions as ratios of sides of triangles, although he did not give the functions their present names. Credit for this is due to Thomas Finck (1583), but Finck's notation was by no means universally accepted at the time. The notation was finally stabilized by the textbooks of Leonhard Euler (1707–1783).

Trigonometry has since evolved from its use by surveyors, navigators, and engineers to present applications involving ocean tides, the rise and fall of food supplies in certain ecologies, brain wave patterns, and many other phenomena.

7.1 Assess Your Understanding

'Are You Prepared?' *Answers are given at the end of these exercises. If you get a wrong answer, read the pages listed in red.*

1. What is the formula for the circumference C of a circle of radius r? What is the formula for the area A of a circle of radius r? (p. 32)

2. If an object has a speed of r feet per second and travels a distance d (in feet) in time t (in seconds), then $d =$ _____. (pp. 138–139)

Concepts and Vocabulary

3. An angle θ is in _____ _____ if its vertex is at the origin of a rectangular coordinate system and its initial side coincides with the positive x-axis.

4. A _____ _____ is a positive angle whose vertex is at the center of a circle.

5. If the radius of a circle is r and the length of the arc subtended by a central angle is also r, then the measure of the angle is 1 _____.
   (a) degree   (b) minute   (c) second   (d) radian

6. On a circle of radius r, a central angle of θ radians subtends an arc of length $s =$ ___; the area of the sector formed by this angle θ is $A =$ _____.

7. $180° =$ _____ radians
   (a) $\dfrac{\pi}{2}$ (b) π (c) $\dfrac{3\pi}{2}$ (d) 2π

8. An object travels on a circle of radius r with constant speed. If s is the distance traveled in time t on the circle and θ is the central angle (in radians) swept out in time t, then the linear speed of the object is $v =$ _____ and the angular speed of the object is $\omega =$ _____.

9. *True or False* The angular speed ω of an object traveling on a circle of radius r is the angle θ (measured in radians) swept out, divided by the elapsed time t.

10. *True or False* For circular motion on a circle of radius r, linear speed equals angular speed divided by r.

Skill Building

In Problems 11–22, draw each angle.

11. $30°$
12. $60°$
13. $135°$
14. $-120°$
15. $450°$
16. $540°$

17. $\dfrac{3\pi}{4}$
18. $\dfrac{4\pi}{3}$
19. $-\dfrac{\pi}{6}$
20. $-\dfrac{2\pi}{3}$
21. $\dfrac{16\pi}{3}$
22. $\dfrac{21\pi}{4}$

In Problems 23–34, convert each angle in degrees to radians. Express your answer as a multiple of π.

23. $30°$
24. $120°$
25. $240°$
26. $330°$
27. $-60°$
28. $-30°$

29. $180°$
30. $270°$
31. $-135°$
32. $-225°$
33. $-90°$
34. $-180°$

In Problems 35–46, convert each angle in radians to degrees.

35. $\dfrac{\pi}{3}$
36. $\dfrac{5\pi}{6}$
37. $-\dfrac{5\pi}{4}$
38. $-\dfrac{2\pi}{3}$
39. $\dfrac{\pi}{2}$
40. 4π

41. $\dfrac{\pi}{12}$
42. $\dfrac{5\pi}{12}$
43. $-\dfrac{\pi}{2}$
44. $-\pi$
45. $-\dfrac{\pi}{6}$
46. $-\dfrac{3\pi}{4}$

In Problems 47–52, convert each angle in degrees to radians. Express your answer in decimal form, rounded to two decimal places.

47. $17°$
48. $73°$
49. $-40°$
50. $-51°$
51. $125°$
52. $350°$

In Problems 53–58, convert each angle in radians to degrees. Express your answer in decimal form, rounded to two decimal places.

53. 3.14
54. 0.75
55. 2
56. 3
57. 6.32
58. $\sqrt{2}$

In Problems 59–64, convert each angle to a decimal in degrees. Round your answer to two decimal places.

59. $40°10'25''$
60. $61°42'21''$
61. $50°14'20''$
62. $73°40'40''$
63. $9°9'9''$
64. $98°22'45''$

In Problems 65–70, convert each angle to $D°M'S''$ form. Round your answer to the nearest second.

65. $40.32°$
66. $61.24°$
67. $18.255°$
68. $29.411°$
69. $19.99°$
70. $44.01°$

In Problems 71–78, s denotes the length of the arc of a circle of radius r subtended by the central angle θ. Find the missing quantity. Round answers to three decimal places.

71. $r = 10$ meters, $\theta = \dfrac{1}{2}$ radian, $s = ?$

72. $r = 6$ feet, $\theta = 2$ radians, $s = ?$

73. $\theta = \dfrac{1}{3}$ radian, $s = 2$ feet, $r = ?$

74. $\theta = \dfrac{1}{4}$ radian, $s = 6$ centimeters, $r = ?$

75. $r = 5$ miles, $s = 3$ miles, $\theta = ?$

76. $r = 6$ meters, $s = 8$ meters, $\theta = ?$

77. $r = 2$ inches, $\theta = 30°$, $s = ?$

78. $r = 3$ meters, $\theta = 120°$, $s = ?$

In Problems 79–86, A denotes the area of the sector of a circle of radius r formed by the central angle θ. Find the missing quantity. Round answers to three decimal places.

79. $r = 10$ meters, $\theta = \dfrac{1}{2}$ radian, $A = ?$

80. $r = 6$ feet, $\theta = 2$ radians, $A = ?$

81. $\theta = \dfrac{1}{3}$ radian, $A = 2$ square feet, $r = ?$

82. $\theta = \dfrac{1}{4}$ radian, $A = 6$ square centimeters, $r = ?$

83. $r = 5$ miles, $A = 3$ square miles, $\theta = ?$

84. $r = 6$ meters, $A = 8$ square meters, $\theta = ?$

85. $r = 2$ inches, $\theta = 30°$, $A = ?$

86. $r = 3$ meters, $\theta = 120°$, $A = ?$

In Problems 87–90, find the length s and the area A. Round answers to three decimal places.

87.
2 ft

88.
4 m

89.
70°
12 yd

90.
50°
9 cm

Applications and Extensions

91. Movement of a Minute Hand The minute hand of a clock is 6 inches long. How far does the tip of the minute hand move in 15 minutes? How far does it move in 25 minutes? Round answers to two decimal places.

92. Movement of a Pendulum A pendulum swings through an angle of 20° each second. If the pendulum is 40 inches long, how far does its tip move each second? Round answers to two decimal places.

93. Area of a Sector Find the area of the sector of a circle of radius 4 meters formed by an angle of 45°. Round the answer to two decimal places.

94. Area of a Sector Find the area of the sector of a circle of radius 3 centimeters formed by an angle of 60°. Round the answer to two decimal places.

95. Watering a Lawn A water sprinkler sprays water over a distance of 30 feet while rotating through an angle of 135°. What area of lawn receives water?

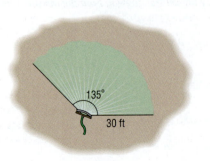
135°
30 ft

96. Designing a Water Sprinkler An engineer is asked to design a water sprinkler that will cover a field of 100 square yards that is in the shape of a sector of a circle of radius 15 yards. Through what angle should the sprinkler rotate?

97. Windshield Wiper The arm and blade of a windshield wiper have a total length of 34 inches. If the blade is 25 inches long and the wiper sweeps out an angle of 120°, how much window area can the blade clean?

98. Windshield Wiper The arm and blade of a windshield wiper have a total length of 30 inches. If the blade is 24 inches long and the wiper sweeps out an angle of 125°, how much window area can the blade clean?

99. Motion on a Circle An object is traveling on a circle with a radius of 5 centimeters. If in 20 seconds a central angle of $\dfrac{1}{3}$ radian is swept out, what is the angular speed of the object? What is its linear speed?

100. Motion on a Circle An object is traveling on a circle with a radius of 2 meters. If in 20 seconds the object travels 5 meters, what is its angular speed? What is its linear speed?

101. Amusement Park Ride A gondola on an amusement park ride, similar to the Spin Cycle at Silverwood Theme Park, spins at a speed of 13 revolutions per minute. If the gondola is 25 feet from the ride's center, what is the linear speed of the gondola in miles per hour?

102. Amusement Park Ride A centrifugal force ride, similar to the Gravitron, spins at a speed of 22 revolutions per minute. If the diameter of the ride is 13 meters, what is the linear speed of the passengers in kilometers per hour?

103. Blu-ray Drive A Blu-ray drive has a maximum speed of 10,000 revolutions per minute. If a Blu-ray disc has a diameter of 12 cm, what is the linear speed, in km/h, of a point 4 cm from the center if the disc is spinning at a rate of 8000 revolutions per minute?

104. DVD Drive A DVD drive has a maximum speed of 7200 revolutions per minute. If a DVD has a diameter of 12 cm, what is the linear speed, in km/h, of a point 5 cm from the disc's center if it is spinning at a rate of 5400 revolutions per minute?

105. Bicycle Wheels The diameter of each wheel of a bicycle is 26 inches. If you are traveling at a speed of 35 miles per hour on this bicycle, through how many revolutions per minute are the wheels turning?

106. Car Wheels The radius of each wheel of a car is 15 inches. If the wheels are turning at the rate of 3 revolutions per second, how fast is the car moving? Express your answer in inches per second and in miles per hour.

In Problems 107–110, the latitude of a location L is the angle formed by a ray drawn from the center of Earth to the equator and a ray drawn from the center of Earth to L. See the figure.

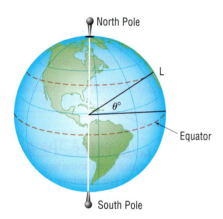

107. Distance between Cities Memphis, Tennessee, is due north of New Orleans, Louisiana. Find the distance between Memphis (35°9′ north latitude) and New Orleans (29°57′ north latitude). Assume that the radius of Earth is 3960 miles.

108. Distance between Cities Charleston, West Virginia, is due north of Jacksonville, Florida. Find the distance between Charleston (38°21′ north latitude) and Jacksonville (30°20′ north latitude). Assume that the radius of Earth is 3960 miles.

109. Linear Speed on Earth Earth rotates on an axis through its poles. The distance from the axis to a location on Earth at 30° north latitude is about 3429.5 miles. Therefore, a location on Earth at 30° north latitude is spinning on a circle of radius 3429.5 miles. Compute the linear speed on the surface of Earth at 30° north latitude.

110. Linear Speed on Earth Earth rotates on an axis through its poles. The distance from the axis to a location on Earth at 40° north latitude is about 3033.5 miles. Therefore, a location on Earth at 40° north latitude is spinning on a circle of radius 3033.5 miles. Compute the linear speed on the surface of Earth at 40° north latitude.

111. Speed of the Moon The mean distance of the moon from Earth is 2.39×10^5 miles. Assuming that the orbit of the moon around Earth is circular and that 1 revolution takes 27.3 days, find the linear speed of the moon. Express your answer in miles per hour.

112. Speed of Earth The mean distance of Earth from the Sun is 9.29×10^7 miles. Assuming that the orbit of Earth around the Sun is circular and that 1 revolution takes 365 days, find the linear speed of Earth. Express your answer in miles per hour.

113. Pulleys Two pulleys, one with radius 2 inches and the other with radius 8 inches, are connected by a belt. (See the figure.) If the 2-inch pulley is caused to rotate at 3 revolutions per minute, determine the revolutions per minute of the 8-inch pulley.

[**Hint:** The linear speeds of the pulleys are the same; both equal the speed of the belt.]

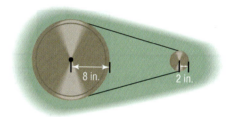

114. Ferris Wheels A neighborhood carnival has a Ferris wheel with a radius of 30 feet. You measure the time it takes for one revolution to be 70 seconds. What is the linear speed (in feet per second) of this Ferris wheel? What is the angular speed in radians per second?

115. Computing the Speed of a River Current To approximate the speed of the current of a river, a circular paddle wheel with radius 4 feet is lowered into the water. If the current causes the wheel to rotate at a speed of 10 revolutions per minute, what is the speed of the current? Express your answer in miles per hour.

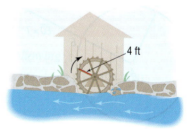

116. Spin Balancing Tires A spin balancer rotates the wheel of a car at 480 revolutions per minute. If the diameter of the wheel is 26 inches, what road speed is being tested? Express your answer in miles per hour. At how many revolutions per minute should the balancer be set to test a road speed of 80 miles per hour?

117. The Cable Cars of San Francisco At the Cable Car Museum you can see the four cable lines that are used to pull cable cars up and down the hills of San Francisco. Each cable travels at a speed of 9.55 miles per hour, driven by a rotating wheel whose diameter is 8.5 feet. How fast is the wheel rotating? Express your answer in revolutions per minute.

Option 2
Using Identities

Begin by finding $\cos \theta$ using the Pythagorean Identity from equation (5).

$$\sin^2 \theta + \cos^2 \theta = 1 \qquad \text{Equation (5)}$$

$$\frac{1}{9} + \cos^2 \theta = 1 \qquad \sin \theta = \frac{1}{3}$$

$$\cos^2 \theta = 1 - \frac{1}{9} = \frac{8}{9}$$

Recall that the trigonometric functions of an acute angle are positive. In particular, $\cos \theta > 0$ for an acute angle θ, so

$$\cos \theta = \sqrt{\frac{8}{9}} = \frac{2\sqrt{2}}{3}$$

Now, use $\sin \theta = \frac{1}{3}$ and $\cos \theta = \frac{2\sqrt{2}}{3}$ and fundamental identities.

$$\tan \theta = \frac{\sin \theta}{\cos \theta} = \frac{\frac{1}{3}}{\frac{2\sqrt{2}}{3}} = \frac{1}{2\sqrt{2}} = \frac{\sqrt{2}}{4} \qquad \cot \theta = \frac{1}{\tan \theta} = \frac{1}{\frac{\sqrt{2}}{4}} = \frac{4}{\sqrt{2}} = 2\sqrt{2}$$

$$\sec \theta = \frac{1}{\cos \theta} = \frac{1}{\frac{2\sqrt{2}}{3}} = \frac{3}{2\sqrt{2}} = \frac{3\sqrt{2}}{4} \qquad \csc \theta = \frac{1}{\sin \theta} = \frac{1}{\frac{1}{3}} = 3$$

●

Finding the Values of the Trigonometric Functions When One Is Known

Given the value of one trigonometric function of an acute angle θ, the exact value of each of the remaining five trigonometric functions of θ can be found in either of two ways.

Option 1 Using the Definition

STEP 1: Draw a right triangle showing the acute angle θ.
STEP 2: Two of the sides can then be assigned values based on the value of the given trigonometric function.
STEP 3: Find the length of the third side by using the Pythagorean Theorem.
STEP 4: Use the definitions in equation (1) to find the value of each of the remaining trigonometric functions.

Option 2 Using Identities

Use appropriately selected identities to find the value of each of the remaining trigonometric functions.

EXAMPLE 5

Given One Value of a Trigonometric Function, Find the Values of the Remaining Trigonometric Functions

Given $\tan \theta = \frac{1}{2}$, θ an acute angle, find the exact value of each of the remaining five trigonometric functions of θ.

Solution
Option 1
Using the Definition

Figure 24 shows a right triangle with acute angle θ, where

$$\tan \theta = \frac{1}{2} = \frac{\text{opposite}}{\text{adjacent}} = \frac{b}{a}$$

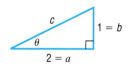

Figure 24 $\tan \theta = \dfrac{1}{2}$

With $b = 1$ and $a = 2$, the hypotenuse c can be found by using the Pythagorean Theorem.

$$c^2 = a^2 + b^2 = 2^2 + 1^2 = 5$$
$$c = \sqrt{5}$$

Now apply the definitions using $a = 2$, $b = 1$, and $c = \sqrt{5}$.

$$\sin \theta = \frac{b}{c} = \frac{1}{\sqrt{5}} = \frac{\sqrt{5}}{5} \qquad \cos \theta = \frac{a}{c} = \frac{2}{\sqrt{5}} = \frac{2\sqrt{5}}{5}$$

$$\csc \theta = \frac{c}{b} = \frac{\sqrt{5}}{1} = \sqrt{5} \qquad \sec \theta = \frac{c}{a} = \frac{\sqrt{5}}{2} \qquad \cot \theta = \frac{a}{b} = \frac{2}{1} = 2$$

Option 2
Using Identities

Because the value of $\tan \theta$ is known, use the Pythagorean Identity that involves $\tan \theta$:

$$\tan^2 \theta + 1 = \sec^2 \theta \qquad \textcolor{teal}{Equation \, (6)}$$

$$\left(\frac{1}{2}\right)^2 + 1 = \sec^2 \theta \qquad \textcolor{teal}{\tan \theta = \frac{1}{2}}$$

$$\sec^2 \theta = \frac{1}{4} + 1 = \frac{5}{4} \qquad \textcolor{teal}{Proceed \, to \, solve \, for \, \sec \theta.}$$

$$\sec \theta = \frac{\sqrt{5}}{2} \qquad \textcolor{teal}{\sec \theta > 0 \, since \, \theta \, is \, acute.}$$

We now know $\tan \theta = \dfrac{1}{2}$ and $\sec \theta = \dfrac{\sqrt{5}}{2}$. Using reciprocal identities,

$$\cos \theta = \frac{1}{\sec \theta} = \frac{1}{\dfrac{\sqrt{5}}{2}} = \frac{2}{\sqrt{5}} = \frac{2\sqrt{5}}{5}$$

$$\cot \theta = \frac{1}{\tan \theta} = \frac{1}{\dfrac{1}{2}} = 2$$

To find $\sin \theta$, use the following reasoning:

Then
$$\tan \theta = \frac{\sin \theta}{\cos \theta}, \quad \text{so} \quad \sin \theta = (\tan \theta)(\cos \theta) = \frac{1}{2} \cdot \frac{2\sqrt{5}}{5} = \frac{\sqrt{5}}{5}$$

$$\csc \theta = \frac{1}{\sin \theta} = \frac{1}{\dfrac{\sqrt{5}}{5}} = \sqrt{5}$$

●

 Now Work PROBLEM 25

4 Use the Complementary Angle Theorem

Two acute angles are called **complementary** if their sum is a right angle, or 90°. Because the sum of the angles of any triangle is 180°, it follows that for a right triangle, the sum of the acute angles is 90°, so the two acute angles are complementary.

Refer now to Figure 25; we have labeled the angle opposite side b as B and the angle opposite side a as A. Notice that side a is adjacent to angle B and is opposite angle A. Similarly, side b is opposite angle B and is adjacent to angle A. As a result,

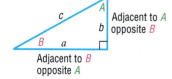

Figure 25

$$\sin B = \frac{b}{c} = \cos A \qquad \cos B = \frac{a}{c} = \sin A \qquad \tan B = \frac{b}{a} = \cot A$$

$$\csc B = \frac{c}{b} = \sec A \qquad \sec B = \frac{c}{a} = \csc A \qquad \cot B = \frac{a}{b} = \tan A$$

(8)

Because of these relationships, the functions sine and cosine, tangent and cotangent, and secant and cosecant are called **cofunctions** of each other. The identities (8) may be expressed in words as follows:

THEOREM

> **Complementary Angle Theorem**
>
> Cofunctions of complementary angles are equal.

Here are examples of this theorem.

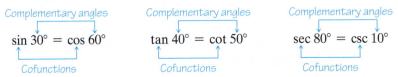

If an angle θ is measured in degrees, we will use the degree symbol when writing a trigonometric function of θ, as, for example, in $\sin 30°$ and $\tan 45°$. If an angle θ is measured in radians, then no symbol is used when writing a trigonometric function of θ, as, for example, in $\cos \pi$ and $\sec \dfrac{\pi}{3}$.

If θ is an acute angle measured in degrees, the angle $90° - \theta$ (or $\dfrac{\pi}{2} - \theta$, if θ is in radians) is the angle complementary to θ. Table 2 restates the preceding theorem on cofunctions.

Table 2

θ (Degrees)	θ (Radians)
$\sin \theta = \cos(90° - \theta)$	$\sin \theta = \cos\left(\dfrac{\pi}{2} - \theta\right)$
$\cos \theta = \sin(90° - \theta)$	$\cos \theta = \sin\left(\dfrac{\pi}{2} - \theta\right)$
$\tan \theta = \cot(90° - \theta)$	$\tan \theta = \cot\left(\dfrac{\pi}{2} - \theta\right)$
$\csc \theta = \sec(90° - \theta)$	$\csc \theta = \sec\left(\dfrac{\pi}{2} - \theta\right)$
$\sec \theta = \csc(90° - \theta)$	$\sec \theta = \csc\left(\dfrac{\pi}{2} - \theta\right)$
$\cot \theta = \tan(90° - \theta)$	$\cot \theta = \tan\left(\dfrac{\pi}{2} - \theta\right)$

The angle θ in Table 2 is acute. We will see later (Section 8.5) that these results are valid for any angle θ.

EXAMPLE 6

Using the Complementary Angle Theorem

(a) $\sin 62° = \cos(90° - 62°) = \cos 28°$

(b) $\tan \dfrac{\pi}{12} = \cot\left(\dfrac{\pi}{2} - \dfrac{\pi}{12}\right) = \cot \dfrac{5\pi}{12}$

(c) $\cos \dfrac{\pi}{4} = \sin\left(\dfrac{\pi}{2} - \dfrac{\pi}{4}\right) = \sin \dfrac{\pi}{4}$

(d) $\csc \dfrac{\pi}{6} = \sec\left(\dfrac{\pi}{2} - \dfrac{\pi}{6}\right) = \sec \dfrac{\pi}{3}$

EXAMPLE 7

Using the Complementary Angle Theorem

Find the exact value of each expression. Do not use a calculator.

(a) $\sec 28° - \csc 62°$

(b) $\dfrac{\sin 35°}{\cos 55°}$

Solution

(a) $\sec 28° - \csc 62° = \csc(90° - 28°) - \csc 62°$

$\qquad\qquad\qquad = \csc 62° - \csc 62° = 0$

(b) $\dfrac{\sin 35°}{\cos 55°} = \dfrac{\cos(90° - 35°)}{\cos 55°} = \dfrac{\cos 55°}{\cos 55°} = 1$

Now Work PROBLEM 43

Historical Feature

he name *sine* for the sine function arose from a medieval confusion. The name comes from the Sanskrit word *jĩva*, (meaning "chord"), first used in India by Araybhata the Elder (AD 510). He really meant half-chord, but abbreviated it. This was brought into Arabic as *jiba*, which was meaningless. Because the proper Arabic word *jaib* would be written the same way (short vowels are not written out in Arabic), *jĩba*, was pronounced as *jaib*, which meant "bosom" or "hollow," and *jaib* remains the Arabic word for *sine* to this day. Scholars translating the Arabic works into Latin found that the word *sinus* also meant "bosom" or "hollow," and from *sinus* we get our *sine*.

The name *tangent*, introduced by Thomas Finck (1583), can be understood by looking at Figure 26. The line segment $\overline{DC}$ is tangent to the circle at C. If $d(O, B) = d(O, C) = 1$, then the length of the line segment $\overline{DC}$ is

$$d(D, C) = \frac{d(D, C)}{1} = \frac{d(D, C)}{d(O, C)} = \tan \alpha$$

The old name for the tangent is *umbra versa* (meaning turned shadow), referring to the use of the tangent in solving height problems with shadows.

The names of the cofunctions came about as follows. If α and β are complementary angles, then $\cos \alpha = \sin \beta$. Because β is the complement of α, it was natural to write the cosine of α as *sin co* α. Probably for reasons involving ease of pronunciation, the *co* migrated to the front, and then cosine received a three-letter abbreviation to match sin, sec, and tan. The two other cofunctions were similarly treated, except that the long forms *cotan* and *cosec* survive to this day in some countries.

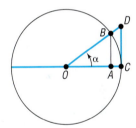

Figure 26

7.2 Assess Your Understanding

'Are You Prepared?' Answers are given at the end of these exercises. If you get a wrong answer, read the pages listed in red.

1. In a right triangle with legs $a = 6$ and $b = 10$, the Pythagorean Theorem tells us that the hypotenuse is $c =$ _____. (pp. 30–31)

2. The value of the function $f(x) = 3x - 7$ at 5 is _____. (pp. 202–205)

Concepts and Vocabulary

3. Two acute angles whose sum is a right angle are called _____ angles.

4. The sine and _____ functions are cofunctions.

5. $\sec 28° = \csc$ ___.

6. Which of the following ratios is used to find the value of $\cot \theta$?

(a) $\dfrac{\text{opposite}}{\text{hypotenuse}}$ (b) $\dfrac{\text{opposite}}{\text{adjacent}}$

(c) $\dfrac{\text{adjacent}}{\text{opposite}}$ (d) $\dfrac{\text{hypotenuse}}{\text{adjacent}}$

7. *True or False* $\tan \theta = \dfrac{\sin \theta}{\cos \theta}$.

8. Which of the following expressions does *not* equal 1?

(a) $\cos^2 \theta - \sin^2 \theta$ (b) $\sec^2 \theta - \tan^2 \theta$

(c) $\csc^2 \theta - \cot^2 \theta$ (d) $\sin^2 \theta + \cos^2 \theta$

9. *True or False* If θ is an acute angle and $\sec \theta = 3$, then $\cos \theta = \dfrac{1}{3}$.

10. *True or False* $\tan \dfrac{\pi}{5} = \cot \dfrac{4\pi}{5}$.

Skill Building

In Problems 11–20, find the value of the six trigonometric functions of the angle θ in each figure.

11.

12.

13.

14.

15.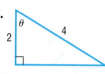

16.

17.

18.

19.

20.

In Problems 21–24, use identities to find the exact value of each of the four remaining trigonometric functions of the acute angle θ.

21. $\sin \theta = \dfrac{1}{2}$, $\cos \theta = \dfrac{\sqrt{3}}{2}$
22. $\sin \theta = \dfrac{\sqrt{3}}{2}$, $\cos \theta = \dfrac{1}{2}$
23. $\sin \theta = \dfrac{2}{3}$, $\cos \theta = \dfrac{\sqrt{5}}{3}$
24. $\sin \theta = \dfrac{1}{3}$, $\cos \theta = \dfrac{2\sqrt{2}}{3}$

In Problems 25–36, use the definition or identities to find the exact value of each of the remaining five trigonometric functions of the acute angle θ.

25. $\sin \theta = \dfrac{\sqrt{2}}{2}$
26. $\cos \theta = \dfrac{\sqrt{2}}{2}$
27. $\cos \theta = \dfrac{1}{3}$
28. $\sin \theta = \dfrac{\sqrt{3}}{4}$

29. $\tan \theta = \dfrac{1}{2}$
30. $\cot \theta = \dfrac{1}{2}$
31. $\sec \theta = 3$
32. $\csc \theta = 5$

33. $\tan \theta = \sqrt{2}$
34. $\sec \theta = \dfrac{5}{3}$
35. $\csc \theta = 2$
36. $\cot \theta = 2$

In Problems 37–54, use Fundamental Identities and/or the Complementary Angle Theorem to find the exact value of each expression. Do not use a calculator.

37. $\sin^2 20° + \cos^2 20°$
38. $\sec^2 28° - \tan^2 28°$
39. $\sin 80° \csc 80°$
40. $\tan 10° \cot 10°$

41. $\tan 50° - \dfrac{\sin 50°}{\cos 50°}$
42. $\cot 25° - \dfrac{\cos 25°}{\sin 25°}$
43. $\sin 38° - \cos 52°$
44. $\tan 12° - \cot 78°$

45. $\dfrac{\cos 10°}{\sin 80°}$
46. $\dfrac{\cos 40°}{\sin 50°}$
47. $1 - \cos^2 20° - \cos^2 70°$
48. $1 + \tan^2 5° - \csc^2 85°$

49. $\tan 20° - \dfrac{\cos 70°}{\cos 20°}$
50. $\cot 40° - \dfrac{\sin 50°}{\sin 40°}$
51. $\tan 35° \cdot \sec 55° \cdot \cos 35°$
52. $\cot 25° \cdot \csc 65° \cdot \sin 25°$

53. $\cos 35° \sin 55° + \cos 55° \sin 35°$
54. $\sec 35° \csc 55° - \tan 35° \cot 55°$

55. Given $\sin 30° = \dfrac{1}{2}$, use trigonometric identities to find the exact value of
  (a) $\cos 60°$ (b) $\cos^2 30°$
  (c) $\csc \dfrac{\pi}{6}$ (d) $\sec \dfrac{\pi}{3}$

56. Given $\sin 60° = \dfrac{\sqrt{3}}{2}$, use trigonometric identities to find the exact value of
  (a) $\cos 30°$ (b) $\cos^2 60°$
  (c) $\sec \dfrac{\pi}{6}$ (d) $\csc \dfrac{\pi}{3}$

57. Given $\tan \theta = 4$, use trigonometric identities to find the exact value of
  (a) $\sec^2 \theta$ (b) $\cot \theta$
  (c) $\cot\left(\dfrac{\pi}{2} - \theta\right)$ (d) $\csc^2 \theta$

58. Given $\sec \theta = 3$, use trigonometric identities to find the exact value of
  (a) $\cos \theta$ (b) $\tan^2 \theta$ (c) $\csc(90° - \theta)$ (d) $\sin^2 \theta$

59. Given $\csc \theta = 4$, use trigonometric identities to find the exact value of
  (a) $\sin \theta$ (b) $\cot^2 \theta$
  (c) $\sec(90° - \theta)$ (d) $\sec^2 \theta$

60. Given $\cot \theta = 2$, use trigonometric identities to find the exact value of
  (a) $\tan \theta$ (b) $\csc^2 \theta$
  (c) $\tan\left(\dfrac{\pi}{2} - \theta\right)$ (d) $\sec^2 \theta$

61. Given the approximation $\sin 38° \approx 0.62$, use trigonometric identities to find the approximate value of
  (a) $\cos 38°$ (b) $\tan 38°$
  (c) $\cot 38°$ (d) $\sec 38°$
  (e) $\csc 38°$ (f) $\sin 52°$
  (g) $\cos 52°$ (h) $\tan 52°$

62. Given the approximation $\cos 21° \approx 0.93$, use trigonometric identities to find the approximate value of
(a) $\sin 21°$ 　　　　　(b) $\tan 21°$
(c) $\cot 21°$ 　　　　　(d) $\sec 21°$
(e) $\csc 21°$ 　　　　　(f) $\sin 69°$
(g) $\cos 69°$ 　　　　　(h) $\tan 69°$

63. If $\sin \theta = 0.3$, find the exact value of $\sin \theta + \cos\left(\dfrac{\pi}{2} - \theta\right)$.

64. If $\tan \theta = 4$, find the exact value of $\tan \theta + \tan\left(\dfrac{\pi}{2} - \theta\right)$.

65. Find an acute angle θ that satisfies the equation
$$\sin \theta = \cos(2\theta + 30°)$$

66. Find an acute angle θ that satisfies the equation
$$\tan \theta = \cot(\theta + 45°)$$

Applications and Extensions

In Problems 67–72, simplify each expression for the given value of x . Assume the angle θ is acute.

67. $\sqrt{25 - x^2}$ if $x = 5\sin\theta$

68. $\sqrt{16 - 9x^2}$ if $x = \dfrac{4}{3}\sin\theta$

69. $\sqrt{49 + x^2}$ if $x = 7\tan\theta$

70. $\sqrt{4 + x^2}$ if $x = -2\tan\theta$

71. $\sqrt{x^2 - 36}$ if $x = -6\sec\theta$

72. $\sqrt{100x^2 - 81}$ if $x = \dfrac{9}{10}\sec\theta$

73. Calculating the Time of a Trip From a parking lot you want to walk to a house on the ocean. The house is located 1500 feet down a paved path that parallels the beach, which is 500 feet wide. Along the path you can walk 300 feet per minute, but in the sand on the beach you can walk only 100 feet per minute. See the illustration.

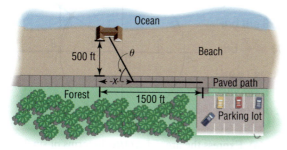

(a) Calculate the time T if you walk 1500 feet along the paved path and then 500 feet in the sand to the house.
(b) Calculate the time T if you walk in the sand directly toward the ocean for 500 feet and then turn left and walk along the beach for 1500 feet to the house.
(c) Express the time T to get from the parking lot to the beachhouse as a function of the angle θ shown in the illustration.
(d) Calculate the time T if you walk directly from the parking lot to the house.
$$\left[\textbf{Hint: } \tan\theta = \frac{500}{1500}\right]$$
(e) Calculate the time T if you walk 1000 feet along the paved path and then walk directly to the house.
(f) Graph $T = T(\theta)$. For what angle θ is T least? What is x for this angle? What is the minimum time?
(g) Explain why $\tan\theta = \dfrac{1}{3}$ gives the smallest angle θ that is possible.

74. Carrying a Ladder around a Corner Two hallways, one of width 3 feet, the other of width 4 feet, meet at a right angle. See the illustration (top, right).
(a) Express the length L of the line segment shown as a function of the angle θ.
(b) Discuss why the length of the longest ladder that can be carried around the corner is equal to the smallest value of L.

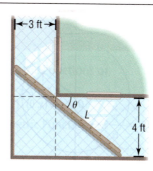

75. Electrical Engineering A resistor and an inductor connected in a series network impede the flow of an alternating current. This impedance Z is determined by the reactance X of the inductor and the resistance R of the resistor. The three quantities, all measured in ohms, can be represented by the sides of a right triangle as illustrated, so $Z^2 = X^2 + R^2$. The angle ϕ is called the **phase angle**. Suppose a series network has an inductive reactance of $X = 400$ ohms and a resistance of $R = 600$ ohms.
(a) Find the impedance Z.
(b) Find the values of the six trigonometric functions of the phase angle ϕ.

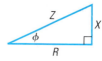

76. Electrical Engineering Refer to Problem 75. A series network has a resistance of $R = 588$ ohms. The phase angle ϕ is such that $\tan\phi = \dfrac{5}{12}$.
(a) Determine the inductive reactance X and the impedance Z.
(b) Determine the values of the remaining five trigonometric functions of the phase angle ϕ.

77. Geometry Suppose that the angle θ is a central angle of a circle of radius 1 (see the figure). Show that
(a) Angle $OAC = \dfrac{\theta}{2}$
(b) $|CD| = \sin\theta$ and $|OD| = \cos\theta$
(c) $\tan\dfrac{\theta}{2} = \dfrac{\sin\theta}{1 + \cos\theta}$

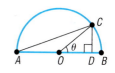

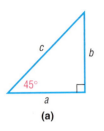

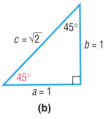

Figure 27 Isosceles right triangle

Then, by the Pythagorean Theorem,

$$c^2 = a^2 + b^2 = 1 + 1 = 2$$

$$c = \sqrt{2}$$

As a result, we have the triangle in Figure 27(b), from which we find

$$\sin \frac{\pi}{4} = \sin 45° = \frac{b}{c} = \frac{1}{\sqrt{2}} = \frac{\sqrt{2}}{2} \qquad \cos \frac{\pi}{4} = \cos 45° = \frac{a}{c} = \frac{1}{\sqrt{2}} = \frac{\sqrt{2}}{2}$$

Using Quotient and Reciprocal Identities,

$$\tan \frac{\pi}{4} = \tan 45° = \frac{\sin 45°}{\cos 45°} = \frac{\frac{\sqrt{2}}{2}}{\frac{\sqrt{2}}{2}} = 1 \qquad \cot \frac{\pi}{4} = \cot 45° = \frac{1}{\tan 45°} = \frac{1}{1} = 1$$

$$\sec \frac{\pi}{4} = \sec 45° = \frac{1}{\cos 45°} = \frac{1}{\frac{1}{\sqrt{2}}} = \sqrt{2} \qquad \csc \frac{\pi}{4} = \csc 45° = \frac{1}{\sin 45°} = \frac{1}{\frac{1}{\sqrt{2}}} = \sqrt{2}$$

EXAMPLE 2

Finding the Exact Value of a Trigonometric Expression

Find the exact value of each expression.

(a) $(\sin 45°)(\tan 45°)$

(b) $\left(\sec \dfrac{\pi}{4}\right)\left(\cot \dfrac{\pi}{4}\right)$

Solution   Use the results obtained in Example 1.

(a) $(\sin 45°)(\tan 45°) = \dfrac{\sqrt{2}}{2} \cdot 1 = \dfrac{\sqrt{2}}{2}$

(b) $\left(\sec \dfrac{\pi}{4}\right)\left(\cot \dfrac{\pi}{4}\right) = \sqrt{2} \cdot 1 = \sqrt{2}$

Now Work PROBLEMS 7 AND 19

2 Find the Exact Values of the Trigonometric Functions of $\dfrac{\pi}{6} = 30°$ and $\dfrac{\pi}{3} = 60°$

EXAMPLE 3

Finding the Exact Values of the Trigonometric Functions of $\dfrac{\pi}{6} = 30°$ and $\dfrac{\pi}{3} = 60°$

Find the exact values of the six trigonometric functions of $\dfrac{\pi}{6} = 30°$ and $\dfrac{\pi}{3} = 60°$.

Solution   Form a right triangle in which one of the angles is $\dfrac{\pi}{6} = 30°$. It then follows that the third angle is $\dfrac{\pi}{3} = 60°$. Figure 28(a) illustrates such a triangle with hypotenuse of length 2. Our problem is to determine a and b.

Begin by placing next to the triangle in Figure 28(a) another triangle congruent to the first, as shown in Figure 28(b). Notice that you now have a triangle whose angles are each 60°. This triangle is therefore equilateral, so each side is of length 2. In particular, the base is $2a = 2$, so $a = 1$. By the Pythagorean Theorem, b satisfies the equation $a^2 + b^2 = c^2$, so

$$a^2 + b^2 = c^2$$
$$1^2 + b^2 = 2^2 \qquad a = 1, c = 2$$
$$b^2 = 4 - 1 = 3$$
$$b = \sqrt{3}$$

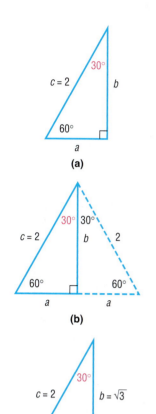

Figure 28 30° – 60° – 90° triangle

Using the triangle in Figure 28(c) and the fact that $\dfrac{\pi}{6} = 30°$ and $\dfrac{\pi}{3} = 60°$ are complementary angles, we find

$$\sin \frac{\pi}{6} = \sin 30° = \frac{\text{opposite}}{\text{hypotenuse}} = \frac{1}{2} \qquad\qquad \cos \frac{\pi}{3} = \cos 60° = \frac{1}{2}$$

$$\cos \frac{\pi}{6} = \cos 30° = \frac{\text{adjacent}}{\text{hypotenuse}} = \frac{\sqrt{3}}{2} \qquad\qquad \sin \frac{\pi}{3} = \sin 60° = \frac{\sqrt{3}}{2}$$

$$\tan \frac{\pi}{6} = \tan 30° = \frac{\sin 30°}{\cos 30°} = \frac{\frac{1}{2}}{\frac{\sqrt{3}}{2}} = \frac{1}{\sqrt{3}} = \frac{\sqrt{3}}{3} \qquad \cot \frac{\pi}{3} = \cot 60° = \frac{\sqrt{3}}{3}$$

$$\csc \frac{\pi}{6} = \csc 30° = \frac{1}{\sin 30°} = \frac{1}{\frac{1}{2}} = 2 \qquad\qquad \sec \frac{\pi}{3} = \sec 60° = 2$$

$$\sec \frac{\pi}{6} = \sec 30° = \frac{1}{\cos 30°} = \frac{1}{\frac{\sqrt{3}}{2}} = \frac{2}{\sqrt{3}} = \frac{2\sqrt{3}}{3} \qquad \csc \frac{\pi}{3} = \csc 60° = \frac{2\sqrt{3}}{3}$$

$$\cot \frac{\pi}{6} = \cot 30° = \frac{1}{\tan 30°} = \frac{1}{\frac{\sqrt{3}}{3}} = \frac{3}{\sqrt{3}} = \sqrt{3} \qquad \tan \frac{\pi}{3} = \tan 60° = \sqrt{3}$$

Table 3 summarizes the information just derived for the angles $\dfrac{\pi}{6} = 30°, \dfrac{\pi}{4} = 45°$, and $\dfrac{\pi}{3} = 60°$. Rather than memorize the entries in Table 3, you can draw the appropriate triangle to determine the values given in the table.

Table 3

θ (Radians)	θ (Degrees)	$\sin \theta$	$\cos \theta$	$\tan \theta$	$\csc \theta$	$\sec \theta$	$\cot \theta$
$\dfrac{\pi}{6}$	30°	$\dfrac{1}{2}$	$\dfrac{\sqrt{3}}{2}$	$\dfrac{\sqrt{3}}{3}$	2	$\dfrac{2\sqrt{3}}{3}$	$\sqrt{3}$
$\dfrac{\pi}{4}$	45°	$\dfrac{\sqrt{2}}{2}$	$\dfrac{\sqrt{2}}{2}$	1	$\sqrt{2}$	$\sqrt{2}$	1
$\dfrac{\pi}{3}$	60°	$\dfrac{\sqrt{3}}{2}$	$\dfrac{1}{2}$	$\sqrt{3}$	$\dfrac{2\sqrt{3}}{3}$	2	$\dfrac{\sqrt{3}}{3}$

EXAMPLE 4

Finding the Exact Value of a Trigonometric Expression

Find the exact value of each expression.

(a) $\sin 45° \cos 30°$ 　　 (b) $\tan \dfrac{\pi}{4} - \sin \dfrac{\pi}{3}$ 　　 (c) $\tan^2 \dfrac{\pi}{6} + \sin^2 \dfrac{\pi}{4}$

Solution

(a) $\sin 45° \cos 30° = \dfrac{\sqrt{2}}{2} \cdot \dfrac{\sqrt{3}}{2} = \dfrac{\sqrt{6}}{4}$

(b) $\tan \dfrac{\pi}{4} - \sin \dfrac{\pi}{3} = 1 - \dfrac{\sqrt{3}}{2} = \dfrac{2 - \sqrt{3}}{2}$

(c) $\tan^2 \dfrac{\pi}{6} + \sin^2 \dfrac{\pi}{4} = \left(\dfrac{\sqrt{3}}{3} \right)^2 + \left(\dfrac{\sqrt{2}}{2} \right)^2 = \dfrac{1}{3} + \dfrac{1}{2} = \dfrac{5}{6}$

Now Work PROBLEMS 11 AND 21

3 Use a Calculator to Approximate the Values of the Trigonometric Functions of Acute Angles

Before getting started, you must first decide whether to enter the angle in the calculator using radians or degrees and then set the calculator to the correct MODE. (Check your instruction manual to find out how your calculator handles degrees and radians.) Your calculator has the keys marked $\boxed{\sin}$, $\boxed{\cos}$, and $\boxed{\tan}$. To find the values of the remaining three trigonometric functions (secant, cosecant, and cotangent), use the reciprocal identities.

WARNING On your calculator the second functions $\sin^{-1}$, $\cos^{-1}$, and $\tan^{-1}$ do not represent the reciprocal of sin, cos, and tan. ∎

$$\sec \theta = \frac{1}{\cos \theta} \qquad \csc \theta = \frac{1}{\sin \theta} \qquad \cot \theta = \frac{1}{\tan \theta}$$

EXAMPLE 5

Using a Calculator to Approximate the Values of Trigonometric Functions

Use a calculator to find the approximate value of each of the following:

(a) $\cos 48°$ (b) $\csc 21°$ (c) $\tan \dfrac{\pi}{12}$

Express your answer rounded to two decimal places.

Solution

(a) First, set the MODE to receive degrees. Rounded to two decimal places,

$$\cos 48° = 0.67$$

(b) Most calculators do not have a csc key. The manufacturers assume that the user knows some trigonometry. To find the value of csc 21°, use the fact that $\csc 21° = \dfrac{1}{\sin 21°}$. Rounded to two decimal places,

$$\csc 21° = 2.79$$

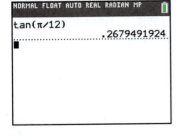

(c) Set the MODE to receive radians. Figure 29 shows the solution using a TI-84 Plus C graphing calculator. Rounded to two decimal places,

$$\tan \frac{\pi}{12} = 0.27$$

Figure 29

✏️ **Now Work** PROBLEM 31

---

4 Model and Solve Applied Problems Involving Right Triangles

Right triangles can be used to model many types of situations, such as the optimal design of a rain gutter.*

EXAMPLE 6

Constructing a Rain Gutter

A rain gutter is to be constructed of aluminum sheets 12 inches wide. After marking off a length of 4 inches from each edge, the sides are bent up at an angle θ. See Figure 30.

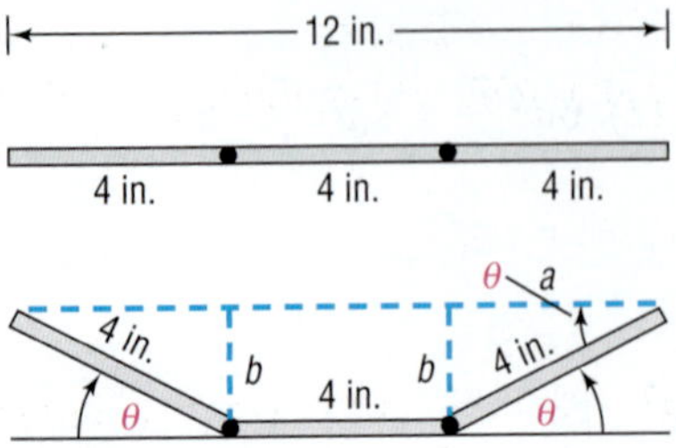

(a) Express the area A of the opening as a function of θ.
    [**Hint:** Let b denote the vertical height of the bend.]

(b) Find the area A of the opening for $\theta = 30°$, $\theta = 45°$, $\theta = 60°$, and $\theta = 75°$.

(c) Graph $A = A(\theta)$. Find the angle θ that makes A largest. (This bend will allow the most water to flow through the gutter.)

Figure 30

---

*In applied problems, it is important that answers be reported with both justifiable accuracy and appropriate significant figures. We shall assume that angles are measured to the nearest tenth and sides are measured to the nearest hundredth, resulting in sides being rounded to two decimal places and angles being rounded to one decimal place.

Solution

Figure 31

(a) Look again at Figure 30. The area A of the opening is the sum of the areas of two congruent right triangles and one rectangle. Look at Figure 31, which shows the triangle on the right in Figure 30 redrawn. Note that

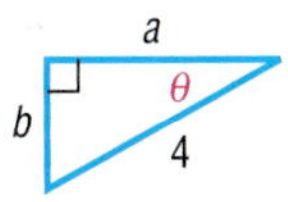

$$\cos\theta = \frac{a}{4}, \quad \text{so} \quad a = 4\cos\theta \qquad \sin\theta = \frac{b}{4}, \quad \text{so} \quad b = 4\sin\theta$$

The area of the triangle is

$$\text{area of triangle} = \frac{1}{2}(\text{base})(\text{height}) = \frac{1}{2}ab = \frac{1}{2}(4\cos\theta)(4\sin\theta) = 8\sin\theta\cos\theta$$

So the area of the two congruent triangles is $16\sin\theta\cos\theta$.
The rectangle has length 4 and height b, so its area is

$$\text{area of rectangle} = 4b = 4(4\sin\theta) = 16\sin\theta$$

The area A of the opening is

$$A = \text{area of the two triangles} + \text{area of the rectangle}$$
$$A(\theta) = 16\sin\theta\cos\theta + 16\sin\theta = 16\sin\theta(\cos\theta + 1)$$

(b) For $\theta = 30°$: $\qquad A(30°) = 16\sin 30°(\cos 30° + 1)$

$$= 16\left(\frac{1}{2}\right)\left(\frac{\sqrt{3}}{2} + 1\right) = 4\sqrt{3} + 8 \approx 14.93$$

The area of the opening for $\theta = 30°$ is about 14.93 square inches.

For $\theta = 45°$: $\qquad A(45°) = 16\sin 45°(\cos 45° + 1)$

$$= 16\left(\frac{\sqrt{2}}{2}\right)\left(\frac{\sqrt{2}}{2} + 1\right) = 8 + 8\sqrt{2} \approx 19.31$$

The area of the opening for $\theta = 45°$ is about 19.31 square inches.

For $\theta = 60°$: $\qquad A(60°) = 16\sin 60°(\cos 60° + 1)$

$$= 16\left(\frac{\sqrt{3}}{2}\right)\left(\frac{1}{2} + 1\right) = 12\sqrt{3} \approx 20.78$$

The area of the opening for $\theta = 60°$ is about 20.78 square inches.

For $\theta = 75°$: $\qquad A(75°) = 16\sin 75°(\cos 75° + 1) \approx 19.45$

The area of the opening for $\theta = 75°$ is about 19.45 square inches.

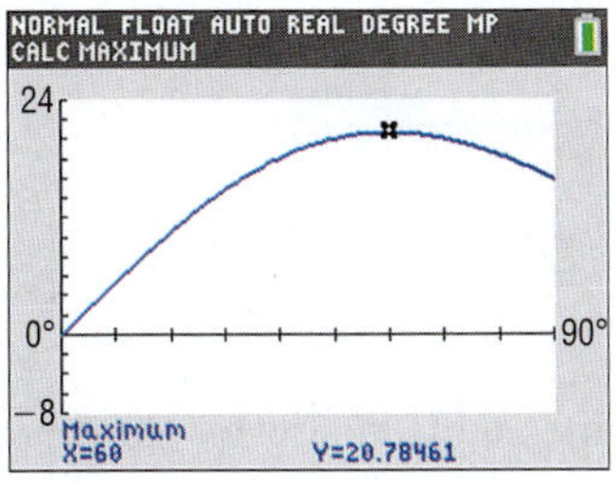

(c) Figure 32 shows the graph of $A = A(\theta)$. Using MAXIMUM, the angle θ that makes A largest is 60°. ●

Figure 32

 **Now Work** PROBLEM 67

In addition to developing models using right triangles, right triangle trigonometry can be used to measure heights and distances that are either awkward or impossible to measure by ordinary means. When using right triangles to solve these problems, pay attention to the known measures. This will indicate which trigonometric function to use. For example, if we know the measure of an angle and the length of the side adjacent to the angle, and we wish to find the length of the opposite side, we would use the tangent function. Do you know why?

| EXAMPLE 7 | **Finding the Width of a River** |

A surveyor can measure the width of a river by setting up a transit at a point C on one side of the river and taking a sighting of a point A on the other side. Refer to Figure 33. After turning through an angle of $90°$ at C, the surveyor walks a distance of 200 meters to point B. Using the transit at B, the angle θ is measured and found to be $20°$. What is the width of the river rounded to the nearest meter?

Solution

We seek the length of side b. We know a and θ, so we use the fact that b is opposite θ and a is adjacent to θ and write

$$\tan \theta = \frac{b}{a}$$

which leads to

$$\tan 20° = \frac{b}{200}$$

$$b = 200 \tan 20° \approx 72.79 \text{ meters}$$

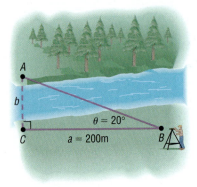

Figure 33

The width of the river is 73 meters, rounded to the nearest meter. ●

Now Work PROBLEM 73

Vertical heights can sometimes be measured using either the *angle of elevation* or the *angle of depression*. If a person is looking up at an object, the acute angle measured from the horizontal to a line of sight to the object is called the **angle of elevation**. See Figure 34(a).

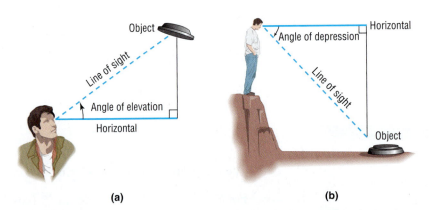

Figure 34   (a)   (b)

If a person is looking down at an object, the acute angle made by the line of sight to the object and the horizontal is called the **angle of depression**. See Figure 34(b).

| EXAMPLE 8 | **Finding the Height of a Cloud** |

Meteorologists find the height of a cloud using an instrument called a **ceilometer**. A ceilometer consists of a **light projector** that directs a vertical light beam up to the cloud base and a **light detector** that scans the cloud to detect the light beam. See Figure 35(a). At Midway Airport in Chicago, a ceilometer was employed to find the height of the cloud cover. It was set up with its light detector 300 feet from its light projector. If the angle of elevation from the light detector to the base of the cloud was $75°$, what was the height of the cloud cover?

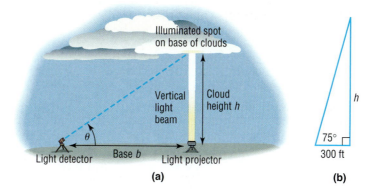

Figure 35

(a)

(b)

Solution   Figure 35(b) illustrates the situation. To determine the height h, use the fact that $\tan 75° = \dfrac{h}{300}$, so

$$h = 300 \tan 75° \approx 1120 \text{ feet}$$

The ceiling (height to the base of the cloud cover) was approximately 1120 feet.   ●

Now Work PROBLEM 75

The idea behind Example 8 can also be used to find the height of an object with a base that is not accessible to the horizontal.

EXAMPLE 9   **Finding the Height of a Statue on a Building**

Adorning the top of the Board of Trade building in Chicago is a statue of Ceres, the Roman goddess of wheat. From street level, two observations are taken 400 feet from the center of the building. The angle of elevation to the base of the statue is found to be 55.1°, and the angle of elevation to the top of the statue is 56.5°. See Figure 36(a). What is the height of the statue?

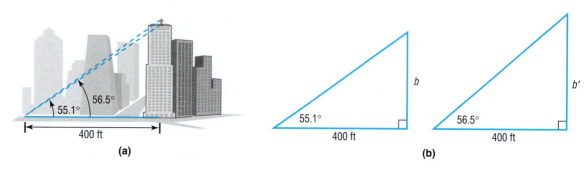

Figure 36

(a)

(b)

Solution   Figure 36(b) shows two triangles that replicate Figure 36(a). The height of the statue of Ceres will be $b' - b$. To find b and b', refer to Figure 36(b).

$$\tan 55.1° = \frac{b}{400} \qquad\qquad \tan 56.5° = \frac{b'}{400}$$

$$b = 400 \tan 55.1° \approx 573.39 \qquad b' = 400 \tan 56.5° \approx 604.33$$

The height of the statue is approximately $604.33 - 573.39 = 30.94$ feet ≈ 31 feet.

●

Now Work PROBLEM 81

7.3 Assess Your Understanding

Concepts and Vocabulary

1. $\tan 45° + \sin 30° = $ ___ .

2. Using a calculator, $\sin 1 = $ _____ , rounded to two decimal places.

3. *True or False* Exact values can be found for the trigonometric functions of 60°.

4. *True or False* Exact values can be found for the sine of any angle.

5. Find the exact value of $\csc \dfrac{\pi}{6}$.

   (a) 1   (b) 2   (c) $\sqrt{2}$ (d) $\dfrac{2\sqrt{3}}{3}$

6. Use a calculator to approximate $\cot 35°$.

   (a) 0.47   (b) 0.70   (c) 1.43   (d) 2.11

Skill Building

7. Write down the exact value of each of the six trigonometric functions of 45°.

8. Write down the exact value of each of the six trigonometric functions of 30° and of 60°.

In Problems 9–18, $f(\theta) = \sin \theta$ and $g(\theta) = \cos \theta$. Find the exact value of each expression if $\theta = 60°$. Do not use a calculator.

9. $f(\theta)$

10. $g(\theta)$

11. $f\left(\dfrac{\theta}{2}\right)$

12. $g\left(\dfrac{\theta}{2}\right)$

13. $[f(\theta)]^2$

14. $[g(\theta)]^2$

15. $2f(\theta)$

16. $2g(\theta)$

17. $\dfrac{f(\theta)}{2}$

18. $\dfrac{g(\theta)}{2}$

In Problems 19–30, find the exact value of each expression. Do not use a calculator.

19. $4 \cos 45° - 2 \sin 45°$

20. $2 \sin 45° + 4 \cos 30°$

21. $6 \tan 45° - 8 \cos 60°$

22. $\sin 30° \cdot \tan 60°$

23. $\sec \dfrac{\pi}{4} + 2 \csc \dfrac{\pi}{3}$

24. $\tan \dfrac{\pi}{4} + \cot \dfrac{\pi}{4}$

25. $\sec^2 \dfrac{\pi}{6} - 4$

26. $4 + \tan^2 \dfrac{\pi}{3}$

27. $\sin^2 30° + \cos^2 60°$

28. $\sec^2 60° - \tan^2 45°$

29. $1 - \cos^2 30° - \cos^2 60°$

30. $1 + \tan^2 30° - \csc^2 45°$

In Problems 31–48, use a calculator to find the approximate value of each expression. Round the answer to two decimal places.

31. $\sin 28°$

32. $\cos 14°$

33. $\tan 21°$

34. $\cot 70°$

35. $\sec 41°$

36. $\csc 55°$

37. $\sin \dfrac{\pi}{10}$

38. $\cos \dfrac{\pi}{8}$

39. $\tan \dfrac{5\pi}{12}$

40. $\cot \dfrac{\pi}{18}$

41. $\sec \dfrac{\pi}{12}$

42. $\csc \dfrac{5\pi}{13}$

43. $\sin 1$

44. $\tan 1$

45. $\sin 1°$

46. $\tan 1°$

47. $\tan 0.3$

48. $\tan 0.1$

Mixed Practice

In Problems 49–58, $f(x) = \sin x$, $g(x) = \cos x$, $h(x) = 2x$, and $p(x) = \dfrac{x}{2}$. Find the value of each of the following:

49. $(f + g)(30°)$

50. $(f - g)(60°)$

51. $(f \cdot g)\left(\dfrac{\pi}{4}\right)$

52. $(f \cdot g)\left(\dfrac{\pi}{3}\right)$

53. $(f \circ h)\left(\dfrac{\pi}{6}\right)$

54. $(g \circ p)(60°)$

55. $(p \circ g)(45°)$

56. $(h \circ f)\left(\dfrac{\pi}{6}\right)$

57. (a) Find $f\left(\dfrac{\pi}{4}\right)$. What point is on the graph of f?

   (b) Assuming $0 < x < \dfrac{\pi}{2}$, f is one-to-one. Use the result of part (a) to find a point on the graph of f^{-1}.

   (c) What point is on the graph of $y = f\left(x + \dfrac{\pi}{4}\right) - 3$ if $x = \dfrac{\pi}{4}$?

58. (a) Find $g\left(\dfrac{\pi}{6}\right)$. What point is on the graph of g?

   (b) Assuming $0 < x < \dfrac{\pi}{2}$, g is one-to-one. Use the result of part (a) to find a point on the graph of g^{-1}.

   (c) What point is on the graph of $y = 2g\left(x - \dfrac{\pi}{6}\right)$ if $x = \dfrac{\pi}{6}$?

Applications and Extensions

59. Geometry A right triangle has a hypotenuse of length 8 inches. If one angle is 35°, find the length of each leg.

60. Geometry A right triangle has a hypotenuse of length 10 centimeters. If one angle is 40°, find the length of each leg.

61. Geometry A right triangle contains a 25° angle.
(a) If one leg is of length 5 inches, what is the length of the hypotenuse?
(b) There are two answers. How is this possible?

62. Geometry A right triangle contains an angle of $\frac{\pi}{8}$ radian.
(a) If one leg is of length 3 meters, what is the length of the hypotenuse?
(b) There are two answers. How is this possible?

Problems 63–66 require the following discussion.

Projectile Motion The path of a projectile fired at an inclination θ to the horizontal with initial speed v_0 is a parabola. See the figure. The **range R** of the projectile—that is, the horizontal distance that the projectile travels—is found by using the function

$$R(\theta) = \frac{2v_0^2 \sin \theta \cos \theta}{g}$$

where $g \approx 32.2$ feet per second per second ≈ 9.8 meters per second per second is the acceleration due to gravity. The maximum height H of the projectile is given by the function

$$H(\theta) = \frac{v_0^2 \sin^2\theta}{2g}$$

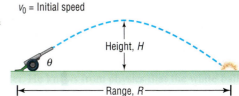

v_0 = Initial speed
Height, H
θ
Range, R

In Problems 63–66, find the range R and maximum height H of the projectile. Round answers to two decimal places.

63. The projectile is fired at an angle of 45° to the horizontal with an initial speed of 100 feet per second.

64. The projectile is fired at an angle of 30° to the horizontal with an initial speed of 150 meters per second.

65. The projectile is fired at an angle of 25° to the horizontal with an initial speed of 500 meters per second.

66. The projectile is fired at an angle of 50° to the horizontal with an initial speed of 200 feet per second.

67. Inclined Plane See the illustration. If friction is ignored, the time t (in seconds) required for a block to slide down an inclined plane is modeled by the function

$$t(\theta) = \sqrt{\frac{2a}{g \sin \theta \cos \theta}}$$

where a is the length (in feet) of the base, and $g \approx 32$ feet per second per second is the acceleration due to gravity. How long does it take a block to slide down an inclined plane with base $a = 10$ feet when
(a) $\theta = 30°$?
(b) $\theta = 45°$?
(c) $\theta = 60°$?

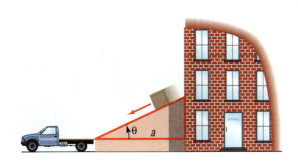

θ a

68. Piston Engines See the illustration. In a certain piston engine, the distance x (in inches) from the center of the drive shaft to the head of the piston is modeled by the function

$$x(\theta) = \cos \theta + \sqrt{16 + 0.5(2 \cos^2\theta - 1)}$$

where θ is the angle between the crank and the path of the piston head. Find x when $\theta = 30°$ and when $\theta = 45°$.

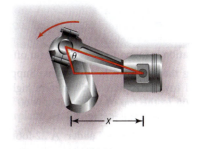

θ
x

69. Find the exact value of $\tan 1° \cdot \tan 2° \cdot \tan 3° \cdots \cdots \tan 89°$.

70. Find the exact value of $\cot 1° \cdot \cot 2° \cdot \cot 3° \cdots \cdots \cot 89°$.

71. Find the exact value of
$\cos 1° \cdot \cos 2° \cdots \cdots \cos 45° \cdot \csc 46° \cdots \cdots \csc 89°$.

72. Find the exact value of
$\sin 1° \cdot \sin 2° \cdots \cdots \sin 45° \cdot \sec 46° \cdots \cdots \sec 89°$.

73. Finding the Width of a Gorge Find the distance from A to C across the gorge illustrated in the figure.

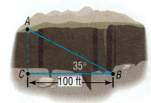

A
$35°$
C 100 ft B

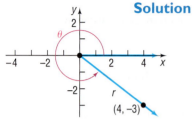

EXAMPLE 1

Finding the Exact Values of the Six Trigonometric Functions of θ, Given a Point on the Terminal Side

Find the exact value of each of the six trigonometric functions of a positive angle θ if $(4, -3)$ is a point on its terminal side.

Solution

Figure 40 illustrates the situation. For the point $(a, b) = (4, -3)$, we have $a = 4$ and $b = -3$. Then $r = \sqrt{a^2 + b^2} = \sqrt{16 + 9} = 5$, so

$$\sin \theta = \frac{b}{r} = -\frac{3}{5} \qquad \cos \theta = \frac{a}{r} = \frac{4}{5} \qquad \tan \theta = \frac{b}{a} = -\frac{3}{4}$$

$$\csc \theta = \frac{r}{b} = -\frac{5}{3} \qquad \sec \theta = \frac{r}{a} = \frac{5}{4} \qquad \cot \theta = \frac{a}{b} = -\frac{4}{3}$$

●

Figure 40

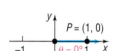 **Now Work** PROBLEM 11

In the next example, we find the exact value of each of the six trigonometric functions at the quadrantal angles $0, \dfrac{\pi}{2}, \pi$, and $\dfrac{3\pi}{2}$.

EXAMPLE 2

Finding the Exact Values of the Six Trigonometric Functions of Quadrantal Angles

Find the exact value of each of the six trigonometric functions of

(a) $\theta = 0 = 0°$ (b) $\theta = \dfrac{\pi}{2} = 90°$ (c) $\theta = \pi = 180°$ (d) $\theta = \dfrac{3\pi}{2} = 270°$

Solution

(a) We can choose any point on the terminal side of $\theta = 0 = 0°$. For convenience, we choose the point $P = (1, 0) = (a, b)$, which is a distance of $r = 1$ unit from the origin. See Figure 41. Then

$$\sin 0 = \sin 0° = \frac{b}{r} = \frac{0}{1} = 0 \qquad \cos 0 = \cos 0° = \frac{a}{r} = \frac{1}{1} = 1$$

$$\tan 0 = \tan 0° = \frac{b}{a} = \frac{0}{1} = 0 \qquad \sec 0 = \sec 0° = \frac{r}{a} = \frac{1}{1} = 1$$

Figure 41 $\theta = 0 = 0°$

Since the y-coordinate of P is 0, $\csc 0$ and $\cot 0$ are not defined.

(b) The point $P = (0, 1) = (a, b)$ is on the terminal side of $\theta = \dfrac{\pi}{2} = 90°$ and is a distance of $r = 1$ unit from the origin. See Figure 42. Then

$$\sin \frac{\pi}{2} = \sin 90° = \frac{b}{r} = \frac{1}{1} = 1 \qquad \cos \frac{\pi}{2} = \cos 90° = \frac{a}{r} = \frac{0}{1} = 0$$

$$\csc \frac{\pi}{2} = \csc 90° = \frac{r}{b} = \frac{1}{1} = 1 \qquad \cot \frac{\pi}{2} = \cot 90° = \frac{a}{b} = \frac{0}{1} = 0$$

Figure 42 $\theta = \dfrac{\pi}{2} = 90°$

Since the x-coordinate of P is 0, $\tan \dfrac{\pi}{2}$ and $\sec \dfrac{\pi}{2}$ are not defined.

(c) The point $P = (-1, 0)$ is on the terminal side of $\theta = \pi = 180°$ and is a distance of $r = 1$ unit from the origin. See Figure 43. Then

$$\sin \pi = \sin 180° = \frac{0}{1} = 0 \qquad \cos \pi = \cos 180° = \frac{-1}{1} = -1$$

$$\tan \pi = \tan 180° = \frac{0}{-1} = 0 \qquad \sec \pi = \sec 180° = \frac{1}{-1} = -1$$

Figure 43 $\theta = \pi = 180°$

Since the y-coordinate of P is 0, $\csc \pi$ and $\cot \pi$ are not defined.

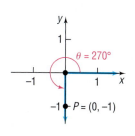

Figure 44 $\theta = \dfrac{3\pi}{2} = 270°$

(d) The point $P = (0, -1)$ is on the terminal side of $\theta = \dfrac{3\pi}{2} = 270°$ and is a distance of $r = 1$ unit from the origin. See Figure 44. Then

$$\sin \frac{3\pi}{2} = \sin 270° = \frac{-1}{1} = -1 \qquad \cos \frac{3\pi}{2} = \cos 270° = \frac{0}{1} = 0$$

$$\csc \frac{3\pi}{2} = \csc 270° = \frac{1}{-1} = -1 \qquad \cot \frac{3\pi}{2} = \cot 270° = \frac{0}{-1} = 0$$

Since the x-coordinate of P is 0, $\tan \dfrac{3\pi}{2}$ and $\sec \dfrac{3\pi}{2}$ are not defined.   ●

Table 4 summarizes the values of the trigonometric functions found in Example 2.

Table 4	θ (Radians)	θ (Degrees)	$\sin \theta$	$\cos \theta$	$\tan \theta$	$\csc \theta$	$\sec \theta$	$\cot \theta$
	0	0°	0	1	0	Not defined	1	Not defined
	$\dfrac{\pi}{2}$	90°	1	0	Not defined	1	Not defined	0
	π	180°	0	−1	0	Not defined	−1	Not defined
	$\dfrac{3\pi}{2}$	270°	−1	0	Not defined	−1	Not defined	0

There is no need to memorize Table 4. To find the value of a trigonometric function of a quadrantal angle, draw the angle and apply the definition, as done in Example 2.

2 Use Coterminal Angles to Find the Exact Value of a Trigonometric Function

DEFINITION

Two angles in standard position are said to be **coterminal** if they have the same terminal side.

See Figure 45.

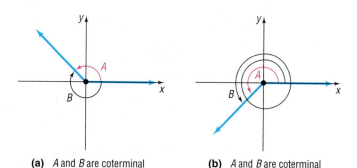

Figure 45 Coterminal angles   **(a)** *A* and *B* are coterminal   **(b)** *A* and *B* are coterminal

For example, the angles 60° and 420° are coterminal, as are the angles −40° and 320°.

In general, if θ is an angle measured in degrees, then $\theta + 360°k$, where k is any integer, is an angle coterminal with θ. If θ is measured in radians, then $\theta + 2\pi k$, where k is any integer, is an angle coterminal with θ.

Because coterminal angles have the same terminal side, it follows that the values of the trigonometric functions of coterminal angles are equal.

Using a Coterminal Angle to Find the Exact Value of a Trigonometric Function

Find the exact value of each of the following:

(a) $\sin 390°$ (b) $\cos 420°$ (c) $\tan \dfrac{9\pi}{4}$ (d) $\sec\left(-\dfrac{7\pi}{4}\right)$ (e) $\csc(-270°)$

Solution

(a) It is best to sketch the angle first. See Figure 46. The angle 390° is coterminal with 30°.

$$\sin 390° = \sin(30° + 360°)$$
$$= \sin 30° = \frac{1}{2}$$

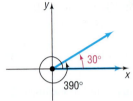

Figure 46

(b) See Figure 47. The angle 420° is coterminal with 60°.

$$\cos 420° = \cos(60° + 360°)$$
$$= \cos 60° = \frac{1}{2}$$

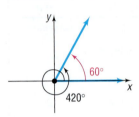

Figure 47

(c) See Figure 48. The angle $\dfrac{9\pi}{4}$ is coterminal with $\dfrac{\pi}{4}$.

$$\tan \frac{9\pi}{4} = \tan\left(\frac{\pi}{4} + 2\pi\right)$$
$$= \tan \frac{\pi}{4} = 1$$

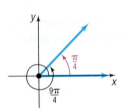

Figure 48

(d) See Figure 49. The angle $-\dfrac{7\pi}{4}$ is coterminal with $\dfrac{\pi}{4}$.

$$\sec\left(-\frac{7\pi}{4}\right) = \sec\left(\frac{\pi}{4} + 2\pi(-1)\right) = \sec \frac{\pi}{4} = \sqrt{2}$$

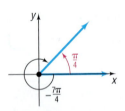

Figure 49

(e) See Figure 50. The angle $-270°$ is coterminal with 90°.

$$\csc(-270°) = \csc(90° + 360°(-1))$$
$$= \csc 90° = 1$$

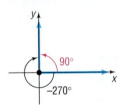

Figure 50

As Example 3 illustrates, the value of a trigonometric function of any angle is equal to the value of the same trigonometric function of an angle θ coterminal to the given angle, where $0° \le \theta < 360°$ (or $0 \le \theta < 2\pi$). Because the angles θ and $\theta + 360°k$ (or $\theta + 2\pi k$), where k is any integer, are coterminal, and because the values of the trigonometric functions are equal for coterminal angles, it follows that

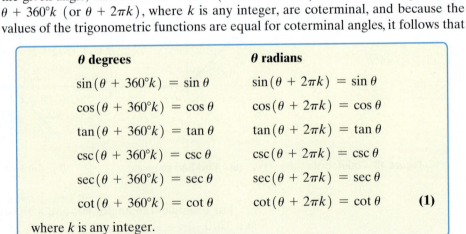

θ degrees	θ radians
$\sin(\theta + 360°k) = \sin\theta$	$\sin(\theta + 2\pi k) = \sin\theta$
$\cos(\theta + 360°k) = \cos\theta$	$\cos(\theta + 2\pi k) = \cos\theta$
$\tan(\theta + 360°k) = \tan\theta$	$\tan(\theta + 2\pi k) = \tan\theta$
$\csc(\theta + 360°k) = \csc\theta$	$\csc(\theta + 2\pi k) = \csc\theta$
$\sec(\theta + 360°k) = \sec\theta$	$\sec(\theta + 2\pi k) = \sec\theta$
$\cot(\theta + 360°k) = \cot\theta$	$\cot(\theta + 2\pi k) = \cot\theta$ **(1)**

where k is any integer.

These formulas show that the values of the trigonometric functions repeat themselves every 360° (or 2π radians).

Now Work PROBLEM 21

3 Determine the Signs of the Trigonometric Functions of an Angle in a Given Quadrant

If θ is not a quadrantal angle, then it will lie in a particular quadrant. In such a case, the signs of the x-coordinate and y-coordinate of a point (a, b) on the terminal side of θ are known. Because $r = \sqrt{a^2 + b^2} > 0$, it follows that we can find the signs of the trigonometric functions of an angle θ if we know in which quadrant θ lies.

For example, if θ lies in quadrant II, as shown in Figure 51, then a point (a, b) on the terminal side of θ has a negative x-coordinate and a positive y-coordinate. Then

$$\sin \theta = \frac{b}{r} > 0 \qquad \cos \theta = \frac{a}{r} < 0 \qquad \tan \theta = \frac{b}{a} < 0$$

$$\csc \theta = \frac{r}{b} > 0 \qquad \sec \theta = \frac{r}{a} < 0 \qquad \cot \theta = \frac{a}{b} < 0$$

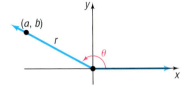

Figure 51
θ in quadrant II, $a < 0, b > 0, r > 0$

Table 5 lists the signs of the six trigonometric functions for each quadrant. Figure 52 provides two illustrations.

Table 5

Quadrant of θ	$\sin \theta, \csc \theta$	$\cos \theta, \sec \theta$	$\tan \theta, \cot \theta$
I	Positive	Positive	Positive
II	Positive	Negative	Negative
III	Negative	Negative	Positive
IV	Negative	Positive	Negative

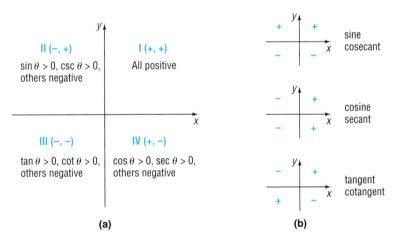

Figure 52 Signs of the trigonometric functions

EXAMPLE 4

Finding the Quadrant in Which an Angle Lies

If $\sin \theta < 0$ and $\cos \theta < 0$, name the quadrant in which the angle θ lies.

Solution

If $\sin \theta < 0$, then θ lies in quadrant III or IV. If $\cos \theta < 0$, then θ lies in quadrant II or III. Therefore, θ lies in quadrant III. ●

 Now Work PROBLEM 33

4 Find the Reference Angle of an Angle

Once we know in which quadrant an angle lies, we know the sign of each trigonometric function of that angle. This information, along with the *reference angle*, will enable us to evaluate the trigonometric functions of such an angle.

DEFINITION

Let θ denote an angle that lies in a quadrant. The acute angle formed by the terminal side of θ and the x-axis is called the **reference angle** for θ.

6 Find the Exact Values of the Trigonometric Functions of an Angle, Given Information about the Functions

EXAMPLE 7

Finding the Exact Values of Trigonometric Functions

Given that $\cos \theta = -\dfrac{2}{3}, \dfrac{\pi}{2} < \theta < \pi$, find the exact value of each of the remaining trigonometric functions.

Solution

The angle θ lies in quadrant II, so $\sin \theta$ and $\csc \theta$ are positive and the other four trigonometric functions are negative. If α is the reference angle for θ, then

$$\cos \alpha = \frac{2}{3} = \frac{\text{adjacent}}{\text{hypotenuse}}.$$

The values of the remaining trigonometric functions of the reference angle α can be found by drawing the appropriate triangle and using the Pythagorean Theorem to determine that the side opposite α is $\sqrt{5}$. Use Figure 63 to obtain

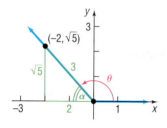

Figure 63 $\cos \alpha = \dfrac{2}{3}$

$$\sin \alpha = \frac{\sqrt{5}}{3} \qquad \cos \alpha = \frac{2}{3} \qquad \tan \alpha = \frac{\sqrt{5}}{2}$$

$$\csc \alpha = \frac{3}{\sqrt{5}} = \frac{3\sqrt{5}}{5} \qquad \sec \alpha = \frac{3}{2} \qquad \cot \alpha = \frac{2}{\sqrt{5}} = \frac{2\sqrt{5}}{5}$$

Now assign the appropriate signs to each of these values to find the values of the trigonometric functions of θ.

$$\sin \theta = \frac{\sqrt{5}}{3} \qquad \cos \theta = -\frac{2}{3} \qquad \tan \theta = -\frac{\sqrt{5}}{2}$$

$$\csc \theta = \frac{3\sqrt{5}}{5} \qquad \sec \theta = -\frac{3}{2} \qquad \cot \theta = -\frac{2\sqrt{5}}{5}$$

● **Now Work PROBLEM 83**

EXAMPLE 8

Finding the Exact Values of Trigonometric Functions

If $\tan \theta = -4$ and $\sin \theta < 0$, find the exact value of each of the remaining trigonometric functions of θ.

Solution

Since $\tan \theta = -4 < 0$ and $\sin \theta < 0$, it follows that θ lies in quadrant IV. If α is the reference angle for θ, then $\tan \alpha = 4 = \dfrac{4}{1} = \dfrac{b}{a}$. With $a = 1$ and $b = 4$, we find $r = \sqrt{1^2 + 4^2} = \sqrt{17}$. See Figure 64. Then

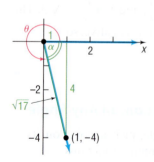

Figure 64 $\tan \alpha = 4$

$$\sin \alpha = \frac{4}{\sqrt{17}} = \frac{4\sqrt{17}}{17} \qquad \cos \alpha = \frac{1}{\sqrt{17}} = \frac{\sqrt{17}}{17} \qquad \tan \alpha = \frac{4}{1} = 4$$

$$\csc \alpha = \frac{\sqrt{17}}{4} \qquad \sec \alpha = \frac{\sqrt{17}}{1} = \sqrt{17} \qquad \cot \alpha = \frac{1}{4}$$

Assign the appropriate sign to each of these to obtain the values of the trigonometric functions of θ.

$$\sin \theta = -\frac{4\sqrt{17}}{17} \qquad \cos \theta = \frac{\sqrt{17}}{17} \qquad \tan \theta = -4$$

$$\csc \theta = -\frac{\sqrt{17}}{4} \qquad \sec \theta = \sqrt{17} \qquad \cot \theta = -\frac{1}{4}$$

● **Now Work PROBLEM 93**

7.4 Assess Your Understanding

Concepts and Vocabulary

1. For an angle θ that lies in quadrant III, the trigonometric functions _____ and _____ are positive.

2. Two angles in standard position that have the same terminal side are _____.

3. The reference angle of $240°$ is ____.

4. *True or False* $\sin 182° = \cos 2°$.

5. *True or False* $\tan \dfrac{\pi}{2}$ is not defined.

6. *True or False* The reference angle of an angle is always an acute angle.

7. What is the reference angle of $600°$?

8. Which function is positive in quadrant IV?
  (a) $\sin \theta$ (b) $\cos \theta$ (c) $\tan \theta$ (d) $\csc \theta$

9. If $0 \le \theta < 2\pi$, for what angles θ, if any, is $\tan \theta$ undefined?

10. What is the reference angle of $\dfrac{13\pi}{3}$?
  (a) $-\dfrac{2\pi}{3}$ (b) $-\dfrac{\pi}{3}$ (c) $\dfrac{2\pi}{3}$ (d) $\dfrac{\pi}{3}$

Skill Building

In Problems 11–20, a point on the terminal side of an angle θ in standard position is given. Find the exact value of each of the six trigonometric functions of θ.

11. $(-3, 4)$

12. $(5, -12)$

13. $(2, -3)$

14. $(-1, -2)$

15. $(-3, -3)$

16. $(2, -2)$

17. $\left(\dfrac{\sqrt{3}}{2}, \dfrac{1}{2}\right)$

18. $\left(-\dfrac{1}{2}, \dfrac{\sqrt{3}}{2}\right)$

19. $\left(\dfrac{\sqrt{2}}{2}, -\dfrac{\sqrt{2}}{2}\right)$

20. $\left(-\dfrac{\sqrt{2}}{2}, -\dfrac{\sqrt{2}}{2}\right)$

In Problems 21–32, use a coterminal angle to find the exact value of each expression. Do not use a calculator.

21. $\sin 405°$

22. $\cos 420°$

23. $\tan 405°$

24. $\sin 390°$

25. $\csc 450°$

26. $\sec 540°$

27. $\cot 390°$

28. $\sec 420°$

29. $\cos \dfrac{33\pi}{4}$

30. $\sin \dfrac{9\pi}{4}$

31. $\tan(21\pi)$

32. $\csc \dfrac{9\pi}{2}$

In Problems 33–40, name the quadrant in which the angle θ lies.

33. $\sin \theta > 0$, $\cos \theta < 0$

34. $\sin \theta < 0$, $\cos \theta > 0$

35. $\sin \theta < 0$, $\tan \theta < 0$

36. $\cos \theta > 0$, $\tan \theta > 0$

37. $\cos \theta > 0$, $\cot \theta < 0$

38. $\sin \theta < 0$, $\cot \theta > 0$

39. $\sec \theta < 0$, $\tan \theta > 0$

40. $\csc \theta > 0$, $\cot \theta < 0$

In Problems 41–58, find the reference angle of each angle.

41. $-30°$

42. $-60°$

43. $120°$

44. $210°$

45. $300°$

46. $330°$

47. $\dfrac{5\pi}{4}$

48. $\dfrac{5\pi}{6}$

49. $\dfrac{8\pi}{3}$

50. $\dfrac{7\pi}{4}$

51. $-135°$

52. $-240°$

53. $-\dfrac{2\pi}{3}$

54. $-\dfrac{7\pi}{6}$

55. $440°$

56. $490°$

57. $\dfrac{15\pi}{4}$

58. $\dfrac{19\pi}{6}$

In Problems 59–82, use the reference angle to find the exact value of each expression. Do not use a calculator.

59. $\sin 150°$

60. $\cos 210°$

61. $\sin 510°$

62. $\cos 600°$

63. $\cos(-45°)$

64. $\sin(-240°)$

65. $\sec 240°$

66. $\csc 300°$

67. $\cot 330°$

68. $\tan 225°$

69. $\sin \dfrac{3\pi}{4}$

70. $\cos \dfrac{2\pi}{3}$

71. $\cos \dfrac{13\pi}{4}$

72. $\tan \dfrac{8\pi}{3}$

73. $\sin\left(-\dfrac{2\pi}{3}\right)$

74. $\cot\left(-\dfrac{\pi}{6}\right)$

75. $\tan \dfrac{14\pi}{3}$

76. $\sec \dfrac{11\pi}{4}$

77. $\sin(8\pi)$

78. $\cos(-2\pi)$

79. $\tan(7\pi)$

80. $\cot(5\pi)$

81. $\sec(-3\pi)$

82. $\csc\left(-\dfrac{5\pi}{2}\right)$

In Problems 83–100, find the exact value of each of the remaining trigonometric functions of θ.

83. $\sin \theta = \dfrac{12}{13}$, θ in quadrant II

84. $\cos \theta = \dfrac{3}{5}$, θ in quadrant IV

85. $\cos \theta = -\dfrac{4}{5}$, θ in quadrant III

86. $\sin \theta = -\dfrac{5}{13}$, θ in quadrant III

87. $\sin \theta = \dfrac{5}{13}$, $90° < \theta < 180°$

88. $\cos \theta = \dfrac{4}{5}$, $270° < \theta < 360°$

89. $\cos \theta = -\dfrac{1}{3}$, $180° < \theta < 270°$

90. $\sin \theta = -\dfrac{2}{3}$, $180° < \theta < 270°$

91. $\sin \theta = \dfrac{2}{3}$, $\tan \theta < 0$

92. $\cos \theta = -\dfrac{1}{4}$, $\tan \theta > 0$

93. $\sec \theta = 2$, $\sin \theta < 0$

94. $\csc \theta = 3$, $\cot \theta < 0$

95. $\tan \theta = \dfrac{3}{4}$, $\sin \theta < 0$

96. $\cot \theta = \dfrac{4}{3}$, $\cos \theta < 0$

97. $\tan \theta = -\dfrac{1}{3}$, $\sin \theta > 0$

98. $\sec \theta = -2$, $\tan \theta > 0$

99. $\csc \theta = -2$, $\tan \theta > 0$

100. $\cot \theta = -2$, $\sec \theta > 0$

101. Find the exact value of $\sin 40° + \sin 130° + \sin 220° + \sin 310°$.

102. Find the exact value of $\tan 40° + \tan 140°$.

Mixed Practice

In Problems 103–106, $f(x) = \sin x, g(x) = \cos x, h(x) = \tan x, F(x) = \csc x, G(x) = \sec x,$ and $H(x) = \cot x.$

103. (a) Find $f(315°)$. What point is on the graph of f?
 (b) Find $G(315°)$. What point is on the graph of G?
 (c) Find $h(315°)$. What point is on the graph of h?

104. (a) Find $g(120°)$. What point is on the graph of g?
 (b) Find $F(120°)$. What point is on the graph of F?
 (c) Find $H(120°)$. What point is on the graph of H?

105. (a) Find $g\left(\dfrac{7\pi}{6}\right)$. What point is on the graph of g?

 (b) Find $F\left(\dfrac{7\pi}{6}\right)$. What point is on the graph of F?

 (c) Find $H(-315°)$. What point is on the graph of H?

106. (a) Find $f\left(\dfrac{7\pi}{4}\right)$. What point is on the graph of f?

 (b) Find $G\left(\dfrac{7\pi}{4}\right)$. What point is on the graph of G?

 (c) Find $F(-225°)$. What point is on the graph of F?

Applications and Extensions

107. If $f(\theta) = \sin \theta = 0.2$, find $f(\theta + \pi)$.

108. If $g(\theta) = \cos \theta = 0.4$, find $g(\theta + \pi)$.

109. If $F(\theta) = \tan \theta = 3$, find $F(\theta + \pi)$.

110. If $G(\theta) = \cot \theta = -2$, find $G(\theta + \pi)$.

111. If $\sin \theta = \dfrac{1}{5}$, find $\csc(\theta + \pi)$.

112. If $\cos \theta = \dfrac{2}{3}$, find $\sec(\theta + \pi)$.

113. Find the exact value of
$\sin 1° + \sin 2° + \sin 3° + \cdots + \sin 358° + \sin 359°$

114. Find the exact value of
$\cos 1° + \cos 2° + \cos 3° + \cdots + \cos 358° + \cos 359°$

115. Projectile Motion An object is propelled upward at an angle θ, $45° < \theta < 90°$, to the horizontal with an initial velocity of v_0 feet per second from the base of a plane that makes an angle of $45°$ with the horizontal. See the illustration.

If air resistance is ignored, the distance R that it travels up the inclined plane is given by the function

$$R(\theta) = \dfrac{v_0^2 \sqrt{2}}{32}\left[\sin(2\theta) - \cos(2\theta) - 1\right]$$

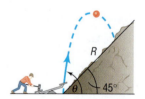

(a) Find the distance R that the object travels along the inclined plane if the initial velocity is 32 feet per second and $\theta = 60°$.

 (b) Graph $R = R(\theta)$ if the initial velocity is 32 feet per second.

(c) What value of θ makes R largest?

Explaining Concepts: Discussion and Writing

116. Give three examples that demonstrate how to use the theorem on reference angles.

117. Write a brief paragraph that explains how to quickly compute the value of the trigonometric functions of $0°, 90°, 180°,$ and $270°$.

118. Explain what a reference angle is. What role does it play in finding the value of a trigonometric function?

Retain Your Knowledge

Problems 119–122 are based on material learned earlier in the course. The purpose of these problems is to keep the material fresh in your mind so that you are better prepared for the final exam.

119. Solve: $25^{x+3} = 5^{x-4}$

120. The demand function for a certain product is $D = 3500 - x^2, 0 \leq x \leq 59.16$, where x is the price per unit in dollars and D is in thousands of units demanded. Determine the number of units demanded if the price per unit is \$32.

121. A function f has an average rate of change of $\dfrac{3}{8}$ over the interval $[0, 12]$. If $f(0) = \dfrac{1}{2}$, find $f(12)$.

122. Given that $(f \circ g)(x) = x^2 - 8x + 19$ and $f(x) = x^2 + 3$, find $g(x)$.

7.5 Unit Circle Approach; Properties of the Trigonometric Functions

PREPARING FOR THIS SECTION *Before getting started, review the following:*

- Unit Circle (Section 2.4, p. 183)
- Functions (Section 3.1, pp. 199–208)
- Even and Odd Functions (Section 3.3, pp. 223–225)

Now Work the **'Are You Prepared?'** problems on page 561.

OBJECTIVES **1** Find the Exact Values of the Trigonometric Functions Using the Unit Circle (p. 553)

2 Know the Domain and Range of the Trigonometric Functions (p. 557)

3 Use Periodic Properties to Find the Exact Values of the Trigonometric Functions (p. 559)

4 Use Even–Odd Properties to Find the Exact Values of the Trigonometric Functions (p. 560)

In this section, we develop important properties of the trigonometric functions. We begin by introducing the trigonometric functions using the unit circle. This approach will lead to the definition given earlier of the trigonometric functions of any angle (page 543).

1 Find the Exact Values of the Trigonometric Functions Using the Unit Circle

Recall that the unit circle is a circle whose radius is 1 and whose center is at the origin of a rectangular coordinate system. Also recall that any circle of radius r has circumference of length $2\pi r$. Therefore, the unit circle (radius = 1) has a circumference of length 2π. So, for 1 revolution around the unit circle, the length of the arc is 2π units.

The following discussion sets the stage for defining the trigonometric functions using the unit circle.

Draw a vertical real number line with the origin of the number line at $(1, 0)$ in the Cartesian plane. Let $t \geq 0$ be any real number and let s be the distance from the origin to t on the real number line. See the red portion of Figure 65(a) on the next page. Now look at the unit circle in Figure 65(a). Beginning at the point $(1, 0)$ on the unit circle, travel $s = t$ units in the counterclockwise direction along the circle to arrive at the point $P = (a, b)$. In this sense, the length $s = t$ units is being **wrapped** around the unit circle.

If $t < 0$, we begin at the point $(1, 0)$ on the unit circle and travel $s = |t|$ units in the clockwise direction to arrive at the point $P = (a, b)$. See Figure 65(b).

DEFINITION

A function f is called **periodic** if there is a positive number p such that whenever θ is in the domain of f, so is $\theta + p$, and

$$f(\theta + p) = f(\theta)$$

If there is a smallest such number p, this smallest value is called the **(fundamental) period** of f.

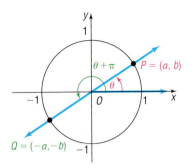

Figure 75

$$\tan\theta = \frac{b}{a} = \frac{-b}{-a} = \tan(\theta + \pi)$$

Based on equation (1), the sine and cosine functions are periodic. In fact, the sine, cosine, secant, and cosecant functions have period 2π. You are asked to prove this in Problems 93 through 96.

The tangent and cotangent functions are periodic with period π. See Figure 75 for a partial justification. You are asked to prove this statement in Problems 97 and 98.

Periodic Properties

$$\sin(\theta + 2\pi) = \sin\theta \qquad \cos(\theta + 2\pi) = \cos\theta \qquad \tan(\theta + \pi) = \tan\theta$$
$$\csc(\theta + 2\pi) = \csc\theta \qquad \sec(\theta + 2\pi) = \sec\theta \qquad \cot(\theta + \pi) = \cot\theta$$

In Words

Tangent and cotangent have period π; the others have period 2π.

Because the sine, cosine, secant, and cosecant functions have period 2π, once their values for $0 \le \theta < 2\pi$ are known, all their values are known. Similarly, since the tangent and cotangent functions have period π, once their values for $0 \le \theta < \pi$ are known, all their values are known.

EXAMPLE 3

Using Periodic Properties to Find Exact Values

Find the exact value of each of the following:

(a) $\sin 420°$ 　　　　 (b) $\tan\dfrac{5\pi}{4}$ 　　　　 (c) $\cos\dfrac{11\pi}{4}$

Solution

(a) $\sin 420° = \sin(60° + 360°) = \sin 60° = \dfrac{\sqrt{3}}{2}$

(b) $\tan\dfrac{5\pi}{4} = \tan\left(\dfrac{\pi}{4} + \pi\right) = \tan\dfrac{\pi}{4} = 1$

(c) $\cos\dfrac{11\pi}{4} = \cos\left(\dfrac{3\pi}{4} + \dfrac{8\pi}{4}\right) = \cos\left(\dfrac{3\pi}{4} + 2\pi\right) = \cos\dfrac{3\pi}{4} = -\dfrac{\sqrt{2}}{2}$ 　　●

The periodic properties of the trigonometric functions will be very helpful when we study their graphs later in this chapter.

Now Work PROBLEMS **23** AND **81**

4 Use Even–Odd Properties to Find the Exact Values of the Trigonometric Functions

Recall that a function f is even if $f(-\theta) = f(\theta)$ for all θ in the domain of f; a function f is odd if $f(-\theta) = -f(\theta)$ for all θ in the domain of f. We will now show that the trigonometric functions sine, tangent, cotangent, and cosecant are odd functions and that the functions cosine and secant are even functions.

THEOREM

Even–Odd Properties

$$\sin(-\theta) = -\sin\theta \qquad \cos(-\theta) = \cos\theta \qquad \tan(-\theta) = -\tan\theta$$
$$\csc(-\theta) = -\csc\theta \qquad \sec(-\theta) = \sec\theta \qquad \cot(-\theta) = -\cot\theta$$

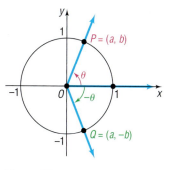

Figure 76

In Words
Cosine and secant are even functions; the others are odd functions.

Proof Let $P = (a, b)$ be the point on the unit circle that corresponds to the angle θ. See Figure 76. The point Q on the unit circle that corresponds to the angle $-\theta$ will have coordinates $(a, -b)$. Using the definition for the trigonometric functions, we have

$$\sin\theta = b \qquad \sin(-\theta) = -b \qquad \cos\theta = a \qquad \cos(-\theta) = a$$

so

$$\sin(-\theta) = -\sin\theta \qquad\qquad\qquad \cos(-\theta) = \cos\theta$$

Now, using these results and some of the Fundamental Identities, we have

$$\tan(-\theta) = \frac{\sin(-\theta)}{\cos(-\theta)} = \frac{-\sin\theta}{\cos\theta} = -\tan\theta$$

$$\cot(-\theta) = \frac{1}{\tan(-\theta)} = \frac{1}{-\tan\theta} = -\cot\theta$$

$$\sec(-\theta) = \frac{1}{\cos(-\theta)} = \frac{1}{\cos\theta} = \sec\theta$$

$$\csc(-\theta) = \frac{1}{\sin(-\theta)} = \frac{1}{-\sin\theta} = -\csc\theta \qquad\blacksquare$$

EXAMPLE 4

Finding Exact Values Using Even–Odd Properties

Find the exact value of each of the following:

(a) $\sin(-45°)$ (b) $\cos(-\pi)$ (c) $\cot\left(-\dfrac{3\pi}{2}\right)$ (d) $\tan\left(-\dfrac{37\pi}{4}\right)$

Solution

(a) $\sin(-45°) = -\sin 45° = -\dfrac{\sqrt{2}}{2}$
  ↑
  Odd function

(b) $\cos(-\pi) = \cos\pi = -1$
  ↑
  Even function

(c) $\cot\left(-\dfrac{3\pi}{2}\right) = -\cot\dfrac{3\pi}{2} = 0$
  ↑
  Odd function

(d) $\tan\left(-\dfrac{37\pi}{4}\right) = -\tan\dfrac{37\pi}{4} = -\tan\left(\dfrac{\pi}{4} + 9\pi\right) = -\tan\dfrac{\pi}{4} = -1$
  ↑   ↑
  Odd function   *Period is π.*

Now Work   PROBLEMS 39 AND 75

7.5 Assess Your Understanding

'Are You Prepared?' *Answers are given at the end of these exercises. If you get a wrong answer, read the pages listed in red.*

1. What is the equation of the unit circle? (p. 183)

2. The domain of the function $f(x) = \dfrac{3x - 6}{x - 4}$ is _____. (pp. 199–208)

3. A function for which $f(x) = f(-x)$ for all x in the domain of f is called a(n) _____ function. (pp. 223–225)

Concepts and Vocabulary

4. The sine, cosine, cosecant, and secant functions have period ____; the tangent and cotangent functions have period ____.

5. Let t be a real number and let $P = (a, b)$ be the point on the unit circle that corresponds to t. Then $\sin t =$ _____ and $\cos t =$ _____.

6. For any angle θ in standard position, let $P = (a, b)$ be any point on the terminal side of θ that is also on the circle $x^2 + y^2 = r^2$. Then $\sin \theta =$ _____ and $\cos \theta =$ _____.

7. If $\sin \theta = 0.2$, then $\sin(-\theta) =$ _____ and $\sin(\theta + 2\pi) =$ _____.

8. *True or False* The only trigonometric functions with a domain of all real numbers are the sine and cosine functions.

9. Which of the following is not in the range of the sine function?
(a) $\dfrac{\pi}{4}$ (b) $\dfrac{3}{2}$ (c) -0.37 (d) -1

10. Which of the following functions is even?
(a) cosine   (b) sine   (c) tangent   (d) cosecant

Skill Building

In Problems 11–16, the point P on the unit circle that corresponds to a real number t is given. Find sin t, cos t, tan t, csc t, sec t, and cot t.

11. $\left(\dfrac{\sqrt{3}}{2}, -\dfrac{1}{2}\right)$ **12.** $\left(-\dfrac{\sqrt{3}}{2}, -\dfrac{1}{2}\right)$ **13.** $\left(-\dfrac{\sqrt{2}}{2}, -\dfrac{\sqrt{2}}{2}\right)$ **14.** $\left(\dfrac{\sqrt{2}}{2}, -\dfrac{\sqrt{2}}{2}\right)$ **15.** $\left(\dfrac{\sqrt{5}}{3}, \dfrac{2}{3}\right)$ **16.** $\left(-\dfrac{\sqrt{5}}{5}, \dfrac{2\sqrt{5}}{5}\right)$

In Problems 17–22, the point P on the circle $x^2 + y^2 = r^2$ that is also on the terminal side of an angle θ in standard position is given. Find $\sin \theta$, $\cos \theta$, $\tan \theta$, $\csc \theta$, $\sec \theta$, and $\cot \theta$.

17. $(3, -4)$ **18.** $(5, -12)$ **19.** $(-2, 3)$ **20.** $(2, -4)$ **21.** $(-1, -1)$ **22.** $(-3, 1)$

In Problems 23–38, use the fact that the trigonometric functions are periodic to find the exact value of each expression. Do not use a calculator.

23. $\sin 405°$ **24.** $\cos 420°$ **25.** $\tan 405°$ **26.** $\sin 390°$

27. $\csc 450°$ **28.** $\sec 540°$ **29.** $\cot 390°$ **30.** $\sec 420°$

31. $\cos \dfrac{33\pi}{4}$ **32.** $\sin \dfrac{9\pi}{4}$ **33.** $\tan(21\pi)$ **34.** $\csc \dfrac{9\pi}{2}$

35. $\sec \dfrac{17\pi}{4}$ **36.** $\cot \dfrac{17\pi}{4}$ **37.** $\tan \dfrac{19\pi}{6}$ **38.** $\sec \dfrac{25\pi}{6}$

In Problems 39–56, use the even–odd properties to find the exact value of each expression. Do not use a calculator.

39. $\sin(-60°)$ **40.** $\cos(-30°)$ **41.** $\tan(-30°)$ **42.** $\sin(-135°)$ **43.** $\sec(-60°)$

44. $\csc(-30°)$ **45.** $\sin(-90°)$ **46.** $\cos(-270°)$ **47.** $\tan\left(-\dfrac{\pi}{4}\right)$ **48.** $\sin(-\pi)$

49. $\cos\left(-\dfrac{\pi}{4}\right)$ **50.** $\sin\left(-\dfrac{\pi}{3}\right)$ **51.** $\tan(-\pi)$ **52.** $\sin\left(-\dfrac{3\pi}{2}\right)$

53. $\csc\left(-\dfrac{\pi}{4}\right)$ **54.** $\sec(-\pi)$ **55.** $\sec\left(-\dfrac{\pi}{6}\right)$ **56.** $\csc\left(-\dfrac{\pi}{3}\right)$

In Problems 57–62, find the exact value of each expression. Do not use a calculator.

57. $\sin(-\pi) + \cos(5\pi)$ **58.** $\tan\left(-\dfrac{5\pi}{6}\right) - \cot \dfrac{7\pi}{2}$ **59.** $\sec(-\pi) + \csc\left(-\dfrac{\pi}{2}\right)$

60. $\tan(-6\pi) + \cos \dfrac{9\pi}{4}$ **61.** $\sin\left(-\dfrac{9\pi}{4}\right) - \tan\left(-\dfrac{9\pi}{4}\right)$ **62.** $\cos\left(-\dfrac{17\pi}{4}\right) - \sin\left(-\dfrac{3\pi}{2}\right)$

63. What is the domain of the sine function?

64. What is the domain of the cosine function?

65. For what numbers θ is $f(\theta) = \tan \theta$ not defined?

66. For what numbers θ is $f(\theta) = \cot \theta$ not defined?

67. For what numbers θ is $f(\theta) = \sec \theta$ not defined?

68. For what numbers θ is $f(\theta) = \csc \theta$ not defined?

69. What is the range of the sine function?

70. What is the range of the cosine function?

71. What is the range of the tangent function?

72. What is the range of the cotangent function?

73. What is the range of the secant function?

74. What is the range of the cosecant function?

75. Is the sine function even, odd, or neither? Is its graph symmetric? With respect to what?

76. Is the cosine function even, odd, or neither? Is its graph symmetric? With respect to what?

77. Is the tangent function even, odd, or neither? Is its graph symmetric? With respect to what?

78. Is the cotangent function even, odd, or neither? Is its graph symmetric? With respect to what?

79. Is the secant function even, odd, or neither? Is its graph symmetric? With respect to what?

80. Is the cosecant function even, odd, or neither? Is its graph symmetric? With respect to what?

81. If $\sin \theta = 0.3$, find the value of
$$\sin \theta + \sin(\theta + 2\pi) + \sin(\theta + 4\pi)$$

82. If $\cos \theta = 0.2$, find the value of
$$\cos \theta + \cos(\theta + 2\pi) + \cos(\theta + 4\pi)$$

83. If $\tan \theta = 3$, find the value of
$$\tan \theta + \tan(\theta + \pi) + \tan(\theta + 2\pi)$$

84. If $\cot \theta = -2$, find the value of
$$\cot \theta + \cot(\theta - \pi) + \cot(\theta - 2\pi)$$

In Problems 85–90, use the periodic and even–odd properties.

85. If $f(x) = \sin x$ and $f(a) = \dfrac{1}{3}$, find the exact value of:

(a) $f(-a)$ (b) $f(a) + f(a + 2\pi) + f(a + 4\pi)$

86. If $f(x) = \cos x$ and $f(a) = \dfrac{1}{4}$, find the exact value of:

(a) $f(-a)$ (b) $f(a) + f(a + 2\pi) + f(a - 2\pi)$

87. If $f(x) = \tan x$ and $f(a) = 2$, find the exact value of:
(a) $f(-a)$ (b) $f(a) + f(a + \pi) + f(a + 2\pi)$

88. If $f(x) = \cot x$ and $f(a) = -3$, find the exact value of:
(a) $f(-a)$ (b) $f(a) + f(a + \pi) + f(a + 4\pi)$

89. If $f(x) = \sec x$ and $f(a) = -4$, find the exact value of:
(a) $f(-a)$ (b) $f(a) + f(a + 2\pi) + f(a + 4\pi)$

90. If $f(x) = \csc x$ and $f(a) = 2$, find the exact value of:
(a) $f(-a)$ (b) $f(a) + f(a + 2\pi) + f(a + 4\pi)$

Applications and Extensions

In Problems 91 and 92, use the figure to approximate the value of the six trigonometric functions at t to the nearest tenth. Then use a calculator to approximate each of the six trigonometric functions at t.

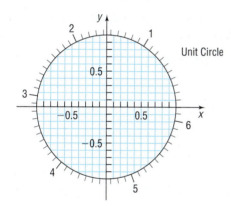

Unit Circle

91. (a) $t = 1$ (b) $t = 5.1$
92. (a) $t = 2$ (b) $t = 4$
93. Show that the period of $f(\theta) = \sin \theta$ is 2π.

[**Hint:** Assume that $0 < p < 2\pi$ exists so that $\sin(\theta + p) = \sin \theta$ for all θ. Let $\theta = 0$ to find p. Then let $\theta = \dfrac{\pi}{2}$ to obtain a contradiction.]

94. Show that the period of $f(\theta) = \cos \theta$ is 2π.

95. Show that the period of $f(\theta) = \sec \theta$ is 2π.
96. Show that the period of $f(\theta) = \csc \theta$ is 2π.
97. Show that the period of $f(\theta) = \tan \theta$ is π.
98. Show that the period of $f(\theta) = \cot \theta$ is π.
99. Show that the range of the tangent function is the set of all real numbers.
100. Show that the range of the cotangent function is the set of all real numbers.
101. If θ, $0 < \theta < \pi$, is the angle between a horizontal ray directed to the right (say, the positive x-axis) and a nonhorizontal, nonvertical line L, show that the slope m of L equals $\tan \theta$. The angle θ is called the **inclination** of L.

[**Hint:** See the illustration, where we have drawn the line M parallel to L and passing through the origin. Use the fact that M intersects the unit circle at the point $(\cos \theta, \sin \theta)$.]

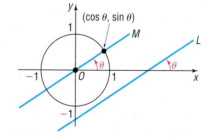

Explaining Concepts: Discussion and Writing

102. Explain how you would find the value of $\sin 390°$ using periodic properties.

103. Explain how you would find the value of $\cos(-45°)$ using even–odd properties.

104. Write down five properties of the tangent function. Explain the meaning of each.

105. Describe your understanding of the meaning of a periodic function.

Retain Your Knowledge

Problems 106–109 are based on material learned earlier in the course. The purpose of these problems is to keep the material fresh in your mind so that you are better prepared for the final exam.

106. Find the real zeros of $f(x) = 3x^2 - 10x + 5$.

107. Find an equation of the line that contains the point $(-3, 7)$ and is perpendicular to the line $y = -4x - 5$.

108. Determine the interest rate required for an investment of $1500 to be worth $1800 after 3 years if interest is compounded quarterly. Round your answer to two decimal places.

109. Find any vertical or horizontal asymptotes for the graph of $f(x) = \dfrac{5x - 2}{x + 3}$.

'Are You Prepared?' Answers

1. $x^2 + y^2 = 1$ **2.** $\{x \mid x \neq 4\}$ **3.** even

7.6 Graphs of the Sine and Cosine Functions*

PREPARING FOR THIS SECTION *Before getting started, review the following:*

- Graphing Techniques: Transformations (Section 3.5, pp. 247–256)

Now Work the 'Are You Prepared?' problems on page 574.

OBJECTIVES **1** Graph Functions of the Form $y = A \sin(\omega x)$ Using Transformations (p. 565)
  2 Graph Functions of the Form $y = A \cos(\omega x)$ Using Transformations (p. 567)
  3 Determine the Amplitude and Period of Sinusoidal Functions (p. 568)
  4 Graph Sinusoidal Functions Using Key Points (p. 570)
  5 Find an Equation for a Sinusoidal Graph (p. 573)

Since we want to graph the trigonometric functions in the *xy*-plane, we shall use the traditional symbols *x* for the independent variable (or argument) and *y* for the dependent variable (or value at *x*) for each function. So the six trigonometric functions can be written as

$$y = f(x) = \sin x \qquad y = f(x) = \cos x \qquad y = f(x) = \tan x$$
$$y = f(x) = \csc x \qquad y = f(x) = \sec x \qquad y = f(x) = \cot x$$

  Here the independent variable *x* represents an angle, measured in radians. In calculus, *x* will usually be treated as a real number. As noted earlier, these are equivalent ways of viewing *x*.

The Graph of the Sine Function $y = \sin x$

Because the sine function has period 2π, it is only necessary to graph $y = \sin x$ on the interval $[0, 2\pi]$. The remainder of the graph will consist of repetitions of this portion of the graph.

To begin, consider Table 7, which lists some points on the graph of $y = \sin x$, $0 \le x \le 2\pi$. As the table shows, the graph of $y = \sin x$, $0 \le x \le 2\pi$, begins at the origin. As *x* increases from 0 to $\dfrac{\pi}{2}$, the value of $y = \sin x$ increases from 0 to 1; as *x* increases from $\dfrac{\pi}{2}$ to π to $\dfrac{3\pi}{2}$, the value of *y* decreases from 1 to 0 to -1;

*For those who wish to include phase shifts here, Section 7.8 can be covered immediately after Section 7.6 without loss of continuity.

Table 7

x	y = sin x	(x, y)
0	0	(0, 0)
$\frac{\pi}{6}$	$\frac{1}{2}$	$\left(\frac{\pi}{6}, \frac{1}{2}\right)$
$\frac{\pi}{2}$	1	$\left(\frac{\pi}{2}, 1\right)$
$\frac{5\pi}{6}$	$\frac{1}{2}$	$\left(\frac{5\pi}{6}, \frac{1}{2}\right)$
π	0	$(\pi, 0)$
$\frac{7\pi}{6}$	$-\frac{1}{2}$	$\left(\frac{7\pi}{6}, -\frac{1}{2}\right)$
$\frac{3\pi}{2}$	-1	$\left(\frac{3\pi}{2}, -1\right)$
$\frac{11\pi}{6}$	$-\frac{1}{2}$	$\left(\frac{11\pi}{6}, -\frac{1}{2}\right)$
2π	0	$(2\pi, 0)$

as x increases from $\frac{3\pi}{2}$ to 2π, the value of y increases from -1 to 0. Plotting the points listed in Table 7 and connecting them with a smooth curve yields the graph shown in Figure 77.

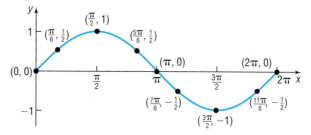

Figure 77 $y = \sin x, 0 \le x \le 2\pi$

The graph in Figure 77 is one period, or **cycle**, of the graph of $y = \sin x$. To obtain a more complete graph of $y = \sin x$, continue the graph in each direction, as shown in Figure 78.

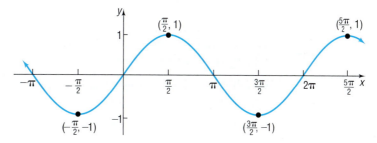

Figure 78 $y = \sin x, -\infty < x < \infty$

The graph of $y = \sin x$ illustrates some of the facts already discussed about the sine function.

> **Properties of the Sine Function $y = \sin x$**
>
> 1. The domain is the set of all real numbers.
> 2. The range consists of all real numbers from -1 to 1, inclusive.
> 3. The sine function is an odd function, as the symmetry of the graph with respect to the origin indicates.
> 4. The sine function is periodic, with period 2π.
> 5. The x-intercepts are $\ldots, -2\pi, -\pi, 0, \pi, 2\pi, 3\pi, \ldots$; the y-intercept is 0.
> 6. The maximum value is 1 and occurs at $x = \ldots, -\frac{3\pi}{2}, \frac{\pi}{2}, \frac{5\pi}{2}, \frac{9\pi}{2}, \ldots$;
>   the minimum value is -1 and occurs at $x = \ldots, -\frac{\pi}{2}, \frac{3\pi}{2}, \frac{7\pi}{2}, \frac{11\pi}{2}, \ldots$.

Now Work PROBLEM 11

1 Graph Functions of the Form $y = A \sin(\omega x)$ Using Transformations

EXAMPLE 1 **Graphing Functions of the Form $y = A \sin(\omega x)$ Using Transformations**

Graph $y = 3 \sin x$ using transformations. Use the graph to determine the domain and the range of the function.

Solution   Figure 79 illustrates the steps.

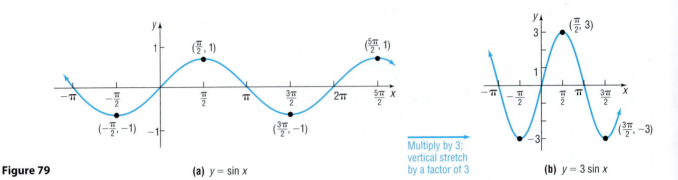

Figure 79

(a) $y = \sin x$

Multiply by 3;
vertical stretch
by a factor of 3

(b) $y = 3 \sin x$

The domain of $y = 3 \sin x$ is the set of all real numbers, or $(-\infty, \infty)$. The range is $\{y | -3 \le y \le 3\}$, or $[-3, 3]$. ●

EXAMPLE 2

Graphing Functions of the Form $y = A \sin(\omega x)$ Using Transformations

Graph $y = -\sin(2x)$ using transformations. Use the graph to determine the domain and the range of the function.

Solution   Figure 80 illustrates the steps.

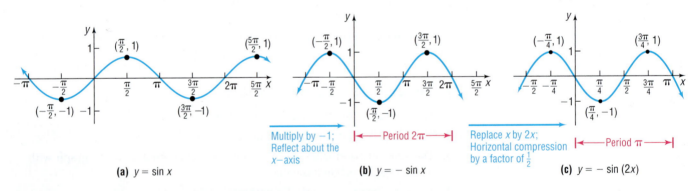

(a) $y = \sin x$

Multiply by -1;
Reflect about the
x–axis

|←— Period 2π —→|

(b) $y = -\sin x$

Replace x by $2x$;
Horizontal compression
by a factor of $\frac{1}{2}$

|←— Period π —→|

(c) $y = -\sin(2x)$

Figure 80

The domain of $y = -\sin(2x)$ is the set of all real numbers, or $(-\infty, \infty)$. The range is $\{y | -1 \le y \le 1\}$, or $[-1, 1]$. ●

Note in Figure 80(c) that the period of the function $y = -\sin(2x)$ is π because of the horizontal compression of the original period 2π by a factor of $\frac{1}{2}$.

➤ **Now Work** PROBLEM 35 USING TRANSFORMATIONS

The Graph of the Cosine Function $y = \cos x$

The cosine function also has period 2π. Proceed as with the sine function by constructing Table 8, which lists some points on the graph of $y = \cos x, 0 \le x \le 2\pi$. As the table shows, the graph of $y = \cos x, 0 \le x \le 2\pi$, begins at the point $(0, 1)$. As x increases from 0 to $\dfrac{\pi}{2}$ to π, the value of y decreases from 1 to 0 to -1; as x increases from π to $\dfrac{3\pi}{2}$ to 2π, the value of y increases from -1 to 0 to 1. As before, plot the points in Table 8 to get one period or cycle of the graph. See Figure 81.

Table 8

x	$y = \cos x$	(x, y)
0	1	$(0, 1)$
$\dfrac{\pi}{3}$	$\dfrac{1}{2}$	$\left(\dfrac{\pi}{3}, \dfrac{1}{2}\right)$
$\dfrac{\pi}{2}$	0	$\left(\dfrac{\pi}{2}, 0\right)$
$\dfrac{2\pi}{3}$	$-\dfrac{1}{2}$	$\left(\dfrac{2\pi}{3}, -\dfrac{1}{2}\right)$
π	-1	$(\pi, -1)$
$\dfrac{4\pi}{3}$	$-\dfrac{1}{2}$	$\left(\dfrac{4\pi}{3}, -\dfrac{1}{2}\right)$
$\dfrac{3\pi}{2}$	0	$\left(\dfrac{3\pi}{2}, 0\right)$
$\dfrac{5\pi}{3}$	$\dfrac{1}{2}$	$\left(\dfrac{5\pi}{3}, \dfrac{1}{2}\right)$
2π	1	$(2\pi, 1)$

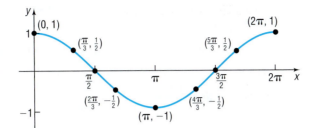

Figure 81 $y = \cos x, \, 0 \leq x \leq 2\pi$

A more complete graph of $y = \cos x$ is obtained by continuing the graph in each direction, as shown in Figure 82.

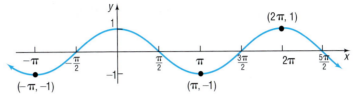

Figure 82 $y = \cos x, \, -\infty < x < \infty$

The graph of $y = \cos x$ illustrates some of the facts already discussed about the cosine function.

Properties of the Cosine Function

1. The domain is the set of all real numbers.
2. The range consists of all real numbers from -1 to 1, inclusive.
3. The cosine function is an even function, as the symmetry of the graph with respect to the y-axis indicates.
4. The cosine function is periodic, with period 2π.
5. The x-intercepts are $\ldots, -\dfrac{3\pi}{2}, -\dfrac{\pi}{2}, \dfrac{\pi}{2}, \dfrac{3\pi}{2}, \dfrac{5\pi}{2}, \ldots$; the y-intercept is 1.
6. The maximum value is 1 and occurs at $x = \ldots, -2\pi, 0, 2\pi, 4\pi, 6\pi, \ldots$; the minimum value is -1 and occurs at $x = \ldots, -\pi, \pi, 3\pi, 5\pi, \ldots$.

2 Graph Functions of the Form $y = A\cos(\omega x)$ Using Transformations

EXAMPLE 3

Graphing Functions of the Form $y = A\cos(\omega x)$ Using Transformations

Graph $y = 2\cos(3x)$ using transformations. Use the graph to determine the domain and the range of the function.

Solution   Figure 83 shows the steps.

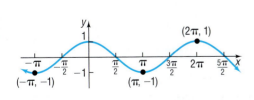

(a) $y = \cos x$

Multiply by 2;
Vertical stretch
by a factor of 2

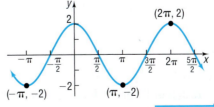

(b) $y = 2\cos x$

Replace x by $3x$;
Horizontal
compression by
a factor of $\frac{1}{3}$

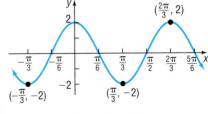

(c) $y = 2\cos(3x)$

Figure 83

The domain of $y = 2 \cos(3x)$ is the set of all real numbers, or $(-\infty, \infty)$. The range is $\{y \mid -2 \le y \le 2\}$, or $[-2, 2]$.

Notice in Figure 83(c) that the period of the function $y = 2 \cos(3x)$ is $\dfrac{2\pi}{3}$ because of the compression of the original period 2π by a factor of $\dfrac{1}{3}$.

Now Work PROBLEM **43** USING TRANSFORMATIONS

Sinusoidal Graphs

Shift the graph of $y = \cos x$ to the right $\dfrac{\pi}{2}$ units to obtain the graph of $y = \cos\left(x - \dfrac{\pi}{2}\right)$. See Figure 84(a). Now look at the graph of $y = \sin x$ in Figure 84(b). Notice that the graph of $y = \sin x$ is the same as the graph of $y = \cos\left(x - \dfrac{\pi}{2}\right)$.

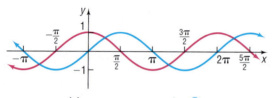
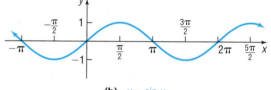

Figure 84   (a) $y = \cos x$ $y = \cos\left(x - \frac{\pi}{2}\right)$ (b) $y = \sin x$

Seeing the Concept

Graph $Y_1 = \sin x$ and $Y_2 = \cos\left(x - \dfrac{\pi}{2}\right)$.

How many graphs do you see?

Based on Figure 84, we conjecture that

$$\sin x = \cos\left(x - \frac{\pi}{2}\right)$$

(We shall prove this fact in Chapter 8.) Because of this relationship, the graphs of functions of the form $y = A \sin(\omega x)$ or $y = A \cos(\omega x)$ are referred to as **sinusoidal graphs**.

3 Determine the Amplitude and Period of Sinusoidal Functions

Figure 85 uses transformations to obtain the graph of $y = 2 \cos x$. Note that the values of $y = 2 \cos x$ lie between -2 and 2, inclusive.

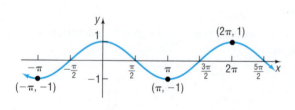

Multiply by 2;
Vertical stretch
by a factor of 2

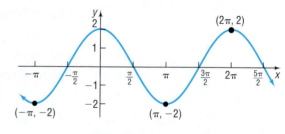

Figure 85   (a) $y = \cos x$ (b) $y = 2 \cos x$

In general, the values of the functions $y = A \sin x$ and $y = A \cos x$, where $A \ne 0$, will always satisfy the inequalities

$$-|A| \le A \sin x \le |A| \quad \text{and} \quad -|A| \le A \cos x \le |A|$$

respectively. The number $|A|$ is called the **amplitude** of $y = A \sin x$ or $y = A \cos x$. See Figure 86.

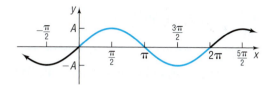

Figure 86 $y = A \sin x, A > 0$; period $= 2\pi$

Figure 87 uses transformations to obtain the graph of $y = \cos(3x)$. Note that the period of this function is $\dfrac{2\pi}{3}$, because of the horizontal compression of the original period 2π by a factor of $\dfrac{1}{3}$.

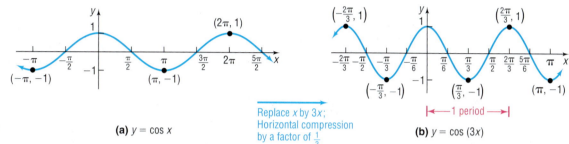

Figure 87

(a) $y = \cos x$

Replace x by $3x$;
Horizontal compression
by a factor of $\frac{1}{3}$

(b) $y = \cos(3x)$

If $\omega > 0$, the functions $y = \sin(\omega x)$ and $y = \cos(\omega x)$ will have period $T = \dfrac{2\pi}{\omega}$. To see why, recall that the graph of $y = \sin(\omega x)$ is obtained from the graph of $y = \sin x$ by performing a horizontal compression or stretch by a factor $\dfrac{1}{\omega}$. This horizontal compression replaces the interval $[0, 2\pi]$, which contains one period of the graph of $y = \sin x$, by the interval $\left[0, \dfrac{2\pi}{\omega}\right]$, which contains one period of the graph of $y = \sin(\omega x)$. So the function $y = \cos(3x)$, graphed in Figure 87(b), with $\omega = 3$, has period $\dfrac{2\pi}{\omega} = \dfrac{2\pi}{3}$.

One period of the graph of $y = \sin(\omega x)$ or $y = \cos(\omega x)$ is called a **cycle**. Figure 88 illustrates the general situation. The blue portion of the graph is one cycle.

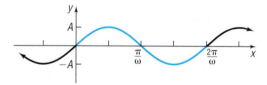

Figure 88 $y = A \sin(\omega x), A > 0, \omega > 0$; period $= \dfrac{2\pi}{\omega}$

NOTE Recall that a function f is even if $f(-x) = f(x)$; a function f is odd if $f(-x) = -f(x)$. Since the sine function is odd, $\sin(-x) = -\sin x$; since the cosine function is even, $\cos(-x) = \cos x$. ∎

When graphing $y = \sin(\omega x)$ or $y = \cos(\omega x)$, we want ω to be positive. To graph either $y = \sin(-\omega x)$, $\omega > 0$, or $y = \cos(-\omega x)$, $\omega > 0$, use the even–odd properties of the sine and cosine functions as follows:

$$\sin(-\omega x) = -\sin(\omega x) \quad \text{and} \quad \cos(-\omega x) = \cos(\omega x)$$

This provides an equivalent form in which the coefficient of x in the argument is positive. For example,

$$\sin(-2x) = -\sin(2x) \quad \text{and} \quad \cos(-\pi x) = \cos(\pi x)$$

Because of this, we can assume that $\omega > 0$.

THEOREM

If $\omega > 0$, the amplitude and period of $y = A \sin(\omega x)$ and $y = A \cos(\omega x)$ are given by

$$\text{Amplitude} = |A| \qquad \text{Period} = T = \frac{2\pi}{\omega} \qquad \textbf{(1)}$$

EXAMPLE 4

Finding the Amplitude and Period of a Sinusoidal Function

Determine the amplitude and period of $y = 3 \sin(4x)$.

Solution

Comparing $y = 3 \sin(4x)$ to $y = A \sin(\omega x)$, note that $A = 3$ and $\omega = 4$. From equation (1),

$$\text{Amplitude} = |A| = 3 \qquad \text{Period} = T = \frac{2\pi}{\omega} = \frac{2\pi}{4} = \frac{\pi}{2}$$

●

Now Work PROBLEM 17

4 Graph Sinusoidal Functions Using Key Points

So far, we have graphed functions of the form $y = A \sin(\omega x)$ or $y = A \cos(\omega x)$ using transformations. We now introduce another method that can be used to graph these functions.

Figure 89 shows one cycle of the graphs of $y = \sin x$ and $y = \cos x$ on the interval $[0, 2\pi]$. Notice that each graph consists of four parts corresponding to the four subintervals:

$$\left[0, \frac{\pi}{2}\right], \quad \left[\frac{\pi}{2}, \pi\right], \quad \left[\pi, \frac{3\pi}{2}\right], \quad \left[\frac{3\pi}{2}, 2\pi\right]$$

Each subinterval is of length $\frac{\pi}{2}$ (the period 2π divided by 4, the number of parts), and the endpoints of these intervals $x = 0, x = \frac{\pi}{2}, x = \pi, x = \frac{3\pi}{2}, x = 2\pi$ give rise to five key points on each graph:

For $y = \sin x$: $(0, 0), \left(\dfrac{\pi}{2}, 1\right), (\pi, 0), \left(\dfrac{3\pi}{2}, -1\right), (2\pi, 0)$

For $y = \cos x$: $(0, 1), \left(\dfrac{\pi}{2}, 0\right), (\pi, -1), \left(\dfrac{3\pi}{2}, 0\right), (2\pi, 1)$

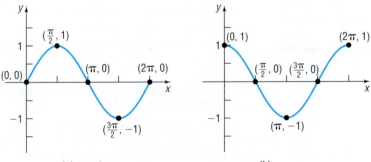

Figure 89

(a) $y = \sin x$

(b) $y = \cos x$

EXAMPLE 5

Graphing a Sinusoidal Function Using Key Points

Graph $y = 3\sin(4x)$ using key points.

Step-by-Step Solution

Step 1: *Determine the amplitude and period of the sinusoidal function.*

Comparing $y = 3\sin(4x)$ to $y = A\sin(\omega x)$, note that $A = 3$ and $\omega = 4$, so the amplitude is $|A| = 3$ and the period is $\dfrac{2\pi}{\omega} = \dfrac{2\pi}{4} = \dfrac{\pi}{2}$. Because the amplitude is 3, the graph of $y = 3\sin(4x)$ will lie between -3 and 3 on the y-axis. Because the period is $\dfrac{\pi}{2}$, one cycle will begin at $x = 0$ and end at $x = \dfrac{\pi}{2}$.

Step 2: *Divide the interval $\left[0, \dfrac{2\pi}{\omega}\right]$ into four subintervals of the same length.*

Divide the interval $\left[0, \dfrac{\pi}{2}\right]$ into four subintervals, each of length $\dfrac{\pi}{2} \div 4 = \dfrac{\pi}{8}$, as follows:

$$\left[0, \frac{\pi}{8}\right] \quad \left[\frac{\pi}{8}, \frac{\pi}{8} + \frac{\pi}{8}\right] = \left[\frac{\pi}{8}, \frac{\pi}{4}\right] \quad \left[\frac{\pi}{4}, \frac{\pi}{4} + \frac{\pi}{8}\right] = \left[\frac{\pi}{4}, \frac{3\pi}{8}\right] \quad \left[\frac{3\pi}{8}, \frac{3\pi}{8} + \frac{\pi}{8}\right] = \left[\frac{3\pi}{8}, \frac{\pi}{2}\right]$$

The endpoints of the subintervals are $0, \dfrac{\pi}{8}, \dfrac{\pi}{4}, \dfrac{3\pi}{8}, \dfrac{\pi}{2}$. These values represent the x-coordinates of the five key points on the graph.

Step 3: *Use the endpoints of the subintervals from Step 2 to obtain five key points on the graph.*

NOTE *The five key points could also be obtained by evaluating $y = 3\sin(4x)$ at each endpoint.* ∎

To obtain the y-coordinates of the five key points of $y = 3\sin(4x)$, multiply the y-coordinates of the five key points for $y = \sin x$ in Figure 89(a) by $A = 3$. The five key points are

$$(0, 0) \quad \left(\frac{\pi}{8}, 3\right) \quad \left(\frac{\pi}{4}, 0\right) \quad \left(\frac{3\pi}{8}, -3\right) \quad \left(\frac{\pi}{2}, 0\right)$$

Step 4: *Plot the five key points and draw a sinusoidal graph to obtain the graph of one cycle. Extend the graph in each direction to make it complete.*

Plot the five key points obtained in Step 3, and fill in the graph of the sine curve as shown in Figure 90(a). Extend the graph in each direction to obtain the complete graph shown in Figure 90(b). Notice that additional key points appear every $\dfrac{\pi}{8}$ radian.

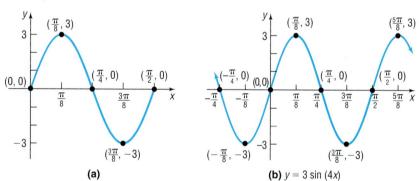

Figure 90

(a)

(b) $y = 3\sin(4x)$

✓**Check:** Graph $y = 3\sin(4x)$ using transformations. Which graphing method do you prefer?

━ **Now Work** PROBLEM 35 USING KEY POINTS

SUMMARY

Steps for Graphing a Sinusoidal Function of the Form $y = A\sin(\omega x)$ or $y = A\cos(\omega x)$ Using Key Points

STEP 1: Determine the amplitude and period of the sinusoidal function.

STEP 2: Divide the interval $\left[0, \dfrac{2\pi}{\omega}\right]$ into four subintervals of the same length.

STEP 3: Use the endpoints of these subintervals to obtain five key points on the graph.

STEP 4: Plot the five key points, and draw a sinusoidal graph to obtain the graph of one cycle. Extend the graph in each direction to make it complete.

EXAMPLE 6

Graphing a Sinusoidal Function Using Key Points

Graph $y = 2 \sin\left(-\dfrac{\pi}{2}x\right)$ using key points.

Solution   Since the sine function is odd, use the equivalent form:

$$y = -2 \sin\left(\dfrac{\pi}{2}x\right)$$

STEP 1: Comparing $y = -2 \sin\left(\dfrac{\pi}{2}x\right)$ to $y = A \sin(\omega x)$, note that $A = -2$ and $\omega = \dfrac{\pi}{2}$. The amplitude is $|A| = |-2| = 2$, and the period is

$$T = \dfrac{2\pi}{\omega} = \dfrac{2\pi}{\dfrac{\pi}{2}} = 4.$$ The graph of $y = -2 \sin\left(\dfrac{\pi}{2}x\right)$ lies between -2 and 2

on the y-axis. One cycle will begin at $x = 0$ and end at $x = 4$.

STEP 2: Divide the interval $[0, 4]$ into four subintervals, each of length $4 \div 4 = 1$. The x-coordinates of the five key points are

0	$0 + 1 = 1$	$1 + 1 = 2$	$2 + 1 = 3$	$3 + 1 = 4$
1st x-coordinate	2nd x-coordinate	3rd x-coordinate	4th x-coordinate	5th x-coordinate

STEP 3: Since $y = -2 \sin\left(\dfrac{\pi}{2}x\right)$, multiply the y-coordinates of the five key points in Figure 89(a) by $A = -2$. The five key points on the graph are

$$(0, 0) \quad (1, -2) \quad (2, 0) \quad (3, 2) \quad (4, 0)$$

STEP 4: Plot these five points, and fill in the graph of the sine function as shown in Figure 91(a). Extend the graph in each direction to obtain Figure 91(b).

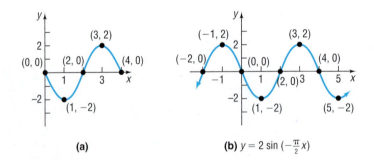

Figure 91   (a)   (b) $y = 2 \sin\left(-\frac{\pi}{2}x\right)$

COMMENT To graph a sinusoidal function of the form $y = A\sin(\omega x)$ or $y = A\cos(\omega x)$ using a graphing utility, use the amplitude to set Y_{min} and Y_{max}, and use the period to set X_{min} and X_{max}. ∎

✓**Check:** Graph $y = 2 \sin\left(-\dfrac{\pi}{2}x\right)$ using transformations. Which graphing method do you prefer?   ●

Now Work   PROBLEM **39** USING KEY POINTS

If the function to be graphed is of the form $y = A \sin(\omega x) + B$ [or $y = A \cos(\omega x) + B$], first graph $y = A \sin(\omega x)$ [or $y = A \cos(\omega x)$] and then use a vertical shift.

EXAMPLE 7

Graphing a Sinusoidal Function Using Key Points

Graph $y = -4 \cos(\pi x) - 2$ using key points. Use the graph to determine the domain and the range of $y = -4 \cos(\pi x) - 2$.

Solution   Begin by graphing the function $y = -4\cos(\pi x)$. Comparing $y = -4\cos(\pi x)$ with $y = A\cos(\omega x)$, note that $A = -4$ and $\omega = \pi$. The amplitude is $|A| = |-4| = 4$, and the period is $T = \dfrac{2\pi}{\omega} = \dfrac{2\pi}{\pi} = 2$.

The graph of $y = -4\cos(\pi x)$ will lie between -4 and 4 on the y-axis. One cycle will begin at $x = 0$ and end at $x = 2$.

Divide the interval $[0, 2]$ into four subintervals, each of length $2 \div 4 = \dfrac{1}{2}$. The x-coordinates of the five key points are

$$\underset{\text{1st }x\text{-coordinate}}{0} \qquad \underset{\text{2nd }x\text{-coordinate}}{0 + \frac{1}{2} = \frac{1}{2}} \qquad \underset{\text{3rd }x\text{-coordinate}}{\frac{1}{2} + \frac{1}{2} = 1} \qquad \underset{\text{4th }x\text{-coordinate}}{1 + \frac{1}{2} = \frac{3}{2}} \qquad \underset{\text{5th }x\text{-coordinate}}{\frac{3}{2} + \frac{1}{2} = 2}$$

Since $y = -4\cos(\pi x)$, multiply the y-coordinates of the five key points of $y = \cos x$ shown in Figure 89(b) by $A = -4$ to obtain the five key points on the graph of $y = -4\cos(\pi x)$:

$$(0, -4) \quad \left(\frac{1}{2}, 0\right) \quad (1, 4) \quad \left(\frac{3}{2}, 0\right) \quad (2, -4)$$

Plot these five points, and fill in the graph of the cosine function as shown in Figure 92(a). Extend the graph in each direction to obtain Figure 92(b), the graph of $y = -4\cos(\pi x)$.

A vertical shift down 2 units gives the graph of $y = -4\cos(\pi x) - 2$, as shown in Figure 92(c).

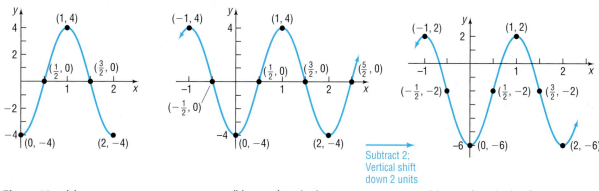

Figure 92   **(a)**   **(b)** $y = -4\cos(\pi x)$ **(c)** $y = -4\cos(\pi x) - 2$

The domain of $y = -4\cos(\pi x) - 2$ is the set of all real numbers, or $(-\infty, \infty)$. The range of $y = -4\cos(\pi x) - 2$ is $\{y \mid -6 \le y \le 2\}$, or $[-6, 2]$.   ●

Now Work   PROBLEM **49**

5 Find an Equation for a Sinusoidal Graph

EXAMPLE 8   ### Finding an Equation for a Sinusoidal Graph

Find an equation for the graph shown in Figure 93.

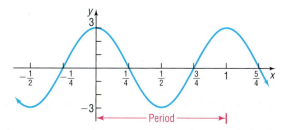

Figure 93

Solution The graph has the characteristics of a cosine function. Do you see why? The maximum value, 3, occurs at $x = 0$. So the equation can be viewed as a cosine function $y = A \cos(\omega x)$ with $A = 3$ and period $T = 1$. Then $\dfrac{2\pi}{\omega} = 1$, so $\omega = 2\pi$. The cosine function whose graph is given in Figure 93 is

$$y = A \cos(\omega x) = 3 \cos(2\pi x)$$

✓ **Check:** Graph $Y_1 = 3 \cos(2\pi x)$ and compare the result with Figure 93. ●

EXAMPLE 9 **Finding an Equation for a Sinusoidal Graph**

Find an equation for the graph shown in Figure 94.

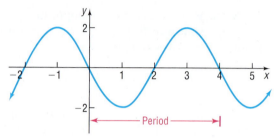

Figure 94

Solution The graph is sinusoidal, with amplitude $|A| = 2$. The period is 4, so $\dfrac{2\pi}{\omega} = 4$, or $\omega = \dfrac{\pi}{2}$. Since the graph passes through the origin, it is easier to view the equation as a sine function,[†] but note that the graph is actually the reflection of a sine function about the x-axis (since the graph is decreasing near the origin). This requires that $A = -2$. The sine function whose graph is given in Figure 94 is

$$y = A \sin(\omega x) = -2 \sin\left(\frac{\pi}{2}x\right)$$

✓ **Check:** Graph $Y_1 = -2 \sin\left(\dfrac{\pi}{2}x\right)$ and compare the result with Figure 94. ●

Now Work PROBLEMS **57** AND **61**

7.6 Assess Your Understanding

'Are You Prepared?' *Answers are given at the end of these exercises. If you get a wrong answer, read the pages listed in red.*

1. Use transformations to graph $y = 3x^2$. (pp. 247–256)

2. Use transformations to graph $y = \sqrt{2x}$. (pp. 247–256)

Concepts and Vocabulary

3. The maximum value of $y = \sin x$, $0 \le x \le 2\pi$, is _____ and occurs at $x = $ _____.

4. The function $y = A \sin(\omega x)$, $A > 0$, has amplitude 3 and period 2; then $A = $ _____ and $\omega = $ _____.

5. The function $y = 3 \cos(6x)$ has amplitude _____ and period _____.

6. *True or False* The graphs of $y = \sin x$ and $y = \cos x$ are identical except for a horizontal shift.

7. *True or False* For $y = 2 \sin(\pi x)$, the amplitude is 2 and the period is $\dfrac{\pi}{2}$.

8. *True or False* The graph of the sine function has infinitely many x-intercepts.

9. One period of the graph of $y = \sin(\omega x)$ or $y = \cos(\omega x)$ is called a(n) _____.
   (a) amplitude   (b) phase shift
   (c) transformation   (d) cycle

10. To graph $y = 3 \sin(-2x)$ using key points, the equivalent form _____ could be graphed instead.
   (a) $y = -3 \sin(-2x)$ (b) $y = -2 \sin(3x)$
   (c) $y = 3 \sin(2x)$ (d) $y = -3 \sin(2x)$

[†]The equation could also be viewed as a cosine function with a horizontal shift, but viewing it as a sine function is easier.

Skill Building

11. $f(x) = \sin x$
(a) What is the y-intercept of the graph of f?
(b) For what numbers x, $-\pi \le x \le \pi$, is the graph of f increasing?
(c) What is the absolute maximum of f?
(d) For what numbers x, $0 \le x \le 2\pi$, does $f(x) = 0$?
(e) For what numbers x, $-2\pi \le x \le 2\pi$, does $f(x) = 1$? Where does $f(x) = -1$?
(f) For what numbers x, $-2\pi \le x \le 2\pi$, does $f(x) = -\dfrac{1}{2}$?
(g) What are the x-intercepts of f?

12. $g(x) = \cos x$
(a) What is the y-intercept of the graph of g?
(b) For what numbers x, $-\pi \le x \le \pi$, is the graph of g decreasing?
(c) What is the absolute minimum of g?
(d) For what numbers x, $0 \le x \le 2\pi$, does $g(x) = 0$?
(e) For what numbers x, $-2\pi \le x \le 2\pi$, does $g(x) = 1$? Where does $g(x) = -1$?
(f) For what numbers x, $-2\pi \le x \le 2\pi$, does $g(x) = \dfrac{\sqrt{3}}{2}$?
(g) What are the x-intercepts of g?

In Problems 13–22, determine the amplitude and period of each function without graphing.

13. $y = 2 \sin x$

14. $y = 3 \cos x$

15. $y = -4 \cos(2x)$

16. $y = -\sin\left(\dfrac{1}{2}x\right)$

17. $y = 6 \sin(\pi x)$

18. $y = -3 \cos(3x)$

19. $y = -\dfrac{1}{2}\cos\left(\dfrac{3}{2}x\right)$

20. $y = \dfrac{4}{3}\sin\left(\dfrac{2}{3}x\right)$

21. $y = \dfrac{5}{3}\sin\left(-\dfrac{2\pi}{3}x\right)$

22. $y = \dfrac{9}{5}\cos\left(-\dfrac{3\pi}{2}x\right)$

In Problems 23–32, match the given function to one of the graphs (A)–(J).

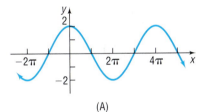

(A)

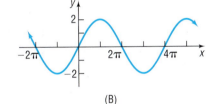

(B)

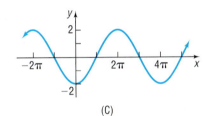

(C)

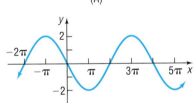

(D)

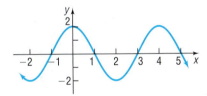

(E)

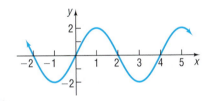

(F)

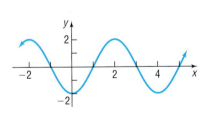

(G)

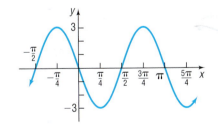

(H)

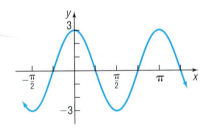

(I)

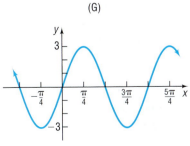

(J)

23. $y = 2 \sin\left(\dfrac{\pi}{2}x\right)$

24. $y = 2 \cos\left(\dfrac{\pi}{2}x\right)$

25. $y = 2 \cos\left(\dfrac{1}{2}x\right)$

26. $y = 3 \cos(2x)$

27. $y = -3 \sin(2x)$

28. $y = 2 \sin\left(\dfrac{1}{2}x\right)$

29. $y = -2 \cos\left(\dfrac{1}{2}x\right)$

30. $y = -2 \cos\left(\dfrac{\pi}{2}x\right)$

31. $y = 3 \sin(2x)$

32. $y = -2 \sin\left(\dfrac{1}{2}x\right)$

In Problems 33–56, graph each function using transformations or the method of key points. Be sure to label key points and show at least two cycles. Use the graph to determine the domain and the range of each function.

33. $y = 4 \cos x$

34. $y = 3 \sin x$

35. $y = -4 \sin x$

36. $y = -3 \cos x$

37. $y = \cos(4x)$

38. $y = \sin(3x)$

39. $y = \sin(-2x)$

40. $y = \cos(-2x)$

41. $y = 2 \sin\left(\dfrac{1}{2}x\right)$

42. $y = 2 \cos\left(\dfrac{1}{4}x\right)$

43. $y = -\dfrac{1}{2} \cos(2x)$

44. $y = -4 \sin\left(\dfrac{1}{8}x\right)$

45. $y = 2 \sin x + 3$

46. $y = 3 \cos x + 2$

47. $y = 5 \cos(\pi x) - 3$

48. $y = 4 \sin\left(\dfrac{\pi}{2}x\right) - 2$

49. $y = -6 \sin\left(\dfrac{\pi}{3}x\right) + 4$

50. $y = -3 \cos\left(\dfrac{\pi}{4}x\right) + 2$

51. $y = 5 - 3 \sin(2x)$

52. $y = 2 - 4 \cos(3x)$

53. $y = \dfrac{5}{3} \sin\left(-\dfrac{2\pi}{3}x\right)$

54. $y = \dfrac{9}{5} \cos\left(-\dfrac{3\pi}{2}x\right)$

55. $y = -\dfrac{3}{2} \cos\left(\dfrac{\pi}{4}x\right) + \dfrac{1}{2}$

56. $y = -\dfrac{1}{2} \sin\left(\dfrac{\pi}{8}x\right) + \dfrac{3}{2}$

In Problems 57–60, write the equation of a sine function that has the given characteristics.

57. Amplitude: 3
Period: π

58. Amplitude: 2
Period: 4π

59. Amplitude: 3
Period: 2

60. Amplitude: 4
Period: 1

In Problems 61–74, find an equation for each graph.

61.

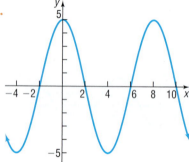

62.

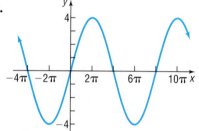

63.

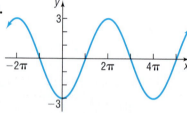

64.

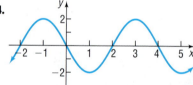

65.

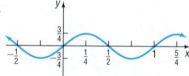

66.

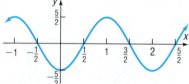

67.

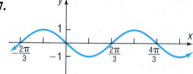

68.

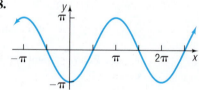

69.

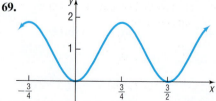

70.

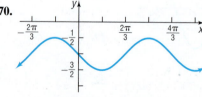

71.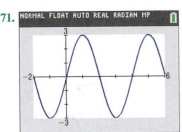

72.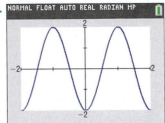

73.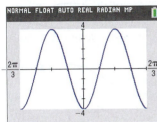

74.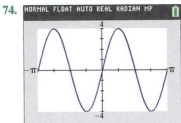

Mixed Practice

In Problems 75–78, find the average rate of change of f from 0 to $\dfrac{\pi}{2}$.

75. $f(x) = \sin x$ **76.** $f(x) = \cos x$ **77.** $f(x) = \sin \dfrac{x}{2}$ **78.** $f(x) = \cos(2x)$

In Problems 79–82, find $(f \circ g)(x)$ and $(g \circ f)(x)$, and graph each of these functions.

79. $f(x) = \sin x$ **80.** $f(x) = \cos x$ **81.** $f(x) = -2x$ **82.** $f(x) = -3x$
    $g(x) = 4x$ $g(x) = \dfrac{1}{2}x$ $g(x) = \cos x$ $g(x) = \sin x$

In Problems 83 and 84, graph each function.

83. $f(x) = \begin{cases} \sin x & 0 \le x < \dfrac{5\pi}{4} \\ \cos x & \dfrac{5\pi}{4} \le x \le 2\pi \end{cases}$

84. $g(x) = \begin{cases} 2\sin x & 0 \le x \le \pi \\ \cos x + 1 & \pi < x \le 2\pi \end{cases}$

Applications and Extensions

85. Alternating Current (ac) Circuits The current I, in amperes, flowing through an ac (alternating current) circuit at time t, in seconds, is

$$I(t) = 220\sin(60\pi t) \qquad t \ge 0$$

What is the period? What is the amplitude? Graph this function over two periods.

86. Alternating Current (ac) Circuits The current I, in amperes, flowing through an ac (alternating current) circuit at time t, in seconds, is

$$I(t) = 120\sin(30\pi t) \qquad t \ge 0$$

What is the period? What is the amplitude? Graph this function over two periods.

87. Alternating Current (ac) Generators The voltage V, in volts, produced by an ac generator at time t, in seconds, is

$$V(t) = 220\sin(120\pi t)$$

(a) What is the amplitude? What is the period?
(b) Graph V over two periods, beginning at $t = 0$.
(c) If a resistance of $R = 10$ ohms is present, what is the current I?
   [**Hint:** Use Ohm's Law, $V = IR$.]
(d) What are the amplitude and period of the current I?
(e) Graph I over two periods, beginning at $t = 0$.

88. Alternating Current (ac) Generators The voltage V, in volts, produced by an ac generator at time t, in seconds, is

$$V(t) = 120\sin(120\pi t)$$

(a) What is the amplitude? What is the period?
(b) Graph V over two periods, beginning at $t = 0$.
(c) If a resistance of $R = 20$ ohms is present, what is the current I?
   [**Hint:** Use Ohm's Law, $V = IR$.]
(d) What are the amplitude and period of the current I?
(e) Graph I over two periods, beginning at $t = 0$.

89. Alternating Current (ac) Generators The voltage V produced by an ac generator is sinusoidal. As a function of time, the voltage V is

$$V(t) = V_0 \sin(2\pi f t)$$

where f is the **frequency**, the number of complete oscillations (cycles) per second. [In the United States and Canada, f is 60 hertz (Hz).] The **power** P delivered to a resistance R at any time t is defined as

$$P(t) = \frac{[V(t)]^2}{R}$$

(a) Show that $P(t) = \dfrac{V_0^2}{R}\sin^2(2\pi f t)$.

(continued on the next page)

(b) The graph of P is shown in the figure. Express P as a sinusoidal function.

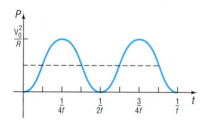

Power in an ac generator

(c) Deduce that

$$\sin^2(2\pi ft) = \frac{1}{2}[1 - \cos(4\pi ft)]$$

90. Bridge Clearance   A one-lane highway runs through a tunnel in the shape of one-half a sine curve cycle. The opening is 28 feet wide at road level and is 15 feet tall at its highest point.

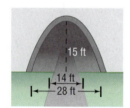

(a) Find an equation for the sine curve that fits the opening. Place the origin at the left end of the opening.
(b) If the road is 14 feet wide with 7-foot shoulders on each side, what is the height of the tunnel at the edge of the road?

Source: en.wikipedia.org/wiki/Interstate_Highway_standards and *Ohio Revised Code*

91. Biorhythms   In the theory of biorhythms, a sine function of the form

$$P(t) = 50\sin(\omega t) + 50$$

is used to measure the percent P of a person's potential at time t, where t is measured in days and $t = 0$ is the day the person is born. Three characteristics are commonly measured:

  Physical potential: period of 23 days
  Emotional potential: period of 28 days
  Intellectual potential: period of 33 days

(a) Find ω for each characteristic.
(b) Using a graphing utility, graph all three functions on the same screen.
(c) Is there a time t when all three characteristics have 100% potential? When is it?
(d) Suppose that you are 20 years old today ($t = 7305$ days). Describe your physical, emotional, and intellectual potential for the next 30 days.

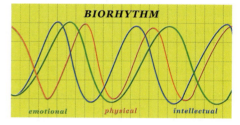

92. Graph $y = |\cos x|$, $-2\pi \le x \le 2\pi$.

93. Graph $y = |\sin x|$, $-2\pi \le x \le 2\pi$.

In Problems 94–97, the graphs of the given pairs of functions intersect infinitely many times. In each problem, find four of these points of intersection.

94. $y = \sin x$
$y = \dfrac{1}{2}$

95. $y = \cos x$
$y = \dfrac{1}{2}$

96. $y = 2\sin x$
$y = -2$

97. $y = \tan x$
$y = 1$

Explaining Concepts: Discussion and Writing

98. Explain how you would scale the x-axis and y-axis before graphing $y = 3\cos(\pi x)$.

99. Explain the term *amplitude* as it relates to the graph of a sinusoidal function.

100. Explain the term *period* as it relates to the graph of a sinusoidal function.

101. Explain how the amplitude and period of a sinusoidal graph are used to establish the scale on each coordinate axis.

102. Find an application in your major field that leads to a sinusoidal graph. Write a summary of your findings.

Retain Your Knowledge

Problems 103–106 are based on material learned earlier in the course. The purpose of these problems is to keep the material fresh in your mind so that you are better prepared for the final exam.

103. If $f(x) = x^2 - 5x + 1$, find $\dfrac{f(x+h) - f(x)}{h}$.

104. Find the vertex of the graph of $g(x) = -3x^2 + 12x - 7$.

105. Find the intercepts of the graph of $h(x) = 3|x+2| - 1$.

106. Solve: $3x - 2(5x + 16) = -3x + 4(8 + x)$

'Are You Prepared?' Answers

1. Vertical stretch by a factor of 3

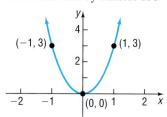

2. Horizontal compression by a factor of $\frac{1}{2}$

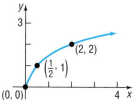

7.7 Graphs of the Tangent, Cotangent, Cosecant, and Secant Functions

PREPARING FOR THIS SECTION *Before getting started, review the following:*

- Vertical Asymptotes (Section 5.2, pp. 345–347)

Now Work the 'Are You Prepared?' problems on page 584.

OBJECTIVES **1** Graph Functions of the Form $y = A \tan(\omega x) + B$ and $y = A \cot(\omega x) + B$ (p. 581)

2 Graph Functions of the Form $y = A \csc(\omega x) + B$ and $y = A \sec(\omega x) + B$ (p. 583)

The Graph of the Tangent Function $y = \tan x$

Because the tangent function has period π, we only need to determine the graph over some interval of length π. The rest of the graph will consist of repetitions of that graph. Because the tangent function is not defined at $\ldots, -\frac{3\pi}{2}, -\frac{\pi}{2}, \frac{\pi}{2}, \frac{3\pi}{2}, \ldots$, we will concentrate on the interval $\left(-\frac{\pi}{2}, \frac{\pi}{2}\right)$, of length π, and construct Table 9, which lists some points on the graph of $y = \tan x$, $-\frac{\pi}{2} < x < \frac{\pi}{2}$. We plot the points in the table and connect them with a smooth curve. See Figure 95 for a partial graph of $y = \tan x$, where $-\frac{\pi}{3} \le x \le \frac{\pi}{3}$.

Table 9

x	$y = \tan x$	(x, y)
$-\frac{\pi}{3}$	$-\sqrt{3} \approx -1.73$	$\left(-\frac{\pi}{3}, -\sqrt{3}\right)$
$-\frac{\pi}{4}$	-1	$\left(-\frac{\pi}{4}, -1\right)$
$-\frac{\pi}{6}$	$-\frac{\sqrt{3}}{3} \approx -0.58$	$\left(-\frac{\pi}{6}, -\frac{\sqrt{3}}{3}\right)$
0	0	$(0, 0)$
$\frac{\pi}{6}$	$\frac{\sqrt{3}}{3} \approx 0.58$	$\left(\frac{\pi}{6}, \frac{\sqrt{3}}{3}\right)$
$\frac{\pi}{4}$	1	$\left(\frac{\pi}{4}, 1\right)$
$\frac{\pi}{3}$	$\sqrt{3} \approx 1.73$	$\left(\frac{\pi}{3}, \sqrt{3}\right)$

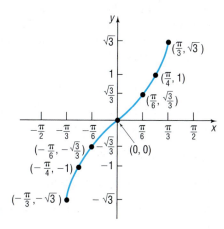

Figure 95 $y = \tan x$, $-\frac{\pi}{3} \le x \le \frac{\pi}{3}$

To complete one period of the graph of $y = \tan x$, we need to investigate the behavior of the function as x approaches $-\dfrac{\pi}{2}$ and $\dfrac{\pi}{2}$. We must be careful, though, because $y = \tan x$ is not defined at these numbers. To determine this behavior, we use the identity

$$\tan x = \frac{\sin x}{\cos x}$$

See Table 10. If x is close to $\dfrac{\pi}{2} \approx 1.5708$ but remains less than $\dfrac{\pi}{2}$, then $\sin x$ will be close to 1, and $\cos x$ will be positive and close to 0. (To see this, refer to the graphs of the sine function and the cosine function.) So the ratio $\dfrac{\sin x}{\cos x}$ will be positive and large. In fact, the closer x gets to $\dfrac{\pi}{2}$, the closer $\sin x$ gets to 1 and $\cos x$ gets to 0, so $\tan x$ approaches ∞ $\left(\lim\limits_{x \to \frac{\pi}{2}^-} \tan x = \infty \right)$. In other words, the vertical line $x = \dfrac{\pi}{2}$ is a vertical asymptote to the graph of $y = \tan x$.

Table 10

x	$\sin x$	$\cos x$	$y = \tan x$
$\dfrac{\pi}{3} \approx 1.05$	$\dfrac{\sqrt{3}}{2}$	$\dfrac{1}{2}$	$\sqrt{3} \approx 1.73$
1.5	0.9975	0.0707	14.1
1.57	0.9999	7.96×10^{-4}	1255.8
1.5707	0.9999	9.6×10^{-5}	10,381
$\dfrac{\pi}{2} \approx 1.5708$	1	0	Undefined

If x is close to $-\dfrac{\pi}{2}$ but remains greater than $-\dfrac{\pi}{2}$, then $\sin x$ is close to -1, and $\cos x$ is positive and close to 0. The ratio $\dfrac{\sin x}{\cos x}$ approaches $-\infty$ $\left(\lim\limits_{x \to -\frac{\pi}{2}^+} \tan x = -\infty \right)$. In other words, the vertical line $x = -\dfrac{\pi}{2}$ is also a vertical asymptote to the graph.

With these observations, one period of the graph can be completed. Obtain the complete graph of $y = \tan x$ by repeating this period, as shown in Figure 96.

✓ **Check:** Graph $Y_1 = \tan x$ and compare the result with Figure 96. Use TRACE to see what happens as x gets close to $\dfrac{\pi}{2}$ but remains less than $\dfrac{\pi}{2}$.

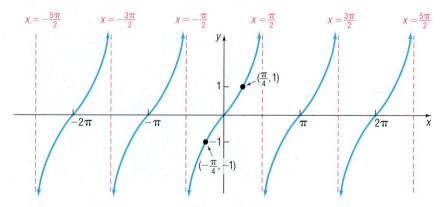

Figure 96 $y = \tan x$, $-\infty < x < \infty$, x not equal to odd multiples of $\dfrac{\pi}{2}$, $-\infty < y < \infty$

The graph of $y = \tan x$ in Figure 96 illustrates the following properties.

> **Properties of the Tangent Function**
>
> 1. The domain is the set of all real numbers, except odd multiples of $\dfrac{\pi}{2}$.
> 2. The range is the set of all real numbers.
> 3. The tangent function is an odd function, as the symmetry of the graph with respect to the origin indicates.
> 4. The tangent function is periodic, with period π.
> 5. The x-intercepts are $\ldots, -2\pi, -\pi, 0, \pi, 2\pi, 3\pi, \ldots$; the y-intercept is 0.
> 6. Vertical asymptotes occur at $x = \ldots, -\dfrac{3\pi}{2}, -\dfrac{\pi}{2}, \dfrac{\pi}{2}, \dfrac{3\pi}{2}, \ldots$.

 Now Work PROBLEMS 7 AND 15

1 Graph Functions of the Form $y = A\tan(\omega x) + B$ and $y = A\cot(\omega x) + B$

For tangent functions, there is no concept of amplitude since the range of the tangent function is $(-\infty, \infty)$. The role of A in $y = A\tan(\omega x) + B$ is to provide the magnitude of the vertical stretch. The period of $y = \tan x$ is π, so the period of $y = A\tan(\omega x) + B$ is $\dfrac{\pi}{\omega}$, caused by the horizontal compression of the graph by a factor of $\dfrac{1}{\omega}$. Finally, the presence of B indicates that a vertical shift is required.

EXAMPLE 1 | **Graphing Functions of the Form $y = A\tan(\omega x) + B$**

Graph $y = 2\tan x - 1$. Use the graph to determine the domain and the range of the function $y = 2\tan x - 1$.

Solution   Figure 97 shows the steps using transformations.

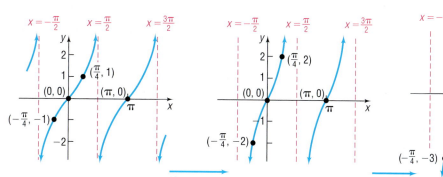

Figure 97

(a) $y = \tan x$

Multiply by 2;
Vertical stretch
by a factor of 2

(b) $y = 2\tan x$

Subtract 1;
Vertically shift
down 1 unit

(c) $y = 2\tan x - 1$

✓ Check: Graph $Y_1 = 2\tan x - 1$ to verify the graph shown in Figure 97(c).

The domain of $y = 2\tan x - 1$ is $\left\{ x \,\middle|\, x \neq \dfrac{k\pi}{2}, k \text{ is an odd integer} \right\}$, and the range is the set of all real numbers, or $(-\infty, \infty)$.   ●

EXAMPLE 2 | **Graphing Functions of the Form $y = A\tan(\omega x) + B$**

Graph $y = 3\tan(2x)$. Use the graph to determine the domain and the range of $y = 3\tan(2x)$.

Solution Figure 98 shows the steps using transformations.

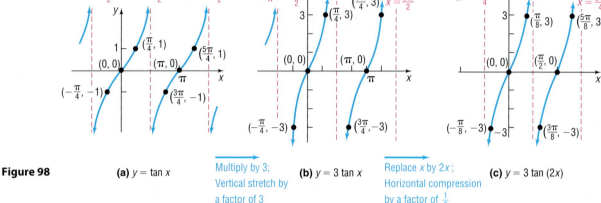

Figure 98   **(a)** $y = \tan x$ Multiply by 3;
Vertical stretch by
a factor of 3   **(b)** $y = 3 \tan x$ Replace x by $2x$;
Horizontal compression
by a factor of $\frac{1}{2}$ **(c)** $y = 3 \tan (2x)$

The domain of $y = 3 \tan (2x)$ is $\left\{ x \mid x \neq \dfrac{k\pi}{4}, k \text{ is an odd integer} \right\}$, and the range is the set of all real numbers, or $(-\infty, \infty)$.

✓ Check: Graph $Y_1 = 3 \tan (2x)$ to verify the graph in Figure 98(c). ●

Note in Figure 98(c) that the period of $y = 3 \tan (2x)$ is $\dfrac{\pi}{2}$ because of the compression of the original period π by a factor of $\dfrac{1}{2}$. Notice that the asymptotes are $x = -\dfrac{\pi}{4}, x = \dfrac{\pi}{4}, x = \dfrac{3\pi}{4}$, and so on, also because of the compression.

- **Now Work** PROBLEM **21**

The Graph of the Cotangent Function $y = \cot x$

The graph of $y = \cot x$ can be obtained in the same manner as the graph of $y = \tan x$. The period of $y = \cot x$ is π. Because the cotangent function is not defined for integer multiples of π, concentrate on the interval $(0, \pi)$. Table 11 lists some points on the graph of $y = \cot x$, $0 < x < \pi$. As x approaches 0 but remains greater than 0, the value of $\cos x$ will be close to 1, and the value of $\sin x$ will be positive and close to 0. The ratio $\dfrac{\cos x}{\sin x} = \cot x$ will be positive and large; so as x approaches 0, with $x > 0$, then $\cot x$ approaches ∞ ($\lim\limits_{x \to 0^+} \cot x = \infty$). Similarly, as x approaches π but remains less than π, the value of $\cos x$ will be close to -1, and the value of $\sin x$ will be positive and close to 0. So the ratio $\dfrac{\cos x}{\sin x} = \cot x$ will be negative and will approach $-\infty$ as x approaches π ($\lim\limits_{x \to \pi^-} \cot x = -\infty$). Figure 99 shows the graph.

Table 11

x	$y = \cot x$	(x, y)
$\dfrac{\pi}{6}$	$\sqrt{3}$	$\left(\dfrac{\pi}{6}, \sqrt{3}\right)$
$\dfrac{\pi}{4}$	1	$\left(\dfrac{\pi}{4}, 1\right)$
$\dfrac{\pi}{3}$	$\dfrac{\sqrt{3}}{3}$	$\left(\dfrac{\pi}{3}, \dfrac{\sqrt{3}}{3}\right)$
$\dfrac{\pi}{2}$	0	$\left(\dfrac{\pi}{2}, 0\right)$
$\dfrac{2\pi}{3}$	$-\dfrac{\sqrt{3}}{3}$	$\left(\dfrac{2\pi}{3}, -\dfrac{\sqrt{3}}{3}\right)$
$\dfrac{3\pi}{4}$	-1	$\left(\dfrac{3\pi}{4}, -1\right)$
$\dfrac{5\pi}{6}$	$-\sqrt{3}$	$\left(\dfrac{5\pi}{6}, -\sqrt{3}\right)$

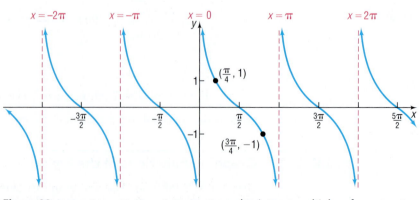

Figure 99 $y = \cot x$, $-\infty < x < \infty$, x not equal to integer multiples of π, $-\infty < y < \infty$

The graph of $y = A \cot(\omega x) + B$ has characteristics similar to those of the tangent function. The cotangent function $y = A \cot(\omega x) + B$ has period $\dfrac{\pi}{\omega}$. The cotangent function has no amplitude. The role of A is to provide the magnitude of the vertical stretch; the presence of B indicates that a vertical shift is required.

   **Now Work** PROBLEM 23

The Graphs of the Cosecant Function and the Secant Function

The cosecant and secant functions, sometimes referred to as **reciprocal functions**, are graphed by making use of the reciprocal identities

$$\csc x = \frac{1}{\sin x} \quad \text{and} \quad \sec x = \frac{1}{\cos x}$$

For example, the value of the cosecant function $y = \csc x$ at a given number x equals the reciprocal of the corresponding value of the sine function, provided that the value of the sine function is not 0. If the value of $\sin x$ is 0, then x is an integer multiple of π. At such numbers, the cosecant function is not defined. In fact, the graph of the cosecant function has vertical asymptotes at integer multiples of π. Figure 100 shows the graph.

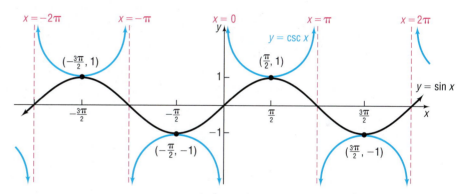

Figure 100 $y = \csc x$, $-\infty < x < \infty$, x not equal to integer multiples of π, $|y| \geq 1$

Using the idea of reciprocals, the graph of $y = \sec x$ can be obtained in a similar manner. See Figure 101.

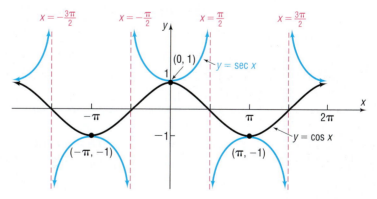

Figure 101 $y = \sec x$, $-\infty < x < \infty$, x not equal to odd multiples of $\dfrac{\pi}{2}$, $|y| \geq 1$

2 Graph Functions of the Form $y = A \csc(\omega x) + B$ and $y = A \sec(\omega x) + B$

The role of A in these functions is to set the range. The range of $y = \csc x$ is $\{y \mid y \leq -1 \text{ or } y \geq 1\}$ or $\{y \mid |y| \geq 1\}$; the range of $y = A \csc x$ is $\{y \mid |y| \geq |A|\}$

because of the vertical stretch of the graph by a factor of $|A|$. Just as with the sine and cosine functions, the period of $y = \csc(\omega x)$ and $y = \sec(\omega x)$ becomes $\dfrac{2\pi}{\omega}$ because of the horizontal compression of the graph by a factor of $\dfrac{1}{\omega}$. The presence of B indicates that a vertical shift is required.

EXAMPLE 3

Graphing Functions of the Form y = A csc(ωx) + B

Graph $y = 2\csc x - 1$. Use the graph to determine the domain and the range of $y = 2\csc x - 1$.

Solution We use transformations. Figure 102 shows the required steps.

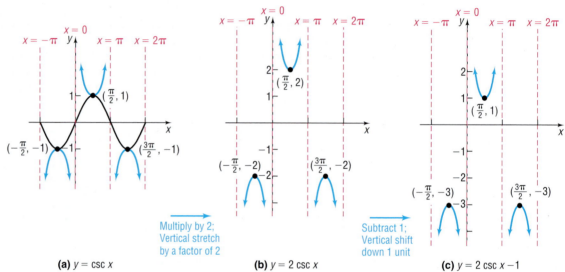

Figure 102   **(a)** $y = \csc x$ **(b)** $y = 2\csc x$ **(c)** $y = 2\csc x - 1$

The domain of $y = 2\csc x - 1$ is $\{x \mid x \neq k\pi, k \text{ is an integer}\}$ and the range is $\{y \mid y \leq -3 \text{ or } y \geq 1\}$, or, using interval notation, $(-\infty, -3] \cup [1, \infty)$.

✓ Check: Graph $Y_1 = 2\csc x - 1$ to verify the graph shown in Figure 102.

Now Work PROBLEM **29**

7.7 Assess Your Understanding

'Are You Prepared?' *Answers are given at the end of these exercises. If you get a wrong answer, read the pages listed in red.*

1. The graph of $y = \dfrac{3x - 6}{x - 4}$ has a vertical asymptote. What is it? (pp. 345–347)

2. **True or False** If $x = 3$ is a vertical asymptote of a rational function R, then $\lim\limits_{x \to 3} |R(x)| = \infty$. (pp. 345–347)

Concepts and Vocabulary

3. The graph of $y = \tan x$ is symmetric with respect to the _____ and has vertical asymptotes at _____.

4. The graph of $y = \sec x$ is symmetric with respect to the _____ and has vertical asymptotes at _____.

5. It is easiest to graph $y = \sec x$ by first sketching the graph of _____.
   (a) $y = \sin x$ (b) $y = \cos x$ (c) $y = \tan x$ (d) $y = \csc x$

6. **True or False** The graphs of $y = \tan x, y = \cot x, y = \sec x$, and $y = \csc x$ each have infinitely many vertical asymptotes.

Skill Building

In Problems 7–16, if necessary, refer to the graphs of the functions to answer each question.

7. What is the y-intercept of $y = \tan x$?

8. What is the y-intercept of $y = \cot x$?

9. What is the y-intercept of $y = \sec x$?

10. What is the y-intercept of $y = \csc x$?

11. For what numbers x, $-2\pi \le x \le 2\pi$, does $\sec x = 1$? For what numbers x does $\sec x = -1$?

12. For what numbers x, $-2\pi \le x \le 2\pi$, does $\csc x = 1$? For what numbers x does $\csc x = -1$?

13. For what numbers x, $-2\pi \le x \le 2\pi$, does the graph of $y = \sec x$ have vertical asymptotes?

14. For what numbers x, $-2\pi \le x \le 2\pi$, does the graph of $y = \csc x$ have vertical asymptotes?

15. For what numbers x, $-2\pi \le x \le 2\pi$, does the graph of $y = \tan x$ have vertical asymptotes?

16. For what numbers x, $-2\pi \le x \le 2\pi$, does the graph of $y = \cot x$ have vertical asymptotes?

In Problems 17–40, graph each function. Be sure to label key points and show at least two cycles. Use the graph to determine the domain and the range of each function.

17. $y = 3 \tan x$

18. $y = -2 \tan x$

19. $y = 4 \cot x$

20. $y = -3 \cot x$

21. $y = \tan\left(\dfrac{\pi}{2}x\right)$

22. $y = \tan\left(\dfrac{1}{2}x\right)$

23. $y = \cot\left(\dfrac{1}{4}x\right)$

24. $y = \cot\left(\dfrac{\pi}{4}x\right)$

25. $y = 2 \sec x$

26. $y = \dfrac{1}{2} \csc x$

27. $y = -3 \csc x$

28. $y = -4 \sec x$

29. $y = 4 \sec\left(\dfrac{1}{2}x\right)$

30. $y = \dfrac{1}{2} \csc(2x)$

31. $y = -2 \csc(\pi x)$

32. $y = -3 \sec\left(\dfrac{\pi}{2}x\right)$

33. $y = \tan\left(\dfrac{1}{4}x\right) + 1$

34. $y = 2 \cot x - 1$

35. $y = \sec\left(\dfrac{2\pi}{3}x\right) + 2$

36. $y = \csc\left(\dfrac{3\pi}{2}x\right)$

37. $y = \dfrac{1}{2} \tan\left(\dfrac{1}{4}x\right) - 2$

38. $y = 3 \cot\left(\dfrac{1}{2}x\right) - 2$

39. $y = 2 \csc\left(\dfrac{1}{3}x\right) - 1$

40. $y = 3 \sec\left(\dfrac{1}{4}x\right) + 1$

Mixed Practice

In Problems 41–44, find the average rate of change of f from 0 to $\dfrac{\pi}{6}$.

41. $f(x) = \tan x$

42. $f(x) = \sec x$

43. $f(x) = \tan(2x)$

44. $f(x) = \sec(2x)$

In Problems 45–48, find $(f \circ g)(x)$ and $(g \circ f)(x)$, and graph each of these functions.

45. $f(x) = \tan x$
  $g(x) = 4x$

46. $f(x) = 2 \sec x$
  $g(x) = \dfrac{1}{2}x$

47. $f(x) = -2x$
  $g(x) = \cot x$

48. $f(x) = \dfrac{1}{2}x$
  $g(x) = 2 \csc x$

In Problems 49 and 50, graph each function.

49. $f(x) = \begin{cases} \tan x & 0 \le x < \dfrac{\pi}{2} \\ 0 & x = \dfrac{\pi}{2} \\ \sec x & \dfrac{\pi}{2} < x \le \pi \end{cases}$

50. $g(x) = \begin{cases} \csc x & 0 < x < \pi \\ 0 & x = \pi \\ \cot x & \pi < x < 2\pi \end{cases}$

Applications and Extensions

51. Carrying a Ladder around a Corner Two hallways, one of width 3 feet, the other of width 4 feet, meet at a right angle. See the illustration.

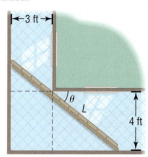

(a) Show that the length L of the ladder shown as a function of the angle θ is

$$L(\theta) = 3\sec\theta + 4\csc\theta$$

 (b) Graph $L = L(\theta), 0 < \theta < \dfrac{\pi}{2}$.

(c) For what value of θ is L the least?

(d) What is the length of the longest ladder that can be carried around the corner? Why is this also the least value of L?

52. A Rotating Beacon Suppose that a fire truck is parked in front of a building as shown in the figure.

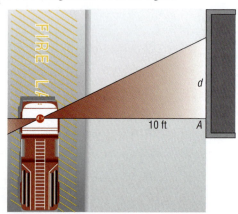

The beacon light on top of the fire truck is located 10 feet from the wall and has a light on each side. If the beacon light rotates 1 revolution every 2 seconds, then a model for determining the distance d, in feet, that the beacon of light is from point A on the wall after t seconds is given by

$$d(t) = |10\tan(\pi t)|$$

(a) Graph $d(t) = |10\tan(\pi t)|$ for $0 \le t \le 2$.

(b) For what values of t is the function undefined? Explain what this means in terms of the beam of light on the wall.

(c) Fill in the following table.

t	0	0.1	0.2	0.3	0.4
$d(t) = 10\tan(\pi t)$					

(d) Compute $\dfrac{d(0.1) - d(0)}{0.1 - 0}$, $\dfrac{d(0.2) - d(0.1)}{0.2 - 0.1}$, and so on, for each consecutive value of t. These are called **first differences**.

(e) Interpret the first differences found in part (d). What is happening to the speed of the beam of light as d increases?

53. Exploration Graph

$$y = \tan x \quad \text{and} \quad y = -\cot\left(x + \frac{\pi}{2}\right)$$

Do you think that $\tan x = -\cot\left(x + \dfrac{\pi}{2}\right)$?

Retain Your Knowledge

Problems 54–57 are based on material learned earlier in the course. The purpose of these problems is to keep the material fresh in your mind so that you are better prepared for the final exam.

54. Factor: $125p^3 - 8q^6$

55. Painting a Room Hazel can paint a room in 2 hours less time than her friend Gwyneth. Working together, they can paint the room in 2.4 hours. How long does it take each woman to paint the room by herself?

56. Solve: $9^{x-1} = 3^{x^2-5}$

57. Use the slope and the y-intercept to graph the linear function $f(x) = \dfrac{1}{4}x - 3$.

'Are You Prepared?' Answers

1. $x = 4$ **2.** True

7.8 Phase Shift; Sinusoidal Curve Fitting

OBJECTIVES 1 Graph Sinusoidal Functions of the Form $y = A \sin(\omega x - \phi) + B$ (p. 587)

2 Build Sinusoidal Models from Data (p. 590)

1 Graph Sinusoidal Functions of the Form $y = A \sin(\omega x - \phi) + B$

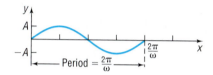

Figure 103 One cycle of $y = A\sin(\omega x)$, $A > 0$, $\omega > 0$

We have seen that the graph of $y = A \sin(\omega x)$, $\omega > 0$, has amplitude $|A|$ and period $T = \dfrac{2\pi}{\omega}$. One cycle can be drawn as x varies from 0 to $\dfrac{2\pi}{\omega}$, or, equivalently, as ωx varies from 0 to 2π. See Figure 103.

Now consider the graph of

$$y = A \sin(\omega x - \phi)$$

which may also be written as

$$y = A \sin\left[\omega\left(x - \frac{\phi}{\omega}\right)\right]$$

NOTE The beginning and end of the period can also be found by solving the inequality:

$$0 \le \omega x - \phi \le 2\pi$$
$$\phi \le \omega x \le 2\pi + \phi$$
$$\frac{\phi}{\omega} \le x \le \frac{2\pi}{\omega} + \frac{\phi}{\omega} \quad ■$$

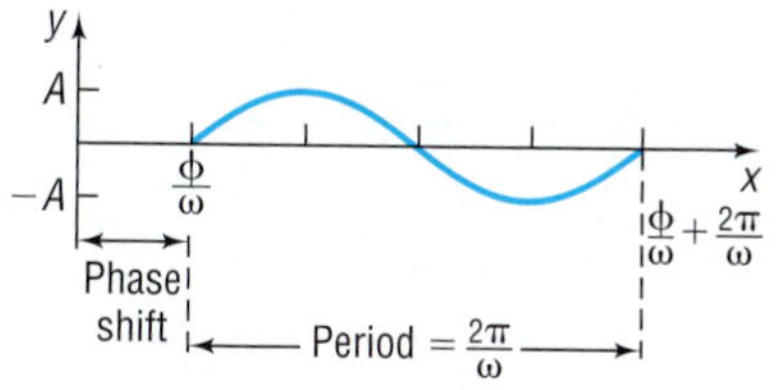

where $\omega > 0$ and ϕ (the Greek letter phi) are real numbers. The graph is a sine curve with amplitude $|A|$. As $\omega x - \phi$ varies from 0 to 2π, one period will be traced out. This period begins when

$$\omega x - \phi = 0 \quad \text{or} \quad x = \frac{\phi}{\omega}$$

and ends when

$$\omega x - \phi = 2\pi \quad \text{or} \quad x = \frac{\phi}{\omega} + \frac{2\pi}{\omega}$$

See Figure 104.

Notice that the graph of $y = A \sin(\omega x - \phi) = A \sin\left[\omega\left(x - \dfrac{\phi}{\omega}\right)\right]$ is the same as the graph of $y = A \sin(\omega x)$, except that it has been shifted $\left|\dfrac{\phi}{\omega}\right|$ units (to the right if $\phi > 0$ and to the left if $\phi < 0$). This number $\dfrac{\phi}{\omega}$ is called the **phase shift** of the graph of $y = A \sin(\omega x - \phi)$.

Figure 104 One cycle of $y = A\sin(\omega x - \phi)$, $A > 0$, $\omega > 0$, $\phi > 0$

For the graphs of $y = A \sin(\omega x - \phi)$ or $y = A \cos(\omega x - \phi)$, $\omega > 0$,

$$\text{Amplitude} = |A| \qquad \text{Period} = T = \frac{2\pi}{\omega} \qquad \text{Phase shift} = \frac{\phi}{\omega}$$

The phase shift is to the left if $\phi < 0$ and to the right if $\phi > 0$.

EXAMPLE 1 **Finding the Amplitude, Period, and Phase Shift of a Sinusoidal Function and Graphing It**

Find the amplitude, period, and phase shift of $y = 3 \sin(2x - \pi)$, and graph the function.

Solution Use the same four steps used to graph sinusoidal functions of the form $y = A \sin(\omega x)$ or $y = A \cos(\omega x)$ given on page 571.

STEP 1: Comparing

$$y = 3 \sin(2x - \pi) = 3 \sin\left[2\left(x - \frac{\pi}{2}\right)\right]$$

to

$$y = A \sin(\omega x - \phi) = A \sin\left[\omega\left(x - \frac{\phi}{\omega}\right)\right]$$

note that $A = 3$, $\omega = 2$, and $\phi = \pi$. The graph is a sine curve with amplitude $|A| = 3$, period $T = \dfrac{2\pi}{\omega} = \dfrac{2\pi}{2} = \pi$, and phase shift $= \dfrac{\phi}{\omega} = \dfrac{\pi}{2}$.

STEP 2: The graph of $y = 3\sin(2x - \pi)$ will lie between -3 and 3 on the y-axis. One cycle will begin at $x = \dfrac{\phi}{\omega} = \dfrac{\pi}{2}$ and end at $x = \dfrac{\phi}{\omega} + \dfrac{2\pi}{\omega} = \dfrac{\pi}{2} + \pi = \dfrac{3\pi}{2}$. To find the five key points, divide the interval $\left[\dfrac{\pi}{2}, \dfrac{3\pi}{2}\right]$ into four subintervals, each of length $\pi \div 4 = \dfrac{\pi}{4}$, by finding the following values of x:

$$\underset{\text{1st } x\text{-coordinate}}{\dfrac{\pi}{2}} \qquad \underset{\text{2nd } x\text{-coordinate}}{\dfrac{\pi}{2} + \dfrac{\pi}{4} = \dfrac{3\pi}{4}} \qquad \underset{\text{3rd } x\text{-coordinate}}{\dfrac{3\pi}{4} + \dfrac{\pi}{4} = \pi} \qquad \underset{\text{4th } x\text{-coordinate}}{\pi + \dfrac{\pi}{4} = \dfrac{5\pi}{4}} \qquad \underset{\text{5th } x\text{-coordinate}}{\dfrac{5\pi}{4} + \dfrac{\pi}{4} = \dfrac{3\pi}{2}}$$

STEP 3: Use these values of x to determine the five key points on the graph:

$$\left(\dfrac{\pi}{2}, 0\right) \quad \left(\dfrac{3\pi}{4}, 3\right) \quad (\pi, 0) \quad \left(\dfrac{5\pi}{4}, -3\right) \quad \left(\dfrac{3\pi}{2}, 0\right)$$

STEP 4: Plot these five points and fill in the graph of the sine function as shown in Figure 105(a). Extend the graph in each direction to obtain Figure 105(b).

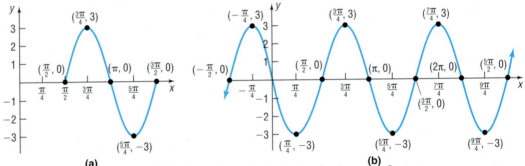

Figure 105
(a)
(b)

NOTE *The interval defining one cycle can also be found by solving the inequality*

$$0 \le 2x - \pi \le 2\pi$$

Then

$$\pi \le 2x \le 3\pi$$
$$\dfrac{\pi}{2} \le x \le \dfrac{3\pi}{2}$$ ∎

The graph of $y = 3\sin(2x - \pi) = 3\sin\left[2\left(x - \dfrac{\pi}{2}\right)\right]$ may also be obtained using transformations. See Figure 106.

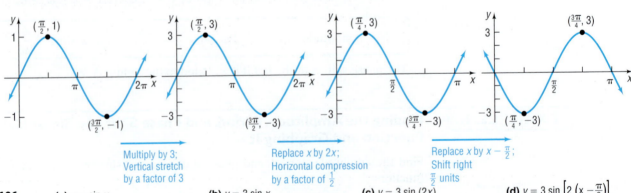

Figure 106
(a) $y = \sin x$
Multiply by 3;
Vertical stretch
by a factor of 3
(b) $y = 3\sin x$
Replace x by $2x$;
Horizontal compression
by a factor of $\frac{1}{2}$
(c) $y = 3\sin(2x)$
Replace x by $x - \frac{\pi}{2}$;
Shift right
$\frac{\pi}{2}$ units
(d) $y = 3\sin\left[2\left(x - \dfrac{\pi}{2}\right)\right]$
$= 3\sin(2x - \pi)$

⬛ ✓**Check:** Graph $Y_1 = 3\sin(2x - \pi)$ using a graphing utility to verify Figure 106(d).

To graph a sinusoidal function of the form $y = A\sin(\omega x - \phi) + B$, first graph the function $y = A\sin(\omega x - \phi)$ and then apply a vertical shift.

EXAMPLE 2

Finding the Amplitude, Period, and Phase Shift of a Sinusoidal Function and Graphing It

Find the amplitude, period, and phase shift of $y = 2\cos(4x + 3\pi) + 1$, and graph the function.

Solution   **STEP 1:** Begin by comparing

$$y = 2\cos(4x + 3\pi) = 2\cos\left[4\left(x + \frac{3\pi}{4}\right)\right]$$

to

$$y = A\cos(\omega x - \phi) = A\cos\left[\omega\left(x - \frac{\phi}{\omega}\right)\right]$$

note that $A = 2$, $\omega = 4$, and $\phi = -3\pi$. The graph is a cosine curve with amplitude $|A| = 2$, period $T = \dfrac{2\pi}{\omega} = \dfrac{2\pi}{4} = \dfrac{\pi}{2}$, and phase shift $= \dfrac{\phi}{\omega} = -\dfrac{3\pi}{4}$.

NOTE The interval defining one cycle can also be found by solving the inequality

$$0 \le 4x + 3\pi \le 2\pi$$

Then

$$-3\pi \le 4x \le -\pi$$
$$-\frac{3\pi}{4} \le x \le -\frac{\pi}{4}$$ ∎

STEP 2: The graph of $y = 2\cos(4x + 3\pi)$ will lie between -2 and 2 on the y-axis. One cycle begins at $x = \dfrac{\phi}{\omega} = -\dfrac{3\pi}{4}$ and ends at $x = \dfrac{\phi}{\omega} + \dfrac{2\pi}{\omega} = -\dfrac{3\pi}{4} + \dfrac{\pi}{2} = -\dfrac{\pi}{4}$. To find the five key points, divide the interval $\left[-\dfrac{3\pi}{4}, -\dfrac{\pi}{4}\right]$ into four subintervals, each of the length $\dfrac{\pi}{2} \div 4 = \dfrac{\pi}{8}$, by finding the following values.

$$-\frac{3\pi}{4} \qquad -\frac{3\pi}{4} + \frac{\pi}{8} = -\frac{5\pi}{8} \qquad -\frac{5\pi}{8} + \frac{\pi}{8} = -\frac{\pi}{2} \qquad -\frac{\pi}{2} + \frac{\pi}{8} = -\frac{3\pi}{8} \qquad -\frac{3\pi}{8} + \frac{\pi}{8} = -\frac{\pi}{4}$$

1st x-coordinate   2nd x-coordinate   3rd x-coordinate   4th x-coordinate   5th x-coordinate

STEP 3: The five key points on the graph of $y = 2\cos(4x + 3\pi)$ are

$$\left(-\frac{3\pi}{4}, 2\right) \quad \left(-\frac{5\pi}{8}, 0\right) \quad \left(-\frac{\pi}{2}, -2\right) \quad \left(-\frac{3\pi}{8}, 0\right) \quad \left(-\frac{\pi}{4}, 2\right)$$

STEP 4: Plot these five points and fill in the graph of the cosine function as shown in Figure 107(a). Extend the graph in each direction to obtain Figure 107(b), the graph of $y = 2\cos(4x + 3\pi)$.

STEP 5: A vertical shift up 1 unit gives the final graph. See Figure 107(c).

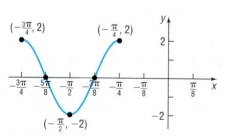

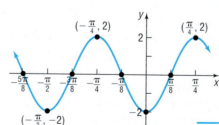

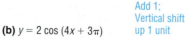

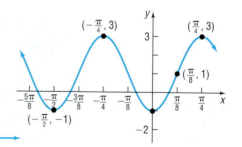

Figure 107   **(a)**   **(b)** $y = 2\cos(4x + 3\pi)$ Add 1; Vertical shift up 1 unit   **(c)** $y = 2\cos(4x + 3\pi) + 1$

The graph of $y = 2\cos(4x + 3\pi) + 1 = 2\cos\left[4\left(x + \dfrac{3\pi}{4}\right)\right] + 1$ may also be obtained using transformations. See Figure 108.

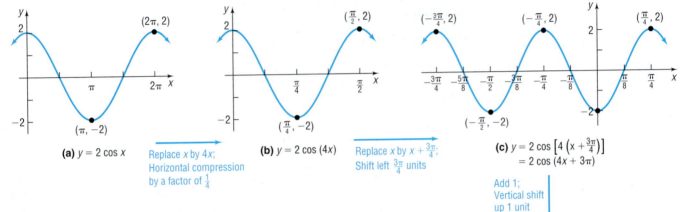

(a) $y = 2\cos x$

Replace x by 4x;
Horizontal compression
by a factor of $\frac{1}{4}$

(b) $y = 2\cos(4x)$

Replace x by $x + \frac{3\pi}{4}$;
Shift left $\frac{3\pi}{4}$ units

(c) $y = 2\cos\left[4\left(x + \frac{3\pi}{4}\right)\right]$
$= 2\cos(4x + 3\pi)$

Add 1;
Vertical shift
up 1 unit

(d) $y = 2\cos(4x + 3\pi) + 1$

Figure 108

Now Work PROBLEM **3**

SUMMARY

Steps for Graphing Sinusoidal Functions $y = A\sin(\omega x - \phi) + B$ or $y = A\cos(\omega x - \phi) + B$

STEP 1: Determine the amplitude $|A|$, period $T = \dfrac{2\pi}{\omega}$, and phase shift $\dfrac{\phi}{\omega}$.

STEP 2: Determine the starting point of one cycle of the graph, $\dfrac{\phi}{\omega}$. Determine the ending point of one cycle of the graph, $\dfrac{\phi}{\omega} + \dfrac{2\pi}{\omega}$. Divide the interval $\left[\dfrac{\phi}{\omega}, \dfrac{\phi}{\omega} + \dfrac{2\pi}{\omega}\right]$ into four subintervals, each of length $\dfrac{2\pi}{\omega} \div 4$.

STEP 3: Use the endpoints of the subintervals to find the five key points on the graph.

STEP 4: Plot the five key points, and connect them with a sinusoidal graph to obtain one cycle of the graph. Extend the graph in each direction to make it complete.

STEP 5: If $B \neq 0$, apply a vertical shift.

2 Build Sinusoidal Models from Data

Scatter diagrams of data sometimes resemble the graph of a sinusoidal function. Let's look at an example.

The data given in Table 12 on the next page represent the average monthly temperatures in Denver, Colorado. Since the data represent *average* monthly temperatures collected over many years, the data will not vary much from year to year and so will essentially repeat each year. In other words, the data are periodic. Figure 109 shows the scatter diagram of these data, where $x = 1$ represents January, $x = 2$ represents February, and so on.

Notice that the scatter diagram looks like the graph of a sinusoidal function. We choose to fit the data to a sine function of the form

$$y = A\sin(\omega x - \phi) + B$$

where A, B, ω, and ϕ are constants.

Table 12

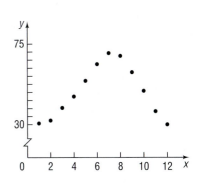

Month, x	Average Monthly Temperature, °F
January, 1	30.7
February, 2	32.5
March, 3	40.4
April, 4	47.4
May, 5	57.1
June, 6	67.4
July, 7	74.2
August, 8	72.5
September, 9	62.4
October, 10	50.9
November, 11	38.3
December, 12	30.0

Source: U.S. National Oceanic and
Atmospheric Administration

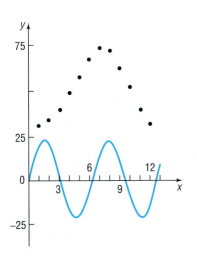

Figure 109 Denver average monthly temperature

EXAMPLE 3

Finding a Sinusoidal Function from Temperature Data

Fit a sine function to the data in Table 12.

Solution

Begin with a scatter diagram of the data for one year. See Figure 110. The data will be fitted to a sine function of the form

$$y = A \sin(\omega x - \phi) + B$$

STEP 1: To find the amplitude A, compute

$$\text{Amplitude} = \frac{\text{largest data value} - \text{smallest data value}}{2}$$

$$= \frac{74.2 - 30.0}{2} = 22.1$$

To see the remaining steps in this process, superimpose the graph of the function $y = 22.1 \sin x$, where x represents months, on the scatter diagram.

Figure 111 shows the two graphs. To fit the data, the graph needs to be shifted vertically, shifted horizontally, and stretched horizontally.

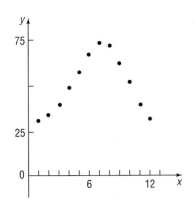

Figure 110

STEP 2: Determine the vertical shift by finding the average of the highest and lowest data values.

$$\text{Vertical shift} = \frac{74.2 + 30.0}{2} = 52.1$$

Now superimpose the graph of $y = 22.1 \sin x + 52.1$ on the scatter diagram. See Figure 112 on the next page.

We see that the graph needs to be shifted horizontally and stretched horizontally.

STEP 3: It is easier to find the horizontal stretch factor first. Since the temperatures repeat every 12 months, the period of the function is $T = 12$. Because

$$T = \frac{2\pi}{\omega} = 12,$$

$$\omega = \frac{2\pi}{12} = \frac{\pi}{6}$$

Figure 111

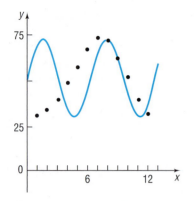

Figure 112

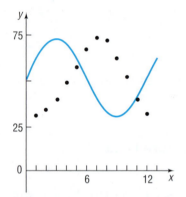

Figure 113

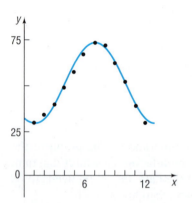

Figure 114

Now superimpose the graph of $y = 22.1 \sin\left(\dfrac{\pi}{6}x\right) + 52.1$ on the scatter diagram. See Figure 113, where it is clear that the graph still needs to be shifted horizontally.

STEP 4: To determine the horizontal shift, use the period $T = 12$ and divide the interval $[0, 12]$ into four subintervals of length $12 \div 4 = 3$:

$$[0, 3], \quad [3, 6], \quad [6, 9], \quad [9, 12]$$

The sine curve is increasing on the interval $(0, 3)$ and is decreasing on the interval $(3, 9)$, so a local maximum occurs at $x = 3$. The data indicate that a maximum occurs at $x = 7$ (corresponding to July's temperature), so the graph of the function must be shifted 4 units to the right by replacing x by $x - 4$. Doing this yields

$$y = 22.1 \sin\left(\dfrac{\pi}{6}(x - 4)\right) + 52.1$$

Multiplying out reveals that a sine function of the form $y = A \sin(\omega x - \phi) + B$ that fits the data is

$$y = 22.1 \sin\left(\dfrac{\pi}{6}x - \dfrac{2\pi}{3}\right) + 52.1$$

The graph of $y = 22.1 \sin\left(\dfrac{\pi}{6}x - \dfrac{2\pi}{3}\right) + 52.1$ and the scatter diagram of the data are shown in Figure 114. ●

The steps to fit a sine function

$$y = A \sin(\omega x - \phi) + B$$

to sinusoidal data follow.

Steps for Fitting a Sine Function $y = A \sin(\omega x - \phi) + B$ to Data

STEP 1: Determine A, the amplitude of the function.

$$\text{Amplitude} = \dfrac{\text{largest data value} - \text{smallest data value}}{2}$$

STEP 2: Determine B, the vertical shift of the function.

$$\text{Vertical shift} = \dfrac{\text{largest data value} + \text{smallest data value}}{2}$$

STEP 3: Determine ω. Since the period T, the time it takes for the data to repeat, is $T = \dfrac{2\pi}{\omega}$, we have

$$\omega = \dfrac{2\pi}{T}$$

STEP 4: Determine the horizontal shift of the function by using the period of the data. Divide the period into four subintervals of equal length. Determine the x-coordinate for the maximum of the sine function and the x-coordinate for the maximum value of the data. Use this information to determine the value of the phase shift, $\dfrac{\phi}{\omega}$.

Now Work PROBLEMS 29(a)–(c)

Let's look at another example. Since the number of hours of sunlight in a day cycles annually, the number of hours of sunlight in a day for a given location can be modeled by a sinusoidal function.

The longest day of the year (in terms of hours of sunlight) occurs on the day of the summer solstice. For locations in the Northern Hemisphere, the summer solstice is the time when the Sun is farthest north. In 2014, the summer solstice occurred on June 21 (the 172nd day of the year) at 6:51 AM EDT. The shortest day of the year occurs on the day of the winter solstice, the time when the Sun is farthest south (for locations in the Northern Hemisphere). In 2014, the winter solstice occurred on December 21 (the 355th day of the year) at 6:03 PM (EST).

EXAMPLE 4   **Finding a Sinusoidal Function for Hours of Daylight**

According to the *Old Farmer's Almanac,* the number of hours of sunlight in Boston on the day of the summer solstice is 15.30, and the number of hours of sunlight on the day of the winter solstice is 9.08.

(a) Find a sinusoidal function of the form $y = A \sin(\omega x - \phi) + B$ that fits the data.

(b) Use the function found in part (a) to predict the number of hours of sunlight in Boston on April 1, the 91st day of the year.

(c) Graph the function found in part (a).

(d) Look up the number of hours of sunlight for April 1 in the *Old Farmer's Almanac* and compare it to the results found in part (b).

Source: The Old Farmer's Almanac, www.almanac.com/rise

Solution   (a) **STEP 1:** Amplitude $= \dfrac{\text{largest data value} - \text{smallest data value}}{2}$

$$= \frac{15.30 - 9.08}{2} = 3.11$$

STEP 2: Vertical shift $= \dfrac{\text{largest data value} + \text{smallest data value}}{2}$

$$= \frac{15.30 + 9.08}{2} = 12.19$$

STEP 3: The data repeat every 365 days. Since $T = \dfrac{2\pi}{\omega} = 365$, we find

$$\omega = \frac{2\pi}{365}$$

So far, we have $y = 3.11 \sin\left(\dfrac{2\pi}{365} x - \phi\right) + 12.19$.

STEP 4: To determine the horizontal shift, use the period $T = 365$ and divide the interval $[0, 365]$ into four subintervals of length $365 \div 4 = 91.25$:

$$[0, 91.25], \quad [91.25, 182.5], \quad [182.5, 273.75], \quad [273.75, 365]$$

The sine curve is increasing on the interval $(0, 91.25)$ and is decreasing on the interval $(91.25, 273.75)$, so a local maximum occurs at $x = 91.25$. Since the maximum occurs on the summer solstice at $x = 172$, we must shift the graph of the function $172 - 91.25 = 80.75$ units to the right by replacing x by $x - 80.75$. Doing this yields

$$y = 3.11 \sin\left(\frac{2\pi}{365} (x - 80.75)\right) + 12.19$$

Next, multiply out to obtain a sine function of the form $y = A \sin(\omega x - \phi) + B$ that fits the data.

$$y = 3.11 \sin\left(\frac{2\pi}{365} x - \frac{323\pi}{730}\right) + 12.19$$

(b) To predict the number of hours of daylight on April 1, let $x = 91$ in the function found in part (a) and obtain

$$y = 3.11 \sin\left(\frac{2\pi}{365} \cdot 91 - \frac{323}{730}\pi\right) + 12.19$$

$$\approx 12.74$$

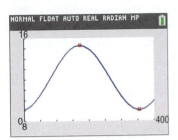

Figure 115

The prediction is that there will be about 12.74 hours = 12 hours 44 minutes of sunlight on April 1 in Boston.

(c) The graph of the function found in part (a) is given in Figure 115.

(d) According to the *Old Farmer's Almanac*, there will be 12 hours 45 minutes of sunlight on April 1 in Boston.   ●

Now Work PROBLEM 35

Certain graphing utilities (such as the TI-83, TI-84 Plus C, and TI-89) have the capability of finding the sine function of best fit for sinusoidal data. At least four data points are required for this process.

EXAMPLE 5

Finding the Sine Function of Best Fit

Use a graphing utility to find the sine function of best fit for the data in Table 12. Graph this function with the scatter diagram of the data.

Solution

Enter the data from Table 12 and execute the SINe REGression program. The result is shown in Figure 116.

The output that the utility provides shows the equation

$$y = a \sin(bx + c) + d$$

The sinusoidal function of best fit is

$$y = 21.43 \sin(0.56x - 2.44) + 51.71$$

where x represents the month and y represents the average temperature.

Figure 117 shows the graph of the sinusoidal function of best fit on the scatter diagram.

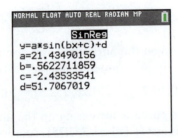

Figure 116

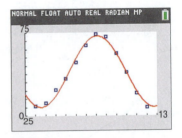

Figure 117

Now Work PROBLEMS 29(d) AND (e)

7.8 Assess Your Understanding

Concepts and Vocabulary

1. For the graph of $y = A \sin(\omega x - \phi)$, the number $\frac{\phi}{\omega}$ is called the _____ _____.

2. *True or False* A graphing utility requires only two data points to find the sine function of best fit.

Skill Building

In Problems 3–14, find the amplitude, period, and phase shift of each function. Graph each function. Be sure to label key points. Show at least two periods.

3. $y = 4 \sin(2x - \pi)$

4. $y = 3 \sin(3x - \pi)$

5. $y = 2 \cos\left(3x + \dfrac{\pi}{2}\right)$

6. $y = 3 \cos(2x + \pi)$

7. $y = -3 \sin\left(2x + \dfrac{\pi}{2}\right)$

8. $y = -2 \cos\left(2x - \dfrac{\pi}{2}\right)$

9. $y = 4 \sin(\pi x + 2) - 5$

10. $y = 2 \cos(2\pi x + 4) + 4$

11. $y = 3 \cos(\pi x - 2) + 5$

12. $y = 2 \cos(2\pi x - 4) - 1$

13. $y = -3 \sin\left(-2x + \dfrac{\pi}{2}\right)$

14. $y = -3 \cos\left(-2x + \dfrac{\pi}{2}\right)$

In Problems 15–18, write the equation of a sine function that has the given characteristics.

15. Amplitude: 2
    Period: π
    Phase shift: $\dfrac{1}{2}$

16. Amplitude: 3
    Period: $\dfrac{\pi}{2}$
    Phase shift: 2

17. Amplitude: 3
    Period: 3π
    Phase shift: $-\dfrac{1}{3}$

18. Amplitude: 2
    Period: π
    Phase shift: -2

Mixed Practice

In Problems 19–26, apply the methods of this and the previous section to graph each function. Be sure to label key points and show at least two periods.

19. $y = 2 \tan(4x - \pi)$

20. $y = \dfrac{1}{2} \cot(2x - \pi)$

21. $y = 3 \csc\left(2x - \dfrac{\pi}{4}\right)$

22. $y = \dfrac{1}{2} \sec(3x - \pi)$

23. $y = -\cot\left(2x + \dfrac{\pi}{2}\right)$

24. $y = -\tan\left(3x + \dfrac{\pi}{2}\right)$

25. $y = -\sec(2\pi x + \pi)$

26. $y = -\csc\left(-\dfrac{1}{2}\pi x + \dfrac{\pi}{4}\right)$

Applications and Extensions

27. **Alternating Current (ac) Circuits** The current I, in amperes, flowing through an ac (alternating current) circuit at time t, in seconds, is

    $$I(t) = 120 \sin\left(30\pi t - \dfrac{\pi}{3}\right) \qquad t \geq 0$$

    What is the period? What is the amplitude? What is the phase shift? Graph this function over two periods.

28. **Alternating Current (ac) Circuits** The current I, in amperes, flowing through an ac (alternating current) circuit at time t, in seconds, is

    $$I(t) = 220 \sin\left(60\pi t - \dfrac{\pi}{6}\right) \qquad t \geq 0$$

    What is the period? What is the amplitude? What is the phase shift? Graph this function over two periods.

29. **Hurricanes** Hurricanes are categorized using the Saffir-Simpson Hurricane Scale, with winds 111–130 miles per hour (mph) corresponding to a category 3 hurricane, winds 131–155 mph corresponding to a category 4 hurricane, and winds in excess of 155 mph corresponding to a category 5 hurricane. The following data represent the number of major hurricanes in the Atlantic Basin (category 3, 4, or 5) each decade from 1921 to 2010.
    (a) Draw a scatter diagram of the data.
    (b) Find a sinusoidal function of the form
    $$y = A \sin(\omega x - \phi) + B \text{ that models the data.}$$

(c) Draw the sinusoidal function found in part (b) on the scatter diagram.
(d) Use a graphing utility to find the sinusoidal function of best fit.
(e) Graph the sinusoidal function of best fit on a scatter diagram of the data.

Decade, x	Major Hurricanes, H
1921–1930, 1	17
1931–1940, 2	16
1941–1950, 3	29
1951–1960, 4	33
1961–1970, 5	27
1971–1980, 6	16
1981–1990, 7	16
1991–2000, 8	27
2001–2010, 9	33

Source: U.S. National Oceanic and Atmospheric Administration

30. **Monthly Temperature** The data on the next page represent the average monthly temperatures for Washington, D.C.
    (a) Draw a scatter diagram of the data for one period.
    (b) Find a sinusoidal function of the form
    $$y = A \sin(\omega x - \phi) + B \text{ that models the data.}$$

(c) Draw the sinusoidal function found in part (b) on the scatter diagram.

(d) Use a graphing utility to find the sinusoidal function of best fit.

(e) Graph the sinusoidal function of best fit on a scatter diagram of the data.

Month, x	Average Monthly Temperature, °F
January, 1	36.0
February, 2	39.0
March, 3	46.8
April, 4	56.8
May, 5	66.0
June, 6	75.2
July, 7	79.8
August, 8	78.1
September, 9	71.0
October, 10	59.5
November, 11	49.6
December, 12	39.7

Source: U.S. National Oceanic and Atmospheric Administration

31. Monthly Temperature The following data represent the average monthly temperatures for Indianapolis, Indiana.

Month, x	Average Monthly Temperature, °F
January, 1	28.1
February, 2	32.1
March, 3	42.2
April, 4	53.0
May, 5	62.7
June, 6	72.0
July, 7	75.4
August, 8	74.2
September, 9	66.9
October, 10	55.0
November, 11	43.6
December, 12	31.6

Source: U.S. National Oceanic and Atmospheric Administration

(a) Draw a scatter diagram of the data for one period.

(b) Find a sinusoidal function of the form

$$y = A \sin(\omega x - \phi) + B$$

that models the data.

(c) Draw the sinusoidal function found in part (b) on the scatter diagram.

(d) Use a graphing utility to find the sinusoidal function of best fit.

(e) Graph the sinusoidal function of best fit on a scatter diagram of the data.

32. Monthly Temperature The following data represent the average monthly temperatures for Baltimore, Maryland.

(a) Draw a scatter diagram of the data for one period.

(b) Find a sinusoidal function of the form

$y = A \sin(\omega x - \phi) + B$ that models the data.

(c) Draw the sinusoidal function found in part (b) on the scatter diagram.

(d) Use a graphing utility to find the sinusoidal function of best fit.

(e) Graph the sinusoidal function of best fit on a scatter diagram of the data.

Month, x	Average Monthly Temperature, °F
January, 1	32.9
February, 2	35.8
March, 3	43.6
April, 4	53.7
May, 5	62.9
June, 6	72.4
July, 7	77.0
August, 8	75.1
September, 9	67.8
October, 10	56.1
November, 11	46.5
December, 12	36.7

Source: U.S. National Oceanic and Atmospheric Administration

33. Tides The length of time between consecutive high tides is 12 hours and 25 minutes. According to the National Oceanic and Atmospheric Administration, on Saturday, April 26, 2014, in Charleston, South Carolina, high tide occurred at 6:30 AM (6.5 hours) and low tide occurred at 12:24 PM (12.4 hours). Water heights are measured as the amounts above or below the mean lower low water. The height of the water at high tide was 5.86 feet, and the height of the water at low tide was −0.38 foot.

(a) Approximately when will the next high tide occur?

(b) Find a sinusoidal function of the form

$$y = A \sin(\omega x - \phi) + B$$

that models the data.

(c) Use the function found in part (b) to predict the height of the water at 3 PM on April 26, 2014.

34. Tides The length of time between consecutive high tides is 12 hours and 25 minutes. According to the National Oceanic and Atmospheric Administration, on Saturday, April 26, 2014, in Sitka, Alaska, high tide occurred at 12:06 AM (0.10 hours) and low tide occurred at 6:24 AM (6.4 hours). Water heights are measured as the amounts above or below the mean lower low water. The height of the water at high tide was 9.97 feet, and the height of the water at low tide was 0.59 feet.

(a) Approximately when will the next high tide occur?

(b) Find a sinusoidal function of the form

$$y = A \sin(\omega x - \phi) + B$$

that models the data.

(c) Use the function found in part (b) to predict the height of the water at 6 PM.

35. Hours of Daylight According to the *Old Farmer's Almanac*, in Miami, Florida, the number of hours of sunlight on the summer solstice of 2014 was 13.75, and the number of hours of sunlight on the winter solstice was 10.52.
(a) Find a sinusoidal function of the form
$$y = A \sin(\omega x - \phi) + B$$
that models the data.
(b) Use the function found in part (a) to predict the number of hours of sunlight on April 1, the 91st day of the year.
(c) Draw a graph of the function found in part (a).
(d) Look up the number of hours of sunlight for April 1 in the *Old Farmer's Almanac*, and compare the actual hours of daylight to the results found in part (b).

36. Hours of Daylight According to the *Old Farmer's Almanac*, in Detroit, Michigan, the number of hours of sunlight on the summer solstice of 2014 was 15.27, and the number of hours of sunlight on the winter solstice was 9.07.
(a) Find a sinusoidal function of the form
$$y = A \sin(\omega x - \phi) + B$$
that models the data.
(b) Use the function found in part (a) to predict the number of hours of sunlight on April 1, the 91st day of the year.
(c) Draw a graph of the function found in part (a).
(d) Look up the number of hours of sunlight for April 1 in the *Old Farmer's Almanac*, and compare the actual hours of daylight to the results found in part (b).

37. Hours of Daylight According to the *Old Farmer's Almanac*, in Anchorage, Alaska, the number of hours of sunlight on the summer solstice of 2014 was 19.37, and the number of hours of sunlight on the winter solstice was 5.45.
(a) Find a sinusoidal function of the form
$$y = A \sin(\omega x - \phi) + B$$
that models the data.
(b) Use the function found in part (a) to predict the number of hours of sunlight on April 1, the 91st day of the year.
(c) Draw a graph of the function found in part (a).
(d) Look up the number of hours of sunlight for April 1 in the *Old Farmer's Almanac*, and compare the actual hours of daylight to the results found in part (b).

38. Hours of Daylight According to the *Old Farmer's Almanac*, in Honolulu, Hawaii, the number of hours of sunlight on the summer solstice of 2014 was 13.42, and the number of hours of sunlight on the winter solstice was 10.83.
(a) Find a sinusoidal function of the form
$$y = A \sin(\omega x - \phi) + B$$
that models the data.
(b) Use the function found in part (a) to predict the number of hours of sunlight on April 1, the 91st day of the year.
(c) Draw a graph of the function found in part (a).
(d) Look up the number of hours of sunlight for April 1 in the *Old Farmer's Almanac*, and compare the actual hours of daylight to the results found in part (b).

Discussion and Writing

39. Explain how the amplitude and period of a sinusoidal graph are used to establish the scale on each coordinate axis.

40. Find an application in your major field that leads to a sinusoidal graph. Write an account of your findings.

Retain Your Knowledge

Problems 41–44 are based on material learned earlier in the course. The purpose of these problems is to keep the material fresh in your mind so that you are better prepared for the final exam.

41. Given $f(x) = \dfrac{4x + 9}{2}$, find $f^{-1}(x)$.

42. Solve: $0.25(0.4x + 0.8) = 3.7 - 1.4x$

43. Multiply: $(8x + 15y)^2$

44. Find the exact distance between the points $(4, -1)$ and $(10, 3)$.

Chapter Review

Things to Know

Definitions

Angle in standard position (p. 506)	Vertex is at the origin; initial side is along the positive x-axis.
1 Degree (1°) (p. 507)	$1° = \dfrac{1}{360}$ revolution
1 Radian (p. 509)	The measure of a central angle of a circle whose rays subtend an arc whose length is equal to the radius of the circle
Acute angle (p. 520)	An angle θ whose measure is $0° < \theta < 90°$ or $0 < \theta < \dfrac{\pi}{2}$, θ in radians
Complementary angles (p. 525)	Two acute angles whose sum is $90°$ $\left(\dfrac{\pi}{2} \text{radians}\right)$
Cofunction (p. 526)	The following pairs of functions are cofunctions of each other: sine and cosine; tangent and cotangent; secant and cosecant.

| Trigonometric functions of any angle (p. 543) | $P = (a, b)$ is a point, not the origin, on the terminal side of an angle θ in standard position a distance r from the origin: |

$$\sin \theta = \frac{b}{r} \qquad \cos \theta = \frac{a}{r} \qquad \tan \theta = \frac{b}{a}, \quad a \neq 0$$

$$\csc \theta = \frac{r}{b}, \quad b \neq 0 \qquad \sec \theta = \frac{r}{a}, \quad a \neq 0 \qquad \cot \theta = \frac{a}{b}, \quad b \neq 0$$

| Reference angle of θ (p. 547) | The acute angle formed by the terminal side of θ and either the positive or negative x-axis |
| Periodic function (p. 560) | $f(\theta + p) = f(\theta)$, for all θ, $p > 0$, where the smallest such p is the fundamental period |

Formulas

1 counterclockwise revolution $= 360°$ (p. 508) $\qquad = 2\pi$ radians (p. 510)	$1° = \dfrac{\pi}{180}$ radian (p. 510); 1 radian $= \dfrac{180}{\pi}$ degrees (p. 510)
Arc length: $s = r\theta$ (p. 509)	θ is measured in radians; s is the length of the arc subtended by the central angle θ of the circle of radius r.
Area of a sector: $A = \dfrac{1}{2}r^2\theta$ (p. 513)	A is the area of the sector of a circle of radius r formed by a central angle of θ radians.
Linear speed: $v = \dfrac{s}{t}$ (p. 514)	v is the linear speed along the circle of radius r; ω is the angular speed (measured in radians per unit time).
Angular speed: $\omega = \dfrac{\theta}{t}$ (p. 511) $\qquad v = r\omega$ (p. 514)	

Table of Values

θ (Radians)	θ (Degrees)	$\sin\theta$	$\cos\theta$	$\tan\theta$	$\csc\theta$	$\sec\theta$	$\cot\theta$
0	0°	0	1	0	Not defined	1	Not defined
$\dfrac{\pi}{6}$	30°	$\dfrac{1}{2}$	$\dfrac{\sqrt{3}}{2}$	$\dfrac{\sqrt{3}}{3}$	2	$\dfrac{2\sqrt{3}}{3}$	$\sqrt{3}$
$\dfrac{\pi}{4}$	45°	$\dfrac{\sqrt{2}}{2}$	$\dfrac{\sqrt{2}}{2}$	1	$\sqrt{2}$	$\sqrt{2}$	1
$\dfrac{\pi}{3}$	60°	$\dfrac{\sqrt{3}}{2}$	$\dfrac{1}{2}$	$\sqrt{3}$	$\dfrac{2\sqrt{3}}{3}$	2	$\dfrac{\sqrt{3}}{3}$
$\dfrac{\pi}{2}$	90°	1	0	Not defined	1	Not defined	0
π	180°	0	-1	0	Not defined	-1	Not defined
$\dfrac{3\pi}{2}$	270°	-1	0	Not defined	-1	Not defined	0

Fundamental identities (p. 522)

Quotient identities: $\tan\theta = \dfrac{\sin\theta}{\cos\theta} \qquad \cot\theta = \dfrac{\cos\theta}{\sin\theta}$ Reciprocal identities: $\csc\theta = \dfrac{1}{\sin\theta} \qquad \sec\theta = \dfrac{1}{\cos\theta} \qquad \cot\theta = \dfrac{1}{\tan\theta}$

Pythagorean identities: $\sin^2\theta + \cos^2\theta = 1 \qquad \tan^2\theta + 1 = \sec^2\theta \qquad \cot^2\theta + 1 = \csc^2\theta$

Properties of the trigonometric functions

| $y = \sin x$ (p. 565) | Domain: $-\infty < x < \infty$
Range: $-1 \le y \le 1$
Periodic: period $= 2\pi$ (360°)
Odd function | |
| $y = \cos x$ (p. 567) | Domain: $-\infty < x < \infty$
Range: $-1 \le y \le 1$
Periodic: period $= 2\pi$ (360°)
Even function | |

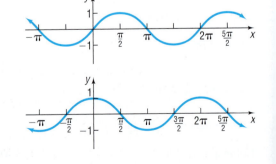

$y = \tan x$ (pp. 579–581)

Domain: $-\infty < x < \infty$, except odd integer multiples of $\dfrac{\pi}{2}$ (90°)

Range: $-\infty < y < \infty$

Periodic: period $= \pi$ (180°)

Odd function

Vertical asymptotes at odd integer multiples of $\dfrac{\pi}{2}$

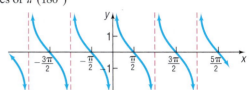

$y = \cot x$ (pp. 582–583)

Domain: $-\infty < x < \infty$, except integer multiples of π (180°)

Range: $-\infty < y < \infty$

Periodic: period $= \pi$ (180°)

Odd function

Vertical asymptotes at integer multiples of π

$y = \csc x$ (p. 583)

Domain: $-\infty < x < \infty$, except integer multiples of π (180°)

Range: $|y| \geq 1$

Periodic: period $= 2\pi$ (360°)

Odd function

Vertical asymptotes at integer multiples of π

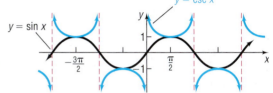

$y = \sec x$ (p. 583)

Domain: $-\infty < x < \infty$, except odd integer multiples of $\dfrac{\pi}{2}$ (90°)

Range: $|y| \geq 1$

Periodic: period $= 2\pi$ (360°)

Even function

Vertical asymptotes at odd integer multiples of $\dfrac{\pi}{2}$

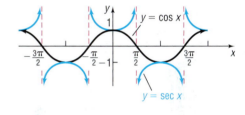

Sinusoidal graphs

$y = A \sin(\omega x) + B, \quad \omega > 0$

$y = A \cos(\omega x) + B, \quad \omega > 0$

$y = A \sin(\omega x - \phi) + B = A \sin\left[\omega\left(x - \dfrac{\phi}{\omega}\right)\right] + B$

$y = A \cos(\omega x - \phi) + B = A \cos\left[\omega\left(x - \dfrac{\phi}{\omega}\right)\right] + B$

Period $= \dfrac{2\pi}{\omega}$ (pp. 570, 587)

Amplitude $= |A|$ (pp. 570, 587)

Phase shift $= \dfrac{\phi}{\omega}$ (p. 587)

Objectives

Section		You should be able to. . .	Example(s)	Review Exercises
7.1	1	Convert between decimal and degree, minute, second measures for angles (p. 508)	2	48
	2	Find the length of an arc of a circle (p. 509)	3	49, 50
	3	Convert from degrees to radians and from radians to degrees (p. 510)	4–6	1–4
	4	Find the area of a sector of a circle (p. 512)	7	49
	5	Find the linear speed of an object traveling in circular motion (p. 513)	8	51, 52
7.2	1	Find the values of trigonometric functions of acute angles (p. 519)	1	41
	2	Use Fundamental Identities (p. 521)	2, 3	11, 12
	3	Find the values of the remaining trigonometric functions, given the value of one of them (p. 523)	4, 5	16
	4	Use the Complementary Angle Theorem (p. 525)	6, 7	13–15

Section		You should be able to ...	Examples	Review Exercises
7.3	1	Find the exact values of the trigonometric functions of $\frac{\pi}{4} = 45°$ (p. 531)	1, 2	5, 6
	2	Find the exact values of the trigonometric functions of $\frac{\pi}{6} = 30°$ and $\frac{\pi}{3} = 60°$ (p. 532)	3, 4	5, 6
	3	Use a calculator to approximate the values of the trigonometric functions of acute angles (p. 534)	5	42
	4	Model and solve applied problems involving right triangles (p. 534)	6–9	53–55
7.4	1	Find the exact values of the trigonometric functions for any angle (p. 543)	1, 2	9, 46
	2	Use coterminal angles to find the exact value of a trigonometric function (p. 545)	3	10
	3	Determine the signs of the trigonometric functions of an angle in a given quadrant (p. 547)	4	43
	4	Find the reference angle of an angle (p. 547)	5	44
	5	Use a reference angle to find the exact value of a trigonometric function (p. 548)	6	7, 8, 10
	6	Find the exact values of the trigonometric functions of an angle, given information about the functions (p. 550)	7, 8	16–23
7.5	1	Find the exact values of the trigonometric functions using the unit circle (p. 553)	1, 2	45, 46
	2	Know the domain and range of the trigonometric functions (p. 557)		47
	3	Use periodic properties to find the exact values of the trigonometric functions (p. 559)	3	10
	4	Use even–odd properties to find the exact values of the trigonometric functions (p. 560)	4	7, 8, 10, 14, 15
7.6	1	Graph functions of the form $y = A \sin(\omega x)$ using transformations (p. 565)	1, 2	24
	2	Graph functions of the form $y = A \cos(\omega x)$ using transformations (p. 567)	3	25
	3	Determine the amplitude and period of sinusoidal functions (p. 568)	4	33, 34
	4	Graph sinusoidal functions using key points (p. 570)	5–7	24, 25, 35
	5	Find an equation for a sinusoidal graph (p. 573)	8, 9	39, 40
7.7	1	Graph functions of the form $y = A \tan(\omega x) + B$ and $y = A \cot(\omega x) + B$ (p. 581)	1, 2	26–28, 32
	2	Graph functions of the form $y = A \csc(\omega x) + B$ and $y = A \sec(\omega x) + B$ (p. 583)	3	29, 30
7.8	1	Graph sinusoidal functions of the form $y = A \sin(\omega x - \phi) + B$ (p. 587)	1, 2	31, 36–38, 56
	2	Build sinusoidal models from data (p. 590)	3–5	57

Review Exercises

In Problems 1 and 2, convert each angle in degrees to radians. Express your answer as a multiple of π.

1. $135°$

2. $15°$

In Problems 3 and 4, convert each angle in radians to degrees.

3. $\dfrac{2\pi}{3}$

4. $-\dfrac{5\pi}{2}$

In Problems 5–15, find the exact value of each expression. Do not use a calculator.

5. $\tan\dfrac{\pi}{4} - \sin\dfrac{\pi}{6}$

6. $3\sin 45° - 4\tan\dfrac{\pi}{6}$

7. $6\cos\dfrac{3\pi}{4} + 2\tan\left(-\dfrac{\pi}{3}\right)$

8. $\sec\left(-\dfrac{\pi}{3}\right) - \cot\left(-\dfrac{5\pi}{4}\right)$

9. $\tan \pi + \sin \pi$

10. $\cos 540° - \tan(-405°)$

11. $\sin^2 20° + \dfrac{1}{\sec^2 20°}$

12. $\sec 50° \cos 50°$

13. $\dfrac{\sin 50°}{\cos 40°}$

14. $\dfrac{\sin(-40°)}{\cos 50°}$

15. $\sin 400° \sec(-50°)$

In Problems 16–23, find the exact value of each of the remaining trigonometric functions.

16. $\sin \theta = \dfrac{4}{5}$, θ is acute

17. $\tan \theta = \dfrac{12}{5}$, $\sin \theta < 0$

18. $\sec \theta = -\dfrac{5}{4}$, $\tan \theta < 0$

19. $\cos \theta = -\dfrac{3}{5}$, θ in quadrant III

20. $\sin \theta = -\dfrac{5}{13}$, $\dfrac{3\pi}{2} < \theta < 2\pi$

21. $\tan \theta = \dfrac{1}{3}$, $180° < \theta < 270°$

22. $\csc \theta = -4$, $\pi < \theta < \dfrac{3\pi}{2}$

23. $\cot \theta = -2$, $\dfrac{\pi}{2} < \theta < \pi$

In Problems 24–32, graph each function. Each graph should contain at least two periods. Use the graph to determine the domain and the range of each function.

24. $y = 2 \sin(4x)$

25. $y = -3 \cos(2x)$

26. $y = \tan(x + \pi)$

27. $y = -2 \tan(3x)$

28. $y = \cot\left(x + \dfrac{\pi}{4}\right)$

29. $y = 4 \sec(2x)$

30. $y = \csc\left(x + \dfrac{\pi}{4}\right)$

31. $y = 4 \sin(2x + 4) - 2$

32. $y = 5 \cot\left(\dfrac{x}{3} - \dfrac{\pi}{4}\right)$

In Problems 33 and 34, determine the amplitude and the period of each function without graphing.

33. $y = \sin(2x)$

34. $y = -2 \cos(3\pi x)$

In Problems 35–38, find the amplitude, period, and phase shift of each function. Graph each function. Show at least two periods.

35. $y = 4 \sin(3x)$

36. $y = -\cos\left(\dfrac{1}{2}x + \dfrac{\pi}{2}\right)$

37. $y = \dfrac{1}{2} \sin\left(\dfrac{3}{2}x - \pi\right)$

38. $y = -\dfrac{2}{3} \cos(\pi x - 6)$

In Problems 39 and 40, find a function whose graph is given.

39.

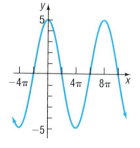

40.

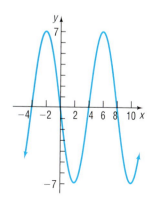

41. Find the value of each of the six trigonometric functions of the acute angle θ in a right triangle where the length of the side opposite θ is 3 and the length of the hypotenuse is 7.

42. Use a calculator to approximate $\sin \dfrac{\pi}{8}$. Use a calculator to approximate $\sec 10°$. Round answers to two decimal places.

43. Determine the signs of the six trigonometric functions of an angle θ whose terminal side is in quadrant III.

44. Find the reference angle of $750°$.

45. Find the exact values of the six trigonometric functions of t if $P = \left(-\dfrac{1}{3}, \dfrac{2\sqrt{2}}{3}\right)$ is the point on the unit circle that corresponds to t.

8 Analytic Trigonometry

Mapping Your Mind

The ability to organize material in your mind is key to understanding. You have been exposed to a lot of concepts at this point in the course, and it is a worthwhile exercise to organize the material. In the past, we might organize material using index cards or an outline. But in today's digital world, we can use interesting software that enables us to digitally organize the material that is in our mind and share it with anyone on the Web.

 —*See the Internet-based Chapter Project I*—

Outline

••• A Look Back

Chapter 6 introduced inverse functions and developed their properties, particularly the relationship between the domain and range of a function and those of its inverse. We learned that the graphs of a function and its inverse are symmetric with respect to the line $y = x$.

Chapter 6 continued by defining two transcendental functions: the exponential function and the inverse of the exponential function, the logarithmic function. Chapter 7 defined six more transcendental functions, the trigonometric functions, and discussed their properties.

A Look Ahead •••

The first two sections of this chapter define the six inverse trigonometric functions and investigate their properties. Section 8.3 discusses equations that contain trigonometric functions. Sections 8.4 through 8.7 continue the derivation of identities. These identities play an important role in calculus, the physical and life sciences, and economics, where they are used to simplify complicated expressions.

8.1 The Inverse Sine, Cosine, and Tangent Functions

PREPARING FOR THIS SECTION *Before getting started, review the following:*

- Inverse Functions (Section 6.2, pp. 411–419)
- Values of the Trigonometric Functions (Section 7.3, pp. 531–534; Section 7.4, pp. 543–550)
- Properties of the Sine, Cosine, and Tangent Functions (Section 7.5, pp. 553–561)
- Graphs of the Sine, Cosine, and Tangent Functions (Sections 7.6, pp. 564–573 and 7.7, pp. 579–582)

✎ **Now Work** the **'Are You Prepared?'** problems on page 616.

OBJECTIVES **1** Find the Exact Value of an Inverse Sine Function (p. 608)

2 Find an Approximate Value of an Inverse Sine Function (p. 609)

3 Use Properties of Inverse Functions to Find Exact Values of Certain Composite Functions (p. 610)

4 Find the Inverse Function of a Trigonometric Function (p. 615)

5 Solve Equations Involving Inverse Trigonometric Functions (p. 616)

In Section 6.2 we discussed inverse functions, and we concluded that if a function is one-to-one, it will have an inverse function. We also observed that if a function is not one-to-one, it may be possible to restrict its domain in some suitable manner so that the restricted function is one-to-one. For example, the function $y = x^2$ is not one-to-one; however, if the domain is restricted to $x \geq 0$, the function is one-to-one.

Other properties of a one-to-one function f and its inverse function f^{-1} that were discussed in Section 6.2 are summarized next.

1. $f^{-1}(f(x)) = x$ for every x in the domain of f, and $f(f^{-1}(x)) = x$ for every x in the domain of f^{-1}.

2. The domain of f = the range of f^{-1}, and the range of f = the domain of f^{-1}.

3. The graph of f and the graph of f^{-1} are reflections of one another about the line $y = x$.

4. If a function $y = f(x)$ has an inverse function, the implicit equation of the inverse function is $x = f(y)$. If we solve this equation for y, we obtain the explicit equation $y = f^{-1}(x)$.

The Inverse Sine Function

Figure 1 shows the graph of $y = \sin x$. Because every horizontal line $y = b$, where b is between -1 and 1, inclusive, intersects the graph of $y = \sin x$ infinitely many times, it follows from the horizontal-line test that the function $y = \sin x$ is not one-to-one.

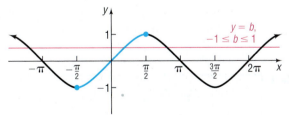

Figure 1 $y = \sin x$, $-\infty < x < \infty$, $-1 \leq y \leq 1$

However, if the domain of $y = \sin x$ is restricted to the interval $\left[-\dfrac{\pi}{2}, \dfrac{\pi}{2}\right]$, the restricted function

$$y = \sin x \qquad -\frac{\pi}{2} \leq x \leq \frac{\pi}{2}$$

is one-to-one and has an inverse function.* See Figure 2.

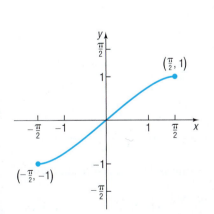

Figure 2
$y = \sin x$, $-\dfrac{\pi}{2} \leq x \leq \dfrac{\pi}{2}$, $-1 \leq y \leq 1$

*Although there are many other ways to restrict the domain and obtain a one-to-one function, mathematicians have agreed to use the interval $\left[-\dfrac{\pi}{2}, \dfrac{\pi}{2}\right]$ to define the inverse of $y = \sin x$.

NOTE Remember, the domain of a function f equals the range of its inverse, function f^{-1}, and the range of a function f equals the domain of its inverse function f^{-1}. Because the restricted domain of the sine function is $\left[-\dfrac{\pi}{2}, \dfrac{\pi}{2}\right]$, the range of inverse sine function is $\left[-\dfrac{\pi}{2}, \dfrac{\pi}{2}\right]$, and because the range of the sine function is $[-1, 1]$, the domain of the inverse sine function is $[-1, 1]$. ∎

An equation for the inverse of $y = f(x) = \sin x$ is obtained by interchanging x and y. The implicit form of the inverse function is $x = \sin y$, $-\dfrac{\pi}{2} \le y \le \dfrac{\pi}{2}$. The explicit form is called the **inverse sine** of x and is symbolized by $y = f^{-1}(x) = \sin^{-1} x$.

$$y = \sin^{-1} x \quad \text{means} \quad x = \sin y$$
$$\text{where} \quad -1 \le x \le 1 \quad \text{and} \quad -\frac{\pi}{2} \le y \le \frac{\pi}{2} \tag{1}$$

Because $y = \sin^{-1} x$ means $x = \sin y$, $y = \sin^{-1} x$ is read as "y is the angle or real number whose sine equals x," or, alternatively, is read "y is the inverse sine of x." Be careful about the notation used. The superscript -1 that appears in $y = \sin^{-1} x$ is not an exponent but the symbol used to denote the inverse function f^{-1} of f. (To avoid confusion, some texts use the notation $y = \text{Arcsin } x$ instead of $y = \sin^{-1} x$.)

The inverse of a function f receives as input an element from the range of f and returns as output an element in the domain of f. The restricted sine function, $y = f(x) = \sin x$, receives as input an angle or real number x in the interval $\left[-\dfrac{\pi}{2}, \dfrac{\pi}{2}\right]$ and outputs a real number in the interval $[-1, 1]$. Therefore, the inverse sine function, $y = \sin^{-1} x$, receives as input a real number in the interval $[-1, 1]$ or $-1 \le x \le 1$, its domain, and outputs an angle or real number in the interval $\left[-\dfrac{\pi}{2}, \dfrac{\pi}{2}\right]$ or $-\dfrac{\pi}{2} \le y \le \dfrac{\pi}{2}$, its range.

The graph of the inverse sine function can be obtained by reflecting the restricted portion of the graph of $y = f(x) = \sin x$ about the line $y = x$, as shown in Figure 3.

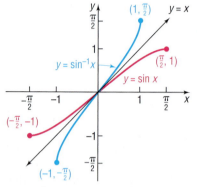

Figure 3
$y = \sin^{-1} x$, $-1 \le x \le 1$, $-\dfrac{\pi}{2} \le y \le \dfrac{\pi}{2}$

✓**Check:** Graph $Y_1 = \sin x$ and $Y_2 = \sin^{-1} x$. Compare the result with Figure 3.

1 Find the Exact Value of an Inverse Sine Function

For some numbers x, it is possible to find the exact value of $y = \sin^{-1} x$.

EXAMPLE 1 **Finding the Exact Value of an Inverse Sine Function**

Find the exact value of: $\sin^{-1} 1$

Solution Let $\theta = \sin^{-1} 1$. Then θ is the angle, $-\dfrac{\pi}{2} \le \theta \le \dfrac{\pi}{2}$, whose sine equals 1.

$$\theta = \sin^{-1} 1 \qquad -\frac{\pi}{2} \le \theta \le \frac{\pi}{2}$$
$$\sin \theta = 1 \qquad -\frac{\pi}{2} \le \theta \le \frac{\pi}{2} \quad \text{By definition of } y = \sin^{-1} x$$

Now look at Table 1 and Figure 4.

Table 1

θ	$-\dfrac{\pi}{2}$	$-\dfrac{\pi}{3}$	$-\dfrac{\pi}{4}$	$-\dfrac{\pi}{6}$	0	$\dfrac{\pi}{6}$	$\dfrac{\pi}{4}$	$\dfrac{\pi}{3}$	$\dfrac{\pi}{2}$
$\sin \theta$	-1	$-\dfrac{\sqrt{3}}{2}$	$-\dfrac{\sqrt{2}}{2}$	$-\dfrac{1}{2}$	0	$\dfrac{1}{2}$	$\dfrac{\sqrt{2}}{2}$	$\dfrac{\sqrt{3}}{2}$	1

The only angle θ in the interval $\left[-\dfrac{\pi}{2}, \dfrac{\pi}{2}\right]$ whose sine is 1 is $\dfrac{\pi}{2}$. (Note that $\sin \dfrac{5\pi}{2}$ also equals 1, but $\dfrac{5\pi}{2}$ lies outside the interval $\left[-\dfrac{\pi}{2}, \dfrac{\pi}{2}\right]$, which is not allowed.) So

$$\sin^{-1} 1 = \frac{\pi}{2}$$

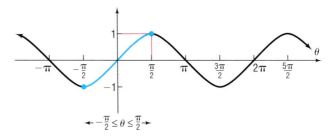

Figure 4

▸ **Now Work** PROBLEM 15

EXAMPLE 2   **Finding the Exact Value of an Inverse Sine Function**

Find the exact value of: $\sin^{-1}\left(-\dfrac{1}{2}\right)$

Solution   Let $\theta = \sin^{-1}\left(-\dfrac{1}{2}\right)$. Then θ is the angle, $-\dfrac{\pi}{2} \le \theta \le \dfrac{\pi}{2}$, whose sine equals $-\dfrac{1}{2}$.

$$\theta = \sin^{-1}\left(-\frac{1}{2}\right) \qquad -\frac{\pi}{2} \le \theta \le \frac{\pi}{2}$$

$$\sin\theta = -\frac{1}{2} \qquad\qquad -\frac{\pi}{2} \le \theta \le \frac{\pi}{2}$$

(Refer to Table 1 and Figure 4, if necessary.) The only angle in the interval $\left[-\dfrac{\pi}{2}, \dfrac{\pi}{2}\right]$ whose sine is $-\dfrac{1}{2}$ is $-\dfrac{\pi}{6}$, so

$$\sin^{-1}\left(-\frac{1}{2}\right) = -\frac{\pi}{6}$$

▸ **Now Work** PROBLEM 21

2 Find an Approximate Value of an Inverse Sine Function

For most numbers x, the value $y = \sin^{-1} x$ must be approximated.

EXAMPLE 3   **Finding an Approximate Value of an Inverse Sine Function**

Find an approximate value of each of the following functions:

(a) $\sin^{-1}\dfrac{1}{3}$ 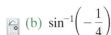 (b) $\sin^{-1}\left(-\dfrac{1}{4}\right)$

Express the answer in radians rounded to two decimal places.

Solution   (a) Because the angle is to be measured in radians, first set the mode of the calculator to radians.* Rounded to two decimal places,

$$\sin^{-1}\frac{1}{3} = 0.34$$

*On most calculators, the inverse sine is obtained by pressing SHIFT or 2nd, followed by sin . On some calculators, $\sin^{-1}$ is pressed first, then $1/3$ is entered; on others, this sequence is reversed. Consult your owner's manual for the correct sequence.

Table 2

θ
0
$\dfrac{\pi}{6}$
$\dfrac{\pi}{4}$
$\dfrac{\pi}{3}$
$\dfrac{\pi}{2}$
$\dfrac{2\pi}{3}$
$\dfrac{3\pi}{4}$
$\dfrac{5\pi}{6}$
π

Figure

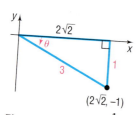

8.2 The Inverse Trigonometric Functions (Continued)

PREPARING FOR THIS SECTION *Before getting started, review the following concepts:*

- Finding Exact Values of the Trigonometric Functions, Given the Value of a Trigonometric Function and the Quadrant of the Angle (Section 7.4, p. 550)
- Graphs of the Secant, Cosecant, and Cotangent Functions (Section 7.7, pp. 582–584)
- Domain and Range of the Secant, Cosecant, and Cotangent Functions (Section 7.5, pp. 557–559)

Now Work the **'Are You Prepared?'** problems on page 623.

OBJECTIVES 1 Find the Exact Value of Expressions Involving the Inverse Sine, Cosine, and Tangent Functions (p. 620)
2 Define the Inverse Secant, Cosecant, and Cotangent Functions (p. 621)
3 Use a Calculator to Evaluate $\sec^{-1} x$, $\csc^{-1} x$, and $\cot^{-1} x$ (p. 622)
4 Write a Trigonometric Expression as an Algebraic Expression (p. 623)

1 Find the Exact Value of Expressions Involving the Inverse Sine, Cosine, and Tangent Functions

EXAMPLE 1

Finding the Exact Value of an Expression Involving Inverse Trigonometric Functions

Find the exact value of: $\sin\left(\tan^{-1}\dfrac{1}{2}\right)$

Solution

Let $\theta = \tan^{-1}\dfrac{1}{2}$. Then $\tan\theta = \dfrac{1}{2}$, where $-\dfrac{\pi}{2} < \theta < \dfrac{\pi}{2}$. We seek $\sin\theta$. Because $\tan\theta > 0$, it follows that $0 < \theta < \dfrac{\pi}{2}$, so θ lies in quadrant I. Figure 15 shows a triangle in quadrant I. Because $\tan\theta = \dfrac{b}{a} = \dfrac{1}{2}$, the side opposite θ is $b = 1$ and the side adjacent to θ is $a = 2$. The hypotenuse of this triangle is found using $a^2 + b^2 = c^2$ or $2^2 + 1^2 = c^2$, so $c = \sqrt{5}$. Then $\sin\theta = \dfrac{b}{c} = \dfrac{1}{\sqrt{5}}$, and

$$\sin\left(\tan^{-1}\dfrac{1}{2}\right) = \sin\theta = \dfrac{1}{\sqrt{5}} = \dfrac{\sqrt{5}}{5}$$

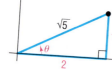

Figure 15 $\tan\theta = \dfrac{1}{2}$

EXAMPLE 2

Finding the Exact Value of an Expression Involving Inverse Trigonometric Functions

Find the exact value of: $\cos\left[\sin^{-1}\left(-\dfrac{1}{3}\right)\right]$

Solution

Let $\theta = \sin^{-1}\left(-\dfrac{1}{3}\right)$. Then $\sin\theta = -\dfrac{1}{3}$ and $-\dfrac{\pi}{2} \leq \theta \leq \dfrac{\pi}{2}$. We seek $\cos\theta$. Because $\sin\theta < 0$, it follows that $-\dfrac{\pi}{2} \leq \theta < 0$, so θ lies in quadrant IV. Figure 16 illustrates $\sin\theta = -\dfrac{1}{3}$ for θ in quadrant IV. Then

$$\cos\left[\sin^{-1}\left(-\dfrac{1}{3}\right)\right] = \cos\theta = \dfrac{2\sqrt{2}}{3}$$

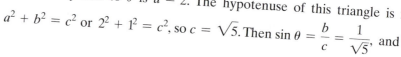

$\cos\theta = \dfrac{a}{r}$

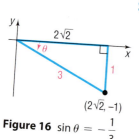

Figure 16 $\sin\theta = -\dfrac{1}{3}$

EXAMPLE 3

Finding the Exact Value of an Expression Involving Inverse Trigonometric Functions

Find the exact value of: $\tan\left[\cos^{-1}\left(-\dfrac{1}{3}\right)\right]$

Solution

Let $\theta = \cos^{-1}\left(-\dfrac{1}{3}\right)$. Then $\cos\theta = -\dfrac{1}{3}$ and $0 \le \theta \le \pi$. We seek $\tan\theta$. Because $\cos\theta < 0$, it follows that $\dfrac{\pi}{2} < \theta \le \pi$, so θ lies in quadrant II. Figure 17 illustrates $\cos\theta = -\dfrac{1}{3}$ for θ in quadrant II. Then

$$\tan\left[\cos^{-1}\left(-\dfrac{1}{3}\right)\right] = \tan\theta = \dfrac{2\sqrt{2}}{-1} = -2\sqrt{2}$$

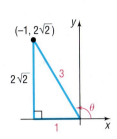

Figure 17 $\cos\theta = -\dfrac{1}{3}$

$(-1, 2\sqrt{2})$

Now Work PROBLEMS **9** AND **27**

2 Define the Inverse Secant, Cosecant, and Cotangent Functions

The inverse secant, inverse cosecant, and inverse cotangent functions are defined as follows:

DEFINITION

$$y = \sec^{-1}x \quad \text{means} \quad x = \sec y$$
$$\text{where} \quad |x| \ge 1 \quad \text{and} \quad 0 \le y \le \pi, \quad y \ne \dfrac{\pi}{2}* \tag{1}$$

$$y = \csc^{-1}x \quad \text{means} \quad x = \csc y$$
$$\text{where} \quad |x| \ge 1 \quad \text{and} \quad -\dfrac{\pi}{2} \le y \le \dfrac{\pi}{2}, \quad y \ne 0^\dagger \tag{2}$$

$$y = \cot^{-1}x \quad \text{means} \quad x = \cot y$$
$$\text{where} \quad -\infty < x < \infty \quad \text{and} \quad 0 < y < \pi \tag{3}$$

Take time to review the graphs of the cotangent, cosecant, and secant functions in Figures 99, 100, and 101 in Section 7.7 to see the basis for these definitions.

EXAMPLE 4

Finding the Exact Value of an Inverse Cosecant Function

Find the exact value of: $\csc^{-1}2$

Solution

Let $\theta = \csc^{-1}2$. Then θ is the angle, $-\dfrac{\pi}{2} \le \theta \le \dfrac{\pi}{2}, \theta \ne 0$, whose cosecant equals 2 $\left(\text{or, equivalently, whose sine equals } \dfrac{1}{2}\right)$.

$$\theta = \csc^{-1}2 \qquad -\dfrac{\pi}{2} \le \theta \le \dfrac{\pi}{2}, \quad \theta \ne 0$$

$$\csc\theta = 2 \qquad -\dfrac{\pi}{2} \le \theta \le \dfrac{\pi}{2}, \quad \theta \ne 0 \quad \sin\theta = \dfrac{1}{2}$$

The only angle θ in the interval $-\dfrac{\pi}{2} \le \theta \le \dfrac{\pi}{2}, \theta \ne 0$, whose cosecant is 2 $\left[\sin\theta = \dfrac{1}{2}\right]$ is $\dfrac{\pi}{6}$, so $\csc^{-1}2 = \dfrac{\pi}{6}$.

Now Work PROBLEM **39**

*Most texts use this definition. A few use the restriction $0 \le y < \dfrac{\pi}{2}, \pi \le y < \dfrac{3\pi}{2}$.

†Most texts use this definition. A few use the restriction $-\pi < y \le -\dfrac{\pi}{2}, 0 < y \le \dfrac{\pi}{2}$.

3 Use a Calculator to Evaluate $\sec^{-1} x$, $\csc^{-1} x$, and $\cot^{-1} x$

Most calculators do not have keys for evaluating the inverse cotangent, cosecant, and secant functions. The easiest way to evaluate them is to convert each to an inverse trigonometric function whose range is the same as the one to be evaluated. In this regard, notice that $y = \cot^{-1} x$ and $y = \sec^{-1} x$, except where undefined, have the same range as $y = \cos^{-1} x$ and that $y = \csc^{-1} x$, except where undefined, has the same range as $y = \sin^{-1} x$.

NOTE Remember that the range of $y = \sin^{-1} x$ is $\left[-\dfrac{\pi}{2}, \dfrac{\pi}{2} \right]$ and that the range of $y = \cos^{-1} x$ is $[0, \pi]$. ∎

EXAMPLE 5

Approximating the Value of Inverse Trigonometric Functions

Use a calculator to approximate each expression in radians rounded to two decimal places.

(a) $\sec^{-1} 3$ (b) $\csc^{-1}(-4)$ (c) $\cot^{-1} \dfrac{1}{2}$ (d) $\cot^{-1}(-2)$

Solution

First, set your calculator to radian mode.

(a) Let $\theta = \sec^{-1} 3$. Then $\sec \theta = 3$ and $0 \le \theta \le \pi, \theta \ne \dfrac{\pi}{2}$. Now find $\cos \theta$ because $y = \cos^{-1} x$ has the same range as $y = \sec^{-1} x$, except where undefined. Because $\sec \theta = \dfrac{1}{\cos \theta} = 3$, this means $\cos \theta = \dfrac{1}{3}$. Then $\theta = \cos^{-1} \dfrac{1}{3}$, and

$$\sec^{-1} 3 = \theta = \cos^{-1} \dfrac{1}{3} \approx 1.23$$
↑
Use a calculator.

(b) Let $\theta = \csc^{-1}(-4)$. Then $\csc \theta = -4, -\dfrac{\pi}{2} \le \theta \le \dfrac{\pi}{2}, \theta \ne 0$. Now find $\sin \theta$ because $y = \sin^{-1} x$ has the same range as $y = \csc^{-1} x$, except where undefined. Because $\csc \theta = \dfrac{1}{\sin \theta} = -4$, this means $\sin \theta = -\dfrac{1}{4}$. Then $\theta = \sin^{-1} \left(-\dfrac{1}{4} \right)$, and

$$\csc^{-1}(-4) = \theta = \sin^{-1} \left(-\dfrac{1}{4} \right) \approx -0.25$$

(c) Let $\theta = \cot^{-1} \dfrac{1}{2}$. Then $\cot \theta = \dfrac{1}{2}, 0 < \theta < \pi$. These facts indicate that θ lies in quadrant I. Now find $\cos \theta$ because $y = \cos^{-1} x$ has the same range as $y = \cot^{-1} x$, except where undefined. Use Figure 18 to find that $\cos \theta = \dfrac{1}{\sqrt{5}}, 0 < \theta < \dfrac{\pi}{2}$. So, $\theta = \cos^{-1} \left(\dfrac{1}{\sqrt{5}} \right)$, and

$$\cot^{-1} \dfrac{1}{2} = \theta = \cos^{-1} \left(\dfrac{1}{\sqrt{5}} \right) \approx 1.11$$

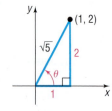

Figure 18
$\cot \theta = \dfrac{1}{2}, 0 < \theta < \pi$

(d) Let $\theta = \cot^{-1}(-2)$. Then $\cot \theta = -2, 0 < \theta < \pi$. These facts indicate that θ lies in quadrant II. Now find $\cos \theta$. Use Figure 19 to find that $\cos \theta = -\dfrac{2}{\sqrt{5}}, \dfrac{\pi}{2} < \theta < \pi$. This means $\theta = \cos^{-1} \left(-\dfrac{2}{\sqrt{5}} \right)$, and

$$\cot^{-1}(-2) = \theta = \cos^{-1} \left(-\dfrac{2}{\sqrt{5}} \right) \approx 2.68$$

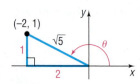

Figure 19
$\cot \theta = -2, 0 < \theta < \pi$

●

━━━━ **Now Work** PROBLEM 45

4 Write a Trigonometric Expression as an Algebraic Expression

| EXAMPLE 6 | **Writing a Trigonometric Expression as an Algebraic Expression** |

Write $\sin(\tan^{-1} u)$ as an algebraic expression containing u.

Solution   Let $\theta = \tan^{-1} u$ so that $\tan \theta = u$, $-\frac{\pi}{2} < \theta < \frac{\pi}{2}$, $-\infty < u < \infty$. This means $\sec \theta > 0$. Then

$$\sin(\tan^{-1} u) = \sin \theta = \sin \theta \cdot \frac{\cos \theta}{\cos \theta} = \tan \theta \cos \theta = \frac{\tan \theta}{\sec \theta} = \frac{\tan \theta}{\sqrt{1 + \tan^2 \theta}} = \frac{u}{\sqrt{1 + u^2}}$$

Multiply by 1: $\frac{\cos \theta}{\cos \theta}$. $\frac{\sin \theta}{\cos \theta} = \tan \theta$ $\sec^2 \theta = 1 + \tan^2 \theta$
$\sec \theta > 0$

Now Work PROBLEM 57

8.2 Assess Your Understanding

'Are You Prepared?' *Answers are given at the end of these exercises. If you get a wrong answer, read the pages listed in* red.

1. What are the domain and the range of $y = \sec x$? (pp. 557–559)

2. *True or False* The graph of $y = \sec x$ is one-to-one on the set $\left[0, \frac{\pi}{2}\right) \cup \left(\frac{\pi}{2}, \pi\right]$. (pp. 582–584)

3. If $\tan \theta = \frac{1}{2}$, $-\frac{\pi}{2} < \theta < \frac{\pi}{2}$, then $\sin \theta = $ _____. (p. 550)

Concepts and Vocabulary

4. $y = \sec^{-1} x$ means _____, where $|x|$ _____ and _____ $\leq y \leq$ __, $y \neq \frac{\pi}{2}$.

5. To find the inverse secant of a real number x such that $|x| \geq 1$, convert the inverse secant to an inverse _____.

6. *True or False* It is impossible to obtain exact values for the inverse secant function.

7. *True or False* $\csc^{-1} 0.5$ is not defined.

8. *True or False* The domain of the inverse cotangent function is the set of real numbers.

Skill Building

In Problems 9–36, find the exact value of each expression.

9. $\cos\left(\sin^{-1}\frac{\sqrt{2}}{2}\right)$

10. $\sin\left(\cos^{-1}\frac{1}{2}\right)$

11. $\tan\left[\cos^{-1}\left(-\frac{\sqrt{3}}{2}\right)\right]$

12. $\tan\left[\sin^{-1}\left(-\frac{1}{2}\right)\right]$

13. $\sec\left(\cos^{-1}\frac{1}{2}\right)$

14. $\cot\left[\sin^{-1}\left(-\frac{1}{2}\right)\right]$

15. $\csc(\tan^{-1} 1)$

16. $\sec\left(\tan^{-1}\sqrt{3}\right)$

17. $\sin[\tan^{-1}(-1)]$

18. $\cos\left[\sin^{-1}\left(-\frac{\sqrt{3}}{2}\right)\right]$

19. $\sec\left[\sin^{-1}\left(-\frac{1}{2}\right)\right]$

20. $\csc\left[\cos^{-1}\left(-\frac{\sqrt{3}}{2}\right)\right]$

21. $\cos^{-1}\left(\sin\frac{5\pi}{4}\right)$

22. $\tan^{-1}\left(\cot\frac{2\pi}{3}\right)$

23. $\sin^{-1}\left[\cos\left(-\frac{7\pi}{6}\right)\right]$

24. $\cos^{-1}\left[\tan\left(-\frac{\pi}{4}\right)\right]$

25. $\tan\left(\sin^{-1}\frac{1}{3}\right)$

26. $\tan\left(\cos^{-1}\frac{1}{3}\right)$

27. $\sec\left(\tan^{-1}\frac{1}{2}\right)$

28. $\cos\left(\sin^{-1}\frac{\sqrt{2}}{3}\right)$

29. $\cot\left[\sin^{-1}\left(-\frac{\sqrt{2}}{3}\right)\right]$

30. $\csc[\tan^{-1}(-2)]$

31. $\sin[\tan^{-1}(-3)]$

32. $\cot\left[\cos^{-1}\left(-\frac{\sqrt{3}}{3}\right)\right]$

33. $\sec\left(\sin^{-1}\frac{2\sqrt{5}}{5}\right)$

34. $\csc\left(\tan^{-1}\frac{1}{2}\right)$

35. $\sin^{-1}\left(\cos\frac{3\pi}{4}\right)$

36. $\cos^{-1}\left(\sin\frac{7\pi}{6}\right)$

In Problems 37–44, find the exact value of each expression.

37. $\cot^{-1}\sqrt{3}$

38. $\cot^{-1} 1$

39. $\csc^{-1}(-1)$

40. $\csc^{-1}\sqrt{2}$

41. $\sec^{-1}\frac{2\sqrt{3}}{3}$

42. $\sec^{-1}(-2)$

43. $\cot^{-1}\left(-\frac{\sqrt{3}}{3}\right)$

44. $\csc^{-1}\left(-\frac{2\sqrt{3}}{3}\right)$

In Problems 45–56, use a calculator to find the value of each expression rounded to two decimal places.

45. $\sec^{-1} 4$
46. $\csc^{-1} 5$
47. $\cot^{-1} 2$
48. $\sec^{-1}(-3)$

49. $\csc^{-1}(-3)$
50. $\cot^{-1}\left(-\dfrac{1}{2}\right)$
51. $\cot^{-1}(-\sqrt{5})$
52. $\cot^{-1}(-8.1)$

53. $\csc^{-1}\left(-\dfrac{3}{2}\right)$
54. $\sec^{-1}\left(-\dfrac{4}{3}\right)$
55. $\cot^{-1}\left(-\dfrac{3}{2}\right)$
56. $\cot^{-1}(-\sqrt{10})$

In Problems 57–66, write each trigonometric expression as an algebraic expression in u.

57. $\cos(\tan^{-1} u)$
58. $\sin(\cos^{-1} u)$
59. $\tan(\sin^{-1} u)$
60. $\tan(\cos^{-1} u)$
61. $\sin(\sec^{-1} u)$

62. $\sin(\cot^{-1} u)$
63. $\cos(\csc^{-1} u)$
64. $\cos(\sec^{-1} u)$
65. $\tan(\cot^{-1} u)$
66. $\tan(\sec^{-1} u)$

Mixed Practice

In Problems 67–78, $f(x) = \sin x$, $-\dfrac{\pi}{2} \le x \le \dfrac{\pi}{2}$; $g(x) = \cos x$, $0 \le x \le \pi$; and $h(x) = \tan x$, $-\dfrac{\pi}{2} < x < \dfrac{\pi}{2}$. Find the exact value of each composite function.

67. $g\left(f^{-1}\left(\dfrac{12}{13}\right)\right)$
68. $f\left(g^{-1}\left(\dfrac{5}{13}\right)\right)$
69. $g^{-1}\left(f\left(-\dfrac{\pi}{4}\right)\right)$
70. $f^{-1}\left(g\left(\dfrac{5\pi}{6}\right)\right)$

71. $h\left(f^{-1}\left(-\dfrac{3}{5}\right)\right)$
72. $h\left(g^{-1}\left(-\dfrac{4}{5}\right)\right)$
73. $g\left(h^{-1}\left(\dfrac{12}{5}\right)\right)$
74. $f\left(h^{-1}\left(\dfrac{5}{12}\right)\right)$

75. $g^{-1}\left(f\left(-\dfrac{\pi}{3}\right)\right)$
76. $g^{-1}\left(f\left(-\dfrac{\pi}{6}\right)\right)$
77. $h\left(g^{-1}\left(-\dfrac{1}{4}\right)\right)$
78. $h\left(f^{-1}\left(-\dfrac{2}{5}\right)\right)$

Applications and Extensions

*Problems 79 and 80 require the following discussion: When granular materials are allowed to fall freely, they form conical (cone-shaped) piles. The naturally occurring angle, measured from the horizontal, at which the loose material comes to rest is called the **angle of repose** and varies for different materials. The angle of repose θ is related to the height h and the base radius r of the conical pile by the equation $\theta = \cot^{-1}\dfrac{r}{h}$. See the illustration.*

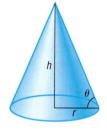

79. Angle of Repose: De-icing Salt Due to potential transportation issues (for example, frozen waterways), de-icing salt used by highway departments in the Midwest must be ordered early and stored for future use. When de-icing salt is stored in a pile 14 feet high, the diameter of the base of the pile is 45 feet.
(a) Find the angle of repose for de-icing salt.
(b) What is the base diameter of a pile that is 17 feet high?
(c) What is the height of a pile that has a base diameter of approximately 122 feet?
Source: Salt Institute, The Salt Storage Handbook, 2013

80. Angle of Repose: Bunker Sand The steepness of sand bunkers on a golf course is affected by the angle of repose of the sand (a larger angle of repose allows for steeper bunkers). A freestanding pile of loose sand from a United States Golf Association (USGA) bunker had a height of 4 feet and a base diameter of approximately 6.68 feet.
(a) Find the angle of repose for USGA bunker sand.
(b) What is the height of such a pile if the diameter of the base is 8 feet?
(c) A 6-foot-high pile of loose Tour Grade 50/50 sand has a base diameter of approximately 8.44 feet. Which type of sand (USGA or Tour Grade 50/50) would be better suited for steep bunkers?
Source: 2004 Annual Report, Purdue University Turfgrass Science Program

81. Artillery A projectile fired into the first quadrant from the origin of a rectangular coordinate system will pass through the point (x, y) at time t according to the relationship $\cot\theta = \dfrac{2x}{2y + gt^2}$, where θ = the angle of elevation of the launcher and g = the acceleration due to gravity = 32.2 feet/second². An artilleryman is firing at an enemy bunker located 2450 feet up the side of a hill that is 6175 feet away. He fires a round, and exactly 2.27 seconds later he scores a direct hit.

(a) What angle of elevation did he use?
(b) If the angle of elevation is also given by $\sec\theta = \dfrac{v_0 t}{x}$, where v_0 is the muzzle velocity of the weapon, find the muzzle velocity of the artillery piece he used.
Source: www.egwald.com/geometry/projectile3d.php

82. Using a graphing utility, graph $y = \cot^{-1} x$.
83. Using a graphing utility, graph $y = \sec^{-1} x$.
84. Using a graphing utility, graph $y = \csc^{-1} x$.

Explaining Concepts: Discussion and Writing

85. Explain in your own words how you would use your calculator to find the value of $\cot^{-1} 10$.

86. Consult three books on calculus, and then write down the definition in each of $y = \sec^{-1} x$ and $y = \csc^{-1} x$. Compare these with the definitions given in this text.

Retain Your Knowledge

Problems 87–90 are based on material learned earlier in the course. The purpose of these problems is to keep the material fresh in your mind so that you are better prepared for the final exam.

87. Find the complex zeros of $f(x) = x^4 + 21x^2 - 100$.

88. Determine algebraically whether $f(x) = x^3 + x^2 - x$ is even, odd, or neither.

89. Convert $315°$ to radians.

90. Find the length of the arc subtended by a central angle of $75°$ on a circle of radius 6 inches. Give both the exact length and an approximation rounded to two decimal places.

'Are You Prepared?' Answers

1. Domain: $\left\{ x \middle| x \neq \text{odd integer multiples of } \dfrac{\pi}{2} \right\}$; range: $\{ y \leq -1 \text{ or } y \geq 1 \}$ **2.** True **3.** $\dfrac{\sqrt{5}}{5}$

8.3 Trigonometric Equations

PREPARING FOR THIS SECTION *Before getting started, review the following:*

- Linear Equations (Section 1.1, pp. 82–85)
- Values of the Trigonometric Functions (Section 7.3, pp. 531–534; Section 7.4, pp. 543–550)
- Quadratic Equations (Section 1.2, pp. 92–99)

- Equations Quadratic in Form (Section 1.4, pp. 114–116)
- Using a Graphing Utility to Solve Equations (Appendix, Section 4, pp. A6–A7)

Now Work the **'Are You Prepared?'** problems on page 630.

OBJECTIVES **1** Solve Equations Involving a Single Trigonometric Function (p. 625)
   2 Solve Trigonometric Equations Using a Calculator (p. 628)
   3 Solve Trigonometric Equations Quadratic in Form (p. 629)
   4 Solve Trigonometric Equations Using Fundamental Identities (p. 629)
   5 Solve Trigonometric Equations Using a Graphing Utility (p. 630)

1 Solve Equations Involving a Single Trigonometric Function

In this section, we discuss **trigonometric equations**—that is, equations involving trigonometric functions that are satisfied only by some values of the variable (or, possibly, are not satisfied by any values of the variable). The values that satisfy the equation are called **solutions** of the equation.

EXAMPLE 1

Checking Whether a Given Number Is a Solution of a Trigonometric Equation

Determine whether $\theta = \dfrac{\pi}{4}$ is a solution of the equation $2 \sin \theta - 1 = 0$. Is $\theta = \dfrac{\pi}{6}$ a solution?

Solution

Replace θ by $\dfrac{\pi}{4}$ in the given equation. The result is

$$2 \sin \frac{\pi}{4} - 1 = 2 \cdot \frac{\sqrt{2}}{2} - 1 = \sqrt{2} - 1 \neq 0$$

Therefore, $\dfrac{\pi}{4}$ is not a solution.

Next replace θ by $\dfrac{\pi}{6}$ in the equation. The result is

$$2 \sin \frac{\pi}{6} - 1 = 2 \cdot \frac{1}{2} - 1 = 0$$

Therefore $\dfrac{\pi}{6}$ is a solution of the given equation. ●

The equation given in Example 1 has other solutions besides $\theta = \dfrac{\pi}{6}$. For example, $\theta = \dfrac{5\pi}{6}$ is also a solution, as is $\theta = \dfrac{13\pi}{6}$. (Check this for yourself.) In fact, the equation has an infinite number of solutions because of the periodicity of the sine function, as can be seen in Figure 20, which shows the graph of $y = 2 \sin x - 1$. Each x-intercept of the graph represents a solution to the equation $2 \sin x - 1 = 0$.

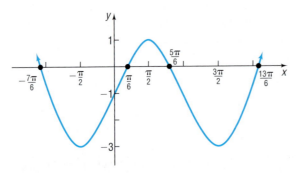

Figure 20 $y = 2 \sin x - 1$

Unless the domain of the variable is restricted, we need to find *all* the solutions of a trigonometric equation. As the next example illustrates, finding all the solutions can be accomplished by first finding solutions over an interval whose length equals the period of the function and then adding multiples of that period to the solutions found.

EXAMPLE 2

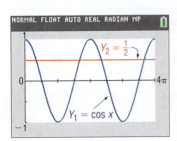

Figure 21

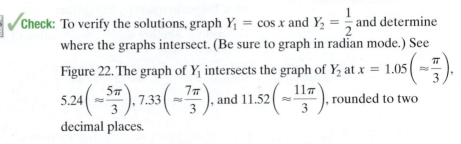

Figure 22

Finding All the Solutions of a Trigonometric Equation

Solve the equation: $\cos \theta = \dfrac{1}{2}$

Give a general formula for all the solutions. List eight of the solutions.

Solution The period of the cosine function is 2π. In the interval $[0, 2\pi)$, there are two angles θ for which $\cos \theta = \dfrac{1}{2}$: $\theta = \dfrac{\pi}{3}$ and $\theta = \dfrac{5\pi}{3}$. See Figure 21. Because the cosine function has period 2π, all the solutions of $\cos \theta = \dfrac{1}{2}$ may be given by the general formula

$$\theta = \frac{\pi}{3} + 2k\pi \quad \text{or} \quad \theta = \frac{5\pi}{3} + 2k\pi \quad k \text{ any integer}$$

Eight of the solutions are

$$\underbrace{-\frac{5\pi}{3}, \quad -\frac{\pi}{3}}_{k=-1}, \quad \underbrace{\frac{\pi}{3}, \quad \frac{5\pi}{3}}_{k=0}, \quad \underbrace{\frac{7\pi}{3}, \quad \frac{11\pi}{3}}_{k=1}, \quad \underbrace{\frac{13\pi}{3}, \quad \frac{17\pi}{3}}_{k=2}$$ ●

✓**Check:** To verify the solutions, graph $Y_1 = \cos x$ and $Y_2 = \dfrac{1}{2}$ and determine where the graphs intersect. (Be sure to graph in radian mode.) See Figure 22. The graph of Y_1 intersects the graph of Y_2 at $x = 1.05 \left(\approx \dfrac{\pi}{3} \right)$, $5.24 \left(\approx \dfrac{5\pi}{3} \right)$, $7.33 \left(\approx \dfrac{7\pi}{3} \right)$, and $11.52 \left(\approx \dfrac{11\pi}{3} \right)$, rounded to two decimal places.

Now Work PROBLEM **37**

In most of the work we do, we shall be interested only in finding solutions of trigonometric equations for $0 \le \theta < 2\pi$.

EXAMPLE 3

Solving a Linear Trigonometric Equation

Solve the equation: $2 \sin \theta + \sqrt{3} = 0$, $\quad 0 \le \theta < 2\pi$

Solution

First solve the equation for $\sin \theta$.

$$2 \sin \theta + \sqrt{3} = 0$$

$$2 \sin \theta = -\sqrt{3} \quad \text{Subtract } \sqrt{3} \text{ from both sides.}$$

$$\sin \theta = -\frac{\sqrt{3}}{2} \quad \text{Divide both sides by 2.}$$

In the interval $[0, 2\pi)$, there are two angles θ for which $\sin \theta = -\frac{\sqrt{3}}{2}$: $\theta = \frac{4\pi}{3}$ and $\theta = \frac{5\pi}{3}$. The solution set is $\left\{\frac{4\pi}{3}, \frac{5\pi}{3}\right\}$.

●

◖▨▨▨ **Now Work** PROBLEM 13

When the argument of the trigonometric function in an equation is a multiple of θ, the general formula must be used to solve the equation.

EXAMPLE 4

Solving a Trigonometric Equation

Solve the equation: $\sin(2\theta) = \frac{1}{2}$, $\quad 0 \le \theta < 2\pi$

Solution

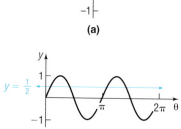

(a)

In the interval $[0, 2\pi)$, the sine function equals $\frac{1}{2}$ at $\frac{\pi}{6}$ and $\frac{5\pi}{6}$. See Figure 23(a). So 2θ must equal $\frac{\pi}{6}$ or $\frac{5\pi}{6}$. Here's the problem, however. The period of $y = \sin(2\theta)$ is $\frac{2\pi}{2} = \pi$. So in the interval $[0, 2\pi)$, the graph of $y = \sin(2\theta)$ will complete two cycles, and the graph of $y = \sin(2\theta)$ will intersect the graph of $y = \frac{1}{2}$ four times. See Figure 23(b). For this reason, there are four solutions to the equation $\sin(2\theta) = \frac{1}{2}$ in $[0, 2\pi)$. To find these solutions, write the general formula that gives all the solutions.

$$2\theta = \frac{\pi}{6} + 2k\pi \quad \text{or} \quad 2\theta = \frac{5\pi}{6} + 2k\pi \quad k \text{ any integer}$$

$$\theta = \frac{\pi}{12} + k\pi \quad \text{or} \quad \theta = \frac{5\pi}{12} + k\pi \quad \text{Divide by 2.}$$

(b)

Figure 23

WARNING In solving a trigonometric equation for θ, $0 \le \theta < 2\pi$, in which the argument is not θ (as in Example 4), you must write down all the solutions first and then list those that are in the interval $[0, 2\pi)$. Otherwise, solutions may be lost. For example, in solving $\sin(2\theta) = \frac{1}{2}$, if you write only the solutions $2\theta = \frac{\pi}{6}$ and $2\theta = \frac{5\pi}{6}$, you will find only $\theta = \frac{\pi}{12}$ and $\theta = \frac{5\pi}{12}$ and miss the other solutions. ∎

Then

$$\theta = \frac{\pi}{12} + (-1)\pi = \frac{-11\pi}{12} \quad k = -1 \qquad \theta = \frac{5\pi}{12} + (-1)\pi = \frac{-7\pi}{12}$$

$$\theta = \frac{\pi}{12} + (0)\pi = \frac{\pi}{12} \quad k = 0 \qquad \theta = \frac{5\pi}{12} + (0)\pi = \frac{5\pi}{12}$$

$$\theta = \frac{\pi}{12} + (1)\pi = \frac{13\pi}{12} \quad k = 1 \qquad \theta = \frac{5\pi}{12} + (1)\pi = \frac{17\pi}{12}$$

$$\theta = \frac{\pi}{12} + (2)\pi = \frac{25\pi}{12} \quad k = 2 \qquad \theta = \frac{5\pi}{12} + (2)\pi = \frac{29\pi}{12}$$

In the interval $[0, 2\pi)$, the solutions of $\sin(2\theta) = \frac{1}{2}$ are $\theta = \frac{\pi}{12}, \theta = \frac{5\pi}{12}, \theta = \frac{13\pi}{12}$, and $\theta = \frac{17\pi}{12}$. The solution set is $\left\{\frac{\pi}{12}, \frac{5\pi}{12}, \frac{13\pi}{12}, \frac{17\pi}{12}\right\}$. This means the graph of $y = \sin(2\theta)$ intersects $y = \frac{1}{2}$ at the points $\left(\frac{\pi}{12}, \frac{1}{2}\right), \left(\frac{5\pi}{12}, \frac{1}{2}\right), \left(\frac{13\pi}{12}, \frac{1}{2}\right)$, and $\left(\frac{17\pi}{12}, \frac{1}{2}\right)$ in the interval $[0, 2\pi)$.

●

| EXAMPLE 9 | **Solving a Trigonometric Equation Using Identities** |

Solve the equation: $\cos^2 \theta + \sin \theta = 2, \quad 0 \leq \theta < 2\pi$

Solution This equation involves two trigonometric functions: sine and cosine. By using a Pythagorean Identity, we can express the equation in terms of just sine functions.

$$\cos^2 \theta + \sin \theta = 2$$

$$(1 - \sin^2 \theta) + \sin \theta = 2 \quad \cos^2 \theta = 1 - \sin^2 \theta$$

$$\sin^2 \theta - \sin \theta + 1 = 0$$

This is a quadratic equation in $\sin \theta$. The discriminant is $b^2 - 4ac = 1 - 4 = -3 < 0$. Therefore, the equation has no real solution. The solution set is the empty set, $\varnothing$. ●

✓Check: Graph $Y_1 = \cos^2 x + \sin x$ and $Y_2 = 2$ to see that the two graphs never intersect, so the equation $Y_1 = Y_2$ has no real solution.

5 Solve Trigonometric Equations Using a Graphing Utility

The techniques introduced in this section apply only to certain types of trigonometric equations. Solutions for other types are usually studied in calculus, using numerical methods.

| EXAMPLE 10 | **Solving a Trigonometric Equation Using a Graphing Utility** |

Solve: $5 \sin x + x = 3$

Express the solution(s) rounded to two decimal places.

Solution This type of trigonometric equation cannot be solved by previous methods. A graphing utility, though, can be used here. Each solution of this equation is the x-coordinate of a point of intersection of the graphs of $Y_1 = 5 \sin x + x$ and $Y_2 = 3$. See Figure 25.

There are three points of intersection; the x-coordinates provide the solutions. Use INTERSECT, to find

$$x = 0.52 \qquad x = 3.18 \qquad x = 5.71$$

The solution set is $\{0.52, 3.18, 5.71\}$. ●

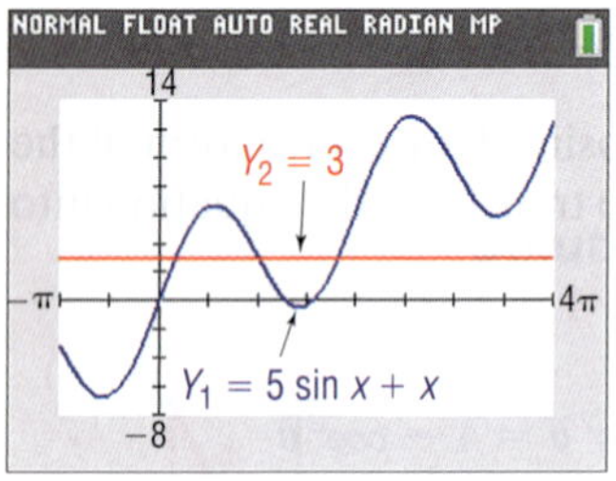

Figure 25

Now Work PROBLEM 83

8.3 Assess Your Understanding

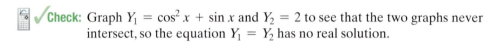

'Are You Prepared?' *Answers are given at the end of these exercises. If you get a wrong answer, read the pages listed in* red.

1. Solve: $3x - 5 = -x + 1$ (pp. 82–85)

2. $\sin\left(\dfrac{\pi}{4}\right) = $ _____ ; $\cos\left(\dfrac{8\pi}{3}\right) = $ _____
(pp. 531–534 and 543–550)

3. Find the real solutions of $4x^2 - x - 5 = 0$. (pp. 92–99)

4. Find the real solutions of $x^2 - x - 1 = 0$. (pp. 92–99)

5. Find the real solutions of $(2x - 1)^2 - 3(2x - 1) - 4 = 0$. (pp. 114–116) _

6. Use a graphing utility to solve $5x^3 - 2 = x - x^2$. Round answers to two decimal places. (pp. A6–A7)

Concepts and Vocabulary

7. *True or False* Most trigonometric equations have unique solutions.

8. *True or False* Two solutions of the equation $\sin \theta = \dfrac{1}{2}$ are $\dfrac{\pi}{6}$ and $\dfrac{5\pi}{6}$.

9. *True or False* The set of all solutions of the equation $\tan \theta = 1$ is given by $\left\{ \theta \middle| \theta = \dfrac{\pi}{4} + k\pi, k \text{ is any integer} \right\}$.

10. *True or False* The equation $\sin \theta = 2$ has a real solution that can be found using a calculator.

11. If all solutions of a trigonometric equation are given by the general formula $\theta = \dfrac{\pi}{6} + 2k\pi$ or $\theta = \dfrac{11\pi}{6} + 2k\pi$, where k is any integer, then which of the following is *not* a solution of the equation?

(a) $\dfrac{35\pi}{6}$ (b) $\dfrac{23\pi}{6}$ (c) $\dfrac{13\pi}{6}$ (d) $\dfrac{7\pi}{6}$

12. Suppose $\theta = \dfrac{\pi}{2}$ is the only solution of a trigonometric equation in the interval $0 \le \theta < 2\pi$. Assuming a period of 2π, which of the following formulas gives all solutions of the equation, where k is any integer?

(a) $\theta = \dfrac{\pi}{2} + 2k\pi$ (b) $\theta = \dfrac{\pi}{2} + k\pi$

(c) $\theta = \dfrac{k\pi}{2}$ (d) $\theta = \dfrac{\pi + k\pi}{2}$

Skill Building

In Problems 13–36, solve each equation on the interval $0 \le \theta < 2\pi$.

13. $2 \sin \theta + 3 = 2$

14. $1 - \cos \theta = \dfrac{1}{2}$

15. $2 \sin \theta + 1 = 0$

16. $\cos \theta + 1 = 0$

17. $\tan \theta + 1 = 0$

18. $\sqrt{3} \cot \theta + 1 = 0$

19. $4 \sec \theta + 6 = -2$

20. $5 \csc \theta - 3 = 2$

21. $3\sqrt{2} \cos \theta + 2 = -1$

22. $4 \sin \theta + 3\sqrt{3} = \sqrt{3}$

23. $4 \cos^2 \theta = 1$

24. $\tan^2 \theta = \dfrac{1}{3}$

25. $2 \sin^2 \theta - 1 = 0$

26. $4 \cos^2 \theta - 3 = 0$

27. $\sin(3\theta) = -1$

28. $\tan \dfrac{\theta}{2} = \sqrt{3}$

29. $\cos(2\theta) = -\dfrac{1}{2}$

30. $\tan(2\theta) = -1$

31. $\sec \dfrac{3\theta}{2} = -2$

32. $\cot \dfrac{2\theta}{3} = -\sqrt{3}$

33. $\cos\left(2\theta - \dfrac{\pi}{2}\right) = -1$

34. $\sin\left(3\theta + \dfrac{\pi}{18}\right) = 1$

35. $\tan\left(\dfrac{\theta}{2} + \dfrac{\pi}{3}\right) = 1$

36. $\cos\left(\dfrac{\theta}{3} - \dfrac{\pi}{4}\right) = \dfrac{1}{2}$

In Problems 37–46, solve each equation. Give a general formula for all the solutions. List six solutions.

37. $\sin \theta = \dfrac{1}{2}$

38. $\tan \theta = 1$

39. $\tan \theta = -\dfrac{\sqrt{3}}{3}$

40. $\cos \theta = -\dfrac{\sqrt{3}}{2}$

41. $\cos \theta = 0$

42. $\sin \theta = \dfrac{\sqrt{2}}{2}$

43. $\cos(2\theta) = -\dfrac{1}{2}$

44. $\sin(2\theta) = -1$

45. $\sin \dfrac{\theta}{2} = -\dfrac{\sqrt{3}}{2}$

46. $\tan \dfrac{\theta}{2} = -1$

In Problems 47–58, use a calculator to solve each equation on the interval $0 \le \theta < 2\pi$. Round answers to two decimal places.

47. $\sin \theta = 0.4$

48. $\cos \theta = 0.6$

49. $\tan \theta = 5$

50. $\cot \theta = 2$

51. $\cos \theta = -0.9$

52. $\sin \theta = -0.2$

53. $\sec \theta = -4$

54. $\csc \theta = -3$

55. $5 \tan \theta + 9 = 0$

56. $4 \cot \theta = -5$

57. $3 \sin \theta - 2 = 0$

58. $4 \cos \theta + 3 = 0$

In Problems 59–82, solve each equation on the interval $0 \le \theta < 2\pi$.

59. $2\cos^2\theta + \cos\theta = 0$

60. $\sin^2\theta - 1 = 0$

61. $2\sin^2\theta - \sin\theta - 1 = 0$

62. $2\cos^2\theta + \cos\theta - 1 = 0$

63. $(\tan\theta - 1)(\sec\theta - 1) = 0$

64. $(\cot\theta + 1)\left(\csc\theta - \dfrac{1}{2}\right) = 0$

65. $\sin^2\theta - \cos^2\theta = 1 + \cos\theta$

66. $\cos^2\theta - \sin^2\theta + \sin\theta = 0$

67. $\sin^2\theta = 6(\cos(-\theta) + 1)$

68. $2\sin^2\theta = 3(1 - \cos(-\theta))$

69. $\cos\theta = -\sin(-\theta)$

70. $\cos\theta - \sin(-\theta) = 0$

71. $\tan\theta = 2\sin\theta$

72. $\tan\theta = \cot\theta$

73. $1 + \sin\theta = 2\cos^2\theta$

74. $\sin^2\theta = 2\cos\theta + 2$

75. $2\sin^2\theta - 5\sin\theta + 3 = 0$

76. $2\cos^2\theta - 7\cos\theta - 4 = 0$

77. $3(1 - \cos\theta) = \sin^2\theta$

78. $4(1 + \sin\theta) = \cos^2\theta$

79. $\tan^2\theta = \dfrac{3}{2}\sec\theta$

80. $\csc^2\theta = \cot\theta + 1$

81. $\sec^2\theta + \tan\theta = 0$

82. $\sec\theta = \tan\theta + \cot\theta$

In Problems 83–94, use a graphing utility to solve each equation. Express the solution(s) rounded to two decimal places.

83. $x + 5\cos x = 0$

84. $x - 4\sin x = 0$

85. $22x - 17\sin x = 3$

86. $19x + 8\cos x = 2$

87. $\sin x + \cos x = x$

88. $\sin x - \cos x = x$

89. $x^2 - 2\cos x = 0$

90. $x^2 + 3\sin x = 0$

91. $x^2 - 2\sin(2x) = 3x$

92. $x^2 = x + 3\cos(2x)$

93. $6\sin x - e^x = 2, \quad x > 0$

94. $4\cos(3x) - e^x = 1, \quad x > 0$

Mixed Practice

95. What are the zeros of $f(x) = 4\sin^2 x - 3$ on the interval $[0, 2\pi]$?

96. What are the zeros of $f(x) = 2\cos(3x) + 1$ on the interval $[0, \pi]$?

97. $f(x) = 3\sin x$
  (a) Find the zeros of f on the interval $[-2\pi, 4\pi]$.
  (b) Graph $f(x) = 3\sin x$ on the interval $[-2\pi, 4\pi]$.
  (c) Solve $f(x) = \dfrac{3}{2}$ on the interval $[-2\pi, 4\pi]$. What points are on the graph of f? Label these points on the graph drawn in part (b).
  (d) Use the graph drawn in part (b) along with the results of part (c) to determine the values of x such that $f(x) > \dfrac{3}{2}$ on the interval $[-2\pi, 4\pi]$.

98. $f(x) = 2\cos x$
  (a) Find the zeros of f on the interval $[-2\pi, 4\pi]$.
  (b) Graph $f(x) = 2\cos x$ on the interval $[-2\pi, 4\pi]$.
  (c) Solve $f(x) = -\sqrt{3}$ on the interval $[-2\pi, 4\pi]$. What points are on the graph of f? Label these points on the graph drawn in part (b).
  (d) Use the graph drawn in part (b) along with the results of part (c) to determine the values of x such that $f(x) < -\sqrt{3}$ on the interval $[-2\pi, 4\pi]$.

99. $f(x) = 4\tan x$
  (a) Solve $f(x) = -4$.
  (b) For what values of x is $f(x) < -4$ on the interval $\left(-\dfrac{\pi}{2}, \dfrac{\pi}{2}\right)$?

100. $f(x) = \cot x$
  (a) Solve $f(x) = -\sqrt{3}$.
  (b) For what values of x is $f(x) > -\sqrt{3}$ on the interval $(0, \pi)$?

101. (a) Graph $f(x) = 3 \sin(2x) + 2$ and $g(x) = \dfrac{7}{2}$ on the same Cartesian plane for the interval $[0, \pi]$.
  (b) Solve $f(x) = g(x)$ on the interval $[0, \pi]$, and label the points of intersection on the graph drawn in part (a).
  (c) Solve $f(x) > g(x)$ on the interval $[0, \pi]$.
  (d) Shade the region bounded by $f(x) = 3 \sin(2x) + 2$ and $g(x) = \dfrac{7}{2}$ between the two points found in part (b) on the graph drawn in part (a).

102. (a) Graph $f(x) = 2 \cos \dfrac{x}{2} + 3$ and $g(x) = 4$ on the same Cartesian plane for the interval $[0, 4\pi]$.
  (b) Solve $f(x) = g(x)$ on the interval $[0, 4\pi]$, and label the points of intersection on the graph drawn in part (a).
  (c) Solve $f(x) < g(x)$ on the interval $[0, 4\pi]$.
  (d) Shade the region bounded by $f(x) = 2 \cos \dfrac{x}{2} + 3$ and $g(x) = 4$ between the two points found in part (b) on the graph drawn in part (a).

103. (a) Graph $f(x) = -4 \cos x$ and $g(x) = 2 \cos x + 3$ on the same Cartesian plane for the interval $[0, 2\pi]$.
  (b) Solve $f(x) = g(x)$ on the interval $[0, 2\pi]$, and label the points of intersection on the graph drawn in part (a).
  (c) Solve $f(x) > g(x)$ on the interval $[0, 2\pi]$.
  (d) Shade the region bounded by $f(x) = -4 \cos x$ and $g(x) = 2 \cos x + 3$ between the two points found in part (b) on the graph drawn in part (a).

104. (a) Graph $f(x) = 2 \sin x$ and $g(x) = -2 \sin x + 2$ on the same Cartesian plane for the interval $[0, 2\pi]$.
  (b) Solve $f(x) = g(x)$ on the interval $[0, 2\pi]$, and label the points of intersection on the graph drawn in part (a).
  (c) Solve $f(x) > g(x)$ on the interval $[0, 2\pi]$.
  (d) Shade the region bounded by $f(x) = 2 \sin x$ and $g(x) = -2 \sin x + 2$ between the two points found in part (b) on the graph drawn in part (a).

Applications and Extensions

105. Blood Pressure Blood pressure is a way of measuring the amount of force exerted on the walls of blood vessels. It is measured using two numbers: systolic (as the heart beats) blood pressure and diastolic (as the heart rests) blood pressure. Blood pressures vary substantially from person to person, but a typical blood pressure is 120/80, which means the systolic blood pressure is 120 mmHg and the diastolic blood pressure is 80 mmHg. Assuming that a person's heart beats 70 times per minute, the blood pressure P of an individual after t seconds can be modeled by the function

$$P(t) = 100 + 20 \sin\left(\frac{7\pi}{3}t\right)$$

  (a) In the interval $[0, 1]$, determine the times at which the blood pressure is 100 mmHg.
  (b) In the interval $[0, 1]$, determine the times at which the blood pressure is 120 mmHg.
  (c) In the interval $[0, 1]$, determine the times at which the blood pressure is between 100 and 105 mmHg.

106. The Ferris Wheel In 1893, George Ferris engineered the Ferris wheel. It was 250 feet in diameter. If a Ferris wheel makes 1 revolution every 40 seconds, then the function

$$h(t) = 125 \sin\left(0.157t - \frac{\pi}{2}\right) + 125$$

represents the height h, in feet, of a seat on the wheel as a function of time t, where t is measured in seconds. The ride begins when $t = 0$.

  (a) During the first 40 seconds of the ride, at what time t is an individual on the Ferris wheel exactly 125 feet above the ground?
  (b) During the first 80 seconds of the ride, at what time t is an individual on the Ferris wheel exactly 250 feet above the ground?
  (c) During the first 40 seconds of the ride, over what interval of time t is an individual on the Ferris wheel more than 125 feet above the ground?

107. Holding Pattern An airplane is asked to stay within a holding pattern near Chicago's O'Hare International

Airport. The function $d(x) = 70 \sin(0.65x) + 150$ represents the distance d, in miles, of the airplane from the airport at time x, in minutes.

  (a) When the plane enters the holding pattern, $x = 0$, how far is it from O'Hare?
  (b) During the first 20 minutes after the plane enters the holding pattern, at what time x is the plane exactly 100 miles from the airport?
  (c) During the first 20 minutes after the plane enters the holding pattern, at what time x is the plane more than 100 miles from the airport?
  (d) While the plane is in the holding pattern, will it ever be within 70 miles of the airport? Why?

108. Projectile Motion A golfer hits a golf ball with an initial velocity of 100 miles per hour. The range R of the ball as a function of the angle θ to the horizontal is given by $R(\theta) = 672 \sin(2\theta)$, where R is measured in feet.

  (a) At what angle θ should the ball be hit if the golfer wants the ball to travel 450 feet (150 yards)?
  (b) At what angle θ should the ball be hit if the golfer wants the ball to travel 540 feet (180 yards)?
  (c) At what angle θ should the ball be hit if the golfer wants the ball to travel at least 480 feet (160 yards)?
  (d) Can the golfer hit the ball 720 feet (240 yards)?

109. Heat Transfer In the study of heat transfer, the equation $x + \tan x = 0$ occurs. Graph $Y_1 = -x$ and $Y_2 = \tan x$ for $x \geq 0$. Conclude that there are an infinite number of points of intersection of these two graphs. Now find the first two positive solutions of $x + \tan x = 0$ rounded to two decimal places.

110. Carrying a Ladder around a Corner Two hallways, one of width 3 feet, the other of width 4 feet, meet at a right angle. See the illustration on page 634. It can be shown that the length L of the ladder as a function of θ is $L(\theta) = 4 \csc \theta + 3 \sec \theta$.

  (a) In calculus, you will be asked to find the length of the longest ladder that can turn the corner by solving the equation

$$3 \sec \theta \tan \theta - 4 \csc \theta \cot \theta = 0 \quad 0° < \theta < 90°$$

Solve this equation for θ.

EXAMPLE 8   **Establishing an Identity**

Establish the identity: $\dfrac{1 - \sin\theta}{\cos\theta} = \dfrac{\cos\theta}{1 + \sin\theta}$

Solution   Start with the left side and multiply the numerator and the denominator by $1 + \sin\theta$. (Alternatively, we could multiply the numerator and the denominator of the right side by $1 - \sin\theta$.)

$$\dfrac{1 - \sin\theta}{\cos\theta} = \dfrac{1 - \sin\theta}{\cos\theta} \cdot \dfrac{1 + \sin\theta}{1 + \sin\theta} \quad \textit{Multiply the numerator and the denominator by } 1 + \sin\theta.$$

$$= \dfrac{1 - \sin^2\theta}{\cos\theta(1 + \sin\theta)}$$

$$= \dfrac{\cos^2\theta}{\cos\theta(1 + \sin\theta)} \quad \textit{1} - \sin^2\theta = \cos^2\theta$$

$$= \dfrac{\cos\theta}{1 + \sin\theta} \quad \textit{Cancel.} \qquad \bullet$$

—✏ **Now Work** PROBLEM 55

Although a lot of practice is the only real way to learn how to establish identities, the following guidelines should prove helpful.

WARNING Be careful not to handle identities to be established as if they were conditional equations. You *cannot* establish an identity by such methods as adding the same expression to each side and obtaining a true statement. This practice is not allowed, because the original statement is precisely the one that you are trying to establish. You do not know until it has been established that it is, in fact, true. ∎

Guidelines for Establishing Identities

1. It is almost always preferable to start with the side containing the more complicated expression.
2. Rewrite sums or differences of quotients as a single quotient.
3. Sometimes it will help to rewrite one side in terms of sine and cosine functions only.
4. Always keep the goal in mind. As you manipulate one side of the expression, keep in mind the form of the expression on the other side.

8.4 Assess Your Understanding

'Are You Prepared?' *Answers are given at the end of these exercises. If you get a wrong answer, read the pages listed in* red.

1. *True or False* $\sin^2\theta = 1 - \cos^2\theta$. (p. 522)

2. *True or False* $\sin(-\theta) + \cos(-\theta) = \cos\theta - \sin\theta$. (p. 560)

Concepts and Vocabulary

3. Suppose that f and g are two functions with the same domain. If $f(x) = g(x)$ for every x in the domain, the equation is called a(n) _____. Otherwise, it is called a(n) _____ equation.

4. $\tan^2\theta - \sec^2\theta =$ _____.

5. $\cos(-\theta) - \cos\theta =$ _____.

6. *True or False* $\sin(-\theta) + \sin\theta = 0$ for any value of θ.

7. *True or False* In establishing an identity, it is often easiest to just multiply both sides by a well-chosen nonzero expression involving the variable.

8. *True or False* $\tan\theta \cdot \cos\theta = \sin\theta$ for any $\theta \neq (2k + 1)\dfrac{\pi}{2}$.

9. Which of the following equations is *not* an identity?
   (a) $\cot^2\theta + 1 = \csc^2\theta$ (b) $\tan(-\theta) = -\tan\theta$
   (c) $\tan\theta = \dfrac{\cos\theta}{\sin\theta}$ (d) $\csc\theta = \dfrac{1}{\sin\theta}$

10. The expression $\dfrac{1}{1 - \sin\theta} + \dfrac{1}{1 + \sin\theta}$ simplifies to which of the following?
    (a) $2\cos^2\theta$ (b) $2\sec^2\theta$ (c) $2\sin^2\theta$ (d) $2\csc^2\theta$

Skill Building

In Problems 11–20, simplify each trigonometric expression by following the indicated direction.

11. Rewrite in terms of sine and cosine functions:

$$\tan \theta \cdot \csc \theta.$$

12. Rewrite in terms of sine and cosine functions:

$$\cot \theta \cdot \sec \theta.$$

13. Multiply $\dfrac{\cos \theta}{1 - \sin \theta}$ by $\dfrac{1 + \sin \theta}{1 + \sin \theta}$.

14. Multiply $\dfrac{\sin \theta}{1 + \cos \theta}$ by $\dfrac{1 - \cos \theta}{1 - \cos \theta}$.

15. Rewrite over a common denominator:

$$\frac{\sin \theta + \cos \theta}{\cos \theta} + \frac{\cos \theta - \sin \theta}{\sin \theta}$$

16. Rewrite over a common denominator:

$$\frac{1}{1 - \cos v} + \frac{1}{1 + \cos v}$$

17. Multiply and simplify: $\dfrac{(\sin \theta + \cos \theta)(\sin \theta + \cos \theta) - 1}{\sin \theta \cos \theta}$

18. Multiply and simplify: $\dfrac{(\tan \theta + 1)(\tan \theta + 1) - \sec^2 \theta}{\tan \theta}$

19. Factor and simplify: $\dfrac{3 \sin^2 \theta + 4 \sin \theta + 1}{\sin^2 \theta + 2 \sin \theta + 1}$

20. Factor and simplify: $\dfrac{\cos^2 \theta - 1}{\cos^2 \theta - \cos \theta}$

In Problems 21–100, establish each identity.

21. $\csc \theta \cdot \cos \theta = \cot \theta$

22. $\sec \theta \cdot \sin \theta = \tan \theta$

23. $1 + \tan^2(-\theta) = \sec^2 \theta$

24. $1 + \cot^2(-\theta) = \csc^2 \theta$

25. $\cos \theta (\tan \theta + \cot \theta) = \csc \theta$

26. $\sin \theta (\cot \theta + \tan \theta) = \sec \theta$

27. $\tan u \cot u - \cos^2 u = \sin^2 u$

28. $\sin u \csc u - \cos^2 u = \sin^2 u$

29. $(\sec \theta - 1)(\sec \theta + 1) = \tan^2 \theta$

30. $(\csc \theta - 1)(\csc \theta + 1) = \cot^2 \theta$

31. $(\sec \theta + \tan \theta)(\sec \theta - \tan \theta) = 1$

32. $(\csc \theta + \cot \theta)(\csc \theta - \cot \theta) = 1$

33. $\cos^2 \theta (1 + \tan^2 \theta) = 1$

34. $(1 - \cos^2 \theta)(1 + \cot^2 \theta) = 1$

35. $(\sin \theta + \cos \theta)^2 + (\sin \theta - \cos \theta)^2 = 2$

36. $\tan^2 \theta \cos^2 \theta + \cot^2 \theta \sin^2 \theta = 1$

37. $\sec^4 \theta - \sec^2 \theta = \tan^4 \theta + \tan^2 \theta$

38. $\csc^4 \theta - \csc^2 \theta = \cot^4 \theta + \cot^2 \theta$

39. $\sec u - \tan u = \dfrac{\cos u}{1 + \sin u}$

40. $\csc u - \cot u = \dfrac{\sin u}{1 + \cos u}$

41. $3 \sin^2 \theta + 4 \cos^2 \theta = 3 + \cos^2 \theta$

42. $9 \sec^2 \theta - 5 \tan^2 \theta = 5 + 4 \sec^2 \theta$

43. $1 - \dfrac{\cos^2 \theta}{1 + \sin \theta} = \sin \theta$

44. $1 - \dfrac{\sin^2 \theta}{1 - \cos \theta} = -\cos \theta$

45. $\dfrac{1 + \tan v}{1 - \tan v} = \dfrac{\cot v + 1}{\cot v - 1}$

46. $\dfrac{\csc v - 1}{\csc v + 1} = \dfrac{1 - \sin v}{1 + \sin v}$

47. $\dfrac{\sec \theta}{\csc \theta} + \dfrac{\sin \theta}{\cos \theta} = 2 \tan \theta$

48. $\dfrac{\csc \theta - 1}{\cot \theta} = \dfrac{\cot \theta}{\csc \theta + 1}$

49. $\dfrac{1 + \sin \theta}{1 - \sin \theta} = \dfrac{\csc \theta + 1}{\csc \theta - 1}$

50. $\dfrac{\cos \theta + 1}{\cos \theta - 1} = \dfrac{1 + \sec \theta}{1 - \sec \theta}$

51. $\dfrac{1 - \sin v}{\cos v} + \dfrac{\cos v}{1 - \sin v} = 2 \sec v$

52. $\dfrac{\cos v}{1 + \sin v} + \dfrac{1 + \sin v}{\cos v} = 2 \sec v$

53. $\dfrac{\sin \theta}{\sin \theta - \cos \theta} = \dfrac{1}{1 - \cot \theta}$

54. $1 - \dfrac{\sin^2 \theta}{1 + \cos \theta} = \cos \theta$

55. $\dfrac{1 - \sin \theta}{1 + \sin \theta} = (\sec \theta - \tan \theta)^2$

56. $\dfrac{1 - \cos \theta}{1 + \cos \theta} = (\csc \theta - \cot \theta)^2$

57. $\dfrac{\cos \theta}{1 - \tan \theta} + \dfrac{\sin \theta}{1 - \cot \theta} = \sin \theta + \cos \theta$

58. $\dfrac{\cot \theta}{1 - \tan \theta} + \dfrac{\tan \theta}{1 - \cot \theta} = 1 + \tan \theta + \cot \theta$

59. $\tan \theta + \dfrac{\cos \theta}{1 + \sin \theta} = \sec \theta$

60. $\dfrac{\sin \theta \cos \theta}{\cos^2 \theta - \sin^2 \theta} = \dfrac{\tan \theta}{1 - \tan^2 \theta}$

61. $\dfrac{\tan \theta + \sec \theta - 1}{\tan \theta - \sec \theta + 1} = \tan \theta + \sec \theta$

62. $\dfrac{\sin \theta - \cos \theta + 1}{\sin \theta + \cos \theta - 1} = \dfrac{\sin \theta + 1}{\cos \theta}$

63. $\dfrac{\tan \theta - \cot \theta}{\tan \theta + \cot \theta} = \sin^2 \theta - \cos^2 \theta$

64. $\dfrac{\sec \theta - \cos \theta}{\sec \theta + \cos \theta} = \dfrac{\sin^2 \theta}{1 + \cos^2 \theta}$

65. $\dfrac{\tan u - \cot u}{\tan u + \cot u} + 1 = 2 \sin^2 u$

66. $\dfrac{\tan u - \cot u}{\tan u + \cot u} + 2 \cos^2 u = 1$

67. $\dfrac{\sec \theta + \tan \theta}{\cot \theta + \cos \theta} = \tan \theta \sec \theta$

68. $\dfrac{\sec \theta}{1 + \sec \theta} = \dfrac{1 - \cos \theta}{\sin^2 \theta}$

69. $\dfrac{1 - \tan^2 \theta}{1 + \tan^2 \theta} + 1 = 2 \cos^2 \theta$

70. $\dfrac{1 - \cot^2 \theta}{1 + \cot^2 \theta} + 2 \cos^2 \theta = 1$

71. $\dfrac{\sec \theta - \csc \theta}{\sec \theta \csc \theta} = \sin \theta - \cos \theta$

72. $\dfrac{\sin^2 \theta - \tan \theta}{\cos^2 \theta - \cot \theta} = \tan^2 \theta$

73. $\sec \theta - \cos \theta = \sin \theta \tan \theta$

74. $\tan \theta + \cot \theta = \sec \theta \csc \theta$

75. $\dfrac{1}{1 - \sin \theta} + \dfrac{1}{1 + \sin \theta} = 2 \sec^2 \theta$

76. $\dfrac{1 + \sin \theta}{1 - \sin \theta} - \dfrac{1 - \sin \theta}{1 + \sin \theta} = 4 \tan \theta \sec \theta$

77. $\dfrac{\sec\theta}{1-\sin\theta} = \dfrac{1+\sin\theta}{\cos^3\theta}$

78. $\dfrac{1+\sin\theta}{1-\sin\theta} = (\sec\theta + \tan\theta)^2$

79. $\dfrac{(\sec v - \tan v)^2 + 1}{\csc v\,(\sec v - \tan v)} = 2\tan v$

80. $\dfrac{\sec^2 v - \tan^2 v + \tan v}{\sec v} = \sin v + \cos v$

81. $\dfrac{\sin\theta + \cos\theta}{\cos\theta} - \dfrac{\sin\theta - \cos\theta}{\sin\theta} = \sec\theta\csc\theta$

82. $\dfrac{\sin\theta + \cos\theta}{\sin\theta} - \dfrac{\cos\theta - \sin\theta}{\cos\theta} = \sec\theta\csc\theta$

83. $\dfrac{\sin^3\theta + \cos^3\theta}{\sin\theta + \cos\theta} = 1 - \sin\theta\cos\theta$

84. $\dfrac{\sin^3\theta + \cos^3\theta}{1 - 2\cos^2\theta} = \dfrac{\sec\theta - \sin\theta}{\tan\theta - 1}$

85. $\dfrac{\cos^2\theta - \sin^2\theta}{1 - \tan^2\theta} = \cos^2\theta$

86. $\dfrac{\cos\theta + \sin\theta - \sin^3\theta}{\sin\theta} = \cot\theta + \cos^2\theta$

87. $\dfrac{(2\cos^2\theta - 1)^2}{\cos^4\theta - \sin^4\theta} = 1 - 2\sin^2\theta$

88. $\dfrac{1 - 2\cos^2\theta}{\sin\theta\cos\theta} = \tan\theta - \cot\theta$

89. $\dfrac{1 + \sin\theta + \cos\theta}{1 + \sin\theta - \cos\theta} = \dfrac{1 + \cos\theta}{\sin\theta}$

90. $\dfrac{1 + \cos\theta + \sin\theta}{1 + \cos\theta - \sin\theta} = \sec\theta + \tan\theta$

91. $(a\sin\theta + b\cos\theta)^2 + (a\cos\theta - b\sin\theta)^2 = a^2 + b^2$

92. $(2a\sin\theta\cos\theta)^2 + a^2(\cos^2\theta - \sin^2\theta)^2 = a^2$

93. $\dfrac{\tan\alpha + \tan\beta}{\cot\alpha + \cot\beta} = \tan\alpha\tan\beta$

94. $(\tan\alpha + \tan\beta)(1 - \cot\alpha\cot\beta) + (\cot\alpha + \cot\beta)(1 - \tan\alpha\tan\beta) = 0$

95. $(\sin\alpha + \cos\beta)^2 + (\cos\beta + \sin\alpha)(\cos\beta - \sin\alpha) = 2\cos\beta(\sin\alpha + \cos\beta)$

96. $(\sin\alpha - \cos\beta)^2 + (\cos\beta + \sin\alpha)(\cos\beta - \sin\alpha) = -2\cos\beta(\sin\alpha - \cos\beta)$

97. $\ln|\sec\theta| = -\ln|\cos\theta|$

98. $\ln|\tan\theta| = \ln|\sin\theta| - \ln|\cos\theta|$

99. $\ln|1 + \cos\theta| + \ln|1 - \cos\theta| = 2\ln|\sin\theta|$

100. $\ln|\sec\theta + \tan\theta| + \ln|\sec\theta - \tan\theta| = 0$

In Problems 101–104, show that the functions f and g are identically equal.

101. $f(x) = \sin x \cdot \tan x \qquad g(x) = \sec x - \cos x$

102. $f(x) = \cos x \cdot \cot x \qquad g(x) = \csc x - \sin x$

103. $f(\theta) = \dfrac{1 - \sin\theta}{\cos\theta} - \dfrac{\cos\theta}{1 + \sin\theta} \qquad g(\theta) = 0$

104. $f(\theta) = \tan\theta + \sec\theta \qquad g(\theta) = \dfrac{\cos\theta}{1 - \sin\theta}$

Applications and Extensions

105. Searchlights A searchlight at the grand opening of a new car dealership casts a spot of light on a wall located 75 meters from the searchlight. The acceleration $\ddot{r}$ of the spot of light is found to be $\ddot{r} = 1200\sec\theta(2\sec^2\theta - 1)$. Show that this is equivalent to $\ddot{r} = 1200\left(\dfrac{1 + \sin^2\theta}{\cos^3\theta}\right)$.

Source: Adapted from Hibbeler, *Engineering Mechanics: Dynamics,* 13th ed., Pearson © 2013.

106. Optical Measurement Optical methods of measurement often rely on the interference of two light waves. If two light waves, identical except for a phase lag, are mixed together, the resulting intensity, or irradiance, is given by

$$I_t = 4A^2\,\dfrac{(\csc\theta - 1)(\sec\theta + \tan\theta)}{\csc\theta\sec\theta}.$$

Show that this is equivalent to $I_t = (2A\cos\theta)^2$.

Source: Experimental Techniques, July/August 2002

Explaining Concepts: Discussion and Writing

107. Write a few paragraphs outlining your strategy for establishing identities.

108. Write down the three Pythagorean Identities.

109. Why do you think it is usually preferable to start with the side containing the more complicated expression when establishing an identity?

110. Make up an identity that is not a basic identity.

Retain Your Knowledge

Problems 111–114 are based on material learned earlier in the course. The purpose of these problems is to keep the material fresh in your mind so that you are better prepared for the final exam.

111. Determine whether $f(x) = -3x^2 + 120x + 50$ has a maximum or a minimum value, and then find the value.

112. Given $f(x) = \dfrac{x + 1}{x - 2}$ and $g(x) = 3x - 4$, find $f \circ g$.

113. Find the exact values of the six trigonometric functions of an angle θ in standard position if $(-12, 5)$ is a point on its terminal side.

114. Find the average rate of change of $f(x) = \cos x$ from 0 to $\dfrac{\pi}{2}$.

'Are You Prepared?' Answers

1. True **2.** True

8.5 Sum and Difference Formulas

PREPARING FOR THIS SECTION *Before getting started, review the following:*

- Distance Formula (Section 2.1, p. 151)
- Values of the Trigonometric Functions (Section 7.3, pp. 531–534; Section 7.4, pp. 543–550)
- Finding Exact Values Given the Value of a Trigonometric Function and the Quadrant of the Angle (Section 7.4, p. 550)

Now Work the 'Are You Prepared?' problems on page 652.

OBJECTIVES 1 Use Sum and Difference Formulas to Find Exact Values (p. 644)
2 Use Sum and Difference Formulas to Establish Identities (p. 647)
3 Use Sum and Difference Formulas Involving Inverse Trigonometric Functions (p. 649)
4 Solve Trigonometric Equations Linear in Sine and Cosine (p. 650)

This section continues the derivation of trigonometric identities by obtaining formulas that involve the sum or the difference of two angles, such as $\cos(\alpha + \beta)$, $\cos(\alpha - \beta)$, and $\sin(\alpha + \beta)$. These formulas are referred to as the **sum and difference formulas**. We begin with the formulas for $\cos(\alpha + \beta)$ and $\cos(\alpha - \beta)$.

THEOREM

Sum and Difference Formulas for the Cosine Function

$$\cos(\alpha + \beta) = \cos\alpha\cos\beta - \sin\alpha\sin\beta \qquad (1)$$

$$\cos(\alpha - \beta) = \cos\alpha\cos\beta + \sin\alpha\sin\beta \qquad (2)$$

In Words
Formula (1) states that the cosine of the sum of two angles equals the cosine of the first angle times the cosine of the second angle minus the sine of the first angle times the sine of the second angle.

Proof We will prove formula (2) first. Although this formula is true for all numbers α and β, we shall assume in our proof that $0 < \beta < \alpha < 2\pi$. Begin with the unit circle and place the angles α and β in standard position, as shown in Figure 26(a). The point P_1 lies on the terminal side of β, so its coordinates are $(\cos\beta, \sin\beta)$; and the point P_2 lies on the terminal side of α, so its coordinates are $(\cos\alpha, \sin\alpha)$.

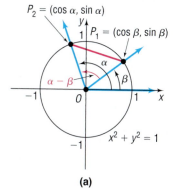

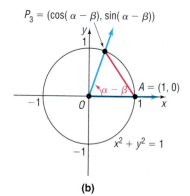

Figure 26

(a)

(b)

Now place the angle $\alpha - \beta$ in standard position, as shown in Figure 26(b). The point A has coordinates $(1, 0)$, and the point P_3 is on the terminal side of the angle $\alpha - \beta$, so its coordinates are $(\cos(\alpha - \beta), \sin(\alpha - \beta))$.

Looking at triangle OP_1P_2 in Figure 26(a) and triangle OAP_3 in Figure 26(b), note that these triangles are congruent. (Do you see why? SAS: two sides and the included angle, $\alpha - \beta$, are equal.) As a result, the unknown side of triangle $OP_1 P_2$ and the unknown side of triangle OAP_3 must be equal; that is,

$$d(A, P_3) = d(P_1, P_2)$$

Now use the distance formula to obtain

$$\sqrt{[\cos(\alpha - \beta) - 1]^2 + [\sin(\alpha - \beta) - 0]^2} = \sqrt{(\cos\alpha - \cos\beta)^2 + (\sin\alpha - \sin\beta)^2} \quad \text{\small{$d(A, P_3) = d(P_1, P_2)$}}$$

$$[\cos(\alpha - \beta) - 1]^2 + \sin^2(\alpha - \beta) = (\cos\alpha - \cos\beta)^2 + (\sin\alpha - \sin\beta)^2 \quad \text{\small{Square both sides.}}$$

$$\cos^2(\alpha - \beta) - 2\cos(\alpha - \beta) + 1 + \sin^2(\alpha - \beta) = \cos^2\alpha - 2\cos\alpha\cos\beta + \cos^2\beta \quad \text{\small{Multiply out the squared terms.}}$$
$$+ \sin^2\alpha - 2\sin\alpha\sin\beta + \sin^2\beta$$

$$2 - 2\cos(\alpha - \beta) = 2 - 2\cos\alpha\cos\beta - 2\sin\alpha\sin\beta \quad \text{\small{Apply a Pythagorean Identity (3 times).}}$$

$$-2\cos(\alpha - \beta) = -2\cos\alpha\cos\beta - 2\sin\alpha\sin\beta \quad \text{\small{Subtract 2 from each side.}}$$

$$\cos(\alpha - \beta) = \cos\alpha\cos\beta + \sin\alpha\sin\beta \quad \text{\small{Divide each side by -2.}}$$

This is formula (2). ∎

The proof of formula (1) follows from formula (2) and the Even–Odd Identities. Because $\alpha + \beta = \alpha - (-\beta)$, it follows that

$$\cos(\alpha + \beta) = \cos[\alpha - (-\beta)]$$

$$= \cos\alpha\cos(-\beta) + \sin\alpha\sin(-\beta) \quad \text{\small{Use formula (2).}}$$

$$= \cos\alpha\cos\beta - \sin\alpha\sin\beta \quad \text{\small{Even–Odd Identities}}$$

1 Use Sum and Difference Formulas to Find Exact Values

One use of formulas (1) and (2) is to obtain the exact value of the cosine of an angle that can be expressed as the sum or difference of angles whose sine and cosine are known exactly.

EXAMPLE 1 **Using the Sum Formula to Find an Exact Value**

Find the exact value of $\cos 75°$.

Solution Because $75° = 45° + 30°$, use formula (1) to obtain

$$\cos 75° = \cos(45° + 30°) = \cos 45° \cos 30° - \sin 45° \sin 30°$$
$$\uparrow$$
$$\text{\small{Formula (1)}}$$

$$= \frac{\sqrt{2}}{2} \cdot \frac{\sqrt{3}}{2} - \frac{\sqrt{2}}{2} \cdot \frac{1}{2} = \frac{1}{4}\left(\sqrt{6} - \sqrt{2}\right)$$ •

EXAMPLE 2 **Using the Difference Formula to Find an Exact Value**

Find the exact value of $\cos\dfrac{\pi}{12}$.

Solution

$$\cos\frac{\pi}{12} = \cos\left(\frac{3\pi}{12} - \frac{2\pi}{12}\right) = \cos\left(\frac{\pi}{4} - \frac{\pi}{6}\right)$$

$$= \cos\frac{\pi}{4}\cos\frac{\pi}{6} + \sin\frac{\pi}{4}\sin\frac{\pi}{6} \quad \text{\small{Use formula (2).}}$$

$$= \frac{\sqrt{2}}{2} \cdot \frac{\sqrt{3}}{2} + \frac{\sqrt{2}}{2} \cdot \frac{1}{2} = \frac{1}{4}\left(\sqrt{6} + \sqrt{2}\right)$$ •

Now Work PROBLEMS 13 AND 19

Another use of formulas (1) and (2) is to establish other identities. Two important identities conjectured earlier, in Section 7.2, are given next.

$$\cos\left(\frac{\pi}{2} - \theta\right) = \sin\theta \qquad\qquad \textbf{(3a)}$$

$$\sin\left(\frac{\pi}{2} - \theta\right) = \cos\theta \qquad\qquad \textbf{(3b)}$$

Proof To prove formula (3a), use the formula for $\cos(\alpha - \beta)$ with $\alpha = \dfrac{\pi}{2}$ and $\beta = \theta$.

$$\begin{aligned}
\cos\left(\frac{\pi}{2} - \theta\right) &= \cos\frac{\pi}{2}\cos\theta + \sin\frac{\pi}{2}\sin\theta \\
&= 0 \cdot \cos\theta + 1 \cdot \sin\theta \\
&= \sin\theta
\end{aligned}$$

To prove formula (3b), make use of the identity (3a) just established.

$$\sin\left(\frac{\pi}{2} - \theta\right) = \underset{\substack{\uparrow \\ \text{Use (3a).}}}{\cos}\left[\frac{\pi}{2} - \left(\frac{\pi}{2} - \theta\right)\right] = \cos\theta \qquad\blacksquare$$

Also, because

$$\cos\left(\frac{\pi}{2} - \theta\right) = \underset{\substack{\uparrow \\ \text{Even Property of Cosine}}}{\cos}\left[-\left(\theta - \frac{\pi}{2}\right)\right] = \cos\left(\theta - \frac{\pi}{2}\right)$$

and because

$$\cos\left(\frac{\pi}{2} - \theta\right) = \underset{\substack{\uparrow \\ \text{(3a)}}}{\sin\theta}$$

it follows that $\cos\left(\theta - \dfrac{\pi}{2}\right) = \sin\theta$. The graphs of $y = \cos\left(\theta - \dfrac{\pi}{2}\right)$ and $y = \sin\theta$ are identical.

Having established the identities in formulas (3a) and (3b), we now can derive the sum and difference formulas for $\sin(\alpha + \beta)$ and $\sin(\alpha - \beta)$.

Proof
$$\begin{aligned}
\sin(\alpha + \beta) &= \cos\left[\frac{\pi}{2} - (\alpha + \beta)\right] && \text{Formula (3a)} \\
&= \cos\left[\left(\frac{\pi}{2} - \alpha\right) - \beta\right] \\
&= \cos\left(\frac{\pi}{2} - \alpha\right)\cos\beta + \sin\left(\frac{\pi}{2} - \alpha\right)\sin\beta && \text{Formula (2)} \\
&= \sin\alpha\cos\beta + \cos\alpha\sin\beta && \text{Formulas (3a) and (3b)}
\end{aligned}$$

$$\begin{aligned}
\sin(\alpha - \beta) &= \sin[\alpha + (-\beta)] \\
&= \sin\alpha\cos(-\beta) + \cos\alpha\sin(-\beta) && \text{Use the sum formula} \\
&&& \text{for sine just obtained.} \\
&= \sin\alpha\cos\beta + \cos\alpha(-\sin\beta) && \text{Even–Odd Identities} \\
&= \sin\alpha\cos\beta - \cos\alpha\sin\beta
\end{aligned}$$

$\blacksquare$

4 Solve Trigonometric Equations Linear in Sine and Cosine

Sometimes it is necessary to square both sides of an equation to obtain expressions that allow the use of identities. Remember, squaring both sides of an equation may introduce extraneous solutions. As a result, apparent solutions must be checked.

EXAMPLE 11   **Solving a Trigonometric Equation Linear in Sine and Cosine**

Solve the equation: $\sin \theta + \cos \theta = 1, \quad 0 \le \theta < 2\pi$

Option 1   Attempts to use available identities do not lead to equations that are easy to solve. (Try it yourself.) So, given the form of this equation, square each side.

$$\sin \theta + \cos \theta = 1$$

$$(\sin \theta + \cos \theta)^2 = 1 \quad \textit{Square each side.}$$

$$\sin^2 \theta + 2 \sin \theta \cos \theta + \cos^2 \theta = 1 \quad \textit{Remove parentheses.}$$

$$2 \sin \theta \cos \theta = 0 \quad \textit{sin}^2 \theta + \cos^2 \theta = 1$$

$$\sin \theta \cos \theta = 0$$

Setting each factor equal to zero leads to

$$\sin \theta = 0 \quad \text{or} \quad \cos \theta = 0$$

The apparent solutions are

$$\theta = 0 \qquad \theta = \pi \qquad \theta = \frac{\pi}{2} \qquad \theta = \frac{3\pi}{2}$$

Because both sides of the original equation were squared, these apparent solutions must be checked to see whether any are extraneous.

$$\theta = 0: \quad \sin 0 + \cos 0 = 0 + 1 = 1 \qquad \textit{A solution}$$

$$\theta = \pi: \quad \sin \pi + \cos \pi = 0 + (-1) = -1 \quad \textit{Not a solution}$$

$$\theta = \frac{\pi}{2}: \quad \sin \frac{\pi}{2} + \cos \frac{\pi}{2} = 1 + 0 = 1 \qquad \textit{A solution}$$

$$\theta = \frac{3\pi}{2}: \quad \sin \frac{3\pi}{2} + \cos \frac{3\pi}{2} = -1 + 0 = -1 \quad \textit{Not a solution}$$

The values $\theta = \pi$ and $\theta = \frac{3\pi}{2}$ are extraneous. The solution set is $\left\{ 0, \frac{\pi}{2} \right\}$.   ●

Option 2   Start with the equation

$$\sin \theta + \cos \theta = 1$$

and divide each side by $\sqrt{2}$. (The reason for this choice will become apparent shortly.) Then

$$\frac{1}{\sqrt{2}} \sin \theta + \frac{1}{\sqrt{2}} \cos \theta = \frac{1}{\sqrt{2}}$$

The left side now resembles the formula for the sine of the sum of two angles, one of which is θ. The other angle is unknown (call it ϕ.) Then

$$\sin(\theta + \phi) = \sin \theta \cos \phi + \cos \theta \sin \phi = \frac{1}{\sqrt{2}} = \frac{\sqrt{2}}{2} \tag{8}$$

where

$$\cos \phi = \frac{1}{\sqrt{2}} = \frac{\sqrt{2}}{2} \qquad \sin \phi = \frac{1}{\sqrt{2}} = \frac{\sqrt{2}}{2} \qquad 0 \le \phi < 2\pi$$

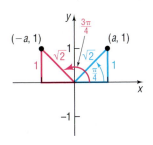

Figure 29

The angle ϕ is therefore $\dfrac{\pi}{4}$. As a result, equation (8) becomes

$$\sin\left(\theta + \frac{\pi}{4}\right) = \frac{\sqrt{2}}{2}$$

In the interval $[0, 2\pi)$, there are two angles whose sine is $\dfrac{\sqrt{2}}{2}$: $\dfrac{\pi}{4}$ and $\dfrac{3\pi}{4}$. See Figure 29. As a result,

$$\theta + \frac{\pi}{4} = \frac{\pi}{4} \quad \text{or} \quad \theta + \frac{\pi}{4} = \frac{3\pi}{4}$$

$$\theta = 0 \quad \text{or} \qquad \theta = \frac{\pi}{2}$$

The solution set is $\left\{0, \dfrac{\pi}{2}\right\}$.

    The second option can be used to solve any linear equation in the variables $\sin\theta$ and $\cos\theta$.

EXAMPLE 12 **Solving a Trigonometric Equation Linear in sin θ and cos θ**

Solve:

$$a\sin\theta + b\cos\theta = c \tag{9}$$

where a, b, and c are constants and either $a \neq 0$ or $b \neq 0$.

Solution Divide each side of equation (9) by $\sqrt{a^2 + b^2}$. Then

$$\frac{a}{\sqrt{a^2 + b^2}}\sin\theta + \frac{b}{\sqrt{a^2 + b^2}}\cos\theta = \frac{c}{\sqrt{a^2 + b^2}} \tag{10}$$

There is a unique angle ϕ, $0 \leq \phi < 2\pi$, for which

$$\cos\phi = \frac{a}{\sqrt{a^2 + b^2}} \quad \text{and} \quad \sin\phi = \frac{b}{\sqrt{a^2 + b^2}} \tag{11}$$

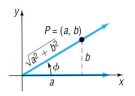

Figure 30

Figure 30 shows the situation for $a > 0$ and $b > 0$. Equation (10) may be written as

$$\sin\theta\cos\phi + \cos\theta\sin\phi = \frac{c}{\sqrt{a^2 + b^2}}$$

or, equivalently,

$$\sin(\theta + \phi) = \frac{c}{\sqrt{a^2 + b^2}} \tag{12}$$

where ϕ satisfies equation (11).

    If $|c| > \sqrt{a^2 + b^2}$, then $\sin(\theta + \phi) > 1$ or $\sin(\theta + \phi) < -1$, and equation (12) has no solution.

    If $|c| \leq \sqrt{a^2 + b^2}$, then the solutions of equation (12) are

$$\theta + \phi = \sin^{-1}\frac{c}{\sqrt{a^2 + b^2}} \quad \text{or} \quad \theta + \phi = \pi - \sin^{-1}\frac{c}{\sqrt{a^2 + b^2}}$$

Because the angle ϕ is determined by equations (11), these give the solutions to equation (9).

Now Work PROBLEM **95**

SUMMARY

Sum and Difference Formulas

$$\cos(\alpha + \beta) = \cos \alpha \cos \beta - \sin \alpha \sin \beta \qquad \cos(\alpha - \beta) = \cos \alpha \cos \beta + \sin \alpha \sin \beta$$

$$\sin(\alpha + \beta) = \sin \alpha \cos \beta + \cos \alpha \sin \beta \qquad \sin(\alpha - \beta) = \sin \alpha \cos \beta - \cos \alpha \sin \beta$$

$$\tan(\alpha + \beta) = \frac{\tan \alpha + \tan \beta}{1 - \tan \alpha \tan \beta} \qquad \tan(\alpha - \beta) = \frac{\tan \alpha - \tan \beta}{1 + \tan \alpha \tan \beta}$$

8.5 Assess Your Understanding

'Are You Prepared?' *Answers are given at the end of these exercises. If you get a wrong answer, read the pages listed in red.*

1. The distance d from the point $(2, -3)$ to the point $(5, 1)$ is _____. (p. 151)

2. If $\sin \theta = \dfrac{4}{5}$ and θ is in quadrant II, then $\cos \theta =$ _____.
   (p. 550)

3. (a) $\sin \dfrac{\pi}{4} \cdot \cos \dfrac{\pi}{3} =$ _____. (pp. 531–534)

   (b) $\tan \dfrac{\pi}{4} - \sin \dfrac{\pi}{6} =$ _____. (pp. 531–534)

4. If $\sin \alpha = -\dfrac{4}{5}, \pi < \alpha < \dfrac{3\pi}{2}$, then $\cos \alpha =$ _____.
   (p. 550)

Concepts and Vocabulary

5. $\cos(\alpha + \beta) = \cos \alpha \cos \beta$ _____ $\sin \alpha \sin \beta$

6. $\sin(\alpha - \beta) = \sin \alpha \cos \beta$ _____ $\cos \alpha \sin \beta$

7. **True or False** $\sin(\alpha + \beta) = \sin \alpha + \sin \beta + 2 \sin \alpha \sin \beta$

8. **True or False** $\tan 75° = \tan 30° + \tan 45°$

9. **True or False** $\cos\left(\dfrac{\pi}{2} - \theta\right) = \cos \theta$

10. **True or False** If $f(x) = \sin x$ and $g(x) = \cos x$, then $g(\alpha + \beta) = g(\alpha)g(\beta) - f(\alpha)f(\beta)$

11. Choose the expression that completes the sum formula for tangent functions: $\tan(\alpha + \beta) =$ _____

    (a) $\dfrac{\tan \alpha + \tan \beta}{1 - \tan \alpha \tan \beta}$ (b) $\dfrac{\tan \alpha - \tan \beta}{1 + \tan \alpha \tan \beta}$

    (c) $\dfrac{\tan \alpha + \tan \beta}{1 + \tan \alpha \tan \beta}$ (d) $\dfrac{\tan \alpha - \tan \beta}{1 - \tan \alpha \tan \beta}$

12. Choose the expression that is equivalent to $\sin 60° \cos 20° + \cos 60° \sin 20°$.

    (a) $\cos 40°$ (b) $\sin 40°$ (c) $\cos 80°$ (d) $\sin 80°$

Skill Building

In Problems 13–24, find the exact value of each expression.

13. $\cos 165°$

14. $\sin 105°$

15. $\tan 15°$

16. $\tan 195°$

17. $\sin \dfrac{5\pi}{12}$

18. $\sin \dfrac{\pi}{12}$

19. $\cos \dfrac{7\pi}{12}$

20. $\tan \dfrac{7\pi}{12}$

21. $\sin \dfrac{17\pi}{12}$

22. $\tan \dfrac{19\pi}{12}$

23. $\sec\left(-\dfrac{\pi}{12}\right)$

24. $\cot\left(-\dfrac{5\pi}{12}\right)$

In Problems 25–34, find the exact value of each expression.

25. $\sin 20° \cos 10° + \cos 20° \sin 10°$

26. $\sin 20° \cos 80° - \cos 20° \sin 80°$

27. $\cos 70° \cos 20° - \sin 70° \sin 20°$

28. $\cos 40° \cos 10° + \sin 40° \sin 10°$

29. $\dfrac{\tan 20° + \tan 25°}{1 - \tan 20° \tan 25°}$

30. $\dfrac{\tan 40° - \tan 10°}{1 + \tan 40° \tan 10°}$

31. $\sin \dfrac{\pi}{12} \cos \dfrac{7\pi}{12} - \cos \dfrac{\pi}{12} \sin \dfrac{7\pi}{12}$

32. $\cos \dfrac{5\pi}{12} \cos \dfrac{7\pi}{12} - \sin \dfrac{5\pi}{12} \sin \dfrac{7\pi}{12}$

33. $\cos \dfrac{\pi}{12} \cos \dfrac{5\pi}{12} + \sin \dfrac{5\pi}{12} \sin \dfrac{\pi}{12}$

34. $\sin \dfrac{\pi}{18} \cos \dfrac{5\pi}{18} + \cos \dfrac{\pi}{18} \sin \dfrac{5\pi}{18}$

In Problems 35–40, find the exact value of each of the following under the given conditions:

(a) $\sin(\alpha + \beta)$ (b) $\cos(\alpha + \beta)$ (c) $\sin(\alpha - \beta)$ (d) $\tan(\alpha - \beta)$

35. $\sin \alpha = \dfrac{3}{5}, 0 < \alpha < \dfrac{\pi}{2}; \quad \cos \beta = \dfrac{2\sqrt{5}}{5}, -\dfrac{\pi}{2} < \beta < 0$

36. $\cos \alpha = \dfrac{\sqrt{5}}{5}, 0 < \alpha < \dfrac{\pi}{2}; \quad \sin \beta = -\dfrac{4}{5}, -\dfrac{\pi}{2} < \beta < 0$

37. $\tan \alpha = -\dfrac{4}{3}, \dfrac{\pi}{2} < \alpha < \pi; \quad \cos \beta = \dfrac{1}{2}, 0 < \beta < \dfrac{\pi}{2}$

38. $\tan \alpha = \dfrac{5}{12}, \pi < \alpha < \dfrac{3\pi}{2}; \quad \sin \beta = -\dfrac{1}{2}, \pi < \beta < \dfrac{3\pi}{2}$

39. $\sin \alpha = \dfrac{5}{13}, -\dfrac{3\pi}{2} < \alpha < -\pi; \quad \tan \beta = -\sqrt{3}, \dfrac{\pi}{2} < \beta < \pi$

40. $\cos \alpha = \dfrac{1}{2}, -\dfrac{\pi}{2} < \alpha < 0; \quad \sin \beta = \dfrac{1}{3}, 0 < \beta < \dfrac{\pi}{2}$

41. If $\sin \theta = \dfrac{1}{3}, \theta$ in quadrant II, find the exact value of:

    (a) $\cos \theta$ (b) $\sin\left(\theta + \dfrac{\pi}{6}\right)$

    (c) $\cos\left(\theta - \dfrac{\pi}{3}\right)$ (d) $\tan\left(\theta + \dfrac{\pi}{4}\right)$

42. If $\cos \theta = \dfrac{1}{4}, \theta$ in quadrant IV, find the exact value of:

    (a) $\sin \theta$ (b) $\sin\left(\theta - \dfrac{\pi}{6}\right)$

    (c) $\cos\left(\theta + \dfrac{\pi}{3}\right)$ (d) $\tan\left(\theta - \dfrac{\pi}{4}\right)$

In Problems 43–48, use the figures to evaluate each function if $f(x) = \sin x$, $g(x) = \cos x$, and $h(x) = \tan x$.

43. $f(\alpha + \beta)$ **44.** $g(\alpha + \beta)$

45. $g(\alpha - \beta)$ **46.** $f(\alpha - \beta)$

47. $h(\alpha + \beta)$ **48.** $h(\alpha - \beta)$

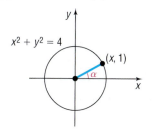

In Problems 49–74, establish each identity.

49. $\sin\left(\dfrac{\pi}{2} + \theta\right) = \cos \theta$

50. $\cos\left(\dfrac{\pi}{2} + \theta\right) = -\sin \theta$

51. $\sin(\pi - \theta) = \sin \theta$

52. $\cos(\pi - \theta) = -\cos \theta$

53. $\sin(\pi + \theta) = -\sin \theta$

54. $\cos(\pi + \theta) = -\cos \theta$

55. $\tan(\pi - \theta) = -\tan \theta$

56. $\tan(2\pi - \theta) = -\tan \theta$

57. $\sin\left(\dfrac{3\pi}{2} + \theta\right) = -\cos \theta$

58. $\cos\left(\dfrac{3\pi}{2} + \theta\right) = \sin \theta$

59. $\sin(\alpha + \beta) + \sin(\alpha - \beta) = 2 \sin \alpha \cos \beta$

60. $\cos(\alpha + \beta) + \cos(\alpha - \beta) = 2 \cos \alpha \cos \beta$

61. $\dfrac{\sin(\alpha + \beta)}{\sin \alpha \cos \beta} = 1 + \cot \alpha \tan \beta$

62. $\dfrac{\sin(\alpha + \beta)}{\cos \alpha \cos \beta} = \tan \alpha + \tan \beta$

63. $\dfrac{\cos(\alpha + \beta)}{\cos \alpha \cos \beta} = 1 - \tan \alpha \tan \beta$

64. $\dfrac{\cos(\alpha - \beta)}{\sin \alpha \cos \beta} = \cot \alpha + \tan \beta$

65. $\dfrac{\sin(\alpha + \beta)}{\sin(\alpha - \beta)} = \dfrac{\tan \alpha + \tan \beta}{\tan \alpha - \tan \beta}$

66. $\dfrac{\cos(\alpha + \beta)}{\cos(\alpha - \beta)} = \dfrac{1 - \tan \alpha \tan \beta}{1 + \tan \alpha \tan \beta}$

67. $\cot(\alpha + \beta) = \dfrac{\cot \alpha \cot \beta - 1}{\cot \beta + \cot \alpha}$

68. $\cot(\alpha - \beta) = \dfrac{\cot \alpha \cot \beta + 1}{\cot \beta - \cot \alpha}$

69. $\sec(\alpha + \beta) = \dfrac{\csc \alpha \csc \beta}{\cot \alpha \cot \beta - 1}$

70. $\sec(\alpha - \beta) = \dfrac{\sec \alpha \sec \beta}{1 + \tan \alpha \tan \beta}$

71. $\sin(\alpha - \beta) \sin(\alpha + \beta) = \sin^2 \alpha - \sin^2 \beta$

72. $\cos(\alpha - \beta) \cos(\alpha + \beta) = \cos^2 \alpha - \sin^2 \beta$

73. $\sin(\theta + k\pi) = (-1)^k \sin \theta, k$ any integer

74. $\cos(\theta + k\pi) = (-1)^k \cos \theta, k$ any integer

In Problems 75–86, find the exact value of each expression.

75. $\sin\left(\sin^{-1}\dfrac{1}{2} + \cos^{-1}0\right)$

76. $\sin\left(\sin^{-1}\dfrac{\sqrt{3}}{2} + \cos^{-1}1\right)$

77. $\sin\left[\sin^{-1}\dfrac{3}{5} - \cos^{-1}\left(-\dfrac{4}{5}\right)\right]$

78. $\sin\left[\sin^{-1}\left(-\dfrac{4}{5}\right) - \tan^{-1}\dfrac{3}{4}\right]$

79. $\cos\left(\tan^{-1}\dfrac{4}{3} + \cos^{-1}\dfrac{5}{13}\right)$

80. $\cos\left[\tan^{-1}\dfrac{5}{12} - \sin^{-1}\left(-\dfrac{3}{5}\right)\right]$

81. $\cos\left(\sin^{-1}\dfrac{5}{13} - \tan^{-1}\dfrac{3}{4}\right)$

82. $\cos\left(\tan^{-1}\dfrac{4}{3} + \cos^{-1}\dfrac{12}{13}\right)$

83. $\tan\left(\sin^{-1}\dfrac{3}{5} + \dfrac{\pi}{6}\right)$

84. $\tan\left(\dfrac{\pi}{4} - \cos^{-1}\dfrac{3}{5}\right)$

85. $\tan\left(\sin^{-1}\dfrac{4}{5} + \cos^{-1}1\right)$

86. $\tan\left(\cos^{-1}\dfrac{4}{5} + \sin^{-1}1\right)$

In Problems 87–92, write each trigonometric expression as an algebraic expression containing u and v. Give the restrictions required on u and v.

87. $\cos\left(\cos^{-1}u + \sin^{-1}v\right)$

88. $\sin\left(\sin^{-1}u - \cos^{-1}v\right)$

89. $\sin\left(\tan^{-1}u - \sin^{-1}v\right)$

90. $\cos\left(\tan^{-1}u + \tan^{-1}v\right)$

91. $\tan\left(\sin^{-1}u - \cos^{-1}v\right)$

92. $\sec\left(\tan^{-1}u + \cos^{-1}v\right)$

In Problems 93–98, solve each equation on the interval $0 \le \theta < 2\pi$.

93. $\sin\theta - \sqrt{3}\cos\theta = 1$

94. $\sqrt{3}\sin\theta + \cos\theta = 1$

95. $\sin\theta + \cos\theta = \sqrt{2}$

96. $\sin\theta - \cos\theta = -\sqrt{2}$

97. $\tan\theta + \sqrt{3} = \sec\theta$

98. $\cot\theta + \csc\theta = -\sqrt{3}$

Applications and Extensions

99. Show that $\sin^{-1}v + \cos^{-1}v = \dfrac{\pi}{2}$.

100. Show that $\tan^{-1}v + \cot^{-1}v = \dfrac{\pi}{2}$.

101. Show that $\tan^{-1}\left(\dfrac{1}{v}\right) = \dfrac{\pi}{2} - \tan^{-1}v$, if $v > 0$.

102. Show that $\cot^{-1}e^v = \tan^{-1}e^{-v}$.

103. Show that $\sin\left(\sin^{-1}v + \cos^{-1}v\right) = 1$.

104. Show that $\cos\left(\sin^{-1}v + \cos^{-1}v\right) = 0$.

---

105. Calculus Show that the difference quotient for $f(x) = \sin x$ is given by

$$\frac{f(x+h) - f(x)}{h} = \frac{\sin(x+h) - \sin x}{h}$$

$$= \cos x \cdot \frac{\sin h}{h} - \sin x \cdot \frac{1 - \cos h}{h}$$

106. Calculus Show that the difference quotient for $f(x) = \cos x$ is given by

$$\frac{f(x+h) - f(x)}{h} = \frac{\cos(x+h) - \cos x}{h}$$

$$= -\sin x \cdot \frac{\sin h}{h} - \cos x \cdot \frac{1 - \cos h}{h}$$

107. One, Two, Three
(a) Show that $\tan\left(\tan^{-1}1 + \tan^{-1}2 + \tan^{-1}3\right) = 0$.
(b) Conclude from part (a) that

$$\tan^{-1}1 + \tan^{-1}2 + \tan^{-1}3 = \pi$$

Source: College Mathematics Journal, Vol. 37, No. 3, May 2006

108. Electric Power In an alternating current (ac) circuit, the instantaneous power p at time t is given by

$$p(t) = V_m I_m \cos\phi \sin^2(\omega t) - V_m I_m \sin\phi \sin(\omega t)\cos(\omega t)$$

Show that this is equivalent to

$$p(t) = V_m I_m \sin(\omega t)\sin(\omega t - \phi)$$

Source: HyperPhysics, hosted by Georgia State University

109. Geometry: Angle between Two Lines Let L_1 and L_2 denote two nonvertical intersecting lines, and let θ denote the acute angle between L_1 and L_2 (see the figure). Show that

$$\tan\theta = \frac{m_2 - m_1}{1 + m_1 m_2}$$

where m_1 and m_2 are the slopes of L_1 and L_2, respectively. [**Hint:** Use the facts that $\tan\theta_1 = m_1$ and $\tan\theta_2 = m_2$.]

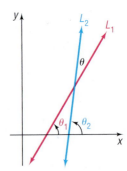

110. If $\alpha + \beta + \gamma = 180°$ and

$$\cot\theta = \cot\alpha + \cot\beta + \cot\gamma \qquad 0 < \theta < 90°$$

show that

$$\sin^3\theta = \sin(\alpha - \theta)\sin(\beta - \theta)\sin(\gamma - \theta)$$

111. If $\tan\alpha = x + 1$ and $\tan\beta = x - 1$, show that

$$2\cot(\alpha - \beta) = x^2$$

Explaining Concepts: Discussion and Writing

112. Discuss the following derivation:

$$\tan\left(\theta + \frac{\pi}{2}\right) = \frac{\tan\theta + \tan\frac{\pi}{2}}{1 - \tan\theta\tan\frac{\pi}{2}} = \frac{\frac{\tan\theta}{\tan\frac{\pi}{2}} + 1}{\frac{1}{\tan\frac{\pi}{2}} - \tan\theta} = \frac{0 + 1}{0 - \tan\theta} = \frac{1}{-\tan\theta} = -\cot\theta$$

Can you justify each step?

113. Explain why formula (7) cannot be used to show that

$$\tan\left(\frac{\pi}{2} - \theta\right) = \cot\theta$$

Establish this identity by using formulas (3a) and (3b).

─ Retain Your Knowledge ──────────

Problems 114–117 are based on material learned earlier in the course. The purpose of these problems is to keep the material fresh in your mind so that you are better prepared for the final exam.

114. Determine the points of intersection of the graphs of $f(x) = x^2 + 5x + 1$ and $g(x) = -2x^2 - 11x - 4$ by solving $f(x) = g(x)$.

115. Convert $\dfrac{17\pi}{6}$ to degrees.

116. Find the area of the sector of a circle of radius 6 centimeters formed by an angle of $45°$. Give both the exact area and an approximation rounded to two decimal places.

117. Given $\tan\theta = -2$, $270° < \theta < 360°$, find the exact value of the remaining five trigonometric functions.

'Are You Prepared?' Answers

1. 5   **2.** $-\dfrac{3}{5}$ **3.** (a) $\dfrac{\sqrt{2}}{4}$ (b) $\dfrac{1}{2}$ **4.** $-\dfrac{3}{5}$

8.6 Double-angle and Half-angle Formulas

OBJECTIVES **1** Use Double-angle Formulas to Find Exact Values (p. 656)
  2 Use Double-angle Formulas to Establish Identities (p. 656)
  3 Use Half-angle Formulas to Find Exact Values (p. 659)

In this section, formulas for $\sin(2\theta), \cos(2\theta), \sin\left(\frac{1}{2}\theta\right)$, and $\cos\left(\frac{1}{2}\theta\right)$ are established in terms of $\sin\theta$ and $\cos\theta$. They are derived using the sum formulas.
In the sum formulas for $\sin(\alpha + \beta)$ and $\cos(\alpha + \beta)$, let $\alpha = \beta = \theta$. Then

$$\sin(\alpha + \beta) = \sin\alpha\cos\beta + \cos\alpha\sin\beta$$
$$\sin(\theta + \theta) = \sin\theta\cos\theta + \cos\theta\sin\theta$$
$$\sin(2\theta) = 2\sin\theta\cos\theta$$

and

$$\cos(\alpha + \beta) = \cos\alpha\cos\beta - \sin\alpha\sin\beta$$
$$\cos(\theta + \theta) = \cos\theta\cos\theta - \sin\theta\sin\theta$$
$$\cos(2\theta) = \cos^2\theta - \sin^2\theta$$

An application of the Pythagorean Identity $\sin^2\theta + \cos^2\theta = 1$ results in two other ways to express $\cos(2\theta)$.

$$\cos(2\theta) = \cos^2\theta - \sin^2\theta = (1 - \sin^2\theta) - \sin^2\theta = 1 - 2\sin^2\theta$$

$$\cos 15° = \cos \frac{30°}{2} = \sqrt{\frac{1 + \cos 30°}{2}}$$

$$= \sqrt{\frac{1 + \sqrt{3}/2}{2}} = \sqrt{\frac{2 + \sqrt{3}}{4}} = \frac{\sqrt{2 + \sqrt{3}}}{2}$$

(b) Use the fact that $\sin(-15°) = -\sin 15°$, and then apply formula (10a).

$$\sin(-15°) = -\sin \frac{30°}{2} = -\sqrt{\frac{1 - \cos 30°}{2}}$$

$$= -\sqrt{\frac{1 - \sqrt{3}/2}{2}} = -\sqrt{\frac{2 - \sqrt{3}}{4}} = -\frac{\sqrt{2 - \sqrt{3}}}{2} \qquad \bullet$$

It is interesting to compare the answer found in Example 6(a) with the answer to Example 2 of Section 8.5. There it was calculated that

$$\cos \frac{\pi}{12} = \cos 15° = \frac{1}{4}\left(\sqrt{6} + \sqrt{2}\right)$$

Based on this and the result of Example 6(a),

$$\frac{1}{4}\left(\sqrt{6} + \sqrt{2}\right) \quad \text{and} \quad \frac{\sqrt{2 + \sqrt{3}}}{2}$$

are equal. (Since each expression is positive, you can verify this equality by squaring each expression.) Two very different-looking, yet correct, answers can be obtained, depending on the approach taken to solve a problem.

Now Work **PROBLEM 21**

EXAMPLE 7 **Finding Exact Values Using Half-angle Formulas**

If $\cos \alpha = -\dfrac{3}{5}$, $\pi < \alpha < \dfrac{3\pi}{2}$, find the exact value of:

(a) $\sin \dfrac{\alpha}{2}$ (b) $\cos \dfrac{\alpha}{2}$ (c) $\tan \dfrac{\alpha}{2}$

Solution   First, observe that if $\pi < \alpha < \dfrac{3\pi}{2}$, then $\dfrac{\pi}{2} < \dfrac{\alpha}{2} < \dfrac{3\pi}{4}$. As a result, $\dfrac{\alpha}{2}$ lies in quadrant II.

(a) Because $\dfrac{\alpha}{2}$ lies in quadrant II, $\sin \dfrac{\alpha}{2} > 0$, so use the $+$ sign in formula (10) to get

$$\sin \frac{\alpha}{2} = \sqrt{\frac{1 - \cos \alpha}{2}} = \sqrt{\frac{1 - \left(-\dfrac{3}{5}\right)}{2}}$$

$$= \sqrt{\frac{\dfrac{8}{5}}{2}} = \sqrt{\frac{4}{5}} = \frac{2}{\sqrt{5}} = \frac{2\sqrt{5}}{5}$$

(b) Because $\dfrac{\alpha}{2}$ lies in quadrant II, $\cos \dfrac{\alpha}{2} < 0$, so use the $-$ sign in formula (11) to get

$$\cos \frac{\alpha}{2} = -\sqrt{\frac{1 + \cos \alpha}{2}} = -\sqrt{\frac{1 + \left(-\dfrac{3}{5}\right)}{2}}$$

$$= -\sqrt{\frac{\dfrac{2}{5}}{2}} = -\frac{1}{\sqrt{5}} = -\frac{\sqrt{5}}{5}$$

(c) Because $\dfrac{\alpha}{2}$ lies in quadrant II, $\tan \dfrac{\alpha}{2} < 0$, so use the $-$ sign in formula (12) to get

$$\tan \frac{\alpha}{2} = -\sqrt{\frac{1 - \cos \alpha}{1 + \cos \alpha}} = -\sqrt{\frac{1 - \left(-\dfrac{3}{5}\right)}{1 + \left(-\dfrac{3}{5}\right)}} = -\sqrt{\frac{\dfrac{8}{5}}{\dfrac{2}{5}}} = -2$$

Another way to solve Example 7(c) is to use the results of parts (a) and (b).

$$\tan \frac{\alpha}{2} = \frac{\sin \dfrac{\alpha}{2}}{\cos \dfrac{\alpha}{2}} = \frac{\dfrac{2\sqrt{5}}{5}}{-\dfrac{\sqrt{5}}{5}} = -2$$

✏️ **Now Work** PROBLEMS 9(C) AND (D)

There is a formula for $\tan \dfrac{\alpha}{2}$ that does not contain $+$ and $-$ signs, making it more useful than formula (12). To derive it, use the formulas

$$1 - \cos \alpha = 2 \sin^2 \frac{\alpha}{2} \quad \textcolor{blue}{\text{Formula (9)}}$$

and

$$\sin \alpha = \sin\left[2\left(\frac{\alpha}{2}\right)\right] = 2 \sin \frac{\alpha}{2} \cos \frac{\alpha}{2} \quad \textcolor{blue}{\text{Double-angle Formula (1)}}$$

Then

$$\frac{1 - \cos \alpha}{\sin \alpha} = \frac{2 \sin^2 \dfrac{\alpha}{2}}{2 \sin \dfrac{\alpha}{2} \cos \dfrac{\alpha}{2}} = \frac{\sin \dfrac{\alpha}{2}}{\cos \dfrac{\alpha}{2}} = \tan \frac{\alpha}{2}$$

Because it also can be shown that

$$\frac{1 - \cos \alpha}{\sin \alpha} = \frac{\sin \alpha}{1 + \cos \alpha}$$

this results in the following two Half-angle Formulas:

Half-angle Formulas for $\tan \dfrac{\alpha}{2}$

$$\tan \frac{\alpha}{2} = \frac{1 - \cos \alpha}{\sin \alpha} = \frac{\sin \alpha}{1 + \cos \alpha} \qquad (13)$$

With this formula, the solution to Example 7(c) can be obtained as follows:

$$\cos \alpha = -\frac{3}{5} \quad \pi < \alpha < \frac{3\pi}{2}$$

$$\sin \alpha = -\sqrt{1 - \cos^2 \alpha} = -\sqrt{1 - \frac{9}{25}} = -\sqrt{\frac{16}{25}} = -\frac{4}{5}$$

Then, by equation (13),

$$\tan \frac{\alpha}{2} = \frac{1 - \cos \alpha}{\sin \alpha} = \frac{1 - \left(-\dfrac{3}{5}\right)}{-\dfrac{4}{5}} = \frac{\dfrac{8}{5}}{-\dfrac{4}{5}} = -2$$

8.6 Assess Your Understanding

Concepts and Vocabulary

1. $\cos(2\theta) = \cos^2\theta - \underline{\hspace{1cm}} = \underline{\hspace{1cm}} - 1 = 1 - \underline{\hspace{1cm}}$.

2. $\sin^2\dfrac{\theta}{2} = \dfrac{\underline{\hspace{1cm}}}{2}$.

3. $\tan\dfrac{\theta}{2} = \dfrac{1 - \cos\theta}{\underline{\hspace{1cm}}}$.

4. *True or False* $\tan(2\theta) = \dfrac{2\tan\theta}{1 - \tan^2\theta}$

5. *True or False* $\sin(2\theta)$ has two equivalent forms:
$$2\sin\theta\cos\theta \quad \text{and} \quad \sin^2\theta - \cos^2\theta$$

6. *True or False* $\tan(2\theta) + \tan(2\theta) = \tan(4\theta)$

7. Choose the expression that completes the Half-angle Formula for cosine functions: $\cos\dfrac{\alpha}{2} = \underline{\hspace{1cm}}$

(a) $\pm\sqrt{\dfrac{1 - \cos\alpha}{2}}$ (b) $\pm\sqrt{\dfrac{1 + \cos\alpha}{2}}$

(c) $\pm\sqrt{\dfrac{\cos\alpha - \sin\alpha}{2}}$ (d) $\pm\sqrt{\dfrac{1 - \cos\alpha}{1 + \cos\alpha}}$

8. If $\sin\alpha = \pm\sqrt{\dfrac{1 - \cos\theta}{2}}$, then which of the following describes how the value of θ is related to the value of α?

(a) $\theta = \alpha$ (b) $\theta = \dfrac{\alpha}{2}$ (c) $\theta = 2\alpha$ (d) $\theta = \alpha^2$

Skill Building

In Problems 9–20, use the information given about the angle θ, $0 \le \theta < 2\pi$, to find the exact value of:

(a) $\sin(2\theta)$ (b) $\cos(2\theta)$ (c) $\sin\dfrac{\theta}{2}$ (d) $\cos\dfrac{\theta}{2}$

9. $\sin\theta = \dfrac{3}{5}$, $0 < \theta < \dfrac{\pi}{2}$

10. $\cos\theta = \dfrac{3}{5}$, $0 < \theta < \dfrac{\pi}{2}$

11. $\tan\theta = \dfrac{4}{3}$, $\pi < \theta < \dfrac{3\pi}{2}$

12. $\tan\theta = \dfrac{1}{2}$, $\pi < \theta < \dfrac{3\pi}{2}$

13. $\cos\theta = -\dfrac{\sqrt{6}}{3}$, $\dfrac{\pi}{2} < \theta < \pi$

14. $\sin\theta = -\dfrac{\sqrt{3}}{3}$, $\dfrac{3\pi}{2} < \theta < 2\pi$

15. $\sec\theta = 3$, $\sin\theta > 0$

16. $\csc\theta = -\sqrt{5}$, $\cos\theta < 0$

17. $\cot\theta = -2$, $\sec\theta < 0$

18. $\sec\theta = 2$, $\csc\theta < 0$

19. $\tan\theta = -3$, $\sin\theta < 0$

20. $\cot\theta = 3$, $\cos\theta < 0$

In Problems 21–30, use the Half-angle Formulas to find the exact value of each expression.

21. $\sin 22.5°$

22. $\cos 22.5°$

23. $\tan\dfrac{7\pi}{8}$

24. $\tan\dfrac{9\pi}{8}$

25. $\cos 165°$

26. $\sin 195°$

27. $\sec\dfrac{15\pi}{8}$

28. $\csc\dfrac{7\pi}{8}$

29. $\sin\left(-\dfrac{\pi}{8}\right)$

30. $\cos\left(-\dfrac{3\pi}{8}\right)$

In Problems 31–42, use the figures to evaluate each function, given that $f(x) = \sin x$, $g(x) = \cos x$, and $h(x) = \tan x$.

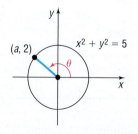

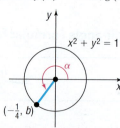

31. $f(2\theta)$

32. $g(2\theta)$

33. $g\left(\dfrac{\theta}{2}\right)$

34. $f\left(\dfrac{\theta}{2}\right)$

35. $h(2\theta)$

36. $h\left(\dfrac{\theta}{2}\right)$

37. $g(2\alpha)$

38. $f(2\alpha)$

39. $f\left(\dfrac{\alpha}{2}\right)$

40. $g\left(\dfrac{\alpha}{2}\right)$

41. $h\left(\dfrac{\alpha}{2}\right)$

42. $h(2\alpha)$

43. Show that $\sin^4\theta = \dfrac{3}{8} - \dfrac{1}{2}\cos(2\theta) + \dfrac{1}{8}\cos(4\theta)$.

44. Show that $\sin(4\theta) = (\cos\theta)(4\sin\theta - 8\sin^3\theta)$.

45. Develop a formula for $\cos(3\theta)$ as a third-degree polynomial in the variable $\cos\theta$.

46. Develop a formula for $\cos(4\theta)$ as a fourth-degree polynomial in the variable $\cos\theta$.

47. Find an expression for $\sin(5\theta)$ as a fifth-degree polynomial in the variable $\sin\theta$.

48. Find an expression for $\cos(5\theta)$ as a fifth-degree polynomial in the variable $\cos\theta$.

In Problems 49–70, establish each identity.

49. $\cos^4\theta - \sin^4\theta = \cos(2\theta)$

50. $\dfrac{\cot\theta - \tan\theta}{\cot\theta + \tan\theta} = \cos(2\theta)$

51. $\cot(2\theta) = \dfrac{\cot^2\theta - 1}{2\cot\theta}$

52. $\cot(2\theta) = \dfrac{1}{2}(\cot\theta - \tan\theta)$

53. $\sec(2\theta) = \dfrac{\sec^2\theta}{2 - \sec^2\theta}$

54. $\csc(2\theta) = \dfrac{1}{2}\sec\theta\csc\theta$

55. $\cos^2(2u) - \sin^2(2u) = \cos(4u)$

56. $(4\sin u\cos u)(1 - 2\sin^2 u) = \sin(4u)$

57. $\dfrac{\cos(2\theta)}{1 + \sin(2\theta)} = \dfrac{\cot\theta - 1}{\cot\theta + 1}$

58. $\sin^2\theta\cos^2\theta = \dfrac{1}{8}[1 - \cos(4\theta)]$

59. $\sec^2\dfrac{\theta}{2} = \dfrac{2}{1 + \cos\theta}$

60. $\csc^2\dfrac{\theta}{2} = \dfrac{2}{1 - \cos\theta}$

61. $\cot^2\dfrac{v}{2} = \dfrac{\sec v + 1}{\sec v - 1}$

62. $\tan\dfrac{v}{2} = \csc v - \cot v$

63. $\cos\theta = \dfrac{1 - \tan^2\dfrac{\theta}{2}}{1 + \tan^2\dfrac{\theta}{2}}$

64. $1 - \dfrac{1}{2}\sin(2\theta) = \dfrac{\sin^3\theta + \cos^3\theta}{\sin\theta + \cos\theta}$

65. $\dfrac{\sin(3\theta)}{\sin\theta} - \dfrac{\cos(3\theta)}{\cos\theta} = 2$

66. $\dfrac{\cos\theta + \sin\theta}{\cos\theta - \sin\theta} - \dfrac{\cos\theta - \sin\theta}{\cos\theta + \sin\theta} = 2\tan(2\theta)$

67. $\tan(3\theta) = \dfrac{3\tan\theta - \tan^3\theta}{1 - 3\tan^2\theta}$

68. $\tan\theta + \tan(\theta + 120°) + \tan(\theta + 240°) = 3\tan(3\theta)$

69. $\ln|\sin\theta| = \dfrac{1}{2}(\ln|1 - \cos(2\theta)| - \ln 2)$

70. $\ln|\cos\theta| = \dfrac{1}{2}(\ln|1 + \cos(2\theta)| - \ln 2)$

In Problems 71–80, solve each equation on the interval $0 \le \theta < 2\pi$.

71. $\cos(2\theta) + 6\sin^2\theta = 4$

72. $\cos(2\theta) = 2 - 2\sin^2\theta$

73. $\cos(2\theta) = \cos\theta$

74. $\sin(2\theta) = \cos\theta$

75. $\sin(2\theta) + \sin(4\theta) = 0$

76. $\cos(2\theta) + \cos(4\theta) = 0$

77. $3 - \sin\theta = \cos(2\theta)$

78. $\cos(2\theta) + 5\cos\theta + 3 = 0$

79. $\tan(2\theta) + 2\sin\theta = 0$

80. $\tan(2\theta) + 2\cos\theta = 0$

Mixed Practice

In Problems 81–92, find the exact value of each expression.

81. $\sin\left(2\sin^{-1}\dfrac{1}{2}\right)$

82. $\sin\left[2\sin^{-1}\dfrac{\sqrt{3}}{2}\right]$

83. $\cos\left(2\sin^{-1}\dfrac{3}{5}\right)$

84. $\cos\left(2\cos^{-1}\dfrac{4}{5}\right)$

85. $\tan\left[2\cos^{-1}\left(-\dfrac{3}{5}\right)\right]$

86. $\tan\left(2\tan^{-1}\dfrac{3}{4}\right)$

87. $\sin\left(2\cos^{-1}\dfrac{4}{5}\right)$

88. $\cos\left[2\tan^{-1}\left(-\dfrac{4}{3}\right)\right]$

89. $\sin^2\left(\dfrac{1}{2}\cos^{-1}\dfrac{3}{5}\right)$

90. $\cos^2\left(\dfrac{1}{2}\sin^{-1}\dfrac{3}{5}\right)$

91. $\sec\left(2\tan^{-1}\dfrac{3}{4}\right)$

92. $\csc\left[2\sin^{-1}\left(-\dfrac{3}{5}\right)\right]$

In Problems 93–95, find the real zeros of each trigonometric function on the interval $0 \le \theta < 2\pi$.

93. $f(x) = \sin(2x) - \sin x$

94. $f(x) = \cos(2x) + \cos x$

95. $f(x) = \cos(2x) + \sin^2 x$

Applications and Extensions

96. Constructing a Rain Gutter A rain gutter is to be constructed of aluminum sheets 12 inches wide. After marking off a length of 4 inches from each edge, the builder bends this length up at an angle θ. See the illustration. The area A of the opening as a function of θ is given by

$$A(\theta) = 16\sin\theta(\cos\theta + 1) \quad 0° < \theta < 90°$$

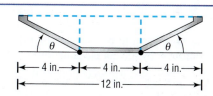

(a) In calculus, you will be asked to find the angle θ that maximizes A by solving the equation

$$\cos(2\theta) + \cos\theta = 0, \quad 0° < \theta < 90°$$

Solve this equation for θ.

(b) What is the maximum area A of the opening?

(c) Graph $A = A(\theta), 0° \le \theta \le 90°$, and find the angle θ that maximizes the area A. Also find the maximum area.

97. Laser Projection In a laser projection system, the **optical angle** or **scanning angle** θ is related to the throw distance D from the scanner to the screen and the projected image width W by the equation

$$D = \frac{\frac{1}{2}W}{\csc\theta - \cot\theta}$$

(a) Show that the projected image width is given by

$$W = 2D\tan\frac{\theta}{2}$$

(b) Find the optical angle if the throw distance is 15 feet and the projected image width is 6.5 feet.

Source: Pangolin Laser Systems, Inc.

98. Product of Inertia The **product of inertia** for an area about inclined axes is given by the formula

$$I_{uv} = I_x\sin\theta\cos\theta - I_y\sin\theta\cos\theta + I_{xy}(\cos^2\theta - \sin^2\theta)$$

Show that this is equivalent to

$$I_{uv} = \frac{I_x - I_y}{2}\sin(2\theta) + I_{xy}\cos(2\theta)$$

Source: Adapted from Hibbeler, *Engineering Mechanics: Statics,* 13th ed., Pearson © 2013.

99. Projectile Motion An object is propelled upward at an angle $\theta, 45° < \theta < 90°$, to the horizontal with an initial velocity of v_0 feet per second from the base of a plane that makes an angle of 45° with the horizontal. See the illustration. If air resistance is ignored, the distance R that it travels up the inclined plane is given by the function

$$R(\theta) = \frac{v_0^2\sqrt{2}}{16}\cos\theta(\sin\theta - \cos\theta)$$

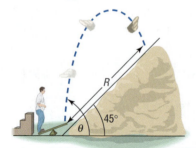

(a) Show that

$$R(\theta) = \frac{v_0^2\sqrt{2}}{32}[\sin(2\theta) - \cos(2\theta) - 1]$$

(b) In calculus, you will be asked to find the angle θ that maximizes R by solving the equation

$$\sin(2\theta) + \cos(2\theta) = 0$$

Solve this equation for θ.

(c) What is the maximum distance R if $v_0 = 32$ feet per second?

(d) Graph $R = R(\theta), 45° \le \theta \le 90°$, and find the angle θ that maximizes the distance R. Also find the maximum distance. Use $v_0 = 32$ feet per second. Compare the results with the answers found in parts (b) and (c).

100. Sawtooth Curve An oscilloscope often displays a sawtooth curve. This curve can be approximated by sinusoidal curves of varying periods and amplitudes. A first approximation to the sawtooth curve is given by

$$y = \frac{1}{2}\sin(2\pi x) + \frac{1}{4}\sin(4\pi x)$$

Show that $y = \sin(2\pi x)\cos^2(\pi x)$.

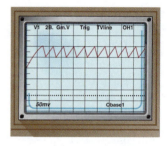

101. Area of an Isosceles Triangle Show that the area A of an isosceles triangle whose equal sides are of length s, and where θ is the angle between them, is

$$A = \frac{1}{2}s^2\sin\theta$$

[Hint: See the illustration. The height h bisects the angle θ and is the perpendicular bisector of the base.]

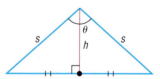

102. Geometry A rectangle is inscribed in a semicircle of radius 1. See the illustration.

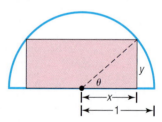

(a) Express the area A of the rectangle as a function of the angle θ shown in the illustration.

(b) Show that $A(\theta) = \sin(2\theta)$.

(c) Find the angle θ that results in the largest area A.

(d) Find the dimensions of this largest rectangle.

103. If $x = 2\tan\theta$, express $\sin(2\theta)$ as a function of x.

104. If $x = 2\tan\theta$, express $\cos(2\theta)$ as a function of x.

105. Find the value of the number C:

$$\frac{1}{2}\sin^2 x + C = -\frac{1}{4}\cos(2x)$$

106. Find the value of the number C:

$$\frac{1}{2}\cos^2 x + C = \frac{1}{4}\cos(2x)$$

△ **107.** If $z = \tan\dfrac{\alpha}{2}$, show that $\sin\alpha = \dfrac{2z}{1+z^2}$.

△ **108.** If $z = \tan\dfrac{\alpha}{2}$, show that $\cos\alpha = \dfrac{1-z^2}{1+z^2}$.

109. Graph $f(x) = \sin^2 x = \dfrac{1-\cos(2x)}{2}$ for $0 \le x \le 2\pi$ by using transformations.

110. Repeat Problem 109 for $g(x) = \cos^2 x$.

111. Use the fact that
$$\cos\frac{\pi}{12} = \frac{1}{4}\left(\sqrt{6}+\sqrt{2}\right)$$
to find $\sin\dfrac{\pi}{24}$ and $\cos\dfrac{\pi}{24}$.

112. Show that
$$\cos\frac{\pi}{8} = \frac{\sqrt{2+\sqrt{2}}}{2}$$
and use it to find $\sin\dfrac{\pi}{16}$ and $\cos\dfrac{\pi}{16}$.

113. Show that
$$\sin^3\theta + \sin^3(\theta + 120°) + \sin^3(\theta + 240°) = -\frac{3}{4}\sin(3\theta)$$

114. If $\tan\theta = a\tan\dfrac{\theta}{3}$, express $\tan\dfrac{\theta}{3}$ in terms of a.

115. For $\cos(2x) + (2m-1)\sin x + m - 1 = 0$, find m such that there is exactly one real solution for x, $-\dfrac{\pi}{2} \le x \le \dfrac{\pi}{2}$.[†]

[†]Courtesy of Joliet Junior College Mathematics Department

Explaining Concepts: Discussion and Writing

116. Go to the library and research Chebyshëv polynomials. Write a report on your findings.

Retain Your Knowledge

Problems 117–120 are based on material learned earlier in the course. The purpose of these problems is to keep the material fresh in your mind so that you are better prepared for the final exam.

117. Find an equation of the line that contains the point $(2, -3)$ and is perpendicular to the line $y = -2x + 9$.

118. Graph $f(x) = -x^2 + 6x + 7$. Label the vertex and any intercepts.

119. Find the exact value of $\sin\left(\dfrac{2\pi}{3}\right) - \cos\left(\dfrac{4\pi}{3}\right)$.

120. Graph $y = -2\cos\left(\dfrac{\pi}{2}x\right)$. Show at least two periods.

8.7 Product-to-Sum and Sum-to-Product Formulas

OBJECTIVES **1** Express Products as Sums (p. 665)
2 Express Sums as Products (p. 666)

1 Express Products as Sums

Sum and difference formulas can be used to derive formulas for writing the products of sines and/or cosines as sums or differences. These identities are usually called the **Product-to-Sum Formulas**.

THEOREM

Product-to-Sum Formulas

$$\sin\alpha\sin\beta = \frac{1}{2}\left[\cos(\alpha-\beta) - \cos(\alpha+\beta)\right] \qquad \textbf{(1)}$$

$$\cos\alpha\cos\beta = \frac{1}{2}\left[\cos(\alpha-\beta) + \cos(\alpha+\beta)\right] \qquad \textbf{(2)}$$

$$\sin\alpha\cos\beta = \frac{1}{2}\left[\sin(\alpha+\beta) + \sin(\alpha-\beta)\right] \qquad \textbf{(3)}$$

These formulas do not have to be memorized. Instead, remember how they are derived. Then, when you want to use them, either look them up or derive them, as needed.

To derive formulas (1) and (2), write down the sum and difference formulas for the cosine:

$$\cos(\alpha - \beta) = \cos \alpha \cos \beta + \sin \alpha \sin \beta \qquad (4)$$

$$\cos(\alpha + \beta) = \cos \alpha \cos \beta - \sin \alpha \sin \beta \qquad (5)$$

To derive formula (1), subtract equation (5) from equation (4) to get

$$\cos(\alpha - \beta) - \cos(\alpha + \beta) = 2 \sin \alpha \sin \beta$$

from which

$$\sin \alpha \sin \beta = \frac{1}{2}[\cos(\alpha - \beta) - \cos(\alpha + \beta)]$$

To derive formula (2), add equations (4) and (5) to get

$$\cos(\alpha - \beta) + \cos(\alpha + \beta) = 2 \cos \alpha \cos \beta$$

from which

$$\cos \alpha \cos \beta = \frac{1}{2}[\cos(\alpha - \beta) + \cos(\alpha + \beta)]$$

To derive Product-to-Sum Formula (3), use the sum and difference formulas for sine in a similar way. (You are asked to do this in Problem 53.)

EXAMPLE 1

Expressing Products as Sums

Express each of the following products as a sum containing only sines or only cosines.

(a) $\sin(6\theta) \sin(4\theta)$ (b) $\cos(3\theta) \cos \theta$ (c) $\sin(3\theta) \cos(5\theta)$

Solution   (a) Use formula (1) to get

$$\sin(6\theta) \sin(4\theta) = \frac{1}{2}[\cos(6\theta - 4\theta) - \cos(6\theta + 4\theta)]$$

$$= \frac{1}{2}[\cos(2\theta) - \cos(10\theta)]$$

(b) Use formula (2) to get

$$\cos(3\theta) \cos \theta = \frac{1}{2}[\cos(3\theta - \theta) + \cos(3\theta + \theta)]$$

$$= \frac{1}{2}[\cos(2\theta) + \cos(4\theta)]$$

(c) Use formula (3) to get

$$\sin(3\theta) \cos(5\theta) = \frac{1}{2}[\sin(3\theta + 5\theta) + \sin(3\theta - 5\theta)]$$

$$= \frac{1}{2}[\sin(8\theta) + \sin(-2\theta)] = \frac{1}{2}[\sin(8\theta) - \sin(2\theta)] \quad \bullet$$

Now Work PROBLEM 7

2 Express Sums as Products

The **Sum-to-Product Formulas** are given next.

THEOREM

Sum-to-Product Formulas

$$\sin \alpha + \sin \beta = 2 \sin \frac{\alpha + \beta}{2} \cos \frac{\alpha - \beta}{2} \qquad (6)$$

$$\sin \alpha - \sin \beta = 2 \sin \frac{\alpha - \beta}{2} \cos \frac{\alpha + \beta}{2} \qquad (7)$$

$$\cos \alpha + \cos \beta = 2 \cos \frac{\alpha + \beta}{2} \cos \frac{\alpha - \beta}{2} \qquad (8)$$

$$\cos \alpha - \cos \beta = -2 \sin \frac{\alpha + \beta}{2} \sin \frac{\alpha - \beta}{2} \qquad (9)$$

Formula (6) is derived here. The derivations of formulas (7) through (9) are left as exercises (see Problems 54 through 56).

Proof

$$2 \sin \frac{\alpha + \beta}{2} \cos \frac{\alpha - \beta}{2} = 2 \cdot \frac{1}{2} \left[\sin\left(\frac{\alpha + \beta}{2} + \frac{\alpha - \beta}{2} \right) + \sin\left(\frac{\alpha + \beta}{2} - \frac{\alpha - \beta}{2} \right) \right]$$

↑
Product-to-Sum Formula (3)

$$= \sin \frac{2\alpha}{2} + \sin \frac{2\beta}{2} = \sin \alpha + \sin \beta \qquad ■$$

| EXAMPLE 2 | **Expressing Sums (or Differences) as Products** |

Express each sum or difference as a product of sines and/or cosines.

(a) $\sin(5\theta) - \sin(3\theta)$ (b) $\cos(3\theta) + \cos(2\theta)$

Solution   (a) Use formula (7) to get

$$\sin(5\theta) - \sin(3\theta) = 2 \sin \frac{5\theta - 3\theta}{2} \cos \frac{5\theta + 3\theta}{2}$$

$$= 2 \sin \theta \cos(4\theta)$$

(b) $\cos(3\theta) + \cos(2\theta) = 2 \cos \dfrac{3\theta + 2\theta}{2} \cos \dfrac{3\theta - 2\theta}{2}$ Formula (8)

$$= 2 \cos \frac{5\theta}{2} \cos \frac{\theta}{2}$$

Now Work PROBLEM 17

8.7 Assess Your Understanding

Skill Building

In Problems 1–6, find the exact value of each expression.

1. $\sin 195° \cdot \cos 75°$

2. $\cos 285° \cdot \cos 195°$

3. $\sin 285° \cdot \sin 75°$

4. $\sin 75° + \sin 15°$

5. $\cos 255° - \cos 195°$

6. $\sin 255° - \sin 15°$

In Problems 7–16, express each product as a sum containing only sines or only cosines.

7. $\sin(4\theta) \sin(2\theta)$ **8.** $\cos(4\theta) \cos(2\theta)$ **9.** $\sin(4\theta) \cos(2\theta)$ **10.** $\sin(3\theta) \sin(5\theta)$ **11.** $\cos(3\theta) \cos(5\theta)$

12. $\sin(4\theta) \cos(6\theta)$ **13.** $\sin \theta \sin(2\theta)$ **14.** $\cos(3\theta) \cos(4\theta)$ **15.** $\sin \dfrac{3\theta}{2} \cos \dfrac{\theta}{2}$ **16.** $\sin \dfrac{\theta}{2} \cos \dfrac{5\theta}{2}$

In Problems 17–24, express each sum or difference as a product of sines and/or cosines.

17. $\sin(4\theta) - \sin(2\theta)$ **18.** $\sin(4\theta) + \sin(2\theta)$ **19.** $\cos(2\theta) + \cos(4\theta)$ **20.** $\cos(5\theta) - \cos(3\theta)$

21. $\sin \theta + \sin(3\theta)$ **22.** $\cos \theta + \cos(3\theta)$ **23.** $\cos \dfrac{\theta}{2} - \cos \dfrac{3\theta}{2}$ **24.** $\sin \dfrac{\theta}{2} - \sin \dfrac{3\theta}{2}$

In Problems 25–42, establish each identity.

25. $\dfrac{\sin \theta + \sin(3\theta)}{2 \sin(2\theta)} = \cos \theta$

26. $\dfrac{\cos \theta + \cos(3\theta)}{2 \cos(2\theta)} = \cos \theta$

27. $\dfrac{\sin(4\theta) + \sin(2\theta)}{\cos(4\theta) + \cos(2\theta)} = \tan(3\theta)$

28. $\dfrac{\cos \theta - \cos(3\theta)}{\sin(3\theta) - \sin \theta} = \tan(2\theta)$

29. $\dfrac{\cos \theta - \cos(3\theta)}{\sin \theta + \sin(3\theta)} = \tan \theta$

30. $\dfrac{\cos \theta - \cos(5\theta)}{\sin \theta + \sin(5\theta)} = \tan(2\theta)$

31. $\sin \theta [\sin \theta + \sin(3\theta)] = \cos \theta [\cos \theta - \cos(3\theta)]$

32. $\sin \theta [\sin(3\theta) + \sin(5\theta)] = \cos \theta [\cos(3\theta) - \cos(5\theta)]$

33. $\dfrac{\sin(4\theta) + \sin(8\theta)}{\cos(4\theta) + \cos(8\theta)} = \tan(6\theta)$

34. $\dfrac{\sin(4\theta) - \sin(8\theta)}{\cos(4\theta) - \cos(8\theta)} = -\cot(6\theta)$

35. $\dfrac{\sin(4\theta) + \sin(8\theta)}{\sin(4\theta) - \sin(8\theta)} = -\dfrac{\tan(6\theta)}{\tan(2\theta)}$

36. $\dfrac{\cos(4\theta) - \cos(8\theta)}{\cos(4\theta) + \cos(8\theta)} = \tan(2\theta) \tan(6\theta)$

37. $\dfrac{\sin \alpha + \sin \beta}{\sin \alpha - \sin \beta} = \tan \dfrac{\alpha + \beta}{2} \cot \dfrac{\alpha - \beta}{2}$

38. $\dfrac{\cos \alpha + \cos \beta}{\cos \alpha - \cos \beta} = -\cot \dfrac{\alpha + \beta}{2} \cot \dfrac{\alpha - \beta}{2}$

39. $\dfrac{\sin \alpha + \sin \beta}{\cos \alpha + \cos \beta} = \tan \dfrac{\alpha + \beta}{2}$

40. $\dfrac{\sin \alpha - \sin \beta}{\cos \alpha - \cos \beta} = -\cot \dfrac{\alpha + \beta}{2}$

41. $1 + \cos(2\theta) + \cos(4\theta) + \cos(6\theta) = 4\cos\theta\cos(2\theta)\cos(3\theta)$

42. $1 - \cos(2\theta) + \cos(4\theta) - \cos(6\theta) = 4\sin\theta\cos(2\theta)\sin(3\theta)$

In Problems 43–46, solve each equation on the interval $0 \le \theta < 2\pi$.

43. $\sin(2\theta) + \sin(4\theta) = 0$

44. $\cos(2\theta) + \cos(4\theta) = 0$

45. $\cos(4\theta) - \cos(6\theta) = 0$

46. $\sin(4\theta) - \sin(6\theta) = 0$

Applications and Extensions

47. Touch-Tone Phones On a Touch-Tone phone, each button produces a unique sound. The sound produced is the sum of two tones, given by

$$y = \sin(2\pi l t) \quad \text{and} \quad y = \sin(2\pi h t)$$

where l and h are the low and high frequencies (cycles per second) shown on the illustration. For example, if you touch 7, the low frequency is $l = 852$ cycles per second and the high frequency is $h = 1209$ cycles per second. The sound emitted when you touch 7 is

$$y = \sin[2\pi(852)t] + \sin[2\pi(1209)t]$$

Touch-Tone phone

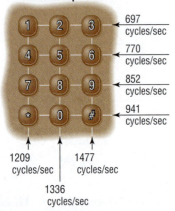

697 cycles/sec
770 cycles/sec
852 cycles/sec
941 cycles/sec

1209 cycles/sec   1477 cycles/sec
1336 cycles/sec

(a) Write this sound as a product of sines and/or cosines.
(b) Determine the maximum value of y.
(c) Graph the sound emitted when 7 is touched.

48. Touch-Tone Phones
(a) Write, as a product of sines and/or cosines, the sound emitted when the # key is touched.
(b) Determine the maximum value of y.
(c) Graph the sound emitted when the # key is touched.

49. Moment of Inertia The moment of inertia I of an object is a measure of how easy it is to rotate the object about some fixed point. In engineering mechanics, it is sometimes

necessary to compute moments of inertia with respect to a set of rotated axes. These moments are given by the equations

$$I_u = I_x \cos^2\theta + I_y \sin^2\theta - 2I_{xy}\sin\theta\cos\theta$$
$$I_v = I_x \sin^2\theta + I_y \cos^2\theta + 2I_{xy}\sin\theta\cos\theta$$

Use Product-to-Sum Formulas to show that

$$I_u = \frac{I_x + I_y}{2} + \frac{I_x - I_y}{2}\cos(2\theta) - I_{xy}\sin(2\theta)$$

and

$$I_v = \frac{I_x + I_y}{2} - \frac{I_x - I_y}{2}\cos(2\theta) + I_{xy}\sin(2\theta)$$

Source: Adapted from Hibbeler, *Engineering Mechanics: Statics,* 13th ed., Pearson © 2013.

50. Projectile Motion The range R of a projectile propelled downward from the top of an inclined plane at an angle θ to the inclined plane is given by

$$R(\theta) = \frac{2v_0^2 \sin\theta \cos(\theta - \phi)}{g \cos^2\phi}$$

where v_0 is the initial velocity of the projectile, ϕ is the angle the plane makes with respect to the horizontal, and g is acceleration due to gravity.

(a) Show that for fixed v_0 and ϕ, the maximum range down the incline is given by $R_{\max} = \dfrac{v_0^2}{g(1 - \sin\phi)}$.

(b) Determine the maximum range if the projectile has an initial velocity of 50 meters/second, the angle of the plane is $\phi = 35°$, and $g = 9.8$ meters/second^2.

51. If $\alpha + \beta + \gamma = \pi$, show that

$$\sin(2\alpha) + \sin(2\beta) + \sin(2\gamma) = 4\sin\alpha\sin\beta\sin\gamma$$

52. If $\alpha + \beta + \gamma = \pi$, show that

$$\tan\alpha + \tan\beta + \tan\gamma = \tan\alpha\tan\beta\tan\gamma$$

53. Derive formula (3).   **54.** Derive formula (7).

55. Derive formula (8).   **56.** Derive formula (9).

Retain Your Knowledge

Problems 57–60 are based on material learned earlier in the course. The purpose of these problems is to keep the material fresh in your mind so that you are better prepared for the final exam.

57. Solve: $27^{x-1} = 9^{x+5}$

58. For $y = 5\cos(4x - \pi)$, find the amplitude, the period, and the phase shift.

59. Find the exact value of $\cos\left(\csc^{-1}\dfrac{7}{5}\right)$.

60. Find the inverse function f^{-1} of $f(x) = 3\sin x - 5$, $-\dfrac{\pi}{2} \le x \le \dfrac{\pi}{2}$. Find the range of f and the domain and range of f^{-1}.

Chapter Review

Things to Know

Definitions of the six inverse trigonometric functions

$y = \sin^{-1} x$ means $x = \sin y$ where $-1 \le x \le 1$, $-\dfrac{\pi}{2} \le y \le \dfrac{\pi}{2}$ (p. 608)

$y = \cos^{-1} x$ means $x = \cos y$ where $-1 \le x \le 1$, $0 \le y \le \pi$ (p. 611)

$y = \tan^{-1} x$ means $x = \tan y$ where $-\infty < x < \infty$, $-\dfrac{\pi}{2} < y < \dfrac{\pi}{2}$ (p. 614)

$y = \sec^{-1} x$ means $x = \sec y$ where $|x| \ge 1$, $0 \le y \le \pi$, $y \ne \dfrac{\pi}{2}$ (p. 621)

$y = \csc^{-1} x$ means $x = \csc y$ where $|x| \ge 1$, $-\dfrac{\pi}{2} \le y \le \dfrac{\pi}{2}$, $y \ne 0$ (p. 621)

$y = \cot^{-1} x$ means $x = \cot y$ where $-\infty < x < \infty$, $0 < y < \pi$ (p. 621)

Sum and Difference Formulas (pp. 643, 646, and 648)

$$\cos(\alpha + \beta) = \cos \alpha \cos \beta - \sin \alpha \sin \beta \qquad \cos(\alpha - \beta) = \cos \alpha \cos \beta + \sin \alpha \sin \beta$$

$$\sin(\alpha + \beta) = \sin \alpha \cos \beta + \cos \alpha \sin \beta \qquad \sin(\alpha - \beta) = \sin \alpha \cos \beta - \cos \alpha \sin \beta$$

$$\tan(\alpha + \beta) = \frac{\tan \alpha + \tan \beta}{1 - \tan \alpha \tan \beta} \qquad \tan(\alpha - \beta) = \frac{\tan \alpha - \tan \beta}{1 + \tan \alpha \tan \beta}$$

Double-angle Formulas (pp. 656 and 657)

$$\sin(2\theta) = 2 \sin \theta \cos \theta \qquad \cos(2\theta) = \cos^2 \theta - \sin^2 \theta \qquad \tan(2\theta) = \frac{2 \tan \theta}{1 - \tan^2 \theta}$$

$$\cos(2\theta) = 2 \cos^2 \theta - 1 \qquad \cos(2\theta) = 1 - 2 \sin^2 \theta$$

Half-angle Formulas (pp. 657, 659, and 661)

$$\sin^2 \frac{\alpha}{2} = \frac{1 - \cos \alpha}{2} \qquad \cos^2 \frac{\alpha}{2} = \frac{1 + \cos \alpha}{2} \qquad \tan^2 \frac{\alpha}{2} = \frac{1 - \cos \alpha}{1 + \cos \alpha}$$

$$\sin \frac{\alpha}{2} = \pm \sqrt{\frac{1 - \cos \alpha}{2}} \qquad \cos \frac{\alpha}{2} = \pm \sqrt{\frac{1 + \cos \alpha}{2}} \qquad \tan \frac{\alpha}{2} = \pm \sqrt{\frac{1 - \cos \alpha}{1 + \cos \alpha}} = \frac{1 - \cos \alpha}{\sin \alpha} = \frac{\sin \alpha}{1 + \cos \alpha}$$

where the $+$ or $-$ sign is determined by the quadrant of $\dfrac{\alpha}{2}$.

Product-to-Sum Formulas (p. 665)

$$\sin \alpha \sin \beta = \frac{1}{2}\left[\cos(\alpha - \beta) - \cos(\alpha + \beta)\right]$$

$$\cos \alpha \cos \beta = \frac{1}{2}\left[\cos(\alpha - \beta) + \cos(\alpha + \beta)\right]$$

$$\sin \alpha \cos \beta = \frac{1}{2}\left[\sin(\alpha + \beta) + \sin(\alpha - \beta)\right]$$

Sum-to-Product Formulas (p. 666)

$$\sin \alpha + \sin \beta = 2 \sin \frac{\alpha + \beta}{2} \cos \frac{\alpha - \beta}{2} \qquad \sin \alpha - \sin \beta = 2 \sin \frac{\alpha - \beta}{2} \cos \frac{\alpha + \beta}{2}$$

$$\cos \alpha + \cos \beta = 2 \cos \frac{\alpha + \beta}{2} \cos \frac{\alpha - \beta}{2} \qquad \cos \alpha - \cos \beta = -2 \sin \frac{\alpha + \beta}{2} \sin \frac{\alpha - \beta}{2}$$

Objectives

Section	You should be able to . . .	Example(s)	Review Exercises
8.1	**1** Find the exact value of an inverse sine function (p. 608)	1, 2, 6, 7, 9	1–6
	2 Find an approximate value of an inverse sine function (p. 609)	3	76–78
	3 Use properties of inverse functions to find exact values of certain composite functions (p. 610)	4, 5, 8	9–17
	4 Find the inverse function of a trigonometric function (p. 615)	10	24, 25
	5 Solve equations involving inverse trigonometric functions (p. 616)	11	84, 85
8.2	**1** Find the exact value of expressions involving the inverse sine, cosine, and tangent functions (p. 620)	1–3	18–21, 23
	2 Define the inverse secant, cosecant, and cotangent functions (p. 621)	4	7, 8, 22
	3 Use a calculator to evaluate $\sec^{-1} x$, $\csc^{-1} x$, and $\cot^{-1} x$ (p. 622)	5	79, 80
	4 Write a trigonometric expression as an algebraic expression (p. 623)	6	26, 27
8.3	**1** Solve equations involving a single trigonometric function (p. 625)	1–5	64–68
	2 Solve trigonometric equations using a calculator (p. 628)	6	69
	3 Solve trigonometric equations quadratic in form (p. 629)	7	72
	4 Solve trigonometric equations using fundamental identities (p. 629)	8, 9	70, 71, 73
	5 Solve trigonometric equations using a graphing utility (p. 630)	10	81–83
8.4	**1** Use algebra to simplify trigonometric expressions (p. 636)	1	28–44
	2 Establish identities (p. 637)	2–8	28–36
8.5	**1** Use sum and difference formulas to find exact values (p. 644)	1–5	45–50, 53–57(a)–(d), 86
	2 Use sum and difference formulas to establish identities (p. 647)	6–8	37, 38
	3 Use sum and difference formulas involving inverse trigonometric functions (p. 649)	9, 10	58–61
	4 Solve trigonometric equations linear in sine and cosine (p. 650)	11, 12	75
8.6	**1** Use Double-angle Formulas to find exact values (p. 656)	1	53–57(e), (f), 62, 63, 87
	2 Use Double-angle Formulas to establish identities (p. 656)	2–5	40, 41, 74
	3 Use Half-angle Formulas to find exact values (p. 659)	6, 7	51, 52, 53–57(g), (h), 86
8.7	**1** Express products as sums (p. 665)	1	42
	2 Express sums as products (p. 666)	2	43, 44

Review Exercises

In Problems 1–8, find the exact value of each expression. Do not use a calculator.

1. $\sin^{-1} 1$ **2.** $\cos^{-1} 0$ **3.** $\tan^{-1} 1$ **4.** $\sin^{-1}\left(-\frac{1}{2}\right)$

5. $\cos^{-1}\left(-\frac{\sqrt{3}}{2}\right)$ **6.** $\tan^{-1}\left(-\sqrt{3}\right)$ **7.** $\sec^{-1}\sqrt{2}$ **8.** $\cot^{-1}(-1)$

In Problems 9–23, find the exact value, if any, of each composite function. If there is no value, say it is "not defined." Do not use a calculator.

9. $\sin^{-1}\left(\sin\frac{3\pi}{8}\right)$ **10.** $\cos^{-1}\left(\cos\frac{3\pi}{4}\right)$ **11.** $\tan^{-1}\left(\tan\frac{2\pi}{3}\right)$ **12.** $\cos^{-1}\left(\cos\frac{15\pi}{7}\right)$

13. $\sin^{-1}\left[\sin\left(-\frac{8\pi}{9}\right)\right]$ **14.** $\sin(\sin^{-1} 0.9)$ **15.** $\cos(\cos^{-1} 0.6)$ **16.** $\tan\left[\tan^{-1} 5\right]$

17. $\cos\left[\cos^{-1}(-1.6)\right]$ **18.** $\sin^{-1}\left(\cos\frac{2\pi}{3}\right)$ **19.** $\cos^{-1}\left(\tan\frac{3\pi}{4}\right)$ **20.** $\tan\left[\sin^{-1}\left(-\frac{\sqrt{3}}{2}\right)\right]$

21. $\sec\left(\tan^{-1}\dfrac{\sqrt{3}}{3}\right)$

22. $\sin\left(\cot^{-1}\dfrac{3}{4}\right)$

23. $\tan\left[\sin^{-1}\left(-\dfrac{4}{5}\right)\right]$

In Problems 24 and 25, find the inverse function f^{-1} of each function f. Find the range of f and the domain and range of f^{-1}.

24. $f(x) = 2\sin(3x)\quad -\dfrac{\pi}{6} \le x \le \dfrac{\pi}{6}$

25. $f(x) = -\cos x + 3\quad 0 \le x \le \pi$

In Problems 26 and 27, write each trigonometric expression as an algebraic expression in u.

26. $\cos(\sin^{-1} u)$

27. $\tan(\csc^{-1} u)$

In Problems 28–44, establish each identity.

28. $\tan\theta\cot\theta - \sin^2\theta = \cos^2\theta$

29. $\sin^2\theta(1 + \cot^2\theta) = 1$

30. $5\cos^2\theta + 3\sin^2\theta = 3 + 2\cos^2\theta$

31. $\dfrac{1 - \cos\theta}{\sin\theta} + \dfrac{\sin\theta}{1 - \cos\theta} = 2\csc\theta$

32. $\dfrac{\cos\theta}{\cos\theta - \sin\theta} = \dfrac{1}{1 - \tan\theta}$

33. $\dfrac{\csc\theta}{1 + \csc\theta} = \dfrac{1 - \sin\theta}{\cos^2\theta}$

34. $\csc\theta - \sin\theta = \cos\theta\cot\theta$

35. $\dfrac{1 - \sin\theta}{\sec\theta} = \dfrac{\cos^3\theta}{1 + \sin\theta}$

36. $\dfrac{1 - 2\sin^2\theta}{\sin\theta\cos\theta} = \cot\theta - \tan\theta$

37. $\dfrac{\cos(\alpha + \beta)}{\cos\alpha\sin\beta} = \cot\beta - \tan\alpha$

38. $\dfrac{\cos(\alpha - \beta)}{\cos\alpha\cos\beta} = 1 + \tan\alpha\tan\beta$

39. $(1 + \cos\theta)\tan\dfrac{\theta}{2} = \sin\theta$

40. $2\cot\theta\cot(2\theta) = \cot^2\theta - 1$

41. $1 - 8\sin^2\theta\cos^2\theta = \cos(4\theta)$

42. $\dfrac{\sin(3\theta)\cos\theta - \sin\theta\cos(3\theta)}{\sin(2\theta)} = 1$

43. $\dfrac{\sin(2\theta) + \sin(4\theta)}{\cos(2\theta) + \cos(4\theta)} = \tan(3\theta)$

44. $\dfrac{\cos(2\theta) - \cos(4\theta)}{\cos(2\theta) + \cos(4\theta)} - \tan\theta\tan(3\theta) = 0$

In Problems 45–52, find the exact value of each expression.

45. $\sin 165°$

46. $\tan 105°$

47. $\cos\dfrac{5\pi}{12}$

48. $\sin\left(-\dfrac{\pi}{12}\right)$

49. $\cos 80° \cos 20° + \sin 80° \sin 20°$

50. $\sin 70° \cos 40° - \cos 70° \sin 40°$

51. $\tan\dfrac{\pi}{8}$

52. $\sin\dfrac{5\pi}{8}$

In Problems 53–57, use the information given about the angles α and β to find the exact value of:

(a) $\sin(\alpha + \beta)$ (b) $\cos(\alpha + \beta)$ (c) $\sin(\alpha - \beta)$ (d) $\tan(\alpha + \beta)$

(e) $\sin(2\alpha)$ (f) $\cos(2\beta)$ (g) $\sin\dfrac{\beta}{2}$ (h) $\cos\dfrac{\alpha}{2}$

53. $\sin\alpha = \dfrac{4}{5}, 0 < \alpha < \dfrac{\pi}{2}; \sin\beta = \dfrac{5}{13}, \dfrac{\pi}{2} < \beta < \pi$

54. $\sin\alpha = -\dfrac{3}{5}, \pi < \alpha < \dfrac{3\pi}{2}; \cos\beta = \dfrac{12}{13}, \dfrac{3\pi}{2} < \beta < 2\pi$

55. $\tan\alpha = \dfrac{3}{4}, \pi < \alpha < \dfrac{3\pi}{2}; \tan\beta = \dfrac{12}{5}, 0 < \beta < \dfrac{\pi}{2}$

56. $\sec\alpha = 2, -\dfrac{\pi}{2} < \alpha < 0; \sec\beta = 3, \dfrac{3\pi}{2} < \beta < 2\pi$

57. $\sin\alpha = -\dfrac{2}{3}, \pi < \alpha < \dfrac{3\pi}{2}; \cos\beta = -\dfrac{2}{3}, \pi < \beta < \dfrac{3\pi}{2}$

In Problems 58–63, find the exact value of each expression.

58. $\cos\left(\sin^{-1}\dfrac{3}{5} - \cos^{-1}\dfrac{1}{2}\right)$

59. $\sin\left(\cos^{-1}\dfrac{5}{13} - \cos^{-1}\dfrac{4}{5}\right)$

60. $\tan\left[\sin^{-1}\left(-\dfrac{1}{2}\right) - \tan^{-1}\dfrac{3}{4}\right]$

61. $\cos\left[\tan^{-1}(-1) + \cos^{-1}\left(-\dfrac{4}{5}\right)\right]$

62. $\sin\left[2\cos^{-1}\left(-\dfrac{3}{5}\right)\right]$

63. $\cos\left(2\tan^{-1}\dfrac{4}{3}\right)$

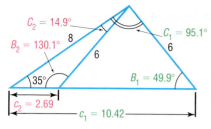

Figure 20(b)

The third side c obeys the Law of Sines, so

$$\frac{\sin A}{a} = \frac{\sin C_1}{c_1} \qquad\qquad \frac{\sin A}{a} = \frac{\sin C_2}{c_2}$$

$$\frac{\sin 35°}{6} = \frac{\sin 95.1°}{c_1} \qquad\qquad \frac{\sin 35°}{6} = \frac{\sin 14.9°}{c_2}$$

$$c_1 = \frac{6 \sin 95.1°}{\sin 35°} \approx 10.42 \qquad c_2 = \frac{6 \sin 14.9°}{\sin 35°} \approx 2.69$$

The two solved triangles are illustrated in Figure 20(b).

Now Work PROBLEMS 25 AND 31

 3 Solve Applied Problems

EXAMPLE 6

Finding the Height of a Mountain

To measure the height of a mountain, a surveyor takes two sightings of the peak at a distance 900 meters apart on a direct line to the mountain.* See Figure 21(a). The first observation results in an angle of elevation of 47°, and the second results in an angle of elevation of 35°. If the transit is 2 meters high, what is the height h of the mountain?

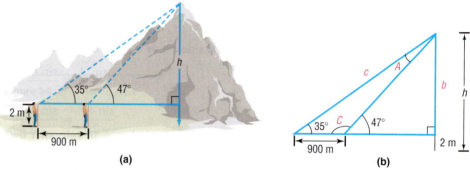

Figure 21 (a) (b)

Solution

Figure 21(b) shows the triangles that replicate the illustration in Figure 21(a). Since $C + 47° = 180°$, this means that $C = 133°$. Also, since $A + C + 35° = 180°$, this means that $A = 180° - 35° - C = 145° - 133° = 12°$. Use the Law of Sines to find c.

$$\frac{\sin A}{a} = \frac{\sin C}{c} \qquad A = 12°, C = 133°, a = 900$$

$$c = \frac{900 \sin 133°}{\sin 12°} \approx 3165.86$$

Using the larger right triangle gives

$$\sin 35° = \frac{b}{c}$$

$$b = 3165.86 \sin 35° \approx 1815.86 \approx 1816 \text{ meters}$$

The height of the peak from ground level is approximately $1816 + 2 = 1818$ meters.

Now Work PROBLEM 37

EXAMPLE 7

Rescue at Sea

Coast Guard Station Zulu is located 120 miles due west of Station X-ray. A ship at sea sends an SOS call that is received by each station. The call to Station Zulu indicates that the bearing of the ship from Zulu is N40°E (40° east of north). The call to Station X-ray indicates that the bearing of the ship from X-ray is N30°W (30° west of north).

(a) How far is each station from the ship?

(b) If a helicopter capable of flying 200 miles per hour is dispatched from the nearest station to the ship, how long will it take to reach the ship?

* For simplicity, assume that these sightings are at the same level.

Solution

(a) Figure 22 illustrates the situation. The angle C is found to be

$$C = 180° - 50° - 60° = 70°$$

The Law of Sines can now be used to find the two distances a and b that are needed.

$$\frac{\sin 50°}{a} = \frac{\sin 70°}{120}$$

$$a = \frac{120 \sin 50°}{\sin 70°} \approx 97.82 \text{ miles}$$

$$\frac{\sin 60°}{b} = \frac{\sin 70°}{120}$$

$$b = \frac{120 \sin 60°}{\sin 70°} \approx 110.59 \text{ miles}$$

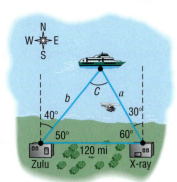

Figure 22

Station Zulu is about 111 miles from the ship, and Station X-ray is about 98 miles from the ship.

(b) The time t needed for the helicopter to reach the ship from Station X-ray is found by using the formula

$$(\text{Rate}, r)(\text{Time}, t) = \text{Distance}, a$$

Then

$$t = \frac{a}{r} = \frac{97.82}{200} \approx 0.49 \text{ hour} \approx 29 \text{ minutes}$$

It will take about 29 minutes for the helicopter to reach the ship. ●

Now Work PROBLEM 47

Proof of the Law of Sines To prove the Law of Sines, construct an altitude of length h from one of the vertices of a triangle. Figure 23(a) shows h for a triangle with three acute angles, and Figure 23(b) shows h for a triangle with an obtuse angle. In each case, the altitude is drawn from the vertex at B. Using either illustration

$$\sin C = \frac{h}{a}$$

from which

$$h = a \sin C \qquad (3)$$

From Figure 23(a), it also follows that

$$\sin A = \frac{h}{c}$$

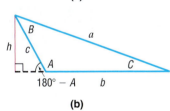

(b)

Figure 23

from which

$$h = c \sin A \qquad (4)$$

From Figure 23(b), it follows that

$$\sin(180° - A) = \sin A = \frac{h}{c}$$

$$\uparrow$$

$$\sin(180° - A) = \sin 180° \cos A - \cos 180° \sin A = \sin A$$

which again gives

$$h = c \sin A$$

Thus, whether the triangle has three acute angles or has two acute angles and one obtuse angle, equations (3) and (4) hold. As a result, the expressions for h in equations (3) and (4) are equal. That is,

$$a \sin C = c \sin A$$

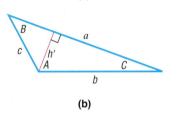

(a)

(b)

Figure 24

from which

$$\frac{\sin A}{a} = \frac{\sin C}{c} \qquad \textbf{(5)}$$

In a similar manner, constructing the altitude h' from the vertex of angle A, as shown in Figure 24, reveals that

$$\sin B = \frac{h'}{c} \quad \text{and} \quad \sin C = \frac{h'}{b}$$

Equating the expressions for h' gives

$$h' = c \sin B = b \sin C$$

from which

$$\frac{\sin B}{b} = \frac{\sin C}{c} \qquad \textbf{(6)}$$

When equations (5) and (6) are combined, the result is equation (1), the Law of Sines. ∎

9.2 Assess Your Understanding

'Are You Prepared?' *Answers are given at the end of these exercises. If you get a wrong answer, read the pages listed in red.*

1. The difference formula for the sine function is
   $\sin(A - B) = $ _____ . (p. 646)

2. If θ is an acute angle, solve the equation $\cos\theta = \dfrac{\sqrt{3}}{2}$. (pp. 625–630)

3. The two triangles shown are similar. Find the missing length. (pp. 30–35)

Concepts and Vocabulary

4. If none of the angles of a triangle is a right angle, the triangle is called _____.
   (a) oblique   (b) obtuse   (c) acute   (d) scalene

5. For a triangle with sides a, b, c and opposite angles A, B, C, the Law of Sines states that _____ .

6. *True or False* An oblique triangle in which two sides and an angle are given always results in at least one triangle.

7. *True or False* The Law of Sines can be used to solve triangles where three sides are known.

8. Triangles for which two sides and the angle opposite one of them are known (SSA) are referred to as the _____ ____ .

Skill Building

In Problems 9–16, solve each triangle.

9.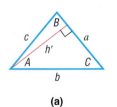

10.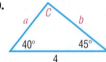

11.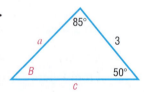

12.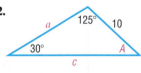

13.

14.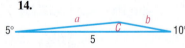

15.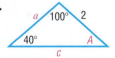

16.

In Problems 17–24, solve each triangle.

17. $A = 40°$, $B = 20°$, $a = 2$

18. $A = 50°$, $C = 20°$, $a = 3$

19. $B = 70°$, $C = 10°$, $b = 5$

20. $A = 70°$, $B = 60°$, $c = 4$

21. $A = 110°$, $C = 30°$, $c = 3$

22. $B = 10°$, $C = 100°$, $b = 2$

23. $A = 40°$, $B = 40°$, $c = 2$

24. $B = 20°$, $C = 70°$, $a = 1$

In Problems 25–36, two sides and an angle are given. Determine whether the given information results in one triangle, two triangles, or no triangle at all. Solve any resulting triangle(s).

25. $a = 3$, $b = 2$, $A = 50°$

26. $b = 4$, $c = 3$, $B = 40°$

27. $b = 5$, $c = 3$, $B = 100°$

28. $a = 2$, $c = 1$, $A = 120°$

29. $a = 4$, $b = 5$, $A = 60°$

30. $b = 2$, $c = 3$, $B = 40°$

31. $b = 4$, $c = 6$, $B = 20°$

32. $a = 3$, $b = 7$, $A = 70°$

33. $a = 2$, $c = 1$, $C = 100°$

34. $b = 4$, $c = 5$, $B = 95°$

35. $a = 2$, $c = 1$, $C = 25°$

36. $b = 4$, $c = 5$, $B = 40°$

Applications and Extensions

37. Finding the Length of a Ski Lift Consult the figure. To find the length of the span of a proposed ski lift from P to Q, a surveyor measures $\angle DPQ$ to be $25°$ and then walks back a distance of 1000 feet to R and measures $\angle PRQ$ to be $15°$. What is the distance from P to Q?

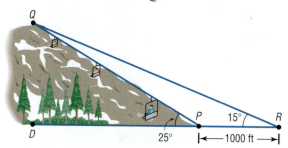

38. Finding the Height of a Mountain Use the illustration in Problem 37 to find the height QD of the mountain.

39. Finding the Height of an Airplane An aircraft is spotted by two observers who are 1000 feet apart. As the airplane passes over the line joining them, each observer takes a sighting of the angle of elevation to the plane, as indicated in the figure. How high is the airplane?

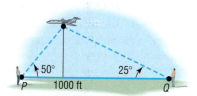

40. Finding the Height of the Bridge over the Royal Gorge The highest bridge in the world is the bridge over the Royal Gorge of the Arkansas River in Colorado. Sightings to the same point at water level directly under the bridge are taken from each side of the 880-foot-long bridge, as indicated in the figure. How high is the bridge?

Source: Guinness Book of World Records

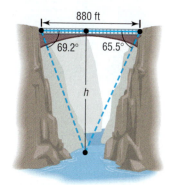

41. Land Dimensions A triangular plot of land has one side along a straight road measuring 200 feet. A second side makes a 50° angle with the road, and the third side makes a 43° angle with the road. How long are the other two sides?

42. Distance between Runners Two runners in a marathon determine that the angles of elevation of a news helicopter covering the race are 38° and 45°. If the helicopter is 1700 feet directly above the finish line, how far apart are the runners?

43. Landscaping Pat needs to determine the height of a tree before cutting it down to be sure that it will not fall on a nearby fence. The angle of elevation of the tree from one position on a flat path from the tree is 30°, and from a second position 40 feet farther along this path it is 20°. What is the height of the tree?

44. Construction A loading ramp 10 feet long that makes an angle of 18° with the horizontal is to be replaced by one that makes an angle of 12° with the horizontal. How long is the new ramp?

45. Commercial Navigation Adam must fly home to St. Louis from a business meeting in Oklahoma City. One flight option flies directly to St. Louis, a distance of about 461.1 miles. A second flight option flies first to Kansas City and then connects to St. Louis. The bearing from Oklahoma City to Kansas City is N29.6°E, and the bearing from Oklahoma City to St. Louis is N57.7°E. The bearing from St. Louis to Oklahoma City is S57.7°W, and the bearing from St. Louis to Kansas City is N79.4°W. How many more frequent flyer miles will Adam receive if he takes the connecting flight rather than the direct flight?

Source: www.landings.com

46. Time Lost to a Navigation Error In attempting to fly from city P to city Q, an aircraft followed a course that was 10° in error, as indicated in the figure. After flying a distance of 50 miles, the pilot corrected the course by turning at point R and flying 300 miles farther. If the constant speed of the aircraft was 250 miles per hour, how much time was lost due to the error?

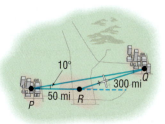

47. Rescue at Sea Coast Guard Station Able is located 150 miles due south of Station Baker. A ship at sea sends an SOS call that is received by each station. The call to Station Able indicates that the ship is located N55°E; the call to Station Baker indicates that the ship is located S60°E.

(continued on next page)

(a) How far is each station from the ship?

(b) If a helicopter capable of flying 200 miles per hour is dispatched from the station nearest the ship, how long will it take to reach the ship?

48. Distance to the Moon At exactly the same time, Tom and Alice measured the angle of elevation to the moon while standing exactly 300 km apart. The angle of elevation to the moon for Tom was 49.8974°, and the angle of elevation to the moon for Alice was 49.9312°. See the figure. To the nearest 1000 km, how far was the moon from Earth when the measurement was obtained?

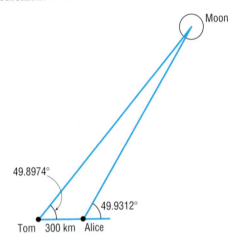

49. Finding the Lean of the Leaning Tower of Pisa The famous Leaning Tower of Pisa was originally 184.5 feet high.* At a distance of 123 feet from the base of the tower, the angle of elevation to the top of the tower is found to be 60°. Find ∠ RPQ indicated in the figure. Also, find the perpendicular distance from R to PQ.

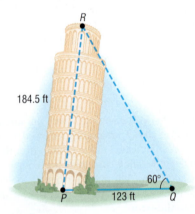

50. Crankshafts on Cars On a certain automobile, the crankshaft is 3 inches long and the connecting rod is 9 inches long (see the figure, top, right). At the time when ∠OPQ is 15°, how far is the piston (P) from the center (O) of the crankshaft?

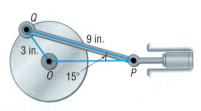

51. Constructing a Highway U.S. 41, a highway whose primary directions are north–south, is being constructed along the west coast of Florida. Near Naples, a bay obstructs the straight path of the road. Since the cost of a bridge is prohibitive, engineers decide to go around the bay. The illustration shows the path that they decide on and the measurements taken. What is the length of highway needed to go around the bay?

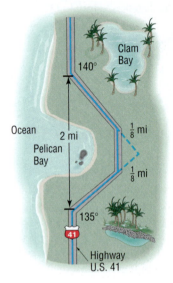

52. Calculating Distances at Sea The navigator of a ship at sea spots two lighthouses that she knows to be 3 miles apart along a straight seashore. She determines that the angles formed between two line-of-sight observations of the lighthouses and the line from the ship directly to shore are 15° and 35°. See the illustration.

(a) How far is the ship from lighthouse P?

(b) How far is the ship from lighthouse Q?

(c) How far is the ship from shore?

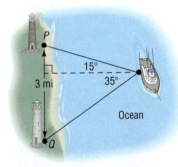

* On February 27, 1964, the government of Italy requested aid in preventing the tower from toppling. A multinational task force of engineers, mathematicians, and historians was assigned and met on the Azores islands to discuss stabilization methods. After over two decades of work on the subject, the tower was closed to the public in January 1990. During the time that the tower was closed, the bells were removed to relieve it of some weight, and cables were cinched around the third level and anchored several hundred meters away. Apartments and houses in the path of the tower were vacated for safety concerns. After a decade of corrective reconstruction and stabilization efforts, the tower was reopened to the public on December 15, 2001. Many methods were proposed to stabilize the tower, including the addition of 800 metric tons of lead counterweights to the raised end of the base. The final solution was to remove 38 cubic meters of soil from underneath the raised end. The tower has been declared stable for at least another 300 years.

Source: http://en.wikipedia.org/wiki/Leaning_Tower_of_Pisa

53. Designing an Awning An awning that covers a sliding glass door that is 88 inches tall forms an angle of 50° with the wall. The purpose of the awning is to prevent sunlight from entering the house when the angle of elevation of the Sun is more than 65°. See the figure. Find the length L of the awning.

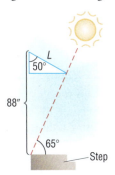

54. Finding Distances A forest ranger is walking on a path inclined at 5° to the horizontal directly toward a 100-foot-tall fire observation tower. The angle of elevation from the path to the top of the tower is 40°. How far is the ranger from the tower at this time?

55. Great Pyramid of Cheops One of the original Seven Wonders of the World, the Great Pyramid of Cheops was built about 2580 BC. Its original height was 480 feet 11 inches, but owing to the loss of its topmost stones, it is now shorter. Find the current height of the Great Pyramid using the information given in the illustration.

Source: Guinness Book of World Records

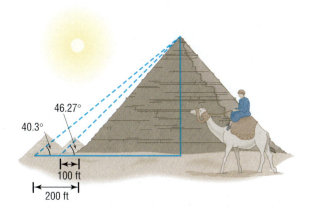

56. Determining the Height of an Aircraft Two sensors are spaced 700 feet apart along the approach to a small airport. When an aircraft is nearing the airport, the angle of elevation from the first sensor to the aircraft is 20°, and from the second sensor to the aircraft it is 15°. Determine how high the aircraft is at this time.

57. Mercury The distance from the Sun to Earth is approximately 149,600,000 kilometers (km). The distance from the Sun to Mercury is approximately 57,910,000 km. The **elongation angle** α is the angle formed between the line of sight from Earth to the Sun and the line of sight from Earth to Mercury.

See the figure. Suppose that the elongation angle for Mercury is 15°. Use this information to find the possible distances between Earth and Mercury.

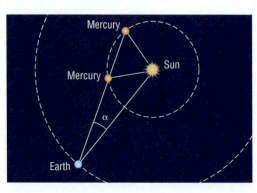

58. Venus The distance from the Sun to Earth is approximately 149,600,000 km. The distance from the Sun to Venus is approximately 108,200,000 km. The elongation angle α is the angle formed between the line of sight from Earth to the Sun and the line of sight from Earth to Venus. Suppose that the elongation angle for Venus is 10°. Use this information to find the possible distances between Earth and Venus.

59. The Original Ferris Wheel George Washington Gale Ferris, Jr., designed the original Ferris wheel for the 1893 World's Columbian Exposition in Chicago, Illinois. The wheel had 36 equally spaced cars each the size of a school bus. The distance between adjacent cars was approximately 22 feet. Determine the diameter of the wheel to the nearest foot.

Source: Carnegie Library of Pittsburgh, www.clpgh.org

60. Mollweide's Formula For any triangle, **Mollweide's Formula** (named after Karl Mollweide, 1774–1825) states that

$$\frac{a + b}{c} = \frac{\cos\left[\frac{1}{2}(A - B)\right]}{\sin\left(\frac{1}{2}C\right)}$$

Derive it.

[**Hint:** Use the Law of Sines and then a Sum-to-Product Formula. Notice that this formula involves all six parts of a triangle. As a result, it is sometimes used to check the solution of a triangle.]

61. Mollweide's Formula Another form of Mollweide's Formula is

$$\frac{a - b}{c} = \frac{\sin\left[\frac{1}{2}(A - B)\right]}{\cos\left(\frac{1}{2}C\right)}$$

Derive it.

62. For any triangle, derive the formula

$$a = b \cos C + c \cos B$$

[**Hint:** Use the fact that $\sin A = \sin(180° - B - C)$.]

63. Law of Tangents For any triangle, derive the **Law of Tangents:**

$$\frac{a - b}{a + b} = \frac{\tan\left[\frac{1}{2}(A - B)\right]}{\tan\left[\frac{1}{2}(A + B)\right]}$$

[**Hint:** Use Mollweide's Formula.]

64. Circumscribing a Triangle Show that

$$\frac{\sin A}{a} = \frac{\sin B}{b} = \frac{\sin C}{c} = \frac{1}{2r}$$

where r is the radius of the circle circumscribing the triangle PQR whose sides are a, b, and c, as shown in the figure.

[**Hint:** Draw the diameter PP'. Then $B = \angle PQR = \angle PP'R$, and angle $\angle PRP' = 90°$.]

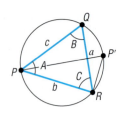

Explaining Concepts: Discussion and Writing

65. Make up three problems involving oblique triangles. One should result in one triangle, the second in two triangles, and the third in no triangle.

66. What do you do first if you are asked to solve a triangle and are given one side and two angles?

67. What do you do first if you are asked to solve a triangle and are given two sides and the angle opposite one of them?

68. Solve Example 6 using right-triangle geometry. Comment on which solution, using the Law of Sines or using right triangles, you prefer. Give reasons.

Retain Your Knowledge

Problems 69–72 are based on material learned earlier in the course. The purpose of these problems is to keep the material fresh in your mind so that you are better prepared for the final exam.

69. Solve: $3x^3 + 4x^2 - 27x - 36 = 0$

70. Find the exact distance between $P_1 = (-1, -7)$ and $P_2 = (2, -1)$. Then approximate the distance to two decimal places.

71. Find the exact value of $\tan\left[\cos^{-1}\left(-\frac{7}{8}\right)\right]$.

72. Graph $y = 4 \sin\left(\frac{1}{2}x\right)$. Show at least two periods.

'Are You Prepared?' Answers

1. $\sin A \cos B - \cos A \sin B$ **2.** $30°$ or $\dfrac{\pi}{6}$ **3.** $\dfrac{15}{2}$

9.3 The Law of Cosines

PREPARING FOR THIS SECTION *Before getting started, review the following:*

- Trigonometric Equations (Section 8.3, pp. 625–630)
- Distance Formula (Section 2.1, p. 151)

Now Work the **'Are You Prepared?'** problems on page 695.

OBJECTIVES **1** Solve SAS Triangles (p. 693)
2 Solve SSS Triangles (p. 694)
3 Solve Applied Problems (p. 694)

In the previous section, the Law of Sines was used to solve Case 1 (SAA or ASA) and Case 2 (SSA) of an oblique triangle. In this section, the Law of Cosines is derived and used to solve Cases 3 and 4.

CASE 3: Two sides and the included angle are known (SAS).
CASE 4: Three sides are known (SSS).

THEOREM

Law of Cosines

For a triangle with sides a, b, c and opposite angles A, B, C, respectively,

$$c^2 = a^2 + b^2 - 2ab \cos C \qquad (1)$$
$$b^2 = a^2 + c^2 - 2ac \cos B \qquad (2)$$
$$a^2 = b^2 + c^2 - 2bc \cos A \qquad (3)$$

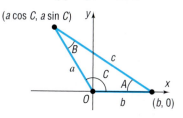

(a) Angle C is acute

(b) Angle C is obtuse

Figure 25

Proof Only formula (1) is proved here. Formulas (2) and (3) may be proved using the same argument.

Begin by strategically placing a triangle on a rectangular coordinate system so that the vertex of angle C is at the origin and side b lies along the positive x-axis. Regardless of whether C is acute, as in Figure 25(a), or obtuse, as in Figure 25(b), the vertex of angle B has coordinates $(a \cos C, a \sin C)$. The vertex of angle A has coordinates $(b, 0)$.

Use the distance formula to compute c^2.

$$
\begin{aligned}
c^2 &= (b - a \cos C)^2 + (0 - a \sin C)^2 \\
&= b^2 - 2ab \cos C + a^2 \cos^2 C + a^2 \sin^2 C \\
&= b^2 - 2ab \cos C + a^2 (\cos^2 C + \sin^2 C) \\
&= a^2 + b^2 - 2ab \cos C
\end{aligned}
$$
∎

Each of formulas (1), (2), and (3) may be stated in words as follows:

THEOREM

Law of Cosines

The square of one side of a triangle equals the sum of the squares of the other two sides, minus twice their product times the cosine of their included angle.

Observe that if the triangle is a right triangle (so that, say, $C = 90°$), formula (1) becomes the familiar Pythagorean Theorem: $c^2 = a^2 + b^2$. The Pythagorean Theorem is a special case of the Law of Cosines!

1 Solve SAS Triangles

The Law of Cosines is used to solve Case 3 (SAS), which applies to triangles for which two sides and the included angle are known.

EXAMPLE 1   **Using the Law of Cosines to Solve an SAS Triangle**

Solve the triangle: $a = 2$, $b = 3$, $C = 60°$

Solution   See Figure 26. Because two sides, a and b, and the included angle, $C = 60°$, are known, the Law of Cosines makes it easy to find the third side, c.

$$
\begin{aligned}
c^2 &= a^2 + b^2 - 2ab \cos C \\
&= 2^2 + 3^2 - 2 \cdot 2 \cdot 3 \cdot \cos 60° \quad \textcolor{blue}{a = 2, b = 3, C = 60°} \\
&= 13 - \left(12 \cdot \frac{1}{2} \right) = 7 \\
c &= \sqrt{7}
\end{aligned}
$$

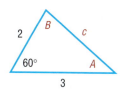

Figure 26

Side c is of length $\sqrt{7}$. To find the angles A and B, either the Law of Sines or the Law of Cosines may be used. It is preferable to use the Law of Cosines because it will lead to an equation with one solution. Using the Law of Sines would lead to an equation with two solutions that would need to be checked to determine which solution fits the given data.* We choose to use formulas (2) and (3) of the Law of Cosines to find A and B.

For A:

$$
\begin{aligned}
a^2 &= b^2 + c^2 - 2bc \cos A \\
2bc \cos A &= b^2 + c^2 - a^2 \\
\cos A &= \frac{b^2 + c^2 - a^2}{2bc} = \frac{9 + 7 - 4}{2 \cdot 3\sqrt{7}} = \frac{12}{6\sqrt{7}} = \frac{2\sqrt{7}}{7}
\end{aligned}
$$

$$
A = \cos^{-1} \frac{2\sqrt{7}}{7} \approx 40.9°
$$

*The Law of Sines can be used if the angle sought is opposite the smaller side, thus ensuring that it is acute. (In Figure 26, use the Law of Sines to find A, the angle opposite the smaller side.)

For *B*:

$$b^2 = a^2 + c^2 - 2ac \cos B$$

$$\cos B = \frac{a^2 + c^2 - b^2}{2ac} = \frac{4 + 7 - 9}{4\sqrt{7}} = \frac{2}{4\sqrt{7}} = \frac{\sqrt{7}}{14}$$

$$B = \cos^{-1} \frac{\sqrt{7}}{14} \approx 79.1°$$

NOTE The angle *B* could also have been found by using the fact that the sum $A + B + C = 180°$, so $B = 180° - 40.9° - 60° = 79.1°$. However, using the Law of Cosines twice allows for a check. ■

Notice that $A + B + C = 40.9° + 79.1° + 60° = 180°$, as required. ●

═══ **Now Work** PROBLEM 9

2 Solve SSS Triangles

The next example illustrates how the Law of Cosines is used when three sides of a triangle are known, Case 4 (SSS).

EXAMPLE 2 **Using the Law of Cosines to Solve an SSS Triangle**

Solve the triangle: $a = 4, b = 3, c = 6$

Solution See Figure 27. To find the angles *A*, *B*, and *C*, proceed as in the solution to Example 1.

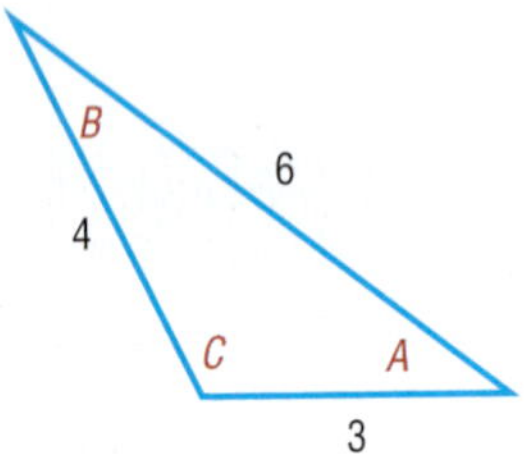

For *A*:

$$\cos A = \frac{b^2 + c^2 - a^2}{2bc} = \frac{9 + 36 - 16}{2 \cdot 3 \cdot 6} = \frac{29}{36}$$

$$A = \cos^{-1} \frac{29}{36} \approx 36.3°$$

For *B*:

$$\cos B = \frac{a^2 + c^2 - b^2}{2ac} = \frac{16 + 36 - 9}{2 \cdot 4 \cdot 6} = \frac{43}{48}$$

$$B = \cos^{-1} \frac{43}{48} \approx 26.4°$$

Figure 27

Now use *A* and *B* to find *C*:

$$C = 180° - A - B \approx 180° - 36.3° - 26.4° = 117.3°$$ ●

═══ **Now Work** PROBLEM 15

3 Solve Applied Problems

EXAMPLE 3 **Correcting a Navigational Error**

A motorized sailboat leaves Naples, Florida, bound for Key West, 150 miles away. Maintaining a constant speed of 15 miles per hour, but encountering heavy crosswinds and strong currents, the crew finds, after 4 hours, that the sailboat is off course by 20°.

(a) How far is the sailboat from Key West at this time?

(b) Through what angle should the sailboat turn to correct its course?

(c) How much time has been added to the trip because of this? (Assume that the speed remains at 15 miles per hour.)

Solution   See Figure 28. With a speed of 15 miles per hour, the sailboat has gone 60 miles after 4 hours. The distance x of the sailboat from Key West is to be found, along with the angle θ that the sailboat should turn through to correct its course.

Figure 28

(a) To find x, use the Law of Cosines, because two sides and the included angle are known.

$$x^2 = 150^2 + 60^2 - 2(150)(60)\cos 20° \approx 9185.53$$
$$x \approx 95.8$$

The sailboat is about 96 miles from Key West.

(b) With all three sides of the triangle now known, use the Law of Cosines again to find the angle A opposite the side of length 150 miles.

$$150^2 = 96^2 + 60^2 - 2(96)(60)\cos A$$
$$9684 = -11{,}520 \cos A$$
$$\cos A \approx -0.8406$$
$$A \approx 147.2°$$

So,

$$\theta = 180° - A \approx 180° - 147.2° = 32.8°$$

The sailboat should turn through an angle of about 33° to correct its course.

(c) The total length of the trip is now $60 + 96 = 156$ miles. The extra 6 miles will only require about 0.4 hour, or 24 minutes, more if the speed of 15 miles per hour is maintained.   ●

 Now Work PROBLEM 45

Historical Feature

he Law of Sines was known vaguely long before it was explicitly stated by Nasir Eddin (about AD 1250). Ptolemy (about AD 150) was aware of it in a form using a chord function instead of the sine function. But it was first clearly stated in Europe by Regiomontanus, writing in 1464.

The Law of Cosines appears first in Euclid's *Elements* (Book II), but in a well-disguised form in which squares built on the sides of triangles are added and a rectangle representing the cosine term is subtracted. It was thus known to all mathematicians because of their familiarity with Euclid's

work. An early modern form of the Law of Cosines, that for finding the angle when the sides are known, was stated by François Viète (in 1593).

The Law of Tangents (see Problem 63 in Section 9.2) has become obsolete. In the past it was used in place of the Law of Cosines, because the Law of Cosines was very inconvenient for calculation with logarithms or slide rules. Mixing of addition and multiplication is now very easy on a calculator, however, and the Law of Tangents has been shelved along with the slide rule.

9.3 Assess Your Understanding

'Are You Prepared?' *Answers are given at the end of these exercises. If you get a wrong answer, read the pages listed in* red.

1. Write the formula for the distance d from $P_1 = (x_1, y_1)$ to $P_2 = (x_2, y_2)$. (p. 151)

2. If θ is an acute angle, solve the equation $\cos \theta = \dfrac{\sqrt{2}}{2}$. (pp. 625–630)

Concepts and Vocabulary

3. If three sides of a triangle are given, the Law of _____ is used to solve the triangle.

4. If one side and two angles of a triangle are given, which law can be used to solve the triangle?
 (a) Law of Sines   (b) Law of Cosines
 (c) Either a or b   (d) The triangle cannot be solved.

5. If two sides and the included angle of a triangle are given, which law can be used to solve the triangle?
 (a) Law of Sines   (b) Law of Cosines
 (c) Either a or b   (d) The triangle cannot be solved.

6. *True or False* Given only the three sides of a triangle, there is insufficient information to solve the triangle.

7. *True or False* The Law of Cosines states that the square of one side of a triangle equals the sum of the squares of the other two sides, minus twice their product.

8. *True or False* A special case of the Law of Cosines is the Pythagorean Theorem.

Skill Building

In Problems 9–16, solve each triangle.

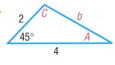

9.

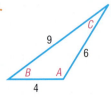

12.

15.
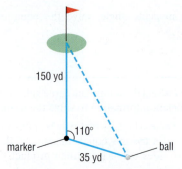

10.
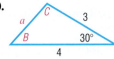

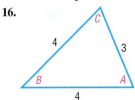

13.

16.
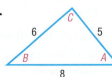

11.

14.

In Problems 17–32, solve each triangle.

17. $a = 3$, $b = 4$, $C = 40°$

20. $a = 6$, $b = 4$, $C = 60°$

23. $a = 2$, $b = 2$, $C = 50°$

26. $a = 4$, $b = 5$, $c = 3$

29. $a = 5$, $b = 8$, $c = 9$

32. $a = 9$, $b = 7$, $c = 10$

18. $a = 2$, $c = 1$, $B = 10°$

21. $a = 3$, $c = 2$, $B = 110°$

24. $a = 3$, $c = 2$, $B = 90°$

27. $a = 2$, $b = 2$, $c = 2$

30. $a = 4$, $b = 3$, $c = 6$

19. $b = 1$, $c = 3$, $A = 80°$

22. $b = 4$, $c = 1$, $A = 120°$

25. $a = 12$, $b = 13$, $c = 5$

28. $a = 3$, $b = 3$, $c = 2$

31. $a = 10$, $b = 8$, $c = 5$

Mixed Practice

In Problems 33–42, solve each triangle using either the Law of Sines or the Law of Cosines.

33. $B = 20°, C = 75°, b = 5$

36. $a = 14, b = 7, A = 85°$

39. $A = 10°, a = 3, b = 10$

42. $a = 10, b = 10, c = 15$

34. $A = 50°, B = 55°, c = 9$

37. $B = 35°, C = 65°, a = 15$

40. $A = 65°, B = 72°, b = 7$

35. $a = 6, b = 8, c = 9$

38. $a = 4, c = 5, B = 55°$

41. $b = 5, c = 12, A = 60°$

Applications and Extensions

43. Distance to the Green A golfer hits an errant tee shot that lands in the rough. A marker in the center of the fairway is 150 yards from the center of the green. While standing on the marker and facing the green, the golfer turns 110° toward his ball. He then paces off 35 yards to his ball. See the figure. How far is the ball from the center of the green?

(a) How far is it directly from Ft. Myers to Orlando?
(b) What bearing should the pilot use to fly directly from Ft. Myers to Orlando?

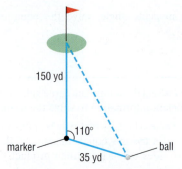

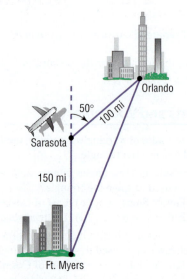

44. Navigation An airplane flies due north from Ft. Myers to Sarasota, a distance of 150 miles, and then turns through an angle of 50° and flies to Orlando, a distance of 100 miles. See the figure.

45. Avoiding a Tropical Storm A cruise ship maintains an average speed of 15 knots in going from San Juan, Puerto Rico, to Barbados, West Indies, a distance of 600 nautical miles. To avoid a tropical storm, the captain heads out of San Juan in a direction of 20° off a direct heading to Barbados. The captain maintains the 15-knot speed for 10 hours, after which time the path to Barbados becomes clear of storms.
(a) Through what angle should the captain turn to head directly to Barbados?
(b) Once the turn is made, how long will it be before the ship reaches Barbados if the same 15-knot speed is maintained?

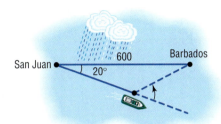

46. Revising a Flight Plan In attempting to fly from Chicago to Louisville, a distance of 330 miles, a pilot inadvertently took a course that was 10° in error, as indicated in the figure.
(a) If the aircraft maintains an average speed of 220 miles per hour, and if the error in direction is discovered after 15 minutes, through what angle should the pilot turn to head toward Louisville?
(b) What new average speed should the pilot maintain so that the total time of the trip is 90 minutes?

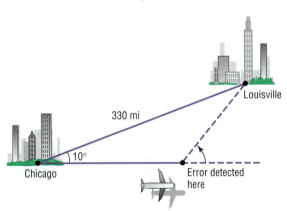

47. Major League Baseball Field A major league baseball diamond is actually a square 90 feet on a side. The pitching rubber is located 60.5 feet from home plate on a line joining home plate and second base.
(a) How far is it from the pitching rubber to first base?
(b) How far is it from the pitching rubber to second base?
(c) If a pitcher faces home plate, through what angle does he need to turn to face first base?

48. Little League Baseball Field According to Little League baseball official regulations, the diamond is a square 60 feet on a side. The pitching rubber is located 46 feet from home plate on a line joining home plate and second base.
(a) How far is it from the pitching rubber to first base?
(b) How far is it from the pitching rubber to second base?
(c) If a pitcher faces home plate, through what angle does he need to turn to face first base?

49. Finding the Length of a Guy Wire The height of a radio tower is 500 feet, and the ground on one side of the tower slopes upward at an angle of 10° (see the figure).
(a) How long should a guy wire be if it is to connect to the top of the tower and be secured at a point on the sloped side 100 feet from the base of the tower?
(b) How long should a second guy wire be if it is to connect to the middle of the tower and be secured at a position 100 feet from the base on the flat side?

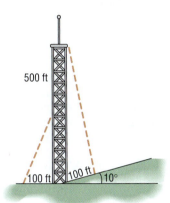

50. Finding the Length of a Guy Wire A radio tower 500 feet high is located on the side of a hill with an inclination to the horizontal of 5°. See the figure. How long should two guy wires be if they are to connect to the top of the tower and be secured at two points 100 feet directly above and directly below the base of the tower?

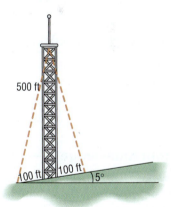

51. Wrigley Field, Home of the Chicago Cubs The distance from home plate to the fence in dead center in Wrigley Field is 400 feet (see the figure). How far is it from the fence in dead center to third base?

52. Little League Baseball The distance from home plate to the fence in dead center at the Oak Lawn Little League field is 280 feet. How far is it from the fence in dead center to third base?

[**Hint:** The distance between the bases in Little League is 60 feet.]

53. Building a Swing Set Clint is building a wooden swing set for his children. Each supporting end of the swing set is to be an A-frame constructed with two 10-foot-long 4 by 4's joined at a 45° angle. To prevent the swing set from tipping over, Clint wants to secure the base of each A-frame to concrete footings. How far apart should the footings for each A-frame be?

54. Rods and Pistons Rod OA rotates about the fixed point O so that point A travels on a circle of radius r. Connected to point A is another rod AB of length $L > 2r$, and point B is connected to a piston. See the figure. Show that the distance x between point O and point B is given by

$$x = r\cos\theta + \sqrt{r^2\cos^2\theta + L^2 - r^2}$$

where θ is the angle of rotation of rod OA.

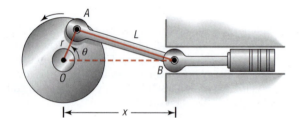

55. Geometry Show that the length d of a chord of a circle of radius r is given by the formula

$$d = 2r\sin\frac{\theta}{2}$$

where θ is the central angle formed by the radii to the ends of the chord. See the figure. Use this result to derive the fact that $\sin\theta < \theta$, where $\theta > 0$ is measured in radians.

56. For any triangle, show that

$$\cos\frac{C}{2} = \sqrt{\frac{s(s-c)}{ab}}$$

where $s = \frac{1}{2}(a+b+c)$.

[**Hint:** Use a Half-angle Formula and the Law of Cosines.]

57. For any triangle, show that

$$\sin\frac{C}{2} = \sqrt{\frac{(s-a)(s-b)}{ab}}$$

where $s = \frac{1}{2}(a+b+c)$.

58. Use the Law of Cosines to prove the identity

$$\frac{\cos A}{a} + \frac{\cos B}{b} + \frac{\cos C}{c} = \frac{a^2+b^2+c^2}{2abc}$$

Explaining Concepts: Discussion and Writing

59. What do you do first if you are asked to solve a triangle and are given two sides and the included angle?

60. What do you do first if you are asked to solve a triangle and are given three sides?

61. Make up an applied problem that requires using the Law of Cosines.

62. Write down your strategy for solving an oblique triangle.

63. State the Law of Cosines in words.

⌐ Retain Your Knowledge

Problems 64–67 are based on material learned earlier in the course. The purpose of these problems is to keep the material fresh in your mind so that you are better prepared for the final exam.

64. Graph: $R(x) = \dfrac{2x+1}{x-3}$

65. Solve $4^x = 3^{x+1}$. If the solution is irrational, express it both in exact form and as a decimal rounded to three places.

66. Given $\tan\theta = -\dfrac{2\sqrt{6}}{5}$ and $\cos\theta = -\dfrac{5}{7}$, find the exact value of each of the four remaining trigonometric functions.

67. Find an equation for the graph.

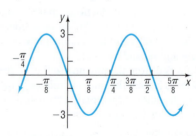

'Are You Prepared?' Answers

1. $d = \sqrt{(x_2-x_1)^2+(y_2-y_1)^2}$ **2.** $\theta = 45°$ or $\dfrac{\pi}{4}$

9.4 Area of a Triangle

PREPARING FOR THIS SECTION *Before getting started, review the following:*

- Geometry Essentials (Chapter R, Review, Section R.3, pp. 30–35)

✎ **Now Work** the **'Are You Prepared?'** problem on page 701.

OBJECTIVES **1** Find the Area of SAS Triangles (p. 699)
               2 Find the Area of SSS Triangles (p. 700)

In this section, several formulas for calculating the area of a triangle are derived. The most familiar of these follows.

THEOREM

NOTE Typically, *A* is used for area. However, because *A* is also used as the measure of an angle, *K* is used here for area to avoid confusion. ∎

The area K of a triangle is

$$K = \frac{1}{2}bh \qquad \text{(1)}$$

where b is the base and h is an altitude drawn to that base.

Proof Look at the triangle in Figure 29. Around the triangle construct a rectangle of height h and base b, as shown in Figure 30.

Triangles 1 and 2 in Figure 30 are equal in area, as are triangles 3 and 4. Consequently, the area of the triangle with base b and altitude h is exactly half the area of the rectangle, which is bh.

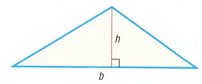

Figure 29

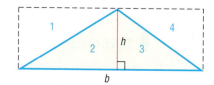

Figure 30 ∎

1 Find the Area of SAS Triangles

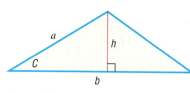

Figure 31

If the base b and the altitude h to that base are known, then the area of such a triangle can be found using formula (1). Usually, though, the information required to use formula (1) is not given. Suppose, for example, that two sides a and b and the included angle C are known. See Figure 31. Then the altitude h can be found by noting that

$$\frac{h}{a} = \sin C$$

so

$$h = a \sin C$$

Using this fact in formula (1) produces

$$K = \frac{1}{2}bh = \frac{1}{2}b(a \sin C) = \frac{1}{2}ab \sin C$$

The area K of the triangle is given by the formula

$$K = \frac{1}{2}ab \sin C \qquad \text{(2)}$$

Dropping altitudes from the other two vertices of the triangle leads to the following corresponding formulas:

$$K = \frac{1}{2}bc \sin A \qquad \text{(3)}$$

$$K = \frac{1}{2}ac \sin B \qquad \text{(4)}$$

It is easiest to remember these formulas by using the following wording:

THEOREM

The area K of a triangle equals one-half the product of two of its sides times the sine of their included angle.

EXAMPLE 1

Finding the Area of an SAS Triangle

Find the area K of the triangle for which $a = 8$, $b = 6$, and $C = 30°$.

Solution   See Figure 32. Use formula (2) to get

$$K = \frac{1}{2}\,ab \sin C = \frac{1}{2}\cdot 8 \cdot 6 \cdot \sin 30° = 12 \text{ square units}$$

Figure 32

━━━━ **Now Work** PROBLEM 7

2 Find the Area of SSS Triangles

If the three sides of a triangle are known, another formula, called **Heron's Formula** (named after Heron of Alexandria), can be used to find the area of a triangle.

THEOREM

Heron's Formula

The area K of a triangle with sides $a, b,$ and c is

$$K = \sqrt{s(s - a)(s - b)(s - c)} \qquad \text{(5)}$$

where $s = \frac{1}{2}(a + b + c)$.

EXAMPLE 2

Finding the Area of an SSS Triangle

Find the area of a triangle whose sides are 4, 5, and 7.

Solution   Let $a = 4$, $b = 5$, and $c = 7$. Then

$$s = \frac{1}{2}(a + b + c) = \frac{1}{2}(4 + 5 + 7) = 8$$

Heron's Formula gives the area K as

$$K = \sqrt{s(s - a)(s - b)(s - c)} = \sqrt{8 \cdot 4 \cdot 3 \cdot 1} = \sqrt{96} = 4\sqrt{6} \text{ square units}$$

━━━━ **Now Work** PROBLEM 13

Proof of Heron's Formula The proof given here uses the Law of Cosines and is quite different from the proof given by Heron.

From the Law of Cosines,

$$c^2 = a^2 + b^2 - 2ab \cos C$$

and the Half-angle Formula,

$$\cos^2 \frac{C}{2} = \frac{1 + \cos C}{2}$$

it follows that

$$\cos^2 \frac{C}{2} = \frac{1 + \cos C}{2} = \frac{1 + \dfrac{a^2 + b^2 - c^2}{2ab}}{2}$$

$$= \frac{a^2 + 2ab + b^2 - c^2}{4ab} = \frac{(a + b)^2 - c^2}{4ab}$$

$$= \frac{(a + b - c)(a + b + c)}{4ab} = \frac{2(s - c) \cdot 2s}{4ab} = \frac{s(s - c)}{ab} \qquad \textbf{(6)}$$

↑
Factor.

↑
$a + b - c = a + b + c - 2c$
$= 2s - 2c = 2(s - c)$

Similarly, using $\sin^2 \dfrac{C}{2} = \dfrac{1 - \cos C}{2}$, it follows that

$$\sin^2 \frac{C}{2} = \frac{(s - a)(s - b)}{ab} \qquad \textbf{(7)}$$

Now use formula (2) for the area.

$$K = \frac{1}{2} ab \sin C$$

$$= \frac{1}{2} ab \cdot 2 \sin \frac{C}{2} \cos \frac{C}{2} \qquad \sin C = \sin\left[2\left(\frac{C}{2}\right)\right] = 2 \sin \frac{C}{2} \cos \frac{C}{2}$$

$$= ab \sqrt{\frac{(s - a)(s - b)}{ab}} \sqrt{\frac{s(s - c)}{ab}} \qquad \text{Use equations (6) and (7).}$$

$$= \sqrt{s(s - a)(s - b)(s - c)} \qquad ■$$

Historical Feature

Heron's Formula (also known as *Hero's Formula*) was first expressed by Heron of Alexandria (first century AD), who had, besides his mathematical talents, a good deal of engineering skills. In various temples his mechanical devices produced effects that seemed supernatural, and visitors presumably were thus moved to generosity. Heron's book *Metrica*, on making such devices, has survived and was discovered in 1896 in the city of Constantinople.

Heron's Formulas for the area of a triangle caused some mild discomfort in Greek mathematics, because a product with two factors was an area and one with three factors was a volume, but four factors seemed contradictory in Heron's time.

9.4 Assess Your Understanding

'Are You Prepared?' *The answer is given at the end of these exercises. If you get the wrong answer, read the page listed in* red.

1. The area K of a triangle whose base is b and whose height is h is _____ . (pp. 30–35)

Concepts and Vocabulary

2. If two sides a and b and the included angle C are known in a triangle, then the area K is found using the formula $K = $ _____ .

3. The area K of a triangle with sides $a, b,$ and c is

   $K = $ _____ , where $s = $ _____ .

4. **True or False** The area of a triangle equals one-half the product of the lengths of two of its sides times the sine of their included angle.

5. Given two sides of a triangle, b and c, and the included angle A, the altitude h from angle B to side b is given by _____ .

   (a) $\frac{1}{2} ab \sin A$ (b) $b \sin A$ (c) $c \sin A$ (d) $\frac{1}{2} bc \sin A$

6. Heron's Formula is used to find the area of _____ triangles.

   (a) ASA (b) SAS (c) SSS (d) AAS

Skill Building

In Problems 7–14, find the area of each triangle. Round answers to two decimal places.

7.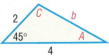

8.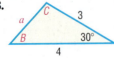

9.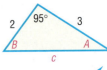

10.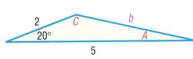

11.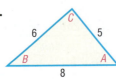

12.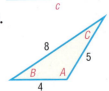

13.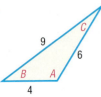

14.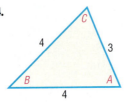

In Problems 15–26, find the area of each triangle. Round answers to two decimal places.

15. $a = 3$, $b = 4$, $C = 40°$ **16.** $a = 2$, $c = 1$, $B = 10°$ **17.** $b = 1$, $c = 3$, $A = 80°$

18. $a = 6$, $b = 4$, $C = 60°$ **19.** $a = 3$, $c = 2$, $B = 110°$ **20.** $b = 4$, $c = 1$, $A = 120°$

21. $a = 12$, $b = 13$, $c = 5$ **22.** $a = 4$, $b = 5$, $c = 3$ **23.** $a = 2$, $b = 2$, $c = 2$

24. $a = 3$, $b = 3$, $c = 2$ **25.** $a = 5$, $b = 8$, $c = 9$ **26.** $a = 4$, $b = 3$, $c = 6$

Applications and Extensions

27. Area of an ASA Triangle If two angles and the included side are given, the third angle is easy to find. Use the Law of Sines to show that the area K of a triangle with side a and angles A, B, and C is

$$K = \frac{a^2 \sin B \sin C}{2 \sin A}$$

28. Area of a Triangle Prove the two other forms of the formula given in Problem 27.

$$K = \frac{b^2 \sin A \sin C}{2 \sin B} \quad \text{and} \quad K = \frac{c^2 \sin A \sin B}{2 \sin C}$$

In Problems 29–34, use the results of Problem 27 or 28 to find the area of each triangle. Round answers to two decimal places.

29. $A = 40°$, $B = 20°$, $a = 2$ **30.** $A = 50°$, $C = 20°$, $a = 3$ **31.** $B = 70°$, $C = 10°$, $b = 5$

32. $A = 70°$, $B = 60°$, $c = 4$ **33.** $A = 110°$, $C = 30°$, $c = 3$ **34.** $B = 10°$, $C = 100°$, $b = 2$

35. Area of a Segment Find the area of the segment (shaded in blue in the figure) of a circle whose radius is 8 feet, formed by a central angle of 70°.

[**Hint:** Subtract the area of the triangle from the area of the sector to obtain the area of the segment.]

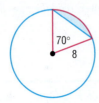

36. Area of a Segment Find the area of the segment of a circle whose radius is 5 inches, formed by a central angle of 40°.

37. Cost of a Triangular Lot The dimensions of a triangular lot are 100 feet by 50 feet by 75 feet. If the price of such land is $3 per square foot, how much does the lot cost?

38. Amount of Material to Make a Tent A cone-shaped tent is made from a circular piece of canvas 24 feet in diameter by removing a sector with central angle 100° and connecting the ends. What is the surface area of the tent?

39. Dimensions of Home Plate The dimensions of home plate at any major league baseball stadium are shown. Find the area of home plate.

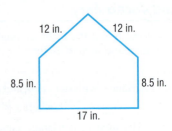

40. Computing Areas See the figure. Find the area of the shaded region enclosed in a semicircle of diameter 10 inches. The length of the chord PQ is 8 inches.

[**Hint:** Triangle PQR is a right triangle.]

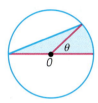

41. Geometry See the figure, which shows a circle of radius r with center at O. Find the area K of the shaded region as a function of the central angle θ.

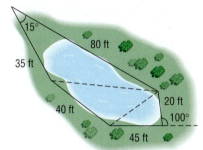

42. Approximating the Area of a Lake To approximate the area of a lake, a surveyor walks around the perimeter of the lake, taking the measurements shown in the illustration. Using this technique, what is the approximate area of the lake?

[**Hint:** Use the Law of Cosines on the three triangles shown, and then find the sum of their areas.]

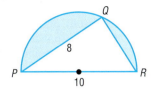

43. The Flatiron Building Completed in 1902 in New York City, the Flatiron Building is triangular shaped and bounded by 22nd Street, Broadway, and 5th Avenue. The building measures approximately 87 feet on the 22nd Street side, 190 feet on the Broadway side, and 173 feet on the 5th Avenue side. Approximate the ground area covered by the building.

Source: Sarah Bradford Landau and Carl W. Condit, *Rise of the New York Skyscraper: 1865–1913.* New Haven, CT: Yale University Press, 1996

44. Bermuda Triangle The Bermuda Triangle is roughly defined by Hamilton, Bermuda; San Juan, Puerto Rico; and Fort Lauderdale, Florida. The distances from Hamilton to Fort Lauderdale, Fort Lauderdale to San Juan, and San Juan to Hamilton are approximately 1028, 1046, and 965 miles, respectively. Ignoring the curvature of Earth, approximate the area of the Bermuda Triangle.

Source: www.worldatlas.com

45. Geometry Refer to the figure. If $|OA| = 1$, show that:

(a) Area $\triangle OAC = \dfrac{1}{2} \sin \alpha \cos \alpha$

(b) Area $\triangle OCB = \dfrac{1}{2} |OB|^2 \sin \beta \cos \beta$

(c) Area $\triangle OAB = \dfrac{1}{2} |OB| \sin(\alpha + \beta)$

(d) $|OB| = \dfrac{\cos \alpha}{\cos \beta}$

(e) $\sin(\alpha + \beta) = \sin \alpha \cos \beta + \cos \alpha \sin \beta$

[**Hint:** Area $\triangle OAB$ = Area $\triangle OAC$ + Area $\triangle OCB$]

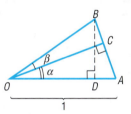

46. Geometry Refer to the figure, in which a unit circle is drawn. The line segment DB is tangent to the circle and θ is acute.

(a) Express the area of $\triangle OBC$ in terms of $\sin \theta$ and $\cos \theta$.

(b) Express the area of $\triangle OBD$ in terms of $\sin \theta$ and $\cos \theta$.

(c) The area of the sector $\overset{\frown}{OBC}$ of the circle is $\dfrac{1}{2}\theta$, where θ is measured in radians. Use the results of parts (a) and (b) and the fact that

$$\text{Area } \triangle OBC < \text{Area } \overset{\frown}{OBC} < \text{Area } \triangle OBD$$

to show that

$$1 < \frac{\theta}{\sin \theta} < \frac{1}{\cos \theta}$$

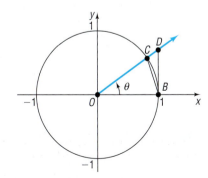

47. The Cow Problem* A cow is tethered to one corner of a square barn, 10 feet by 10 feet, with a rope 100 feet long. What is the maximum grazing area for the cow? [See the illustration that follows.]

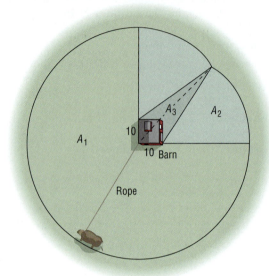

48. Another Cow Problem If the barn in Problem 47 is rectangular, 10 feet by 20 feet, what is the maximum grazing area for the cow?

49. Perfect Triangles A *perfect triangle* is one having integers for sides for which the area is numerically equal to the perimeter. Show that the triangles with the given side lengths are perfect.

(a) $9, 10, 17$ (b) $6, 25, 29$

Source: M.V. Bonsangue, G. E. Gannon, E. Buchman, and N. Gross, "In Search of Perfect Triangles," *Mathematics Teacher*, Vol. 92, No. 1, 1999: 56–61

50. If h_1, h_2, and h_3 are the altitudes dropped from P, Q, and R, respectively, in a triangle (see the figure), show that

$$\frac{1}{h_1} + \frac{1}{h_2} + \frac{1}{h_3} = \frac{s}{K}$$

where K is the area of the triangle and $s = \frac{1}{2}(a + b + c)$.

[**Hint:** $h_1 = \dfrac{2K}{a}$.]

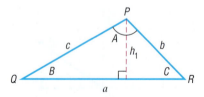

51. Show that a formula for the altitude h from a vertex to the opposite side a of a triangle is

$$h = \frac{a \sin B \sin C}{\sin A}$$

52. A triangle has vertices $A(0, 0)$, $B(1, 0)$, and C, where C is the point on the unit circle corresponding to an angle of $105°$ when it is drawn in standard position. Find the area of the triangle. State the answer in complete simplified form with a rationalized denominator.

Inscribed Circle *For Problems 53–56, the lines that bisect each angle of a triangle meet in a single point O, and the perpendicular distance r from O to each side of the triangle is the same. The circle with center at O and radius r is called the inscribed circle of the triangle (see the figure).*

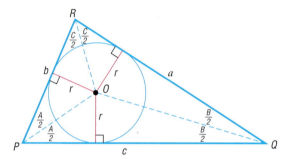

53. Apply the formula from Problem 51 to triangle OPQ to show that

$$r = \frac{c \sin \dfrac{A}{2} \sin \dfrac{B}{2}}{\cos \dfrac{C}{2}}$$

54. Use the result of Problem 53 and the results of Problems 56 and 57 in Section 9.3 to show that

$$\cot \frac{C}{2} = \frac{s - c}{r}$$

where $s = \frac{1}{2}(a + b + c)$.

55. Show that

$$\cot \frac{A}{2} + \cot \frac{B}{2} + \cot \frac{C}{2} = \frac{s}{r}$$

56. Show that the area K of triangle PQR is $K = rs$, where $s = \frac{1}{2}(a + b + c)$. Then show that

$$r = \sqrt{\frac{(s - a)(s - b)(s - c)}{s}}$$

Explaining Concepts: Discussion and Writing

57. What do you do first if you are asked to find the area of a triangle and are given two sides and the included angle?

58. What do you do first if you are asked to find the area of a triangle and are given three sides?

59. State the area of an SAS triangle in words.

---

* Suggested by Professor Teddy Koukounas of Suffolk Community College, who learned of it from an old farmer in Virginia. Solution provided by Professor Kathleen Miranda of SUNY at Old Westbury.

Retain Your Knowledge

Problems 60–63 are based on material learned earlier in the course. The purpose of these problems is to keep the material fresh in your mind so that you are better prepared for the final exam.

60. Without graphing, determine whether the quadratic function $f(x) = -3x^2 + 12x + 5$ has a maximum value or a minimum value, and then find the value.

61. Solve the inequality: $\dfrac{x + 1}{x^2 - 9} \le 0$

62. $P = \left(-\dfrac{\sqrt{7}}{3}, \dfrac{\sqrt{2}}{3} \right)$ is the point on the unit circle that corresponds to a real number t. Find the exact values of the six trigonometric functions of t.

63. Establish the identity: $\csc \theta - \sin \theta = \cos \theta \cot \theta$

'Are You Prepared?' Answer

1. $K = \dfrac{1}{2}bh$

9.5 Simple Harmonic Motion; Damped Motion; Combining Waves

PREPARING FOR THIS SECTION *Before getting started, review the following:*

- Sinusoidal Graphs (Section 7.6, pp. 568–574)

 Now Work the 'Are You Prepared?' problem on page 711.

OBJECTIVES 1 Build a Model for an Object in Simple Harmonic Motion (p. 705)
2 Analyze Simple Harmonic Motion (p. 707)
3 Analyze an Object in Damped Motion (p. 708)
4 Graph the Sum of Two Functions (p. 709)

1 Build a Model for an Object in Simple Harmonic Motion

Many physical phenomena can be described as simple harmonic motion. Radio and television waves, light waves, sound waves, and water waves exhibit motion that is simple harmonic.

The swinging of a pendulum, the vibrations of a tuning fork, and the bobbing of a weight attached to a coiled spring are examples of vibrational motion. In this type of motion, an object swings back and forth over the same path. In Figure 33, the point B is the **equilibrium (rest) position** of the vibrating object. The **amplitude** is the distance from the object's rest position to its point of greatest displacement (either point A or point C in Figure 33). The **period** is the time required to complete one vibration—that is, the time it takes to go from, say, point A through B to C and back to A.

Tuning fork

Simple harmonic motion is a special kind of vibrational motion in which the acceleration a of the object is directly proportional to the negative of its displacement d from its rest position. That is, $a = -kd, k > 0$.

For example, when the mass hanging from the spring in Figure 33 is pulled down from its rest position B to the point C, the force of the spring tries to restore the mass to its rest position. Assuming that there is no frictional force* to retard the motion, the amplitude will remain constant. The force increases in direct proportion to the distance that the mass is pulled from its rest position. Since the force increases directly, the acceleration of the mass of the object must do likewise, because (by Newton's Second Law of Motion) force is directly proportional to acceleration. As a result, the acceleration of the object varies directly with its displacement, and the motion is an example of simple harmonic motion.

Figure 33 Coiled spring

*If friction is present, the amplitude will decrease with time to 0. This type of motion is an example of **damped motion**, which is discussed later in this section.

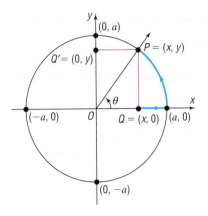

Figure 34

Simple harmonic motion is related to circular motion. To see this relationship, consider a circle of radius a, with center at $(0,0)$. See Figure 34. Suppose that an object initially placed at $(a, 0)$ moves counterclockwise around the circle at a constant angular speed ω. Suppose further that after time t has elapsed the object is at the point $P = (x, y)$ on the circle. The angle θ, in radians, swept out by the ray $\overrightarrow{OP}$ in this time t is

$$\theta = \omega t \qquad \omega = \frac{\theta}{t}$$

The coordinates of the point P at time t are

$$x = a \cos \theta = a \cos(\omega t)$$
$$y = a \sin \theta = a \sin(\omega t)$$

Corresponding to each position $P = (x, y)$ of the object moving about the circle, there is the point $Q = (x, 0)$, called the **projection of P on the x-axis**. As P moves around the circle at a constant rate, the point Q moves back and forth between the points $(a, 0)$ and $(-a, 0)$ along the x-axis with a motion that is simple harmonic. Similarly, for each point P there is a point $Q' = (0, y)$, called the **projection of P on the y-axis**. As P moves around the circle, the point Q' moves back and forth between the points $(0, a)$ and $(0, -a)$ on the y-axis with a motion that is simple harmonic. Simple harmonic motion can be described as the projection of constant circular motion on a coordinate axis.

To put it another way, again consider a mass hanging from a spring where the mass is pulled down from its rest position to the point C and then released. See Figure 35(a). The graph shown in Figure 35(b) describes the displacement d of the object from its rest position as a function of time t, assuming that no frictional force is present.

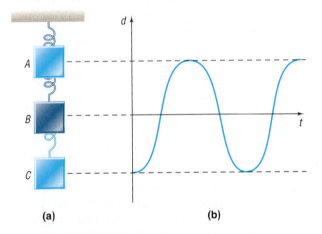

Figure 35 (a) (b)

THEOREM

Simple Harmonic Motion

An object that moves on a coordinate axis so that the displacement d from its rest position at time t is given by either

$$d = a \cos(\omega t) \quad \text{or} \quad d = a \sin(\omega t)$$

where a and $\omega > 0$ are constants, moves with simple harmonic motion. The motion has amplitude $|a|$ and period $\dfrac{2\pi}{\omega}$.

The **frequency** f of an object in simple harmonic motion is the number of oscillations per unit time. Since the period is the time required for one oscillation, it follows that the frequency is the reciprocal of the period; that is,

$$f = \frac{\omega}{2\pi} \qquad \omega > 0$$

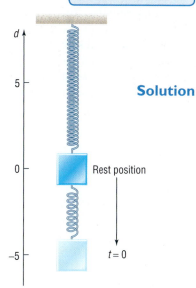

Figure 36

NOTE In the solution to Example 1, $a = -5$ because the object is initially pulled down. (If the initial direction is up, then use $a = 5$.)

EXAMPLE 1

Build a Model for an Object in Harmonic Motion

Suppose that an object attached to a coiled spring is pulled down a distance of 5 inches from its rest position and then released. If the time for one oscillation is 3 seconds, develop a model that relates the displacement d of the object from its rest position after time t (in seconds). Assume no friction.

Solution

The motion of the object is simple harmonic. See Figure 36. When the object is released $(t = 0)$, the displacement of the object from the rest position is -5 units (since the object was pulled down). Because $d = -5$ when $t = 0$, it is easier to use the cosine function*

$$d = a \cos(\omega t)$$

to describe the motion. Now the amplitude is $\left|-5\right| = 5$ and the period is 3, so

$$a = -5 \quad \text{and} \quad \frac{2\pi}{\omega} = \text{period} = 3, \quad \text{so } \omega = \frac{2\pi}{3}$$

An equation that models the motion of the object is

$$d = -5 \cos\left[\frac{2\pi}{3} t\right]$$

━━━ **Now Work** PROBLEM 5

■ **2 Analyze Simple Harmonic Motion**

EXAMPLE 2

Analyzing the Motion of an Object

Suppose that the displacement d (in meters) of an object at time t (in seconds) satisfies the equation

$$d = 10 \sin(5t)$$

(a) Describe the motion of the object.
(b) What is the maximum displacement from its resting position?
(c) What is the time required for one oscillation?
(d) What is the frequency?

Solution

Observe that the given equation is of the form

$$d = a \sin(\omega t) \quad d = 10 \sin(5t)$$

where $a = 10$ and $\omega = 5$.

(a) The motion is simple harmonic.
(b) The maximum displacement of the object from its resting position is the amplitude: $\left|a\right| = 10$ meters.
(c) The time required for one oscillation is the period:

$$\text{Period} = \frac{2\pi}{\omega} = \frac{2\pi}{5} \text{ seconds}$$

(d) The frequency is the reciprocal of the period.

$$\text{Frequency} = f = \frac{5}{2\pi} \text{ oscillation per second}$$

━━━ **Now Work** PROBLEM 13

* No phase shift is required when a cosine function is used.

3 Analyze an Object in Damped Motion

Most physical phenomena are affected by friction or other resistive forces. These forces remove energy from a moving system and thereby damp its motion. For example, when a mass hanging from a spring is pulled down a distance a and released, the friction in the spring causes the distance that the mass moves from its at-rest position to decrease over time. As a result, the amplitude of any real oscillating spring or swinging pendulum decreases with time due to air resistance, friction, and so forth. See Figure 37.

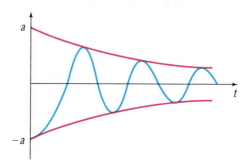

Figure 37

A model that describes this phenomenon maintains a sinusoidal component, but the amplitude of this component will decrease with time to account for the damping effect. In addition, the period of the oscillating component will be affected by the damping. The next result, from physics, describes damped motion.

THEOREM

Damped Motion

The displacement d of an oscillating object from its at-rest position at time t is given by

$$d(t) = ae^{-bt/(2m)} \cos\left(\sqrt{\omega^2 - \frac{b^2}{4m^2}}\, t\right)$$

where b is the **damping factor** or **damping coefficient** and m is the mass of the oscillating object. Here $|a|$ is the displacement at $t = 0$, and $\dfrac{2\pi}{\omega}$ is the period under simple harmonic motion (no damping).

Notice that for $b = 0$ (zero damping), we have the formula for simple harmonic motion with amplitude $|a|$ and period $\dfrac{2\pi}{\omega}$.

EXAMPLE 3

Analyzing a Damped Vibration Curve

Analyze the damped vibration curve

$$d(t) = e^{-t/\pi} \cos t \quad t \geq 0$$

Solution

The displacement d is the product of $y = e^{-t/\pi}$ and $y = \cos t$. Using properties of absolute value and the fact that $|\cos t| \leq 1$, it follows that

$$|d(t)| = \left|e^{-t/\pi} \cos t\right| = \left|e^{-t/\pi}\right| |\cos t| \leq \left|e^{-t/\pi}\right| = e^{-t/\pi}$$

$$\underset{\substack{\uparrow \\ e^{-t/\pi} > 0}}{}$$

As a result,

$$-e^{-t/\pi} \leq d(t) \leq e^{-t/\pi}$$

This means that the graph of d will lie between the graphs of $y = e^{-t/\pi}$ and $y = -e^{-t/\pi}$, called the **bounding curves** of d.

Also, the graph of d will touch these graphs when $|\cos t| = 1$; that is, when $t = 0, \pi, 2\pi$, and so on. The x-intercepts of the graph of d occur when $\cos t = 0$, that is, at $\dfrac{\pi}{2}, \dfrac{3\pi}{2}, \dfrac{5\pi}{2}$, and so on. See Table 1.

Table 1

t	0	$\dfrac{\pi}{2}$	π	$\dfrac{3\pi}{2}$	2π
$e^{-t/\pi}$	1	$e^{-1/2}$	e^{-1}	$e^{-3/2}$	e^{-2}
$\cos t$	1	0	-1	0	1
$d(t) = e^{-t/\pi} \cos t$	1	0	$-e^{-1}$	0	e^{-2}
Point on graph of d	$(0, 1)$	$\left(\dfrac{\pi}{2}, 0\right)$	$(\pi, -e^{-1})$	$\left(\dfrac{3\pi}{2}, 0\right)$	$(2\pi, e^{-2})$

The graphs of $y = \cos t$, $y = e^{-t/\pi}$, $y = -e^{-t/\pi}$, and $d(t) = e^{-t/\pi} \cos t$ are shown in in Figure 38.

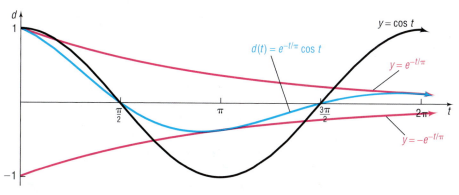

Figure 38

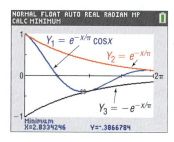

Figure 39

Exploration

Graph $Y_1 = e^{-x/\pi} \cos x$, along with $Y_2 = e^{-x/\pi}$ and $Y_3 = -e^{-x/\pi}$, for $0 \le x \le 2\pi$. Determine where Y_1 has its first turning point (local minimum). Compare this to where Y_1 intersects Y_3.

Result Figure 39 shows the graphs of $Y_1 = e^{-x/\pi} \cos x$, $Y_2 = e^{-x/\pi}$, and $Y_3 = -e^{-x/\pi}$. Using MINIMUM, the first turning point occurs at $x \approx 2.83$; Y_1 INTERSECTS Y_3 at $x = \pi \approx 3.14$.

 Now Work PROBLEM 21

4 Graph the Sum of Two Functions

Many physical and biological applications require the graph of the sum of two functions, such as

$$f(x) = x + \sin x \quad \text{or} \quad g(x) = \sin x + \cos(2x)$$

For example, if two tones are emitted, the sound produced is the sum of the waves produced by the two tones. See Problem 57 for an explanation of Touch-Tone phones.

To graph the sum of two (or more) functions, use the method of adding y-coordinates described next.

EXAMPLE 4

Graphing the Sum of Two Functions

Use the method of adding y-coordinates to graph $f(x) = x + \sin x$.

Solution First, graph the component functions,

$$y = f_1(x) = x \qquad y = f_2(x) = \sin x$$

on the same coordinate system. See Figure 40(a). Now, select several values of x say $x = 0$, $x = \dfrac{\pi}{2}$, $x = \pi$, $x = \dfrac{3\pi}{2}$, and $x = 2\pi$, and use them to compute $f(x) = f_1(x) + f_2(x)$. Table 2 shows the computations. Plot these points and connect them to get the graph, as shown in Figure 40(b).

Table 2

x	0	$\dfrac{\pi}{2}$	π	$\dfrac{3\pi}{2}$	2π
$y = f_1(x) = x$	0	$\dfrac{\pi}{2}$	π	$\dfrac{3\pi}{2}$	2π
$y = f_2(x) = \sin x$	0	1	0	-1	0
$f(x) = x + \sin x$	0	$\dfrac{\pi}{2} + 1 \approx 2.57$	π	$\dfrac{3\pi}{2} - 1 \approx 3.71$	2π
Point on graph of f	$(0, 0)$	$\left(\dfrac{\pi}{2}, 2.57\right)$	(π, π)	$\left(\dfrac{3\pi}{2}, 3.71\right)$	$(2\pi, 2\pi)$

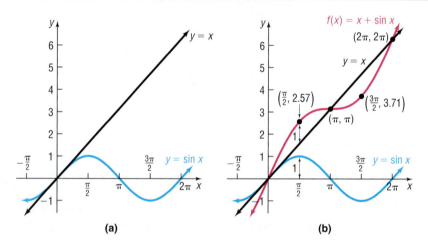

Figure 40
(a)   (b)

In Figure 40(b), notice that the graph of $f(x) = x + \sin x$ intersects the line $y = x$ whenever $\sin x = 0$. Also, notice that the graph of f is not periodic. ●

✓ **Check:** Graph $Y_1 = x$, $Y_2 = \sin x$, and $Y_3 = x + \sin x$ and compare the result with Figure 40(b). Use INTERSECT to verify that the graphs of Y_1 and Y_3 intersect when $\sin x = 0$.

The next example shows a periodic graph.

EXAMPLE 5   **Graphing the Sum of Two Sinusoidal Functions**

Use the method of adding y-coordinates to graph

$$f(x) = \sin x + \cos(2x)$$

Solution   Table 3 shows the steps for computing several points on the graph of f. Figure 41 on page 711 illustrates the graphs of the component functions, $y = f_1(x) = \sin x$ and $y = f_2(x) = \cos(2x)$, and the graph of $f(x) = \sin x + \cos(2x)$, which is shown in red.

Table 3

x	$-\dfrac{\pi}{2}$	0	$\dfrac{\pi}{2}$	π	$\dfrac{3\pi}{2}$	2π
$y = f_1(x) = \sin x$	-1	0	1	0	-1	0
$y = f_2(x) = \cos(2x)$	-1	1	-1	1	-1	1
$f(x) = \sin x + \cos(2x)$	-2	1	0	1	-2	1
Point on graph of f	$\left(-\dfrac{\pi}{2}, -2\right)$	$(0, 1)$	$\left(\dfrac{\pi}{2}, 0\right)$	$(\pi, 1)$	$\left(\dfrac{3\pi}{2}, -2\right)$	$(2\pi, 1)$

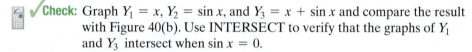

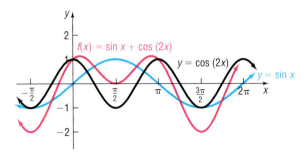

Figure 41

Notice that f is periodic, with period 2π.

✓ **Check:** Graph $Y_1 = \sin x$, $Y_2 = \cos(2x)$, and $Y_3 = \sin x + \cos(2x)$ and compare the result with Figure 41.

Now Work PROBLEM 25

9.5 Assess Your Understanding

'Are You Prepared?' *The answers are given at the end of these exercises. If you get a wrong answer, read the pages listed in red.*

1. The amplitude A and period T of $f(x) = 5\sin(4x)$ are ____ and ____ . (pp. 568–574)

Concepts and Vocabulary

2. The motion of an object obeys the equation $d = 4\cos(6t)$. Such motion is described as _____ _____ . The number 4 is called the _____ .

3. When a mass hanging from a spring is pulled down and then released, the motion is called _____ _____ if there is

   no frictional force to retard the motion, and the motion is called _____ if there is such friction.

4. *True or False* If the distance d of an object from its rest position at time t is given by a sinusoidal graph, the motion of the object is simple harmonic motion.

Skill Building

In Problems 5–8, an object attached to a coiled spring is pulled down a distance a from its rest position and then released. Assuming that the motion is simple harmonic with period T, write an equation that relates the displacement d of the object from its rest position after t seconds. Also assume that the positive direction of the motion is up.

5. $a = 5$; $T = 2$ seconds

6. $a = 10$; $T = 3$ seconds

7. $a = 6$; $T = \pi$ seconds

8. $a = 4$; $T = \dfrac{\pi}{2}$ seconds

9. Rework Problem 5 under the same conditions, except that at time $t = 0$, the object is at its resting position and moving down.

10. Rework Problem 6 under the same conditions, except that at time $t = 0$, the object is at its resting position and moving down.

11. Rework Problem 7 under the same conditions, except that at time $t = 0$, the object is at its resting position and moving down.

12. Rework Problem 8 under the same conditions, except that at time $t = 0$, the object is at its resting position and moving down.

In Problems 13–20, the displacement d (in meters) of an object at time t (in seconds) is given.
   (a) Describe the motion of the object.
   (b) What is the maximum displacement from its resting position?
   (c) What is the time required for one oscillation?
   (d) What is the frequency?

13. $d = 5\sin(3t)$

14. $d = 4\sin(2t)$

15. $d = 6\cos(\pi t)$

16. $d = 5\cos\left(\dfrac{\pi}{2}t\right)$

17. $d = -3\sin\left(\dfrac{1}{2}t\right)$

18. $d = -2\cos(2t)$

19. $d = 6 + 2\cos(2\pi t)$

20. $d = 4 + 3\sin(\pi t)$

Chapter Projects

I. Spherical Trigonometry When the distance between two locations on the surface of Earth is small, we can compute the distance in statutory miles. Using this assumption, we can use the Law of Sines and the Law of Cosines to approximate distances and angles. However, if you look at a globe, you notice that Earth is a sphere, so as the distance between two points on its surface increases, the linear distance is less accurate because of curvature. Under this circumstance, we need to take into account the curvature of Earth when using the Law of Sines and the Law of Cosines.

1. See the figure. The points A, B and C are the vertices of a spherical triangle with sides a, b, and c, a three-sided figure drawn on the surface of a sphere with center at the point O. Then connect each vertex by a radius to the center O of the sphere. Now draw tangent lines to the sides a and b of the triangle that go through C. Extend the lines OA and OB to intersect the tangent lines at P and Q, respectively. List the plane right triangles. Determine the measures of the central angles.

2. Apply the Law of Cosines to triangles OPQ and CPQ to find two expressions for the length of PQ.

3. Subtract the expressions in part (2) from each other. Solve for the term containing cos c.

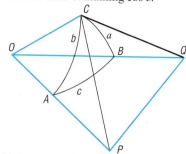

4. Use the Pythagorean Theorem to find another value for $OQ^2 - CQ^2$ and $OP^2 - CP^2$. Now solve for cos c.

5. Replacing the ratios in part (4) by the cosines of the sides of the spherical triangle, you should now have the Law of Cosines for spherical triangles:

$$\cos c = \cos a \cos b + \sin a \sin b \cos C$$

Source: For the spherical Law of Cosines, see *Mathematics from the Birth of Numbers* by Jan Gullberg. W. W. Norton & Co., Publishers, 1996, pp. 491–494.

II. The Lewis and Clark Expedition Lewis and Clark followed several rivers in their trek from what is now Great Falls, Montana, to the Pacific coast. First, they went down the Missouri and Jefferson rivers from Great Falls to Lemhi, Idaho. Because the two cities are at different longitudes and different latitudes, we must account for the curvature of Earth when computing the distance that they traveled. Assume that the radius of Earth is 3960 miles.

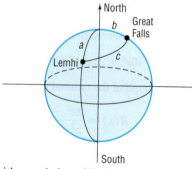

1. Great Falls is at approximately 47.5°N and 111.3°W. Lemhi is at approximately 45.5°N and 113.5°W. (We will assume that

the rivers flow straight from Great Falls to Lemhi on the surface of Earth.) This line is called a geodesic line. Apply the Law of Cosines for a spherical triangle to find the angle between Great Falls and Lemhi. (The central angles are found by using the differences in the latitudes and longitudes of the towns. See the diagram.) Then find the length of the arc joining the two towns. (Recall $s = r\theta$.)

2. From Lemhi, they went up the Bitteroot River and the Snake River to what is now Lewiston and Clarkston on the border of Idaho and Washington. Although this is not really a side to a triangle, we will make a side that goes from Lemhi to Lewiston and Clarkston. If Lewiston and Clarkston are at about 46.5°N 117.0°W, find the distance from Lemhi using the Law of Cosines for a spherical triangle and the arc length.

3. How far did the explorers travel just to get that far?

4. Draw a plane triangle connecting the three towns. If the distance from Lewiston to Great Falls is 282 miles and the angle at Great Falls is 42° and the angle at Lewiston is 48.5°, find the distance from Great Falls to Lemhi and from Lemhi to Lewiston. How do these distances compare with the ones computed in parts (1) and (2)?

Source: For Lewis and Clark Expedition: *American Journey: The Quest for Liberty to 1877, Texas Edition.* Prentice Hall, 1992, p. 345.

Source: For map coordinates: *National Geographic Atlas of the World,* published by National Geographic Society, 1981, pp. 74–75.

Citation: Used with permission of *Technology Review*, from W. Roush, "From Lewis and Clark to Landsat: David Rumsey's Digital Maps Marry Past and Present," 108, no. 7, © 2005; permission conveyed through Copyright Clearance Center, Inc.

The following projects are available at the Instructor's Resource Center (IRC).

III. Project at Motorola: *How Can You Build or Analyze a Vibration Profile?* Fourier functions not only are important to analyze vibrations but also are what a mathematician would call interesting. Complete the project to see why.

IV. Leaning Tower of Pisa Trigonometry is used to analyze the apparent height and tilt of the Leaning Tower of Pisa.

V. Locating Lost Treasure Clever treasure seekers who know the Law of Sines are able to find a buried treasure efficiently.

VI. Jacob's Field Angles of elevation and the Law of Sines are used to determine the height of the stadium wall and the distance from home plate to the top of the wall.

Polar Coordinates; Vectors

How Airplanes Fly

Four aerodynamic forces act on an airplane in flight: **lift**, **drag**, **thrust**, and **weight** (gravity).

Drag is the resistance of air molecules hitting the airplane (the *backward* force), thrust is the power of the airplane's engine (the *forward* force), lift is the *upward* force, and weight is the *downward* force. So for airplanes to fly and stay airborne, the thrust must be greater than the drag, and the lift must be greater than the weight (*so, drag opposes thrust, and lift opposes weight*).

This is certainly the case when an airplane takes off or climbs. However, when it is in straight and level flight, the opposing forces of lift, and weight are balanced. During a descent, weight exceeds lift, and to slow the airplane, drag has to overcome thrust.

Thrust is generated by the airplane's engine (propeller or jet), weight is created by the natural force of gravity acting on the airplane, and drag comes from friction as the plane moves through air molecules. Drag is also a *reaction* to lift, and this lift must be generated by the airplane in flight. This is done by the **wings** of the airplane.

A cross section of a typical airplane wing shows the top surface to be more curved than the bottom surface. This shaped profile is called an **airfoil** (or aerofoil), and the shape is used because an airfoil generates significantly more lift than opposing drag. In other words, it is very **efficient** at generating lift.

During flight, air naturally flows over and beneath the wing and is deflected upward over the top surface and downward beneath the lower surface. Any difference in deflection causes a difference in air pressure (pressure gradient), and because of the airfoil shape, the pressure of the deflected air is lower above the airfoil than below it. As a result the wing is "pushed" upward by the higher pressure beneath, or, you can argue, it is "sucked" upward by the lower pressure above.

Source: Adapted from Pete Carpenter. How Airplanes Fly—The Basic Principles of Flight
http://www.rc-airplane-world.com/how-airplanes-fly.html,
accessed June 2014. © rc-airplane-world.com

—See Chapter Project I—

••• A Look Back, A Look Ahead •••

This chapter is in two parts: Polar Coordinates (Sections 10.1–10.3) and Vectors (Sections 10.4–10.5). They are independent of each other and may be covered in either order.

Sections 10.1–10.3: In Chapter 2 we introduced rectangular coordinates (the *xy*-plane) and discussed the graph of an equation in two variables involving *x* and *y*. In Sections 10.1 and 10.2, we introduce polar coordinates, an alternative to rectangular coordinates, and discuss graphing equations that involve polar coordinates. In Section 6.3, we discussed raising a real number to a real power. In Section 10.3, we extend this idea by raising a complex number to a real power. As it turns out, polar coordinates are useful for the discussion.

Sections 10.4–10.5: We have seen in many chapters that we are often required to solve an equation to obtain a solution to applied problems. In the last two sections of this chapter, we develop the notion of a vector and show how it can be used to model applied problems in physics and engineering.

Outline

10.1 Polar Coordinates

PREPARING FOR THIS SECTION *Before getting started, review the following:*

- Rectangular Coordinates (Section 2.1, pp. 150–154)
- Definition of the Trigonometric Functions (Section 7.4, pp. 543–545)
- Inverse Tangent Function (Section 8.1, pp. 613–616)
- Completing the Square (Chapter R, Review, Section R.5, p. 56)

Now Work the 'Are You Prepared?' problems on page 727.

OBJECTIVES 1 Plot Points Using Polar Coordinates (p. 720)
2 Convert from Polar Coordinates to Rectangular Coordinates (p. 722)
3 Convert from Rectangular Coordinates to Polar Coordinates (p. 724)
4 Transform Equations between Polar and Rectangular Forms (p. 726)

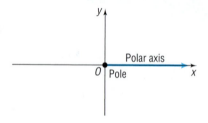

Figure 1

So far, we have always used a system of rectangular coordinates to plot points in the plane. Now we are ready to describe another system, called *polar coordinates*. In many instances, polar coordinates offer certain advantages over rectangular coordinates.

In a rectangular coordinate system, you will recall, a point in the plane is represented by an ordered pair of numbers (x, y), where x and y equal the signed distances of the point from the y-axis and the x-axis, respectively. In a polar coordinate system, we select a point, called the **pole**, and then a ray with vertex at the pole, called the **polar axis**. See Figure 1. Comparing the rectangular and polar coordinate systems, note that the origin in rectangular coordinates coincides with the pole in polar coordinates, and the positive x-axis in rectangular coordinates coincides with the polar axis in polar coordinates.

1 Plot Points Using Polar Coordinates

A point P in a polar coordinate system is represented by an ordered pair of numbers (r, θ). If $r > 0$, then r is the distance of the point from the pole; θ is an angle (in degrees or radians) formed by the polar axis and a ray from the pole through the point. We call the ordered pair (r, θ) the **polar coordinates** of the point. See Figure 2.

As an example, suppose that a point P has polar coordinates $\left(2, \dfrac{\pi}{4}\right)$. Locate P by first drawing an angle of $\dfrac{\pi}{4}$ radian, placing its vertex at the pole and its initial side along the polar axis. Then go out a distance of 2 units along the terminal side of the angle to reach the point P. See Figure 3.

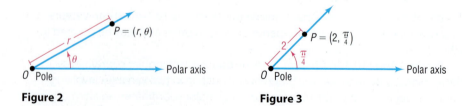

Figure 2   **Figure 3**

In using polar coordinates (r, θ), it is possible for r to be negative. When this happens, instead of the point being on the terminal side of θ, it is on the ray from the pole extending in the direction *opposite* the terminal side of θ at a distance $|r|$ units from the pole. See Figure 4 for an illustration.

For example, to plot the point $\left(-3, \dfrac{2\pi}{3}\right)$, use the ray in the opposite direction of $\dfrac{2\pi}{3}$ and go out $|-3| = 3$ units along that ray. See Figure 5.

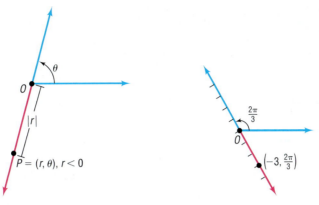

Figure 4   Figure 5

EXAMPLE 1   **Plotting Points Using Polar Coordinates**

Plot the points with the following polar coordinates:

(a) $\left(3, \dfrac{5\pi}{3}\right)$ (b) $\left(2, -\dfrac{\pi}{4}\right)$ (c) $(3, 0)$ (d) $\left(-2, \dfrac{\pi}{4}\right)$

Solution   Figure 6 shows the points.

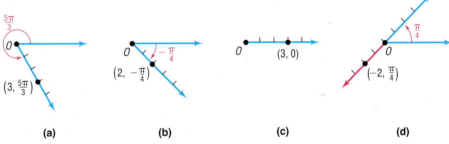

Figure 6   (a)   (b)   (c)   (d)

Now Work PROBLEMS **11, 19, AND 29**

Recall that an angle measured counterclockwise is positive and an angle measured clockwise is negative. This convention has some interesting consequences related to polar coordinates.

EXAMPLE 2   **Finding Several Polar Coordinates of a Single Point**

Consider again the point P with polar coordinates $\left(2, \dfrac{\pi}{4}\right)$, as shown in Figure 7(a). Because $\dfrac{\pi}{4}, \dfrac{9\pi}{4}$, and $-\dfrac{7\pi}{4}$ all have the same terminal side, this point P also can be located by using the polar coordinates $\left(2, \dfrac{9\pi}{4}\right)$ or the polar coordinates $\left(2, -\dfrac{7\pi}{4}\right)$, as shown in Figures 7(b) and (c). The point $\left(2, \dfrac{\pi}{4}\right)$ can also be represented by the polar coordinates $\left(-2, \dfrac{5\pi}{4}\right)$. See Figure 7(d).

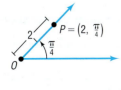 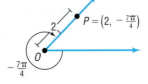 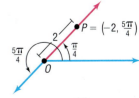

Figure 7   (a)   (b)   (c)   (d)

EXAMPLE 3 **Finding Other Polar Coordinates of a Given Point**

Plot the point P with polar coordinates $\left(3, \dfrac{\pi}{6}\right)$, and find other polar coordinates (r, θ) of this same point for which:

(a) $r > 0, \quad 2\pi \le \theta < 4\pi$ (b) $r < 0, \quad 0 \le \theta < 2\pi$

(c) $r > 0, \quad -2\pi \le \theta < 0$

Solution The point $\left(3, \dfrac{\pi}{6}\right)$ is plotted in Figure 8.

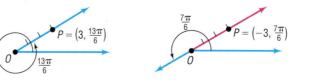

Figure 8

(a) Add 1 revolution (2π radians) to the angle $\dfrac{\pi}{6}$ to get

$$P = \left(3, \frac{\pi}{6} + 2\pi\right) = \left(3, \frac{13\pi}{6}\right).$$

See Figure 9.

(b) Add $\dfrac{1}{2}$ revolution (π radians) to the angle $\dfrac{\pi}{6}$, and replace 3 by -3 to get

$$P = \left(-3, \frac{\pi}{6} + \pi\right) = \left(-3, \frac{7\pi}{6}\right). \text{ See Figure 10.}$$

(c) Subtract 2π from the angle $\dfrac{\pi}{6}$ to get $P = \left(3, \dfrac{\pi}{6} - 2\pi\right) = \left(3, -\dfrac{11\pi}{6}\right)$. See Figure 11.

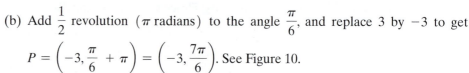

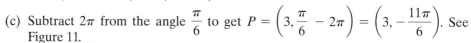

Figure 9   **Figure 10**   **Figure 11**

These examples show a major difference between rectangular coordinates and polar coordinates. A point has exactly one pair of rectangular coordinates; however, a point has infinitely many pairs of polar coordinates.

SUMMARY

A point with polar coordinates (r, θ), θ in radians, can also be represented by either of the following:

$$(r, \theta + 2\pi k) \quad \text{or} \quad (-r, \theta + \pi + 2\pi k) \qquad k \text{ any integer}$$

The polar coordinates of the pole are $(0, \theta)$, where θ can be any angle.

Now Work PROBLEM 33

2 Convert from Polar Coordinates to Rectangular Coordinates

Sometimes it is necessary to convert coordinates or equations in rectangular form to polar form, and vice versa. To do this, recall that the origin in rectangular coordinates is the pole in polar coordinates and that the positive x-axis in rectangular coordinates is the polar axis in polar coordinates.

THEOREM

Conversion from Polar Coordinates to Rectangular Coordinates

If P is a point with polar coordinates (r, θ), the rectangular coordinates (x, y) of P are given by

$$x = r \cos \theta \qquad y = r \sin \theta \tag{1}$$

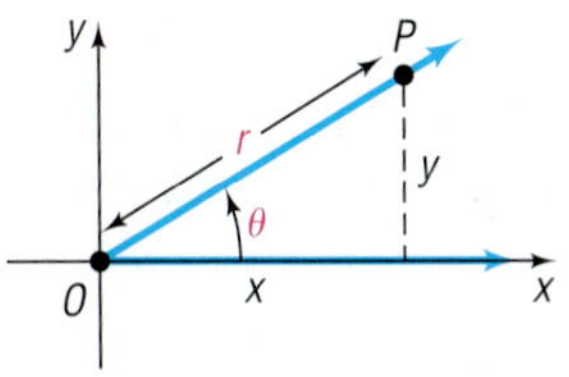

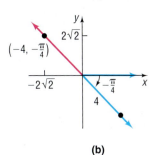

Figure 12

Proof Suppose that P has the polar coordinates (r, θ). We seek the rectangular coordinates (x, y) of P. Refer to Figure 12.

If $r = 0$, then, regardless of θ, the point P is the pole, for which the rectangular coordinates are $(0, 0)$. Formula (1) is valid for $r = 0$.

If $r > 0$, the point P is on the terminal side of θ, and $r = d(O, P) = \sqrt{x^2 + y^2}$. Because

$$\cos \theta = \frac{x}{r} \qquad \sin \theta = \frac{y}{r}$$

this means

$$x = r \cos \theta \qquad y = r \sin \theta$$

If $r < 0$ and θ is in radians, the point $P = (r, \theta)$ can be represented as $(-r, \pi + \theta)$, where $-r > 0$. Because

$$\cos(\pi + \theta) = -\cos \theta = \frac{x}{-r} \qquad \sin(\pi + \theta) = -\sin \theta = \frac{y}{-r}$$

this means

$$x = r \cos \theta \qquad y = r \sin \theta \qquad\qquad ■$$

EXAMPLE 4

Converting from Polar Coordinates to Rectangular Coordinates

Find the rectangular coordinates of the points with the following polar coordinates:

(a) $\left(6, \dfrac{\pi}{6} \right)$ (b) $\left(-4, -\dfrac{\pi}{4} \right)$

Solution Use formula (1): $x = r \cos \theta$ and $y = r \sin \theta$.

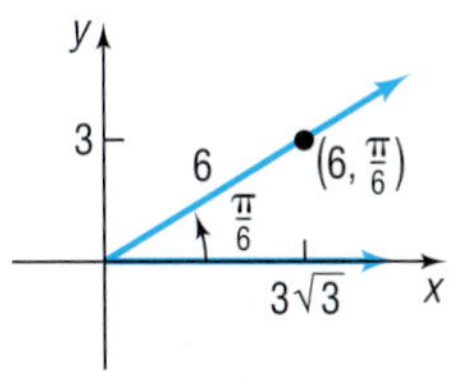

(a) Figure 13(a) shows $\left(6, \dfrac{\pi}{6} \right)$ plotted. Notice that $\left(6, \dfrac{\pi}{6} \right)$ lies in quadrant I of the rectangular coordinate system. So both the x-coordinate and the y-coordinate will be positive. Substituting $r = 6$ and $\theta = \dfrac{\pi}{6}$ gives

$$x = r \cos \theta = 6 \cos \frac{\pi}{6} = 6 \cdot \frac{\sqrt{3}}{2} = 3\sqrt{3}$$

$$y = r \sin \theta = 6 \sin \frac{\pi}{6} = 6 \cdot \frac{1}{2} = 3$$

The rectangular coordinates of the point $\left(6, \dfrac{\pi}{6} \right)$ are $\left(3\sqrt{3}, 3 \right)$, which lies in quadrant I, as expected.

(b) Figure 13(b) shows $\left(-4, -\dfrac{\pi}{4} \right)$ plotted. Notice that $\left(-4, -\dfrac{\pi}{4} \right)$ lies in quadrant II of the rectangular coordinate system. Substituting $r = -4$ and $\theta = -\dfrac{\pi}{4}$ gives

$$x = r \cos \theta = -4 \cos\left(-\frac{\pi}{4} \right) = -4 \cdot \frac{\sqrt{2}}{2} = -2\sqrt{2}$$

$$y = r \sin \theta = -4 \sin\left(-\frac{\pi}{4} \right) = -4\left(-\frac{\sqrt{2}}{2} \right) = 2\sqrt{2}$$

The rectangular coordinates of the point $\left(-4, -\dfrac{\pi}{4} \right)$ are $\left(-2\sqrt{2}, 2\sqrt{2} \right)$, which lies in quadrant II, as expected.

(a)

(b)

Figure 13

COMMENT Many calculators have the capability of converting from polar coordinates to rectangular coordinates. Consult your owner's manual for the proper keystrokes. In most cases this procedure is tedious, so you will probably find that using formula (1) is faster. ■

Now Work PROBLEMS **41** AND **53**

3 Convert from Rectangular Coordinates to Polar Coordinates

Converting from rectangular coordinates (x, y) to polar coordinates (r, θ) is a little more complicated. Notice that each solution begins by plotting the given rectangular coordinates.

EXAMPLE 5   **How to Convert from Rectangular Coordinates to Polar Coordinates with the Point on a Coordinate Axis**

Find polar coordinates of a point whose rectangular coordinates are $(0, 3)$.

Step-by-Step Solution

Step 1: Plot the point (x, y) and note the quadrant the point lies in or the coordinate axis the point lies on.

Plot the point $(0, 3)$ in a rectangular coordinate system. See Figure 14. The point lies on the positive y-axis.

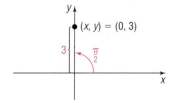

Figure 14

Step 2: Determine the distance r from the origin to the point.

The point $(0, 3)$ lies on the y-axis a distance of 3 units from the origin (pole), so $r = 3$.

Step 3: Determine θ.

A ray with vertex at the pole through $(0, 3)$ forms an angle $\theta = \dfrac{\pi}{2}$ with the polar axis.

Polar coordinates for this point can be given by $\left(3, \dfrac{\pi}{2}\right)$. Other possible representations include $\left(-3, -\dfrac{\pi}{2}\right)$ and $\left(3, \dfrac{5\pi}{2}\right)$. •

Figure 15 shows polar coordinates of points that lie on either the x-axis or the y-axis. In each illustration, $a > 0$.

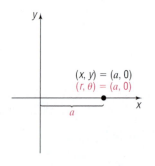 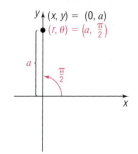 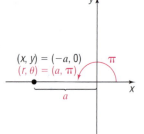 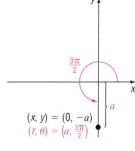

Figure 15   (a) $(x, y) = (a, 0), a > 0$ (b) $(x, y) = (0, a), a > 0$ (c) $(x, y) = (-a, 0), a > 0$ (d) $(x, y) = (0, -a), a > 0$

 **Now Work** PROBLEM 57

EXAMPLE 6   **How to Convert from Rectangular Coordinates to Polar Coordinates with the Point in a Quadrant**

Find the polar coordinates of a point whose rectangular coordinates are $(2, -2)$.

Step-by-Step Solution

Step 1: Plot the point (x, y) and note the quadrant the point lies in or the coordinate axis the point lies on.

Plot the point $(2, -2)$ in a rectangular coordinate system. See Figure 16. The point lies in quadrant IV.

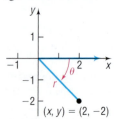

Figure 16

Step 2: Determine the distance r from the origin to the point using $r = \sqrt{x^2 + y^2}$.

$$r = \sqrt{x^2 + y^2} = \sqrt{(2)^2 + (-2)^2} = \sqrt{8} = 2\sqrt{2}$$

Step 3: Determine θ.

Find θ by recalling that $\tan\theta = \dfrac{y}{x}$, so $\theta = \tan^{-1}\dfrac{y}{x}, -\dfrac{\pi}{2} < \theta < \dfrac{\pi}{2}$. Because $(2, -2)$ lies in quadrant IV, this means that $-\dfrac{\pi}{2} < \theta < 0$. As a result,

$$\theta = \tan^{-1}\frac{y}{x} = \tan^{-1}\left(\frac{-2}{2}\right) = \tan^{-1}(-1) = -\frac{\pi}{4}$$

 COMMENT Many calculators have the capability of converting from rectangular coordinates to polar coordinates. Consult your owner's manual for the proper keystrokes. ∎

A set of polar coordinates for the point $(2, -2)$ is $\left(2\sqrt{2}, -\dfrac{\pi}{4}\right)$. Other possible representations include $\left(2\sqrt{2}, \dfrac{7\pi}{4}\right)$ and $\left(-2\sqrt{2}, \dfrac{3\pi}{4}\right)$. ●

EXAMPLE 7

Converting from Rectangular Coordinates to Polar Coordinates

Find polar coordinates of a point whose rectangular coordinates are $\left(-1, -\sqrt{3}\right)$.

Solution

STEP 1: See Figure 17. The point lies in quadrant III.

STEP 2: The distance r from the origin to the point $\left(-1, -\sqrt{3}\right)$ is

$$r = \sqrt{(-1)^2 + \left(-\sqrt{3}\right)^2} = \sqrt{4} = 2$$

STEP 3: To find θ, use $\alpha = \tan^{-1}\dfrac{y}{x} = \tan^{-1}\dfrac{-\sqrt{3}}{-1} = \tan^{-1}\sqrt{3}, -\dfrac{\pi}{2} < \alpha < \dfrac{\pi}{2}$.

Since the point $\left(-1, -\sqrt{3}\right)$ lies in quadrant III and the inverse tangent function gives an angle in quadrant I, add π to the result to obtain an angle in quadrant III.

$$\theta = \pi + \tan^{-1}\left(\frac{-\sqrt{3}}{-1}\right) = \pi + \tan^{-1}\sqrt{3} = \pi + \frac{\pi}{3} = \frac{4\pi}{3}$$

A set of polar coordinates for this point is $\left(2, \dfrac{4\pi}{3}\right)$. Other possible representations include $\left(-2, \dfrac{\pi}{3}\right)$ and $\left(2, -\dfrac{2\pi}{3}\right)$. ●

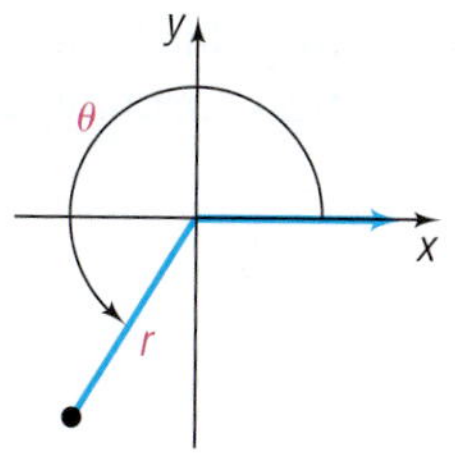

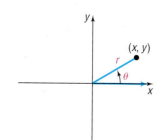

$(x, y) = (-1, -\sqrt{3})$

Figure 17

Figure 18 shows how to find polar coordinates of a point that lies in a quadrant when its rectangular coordinates (x, y) are given.

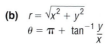
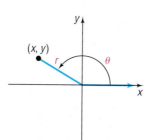
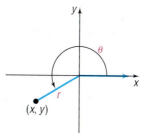

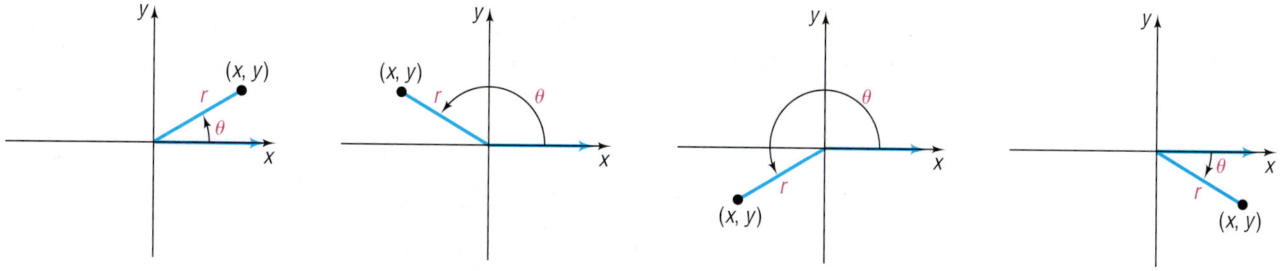

Figure 18 (a) $r = \sqrt{x^2 + y^2}$
$\theta = \tan^{-1}\dfrac{y}{x}$

(b) $r = \sqrt{x^2 + y^2}$
$\theta = \pi + \tan^{-1}\dfrac{y}{x}$

(c) $r = \sqrt{x^2 + y^2}$
$\theta = \pi + \tan^{-1}\dfrac{y}{x}$

(d) $r = \sqrt{x^2 + y^2}$
$\theta = \tan^{-1}\dfrac{y}{x}$

The preceding discussion provides the formulas

$$r^2 = x^2 + y^2 \qquad \tan\theta = \frac{y}{x} \qquad \text{if } x \neq 0 \qquad (2)$$

To use formula (2) effectively, follow these steps:

Steps for Converting from Rectangular to Polar Coordinates

STEP 1: Always plot the point (x, y) first, as shown in Examples 5, 6, and 7. Note the quadrant the point lies in or the coordinate axis the point lies on.

STEP 2: If $x = 0$ or $y = 0$, use your illustration to find r. If $x \neq 0$ and $y \neq 0$, then $r = \sqrt{x^2 + y^2}$.

STEP 3: Find θ. If $x = 0$ or $y = 0$, use your illustration to find θ. If $x \neq 0$ and $y \neq 0$, note the quadrant in which the point lies.

$$\text{Quadrant I or IV:} \quad \theta = \tan^{-1} \frac{y}{x}$$

$$\text{Quadrant II or III:} \quad \theta = \pi + \tan^{-1} \frac{y}{x}$$

 **Now Work** PROBLEM 61

4 Transform Equations between Polar and Rectangular Forms

Formulas (1) and (2) can also be used to transform equations from polar form to rectangular form, and vice versa. Two common techniques for transforming an equation from polar form to rectangular form are

1. Multiplying both sides of the equation by r
2. Squaring both sides of the equation

EXAMPLE 8

Transforming an Equation from Polar to Rectangular Form

Transform the equation $r = 6 \cos \theta$ from polar coordinates to rectangular coordinates, and identify the graph.

Solution

Multiplying each side by r makes it easier to apply formulas (1) and (2).

$$r = 6 \cos \theta$$
$$r^2 = 6r \cos \theta \qquad \text{\textit{Multiply each side by r.}}$$
$$x^2 + y^2 = 6x \qquad \text{\textit{$r^2 = x^2 + y^2$; $x = r \cos \theta$}}$$

This is the equation of a circle. Complete the square to obtain the standard form of the equation.

$$x^2 + y^2 = 6x$$
$$(x^2 - 6x) + y^2 = 0 \qquad \text{\textit{General form}}$$
$$(x^2 - 6x + 9) + y^2 = 9 \qquad \text{\textit{Complete the square in x.}}$$
$$(x - 3)^2 + y^2 = 9 \qquad \text{\textit{Factor.}}$$

This is the standard form of the equation of a circle with center $(3, 0)$ and radius 3.

●

Now Work PROBLEM 77

EXAMPLE 9

Transforming an Equation from Rectangular to Polar Form

Transform the equation $4xy = 9$ from rectangular coordinates to polar coordinates.

Solution   Use formula (1): $x = r \cos \theta$ and $y = r \sin \theta$.

$$4xy = 9$$

$$4(r \cos \theta)(r \sin \theta) = 9 \quad \textcolor{orange}{x = r\cos\theta, y = r\sin\theta}$$

$$4r^2 \cos \theta \sin \theta = 9$$

This is the polar form of the equation. It can be simplified as follows:

$$2r^2 (2 \sin \theta \cos \theta) = 9 \quad \textcolor{orange}{\text{Factor out } 2r^2.}$$

$$2r^2 \sin(2\theta) = 9 \quad \textcolor{orange}{\text{Double-angle Formula}}$$

Now Work PROBLEM 71

10.1 Assess Your Understanding

'Are You Prepared?' *Answers are given at the end of these exercises. If you get a wrong answer, read the pages listed in red.*

1. Plot the point whose rectangular coordinates are $(3, -1)$. What quadrant does the point lie in? (pp. 150–154)

2. To complete the square of $x^2 + 6x$, add _____ . (p. 56)

3. If $P = (a, b)$ is a point on the terminal side of the angle θ at a distance r from the origin, then $\tan \theta =$ _____ . (pp. 543–545)

4. $\tan^{-1}(-1) =$ _____ . (pp. 613–616)

Concepts and Vocabulary

5. The origin in rectangular coordinates coincides with the _____ in polar coordinates; the positive x-axis in rectangular coordinates coincides with the _____ _____ in polar coordinates.

6. If P is a point with polar coordinates (r, θ), the rectangular coordinates (x, y) of P are given by $x =$ _____ and $y =$ _____ .

7. For the point with polar coordinates $\left(1, -\dfrac{\pi}{2}\right)$, which of the following best describes the location of the point in a rectangular coordinate system?
   (a) in quadrant IV (b) on the y-axis
   (c) in quadrant II (d) on the x-axis

8. The point $\left(5, \dfrac{\pi}{6}\right)$ can also be represented by which of the following polar coordinates?
   (a) $\left(5, -\dfrac{\pi}{6}\right)$ (b) $\left(-5, \dfrac{13\pi}{6}\right)$
   (c) $\left(5, -\dfrac{5\pi}{6}\right)$ (d) $\left(-5, \dfrac{7\pi}{6}\right)$

9. **True or False** In the polar coordinates (r, θ), r can be negative.

10. **True or False** The polar coordinates of a point are unique.

Skill Building

In Problems 11–18, match each point in polar coordinates with either A, B, C, or D on the graph.

11. $\left(2, -\dfrac{11\pi}{6}\right)$ 12. $\left(-2, -\dfrac{\pi}{6}\right)$ 13. $\left(-2, \dfrac{\pi}{6}\right)$ 14. $\left(2, \dfrac{7\pi}{6}\right)$

15. $\left(2, \dfrac{5\pi}{6}\right)$ 16. $\left(-2, \dfrac{5\pi}{6}\right)$ 17. $\left(-2, \dfrac{7\pi}{6}\right)$ 18. $\left(2, \dfrac{11\pi}{6}\right)$

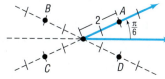

In Problems 19–32, plot each point given in polar coordinates.

19. $(3, 90°)$ 20. $(4, 270°)$ 21. $(-2, 0)$ 22. $(-3, \pi)$ 23. $\left(6, \dfrac{\pi}{6}\right)$

24. $\left(5, \dfrac{5\pi}{3}\right)$ 25. $(-2, 135°)$ 26. $(-3, 120°)$ 27. $\left(4, -\dfrac{2\pi}{3}\right)$ 28. $\left(2, -\dfrac{5\pi}{4}\right)$

29. $\left(-1, -\dfrac{\pi}{3}\right)$ 30. $\left(-3, -\dfrac{3\pi}{4}\right)$ 31. $(-2, -\pi)$ 32. $\left(-3, -\dfrac{\pi}{2}\right)$

In Problems 33–40, plot each point given in polar coordinates, and find other polar coordinates (r, θ) of the point for which:
   (a) $r > 0$, $-2\pi \le \theta < 0$ (b) $r < 0$, $0 \le \theta < 2\pi$ (c) $r > 0$, $2\pi \le \theta < 4\pi$

33. $\left(5, \dfrac{2\pi}{3}\right)$ 34. $\left(4, \dfrac{3\pi}{4}\right)$ 35. $(-2, 3\pi)$ 36. $(-3, 4\pi)$

37. $\left(1, \dfrac{\pi}{2}\right)$ 38. $(2, \pi)$ 39. $\left(-3, -\dfrac{\pi}{4}\right)$ 40. $\left(-2, -\dfrac{2\pi}{3}\right)$

In Problems 41–56, polar coordinates of a point are given. Find the rectangular coordinates of each point.

41. $\left(3, \dfrac{\pi}{2}\right)$
42. $\left(4, \dfrac{3\pi}{2}\right)$
43. $(-2, 0)$
44. $(-3, \pi)$

45. $(6, 150°)$
46. $(5, 300°)$
47. $\left(-2, \dfrac{3\pi}{4}\right)$
48. $\left(-2, \dfrac{2\pi}{3}\right)$

49. $\left(-1, -\dfrac{\pi}{3}\right)$
50. $\left(-3, -\dfrac{3\pi}{4}\right)$
51. $(-2, -180°)$
52. $(-3, -90°)$

53. $(7.5, 110°)$
54. $(-3.1, 182°)$
55. $(6.3, 3.8)$
56. $(8.1, 5.2)$

In Problems 57–68, the rectangular coordinates of a point are given. Find polar coordinates for each point.

57. $(3, 0)$
58. $(0, 2)$
59. $(-1, 0)$
60. $(0, -2)$

61. $(1, -1)$
62. $(-3, 3)$
63. $\left(\sqrt{3}, 1\right)$
64. $\left(-2, -2\sqrt{3}\right)$

65. $(1.3, -2.1)$
66. $(-0.8, -2.1)$
67. $(8.3, 4.2)$
68. $(-2.3, 0.2)$

In Problems 69–76, the letters x and y represent rectangular coordinates. Write each equation using polar coordinates (r, θ).

69. $2x^2 + 2y^2 = 3$
70. $x^2 + y^2 = x$
71. $x^2 = 4y$
72. $y^2 = 2x$

73. $2xy = 1$
74. $4x^2 y = 1$
75. $x = 4$
76. $y = -3$

In Problems 77–84, the letters r and θ represent polar coordinates. Write each equation using rectangular coordinates (x, y).

77. $r = \cos \theta$
78. $r = \sin \theta + 1$
79. $r^2 = \cos \theta$
80. $r = \sin \theta - \cos \theta$

81. $r = 2$
82. $r = 4$
83. $r = \dfrac{4}{1 - \cos \theta}$
84. $r = \dfrac{3}{3 - \cos \theta}$

Applications and Extensions

85. Chicago In Chicago, the road system is set up like a Cartesian plane, where streets are indicated by the number of blocks they are from Madison Street and State Street. For example, Wrigley Field in Chicago is located at 1060 West Addison, which is 10 blocks west of State Street and 36 blocks north of Madison Street. Treat the intersection of Madison Street and State Street as the origin of a coordinate system, with east being the positive *x*-axis.
 (a) Write the location of Wrigley Field using rectangular coordinates.
 (b) Write the location of Wrigley Field using polar coordinates. Use the east direction for the polar axis. Express θ in degrees.
 (c) U.S. Cellular Field, home of the White Sox, is located at 35th and Princeton, which is 3 blocks west of State Street and 35 blocks south of Madison. Write the location of U.S. Cellular Field using rectangular coordinates.
 (d) Write the location of U.S. Cellular Field using polar coordinates. Use the east direction for the polar axis. Express θ in degrees.

86. Show that the formula for the distance *d* between two points $P_1 = (r_1, \theta_1)$ and $P_2 = (r_2, \theta_2)$ is
$$d = \sqrt{r_1^2 + r_2^2 - 2r_1 r_2 \cos(\theta_2 - \theta_1)}$$

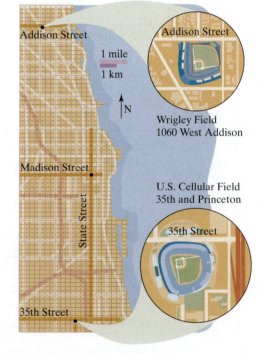

City of Chicago, Illinois

Addison Street

1 mile
1 km

N

Addison Street

Wrigley Field
1060 West Addison

Madison Street

State Street

U.S. Cellular Field
35th and Princeton

35th Street

35th Street

Explaining Concepts: Discussion and Writing

87. In converting from polar coordinates to rectangular coordinates, what formulas will you use?

88. Explain how to convert from rectangular coordinates to polar coordinates.

89. Is the street system in your town based on a rectangular coordinate system, a polar coordinate system, or some other system? Explain.

Retain Your Knowledge

Problems 90–93 are based on material learned earlier in the course. The purpose of these problems is to keep the material fresh in your mind so that you are better prepared for the final exam.

90. Solve: $\log_4(x + 3) - \log_4(x - 1) = 2$

91. Use Descartes' Rule of Signs to determine the possible number of positive or negative real zeros for the function $f(x) = -2x^3 + 6x^2 - 7x - 8$.

92. Find the midpoint of the line segment connecting the points $(-3, 7)$ and $\left(\frac{1}{2}, 2\right)$.

93. Given that the point $(3, 8)$ is on the graph of $y = f(x)$, what is the corresponding point on the graph of $y = -2f(x + 3) + 5$?

'Are You Prepared?' Answers

1.

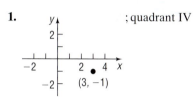

; quadrant IV   **2.** 9   **3.** $\frac{b}{a}$ **4.** $-\frac{\pi}{4}$

10.2 Polar Equations and Graphs

PREPARING FOR THIS SECTION *Before getting started, review the following:*

- Symmetry (Section 2.2, pp. 160–162)
- Circles (Section 2.4, pp. 182–185)
- Even–Odd Properties of Trigonometric Functions (Section 7.5, pp. 560–561)

- Difference Formulas for Sine and Cosine (Section 8.5, pp. 643 and 646)
- Values of the Sine and Cosine Functions at Certain Angles (Section 7.3, pp. 531–537, Section 7.4, pp. 543–550)

Now Work the **'Are You Prepared?'** problems on page 741.

OBJECTIVES **1** Identify and Graph Polar Equations by Converting to Rectangular Equations (p. 730)

2 Test Polar Equations for Symmetry (p. 733)

3 Graph Polar Equations by Plotting Points (p. 734)

Just as a rectangular grid may be used to plot points given by rectangular coordinates, such as the points $(-3, 1)$ and $(1, 2)$ shown in Figure 19(a), a grid consisting of concentric circles (with centers at the pole) and rays (with vertices at the pole) can be used to plot points given by polar coordinates, such as the points $\left(4, \frac{5\pi}{4}\right)$ and $\left(2, \frac{\pi}{4}\right)$ shown in Figure 19(b). Such **polar grids** are used to graph *polar equations*.

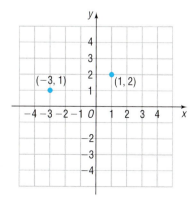

Figure 19   **(a)** Rectangular grid   **(b)** Polar grid

EXAMPLE 5

Identifying and Graphing a Polar Equation (Circle)

Identify and graph the equation: $r = 4 \sin \theta$

Solution

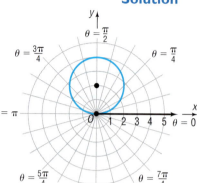

To transform the equation to rectangular coordinates, multiply each side by r.

$$r^2 = 4r \sin \theta$$

Now use the facts that $r^2 = x^2 + y^2$ and $y = r \sin \theta$. Then

$$x^2 + y^2 = 4y$$
$$x^2 + (y^2 - 4y) = 0$$
$$x^2 + (y^2 - 4y + 4) = 4 \qquad \text{Complete the square in } y.$$
$$x^2 + (y - 2)^2 = 4 \qquad \text{Factor.}$$

This is the standard equation of a circle with center at $(0, 2)$ in rectangular coordinates and radius 2. See Figure 24.

Figure 24
$r = 4 \sin \theta$, or $x^2 + (y - 2)^2 = 4$

EXAMPLE 6

Identifying and Graphing a Polar Equation (Circle)

Identify and graph the equation: $r = -2 \cos \theta$

Solution

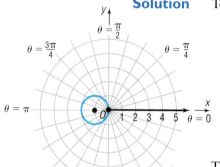

To transform the equation to rectangular coordinates, multiply each side by r.

$$r^2 = -2r \cos \theta \qquad \text{Multiply both sides by } r.$$
$$x^2 + y^2 = -2x \qquad r^2 = x^2 + y^2; \quad x = r \cos \theta$$
$$x^2 + 2x + y^2 = 0$$
$$(x^2 + 2x + 1) + y^2 = 1 \qquad \text{Complete the square in } x.$$
$$(x + 1)^2 + y^2 = 1 \qquad \text{Factor.}$$

This is the standard equation of a circle with center at $(-1, 0)$ in rectangular coordinates and radius 1. See Figure 25.

Figure 25
$r = -2 \cos \theta$, or $(x + 1)^2 + y^2 = 1$

Exploration

Using a square screen, graph $r_1 = \sin \theta$, $r_2 = 2 \sin \theta$, and $r_3 = 3 \sin \theta$. Do you see the pattern? Clear the screen and graph $r_1 = -\sin \theta$, $r_2 = -2 \sin \theta$, and $r_3 = -3 \sin \theta$. Do you see the pattern? Clear the screen and graph $r_1 = \cos \theta$, $r_2 = 2 \cos \theta$, and $r_3 = 3 \cos \theta$. Do you see the pattern? Clear the screen and graph $r_1 = -\cos \theta$, $r_2 = -2 \cos \theta$, and $r_3 = -3 \cos \theta$. Do you see the pattern?

Based on Examples 5 and 6 and the Exploration above, we are led to the following results. (The proofs are left as exercises. See Problems 85–88.)

THEOREM

Let a be a positive real number. Then

Equation	Description
(a) $r = 2a \sin \theta$	Circle: radius a; center at $(0, a)$ in rectangular coordinates
(b) $r = -2a \sin \theta$	Circle: radius a; center at $(0, -a)$ in rectangular coordinates
(c) $r = 2a \cos \theta$	Circle: radius a; center at $(a, 0)$ in rectangular coordinates
(d) $r = -2a \cos \theta$	Circle: radius a; center at $(-a, 0)$ in rectangular coordinates

Each circle passes through the pole.

Now Work PROBLEM 23

The method of converting a polar equation to an identifiable rectangular equation to obtain the graph is not always helpful, nor is it always necessary. Usually, a table is created that lists several points on the graph. By checking for symmetry, it may be possible to reduce the number of points needed to draw the graph.

2 Test Polar Equations for Symmetry

In polar coordinates, the points (r, θ) and $(r, -\theta)$ are symmetric with respect to the polar axis (and to the x-axis). See Figure 26(a). The points (r, θ) and $(r, \pi - \theta)$ are symmetric with respect to the line $\theta = \dfrac{\pi}{2}$ (the y-axis). See Figure 26(b). The points (r, θ) and $(-r, \theta)$ are symmetric with respect to the pole (the origin). See Figure 26(c).

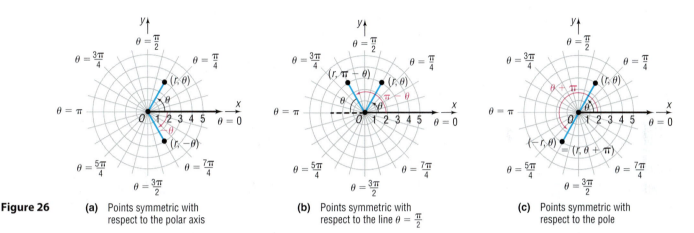

Figure 26
(a) Points symmetric with respect to the polar axis

(b) Points symmetric with respect to the line $\theta = \frac{\pi}{2}$

(c) Points symmetric with respect to the pole

The following tests are a consequence of these observations.

THEOREM

Tests for Symmetry

Symmetry with Respect to the Polar Axis (x-Axis)

In a polar equation, replace θ by $-\theta$. If an equivalent equation results, the graph is symmetric with respect to the polar axis.

Symmetry with Respect to the Line $\theta = \dfrac{\pi}{2}$ (y-Axis)

In a polar equation, replace θ by $\pi - \theta$. If an equivalent equation results, the graph is symmetric with respect to the line $\theta = \dfrac{\pi}{2}$.

Symmetry with Respect to the Pole (Origin)

In a polar equation, replace r by $-r$ or θ by $\theta + \pi$. If an equivalent equation results, the graph is symmetric with respect to the pole.

The three tests for symmetry given here are *sufficient* conditions for symmetry, but they are not *necessary* conditions. That is, an equation may fail these tests and still have a graph that is symmetric with respect to the polar axis, the line $\theta = \dfrac{\pi}{2}$, or the pole. For example, the graph of $r = \sin(2\theta)$ turns out to be symmetric with respect to the polar axis, the line $\theta = \dfrac{\pi}{2}$, and the pole, but only the test for symmetry with respect to the pole (replace θ by $\theta + \pi$) works. See also Problems 89–91.

3 Graph Polar Equations by Plotting Points

EXAMPLE 7

Graphing a Polar Equation (Cardioid)

Graph the equation: $r = 1 - \sin\theta$

Solution   Check for symmetry first.

Polar Axis: Replace θ by $-\theta$. The result is

$$r = 1 - \sin(-\theta) = 1 + \sin\theta \quad \text{\small $\sin(-\theta) = -\sin\theta$}$$

The test fails, so the graph may or may not be symmetric with respect to the polar axis.

The Line $\theta = \dfrac{\pi}{2}$: Replace θ by $\pi - \theta$. The result is

$$r = 1 - \sin(\pi - \theta) = 1 - (\sin\pi\cos\theta - \cos\pi\sin\theta)$$
$$= 1 - [0\cdot\cos\theta - (-1)\sin\theta] = 1 - \sin\theta$$

The test is satisfied, so the graph is symmetric with respect to the line $\theta = \dfrac{\pi}{2}$.

The Pole: Replace r by $-r$. Then the result is $-r = 1 - \sin\theta$, so $r = -1 + \sin\theta$. The test fails. Replace θ by $\theta + \pi$. The result is

$$r = 1 - \sin(\theta + \pi)$$
$$= 1 - [\sin\theta\cos\pi + \cos\theta\sin\pi]$$
$$= 1 - [\sin\theta\cdot(-1) + \cos\theta\cdot 0]$$
$$= 1 + \sin\theta$$

This test also fails, so the graph may or may not be symmetric with respect to the pole.

Next, identify points on the graph by assigning values to the angle θ and calculating the corresponding values of r. Due to the periodicity of the sine function and the symmetry with respect to the line $\theta = \dfrac{\pi}{2}$, just assign values to θ from $-\dfrac{\pi}{2}$ to $\dfrac{\pi}{2}$, as given in Table 1.

Now plot the points (r, θ) from Table 1 and trace out the graph, beginning at the point $\left(2, -\dfrac{\pi}{2}\right)$ and ending at the point $\left(0, \dfrac{\pi}{2}\right)$. Then reflect this portion of the graph about the line $\theta = \dfrac{\pi}{2}$ (the y-axis) to obtain the complete graph. See Figure 27.

Table 1

θ	$r = 1 - \sin\theta$
$-\dfrac{\pi}{2}$	$1 - (-1) = 2$
$-\dfrac{\pi}{3}$	$1 - \left(-\dfrac{\sqrt{3}}{2}\right) \approx 1.87$
$-\dfrac{\pi}{6}$	$1 - \left(-\dfrac{1}{2}\right) = \dfrac{3}{2}$
0	$1 - 0 = 1$
$\dfrac{\pi}{6}$	$1 - \dfrac{1}{2} = \dfrac{1}{2}$
$\dfrac{\pi}{3}$	$1 - \dfrac{\sqrt{3}}{2} \approx 0.13$
$\dfrac{\pi}{2}$	$1 - 1 = 0$

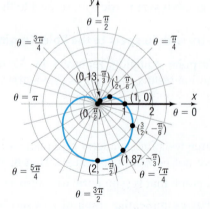

Figure 27 $r = 1 - \sin\theta$

The curve in Figure 27 is an example of a *cardioid* (a heart-shaped curve).

Exploration

Graph $r_1 = 1 + \sin\theta$. Clear the screen and graph $r_1 = 1 - \cos\theta$. Clear the screen and graph $r_1 = 1 + \cos\theta$. Do you see a pattern?

DEFINITION

Cardioids are characterized by equations of the form

$$r = a(1 + \cos \theta) \qquad r = a(1 + \sin \theta)$$
$$r = a(1 - \cos \theta) \qquad r = a(1 - \sin \theta)$$

where $a > 0$. The graph of a cardioid passes through the pole.

━━━━━━ **Now Work** PROBLEM **39**

EXAMPLE 8

Graphing a Polar Equation (Limaçon without an Inner Loop)

Graph the equation: $r = 3 + 2 \cos \theta$

Solution

Check for symmetry first.

Polar Axis: Replace θ by $-\theta$. The result is

$$r = 3 + 2 \cos(-\theta) = 3 + 2 \cos \theta \qquad \cos(-\theta) = \cos \theta$$

The test is satisfied, so the graph is symmetric with respect to the polar axis.

The Line $\theta = \dfrac{\pi}{2}$: Replace θ by $\pi - \theta$. The result is

$$r = 3 + 2 \cos(\pi - \theta) = 3 + 2(\cos \pi \cos \theta + \sin \pi \sin \theta)$$
$$= 3 - 2 \cos \theta$$

The test fails, so the graph may or may not be symmetric with respect to the line $\theta = \dfrac{\pi}{2}$.

The Pole: Replace r by $-r$. The test fails, so the graph may or may not be symmetric with respect to the pole. Replace θ by $\theta + \pi$. The test fails, so the graph may or may not be symmetric with respect to the pole.

Next, identify points on the graph by assigning values to the angle θ and calculating the corresponding values of r. Due to the periodicity of the cosine function and the symmetry with respect to the polar axis, just assign values to θ from 0 to π, as given in Table 2.

Now plot the points (r, θ) from Table 2 and trace out the graph, beginning at the point $(5, 0)$ and ending at the point $(1, \pi)$. Then reflect this portion of the graph about the polar axis (the x-axis) to obtain the complete graph. See Figure 28.

Table 2

θ	$r = 3 + 2\cos \theta$
0	$3 + 2(1) = 5$
$\dfrac{\pi}{6}$	$3 + 2\left(\dfrac{\sqrt{3}}{2}\right) \approx 4.73$
$\dfrac{\pi}{3}$	$3 + 2\left(\dfrac{1}{2}\right) = 4$
$\dfrac{\pi}{2}$	$3 + 2(0) = 3$
$\dfrac{2\pi}{3}$	$3 + 2\left(-\dfrac{1}{2}\right) = 2$
$\dfrac{5\pi}{6}$	$3 + 2\left(-\dfrac{\sqrt{3}}{2}\right) \approx 1.27$
π	$3 + 2(-1) = 1$

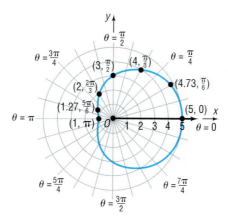

Figure 28 $r = 3 + 2\cos \theta$

Exploration

Graph $r_1 = 3 - 2 \cos \theta$. Clear the screen and graph $r_1 = 3 + 2 \sin \theta$. Clear the screen and graph $r_1 = 3 - 2 \sin \theta$. Do you see a pattern?

The curve in Figure 28 is an example of a *limaçon* (a French word for *snail*) *without an inner loop*.

DEFINITION

> **Limaçons without an inner loop** are characterized by equations of the form
>
> $$r = a + b \cos \theta \qquad r = a + b \sin \theta$$
> $$r = a - b \cos \theta \qquad r = a - b \sin \theta$$

where $a > 0$, $b > 0$, and $a > b$. The graph of a limaçon without an inner loop does not pass through the pole.

━━━━━━ **Now Work** PROBLEM 45

EXAMPLE 9

Graphing a Polar Equation (Limaçon with an Inner Loop)

Graph the equation: $r = 1 + 2 \cos \theta$

Solution

First, check for symmetry.

Polar Axis: Replace θ by $-\theta$. The result is

$$r = 1 + 2 \cos(-\theta) = 1 + 2 \cos \theta$$

The test is satisfied, so the graph is symmetric with respect to the polar axis.

The Line $\theta = \dfrac{\pi}{2}$: Replace θ by $\pi - \theta$. The result is

$$r = 1 + 2 \cos(\pi - \theta) = 1 + 2(\cos \pi \cos \theta + \sin \pi \sin \theta)$$
$$= 1 - 2 \cos \theta$$

The test fails, so the graph may or may not be symmetric with respect to the line $\theta = \dfrac{\pi}{2}$.

The Pole: Replace r by $-r$. The test fails, so the graph may or may not be symmetric with respect to the pole. Replace θ by $\theta + \pi$. The test fails, so the graph may or may not be symmetric with respect to the pole.

Next, identify points on the graph of $r = 1 + 2 \cos \theta$ by assigning values to the angle θ and calculating the corresponding values of r. Due to the periodicity of the cosine function and the symmetry with respect to the polar axis, just assign values to θ from 0 to π, as given in Table 3.

Now plot the points (r, θ) from Table 3, beginning at $(3, 0)$ and ending at $(-1, \pi)$. See Figure 29(a). Finally, reflect this portion of the graph about the polar axis (the x-axis) to obtain the complete graph. See Figure 29(b).

Table 3

θ	$r = 1 + 2 \cos \theta$
0	$1 + 2(1) = 3$
$\dfrac{\pi}{6}$	$1 + 2\left(\dfrac{\sqrt{3}}{2}\right) \approx 2.73$
$\dfrac{\pi}{3}$	$1 + 2\left(\dfrac{1}{2}\right) = 2$
$\dfrac{\pi}{2}$	$1 + 2(0) = 1$
$\dfrac{2\pi}{3}$	$1 + 2\left(-\dfrac{1}{2}\right) = 0$
$\dfrac{5\pi}{6}$	$1 + 2\left(-\dfrac{\sqrt{3}}{2}\right) \approx -0.73$
π	$1 + 2(-1) = -1$

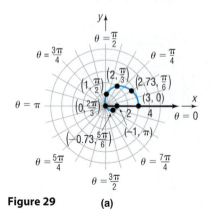

(a)

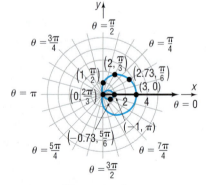

(b) $r = 1 + 2 \cos \theta$

Figure 29

The curve in Figure 29(b) is an example of a *limaçon with an inner loop*.

Exploration

Graph $r_1 = 1 - 2 \cos \theta$. Clear the screen and graph $r_1 = 1 + 2 \sin \theta$. Clear the screen and graph $r_1 = 1 - 2 \sin \theta$. Do you see a pattern?

DEFINITION

Limaçons with an inner loop are characterized by equations of the form

$$r = a + b \cos \theta \qquad r = a + b \sin \theta$$
$$r = a - b \cos \theta \qquad r = a - b \sin \theta$$

where $a > 0, b > 0$, and $a < b$. The graph of a limaçon with an inner loop passes through the pole twice.

──── **Now Work** PROBLEM 47

EXAMPLE 10

Graphing a Polar Equation (Rose)

Graph the equation: $r = 2 \cos(2\theta)$

Solution

Check for symmetry.

Polar Axis: Replace θ by $-\theta$. The result is

$$r = 2 \cos[2(-\theta)] = 2 \cos(2\theta)$$

The test is satisfied, so the graph is symmetric with respect to the polar axis.

The Line $\theta = \dfrac{\pi}{2}$: Replace θ by $\pi - \theta$. The result is

$$r = 2 \cos[2(\pi - \theta)] = 2 \cos(2\pi - 2\theta) = 2 \cos(2\theta)$$

The test is satisfied, so the graph is symmetric with respect to the line $\theta = \dfrac{\pi}{2}$.

The Pole: Since the graph is symmetric with respect to both the polar axis and the line $\theta = \dfrac{\pi}{2}$, it must be symmetric with respect to the pole.

Next, construct Table 4. Because of the periodicity of the cosine function and the symmetry with respect to the polar axis, the line $\theta = \dfrac{\pi}{2}$, and the pole, consider only values of θ from 0 to $\dfrac{\pi}{2}$.

Plot and connect these points as shown in Figure 30(a). Finally, because of symmetry, reflect this portion of the graph first about the polar axis (the x-axis) and then about the line $\theta = \dfrac{\pi}{2}$ (the y-axis) to obtain the complete graph. See Figure 30(b).

Table 4

θ	$r = 2 \cos(2\theta)$
0	$2(1) = 2$
$\dfrac{\pi}{6}$	$2\left(\dfrac{1}{2}\right) = 1$
$\dfrac{\pi}{4}$	$2(0) = 0$
$\dfrac{\pi}{3}$	$2\left(-\dfrac{1}{2}\right) = -1$
$\dfrac{\pi}{2}$	$2(-1) = -2$

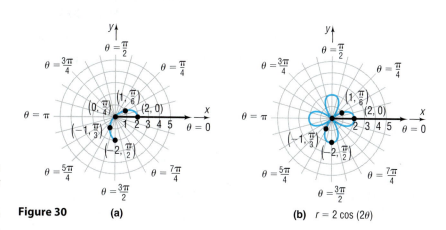

Figure 30   **(a)**   **(b)** $r = 2 \cos(2\theta)$

Exploration

Graph $r_1 = 2 \cos(4\theta)$; clear the screen and graph $r_1 = 2 \cos(6\theta)$. How many petals did each of these graphs have?

Clear the screen and graph, in order, each on a clear screen, $r_1 = 2 \cos(3\theta)$, $r_1 = 2 \cos(5\theta)$, and $r_1 = 2 \cos(7\theta)$. What do you notice about the number of petals?

The curve in Figure 30(b) is called a *rose* with four petals.

DEFINITION

> **Rose** curves are characterized by equations of the form
>
> $$r = a\cos(n\theta) \qquad r = a\sin(n\theta) \qquad a \neq 0$$
>
> and have graphs that are rose shaped. If $n \neq 0$ is even, the rose has $2n$ petals; if $n \neq \pm 1$ is odd, the rose has n petals.

Now Work PROBLEM 51

EXAMPLE 11

Graphing a Polar Equation (Lemniscate)

Graph the equation: $r^2 = 4\sin(2\theta)$

Solution

We leave it to you to verify that the graph is symmetric with respect to the pole. Because of the symmetry with respect to the pole, consider only those values of θ between $\theta = 0$ and $\theta = \pi$. Note that there are no points on the graph for $\frac{\pi}{2} < \theta < \pi$ (quadrant II), since $r^2 < 0$ for such values. Table 5 lists points on the graph for values of $\theta = 0$ through $\theta = \frac{\pi}{2}$. The points from Table 5 where $r \geq 0$ are plotted in Figure 31(a). The remaining points on the graph may be obtained by using symmetry. Figure 31(b) shows the final graph drawn.

Table 5

θ	$r^2 = 4\sin(2\theta)$	r
0	$4(0) = 0$	0
$\frac{\pi}{6}$	$4\left(\frac{\sqrt{3}}{2}\right) = 2\sqrt{3}$	± 1.9
$\frac{\pi}{4}$	$4(1) = 4$	± 2
$\frac{\pi}{3}$	$4\left(\frac{\sqrt{3}}{2}\right) = 2\sqrt{3}$	± 1.9
$\frac{\pi}{2}$	$4(0) = 0$	0

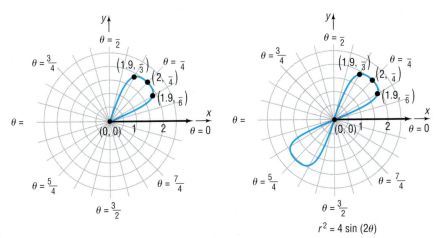

Figure 31

The curve in Figure 31(b) is an example of a *lemniscate* (from the Greek word for *ribbon*).

DEFINITION

> **Lemniscates** are characterized by equations of the form
>
> $$r^2 = a^2\sin(2\theta) \qquad r^2 = a^2\cos(2\theta)$$
>
> where $a \neq 0$, and have graphs that are propeller shaped.

Now Work PROBLEM 55

EXAMPLE 12

Graphing a Polar Equation (Spiral)

Graph the equation: $r = e^{\theta/5}$

Solution

The tests for symmetry with respect to the pole, the polar axis, and the line $\theta = \frac{\pi}{2}$ fail. Furthermore, there is no number θ for which $r = 0$, so the graph does not pass through the pole. Observe that r is positive for all θ, r increases as θ increases, $r \to 0$

Table 6

θ	$r = e^{\theta/5}$
$-\dfrac{3\pi}{2}$	0.39
$-\pi$	0.53
$-\dfrac{\pi}{2}$	0.73
$-\dfrac{\pi}{4}$	0.85
0	1
$\dfrac{\pi}{4}$	1.17
$\dfrac{\pi}{2}$	1.37
π	1.87
$\dfrac{3\pi}{2}$	2.57
2π	3.51

as $\theta \to -\infty$, and $r \to \infty$ as $\theta \to \infty$. With the help of a calculator, the values in Table 6 can be obtained. See Figure 32.

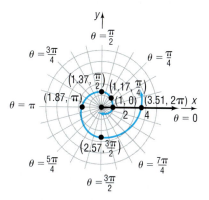

Figure 32 $r = e^{\theta/5}$

The curve in Figure 32 is called a **logarithmic spiral**, since its equation may be written as $\theta = 5 \ln r$ and it spirals infinitely both toward the pole and away from it.

Classification of Polar Equations

The equations of some lines and circles in polar coordinates and their corresponding equations in rectangular coordinates are given in Table 7. Also included are the names and graphs of a few of the more frequently encountered polar equations.

Table 7

Lines			
Description	Line passing through the pole making an angle α with the polar axis	Vertical line	Horizontal line
Rectangular equation	$y = (\tan \alpha)x$	$x = a$	$y = b$
Polar equation	$\theta = \alpha$	$r \cos \theta = a$	$r \sin \theta = b$
Typical graph			

Circles			
Description	Center at the pole, radius a	Passing through the pole, tangent to the line $\theta = \dfrac{\pi}{2}$, center on the polar axis, radius a	Passing through the pole, tangent to the polar axis, center on the line $\theta = \dfrac{\pi}{2}$, radius a
Rectangular equation	$x^2 + y^2 = a^2, \ a > 0$	$x^2 + y^2 = \pm 2ax, \ a > 0$	$x^2 + y^2 = \pm 2ay, \ a > 0$
Polar equation	$r = a, \ a > 0$	$r = \pm 2a \cos \theta, \ a > 0$	$r = \pm 2a \sin \theta, \ a > 0$
Typical graph			

(continued)

10.3 The Complex Plane; De Moivre's Theorem

PREPARING FOR THIS SECTION *Before getting started, review the following:*

- Complex Numbers (Section 1.3, pp. 104–109)
- Values of the Sine and Cosine Functions at Certain Angles (Section 7.3, pp. 531–537; Section 7.4, pp. 543–550)
- Sum and Difference Formulas for Sine and Cosine (Section 8.5, pp. 643 and 646)

Now Work the **'Are You Prepared?'** problems on page 750.

OBJECTIVES 1 Plot Points in the Complex Plane (p. 744)
2 Convert a Complex Number between Rectangular Form and Polar Form (p. 745)
3 Find Products and Quotients of Complex Numbers in Polar Form (p. 746)
4 Use De Moivre's Theorem (p. 747)
5 Find Complex Roots (p. 748)

1 Plot Points in the Complex Plane

Complex numbers are discussed in Chapter 1, Section 1.3. In that discussion, we were not prepared to give a geometric interpretation of a complex number. Now we are ready.

A complex number $z = x + yi$ can be interpreted geometrically as the point (x, y) in the xy-plane. Each point in the plane corresponds to a complex number, and conversely, each complex number corresponds to a point in the plane. The collection of such points is referred to as the **complex plane**. The x-axis is referred to as the **real axis**, because any point that lies on the real axis is of the form $z = x + 0i = x$, a real number. The y-axis is called the **imaginary axis**, because any point that lies on it is of the form $z = 0 + yi = yi$, a pure imaginary number. See Figure 34.

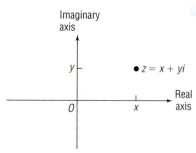

Figure 34 Complex plane

EXAMPLE 1

Plotting a Point in the Complex Plane

Plot the point corresponding to $z = \sqrt{3} - i$ in the complex plane.

Solution The point corresponding to $z = \sqrt{3} - i$ has the rectangular coordinates $(\sqrt{3}, -1)$. This point, located in quadrant IV, is plotted in Figure 35.

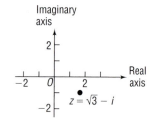

Figure 35

DEFINITION

Let $z = x + yi$ be a complex number. The **magnitude** or **modulus** of z, denoted by $|z|$, is defined as the distance from the origin to the point (x, y). That is,

$$|z| = \sqrt{x^2 + y^2} \tag{1}$$

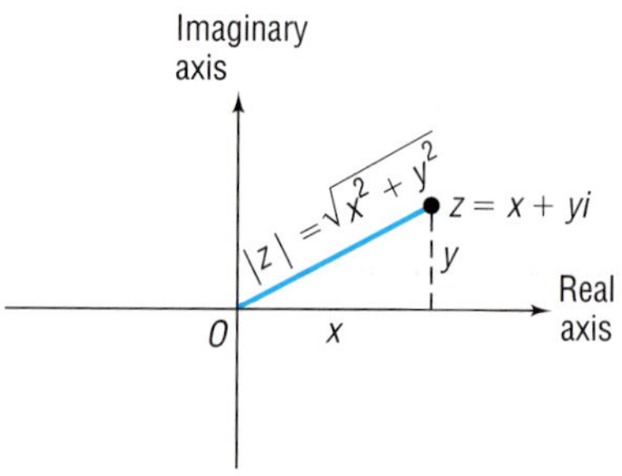

See Figure 36 for an illustration.

This definition for $|z|$ is consistent with the definition for the absolute value of a real number: If $z = x + yi$ is real, then $z = x + 0i$ and

$$|z| = \sqrt{x^2 + 0^2} = \sqrt{x^2} = |x|$$

For this reason, the magnitude of z is sometimes called the **absolute value of z**.

Figure 36

Recall that if $z = x + yi$, then its **conjugate**, denoted by $\bar{z}$, is $\bar{z} = x - yi$. Because $z\bar{z} = x^2 + y^2$, which is a nonnegative real number, it follows from equation (1) that the magnitude of z can be written as

$$|z| = \sqrt{z\bar{z}} \tag{2}$$

2 Convert a Complex Number between Rectangular Form and Polar Form

When a complex number is written in the standard form $z = x + yi$, it is in **rectangular**, or **Cartesian**, **form**, because (x, y) are the rectangular coordinates of the corresponding point in the complex plane. Suppose that (r, θ) are polar coordinates of this point. Then

$$x = r \cos \theta \qquad y = r \sin \theta \tag{3}$$

DEFINITION

If $r \geq 0$ and $0 \leq \theta < 2\pi$, the complex number $z = x + yi$ may be written in **polar form** as

$$z = x + yi = (r \cos \theta) + (r \sin \theta)i = r(\cos \theta + i \sin \theta) \tag{4}$$

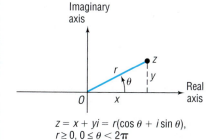

Imaginary axis

$z = x + yi = r(\cos\theta + i\sin\theta)$,
$r \geq 0, 0 \leq \theta < 2\pi$

Figure 37

See Figure 37.

If $z = r(\cos \theta + i \sin \theta)$ is the polar form of a complex number,* the angle θ, $0 \leq \theta < 2\pi$, is called the **argument of z**.

Also, because $r \geq 0$, we have $r = \sqrt{x^2 + y^2}$. From equation (1), it follows that the magnitude of $z = r(\cos \theta + i \sin \theta)$ is

$$|z| = r$$

EXAMPLE 2

Writing a Complex Number in Polar Form

Write an expression for $z = \sqrt{3} - i$ in polar form.

Solution

The point, located in quadrant IV, is plotted in Figure 35. Because $x = \sqrt{3}$ and $y = -1$, it follows that

$$r = \sqrt{x^2 + y^2} = \sqrt{(\sqrt{3})^2 + (-1)^2} = \sqrt{4} = 2$$

so

$$\sin \theta = \frac{y}{r} = \frac{-1}{2} \qquad \cos \theta = \frac{x}{r} = \frac{\sqrt{3}}{2} \qquad 0 \leq \theta < 2\pi$$

The angle $\theta, 0 \leq \theta < 2\pi$, that satisfies both equations is $\theta = \dfrac{11\pi}{6}$. With $\theta = \dfrac{11\pi}{6}$ and $r = 2$, the polar form of $z = \sqrt{3} - i$ is

$$z = r(\cos \theta + i \sin \theta) = 2\left(\cos \frac{11\pi}{6} + i \sin \frac{11\pi}{6}\right)$$

●

 Now Work PROBLEM 13

EXAMPLE 3

Plotting a Point in the Complex Plane and Converting from Polar to Rectangular Form

Plot the point corresponding to $z = 2(\cos 30° + i \sin 30°)$ in the complex plane, and write an expression for z in rectangular form.

*Some texts abbreviate the polar form using $z = r(\cos \theta + i \sin \theta) = r \operatorname{cis} \theta$.

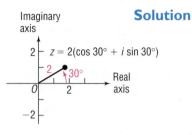

Figure 38 $z = \sqrt{3} + i$

Solution

To plot the complex number $z = 2(\cos 30° + i \sin 30°)$, plot the point whose polar coordinates are $(r, \theta) = (2, 30°)$, as shown in Figure 38. In rectangular form,

$$z = 2(\cos 30° + i \sin 30°) = 2\left(\frac{\sqrt{3}}{2} + \frac{1}{2}i\right) = \sqrt{3} + i$$

●

✏ **Now Work** PROBLEM 25

3 Find Products and Quotients of Complex Numbers in Polar Form

The polar form of a complex number provides an alternative method for finding products and quotients of complex numbers.

THEOREM

Let $z_1 = r_1(\cos \theta_1 + i \sin \theta_1)$ and $z_2 = r_2(\cos \theta_2 + i \sin \theta_2)$ be two complex numbers. Then

$$z_1 z_2 = r_1 r_2 \left[\cos(\theta_1 + \theta_2) + i \sin(\theta_1 + \theta_2)\right] \qquad (5)$$

If $z_2 \neq 0$, then

$$\frac{z_1}{z_2} = \frac{r_1}{r_2}\left[\cos(\theta_1 - \theta_2) + i \sin(\theta_1 - \theta_2)\right] \qquad (6)$$

> **In Words**
> The magnitude of a complex number z is r, and its argument is θ, so when
> $$z = r(\cos\theta + i\sin\theta)$$
> the magnitude of the product (quotient) of two complex numbers equals the product (quotient) of their magnitudes; the argument of the product (quotient) of two complex numbers is determined by the sum (difference) of their arguments.

Proof We will prove formula (5). The proof of formula (6) is left as an exercise (see Problem 68).

$$z_1 z_2 = \left[r_1(\cos \theta_1 + i \sin \theta_1)\right]\left[r_2(\cos \theta_2 + i \sin \theta_2)\right]$$
$$= r_1 r_2\left[(\cos \theta_1 + i \sin \theta_1)(\cos \theta_2 + i \sin \theta_2)\right]$$
$$= r_1 r_2\left[(\cos \theta_1 \cos \theta_2 - \sin \theta_1 \sin \theta_2) + i(\sin \theta_1 \cos \theta_2 + \cos \theta_1 \sin \theta_2)\right]$$
$$= r_1 r_2\left[\cos(\theta_1 + \theta_2) + i \sin(\theta_1 + \theta_2)\right] \qquad ■$$

Let's look at an example of how this theorem can be used.

EXAMPLE 4

Finding Products and Quotients of Complex Numbers in Polar Form

If $z = 3(\cos 20° + i \sin 20°)$ and $w = 5(\cos 100° + i \sin 100°)$, find the following (leave your answers in polar form).

(a) zw (b) $\dfrac{z}{w}$

Solution

(a) $zw = \left[3(\cos 20° + i \sin 20°)\right]\left[5(\cos 100° + i \sin 100°)\right]$
$$= (3 \cdot 5)\left[\cos(20° + 100°) + i \sin(20° + 100°)\right] \qquad \text{Apply equation (5).}$$
$$= 15(\cos 120° + i \sin 120°)$$

(b) $\dfrac{z}{w} = \dfrac{3(\cos 20° + i \sin 20°)}{5(\cos 100° + i \sin 100°)}$

$$= \frac{3}{5}\left[\cos(20° - 100°) + i \sin(20° - 100°)\right] \quad \text{Apply equation (6).}$$

$$= \frac{3}{5}\left[\cos(-80°) + i \sin(-80°)\right]$$

$$= \frac{3}{5}(\cos 280° + i \sin 280°) \qquad \text{The argument must lie between } 0° \text{ and } 360°.$$

●

✏ **Now Work** PROBLEM 35

4 Use De Moivre's Theorem

De Moivre's Theorem, stated by Abraham De Moivre (1667–1754) in 1730, but already known to many people by 1710, is important for the following reason: The fundamental processes of algebra are the four operations of addition, subtraction, multiplication, and division, together with powers and the extraction of roots. De Moivre's Theorem allows the last two fundamental algebraic operations to be applied to complex numbers.

De Moivre's Theorem, in its most basic form, is a formula for raising a complex number z to the power n, where $n \geq 1$ is a positive integer. Let's try to conjecture the form of the result.

Let $z = r(\cos\theta + i\sin\theta)$ be a complex number. Then equation (5) yields

$n = 2$: $z^2 = r^2\left[\cos(2\theta) + i\sin(2\theta)\right]$ Equation (5)

$n = 3$: $z^3 = z^2 \cdot z$
$= \{r^2\left[\cos(2\theta) + i\sin(2\theta)\right]\}\left[r(\cos\theta + i\sin\theta)\right]$
$= r^3\left[\cos(3\theta) + i\sin(3\theta)\right]$ Equation (5)

$n = 4$: $z^4 = z^3 \cdot z$
$= \{r^3\left[\cos(3\theta) + i\sin(3\theta)\right]\}\left[r(\cos\theta + i\sin\theta)\right]$
$= r^4\left[\cos(4\theta) + i\sin(4\theta)\right]$ Equation (5)

Do you see the pattern?

THEOREM

> **De Moivre's Theorem**
>
> If $z = r(\cos\theta + i\sin\theta)$ is a complex number, then
>
> $$z^n = r^n\left[\cos(n\theta) + i\sin(n\theta)\right] \qquad (7)$$
>
> where $n \geq 1$ is a positive integer.

The proof of De Moivre's Theorem requires mathematical induction (which is not discussed until Section 13.4), so it is omitted here. The theorem is actually true for all integers, n. You are asked to prove this in Problem 69.

EXAMPLE 5

Using De Moivre's Theorem

Write $\left[2(\cos 20° + i\sin 20°)\right]^3$ in the standard form $a + bi$.

Solution $\left[2(\cos 20° + i\sin 20°)\right]^3 = 2^3\left[\cos(3\cdot 20°) + i\sin(3\cdot 20°)\right]$ Apply De Moivre's Theorem.
$= 8(\cos 60° + i\sin 60°)$
$= 8\left(\dfrac{1}{2} + \dfrac{\sqrt{3}}{2}i\right) = 4 + 4\sqrt{3}i$ ●

Now Work PROBLEM 43

EXAMPLE 6

Using De Moivre's Theorem

Write $(1 + i)^5$ in the standard form $a + bi$.

Solution   To apply De Moivre's Theorem, first write the complex number in polar form. Since the magnitude of $1 + i$ is $\sqrt{1^2 + 1^2} = \sqrt{2}$, begin by writing

$$1 + i = \sqrt{2}\left(\dfrac{1}{\sqrt{2}} + \dfrac{1}{\sqrt{2}}i\right) = \sqrt{2}\left(\cos\dfrac{\pi}{4} + i\sin\dfrac{\pi}{4}\right)$$

NOTE In the solution of Example 6, the approach used in Example 2 could also be used to write $1 + i$ in polar form. ∎

Now

$$(1 + i)^5 = \left[\sqrt{2} \left(\cos \frac{\pi}{4} + i \sin \frac{\pi}{4} \right) \right]^5$$

$$= (\sqrt{2})^5 \left[\cos \left(5 \cdot \frac{\pi}{4} \right) + i \sin \left(5 \cdot \frac{\pi}{4} \right) \right]$$

$$= 4\sqrt{2} \left(\cos \frac{5\pi}{4} + i \sin \frac{5\pi}{4} \right)$$

$$= 4\sqrt{2} \left[-\frac{1}{\sqrt{2}} + \left(-\frac{1}{\sqrt{2}} \right) i \right] = -4 - 4i \qquad \bullet$$

5 Find Complex Roots

Let w be a given complex number, and let $n \geq 2$ denote a positive integer. Any complex number z that satisfies the equation

$$z^n = w$$

is a **complex nth root** of w. In keeping with previous usage, if $n = 2$, the solutions of the equation $z^2 = w$ are called **complex square roots** of w, and if $n = 3$, the solutions of the equation $z^3 = w$ are called **complex cube roots** of w.

THEOREM

Finding Complex Roots

Let $w = r(\cos \theta_0 + i \sin \theta_0)$ be a complex number, and let $n \geq 2$ be an integer. If $w \neq 0$, there are n distinct complex nth roots of w, given by the formula

$$z_k = \sqrt[n]{r} \left[\cos \left(\frac{\theta_0}{n} + \frac{2k\pi}{n} \right) + i \sin \left(\frac{\theta_0}{n} + \frac{2k\pi}{n} \right) \right] \qquad \textbf{(8)}$$

where $k = 0, 1, 2, \ldots, n - 1$.

Proof (Outline) We will not prove this result in its entirety. Instead, we shall show only that each z_k in equation (8) satisfies the equation $z_k^n = w$, proving that each z_k is a complex nth root of w.

$$z_k^n = \left\{ \sqrt[n]{r} \left[\cos \left(\frac{\theta_0}{n} + \frac{2k\pi}{n} \right) + i \sin \left(\frac{\theta_0}{n} + \frac{2k\pi}{n} \right) \right] \right\}^n$$

$$= (\sqrt[n]{r})^n \left\{ \cos \left[n \left(\frac{\theta_0}{n} + \frac{2k\pi}{n} \right) \right] + i \sin \left[n \left(\frac{\theta_0}{n} + \frac{2k\pi}{n} \right) \right] \right\} \quad \text{Apply De Moivre's Theorem.}$$

$$= r[\cos (\theta_0 + 2k\pi) + i \sin (\theta_0 + 2k\pi)] \quad \text{Simplify.}$$

$$= r(\cos \theta_0 + i \sin \theta_0) = w \quad \text{Periodic Property}$$

So each z_k, $k = 0, 1, \ldots, n - 1$, is a complex nth root of w. To complete the proof, we would need to show that each z_k, $k = 0, 1, \ldots, n - 1$, is, in fact, distinct and that there are no complex nth roots of w other than those given by equation (8). ∎

EXAMPLE 7

Finding Complex Cube Roots

Find the complex cube roots of $-1 + \sqrt{3}i$. Leave your answers in polar form, with the argument in degrees.

Solution

First, express $-1 + \sqrt{3}i$ in polar form using degrees.

$$-1 + \sqrt{3}i = 2 \left(-\frac{1}{2} + \frac{\sqrt{3}}{2} i \right) = 2 (\cos 120° + i \sin 120°)$$

The three complex cube roots of $-1 + \sqrt{3}i = 2(\cos 120° + i \sin 120°)$ are

$$z_k = \sqrt[3]{2}\left[\cos\left(\frac{120°}{3} + \frac{360°k}{3}\right) + i\sin\left(\frac{120°}{3} + \frac{360°k}{3}\right)\right]$$

$$= \sqrt[3]{2}\left[\cos(40° + 120°k) + i\sin(40° + 120°k)\right] \qquad k = 0, 1, 2$$

so

$$z_0 = \sqrt[3]{2}\left[\cos(40° + 120°\cdot 0) + i\sin(40° + 120°\cdot 0)\right] = \sqrt[3]{2}\,(\cos 40° + i\sin 40°)$$

$$z_1 = \sqrt[3]{2}\left[\cos(40° + 120°\cdot 1) + i\sin(40° + 120°\cdot 1)\right] = \sqrt[3]{2}\,(\cos 160° + i\sin 160°)$$

$$z_2 = \sqrt[3]{2}\left[\cos(40° + 120°\cdot 2) + i\sin(40° + 120°\cdot 2)\right] = \sqrt[3]{2}\,(\cos 280° + i\sin 280°)$$

 **WARNING** Most graphing utilities will provide only the answer z_0 to the calculation $(-1 + \sqrt{3}\,i) \wedge (1/3)$. The paragraph following Example 7 explains how to obtain z_1 and z_2 from z_0. ∎

Notice that all of the three complex roots of $-1 + \sqrt{3}i$ have the same magnitude, $\sqrt[3]{2}$. This means that the points corresponding to each cube root lie the same distance from the origin; that is, the three points lie on a circle with center at the origin and radius $\sqrt[3]{2}$. Furthermore, the arguments of these cube roots are 40°, 160°, and 280°, the difference of consecutive pairs being $120° = \dfrac{360°}{3}$. This means that the three points are equally spaced on the circle, as shown in Figure 39. These results are not coincidental. In fact, you are asked to show that these results hold for complex nth roots in Problems 65 through 67.

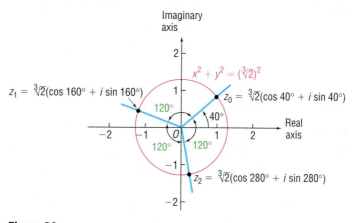

Figure 39

Now Work PROBLEM **55**

Historical Feature

The Babylonians, Greeks, and Arabs considered square roots of negative quantities to be impossible and equations with complex solutions to be unsolvable. The first hint that there was some connection between real solutions of equations and complex numbers came when Girolamo Cardano (1501–1576) and Tartaglia (1499–1557) found *real* roots of cubic equations by taking cube roots of *complex* quantities. For centuries thereafter,

John Wallis

mathematicians worked with complex numbers without much belief in their actual existence. In 1673, John Wallis appears to have been the first to suggest the graphical representation of complex numbers, a truly significant idea that was not pursued further until about 1800. Several people, including Karl Friedrich Gauss (1777–1855), then rediscovered the idea, and graphical representation helped to establish complex numbers as equal members of the number family. In practical applications, complex numbers have found their greatest uses in the study of alternating current, where they are a commonplace tool, and in the field of subatomic physics.

Historical Problems

1. The quadratic formula works perfectly well if the coefficients are complex numbers. Solve the following.
   (a) $z^2 - (2 + 5i)z - 3 + 5i = 0$ **(b)** $z^2 - (1 + i)z - 2 - i = 0$

10.3 Assess Your Understanding

'Are You Prepared?' *Answers are given at the end of these exercises. If you get a wrong answer, read the pages listed in* red.

1. The conjugate of $-4 - 3i$ is _____ . (pp. 104–109)

2. The sum formula for the sine function is
$\sin(A + B) = $ _____ . (p. 646)

3. The sum formula for the cosine function is
$\cos(A + B) = $ _____ . (p. 643)

4. $\sin 120° = $ _____ ; $\cos 240° = $ _____ . (pp. 543–550)

Concepts and Vocabulary

5. In the complex plane, the x-axis is referred to as the _____ axis, and the y-axis is called the _____ axis.

6. When a complex number z is written in the polar form $z = r(\cos\theta + i\sin\theta)$, the nonnegative number r is the _____ or _____ of z, and the angle θ, $0 \le \theta < 2\pi$, is the _____ of z.

7. Let $z_1 = r_1(\cos\theta_1 + i\sin\theta_1)$ and $z_2 = r_2(\cos\theta_2 + i\sin\theta_2)$ be two complex numbers. Then
$z_1 z_2 = $ ____ $[\cos ($ _____ $) + i\sin ($ _____ $)]$.

8. If $z = r(\cos\theta + i\sin\theta)$ is a complex number, then $z^n = $ ____ $[\cos($ ___ $) + i\sin($ ___ $)]$.

9. Every nonzero complex number will have exactly _____ distinct complex cube roots.

10. **True or False** The polar form of a nonzero complex number is unique.

11. If $z = x + yi$ is a complex number, then $|z|$ equals which of the following?
(a) $x^2 + y^2$ (b) $|x| + |y|$
(c) $\sqrt{x^2 + y^2}$ (d) $\sqrt{|x| + |y|}$

12. If $z_1 = r_1(\cos\theta_1 + i\sin\theta_1)$ and $z_2 = r_2(\cos\theta_2 + i\sin\theta_2)$ are complex numbers, then $\dfrac{z_1}{z_2}$, $z_2 \ne 0$, equals which of the following?

(a) $\dfrac{r_1}{r_2}[\cos(\theta_1 - \theta_2) + i\sin(\theta_1 - \theta_2)]$

(b) $\dfrac{r_1}{r_2}\left[\cos\left(\dfrac{\theta_1}{\theta_2}\right) + i\sin\left(\dfrac{\theta_1}{\theta_2}\right)\right]$

(c) $\dfrac{r_1}{r_2}[\cos(\theta_1 + \theta_2) - i\sin(\theta_1 + \theta_2)]$

(d) $\dfrac{r_1}{r_2}\left[\cos\left(\dfrac{\theta_1}{\theta_2}\right) - i\sin\left(\dfrac{\theta_1}{\theta_2}\right)\right]$

Skill Building

In Problems 13–24, plot each complex number in the complex plane and write it in polar form. Express the argument in degrees.

13. $1 + i$
14. $-1 + i$
15. $\sqrt{3} - i$
16. $1 - \sqrt{3}i$
17. $-3i$
18. -2

19. $4 - 4i$
20. $9\sqrt{3} + 9i$
21. $3 - 4i$
22. $2 + \sqrt{3}i$
23. $-2 + 3i$
24. $\sqrt{5} - i$

In Problems 25–34, write each complex number in rectangular form.

25. $2(\cos 120° + i\sin 120°)$

26. $3(\cos 210° + i\sin 210°)$

27. $4\left(\cos\dfrac{7\pi}{4} + i\sin\dfrac{7\pi}{4}\right)$

28. $2\left(\cos\dfrac{5\pi}{6} + i\sin\dfrac{5\pi}{6}\right)$

29. $3\left(\cos\dfrac{3\pi}{2} + i\sin\dfrac{3\pi}{2}\right)$

30. $4\left(\cos\dfrac{\pi}{2} + i\sin\dfrac{\pi}{2}\right)$

31. $0.2(\cos 100° + i\sin 100°)$

32. $0.4(\cos 200° + i\sin 200°)$

33. $2\left(\cos\dfrac{\pi}{18} + i\sin\dfrac{\pi}{18}\right)$

34. $3\left(\cos\dfrac{\pi}{10} + i\sin\dfrac{\pi}{10}\right)$

In Problems 35–42, find zw and $\dfrac{z}{w}$. Leave your answers in polar form.

35. $z = 2(\cos 40° + i\sin 40°)$
$w = 4(\cos 20° + i\sin 20°)$

36. $z = \cos 120° + i\sin 120°$
$w = \cos 100° + i\sin 100°$

37. $z = 3(\cos 130° + i\sin 130°)$
$w = 4(\cos 270° + i\sin 270°)$

38. $z = 2(\cos 80° + i\sin 80°)$
$w = 6(\cos 200° + i\sin 200°)$

39. $z = 2\left(\cos\dfrac{\pi}{8} + i\sin\dfrac{\pi}{8}\right)$
$w = 2\left(\cos\dfrac{\pi}{10} + i\sin\dfrac{\pi}{10}\right)$

40. $z = 4\left(\cos\dfrac{3\pi}{8} + i\sin\dfrac{3\pi}{8}\right)$
$w = 2\left(\cos\dfrac{9\pi}{16} + i\sin\dfrac{9\pi}{16}\right)$

41. $z = 2 + 2i$
$w = \sqrt{3} - i$

42. $z = 1 - i$
$w = 1 - \sqrt{3}i$

In Problems 43–54, write each expression in the standard form $a + bi$.

43. $[4(\cos 40° + i\sin 40°)]^3$

44. $[3(\cos 80° + i\sin 80°)]^3$

45. $\left[2\left(\cos\dfrac{\pi}{10} + i\sin\dfrac{\pi}{10}\right)\right]^5$

46. $\left[\sqrt{2}\left(\cos\dfrac{5\pi}{16} + i\sin\dfrac{5\pi}{16}\right)\right]^4$

47. $\left[\sqrt{3}\,(\cos 10° + i\sin 10°)\right]^6$

48. $\left[\dfrac{1}{2}(\cos 72° + i\sin 72°)\right]^5$

49. $\left[\sqrt{5}\left(\cos\dfrac{3\pi}{16}+i\sin\dfrac{3\pi}{16}\right)\right]^4$

50. $\left[\sqrt{3}\left(\cos\dfrac{5\pi}{18}+i\sin\dfrac{5\pi}{18}\right)\right]^6$

51. $(1-i)^5$

52. $\left(\sqrt{3}-i\right)^6$

53. $\left(\sqrt{2}-i\right)^6$

54. $\left(1-\sqrt{5}i\right)^8$

In Problems 55–62, find all the complex roots. Leave your answers in polar form with the argument in degrees.

55. The complex cube roots of $1+i$

56. The complex fourth roots of $\sqrt{3}-i$

57. The complex fourth roots of $4-4\sqrt{3}i$

58. The complex cube roots of $-8-8i$

59. The complex fourth roots of $-16i$

60. The complex cube roots of -8

61. The complex fifth roots of i

62. The complex fifth roots of $-i$

Applications and Extensions

63. Find the four complex fourth roots of unity (1) and plot them.

64. Find the six complex sixth roots of unity (1) and plot them.

65. Show that each complex nth root of a nonzero complex number w has the same magnitude.

66. Use the result of Problem 65 to draw the conclusion that each complex nth root lies on a circle with center at the origin. What is the radius of this circle?

67. Refer to Problem 66. Show that the complex nth roots of a nonzero complex number w are equally spaced on the circle.

68. Prove formula (6).

69. Prove that De Moivre's Theorem is true for *all* integers n by assuming it is true for integers $n \geq 1$ and then showing it is true for 0 and for negative integers.

Hint: Multiply the numerator and the denominator by the conjugate of the denominator, and use even-odd properties.

70. Mandelbrot Sets
(a) Consider the expression $a_n = (a_{n-1})^2 + z$, where z is some complex number (called the **seed**) and $a_0 = z$. Compute $a_1\,(=a_0^2 + z)$, $a_2\,(=a_1^2 + z)$, $a_3\,(=a_2^2 + z)$, a_4, a_5, and a_6 for the following seeds: $z_1 = 0.1 - 0.4i$, $z_2 = 0.5 + 0.8i$, $z_3 = -0.9 + 0.7i$, $z_4 = -1.1 + 0.1i$, $z_5 = 0 - 1.3i$, and $z_6 = 1 + 1i$.
(b) The dark portion of the graph represents the set of all values $z = x + yi$ that are in the Mandelbrot set.

Determine which complex numbers in part (a) are in this set by plotting them on the graph. Do the complex numbers that are not in the Mandelbrot set have any common characteristics regarding the values of a_6 found in part (a)?

(c) Compute $|z| = \sqrt{x^2 + y^2}$ for each of the complex numbers in part (a). Now compute $|a_6|$ for each of the complex numbers in part (a). For which complex numbers is $|a_6| \leq |z|$ and $|z| \leq 2$? Conclude that the criterion for a complex number to be in the Mandelbrot set is that $|a_n| \leq |z|$ and $|z| \leq 2$.

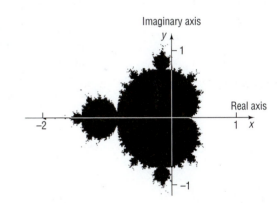

Retain Your Knowledge

Problems 71–74 are based on material learned earlier in the course. The purpose of these problems is to keep the material fresh in your mind so that you are better prepared for the final exam.

71. Find the area of the triangle with $a = 8$, $b = 11$, and $C = 113°$.

72. Convert $240°$ to radians. Express your answer as a multiple of π.

73. Simplify: $\sqrt[3]{24x^2y^5}$

74. Determine whether $f(x) = 5x^2 - 12x + 4$ has a maximum value or a minimum value, and then find the value.

'Are You Prepared?' Answers

1. $-4 + 3i$

2. $\sin A \cos B + \cos A \sin B$

3. $\cos A \cos B - \sin A \sin B$

4. $\dfrac{\sqrt{3}}{2}; -\dfrac{1}{2}$

10.4 Vectors

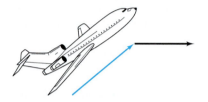

Figure 40

In simple terms, a **vector** (derived from the Latin *vehere,* meaning "to carry") is a quantity that has both magnitude and direction. It is customary to represent a vector by using an arrow. The length of the arrow represents the **magnitude** of the vector, and the arrowhead indicates the **direction** of the vector.

Many quantities in physics can be represented by vectors. For example, the velocity of an aircraft can be represented by an arrow that points in the direction of movement; the length of the arrow represents the speed. If the aircraft speeds up, we lengthen the arrow; if the aircraft changes direction, we introduce an arrow in the new direction. See Figure 40. Based on this representation, it is not surprising that vectors and *directed line segments* are somehow related.

Geometric Vectors

If P and Q are two distinct points in the xy-plane, there is exactly one line containing both P and Q [Figure 41(a)]. The points on that part of the line that joins P to Q, including P and Q, form what is called the **line segment** $\overline{PQ}$ [Figure 41(b)]. Ordering the points so that they proceed from P to Q results in a **directed line segment** from P to Q, or a **geometric vector**, which is denoted by $\overrightarrow{PQ}$. In a directed line segment $\overrightarrow{PQ}$, P is called the **initial point** and Q the **terminal point**, as indicated in Figure 41(c).

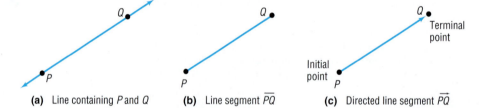

Figure 41   **(a)** Line containing P and Q **(b)** Line segment $\overline{PQ}$ **(c)** Directed line segment $\overrightarrow{PQ}$

The magnitude of the directed line segment $\overrightarrow{PQ}$ is the distance from the point P to the point Q; that is, it is the length of the line segment. The direction of $\overrightarrow{PQ}$ is from P to Q. If a vector $\mathbf{v}$* has the same magnitude and the same direction as the directed line segment $\overrightarrow{PQ}$, write

$$\mathbf{v} = \overrightarrow{PQ}$$

The vector $\mathbf{v}$ whose magnitude is 0 is called the **zero vector**, **0**. The zero vector is assigned no direction.

Two vectors $\mathbf{v}$ and $\mathbf{w}$ are **equal**, written

$$\mathbf{v} = \mathbf{w}$$

if they have the same magnitude and the same direction.

For example, the three vectors shown in Figure 42 have the same magnitude and the same direction, so they are equal, even though they have different initial points and different terminal points. As a result, it is useful to think of a vector simply as an arrow, keeping in mind that two arrows (vectors) are equal if they have the same direction and the same magnitude (length).

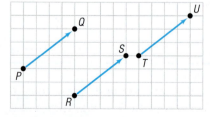

Figure 42 Equal vectors

*Boldface letters will be used to denote vectors, to distinguish them from numbers. For handwritten work, an arrow is placed over the letter to signify a vector. For example, write a vector by hand as $\vec{v}$.

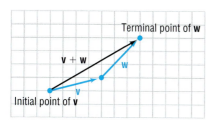

Figure 43 Adding vectors

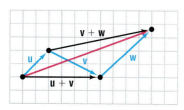

Figure 44 v + w = w + v

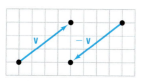

Figure 45
(u + v) + w = u + (v + w)

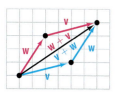

Figure 46 Opposite vectors

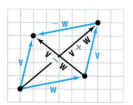

Figure 47

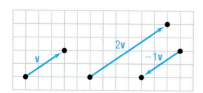

Figure 48 Scalar multiples

Adding Vectors Geometrically

The **sum v + w** of two vectors is defined as follows: Position the vectors **v** and **w** so that the terminal point of **v** coincides with the initial point of **w**, as shown in Figure 43. The vector **v + w** is then the unique vector whose initial point coincides with the initial point of **v** and whose terminal point coincides with the terminal point of **w**.

Vector addition is **commutative**. That is, if **v** and **w** are any two vectors, then

$$\mathbf{v} + \mathbf{w} = \mathbf{w} + \mathbf{v}$$

Figure 44 illustrates this fact. (Observe that the commutative property is another way of saying that opposite sides of a parallelogram are equal and parallel.)

Vector addition is also **associative**. That is, if **u**, **v**, and **w** are vectors, then

$$\mathbf{u} + (\mathbf{v} + \mathbf{w}) = (\mathbf{u} + \mathbf{v}) + \mathbf{w}$$

Figure 45 illustrates the associative property for vectors.

The zero vector **0** has the property that

$$\mathbf{v} + \mathbf{0} = \mathbf{0} + \mathbf{v} = \mathbf{v}$$

for any vector **v**.

If **v** is a vector, then −**v** is the vector that has the same magnitude as **v**, but whose direction is opposite to **v**, as shown in Figure 46.
Furthermore,

$$\mathbf{v} + (-\mathbf{v}) = \mathbf{0}$$

If **v** and **w** are two vectors, then the **difference v − w** is defined as

$$\mathbf{v} - \mathbf{w} = \mathbf{v} + (-\mathbf{w})$$

Figure 47 illustrates the relationships among **v**, **w**, **v + w**, and **v − w**.

Multiplying Vectors by Numbers Geometrically

When dealing with vectors, real numbers are referred to as **scalars**. Scalars are quantities that have only magnitude. Examples of scalar quantities from physics are temperature, speed, and time. We now define how to multiply a vector by a scalar.

DEFINITION

If α is a scalar and **v** is a vector, the **scalar multiple** $\alpha \mathbf{v}$ is defined as follows:

1. If $\alpha > 0$, $\alpha \mathbf{v}$ is the vector whose magnitude is α times the magnitude of **v** and whose direction is the same as that of **v**.
2. If $\alpha < 0$, $\alpha \mathbf{v}$ is the vector whose magnitude is $|\alpha|$ times the magnitude of **v** and whose direction is opposite that of **v**.
3. If $\alpha = 0$ or if **v = 0**, then $\alpha \mathbf{v} = \mathbf{0}$.

See Figure 48 for some illustrations.
For example, if **a** is the acceleration of an object of mass m due to a force **F** being exerted on it, then, by Newton's second law of motion, **F** = m**a**. Here, m**a** is the product of the scalar m and the vector **a**.

Scalar multiples have the following properties:

$$0\mathbf{v} = \mathbf{0} \qquad 1\mathbf{v} = \mathbf{v} \qquad -1\mathbf{v} = -\mathbf{v}$$

$$(\alpha + \beta)\mathbf{v} = \alpha\mathbf{v} + \beta\mathbf{v} \qquad \alpha(\mathbf{v} + \mathbf{w}) = \alpha\mathbf{v} + \alpha\mathbf{w}$$

$$\alpha(\beta\mathbf{v}) = (\alpha\beta)\mathbf{v}$$

1 Graph Vectors

EXAMPLE 1

Graphing Vectors

Use the vectors illustrated in Figure 49 to graph each of the following vectors:

(a) $\mathbf{v} - \mathbf{w}$ (b) $2\mathbf{v} + 3\mathbf{w}$ (c) $2\mathbf{v} - \mathbf{w} + \mathbf{u}$

Solution Figure 50 shows each graph.

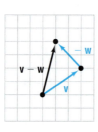

Figure 49

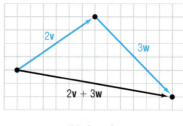

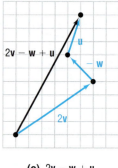

Figure 50   (a) $\mathbf{v} - \mathbf{w}$ (b) $2\mathbf{v} + 3\mathbf{w}$ (c) $2\mathbf{v} - \mathbf{w} + \mathbf{u}$ ●

Now Work PROBLEMS **11** AND **13**

Magnitude of Vectors

The symbol $\|\mathbf{v}\|$ represents the **magnitude** of a vector $\mathbf{v}$. Since $\|\mathbf{v}\|$ equals the length of a directed line segment, it follows that $\|\mathbf{v}\|$ has the following properties:

THEOREM

Properties of $\|\mathbf{v}\|$

If $\mathbf{v}$ is a vector and if α is a scalar, then

(a) $\|\mathbf{v}\| \geq 0$ (b) $\|\mathbf{v}\| = 0$ if and only if $\mathbf{v} = \mathbf{0}$

(c) $\|-\mathbf{v}\| = \|\mathbf{v}\|$ (d) $\|\alpha\mathbf{v}\| = |\alpha|\|\mathbf{v}\|$

Property (a) is a consequence of the fact that distance is a nonnegative number. Property (b) follows because the length of the directed line segment $\overrightarrow{PQ}$ is positive unless P and Q are the same point, in which case the length is 0. Property (c) follows because the length of the line segment $\overline{PQ}$ equals the length of the line segment $\overline{QP}$. Property (d) is a direct consequence of the definition of a scalar multiple.

DEFINITION

A vector $\mathbf{u}$ for which $\|\mathbf{u}\| = 1$ is called a **unit vector**.

2 Find a Position Vector

To compute the magnitude and direction of a vector, an algebraic way of representing vectors is needed.

DEFINITION

An **algebraic vector** $\mathbf{v}$ is represented as

$$\mathbf{v} = \langle a, b \rangle$$

where a and b are real numbers (scalars) called the **components** of the vector $\mathbf{v}$.

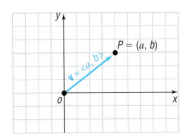

Figure 51 Position vector **v**

A rectangular coordinate system is used to represent algebraic vectors in the plane. If $\mathbf{v} = \langle a, b \rangle$ is an algebraic vector whose initial point is at the origin, then $\mathbf{v}$ is called a **position vector**. See Figure 51. Notice that the terminal point of the position vector $\mathbf{v} = \langle a, b \rangle$ is $P = (a, b)$.

The next result states that any vector whose initial point is not at the origin is equal to a unique position vector.

THEOREM

Suppose that $\mathbf{v}$ is a vector with initial point $P_1 = (x_1, y_1)$, not necessarily the origin, and terminal point $P_2 = (x_2, y_2)$. If $\mathbf{v} = \overrightarrow{P_1 P_2}$, then $\mathbf{v}$ is equal to the position vector

$$\mathbf{v} = \langle x_2 - x_1, y_2 - y_1 \rangle \qquad (1)$$

In Words

An algebraic vector represents "driving directions" to get from the initial point to the terminal point of a vector. So if $\mathbf{v} = \langle 5, 4 \rangle$, travel 5 units right and 4 units up from the initial point to arrive at the terminal point.

To see why this is true, look at Figure 52.

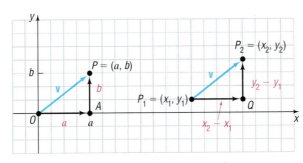

Figure 52 $\mathbf{v} = \langle a, b \rangle = \langle x_2 - x_1, y_2 - y_1 \rangle$

Triangle OPA and triangle P_1P_2Q are congruent. [Do you see why? The line segments have the same magnitude, so $d(O, P) = d(P_1, P_2)$; and they have the same direction, so $\angle POA = \angle P_2P_1Q$. Since the triangles are right triangles, we have angle–side–angle.] It follows that corresponding sides are equal. As a result, $x_2 - x_1 = a$ and $y_2 - y_1 = b$, so $\mathbf{v}$ may be written as

$$\mathbf{v} = \langle a, b \rangle = \langle x_2 - x_1, y_2 - y_1 \rangle$$

Because of this result, any algebraic vector can be replaced by a unique position vector, and vice versa. This flexibility is one of the main reasons for the wide use of vectors.

EXAMPLE 2

Finding a Position Vector

Find the position vector of the vector $\mathbf{v} = \overrightarrow{P_1P_2}$ if $P_1 = (-1, 2)$ and $P_2 = (4, 6)$.

Solution By equation (1), the position vector equal to $\mathbf{v}$ is

$$\mathbf{v} = \langle 4 - (-1), 6 - 2 \rangle = \langle 5, 4 \rangle$$

See Figure 53.

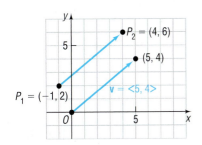

Figure 53

Two position vectors **v** and **w** are equal if and only if the terminal point of **v** is the same as the terminal point of **w**. This leads to the following result:

THEOREM

Equality of Vectors

Two vectors **v** and **w** are equal if and only if their corresponding components are equal. That is,

$$\text{If } \mathbf{v} = \langle a_1, b_1 \rangle \quad \text{and} \quad \mathbf{w} = \langle a_2, b_2 \rangle$$
$$\text{then} \quad \mathbf{v} = \mathbf{w} \text{ if and only if } \quad a_1 = a_2 \quad \text{and} \quad b_1 = b_2.$$

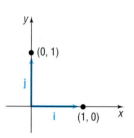

Figure 54 Unit vectors **i** and **j**

We now present an alternative representation of a vector in the plane that is common in the physical sciences. Let **i** denote the unit vector whose direction is along the positive x-axis; let **j** denote the unit vector whose direction is along the positive y-axis. Then $\mathbf{i} = \langle 1, 0 \rangle$ and $\mathbf{j} = \langle 0, 1 \rangle$, as shown in Figure 54. Any vector $\mathbf{v} = \langle a, b \rangle$ can be written using the unit vectors **i** and **j** as follows:

$$\mathbf{v} = \langle a, b \rangle = a\langle 1, 0 \rangle + b\langle 0, 1 \rangle = a\mathbf{i} + b\mathbf{j}$$

The quantities a and b are called the **horizontal** and **vertical components** of **v**, respectively. For example, if $\mathbf{v} = \langle 5, 4 \rangle = 5\mathbf{i} + 4\mathbf{j}$, then 5 is the horizontal component and 4 is the vertical component.

✏️ **Now Work** PROBLEM 31

3 Add and Subtract Vectors Algebraically

The sum, difference, scalar multiple, and magnitude of algebraic vectors are defined in terms of their components.

DEFINITION

Let $\mathbf{v} = a_1\mathbf{i} + b_1\mathbf{j} = \langle a_1, b_1 \rangle$ and $\mathbf{w} = a_2\mathbf{i} + b_2\mathbf{j} = \langle a_2, b_2 \rangle$ be two vectors, and let α be a scalar. Then

$$\mathbf{v} + \mathbf{w} = (a_1 + a_2)\mathbf{i} + (b_1 + b_2)\mathbf{j} = \langle a_1 + a_2, b_1 + b_2 \rangle \tag{2}$$
$$\mathbf{v} - \mathbf{w} = (a_1 - a_2)\mathbf{i} + (b_1 - b_2)\mathbf{j} = \langle a_1 - a_2, b_1 - b_2 \rangle \tag{3}$$
$$\alpha\mathbf{v} = (\alpha a_1)\mathbf{i} + (\alpha b_1)\mathbf{j} = \langle \alpha a_1, \alpha b_1 \rangle \tag{4}$$
$$\|\mathbf{v}\| = \sqrt{a_1^2 + b_1^2} \tag{5}$$

In Words

To add two vectors, add corresponding components. To subtract two vectors, subtract corresponding components.

These definitions are compatible with the geometric definitions given earlier in this section. See Figure 55.

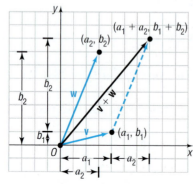

(a) Illustration of property (2)

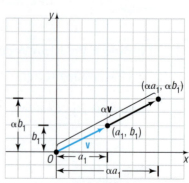

(b) Illustration of property (4), $\alpha > 0$

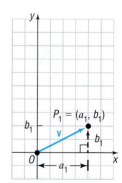

(c) Illustration of property (5): $\|\mathbf{v}\|$ = Distance from O to P_1 $\|\mathbf{v}\| = \sqrt{a_1^2 + b_1^2}$

Figure 55

EXAMPLE 3

Adding and Subtracting Vectors

If $\mathbf{v} = 2\mathbf{i} + 3\mathbf{j} = \langle 2, 3 \rangle$ and $\mathbf{w} = 3\mathbf{i} - 4\mathbf{j} = \langle 3, -4 \rangle$, find:

(a) $\mathbf{v} + \mathbf{w}$ (b) $\mathbf{v} - \mathbf{w}$

Solution

(a) $\mathbf{v} + \mathbf{w} = (2\mathbf{i} + 3\mathbf{j}) + (3\mathbf{i} - 4\mathbf{j}) = (2 + 3)\mathbf{i} + (3 - 4)\mathbf{j} = 5\mathbf{i} - \mathbf{j}$

or

$\mathbf{v} + \mathbf{w} = \langle 2, 3 \rangle + \langle 3, -4 \rangle = \langle 2 + 3, 3 + (-4) \rangle = \langle 5, -1 \rangle$

(b) $\mathbf{v} - \mathbf{w} = (2\mathbf{i} + 3\mathbf{j}) - (3\mathbf{i} - 4\mathbf{j}) = (2 - 3)\mathbf{i} + [3 - (-4)]\mathbf{j} = -\mathbf{i} + 7\mathbf{j}$

or

$\mathbf{v} - \mathbf{w} = \langle 2, 3 \rangle - \langle 3, -4 \rangle = \langle 2 - 3, 3 - (-4) \rangle = \langle -1, 7 \rangle$ ●

4 Find a Scalar Multiple and the Magnitude of a Vector

EXAMPLE 4

Finding Scalar Multiples and Magnitudes of Vectors

If $\mathbf{v} = 2\mathbf{i} + 3\mathbf{j} = \langle 2, 3 \rangle$ and $\mathbf{w} = 3\mathbf{i} - 4\mathbf{j} = \langle 3, -4 \rangle$, find:

(a) $3\mathbf{v}$ (b) $2\mathbf{v} - 3\mathbf{w}$ (c) $\|\mathbf{v}\|$

Solution

(a) $3\mathbf{v} = 3(2\mathbf{i} + 3\mathbf{j}) = 6\mathbf{i} + 9\mathbf{j}$

or

$3\mathbf{v} = 3\langle 2, 3 \rangle = \langle 6, 9 \rangle$

(b) $2\mathbf{v} - 3\mathbf{w} = 2(2\mathbf{i} + 3\mathbf{j}) - 3(3\mathbf{i} - 4\mathbf{j}) = 4\mathbf{i} + 6\mathbf{j} - 9\mathbf{i} + 12\mathbf{j}$

$= -5\mathbf{i} + 18\mathbf{j}$

or

$2\mathbf{v} - 3\mathbf{w} = 2\langle 2, 3 \rangle - 3\langle 3, -4 \rangle = \langle 4, 6 \rangle - \langle 9, -12 \rangle$

$= \langle 4 - 9, 6 - (-12) \rangle = \langle -5, 18 \rangle$

(c) $\|\mathbf{v}\| = \|2\mathbf{i} + 3\mathbf{j}\| = \sqrt{2^2 + 3^2} = \sqrt{13}$ ●

 Now Work PROBLEMS 37 AND 43

For the remainder of the section, we will express a vector $\mathbf{v}$ in the form $a\mathbf{i} + b\mathbf{j}$.

5 Find a Unit Vector

Recall that a unit vector $\mathbf{u}$ is a vector for which $\|\mathbf{u}\| = 1$. In many applications, it is useful to be able to find a unit vector $\mathbf{u}$ that has the same direction as a given vector $\mathbf{v}$.

THEOREM

Unit Vector in the Direction of v

For any nonzero vector $\mathbf{v}$, the vector

$$\mathbf{u} = \frac{\mathbf{v}}{\|\mathbf{v}\|} \tag{6}$$

is a unit vector that has the same direction as $\mathbf{v}$.

Proof Let $\mathbf{v} = a\mathbf{i} + b\mathbf{j}$. Then $\|\mathbf{v}\| = \sqrt{a^2 + b^2}$ and

$$\mathbf{u} = \frac{\mathbf{v}}{\|\mathbf{v}\|} = \frac{a\mathbf{i} + b\mathbf{j}}{\sqrt{a^2 + b^2}} = \frac{a}{\sqrt{a^2 + b^2}}\mathbf{i} + \frac{b}{\sqrt{a^2 + b^2}}\mathbf{j}$$

10.4 Assess Your Understanding

Concepts and Vocabulary

1. A _____ is a quantity that has both magnitude and direction.

2. If **v** is a vector, then **v** + (−**v**) = ____.

3. A vector **u** for which ‖**u**‖ = 1 is called a(n) _____ vector.

4. If **v** = <a, b> is an algebraic vector whose initial point is the origin, then **v** is called a(n) _____ vector.

5. If **v** = a**i** + b**j**, then a is called the _____ component of **v** and b is called the _____ component of **v**.

6. If **F**₁ and **F**₂ are two forces simultaneously acting on an object, the vector sum **F**₁ + **F**₂ is called the _____ force.

7. *True or False* Force is an example of a vector.

8. *True or False* Mass is an example of a vector.

9. If **v** is a vector with initial point (x_1, y_1) and terminal point (x_2, y_2), then which of the following is the position vector that equals **v**?
   (a) $\langle x_2 - x_1, y_2 - y_1 \rangle$ (b) $\langle x_1 - x_2, y_1 - y_2 \rangle$
   (c) $\left\langle \frac{x_2 - x_1}{2}, \frac{y_2 - y_1}{2} \right\rangle$ (d) $\left\langle \frac{x_1 + x_2}{2}, \frac{y_1 + y_2}{2} \right\rangle$

10. If **v** is a nonzero vector with direction angle α, 0° ≤ α < 360°, between **v** and **i**, then **v** equals which of the following?
   (a) ‖**v**‖(cos α**i** − sin α**j**)   (b) ‖**v**‖(cos α**i** + sin α**j**)
   (c) ‖**v**‖(sin α**i** − cos α**j**)   (d) ‖**v**‖(sin α**i** + cos α**j**)

Skill Building

In Problems 11–18, use the vectors in the figure at the right to graph each of the following vectors.

11. **v** + **w**   12. **u** + **v**

13. 3**v**   14. 2**w**

15. **v** − **w**   16. **u** − **v**

17. 3**v** + **u** − 2**w**   18. 2**u** − 3**v** + **w**

In Problems 19–26, use the figure at the right. Determine whether each statement given is true or false.

19. **A** + **B** = **F**   20. **K** + **G** = **F**

21. **C** = **D** − **E** + **F**   22. **G** + **H** + **E** = **D**

23. **E** + **D** = **G** + **H**   24. **H** − **C** = **G** − **F**

25. **A** + **B** + **K** + **G** = **0**   26. **A** + **B** + **C** + **H** + **G** = **0**

27. If ‖**v**‖ = 4, what is ‖3**v**‖?   28. If ‖**v**‖ = 2, what is ‖−4**v**‖?

*In Problems 29–36, the vector **v** has initial point P and terminal point Q. Write **v** in the form a**i** + b**j**; that is, find its position vector.*

29. P = (0, 0); Q = (3, 4)   30. P = (0, 0); Q = (−3, −5)

31. P = (3, 2); Q = (5, 6)   32. P = (−3, 2); Q = (6, 5)

33. P = (−2, −1); Q = (6, −2)   34. P = (−1, 4); Q = (6, 2)

35. P = (1, 0); Q = (0, 1)   36. P = (1, 1); Q = (2, 2)

*In Problems 37–42, find ‖**v**‖.*

37. **v** = 3**i** − 4**j**   38. **v** = −5**i** + 12**j**   39. **v** = **i** − **j**

40. **v** = −**i** − **j**   41. **v** = −2**i** + 3**j**   42. **v** = 6**i** + 2**j**

*In Problems 43–48, find each quantity if **v** = 3**i** − 5**j** and **w** = −2**i** + 3**j**.*

43. 2**v** + 3**w**   44. 3**v** − 2**w**   45. ‖**v** − **w**‖

46. ‖**v** + **w**‖   47. ‖**v**‖ − ‖**w**‖   48. ‖**v**‖ + ‖**w**‖

*In Problems 49–54, find the unit vector in the same direction as **v**.*

49. **v** = 5**i**   50. **v** = −3**j**   51. **v** = 3**i** − 4**j**

52. **v** = −5**i** + 12**j**   53. **v** = **i** − **j**   54. **v** = 2**i** − **j**

55. Find a vector **v** whose magnitude is 4 and whose component in the **i** direction is twice the component in the **j** direction.

56. Find a vector **v** whose magnitude is 3 and whose component in the **i** direction is equal to the component in the **j** direction.

57. If **v** = 2**i** − **j** and **w** = x**i** + 3**j**, find all numbers x for which ‖**v** + **w**‖ = 5.

58. If P = (−3, 1) and Q = (x, 4), find all numbers x such that the vector represented by $\overrightarrow{PQ}$ has length 5.

*In Problems 59–64, write the vector **v** in the form a**i** + b**j**, given its magnitude ‖**v**‖ and the angle α it makes with the positive x-axis.*

59. ‖**v**‖ = 5, α = 60°   **60.** ‖**v**‖ = 8, α = 45°   **61.** ‖**v**‖ = 14, α = 120°

62. ‖**v**‖ = 3, α = 240°   **63.** ‖**v**‖ = 25, α = 330°   **64.** ‖**v**‖ = 15, α = 315°

*In Problems 65–72, find the direction angle of **v**.*

65. **v** = 3**i** + 3**j**   **66.** **v** = **i** + √3**j**   **67.** **v** = −3√3**i** + 3**j**   **68.** **v** = −5**i** − 5**j**

69. **v** = 4**i** − 2**j**   **70.** **v** = 6**i** − 4**j**   **71.** **v** = −**i** − 5**j**   **72.** **v** = −**i** + 3**j**

Applications and Extensions

73. Force Vectors A child pulls a wagon with a force of 40 pounds. The handle of the wagon makes an angle of 30° with the ground. Express the force vector **F** in terms of **i** and **j**.

74. Force Vectors A man pushes a wheelbarrow up an incline of 20° with a force of 100 pounds. Express the force vector **F** in terms of **i** and **j**.

75. Resultant Force Two forces of magnitude 40 newtons (N) and 60 N act on an object at angles of 30° and −45° with the positive x-axis, as shown in the figure. Find the direction and magnitude of the resultant force; that is, find **F**₁ + **F**₂.

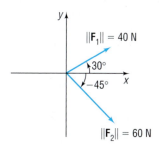

76. Resultant Force Two forces of magnitude 30 newtons (N) and 70 N act on an object at angles of 45° and 120° with the positive x-axis, as shown in the figure. Find the direction and magnitude of the resultant force; that is, find **F**₁ + **F**₂.

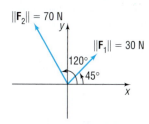

77. Finding the Actual Speed and Direction of an Aircraft A Boeing 747 jumbo jet maintains a constant airspeed of 550 miles per hour (mph) headed due north. The jet stream is 100 mph in the northeasterly direction.
(a) Express the velocity **v**ₐ of the 747 relative to the air and the velocity **v**_w of the jet stream in terms of **i** and **j**.
(b) Find the velocity of the 747 relative to the ground.
(c) Find the actual speed and direction of the 747 relative to the ground.

78. Finding the Actual Speed and Direction of an Aircraft An Airbus A320 jet maintains a constant airspeed of 500 mph headed due west. The jet stream is 100 mph in the south-easterly direction.
(a) Express the velocity **v**ₐ of the A320 relative to the air and the velocity **v**_w of the jet stream in terms of **i** and **j**.
(b) Find the velocity of the A320 relative to the ground.
(c) Find the actual speed and direction of the A320 relative to the ground.

79. Ground Speed and Direction of an Airplane An airplane has an airspeed of 500 kilometers per hour (km/h) bearing N45°E. The wind velocity is 60 km/h in the direction N30°W. Find the resultant vector representing the path of the plane relative to the ground. What is the groundspeed of the plane? What is its direction?

80. Ground Speed and Direction of an Airplane An airplane has an airspeed of 600 km/h bearing S30°E. The wind velocity is 40 km/h in the direction S45°E. Find the resultant vector representing the path of the plane relative to the ground. What is the groundspeed of the plane? What is its direction?

81. Weight of a Boat A magnitude of 700 pounds of force is required to hold a boat and its trailer in place on a ramp whose incline is 10° to the horizontal. What is the combined weight of the boat and its trailer?

82. Weight of a Car A magnitude of 1200 pounds of force is required to prevent a car from rolling down a hill whose incline is 15° to the horizontal. What is the weight of the car?

83. Correct Direction for Crossing a River A river has a constant current of 3 km/h. At what angle to a boat dock should a motorboat capable of maintaining a constant speed of 20 km/h be headed in order to reach a point directly opposite the dock? If the river is $\frac{1}{2}$ kilometer wide, how long will it take to cross?

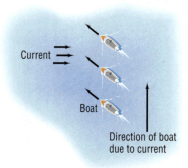

84. Finding the Correct Compass Heading The pilot of an aircraft wishes to head directly east but is faced with a wind speed of 40 mph from the northwest. If the pilot maintains an airspeed of 250 mph, what compass heading should be maintained to head directly east? What is the actual speed of the aircraft?

85. Charting a Course A helicopter pilot needs to travel to a regional airport 25 miles away. She flies at an actual heading of N16.26°E with an airspeed of 120 mph, and there is a wind blowing directly east at 20 mph.
(a) Determine the compass heading that the pilot needs to reach her destination.
(b) How long will it take her to reach her destination? Round to the nearest minute.

86. Crossing a River A captain needs to pilot a boat across a river that is 2 km wide. The current in the river is 2 km/h and the speed of the boat in still water is 10 km/h. The desired landing point on the other side is 1 km upstream.
(a) Determine the direction in which the captain should aim the boat.
(b) How long will the trip take?

87. Static Equilibrium A weight of 1000 pounds is suspended from two cables, as shown in the figure. What are the tensions in the two cables?

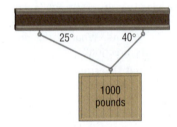

88. Static Equilibrium A weight of 800 pounds is suspended from two cables, as shown in the figure. What are the tensions in the two cables?

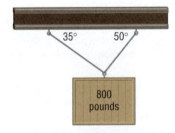

89. Static Equilibrium A tightrope walker located at a certain point deflects the rope as indicated in the figure. If the weight of the tightrope walker is 150 pounds, how much tension is in each part of the rope?

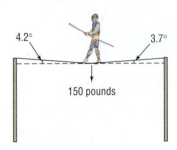

90. Static Equilibrium Repeat Problem 89 if the angle on the left is 3.8°, the angle on the right is 2.6°, and the weight of the tightrope walker is 135 pounds.

91. Static Friction A 20-pound box sits at rest on a horizontal surface, and there is friction between the box and the surface. One side of the surface is raised slowly to create a ramp. The friction force $\mathbf{f}$ opposes the direction of motion and is proportional to the normal force $\mathbf{F_N}$ exerted by the surface on the box. The proportionality constant is called the **coefficient of friction**, μ. When the angle of the ramp, θ, reaches 20°, the box begins to slide. Find the value of μ to two decimal places.

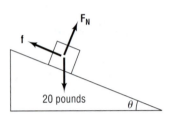

92. Inclined Ramp A 2-pound weight is attached to a 3-pound weight by a rope that passes over an ideal pulley. The smaller weight hangs vertically, while the larger weight sits on a frictionless inclined ramp with angle θ. The rope exerts a tension force $\mathbf{T}$ on both weights along the direction of the rope. Find the angle measure for θ that is needed to keep the larger weight from sliding down the ramp. Round your answer to the nearest tenth of a degree.

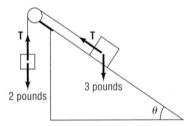

93. Inclined Ramp A box sitting on a horizontal surface is attached to a second box sitting on an inclined ramp by a rope that passes over an ideal pulley. The rope exerts a tension force $\mathbf{T}$ on both weights along the direction of the rope, and the coefficient of friction between the surface and boxes is 0.6 (see Problems 91 and 92). If the box on the right weighs 100 pounds and the angle of the ramp is 35°, how much must the box on the left weigh for the system to be in static equilibrium? Round your answer to two decimal places.

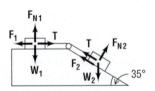

94. Muscle Force Two muscles exert force on a bone at the same point. The first muscle exerts a force of 800 N at a 10° angle with the bone. The second muscle exerts a force of 710 N at a 35° angle with the bone. What are the direction and magnitude of the resulting force on the bone?

95. Truck Pull At a county fair truck pull, two pickup trucks are attached to the back end of a monster truck as illustrated in the figure. One of the pickups pulls with a force of 2000 pounds, and the other pulls with a force of 3000 pounds. There is an angle of 45° between them. With how much force must the monster truck pull in order to remain unmoved?

[**Hint:** Find the resultant force of the two trucks.]

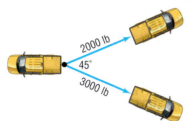

96. Removing a Stump A farmer wishes to remove a stump from a field by pulling it out with his tractor. Having removed many stumps before, he estimates that he will need 6 tons (12,000 pounds) of force to remove the stump. However, his tractor is only capable of pulling with a force of 7000 pounds, so he asks his neighbor to help. His neighbor's tractor can pull with a force of 5500 pounds. They attach the two tractors to the stump with a 40° angle between the forces, as shown in the figure.

(a) Assuming the farmer's estimate of a needed 6-ton force is correct, will the farmer be successful in removing the stump?

(b) Had the farmer arranged the tractors with a 25° angle between the forces, would he have been successful in removing the stump?

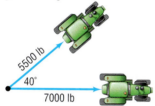

97. Computer Graphics The field of computer graphics utilizes vectors to compute translations of points. For example, if the point $(-3, 2)$ is to be translated by $\mathbf{v} = \langle 5, 2 \rangle$, then the new location will be $\mathbf{u'} = \mathbf{u} + \mathbf{v} = \langle -3, 2 \rangle + \langle 5, 2 \rangle = \langle 2, 4 \rangle$.

As illustrated in the figure, the point $(-3, 2)$ is translated to $(2, 4)$ by $\mathbf{v}$.

(a) Determine the new coordinates of $(3, -1)$ if it is translated by $\mathbf{v} = \langle -4, 5 \rangle$.

(b) Illustrate this translation graphically.

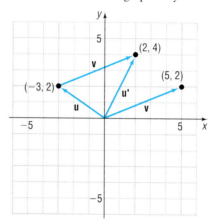

Source: Phil Dadd. Vectors and Matrices: A Primer. www.gamedev.net/reference/articles/article1832.asp

98. Computer Graphics Refer to Problem 97. The points $(-3, 0)$, $(-1, -2)$, $(3, 1)$, and $(1, 3)$ are the vertices of a parallelogram $ABCD$.

(a) Find the new vertices of a parallelogram $A'B'C'D'$ if it is translated by $\mathbf{v} = \langle 3, -2 \rangle$.

(b) Find the new vertices of a parallelogram $A'B'C'D'$ if it is translated by $-\dfrac{1}{2}\mathbf{v}$.

99. Static Equilibrium Show on the following graph the force needed for the object at P to be in static equilibrium.

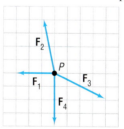

Explaining Concepts: Discussion and Writing

100. Explain in your own words what a vector is. Give an example of a vector.

101. Write a brief paragraph comparing the algebra of complex numbers and the algebra of vectors.

102. Explain the difference between an algebraic vector and a position vector.

Retain Your Knowledge

Problems 103–106 are based on material learned earlier in the course. The purpose of these problems is to keep the material fresh in your mind so that you are better prepared for the final exam.

103. Solve: $\sqrt[3]{x - 2} = 3$.

104. Factor $-3x^3 + 12x^2 + 36x$ completely.

105. Find the exact value of $\tan\left[\cos^{-1}\left(\dfrac{1}{2}\right)\right]$.

106. Find the amplitude, period, and phase shift of $y = \dfrac{3}{2}\cos(6x + 3\pi)$. Graph the function, showing at least two periods.

Determine the vector projection of $\mathbf{F}_g$ onto $\mathbf{w}$, which is the force parallel to the hill. The vector $\mathbf{w}$ is given by

$$\mathbf{w} = \cos 20°\mathbf{i} + \sin 20°\mathbf{j}$$

The vector projection of $\mathbf{F}_g$ onto $\mathbf{w}$ is

$$\mathbf{v} = \frac{\mathbf{F}_g \cdot \mathbf{w}}{\|\mathbf{w}\|^2} \mathbf{w}$$

$$= \frac{-100(\sin 20°)}{\left(\sqrt{\cos^2 20° + \sin^2 20°}\right)^2}(\cos 20°\mathbf{i} + \sin 20°\mathbf{j})$$

$$= -34.2(\cos 20°\mathbf{i} + \sin 20°\mathbf{j})$$

The magnitude of $\mathbf{v}$ is 34.2 pounds, so the magnitude of the force required to keep the wagon from rolling down the hill is 34.2 pounds.

6 Compute Work

In elementary physics, the **work** W done by a constant force $\mathbf{F}$ in moving an object from a point A to a point B is defined as

$$W = (\text{magnitude of force})(\text{distance}) = \|\mathbf{F}\|\|\overrightarrow{AB}\|$$

Work is commonly measured in foot-pounds or in newton-meters (joules).

In this definition, it is assumed that the force $\mathbf{F}$ is applied along the line of motion. If the constant force $\mathbf{F}$ is not along the line of motion, but instead is at an angle θ to the direction of the motion, as illustrated in Figure 72, then the **work** W **done by** $\mathbf{F}$ in moving an object from A to B is defined as

$$W = \mathbf{F} \cdot \overrightarrow{AB} \qquad (12)$$

This definition is compatible with the force-times-distance definition, since

$$W = (\text{amount of force in the direction of } \overrightarrow{AB})(\text{distance})$$

$$= \|\text{projection of } \mathbf{F} \text{ on } AB\|\|\overrightarrow{AB}\| = \frac{\mathbf{F} \cdot \overrightarrow{AB}}{\|\overrightarrow{AB}\|^2}\|\overrightarrow{AB}\|\|\overrightarrow{AB}\| = \mathbf{F} \cdot \overrightarrow{AB}$$

Use formula (10).

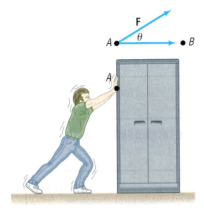

Figure 72

| EXAMPLE 7 | **Computing Work** |

A girl is pulling a wagon with a force of 50 pounds. How much work is done in moving the wagon 100 feet if the handle makes an angle of 30° with the ground? See Figure 73(a).

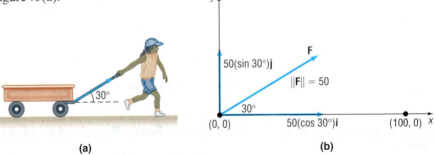

Figure 73   (a)   (b)

Solution   Position the vectors in a coordinate system in such a way that the wagon is moved from $(0,0)$ to $(100,0)$. The motion is from $A = (0,0)$ to $B = (100,0)$, so $\overrightarrow{AB} = 100\mathbf{i}$. The force vector $\mathbf{F}$, as shown in Figure 73(b), is

$$\mathbf{F} = 50(\cos 30°\mathbf{i} + \sin 30°\mathbf{j}) = 50\left(\frac{\sqrt{3}}{2}\mathbf{i} + \frac{1}{2}\mathbf{j}\right) = 25(\sqrt{3}\mathbf{i} + \mathbf{j})$$

By formula (12), the work done is

$$W = \mathbf{F} \cdot \overrightarrow{AB} = 25(\sqrt{3}\mathbf{i} + \mathbf{j}) \cdot 100\mathbf{i} = 2500\sqrt{3} \text{ foot-pounds}$$

 **Now Work** PROBLEM 29

Historical Feature

We stated in the Historical Feature in Section 10.4 that complex numbers were used as vectors in the plane before the general notion of a vector was clarified. Suppose that we make the correspondence

$$\text{Vector} \leftrightarrow \text{Complex number}$$
$$a\mathbf{i} + b\mathbf{j} \leftrightarrow a + bi$$
$$c\mathbf{i} + d\mathbf{j} \leftrightarrow c + di$$

Show that

$$(a\mathbf{i} + b\mathbf{j}) \cdot (c\mathbf{i} + d\mathbf{j}) = \text{real part } [(\overline{a + bi})(c + di)]$$

This is how the dot product was found originally. The imaginary part is also interesting. It is a determinant (see Section 12.3) and represents the area of the parallelogram whose edges are the vectors. This is close to some of Hermann Grassmann's ideas and is also connected with the scalar triple product of three-dimensional vectors.

10.5 Assess Your Understanding

'Are You Prepared?' *The answer is given at the end of these exercises. If you get the wrong answer, read the page listed in* red.

1. In a triangle with sides a, b, c and angles A, B, C, the Law of Cosines states that _____. (p. 692)

Concepts and Vocabulary

2. If $\mathbf{v} = a_1\mathbf{i} + b_1\mathbf{j}$ and $\mathbf{w} = a_2\mathbf{i} + b_2\mathbf{j}$ are two vectors, then the _____ _____ is defined as $\mathbf{v} \cdot \mathbf{w} = a_1a_2 + b_1b_2$.

3. If $\mathbf{v} \cdot \mathbf{w} = 0$, then the two vectors $\mathbf{v}$ and $\mathbf{w}$ are _____.

4. If $\mathbf{v} = 3\mathbf{w}$, then the two vectors $\mathbf{v}$ and $\mathbf{w}$ are _____.

5. *True or False* Given two nonzero vectors $\mathbf{v}$ and $\mathbf{w}$, it is always possible to decompose $\mathbf{v}$ into two vectors, one parallel to $\mathbf{w}$ and the other orthogonal to $\mathbf{w}$.

6. *True or False* Work is a physical example of a vector.

7. The angle θ, $0 \leq \theta \leq \pi$, between two nonzero vectors $\mathbf{u}$ and $\mathbf{v}$ can be found using which of the following formulas?

    (a) $\sin \theta = \dfrac{\|\mathbf{u}\|}{\|\mathbf{v}\|}$ (b) $\cos \theta = \dfrac{\|\mathbf{u}\|}{\|\mathbf{v}\|}$

    (c) $\sin \theta = \dfrac{\mathbf{u} \cdot \mathbf{v}}{\|\mathbf{u}\| \|\mathbf{v}\|}$ (d) $\cos \theta = \dfrac{\mathbf{u} \cdot \mathbf{v}}{\|\mathbf{u}\| \|\mathbf{v}\|}$

8. If two nonzero vectors $\mathbf{v}$ and $\mathbf{w}$ are orthogonal, then the angle between them has which of the following measures?

    (a) π (b) $\dfrac{\pi}{2}$ (c) $\dfrac{3\pi}{2}$ (d) 2π

Skill Building

In Problems 9–18, (a) find the dot product $\mathbf{v} \cdot \mathbf{w}$*; (b) find the angle between* $\mathbf{v}$ *and* $\mathbf{w}$*; (c) state whether the vectors are parallel, orthogonal, or neither.*

9. $\mathbf{v} = \mathbf{i} - \mathbf{j}, \quad \mathbf{w} = \mathbf{i} + \mathbf{j}$

10. $\mathbf{v} = \mathbf{i} + \mathbf{j}, \quad \mathbf{w} = -\mathbf{i} + \mathbf{j}$

11. $\mathbf{v} = 2\mathbf{i} + \mathbf{j}, \quad \mathbf{w} = \mathbf{i} - 2\mathbf{j}$

12. $\mathbf{v} = 2\mathbf{i} + 2\mathbf{j}, \quad \mathbf{w} = \mathbf{i} + 2\mathbf{j}$

13. $\mathbf{v} = \sqrt{3}\mathbf{i} - \mathbf{j}, \quad \mathbf{w} = \mathbf{i} + \mathbf{j}$

14. $\mathbf{v} = \mathbf{i} + \sqrt{3}\mathbf{j}, \quad \mathbf{w} = \mathbf{i} - \mathbf{j}$

15. $\mathbf{v} = 3\mathbf{i} + 4\mathbf{j}, \quad \mathbf{w} = -6\mathbf{i} - 8\mathbf{j}$

16. $\mathbf{v} = 3\mathbf{i} - 4\mathbf{j}, \quad \mathbf{w} = 9\mathbf{i} - 12\mathbf{j}$

17. $\mathbf{v} = 4\mathbf{i}, \quad \mathbf{w} = \mathbf{j}$

18. $\mathbf{v} = \mathbf{i}, \quad \mathbf{w} = -3\mathbf{j}$

19. Find a so that the vectors $\mathbf{v} = \mathbf{i} - a\mathbf{j}$ and $\mathbf{w} = 2\mathbf{i} + 3\mathbf{j}$ are orthogonal.

20. Find b so that the vectors $\mathbf{v} = \mathbf{i} + \mathbf{j}$ and $\mathbf{w} = \mathbf{i} + b\mathbf{j}$ are orthogonal.

In Problems 21–26, decompose $\mathbf{v}$ *into two vectors* $\mathbf{v}_1$ *and* $\mathbf{v}_2$*, where* $\mathbf{v}_1$ *is parallel to* $\mathbf{w}$*, and* $\mathbf{v}_2$ *is orthogonal to* $\mathbf{w}$*.*

21. $\mathbf{v} = 2\mathbf{i} - 3\mathbf{j}, \quad \mathbf{w} = \mathbf{i} - \mathbf{j}$

22. $\mathbf{v} = -3\mathbf{i} + 2\mathbf{j}, \quad \mathbf{w} = 2\mathbf{i} + \mathbf{j}$

23. $\mathbf{v} = \mathbf{i} - \mathbf{j}, \quad \mathbf{w} = -\mathbf{i} - 2\mathbf{j}$

24. $\mathbf{v} = 2\mathbf{i} - \mathbf{j}, \quad \mathbf{w} = \mathbf{i} - 2\mathbf{j}$

25. $\mathbf{v} = 3\mathbf{i} + \mathbf{j}, \quad \mathbf{w} = -2\mathbf{i} - \mathbf{j}$

26. $\mathbf{v} = \mathbf{i} - 3\mathbf{j}, \quad \mathbf{w} = 4\mathbf{i} - \mathbf{j}$

Applications and Extensions

27. Given vectors $\mathbf{u} = \mathbf{i} + 5\mathbf{j}$ and $\mathbf{v} = 4\mathbf{i} + y\mathbf{j}$, find y so that the angle between the vectors is 60°.[†]

28. Given vectors $\mathbf{u} = x\mathbf{i} + 2\mathbf{j}$ and $\mathbf{v} = 7\mathbf{i} - 3\mathbf{j}$, find x so that the angle between the vectors is 30°.

29. **Computing Work** Find the work done by a force of 3 pounds acting in the direction 60° to the horizontal in moving an object 6 feet from $(0, 0)$ to $(6, 0)$.

30. **Computing Work** A wagon is pulled horizontally by exerting a force of 20 pounds on the handle at an angle of 30° with the horizontal. How much work is done in moving the wagon 100 feet?

31. **Solar Energy** The amount of energy collected by a solar panel depends on the intensity of the sun's rays and the area of the panel. Let the vector **I** represent the intensity, in watts

[†]Courtesy of the Joliet Junior College Mathematics Department

per square centimeter, having the direction of the sun's rays. Let the vector **A** represent the area, in square centimeters, whose direction is the orientation of a solar panel. See the figure. The total number of watts collected by the panel is given by $W = |\mathbf{I} \cdot \mathbf{A}|$.

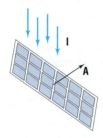

Suppose that $\mathbf{I} = \langle -0.02, -0.01 \rangle$ and $\mathbf{A} = \langle 300, 400 \rangle$.
(a) Find $\|\mathbf{I}\|$ and $\|\mathbf{A}\|$, and interpret the meaning of each.
(b) Compute W and interpret its meaning.
(c) If the solar panel is to collect the maximum number of watts, what must be true about **I** and **A**?

32. Rainfall Measurement Let the vector **R** represent the amount of rainfall, in inches, whose direction is the inclination of the rain to a rain gauge. Let the vector **A** represent the area, in square inches, whose direction is the orientation of the opening of the rain gauge. See the figure. The volume of rain collected in the gauge, in cubic inches, is given by $V = |\mathbf{R} \cdot \mathbf{A}|$, even when the rain falls in a slanted direction or the gauge is not perfectly vertical.

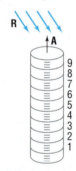

Suppose that $\mathbf{R} = \langle 0.75, -1.75 \rangle$ and $\mathbf{A} = \langle 0.3, 1 \rangle$.
(a) Find $\|\mathbf{R}\|$ and $\|\mathbf{A}\|$, and interpret the meaning of each.
(b) Compute V and interpret its meaning.
(c) If the gauge is to collect the maximum volume of rain, what must be true about **R** and **A**?

33. Braking Load A Toyota Sienna with a gross weight of 5300 pounds is parked on a street with an 8° grade. See the figure. Find the magnitude of the force required to keep the Sienna from rolling down the hill. What is the magnitude of the force perpendicular to the hill?

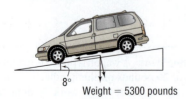

Weight = 5300 pounds

34. Braking Load A Chevrolet Silverado with a gross weight of 4500 pounds is parked on a street with a 10° grade. Find the magnitude of the force required to keep the Silverado from rolling down the hill. What is the magnitude of the force perpendicular to the hill?

35. Ramp Angle Billy and Timmy are using a ramp to load furniture into a truck. While rolling a 250-pound piano up the ramp, they discover that the truck is too full of other furniture for the piano to fit. Timmy holds the piano in place on the ramp while Billy repositions other items to make room for it in the truck. If the angle of inclination of the ramp is 20°, how many pounds of force must Timmy exert to hold the piano in position?

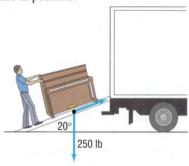

250 lb

36. Incline Angle A bulldozer exerts 1000 pounds of force to prevent a 5000-pound boulder from rolling down a hill. Determine the angle of inclination of the hill.

37. Find the acute angle that a constant unit force vector makes with the positive x-axis if the work done by the force in moving a particle from $(0, 0)$ to $(4, 0)$ equals 2.

38. Prove the distributive property:

$$\mathbf{u} \cdot (\mathbf{v} + \mathbf{w}) = \mathbf{u} \cdot \mathbf{v} + \mathbf{u} \cdot \mathbf{w}$$

39. Prove property (5): $\mathbf{0} \cdot \mathbf{v} = 0$.

40. If **v** is a unit vector and the angle between **v** and **i** is α, show that $\mathbf{v} = \cos \alpha \mathbf{i} + \sin \alpha \mathbf{j}$.

41. Suppose that **v** and **w** are unit vectors. If the angle between **v** and **i** is α and the angle between **w** and **i** is β, use the idea of the dot product $\mathbf{v} \cdot \mathbf{w}$ to prove that

$$\cos (\alpha - \beta) = \cos \alpha \cos \beta + \sin \alpha \sin \beta$$

42. Show that the projection of **v** onto **i** is $(\mathbf{v} \cdot \mathbf{i})\mathbf{i}$. Then show that we can always write a vector **v** as

$$\mathbf{v} = (\mathbf{v} \cdot \mathbf{i})\mathbf{i} + (\mathbf{v} \cdot \mathbf{j})\mathbf{j}$$

43. (a) If **u** and **v** have the same magnitude, show that $\mathbf{u} + \mathbf{v}$ and $\mathbf{u} - \mathbf{v}$ are orthogonal.
(b) Use this to prove that an angle inscribed in a semicircle is a right angle (see the figure).

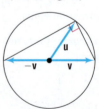

44. Let **v** and **w** denote two nonzero vectors. Show that the vector $\mathbf{v} - \alpha \mathbf{w}$ is orthogonal to **w** if $\alpha = \dfrac{\mathbf{v} \cdot \mathbf{w}}{\|\mathbf{w}\|^2}$.

45. Let **v** and **w** denote two nonzero vectors. Show that the vectors $\|\mathbf{w}\|\mathbf{v} + \|\mathbf{v}\|\mathbf{w}$ and $\|\mathbf{w}\|\mathbf{v} - \|\mathbf{v}\|\mathbf{w}$ are orthogonal.

46. In the definition of work given in this section, what is the work done if **F** is orthogonal to $\overrightarrow{AB}$?

47. Prove the **polarization identity**,

$$\|\mathbf{u} + \mathbf{v}\|^2 - \|\mathbf{u} - \mathbf{v}\|^2 = 4(\mathbf{u} \cdot \mathbf{v})$$

Explaining Concepts: Discussion and Writing

48. Create an application (different from any found in the text) that requires a dot product.

Retain Your Knowledge

Problems 49–52 are based on material learned earlier in the course. The purpose of these problems is to keep the material fresh in your mind so that you are better prepared for the final exam.

49. Find the average rate of change of $f(x) = x^3 - 5x^2 + 27$ from -3 to 2.

50. Find the exact value of $5 \cos 60° + 2 \tan \dfrac{\pi}{4}$. Do not use a calculator.

51. Establish the identity: $(1 - \sin^2 \theta)(1 + \tan^2 \theta) = 1$

52. Volume of a Box An open-top box is made from a sheet of metal by cutting squares from each corner and folding up the sides. The sheet has a length of 19 inches and a width of 13 inches. If x is the length of one side of the square to be cut out, write a function, $V(x)$, for the volume of the box in terms of x.

'Are You Prepared?' Answer

1. $c^2 = a^2 + b^2 - 2ab \cos C$

Chapter Review

Things to Know

Polar Coordinates (pp. 720–727)

Relationship between polar coordinates (r, θ) and rectangular coordinates (x, y) (pp. 722 and 725)	$x = r \cos \theta, \ y = r \sin \theta$ $r^2 = x^2 + y^2, \ \tan \theta = \dfrac{y}{x}, \ x \neq 0$

Complex Numbers and De Moivre's Theorem (pp. 744–749)

Polar form of a complex number (p. 745)	If $z = x + yi$, then $z = r(\cos \theta + i \sin \theta)$, where $r = |z| = \sqrt{x^2 + y^2}$, $\sin \theta = \dfrac{y}{r}$, $\cos \theta = \dfrac{x}{r}$, $0 \le \theta < 2\pi$.
De Moivre's Theorem (p. 747)	If $z = r(\cos \theta + i \sin \theta)$, then $z^n = r^n [\cos(n\theta) + i \sin(n\theta)]$, where $n \ge 1$ is a positive integer.
nth root of a complex number $w = r(\cos \theta_0 + i \sin \theta_0)$ (p. 748)	$z_k = \sqrt[n]{r}\left[\cos\left(\dfrac{\theta_0}{n} + \dfrac{2k\pi}{n}\right) + i \sin\left(\dfrac{\theta_0}{n} + \dfrac{2k\pi}{n}\right)\right]$, $k = 0, \ldots, n-1$, where $n \ge 2$ is an integer

Vectors (pp. 752–761)

	A quantity having magnitude and direction; equivalent to a directed line segment $\overrightarrow{PQ}$
Position vector (p. 755)	A vector whose initial point is at the origin
Unit vector (pp. 754 and 757)	A vector whose magnitude is 1
Direction angle of a vector $\mathbf{v}$ (p. 758)	The angle α, $0° \le \alpha < 360°$, between $\mathbf{i}$ and $\mathbf{v}$
Dot product (p. 766)	If $\mathbf{v} = a_1\mathbf{i} + b_1\mathbf{j}$ and $\mathbf{w} = a_2\mathbf{i} + b_2\mathbf{j}$, then $\mathbf{v} \cdot \mathbf{w} = a_1 a_2 + b_1 b_2$.
Angle θ between two nonzero vectors $\mathbf{u}$ and $\mathbf{v}$ (p. 767)	$\cos \theta = \dfrac{\mathbf{u} \cdot \mathbf{v}}{\|\mathbf{u}\| \|\mathbf{v}\|}, 0 \le \theta \le \pi$

Objectives

Section		You should be able to . . .	Example(s)	Review Exercises
10.1	**1**	Plot points using polar coordinates (p. 720)	1–3	1–3
	2	Convert from polar coordinates to rectangular coordinates (p. 722)	4	1–3
	3	Convert from rectangular coordinates to polar coordinates (p. 724)	5–7	4–6
	4	Transform equations between polar and rectangular forms (p. 726)	8, 9	7(a)–10(a)
10.2	**1**	Identify and graph polar equations by converting to rectangular equations (p. 730)	1–6	7(b)–10(b)
	2	Test polar equations for symmetry (p. 733)	7–10	11–13
	3	Graph polar equations by plotting points (p. 734)	7–12	11–13

Section	You should be able to . . .	Example(s)	Review Exercises
10.3	**1** Plot points in the complex plane (p. 744)	1	16–18
	2 Convert a complex number between rectangular form and polar form (p. 745)	2, 3	14–18
	3 Find products and quotients of complex numbers in polar form (p. 746)	4	19–21
	4 Use De Moivre's Theorem (p. 747)	5, 6	22–25
	5 Find complex roots (p. 748)	7	26
10.4	**1** Graph vectors (p. 754)	1	27, 28
	2 Find a position vector (p. 754)	2	29, 30
	3 Add and subtract vectors algebraically (p. 756)	3	31
	4 Find a scalar multiple and the magnitude of a vector (p. 757)	4	29, 30, 32–34
	5 Find a unit vector (p. 757)	5	35
	6 Find a vector from its direction and magnitude (p. 758)	6	36, 37
	7 Model with vectors (p. 759)	8–10	45, 46
10.5	**1** Find the dot product of two vectors (p. 766)	1	38, 39
	2 Find the angle between two vectors (p. 767)	2	38, 39
	3 Determine whether two vectors are parallel (p. 768)	3	40–42
	4 Determine whether two vectors are orthogonal (p. 768)	4	40–42
	5 Decompose a vector into two orthogonal vectors (p. 768)	5, 6	43, 44, 48
	6 Compute work (p. 770)	7	47

Review Exercises

In Problems 1–3, plot each point given in polar coordinates, and find its rectangular coordinates.

1. $\left(3, \dfrac{\pi}{6}\right)$

2. $\left(-2, \dfrac{4\pi}{3}\right)$

3. $\left(-3, -\dfrac{\pi}{2}\right)$

In Problems 4–6, the rectangular coordinates of a point are given. Find two pairs of polar coordinates (r, θ) for each point, one with $r > 0$ and the other with $r < 0$. Express θ in radians.

4. $(-3, 3)$

5. $(0, -2)$

6. $(3, 4)$

In Problems 7–10, the variables r and θ represent polar coordinates. (a) Write each polar equation as an equation in rectangular coordinates (x, y). (b) Identify the equation and graph it.

7. $r = 2 \sin \theta$

8. $r = 5$

9. $\theta = \dfrac{\pi}{4}$

10. $r^2 + 4r \sin \theta - 8r \cos \theta = 5$

In Problems 11–13, graph each polar equation. Be sure to test for symmetry.

11. $r = 4 \cos \theta$

12. $r = 3 - 3 \sin \theta$

13. $r = 4 - \cos \theta$

In Problems 14 and 15, write each complex number in polar form. Express each argument in degrees.

14. $-1 - i$

15. $4 - 3i$

In Problems 16–18, write each complex number in the standard form $a + bi$, and plot each in the complex plane.

16. $2(\cos 150° + i \sin 150°)$

17. $3\left(\cos \dfrac{2\pi}{3} + i \sin \dfrac{2\pi}{3}\right)$

18. $0.1(\cos 350° + i \sin 350°)$

In Problems 19–21, find zw and $\dfrac{z}{w}$. Leave your answers in polar form.

19. $z = \cos 80° + i \sin 80°$
$w = \cos 50° + i \sin 50°$

20. $z = 3\left(\cos \dfrac{9\pi}{5} + i \sin \dfrac{9\pi}{5}\right)$
$w = 2\left(\cos \dfrac{\pi}{5} + i \sin \dfrac{\pi}{5}\right)$

21. $z = 5(\cos 10° + i \sin 10°)$
$w = \cos 355° + i \sin 355°$

In Problems 22–25, write each expression in the standard form $a + bi$.

22. $[3(\cos 20° + i \sin 20°)]^3$

23. $\left[\sqrt{2}\left(\cos \dfrac{5\pi}{8} + i \sin \dfrac{5\pi}{8}\right)\right]^4$

24. $\left(1 - \sqrt{3}i\right)^6$

25. $(3 + 4i)^4$

26. Find all the complex cube roots of 27.

In Problems 27 and 28, use the figure to graph each of the following:

27. $\mathbf{u} + \mathbf{v}$

28. $2\mathbf{u} + 3\mathbf{v}$

In Problems 29 and 30, the vector $\mathbf{v}$ is represented by the directed line segment $\overrightarrow{PQ}$. Write $\mathbf{v}$ in the form $a\mathbf{i} + b\mathbf{j}$ and find $\|\mathbf{v}\|$.

29. $P = (1, -2)$; $Q = (3, -6)$

30. $P = (0, -2)$; $Q = (-1, 1)$

In Problems 31–35, use the vectors $\mathbf{v} = -2\mathbf{i} + \mathbf{j}$ and $\mathbf{w} = 4\mathbf{i} - 3\mathbf{j}$ to find:

31. $\mathbf{v} + \mathbf{w}$ **32.** $4\mathbf{v} - 3\mathbf{w}$ **33.** $\|\mathbf{v}\|$ **34.** $\|\mathbf{v}\| + \|\mathbf{w}\|$

35. A unit vector in the same direction as $\mathbf{v}$.

36. Find the vector $\mathbf{v}$ in the xy-plane with magnitude 3 if the direction angle of $\mathbf{v}$ is $60°$.

37. Find the direction angle α of $\mathbf{v} = -\mathbf{i} + \sqrt{3}\,\mathbf{j}$.

In Problems 38 and 39, find the dot product $\mathbf{v} \cdot \mathbf{w}$ and the angle between $\mathbf{v}$ and $\mathbf{w}$.

38. $\mathbf{v} = -2\mathbf{i} + \mathbf{j}$, $\mathbf{w} = 4\mathbf{i} - 3\mathbf{j}$

39. $\mathbf{v} = \mathbf{i} - 3\mathbf{j}$, $\mathbf{w} = -\mathbf{i} + \mathbf{j}$

In Problems 40–42, determine whether $\mathbf{v}$ and $\mathbf{w}$ are parallel, orthogonal, or neither.

40. $\mathbf{v} = 2\mathbf{i} + 3\mathbf{j}$; $\mathbf{w} = -4\mathbf{i} - 6\mathbf{j}$

41. $\mathbf{v} = -2\mathbf{i} + 2\mathbf{j}$; $\mathbf{w} = -3\mathbf{i} + 2\mathbf{j}$

42. $\mathbf{v} = 3\mathbf{i} - 2\mathbf{j}$; $\mathbf{w} = 4\mathbf{i} + 6\mathbf{j}$

In Problems 43 and 44, decompose $\mathbf{v}$ into two vectors, one parallel to $\mathbf{w}$ and the other orthogonal to $\mathbf{w}$.

43. $\mathbf{v} = 2\mathbf{i} + \mathbf{j}$; $\mathbf{w} = -4\mathbf{i} + 3\mathbf{j}$

44. $\mathbf{v} = 2\mathbf{i} + 3\mathbf{j}$; $\mathbf{w} = 3\mathbf{i} + \mathbf{j}$

---

45. Actual Speed and Direction of a Swimmer A swimmer can maintain a constant speed of 5 miles per hour. If the swimmer heads directly across a river that has a current moving at the rate of 2 miles per hour, what is the actual speed of the swimmer? (See the figure.) If the river is 1 mile wide, how far downstream will the swimmer end up from the point directly across the river from the starting point?

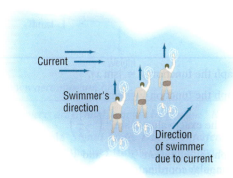

Current

Swimmer's direction

Direction of swimmer due to current

46. Static Equilibrium A weight of 2000 pounds is suspended from two cables, as shown in the figure. What are the tensions in the two cables?

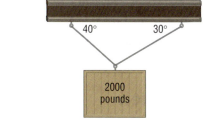

$40°$ $30°$

2000 pounds

47. Computing Work Find the work done by a force of 5 pounds acting in the direction $60°$ to the horizontal in moving an object 20 feet from $(0, 0)$ to $(20, 0)$.

48. Braking Load A moving van with a gross weight of 8000 pounds is parked on a street with a $5°$ grade. Find the magnitude of the force required to keep the van from rolling down the hill. What is the magnitude of the force perpendicular to the hill?

Recall that we obtained equation (2) after placing the focus on the positive x-axis. If the focus is placed on the negative x-axis, positive y-axis, or negative y-axis, a different form of the equation for the parabola results. The four forms of the equation of a parabola with vertex at $(0, 0)$ and focus on a coordinate axis a distance a from $(0, 0)$ are given in Table 1, and their graphs are given in Figure 7. Notice that each graph is symmetric with respect to its axis of symmetry.

Table 1

Equations of a Parabola: Vertex at (0, 0); Focus on an Axis; $a > 0$				
Vertex	**Focus**	**Directrix**	**Equation**	**Description**
$(0, 0)$	$(a, 0)$	$x = -a$	$y^2 = 4ax$	Axis of symmetry is the x-axis, opens right
$(0, 0)$	$(-a, 0)$	$x = a$	$y^2 = -4ax$	Axis of symmetry is the x-axis, opens left
$(0, 0)$	$(0, a)$	$y = -a$	$x^2 = 4ay$	Axis of symmetry is the y-axis, opens up
$(0, 0)$	$(0, -a)$	$y = a$	$x^2 = -4ay$	Axis of symmetry is the y-axis, opens down

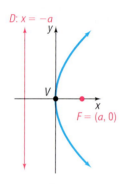 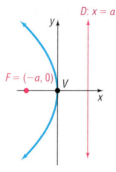 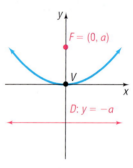 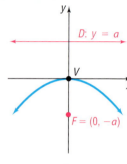

Figure 7   **(a)** $y^2 = 4ax$ **(b)** $y^2 = -4ax$ **(c)** $x^2 = 4ay$ **(d)** $x^2 = -4ay$

EXAMPLE 3

Analyzing the Equation of a Parabola

Analyze the equation: $x^2 = -12y$

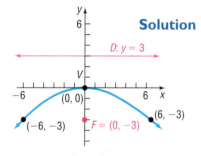

Solution   The equation $x^2 = -12y$ is of the form $x^2 = -4ay$, with $a = 3$. Consequently, the graph of the equation is a parabola with vertex at $(0, 0)$, focus at $(0, -3)$, and directrix the line $y = 3$. The parabola opens down, and its axis of symmetry is the y-axis. To obtain the points defining the latus rectum, let $y = -3$. Then $x^2 = 36$, so $x = \pm 6$. The points $(-6, -3)$ and $(6, -3)$ determine the latus rectum. See Figure 8 for the graph. ●

Figure 8 $x^2 = -12y$

🖉 **Now Work** PROBLEM **41**

EXAMPLE 4

Finding the Equation of a Parabola

Find the equation of the parabola with focus at $(0, 4)$ and directrix the line $y = -4$. Graph the equation.

Solution   A parabola whose focus is at $(0, 4)$ and whose directrix is the horizontal line $y = -4$ will have its vertex at $(0, 0)$. (Do you see why? The vertex is midway between the focus and the directrix.) Since the focus is on the positive y-axis at $(0, 4)$, the equation of this parabola is of the form $x^2 = 4ay$, with $a = 4$. That is,

$$x^2 = 4ay = 4(4)y = 16y$$
$$\underset{a = 4}{\uparrow}$$

Letting $y = 4$ yields $x^2 = 64$, so $x = \pm 8$. The points $(8, 4)$ and $(-8, 4)$ determine the latus rectum. Figure 9 shows the graph of $x^2 = 16y$. ●

Figure 9 $x^2 = 16y$

EXAMPLE 5   **Finding the Equation of a Parabola**

Find the equation of a parabola with vertex at $(0,0)$ if its axis of symmetry is the x-axis and its graph contains the point $\left(-\dfrac{1}{2}, 2\right)$. Find its focus and directrix, and graph the equation.

Solution   The vertex is at the origin, the axis of symmetry is the x-axis, and the graph contains a point in the second quadrant, so the parabola opens to the left. From Table 1, note that the form of the equation is

$$y^2 = -4ax$$

Because the point $\left(-\dfrac{1}{2}, 2\right)$ is on the parabola, the coordinates $x = -\dfrac{1}{2}, y = 2$ must satisfy $y^2 = -4ax$. Substituting $x = -\dfrac{1}{2}$ and $y = 2$ into this equation leads to

$$4 = -4a\left(-\frac{1}{2}\right) \qquad y^2 = -4ax; x = -\frac{1}{2}, y = 2$$

$$a = 2$$

The equation of the parabola is

$$y^2 = -4(2)x = -8x$$

The focus is at $(-2, 0)$ and the directrix is the line $x = 2$. Letting $x = -2$ gives $y^2 = 16$, so $y = \pm 4$. The points $(-2, 4)$ and $(-2, -4)$ determine the latus rectum. See Figure 10.

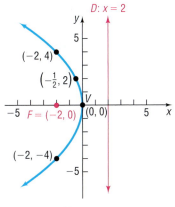

Figure 10 $y^2 = -8x$

Now Work PROBLEM 29

2 Analyze Parabolas with Vertex at (*h*, *k*)

If a parabola with vertex at the origin and axis of symmetry along a coordinate axis is shifted horizontally h units and then vertically k units, the result is a parabola with vertex at (h, k) and axis of symmetry parallel to a coordinate axis. The equations of such parabolas have the same forms as those in Table 1, but with x replaced by $x - h$ (the horizontal shift) and y replaced by $y - k$ (the vertical shift). Table 2 gives the forms of the equations of such parabolas. Figures 11(a)–(d) on page 784 illustrate the graphs for $h > 0, k > 0$.

NOTE It is not recommended that Table 2 be memorized. Rather, use transformations (shift horizontally h units, vertically k units), along with the fact that a represents the distance from the vertex to the focus, to determine the various components of a parabola. It is also helpful to remember that parabolas of the form "$x^2 = $" open up or down, while parabolas of the form "$y^2 = $" open left or right. ∎

Table 2

Equations of a Parabola: Vertex at (*h*, *k*); Axis of Symmetry Parallel to a Coordinate Axis; *a* > 0				
Vertex	Focus	Directrix	Equation	Description
(h, k)	$(h + a, k)$	$x = h - a$	$(y - k)^2 = 4a(x - h)$	Axis of symmetry is parallel to the x-axis, opens right
(h, k)	$(h - a, k)$	$x = h + a$	$(y - k)^2 = -4a(x - h)$	Axis of symmetry is parallel to the x-axis, opens left
(h, k)	$(h, k + a)$	$y = k - a$	$(x - h)^2 = 4a(y - k)$	Axis of symmetry is parallel to the y-axis, opens up
(h, k)	$(h, k - a)$	$y = k + a$	$(x - h)^2 = -4a(y - k)$	Axis of symmetry is parallel to the y-axis, opens down

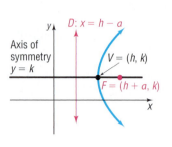

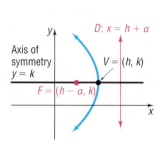

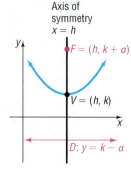

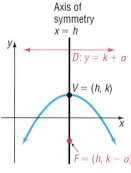

(a) $(y - k)^2 = 4a(x - h)$ **(b)** $(y - k)^2 = -4a(x - h)$ **(c)** $(x - h)^2 = 4a(y - k)$ **(d)** $(x - h)^2 = -4a(y - k)$

Figure 11

EXAMPLE 6   **Finding the Equation of a Parabola, Vertex Not at the Origin**

Find an equation of the parabola with vertex at $(-2, 3)$ and focus at $(0, 3)$. Graph the equation.

Solution   The vertex $(-2, 3)$ and focus $(0, 3)$ both lie on the horizontal line $y = 3$ (the axis of symmetry). The distance a from the vertex $(-2, 3)$ to the focus $(0, 3)$ is $a = 2$. Also, because the focus lies to the right of the vertex, the parabola opens to the right. Consequently, the form of the equation is

$$(y - k)^2 = 4a(x - h)$$

where $(h, k) = (-2, 3)$ and $a = 2$. Therefore, the equation is

$$(y - 3)^2 = 4 \cdot 2[x - (-2)]$$
$$(y - 3)^2 = 8(x + 2)$$

To find the points that define the latus rectum, let $x = 0$, so that $(y - 3)^2 = 16$. Then $y - 3 = \pm 4$, so $y = -1$ or $y = 7$. The points $(0, -1)$ and $(0, 7)$ determine the latus rectum; the line $x = -4$ is the directrix. See Figure 12.   ●

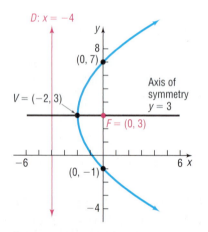

Figure 12 $(y - 3)^2 = 8(x + 2)$

 **Now Work** PROBLEM 31

Polynomial equations define parabolas whenever they involve two variables that are quadratic in one variable and linear in the other.

EXAMPLE 7   **Analyzing the Equation of a Parabola**

Analyze the equation: $x^2 + 4x - 4y = 0$

Solution   To analyze the equation $x^2 + 4x - 4y = 0$, complete the square involving the variable x.

$$x^2 + 4x - 4y = 0$$
$$x^2 + 4x = 4y \qquad \text{Isolate the terms involving } x \text{ on the left side.}$$
$$x^2 + 4x + 4 = 4y + 4 \qquad \text{Complete the square on the left side.}$$
$$(x + 2)^2 = 4(y + 1) \qquad \text{Factor.}$$

This equation is of the form $(x - h)^2 = 4a(y - k)$, with $h = -2, k = -1$, and $a = 1$. The graph is a parabola with vertex at $(h, k) = (-2, -1)$ that opens up. The focus is at $(-2, 0)$, and the directrix is the line $y = -2$. See Figure 13.   ●

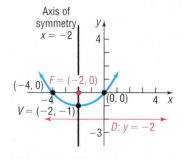

Figure 13 $x^2 + 4x - 4y = 0$

 **Now Work** PROBLEM 49

3 Solve Applied Problems Involving Parabolas

Parabolas find their way into many applications. For example, as discussed in Section 4.4, suspension bridges have cables in the shape of a parabola. Another property of parabolas that is used in applications is their reflecting property.

Suppose that a mirror is shaped like a **paraboloid of revolution**, a surface formed by rotating a parabola about its axis of symmetry. If a light (or any other emitting source) is placed at the focus of the parabola, all the rays emanating from the light will reflect off the mirror in lines parallel to the axis of symmetry. This principle is used in the design of searchlights, flashlights, certain automobile headlights, and other such devices. See Figure 14.

Conversely, suppose that rays of light (or other signals) emanate from a distant source so that they are essentially parallel. When these rays strike the surface of a parabolic mirror whose axis of symmetry is parallel to these rays, they are reflected to a single point at the focus. This principle is used in the design of some solar energy devices, satellite dishes, and the mirrors used in some types of telescopes. See Figure 15.

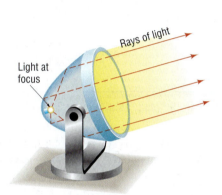

Figure 14 Searchlight

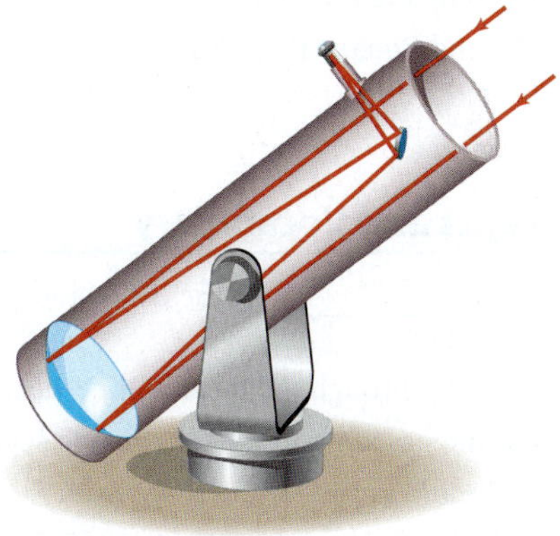

Figure 15 Telescope

EXAMPLE 8

Satellite Dish

A satellite dish is shaped like a paraboloid of revolution. The signals that emanate from a satellite strike the surface of the dish and are reflected to a single point, where the receiver is located. If the dish is 8 feet across at its opening and 3 feet deep at its center, at what position should the receiver be placed? That is, where is the focus?

Solution

Figure 16(a) shows the satellite dish. On a rectangular coordinate system, draw the parabola used to form the dish so that the vertex of the parabola is at the origin and its focus is on the positive y-axis. See Figure 16(b).

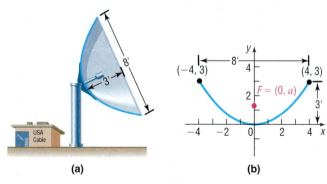

(a)

(b)

Figure 16

The form of the equation of the parabola is

$$x^2 = 4ay$$

and its focus is at $(0, a)$. Since $(4, 3)$ is a point on the graph, this gives

$$4^2 = 4a(3) \quad \text{\color{blue}{$x^2 = 4ay; x = 4, y = 3$}}$$

$$a = \frac{4}{3} \quad \text{\color{blue}{Solve for a.}}$$

The receiver should be located $1\frac{1}{3}$ feet (1 foot, 4 inches) from the base of the dish, along its axis of symmetry.

 **Now Work** PROBLEM 65

11.2 Assess Your Understanding

'Are You Prepared?' *Answers are given at the end of these exercises. If you get a wrong answer, read the pages listed in* red.

1. The formula for the distance d from $P_1 = (x_1, y_1)$ to $P_2 = (x_2, y_2)$ is $d =$ _____. (p. 151)

2. To complete the square of $x^2 - 4x$, add ____ . (p. 56)

3. Use the Square Root Method to find the real solutions of $(x + 4)^2 = 9$. (pp. 94–95)

4. The point that is symmetric with respect to the x-axis to the point $(-2, 5)$ is _____. (pp. 160–162)

5. To graph $y = (x - 3)^2 + 1$, shift the graph of $y = x^2$ to the right _____ units and then _____ 1 unit. (pp. 247–256)

Concepts and Vocabulary

6. A(n) _____ is the collection of all points in the plane that are the same distance from a fixed point as they are from a fixed line.

7. The line through the focus and perpendicular to the directrix is called the _____ of the parabola.

8. For the parabola $y^2 = 4ax$, the line segment joining the two points $(a, 2a)$ and $(a, -2a)$ is called the _____ .

Answer Problems 9–12 using the figure.

9. If $a > 0$, the equation of the parabola is of the form
(a) $(y - k)^2 = 4a(x - h)$ (b) $(y - k)^2 = -4a(x - h)$
(c) $(x - h)^2 = 4a(y - k)$ (d) $(x - h)^2 = -4a(y - k)$

10. The coordinates of the vertex are _____ .

11. If $a = 4$, then the coordinates of the focus are _____ .
(a) $(-1, 2)$ (b) $(3, -2)$ (c) $(7, 2)$ (d) $(3, 6)$

12. If $a = 4$, then the equation of the directrix is _____ .
(a) $x = -3$ (b) $x = 3$ (c) $y = -2$ (d) $y = 2$

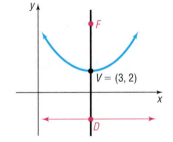

Skill Building

In Problems 13–20, the graph of a parabola is given. Match each graph to its equation.

(A) $y^2 = 4x$ (C) $y^2 = -4x$ (E) $(y - 1)^2 = 4(x - 1)$ (G) $(y - 1)^2 = -4(x - 1)$
(B) $x^2 = 4y$ (D) $x^2 = -4y$ (F) $(x + 1)^2 = 4(y + 1)$ (H) $(x + 1)^2 = -4(y + 1)$

13.
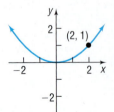

14.
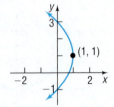

15.
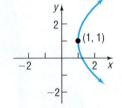

16.
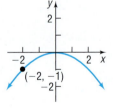

17.
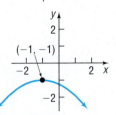

18.
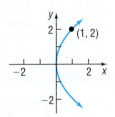

19.
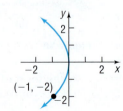

20.
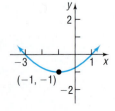

In Problems 21–38, find the equation of the parabola described. Find the two points that define the latus rectum, and graph the equation.

21. Focus at $(4, 0)$; vertex at $(0, 0)$

22. Focus at $(0, 2)$; vertex at $(0, 0)$

23. Focus at $(0, -3)$; vertex at $(0, 0)$

24. Focus at $(-4, 0)$; vertex at $(0, 0)$

25. Focus at $(-2, 0)$; directrix the line $x = 2$

26. Focus at $(0, -1)$; directrix the line $y = 1$

27. Directrix the line $y = -\dfrac{1}{2}$; vertex at $(0, 0)$

28. Directrix the line $x = -\dfrac{1}{2}$; vertex at $(0, 0)$

29. Vertex at $(0, 0)$; axis of symmetry the y-axis; containing the point $(2, 3)$

30. Vertex at $(0, 0)$; axis of symmetry the x-axis; containing the point $(2, 3)$

31. Vertex at $(2, -3)$; focus at $(2, -5)$

32. Vertex at $(4, -2)$; focus at $(6, -2)$

33. Vertex at $(-1, -2)$; focus at $(0, -2)$

34. Vertex at $(3, 0)$; focus at $(3, -2)$

35. Focus at $(-3, 4)$; directrix the line $y = 2$

36. Focus at $(2, 4)$; directrix the line $x = -4$

37. Focus at $(-3, -2)$; directrix the line $x = 1$

38. Focus at $(-4, 4)$; directrix the line $y = -2$

In Problems 39–56, find the vertex, focus, and directrix of each parabola. Graph the equation.

39. $x^2 = 4y$

40. $y^2 = 8x$

41. $y^2 = -16x$

42. $x^2 = -4y$

43. $(y - 2)^2 = 8(x + 1)$

44. $(x + 4)^2 = 16(y + 2)$

45. $(x - 3)^2 = -(y + 1)$

46. $(y + 1)^2 = -4(x - 2)$

47. $(y + 3)^2 = 8(x - 2)$

48. $(x - 2)^2 = 4(y - 3)$

49. $y^2 - 4y + 4x + 4 = 0$

50. $x^2 + 6x - 4y + 1 = 0$

51. $x^2 + 8x = 4y - 8$

52. $y^2 - 2y = 8x - 1$

53. $y^2 + 2y - x = 0$

54. $x^2 - 4x = 2y$

55. $x^2 - 4x = y + 4$

56. $y^2 + 12y = -x + 1$

In Problems 57–64, write an equation for each parabola.

57.
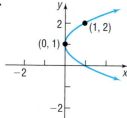

58.
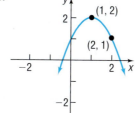

59.
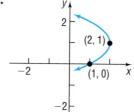

60.
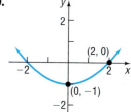

61.
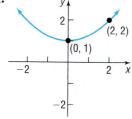

62.
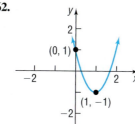

63.
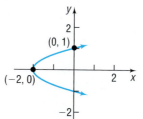

64.
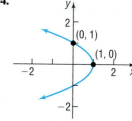

Applications and Extensions

65. Satellite Dish A satellite dish is shaped like a paraboloid of revolution. The signals that emanate from a satellite strike the surface of the dish and are reflected to a single point, where the receiver is located. If the dish is 10 feet across at its opening and 4 feet deep at its center, at what position should the receiver be placed?

66. Constructing a TV Dish A cable TV receiving dish is in the shape of a paraboloid of revolution. Find the location of the receiver, which is placed at the focus, if the dish is 6 feet across at its opening and 2 feet deep.

67. Constructing a Flashlight The reflector of a flashlight is in the shape of a paraboloid of revolution. Its diameter is 4 inches and its depth is 1 inch. How far from the vertex should the light bulb be placed so that the rays will be reflected parallel to the axis?

68. Constructing a Headlight A sealed-beam headlight is in the shape of a paraboloid of revolution. The bulb, which is placed at the focus, is 1 inch from the vertex. If the depth is to be 2 inches, what is the diameter of the headlight at its opening?

69. Suspension Bridge The cables of a suspension bridge are in the shape of a parabola, as shown in the figure. The towers supporting the cable are 600 feet apart and 80 feet high. If the cables touch the road surface midway between the towers, what is the height of the cable from the road at a point 150 feet from the center of the bridge?

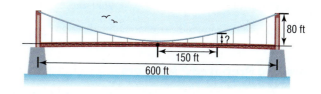

70. Suspension Bridge The cables of a suspension bridge are in the shape of a parabola. The towers supporting the cable are 400 feet apart and 100 feet high. If the cables are at a height of 10 feet midway between the towers, what is the height of the cable at a point 50 feet from the center of the bridge?

71. Searchlight A searchlight is shaped like a paraboloid of revolution. If the light source is located 2 feet from the base along the axis of symmetry and the opening is 5 feet across, how deep should the searchlight be?

72. Searchlight A searchlight is shaped like a paraboloid of revolution. If the light source is located 2 feet from the base along the axis of symmetry and the depth of the searchlight is 4 feet, what should the width of the opening be?

73. Solar Heat A mirror is shaped like a paraboloid of revolution and will be used to concentrate the rays of the sun at its focus, creating a heat source. See the figure. If the mirror is 20 feet across at its opening and is 6 feet deep, where will the heat source be concentrated?

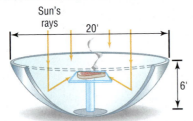

74. Reflecting Telescope A reflecting telescope contains a mirror shaped like a paraboloid of revolution. If the mirror is 4 inches across at its opening and is 3 inches deep, where will the collected light be concentrated?

75. Parabolic Arch Bridge A bridge is built in the shape of a parabolic arch. The bridge has a span of 120 feet and a maximum height of 25 feet. See the illustration. Choose a suitable rectangular coordinate system and find the height of the arch at distances of 10, 30, and 50 feet from the center.

76. Parabolic Arch Bridge A bridge is to be built in the shape of a parabolic arch and is to have a span of 100 feet. The height of the arch a distance of 40 feet from the center is to be 10 feet. Find the height of the arch at its center.

77. Gateway Arch The Gateway Arch in St. Louis is often mistaken to be parabolic in shape. In fact, it is a *catenary*, which has a more complicated formula than a parabola. The Arch is 630 feet high and 630 feet wide at its base.
   (a) Find the equation of a parabola with the same dimensions. Let x equal the horizontal distance from the center of the arch.
   (b) The table below gives the height of the Arch at various widths; find the corresponding heights for the parabola found in (a).

Width (ft)	Height (ft)
567	100
478	312.5
308	525

   (c) Do the data support the notion that the Arch is in the shape of a parabola?

Source: gatewayarch.com

78. Show that an equation of the form

$$Ax^2 + Ey = 0 \qquad A \neq 0, E \neq 0$$

is the equation of a parabola with vertex at $(0, 0)$ and axis of symmetry the y-axis. Find its focus and directrix.

79. Show that an equation of the form

$$Cy^2 + Dx = 0 \qquad C \neq 0, D \neq 0$$

is the equation of a parabola with vertex at $(0, 0)$ and axis of symmetry the x-axis. Find its focus and directrix.

80. Show that the graph of an equation of the form

$$Ax^2 + Dx + Ey + F = 0 \qquad A \neq 0$$

   (a) Is a parabola if $E \neq 0$.
   (b) Is a vertical line if $E = 0$ and $D^2 - 4AF = 0$.
   (c) Is two vertical lines if $E = 0$ and $D^2 - 4AF > 0$.
   (d) Contains no points if $E = 0$ and $D^2 - 4AF < 0$.

81. Show that the graph of an equation of the form

$$Cy^2 + Dx + Ey + F = 0 \qquad C \neq 0$$

   (a) Is a parabola if $D \neq 0$.
   (b) Is a horizontal line if $D = 0$ and $E^2 - 4CF = 0$.
   (c) Is two horizontal lines if $D = 0$ and $E^2 - 4CF > 0$.
   (d) Contains no points if $D = 0$ and $E^2 - 4CF < 0$.

Retain Your Knowledge

Problems 82–85 are based on material learned earlier in the course. The purpose of these problems is to keep the material fresh in your mind so that you are better prepared for the final exam.

82. For $x = 9y^2 - 36$, list the intercepts and test for symmetry.

83. Solve: $4^{x+1} = 8^{x-1}$

84. Given $\tan \theta = -\dfrac{5}{8}, \dfrac{\pi}{2} < \theta < \pi$, find the exact value of each of the remaining trigonometric functions.

85. Find the exact value: $\tan\left[\cos^{-1}\left(-\dfrac{3}{7}\right)\right]$

'Are You Prepared?' Answers

1. $\sqrt{(x_2 - x_1)^2 + (y_2 - y_1)^2}$ **2.** 4   **3.** $\{-7, -1\}$ **4.** $(-2, -5)$ **5.** 3; up

11.3 The Ellipse

PREPARING FOR THIS SECTION *Before getting started, review the following:*

- Distance Formula (Section 2.1, p. 151)
- Completing the Square (Chapter R, Review, Section R.5, p. 56)
- Intercepts (Section 2.2, pp. 159–160)

- Symmetry (Section 2.2, pp. 160–162)
- Circles (Section 2.4, pp. 182–185)
- Graphing Techniques: Transformations (Section 3.5, pp. 247–256)

Now Work the **'Are You Prepared?'** problems on page 796.

OBJECTIVES **1** Analyze Ellipses with Center at the Origin (p. 789)
             2 Analyze Ellipses with Center at (h, k) (p. 793)
             3 Solve Applied Problems Involving Ellipses (p. 795)

DEFINITION

An **ellipse** is the collection of all points in the plane the sum of whose distances from two fixed points, called the **foci**, is a constant.

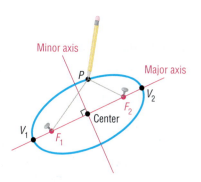

Figure 17 Ellipse

The definition contains within it a physical means for drawing an ellipse. Find a piece of string (the length of this string is the constant referred to in the definition). Then take two thumbtacks (the foci) and stick them into a piece of cardboard so that the distance between them is less than the length of the string. Now attach the ends of the string to the thumbtacks and, using the point of a pencil, pull the string taut. See Figure 17. Keeping the string taut, rotate the pencil around the two thumbtacks. The pencil traces out an ellipse, as shown in Figure 17.

In Figure 17, the foci are labeled F_1 and F_2. The line containing the foci is called the **major axis**. The midpoint of the line segment joining the foci is the **center** of the ellipse. The line through the center and perpendicular to the major axis is the **minor axis**.

The two points of intersection of the ellipse and the major axis are the **vertices**, V_1 and V_2, of the ellipse. The distance from one vertex to the other is the **length of the major axis**. The ellipse is symmetric with respect to its major axis, with respect to its minor axis, and with respect to its center.

1 Analyze Ellipses with Center at the Origin

With these ideas in mind, we are ready to find the equation of an ellipse in a rectangular coordinate system. First, place the center of the ellipse at the origin. Second, position the ellipse so that its major axis coincides with a coordinate axis, say the x-axis, as shown in Figure 18. If c is the distance from the center to a focus, one focus will be at $F_1 = (-c, 0)$ and the other at $F_2 = (c, 0)$. As we shall see, it is

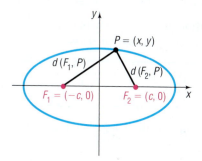

Figure 18

convenient to let $2a$ denote the constant distance referred to in the definition. Then, if $P = (x, y)$ is any point on the ellipse,

$$d(F_1, P) + d(F_2, P) = 2a$$ The sum of the distances from P to the foci equals a constant, $2a$.

$$\sqrt{(x + c)^2 + y^2} + \sqrt{(x - c)^2 + y^2} = 2a$$ Use the Distance Formula.

$$\sqrt{(x + c)^2 + y^2} = 2a - \sqrt{(x - c)^2 + y^2}$$ Isolate one radical.

$$(x + c)^2 + y^2 = 4a^2 - 4a\sqrt{(x - c)^2 + y^2} + (x - c)^2 + y^2$$ Square both sides.

$$x^2 + 2cx + c^2 + y^2 = 4a^2 - 4a\sqrt{(x - c)^2 + y^2} + x^2 - 2cx + c^2 + y^2$$ Multiply out.

$$4cx - 4a^2 = -4a\sqrt{(x - c)^2 + y^2}$$ Simplify; isolate the radical.

$$cx - a^2 = -a\sqrt{(x - c)^2 + y^2}$$ Divide each side by 4.

$$(cx - a^2)^2 = a^2[(x - c)^2 + y^2]$$ Square both sides again.

$$c^2x^2 - 2a^2cx + a^4 = a^2(x^2 - 2cx + c^2 + y^2)$$ Multiply out.

$$(c^2 - a^2)x^2 - a^2y^2 = a^2c^2 - a^4$$ Rearrange the terms.

$$(a^2 - c^2)x^2 + a^2y^2 = a^2(a^2 - c^2)$$ Multiply each side by -1; **(1)** factor a^2 on the right side.

To obtain points on the ellipse off the x-axis, it must be that $a > c$. To see why, look again at Figure 18. Then

$$d(F_1, P) + d(F_2, P) > d(F_1, F_2)$$ The sum of the lengths of two sides of a triangle is greater than the length of the third side.

$$2a > 2c$$ $d(F_1, P) + d(F_2, P) = 2a$, $d(F_1, F_2) = 2c$

$$a > c$$

Because $a > c > 0$, this means $a^2 > c^2$, so $a^2 - c^2 > 0$. Let $b^2 = a^2 - c^2$, $b > 0$. Then $a > b$ and equation (1) can be written as

$$b^2x^2 + a^2y^2 = a^2b^2$$

$$\frac{x^2}{a^2} + \frac{y^2}{b^2} = 1$$ Divide each side by a^2b^2.

As you can verify, the graph of this equation has symmetry with respect to the x-axis, the y-axis, and the origin.

Because the major axis is the x-axis, find the vertices of this ellipse by letting $y = 0$. The vertices satisfy the equation $\frac{x^2}{a^2} = 1$, the solutions of which are $x = \pm a$. Consequently, the vertices of this ellipse are $V_1 = (-a, 0)$ and $V_2 = (a, 0)$. The y-intercepts of the ellipse, found by letting $x = 0$, have coordinates $(0, -b)$ and $(0, b)$. The four intercepts, $(a, 0)$, $(-a, 0)$, $(0, b)$, and $(0, -b)$, are used to graph the ellipse.

THEOREM

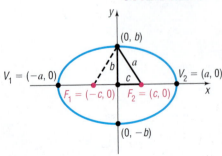

Equation of an Ellipse: Center at (0, 0); Major Axis along the x-Axis

An equation of the ellipse with center at $(0, 0)$, foci at $(-c, 0)$ and $(c, 0)$, and vertices at $(-a, 0)$ and $(a, 0)$ is

$$\frac{x^2}{a^2} + \frac{y^2}{b^2} = 1 \qquad \text{where } a > b > 0 \text{ and } b^2 = a^2 - c^2 \qquad \textbf{(2)}$$

The major axis is the x-axis. See Figure 19.

Figure 19

Notice in Figure 19 the points $(0,0)$, $(c,0)$, and $(0,b)$ form a right triangle. Because $b^2 = a^2 - c^2$ (or $b^2 + c^2 = a^2$), the distance from the focus at $(c,0)$ to the point $(0,b)$ is a.

This can be seen another way. Look at the two right triangles in Figure 19. They are congruent. Do you see why? Because the sum of the distances from the foci to a point on the ellipse is $2a$, it follows that the distance from $(c,0)$ to $(0,b)$ is a.

EXAMPLE 1

Finding an Equation of an Ellipse

Find an equation of the ellipse with center at the origin, one focus at $(3,0)$, and a vertex at $(-4,0)$. Graph the equation.

Solution

The ellipse has its center at the origin, and since the given focus and vertex lie on the x-axis, the major axis is the x-axis. The distance from the center, $(0,0)$, to one of the foci, $(3,0)$, is $c = 3$. The distance from the center, $(0,0)$, to one of the vertices, $(-4,0)$, is $a = 4$. From equation (2), it follows that

$$b^2 = a^2 - c^2 = 16 - 9 = 7$$

so an equation of the ellipse is

$$\frac{x^2}{16} + \frac{y^2}{7} = 1$$

Figure 20 shows the graph.

In Figure 20, the intercepts of the equation are used to graph the ellipse. Following this practice will make it easier for you to obtain an accurate graph of an ellipse.

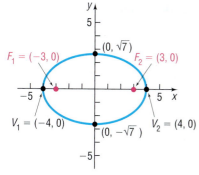

Figure 20 $\dfrac{x^2}{16} + \dfrac{y^2}{7} = 1$

Now Work PROBLEM 27

COMMENT The intercepts of the ellipse also provide information about how to set the viewing rectangle for graphing an ellipse. To graph the ellipse

$$\frac{x^2}{16} + \frac{y^2}{7} = 1$$

discussed in Example 1, set the viewing rectangle using a square screen that includes the intercepts, perhaps $-4.8 \leq x \leq 4.8$, $-3 \leq y \leq 3$. Then solve the equation for y:

$$\frac{x^2}{16} + \frac{y^2}{7} = 1$$

$$\frac{y^2}{7} = 1 - \frac{x^2}{16} \qquad \text{Subtract } \frac{x^2}{16} \text{ from each side.}$$

$$y^2 = 7\left(1 - \frac{x^2}{16}\right) \qquad \text{Multiply both sides by 7.}$$

$$y = \pm\sqrt{7\left(1 - \frac{x^2}{16}\right)} \qquad \text{Take the square root of each side.}$$

Now graph the two functions

$$Y_1 = \sqrt{7\left(1 - \frac{x^2}{16}\right)} \quad \text{and} \quad Y_2 = -\sqrt{7\left(1 - \frac{x^2}{16}\right)}$$

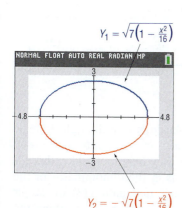

Figure 21

Figure 21 shows the result.

An equation of the form of equation (2), with $a^2 > b^2$, is the equation of an ellipse with center at the origin, foci on the x-axis at $(-c, 0)$ and $(c, 0)$, where $c^2 = a^2 - b^2$, and major axis along the x-axis.

For the remainder of this section, the direction **"Analyze the equation"** will mean to find the center, major axis, foci, and vertices of the ellipse and graph it.

EXAMPLE 2

Analyzing the Equation of an Ellipse

Analyze the equation: $\dfrac{x^2}{25} + \dfrac{y^2}{9} = 1$

Solution

The equation is of the form of equation (2), with $a^2 = 25$ and $b^2 = 9$. The equation is that of an ellipse with center $(0, 0)$ and major axis along the x-axis. The vertices are at $(\pm a, 0) = (\pm 5, 0)$. Because $b^2 = a^2 - c^2$, this means

$$c^2 = a^2 - b^2 = 25 - 9 = 16$$

The foci are at $(\pm c, 0) = (\pm 4, 0)$. Figure 22 shows the graph.

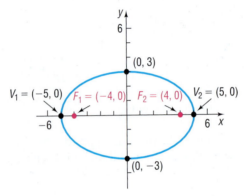

Figure 22 $\dfrac{x^2}{25} + \dfrac{y^2}{9} = 1$

Now Work PROBLEM 17

If the major axis of an ellipse with center at $(0, 0)$ lies on the y-axis, the foci are at $(0, -c)$ and $(0, c)$. Using the same steps as before, the definition of an ellipse leads to the following result.

THEOREM

Equation of an Ellipse: Center at (0, 0); Major Axis along the y-Axis

An equation of the ellipse with center at $(0, 0)$, foci at $(0, -c)$ and $(0, c)$, and vertices at $(0, -a)$ and $(0, a)$ is

$$\frac{x^2}{b^2} + \frac{y^2}{a^2} = 1 \qquad \text{where } a > b > 0 \text{ and } b^2 = a^2 - c^2 \qquad \textbf{(3)}$$

The major axis is the y-axis.

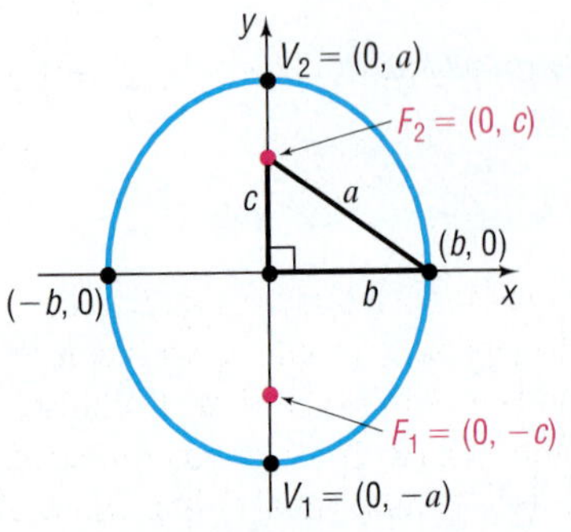

Figure 23 illustrates the graph of such an ellipse. Again, notice the right triangle formed by the points at $(0, 0)$, $(b, 0)$, and $(0, c)$, so that $a^2 = b^2 + c^2$ (or $b^2 = a^2 - c^2$).

Look closely at equations (2) and (3). Although they may look alike, there is a difference! In equation (2), the larger number, a^2, is in the denominator of the x^2-term, so the major axis of the ellipse is along the x-axis. In equation (3), the larger number, a^2, is in the denominator of the y^2-term, so the major axis is along the y-axis.

Figure 23 $\dfrac{x^2}{b^2} + \dfrac{y^2}{a^2} = 1, a > b > 0$

EXAMPLE 3

Analyzing the Equation of an Ellipse

Analyze the equation: $9x^2 + y^2 = 9$

Solution

To put the equation in proper form, divide each side by 9.

$$x^2 + \frac{y^2}{9} = 1$$

The larger denominator, 9, is in the y^2-term so, based on equation (3), this is the equation of an ellipse with center at the origin and major axis along the y-axis. Also, $a^2 = 9, b^2 = 1$, and $c^2 = a^2 - b^2 = 9 - 1 = 8$. The vertices are at $(0, \pm a) = (0, \pm 3)$, and the foci are at $(0, \pm c) = (0, \pm 2\sqrt{2})$. The x-intercepts are at $(\pm b, 0) = (\pm 1, 0)$. Figure 24 shows the graph. ●

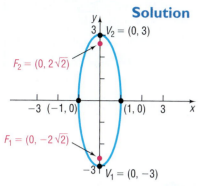

Figure 24 $9x^2 + y^2 = 9$

Now Work PROBLEM 21

EXAMPLE 4

Finding an Equation of an Ellipse

Find an equation of the ellipse having one focus at $(0, 2)$ and vertices at $(0, -3)$ and $(0, 3)$. Graph the equation.

Solution

Plot the given focus and vertices, and note that the major axis is the y-axis. Because the vertices are at $(0, -3)$ and $(0, 3)$, the center of this ellipse is at their midpoint, the origin. The distance from the center, $(0, 0)$, to the given focus, $(0, 2)$, is $c = 2$. The distance from the center, $(0, 0)$, to one of the vertices, $(0, 3)$, is $a = 3$. So $b^2 = a^2 - c^2 = 9 - 4 = 5$. The form of the equation of this ellipse is given by equation (3).

$$\frac{x^2}{b^2} + \frac{y^2}{a^2} = 1$$

$$\frac{x^2}{5} + \frac{y^2}{9} = 1$$

Figure 25 shows the graph. ●

Figure 25 $\dfrac{x^2}{5} + \dfrac{y^2}{9} = 1$

Now Work PROBLEM 29

A circle may be considered a special kind of ellipse. To see why, let $a = b$ in equation (2) or (3). Then

$$\frac{x^2}{a^2} + \frac{y^2}{a^2} = 1$$

$$x^2 + y^2 = a^2$$

This is the equation of a circle with center at the origin and radius a. The value of c is

$$c^2 = a^2 - b^2 = 0$$

$$a = b$$

This indicates that the closer the two foci of an ellipse are to the center, the more the ellipse will look like a circle.

2 Analyze Ellipses with Center at (*h, k*)

If an ellipse with center at the origin and major axis coinciding with a coordinate axis is shifted horizontally h units and then vertically k units, the result is an ellipse with center at (h, k) and major axis parallel to a coordinate axis. The equations of such ellipses have the same forms as those given in equations (2) and (3), except that x is replaced by $x - h$ (the horizontal shift) and y is replaced by $y - k$ (the vertical shift). Table 3 (on the next page) gives the forms of the equations of such ellipses, and Figure 26 shows their graphs.

Table 3

Equations of an Ellipse: Center at (h, k); Major Axis Parallel to a Coordinate Axis				
Center	**Major Axis**	**Foci**	**Vertices**	**Equation**
(h, k)	Parallel to the x-axis	$(h + c, k)$	$(h + a, k)$	$\dfrac{(x - h)^2}{a^2} + \dfrac{(y - k)^2}{b^2} = 1$
		$(h - c, k)$	$(h - a, k)$	$a > b > 0$ and $b^2 = a^2 - c^2$
(h, k)	Parallel to the y-axis	$(h, k + c)$	$(h, k + a)$	$\dfrac{(x - h)^2}{b^2} + \dfrac{(y - k)^2}{a^2} = 1$
		$(h, k - c)$	$(h, k - a)$	$a > b > 0$ and $b^2 = a^2 - c^2$

NOTE It is not recommended that Table 3 be memorized. Rather, use transformations (shift horizontally h units, vertically k units), along with the fact that a represents the distance from the center to the vertices, c represents the distance from the center to the foci, and $b^2 = a^2 - c^2$ (or $c^2 = a^2 - b^2$). ∎

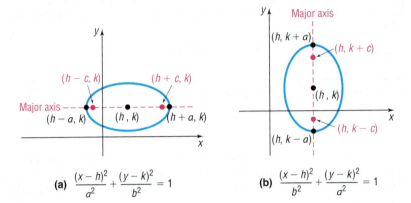

Figure 26

(a) $\dfrac{(x - h)^2}{a^2} + \dfrac{(y - k)^2}{b^2} = 1$

(b) $\dfrac{(x - h)^2}{b^2} + \dfrac{(y - k)^2}{a^2} = 1$

EXAMPLE 5

Finding an Equation of an Ellipse, Center Not at the Origin

Find an equation for the ellipse with center at $(2, -3)$, one focus at $(3, -3)$, and one vertex at $(5, -3)$. Graph the equation.

Solution The center is at $(h, k) = (2, -3)$, so $h = 2$ and $k = -3$. Plot the center, focus, and vertex, and note that the points all lie on the line $y = -3$. Therefore, the major axis is parallel to the x-axis. The distance from the center $(2, -3)$ to a focus $(3, -3)$ is $c = 1$; the distance from the center $(2, -3)$ to a vertex $(5, -3)$ is $a = 3$. Then $b^2 = a^2 - c^2 = 9 - 1 = 8$. The form of the equation is

$$\frac{(x - h)^2}{a^2} + \frac{(y - k)^2}{b^2} = 1 \quad h = 2, k = -3, a = 3, b = 2\sqrt{2}$$

$$\frac{(x - 2)^2}{9} + \frac{(y + 3)^2}{8} = 1$$

To graph the equation, use the center $(h, k) = (2, -3)$ to locate the vertices. The major axis is parallel to the x-axis, so the vertices are $a = 3$ units left and right of the center $(2, -3)$. Therefore, the vertices are

$$V_1 = (2 - 3, -3) = (-1, -3) \quad \text{and} \quad V_2 = (2 + 3, -3) = (5, -3)$$

Since $c = 1$ and the major axis is parallel to the x-axis, the foci are 1 unit left and right of the center. Therefore, the foci are

$$F_1 = (2 - 1, -3) = (1, -3) \quad \text{and} \quad F_2 = (2 + 1, -3) = (3, -3)$$

Finally, use the value of $b = 2\sqrt{2}$ to find the two points above and below the center.

$$\left(2, -3 - 2\sqrt{2}\right) \quad \text{and} \quad \left(2, -3 + 2\sqrt{2}\right)$$

Figure 27 shows the graph. ●

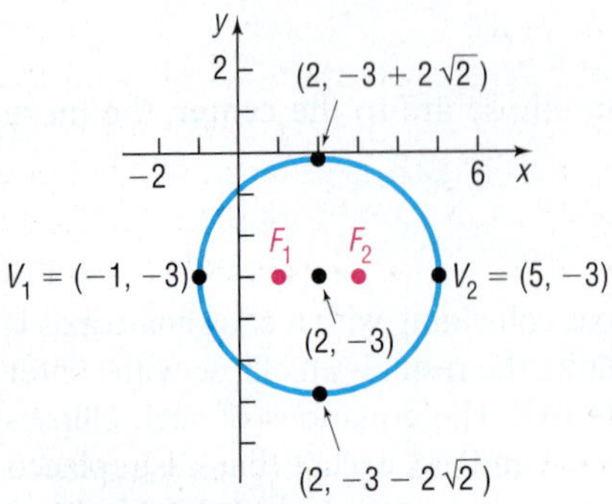

Figure 27 $\dfrac{(x - 2)^2}{9} + \dfrac{(y + 3)^2}{8} = 1$

Now Work PROBLEM 55

| EXAMPLE 6 | **Analyzing the Equation of an Ellipse** |

Analyze the equation: $4x^2 + y^2 - 8x + 4y + 4 = 0$

Solution Complete the squares in x and in y.

$$4x^2 + y^2 - 8x + 4y + 4 = 0$$
$$4x^2 - 8x + y^2 + 4y = -4 \quad \text{Group like variables; place the constant on the right side.}$$
$$4(x^2 - 2x) + (y^2 + 4y) = -4 \quad \text{Factor out 4 from the first two terms.}$$
$$4(x^2 - 2x + 1) + (y^2 + 4y + 4) = -4 + 4 + 4 \quad \text{Complete each square.}$$
$$4(x - 1)^2 + (y + 2)^2 = 4 \quad \text{Factor.}$$
$$(x - 1)^2 + \frac{(y + 2)^2}{4} = 1 \quad \text{Divide each side by 4.}$$

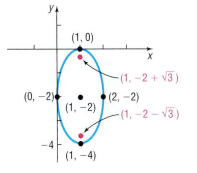

This is the equation of an ellipse with center at $(1, -2)$ and major axis parallel to the y-axis. Since $a^2 = 4$ and $b^2 = 1$, we have $c^2 = a^2 - b^2 = 4 - 1 = 3$. The vertices are at $(h, k \pm a) = (1, -2 \pm 2)$, or $(1, -4)$ and $(1, 0)$. The foci are at $(h, k \pm c) = (1, -2 \pm \sqrt{3})$, or $(1, -2 - \sqrt{3})$ and $(1, -2 + \sqrt{3})$. Figure 28 shows the graph. ●

Figure 28
$4x^2 + y^2 - 8x + 4y + 4 = 0$

 Now Work PROBLEM 47

3 Solve Applied Problems Involving Ellipses

Ellipses are found in many applications in science and engineering. For example, the orbits of the planets around the Sun are elliptical, with the Sun's position at a focus. See Figure 29.

Figure 29

Stone and concrete bridges are often shaped as semielliptical arches. Elliptical gears are used in machinery when a variable rate of motion is required.

Ellipses also have an interesting reflection property. If a source of light (or sound) is placed at one focus, the waves transmitted by the source will reflect off the ellipse and concentrate at the other focus. This is the principle behind *whispering galleries*, which are rooms designed with elliptical ceilings. A person standing at one focus of the ellipse can whisper and be heard by a person standing at the other focus, because all the sound waves that reach the ceiling are reflected to the other person.

| EXAMPLE 7 | **A Whispering Gallery** |

The whispering gallery in the Museum of Science and Industry in Chicago is 47.3 feet long. The distance from the center of the room to the foci is 20.3 feet. Find an equation that describes the shape of the room. How high is the room at its center?

Source: *Chicago Museum of Science and Industry Web site; www.msichicago.org*

Solution Set up a rectangular coordinate system so that the center of the ellipse is at the origin and the major axis is along the *x*-axis. The equation of the ellipse is

$$\frac{x^2}{a^2} + \frac{y^2}{b^2} = 1$$

Since the length of the room is 47.3 feet, the distance from the center of the room to each vertex (the end of the room) will be $\frac{47.3}{2} = 23.65$ feet; so $a = 23.65$ feet. The distance from the center of the room to each focus is $c = 20.3$ feet. See Figure 30. Because $b^2 = a^2 - c^2$, this means that $b^2 = 23.65^2 - 20.3^2 = 147.2325$. An equation that describes the shape of the room is given by

$$\frac{x^2}{23.65^2} + \frac{y^2}{147.2325} = 1$$

Figure 30

The height of the room at its center is $b = \sqrt{147.2325} \approx 12.1$ feet. ●

➤ **Now Work** PROBLEM 71

11.3 Assess Your Understanding

'Are You Prepared?' *Answers are given at the end of these exercises. If you get a wrong answer, read the pages listed in* red.

1. The distance d from $P_1 = (2, -5)$ to $P_2 = (4, -2)$ is $d = \underline{\hspace{1cm}}$. (p. 151)

2. To complete the square of $x^2 - 3x$, add $\underline{\hspace{1cm}}$. (p. 56)

3. Find the intercepts of the equation $y^2 = 16 - 4x^2$. (pp. 159–160)

4. The point that is symmetric with respect to the *y*-axis to the point $(-2, 5)$ is $\underline{\hspace{1cm}}$. (pp. 160–162)

5. To graph $y = (x + 1)^2 - 4$, shift the graph of $y = x^2$ to the (left/right) $\underline{\hspace{1cm}}$ unit(s) and then (up/down) $\underline{\hspace{1cm}}$ unit(s). (pp. 247–256)

6. The standard equation of a circle with center at $(2, -3)$ and radius 1 is $\underline{\hspace{1cm}}$. (pp. 182–185)

Concepts and Vocabulary

7. A(n) $\underline{\hspace{1cm}}$ is the collection of all points in the plane the sum of whose distances from two fixed points is a constant.

8. For an ellipse, the foci lie on a line called the $\underline{\hspace{1cm}}$.
   (a) minor axis   (b) major axis
   (c) directrix   (d) latus rectum

9. For the ellipse $\frac{x^2}{4} + \frac{y^2}{25} = 1$, the vertices are the points $\underline{\hspace{1cm}}$ and $\underline{\hspace{1cm}}$.

10. For the ellipse $\frac{x^2}{25} + \frac{y^2}{9} = 1$, the value of a is $\underline{\hspace{1cm}}$, the value of b is $\underline{\hspace{1cm}}$, and the major axis is the $\underline{\hspace{1cm}}$-axis.

11. If the center of an ellipse is $(2, -3)$, the major axis is parallel to the *x*-axis, and the distance from the center of the ellipse to its vertices is $a = 4$ units, then the coordinates of the vertices are $\underline{\hspace{1cm}}$ and $\underline{\hspace{1cm}}$.

12. If the foci of an ellipse are $(-4, 4)$ and $(6, 4)$, then the coordinates of the center of the ellipse are $\underline{\hspace{1cm}}$.
    (a) $(1, 4)$ (b) $(4, 1)$
    (c) $(1, 0)$ (d) $(5, 4)$

Skill Building

In Problems 13–16, the graph of an ellipse is given. Match each graph to its equation.

(A) $\frac{x^2}{4} + y^2 = 1$ (B) $x^2 + \frac{y^2}{4} = 1$ (C) $\frac{x^2}{16} + \frac{y^2}{4} = 1$ (D) $\frac{x^2}{4} + \frac{y^2}{16} = 1$

13.
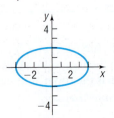

14.
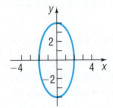

15.
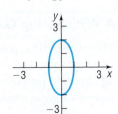

16.
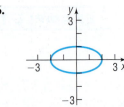

In Problems 17–26, analyze each equation. That is, find the center, vertices, and foci of each ellipse and graph it.

17. $\dfrac{x^2}{25} + \dfrac{y^2}{4} = 1$ **18.** $\dfrac{x^2}{9} + \dfrac{y^2}{4} = 1$ **19.** $\dfrac{x^2}{9} + \dfrac{y^2}{25} = 1$ **20.** $x^2 + \dfrac{y^2}{16} = 1$ **21.** $4x^2 + y^2 = 16$

22. $x^2 + 9y^2 = 18$ **23.** $4y^2 + x^2 = 8$ **24.** $4y^2 + 9x^2 = 36$ **25.** $x^2 + y^2 = 16$ **26.** $x^2 + y^2 = 4$

In Problems 27–38, find an equation for each ellipse. Graph the equation.

27. Center at $(0,0)$; focus at $(3,0)$; vertex at $(5,0)$

28. Center at $(0,0)$; focus at $(-1,0)$; vertex at $(3,0)$

29. Center at $(0,0)$; focus at $(0,-4)$; vertex at $(0,5)$

30. Center at $(0,0)$; focus at $(0,1)$; vertex at $(0,-2)$

31. Foci at $(\pm2,0)$; length of the major axis is 6

32. Foci at $(0,\pm2)$; length of the major axis is 8

33. Focus at $(-4,0)$; vertices at $(\pm5,0)$

34. Focus at $(0,-4)$; vertices at $(0,\pm8)$

35. Foci at $(0,\pm3)$; x-intercepts are ±2

36. Vertices at $(\pm4,0)$; y-intercepts are ±1

37. Center at $(0,0)$; vertex at $(0,4)$; $b = 1$

38. Vertices at $(\pm5,0)$; $c = 2$

In Problems 39–42, write an equation for each ellipse.

39.
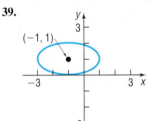

40.
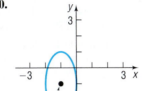

41.
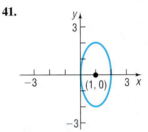

42.
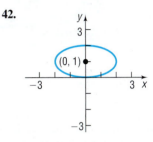

In Problems 43–54, analyze each equation; that is, find the center, foci, and vertices of each ellipse. Graph each equation.

43. $\dfrac{(x-3)^2}{4} + \dfrac{(y+1)^2}{9} = 1$ **44.** $\dfrac{(x+4)^2}{9} + \dfrac{(y+2)^2}{4} = 1$ **45.** $(x+5)^2 + 4(y-4)^2 = 16$

46. $9(x-3)^2 + (y+2)^2 = 18$ **47.** $x^2 + 4x + 4y^2 - 8y + 4 = 0$ **48.** $x^2 + 3y^2 - 12y + 9 = 0$

49. $2x^2 + 3y^2 - 8x + 6y + 5 = 0$ **50.** $4x^2 + 3y^2 + 8x - 6y = 5$ **51.** $9x^2 + 4y^2 - 18x + 16y - 11 = 0$

52. $x^2 + 9y^2 + 6x - 18y + 9 = 0$ **53.** $4x^2 + y^2 + 4y = 0$ **54.** $9x^2 + y^2 - 18x = 0$

In Problems 55–64, find an equation for each ellipse. Graph the equation.

55. Center at $(2,-2)$; vertex at $(7,-2)$; focus at $(4,-2)$

56. Center at $(-3,1)$; vertex at $(-3,3)$; focus at $(-3,0)$

57. Vertices at $(4,3)$ and $(4,9)$; focus at $(4,8)$

58. Foci at $(1,2)$ and $(-3,2)$; vertex at $(-4,2)$

59. Foci at $(5,1)$ and $(-1,1)$; length of the major axis is 8

60. Vertices at $(2,5)$ and $(2,-1)$; $c = 2$

61. Center at $(1,2)$; focus at $(4,2)$; contains the point $(1,3)$

62. Center at $(1,2)$; focus at $(1,4)$; contains the point $(2,2)$

63. Center at $(1,2)$; vertex at $(4,2)$; contains the point $(1,5)$

64. Center at $(1,2)$; vertex at $(1,4)$; contains the point $(1 + \sqrt{3}, 3)$

In Problems 65–68, graph each function. Be sure to label all the intercepts. **[Hint:** *Notice that each function is half an ellipse.*]

65. $f(x) = \sqrt{16 - 4x^2}$ **66.** $f(x) = \sqrt{9 - 9x^2}$ **67.** $f(x) = -\sqrt{64 - 16x^2}$ **68.** $f(x) = -\sqrt{4 - 4x^2}$

Applications and Extensions

69. Semielliptical Arch Bridge An arch in the shape of the upper half of an ellipse is used to support a bridge that is to span a river 20 meters wide. The center of the arch is 6 meters above the center of the river. See the figure. Write an equation for the ellipse in which the x-axis coincides with the water level and the y-axis passes through the center of the arch.

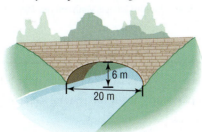

70. Semielliptical Arch Bridge The arch of a bridge is a semiellipse with a horizontal major axis. The span is 30 feet, and the top of the arch is 10 feet above the major axis. The roadway is horizontal and is 2 feet above the top of the arch. Find the vertical distance from the roadway to the arch at 5-foot intervals along the roadway.

71. Whispering Gallery A hall 100 feet in length is to be designed as a whispering gallery. If the foci are located 25 feet from the center, how high will the ceiling be at the center?

72. Whispering Gallery Jim, standing at one focus of a whispering gallery, is 6 feet from the nearest wall. His friend is standing at the other focus, 100 feet away. What is the length of this whispering gallery? How high is its elliptical ceiling at the center?

73. **Semielliptical Arch Bridge** A bridge is built in the shape of a semielliptical arch. The bridge has a span of 120 feet and a maximum height of 25 feet. Choose a suitable rectangular coordinate system and find the height of the arch at distances of 10, 30, and 50 feet from the center.

74. **Semielliptical Arch Bridge** A bridge is to be built in the shape of a semielliptical arch and is to have a span of 100 feet. The height of the arch, at a distance of 40 feet from the center, is to be 10 feet. Find the height of the arch at its center.

75. **Racetrack Design** Consult the figure. A racetrack is in the shape of an ellipse 100 feet long and 50 feet wide. What is the width 10 feet from a vertex?

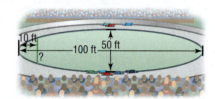

76. **Semielliptical Arch Bridge** An arch for a bridge over a highway is in the form of half an ellipse. The top of the arch is 20 feet above the ground level (the major axis). The highway has four lanes, each 12 feet wide; a center safety strip 8 feet wide; and two side strips, each 4 feet wide. What should the span of the bridge be (the length of its major axis) if the height 28 feet from the center is to be 13 feet?

77. **Installing a Vent Pipe** A homeowner is putting in a fireplace that has a 4-inch-radius vent pipe. He needs to cut an elliptical hole in his roof to accommodate the pipe. If the pitch of his roof is $\dfrac{5}{4}$ (a rise of 5, run of 4), what are the dimensions of the hole?

Source: www.doe.virginia.gov

78. **Volume of a Football** A football is in the shape of a **prolate spheroid**, which is simply a solid obtained by rotating an ellipse $\left(\dfrac{x^2}{a^2} + \dfrac{y^2}{b^2} = 1\right)$ about its major axis. An inflated NFL football averages 11.125 inches in length and 28.25 inches in center circumference. If the volume of a prolate spheroid is $\dfrac{4}{3}\pi ab^2$, how much air does the football contain? (Neglect material thickness).

Source: www.answerbag.com

*In Problems 79–83, use the fact that the orbit of a planet about the Sun is an ellipse, with the Sun at one focus. The **aphelion** of a planet is its greatest distance from the Sun, and the **perihelion** is its shortest distance. The **mean distance** of a planet from the Sun is the length of the semimajor axis of the elliptical orbit. See the illustration.*

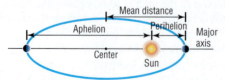

79. **Earth** The mean distance of Earth from the Sun is 93 million miles. If the aphelion of Earth is 94.5 million miles, what is the perihelion? Write an equation for the orbit of Earth around the Sun.

80. **Mars** The mean distance of Mars from the Sun is 142 million miles. If the perihelion of Mars is 128.5 million miles, what is the aphelion? Write an equation for the orbit of Mars about the Sun.

81. **Jupiter** The aphelion of Jupiter is 507 million miles. If the distance from the center of its elliptical orbit to the Sun is 23.2 million miles, what is the perihelion? What is the mean distance? Write an equation for the orbit of Jupiter around the Sun.

82. **Pluto** The perihelion of Pluto is 4551 million miles, and the distance from the center of its elliptical orbit to the Sun is 897.5 million miles. Find the aphelion of Pluto. What is the mean distance of Pluto from the Sun? Write an equation for the orbit of Pluto about the Sun.

83. **Elliptical Orbit** A planet orbits a star in an elliptical orbit with the star located at one focus. The perihelion of the planet is 5 million miles. The **eccentricity** e of a conic section is $e = \dfrac{c}{a}$. If the eccentricity of the orbit is 0.75, find the aphelion of the planet.[†]

84. A rectangle is inscribed in an ellipse with major axis of length 14 meters and minor axis of length 4 meters. Find the maximum area of a rectangle inscribed in the ellipse. Round your answer to two decimal places.[†]

85. Let D be the line given by the equation $x + 5 = 0$. Let E be the conic section given by the equation $x^2 + 5y^2 = 20$. Let the point C be the vertex of E with the smaller x-coordinate, and let B be the endpoint of the minor axis of E with the larger y-coordinate. Determine the exact y-coordinate of the point M on D that is equidistant from points B and C.[†]

86. Show that an equation of the form
$$Ax^2 + Cy^2 + F = 0 \qquad A \neq 0, C \neq 0, F \neq 0$$
where A and C are of the same sign and F is of opposite sign,
(a) is the equation of an ellipse with center at $(0,0)$ if $A \neq C$.
(b) is the equation of a circle with center $(0,0)$ if $A = C$.

87. Show that the graph of an equation of the form
$$Ax^2 + Cy^2 + Dx + Ey + F = 0 \qquad A \neq 0, C \neq 0$$
where A and C are of the same sign,
(a) is an ellipse if $\dfrac{D^2}{4A} + \dfrac{E^2}{4C} - F$ is the same sign as A.
(b) is a point if $\dfrac{D^2}{4A} + \dfrac{E^2}{4C} - F = 0$.
(c) contains no points if $\dfrac{D^2}{4A} + \dfrac{E^2}{4C} - F$ is of opposite sign to A.

[†]Courtesy of the Joliet Junior College Mathematics Department

Discussion and Writing

88. The **eccentricity** e of an ellipse is defined as the number $\dfrac{c}{a}$, where a is the distance of a vertex from the center and c is the distance of a focus from the center. Because $a > c$, it follows that $e < 1$. Write a brief paragraph about the general shape of each of the following ellipses. Be sure to justify your conclusions.

(a) Eccentricity close to 0
(b) Eccentricity = 0.5
(c) Eccentricity close to 1

Retain Your Knowledge

Problems 89–92 are based on material learned earlier in the course. The purpose of these problems is to keep the material fresh in your mind so that you are better prepared for the final exam.

89. Find the zeros of the quadratic function $f(x) = (x - 5)^2 - 12$. What are the *x*-intercepts, if any, of the graph of the function?

90. Find the domain of the rational function $f(x) = \dfrac{2x - 3}{x - 5}$.
Find any horizontal, vertical, or oblique asymptotes.

91. Find the work done by a force of 80 pounds acting in the direction of 50° to the horizontal in moving an object 12 feet from $(0,0)$ to $(12,0)$. Round to one decimal place.

92. Solve the right triangle shown.

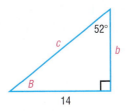

'Are You Prepared?' Answers

1. $\sqrt{13}$ **2.** $\dfrac{9}{4}$ **3.** $(-2,0), (2,0), (0,-4), (0,4)$ **4.** $(2,5)$ **5.** left; 1; down; 4   **6.** $(x - 2)^2 + (y + 3)^2 = 1$

11.4 The Hyperbola

PREPARING FOR THIS SECTION *Before getting started, review the following:*

- Distance Formula (Section 2.1, p. 151)
- Completing the Square (Chapter R, Section R.5, p. 56)
- Intercepts (Section 2.2, pp. 159–160)
- Symmetry (Section 2.2, pp. 160–162)
- Asymptotes (Section 5.2, pp. 345–350)
- Graphing Techniques: Transformations (Section 3.5, pp. 247–256)
- Square Root Method (Section 1.2, pp. 94–95)

✏️ **Now Work** the **'Are You Prepared?'** problems on page 809.

OBJECTIVES   **1** Analyze Hyperbolas with Center at the Origin (p. 799)
   2 Find the Asymptotes of a Hyperbola (p. 804)
   3 Analyze Hyperbolas with Center at (h, k) (p. 806)
   4 Solve Applied Problems Involving Hyperbolas (p. 807)

DEFINITION

A **hyperbola** is the collection of all points in the plane the difference of whose distances from two fixed points, called the **foci**, is a constant.

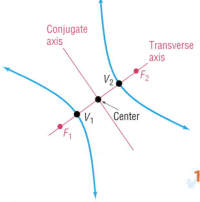

Figure 31 Hyperbola

Figure 31 illustrates a hyperbola with foci F_1 and F_2. The line containing the foci is called the **transverse axis**. The midpoint of the line segment joining the foci is the **center** of the hyperbola. The line through the center and perpendicular to the transverse axis is the **conjugate axis**. The hyperbola consists of two separate curves, called **branches**, that are symmetric with respect to the transverse axis, conjugate axis, and center. The two points of intersection of the hyperbola and the transverse axis are the **vertices**, V_1 and V_2, of the hyperbola.

1 Analyze Hyperbolas with Center at the Origin

With these ideas in mind, we are now ready to find the equation of a hyperbola in the rectangular coordinate system. First, place the center at the origin. Next, position

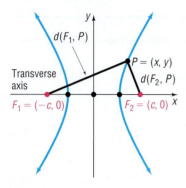

Figure 32
$d(F_1, P) - d(F_2, P) = \pm 2a$

the hyperbola so that its transverse axis coincides with a coordinate axis. Suppose that the transverse axis coincides with the x-axis, as shown in Figure 32.

If c is the distance from the center to a focus, one focus will be at $F_1 = (-c, 0)$ and the other at $F_2 = (c, 0)$. Now we let the constant difference of the distances from any point $P = (x, y)$ on the hyperbola to the foci F_1 and F_2 be denoted by $\pm 2a$, where $a > 0$. (If P is on the right branch, the $+$ sign is used; if P is on the left branch, the $-$ sign is used.) The coordinates of P must satisfy the equation

$$d(F_1, P) - d(F_2, P) = \pm 2a \qquad \text{\textit{Difference of the distances from } P \text{ \textit{to the foci equals }} \pm 2a.}$$

$$\sqrt{(x + c)^2 + y^2} - \sqrt{(x - c)^2 + y^2} = \pm 2a \qquad \text{\textit{Use the Distance Formula.}}$$

$$\sqrt{(x + c)^2 + y^2} = \pm 2a + \sqrt{(x - c)^2 + y^2} \qquad \text{\textit{Isolate one radical.}}$$

$$(x + c)^2 + y^2 = 4a^2 \pm 4a\sqrt{(x - c)^2 + y^2} \qquad \text{\textit{Square both sides.}}$$
$$+ (x - c)^2 + y^2$$

Next multiply out.

$$x^2 + 2cx + c^2 + y^2 = 4a^2 \pm 4a\sqrt{(x - c)^2 + y^2} + x^2 - 2cx + c^2 + y^2$$

$$4cx - 4a^2 = \pm 4a\sqrt{(x - c)^2 + y^2} \qquad \text{\textit{Simplify; isolate the radical.}}$$

$$cx - a^2 = \pm a\sqrt{(x - c)^2 + y^2} \qquad \text{\textit{Divide each side by 4.}}$$

$$(cx - a^2)^2 = a^2[(x - c)^2 + y^2] \qquad \text{\textit{Square both sides.}}$$

$$c^2x^2 - 2ca^2x + a^4 = a^2(x^2 - 2cx + c^2 + y^2) \qquad \text{\textit{Multiply out.}}$$

$$c^2x^2 + a^4 = a^2x^2 + a^2c^2 + a^2y^2 \qquad \text{\textit{Distribute and simplify.}}$$

$$(c^2 - a^2)x^2 - a^2y^2 = a^2c^2 - a^4 \qquad \text{\textit{Rearrange terms.}}$$

$$(c^2 - a^2)x^2 - a^2y^2 = a^2(c^2 - a^2) \qquad \text{\textit{Factor } a^2 \text{ \textit{on the right side.}}} \quad \textbf{(1)}$$

To obtain points on the hyperbola off the x-axis, it must be that $a < c$. To see why, look again at Figure 32.

$$d(F_1, P) < d(F_2, P) + d(F_1, F_2) \qquad \text{\textit{Use triangle } F_1PF_2.}$$

$$d(F_1, P) - d(F_2, P) < d(F_1, F_2)$$

$$2a < 2c \qquad \text{\textit{P is on the right branch, so}}$$
$$\text{\textit{d}(F_1, P) - d(F_2, P) = 2a;}$$
$$\text{\textit{d}(F_1, F_2) = 2c.}$$

$$a < c$$

Since $a < c$, we also have $a^2 < c^2$, so $c^2 - a^2 > 0$. Let $b^2 = c^2 - a^2$, $b > 0$. Then equation (1) can be written as

$$b^2x^2 - a^2y^2 = a^2b^2$$

$$\frac{x^2}{a^2} - \frac{y^2}{b^2} = 1 \qquad \text{\textit{Divide each side by } a^2b^2.}$$

To find the vertices of the hyperbola defined by this equation, let $y = 0$. The vertices satisfy the equation $\frac{x^2}{a^2} = 1$, the solutions of which are $x = \pm a$. Consequently, the vertices of the hyperbola are $V_1 = (-a, 0)$ and $V_2 = (a, 0)$. Notice that the distance from the center $(0, 0)$ to either vertex is a.

THEOREM

Equation of a Hyperbola: Center at (0, 0); Transverse Axis along the x-Axis

An equation of the hyperbola with center at $(0, 0)$, foci at $(-c, 0)$ and $(c, 0)$, and vertices at $(-a, 0)$ and $(a, 0)$ is

$$\frac{x^2}{a^2} - \frac{y^2}{b^2} = 1 \qquad \text{where } b^2 = c^2 - a^2 \qquad (2)$$

The transverse axis is the x-axis.

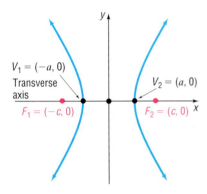

Figure 33
$\dfrac{x^2}{a^2} - \dfrac{y^2}{b^2} = 1, \quad b^2 = c^2 - a^2$

See Figure 33. As you can verify, the hyperbola defined by equation (2) is symmetric with respect to the x-axis, y-axis, and origin. To find the y-intercepts, if any, let $x = 0$ in equation (2). This results in the equation $\dfrac{y^2}{b^2} = -1$, which has no real solution, so the hyperbola defined by equation (2) has no y-intercepts. In fact, since $\dfrac{x^2}{a^2} - 1 = \dfrac{y^2}{b^2} \geq 0$, it follows that $\dfrac{x^2}{a^2} \geq 1$. There are no points on the graph for $-a < x < a$.

EXAMPLE 1

Finding and Graphing an Equation of a Hyperbola

Find an equation of the hyperbola with center at the origin, one focus at $(3, 0)$, and one vertex at $(-2, 0)$. Graph the equation.

Solution

The hyperbola has its center at the origin. Plot the center, focus, and vertex. Since they all lie on the x-axis, the transverse axis coincides with the x-axis. One focus is at $(c, 0) = (3, 0)$, so $c = 3$. One vertex is at $(-a, 0) = (-2, 0)$, so $a = 2$. From equation (2), it follows that $b^2 = c^2 - a^2 = 9 - 4 = 5$, so an equation of the hyperbola is

$$\frac{x^2}{4} - \frac{y^2}{5} = 1$$

To graph a hyperbola, it is helpful to locate and plot other points on the graph. For example, to find the points above and below the foci, let $x = \pm 3$. Then

$$\frac{x^2}{4} - \frac{y^2}{5} = 1$$

$$\frac{(\pm 3)^2}{4} - \frac{y^2}{5} = 1 \qquad x = \pm 3$$

$$\frac{9}{4} - \frac{y^2}{5} = 1$$

$$\frac{y^2}{5} = \frac{5}{4}$$

$$y^2 = \frac{25}{4}$$

$$y = \pm \frac{5}{2}$$

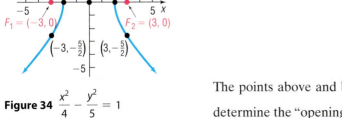

Figure 34 $\dfrac{x^2}{4} - \dfrac{y^2}{5} = 1$

The points above and below the foci are $\left(\pm 3, \dfrac{5}{2} \right)$ and $\left(\pm 3, -\dfrac{5}{2} \right)$. These points determine the "opening" of the hyperbola. See Figure 34. ●

Look at the equations of the hyperbolas in Examples 2 and 4. For the hyperbola in Example 2, $a^2 = 16$ and $b^2 = 4$, so $a > b$; for the hyperbola in Example 4, $a^2 = 4$ and $b^2 = 5$, so $a < b$. This indicates that for hyperbolas, there are no requirements involving the relative sizes of a and b. Contrast this situation to the case of an ellipse, in which the relative sizes of a and b dictate which axis is the major axis. Hyperbolas have another feature to distinguish them from ellipses and parabolas: hyperbolas have asymptotes.

2 Find the Asymptotes of a Hyperbola

Recall from Section 5.2 that a horizontal or oblique asymptote of a graph is a line with the property that the distance from the line to points on the graph approaches 0 as $x \to -\infty$ or as $x \to \infty$. Asymptotes provide information about the end behavior of the graph of a hyperbola.

THEOREM

Asymptotes of a Hyperbola

The hyperbola $\dfrac{x^2}{a^2} - \dfrac{y^2}{b^2} = 1$ has the two oblique asymptotes

$$y = \frac{b}{a}x \quad \text{and} \quad y = -\frac{b}{a}x \tag{4}$$

Proof Begin by solving for y in the equation of the hyperbola.

$$\frac{x^2}{a^2} - \frac{y^2}{b^2} = 1$$

$$\frac{y^2}{b^2} = \frac{x^2}{a^2} - 1$$

$$y^2 = b^2\left(\frac{x^2}{a^2} - 1\right)$$

Since $x \neq 0$, the right side can be rearranged in the form

$$y^2 = \frac{b^2 x^2}{a^2}\left(1 - \frac{a^2}{x^2}\right)$$

$$y = \pm\frac{bx}{a}\sqrt{1 - \frac{a^2}{x^2}}$$

Now, as $x \to -\infty$ or as $x \to \infty$, the term $\dfrac{a^2}{x^2}$ approaches 0, so the expression under the radical approaches 1. So as $x \to -\infty$ or as $x \to \infty$, the value of y approaches $\pm\dfrac{bx}{a}$; that is, the graph of the hyperbola approaches the lines

$$y = -\frac{b}{a}x \quad \text{and} \quad y = \frac{b}{a}x$$

These lines are oblique asymptotes of the hyperbola. ∎

The asymptotes of a hyperbola are not part of the hyperbola, but they do serve as a guide for graphing a hyperbola. For example, suppose that we want to graph the equation

$$\frac{x^2}{a^2} - \frac{y^2}{b^2} = 1$$

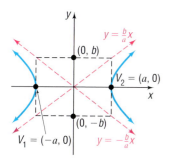

Figure 39 $\dfrac{x^2}{a^2} - \dfrac{y^2}{b^2} = 1$

Begin by plotting the vertices $(-a, 0)$ and $(a, 0)$. Then plot the points $(0, -b)$ and $(0, b)$ and use these four points to construct a rectangle, as shown in Figure 39. The diagonals of this rectangle have slopes $\dfrac{b}{a}$ and $-\dfrac{b}{a}$, and their extensions are the asymptotes of the hyperbola, $y = \dfrac{b}{a}x$ and $y = -\dfrac{b}{a}x$. If we graph the asymptotes, we can use them to establish the "opening" of the hyperbola and avoid plotting other points.

THEOREM

Asymptotes of a Hyperbola

The hyperbola $\dfrac{y^2}{a^2} - \dfrac{x^2}{b^2} = 1$ has the two oblique asymptotes

$$y = \frac{a}{b}x \quad \text{and} \quad y = -\frac{a}{b}x \qquad (5)$$

You are asked to prove this result in Problem 84.

For the remainder of this section, the direction **"Analyze the equation"** will mean to find the center, transverse axis, vertices, foci, and asymptotes of the hyperbola and graph it.

EXAMPLE 5 | Analyzing the Equation of a Hyperbola

Analyze the equation: $\dfrac{y^2}{4} - x^2 = 1$

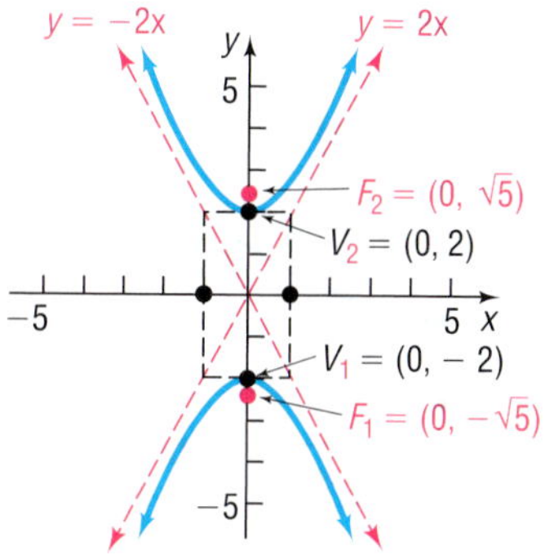

Solution Since the x^2-term is subtracted from the y^2-term, the equation is of the form of equation (3) and is a hyperbola with center at the origin and transverse axis along the y-axis. Comparing this equation to equation (3), note that $a^2 = 4$, $b^2 = 1$, and $c^2 = a^2 + b^2 = 5$. The vertices are at $(0, \pm a) = (0, \pm 2)$, and the foci are at $(0, \pm c) = (0, \pm \sqrt{5})$. Using equation (5) with $a = 2$ and $b = 1$, the asymptotes are the lines $y = \dfrac{a}{b}x = 2x$ and $y = -\dfrac{a}{b}x = -2x$. Form the rectangle containing the points $(0, \pm a) = (0, \pm 2)$ and $(\pm b, 0) = (\pm 1, 0)$. The extensions of the diagonals of this rectangle are the asymptotes. Now graph the asymptotes and the hyperbola. See Figure 40. ●

Figure 40 $\dfrac{y^2}{4} - x^2 = 1$

EXAMPLE 6 | Analyzing the Equation of a Hyperbola

Analyze the equation: $9x^2 - 4y^2 = 36$

Solution Divide each side of the equation by 36 to put the equation in proper form.

$$\frac{x^2}{4} - \frac{y^2}{9} = 1$$

The center of the hyperbola is the origin. Since the y^2-term is subtracted from the x^2-term, the transverse axis is along the x-axis, and the vertices and foci will lie on the x-axis. Using equation (2), note that $a^2 = 4$, $b^2 = 9$, and $c^2 = a^2 + b^2 = 13$. The vertices are $a = 2$ units left and right of the center at $(\pm a, 0) = (\pm 2, 0)$, the foci

Substituting these values into the original equation and simplifying gives

$$x^2 + \sqrt{3}\,xy + 2y^2 - 10 = 0$$

$$\frac{1}{4}(x' - \sqrt{3}\,y')^2 + \sqrt{3}\left[\frac{1}{2}(x' - \sqrt{3}\,y')\right]\left[\frac{1}{2}(\sqrt{3}\,x' + y')\right] + 2\left[\frac{1}{4}(\sqrt{3}\,x' + y')^2\right] = 10$$

Multiply both sides by 4 and expand to obtain

$$x'^2 - 2\sqrt{3}\,x'y' + 3y'^2 + \sqrt{3}\left(\sqrt{3}\,x'^2 - 2x'y' - \sqrt{3}\,y'^2\right) + 2\left(3x'^2 + 2\sqrt{3}\,x'y' + y'^2\right) = 40$$

$$10x'^2 + 2y'^2 = 40$$

$$\frac{x'^2}{4} + \frac{y'^2}{20} = 1$$

This is the equation of an ellipse with center at $(0,0)$ and major axis along the y'-axis. The vertices are at $(0, \pm 2\sqrt{5})$ on the y'-axis. See Figure 49 for the graph. ●

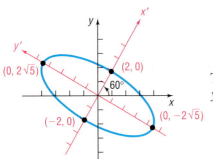

Figure 49 $\dfrac{x'^2}{4} + \dfrac{y'^2}{20} = 1$

---- **Now Work** PROBLEM 31

In Example 3, the acute angle θ through which to rotate the axes was easy to find because of the numbers used in the given equation. In general, the equation $\cot(2\theta) = \dfrac{A - C}{B}$ will not have such a "nice" solution. As the next example shows, we can still find the appropriate rotation formulas without using a calculator approximation by applying Half-angle Formulas.

EXAMPLE 4

Analyzing an Equation Using a Rotation of Axes

Analyze the equation: $\quad 4x^2 - 4xy + y^2 + 5\sqrt{5}\,x + 5 = 0$

Solution

Letting $A = 4$, $B = -4$, and $C = 1$ in equation (7), the appropriate angle θ through which to rotate the axes satisfies

$$\cot(2\theta) = \frac{A - C}{B} = \frac{3}{-4} = -\frac{3}{4}$$

To use the rotation formulas (5), we need to know the values of $\sin\theta$ and $\cos\theta$. Because we seek an acute angle θ, we know that $\sin\theta > 0$ and $\cos\theta > 0$. Use the Half-angle Formulas in the form

$$\sin\theta = \sqrt{\frac{1 - \cos(2\theta)}{2}} \qquad \cos\theta = \sqrt{\frac{1 + \cos(2\theta)}{2}}$$

Now we need to find the value of $\cos(2\theta)$. Because $\cot(2\theta) = -\dfrac{3}{4}$, then $90° < 2\theta < 180°$ (Do you know why?), so $\cos(2\theta) = -\dfrac{3}{5}$. Then

$$\sin\theta = \sqrt{\frac{1 - \cos(2\theta)}{2}} = \sqrt{\frac{1 - \left(-\dfrac{3}{5}\right)}{2}} = \sqrt{\frac{4}{5}} = \frac{2}{\sqrt{5}} = \frac{2\sqrt{5}}{5}$$

$$\cos\theta = \sqrt{\frac{1 + \cos(2\theta)}{2}} = \sqrt{\frac{1 + \left(-\dfrac{3}{5}\right)}{2}} = \sqrt{\frac{1}{5}} = \frac{1}{\sqrt{5}} = \frac{\sqrt{5}}{5}$$

With these values, the rotation formulas (5) are

$$x = \frac{\sqrt{5}}{5}x' - \frac{2\sqrt{5}}{5}y' = \frac{\sqrt{5}}{5}(x' - 2y')$$

$$y = \frac{2\sqrt{5}}{5}x' + \frac{\sqrt{5}}{5}y' = \frac{\sqrt{5}}{5}(2x' + y')$$

Substituting these values in the original equation and simplifying gives

$$4x^2 - 4xy + y^2 + 5\sqrt{5}x + 5 = 0$$

$$4\left[\frac{\sqrt{5}}{5}(x' - 2y')\right]^2 - 4\left[\frac{\sqrt{5}}{5}(x' - 2y')\right]\left[\frac{\sqrt{5}}{5}(2x' + y')\right]$$

$$+ \left[\frac{\sqrt{5}}{5}(2x' + y')\right]^2 + 5\sqrt{5}\left[\frac{\sqrt{5}}{5}(x' - 2y')\right] = -5$$

Multiply both sides by 5 and expand to obtain

$$4(x'^2 - 4x'y' + 4y'^2) - 4(2x'^2 - 3x'y' - 2y'^2)$$

$$+ 4x'^2 + 4x'y' + y'^2 + 25(x' - 2y') = -25$$

$$25y'^2 - 50y' + 25x' = -25 \quad \textcolor{teal}{\textit{Combine like terms.}}$$

$$y'^2 - 2y' + x' = -1 \quad \textcolor{teal}{\textit{Divide by 25.}}$$

$$y'^2 - 2y' + 1 = -x' \quad \textcolor{teal}{\textit{Complete the square in } y'.}$$

$$(y' - 1)^2 = -x'$$

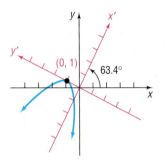

Figure 50 $(y' - 1)^2 = -x'$

This is the equation of a parabola with vertex at $(0, 1)$ in the $x'y'$-plane. The axis of symmetry is parallel to the x'-axis. Use a calculator to solve $\sin\theta = \dfrac{2\sqrt{5}}{5}$, and find that $\theta \approx 63.4°$. See Figure 50 for the graph. •

Now Work PROBLEM 37

4 Identify Conics without a Rotation of Axes

Suppose that we are required only to identify (rather than analyze) the graph of an equation of the form

$$Ax^2 + Bxy + Cy^2 + Dx + Ey + F = 0 \qquad B \neq 0 \qquad \textbf{(8)}$$

Applying the rotation formulas (5) to this equation gives an equation of the form

$$A'x'^2 + B'x'y' + C'y'^2 + D'x' + E'y' + F' = 0 \qquad \textbf{(9)}$$

where A', B', C', D', E', and F' can be expressed in terms of A, B, C, D, E, F and the angle θ of rotation (see Problem 53). It can be shown that the value of $B^2 - 4AC$ in equation (8) and the value of $B'^2 - 4A'C'$ in equation (9) are equal no matter what angle θ of rotation is chosen (see Problem 55). In particular, if the angle θ of rotation satisfies equation (7), then $B' = 0$ in equation (9), and $B^2 - 4AC = -4A'C'$. Since equation (9) then has the form of equation (2),

$$A'x'^2 + C'y'^2 + D'x' + E'y' + F' = 0$$

we can identify its graph without completing the squares, as we did in the beginning of this section. In fact, now we can identify the conic described by any equation of the form of equation (8) without a rotation of axes.

THEOREM

Identifying Conics without a Rotation of Axes

Except for degenerate cases, the equation

$$Ax^2 + Bxy + Cy^2 + Dx + Ey + F = 0$$

(a) Defines a parabola if $B^2 - 4AC = 0$.
(b) Defines an ellipse (or a circle) if $B^2 - 4AC < 0$.
(c) Defines a hyperbola if $B^2 - 4AC > 0$.

You are asked to prove this theorem in Problem 56.

EXAMPLE 5

Identifying a Conic without a Rotation of Axes

Identify the graph of the equation: $8x^2 - 12xy + 17y^2 - 4\sqrt{5}x - 2\sqrt{5}y - 15 = 0$

Solution

Here $A = 8$, $B = -12$, and $C = 17$, so $B^2 - 4AC = -400$. Since $B^2 - 4AC < 0$, the equation defines an ellipse. ●

 **Now Work** PROBLEM 43

11.5 Assess Your Understanding

'Are You Prepared?' *Answers are given at the end of these exercises. If you get a wrong answer, read the pages listed in* red.

1. The sum formula for the sine function is $\sin(A + B) = \underline{\hspace{1cm}}$. (p. 646)

2. The Double-angle Formula for the sine function is $\sin(2\theta) = \underline{\hspace{1cm}}$. (p. 656)

3. If θ is acute, the Half-angle Formula for the sine function is $\sin\dfrac{\theta}{2} = \underline{\hspace{1cm}}$. (p. 659)

4. If θ is acute, the Half-angle Formula for the cosine function is $\cos\dfrac{\theta}{2} = \underline{\hspace{1cm}}$. (p. 659)

Concepts and Vocabulary

5. To transform the equation
$$Ax^2 + Bxy + Cy^2 + Dx + Ey + F = 0 \qquad B \neq 0$$
into one in x' and y' without an $x'y'$-term, rotate the axes through an acute angle θ that satisfies the equation $\underline{\hspace{1cm}}$.

6. Except for degenerate cases, the equation
$$Ax^2 + Bxy + Cy^2 + Dx + Ey + F = 0$$
defines a(n) $\underline{\hspace{1cm}}$ if $B^2 - 4AC = 0$.

7. Except for degenerate cases, the equation
$$Ax^2 + Bxy + Cy^2 + Dx + Ey + F = 0$$
defines an ellipse if $\underline{\hspace{1cm}}$.

8. **True or False** The equation $ax^2 + 6y^2 - 12y = 0$ defines an ellipse if $a > 0$.

9. **True or False** The equation $3x^2 + Bxy + 12y^2 = 10$ defines a parabola if $B = -12$.

10. **True or False** To eliminate the xy-term from the equation $x^2 - 2xy + y^2 - 2x + 3y + 5 = 0$, rotate the axes through an angle θ, where $\cot\theta = B^2 - 4AC$.

Skill Building

In Problems 11–20, identify the graph of each equation without completing the squares.

11. $x^2 + 4x + y + 3 = 0$

12. $2y^2 - 3y + 3x = 0$

13. $6x^2 + 3y^2 - 12x + 6y = 0$

14. $2x^2 + y^2 - 8x + 4y + 2 = 0$

15. $3x^2 - 2y^2 + 6x + 4 = 0$

16. $4x^2 - 3y^2 - 8x + 6y + 1 = 0$

17. $2y^2 - x^2 - y + x = 0$

18. $y^2 - 8x^2 - 2x - y = 0$

19. $x^2 + y^2 - 8x + 4y = 0$

20. $2x^2 + 2y^2 - 8x + 8y = 0$

In Problems 21–30, determine the appropriate rotation formulas to use so that the new equation contains no xy-term.

21. $x^2 + 4xy + y^2 - 3 = 0$

22. $x^2 - 4xy + y^2 - 3 = 0$

23. $5x^2 + 6xy + 5y^2 - 8 = 0$

24. $3x^2 - 10xy + 3y^2 - 32 = 0$

25. $13x^2 - 6\sqrt{3}xy + 7y^2 - 16 = 0$

26. $11x^2 + 10\sqrt{3}xy + y^2 - 4 = 0$

27. $4x^2 - 4xy + y^2 - 8\sqrt{5}x - 16\sqrt{5}y = 0$

28. $x^2 + 4xy + 4y^2 + 5\sqrt{5}y + 5 = 0$

29. $25x^2 - 36xy + 40y^2 - 12\sqrt{13}x - 8\sqrt{13}y = 0$

30. $34x^2 - 24xy + 41y^2 - 25 = 0$

In Problems 31–42, rotate the axes so that the new equation contains no xy-term. Analyze and graph the new equation. Refer to Problems 21–30 for Problems 31–40.

31. $x^2 + 4xy + y^2 - 3 = 0$

32. $x^2 - 4xy + y^2 - 3 = 0$

33. $5x^2 + 6xy + 5y^2 - 8 = 0$

34. $3x^2 - 10xy + 3y^2 - 32 = 0$

35. $13x^2 - 6\sqrt{3}xy + 7y^2 - 16 = 0$

36. $11x^2 + 10\sqrt{3}xy + y^2 - 4 = 0$

37. $4x^2 - 4xy + y^2 - 8\sqrt{5}x - 16\sqrt{5}y = 0$

38. $x^2 + 4xy + 4y^2 + 5\sqrt{5}y + 5 = 0$

39. $25x^2 - 36xy + 40y^2 - 12\sqrt{13}x - 8\sqrt{13}y = 0$

40. $34x^2 - 24xy + 41y^2 - 25 = 0$

41. $16x^2 + 24xy + 9y^2 - 130x + 90y = 0$

42. $16x^2 + 24xy + 9y^2 - 60x + 80y = 0$

In Problems 43–52, identify the graph of each equation without applying a rotation of axes.

43. $x^2 + 3xy - 2y^2 + 3x + 2y + 5 = 0$

44. $2x^2 - 3xy + 4y^2 + 2x + 3y - 5 = 0$

45. $x^2 - 7xy + 3y^2 - y - 10 = 0$

46. $2x^2 - 3xy + 2y^2 - 4x - 2 = 0$

47. $9x^2 + 12xy + 4y^2 - x - y = 0$

48. $10x^2 + 12xy + 4y^2 - x - y + 10 = 0$

49. $10x^2 - 12xy + 4y^2 - x - y - 10 = 0$

50. $4x^2 + 12xy + 9y^2 - x - y = 0$

51. $3x^2 - 2xy + y^2 + 4x + 2y - 1 = 0$

52. $3x^2 + 2xy + y^2 + 4x - 2y + 10 = 0$

Applications and Extensions

In Problems 53–56, apply the rotation formulas (5) to

$$Ax^2 + Bxy + Cy^2 + Dx + Ey + F = 0$$

to obtain the equation

$$A'x'^2 + B'x'y' + C'y'^2 + D'x' + E'y' + F' = 0$$

53. Express A', B', C', D', E', and F' in terms of A, B, C, D, E, F, and the angle θ of rotation.
   [Hint: Refer to equation (6).]

54. Show that $A + C = A' + C'$, which proves that $A + C$ is **invariant**; that is, its value does not change under a rotation of axes.

55. Refer to Problem 54. Show that $B^2 - 4AC$ is invariant.

56. Prove that, except for degenerate cases, the equation

$$Ax^2 + Bxy + Cy^2 + Dx + Ey + F = 0$$

   (a) Defines a parabola if $B^2 - 4AC = 0$.
   (b) Defines an ellipse (or a circle) if $B^2 - 4AC < 0$.
   (c) Defines a hyperbola if $B^2 - 4AC > 0$.

57. Use the rotation formulas (5) to show that distance is invariant under a rotation of axes. That is, show that the distance from $P_1 = (x_1, y_1)$ to $P_2 = (x_2, y_2)$ in the xy-plane equals the distance from $P_1 = (x'_1, y'_1)$ to $P_2 = (x'_2, y'_2)$ in the $x'y'$-plane.

58. Show that the graph of the equation $x^{1/2} + y^{1/2} = a^{1/2}$ is part of the graph of a parabola.

Explaining Concepts: Discussion and Writing

59. Formulate a strategy for analyzing and graphing an equation of the form

$$Ax^2 + Cy^2 + Dx + Ey + F = 0$$

60. Explain how your strategy presented in Problem 59 changes if the equation is of the form

$$Ax^2 + Bxy + Cy^2 + Dx + Ey + F = 0$$

Retain Your Knowledge

Problems 61–64 are based on material learned earlier in the course. The purpose of these problems is to keep the material fresh in your mind so that you are better prepared for the final exam.

61. Solve the triangle described: $a = 7, b = 9$, and $c = 11$.

62. Find the area of the triangle described: $a = 14$, $b = 11$, and $C = 30°$.

63. Transform the equation $xy = 1$ from rectangular coordinates to polar coordinates.

64. Write the complex number $2 - 5i$ in polar form.

'Are You Prepared?' Answers

1. $\sin A \cos B + \cos A \sin B$ **2.** $2 \sin \theta \cos \theta$ **3.** $\sqrt{\dfrac{1 - \cos \theta}{2}}$ **4.** $\sqrt{\dfrac{1 + \cos \theta}{2}}$

Solution

(a) We have $v_0 = 150$ ft/sec, $\theta = 30°$, $h = 0$ ft (the ball is on the ground), and $g = 32$ ft/sec^2 (since the units are in feet and seconds). Substitute these values into equations (2) to get

$$x = (v_0 \cos \theta)t = (150 \cos 30°)t = 75\sqrt{3}t$$

$$y = -\frac{1}{2}gt^2 + (v_0 \sin \theta)t + h = -\frac{1}{2}(32)t^2 + (150 \sin 30°)t + 0$$

$$= -16t^2 + 75t$$

(b) To determine the length of time that the ball was in the air, solve the equation $y = 0$.

$$-16t^2 + 75t = 0$$

$$t(-16t + 75) = 0$$

$$t = 0 \text{ sec} \quad \text{or} \quad t = \frac{75}{16} = 4.6875 \text{ sec}$$

The ball struck the ground after 4.6875 seconds.

(c) Notice that the height y of the ball is a quadratic function of t, so the maximum height of the ball can be found by determining the vertex of $y = -16t^2 + 75t$. The value of t at the vertex is

$$t = \frac{-b}{2a} = \frac{-75}{-32} = 2.34375 \text{ sec}$$

The ball was at its maximum height after 2.34375 seconds. The maximum height of the ball is found by evaluating the function y at $t = 2.34375$ seconds.

$$\text{Maximum height} = -16(2.34375)^2 + 75(2.34375) \approx 87.89 \text{ feet}$$

(d) Since the ball was in the air for 4.6875 seconds, the horizontal distance that the ball traveled is

$$x = (75\sqrt{3})4.6875 \approx 608.92 \text{ feet}$$

 (e) Enter the equations from part (a) into a graphing utility with Tmin = 0, Tmax = 4.7, and Tstep = 0.1. Use ZOOM-SQUARE to avoid any distortion to the angle of elevation. See Figure 62. ●

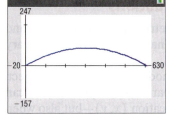

Figure 62

Exploration

Simulate the motion of a ball thrown straight up with an initial speed of 100 feet per second from a height of 5 feet above the ground. Use PARametric mode with Tmin = 0, Tmax = 6.5, Tstep = 0.1, Xmin = 0, Xmax = 5, Ymin = 0, and Ymax = 180. What happens to the speed with which the graph is drawn as the ball goes up and then comes back down? How do you interpret this physically? Repeat the experiment using other values for Tstep. How does this affect the experiment?

[**Hint:** In the projectile motion equations, let $\theta = 90°$, $v_0 = 100$, $h = 5$, and $g = 32$. Use $x = 3$ instead of $x = 0$ to see the vertical motion better.]

Result In Figure 63(a) the ball is going up. In Figure 63(b) the ball is near its highest point. Finally, in Figure 63(c), the ball is coming back down.

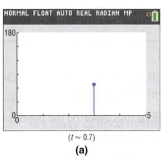

($t \approx 0.7$)

(a)

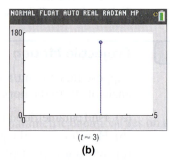

($t \approx 3$)

(b)

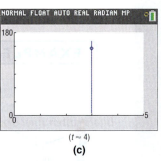

($t \approx 4$)

(c)

Figure 63

Notice that as the ball goes up, its speed decreases, until at the highest point it is zero. Then the speed increases as the ball comes back down.

Now Work PROBLEM 49

A graphing utility can be used to simulate other kinds of motion as well. Let's work again Example 5 from Section 1.7.

EXAMPLE 5

Simulating Motion

Tanya, who is a long-distance runner, runs at an average speed of 8 miles per hour. Two hours after Tanya leaves your house, you leave in your Honda and follow the same route. If your average speed is 40 miles per hour, how long will it be before you catch up to Tanya? See Figure 64. Use a simulation of the two motions to verify the answer.

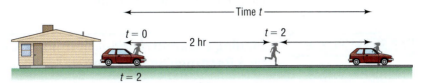

Figure 64

Solution

Begin with two sets of parametric equations: one to describe Tanya's motion, the other to describe the motion of the Honda. We choose time $t = 0$ to be when Tanya leaves the house. If we choose $y_1 = 2$ as Tanya's path, then we can use $y_2 = 4$ as the parallel path of the Honda. The horizontal distances traversed in time t (Distance = Rate $\times$ Time) are

$$\text{Tanya:} \quad x_1 = 8t \qquad \text{Honda:} \quad x_2 = 40(t - 2)$$

The Honda catches up to Tanya when $x_1 = x_2$.

$$8t = 40(t - 2)$$
$$8t = 40t - 80$$
$$-32t = -80$$
$$t = \frac{-80}{-32} = 2.5$$

The Honda catches up to Tanya 2.5 hours after Tanya leaves the house.

In PARametric mode with Tstep $= 0.01$, simultaneously graph

$$\text{Tanya:} \quad x_1 = 8t \qquad \text{Honda:} \quad x_2 = 40(t - 2)$$
$$y_1 = 2 \qquad \qquad \quad y_2 = 4$$

for $0 \le t \le 3$.

Figure 65 shows the relative positions of Tanya and the Honda for $t = 0$, $t = 2$, $t = 2.25$, $t = 2.5$, and $t = 2.75$.

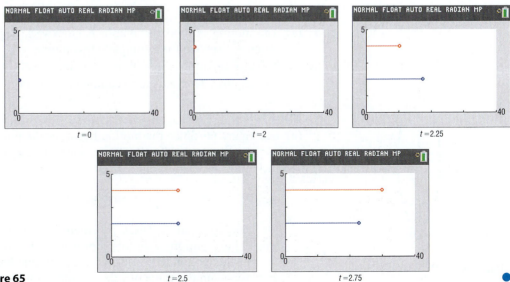

Figure 65

4 Find Parametric Equations for Curves Defined by Rectangular Equations

We now take up the question of how to find parametric equations of a given curve. If a curve is defined by the equation $y = f(x)$, where f is a function, one way of finding parametric equations is to let $x = t$. Then $y = f(t)$ and

$$x = t \quad y = f(t) \qquad t \text{ in the domain of } f$$

are parametric equations of the curve.

EXAMPLE 6

Finding Parametric Equations for a Curve Defined by a Rectangular Equation

Find two different pairs of parametric equations for the equation $y = x^2 - 4$.

Solution   For the first pair of parametric equations, let $x = t$. Then the parametric equations are

$$x = t \quad y = t^2 - 4 \qquad -\infty < t < \infty$$

A second pair of parametric equations is found by letting $x = t^3$. Then the parametric equations become

$$x = t^3 \quad y = t^6 - 4 \qquad -\infty < t < \infty \qquad \bullet$$

Care must be taken when using the second approach in Example 6, since the substitution for x must be a function that allows x to take on all the values stipulated by the domain of f. For example, letting $x = t^2$ so that $y = t^4 - 4$ does not result in equivalent parametric equations for $y = x^2 - 4$, since only points for which $x \geq 0$ are obtained; yet the domain of $y = x^2 - 4$ is $\{x \mid x \text{ is any real number}\}$.

 Now Work PROBLEM 33

EXAMPLE 7

Finding Parametric Equations for an Object in Motion

Find parametric equations for the ellipse

$$x^2 + \frac{y^2}{9} = 1$$

where the parameter t is time (in seconds) and

(a) The motion around the ellipse is clockwise, begins at the point $(0, 3)$, and requires 1 second for a complete revolution.
(b) The motion around the ellipse is counterclockwise, begins at the point $(1, 0)$, and requires 2 seconds for a complete revolution.

Solution

(a) See Figure 66. Since the motion begins at the point $(0, 3)$, we want $x = 0$ and $y = 3$ when $t = 0$. Furthermore, since the given equation is an ellipse, begin by letting

$$x = \sin(\omega t) \quad y = 3\cos(\omega t)$$

for some constant ω. These parametric equations satisfy the equation of the ellipse. Furthermore, with this choice, when $t = 0$ we have $x = 0$ and $y = 3$.

For the motion to be clockwise, the motion has to begin with the value of x increasing and the value of y decreasing as t increases. This requires that $\omega > 0$.

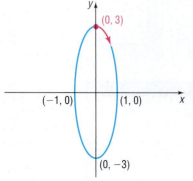

Figure 66 $x^2 + \dfrac{y^2}{9} = 1$

[Do you know why? If $\omega > 0$, then $x = \sin(\omega t)$ is increasing when $t > 0$ is near zero, and $y = 3\cos(\omega t)$ is decreasing when $t > 0$ is near zero.] See the red part of the graph in Figure 66.

Finally, since 1 revolution requires 1 second, the period $\dfrac{2\pi}{\omega} = 1$, so $\omega = 2\pi$. Parametric equations that satisfy the conditions stipulated are

$$x = \sin(2\pi t) \qquad y = 3\cos(2\pi t) \qquad 0 \le t \le 1 \qquad \textbf{(3)}$$

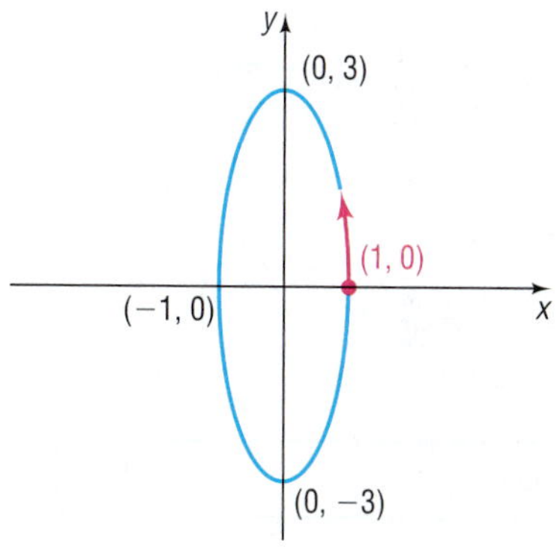

(b) See Figure 67. Since the motion begins at the point $(1, 0)$, we want $x = 1$ and $y = 0$ when $t = 0$. The given equation is an ellipse, so begin by letting

$$x = \cos(\omega t) \qquad y = 3\sin(\omega t)$$

for some constant ω. These parametric equations satisfy the equation of the ellipse. Furthermore, with this choice, when $t = 0$ we have $x = 1$ and $y = 0$.

For the motion to be counterclockwise, the motion has to begin with the value of x decreasing and the value of y increasing as t increases. This requires that $\omega > 0$. (Do you know why?) Finally, since 1 revolution requires 2 seconds, the period is $\dfrac{2\pi}{\omega} = 2$, so $\omega = \pi$. The parametric equations that satisfy the conditions stipulated are

$$x = \cos(\pi t) \qquad y = 3\sin(\pi t) \qquad 0 \le t \le 2 \qquad \textbf{(4)}$$

Either equations (3) or equations (4) can serve as parametric equations for the ellipse $x^2 + \dfrac{y^2}{9} = 1$ given in Example 7. The direction of the motion, the beginning point, and the time for 1 revolution give a particular parametric representation.

Now Work PROBLEM 39

Figure 67 $x^2 + \dfrac{y^2}{9} = 1$

The Cycloid

Suppose that a circle of radius a rolls along a horizontal line without slipping. As the circle rolls along the line, a point P on the circle will trace out a curve called a **cycloid** (see Figure 68). We now seek parametric equations* for a cycloid.

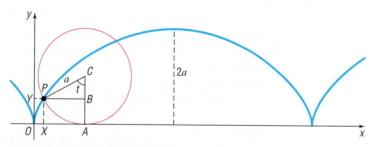

Figure 68 Cycloid

We begin with a circle of radius a and take the fixed line on which the circle rolls as the x-axis. Let the origin be one of the points at which the point P comes in contact with the x-axis. Figure 68 illustrates the position of this point P after the circle has rolled somewhat. The angle t (in radians) measures the angle through which the circle has rolled.

---

*Any attempt to derive the rectangular equation of a cycloid would soon demonstrate how complicated the task is.

Since we require no slippage, it follows that

$$\text{Arc } AP = d(O, A)$$

The length of the arc AP is given by $s = r\theta$, where $r = a$ and $\theta = t$ radians. Then

$$at = d(O, A) \quad \textcolor{blue}{s = r\theta, \text{where } r = a \text{ and } \theta = t}$$

The x-coordinate of the point P is

$$d(O, X) = d(O, A) - d(X, A) = at - a \sin t = a(t - \sin t)$$

The y-coordinate of the point P is

$$d(O, Y) = d(A, C) - d(B, C) = a - a \cos t = a(1 - \cos t)$$

Exploration

Graph $x = t - \sin t, y = 1 - \cos t,$ $0 \le t \le 3\pi$, using your graphing utility with Tstep $= \dfrac{\pi}{36}$ and a square screen. Compare your results with Figure 68.

The parametric equations of the cycloid are

$$x = a(t - \sin t) \qquad y = a(1 - \cos t) \tag{5}$$

Applications to Mechanics

If a is negative in equation (5), we obtain an inverted cycloid, as shown in Figure 69(a). The inverted cycloid occurs as a result of some remarkable applications in the field of mechanics. We shall mention two of them: the *brachistochrone* and the *tautochrone*.*

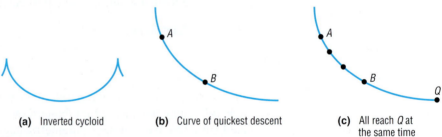

(a) Inverted cycloid   (b) Curve of quickest descent   (c) All reach Q at the same time

Figure 69

The **brachistochrone** is the curve of quickest descent. If a particle is constrained to follow some path from one point A to a lower point B (not on the same vertical line) and is acted on only by gravity, the time needed to make the descent is least if the path is an inverted cycloid. See Figure 69(b). This remarkable discovery, which has been attributed to many famous mathematicians (including Johann Bernoulli and Blaise Pascal), was a significant step in creating the branch of mathematics known as the *calculus of variations*.

To define the **tautochrone**, let Q be the lowest point on an inverted cycloid. If several particles placed at various positions on an inverted cycloid simultaneously begin to slide down the cycloid, they will reach the point Q at the same time, as indicated in Figure 69(c). The tautochrone property of the cycloid was used by Christiaan Huygens (1629–1695), the Dutch mathematician, physicist, and astronomer, to construct a pendulum clock with a bob that swings along a cycloid (see Figure 70). In Huygens's clock, the bob was made to swing along a cycloid by suspending the bob on a thin wire constrained by two plates shaped like cycloids. In a clock of this design, the period of the pendulum is independent of its amplitude.

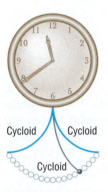

Figure 70

* In Greek, *brachistochrone* means "the shortest time" and *tautochrone* means "equal time."

11.7 Assess Your Understanding

'Are You Prepared?' *The answer is given at the end of these exercises. If you get a wrong answer, read the pages listed in* red.

1. The function $f(x) = 3\sin(4x)$ has amplitude _____ and period _____. (pp. 568–570)

Concepts and Vocabulary

2. Let $x = f(t)$ and $y = g(t)$, where f and g are two functions whose common domain is some interval I. The collection of points defined by $(x, y) = (f(t), g(t))$ is called a(n) _____ _____. The variable t is called a(n) _____.

3. The parametric equations $x = 2\sin t, y = 3\cos t$ define a(n) _____.

4. If a circle rolls along a horizontal line without slippage, a fixed point P on the circle will trace out a curve called a(n) _____.

5. *True or False* Parametric equations defining a curve are unique.

6. *True or False* Curves defined using parametric equations have an orientation.

Skill Building

In Problems 7–26, graph the curve whose parametric equations are given, and show its orientation. Find the rectangular equation of each curve.

7. $x = 3t + 2, \quad y = t + 1; \quad 0 \le t \le 4$

8. $x = t - 3, \quad y = 2t + 4; \quad 0 \le t \le 2$

9. $x = t + 2, \quad y = \sqrt{t}; \quad t \ge 0$

10. $x = \sqrt{2t}, \quad y = 4t; \quad t \ge 0$

11. $x = t^2 + 4, \quad y = t^2 - 4; \quad -\infty < t < \infty$

12. $x = \sqrt{t} + 4, \quad y = \sqrt{t} - 4; \quad t \ge 0$

13. $x = 3t^2, \quad y = t + 1; \quad -\infty < t < \infty$

14. $x = 2t - 4, \quad y = 4t^2; \quad -\infty < t < \infty$

15. $x = 2e^t, \quad y = 1 + e^t; \quad t \ge 0$

16. $x = e^t, \quad y = e^{-t}; \quad t \ge 0$

17. $x = \sqrt{t}, \quad y = t^{3/2}; \quad t \ge 0$

18. $x = t^{3/2} + 1, \quad y = \sqrt{t}; \quad t \ge 0$

19. $x = 2\cos t, \quad y = 3\sin t; \quad 0 \le t \le 2\pi$

20. $x = 2\cos t, \quad y = 3\sin t; \quad 0 \le t \le \pi$

21. $x = 2\cos t, \quad y = 3\sin t; \quad -\pi \le t \le 0$

22. $x = 2\cos t, \quad y = \sin t; \quad 0 \le t \le \dfrac{\pi}{2}$

23. $x = \sec t, \quad y = \tan t; \quad 0 \le t \le \dfrac{\pi}{4}$

24. $x = \csc t, \quad y = \cot t; \quad \dfrac{\pi}{4} \le t \le \dfrac{\pi}{2}$

25. $x = \sin^2 t, \quad y = \cos^2 t; \quad 0 \le t \le 2\pi$

26. $x = t^2, \quad y = \ln t; \quad t > 0$

In Problems 27–34, find two different pairs of parametric equations for each rectangular equation.

27. $y = 4x - 1$

28. $y = -8x + 3$

29. $y = x^2 + 1$

30. $y = -2x^2 + 1$

31. $y = x^3$

32. $y = x^4 + 1$

33. $x = y^{3/2}$

34. $x = \sqrt{y}$

In Problems 35–38, find parametric equations that define the curve shown.

35.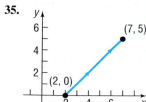

36.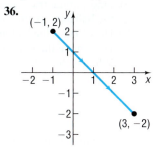

37.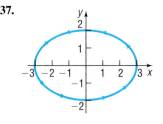

38.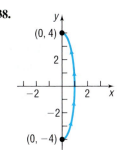

In Problems 39–42, find parametric equations for an object that moves along the ellipse $\dfrac{x^2}{4} + \dfrac{y^2}{9} = 1$ *with the motion described.*

39. The motion begins at $(2, 0)$, is clockwise, and requires 2 seconds for a complete revolution.

40. The motion begins at $(0, 3)$, is counterclockwise, and requires 1 second for a complete revolution.

41. The motion begins at $(0, 3)$, is clockwise, and requires 1 second for a complete revolution.

42. The motion begins at $(2, 0)$, is counterclockwise, and requires 3 seconds for a complete revolution.

side of equation (3) by 6, and add equations (2) and (3). The result is the new equation (3).

$$-7y + 6z = 20 \quad (2)$$
$$-4y - z = 7 \quad (3) \quad \text{Multiply by 6.}$$

$$-7y + 6z = 20 \quad (2)$$
$$\underline{-24y - 6z = 42 \quad (3)}$$
$$-31y = 62 \quad \text{Add}$$

$$\begin{cases} x + y - z = -1 \quad (1) \\ -7y + 6z = 20 \quad (2) \\ -31y = 62 \quad (3) \end{cases}$$

Now solve equation (3) for y by dividing both sides of the equation by -31.

$$\begin{cases} x + y - z = -1 \quad (1) \\ -7y + 6z = 20 \quad (2) \\ y = -2 \quad (3) \end{cases}$$

Back-substitute $y = -2$ in equation (2) and solve for z.

$$-7y + 6z = 20 \quad (2)$$
$$-7(-2) + 6z = 20 \quad \text{Substitute } y = -2 \text{ in (2).}$$
$$6z = 6 \quad \text{Subtract 14 from both sides of the equation.}$$
$$z = 1 \quad \text{Divide both sides of the equation by 6.}$$

Finally, back-substitute $y = -2$ and $z = 1$ in equation (1) and solve for x.

$$x + y - z = -1 \quad (1)$$
$$x + (-2) - 1 = -1 \quad \text{Substitute } y = -2 \text{ and } z = 1 \text{ in (1).}$$
$$x - 3 = -1 \quad \text{Simplify.}$$
$$x = 2 \quad \text{Add 3 to both sides.}$$

The solution of the original system is $x = 2, y = -2, z = 1$, or, using an ordered triplet, $(2, -2, 1)$. You should check this solution. ●

Look back over the solution given in Example 9. Note the pattern of removing one of the variables from two of the equations, followed by solving this system of two equations and two unknowns. Although which variables to remove is your choice, the methodology remains the same for all systems.

Now Work PROBLEM 45

6 Identify Inconsistent Systems of Equations Containing Three Variables

EXAMPLE 10

Identify an Inconsistent System of Linear Equations

Solve: $\begin{cases} 2x + y - z = -2 \quad (1) \\ x + 2y - z = -9 \quad (2) \\ x - 4y + z = 1 \quad (3) \end{cases}$

Solution

Our strategy is the same as in Example 9. However, in this system, it seems easiest to eliminate the variable z first. Do you see why?

Multiply each side of equation (1) by -1, and add the result to equation (2). Also, add equations (2) and (3). Here equation (2) is used twice and becomes equation (1) in the new system.

$$2x + y - z = -2 \quad (1) \quad \text{Multiply by } -1.$$
$$x + 2y - z = -9 \quad (2)$$

$$-2x - y + z = 2 \quad (1)$$
$$\underline{x + 2y - z = -9 \quad (2)}$$
$$-x + y = -7 \quad \text{Add}$$

$$x + 2y - z = -9 \quad (2)$$
$$\underline{x - 4y + z = 1 \quad (3)}$$
$$2x - 2y = -8 \quad \text{Add}$$

$$\begin{cases} x + 2y - z = -9 \quad (1) \\ -x + y = -7 \quad (2) \\ 2x - 2y = -8 \quad (3) \end{cases}$$

Now concentrate on the new equations (2) and (3), treating them as a system of two equations containing two variables. Multiply each side of equation (2) by 2, and add the result to equation (3).

$$
\begin{aligned}
-x + y &= -7 \quad (2) \\
2x - 2y &= -8 \quad (3)
\end{aligned}
\quad \text{Multiply by 2.} \quad
\begin{aligned}
-2x + 2y &= -14 \quad (2) \\
\underline{2x - 2y} &= \underline{-8} \quad (3) \\
0 &= -22 \quad \text{Add}
\end{aligned}
\quad\longrightarrow\quad
\begin{cases}
x + 2y - z = -9 & (1) \\
-x + y \quad\ = -7 & (2) \\
\quad\quad\ 0 = -22 & (3)
\end{cases}
$$

Equation (3) has no solution, so the system is inconsistent. ●

7 Express the Solution of a System of Dependent Equations Containing Three Variables

EXAMPLE 11 **Solving a System of Dependent Equations**

Solve: $\begin{cases} x - 2y - z = 8 & (1) \\ 2x - 3y + z = 23 & (2) \\ 4x - 5y + 5z = 53 & (3) \end{cases}$

Solution Our plan is to eliminate x from equations (2) and (3). Multiply each side of equation (1) by -2, and add the result to equation (2). Also, multiply each side of equation (1) by -4, and add the result to equation (3).

$$
\begin{aligned}
x - 2y - z &= 8 \quad (1) \\
2x - 3y + z &= 23 \quad (2)
\end{aligned}
\quad \text{Multiply by } -2. \quad
\begin{aligned}
-2x + 4y + 2z &= -16 \quad (1) \\
\underline{2x - 3y + z} &= \underline{23} \quad (2) \\
y + 3z &= 7 \quad \text{Add}
\end{aligned}
$$

$$
\begin{aligned}
x - 2y - z &= 8 \quad (1) \\
4x - 5y + 5z &= 53 \quad (3)
\end{aligned}
\quad \text{Multiply by } -4. \quad
\begin{aligned}
-4x + 8y + 4z &= -32 \quad (1) \\
\underline{4x - 5y + 5z} &= \underline{53} \quad (3) \\
3y + 9z &= 21 \quad \text{Add}
\end{aligned}
$$

$$
\begin{cases}
x - 2y - z = 8 & (1) \\
y + 3z = 7 & (2) \\
3y + 9z = 21 & (3)
\end{cases}
$$

Treat equations (2) and (3) as a system of two equations containing two variables, and eliminate the variable y by multiplying both sides of equation (2) by -3 and adding the result to equation (3).

$$
\begin{aligned}
y + 3z &= 7 \quad (2) \\
3y + 9z &= 21 \quad (3)
\end{aligned}
\quad \text{Multiply by } -3. \quad
\begin{aligned}
-3y - 9z &= -21 \\
\underline{3y + 9z} &= \underline{21} \quad \text{Add} \\
0 &= 0
\end{aligned}
\quad\longrightarrow\quad
\begin{cases}
x - 2y - z = 8 & (1) \\
y + 3z = 7 & (2) \\
0 = 0 & (3)
\end{cases}
$$

The original system is equivalent to a system containing two equations, so the equations are dependent and the system has infinitely many solutions. If we solve equation (2) for y, we can express y in terms of z as $y = -3z + 7$. Substitute this expression into equation (1) to determine x in terms of z.

$$
\begin{aligned}
x - 2y - z &= 8 && (1) \\
x - 2(-3z + 7) - z &= 8 && \text{Substitute } y = -3z + 7 \text{ in (1).} \\
x + 6z - 14 - z &= 8 && \text{Multiply out.} \\
x + 5z &= 22 && \text{Combine like terms.} \\
x &= -5z + 22 && \text{Solve for } x.
\end{aligned}
$$

We will write the solution to the system as

$$
\begin{cases}
x = -5z + 22 \\
y = -3z + 7
\end{cases}
\quad \text{where } z \text{ can be any real number.}
$$

Finally, back-substitute $y = 2$ and $z = -2$ into the equation $x + y + z = 1$ (from the first row) and obtain

$$x + y + z = 1$$
$$x + 2 + (-2) = 1 \quad y = 2, z = -2$$
$$x = 1 \quad \text{Solve for x.}$$

The solution of the system is $x = 1, y = 2, z = -2$, or, using an ordered triplet, $(1, 2, -2)$. ●

In Words

To obtain an augmented matrix in row echelon form:

• Add rows, interchange rows, or multiply a row by a nonzero constant.
• Work from top to bottom and left to right.
• Get 1's in the main diagonal with 0's below the 1's.
• Once the entry in row 1, column 1 is 1 with 0's below it, do not use row 1 in your row operations.
• Once the entries in row 1, column 1 and row 2, column 2 are 1 with 0's below, do not use row 1 or 2 in your row operations (and so on).

Matrix Method for Solving a System of Linear Equations (Row Echelon Form)

STEP 1: Write the augmented matrix that represents the system.

STEP 2: Perform row operations to get the entry 1 in row 1, column 1.

STEP 3: Perform row operations that leave the entry 1 in row 1, column 1 unchanged, while causing 0's to appear below it in column 1.

STEP 4: Perform row operations to get the entry 1 in row 2, column 2, but leave the entries in columns to the left unchanged. If it is impossible to place a 1 in row 2, column 2, proceed to place a 1 in row 2, column 3. Once a 1 is in place, perform row operations to place 0's below it. (Place any rows that contain only 0's on the left side of the vertical bar, at the bottom of the matrix.)

STEP 5: Now repeat Step 4 to get a 1 in the next row, but one column to the right. Continue until the bottom row or the vertical bar is reached.

STEP 6: The matrix that results is the row echelon form of the augmented matrix. Analyze the system of equations corresponding to it to solve the original system.

EXAMPLE 6 **Solving a System of Linear Equations Using Matrices (Row Echelon Form)**

Solve: $\begin{cases} x - y + z = 8 & (1) \\ 2x + 3y - z = -2 & (2) \\ 3x - 2y - 9z = 9 & (3) \end{cases}$

Solution

STEP 1: The augmented matrix of the system is

$$\begin{bmatrix} 1 & -1 & 1 & | & 8 \\ 2 & 3 & -1 & | & -2 \\ 3 & -2 & -9 & | & 9 \end{bmatrix}$$

STEP 2: Because the entry 1 is already present in row 1, column 1, go to Step 3.

STEP 3: Perform the row operations $R_2 = -2r_1 + r_2$ and $R_3 = -3r_1 + r_3$. Each of these leaves the entry 1 in row 1, column 1 unchanged, while causing 0's to appear under it.

$$\begin{bmatrix} 1 & -1 & 1 & | & 8 \\ 2 & 3 & -1 & | & -2 \\ 3 & -2 & -9 & | & 9 \end{bmatrix} \rightarrow \begin{bmatrix} 1 & -1 & 1 & | & 8 \\ 0 & 5 & -3 & | & -18 \\ 0 & 1 & -12 & | & -15 \end{bmatrix}$$

$$R_2 = -2r_1 + r_2$$
$$R_3 = -3r_1 + r_3$$

STEP 4: The easiest way to obtain the entry 1 in row 2, column 2 without altering column 1 is to interchange rows 2 and 3 (another way would be to multiply row 2 by $\frac{1}{5}$, but this introduces fractions).

$$\begin{bmatrix} 1 & -1 & 1 & | & 8 \\ 0 & 1 & -12 & | & -15 \\ 0 & 5 & -3 & | & -18 \end{bmatrix}$$

To get a 0 under the 1 in row 2, column 2, perform the row operation $R_3 = -5r_2 + r_3$.

$$\begin{bmatrix} 1 & -1 & 1 & | & 8 \\ 0 & 1 & -12 & | & -15 \\ 0 & 5 & -3 & | & -18 \end{bmatrix} \rightarrow \begin{bmatrix} 1 & -1 & 1 & | & 8 \\ 0 & 1 & -12 & | & -15 \\ 0 & 0 & 57 & | & 57 \end{bmatrix}$$

$$\uparrow R_3 = -5r_2 + r_3$$

STEP 5: Continuing, obtain a 1 in row 3, column 3 by using $R_3 = \frac{1}{57}r_3$.

$$\begin{bmatrix} 1 & -1 & 1 & | & 8 \\ 0 & 1 & -12 & | & -15 \\ 0 & 0 & 57 & | & 57 \end{bmatrix} \rightarrow \begin{bmatrix} 1 & -1 & 1 & | & 8 \\ 0 & 1 & -12 & | & -15 \\ 0 & 0 & 1 & | & 1 \end{bmatrix}$$

$$\uparrow R_3 = \frac{1}{57}r_3$$

STEP 6: The matrix on the right is the row echelon form of the augmented matrix. The system of equations represented by the matrix in row echelon form is

$$\begin{cases} x - y + z = 8 & (1) \\ y - 12z = -15 & (2) \\ z = 1 & (3) \end{cases}$$

Using $z = 1$, back-substitute to get

$$\begin{cases} x - y + 1 = 8 & (1) \\ y - 12(1) = -15 & (2) \end{cases} \xrightarrow{\text{Simplify.}} \begin{cases} x - y = 7 & (1) \\ y = -3 & (2) \end{cases}$$

We get $y = -3$ and, back-substituting into $x - y = 7$, we find that $x = 4$. The solution of the system is $x = 4, y = -3, z = 1$, or, using an ordered triplet, $(4, -3, 1)$. ●

Sometimes it is advantageous to write a matrix in **reduced row echelon form**. In this form, row operations are used to obtain entries that are 0 above (as well as below) the leading 1 in a row. For example, the row echelon form obtained in the solution to Example 6 is

$$\begin{bmatrix} 1 & -1 & 1 & | & 8 \\ 0 & 1 & -12 & | & -15 \\ 0 & 0 & 1 & | & 1 \end{bmatrix}$$

To write this matrix in reduced row echelon form, proceed as follows:

$$\begin{bmatrix} 1 & -1 & 1 & | & 8 \\ 0 & 1 & -12 & | & -15 \\ 0 & 0 & 1 & | & 1 \end{bmatrix} \rightarrow \begin{bmatrix} 1 & 0 & -11 & | & -7 \\ 0 & 1 & -12 & | & -15 \\ 0 & 0 & 1 & | & 1 \end{bmatrix} \rightarrow \begin{bmatrix} 1 & 0 & 0 & | & 4 \\ 0 & 1 & 0 & | & -3 \\ 0 & 0 & 1 & | & 1 \end{bmatrix}$$

$$\uparrow R_1 = r_2 + r_1 \qquad \uparrow \begin{array}{l} R_1 = 11r_3 + r_1 \\ R_2 = 12r_3 + r_2 \end{array}$$

The matrix is now written in reduced row echelon form. The advantage of writing the matrix in this form is that the solution to the system, $x = 4$, $y = -3, z = 1$, is readily found, without the need to back-substitute. Another advantage will be seen in Section 12.4, where the inverse of a matrix is discussed. The method used to write a matrix in reduced row echelon form is called **Gauss-Jordan elimination**.

Now Work PROBLEMS 39 AND 49

The matrix method for solving a system of linear equations also identifies systems that have infinitely many solutions and systems that are inconsistent.

EXAMPLE 7

Solving a Dependent System of Linear Equations Using Matrices

Solve: $\begin{cases} 6x - y - z = 4 & (1) \\ -12x + 2y + 2z = -8 & (2) \\ 5x + y - z = 3 & (3) \end{cases}$

Solution
Start with the augmented matrix of the system and proceed to obtain a 1 in row 1, column 1 with 0's below.

$$\begin{bmatrix} 6 & -1 & -1 & | & 4 \\ -12 & 2 & 2 & | & -8 \\ 5 & 1 & -1 & | & 3 \end{bmatrix} \rightarrow \begin{bmatrix} 1 & -2 & 0 & | & 1 \\ -12 & 2 & 2 & | & -8 \\ 5 & 1 & -1 & | & 3 \end{bmatrix} \rightarrow \begin{bmatrix} 1 & -2 & 0 & | & 1 \\ 0 & -22 & 2 & | & 4 \\ 0 & 11 & -1 & | & -2 \end{bmatrix}$$

$$\underset{R_1 = -1r_3 + r_1}{} \qquad \underset{\substack{R_2 = 12r_1 + r_2 \\ R_3 = -5r_1 + r_3}}{}$$

Obtaining a 1 in row 2, column 2 without altering column 1 can be accomplished by $R_2 = -\dfrac{1}{22}r_2$, by $R_3 = \dfrac{1}{11}r_3$ and interchanging rows 2 and 3, or by $R_2 = \dfrac{23}{11}r_3 + r_2$. We shall use the first of these.

$$\begin{bmatrix} 1 & -2 & 0 & | & 1 \\ 0 & -22 & 2 & | & 4 \\ 0 & 11 & -1 & | & -2 \end{bmatrix} \rightarrow \begin{bmatrix} 1 & -2 & 0 & | & 1 \\ 0 & 1 & -\frac{1}{11} & | & -\frac{2}{11} \\ 0 & 11 & -1 & | & -2 \end{bmatrix} \rightarrow \begin{bmatrix} 1 & -2 & 0 & | & 1 \\ 0 & 1 & -\frac{1}{11} & | & -\frac{2}{11} \\ 0 & 0 & 0 & | & 0 \end{bmatrix}$$

$$\underset{R_2 = -\frac{1}{22}r_2}{} \qquad \underset{R_3 = -11r_2 + r_3}{}$$

This matrix is in row echelon form. Because the bottom row consists entirely of 0's, the system actually consists of only two equations.

$$\begin{cases} x - 2y = 1 & (1) \\ y - \dfrac{1}{11}z = -\dfrac{2}{11} & (2) \end{cases}$$

To make it easier to write down some of the solutions, we express both x and y in terms of z.

From the second equation, $y = \dfrac{1}{11}z - \dfrac{2}{11}$. Now back-substitute this solution for y into the first equation to get

$$x = 2y + 1 = 2\left(\dfrac{1}{11}z - \dfrac{2}{11}\right) + 1 = \dfrac{2}{11}z + \dfrac{7}{11}$$

The original system is equivalent to the system

$$\begin{cases} x = \dfrac{2}{11}z + \dfrac{7}{11} & \text{(1)} \\[2mm] y = \dfrac{1}{11}z - \dfrac{2}{11} & \text{(2)} \end{cases} \quad \text{where } z \text{ can be any real number}$$

Let's look at the situation. The original system of three equations is equivalent to a system containing two equations. This means that any values of x, y, z that satisfy both

$$x = \frac{2}{11}z + \frac{7}{11} \quad \text{and} \quad y = \frac{1}{11}z - \frac{2}{11}$$

will be solutions. For example, $z = 0, x = \dfrac{7}{11}, y = -\dfrac{2}{11}; z = 1, x = \dfrac{9}{11}, y = -\dfrac{1}{11};$ and $z = -1, x = \dfrac{5}{11}, y = -\dfrac{3}{11}$ are some of the solutions of the original system. There are, in fact, infinitely many values of $x, y,$ and z for which the two equations are satisfied. That is, the original system has infinitely many solutions. We will write the solution of the original system as

$$\begin{cases} x = \dfrac{2}{11}z + \dfrac{7}{11} \\[2mm] y = \dfrac{1}{11}z - \dfrac{2}{11} \end{cases} \quad \text{where } z \text{ can be any real number}$$

or, using ordered triplets, as

$$\left\{ (x, y, z) \,\middle|\, x = \frac{2}{11}z + \frac{7}{11}, y = \frac{1}{11}z - \frac{2}{11}, z \text{ any real number} \right\}$$

We can also find the solution by writing the augmented matrix in reduced row echelon form. Starting with the row echelon form, we have

$$\begin{bmatrix} 1 & -2 & 0 & \bigm| & 1 \\[1mm] 0 & 1 & -\dfrac{1}{11} & \bigm| & -\dfrac{2}{11} \\[2mm] 0 & 0 & 0 & \bigm| & 0 \end{bmatrix} \rightarrow \begin{bmatrix} 1 & 0 & -\dfrac{2}{11} & \bigm| & \dfrac{7}{11} \\[2mm] 0 & 1 & -\dfrac{1}{11} & \bigm| & -\dfrac{2}{11} \\[2mm] 0 & 0 & 0 & \bigm| & 0 \end{bmatrix}$$
$$\uparrow$$
$$R_1 = 2r_2 + r_1$$

The matrix on the right is in reduced row echelon form. The corresponding system of equations is

$$\begin{cases} x - \dfrac{2}{11}z = \dfrac{7}{11} & \text{(1)} \\[2mm] y - \dfrac{1}{11}z = -\dfrac{2}{11} & \text{(2)} \end{cases} \quad \text{where } z \text{ can be any real number}$$

or, equivalently,

$$\begin{cases} x = \dfrac{2}{11}z + \dfrac{7}{11} \\[2mm] y = \dfrac{1}{11}z - \dfrac{2}{11} \end{cases} \quad \text{where } z \text{ can be any real number}$$

Now Work PROBLEM 55

The matrix is now in row echelon form. The final matrix represents the system

$$\begin{cases} n + b + c = 600{,}000 & (1) \\ b + 1.5c = 350{,}000 & (2) \\ c = 100{,}000 & (3) \end{cases}$$

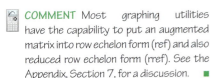

COMMENT Most graphing utilities have the capability to put an augmented matrix into row echelon form (ref) and also reduced row echelon form (rref). See the Appendix, Section 7, for a discussion. ■

From equation (3), we determine that Adam and Michelle should invest $100,000 in corporate bonds. Back-substitute $100,000 into equation (2) to find that $b = 200{,}000$, so Adam and Michelle should invest $200,000 in Treasury bonds. Back-substitute these values into equation (1) and find that $n = \$300{,}000$, so $300,000 should be invested in Treasury notes. ●

12.2 Assess Your Understanding

Concepts and Vocabulary

1. An m by n rectangular array of numbers is called a(n) _____.

2. The matrix used to represent a system of linear equations is called a(n) _____ matrix.

3. The notation a_{35} refers to the entry in the _____ row and _____ column of a matrix.

4. *True or False* The matrix $\begin{bmatrix} 1 & 3 & | & -2 \\ 0 & 1 & | & 5 \\ 0 & 0 & | & 0 \end{bmatrix}$ is in row echelon form.

5. Which of the following matrices is in reduced row echelon form?

(a) $\begin{bmatrix} 1 & 2 & | & 9 \\ 3 & -1 & | & -1 \end{bmatrix}$ (b) $\begin{bmatrix} 1 & 0 & | & 1 \\ 0 & 1 & | & 4 \end{bmatrix}$

(c) $\begin{bmatrix} 1 & 2 & | & 9 \\ 0 & 0 & | & 28 \end{bmatrix}$ (d) $\begin{bmatrix} 1 & 2 & | & 9 \\ 0 & 1 & | & 4 \end{bmatrix}$

6. Which of the following statements accurately describes the system represented by the matrix $\begin{bmatrix} 1 & 5 & -2 & | & 3 \\ 0 & 1 & 3 & | & -2 \\ 0 & 0 & 0 & | & 5 \end{bmatrix}$?

(a) The system has one solution.
(b) The system has infinitely many solutions.
(c) The system has no solution.
(d) The number of solutions cannot be determined.

Skill Building

In Problems 7–18, write the augmented matrix of the given system of equations.

7. $\begin{cases} x - 5y = 5 \\ 4x + 3y = 6 \end{cases}$

8. $\begin{cases} 3x + 4y = 7 \\ 4x - 2y = 5 \end{cases}$

9. $\begin{cases} 2x + 3y - 6 = 0 \\ 4x - 6y + 2 = 0 \end{cases}$

10. $\begin{cases} 9x - y = 0 \\ 3x - y - 4 = 0 \end{cases}$

11. $\begin{cases} 0.01x - 0.03y = 0.06 \\ 0.13x + 0.10y = 0.20 \end{cases}$

12. $\begin{cases} \dfrac{4}{3}x - \dfrac{3}{2}y = \dfrac{3}{4} \\ -\dfrac{1}{4}x + \dfrac{1}{3}y = \dfrac{2}{3} \end{cases}$

13. $\begin{cases} x - y + z = 10 \\ 3x + 3y = 5 \\ x + y + 2z = 2 \end{cases}$

14. $\begin{cases} 5x - y - z = 0 \\ x + y = 5 \\ 2x - 3z = 2 \end{cases}$

15. $\begin{cases} x + y - z = 2 \\ 3x - 2y = 2 \\ 5x + 3y - z = 1 \end{cases}$

16. $\begin{cases} 2x + 3y - 4z = 0 \\ x - 5z + 2 = 0 \\ x + 2y - 3z = -2 \end{cases}$

17. $\begin{cases} x - y - z = 10 \\ 2x + y + 2z = -1 \\ -3x + 4y = 5 \\ 4x - 5y + z = 0 \end{cases}$

18. $\begin{cases} x - y + 2z - w = 5 \\ x + 3y - 4z + 2w = 2 \\ 3x - y - 5z - w = -1 \end{cases}$

In Problems 19–26, write the system of equations corresponding to each augmented matrix. Then perform the indicated row operation(s) on the given augmented matrix.

19. $\begin{bmatrix} 1 & -3 & | & -2 \\ 2 & -5 & | & 5 \end{bmatrix}$ $R_2 = -2r_1 + r_2$

20. $\begin{bmatrix} 1 & -3 & | & -3 \\ 2 & -5 & | & -4 \end{bmatrix}$ $R_2 = -2r_1 + r_2$

21. $\begin{bmatrix} 1 & -3 & 4 & | & 3 \\ 3 & -5 & 6 & | & 6 \\ -5 & 3 & 4 & | & 6 \end{bmatrix}$ $\begin{array}{l} R_2 = -3r_1 + r_2 \\ R_3 = 5r_1 + r_3 \end{array}$

22. $\begin{bmatrix} 1 & -3 & 3 & | & -5 \\ -4 & -5 & -3 & | & -5 \\ -3 & -2 & 4 & | & 6 \end{bmatrix}$ $\begin{array}{l} R_2 = 4r_1 + r_2 \\ R_3 = 3r_1 + r_3 \end{array}$

23. $\begin{bmatrix} 1 & -3 & 2 & | & -6 \\ 2 & -5 & 3 & | & -4 \\ -3 & -6 & 4 & | & 6 \end{bmatrix}$ $\begin{array}{l} R_2 = -2r_1 + r_2 \\ R_3 = 3r_1 + r_3 \end{array}$

24. $\begin{bmatrix} 1 & -3 & -4 & | & -6 \\ 6 & -5 & 6 & | & -6 \\ -1 & 1 & 4 & | & 6 \end{bmatrix}$ $\begin{array}{l} R_2 = -6r_1 + r_2 \\ R_3 = r_1 + r_3 \end{array}$

25. $\begin{bmatrix} 5 & -3 & 1 & | & -2 \\ 2 & -5 & 6 & | & -2 \\ -4 & 1 & 4 & | & 6 \end{bmatrix}$ $\begin{aligned} R_1 &= -2r_2 + r_1 \\ R_3 &= 2r_2 + r_3 \end{aligned}$

26. $\begin{bmatrix} 4 & -3 & -1 & | & 2 \\ 3 & -5 & 2 & | & 6 \\ -3 & -6 & 4 & | & 6 \end{bmatrix}$ $\begin{aligned} R_1 &= -r_2 + r_1 \\ R_3 &= r_2 + r_3 \end{aligned}$

In Problems 27–38, the reduced row echelon form of a system of linear equations is given. Write the system of equations corresponding to the given matrix. Use x, y; or x, y, z; or x_1, x_2, x_3, x_4 as variables. Determine whether the system is consistent or inconsistent. If it is consistent, give the solution.

27. $\begin{bmatrix} 1 & 0 & | & 5 \\ 0 & 1 & | & -1 \end{bmatrix}$

28. $\begin{bmatrix} 1 & 0 & | & -4 \\ 0 & 1 & | & 0 \end{bmatrix}$

29. $\begin{bmatrix} 1 & 0 & 0 & | & 1 \\ 0 & 1 & 0 & | & 2 \\ 0 & 0 & 0 & | & 3 \end{bmatrix}$

30. $\begin{bmatrix} 1 & 0 & 0 & | & 0 \\ 0 & 1 & 0 & | & 0 \\ 0 & 0 & 0 & | & 2 \end{bmatrix}$

31. $\begin{bmatrix} 1 & 0 & 2 & | & -1 \\ 0 & 1 & -4 & | & -2 \\ 0 & 0 & 0 & | & 0 \end{bmatrix}$

32. $\begin{bmatrix} 1 & 0 & 4 & | & 4 \\ 0 & 1 & 3 & | & 2 \\ 0 & 0 & 0 & | & 0 \end{bmatrix}$

33. $\begin{bmatrix} 1 & 0 & 0 & 0 & | & 1 \\ 0 & 1 & 0 & 1 & | & 2 \\ 0 & 0 & 1 & 2 & | & 3 \end{bmatrix}$

34. $\begin{bmatrix} 1 & 0 & 0 & 0 & | & 1 \\ 0 & 1 & 0 & 2 & | & 2 \\ 0 & 0 & 1 & 3 & | & 0 \end{bmatrix}$

35. $\begin{bmatrix} 1 & 0 & 0 & 4 & | & 2 \\ 0 & 1 & 1 & 3 & | & 3 \\ 0 & 0 & 0 & 0 & | & 0 \end{bmatrix}$

36. $\begin{bmatrix} 1 & 0 & 0 & 0 & | & 1 \\ 0 & 1 & 0 & 0 & | & 2 \\ 0 & 0 & 1 & 2 & | & 3 \end{bmatrix}$

37. $\begin{bmatrix} 1 & 0 & 0 & 1 & | & -2 \\ 0 & 1 & 0 & 2 & | & 2 \\ 0 & 0 & 1 & -1 & | & 0 \\ 0 & 0 & 0 & 0 & | & 0 \end{bmatrix}$

38. $\begin{bmatrix} 1 & 0 & 0 & 0 & | & 1 \\ 0 & 1 & 0 & 0 & | & 2 \\ 0 & 0 & 1 & 0 & | & 3 \\ 0 & 0 & 0 & 1 & | & 0 \end{bmatrix}$

In Problems 39–74, solve each system of equations using matrices (row operations). If the system has no solution, say that it is inconsistent.

39. $\begin{cases} x + y = 8 \\ x - y = 4 \end{cases}$

40. $\begin{cases} x + 2y = 5 \\ x + y = 3 \end{cases}$

41. $\begin{cases} 2x - 4y = -2 \\ 3x + 2y = 3 \end{cases}$

42. $\begin{cases} 3x + 3y = 3 \\ 4x + 2y = \dfrac{8}{3} \end{cases}$

43. $\begin{cases} x + 2y = 4 \\ 2x + 4y = 8 \end{cases}$

44. $\begin{cases} 3x - y = 7 \\ 9x - 3y = 21 \end{cases}$

45. $\begin{cases} 2x + 3y = 6 \\ x - y = \dfrac{1}{2} \end{cases}$

46. $\begin{cases} \dfrac{1}{2}x + y = -2 \\ x - 2y = 8 \end{cases}$

47. $\begin{cases} 3x - 5y = 3 \\ 15x + 5y = 21 \end{cases}$

48. $\begin{cases} 2x - y = -1 \\ x + \dfrac{1}{2}y = \dfrac{3}{2} \end{cases}$

49. $\begin{cases} x - y = 6 \\ 2x - 3z = 16 \\ 2y + z = 4 \end{cases}$

50. $\begin{cases} 2x + y = -4 \\ -2y + 4z = 0 \\ 3x - 2z = -11 \end{cases}$

51. $\begin{cases} x - 2y + 3z = 7 \\ 2x + y + z = 4 \\ -3x + 2y - 2z = -10 \end{cases}$

52. $\begin{cases} 2x + y - 3z = 0 \\ -2x + 2y + z = -7 \\ 3x - 4y - 3z = 7 \end{cases}$

53. $\begin{cases} 2x - 2y - 2z = 2 \\ 2x + 3y + z = 2 \\ 3x + 2y = 0 \end{cases}$

54. $\begin{cases} 2x - 3y - z = 0 \\ -x + 2y + z = 5 \\ 3x - 4y - z = 1 \end{cases}$

55. $\begin{cases} -x + y + z = -1 \\ -x + 2y - 3z = -4 \\ 3x - 2y - 7z = 0 \end{cases}$

56. $\begin{cases} 2x - 3y - z = 0 \\ 3x + 2y + 2z = 2 \\ x + 5y + 3z = 2 \end{cases}$

57. $\begin{cases} 2x - 2y + 3z = 6 \\ 4x - 3y + 2z = 0 \\ -2x + 3y - 7z = 1 \end{cases}$

58. $\begin{cases} 3x - 2y + 2z = 6 \\ 7x - 3y + 2z = -1 \\ 2x - 3y + 4z = 0 \end{cases}$

59. $\begin{cases} x + y - z = 6 \\ 3x - 2y + z = -5 \\ x + 3y - 2z = 14 \end{cases}$

60. $\begin{cases} x - y + z = -4 \\ 2x - 3y + 4z = -15 \\ 5x + y - 2z = 12 \end{cases}$

61. $\begin{cases} x + 2y - z = -3 \\ 2x - 4y + z = -7 \\ -2x + 2y - 3z = 4 \end{cases}$

62. $\begin{cases} x + 4y - 3z = -8 \\ 3x - y + 3z = 12 \\ x + y + 6z = 1 \end{cases}$

Because $D \neq 0$, proceed to find the values of D_x, D_y, and D_z. To find D_x, replace the coefficients of x in D with the constants and then evaluate the determinant.

$$D_x = \begin{vmatrix} 3 & 1 & -1 \\ -3 & 2 & 4 \\ 4 & -2 & -3 \end{vmatrix} = (-1)^{1+1} \cdot 3 \cdot \begin{vmatrix} 2 & 4 \\ -2 & -3 \end{vmatrix} + (-1)^{1+2} \cdot 1 \cdot \begin{vmatrix} -3 & 4 \\ 4 & -3 \end{vmatrix} + (-1)^{1+3}(-1)\begin{vmatrix} -3 & 2 \\ 4 & -2 \end{vmatrix}$$

$$= 3(2) - 1(-7) + (-1)(-2) = 15$$

$$D_y = \begin{vmatrix} 2 & 3 & -1 \\ -1 & -3 & 4 \\ 1 & 4 & -3 \end{vmatrix} = (-1)^{1+1} \cdot 2 \cdot \begin{vmatrix} -3 & 4 \\ 4 & -3 \end{vmatrix} + (-1)^{1+2} \cdot 3 \cdot \begin{vmatrix} -1 & 4 \\ 1 & -3 \end{vmatrix} + (-1)^{1+3}(-1)\begin{vmatrix} -1 & -3 \\ 1 & 4 \end{vmatrix}$$

$$= 2(-7) - 3(-1) + (-1)(-1) = -10$$

$$D_z = \begin{vmatrix} 2 & 1 & 3 \\ -1 & 2 & -3 \\ 1 & -2 & 4 \end{vmatrix} = (-1)^{1+1} \cdot 2 \cdot \begin{vmatrix} 2 & -3 \\ -2 & 4 \end{vmatrix} + (-1)^{1+2} \cdot 1 \cdot \begin{vmatrix} -1 & -3 \\ 1 & 4 \end{vmatrix} + (-1)^{1+3} \cdot 3 \cdot \begin{vmatrix} -1 & 2 \\ 1 & -2 \end{vmatrix}$$

$$= 2(2) - 1(-1) + 3(0) = 5$$

As a result,

$$x = \frac{D_x}{D} = \frac{15}{5} = 3 \qquad y = \frac{D_y}{D} = \frac{-10}{5} = -2 \qquad z = \frac{D_z}{D} = \frac{5}{5} = 1$$

The solution is $x = 3$, $y = -2$, $z = 1$, or, using an ordered triplet, $(3, -2, 1)$. ●

Cramer's Rule cannot be used when the determinant of the coefficients of the variables, D, is 0. But can anything be learned about the system other than it is not a consistent and independent system if $D = 0$? The answer is yes!

Cramer's Rule with Inconsistent or Dependent Systems

- If $D = 0$ and at least one of the determinants D_x, D_y, or D_z is different from 0, then the system is inconsistent and the solution set is $\varnothing$, or { }.
- If $D = 0$ and all the determinants D_x, D_y, and D_z equal 0, then the system is consistent and dependent, so there are infinitely many solutions. The system must be solved using row reduction techniques.

━━━━ **Now Work** PROBLEM 33

5 Know Properties of Determinants

Determinants have several properties that are sometimes helpful for obtaining their value. We list some of them here.

THEOREM

The value of a determinant changes sign if any two rows (or any two columns) are interchanged.

(11)

Proof for 2 by 2 Determinants

$$\begin{vmatrix} a & b \\ c & d \end{vmatrix} = ad - bc \quad \text{and} \quad \begin{vmatrix} c & d \\ a & b \end{vmatrix} = bc - ad = -(ad - bc) \quad \blacksquare$$

EXAMPLE 6　**Demonstrating Theorem (11)**

$$\begin{vmatrix} 3 & 4 \\ 1 & 2 \end{vmatrix} = 6 - 4 = 2 \qquad \begin{vmatrix} 1 & 2 \\ 3 & 4 \end{vmatrix} = 4 - 6 = -2$$

●

THEOREM

If all the entries in any row (or any column) equal 0, the value of the determinant is 0.

(12)

Proof Expand across the row (or down the column) containing the 0's. ∎

THEOREM

> If any two rows (or any two columns) of a determinant have corresponding entries that are equal, the value of the determinant is 0. **(13)**

In Problem 68, you are asked to prove this result for a 3 by 3 determinant in which the entries in column 1 equal the entries in column 3.

EXAMPLE 7

Demonstrating Theorem (13)

$$\begin{vmatrix} 1 & 2 & 3 \\ 1 & 2 & 3 \\ 4 & 5 & 6 \end{vmatrix} = (-1)^{1+1} \cdot 1 \cdot \begin{vmatrix} 2 & 3 \\ 5 & 6 \end{vmatrix} + (-1)^{1+2} \cdot 2 \cdot \begin{vmatrix} 1 & 3 \\ 4 & 6 \end{vmatrix} + (-1)^{1+3} \cdot 3 \cdot \begin{vmatrix} 1 & 2 \\ 4 & 5 \end{vmatrix}$$

$$= 1(-3) - 2(-6) + 3(-3) = -3 + 12 - 9 = 0$$
●

THEOREM

> If any row (or any column) of a determinant is multiplied by a nonzero number k, the value of the determinant is also changed by a factor of k. **(14)**

In Problem 67, you are asked to prove this result for a 3 by 3 determinant using row 2.

EXAMPLE 8

Demonstrating Theorem (14)

$$\begin{vmatrix} 1 & 2 \\ 4 & 6 \end{vmatrix} = 6 - 8 = -2$$

$$\begin{vmatrix} k & 2k \\ 4 & 6 \end{vmatrix} = 6k - 8k = -2k = k(-2) = k\begin{vmatrix} 1 & 2 \\ 4 & 6 \end{vmatrix}$$
●

THEOREM

> If the entries of any row (or any column) of a determinant are multiplied by a nonzero number k and the result is added to the corresponding entries of another row (or column), the value of the determinant remains unchanged. **(15)**

In Problem 69, you are asked to prove this result for a 3 by 3 determinant using rows 1 and 2.

EXAMPLE 9

Demonstrating Theorem (15)

$$\begin{vmatrix} 3 & 4 \\ 5 & 2 \end{vmatrix} = -14 \qquad \begin{vmatrix} 3 & 4 \\ 5 & 2 \end{vmatrix} \rightarrow \begin{vmatrix} -7 & 0 \\ 5 & 2 \end{vmatrix} = -14$$

Multiply row 2 by −2 and add to row 1.
●

Now Work PROBLEM 45

12.3 Assess Your Understanding

Concepts and Vocabulary

1. $D = \begin{vmatrix} a & b \\ c & d \end{vmatrix} = $ _____.

2. Using Cramer's Rule, the value of x that satisfies the system of equations $\begin{cases} 2x + 3y = 5 \\ x - 4y = -3 \end{cases}$ is $x = \dfrac{}{\begin{vmatrix} 2 & 3 \\ 1 & -4 \end{vmatrix}}$.

3. *True or False* A determinant can never equal 0.

4. *True or False* When using Cramer's Rule, if $D = 0$, then the system of linear equations is inconsistent.

5. *True or False* If any row (or any column) of a determinant is multiplied by a nonzero number k, the value of the determinant remains unchanged.

6. If any two rows of a determinant are interchanged, its value is best described by which of the following?
   (a) changes sign   (b) becomes zero   (c) remains the same
   (d) no longer relates to the original value

Skill Building

In Problems 7–14, find the value of each determinant.

7. $\begin{vmatrix} 6 & 4 \\ -1 & 3 \end{vmatrix}$

8. $\begin{vmatrix} 8 & -3 \\ 4 & 2 \end{vmatrix}$

9. $\begin{vmatrix} -3 & -1 \\ 4 & 2 \end{vmatrix}$

10. $\begin{vmatrix} -4 & 2 \\ -5 & 3 \end{vmatrix}$

11. $\begin{vmatrix} 3 & 4 & 2 \\ 1 & -1 & 5 \\ 1 & 2 & -2 \end{vmatrix}$

12. $\begin{vmatrix} 1 & 3 & -2 \\ 6 & 1 & -5 \\ 8 & 2 & 3 \end{vmatrix}$

13. $\begin{vmatrix} 4 & -1 & 2 \\ 6 & -1 & 0 \\ 1 & -3 & 4 \end{vmatrix}$

14. $\begin{vmatrix} 3 & -9 & 4 \\ 1 & 4 & 0 \\ 8 & -3 & 1 \end{vmatrix}$

In Problems 15–42, solve each system of equations using Cramer's Rule if it is applicable. If Cramer's Rule is not applicable, say so.

15. $\begin{cases} x + y = 8 \\ x - y = 4 \end{cases}$

16. $\begin{cases} x + 2y = 5 \\ x - y = 3 \end{cases}$

17. $\begin{cases} 5x - y = 13 \\ 2x + 3y = 12 \end{cases}$

18. $\begin{cases} x + 3y = 5 \\ 2x - 3y = -8 \end{cases}$

19. $\begin{cases} 3x = 24 \\ x + 2y = 0 \end{cases}$

20. $\begin{cases} 4x + 5y = -3 \\ -2y = -4 \end{cases}$

21. $\begin{cases} 3x - 6y = 24 \\ 5x + 4y = 12 \end{cases}$

22. $\begin{cases} 2x + 4y = 16 \\ 3x - 5y = -9 \end{cases}$

23. $\begin{cases} 3x - 2y = 4 \\ 6x - 4y = 0 \end{cases}$

24. $\begin{cases} -x + 2y = 5 \\ 4x - 8y = 6 \end{cases}$

25. $\begin{cases} 2x - 4y = -2 \\ 3x + 2y = 3 \end{cases}$

26. $\begin{cases} 3x + 3y = 3 \\ 4x + 2y = \dfrac{8}{3} \end{cases}$

27. $\begin{cases} 2x - 3y = -1 \\ 10x + 10y = 5 \end{cases}$

28. $\begin{cases} 3x - 2y = 0 \\ 5x + 10y = 4 \end{cases}$

29. $\begin{cases} 2x + 3y = 6 \\ x - y = \dfrac{1}{2} \end{cases}$

30. $\begin{cases} \dfrac{1}{2}x + y = -2 \\ x - 2y = 8 \end{cases}$

31. $\begin{cases} 3x - 5y = 3 \\ 15x + 5y = 21 \end{cases}$

32. $\begin{cases} 2x - y = -1 \\ x + \dfrac{1}{2}y = \dfrac{3}{2} \end{cases}$

33. $\begin{cases} x + y - z = 6 \\ 3x - 2y + z = -5 \\ x + 3y - 2z = 14 \end{cases}$

34. $\begin{cases} x - y + z = -4 \\ 2x - 3y + 4z = -15 \\ 5x + y - 2z = 12 \end{cases}$

35. $\begin{cases} x + 2y - z = -3 \\ 2x - 4y + z = -7 \\ -2x + 2y - 3z = 4 \end{cases}$

36. $\begin{cases} x + 4y - 3z = -8 \\ 3x - y + 3z = 12 \\ x + y + 6z = 1 \end{cases}$

37. $\begin{cases} x - 2y + 3z = 1 \\ 3x + y - 2z = 0 \\ 2x - 4y + 6z = 2 \end{cases}$

38. $\begin{cases} x - y + 2z = 5 \\ 3x + 2y = 4 \\ -2x + 2y - 4z = -10 \end{cases}$

39. $\begin{cases} x + 2y - z = 0 \\ 2x - 4y + z = 0 \\ -2x + 2y - 3z = 0 \end{cases}$

40. $\begin{cases} x + 4y - 3z = 0 \\ 3x - y + 3z = 0 \\ x + y + 6z = 0 \end{cases}$

41. $\begin{cases} x - 2y + 3z = 0 \\ 3x + y - 2z = 0 \\ 2x - 4y + 6z = 0 \end{cases}$

42. $\begin{cases} x - y + 2z = 0 \\ 3x + 2y = 0 \\ -2x + 2y - 4z = 0 \end{cases}$

In Problems 43–50, use properties of determinants to find the value of each determinant if it is known that

$$\begin{vmatrix} x & y & z \\ u & v & w \\ 1 & 2 & 3 \end{vmatrix} = 4$$

43. $\begin{vmatrix} 1 & 2 & 3 \\ u & v & w \\ x & y & z \end{vmatrix}$

44. $\begin{vmatrix} x & y & z \\ u & v & w \\ 2 & 4 & 6 \end{vmatrix}$

45. $\begin{vmatrix} x & y & z \\ -3 & -6 & -9 \\ u & v & w \end{vmatrix}$

46. $\begin{vmatrix} 1 & 2 & 3 \\ x - u & y - v & z - w \\ u & v & w \end{vmatrix}$

47. $\begin{vmatrix} 1 & 2 & 3 \\ x - 3 & y - 6 & z - 9 \\ 2u & 2v & 2w \end{vmatrix}$

48. $\begin{vmatrix} x & y & z - x \\ u & v & w - u \\ 1 & 2 & 2 \end{vmatrix}$

49. $\begin{vmatrix} 1 & 2 & 3 \\ 2x & 2y & 2z \\ u - 1 & v - 2 & w - 3 \end{vmatrix}$

50. $\begin{vmatrix} x + 3 & y + 6 & z + 9 \\ 3u - 1 & 3v - 2 & 3w - 3 \\ 1 & 2 & 3 \end{vmatrix}$

Mixed Practice

In Problems 51–56, solve for x.

51. $\begin{vmatrix} x & x \\ 4 & 3 \end{vmatrix} = 5$

52. $\begin{vmatrix} x & 1 \\ 3 & x \end{vmatrix} = -2$

53. $\begin{vmatrix} x & 1 & 1 \\ 4 & 3 & 2 \\ -1 & 2 & 5 \end{vmatrix} = 2$

54. $\begin{vmatrix} 3 & 2 & 4 \\ 1 & x & 5 \\ 0 & 1 & -2 \end{vmatrix} = 0$

55. $\begin{vmatrix} x & 2 & 3 \\ 1 & x & 0 \\ 6 & 1 & -2 \end{vmatrix} = 7$

56. $\begin{vmatrix} x & 1 & 2 \\ 1 & x & 3 \\ 0 & 1 & 2 \end{vmatrix} = -4x$

Applications and Extensions

57. Geometry: Equation of a Line An equation of the line containing the two points (x_1, y_1) and (x_2, y_2) may be expressed as the determinant

$$\begin{vmatrix} x & y & 1 \\ x_1 & y_1 & 1 \\ x_2 & y_2 & 1 \end{vmatrix} = 0$$

Prove this result by expanding the determinant and comparing the result to the two-point form of the equation of a line.

58. Geometry: Collinear Points Using the result obtained in Problem 57, show that three distinct points (x_1, y_1), (x_2, y_2), and (x_3, y_3) are collinear (lie on the same line) if and only if

$$\begin{vmatrix} x_1 & y_1 & 1 \\ x_2 & y_2 & 1 \\ x_3 & y_3 & 1 \end{vmatrix} = 0$$

59. Geometry: Area of a Triangle A triangle has vertices (x_1, y_1), (x_2, y_2), and (x_3, y_3). The area of the triangle is given by the absolute value of D, where $D = \dfrac{1}{2} \begin{vmatrix} x_1 & x_2 & x_3 \\ y_1 & y_2 & y_3 \\ 1 & 1 & 1 \end{vmatrix}$.

Use this formula to find the area of a triangle with vertices $(2, 3)$, $(5, 2)$, and $(6, 5)$.

60. Geometry: Area of a Polygon The formula from Problem 59 can be used to find the area of a polygon. To do so, divide the polygon into non-overlapping triangular regions and find the sum of the areas. Use this approach to find the area of the given polygon.

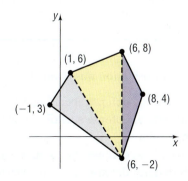

61. Geometry: Area of a Polygon Another approach for finding the area of a polygon by using determinants is to use the formula

$$A = \frac{1}{2}\left(\begin{vmatrix} x_1 & y_1 \\ x_2 & y_2 \end{vmatrix} + \begin{vmatrix} x_2 & y_2 \\ x_3 & y_3 \end{vmatrix} + \begin{vmatrix} x_3 & y_3 \\ x_4 & y_4 \end{vmatrix} + \cdots + \begin{vmatrix} x_n & y_n \\ x_1 & y_1 \end{vmatrix} \right)$$

where $(x_1, y_1), (x_2, y_2), \ldots, (x_n, y_n)$ are the n corner points in counterclockwise order. Use this formula to compute the area of the polygon from Problem 60 again. Which method do you prefer?

62. Show that the formula in Problem 61 yields the same result as the formula using Problem 59.

63. Geometry: Equation of a Circle An equation of the circle containing the distinct points (x_1, y_1), (x_2, y_2), and (x_3, y_3) can be found using the following equation.

$$\begin{vmatrix} 1 & 1 & 1 & 1 \\ x & x_1 & x_2 & x_3 \\ y & y_1 & y_2 & y_3 \\ x^2 + y^2 & x_1^2 + y_1^2 & x_2^2 + y_2^2 & x_3^2 + y_3^2 \end{vmatrix} = 0$$

Find the equation of the circle containing the points $(7, -5), (3, 3)$, and $(6, 2)$. Write the equation in standard form.

64. Show that $\begin{vmatrix} x^2 & x & 1 \\ y^2 & y & 1 \\ z^2 & z & 1 \end{vmatrix} = (y - z)(x - y)(x - z)$.

65. Complete the proof of Cramer's Rule for two equations containing two variables.

[Hint: In system (5), page 876, if $a = 0$, then $b \neq 0$ and $c \neq 0$, since $D = -bc \neq 0$. Now show that equation (6) provides a solution of the system when $a = 0$. Then three cases remain: $b = 0$, $c = 0$, and $d = 0$.]

66. Interchange columns 1 and 3 of a 3 by 3 determinant. Show that the value of the new determinant is -1 times the value of the original determinant.

67. Multiply each entry in row 2 of a 3 by 3 determinant by the number $k, k \neq 0$. Show that the value of the new determinant is k times the value of the original determinant.

68. Prove that a 3 by 3 determinant in which the entries in column 1 equal those in column 3 has the value 0.

69. Prove that if row 2 of a 3 by 3 determinant is multiplied by $k, k \neq 0$, and the result is added to the entries in row 1, there is no change in the value of the determinant.

Retain Your Knowledge

Problems 70–73 are based on material learned earlier in the course. The purpose of these problems is to keep the material fresh in your mind so that you are better prepared for the final exam.

70. For the points $P = (-4, 3)$ and $Q = (5, -1)$ write the vector **v** represented by the directed line segment $\overrightarrow{PQ}$ in the form $a\mathbf{i} + b\mathbf{j}$ and find $\|\mathbf{v}\|$.

71. List the potential rational zeros of the polynomial function $P(x) = 2x^3 - 5x^2 + x - 10$.

72. Graph $f(x) = (x + 1)^2 - 4$ using transformations (shifting, compressing, stretching, and/or reflecting).

73. Find the exact value of $\tan 42° - \cot 48°$ without using a calculator.

12.4 Matrix Algebra

OBJECTIVES
1. Find the Sum and Difference of Two Matrices (p. 885)
2. Find Scalar Multiples of a Matrix (p. 887)
3. Find the Product of Two Matrices (p. 888)
4. Find the Inverse of a Matrix (p. 892)
5. Solve a System of Linear Equations Using an Inverse Matrix (p. 896)

Section 12.2 defined a matrix as a rectangular array of real numbers and used an augmented matrix to represent a system of linear equations. There is, however, a branch of mathematics, called **linear algebra**, that deals with matrices in such a way that an algebra of matrices is permitted. This section provides a survey of how this **matrix algebra** is developed.

Before getting started, recall the definition of a matrix.

DEFINITION

A **matrix** is defined as a rectangular array of numbers:

$$
\begin{array}{c}
\\
\text{Row 1} \\
\text{Row 2} \\
\vdots \\
\text{Row } i \\
\vdots \\
\text{Row } m
\end{array}
\begin{array}{ccccc}
\text{Column 1} & \text{Column 2} & \text{Column } j & & \text{Column } n \\
\begin{bmatrix}
a_{11} & a_{12} & \cdots & a_{1j} & \cdots & a_{1n} \\
a_{21} & a_{22} & \cdots & a_{2j} & \cdots & a_{2n} \\
\vdots & \vdots & & \vdots & & \vdots \\
a_{i1} & a_{i2} & \cdots & a_{ij} & \cdots & a_{in} \\
\vdots & \vdots & & \vdots & & \vdots \\
a_{m1} & a_{m2} & \cdots & a_{mj} & \cdots & a_{mn}
\end{bmatrix}
\end{array}
$$

Each number a_{ij} of the matrix has two indexes: the **row index** i and the **column index** j. The matrix shown here has m rows and n columns. The numbers a_{ij} are usually referred to as the **entries** of the matrix. For example, a_{23} refers to the entry in the second row, third column.

EXAMPLE 1

Arranging Data in a Matrix

In a survey of 900 people, the following information was obtained:

200 males	Thought federal defense spending was too high
150 males	Thought federal defense spending was too low
45 males	Had no opinion
315 females	Thought federal defense spending was too high
125 females	Thought federal defense spending was too low
65 females	Had no opinion

We can arrange these data in a rectangular array as follows:

	Too High	Too Low	No Opinion
Male	200	150	45
Female	315	125	65

or as the matrix

$$\begin{bmatrix} 200 & 150 & 45 \\ 315 & 125 & 65 \end{bmatrix}$$

This matrix has two rows (representing male and female) and three columns (representing "too high," "too low," and "no opinion"). ●

The matrix developed in Example 1 has 2 rows and 3 columns. In general, a matrix with m rows and n columns is called an ***m* by *n* matrix**. The matrix developed in Example 1 is a 2 by 3 matrix and contains $2 \cdot 3 = 6$ entries. An m by n matrix will contain $m \cdot n$ entries.

If an m by n matrix has the same number of rows as columns, that is, if $m = n$, then the matrix is a **square matrix**.

EXAMPLE 2 **Examples of Matrices**

(a) $\begin{bmatrix} 5 & 0 \\ -6 & 1 \end{bmatrix}$ A 2 by 2 square matrix   (b) $\begin{bmatrix} 1 & 0 & 3 \end{bmatrix}$ A 1 by 3 matrix

(c) $\begin{bmatrix} 6 & -2 & 4 \\ 4 & 3 & 5 \\ 8 & 0 & 1 \end{bmatrix}$ A 3 by 3 square matrix ●

1 Find the Sum and Difference of Two Matrices

We begin our discussion of matrix algebra by first defining equivalent matrices and then defining the operations of addition and subtraction. It is important to note that these definitions require both matrices to have the same number of rows *and* the same number of columns as a condition for equality and for addition and subtraction.

Matrices usually are represented by capital letters, such as A, B, and C.

DEFINITION

Two matrices A and B are **equal**, written as

$$A = B$$

provided that A and B have the same number of rows and the same number of columns and each entry a_{ij} in A is equal to the corresponding entry b_{ij} in B.

For example,

$$\begin{bmatrix} 2 & 1 \\ 0.5 & -1 \end{bmatrix} = \begin{bmatrix} \sqrt{4} & 1 \\ \frac{1}{2} & -1 \end{bmatrix} \quad \text{and} \quad \begin{bmatrix} 3 & 2 & 1 \\ 0 & 1 & -2 \end{bmatrix} = \begin{bmatrix} \sqrt{9} & \sqrt{4} & 1 \\ 0 & 1 & \sqrt[3]{-8} \end{bmatrix}$$

$$\begin{bmatrix} 4 & 1 \\ 6 & 1 \end{bmatrix} \neq \begin{bmatrix} 4 & 0 \\ 6 & 1 \end{bmatrix}$$ Because the entries in row 1, column 2 are not equal

$$\begin{bmatrix} 4 & 1 & 2 \\ 6 & 1 & 2 \end{bmatrix} \neq \begin{bmatrix} 4 & 1 & 2 & 3 \\ 6 & 1 & 2 & 4 \end{bmatrix}$$ Because the matrix on the left has 3 columns and the matrix on the right has 4 columns

Check: Enter the matrices A, B, and C into a graphing utility. Then find $4A$, $\frac{1}{3}C$, and $3A - 2B$.

Now Work PROBLEM 13

Some of the algebraic properties of scalar multiplication are listed next. Let h and k be real numbers, and let A and B be m by n matrices. Then

> **Properties of Scalar Multiplication**
>
> $$k(hA) = (kh)A$$
> $$(k + h)A = kA + hA$$
> $$k(A + B) = kA + kB$$

3 Find the Product of Two Matrices

Unlike the straightforward definition for adding two matrices, the definition for multiplying two matrices is not what might be expected. In preparation for this definition, we need the following definitions:

DEFINITION

A **row vector** R is a 1 by n matrix

$$R = \begin{bmatrix} r_1 & r_2 & \cdots & r_n \end{bmatrix}$$

A **column vector** C is an n by 1 matrix

$$C = \begin{bmatrix} c_1 \\ c_2 \\ \vdots \\ c_n \end{bmatrix}$$

The **product** RC of R times C is defined as the number

> $$RC = \begin{bmatrix} r_1 & r_2 & \cdots & r_n \end{bmatrix} \begin{bmatrix} c_1 \\ c_2 \\ \vdots \\ c_n \end{bmatrix} = r_1 c_1 + r_2 c_2 + \cdots + r_n c_n$$

Notice that a row vector and a column vector can be multiplied only if they contain the same number of entries.

EXAMPLE 6

The Product of a Row Vector and a Column Vector

If $R = \begin{bmatrix} 3 & -5 & 2 \end{bmatrix}$ and $C = \begin{bmatrix} 3 \\ 4 \\ -5 \end{bmatrix}$, then

$$RC = \begin{bmatrix} 3 & -5 & 2 \end{bmatrix} \begin{bmatrix} 3 \\ 4 \\ -5 \end{bmatrix} = 3 \cdot 3 + (-5)4 + 2(-5) = 9 - 20 - 10 = -21$$

EXAMPLE 7

Using Matrices to Compute Revenue

A clothing store sells men's shirts for $40, silk ties for $20, and wool suits for $400. Last month, the store had sales consisting of 100 shirts, 200 ties, and 50 suits. What was the total revenue due to these sales?

Solution

Set up a row vector R to represent the prices of these three items and a column vector C to represent the corresponding number of items sold. Then

$$
\begin{array}{cc}
\begin{array}{c} \text{Prices} \\ \text{Shirts Ties Suits} \end{array} & \begin{array}{c} \text{Number} \\ \text{sold} \end{array} \\
R = \begin{bmatrix} 40 & 20 & 400 \end{bmatrix} & C = \begin{bmatrix} 100 \\ 200 \\ 50 \end{bmatrix} \begin{array}{c} \text{Shirts} \\ \text{Ties} \\ \text{Suits} \end{array}
\end{array}
$$

The total revenue obtained is the product RC. That is,

$$
RC = \begin{bmatrix} 40 & 20 & 400 \end{bmatrix} \begin{bmatrix} 100 \\ 200 \\ 50 \end{bmatrix}
$$

$$
= \underbrace{40 \cdot 100}_{\text{Shirt revenue}} + \underbrace{20 \cdot 200}_{\text{Tie revenue}} + \underbrace{400 \cdot 50}_{\text{Suit revenue}} = \underbrace{\$28{,}000}_{\text{Total revenue}}
$$

The definition for multiplying two matrices is based on the definition of a row vector times a column vector.

DEFINITION

Let A denote an m by r matrix and let B denote an r by n matrix. The **product** AB is defined as the m by n matrix whose entry in row i, column j is the product of the ith row of A and the jth column of B.

The definition of the product AB of two matrices A and B, in this order, requires that the number of columns of A equals the number of rows of B; otherwise, no product is defined.

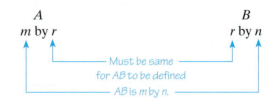

In Words
To find the product AB, the number of columns in A must equal the number of rows in B.

An example will help to clarify the definition.

EXAMPLE 8

Multiplying Two Matrices

Find the product AB if

$$
A = \begin{bmatrix} 2 & 4 & -1 \\ 5 & 8 & 0 \end{bmatrix} \quad \text{and} \quad B = \begin{bmatrix} 2 & 5 & 1 & 4 \\ 4 & 8 & 0 & 6 \\ -3 & 1 & -2 & -1 \end{bmatrix}
$$

Solution

First, observe that A is 2 by 3 and B is 3 by 4. The number of columns in A equals the number of rows in B, so the product AB is defined and will be a 2 by 4 matrix.

Suppose we want the entry in row 2, column 3 of AB. To find it, find the product of the row vector from row 2 of A and the column vector from column 3 of B.

Column 3 of B

$$\text{Row 2 of } A \begin{bmatrix} 5 & 8 & 0 \end{bmatrix} \begin{bmatrix} 1 \\ 0 \\ -2 \end{bmatrix} = 5 \cdot 1 + 8 \cdot 0 + 0(-2) = 5$$

So far, we have

Column 3
↓

$$AB = \begin{bmatrix} \_\_ & \_\_ & 5 & \_\_ \\ \end{bmatrix} \quad \leftarrow \text{Row 2}$$

Now, to find the entry in row 1, column 4 of AB, find the product of row 1 of A and column 4 of B.

Column 4 of B

$$\text{Row 1 of } A \begin{bmatrix} 2 & 4 & -1 \end{bmatrix} \begin{bmatrix} 4 \\ 6 \\ -1 \end{bmatrix} = 2 \cdot 4 + 4 \cdot 6 + (-1)(-1) = 33$$

Continuing in this fashion, we find AB.

$$AB = \begin{bmatrix} 2 & 4 & -1 \\ 5 & 8 & 0 \end{bmatrix} \begin{bmatrix} 2 & 5 & 1 & 4 \\ 4 & 8 & 0 & 6 \\ -3 & 1 & -2 & -1 \end{bmatrix}$$

$$= \begin{bmatrix} \text{Row 1 of } A & \text{Row 1 of } A & \text{Row 1 of } A & \text{Row 1 of } A \\ \text{times} & \text{times} & \text{times} & \text{times} \\ \text{column 1 of } B & \text{column 2 of } B & \text{column 3 of } B & \text{column 4 of } B \\ & & & \\ \text{Row 2 of } A & \text{Row 2 of } A & \text{Row 2 of } A & \text{Row 2 of } A \\ \text{times} & \text{times} & \text{times} & \text{times} \\ \text{column 1 of } B & \text{column 2 of } B & \text{column 3 of } B & \text{column 4 of } B \end{bmatrix}$$

$$= \begin{bmatrix} 2 \cdot 2 + 4 \cdot 4 + (-1)(-3) & 2 \cdot 5 + 4 \cdot 8 + (-1)1 & 2 \cdot 1 + 4 \cdot 0 + (-1)(-2) & 33 \, (\text{from earlier}) \\ 5 \cdot 2 + 8 \cdot 4 + 0(-3) & 5 \cdot 5 + 8 \cdot 8 + 0 \cdot 1 & 5 \, (\text{from earlier}) & 5 \cdot 4 + 8 \cdot 6 + 0(-1) \end{bmatrix}$$

$$= \begin{bmatrix} 23 & 41 & 4 & 33 \\ 42 & 89 & 5 & 68 \end{bmatrix}$$

●

✓ **Check:** Enter the matrices A and B. Then find AB. (See what happens if you try to find BA.)

Now Work PROBLEM 25

Notice that for the matrices given in Example 8, the product BA is not defined because B is 3 by 4 and A is 2 by 3.

EXAMPLE 9 **Multiplying Two Matrices**

If

$$A = \begin{bmatrix} 2 & 1 & 3 \\ 1 & -1 & 0 \end{bmatrix} \quad \text{and} \quad B = \begin{bmatrix} 1 & 0 \\ 2 & 1 \\ 3 & 2 \end{bmatrix}$$

find: (a) AB (b) BA

Solution   (a) $AB = \begin{bmatrix} 2 & 1 & 3 \\ 1 & -1 & 0 \end{bmatrix} \begin{bmatrix} 1 & 0 \\ 2 & 1 \\ 3 & 2 \end{bmatrix} = \begin{bmatrix} 13 & 7 \\ -1 & -1 \end{bmatrix}$

                           2 by 3   3 by 2   2 by 2

(b) $BA = \begin{bmatrix} 1 & 0 \\ 2 & 1 \\ 3 & 2 \end{bmatrix} \begin{bmatrix} 2 & 1 & 3 \\ 1 & -1 & 0 \end{bmatrix} = \begin{bmatrix} 2 & 1 & 3 \\ 5 & 1 & 6 \\ 8 & 1 & 9 \end{bmatrix}$

                   3 by 2   2 by 3   3 by 3   ●

Notice in Example 9 that AB is 2 by 2 and BA is 3 by 3. It is possible for both AB and BA to be defined and yet be unequal. In fact, even if A and B are both n by n matrices so that AB and BA are each defined and n by n, AB and BA will usually be unequal.

EXAMPLE 10   **Multiplying Two Square Matrices**

If

$$A = \begin{bmatrix} 2 & 1 \\ 0 & 4 \end{bmatrix} \quad \text{and} \quad B = \begin{bmatrix} -3 & 1 \\ 1 & 2 \end{bmatrix}$$

find: (a) AB (b) BA

Solution   (a) $AB = \begin{bmatrix} 2 & 1 \\ 0 & 4 \end{bmatrix} \begin{bmatrix} -3 & 1 \\ 1 & 2 \end{bmatrix} = \begin{bmatrix} -5 & 4 \\ 4 & 8 \end{bmatrix}$

(b) $BA = \begin{bmatrix} -3 & 1 \\ 1 & 2 \end{bmatrix} \begin{bmatrix} 2 & 1 \\ 0 & 4 \end{bmatrix} = \begin{bmatrix} -6 & 1 \\ 2 & 9 \end{bmatrix}$ ●

The preceding examples demonstrate that an important property of real numbers, the commutative property of multiplication, is not shared by matrices. In general:

THEOREM   Matrix multiplication is not commutative.

Now Work PROBLEMS **15 AND 17**

Next, consider two of the properties of real numbers that *are* shared by matrices. Assuming that each product and sum is defined, the following is true:

Associative Property of Matrix Multiplication

$$A(BC) = (AB)C$$

Distributive Property

$$A(B + C) = AB + AC$$

For an n by n square matrix, the entries located in row i, column i, $1 \leq i \leq n$, are called the **diagonal entries** or **the main diagonal**. The n by n square matrix whose diagonal entries are 1's, and all other entries are 0's, is called the **identity matrix I_n**. For example,

$$I_2 = \begin{bmatrix} 1 & 0 \\ 0 & 1 \end{bmatrix} \qquad I_3 = \begin{bmatrix} 1 & 0 & 0 \\ 0 & 1 & 0 \\ 0 & 0 & 1 \end{bmatrix}$$

and so on.

EXAMPLE 11

Multiplication with an Identity Matrix

Let

$$A = \begin{bmatrix} -1 & 2 & 0 \\ 0 & 1 & 3 \end{bmatrix} \quad \text{and} \quad B = \begin{bmatrix} 3 & 2 \\ 4 & 6 \\ 5 & 2 \end{bmatrix}$$

Find: (a) AI_3 (b) $I_2 A$ (c) BI_2

Solution

(a) $AI_3 = \begin{bmatrix} -1 & 2 & 0 \\ 0 & 1 & 3 \end{bmatrix} \begin{bmatrix} 1 & 0 & 0 \\ 0 & 1 & 0 \\ 0 & 0 & 1 \end{bmatrix} = \begin{bmatrix} -1 & 2 & 0 \\ 0 & 1 & 3 \end{bmatrix} = A$

(b) $I_2 A = \begin{bmatrix} 1 & 0 \\ 0 & 1 \end{bmatrix} \begin{bmatrix} -1 & 2 & 0 \\ 0 & 1 & 3 \end{bmatrix} = \begin{bmatrix} -1 & 2 & 0 \\ 0 & 1 & 3 \end{bmatrix} = A$

(c) $BI_2 = \begin{bmatrix} 3 & 2 \\ 4 & 6 \\ 5 & 2 \end{bmatrix} \begin{bmatrix} 1 & 0 \\ 0 & 1 \end{bmatrix} = \begin{bmatrix} 3 & 2 \\ 4 & 6 \\ 5 & 2 \end{bmatrix} = B$ ●

Example 11 demonstrates the following property:

Identity Property

If A is an m by n matrix, then

$$I_m A = A \quad \text{and} \quad AI_n = A$$

If A is an n by n square matrix,

$$AI_n = I_n A = A$$

An identity matrix has properties similar to those of the real number 1. In other words, the identity matrix is a multiplicative identity in matrix algebra.

4 Find the Inverse of a Matrix

DEFINITION

Let A be a square n by n matrix. If there exists an n by n matrix A^{-1} (read as "A inverse") for which

$$AA^{-1} = A^{-1}A = I_n$$

then A^{-1} is called the **inverse** of the matrix A.

NOTE If the determinant of A is zero, A is singular. (Refer to Section 12.3.) ∎

Not every square matrix has an inverse. When a matrix A does have an inverse A^{-1}, then A is said to be **nonsingular**. If a matrix A has no inverse, it is called **singular**.

EXAMPLE 12 **Multiplying a Matrix by Its Inverse**

Show that the inverse of

$$A = \begin{bmatrix} 3 & 1 \\ 2 & 1 \end{bmatrix} \quad \text{is} \quad A^{-1} = \begin{bmatrix} 1 & -1 \\ -2 & 3 \end{bmatrix}$$

Solution We need to show that $AA^{-1} = A^{-1}A = I_2$.

$$AA^{-1} = \begin{bmatrix} 3 & 1 \\ 2 & 1 \end{bmatrix}\begin{bmatrix} 1 & -1 \\ -2 & 3 \end{bmatrix} = \begin{bmatrix} 1 & 0 \\ 0 & 1 \end{bmatrix} = I_2$$

$$A^{-1}A = \begin{bmatrix} 1 & -1 \\ -2 & 3 \end{bmatrix}\begin{bmatrix} 3 & 1 \\ 2 & 1 \end{bmatrix} = \begin{bmatrix} 1 & 0 \\ 0 & 1 \end{bmatrix} = I_2$$

The following shows one way to find the inverse of

$$A = \begin{bmatrix} 3 & 1 \\ 2 & 1 \end{bmatrix}$$

Suppose that A^{-1} is given by

$$A^{-1} = \begin{bmatrix} x & y \\ z & w \end{bmatrix} \tag{1}$$

where x, y, z, and w are four variables. Based on the definition of an inverse, if A has an inverse, then

$$AA^{-1} = I_2$$

$$\begin{bmatrix} 3 & 1 \\ 2 & 1 \end{bmatrix}\begin{bmatrix} x & y \\ z & w \end{bmatrix} = \begin{bmatrix} 1 & 0 \\ 0 & 1 \end{bmatrix}$$

$$\begin{bmatrix} 3x + z & 3y + w \\ 2x + z & 2y + w \end{bmatrix} = \begin{bmatrix} 1 & 0 \\ 0 & 1 \end{bmatrix}$$

Because corresponding entries must be equal, it follows that this matrix equation is equivalent to two systems of linear equations.

$$\begin{cases} 3x + z = 1 \\ 2x + z = 0 \end{cases} \qquad \begin{cases} 3y + w = 0 \\ 2y + w = 1 \end{cases}$$

The augmented matrix of each system is

$$\begin{bmatrix} 3 & 1 & | & 1 \\ 2 & 1 & | & 0 \end{bmatrix} \qquad \begin{bmatrix} 3 & 1 & | & 0 \\ 2 & 1 & | & 1 \end{bmatrix} \tag{2}$$

The usual procedure would be to transform each augmented matrix into reduced row echelon form. Notice, though, that the left sides of the augmented matrices are equal, so the same row operations (see Section 12.2) can be used to reduce each one. It is more efficient to combine the two augmented matrices (2) into a single matrix, as shown next.

$$\begin{bmatrix} 3 & 1 & | & 1 & 0 \\ 2 & 1 & | & 0 & 1 \end{bmatrix}$$

Next, use row operations to transform the matrix into reduced row echelon form.

$$\begin{bmatrix} 3 & 1 & | & 1 & 0 \\ 2 & 1 & | & 0 & 1 \end{bmatrix} \rightarrow \begin{bmatrix} 1 & 0 & | & 1 & -1 \\ 2 & 1 & | & 0 & 1 \end{bmatrix}$$

$$\uparrow$$
$$R_1 = -1r_2 + r_1$$

$$\rightarrow \begin{bmatrix} 1 & 0 & | & 1 & -1 \\ 0 & 1 & | & -2 & 3 \end{bmatrix} \tag{3}$$

$$\uparrow$$
$$R_2 = -2r_1 + r_2$$

Matrix (3) is in reduced row echelon form.

Now reverse the earlier step of combining the two augmented matrices in (2), and write the single matrix (3) as two augmented matrices.

$$\left[\begin{array}{cc|c} 1 & 0 & 1 \\ 0 & 1 & -2 \end{array}\right] \quad \text{and} \quad \left[\begin{array}{cc|c} 1 & 0 & -1 \\ 0 & 1 & 3 \end{array}\right]$$

The conclusion from these matrices is that $x = 1, z = -2$, and $y = -1, w = 3$. Substituting these values into matrix (1) results in

$$A^{-1} = \left[\begin{array}{cc} 1 & -1 \\ -2 & 3 \end{array}\right]$$

Notice in display (3) that the 2 by 2 matrix to the right of the vertical bar is, in fact, the inverse of A. Also notice that the identity matrix I_2 is the matrix that appears to the left of the vertical bar. These observations and the procedures used to get display (3) will work in general.

> **In Words**
>
> If A is nonsingular, begin with the matrix $[A \,|\, I_n]$, and after transforming it into reduced row echelon form, you end up with the matrix $[I_n \,|\, A^{-1}]$.

Procedure for Finding the Inverse of a Nonsingular Matrix*

To find the inverse of an n by n nonsingular matrix A, proceed as follows:

STEP 1: Form the matrix $[A \,|\, I_n]$.

STEP 2: Transform the matrix $[A \,|\, I_n]$ into reduced row echelon form.

STEP 3: The reduced row echelon form of $[A \,|\, I_n]$ will contain the identity matrix I_n on the left of the vertical bar; the n by n matrix on the right of the vertical bar is the inverse of A.

EXAMPLE 13

Finding the Inverse of a Matrix

The matrix

$$A = \left[\begin{array}{ccc} 1 & 1 & 0 \\ -1 & 3 & 4 \\ 0 & 4 & 3 \end{array}\right]$$

is nonsingular. Find its inverse.

Solution

First, form the matrix

$$[A \,|\, I_3] = \left[\begin{array}{ccc|ccc} 1 & 1 & 0 & 1 & 0 & 0 \\ -1 & 3 & 4 & 0 & 1 & 0 \\ 0 & 4 & 3 & 0 & 0 & 1 \end{array}\right]$$

Next, use row operations to transform $[A \,|\, I_3]$ into reduced row echelon form.

$$\left[\begin{array}{ccc|ccc} 1 & 1 & 0 & 1 & 0 & 0 \\ -1 & 3 & 4 & 0 & 1 & 0 \\ 0 & 4 & 3 & 0 & 0 & 1 \end{array}\right] \rightarrow \left[\begin{array}{ccc|ccc} 1 & 1 & 0 & 1 & 0 & 0 \\ 0 & 4 & 4 & 1 & 1 & 0 \\ 0 & 4 & 3 & 0 & 0 & 1 \end{array}\right] \rightarrow \left[\begin{array}{ccc|ccc} 1 & 1 & 0 & 1 & 0 & 0 \\ 0 & 1 & 1 & \frac{1}{4} & \frac{1}{4} & 0 \\ 0 & 4 & 3 & 0 & 0 & 1 \end{array}\right]$$

$$\uparrow R_2 = r_1 + r_2 \qquad\qquad \uparrow R_2 = \frac{1}{4}r_2$$

$$\rightarrow \left[\begin{array}{ccc|ccc} 1 & 0 & -1 & \frac{3}{4} & -\frac{1}{4} & 0 \\ 0 & 1 & 1 & \frac{1}{4} & \frac{1}{4} & 0 \\ 0 & 0 & -1 & -1 & -1 & 1 \end{array}\right] \rightarrow \left[\begin{array}{ccc|ccc} 1 & 0 & -1 & \frac{3}{4} & -\frac{1}{4} & 0 \\ 0 & 1 & 1 & \frac{1}{4} & \frac{1}{4} & 0 \\ 0 & 0 & 1 & 1 & 1 & -1 \end{array}\right] \rightarrow \left[\begin{array}{ccc|ccc} 1 & 0 & 0 & \frac{7}{4} & \frac{3}{4} & -1 \\ 0 & 1 & 0 & -\frac{3}{4} & -\frac{3}{4} & 1 \\ 0 & 0 & 1 & 1 & 1 & -1 \end{array}\right]$$

$$\uparrow \begin{array}{l} R_1 = -1r_2 + r_1 \\ R_3 = -4r_2 + r_3 \end{array} \qquad \uparrow R_3 = -1r_3 \qquad\qquad \uparrow \begin{array}{l} R_1 = r_3 + r_1 \\ R_2 = -1r_3 + r_2 \end{array}$$

*For 2×2 matrices there is a simple formula that can be used. See Problem 89.

The matrix $[A \,|\, I_3]$ is now in reduced row echelon form, and the identity matrix I_3 is on the left of the vertical bar. The inverse of A is

$$A^{-1} = \begin{bmatrix} \dfrac{7}{4} & \dfrac{3}{4} & -1 \\[2mm] -\dfrac{3}{4} & -\dfrac{3}{4} & 1 \\[2mm] 1 & 1 & -1 \end{bmatrix}$$

You should verify that this is the correct inverse by showing that

$$AA^{-1} = A^{-1}A = I_3 \qquad \bullet$$

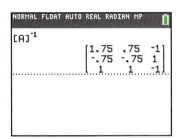

Figure 8 Inverse matrix

✓ **Check:** Enter the matrix A into a graphing utility. Figure 8 shows A^{-1}.

━━━ **Now Work** PROBLEM 33

If transforming the matrix $[A \,|\, I_n]$ into reduced row echelon form does not result in the identity matrix I_n to the left of the vertical bar, then A is singular and has no inverse.

EXAMPLE 14 **Showing That a Matrix Has No Inverse**

Show that the matrix $A = \begin{bmatrix} 4 & 6 \\ 2 & 3 \end{bmatrix}$ has no inverse.

Solution Begin by writing the matrix $[A \,|\, I_2]$.

$$[A \,|\, I_2] = \begin{bmatrix} 4 & 6 & | & 1 & 0 \\ 2 & 3 & | & 0 & 1 \end{bmatrix}$$

Then use row operations to transform $[A \,|\, I_2]$ into reduced row echelon form.

$$[A \,|\, I_2] = \begin{bmatrix} 4 & 6 & | & 1 & 0 \\ 2 & 3 & | & 0 & 1 \end{bmatrix} \rightarrow \begin{bmatrix} 1 & \dfrac{3}{2} & | & \dfrac{1}{4} & 0 \\[2mm] 2 & 3 & | & 0 & 1 \end{bmatrix} \rightarrow \begin{bmatrix} 1 & \dfrac{3}{2} & | & \dfrac{1}{4} & 0 \\[2mm] 0 & 0 & | & -\dfrac{1}{2} & 1 \end{bmatrix}$$

$$\uparrow \qquad\qquad\qquad \uparrow$$
$$R_1 = \dfrac{1}{4}r_1 \qquad\qquad R_2 = -2r_1 + r_2$$

The matrix $[A \,|\, I_2]$ is sufficiently reduced to see that the identity matrix cannot appear to the left of the vertical bar, so A is singular and has no inverse. $\qquad \bullet$

It can be shown that if the determinant of a matrix is 0, the matrix is singular. For example, the determinant of matrix A from Example 14 is

$$\begin{vmatrix} 4 & 6 \\ 2 & 3 \end{vmatrix} = 4 \cdot 3 - 6 \cdot 2 = 0$$

✓ **Check:** Enter the matrix A. Try to find its inverse. What happens?

━━━ **Now Work** PROBLEM 61

2 Decompose $\dfrac{P}{Q}$ Where Q Has Repeated Linear Factors

Case 2: Q has repeated linear factors.

If the polynomial Q has a repeated linear factor, say $(x - a)^n$, $n \geq 2$ an integer, then, in the partial fraction decomposition of $\dfrac{P}{Q}$, allow for the terms

$$\frac{A_1}{x - a} + \frac{A_2}{(x - a)^2} + \cdots + \frac{A_n}{(x - a)^n}$$

where the numbers $A_1, A_2, \ldots, A_n$ are to be determined.

EXAMPLE 2

Repeated Linear Factors

Find the partial fraction decomposition of $\dfrac{x + 2}{x^3 - 2x^2 + x}$.

Solution

First, factor the denominator,

$$x^3 - 2x^2 + x = x(x^2 - 2x + 1) = x(x - 1)^2$$

and notice that the denominator has the nonrepeated linear factor x and the repeated linear factor $(x - 1)^2$. By Case 1, the term $\dfrac{A}{x}$ must be in the decomposition; and by Case 2, the terms $\dfrac{B}{x - 1} + \dfrac{C}{(x - 1)^2}$ must be in the decomposition.

Now write

$$\frac{x + 2}{x^3 - 2x^2 + x} = \frac{A}{x} + \frac{B}{x - 1} + \frac{C}{(x - 1)^2} \qquad \textbf{(4)}$$

Again, clear fractions by multiplying each side by $x^3 - 2x^2 + x = x(x - 1)^2$. The result is the identity

$$x + 2 = A(x - 1)^2 + Bx(x - 1) + Cx \qquad \textbf{(5)}$$

Let $x = 0$ in this expression and the terms containing B and C drop out, leaving $2 = A(-1)^2$, or $A = 2$. Similarly, let $x = 1$, and the terms containing A and B drop out, leaving $3 = C$. Then equation (5) becomes

$$x + 2 = 2(x - 1)^2 + Bx(x - 1) + 3x$$

Let $x = 2$ (any choice other than 0 or 1 will work as well). The result is

$$4 = 2(1)^2 + B(2)(1) + 3(2)$$
$$4 = 2 + 2B + 6$$
$$2B = -4$$
$$B = -2$$

Therefore, $A = 2$, $B = -2$, and $C = 3$.

From equation (4), the partial fraction decomposition is

$$\frac{x + 2}{x^3 - 2x^2 + x} = \frac{2}{x} + \frac{-2}{x - 1} + \frac{3}{(x - 1)^2}$$

| EXAMPLE 3 | **Repeated Linear Factors** |

Find the partial fraction decomposition of $\dfrac{x^3 - 8}{x^2(x - 1)^3}$.

Solution The denominator contains the repeated linear factors x^2 and $(x - 1)^3$. The partial fraction decomposition takes the form

$$\frac{x^3 - 8}{x^2(x - 1)^3} = \frac{A}{x} + \frac{B}{x^2} + \frac{C}{x - 1} + \frac{D}{(x - 1)^2} + \frac{E}{(x - 1)^3} \qquad (6)$$

As before, clear fractions and obtain the identity

$$x^3 - 8 = Ax(x - 1)^3 + B(x - 1)^3 + Cx^2(x - 1)^2 + Dx^2(x - 1) + Ex^2 \quad (7)$$

Let $x = 0$. (Do you see why this choice was made?) Then

$$-8 = B(-1)$$
$$B = 8$$

Let $x = 1$ in equation (7). Then

$$-7 = E$$

Use $B = 8$ and $E = -7$ in equation (7), and collect like terms.

$$x^3 - 8 = Ax(x - 1)^3 + 8(x - 1)^3$$
$$+ Cx^2(x - 1)^2 + Dx^2(x - 1) - 7x^2$$
$$x^3 - 8 - 8(x^3 - 3x^2 + 3x - 1) + 7x^2 = Ax(x - 1)^3$$
$$+ Cx^2(x - 1)^2 + Dx^2(x - 1)$$
$$-7x^3 + 31x^2 - 24x = x(x - 1)[A(x - 1)^2 + Cx(x - 1) + Dx]$$
$$x(x - 1)(-7x + 24) = x(x - 1)[A(x - 1)^2 + Cx(x - 1) + Dx]$$
$$-7x + 24 = A(x - 1)^2 + Cx(x - 1) + Dx \qquad (8)$$

Now work with equation (8). Let $x = 0$. Then

$$24 = A$$

Let $x = 1$ in equation (8). Then

$$17 = D$$

Use $A = 24$ and $D = 17$ in equation (8).

$$-7x + 24 = 24(x - 1)^2 + Cx(x - 1) + 17x$$

Let $x = 2$ and simplify.

$$-14 + 24 = 24 + C(2) + 34$$
$$-48 = 2C$$
$$-24 = C$$

The numbers $A, B, C, D,$ and E are all now known. So, from equation (6),

$$\frac{x^3 - 8}{x^2(x - 1)^3} = \frac{24}{x} + \frac{8}{x^2} + \frac{-24}{x - 1} + \frac{17}{(x - 1)^2} + \frac{-7}{(x - 1)^3}$$

●

Now Work Example 3 by solving the system of five equations containing five variables that the expansion of equation (7) leads to.

Now Work PROBLEM 19

The final two cases involve irreducible quadratic factors. A quadratic factor is irreducible if it cannot be factored into linear factors with real coefficients. A quadratic expression $ax^2 + bx + c$ is irreducible whenever $b^2 - 4ac < 0$. For example, $x^2 + x + 1$ and $x^2 + 4$ are irreducible.

3 Decompose $\dfrac{P}{Q}$ Where Q Has a Nonrepeated Irreducible Quadratic Factor

Case 3: Q contains a nonrepeated irreducible quadratic factor.

If Q contains a nonrepeated irreducible quadratic factor of the form $ax^2 + bx + c$, then, in the partial fraction decomposition of $\dfrac{P}{Q}$, allow for the term

$$\frac{Ax + B}{ax^2 + bx + c}$$

where the numbers A and B are to be determined.

EXAMPLE 4

Nonrepeated Irreducible Quadratic Factor

Find the partial fraction decomposition of $\dfrac{3x - 5}{x^3 - 1}$.

Solution Factor the denominator,

$$x^3 - 1 = (x - 1)(x^2 + x + 1)$$

Notice that it has a nonrepeated linear factor $x - 1$ and a nonrepeated irreducible quadratic factor $x^2 + x + 1$. Allow for the term $\dfrac{A}{x - 1}$ by Case 1, and allow for the term $\dfrac{Bx + C}{x^2 + x + 1}$ by Case 3. Then

$$\frac{3x - 5}{x^3 - 1} = \frac{A}{x - 1} + \frac{Bx + C}{x^2 + x + 1} \qquad \textbf{(9)}$$

Multiply each side of equation (9) by $x^3 - 1 = (x - 1)(x^2 + x + 1)$ to obtain the identity

$$3x - 5 = A(x^2 + x + 1) + (Bx + C)(x - 1) \qquad \textbf{(10)}$$

Expand the identity in (10) to obtain

$$3x - 5 = (A + B)x^2 + (A - B + C)x + (A - C)$$

This identity leads to the system of equations

$$\begin{cases} A + B & = & 0 & \text{(1)} \\ A - B + C & = & 3 & \text{(2)} \\ A \quad\;\; - C & = & -5 & \text{(3)} \end{cases}$$

The solution of this system is $A = -\dfrac{2}{3}$, $B = \dfrac{2}{3}$, $C = \dfrac{13}{3}$. Then, from equation (9),

$$\frac{3x - 5}{x^3 - 1} = \frac{-\dfrac{2}{3}}{x - 1} + \frac{\dfrac{2}{3}x + \dfrac{13}{3}}{x^2 + x + 1}$$

➤ **Now Work** Example **4** using equation **(10)** and assigning values to x.

➤ **Now Work** PROBLEM **21**

4 Decompose $\dfrac{P}{Q}$ Where Q Has a Repeated Irreducible Quadratic Factor

Case 4: **Q contains a repeated irreducible quadratic factor.**

If the polynomial Q contains a repeated irreducible quadratic factor $(ax^2 + bx + c)^n$, $n \geq 2$, n an integer, then, in the partial fraction decomposition of $\dfrac{P}{Q}$, allow for the terms

$$\frac{A_1 x + B_1}{ax^2 + bx + c} + \frac{A_2 x + B_2}{(ax^2 + bx + c)^2} + \cdots + \frac{A_n x + B_n}{(ax^2 + bx + c)^n}$$

where the numbers $A_1, B_1, A_2, B_2, \ldots, A_n, B_n$ are to be determined.

EXAMPLE 5

Repeated Irreducible Quadratic Factor

Find the partial fraction decomposition of $\dfrac{x^3 + x^2}{(x^2 + 4)^2}$.

Solution

The denominator contains the repeated irreducible quadratic factor $(x^2 + 4)^2$, so write

$$\frac{x^3 + x^2}{(x^2 + 4)^2} = \frac{Ax + B}{x^2 + 4} + \frac{Cx + D}{(x^2 + 4)^2} \qquad \textbf{(11)}$$

Clear fractions to obtain

$$x^3 + x^2 = (Ax + B)(x^2 + 4) + Cx + D$$

Collecting like terms yields the identity

$$x^3 + x^2 = Ax^3 + Bx^2 + (4A + C)x + 4B + D$$

Equating coefficients results in the system

$$\begin{cases} A = 1 \\ B = 1 \\ 4A + C = 0 \\ 4B + D = 0 \end{cases}$$

The solution is $A = 1$, $B = 1$, $C = -4$, $D = -4$. From equation (11),

$$\frac{x^3 + x^2}{(x^2 + 4)^2} = \frac{x + 1}{x^2 + 4} + \frac{-4x - 4}{(x^2 + 4)^2}$$

➤ **Now Work** PROBLEM **35**

12.5 Assess Your Understanding

'Are You Prepared?' *Answers are given at the end of these exercises. If you get a wrong answer, read the pages listed in* red.

1. **True or False** The equation $(x - 1)^2 - 1 = x(x - 2)$ is an example of an identity. (p. 82)

2. **True or False** The rational expression $\dfrac{5x^2 - 1}{x^3 + 1}$ is proper. (p. 348)

3. Factor completely: $3x^4 + 6x^3 + 3x^2$ (pp. 49–55)

4. **True or False** Every polynomial with real numbers as coefficients can be factored into products of linear and/or irreducible quadratic factors. (p. 392)

Skill Building

In Problems 5–12, tell whether the given rational expression is proper or improper. If improper, rewrite it as the sum of a polynomial and a proper rational expression.

5. $\dfrac{x}{x^2 - 1}$

6. $\dfrac{5x + 2}{x^3 - 1}$

7. $\dfrac{x^2 + 5}{x^2 - 4}$

8. $\dfrac{3x^2 - 2}{x^2 - 1}$

9. $\dfrac{5x^3 + 2x - 1}{x^2 - 4}$

10. $\dfrac{3x^4 + x^2 - 2}{x^3 + 8}$

11. $\dfrac{x(x - 1)}{(x + 4)(x - 3)}$

12. $\dfrac{2x(x^2 + 4)}{x^2 + 1}$

In Problems 13–46, find the partial fraction decomposition of each rational expression.

13. $\dfrac{4}{x(x - 1)}$

14. $\dfrac{3x}{(x + 2)(x - 1)}$

15. $\dfrac{1}{x(x^2 + 1)}$

16. $\dfrac{1}{(x + 1)(x^2 + 4)}$

17. $\dfrac{x}{(x - 1)(x - 2)}$

18. $\dfrac{3x}{(x + 2)(x - 4)}$

19. $\dfrac{x^2}{(x - 1)^2(x + 1)}$

20. $\dfrac{x + 1}{x^2(x - 2)}$

21. $\dfrac{1}{x^3 - 8}$

22. $\dfrac{2x + 4}{x^3 - 1}$

23. $\dfrac{x^2}{(x - 1)^2(x + 1)^2}$

24. $\dfrac{x + 1}{x^2(x - 2)^2}$

25. $\dfrac{x - 3}{(x + 2)(x + 1)^2}$

26. $\dfrac{x^2 + x}{(x + 2)(x - 1)^2}$

27. $\dfrac{x + 4}{x^2(x^2 + 4)}$

28. $\dfrac{10x^2 + 2x}{(x - 1)^2(x^2 + 2)}$

29. $\dfrac{x^2 + 2x + 3}{(x + 1)(x^2 + 2x + 4)}$

30. $\dfrac{x^2 - 11x - 18}{x(x^2 + 3x + 3)}$

31. $\dfrac{x}{(3x - 2)(2x + 1)}$

32. $\dfrac{1}{(2x + 3)(4x - 1)}$

33. $\dfrac{x}{x^2 + 2x - 3}$

34. $\dfrac{x^2 - x - 8}{(x + 1)(x^2 + 5x + 6)}$

35. $\dfrac{x^2 + 2x + 3}{(x^2 + 4)^2}$

36. $\dfrac{x^3 + 1}{(x^2 + 16)^2}$

37. $\dfrac{7x + 3}{x^3 - 2x^2 - 3x}$

38. $\dfrac{x^3 + 1}{x^5 - x^4}$

39. $\dfrac{x^2}{x^3 - 4x^2 + 5x - 2}$

40. $\dfrac{x^2 + 1}{x^3 + x^2 - 5x + 3}$

41. $\dfrac{x^3}{(x^2 + 16)^3}$

42. $\dfrac{x^2}{(x^2 + 4)^3}$

43. $\dfrac{4}{2x^2 - 5x - 3}$

44. $\dfrac{4x}{2x^2 + 3x - 2}$

45. $\dfrac{2x + 3}{x^4 - 9x^2}$

46. $\dfrac{x^2 + 9}{x^4 - 2x^2 - 8}$

Mixed Practice

In Problems 47–54, use the division algorithm to rewrite each improper rational expression as the sum of a polynomial and a proper rational expression. Find the partial fraction decomposition of the proper rational expression. Finally, express the improper rational expression as the sum of a polynomial and the partial fraction decomposition.

47. $\dfrac{x^3 + x^2 - 3}{x^2 + 3x - 4}$

48. $\dfrac{x^3 - 3x^2 + 1}{x^2 + 5x + 6}$

49. $\dfrac{x^3}{x^2 + 1}$

50. $\dfrac{x^3 + x}{x^2 + 4}$

51. $\dfrac{x^4 - 5x^2 + x - 4}{x^2 + 4x + 4}$

52. $\dfrac{x^4 + x^3 - x + 2}{x^2 - 2x + 1}$

53. $\dfrac{x^5 + x^4 - x^2 + 2}{x^4 - 2x^2 + 1}$

54. $\dfrac{x^5 - x^3 + x^2 + 1}{x^4 + 6x^2 + 9}$

55. Credit Card Balance Nick has a credit card balance of $4200. If the credit card company charges 18% interest compound daily, and Nick does not make any payments on the account, how long will it take for his balance to double? Round to two decimal places.

56. Given $f(x) = x + 4$ and $g(x) = x^2 - 3x$, find $(g \circ f)(-3)$.

57. Find the exact value of $\sec 52° \cos 308°$.

58. Plot the point given by the polar coordinates $\left(-1, \dfrac{5\pi}{4}\right)$ and find its rectangular coordinates.

'Are You Prepared?' Answers

1. True **2.** True **3.** $3x^2(x+1)^2$ **4.** True

12.6 Systems of Nonlinear Equations

PREPARING FOR THIS SECTION *Before getting started, review the following:*

- Lines (Section 2.3, pp. 167–177)
- Circles (Section 2.4, pp. 182–185)
- Parabolas (Section 11.2, pp. 780–784)
- Ellipses (Section 11.3, pp. 789–795)
- Hyperbolas (Section 11.4, pp. 799–807)

Now Work the 'Are You Prepared?' problems on page 914.

OBJECTIVES 1 Solve a System of Nonlinear Equations Using Substitution (p. 909)
2 Solve a System of Nonlinear Equations Using Elimination (p. 910)

1 Solve a System of Nonlinear Equations Using Substitution

In Section 12.1, we observed that the solution to a system of linear equations could be found geometrically by determining the point(s) of intersection (if any) of the equations in the system. Similarly, in solving systems of nonlinear equations, the solution(s) also represent(s) the point(s) of intersection (if any) of the graphs of the equations.

There is no general method for solving a system of nonlinear equations. Sometimes substitution is best; other times elimination is best; and sometimes neither of these methods works. Experience and a certain degree of imagination are your allies here.

Before we begin, two comments are in order.

- If the system contains two variables and if the equations in the system are easy to graph, then graph them. By graphing each equation in the system, you can get an idea of how many solutions a system has and approximately where they are located.

- Extraneous solutions can creep in when solving nonlinear systems, so it is imperative to check all apparent solutions.

EXAMPLE 1 **Solving a System of Nonlinear Equations Using Substitution**

Solve the following system of equations:

$$\begin{cases} 3x - y = -2 & (1) \\ 2x^2 - y = 0 & (2) \end{cases}$$

Solution First, notice that the system contains two variables and that we know how to graph each equation. Equation (1) is the line $y = 3x + 2$, and equation (2) is the parabola $y = 2x^2$. See Figure 9 on the next page. The system apparently has two solutions.

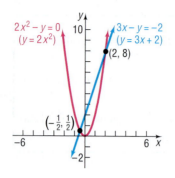

Figure 9

To use substitution to solve the system, we choose to solve equation (1) for y.

$$3x - y = -2 \qquad \text{Equation (1)}$$
$$y = 3x + 2$$

Substitute this expression for y in equation (2). The result is an equation containing just the variable x, which we can then solve.

$$2x^2 - y = 0 \qquad \text{Equation (2)}$$
$$2x^2 - (3x + 2) = 0 \qquad \text{Substitute } 3x + 2 \text{ for } y.$$
$$2x^2 - 3x - 2 = 0 \qquad \text{Remove parentheses.}$$
$$(2x + 1)(x - 2) = 0 \qquad \text{Factor.}$$
$$2x + 1 = 0 \quad \text{or} \quad x - 2 = 0 \qquad \text{Apply the Zero-Product Property.}$$
$$x = -\frac{1}{2} \quad \text{or} \quad x = 2$$

Use these values for x in $y = 3x + 2$ to find

$$y = 3\left(-\frac{1}{2}\right) + 2 = \frac{1}{2} \quad \text{or} \quad y = 3(2) + 2 = 8$$

The apparent solutions are $x = -\frac{1}{2}, y = \frac{1}{2}$ and $x = 2, y = 8$.

✓**Check:** For $x = -\frac{1}{2}, y = \frac{1}{2}$,

$$\begin{cases} 3\left(-\frac{1}{2}\right) - \frac{1}{2} = -\frac{3}{2} - \frac{1}{2} = -2 & (1) \\ 2\left(-\frac{1}{2}\right)^2 - \frac{1}{2} = 2\left(\frac{1}{4}\right) - \frac{1}{2} = 0 & (2) \end{cases}$$

For $x = 2, y = 8$,

$$\begin{cases} 3(2) - 8 = 6 - 8 = -2 & (1) \\ 2(2)^2 - 8 = 2(4) - 8 = 0 & (2) \end{cases}$$

Each solution checks. The graphs of the two equations intersect at the points $\left(-\frac{1}{2}, \frac{1}{2}\right)$ and $(2, 8)$, as shown in Figure 9. ●

✓**Check:** Graph $3x - y = -2$ ($Y_1 = 3x + 2$) and $2x^2 - y = 0$ ($Y_2 = 2x^2$), and compare what you see with Figure 9. Use INTERSECT (twice) to find the two points of intersection.

Now Work PROBLEM 15 USING SUBSTITUTION

2 Solve a System of Nonlinear Equations Using Elimination

EXAMPLE 2 **Solving a System of Nonlinear Equations Using Elimination**

Solve: $\begin{cases} x^2 + y^2 = 13 & (1) \quad \text{A circle} \\ x^2 - y = 7 & (2) \quad \text{A parabola} \end{cases}$

Solution First graph each equation, as shown in Figure 10. Based on the graph, we expect four solutions. Notice that subtracting equation (2) from equation (1) eliminates the variable x.

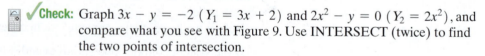

$$\begin{cases} x^2 + y^2 = 13 \\ x^2 - y = 7 \end{cases} \qquad \text{Subtract.}$$
$$y^2 + y = 6$$

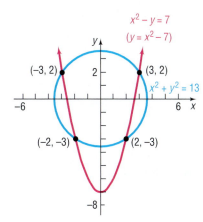

Figure 10

This quadratic equation in y can be solved by factoring.

$$y^2 + y - 6 = 0$$
$$(y + 3)(y - 2) = 0$$
$$y = -3 \quad \text{or} \quad y = 2$$

Use these values for y in equation (2) to find x.

If $y = 2$, then $x^2 = y + 7 = 9$, so $x = 3$ or -3.

If $y = -3$, then $x^2 = y + 7 = 4$, so $x = 2$ or -2.

There are four solutions: $x = 3, y = 2$; $x = -3, y = 2$; $x = 2, y = -3$; and $x = -2, y = -3$.

You should verify that, in fact, these four solutions also satisfy equation (1), so all four are solutions of the system. The four points, $(3, 2)$, $(-3, 2)$, $(2, -3)$, and $(-2, -3)$, are the points of intersection of the graphs. Look again at Figure 10.

✔ Check: Graph $x^2 + y^2 = 13$ and $x^2 - y = 7$. (Remember that to graph $x^2 + y^2 = 13$ requires two functions: $Y_1 = \sqrt{13 - x^2}$ and $Y_2 = -\sqrt{13 - x^2}$.) Use INTERSECT to find the points of intersection.

Now Work PROBLEM **13** USING ELIMINATION

EXAMPLE 3

Figure 11

Solving a System of Nonlinear Equations

Solve: $\begin{cases} x^2 - y^2 = 4 & (1) \quad \text{A hyperbola} \\ y = x^2 & (2) \quad \text{A parabola} \end{cases}$

Solution Either substitution or elimination can be used here. To use substitution, replace x^2 by y in equation (1).

$$x^2 - y^2 = 4 \quad \text{Equation (1)}$$
$$y - y^2 = 4 \quad y = x^2$$
$$y^2 - y + 4 = 0 \quad \text{Place in standard form.}$$

This is a quadratic equation. Its discriminant is $(-1)^2 - 4 \cdot 1 \cdot 4 = 1 - 16 = -15 < 0$. The equation has no real solutions, so the system is inconsistent. The graphs of these two equations do not intersect. See Figure 11.

EXAMPLE 4

Solving a System of Nonlinear Equations Using Elimination

Solve: $\begin{cases} x^2 + x + y^2 - 3y + 2 = 0 & (1) \\ x + 1 + \dfrac{y^2 - y}{x} = 0 & (2) \end{cases}$

Solution First, multiply equation (2) by x to eliminate the fraction. The result is an equivalent system because x cannot be 0. [Look at equation (2) to see why.]

$$\begin{cases} x^2 + x + y^2 - 3y + 2 = 0 & (1) \\ x^2 + x + y^2 - y = 0 & (2) \quad x \neq 0 \end{cases}$$

Now subtract equation (2) from equation (1) to eliminate x. The result is

$$-2y + 2 = 0$$
$$y = 1 \quad \text{Solve for } y.$$

To find x, back-substitute $y = 1$ in equation (1).

$$
\begin{aligned}
x^2 + x + y^2 - 3y + 2 &= 0 && \text{Equation (1)} \\
x^2 + x + 1 - 3 + 2 &= 0 && \text{Substitute 1 for } y. \\
x^2 + x &= 0 && \text{Simplify.} \\
x(x + 1) &= 0 && \text{Factor.} \\
x = 0 \quad \text{or} \quad x &= -1 && \text{Apply the Zero-Product Property.}
\end{aligned}
$$

Because x cannot be 0, the value $x = 0$ is extraneous, so discard it.

✓ **Check:** Check $x = -1$, $y = 1$:

$$
\begin{cases}
(-1)^2 + (-1) + 1^2 - 3(1) + 2 = 1 - 1 + 1 - 3 + 2 = 0 & (1) \\
-1 + 1 + \dfrac{1^2 - 1}{-1} = 0 + \dfrac{0}{-1} = 0 & (2)
\end{cases}
$$

The solution is $x = -1$, $y = 1$. The point of intersection of the graphs of the equations is $(-1, 1)$. ●

In Problem 55 you are asked to graph the equations given in Example 4. Be sure to show holes in the graph of equation (2) for $x = 0$.

✏ **Now Work** PROBLEMS 29 AND 49

EXAMPLE 5

Solving a System of Nonlinear Equations

Solve: $\begin{cases} 3xy - 2y^2 = -2 & (1) \\ 9x^2 + 4y^2 = 10 & (2) \end{cases}$

Solution Multiply equation (1) by 2, and add the result to equation (2), to eliminate the y^2 terms.

$$
\begin{cases}
6xy - 4y^2 = -4 & (1) \\
9x^2 + 4y^2 = 10 & (2)
\end{cases}
$$
$$
\begin{aligned}
9x^2 + 6xy &= 6 && \text{Add.} \\
3x^2 + 2xy &= 2 && \text{Divide each side by 3.}
\end{aligned}
$$

Since $x \neq 0$ (do you see why?), solve for y in this equation to get

$$
y = \frac{2 - 3x^2}{2x} \qquad x \neq 0
$$

Now substitute for y in equation (2) of the system.

$$
\begin{aligned}
9x^2 + 4y^2 &= 10 && \text{Equation (2)} \\
9x^2 + 4\left(\frac{2 - 3x^2}{2x}\right)^2 &= 10 && \text{Substitute } y = \frac{2 - 3x^2}{2x}. \\
9x^2 + \frac{4 - 12x^2 + 9x^4}{x^2} &= 10 && \text{Expand and simplify.} \\
9x^4 + 4 - 12x^2 + 9x^4 &= 10x^2 && \text{Multiply both sides by } x^2. \\
18x^4 - 22x^2 + 4 &= 0 && \text{Subtract } 10x^2 \text{ from both sides.} \\
9x^4 - 11x^2 + 2 &= 0 && \text{Divide both sides by 2.}
\end{aligned}
$$

This quadratic equation (in x^2) can be factored:

$$(9x^2 - 2)(x^2 - 1) = 0$$

$$9x^2 - 2 = 0 \quad\quad \text{or} \quad x^2 - 1 = 0$$

$$x^2 = \frac{2}{9} \quad\quad \text{or} \quad\quad x^2 = 1$$

$$x = \pm\sqrt{\frac{2}{9}} = \pm\frac{\sqrt{2}}{3} \quad \text{or} \quad\quad x = \pm 1$$

To find y, use the fact that $y = \dfrac{2 - 3x^2}{2x}$.

$$\text{If } x = \frac{\sqrt{2}}{3}: \quad y = \frac{2 - 3x^2}{2x} = \frac{2 - \dfrac{2}{3}}{2\left(\dfrac{\sqrt{2}}{3}\right)} = \frac{4}{2\sqrt{2}} = \sqrt{2}$$

$$\text{If } x = -\frac{\sqrt{2}}{3}: \quad y = \frac{2 - 3x^2}{2x} = \frac{2 - \dfrac{2}{3}}{2\left(-\dfrac{\sqrt{2}}{3}\right)} = \frac{4}{-2\sqrt{2}} = -\sqrt{2}$$

$$\text{If } x = 1: \quad y = \frac{2 - 3x^2}{2x} = \frac{2 - 3(1)^2}{2} = -\frac{1}{2}$$

$$\text{If } x = -1: \quad y = \frac{2 - 3x^2}{2x} = \frac{2 - 3(-1)^2}{-2} = \frac{1}{2}$$

The system has four solutions: $\left(\dfrac{\sqrt{2}}{3}, \sqrt{2}\right), \left(-\dfrac{\sqrt{2}}{3}, -\sqrt{2}\right), \left(1, -\dfrac{1}{2}\right), \left(-1, \dfrac{1}{2}\right)$.
Check them for yourself. ●

Now Work PROBLEM **47**

The next example illustrates an imaginative solution to a system of nonlinear equations.

| EXAMPLE 6 | **Running a Long-Distance Race** |

In a 50-mile race, the winner crosses the finish line 1 mile ahead of the second-place runner and 4 miles ahead of the third-place runner. Assuming that each runner maintains a constant speed throughout the race, by how many miles does the second-place runner beat the third-place runner?

3 miles   1 mile

Solution   Let v_1, v_2, and v_3 denote the speeds of the first-, second-, and third-place runners, respectively. Let t_1 and t_2 denote the times (in hours) required for the first-place runner and the second-place runner to finish the race. Then the following system of equations results:

$$\begin{cases} 50 = v_1 t_1 & (1) \quad \textit{First-place runner goes 50 miles in } t_1 \textit{ hours.} \\ 49 = v_2 t_1 & (2) \quad \textit{Second-place runner goes 49 miles in } t_1 \textit{ hours.} \\ 46 = v_3 t_1 & (3) \quad \textit{Third-place runner goes 46 miles in } t_1 \textit{ hours.} \\ 50 = v_2 t_2 & (4) \quad \textit{Second-place runner goes 50 miles in } t_2 \textit{ hours.} \end{cases}$$

We seek the distance d of the third-place runner from the finish at time t_2. At time t_2, the third-place runner has gone a distance of $v_3 t_2$ miles, so the distance d remaining is $50 - v_3 t_2$. Now

$$d = 50 - v_3 t_2$$

$$= 50 - v_3 \left(t_1 \cdot \frac{t_2}{t_1} \right)$$

$$= 50 - (v_3 t_1) \cdot \frac{t_2}{t_1}$$

$$= 50 - 46 \cdot \frac{\dfrac{50}{v_2}}{\dfrac{50}{v_1}} \qquad \left\{ \begin{array}{l} \text{From (3), } v_3 t_1 = 46 \\ \text{From (4), } t_2 = \dfrac{50}{v_2} \\ \text{From (1), } t_1 = \dfrac{50}{v_1} \end{array} \right.$$

$$= 50 - 46 \cdot \frac{v_1}{v_2}$$

$$= 50 - 46 \cdot \frac{50}{49} \qquad \text{From the quotient of (1) and (2).}$$

$$\approx 3.06 \text{ miles}$$

Historical Feature

I n the beginning of this section, it was stated that imagination and experience are important in solving systems of nonlinear equations. Indeed, these kinds of problems lead into some of the deepest and most difficult parts of modern mathematics. Look again at the graphs in Examples 1 and 2 of this section (Figures 9 and 10). Example 1 has two solutions, and Example 2 has four solutions. We might conjecture that the number of solutions is equal to the product of the degrees of the equations involved. This conjecture was indeed made by Étienne Bézout (1730–1783), but working out the details took about 150 years. It turns out that arriving at the correct number of intersections requires counting not only the complex number intersections, but also those intersections that, in a certain sense, lie at infinity. For example, a parabola and a line lying on the axis of the parabola intersect at the vertex and at infinity. This topic is part of the study of algebraic geometry.

Historical Problem

A papyrus dating back to 1950 BC contains the following problem: "A given surface area of 100 units of area shall be represented as the sum of two squares whose sides are to each other as 1 is to $\frac{3}{4}$."

Solve for the sides by solving the system of equations

$$\begin{cases} x^2 + y^2 = 100 \\ \quad x = \dfrac{3}{4} y \end{cases}$$

12.6 Assess Your Understanding

'Are You Prepared?' *Answers are given at the end of these exercises. If you get a wrong answer, read the pages listed in* red.

1. Graph the equation: $y = 3x + 2$ (pp. 167–177)
2. Graph the equation: $y + 4 = x^2$ (pp. 780–784)
3. Graph the equation: $y^2 = x^2 - 1$ (pp. 799–807)
4. Graph the equation: $x^2 + 4y^2 = 4$ (pp. 789–795)

Skill Building

In Problems 5–24, graph each equation of the system. Then solve the system to find the points of intersection.

5. $\begin{cases} y = x^2 + 1 \\ y = x + 1 \end{cases}$

6. $\begin{cases} y = x^2 + 1 \\ y = 4x + 1 \end{cases}$

7. $\begin{cases} y = \sqrt{36 - x^2} \\ y = 8 - x \end{cases}$

8. $\begin{cases} y = \sqrt{4 - x^2} \\ y = 2x + 4 \end{cases}$

9. $\begin{cases} y = \sqrt{x} \\ y = 2 - x \end{cases}$

10. $\begin{cases} y = \sqrt{x} \\ y = 6 - x \end{cases}$

11. $\begin{cases} x = 2y \\ x = y^2 - 2y \end{cases}$

12. $\begin{cases} y = x - 1 \\ y = x^2 - 6x + 9 \end{cases}$

13. $\begin{cases} x^2 + y^2 = 4 \\ x^2 + 2x + y^2 = 0 \end{cases}$

14. $\begin{cases} x^2 + y^2 = 8 \\ x^2 + y^2 + 4y = 0 \end{cases}$

15. $\begin{cases} y = 3x - 5 \\ x^2 + y^2 = 5 \end{cases}$

16. $\begin{cases} x^2 + y^2 = 10 \\ y = x + 2 \end{cases}$

17. $\begin{cases} x^2 + y^2 = 4 \\ y^2 - x = 4 \end{cases}$ **18.** $\begin{cases} x^2 + y^2 = 16 \\ x^2 - 2y = 8 \end{cases}$ **19.** $\begin{cases} xy = 4 \\ x^2 + y^2 = 8 \end{cases}$ **20.** $\begin{cases} x^2 = y \\ xy = 1 \end{cases}$

21. $\begin{cases} x^2 + y^2 = 4 \\ y = x^2 - 9 \end{cases}$ **22.** $\begin{cases} xy = 1 \\ y = 2x + 1 \end{cases}$ **23.** $\begin{cases} y = x^2 - 4 \\ y = 6x - 13 \end{cases}$ **24.** $\begin{cases} x^2 + y^2 = 10 \\ xy = 3 \end{cases}$

In Problems 25–54, solve each system. Use any method you wish.

25. $\begin{cases} 2x^2 + y^2 = 18 \\ xy = 4 \end{cases}$ **26.** $\begin{cases} x^2 - y^2 = 21 \\ x + y = 7 \end{cases}$ **27.** $\begin{cases} y = 2x + 1 \\ 2x^2 + y^2 = 1 \end{cases}$

28. $\begin{cases} x^2 - 4y^2 = 16 \\ 2y - x = 2 \end{cases}$ **29.** $\begin{cases} x + y + 1 = 0 \\ x^2 + y^2 + 6y - x = -5 \end{cases}$ **30.** $\begin{cases} 2x^2 - xy + y^2 = 8 \\ xy = 4 \end{cases}$

31. $\begin{cases} 4x^2 - 3xy + 9y^2 = 15 \\ 2x + 3y = 5 \end{cases}$ **32.** $\begin{cases} 2y^2 - 3xy + 6y + 2x + 4 = 0 \\ 2x - 3y + 4 = 0 \end{cases}$ **33.** $\begin{cases} x^2 - 4y^2 + 7 = 0 \\ 3x^2 + y^2 = 31 \end{cases}$

34. $\begin{cases} 3x^2 - 2y^2 + 5 = 0 \\ 2x^2 - y^2 + 2 = 0 \end{cases}$ **35.** $\begin{cases} 7x^2 - 3y^2 + 5 = 0 \\ 3x^2 + 5y^2 = 12 \end{cases}$ **36.** $\begin{cases} x^2 - 3y^2 + 1 = 0 \\ 2x^2 - 7y^2 + 5 = 0 \end{cases}$

37. $\begin{cases} x^2 + 2xy = 10 \\ 3x^2 - xy = 2 \end{cases}$ **38.** $\begin{cases} 5xy + 13y^2 + 36 = 0 \\ xy + 7y^2 = 6 \end{cases}$ **39.** $\begin{cases} 2x^2 + y^2 = 2 \\ x^2 - 2y^2 + 8 = 0 \end{cases}$

40. $\begin{cases} y^2 - x^2 + 4 = 0 \\ 2x^2 + 3y^2 = 6 \end{cases}$ **41.** $\begin{cases} x^2 + 2y^2 = 16 \\ 4x^2 - y^2 = 24 \end{cases}$ **42.** $\begin{cases} 4x^2 + 3y^2 = 4 \\ 2x^2 - 6y^2 = -3 \end{cases}$

43. $\begin{cases} \dfrac{5}{x^2} - \dfrac{2}{y^2} + 3 = 0 \\ \dfrac{3}{x^2} + \dfrac{1}{y^2} = 7 \end{cases}$ **44.** $\begin{cases} \dfrac{2}{x^2} - \dfrac{3}{y^2} + 1 = 0 \\ \dfrac{6}{x^2} - \dfrac{7}{y^2} + 2 = 0 \end{cases}$ **45.** $\begin{cases} \dfrac{1}{x^4} + \dfrac{6}{y^4} = 6 \\ \dfrac{2}{x^4} - \dfrac{2}{y^4} = 19 \end{cases}$

46. $\begin{cases} \dfrac{1}{x^4} - \dfrac{1}{y^4} = 1 \\ \dfrac{1}{x^4} + \dfrac{1}{y^4} = 4 \end{cases}$ **47.** $\begin{cases} x^2 - 3xy + 2y^2 = 0 \\ x^2 + xy = 6 \end{cases}$ **48.** $\begin{cases} x^2 - xy - 2y^2 = 0 \\ xy + x + 6 = 0 \end{cases}$

49. $\begin{cases} y^2 + y + x^2 - x - 2 = 0 \\ y + 1 + \dfrac{x - 2}{y} = 0 \end{cases}$ **50.** $\begin{cases} x^3 - 2x^2 + y^2 + 3y - 4 = 0 \\ x - 2 + \dfrac{y^2 - y}{x^2} = 0 \end{cases}$ **51.** $\begin{cases} \log_x y = 3 \\ \log_x(4y) = 5 \end{cases}$

52. $\begin{cases} \log_x(2y) = 3 \\ \log_x(4y) = 2 \end{cases}$ **53.** $\begin{cases} \ln x = 4 \ln y \\ \log_3 x = 2 + 2\log_3 y \end{cases}$ **54.** $\begin{cases} \ln x = 5 \ln y \\ \log_2 x = 3 + 2\log_2 y \end{cases}$

55. Graph the equations given in Example 4. **56.** Graph the equations given in Problem 49.

In Problems 57–64, use a graphing utility to solve each system of equations. Express the solution(s) rounded to two decimal places.

57. $\begin{cases} y = x^{2/3} \\ y = e^{-x} \end{cases}$ **58.** $\begin{cases} y = x^{3/2} \\ y = e^{-x} \end{cases}$ **59.** $\begin{cases} x^2 + y^3 = 2 \\ x^3 y = 4 \end{cases}$ **60.** $\begin{cases} x^3 + y^2 = 2 \\ x^2 y = 4 \end{cases}$

61. $\begin{cases} x^4 + y^4 = 12 \\ xy^2 = 2 \end{cases}$ **62.** $\begin{cases} x^4 + y^4 = 6 \\ xy = 1 \end{cases}$ **63.** $\begin{cases} xy = 2 \\ y = \ln x \end{cases}$ **64.** $\begin{cases} x^2 + y^2 = 4 \\ y = \ln x \end{cases}$

Mixed Practice

In Problems 65–70, graph each equation and find the point(s) of intersection, if any.

65. The line $x + 2y = 0$ and the circle $(x - 1)^2 + (y - 1)^2 = 5$

66. The line $x + 2y + 6 = 0$ and the circle $(x + 1)^2 + (y + 1)^2 = 5$

67. The circle $(x - 1)^2 + (y + 2)^2 = 4$ and the parabola $y^2 + 4y - x + 1 = 0$

68. The circle $(x + 2)^2 + (y - 1)^2 = 4$ and the parabola $y^2 - 2y - x - 5 = 0$

69. $y = \dfrac{4}{x - 3}$ and the circle $x^2 - 6x + y^2 + 1 = 0$

70. $y = \dfrac{4}{x + 2}$ and the circle $x^2 + 4x + y^2 - 4 = 0$

| EXAMPLE 4 | **Graphing Linear Inequalities** |

Graph: (a) $y < 2$ (b) $y \geq 2x$

Solution

(a) Points on the horizontal line $y = 2$ are not part of the graph of the inequality, so the graph is shown as a dashed line. Since $(0, 0)$ satisfies the inequality, the graph consists of the half-plane below the line $y = 2$. See Figure 15.

(b) Points on the line $y = 2x$ are part of the graph of the inequality, so the graph is shown as a solid line. Use $(3, 0)$ as a test point. It does not satisfy the inequality $[0 < 2 \cdot 3]$. Points in the half-plane on the opposite side of $(3, 0)$ satisfy the inequality. See Figure 16.

COMMENT A graphing utility can be used to graph inequalities. To see how, read Section 6 in the Appendix. ■

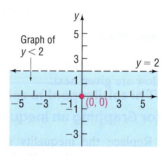

Figure 15 $y < 2$

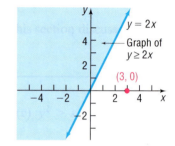

Figure 16 $y \geq 2x$

●

Now Work PROBLEM 13

2 Graph a System of Inequalities

The **graph of a system of inequalities** in two variables x and y is the set of all points (x, y) that simultaneously satisfy *each* inequality in the system. The graph of a system of inequalities can be obtained by graphing each inequality individually and then determining where, if at all, they intersect.

| EXAMPLE 5 | **Graphing a System of Linear Inequalities** |

Graph the system: $\begin{cases} x + y \geq 2 \\ 2x - y \leq 4 \end{cases}$

Solution

Begin by graphing the lines $x + y = 2$ and $2x - y = 4$ using a solid line since both inequalities are nonstrict. Use the test point $(0, 0)$ on each inequality. For example, $(0, 0)$ does not satisfy $x + y \geq 2$, so shade above the line $x + y = 2$. See Figure 17(a). Also, $(0, 0)$ does satisfy $2x - y \leq 4$, so shade above the line $2x - y = 4$. See Figure 17(b). The intersection of the shaded regions (in purple) gives the result presented in Figure 17(c).

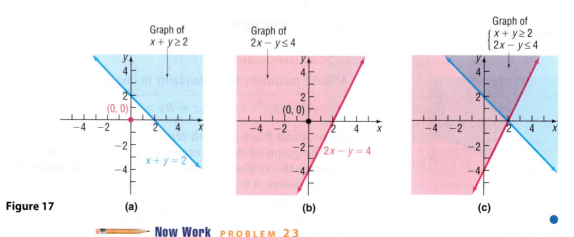

Figure 17   (a)   (b)   (c)

Now Work PROBLEM 23

●

EXAMPLE 6

Graphing a System of Linear Inequalities

Graph the system: $\begin{cases} x + y \le 2 \\ x + y \ge 0 \end{cases}$

Solution   See Figure 18. The overlapping purple-shaded region between the two boundary lines is the graph of the system.

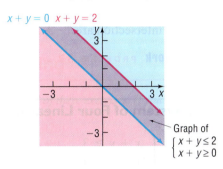

$x + y = 0$ $x + y = 2$

Graph of
$\begin{cases} x + y \le 2 \\ x + y \ge 0 \end{cases}$

Figure 18

•

Now Work PROBLEM 29

EXAMPLE 7

Graphing a System of Linear Inequalities

Graph the systems:

(a) $\begin{cases} 2x - y \ge 0 \\ 2x - y \ge 2 \end{cases}$

(b) $\begin{cases} x + 2y \le 2 \\ x + 2y \ge 6 \end{cases}$

Solution   (a) See Figure 19. The overlapping purple-shaded region is the graph of the system. Note that the graph of the system is identical to the graph of the single inequality $2x - y \ge 2$.

(b) See Figure 20. Here, because no overlapping region results, there are no points in the xy-plane that simultaneously satisfy each inequality. The system has no solution.

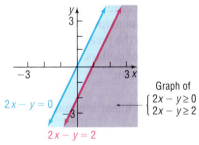

$2x - y = 0$

Graph of
$\begin{cases} 2x - y \ge 0 \\ 2x - y \ge 2 \end{cases}$

$2x - y = 2$

$x + 2y = 6$

$x + 2y = 2$

Figure 19   **Figure 20**

•

EXAMPLE 8

Graphing a System of Nonlinear Inequalities

Graph the region below the graph of $x + y = 2$ and above the graph of $y = x^2 - 4$ by graphing the system

$$\begin{cases} y \ge x^2 - 4 \\ x + y \le 2 \end{cases}$$

Label all points of intersection.

Solution   Figure 21 on the next page shows the graph of the region above the graph of the parabola $y = x^2 - 4$ and below the graph of the line $x + y = 2$. The points of intersection are found by solving the system of equations

$$\begin{cases} y = x^2 - 4 \\ x + y = 2 \end{cases}$$

discussion to problems containing only two variables, because we can solve such problems using graphing techniques.*

1 Set Up a Linear Programming Problem

Let's begin by returning to Example 10 of the previous section.

EXAMPLE 1

Financial Planning

A retired couple has up to $25,000 to invest. As their financial adviser, you recommend that they place at least $15,000 in Treasury bills yielding 2% and at most $5000 in corporate bonds yielding 3%. Develop a model that can be used to determine how much money they should place in each investment so that income is maximized.

Solution   The problem is typical of a *linear programming problem.* The problem requires that a certain linear expression, the income, be maximized. If I represents income, x the amount invested in Treasury bills at 2%, and y the amount invested in corporate bonds at 3%, then

$$I = 0.02x + 0.03y$$

Assume, as before, that I, x, and y are in thousands of dollars.

The linear expression $I = 0.02x + 0.03y$ is called the **objective function**. Further, the problem requires that the maximum income be achieved under certain conditions, or **constraints**, each of which is a linear inequality involving the variables. (See Example 10 in Section 12.7.) The linear programming problem may be modeled as

$$\text{Maximize} \quad I = 0.02x + 0.03y$$

subject to the conditions that

$$\begin{cases} x \geq \ 0 \\ y \geq \ 0 \\ x + y \leq 25 \\ x \geq 15 \\ y \leq 5 \end{cases}$$

In general, every linear programming problem has two components:

1. A linear objective function that is to be maximized or minimized
2. A collection of linear inequalities that must be satisfied simultaneously

DEFINITION

A **linear programming problem** in two variables x and y consists of maximizing (or minimizing) a linear objective function

$$z = Ax + By \qquad A \text{ and } B \text{ are real numbers, not both } 0$$

subject to certain conditions, or constraints, expressible as linear inequalities in x and y.

2 Solve a Linear Programming Problem

To maximize (or minimize) the quantity $z = Ax + By$, we need to identify points (x, y) that make the expression for z the largest (or smallest) possible. But not all points (x, y) are eligible; only those that also satisfy each linear inequality (constraint) can be used. Each point (x, y) that satisfies the system of linear inequalities (the constraints) is a **feasible point**. Linear programming problems seek the feasible point(s) that maximizes (or minimizes) the objective function.

Look again at the linear programming problem in Example 1.

*The **simplex method** is a way to solve linear programming problems involving many inequalities and variables. This method was developed by George Dantzig in 1946 and is particularly well suited for computerization. In 1984, Narendra Karmarkar of Bell Laboratories discovered a way of solving large linear programming problems that improves on the simplex method.

EXAMPLE 2   **Analyzing a Linear Programming Problem**

Consider the linear programming problem

$$\text{Maximize} \qquad I = 0.02x + 0.03y$$

subject to the conditions that

$$\begin{cases} x \geq 0 \\ y \geq 0 \\ x + y \leq 25 \\ x \geq 15 \\ y \leq 5 \end{cases}$$

Graph the constraints. Then graph the objective function for $I = 0, 0.3, 0.45, 0.55,$ and 0.6.

Solution   Figure 24 shows the graph of the constraints. We superimpose on this graph the graph of the objective function for the given values of I.

For $I = 0$, the objective function is the line $0 = 0.02x + 0.03y$.

For $I = 0.3$, the objective function is the line $0.3 = 0.02x + 0.03y$.

For $I = 0.45$, the objective function is the line $0.45 = 0.02x + 0.03y$.

For $I = 0.55$, the objective function is the line $0.55 = 0.02x + 0.03y$.

For $I = 0.6$, the objective function is the line $0.6 = 0.02x + 0.03y$.

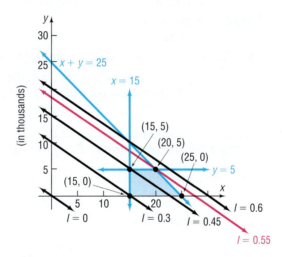

Figure 24   ●

DEFINITION   A **solution** to a linear programming problem consists of a feasible point that maximizes (or minimizes) the objective function, together with the corresponding value of the objective function.

One condition for a linear programming problem in two variables to have a solution is that the graph of the feasible points be bounded. (Refer to page 922.)

If none of the feasible points maximizes (or minimizes) the objective function or if there are no feasible points, the linear programming problem has no solution.

Consider the linear programming problem posed in Example 2, and look again at Figure 24. The feasible points are the points that lie in the shaded region. For example, $(20, 3)$ is a feasible point, as are $(15, 5), (20, 5), (18, 4)$, and so on. To find the solution of the problem requires finding a feasible point (x, y) that makes $I = 0.02x + 0.03y$ as large as possible. Notice that as I increases in value from $I = 0$ to $I = 0.3$ to $I = 0.45$ to $I = 0.55$ to $I = 0.6$, the result is a collection of parallel lines. Further, notice that the largest value of I that can be obtained using feasible points is $I = 0.55$, which corresponds to the line $0.55 = 0.02x + 0.03y$. Any larger value of I results in a line that does not pass through any feasible points. Finally, notice that the feasible point that yields $I = 0.55$ is the point $(20, 5)$, a corner point. These observations form the basis of the following results, which are stated without proof.

THEOREM

Location of the Solution of a Linear Programming Problem

If a linear programming problem has a solution, it is located at a corner point of the graph of the feasible points.

If a linear programming problem has multiple solutions, at least one of them is located at a corner point of the graph of the feasible points.

In either case, the corresponding value of the objective function is unique.

We shall not consider linear programming problems that have no solution. As a result, we can outline the procedure for solving a linear programming problem as follows:

Procedure for Solving a Linear Programming Problem

STEP 1: Write an expression for the quantity to be maximized (or minimized). This expression is the objective function.

STEP 2: Write all the constraints as a system of linear inequalities, and graph the system.

STEP 3: List the corner points of the graph of the feasible points.

STEP 4: List the corresponding values of the objective function at each corner point. The largest (or smallest) of these is the solution.

EXAMPLE 3

Solving a Minimum Linear Programming Problem

Minimize the expression

$$z = 2x + 3y$$

subject to the constraints

$$y \le 5 \qquad x \le 6 \qquad x + y \ge 2 \qquad x \ge 0 \qquad y \ge 0$$

Solution **STEP 1:** The objective function is $z = 2x + 3y$.

STEP 2: We seek the smallest value of z that can occur if x and y are solutions of the system of linear inequalities

$$\begin{cases} y \le 5 \\ x \le 6 \\ x + y \ge 2 \\ x \ge 0 \\ y \ge 0 \end{cases}$$

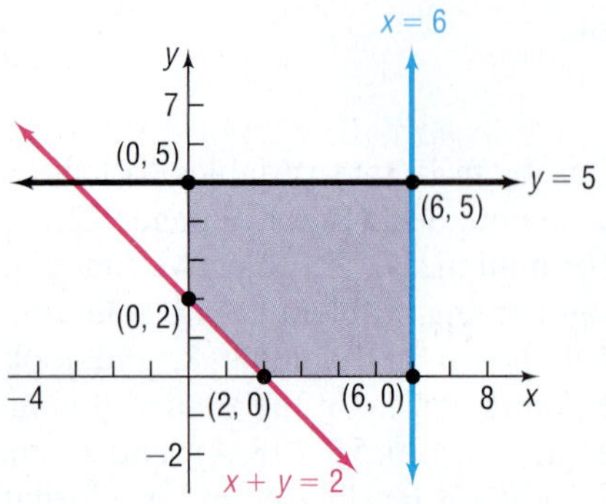

Figure 25

STEP 3: The graph of this system (the set of feasible points) is shown as the shaded region in Figure 25. The corner points have also been plotted.

STEP 4: Table 1 lists the corner points and the corresponding values of the objective function. From the table, the minimum value of z is 4, and it occurs at the point $(2, 0)$.

Table 1

Corner Point (x, y)	Value of the Objective Function $z = 2x + 3y$
$(0, 2)$	$z = 2(0) + 3(2) = 6$
$(0, 5)$	$z = 2(0) + 3(5) = 15$
$(6, 5)$	$z = 2(6) + 3(5) = 27$
$(6, 0)$	$z = 2(6) + 3(0) = 12$
$(2, 0)$	$z = 2(2) + 3(0) = 4$

Now Work PROBLEMS 5 AND 11

EXAMPLE 4

Maximizing Profit

At the end of every month, after filling orders for its regular customers, a coffee company has some pure Colombian coffee and some special-blend coffee remaining. The practice of the company has been to package a mixture of the two coffees into 1-pound (lb) packages as follows: a low-grade mixture containing 4 ounces (oz) of Colombian coffee and 12 oz of special-blend coffee, and a high-grade mixture containing 8 oz of Colombian and 8 oz of special-blend coffee. A profit of $0.30 per package is made on the low-grade mixture, whereas a profit of $0.40 per package is made on the high-grade mixture. This month, 120 lb of special-blend coffee and 100 lb of pure Colombian coffee remain. How many packages of each mixture should be prepared to achieve a maximum profit? Assume that all packages prepared can be sold.

Solution

STEP 1: Begin by assigning symbols for the two variables.

x = Number of packages of the low-grade mixture

y = Number of packages of the high-grade mixture

If P denotes the profit, then

$$P = \$0.30x + \$0.40y \quad \textit{Objective function}$$

STEP 2: The goal is to maximize P subject to certain constraints on x and y. Because x and y represent numbers of packages, the only meaningful values for x and y are nonnegative integers. This yields the two constraints

$$x \geq 0 \quad y \geq 0 \quad \textit{Nonnegative constraints}$$

There is only so much of each type of coffee available. For example, the total amount of Colombian coffee used in the two mixtures cannot exceed 100 lb, or 1600 oz. Because 4 oz are used in each low-grade package and 8 oz are used in each high-grade package, this leads to the constraint

$$4x + 8y \leq 1600 \quad \textit{Colombian coffee constraint}$$

Similarly, the supply of 120 lb, or 1920 oz, of special-blend coffee leads to the constraint

$$12x + 8y \leq 1920 \quad \textit{Special-blend coffee constraint}$$

The linear programming problem may be stated as

$$\text{Maximize} \quad P = 0.3x + 0.4y$$

subject to the constraints

$$x \geq 0 \quad y \geq 0 \quad 4x + 8y \leq 1600 \quad 12x + 8y \leq 1920$$

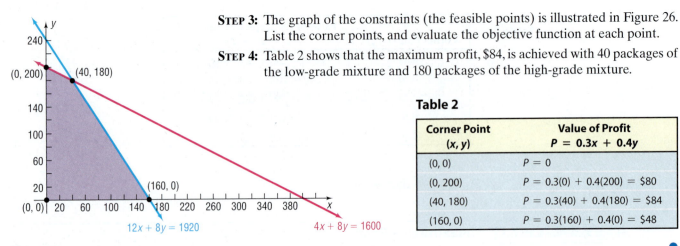

STEP 3: The graph of the constraints (the feasible points) is illustrated in Figure 26. List the corner points, and evaluate the objective function at each point.

STEP 4: Table 2 shows that the maximum profit, $84, is achieved with 40 packages of the low-grade mixture and 180 packages of the high-grade mixture.

Table 2

Corner Point (x, y)	Value of Profit $P = 0.3x + 0.4y$
(0, 0)	$P = 0$
(0, 200)	$P = 0.3(0) + 0.4(200) = \$80$
(40, 180)	$P = 0.3(40) + 0.4(180) = \$84$
(160, 0)	$P = 0.3(160) + 0.4(0) = \$48$

Figure 26

Now Work PROBLEM 19

12.8 Assess Your Understanding

Concepts and Vocabulary

1. A linear programming problem requires that a linear expression, called the _____ _____, be maximized or minimized.

2. *True or False* If a linear programming problem has a solution, it is located at a corner point of the graph of the feasible points.

Skill Building

In Problems 3–8, find the maximum and minimum value of the given objective function of a linear programming problem. The figure illustrates the graph of the feasible points.

3. $z = x + y$

4. $z = 2x + 3y$

5. $z = x + 10y$

6. $z = 10x + y$

7. $z = 5x + 7y$

8. $z = 7x + 5y$

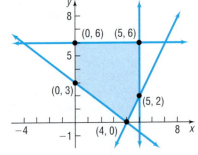

In Problems 9–18, solve each linear programming problem.

9. Maximize $z = 2x + y$ subject to $x \geq 0$, $y \geq 0$, $x + y \leq 6$, $x + y \geq 1$

10. Maximize $z = x + 3y$ subject to $x \geq 0$, $y \geq 0$, $x + y \geq 3$, $x \leq 5$, $y \leq 7$

11. Minimize $z = 2x + 5y$ subject to $x \geq 0$, $y \geq 0$, $x + y \geq 2$, $x \leq 5$, $y \leq 3$

12. Minimize $z = 3x + 4y$ subject to $x \geq 0$, $y \geq 0$, $2x + 3y \geq 6$, $x + y \leq 8$

13. Maximize $z = 3x + 5y$ subject to $x \geq 0$, $y \geq 0$, $x + y \geq 2$, $2x + 3y \leq 12$, $3x + 2y \leq 12$

14. Maximize $z = 5x + 3y$ subject to $x \geq 0$, $y \geq 0$, $x + y \geq 2$, $x + y \leq 8$, $2x + y \leq 10$

15. Minimize $z = 5x + 4y$ subject to $x \geq 0$, $y \geq 0$, $x + y \geq 2$, $2x + 3y \leq 12$, $3x + y \leq 12$

16. Minimize $z = 2x + 3y$ subject to $x \geq 0$, $y \geq 0$, $x + y \geq 3$, $x + y \leq 9$, $x + 3y \geq 6$

17. Maximize $z = 5x + 2y$ subject to $x \geq 0$, $y \geq 0$, $x + y \leq 10$, $2x + y \geq 10$, $x + 2y \geq 10$

18. Maximize $z = 2x + 4y$ subject to $x \geq 0$, $y \geq 0$, $2x + y \geq 4$, $x + y \leq 9$

Applications and Extensions

19. Maximizing Profit A manufacturer of skis produces two types: downhill and cross-country. Use the following table to determine how many of each kind of ski should be produced to achieve a maximum profit. What is the maximum profit? What would the maximum profit be if the time available for manufacturing were increased to 48 hours?

	Downhill	Cross-country	Time Available
Manufacturing time per ski	2 hours	1 hour	40 hours
Finishing time per ski	1 hour	1 hour	32 hours
Profit per ski	$70	$50	

20. Farm Management A farmer has 70 acres of land available for planting either soybeans or wheat. The cost of preparing the soil, the workdays required, and the expected profit per acre planted for each type of crop are given in the following table.

	Soybeans	Wheat
Preparation cost per acre	$60	$30
Workdays required per acre	3	4
Profit per acre	$180	$100

The farmer cannot spend more than $1800 in preparation costs and cannot use a total of more than 120 workdays. How many acres of each crop should be planted to maximize the profit? What is the maximum profit? What is the maximum profit if the farmer is willing to spend no more than $2400 on preparation?

21. Banquet Seating A banquet hall offers two types of tables for rent: 6-person rectangular tables at a cost of $28 each and 10-person round tables at a cost of $52 each. Kathleen would like to rent the hall for a wedding banquet and needs tables for 250 people. The hall can have a maximum of 35 tables, and the hall has only 15 rectangular tables available. How many of each type of table should be rented to minimize cost and what is the minimum cost?

Source: facilities.princeton.edu

22. Spring Break The student activities department of a community college plans to rent buses and vans for a spring-break trip. Each bus has 40 regular seats and 1 special seat designed to accommodate travelers with disabilities. Each van has 8 regular seats and 3 special seats. The rental cost is $350 for each van and $975 for each bus. If 320 regular and 36 special seats are required for the trip, how many vehicles of each type should be rented to minimize cost?

Source: www.busrates.com

23. Return on Investment An investment broker is instructed by her client to invest up to $20,000, some in a junk bond yielding 9% per annum and some in Treasury bills yielding 7% per annum. The client wants to invest at least $8000 in T-bills and no more than $12,000 in the junk bond.
(a) How much should the broker recommend that the client place in each investment to maximize income if

the client insists that the amount invested in T-bills must equal or exceed the amount placed in the junk bond?
(b) How much should the broker recommend that the client place in each investment to maximize income if the client insists that the amount invested in T-bills must not exceed the amount placed in the junk bond?

24. Production Scheduling In a factory, machine 1 produces 8-inch (in.) pliers at the rate of 60 units per hour (h) and 6-in. pliers at the rate of 70 units/h. Machine 2 produces 8-in. pliers at the rate of 40 units/h and 6-in. pliers at the rate of 20 units/h. It costs $50/h to operate machine 1, and machine 2 costs $30/h to operate. The production schedule requires that at least 240 units of 8-in. pliers and at least 140 units of 6-in. pliers be produced during each 10-h day. Which combination of machines will cost the least money to operate?

25. Managing a Meat Market A meat market combines ground beef and ground pork in a single package for meat loaf. The ground beef is 75% lean (75% beef, 25% fat) and costs the market $0.75 per pound (lb). The ground pork is 60% lean and costs the market $0.45/lb. The meat loaf must be at least 70% lean. If the market wants to use at least 50 lb of its available pork, but no more than 200 lb of its available ground beef, how much ground beef should be mixed with ground pork so that the cost is minimized?

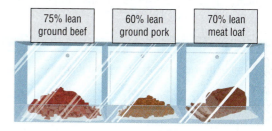

26. Ice Cream The Mom and Pop Ice Cream Company makes two kinds of chocolate ice cream: regular and premium. The properties of 1 gallon (gal) of each type are shown in the table:

	Regular	Premium
Flavoring	24 oz	20 oz
Milk-fat products	12 oz	20 oz
Shipping weight	5 lbs	6 lbs
Profit	$0.75	$0.90

In addition, current commitments require the company to make at least 1 gal of premium for every 4 gal of regular. Each day, the company has available 725 pounds (lb) of flavoring and 425 lb of milk-fat products. If the company can ship no more than 3000 lb of product per day, how many gallons of each type should be produced daily to maximize profit?

Source: www.scitoys.com/ingredients/ice_cream.html

27. Maximizing Profit on Ice Skates A factory manufactures two kinds of ice skates: racing skates and figure skates. The racing skates require 6 work-hours in the fabrication department, whereas the figure skates require 4 work-hours there. The racing skates require 1 work-hour in the finishing department, whereas the figure skates require 2 work-hours there. The fabricating department has available at most 120 work-hours per day, and the finishing department has no more than 40 work-hours per day available. If the profit on each racing skate is $10 and the profit on each figure skate is $12, how many of each should be manufactured each day to maximize profit? (Assume that all skates made are sold.)

28. **Financial Planning** A retired couple have up to $50,000 to place in fixed-income securities. Their financial adviser suggests two securities to them: one is an AAA bond that yields 8% per annum; the other is a certificate of deposit (CD) that yields 4%. After careful consideration of the alternatives, the couple decide to place at most $20,000 in the AAA bond and at least $15,000 in the CD. They also instruct the financial adviser to place at least as much in the CD as in the AAA bond. How should the financial adviser proceed to maximize the return on their investment?

29. **Product Design** An entrepreneur is having a design group produce at least six samples of a new kind of fastener that he wants to market. It costs $9.00 to produce each metal fastener and $4.00 to produce each plastic fastener. He wants to have at least two of each version of the fastener and needs to have all the samples 24 hours (h) from now. It takes 4 h to produce each metal sample and 2 h to produce each plastic sample. To minimize the cost of the samples, how many of each kind should the entrepreneur order? What will be the cost of the samples?

30. **Animal Nutrition** Kevin's dog Amadeus likes two kinds of canned dog food. Gourmet Dog costs 40 cents a can and has 20 units of a vitamin complex; the calorie content is 75 calories. Chow Hound costs 32 cents a can and has 35 units of vitamins and 50 calories. Kevin likes Amadeus to have at least 1175 units of vitamins a month and at least 2375 calories during the same time period. Kevin has space to store only 60 cans of dog food at a time. How much of each kind of dog food should Kevin buy each month to minimize his cost?

31. **Airline Revenue** An airline has two classes of service: first class and coach. Management's experience has been that each aircraft should have at least 8 but no more than 16 first-class seats and at least 80 but no more than 120 coach seats.
   (a) If management decides that the ratio of first class to coach seats should never exceed 1:12, with how many of each type of seat should an aircraft be configured to maximize revenue?
   (b) If management decides that the ratio of first class to coach seats should never exceed 1:8, with how many of each type of seat should an aircraft be configured to maximize revenue?
   (c) If you were management, what would you do? [**Hint:** Assume that the airline charges $C for a coach seat and $F for a first-class seat; $C > 0$, $F > C$.]

Explaining Concepts: Discussion and Writing

32. Explain in your own words what a linear programming problem is and how it can be solved.

Retain Your Knowledge

Problems 33–36 are based on material learned earlier in the course. The purpose of these problems is to keep the material fresh in your mind so that you are better prepared for the final exam.

33. Solve: $2m^{2/5} - m^{1/5} = 1$

34. Graph $y = -\tan\left(x - \dfrac{\pi}{2}\right)$ for at least two periods. Use the graph to determine the domain and range.

35. **Radioactive Decay** The half-life of titanium-44 is 63 years. How long will it take 200 grams to decay to 75 grams? Round to one decimal place.

36. Find the equation of the line that is parallel to $y = 3x + 11$ and passes through the point $(-2, 1)$.

Chapter Review

Things to Know

Systems of equations (pp. 844–854)

Systems with no solutions are inconsistent.   Systems with a solution are consistent.

Consistent systems of linear equations have either a unique solution (independent) or an infinite number of solutions (dependent).

Matrix (p. 859)   Rectangular array of numbers, called entries

Augmented matrix (p. 859)

Row operations (p. 861)

Row echelon form (p. 862)

Reduced row echelon form (p. 865)

Determinants and Cramer's Rule (pp. 874–881)

Matrix Algebra (pp. 884–896)

m by n matrix (p. 885)   Matrix with m rows and n columns

Identity matrix I_n (p. 892)   An n by n square matrix whose diagonal entries are 1's, while all other entries are 0's

Inverse of a matrix (p. 892) A^{-1} is the inverse of A if $AA^{-1} = A^{-1}A = I_n$.

Nonsingular matrix (p. 892) A square matrix that has an inverse

Linear programming problem (p. 926)

Maximize (or minimize) a linear objective function, $z = Ax + By$, subject to certain conditions, or constraints, expressible as linear inequalities in x and y. A feasible point (x, y) is a point that satisfies the constraints (linear inequalities) of a linear programming problem.

Location of the solution of a linear programming problem (p. 928)

If a linear programming problem has a solution, it is located at a corner point of the graph of the feasible points. If a linear programming problem has multiple solutions, at least one of them is located at a corner point of the graph of the feasible points. In either case, the corresponding value of the objective function is unique.

Objectives

Section		You should be able to ...	Example(s)	Review Exercises
12.1	1	Solve systems of equations by substitution (p. 846)	4	1–7, 56, 59
	2	Solve systems of equations by elimination (p. 847)	5, 6	1–7, 56, 59
	3	Identify inconsistent systems of equations containing two variables (p. 849)	7	5, 54
	4	Express the solution of a system of dependent equations containing two variables (p. 849)	8	7, 53
	5	Solve systems of three equations containing three variables (p. 850)	9, 12	8–10, 55, 57, 60
	6	Identify inconsistent systems of equations containing three variables (p. 852)	10	10
	7	Express the solution of a system of dependent equations containing three variables (p. 853)	11	9
12.2	1	Write the augmented matrix of a system of linear equations (p. 859)	1	20–25
	2	Write the system of equations from the augmented matrix (p. 860)	2	11, 12
	3	Perform row operations on a matrix (p. 861)	3, 4	20–25
	4	Solve a system of linear equations using matrices (p. 862)	5–10	20–25
12.3	1	Evaluate 2 by 2 determinants (p. 874)	1	26
	2	Use Cramer's Rule to solve a system of two equations containing two variables (p. 875)	2	29, 30
	3	Evaluate 3 by 3 determinants (p. 877)	4	27, 28
	4	Use Cramer's Rule to solve a system of three equations containing three variables (p. 879)	5	31
	5	Know properties of determinants (p. 880)	6–9	32, 33
12.4	1	Find the sum and difference of two matrices (p. 885)	3, 4	13
	2	Find scalar multiples of a matrix (p. 887)	5	14
	3	Find the product of two matrices (p. 888)	6–11	15, 16
	4	Find the inverse of a matrix (p. 892)	12–14	17–19
	5	Solve a system of linear equations using an inverse matrix (p. 896)	15	20–25
12.5	1	Decompose $\dfrac{P}{Q}$ where Q has only nonrepeated linear factors (p. 902)	1	34
	2	Decompose $\dfrac{P}{Q}$ where Q has repeated linear factors (p. 904)	2, 3	35
	3	Decompose $\dfrac{P}{Q}$ where Q has a nonrepeated irreducible quadratic factor (p. 906)	4	36, 38
	4	Decompose $\dfrac{P}{Q}$ where Q has a repeated irreducible quadratic factor (p. 907)	5	37
12.6	1	Solve a system of nonlinear equations using substitution (p. 909)	1, 3	39–43
	2	Solve a system of nonlinear equations using elimination (p. 910)	2, 4, 5	39–43
12.7	1	Graph an inequality (p. 918)	2–4	44, 45
	2	Graph a system of inequalities (p. 920)	5–10	46–50, 58
12.8	1	Set up a linear programming problem (p. 926)	1	61
	2	Solve a linear programming problem (p. 926)	2–4	51, 52, 61

Review Exercises

In Problems 1–10, solve each system of equations using the method of substitution or the method of elimination. If the system has no solution, say that it is inconsistent.

1. $\begin{cases} 2x - y = 5 \\ 5x + 2y = 8 \end{cases}$

2. $\begin{cases} 3x - 4y = 4 \\ x - 3y = \dfrac{1}{2} \end{cases}$

3. $\begin{cases} x - 2y - 4 = 0 \\ 3x + 2y - 4 = 0 \end{cases}$

4. $\begin{cases} y = 2x - 5 \\ x = 3y + 4 \end{cases}$

5. $\begin{cases} x - 3y + 4 = 0 \\ \dfrac{1}{2}x - \dfrac{3}{2}y + \dfrac{4}{3} = 0 \end{cases}$

6. $\begin{cases} 2x + 3y - 13 = 0 \\ 3x - 2y = 0 \end{cases}$

7. $\begin{cases} 2x + 5y = 10 \\ 4x + 10y = 20 \end{cases}$

8. $\begin{cases} x + 2y - z = 6 \\ 2x - y + 3z = -13 \\ 3x - 2y + 3z = -16 \end{cases}$

9. $\begin{cases} 2x - 4y + z = -15 \\ x + 2y - 4z = 27 \\ 5x - 6y - 2z = -3 \end{cases}$

10. $\begin{cases} x - 4y + 3z = 15 \\ -3x + y - 5z = -5 \\ -7x - 5y - 9z = 10 \end{cases}$

In Problems 11 and 12, write the system of equations that corresponds to the given augmented matrix.

11. $\begin{bmatrix} 3 & 2 & | & 8 \\ 1 & 4 & | & -1 \end{bmatrix}$

12. $\begin{bmatrix} 1 & 2 & 5 & | & -2 \\ 5 & 0 & -3 & | & 8 \\ 2 & -1 & 0 & | & 0 \end{bmatrix}$

In Problems 13–16, use the following matrices to compute each expression.

$$A = \begin{bmatrix} 1 & 0 \\ 2 & 4 \\ -1 & 2 \end{bmatrix} \quad B = \begin{bmatrix} 4 & -3 & 0 \\ 1 & 1 & -2 \end{bmatrix} \quad C = \begin{bmatrix} 3 & -4 \\ 1 & 5 \\ 5 & 2 \end{bmatrix}$$

13. $A + C$

14. $6A$

15. AB

16. BC

In Problems 17–19, find the inverse, if there is one, of each matrix. If there is no inverse, say that the matrix is singular.

17. $\begin{bmatrix} 4 & 6 \\ 1 & 3 \end{bmatrix}$

18. $\begin{bmatrix} 1 & 3 & 3 \\ 1 & 2 & 1 \\ 1 & -1 & 2 \end{bmatrix}$

19. $\begin{bmatrix} 4 & -8 \\ -1 & 2 \end{bmatrix}$

In Problems 20–25, solve each system of equations using matrices. If the system has no solution, say that it is inconsistent.

20. $\begin{cases} 3x - 2y = 1 \\ 10x + 10y = 5 \end{cases}$

21. $\begin{cases} 5x - 6y - 3z = 6 \\ 4x - 7y - 2z = -3 \\ 3x + y - 7z = 1 \end{cases}$

22. $\begin{cases} 2x + y + z = 5 \\ 4x - y - 3z = 1 \\ 8x + y - z = 5 \end{cases}$

23. $\begin{cases} x - 2z = 1 \\ 2x + 3y = -3 \\ 4x - 3y - 4z = 3 \end{cases}$

24. $\begin{cases} x - y + z = 0 \\ x - y - 5z - 6 = 0 \\ 2x - 2y + z - 1 = 0 \end{cases}$

25. $\begin{cases} x - y - z - t = 1 \\ 2x + y - z + 2t = 3 \\ x - 2y - 2z - 3t = 0 \\ 3x - 4y + z + 5t = -3 \end{cases}$

In Problems 26–28, find the value of each determinant.

26. $\begin{vmatrix} 3 & 4 \\ 1 & 3 \end{vmatrix}$

27. $\begin{vmatrix} 1 & 4 & 0 \\ -1 & 2 & 6 \\ 4 & 1 & 3 \end{vmatrix}$

28. $\begin{vmatrix} 2 & 1 & -3 \\ 5 & 0 & 1 \\ 2 & 6 & 0 \end{vmatrix}$

In Problems 29–31, use Cramer's Rule, if applicable, to solve each system.

29. $\begin{cases} x - 2y = 4 \\ 3x + 2y = 4 \end{cases}$

30. $\begin{cases} 2x + 3y - 13 = 0 \\ 3x - 2y = 0 \end{cases}$

31. $\begin{cases} x + 2y - z = 6 \\ 2x - y + 3z = -13 \\ 3x - 2y + 3z = -16 \end{cases}$

In Problems 32 and 33, use properties of determinants to find the value of each determinant if it is known that $\begin{vmatrix} x & y \\ a & b \end{vmatrix} = 8.$

32. $\begin{vmatrix} 2x & y \\ 2a & b \end{vmatrix}$

33. $\begin{vmatrix} y & x \\ b & a \end{vmatrix}$

In Problems 34–38, find the partial fraction decomposition of each rational expression.

34. $\dfrac{6}{x(x-4)}$ **35.** $\dfrac{x-4}{x^2(x-1)}$ **36.** $\dfrac{x}{(x^2+9)(x+1)}$ **37.** $\dfrac{x^3}{(x^2+4)^2}$ **38.** $\dfrac{x^2}{(x^2+1)(x^2-1)}$

In Problems 39–43, solve each system of equations.

39. $\begin{cases} 2x+y+3=0 \\ x^2+y^2=5 \end{cases}$ **40.** $\begin{cases} 2xy+y^2=10 \\ 3y^2-xy=2 \end{cases}$ **41.** $\begin{cases} x^2+y^2=6y \\ x^2=3y \end{cases}$

42. $\begin{cases} 3x^2+4xy+5y^2=8 \\ x^2+3xy+2y^2=0 \end{cases}$ **43.** $\begin{cases} x^2-3x+y^2+y=-2 \\ \dfrac{x^2-x}{y}+y+1=0 \end{cases}$

In Problems 44 and 45 graph each inequality.

44. $3x+4y\le 12$ **45.** $y\le x^2$

In Problems 46–48, graph each system of inequalities. Tell whether the graph is bounded or unbounded, and label the corner points.

46. $\begin{cases} -2x+y\le 2 \\ x+y\ge 2 \end{cases}$ **47.** $\begin{cases} x\ge 0 \\ y\ge 0 \\ x+y\le 4 \\ 2x+3y\le 6 \end{cases}$ **48.** $\begin{cases} x\ge 0 \\ y\ge 0 \\ 2x+y\le 8 \\ x+2y\ge 2 \end{cases}$

In Problems 49 and 50, graph each system of inequalities.

49. $\begin{cases} x^2+y^2\le 16 \\ x+y\ge 2 \end{cases}$ **50.** $\begin{cases} y\le x^2 \\ xy\le 4 \end{cases}$

In Problems 51 and 52, solve each linear programming problem.

51. Maximize $z=3x+4y$ subject to $x\ge 0, y\ge 0, 3x+2y\ge 6, x+y\le 8$
52. Minimize $z=3x+5y$ subject to $x\ge 0, y\ge 0, x+y\ge 1, 3x+2y\le 12, x+3y\le 12$

---

53. Find A so that the system of equations has infinitely many solutions.
$$\begin{cases} 2x+5y=5 \\ 4x+10y=A \end{cases}$$

54. Find A so that the system in Problem 53 is inconsistent.

55. Curve Fitting Find the quadratic function $y=ax^2+bx+c$ that passes through the three points $(0,1),(1,0)$, and $(-2,1)$.

56. Blending Coffee A coffee distributor is blending a new coffee that will cost $6.90 per pound. It will consist of a blend of $6.00-per-pound coffee and $9.00-per-pound coffee. What amounts of each type of coffee should be mixed to achieve the desired blend?

[**Hint:** Assume that the weight of the blended coffee is 100 pounds.]

$6.00/lb $6.90/lb $9.00/lb

57. Cookie Orders A cookie company makes three kinds of cookies (oatmeal raisin, chocolate chip, and shortbread) packaged in small, medium, and large boxes. The small box contains 1 dozen oatmeal raisin and 1 dozen chocolate chip; the medium box has 2 dozen oatmeal raisin, 1 dozen chocolate chip, and 1 dozen shortbread; the large box contains 2 dozen oatmeal raisin, 2 dozen chocolate chip, and 3 dozen shortbread. If you require exactly 15 dozen oatmeal raisin, 10 dozen chocolate chip, and 11 dozen shortbread, how many of each size box should you buy?

58. Mixed Nuts A store that specializes in selling nuts has available 72 pounds (lb) of cashews and 120 lb of peanuts. These are to be mixed in 12-ounce (oz) packages as follows: a lower-priced package containing 8 oz of peanuts and 4 oz of cashews, and a quality package containing 6 oz of peanuts and 6 oz of cashews.
(a) Use x to denote the number of lower-priced packages, and use y to denote the number of quality packages. Write a system of linear inequalities that describes the possible numbers of each kind of package.
(b) Graph the system and label the corner points.

59. Determining the Speed of the Current of the Aguarico River On a recent trip to the Cuyabeno Wildlife Reserve in the Amazon region of Ecuador, Mike took a 100-kilometer trip by speedboat down the Aguarico River from Chiritza to the Flotel Orellana. As Mike watched the Amazon unfold, he wondered how fast the speedboat was going and how fast the current of the white-water Aguarico River was. Mike timed the trip downstream at 2.5 hours and the return trip at 3 hours. What were the two speeds?

60. Constant Rate Jobs If Bruce and Bryce work together for 1 hour and 20 minutes, they will finish a certain job. If Bryce and Marty work together for 1 hour and 36 minutes, the same job can be finished. If Marty and Bruce work together, they can complete this job in 2 hours and 40 minutes. How long would it take each of them, working alone, to finish the job?

61. Minimizing Production Cost A factory produces gasoline engines and diesel engines. Each week the factory is obligated to deliver at least 20 gasoline engines and at least 15 diesel engines. Due to physical limitations, however, the factory cannot make more than 60 gasoline engines or more than 40 diesel engines in any given week. Finally, to prevent layoffs, a total of at least 50 engines must be produced. If gasoline engines cost $450 each to produce and diesel engines cost $550 each to produce, how many of each should be produced per week to minimize the cost? What is the excess capacity of the factory? That is, how many of each kind of engine are being produced in excess of the number that the factory is obligated to deliver?

62. Describe four ways of solving a system of three linear equations containing three variables. Which method do you prefer? Why?

CHAPTER
Chapter Test

The Chapter Test Prep Videos are step-by-step solutions available in **MyMathLab**, or on this text's You Tube™ Channel. Flip back to the Resources for Success page for a link to this text's YouTube channel.

In Problems 1–4, solve each system of equations using the method of substitution or the method of elimination. If the system has no solution, say that it is inconsistent.

1. $\begin{cases} -2x + y = -7 \\ 4x + 3y = 9 \end{cases}$

2. $\begin{cases} \dfrac{1}{3}x - 2y = 1 \\ 5x - 30y = 18 \end{cases}$

3. $\begin{cases} x - y + 2z = 5 \\ 3x + 4y - z = -2 \\ 5x + 2y + 3z = 8 \end{cases}$

4. $\begin{cases} 3x + 2y - 8z = -3 \\ -x - \dfrac{2}{3}y + z = 1 \\ 6x - 3y + 15z = 8 \end{cases}$

5. Write the augmented matrix corresponding to the system of equations: $\begin{cases} 4x - 5y + z = 0 \\ -2x - y + 6 = -19 \\ x + 5y - 5z = 10 \end{cases}$

6. Write the system of equations corresponding to the augmented matrix: $\left[\begin{array}{ccc|c} 3 & 2 & 4 & -6 \\ 1 & 0 & 8 & 2 \\ -2 & 1 & 3 & -11 \end{array} \right]$

In Problems 7–10, use the given matrices to compute each expression.

$$A = \begin{bmatrix} 1 & -1 \\ 0 & -4 \\ 3 & 2 \end{bmatrix} \quad B = \begin{bmatrix} 1 & -2 & 5 \\ 0 & 3 & 1 \end{bmatrix} \quad C = \begin{bmatrix} 4 & 6 \\ 1 & -3 \\ -1 & 8 \end{bmatrix}$$

7. $2A + C$ **8.** $A - 3C$ **9.** CB **10.** BA

In Problems 11 and 12, find the inverse of each nonsingular matrix.

11. $A = \begin{bmatrix} 3 & 2 \\ 5 & 4 \end{bmatrix}$

12. $B = \begin{bmatrix} 1 & -1 & 1 \\ 2 & 5 & -1 \\ 2 & 3 & 0 \end{bmatrix}$

In Problems 13–16, solve each system of equations using matrices. If the system has no solution, say that it is inconsistent.

13. $\begin{cases} 6x + 3y = 12 \\ 2x - y = -2 \end{cases}$

14. $\begin{cases} x + \dfrac{1}{4}y = 7 \\ 8x + 2y = 56 \end{cases}$

15. $\begin{cases} x + 2y + 4z = -3 \\ 2x + 7y + 15z = -12 \\ 4x + 7y + 13z = -10 \end{cases}$

16. $\begin{cases} 2x + 2y - 3z = 5 \\ x - y + 2z = 8 \\ 3x + 5y - 8z = -2 \end{cases}$

In Problems 17 and 18, find the value of each determinant.

17. $\begin{vmatrix} -2 & 5 \\ 3 & 7 \end{vmatrix}$

18. $\begin{vmatrix} 2 & -4 & 6 \\ 1 & 4 & 0 \\ -1 & 2 & -4 \end{vmatrix}$

In Problems 19 and 20, use Cramer's Rule, if possible, to solve each system.

19. $\begin{cases} 4x + 3y = -23 \\ 3x - 5y = 19 \end{cases}$

20. $\begin{cases} 4x - 3y + 2z = 15 \\ -2x + y - 3z = -15 \\ 5x - 5y + 2z = 18 \end{cases}$

In Problems 21 and 22, solve each system of equations.

21. $\begin{cases} 3x^2 + y^2 = 12 \\ y^2 = 9x \end{cases}$

22. $\begin{cases} 2y^2 - 3x^2 = 5 \\ y - x = 1 \end{cases}$

23. Graph the system of inequalities: $\begin{cases} x^2 + y^2 \leq 100 \\ 4x - 3y \geq 0 \end{cases}$

In Problems 24 and 25, find the partial fraction decomposition of each rational expression.

24. $\dfrac{3x + 7}{(x + 3)^2}$

25. $\dfrac{4x^2 - 3}{x(x^2 + 3)^2}$

26. Graph the system of inequalities. Tell whether the graph is bounded or unbounded, and label all corner points.

$$\begin{cases} x \ge 0 \\ y \ge 0 \\ x + 2y \ge 8 \\ 2x - 3y \ge 2 \end{cases}$$

27. Maximize $z = 5x + 8y$ subject to the constraints $x \ge 0$, $2x + y \le 8$, and $x - 3y \le -3$.

28. Megan went clothes shopping and bought 2 pairs of flare jeans, 2 camisoles, and 4 T-shirts for $90.00. At the same store, Paige bought one pair of flare jeans and 3 T-shirts for $42.50, while Kara bought 1 pair of flare jeans, 3 camisoles, and 2 T-shirts for $62.00. Determine the price of each clothing item.

Cumulative Review

In Problems 1–6, solve each equation.

1. $2x^2 - x = 0$

2. $\sqrt{3x + 1} = 4$

3. $2x^3 - 3x^2 - 8x - 3 = 0$

4. $3^x = 9^{x+1}$

5. $\log_3(x - 1) + \log_3(2x + 1) = 2$

6. $3^x = e$

7. Determine whether the function $g(x) = \dfrac{2x^3}{x^4 + 1}$ is even, odd, or neither. Is the graph of g symmetric with respect to the x-axis, y-axis, or origin?

8. Find the center and radius of the circle
$$x^2 + y^2 - 2x + 4y - 11 = 0$$
Graph the circle.

9. Graph $f(x) = 3^{x-2} + 1$ using transformations. What are the domain, range, and horizontal asymptote of f?

10. The function $f(x) = \dfrac{5}{x + 2}$ is one-to-one. Find f^{-1}. Find the domain and the range of f and the domain and the range of f^{-1}.

11. Graph each equation.
(a) $y = 3x + 6$
(b) $x^2 + y^2 = 4$
(c) $y = x^3$
(d) $y = \dfrac{1}{x}$
(e) $y = \sqrt{x}$
(f) $y = e^x$
(g) $y = \ln x$
(h) $2x^2 + 5y^2 = 1$
(i) $x^2 - 3y^2 = 1$
(j) $x^2 - 2x - 4y + 1 = 0$

12. $f(x) = x^3 - 3x + 5$
(a) Using a graphing utility, graph f and approximate the zero(s) of f.
(b) Using a graphing utility, approximate the local maxima and the local minima.
(c) Determine the intervals on which f is increasing.

Chapter Projects

I. Markov Chains A **Markov chain** (or process) is one in which future outcomes are determined by a current state. Future outcomes are based on probabilities. The probability of moving to a certain state depends only on the state previously occupied and does not vary with time. An example of a Markov chain is the maximum education achieved by children based on the highest educational level attained by their parents, where the states are (1) earned college degree, (2) high school diploma only, (3) elementary school only. If p_{ij} is the probability of moving from state i to state j, the **transition matrix** is the $m \times m$ matrix

$$P = \begin{bmatrix} p_{11} & p_{12} & \cdots & p_{1m} \\ \vdots & \vdots & & \vdots \\ p_{m1} & p_{m2} & \cdots & p_{mm} \end{bmatrix}$$

The table represents the probabilities for the highest educational level of children based on the highest educational level of their parents. For example, the table shows that the probability p_{21} is 40% that parents with a high-school education (row 2) will have children with a college education (column 1).

Highest Educational Level of Parents	Maximum Education That Children Achieve		
	College	High School	Elementary
College	80%	18%	2%
High school	40%	50%	10%
Elementary	20%	60%	20%

1. Convert the percentages to decimals.

2. What is the transition matrix?

3. Sum across the rows. What do you notice? Why do you think that you obtained this result?

4. If P is the transition matrix of a Markov chain, the (i, j)th entry of P^n (nth power of P) gives the probability of passing from state i to state j in n stages. What is the probability that the grandchild of a college graduate is a college graduate?

5. What is the probability that the grandchild of a high school graduate finishes college?

6. The row vector $v^{(0)} = \begin{bmatrix} 0.317 & 0.565 & 0.118 \end{bmatrix}$ represents the proportion of the U.S. population 25 years or older that has college, high school, and elementary school, respectively, as the highest educational level in 2013.* In a Markov chain the probability distribution $v^{(k)}$ after k stages is $v^{(k)} = v^{(0)}P^k$, where P^k is the kth power of the transition matrix. What will be the distribution of highest educational attainment of the grandchildren of the current population?

7. Calculate $P^3, P^4, P^5, \ldots$. Continue until the matrix does not change. This is called the long-run or steady-state distribution. What is the long-run distribution of highest educational attainment of the population?

   Source: U.S. Census Bureau.

The following projects are available at the Instructor's Resource Center (IRC).

II. Project at Motorola: *Error Control Coding* The high-powered engineering needed to ensure that wireless communications are transmitted correctly is analyzed using matrices to control coding errors.

III. Using Matrices to Find the Line of Best Fit Have you wondered how our calculators get a line of best fit? See how to find the line by solving a matrix equation.

IV. CBL Experiment Simulate two people walking toward each other at a constant rate. Then solve the resulting system of equations to determine when and where they will meet.

Sequences; Induction; the Binomial Theorem

13

UN Projects World Population Will Hit 9.6 Billion by 2050

The population of the planet is expected to reach 9.6 billion by 2050, according to a new UN report—a slightly larger number than anticipated, because fertility projections have increased in nations where women have the most children.

More than half of this projected demographic growth will be in Africa, which continues to add people even as population growth in the world at large slows down.

"Although population growth has slowed for the world as a whole, this report reminds us that some developing countries, especially in Africa, are still growing rapidly," said Under-Secretary-General for Economic and Social Affairs Wu Hongbo, in the UN report.

World Population Prospects: The 2012 Revision notes that the population in the developed regions of the world should remain stable at around 1.3 billion until 2050 thanks to trends of low fertility, while the world's 49 least-developed countries are projected to double in population size.

The new figures do *not* mean that world population growth has started to speed up again. As countries industrialize, they tend to undergo "demographic transition," wherein high death and birth rates are slowly replaced with low birth and death rates. It's a demographic shift often helped along by an increase in the granting of rights for women.

Some experts suspect that the world population will plateau around 2060, and there's a possibility that—after a transitional period of a higher death rate than birth rate—world birth and death rates could actually even out, keeping the human population stable.

Source: Faine Greenwood, June 14, 2013, 11:36. GlobalPost® (www.globalpost.com/dispatch/news/science/130614/un-projects-world-population-will-hit-96-billion-2050)

 —See the Internet-based Chapter Project I—

••• A Look Back, A Look Ahead •••

This chapter may be divided into three independent parts: Sections 13.1–13.3, Section 13.4, and Section 13.5.

In Chapter 3, we defined a function and its domain, which was usually some set of real numbers. In Sections 13.1–13.3, we discuss a sequence, which is a function whose domain is the set of positive integers.

Throughout this text, where it seemed appropriate, we have given proofs of many of the results. In Section 13.4, a technique for proving theorems involving natural numbers is discussed.

In Chapter R, Review, Section R.4, there are formulas for expanding $(x + a)^2$ and $(x + a)^3$. In Section 13.5, we discuss the Binomial Theorem, a formula for the expansion of $(x + a)^n$, where n is any positive integer.

The topics introduced in this chapter are covered in more detail in courses titled *Discrete Mathematics*. Applications of these topics can be found in the fields of computer science, engineering, business and economics, the social sciences, and the physical and biological sciences.

Outline

13.1 Sequences

PREPARING FOR THIS SECTION *Before getting started, review the following:*

- Functions (Section 3.1, pp. 199–204)

Now Work the **'Are You Prepared?'** problems on page 946.

> **OBJECTIVES** 1 Write the First Several Terms of a Sequence (p. 940)
> 2 Write the Terms of a Sequence Defined by a Recursive Formula (p. 943)
> 3 Use Summation Notation (p. 944)
> 4 Find the Sum of a Sequence (p. 945)

When you hear the word *sequence* as it is used in the phrase "a sequence of events," you probably think of a collection of events, one of which happens first, another second, and so on. In mathematics, the word *sequence* also refers to outcomes that are first, second, and so on.

DEFINITION

> A **sequence** is a function whose domain is the set of positive integers.

In a sequence, then, the inputs are $1, 2, 3, \ldots$. Because a sequence is a function, it will have a graph. Figure 1(a) shows the graph of the function $f(x) = \dfrac{1}{x}, x > 0$. If all the points on this graph were removed except those whose x-coordinates are positive integers—that is, if all points were removed except $(1, 1)$, $\left(2, \dfrac{1}{2}\right)$, $\left(3, \dfrac{1}{3}\right)$, and so on—the remaining points would be the graph of the sequence $f(n) = \dfrac{1}{n}$, as shown in Figure 1(b). Note that n is used to represent the independent variable in a sequence. This serves to remind us that n is a positive integer.

Figure 1

(a) $f(x) = \frac{1}{x}, x > 0$ (b) $f(n) = \frac{1}{n}, n$ a positive integer

1 Write the First Several Terms of a Sequence

A sequence is usually represented by listing its values in order. For example, the sequence whose graph is given in Figure 1(b) might be represented as

$$f(1), f(2), f(3), f(4), \ldots \quad \text{or} \quad 1, \frac{1}{2}, \frac{1}{3}, \frac{1}{4}, \ldots$$

The list never ends, as the ellipsis indicates. The numbers in this ordered list are called the **terms** of the sequence.

In dealing with sequences, subscripted letters are used such as a_1 to represent the first term, a_2 for the second term, a_3 for the third term, and so on.

For the sequence $f(n) = \dfrac{1}{n}$, this means

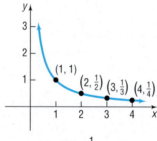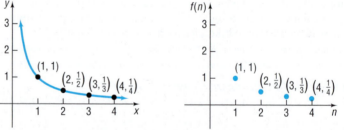

$$\underbrace{a_1 = f(1) = 1}_{\text{first term}} \quad \underbrace{a_2 = f(2) = \frac{1}{2}}_{\text{second term}} \quad \underbrace{a_3 = f(3) = \frac{1}{3}}_{\text{third term}} \quad \underbrace{a_4 = f(4) = \frac{1}{4}}_{\text{fourth term}} \ldots \quad \underbrace{a_n = f(n) = \frac{1}{n}}_{\text{nth term}} \ldots$$

In other words, the traditional function notation $f(n)$ is typically not used for sequences. For this particular sequence, we have a rule for the nth term, which is $a_n = \dfrac{1}{n}$, so it is easy to find any term of the sequence.

When a formula for the nth term (sometimes called the **general term**) of a sequence is known, the entire sequence can be represented by placing braces around the formula for the nth term.

For example, the sequence whose nth term is $b_n = \left(\dfrac{1}{2}\right)^n$ may be represented as

$$\{b_n\} = \left\{\left(\frac{1}{2}\right)^n\right\}$$

or by

$$b_1 = \frac{1}{2}, \quad b_2 = \frac{1}{4}, \quad b_3 = \frac{1}{8}, \dots, \quad b_n = \left(\frac{1}{2}\right)^n, \dots$$

EXAMPLE 1   **Writing the First Several Terms of a Sequence**

Write down the first six terms of the following sequence and graph it.

$$\{a_n\} = \left\{\frac{n-1}{n}\right\}$$

Solution   The first six terms of the sequence are

$$a_1 = \frac{1-1}{1} = 0, \quad a_2 = \frac{2-1}{2} = \frac{1}{2}, \quad a_3 = \frac{3-1}{3} = \frac{2}{3}, \quad a_4 = \frac{3}{4}, \quad a_5 = \frac{4}{5}, \quad a_6 = \frac{5}{6}$$

See Figure 2 for the graph.   ●

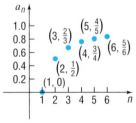

Figure 2 $\{a_n\} = \left\{\dfrac{n-1}{n}\right\}$

📱 COMMENT Graphing utilities can be used to write the terms of a sequence and graph them. Figure 3 shows the sequence given in Example 1 generated on a TI-84 Plus C graphing calculator. The first few terms of the sequence are shown on the viewing window. Press the right arrow key to scroll right to see the remaining terms of the sequence. Figure 4 shows a graph of the sequence. Note that the first term of the sequence is barely visible since it lies on the x-axis. TRACEing the graph will enable you to see the terms of the sequence. The TABLE feature can also be used to generate the terms of the sequence. See Table 1.

Table 1

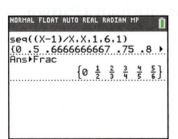

Figure 3   **Figure 4**   ■

✏️ **Now Work** PROBLEM 17

EXAMPLE 2   **Writing the First Several Terms of a Sequence**

Write down the first six terms of the following sequence and graph it.

$$\{b_n\} = \left\{(-1)^{n+1}\left(\frac{2}{n}\right)\right\}$$

Solution   (a) Formula (2) states that $a_n = a_1 + (n-1)d$. As a result,

$$\begin{cases} a_8 = a_1 + 7d = 75 \\ a_{20} = a_1 + 19d = 39 \end{cases}$$

This is a system of two linear equations containing two variables, a_1 and d, which can be solved by elimination. Subtracting the second equation from the first gives

$$-12d = 36$$
$$d = -3$$

With $d = -3$, use $a_1 + 7d = 75$ to find that $a_1 = 75 - 7d = 75 - 7(-3) = 96$. The first term is $a_1 = 96$, and the common difference is $d = -3$.

(b) Using formula (1), a recursive formula for this sequence is

$$a_1 = 96 \qquad a_n = a_{n-1} - 3$$

(c) Using formula (2), a formula for the nth term of the sequence $\{a_n\}$ is

$$a_n = a_1 + (n-1)d = 96 + (n-1)(-3) = 99 - 3n \qquad \bullet$$

Exploration

Graph the recursive formula from Example 5, $a_1 = 96$, $a_n = a_{n-1} - 3$, using a graphing utility. Conclude that the graph of the recursive formula behaves like the graph of a linear function. How is d, the common difference, related to m, the slope of a line?

 Now Work PROBLEMS **17** AND **31**

3 Find the Sum of an Arithmetic Sequence

The next result gives two formulas for finding the sum of the first n terms of an arithmetic sequence.

THEOREM

> **Sum of the First n Terms of an Arithmetic Sequence**
>
> Let $\{a_n\}$ be an arithmetic sequence with first term a_1 and common difference d. The sum S_n of the first n terms of $\{a_n\}$ may be found in two ways:
>
> $$S_n = a_1 + a_2 + a_3 + \cdots + a_n$$
>
> $$= \frac{n}{2}[2a_1 + (n-1)d] \qquad (3)$$
>
> $$= \frac{n}{2}(a_1 + a_n) \qquad (4)$$

Proof

$$\begin{aligned} S_n &= a_1 + a_2 + a_3 + \cdots + a_n && \text{Sum of first } n \text{ terms} \\ &= a_1 + (a_1 + d) + (a_1 + 2d) + \cdots + [a_1 + (n-1)d] && \text{Formula (2)} \\ &= \underbrace{(a_1 + a_1 + \cdots + a_1)}_{n \text{ terms}} + [d + 2d + \cdots + (n-1)d] && \text{Rearrange terms.} \\ &= na_1 + d[1 + 2 + \cdots + (n-1)] \\ &= na_1 + d\left[\frac{(n-1)n}{2}\right] && \text{Formula 6, Section 13.1} \\ &= na_1 + \frac{n}{2}(n-1)d \\ &= \frac{n}{2}[2a_1 + (n-1)d] && \text{Factor out } \frac{n}{2}; \text{ this is formula (3).} \\ &= \frac{n}{2}[a_1 + a_1 + (n-1)d] \\ &= \frac{n}{2}(a_1 + a_n) && \text{Use formula (2); this is formula (4).} \qquad \blacksquare \end{aligned}$$

There are two ways to find the sum of the first n terms of an arithmetic sequence. Notice that formula (3) involves the first term and common difference, whereas formula (4) involves the first term and the nth term. Use whichever form is easier.

EXAMPLE 6 | **Finding the Sum of an Arithmetic Sequence**

Find the sum S_n of the first n terms of the sequence $\{a_n\} = \{3n + 5\}$; that is, find

$$8 + 11 + 14 + \cdots + (3n + 5) = \sum_{k=1}^{n} (3k + 5)$$

Solution The sequence $\{a_n\} = \{3n + 5\}$ is an arithmetic sequence with first term $a_1 = 8$ and nth term $a_n = 3n + 5$. To find the sum S_n, use formula (4).

$$S_n = \sum_{k=1}^{n} (3k + 5) = \underset{\underset{S_n = \frac{n}{2}(a_1 + a_n)}{\uparrow}}{\frac{n}{2}[8 + (3n + 5)]} = \frac{n}{2}(3n + 13)$$

● **Now Work** PROBLEM **39**

EXAMPLE 7 | **Finding the Sum of an Arithmetic Sequence**

Find the sum: $60 + 64 + 68 + 72 + \cdots + 120$

Solution This is the sum S_n of an arithmetic sequence $\{a_n\}$ whose first term is $a_1 = 60$ and whose common difference is $d = 4$. The nth term is $a_n = 120$. Use formula (2) to find n.

$$a_n = a_1 + (n - 1)d \qquad \text{Formula (2)}$$
$$120 = 60 + (n - 1) \cdot 4 \qquad a_n = 120, a_1 = 60, d = 4$$
$$60 = 4(n - 1) \qquad \text{Simplify.}$$
$$15 = n - 1 \qquad \text{Simplify.}$$
$$n = 16 \qquad \text{Solve for } n.$$

Now use formula (4) to find the sum S_{16}.

$$60 + 64 + 68 + \cdots + 120 = S_{16} = \underset{\underset{S_n = \frac{n}{2}(a_1 + a_n)}{\uparrow}}{\frac{16}{2}(60 + 120)} = 1440$$

● **Now Work** PROBLEM **43**

EXAMPLE 8 | **Creating a Floor Design**

A ceramic tile floor is designed in the shape of a trapezoid 20 feet wide at the base and 10 feet wide at the top. See Figure 7. The tiles, which measure 12 inches by 12 inches, are to be placed so that each successive row contains one fewer tile than the preceding row. How many tiles will be required?

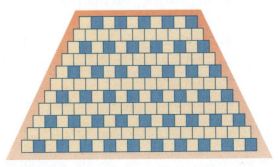

Figure 7

Solution The bottom row requires 20 tiles and the top row, 10 tiles. Since each successive row requires one fewer tile, the total number of tiles required is

$$S = 20 + 19 + 18 + \cdots + 11 + 10$$

This is the sum of an arithmetic sequence; the common difference is -1. The number of terms to be added is $n = 11$, with the first term $a_1 = 20$ and the last term $a_{11} = 10$. The sum S is

$$S = \frac{n}{2}(a_1 + a_{11}) = \frac{11}{2}(20 + 10) = 165$$

In all, 165 tiles will be required.   ●

13.2 Assess Your Understanding

Concepts and Vocabulary

1. In a(n) _____ sequence, the difference between successive terms is a constant.

2. **True or False** For an arithmetic sequence $\{a_n\}$ whose first term is a_1 and whose common difference is d, the nth term is determined by the formula $a_n = a_1 + nd$.

3. If the 5th term of an arithmetic sequence is 12 and the common difference is 5, then the 6th term of the sequence is ____ .

4. **True or False** The sum S_n of the first n terms of an arithmetic sequence $\{a_n\}$ whose first term is a_1 can be found using the formula $S_n = \frac{n}{2}(a_1 + a_n)$.

5. An arithmetic sequence can always be expressed as a(n) _____ sequence.
   (a) Fibonacci   (b) alternating
   (c) geometric   (d) recursive

6. If $a_n = -2n + 7$ is the nth term of an arithmetic sequence, the first term is _____.
   (a) -2 (b) 0
   (c) 5   (d) 7

Skill Building

In Problems 7–16, show that each sequence is arithmetic. Find the common difference, and write out the first four terms.

7. $\{s_n\} = \{n + 4\}$

8. $\{s_n\} = \{n - 5\}$

9. $\{a_n\} = \{2n - 5\}$

10. $\{b_n\} = \{3n + 1\}$

11. $\{c_n\} = \{6 - 2n\}$

12. $\{a_n\} = \{4 - 2n\}$

13. $\{t_n\} = \left\{\frac{1}{2} - \frac{1}{3}n\right\}$

14. $\{t_n\} = \left\{\frac{2}{3} + \frac{n}{4}\right\}$

15. $\{s_n\} = \{\ln 3^n\}$

16. $\{s_n\} = \{e^{\ln n}\}$

In Problems 17–24, find the nth term of the arithmetic sequence $\{a_n\}$ whose initial term a and common difference d are given. What is the 51st term?

17. $a_1 = 2; \quad d = 3$

18. $a_1 = -2; \quad d = 4$

19. $a_1 = 5; \quad d = -3$

20. $a_1 = 6; \quad d = -2$

21. $a_1 = 0; \quad d = \frac{1}{2}$

22. $a_1 = 1; \quad d = -\frac{1}{3}$

23. $a_1 = \sqrt{2}; \quad d = \sqrt{2}$

24. $a_1 = 0; \quad d = \pi$

In Problems 25–30, find the indicated term in each arithmetic sequence.

25. 100th term of $2, 4, 6, \ldots$

26. 80th term of $-1, 1, 3, \ldots$

27. 90th term of $1, -2, -5, \ldots$

28. 80th term of $5, 0, -5, \ldots$

29. 80th term of $2, \frac{5}{2}, 3, \frac{7}{2}, \ldots$

30. 70th term of $2\sqrt{5}, 4\sqrt{5}, 6\sqrt{5}, \ldots$

In Problems 31–38, find the first term and the common difference of the arithmetic sequence described. Give a recursive formula for the sequence. Find a formula for the nth term.

31. 8th term is 8; 20th term is 44

32. 4th term is 3; 20th term is 35

33. 9th term is -5; 15th term is 31

34. 8th term is 4; 18th term is -96

35. 15th term is 0; 40th term is -50

36. 5th term is -2; 13th term is 30

37. 14th term is -1; 18th term is -9

38. 12th term is 4; 18th term is 28

In Problems 39–56, find each sum.

39. $1 + 3 + 5 + \cdots + (2n - 1)$

40. $2 + 4 + 6 + \cdots + 2n$

41. $7 + 12 + 17 + \cdots + (2 + 5n)$

42. $-1 + 3 + 7 + \cdots + (4n - 5)$

43. $2 + 4 + 6 + \cdots + 70$

44. $1 + 3 + 5 + \cdots + 59$

45. $5 + 9 + 13 + \cdots + 49$

46. $2 + 5 + 8 + \cdots + 41$

47. $73 + 78 + 83 + 88 + \cdots + 558$

48. $7 + 1 - 5 - 11 - \cdots - 299$

49. $4 + 4.5 + 5 + 5.5 + \cdots + 100$

50. $8 + 8\frac{1}{4} + 8\frac{1}{2} + 8\frac{3}{4} + 9 + \cdots + 50$

51. $\sum_{n=1}^{80} (2n - 5)$

52. $\sum_{n=1}^{90} (3 - 2n)$

53. $\sum_{n=1}^{100} \left(6 - \frac{1}{2}n\right)$

54. $\sum_{n=1}^{80} \left(\frac{1}{3}n + \frac{1}{2}\right)$

55. The sum of the first 120 terms of the sequence
$14, 16, 18, 20, \ldots$

56. The sum of the first 46 terms of the sequence
$2, -1, -4, -7, \ldots$

Applications and Extensions

57. Find x so that $x + 3, 2x + 1$, and $5x + 2$ are consecutive terms of an arithmetic sequence.

58. Find x so that $2x, 3x + 2$, and $5x + 3$ are consecutive terms of an arithmetic sequence.

59. How many terms must be added in an arithmetic sequence whose first term is 11 and whose common difference is 3 to obtain a sum of 1092?

60. How many terms must be added in an arithmetic sequence whose first term is 78 and whose common difference is -4 to obtain a sum of 702?

61. Drury Lane Theater The Drury Lane Theater has 25 seats in the first row and 30 rows in all. Each successive row contains one additional seat. How many seats are in the theater?

62. Football Stadium The corner section of a football stadium has 15 seats in the first row and 40 rows in all. Each successive row contains two additional seats. How many seats are in this section?

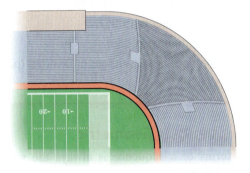

63. Creating a Mosaic A mosaic is designed in the shape of an equilateral triangle, 20 feet on each side. Each tile in the mosaic is in the shape of an equilateral triangle, 12 inches to a side. The tiles are to alternate in color as shown in the illustration. How many tiles of each color will be required?

64. Constructing a Brick Staircase A brick staircase has a total of 30 steps. The bottom step requires 100 bricks. Each successive step requires two fewer bricks than the prior step.
(a) How many bricks are required for the top step?
(b) How many bricks are required to build the staircase?

65. Cooling Air As a parcel of air rises (for example, as it is pushed over a mountain), it cools at the *dry adiabatic lapse rate* of 5.5°F per 1000 feet until it reaches its dew point. If the ground temperature is 67°F, write a formula for the sequence of temperatures, $\{T_n\}$, of a parcel of air that has risen n thousand feet. What is the temperature of a parcel of air if it has risen 5000 feet?

Source: National Aeronautics and Space Administration

66. Citrus Ladders Ladders used by fruit pickers are typically tapered with a wide bottom for stability and a narrow top for ease of picking. If the bottom rung of such a ladder is 49 inches wide and the top rung is 24 inches wide, how many rungs does the ladder have if each rung is 2.5 inches shorter than the one below it? How much material would be needed to make the rungs for the ladder described?

Source: www.stokesladders.com

67. Seats in an Amphitheater An outdoor amphitheater has 35 seats in the first row, 37 in the second row, 39 in the third row, and so on. There are 27 rows altogether. How many can the amphitheater seat?

68. Stadium Construction How many rows are in the corner section of a stadium containing 2040 seats if the first row has 10 seats and each successive row has 4 additional seats?

69. Salary If you take a job with a starting salary of $35,000 per year and a guaranteed raise of $1400 per year, how many years will it be before your aggregate salary is $280,000?

[**Hint:** Remember that your aggregate salary after 2 years is $35,000 + ($35,000 + $1400).]

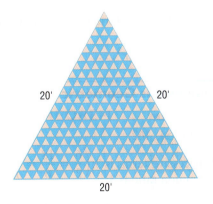

Explaining Concepts: Discussion and Writing

70. Make up an arithmetic sequence. Give it to a friend and ask for its 20th term.

71. Describe the similarities and differences between arithmetic sequences and linear functions.

Solution

(a) The first term of this geometric sequence is $a_1 = 10$, and the common ratio is

$$\frac{9}{10} . \text{(Use } \frac{9}{10} \text{ or } \frac{\frac{81}{10}}{9} = \frac{9}{10} \text{ or any two successive terms.) Then, by formula (2), the}$$

nth term is

$$a_n = 10\left(\frac{9}{10}\right)^{n-1} \qquad a_n = a_1 r^{n-1}; a_1 = 10, r = \frac{9}{10}$$

(b) The 9th term is

$$a_9 = 10\left(\frac{9}{10}\right)^{9-1} = 10\left(\frac{9}{10}\right)^{8} = 4.3046721$$

(c) The first term in the sequence is 10, and the common ratio is $r = \dfrac{9}{10}$. Using formula (1), the recursive formula is $a_1 = 10, \ a_n = \dfrac{9}{10} a_{n-1}$. ●

 **Now Work** PROBLEMS 19, 27, AND 35

Exploration

Use a graphing utility to find the ninth term of the sequence given in Example 4. Use it to find the 20th and 50th terms. Now use a graphing utility to graph the recursive formula found in Example 4(c). Conclude that the graph of the recursive formula behaves like the graph of an exponential function. How is r, the common ratio, related to a, the base of the exponential function $y = a^x$?

3 Find the Sum of a Geometric Sequence

THEOREM

Sum of the First n Terms of a Geometric Sequence

Let $\{a_n\}$ be a geometric sequence with first term a_1 and common ratio r, where $r \neq 0, r \neq 1$. The sum S_n of the first n terms of $\{a_n\}$ is

$$S_n = a_1 + a_1 r + a_1 r^2 + \cdots + a_1 r^{n-1} = \sum_{k=1}^{n} a_1 r^{k-1}$$

$$= a_1 \cdot \frac{1 - r^n}{1 - r} \qquad r \neq 0, 1 \tag{3}$$

Proof The sum S_n of the first n terms of $\{a_n\} = \{a_1 r^{n-1}\}$ is

$$S_n = a_1 + a_1 r + \cdots + a_1 r^{n-1} \tag{4}$$

Multiply each side by r to obtain

$$r S_n = a_1 r + a_1 r^2 + \cdots + a_1 r^{n} \tag{5}$$

Now, subtract (5) from (4). The result is

$$S_n - r S_n = a_1 - a_1 r^n$$

$$(1 - r) S_n = a_1 (1 - r^n)$$

Since $r \neq 1$, solve for S_n.

$$S_n = a_1 \cdot \frac{1 - r^n}{1 - r} \qquad\qquad ∎$$

EXAMPLE 5

Finding the Sum of the First n Terms of a Geometric Sequence

Find the sum S_n of the first n terms of the sequence $\left\{\left(\dfrac{1}{2}\right)^n\right\}$; that is, find

$$\frac{1}{2} + \frac{1}{4} + \frac{1}{8} + \cdots + \left(\frac{1}{2}\right)^n = \sum_{k=1}^{n} \frac{1}{2}\left(\frac{1}{2}\right)^{k-1}$$

Solution The sequence $\left\{ \left(\dfrac{1}{2}\right)^n \right\}$ is a geometric sequence with $a_1 = \dfrac{1}{2}$ and $r = \dfrac{1}{2}$. Use formula (3) to get

$$S_n = \sum_{k=1}^{n} \frac{1}{2}\left(\frac{1}{2}\right)^{k-1} = \frac{1}{2} + \frac{1}{4} + \frac{1}{8} + \cdots + \left(\frac{1}{2}\right)^n$$

$$= \frac{1}{2}\left[\frac{1 - \left(\dfrac{1}{2}\right)^n}{1 - \dfrac{1}{2}}\right] \qquad \text{Formula (3); } a_1 = \frac{1}{2}, r = \frac{1}{2}$$

$$= \frac{1}{2}\left[\frac{1 - \left(\dfrac{1}{2}\right)^n}{\dfrac{1}{2}}\right]$$

$$= 1 - \left(\frac{1}{2}\right)^n \qquad\qquad \bullet$$

 Now Work PROBLEM 41

EXAMPLE 6

Using a Graphing Utility to Find the Sum of a Geometric Sequence

Use a graphing utility to find the sum of the first 15 terms of the sequence $\left\{ \left(\dfrac{1}{3}\right)^n \right\}$; that is, find

$$\frac{1}{3} + \frac{1}{9} + \frac{1}{27} + \cdots + \left(\frac{1}{3}\right)^{15} = \sum_{k=1}^{15} \frac{1}{3}\left(\frac{1}{3}\right)^{k-1}$$

Solution Figure 8 shows the result using a TI-84 Plus C graphing calculator. The sum of the first 15 terms of the sequence $\left\{ \left(\dfrac{1}{3}\right)^n \right\}$ is approximately 0.4999999652. $\bullet$

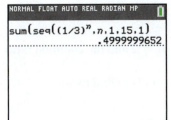

Figure 8

 Now Work PROBLEM 47

4 Determine Whether a Geometric Series Converges or Diverges

DEFINITION

An infinite sum of the form

$$a_1 + a_1 r + a_1 r^2 + \cdots + a_1 r^{n-1} + \cdots$$

with first term a_1 and common ratio r, is called an **infinite geometric series** and is denoted by

$$\sum_{k=1}^{\infty} a_1 r^{k-1}$$

Based on formula (3), the sum S_n of the first n terms of a geometric series is

$$S_n = a_1 \cdot \frac{1 - r^n}{1 - r} = \frac{a_1}{1 - r} - \frac{a_1 r^n}{1 - r} \qquad\qquad (6)$$

NOTE In calculus, limit notation is used, and the sum is written

$$L = \lim_{n \to \infty} S_n = \lim_{n \to \infty} \sum_{k=1}^{n} a_1 r^{k-1} = \sum_{k=1}^{\infty} a_1 r^{k-1}$$

∎

If this finite sum S_n approaches a number L as $n \to \infty$, then the infinite geometric series $\sum_{k=1}^{\infty} a_1 r^{k-1}$ **converges** to L and L is called the **sum of the infinite geometric series**. The sum is written as

$$L = \sum_{k=1}^{\infty} a_1 r^{k-1}$$

A series that does not converge is called a **divergent series**.

THEOREM

Convergence of an Infinite Geometric Series

If $|r| < 1$, the infinite geometric series $\sum_{k=1}^{\infty} a_1 r^{k-1}$ converges. Its sum is

$$\sum_{k=1}^{\infty} a_1 r^{k-1} = \frac{a_1}{1 - r} \qquad (7)$$

Intuitive Proof Since $|r| < 1$, it follows that $|r^n|$ approaches 0 as $n \to \infty$. Then, based on formula (6), the term $\dfrac{a_1 r^n}{1 - r}$ approaches 0, so the sum S_n approaches $\dfrac{a_1}{1 - r}$ as $n \to \infty$. ∎

EXAMPLE 7

Determining Whether a Geometric Series Converges or Diverges

Determine whether the geometric series

$$\sum_{k=1}^{\infty} 2\left(\frac{2}{3}\right)^{k-1} = 2 + \frac{4}{3} + \frac{8}{9} + \cdots$$

converges or diverges. If it converges, find its sum.

Solution

Comparing $\sum_{k=1}^{\infty} 2\left(\dfrac{2}{3}\right)^{k-1}$ to $\sum_{k=1}^{\infty} a_1 r^{k-1}$, the first term is $a_1 = 2$ and the common ratio is $r = \dfrac{2}{3}$. Since $|r| < 1$, the series converges. Use formula (7) to find its sum:

$$\sum_{k=1}^{\infty} 2\left(\frac{2}{3}\right)^{k-1} = 2 + \frac{4}{3} + \frac{8}{9} + \cdots = \frac{2}{1 - \dfrac{2}{3}} = 6$$

●

 Now Work PROBLEM 53

EXAMPLE 8

Repeating Decimals

Show that the repeating decimal $0.999\ldots$ equals 1.

Solution

The decimal $0.999\ldots = 0.9 + 0.09 + 0.009 + \cdots = \dfrac{9}{10} + \dfrac{9}{100} + \dfrac{9}{1000} + \cdots$ is an infinite geometric series. Write it in the form $\sum_{k=1}^{\infty} a_1 r^{k-1}$ and use formula (7).

$$0.999\ldots = \frac{9}{10} + \frac{9}{100} + \frac{9}{1000} + \cdots = \sum_{k=1}^{\infty} \frac{9}{10^k} = \sum_{k=1}^{\infty} \frac{9}{10 \cdot 10^{k-1}} = \sum_{k=1}^{\infty} \frac{9}{10}\left(\frac{1}{10}\right)^{k-1}$$

Compare this series to $\displaystyle\sum_{k=1}^{\infty} a_1 r^{k-1}$ and note that $a_1 = \dfrac{9}{10}$ and $r = \dfrac{1}{10}$. Since $|r| < 1$, the series converges and its sum is

$$0.999\ldots = \frac{\dfrac{9}{10}}{1 - \dfrac{1}{10}} = \frac{\dfrac{9}{10}}{\dfrac{9}{10}} = 1$$

The repeating decimal $0.999\ldots$ equals 1.

EXAMPLE 9   **Pendulum Swings**

Initially, a pendulum swings through an arc of 18 inches. See Figure 9. On each successive swing, the length of the arc is 0.98 of the previous length.

(a) What is the length of the arc of the 10th swing?
(b) On which swing is the length of the arc first less than 12 inches?
(c) After 15 swings, what total distance will the pendulum have swung?
(d) When it stops, what total distance will the pendulum have swung?

Solution

(a) The length of the first swing is 18 inches.
The length of the second swing is $0.98(18)$ inches.
The length of the third swing is $0.98(0.98)(18) = 0.98^2(18)$ inches.
The length of the arc of the 10th swing is

$$(0.98)^9(18) \approx 15.007 \text{ inches}$$

(b) The length of the arc of the nth swing is $(0.98)^{n-1}(18)$. For this to be exactly 12 inches requires that

$$(0.98)^{n-1}(18) = 12$$
$$(0.98)^{n-1} = \frac{12}{18} = \frac{2}{3} \qquad \textit{Divide both sides by 18.}$$
$$n - 1 = \log_{0.98}\left(\frac{2}{3}\right) \qquad \textit{Express as a logarithm.}$$
$$n = 1 + \frac{\ln\left(\dfrac{2}{3}\right)}{\ln 0.98} \approx 1 + 20.07 = 21.07 \quad \begin{array}{l}\textit{Solve for n; use the Change}\\\textit{of Base Formula.}\end{array}$$

The length of the arc of the pendulum exceeds 12 inches on the 21st swing and is first less than 12 inches on the 22nd swing.

(c) After 15 swings, the pendulum will have swung the following total distance L:

$$L = \underset{\text{1st}}{18} + \underset{\text{2nd}}{0.98(18)} + \underset{\text{3rd}}{(0.98)^2(18)} + \underset{\text{4th}}{(0.98)^3(18)} + \cdots + \underset{\text{15th}}{(0.98)^{14}(18)}$$

This is the sum of a geometric sequence. The common ratio is 0.98; the first term is 18. The sum has 15 terms, so

$$L = 18 \cdot \frac{1 - 0.98^{15}}{1 - 0.98} \approx 18(13.07) \approx 235.3 \text{ inches}$$

The pendulum will have swung through approximately 235.3 inches after 15 swings.

(d) When the pendulum stops, it will have swung the following total distance T:

$$T = 18 + 0.98(18) + (0.98)^2(18) + (0.98)^3(18) + \cdots$$

This is the sum of an infinite geometric series. The common ratio is $r = 0.98$; the first term is $a_1 = 18$. Since $|r| < 1$, the series converges. Its sum is

$$T = \frac{a_1}{1 - r} = \frac{18}{1 - 0.98} = 900$$

The pendulum will have swung a total of 900 inches when it finally stops.

 Now Work PROBLEM **87**

Figure 9

EXAMPLE 3

Expanding a Binomial

Expand $(2y - 3)^4$ using the Binomial Theorem.

Solution

First, rewrite the expression $(2y - 3)^4$ as $[2y + (-3)]^4$. Now use the Binomial Theorem with $n = 4$, $x = 2y$, and $a = -3$.

$$[2y + (-3)]^4 = \binom{4}{0}(2y)^4 + \binom{4}{1}(-3)(2y)^3 + \binom{4}{2}(-3)^2(2y)^2$$

$$+ \binom{4}{3}(-3)^3(2y) + \binom{4}{4}(-3)^4$$

$$= 1 \cdot 16y^4 + 4(-3)8y^3 + 6 \cdot 9 \cdot 4y^2 + 4(-27)2y + 1 \cdot 81$$

Use row $n = 4$ of the Pascal triangle or formula (1) for $\binom{n}{j}$.

$$= 16y^4 - 96y^3 + 216y^2 - 216y + 81$$

In this expansion, note that the signs alternate because $a = -3 < 0$. ●

> **Now Work** PROBLEM 21

EXAMPLE 4

Finding a Particular Coefficient in a Binomial Expansion

Find the coefficient of y^8 in the expansion of $(2y + 3)^{10}$.

Solution

Write out the expansion using the Binomial Theorem.

$$(2y + 3)^{10} = \binom{10}{0}(2y)^{10} + \binom{10}{1}(2y)^9(3)^1 + \binom{10}{2}(2y)^8(3)^2 + \binom{10}{3}(2y)^7(3)^3$$

$$+ \binom{10}{4}(2y)^6(3)^4 + \cdots + \binom{10}{9}(2y)(3)^9 + \binom{10}{10}(3)^{10}$$

From the third term in the expansion, the coefficient of y^8 is

$$\binom{10}{2}(2)^8(3)^2 = \frac{10!}{2!\,8!} \cdot 2^8 \cdot 9 = \frac{10 \cdot 9 \cdot 8!}{2 \cdot 8!} \cdot 2^8 \cdot 9 = 103,680$$ ●

As this solution demonstrates, the Binomial Theorem can be used to find a particular term in an expansion without writing the entire expansion.

Based on the expansion of $(x + a)^n$, the term containing x^j is

$$\binom{n}{n - j}a^{n-j}x^j \qquad (3)$$

Example 4 can be solved by using formula (3) with $n = 10$, $a = 3$, $x = 2y$, and $j = 8$. Then the term containing y^8 is

$$\binom{10}{10 - 8}3^{10-8}(2y)^8 = \binom{10}{2} \cdot 3^2 \cdot 2^8 \cdot y^8 = \frac{10!}{2!\,8!} \cdot 9 \cdot 2^8 y^8$$

$$= \frac{10 \cdot 9 \cdot 8!}{2 \cdot 8!} \cdot 9 \cdot 2^8\, y^8 = 103,680y^8$$

| EXAMPLE 5 | **Finding a Particular Term in a Binomial Expansion** |

Find the 6th term in the expansion of $(x + 2)^9$.

Solution A Expand using the Binomial Theorem until the 6th term is reached.

$$(x + 2)^9 = \binom{9}{0}x^9 + \binom{9}{1}x^8 \cdot 2 + \binom{9}{2}x^7 \cdot 2^2 + \binom{9}{3}x^6 \cdot 2^3 + \binom{9}{4}x^5 \cdot 2^4$$
$$+ \binom{9}{5}x^4 \cdot 2^5 + \cdots$$

The 6th term is

$$\binom{9}{5}x^4 \cdot 2^5 = \frac{9!}{5! \, 4!} \cdot x^4 \cdot 32 = 4032x^4$$

Solution B The 6th term in the expansion of $(x + 2)^9$, which has 10 terms total, contains x^4. (Do you see why?) By formula (3), the 6th term is

$$\binom{9}{9 - 4}2^{9-4}x^4 = \binom{9}{5}2^5 x^4 = \frac{9!}{5! \, 4!} \cdot 32x^4 = 4032x^4$$

 Now Work PROBLEMS 29 AND 35

The following theorem shows that the *triangular addition* feature of the Pascal triangle illustrated in Figure 12 always works.

THEOREM If n and j are integers with $1 \le j \le n$, then

$$\binom{n}{j - 1} + \binom{n}{j} = \binom{n + 1}{j} \tag{4}$$

Proof

$$\binom{n}{j - 1} + \binom{n}{j} = \frac{n!}{(j - 1)! \, [n - (j - 1)]!} + \frac{n!}{j! \, (n - j)!}$$

$$= \frac{n!}{(j - 1)! \, (n - j + 1)!} + \frac{n!}{j! \, (n - j)!}$$

$$= \frac{jn!}{j(j - 1)! \, (n - j + 1)!} + \frac{(n - j + 1)n!}{j! \, (n - j + 1)(n - j)!} \qquad \text{Multiply the first term by } \frac{j}{j} \text{ and}$$
$$\text{the second term by } \frac{n - j + 1}{n - j + 1}$$

$$= \frac{jn!}{j! \, (n - j + 1)!} + \frac{(n - j + 1)n!}{j! \, (n - j + 1)!} \qquad \text{to make the denominators equal.}$$

$$= \frac{jn! + (n - j + 1)n!}{j! \, (n - j + 1)!}$$

$$= \frac{n!(j + n - j + 1)}{j! \, (n - j + 1)!}$$

$$= \frac{n!(n + 1)}{j! \, (n - j + 1)!} = \frac{(n + 1)!}{j! \, [(n + 1) - j]!} = \binom{n + 1}{j} \qquad \blacksquare$$

Historical Feature

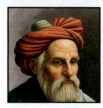

*Omar Khayyám
(1048–1131)*

The case $n = 2$ of the Binomial Theorem, $(a + b)^2$, was known to Euclid in 300 BC, but the general law seems to have been discovered by the Persian mathematician and astronomer Omar Khayyám (1048–1131), who is also well known as the author of the *Rubáiyát*, a collection of four-line poems making observations on the human condition. Omar Khayyám did not state the Binomial Theorem explicitly, but he claimed to have a method for extracting third, fourth, fifth roots, and so on. A little study shows that one must know the Binomial Theorem to create such a method.

The heart of the Binomial Theorem is the formula for the numerical coefficients, and, as we saw, they can be written in a symmetric triangular form. The Pascal triangle appears first in the books of Yang Hui (about 1270) and Chu Shih-chieh (1303). Pascal's name is attached to the triangle because of the many applications he made of it, especially to counting and probability. In establishing these results, he was one of the earliest users of mathematical induction.

Many people worked on the proof of the Binomial Theorem, which was finally completed for all n (including complex numbers) by Niels Abel (1802–1829).

13.5 Assess Your Understanding

Concepts and Vocabulary

1. The _____ _____ is a triangular display of the binomial coefficients.

2. $\binom{n}{0} =$ ____ and $\binom{n}{1} =$ ____ .

3. *True or False* $\binom{n}{j} = \dfrac{j!}{(n-j)!\, n!}$

4. The _____ _____ can be used to expand expressions like $(2x + 3)^6$.

Skill Building

In Problems 5–16, evaluate each expression.

5. $\binom{5}{3}$

6. $\binom{7}{3}$

7. $\binom{7}{5}$

8. $\binom{9}{7}$

9. $\binom{50}{49}$

10. $\binom{100}{98}$

11. $\binom{1000}{1000}$

12. $\binom{1000}{0}$

13. $\binom{55}{23}$

14. $\binom{60}{20}$

15. $\binom{47}{25}$

16. $\binom{37}{19}$

In Problems 17–28, expand each expression using the Binomial Theorem.

17. $(x + 1)^5$

18. $(x - 1)^5$

19. $(x - 2)^6$

20. $(x + 3)^5$

21. $(3x + 1)^4$

22. $(2x + 3)^5$

23. $(x^2 + y^2)^5$

24. $(x^2 - y^2)^6$

25. $(\sqrt{x} + \sqrt{2}\,)^6$

26. $(\sqrt{x} - \sqrt{3}\,)^4$

27. $(ax + by)^5$

28. $(ax - by)^4$

In Problems 29–42, use the Binomial Theorem to find the indicated coefficient or term.

29. The coefficient of x^6 in the expansion of $(x + 3)^{10}$

30. The coefficient of x^3 in the expansion of $(x - 3)^{10}$

31. The coefficient of x^7 in the expansion of $(2x - 1)^{12}$

32. The coefficient of x^3 in the expansion of $(2x + 1)^{12}$

33. The coefficient of x^7 in the expansion of $(2x + 3)^9$

34. The coefficient of x^2 in the expansion of $(2x - 3)^9$

35. The 5th term in the expansion of $(x + 3)^7$

36. The 3rd term in the expansion of $(x - 3)^7$

37. The 3rd term in the expansion of $(3x - 2)^9$

38. The 6th term in the expansion of $(3x + 2)^8$

39. The coefficient of x^0 in the expansion of $\left(x^2 + \dfrac{1}{x}\right)^{12}$

40. The coefficient of x^0 in the expansion of $\left(x - \dfrac{1}{x^2}\right)^9$

41. The coefficient of x^4 in the expansion of $\left(x - \dfrac{2}{\sqrt{x}}\right)^{10}$

42. The coefficient of x^2 in the expansion of $\left(\sqrt{x} + \dfrac{3}{\sqrt{x}}\right)^8$

Applications and Extensions

43. Use the Binomial Theorem to find the numerical value of $(1.001)^5$ correct to five decimal places.
[**Hint:** $(1.001)^5 = (1 + 10^{-3})^5$]

44. Use the Binomial Theorem to find the numerical value of $(0.998)^6$ correct to five decimal places.

45. Show that $\dbinom{n}{n-1} = n$ and $\dbinom{n}{n} = 1$.

46. Show that if n and j are integers with $0 \le j \le n$, then,

$$\binom{n}{j} = \binom{n}{n-j}$$

Conclude that the Pascal triangle is symmetric with respect to a vertical line drawn from the topmost entry.

47. If n is a positive integer, show that

$$\binom{n}{0} + \binom{n}{1} + \cdots + \binom{n}{n} = 2^n$$

[**Hint:** $2^n = (1 + 1)^n$; now use the Binomial Theorem.]

48. If n is a positive integer, show that

$$\binom{n}{0} - \binom{n}{1} + \binom{n}{2} - \cdots + (-1)^n\binom{n}{n} = 0$$

49. $\dbinom{5}{0}\left(\dfrac{1}{4}\right)^5 + \dbinom{5}{1}\left(\dfrac{1}{4}\right)^4\left(\dfrac{3}{4}\right) + \dbinom{5}{2}\left(\dfrac{1}{4}\right)^3\left(\dfrac{3}{4}\right)^2$

$+ \dbinom{5}{3}\left(\dfrac{1}{4}\right)^2\left(\dfrac{3}{4}\right)^3 + \dbinom{5}{4}\left(\dfrac{1}{4}\right)\left(\dfrac{3}{4}\right)^4 + \dbinom{5}{5}\left(\dfrac{3}{4}\right)^5 = ?$

50. Stirling's Formula An approximation for $n!$, when n is large, is given by

$$n! \approx \sqrt{2n\pi}\left(\dfrac{n}{e}\right)^n\left(1 + \dfrac{1}{12n-1}\right)$$

Calculate 12!, 20!, and 25! on your calculator. Then use Stirling's formula to approximate 12!, 20!, and 25!.

Retain Your Knowledge

Problems 51–54 are based on material learned earlier in the course. The purpose of these problems is to keep the material fresh in your mind so that you are better prepared for the final exam.

51. Solve $6^x = 5^{x+1}$. Express the answer both in exact form and as a decimal rounded to three decimal places.

52. For $\mathbf{v} = 2\mathbf{i} + 3\mathbf{j}$ and $\mathbf{w} = 3\mathbf{i} - 2\mathbf{j}$:
(a) Find the dot product $\mathbf{v} \cdot \mathbf{w}$.
(b) Find the angle between $\mathbf{v}$ and $\mathbf{w}$.
(c) Are the vectors parallel, orthogonal, or neither?

53. Solve the system of equations:

$$\begin{cases} x - y - z = 0 \\ 2x + y + 3z = -1 \\ 4x + 2y - z = 12 \end{cases}$$

54. Graph the system of inequalities. Tell whether the graph is bounded or unbounded, and label the corner points.

$$\begin{cases} x \ge 0 \\ y \ge 0 \\ x + y \le 6 \\ 2x + y \le 10 \end{cases}$$

Chapter Review

Things to Know

Sequence (p. 940)	A function whose domain is the set of positive integers
Factorials (p. 942)	$0! = 1, 1! = 1, n! = n(n-1)\cdots\cdot 3 \cdot 2 \cdot 1$ if $n \ge 2$ is an integer
Arithmetic sequence (pp. 950 and 951)	$a_1 = a,\ a_n = a_{n-1} + d$, where $a_1 = a = $ first term, $d = $ common difference $a_n = a_1 + (n-1)d$
Sum of the first n terms of an arithmetic sequence (p. 952)	$S_n = \dfrac{n}{2}[2a_1 + (n-1)d] = \dfrac{n}{2}(a_1 + a_n)$
Geometric sequence (pp. 956 and 957)	$a_1 = a,\ \ a_n = ra_{n-1}$, where $a_1 = a = $ first term, $r = $ common ratio $a_n = a_1 r^{n-1}\quad r \ne 0$
Sum of the first n terms of a geometric sequence (p. 958)	$S_n = a_1\dfrac{1 - r^n}{1 - r}\quad r \ne 0, 1$
Infinite geometric series (p. 959)	$a_1 + a_1 r + \cdots + a_1 r^{n-1} + \cdots = \displaystyle\sum_{k=1}^{\infty} a_1 r^{k-1}$
Sum of a convergent infinite geometric series (p. 960)	If $\lvert r \rvert < 1$, $\displaystyle\sum_{k=1}^{\infty} a_1 r^{k-1} = \dfrac{a_1}{1 - r}$

Amount of an annuity (962)	$A = P \dfrac{(1 + i)^n - 1}{i}$, where $P =$ the deposit (in dollars) made at the end of each payment period, $i =$ interest rate per payment period, and $A =$ the amount of the annuity after n deposits.
Principle of Mathematical Induction (p. 968)	If the following two conditions are satisfied,
	Condition I: The statement is true for the natural number 1.
	Condition II: If the statement is true for some natural number k, it is also true for $k + 1$
	then the statement is true for all natural numbers.
Binomial coefficient (p. 972)	$\dbinom{n}{j} = \dfrac{n!}{j!\,(n - j)!}$
The Pascal triangle (p. 973)	See Figure 12.
Binomial Theorem (p. 973)	$(x + a)^n = \dbinom{n}{0}x^n + \dbinom{n}{1}ax^{n-1} + \cdots + \dbinom{n}{j}a^j x^{n-j} + \cdots + \dbinom{n}{n}a^n = \sum\limits_{j=0}^{n}\dbinom{n}{j}x^{n-j}a^j$

Objectives

Section		You should be able to...	Example(s)	Review Exercises
13.1	1	Write the first several terms of a sequence (p. 940)	1–4	1, 2
	2	Write the terms of a sequence defined by a recursive formula (p. 943)	5, 6	3, 4
	3	Use summation notation (p. 944)	7, 8	5, 6
	4	Find the sum of a sequence (p. 945)	9	13, 14
13.2	1	Determine whether a sequence is arithmetic (p. 950)	1–3	7–12
	2	Find a formula for an arithmetic sequence (p. 951)	4, 5	17, 19–21, 34(a)
	3	Find the sum of an arithmetic sequence (p. 952)	6–8	7, 10, 14, 34(b), 35
13.3	1	Determine whether a sequence is geometric (p. 956)	1–3	7–12
	2	Find a formula for a geometric sequence (p. 957)	4	11, 18, 36(a)–(c), 38
	3	Find the sum of a geometric sequence (p. 958)	5, 6	9, 11, 15, 16
	4	Determine whether a geometric series converges or diverges (p. 959)	7–9	22–25, 36(d)
	5	Solve annuity problems (p. 962)	10, 11	37
13.4	1	Prove statements using mathematical induction (p. 967)	1–4	26–28
13.5	1	Evaluate $\dbinom{n}{j}$ (p. 971)	1	29
	2	Use the Binomial Theorem (p. 973)	2–5	30–33

Review Exercises

In Problems 1–4, write down the first five terms of each sequence.

1. $\{a_n\} = \left\{(-1)^n\left(\dfrac{n+3}{n+2}\right)\right\}$ **2.** $\{c_n\} = \left\{\dfrac{2^n}{n^2}\right\}$ **3.** $a_1 = 3;\ a_n = \dfrac{2}{3}a_{n-1}$ **4.** $a_1 = 2;\ a_n = 2 - a_{n-1}$

5. Write out $\sum\limits_{k=1}^{4}(4k + 2)$.

6. Express $1 - \dfrac{1}{2} + \dfrac{1}{3} - \dfrac{1}{4} + \cdots + \dfrac{1}{13}$ using summation notation.

In Problems 7–12, determine whether the given sequence is arithmetic, geometric, or neither. If the sequence is arithmetic, find the common difference and the sum of the first n terms. If the sequence is geometric, find the common ratio and the sum of the first n terms.

7. $\{a_n\} = \{n + 5\}$ **8.** $\{c_n\} = \{2n^3\}$ **9.** $\{s_n\} = \{2^{3n}\}$

10. $0, 4, 8, 12, \ldots$ **11.** $3, \dfrac{3}{2}, \dfrac{3}{4}, \dfrac{3}{8}, \dfrac{3}{16}, \ldots$ **12.** $\dfrac{2}{3}, \dfrac{3}{4}, \dfrac{4}{5}, \dfrac{5}{6}, \ldots$

In Problems 13–16, find each sum.

13. $\displaystyle\sum_{k=1}^{30} (k^2 + 2)$

14. $\displaystyle\sum_{k=1}^{40} (-2k + 8)$

15. $\displaystyle\sum_{k=1}^{7} \left(\frac{1}{3}\right)^k$

16. $\displaystyle\sum_{k=1}^{10} (-2)^k$

In Problems 17–19, find the indicated term in each sequence. [**Hint:** *Find the general term first.*]

17. 9th term of $3, 7, 11, 15, \ldots$

18. 11th term of $1, \dfrac{1}{10}, \dfrac{1}{100}, \ldots$

19. 9th term of $\sqrt{2}, 2\sqrt{2}, 3\sqrt{2}, \ldots$

In Problems 20 and 21, find a general formula for each arithmetic sequence.

20. 7th term is 31; 20th term is 96

21. 10th term is 0; 18th term is 8

In Problems 22–25, determine whether each infinite geometric series converges or diverges. If it converges, find its sum.

22. $3 + 1 + \dfrac{1}{3} + \dfrac{1}{9} + \cdots$

23. $2 - 1 + \dfrac{1}{2} - \dfrac{1}{4} + \cdots$

24. $\dfrac{1}{2} + \dfrac{3}{4} + \dfrac{9}{8} + \cdots$

25. $\displaystyle\sum_{k=1}^{\infty} 4\left(\frac{1}{2}\right)^{k-1}$

In Problems 26–28, use the Principle of Mathematical Induction to show that the given statement is true for all natural numbers.

26. $3 + 6 + 9 + \cdots + 3n = \dfrac{3n}{2}(n + 1)$

27. $2 + 6 + 18 + \cdots + 2 \cdot 3^{n-1} = 3^n - 1$

28. $1^2 + 4^2 + 7^2 + \cdots + (3n - 2)^2 = \dfrac{1}{2}n(6n^2 - 3n - 1)$

29. Evaluate: $\dbinom{5}{2}$

In Problems 30 and 31, expand each expression using the Binomial Theorem.

30. $(x + 2)^5$

31. $(3x - 4)^4$

---

32. Find the coefficient of x^7 in the expansion of $(x + 2)^9$

33. Find the coefficient of x^2 in the expansion of $(2x + 1)^7$.

34. **Constructing a Brick Staircase** A brick staircase has a total of 25 steps. The bottom step requires 80 bricks. Each step thereafter requires three fewer bricks than the prior step.
 (a) How many bricks are required for the top step?
 (b) How many bricks are required to build the staircase?

35. **Creating a Floor Design** A mosaic tile floor is designed in the shape of a trapezoid 30 feet wide at the base and 15 feet wide at the top. The tiles, 12 inches by 12 inches, are to be placed so that each successive row contains one fewer tile than the row below. How many tiles will be required?

36. **Bouncing Balls** A ball is dropped from a height of 20 feet. Each time it strikes the ground, it bounces up to three-quarters of the height of the previous bounce.

 (a) What height will the ball bounce up to after it strikes the ground for the 3rd time?
 (b) How high will it bounce after it strikes the ground for the nth time?
 (c) How many times does the ball need to strike the ground before its bounce is less than 6 inches?
 (d) What total distance does the ball travel before it stops bouncing?

37. **Retirement Planning** Chris gets paid once a month and contributes \$200 each pay period into his 401(k). If Chris plans on retiring in 20 years, what will be the value of his 401(k) if the per annum rate of return of the 401(k) is 10% compounded monthly?

38. **Salary Increases** Your friend has just been hired at an annual salary of \$50,000. If she expects to receive annual increases of 4%, what will be her salary as she begins her 5th year?

Chapter Test

The Chapter Test Prep Videos are step-by-step solutions available in **MyMathLab**, or on this text's **YouTube** Channel. Flip back to the Resources for Success page for a link to this text's YouTube channel.

In Problems 1 and 2, write down the first five terms of each sequence.

1. $\{s_n\} = \left\{\dfrac{n^2 - 1}{n + 8}\right\}$

2. $a_1 = 4, a_n = 3a_{n-1} + 2$

In Problems 3 and 4, write out each sum. Evaluate each sum.

3. $\displaystyle\sum_{k=1}^{3} (-1)^{k+1}\left(\dfrac{k+1}{k^2}\right)$

4. $\displaystyle\sum_{k=1}^{4}\left[\left(\dfrac{2}{3}\right)^k - k\right]$

5. Write the following sum using summation notation.
$$-\dfrac{2}{5} + \dfrac{3}{6} - \dfrac{4}{7} + \cdots + \dfrac{11}{14}$$

In Problems 6–11, determine whether the given sequence is arithmetic, geometric, or neither. If the sequence is arithmetic, find the common difference and the sum of the first n terms. If the sequence is geometric, find the common ratio and the sum of the first n terms.

6. $6, 12, 36, 144, \ldots$

7. $\left\{-\dfrac{1}{2} \cdot 4^n\right\}$

8. $-2, -10, -18, -26, \ldots$

9. $\left\{-\dfrac{n}{2} + 7\right\}$

10. $25, 10, 4, \dfrac{8}{5}, \ldots$

11. $\left\{\dfrac{2n - 3}{2n + 1}\right\}$

12. Determine whether the infinite geometric series
$$256 - 64 + 16 - 4 + \cdots$$
converges or diverges. If it converges, find its sum.

13. Expand $(3m + 2)^5$ using the Binomial Theorem.

14. Use the Principle of Mathematical Induction to show that the given statement is true for all natural numbers.
$$\left(1 + \dfrac{1}{1}\right)\left(1 + \dfrac{1}{2}\right)\left(1 + \dfrac{1}{3}\right)\cdots\left(1 + \dfrac{1}{n}\right) = n + 1$$

15. A new car sold for $31,000. If the vehicle loses 15% of its value each year, how much will it be worth after 10 years?

16. A weightlifter begins his routine by benching 100 pounds and increases the weight by 30 pounds for each set. If he does 10 repetitions in each set, what is the total weight lifted after 5 sets?

Cumulative Review

1. Find all the solutions, real and complex, of the equation
$$|x^2| = 9$$

2. (a) Graph the circle $x^2 + y^2 = 100$ and the parabola $y = 3x^2$.

  (b) Solve the system of equations: $\begin{cases} x^2 + y^2 = 100 \\ y = 3x^2 \end{cases}$

  (c) Where do the circle and the parabola intersect?

3. Solve the equation: $2e^x = 5$

4. Find an equation of the line with slope 5 and x-intercept 2.

5. Find the standard equation of the circle whose center is the point $(-1, 2)$ if $(3, 5)$ is a point on the circle.

6. $f(x) = \dfrac{3x}{x - 2}$ and $g(x) = 2x + 1$

Find:

  (a) $(f \circ g)(2)$

  (b) $(g \circ f)(4)$

  (c) $(f \circ g)(x)$

  (d) The domain of $(f \circ g)(x)$

  (e) $(g \circ f)(x)$

  (f) The domain of $(g \circ f)(x)$

  (g) The function g^{-1} and its domain

  (h) The function f^{-1} and its domain

7. Find the equation of an ellipse with center at the origin, a focus at $(0, 3)$, and a vertex at $(0, 4)$.

8. Find the equation of a parabola with vertex at $(-1, 2)$ and focus at $(-1, 3)$.

9. Find the polar equation of a circle with center at $(0, 4)$ that passes through the pole. What is the rectangular equation?

10. Solve the equation
$$2 \sin^2 x - \sin x - 3 = 0, \quad 0 \le x < 2\pi$$

11. Find the exact value of $\cos^{-1}(-0.5)$.

12. If $\sin \theta = \dfrac{1}{4}$ and θ is in the second quadrant, find:

  (a) $\cos \theta$

  (b) $\tan \theta$

  (c) $\sin(2\theta)$

  (d) $\cos(2\theta)$

  (e) $\sin\left(\dfrac{1}{2}\theta\right)$

Chapter Projects

Internet-based Project

I. Population Growth The size of the population of the United States essentially depends on its current population, the birth and death rates of the population, and immigration. Let b represent the birth rate of the U.S. population, and let d represent its death rate. Then $r = b - d$ represents the growth rate of the population, where r varies from year to year. The U.S. population after n years can be modeled using the recursive function

$$p_n = (1 + r)p_{n-1} + I$$

where I represents net immigration into the United States.

1. Using data from the CIA World Factbook at *www.cia.gov/cia/publications/factbook/index.html*, determine the birth and death rates in the United States for the most recent year that data are available. Birth rates

and death rates are given as the number of live births per 1000 population. Each must be computed as the number of births (deaths) per individual. For example, in 2014, the birth rate was 13.42 per 1000 and the death rate was 8.15 per 1000, so

$$b = \frac{13.42}{1000} = 0.01342, \text{ while } d = \frac{8.15}{1000} = 0.00815.$$

Next, using data from the Immigration and Naturalization Service at *www.fedstats.gov*, determine the net immigration into the United States for the same year used to obtain b and d.

2. Determine the value of r, the growth rate of the population.

3. Find a recursive formula for the population of the United States.

4. Use the recursive formula to predict the population of the United States in the following year. In other words, if data are available up to the year 2014, predict the U.S. population in 2015.

5. Does your prediction seem reasonable? Explain.

6. Repeat Problems 1–5 for Uganda using the CIA World Factbook (in 2014, the birth rate was 47.17 per 1000 and the death rate was 10.97 per 1000).

7. Do your results for the United States (a developed country) and Uganda (a developing country) seem in line with the article in the chapter opener? Explain.

8. Do you think the recursive formula found in Problem 3 will be useful in predicting future populations? Why or why not?

The following projects are available at the Instructor's Resource Center (IRC):

II. Project at Motorola *Digital Wireless Communication* Cell phones take speech and change it into digital code using only zeros and ones. See how the code length can be modeled using a mathematical sequence.

III. Economics Economists use the current price of a good and a recursive model to predict future consumer demand and to determine future production.

IV. Standardized Tests Many tests of intelligence, aptitude, and achievement contain questions asking for the terms of a mathematical sequence.

14 Counting and Probability

Purchasing a Lottery Ticket

In recent years, the jackpot prizes for the nation's two major multistate lotteries, Mega Millions and Powerball, have climbed to all-time highs. This has happened since Powerball (in January 2012) and Mega Millions (in October 2013) made it more difficult to win their top prizes. The probability of winning the Mega Millions jackpot is now about 1 in 259 million, and the probability for Powerball is about 1 in 175 million.

With such improbable chances of winning the jackpots, one might wonder if there *ever* comes a point when purchasing a lottery ticket is worthwhile. One important consideration in making this determination is the **expected profit**. For a game of chance, the expected profit is a measure of how much a player will profit (or lose) if she or he plays the game a large number of times.

The Internet-based project at the end of this chapter explores the expected profits from playing Mega Millions and Powerball and examines how the expected profit is related to the jackpot amounts.

—See Chapter Project I—

Outline

••• A Look Back

We introduced sets in Chapter R, Review, and have been using them to represent solutions of equations and inequalities and to represent the domain and range of functions.

A Look Ahead •••

Here we discuss methods for counting the number of elements in a set and consider the role of sets in probability.

14.1 Counting

PREPARING FOR THIS SECTION *Before getting started, review the following:*

- Sets (Chapter R, Review, Section R.1, pp. 2–3)

✎ **Now Work** the '**Are You Prepared?**' problems on page 987.

OBJECTIVES 1 Find All the Subsets of a Set (p. 983)
2 Count the Number of Elements in a Set (p. 983)
3 Solve Counting Problems Using the Multiplication Principle (p. 985)

Counting plays a major role in many diverse areas, such as probability, statistics, and computer science; counting techniques are a part of a branch of mathematics called **combinatorics**.

1 Find All the Subsets of a Set

We begin by reviewing the ways in which two sets can be compared.

If two sets A and B have precisely the same elements, we say that A and B are **equal** and write $A = B$.

If each element of a set A is also an element of a set B, we say that A is a **subset** of B and write $A \subseteq B$.

If $A \subseteq B$ and $A \neq B$, we say that A is a **proper subset** of B and write $A \subset B$.

If $A \subseteq B$, every element in set A is also in set B, but B may or may not have additional elements. If $A \subset B$, every element in A is also in B, and B has at least one element not found in A.

Finally, we agree that the empty set, $\varnothing$, is a subset of every set; that is,

$$\varnothing \subseteq A \qquad \text{for any set } A$$

EXAMPLE 1 **Finding All the Subsets of a Set**

Write down all the subsets of the set $\{a, b, c\}$.

Solution To organize the work, write down all the subsets with no elements, then those with one element, then those with two elements, and finally those with three elements. This gives all the subsets. Do you see why?

0 Elements	1 Element	2 Elements	3 Elements
$\varnothing$	$\{a\}, \{b\}, \{c\}$	$\{a, b\}, \{b, c\}, \{a, c\}$	$\{a, b, c\}$

●

✎ **Now Work** PROBLEM 9

2 Count the Number of Elements in a Set

As you count the number of students in a classroom or the number of pennies in your pocket, what you are really doing is matching, on a one-to-one basis, each object to be counted with the set of counting numbers, 1, 2, 3, ..., n, for some number n. If a set A matched up in this fashion with the set $\{1, 2, ..., 25\}$, you would conclude that there are 25 elements in the set A. The notation $n(A) = 25$ is used to indicate that there are 25 elements in the set A.

Because the empty set has no elements, we write

$$n(\varnothing) = 0$$

If the number of elements in a set is a nonnegative integer, the set is **finite.** Otherwise, it is **infinite**. We shall concern ourselves only with finite sets.

Look again at Example 1. A set with 3 elements has $2^3 = 8$ subsets. This result can be generalized.

In Words
The notation $n(A)$ means "the number of elements in set A."

If A is a set with n elements, then A has 2^n subsets.

For example, the set $\{a, b, c, d, e\}$ has $2^5 = 32$ subsets.

EXAMPLE 2 **Analyzing Survey Data**

In a survey of 100 college students, 35 were registered in College Algebra, 52 were registered in Computer Science I, and 18 were registered in both courses.

(a) How many students were registered in College Algebra or Computer Science I?
(b) How many were registered in neither course?

Solution   (a) First, let A = set of students in College Algebra

$$B = \text{set of students in Computer Science I}$$

Then the given information tells us that

$$n(A) = 35 \qquad n(B) = 52 \qquad n(A \cap B) = 18$$

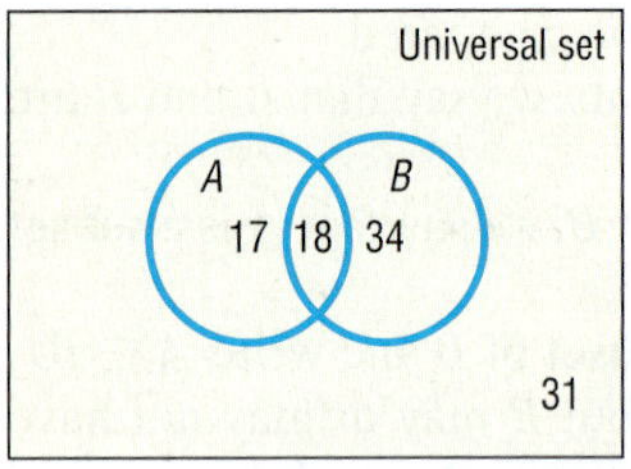

Refer to Figure 1. Since $n(A \cap B) = 18$, the common part of the circles representing set A and set B has 18 elements. In addition, the remaining portion of the circle representing set A will have $35 - 18 = 17$ elements. Similarly, the remaining portion of the circle representing set B has $52 - 18 = 34$ elements. This means that $17 + 18 + 34 = 69$ students were registered in College Algebra or Computer Science I.

(b) Since 100 students were surveyed, it follows that $100 - 69 = 31$ were registered in neither course. ●

Figure 1

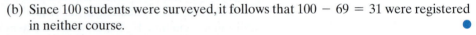

Now Work PROBLEMS 17 AND 27

The solution to Example 2 contains the basis for a general counting formula. If we count the elements in each of two sets A and B, we necessarily count twice any elements that are in both A and B—that is, those elements in $A \cap B$. To count correctly the elements that are in A or B—that is, to find $n(A \cup B)$—we need to subtract those in $A \cap B$ from $n(A) + n(B)$.

THEOREM **Counting Formula**

If A and B are finite sets,

$$n(A \cup B) = n(A) + n(B) - n(A \cap B) \tag{1}$$

Refer to Example 2. Using formula (1), we have

$$n(A \cup B) = n(A) + n(B) - n(A \cap B)$$
$$= 35 + 52 - 18$$
$$= 69$$

There are 69 students registered in College Algebra or Computer Science I.

A special case of the counting formula (1) occurs if A and B have no elements in common. In this case, $A \cap B = \varnothing$, so $n(A \cap B) = 0$.

THEOREM **Addition Principle of Counting**

If two sets A and B have no elements in common, that is,

$$\text{if } A \cap B = \varnothing, \text{ then } n(A \cup B) = n(A) + n(B) \tag{2}$$

Formula (2) can be generalized.

THEOREM

General Addition Principle of Counting

If, for n sets $A_1, A_2, \ldots, A_n$, no two have elements in common, then

$$n(A_1 \cup A_2 \cup \cdots \cup A_n) = n(A_1) + n(A_2) + \cdots + n(A_n) \qquad \textbf{(3)}$$

EXAMPLE 3

Counting

Table 1 lists the level of education for all United States residents 25 years of age or older in 2013.

Table 1

Level of Education	Number of U.S. Residents at Least 25 Years Old
Not a high school graduate	24,517,000
High school graduate	61,704,000
Some college, but no degree	34,805,000
Associate's degree	20,367,000
Bachelor's degree	41,575,000
Advanced degree	23,931,000

Source: U.S. Census Bureau

(a) How many U.S. residents 25 years of age or older had an associate's degree or a bachelor's degree?

(b) How many U.S. residents 25 years of age or older had an associate's degree, a bachelor's degree, or an advanced degree?

Solution   Let A represent the set of associate's degree holders, B represent the set of bachelor's degree holders, and C represent the set of advanced degree holders. No two of the sets A, B, and C have elements in common (although the holder of an advanced degree certainly also holds a bachelor's degree, the individual would be part of the set for which the highest degree has been conferred). Then

$$n(A) = 20{,}367{,}000 \quad n(B) = 41{,}575{,}000 \quad n(C) = 23{,}931{,}000$$

(a) Using formula (2),

$$n(A \cup B) = n(A) + n(B) = 20{,}367{,}000 + 41{,}575{,}000 = 61{,}942{,}000$$

There were 61,942,000 U.S. residents 25 years of age or older who had an associate's degree or a bachelor's degree.

(b) Using formula (3),

$$n(A \cup B \cup C) = n(A) + n(B) + n(C)$$
$$= 20{,}367{,}000 + 41{,}575{,}000 + 23{,}931{,}000$$
$$= 85{,}873{,}000$$

There were 85,873,000 U.S. residents 25 years of age or older who had an associate's degree, a bachelor's degree, or an advanced degree. ●

 Now Work PROBLEM 31

3 Solve Counting Problems Using the Multiplication Principle

EXAMPLE 4

Counting the Number of Possible Meals

The fixed-price dinner at Mabenka Restaurant provides the following choices:

Appetizer: soup or salad

Entrée: baked chicken, broiled beef patty, beef liver, or roast beef au jus

Dessert: ice cream or cheese cake

How many different meals can be ordered?

Solution   Ordering such a meal requires three separate decisions:

Choose an Appetizer	**Choose an Entrée**	**Choose a Dessert**
2 choices	4 choices	2 choices

Look at the **tree diagram** in Figure 2. Note that for each choice of appetizer, there are 4 choices of entrées. And for each of these $2 \cdot 4 = 8$ choices, there are 2 choices for dessert. A total of

$$2 \cdot 4 \cdot 2 = 16$$

different meals can be ordered.

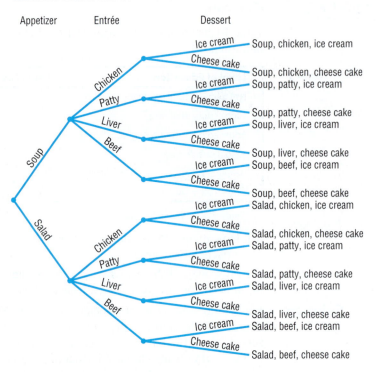

Figure 2

Example 4 demonstrates a general principle of counting.

THEOREM

Multiplication Principle of Counting

If a task consists of a sequence of choices in which there are p selections for the first choice, q selections for the second choice, r selections for the third choice, and so on, the task of making these selections can be done in

$$p \cdot q \cdot r \cdot \ldots$$

different ways.

EXAMPLE 5   **Forming Codes**

How many two-symbol code words can be formed if the first symbol is an uppercase letter and the second symbol is a digit?

Solution   It sometimes helps to begin by listing some of the possibilities. The code consists of an uppercase letter followed by a digit, so some possibilities are A1, A2, B3, X0, and so on. The task consists of making two selections: The first selection requires choosing an uppercase letter (26 choices), and the second task requires choosing a digit (10 choices). By the Multiplication Principle, there are

$$26 \cdot 10 = 260$$

different code words of the type described.

Now Work PROBLEM 23

14.1 Assess Your Understanding

1. The _____ of A and B consists of all elements in either A or B or both. (pp. 2–3)

2. The _____ of A with B consists of all elements in both A and B. (pp. 2–3)

3. **True or False** The intersection of two sets is always a subset of their union. (pp. 2–3)

4. **True or False** If A is a set, the complement of A is the set of all the elements in the universal set that are not in A. (pp. 2–3)

Concepts and Vocabulary

5. If each element of a set A is also an element of a set B, we say that A is a _____ of B and write A _____ B.

6. If the number of elements in a set is a nonnegative integer, we say that the set is _____.

7. The Counting Formula states that if A and B are finite sets, then $n(A \cup B) = $ _____.

8. **True or False** If a task consists of a sequence of three choices in which there are p selections for the first choice, q selections for the second choice, and r selections for the third choice, then the task of making these selections can be done in $p \cdot q \cdot r$ different ways.

Skill Building

9. Write down all the subsets of $\{a, b, c, d\}$.

10. Write down all the subsets of $\{a, b, c, d, e\}$.

11. If $n(A) = 15$, $n(B) = 20$, and $n(A \cap B) = 10$, find $n(A \cup B)$.

12. If $n(A) = 30$, $n(B) = 40$, and $n(A \cup B) = 45$, find $n(A \cap B)$.

13. If $n(A \cup B) = 50$, $n(A \cap B) = 10$, and $n(B) = 20$, find $n(A)$.

14. If $n(A \cup B) = 60$, $n(A \cap B) = 40$, and $n(A) = n(B)$, find $n(A)$.

In Problems 15–22, use the information given in the figure.

15. How many are in set A?

16. How many are in set B?

17. How many are in A or B?

18. How many are in A and B?

19. How many are in A but not C?

20. How many are not in A?

21. How many are in A and B and C?

22. How many are in A or B or C?

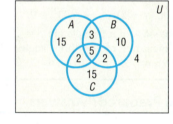

Applications and Extensions

23. **Shirts and Ties** A man has 5 shirts and 3 ties. How many different shirt-and-tie arrangements can he wear?

24. **Blouses and Skirts** A woman has 5 blouses and 8 skirts. How many different outfits can she wear?

25. **Four-digit Numbers** How many four-digit numbers can be formed using the digits 0, 1, 2, 3, 4, 5, 6, 7, 8, and 9 if the first digit cannot be 0? Repeated digits are allowed.

26. **Five-digit Numbers** How many five-digit numbers can be formed using the digits 0, 1, 2, 3, 4, 5, 6, 7, 8, and 9 if the first digit cannot be 0 or 1? Repeated digits are allowed.

27. **Analyzing Survey Data** In a consumer survey of 500 people, 200 indicated that they would be buying a major appliance within the next month, 150 indicated that they would buy a car, and 25 said that they would purchase both a major appliance and a car. How many will purchase neither? How many will purchase only a car?

28. **Analyzing Survey Data** In a student survey, 200 indicated that they would attend Summer Session I, and 150 indicated Summer Session II. If 75 students plan to attend both summer sessions, and 275 indicated that they would attend neither session, how many students participated in the survey?

29. **Analyzing Survey Data** In a survey of 100 investors in the stock market,

    50 owned shares in IBM
    40 owned shares in AT&T
    45 owned shares in GE
    20 owned shares in both IBM and GE
    15 owned shares in both AT&T and GE
    20 owned shares in both IBM and AT&T
    5 owned shares in all three

  (a) How many of the investors surveyed did not have shares in any of the three companies?

  (b) How many owned just IBM shares?

  (c) How many owned just GE shares?

  (d) How many owned neither IBM nor GE?

  (e) How many owned either IBM or AT&T but no GE?

30. **Classifying Blood Types** Human blood is classified as either Rh+ or Rh−. Blood is also classified by type: A, if it contains an A antigen but not a B antigen; B, if it contains a B antigen but not an A antigen; AB, if it contains both A and B antigens; and O, if it contains neither antigen. Draw a Venn diagram illustrating the various blood types. Based on this classification, how many different kinds of blood are there?

31. Demographics   The following data represent the marital status of males 18 years old and older in the U.S. in 2013.

Marital Status	Number (in millions)
Married	65.3
Widowed	3.1
Divorced	10.9
Never married	35.0

Source: Current Population Survey

(a) Determine the number of males 18 years old and older who are widowed or divorced.

(b) Determine the number of males 18 years old and older who are married, widowed, or divorced.

32. Demographics   The following data represent the marital status of females 18 years old and older in the U.S. in 2013.

Marital Status	Number (in millions)
Married	66.2
Widowed	11.2
Divorced	14.4
Never married	30.5

Source: Current Population Survey

(a) Determine the number of females 18 years old and older who are widowed or divorced.

(b) Determine the number of females 18 years old and older who are married, widowed, or divorced.

33. Stock Portfolios   As a financial planner, you are asked to select one stock each from the following groups: 8 Dow Jones stocks, 15 NASDAQ stocks, and 4 global stocks. How many different portfolios are possible?

Explaining Concepts: Discussion and Writing

34. Make up a problem different from any found in the text that requires the addition principle of counting to solve. Give it to a friend to solve and critique.

35. Investigate the notion of counting as it relates to infinite sets. Write an essay on your findings.

Retain Your Knowledge

Problems 36–39 are based on material learned earlier in the course. The purpose of these problems is to keep the material fresh in your mind so that you are better prepared for the final exam.

36. Graph $(x - 2)^2 + (y + 1)^2 = 9$.

37. If the sides of a triangle are $a = 2$, $b = 2$, and $c = 3$, find the measures of the three angles. Round to the nearest tenth.

38. Find all the real zeros of the function
$$f(x) = (x - 2)(x^2 - 3x - 10).$$

39. Solve: $\log_3 x + \log_3 2 = -2$

'Are You Prepared?' Answers

1. union   **2.** intersection   **3.** True   **4.** True

14.2 Permutations and Combinations

PREPARING FOR THIS SECTION   *Before getting started, review the following:*

- Factorial (Section 13.1, p. 942)

Now Work the 'Are You Prepared?' problems on page 994.

OBJECTIVES   **1** Solve Counting Problems Using Permutations Involving *n* Distinct Objects (p. 988)

2 Solve Counting Problems Using Combinations (p. 991)

3 Solve Counting Problems Using Permutations Involving *n* Nondistinct Objects (p. 993)

1 Solve Counting Problems Using Permutations Involving *n* Distinct Objects

DEFINITION   A **permutation** is an ordered arrangement of *r* objects chosen from *n* objects.

Three types of permutations are discussed:

1. The n objects are distinct (different), and repetition is allowed in the selection of r of them. [Distinct, with repetition]

2. The n objects are distinct (different), and repetition is not allowed in the selection of r of them, where $r \leq n$. [Distinct, without repetition]

3. The n objects are not distinct, and all of them are used in the arrangement. [Not distinct]

We take up the first two types here and deal with the third type at the end of this section.

The first type of permutation (n distinct objects, repetition allowed) is handled using the Multiplication Principle.

EXAMPLE 1

Counting Airport Codes [Permutation: Distinct, with Repetition]

The International Airline Transportation Association (IATA) assigns three-letter codes to represent airport locations. For example, the airport code for Ft. Lauderdale, Florida, is FLL. Notice that repetition is allowed in forming this code. How many airport codes are possible?

Solution

An airport code is formed by choosing 3 letters from 26 letters and arranging them in order. In the ordered arrangement, a letter may be repeated. This is an example of a permutation with repetition in which 3 objects are chosen from 26 distinct objects.

The task of counting the number of such arrangements consists of making three selections. Each selection requires choosing a letter of the alphabet (26 choices). By the Multiplication Principle, there are

$$26 \cdot 26 \cdot 26 = 26^3 = 17{,}576$$

possible airport codes.   ●

The solution given to Example 1 can be generalized.

THEOREM

Permutations: Distinct Objects with Repetition

The number of ordered arrangements of r objects chosen from n objects, in which the n objects are distinct and repetition is allowed, is n^r.

 Now Work PROBLEM 33

Now let's consider permutations in which the objects are distinct and repetition is not allowed.

EXAMPLE 2

Forming Codes [Permutation: Distinct, without Repetition]

Suppose that a three-letter code is to be formed using any of the 26 uppercase letters of the alphabet, but no letter is to be used more than once. How many different three-letter codes are there?

Solution

Some of the possibilities are ABC, ABD, ABZ, ACB, CBA, and so on. The task consists of making three selections. The first selection requires choosing from 26 letters. Since no letter can be used more than once, the second selection requires choosing from 25 letters. The third selection requires choosing from 24 letters. (Do you see why?) By the Multiplication Principle, there are

$$26 \cdot 25 \cdot 24 = 15{,}600$$

different three-letter codes with no letter repeated.   ●

For the second type of permutation, we introduce the following notation.

> The notation **P(n, r)** represents the number of ordered arrangements of r objects chosen from n distinct objects, where $r \leq n$ and repetition is not allowed.

For example, the question posed in Example 2 asks for the number of ways in which the 26 letters of the alphabet can be arranged, in order, using three nonrepeated letters. The answer is

$$P(26, 3) = 26 \cdot 25 \cdot 24 = 15{,}600$$

EXAMPLE 3

Lining People Up

In how many ways can 5 people be lined up?

Solution

The 5 people are distinct. Once a person is in line, that person will not be repeated elsewhere in the line; and, in lining people up, order is important. This is a permutation of 5 objects taken 5 at a time, so 5 people can be lined up in

$$P(5, 5) = \underbrace{5 \cdot 4 \cdot 3 \cdot 2 \cdot 1}_{\text{5 factors}} = 120 \text{ ways}$$

●

 Now Work PROBLEM **35**

To arrive at a formula for $P(n, r)$, note that the task of obtaining an ordered arrangement of n objects in which only $r \leq n$ of them are used, without repeating any of them, requires making r selections. For the first selection, there are n choices; for the second selection, there are $n - 1$ choices; for the third selection, there are $n - 2$ choices; ...; for the rth selection, there are $n - (r - 1)$ choices. By the Multiplication Principle, this means

$$
\begin{aligned}
P(n, r) &= \overset{\text{1st}}{n} \cdot \overset{\text{2nd}}{(n-1)} \cdot \overset{\text{3rd}}{(n-2)} \cdot \,\cdots\, \cdot \overset{\text{rth}}{[n-(r-1)]} \\
&= n \cdot (n-1) \cdot (n-2) \cdot \,\cdots\, \cdot (n-r+1)
\end{aligned}
$$

This formula for $P(n, r)$ can be compactly written using factorial notation.*

$$P(n, r) = n \cdot (n - 1) \cdot (n - 2) \cdot \,\cdots\, \cdot (n - r + 1)$$

$$= n \cdot (n - 1) \cdot (n - 2) \cdot \,\cdots\, \cdot (n - r + 1) \cdot \frac{(n - r) \cdot \,\cdots\, \cdot 3 \cdot 2 \cdot 1}{(n - r) \cdot \,\cdots\, \cdot 3 \cdot 2 \cdot 1} = \frac{n!}{(n - r)!}$$

THEOREM

> **Permutations of r Objects Chosen from n Distinct Objects without Repetition**
>
> The number of arrangements of n objects using $r \leq n$ of them, in which
>
> 1. the n objects are distinct,
> 2. once an object is used it cannot be repeated, and
> 3. order is important,
>
> is given by the formula
>
> $$P(n, r) = \frac{n!}{(n - r)!} \qquad (1)$$

*Recall that $0! = 1$, $1! = 1$, $2! = 2 \cdot 1, \ldots, n! = n(n - 1) \cdot \ldots \cdot 3 \cdot 2 \cdot 1$.

EXAMPLE 4

Computing Permutations

Evaluate: (a) $P(7, 3)$ (b) $P(6, 1)$ (c) $P(52, 5)$

Solution

Parts (a) and (b) are each worked two ways.

(a) $P(7, 3) = \underbrace{7 \cdot 6 \cdot 5}_{3\text{ factors}} = 210$

or

$$P(7, 3) = \frac{7!}{(7-3)!} = \frac{7!}{4!} = \frac{7 \cdot 6 \cdot 5 \cdot 4!}{4!} = 210$$

(b) $P(6, 1) = \underbrace{6}_{1\text{ factor}} = 6$

or

$$P(6, 1) = \frac{6!}{(6-1)!} = \frac{6!}{5!} = \frac{6 \cdot 5!}{5!} = 6$$

(c) Figure 3 shows the solution using a TI-84 Plus C graphing calculator. So

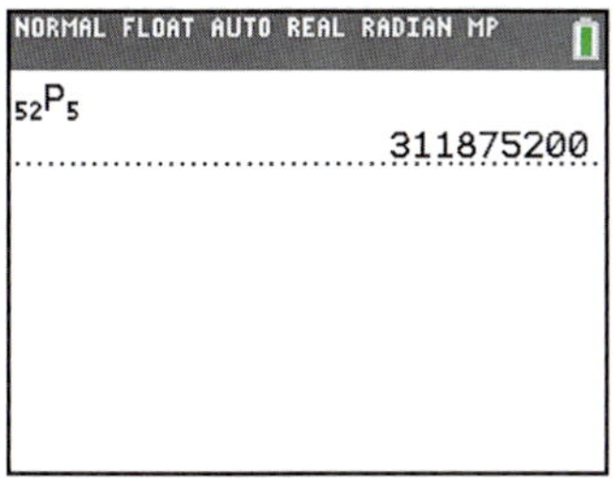

$$P(52, 5) = 311,875,200$$

```
NORMAL FLOAT AUTO REAL RADIAN MP
52P5
                    311875200
```

Figure 3 $P(52, 5)$

 Now Work PROBLEM 7

EXAMPLE 5

The Birthday Problem

All we know about Shannon, Patrick, and Ryan is that they have different birthdays. If all the possible ways this could occur were listed, how many would there be? Assume that there are 365 days in a year.

Solution

This is an example of a permutation in which 3 birthdays are selected from a possible 365 days, and no birthday may repeat itself. The number of ways this can occur is

$$P(365, 3) = \frac{365!}{(365-3)!} = \frac{365 \cdot 364 \cdot 363 \cdot 362!}{362!} = 365 \cdot 364 \cdot 363 = 48{,}228{,}180$$

There are 48,228,180 ways in which three people can all have different birthdays.

 Now Work PROBLEM 47

2 Solve Counting Problems Using Combinations

In a permutation, order is important. For example, the arrangements ABC, CAB, BAC, ... are considered different arrangements of the letters A, B, and C. In many situations, though, order is unimportant. For example, in the card game of poker, the order in which the cards are received does not matter; it is the *combination* of the cards that matters.

DEFINITION

A **combination** is an arrangement, without regard to order, of r objects selected from n distinct objects without repetition, where $r \le n$. The notation $C(n, r)$ represents the number of combinations of n distinct objects using r of them.

EXAMPLE 6

Listing Combinations

List all the combinations of the 4 objects a, b, c, d taken 2 at a time. What is $C(4, 2)$?

Solution

One combination of a, b, c, d taken 2 at a time is

ab

EXAMPLE 2

Determining Probability Models

In a bag of M&Ms,™ the candies are colored red, green, blue, brown, yellow, and orange. A candy is drawn from the bag and the color is recorded. The sample space of this experiment is { red, green, blue, brown, yellow, orange }. Determine which of the following are probability models.

(a)

Outcome	Probability
red	0.3
green	0.15
blue	0
brown	0.15
yellow	0.2
orange	0.2

(b)

Outcome	Probability
red	0.1
green	0.1
blue	0.1
brown	0.4
yellow	0.2
orange	0.3

(c)

Outcome	Probability
red	0.3
green	−0.3
blue	0.2
brown	0.4
yellow	0.2
orange	0.2

(d)

Outcome	Probability
red	0
green	0
blue	0
brown	0
yellow	1
orange	0

Solution

(a) This model is a probability model because all the outcomes have probabilities that are nonnegative, and the sum of the probabilities is 1.

(b) This model is not a probability model because the sum of the probabilities is not 1.

(c) This model is not a probability model because $P(\text{green})$ is less than 0. Remember that all probabilities must be nonnegative.

(d) This model is a probability model because all the outcomes have probabilities that are nonnegative, and the sum of the probabilities is 1. Notice that $P(\text{yellow}) = 1$, meaning that this outcome will occur with 100% certainty each time that the experiment is repeated. This means that the bag of M&Ms™ contains only yellow candies. ●

 Now Work PROBLEM 7

EXAMPLE 3

Constructing a Probability Model

An experiment consists of rolling a fair die once. A die is a cube with each face having 1, 2, 3, 4, 5, or 6 dots on it. See Figure 5. Construct a probability model for this experiment.

Solution

A sample space S consists of all the possibilities that can occur. Because rolling the die will result in one of six faces showing, the sample space S consists of

$$S = \{1, 2, 3, 4, 5, 6\}$$

Because the die is fair, one face is no more likely to occur than another. As a result, our assignment of probabilities is

$$P(1) = \frac{1}{6} \qquad P(2) = \frac{1}{6}$$

$$P(3) = \frac{1}{6} \qquad P(4) = \frac{1}{6}$$

$$P(5) = \frac{1}{6} \qquad P(6) = \frac{1}{6}$$

Figure 5 A six-sided die

Now suppose that a die is loaded (weighted) so that the probability assignments are

$$P(1) = 0 \quad P(2) = 0 \quad P(3) = \frac{1}{3} \quad P(4) = \frac{2}{3} \quad P(5) = 0 \quad P(6) = 0$$

This assignment would be made if the die were loaded so that only a 3 or 4 could occur and the 4 was twice as likely as the 3 to occur. This assignment is consistent with the definition, since each assignment is nonnegative, and the sum of all the probability assignments equals 1.

 Now Work PROBLEM 23

EXAMPLE 4

Constructing a Probability Model

An experiment consists of tossing a coin. The coin is weighted so that heads (H) is three times as likely to occur as tails (T). Construct a probability model for this experiment.

Solution

The sample space S is $S = \{H, T\}$. If x denotes the probability that a tail occurs,

$$P(T) = x \quad \text{and} \quad P(H) = 3x$$

The sum of the probabilities of the possible outcomes must equal 1, so

$$P(T) + P(H) = x + 3x = 1$$
$$4x = 1$$
$$x = \frac{1}{4}$$

Assign the probabilities

$$P(T) = \frac{1}{4} \qquad P(H) = \frac{3}{4}$$

●

 Now Work PROBLEM 27

In working with probability models, the term **event** is used to describe a set of possible outcomes of the experiment. An event E is some subset of the sample space S. The **probability of an event** $E, E \neq \emptyset$, denoted by $P(E)$, is defined as the sum of the probabilities of the outcomes in E. We can also think of the probability of an event E as the likelihood that the event E occurs. If $E = \emptyset$, then $P(E) = 0$; if $E = S$, then $P(E) = P(S) = 1$.

In Words
$P(S) = 1$ means that one of the outcomes in the sample space must occur in an experiment.

2 Compute Probabilities of Equally Likely Outcomes

When the same probability is assigned to each outcome of the sample space, the experiment is said to have **equally likely outcomes**.

THEOREM

Probability for Equally Likely Outcomes

If an experiment has n equally likely outcomes, and if the number of ways in which an event E can occur is m, then the probability of E is

$$P(E) = \frac{\text{Number of ways that } E \text{ can occur}}{\text{Number of possible outcomes}} = \frac{m}{n} \qquad (3)$$

If S is the sample space of this experiment,

$$P(E) = \frac{n(E)}{n(S)} \qquad (4)$$

In Problems 45–48, find the probability of the indicated event if
P(A) = 0.25 and P(B) = 0.45.

45. $P(A \cup B)$ if $P(A \cap B) = 0.15$

46. $P(A \cap B)$ if $P(A \cup B) = 0.6$

47. $P(A \cup B)$ if A, B are mutually exclusive

48. $P(A \cap B)$ if A, B are mutually exclusive

49. If $P(A) = 0.60, P(A \cup B) = 0.85$, and $P(A \cap B) = 0.05$, find $P(B)$.

50. If $P(B) = 0.30, P(A \cup B) = 0.65$, and $P(A \cap B) = 0.15$, find $P(A)$.

51. Automobile Theft According to the Insurance Information Institute, in 2012 there was a 12% probability that an automobile theft in the United States would be cleared by arrests. If an automobile theft case from 2012 is randomly selected, what is the probability that it was not cleared by an arrest?

52. Pet Ownership According to the American Pet Products Manufacturers Association's *2013–2014 National Pet Owners Survey*, there is a 68% probability that a U.S. household owns a pet. If a U.S. household is randomly selected, what is the probability that it does not own a pet?

53. Cat Ownership According to the American Pet Products Manufacturers Association's *2013–2014 National Pet Owners Survey*, there is a 37% probability that a U.S. household owns a cat. If a U.S. household is randomly selected, what is the probability that it does not own a cat?

54. Doctorate Degrees According to the National Science Foundation, in 2012 there was a 16.5% probability that a doctoral degree awarded at a U.S. university was awarded in engineering. If a 2012 U.S. doctoral recipient is randomly selected, what is the probability that his or her degree was not in engineering?

55. Online Gambling According to a Casino FYI survey, 6.4% of U.S. adults admitted to having spent money gambling online. If a U.S. adult is selected at random, what is the probability that he or she has never spent any money gambling online?

56. Girl Scout Cookies According to the Girl Scouts of America, 9% of all Girl Scout cookies sold are shortbread/trefoils. If a box of Girl Scout cookies is selected at random, what is the probability that it does not contain shortbread/trefoils?

For Problems 57–60, a golf ball is selected at random from a container. If the container has 9 white balls, 8 green balls, and 3 orange balls, find the probability of each event.

57. The golf ball is white or green.

58. The golf ball is white or orange.

59. The golf ball is not white.

60. The golf ball is not green.

61. On *The Price Is Right*, there is a game in which a bag is filled with 3 strike chips and 5 numbers. Let's say that the numbers in the bag are 0, 1, 3, 6, and 9. What is the probability of selecting a strike chip or the number 1?

62. Another game on *The Price Is Right* requires the contestant to spin a wheel with the numbers 5, 10, 15, 20, …, 100. What is the probability that the contestant spins 100 or 30?

Problems 63–66 are based on a survey of annual incomes in 100 households. The following table gives the data.

Income	$0–24,999	$25,000–49,999	$50,000–74,999	$75,000–99,999	$100,000 or more
Number of households	24	24	18	12	22

63. What is the probability that a household has an annual income of $75,000 or more?

64. What is the probability that a household has an annual income between $25,000 and $74,999, inclusive?

65. What is the probability that a household has an annual income of less than $50,000?

66. What is the probability that a household has an annual income of $50,000 or more?

67. Surveys In a survey about the number of TV sets in a house, the following probability table was constructed:

Number of TV sets	0	1	2	3	4 or more
Probability	0.05	0.24	0.33	0.21	0.17

Find the probability of a house having:
(a) 1 or 2 TV sets
(b) 1 or more TV sets
(c) 3 or fewer TV sets
(d) 3 or more TV sets

(e) Fewer than 2 TV sets
(f) Fewer than 1 TV set
(g) 1, 2, or 3 TV sets
(h) 2 or more TV sets

68. Checkout Lines Through observation, it has been determined that the probability for a given number of people waiting in line at the "5 items or less" checkout register of a supermarket is as follows:

Number waiting in line	0	1	2	3	4 or more
Probability	0.10	0.15	0.20	0.24	0.31

Find the probability of:
(a) At most 2 people in line
(b) At least 2 people in line
(c) At least 1 person in line

69. In a certain Algebra and Trigonometry class, there are 18 freshmen and 15 sophomores. Of the 18 freshmen, 10 are male, and of the 15 sophomores, 8 are male. Find the probability that a randomly selected student is:

(a) A freshman or female
(b) A sophomore or male

70. The faculty of the mathematics department at Joliet Junior College is composed of 4 females and 9 males. Of the 4 females, 2 are under age 40, and of the 9 males, 3 are under age 40. Find the probability that a randomly selected faculty member is:

(a) Female or under age 40

(b) Male or over age 40

71. Birthday Problem What is the probability that at least 2 people in a group of 12 people have the same birthday? Assume that there are 365 days in a year.

72. Birthday Problem What is the probability that at least 2 people in a group of 35 people have the same birthday? Assume that there are 365 days in a year.

73. Winning a Lottery Powerball is a multistate lottery in which 5 white balls from a drum with 59 balls and 1 red ball from a drum with 35 balls are selected. For a $2 ticket, players get one chance at winning the jackpot by matching all 6 numbers. What is the probability of selecting the winning numbers on a $2 play?

Retain Your Knowledge

Problems 74–77 are based on material learned earlier in the course. The purpose of these problems is to keep the material fresh in your mind so that you are better prepared for the final exam.

74. To graph $g(x) = |x + 2| - 3$, shift the graph of $f(x) = |x|$ _____ units _____ and then _____ units _____.
_{number} _{left/right} _{number} _{up/down}

75. Find the rectangular coordinates of the point whose polar coordinates are $\left(6, \dfrac{2\pi}{3}\right)$.

76. Solve: $\log_5 (x + 3) = 2$

77. Solve the given system using matrices.
$$\begin{cases} 3x + y + 2z = 1 \\ 2x - 2y + 5z = 5 \\ x + 3y + 2z = -9 \end{cases}$$

Chapter Review

Things to Know

Counting formula (p. 984) — $n(A \cup B) = n(A) + n(B) - n(A \cap B)$

Addition Principle of Counting (p. 984) — If $A \cap B = \varnothing$, then $n(A \cup B) = n(A) + n(B)$.

Multiplication Principle of Counting (p. 986) — If a task consists of a sequence of choices in which there are p selections for the first choice, q selections for the second choice, and so on, the task of making these selections can be done in $p \cdot q \cdot \cdots$ different ways.

Permutation (p. 988) — An ordered arrangement of r objects chosen from n objects

Number of permutations: Distinct, with repetition (p. 989) — n^r

The n objects are distinct (different), and repetition is allowed in the selection of r of them.

Number of permutations: Distinct, without repetition (p. 990) — $P(n, r) = n(n - 1) \cdot \cdots \cdot [n - (r - 1)] = \dfrac{n!}{(n - r)!}$

The n objects are distinct (different), and repetition is not allowed in the selection of r of them, where $r \leq n$.

Combination (p. 991) — An arrangement, without regard to order, of r objects selected from n distinct objects, where $r \leq n$

Number of combinations (p. 992) — $C(n, r) = \dfrac{P(n, r)}{r!} = \dfrac{n!}{(n - r)! \, r!}$

Number of permutations: Not distinct (p. 994) — $\dfrac{n!}{n_1! n_2! \cdots n_k!}$

The number of permutations of n objects of which n_1 are of one kind, n_2 are of a second kind, ..., and n_k are of a kth kind, where $n = n_1 + n_2 + \cdots + n_k$

Sample space (p. 997) — Set whose elements represent the possible outcomes that can occur as a result of an experiment

Probability (p. 997) — A nonnegative number assigned to each outcome of a sample space; the sum of all the probabilities of the outcomes equals 1.

Appendix
Graphing Utilities

Outline

1 The Viewing Rectangle

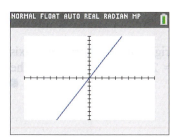

Figure 1 $y = 2x$

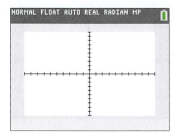

Figure 2 Viewing window

All graphing utilities (that is, all graphing calculators and all computer software graphing packages) graph equations by plotting points on a screen. The screen itself actually consists of small rectangles called **pixels**. The more pixels the screen has, the better the resolution. Most graphing calculators have 50 to 100 pixels per inch; most smartphones have 300 to 450 pixels per inch. When a point to be plotted lies inside a pixel, the pixel is turned on (lights up). The graph of an equation is a collection of pixels. Figure 1 shows how the graph of $y = 2x$ looks on a TI-84 Plus C graphing calculator.

The screen of a graphing utility will display the coordinate axes of a rectangular coordinate system. However, the scale must be set on each axis. The smallest and largest values of x and y to be included in the graph must also be set. This is called **setting the viewing rectangle** or **viewing window**. Figure 2 shows a typical viewing window.

To select the viewing window, values must be given to the following expressions:

Xmin:	the smallest value of x
Xmax:	the largest value of x
Xscl:	the number of units per tick mark on the x-axis
Ymin:	the smallest value of y
Ymax:	the largest value of y
Yscl:	the number of units per tick mark on the y-axis

Figure 3 illustrates these settings and their relation to the Cartesian coordinate system.

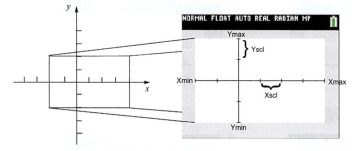

Figure 3

If the scale used on each axis is known, the minimum and maximum values of x and y shown on the screen can be determined by counting the tick marks. Look again at Figure 2. For a scale of 1 on each axis, the minimum and maximum values of x are -10 and 10, respectively; the minimum and maximum values of y are

A1

Graphing Parametric Equations Using a Graphing Utility

STEP 1: Set the mode to PARAMETRIC. Enter $x(t)$ and $y(t)$.

STEP 2: Select the viewing window. In addition to setting Xmin, Xmax, Xscl, and so on, the viewing window in parametric mode requires setting minimum and maximum values for the parameter t and an increment setting for t (Tstep).

STEP 3: Graph.

EXAMPLE 1

Graphing a Curve Defined by Parametric Equations Using a Graphing Utility

Graph the curve defined by the parametric equations

$$x = 3t^2 \qquad y = 2t \qquad -2 \le t \le 2$$

Solution **STEP 1:** Enter the equations $x(t) = 3t^2$, $y(t) = 2t$ with the graphing utility in PARAMETRIC mode.

STEP 2: Select the viewing window. The interval is $-2 \le t \le 2$, so select the following square viewing window:

Tmin $= -2$	Xmin $= 0$	Ymin $= -5$
Tmax $= 2$	Xmax $= 16$	Ymax $= 5$
Tstep $= 0.1$	Xscl $= 1$	Yscl $= 1$

Choose Tmin $= -2$ and Tmax $= 2$ because $-2 \le t \le 2$. Finally, the choice for Tstep will determine the number of points that the graphing utility will plot. For example, with Tstep at 0.1, the graphing utility will evaluate x and y at $t = -2, -1.9, -1.8$, and so on. The smaller the Tstep, the more points the graphing utility will plot. Experiment with different values of Tstep to see how the graph is affected.

STEP 3: Graph. Watch the direction in which the graph is drawn. This direction shows the orientation of the curve.

The graph shown in Figure 22 is complete. ●

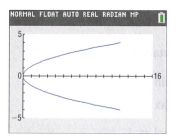

Figure 22
$x = 3t^2, y = 2t, -2 \le t \le 2$

Exploration

Graph the following parametric equations using a graphing utility with Xmin $= 0$, Xmax $= 16$, Ymin $= -5$, Ymax $= 5$, and Tstep $= 0.1$.

1. $x = \dfrac{3t^2}{4}$, $y = t$, $-4 \le t \le 4$

2. $x = 3t^2 + 12t + 12$, $y = 2t + 4$, $-4 \le t \le 0$

3. $x = 3t^{2/3}$, $y = 2\sqrt[3]{t}$, $-8 \le t \le 8$

Compare these graphs to Figure 22. Conclude that parametric equations defining a curve are not unique; that is, different parametric equations can represent the same graph.

Exploration

In FUNCTION mode, graph $x = \dfrac{3y^2}{4}$ $\left(Y_1 = \sqrt{\dfrac{4x}{3}} \text{ and } Y_2 = -\sqrt{\dfrac{4x}{3}} \right)$ with Xmin $= 0$, Xmax $= 16$, Ymin $= -5$, Ymax $= 5$. Compare this graph with Figure 22. Why do the graphs differ?

Answers

CHAPTER R Review

R.1 Assess Your Understanding (*page 15*)

1. rational **2.** 31 **3.** Distributive **4.** $5(x + 3) = 6$ **5.** a **6.** b **7.** T **8.** F **9.** F **10.** T **11.** $\{1, 2, 3, 4, 5, 6, 7, 8, 9\}$ **13.** $\{4\}$ **15.** $\{1, 3, 4, 6\}$

17. $\{0, 2, 6, 7, 8\}$ **19.** $\{0, 1, 2, 3, 5, 6, 7, 8, 9\}$ **21.** $\{0, 1, 2, 3, 5, 6, 7, 8, 9\}$ **23.** **(a)** $\{2, 5\}$ **(b)** $\{-6, 2, 5\}$ **(c)** $\left\{-6, \frac{1}{2}, -1.333\cdots, 2, 5\right\}$ **(d)** $\{\pi\}$

(e) $\left\{-6, \frac{1}{2}, -1.333..., \pi, 2, 5\right\}$ **25.** **(a)** $\{1\}$ **(b)** $\{0, 1\}$ **(c)** $\left\{0, 1, \frac{1}{2}, \frac{1}{3}, \frac{1}{4}\right\}$ **(d)** None **(e)** $\left\{0, 1, \frac{1}{2}, \frac{1}{3}, \frac{1}{4}\right\}$ **27.** **(a)** None **(b)** None **(c)** None

(d) $\left\{\sqrt{2}, \pi, \sqrt{2} + 1, \pi + \frac{1}{2}\right\}$ **(e)** $\left\{\sqrt{2}, \pi, \sqrt{2} + 1, \pi + \frac{1}{2}\right\}$ **29.** **(a)** 18.953 **(b)** 18.952 **31.** **(a)** 28.653 **(b)** 28.653 **33.** **(a)** 0.063 **(b)** 0.062

35. (a) 9.999 **(b)** 9.998 **37. (a)** 0.429 **(b)** 0.428 **39. (a)** 34.733 **(b)** 34.733 **41.** $3 + 2 = 5$ **43.** $x + 2 = 3 \cdot 4$ **45.** $3y = 1 + 2$

47. $x - 2 = 6$ **49.** $\frac{x}{2} = 6$ **51.** 7 **53.** 6 **55.** 1 **57.** $\frac{13}{3}$ **59.** -11 **61.** 11 **63.** -4 **65.** 1 **67.** 6 **69.** $\frac{2}{7}$ **71.** $\frac{4}{45}$ **73.** $\frac{23}{20}$ **75.** $\frac{79}{30}$ **77.** $\frac{13}{36}$

79. $-\frac{16}{45}$ **81.** $\frac{1}{60}$ **83.** $\frac{15}{22}$ **85.** 1 **87.** $\frac{15}{8}$ **89.** $6x + 24$ **91.** $x^2 - 4x$ **93.** $\frac{3}{2}x - 1$ **95.** $x^2 + 6x + 8$ **97.** $x^2 - x - 2$ **99.** $x^2 - 10x + 16$

101. $2x + 3x = (2 + 3)x = 5x$ **103.** $2(3 \cdot 4) = 2 \cdot 12 = 24$; $(2 \cdot 3) \cdot (2 \cdot 4) = 6 \cdot 8 = 48$ **105.** No; $2 - 3 \neq 3 - 2$ **107.** No; $\frac{2}{3} \neq \frac{3}{2}$

109. Symmetric Property **111.** No; no

R.2 Assess Your Understanding (*page 26*)

1. variable **2.** origin **3.** strict **4.** base; exponent, or power **5.** 1.2345678×10^3 **6.** d **7.** b **8.** T **9.** T **10.** F **11.** F **12.** F

13. **15.** $>$ **17.** $>$ **19.** $>$ **21.** $=$ **23.** $<$ **25.** $x > 0$ **27.** $x < 2$ **29.** $x \leq 1$

31. **33.** **35.** 1 **37.** 2 **39.** 6 **41.** 4 **43.** -28 **45.** $\frac{4}{5}$ **47.** 0 **49.** 1 **51.** 5 **53.** 1

55. 22 **57.** 2 **59.** $x = 0$ **61.** $x = 3$ **63.** None **65.** $x = 0, x = 1, x = -1$ **67.** $\{x | x \neq 5\}$ **69.** $\{x | x \neq -4\}$ **71.** $0°C$ **73.** $25°C$ **75.** 16

77. $\frac{1}{16}$ **79.** $\frac{1}{9}$ **81.** 9 **83.** 5 **85.** 4 **87.** $64x^6$ **89.** $\frac{x^4}{y^2}$ **91.** $\frac{x}{y}$ **93.** $-\frac{8x^3 z}{9y}$ **95.** $\frac{16x^2}{9y^2}$ **97.** -4 **99.** 5 **101.** 4 **103.** 2 **105.** $\sqrt{5}$ **107.** $\frac{1}{2}$ **109.** 10; 0

111. 81 **113.** 304,006.671 **115.** 0.004 **117.** 481.890 **119.** 0.000 **121.** 4.542×10^2 **123.** 1.3×10^{-2} **125.** 3.2155×10^4 **127.** 4.23×10^{-4}

129. 61,500 **131.** 0.001214 **133.** 110,000,000 **135.** 0.081 **137.** $A = lw$ **139.** $C = \pi d$ **141.** $A = \frac{\sqrt{3}}{4}x^2$ **143.** $V = \frac{4}{3}\pi r^3$ **145.** $V = x^3$

147. (a) $6000 **(b)** $8000 **149.** $|x - 4| \geq 6$ **151. (a)** $2 \leq 5$ **(b)** $6 > 5$ **153. (a)** Yes **(b)** No **155.** 400,000,000 m **157.** 0.0000005 m

159. 5×10^{-4} in. **161.** 5.865696×10^{12} mi **163.** No; $\frac{1}{3}$ is larger; 0.000333... **165.** No

R.3 Assess Your Understanding (*page 36*)

1. right; hypotenuse **2.** $A = \frac{1}{2}bh$ **3.** $C = 2\pi r$ **4.** similar **5.** c **6.** b **7.** T **8.** T **9.** F **10.** T **11.** T **12.** F **13.** 13 **15.** 26 **17.** 25

19. Right triangle; 5 **21.** Not a right triangle **23.** Right triangle; 25 **25.** Not a right triangle **27.** 8 in.^{2} **29.** 4 in.^{2} **31.** $A = 25\pi$ m^2; $C = 10\pi$ m

33. $V = 224$ ft^3; $S = 232$ ft^2 **35.** $V = \frac{256}{3}\pi$ cm^3; $S = 64\pi$ cm^2 **37.** $V = 648\pi$ in.^{3}; $S = 306\pi$ in.^{2} **39.** π square units **41.** 2π square units

43. $x = 4$ units; $A = 90°$; $B = 60°$; $C = 30°$ **45.** $x = 67.5$ units; $A = 60°$; $B = 95°$; $C = 25°$ **47.** About 16.8 ft **49.** 64 ft^2

51. $24 + 2\pi \approx 30.28$ ft^2; $16 + 2\pi \approx 22.28$ ft **53.** 160 paces **55.** About 5.477 mi **57.** From 100 ft: 12.2 mi; From 150 ft: 15.0 mi

R.4 Assess Your Understanding (*page 47*)

1. 4; 3 **2.** $x^4 - 16$ **3.** $x^3 - 8$ **4.** a **5.** c **6.** F **7.** T **8.** F **9.** Monomial; variable: x; coefficient: 2; degree: 3 **11.** Not a monomial; the exponent of the variable is not a nonnegative integer **13.** Monomial; variables: x, y; coefficient: -2; degree: 3 **15.** Not a monomial; the exponent of one of the variables is not a nonnegative integer **17.** Not a monomial; it has more than one term **19.** Yes; 2 **21.** Yes; 0 **23.** No; the exponent of the variable of one of the terms is not a nonnegative integer **25.** Yes; 3 **27.** No; the polynomial of the denominator has a degree greater than 0 **29.** $x^2 + 7x + 2$

31. $x^3 - 4x^2 + 9x + 7$ **33.** $6x^5 + 5x^4 + 3x^2 + x$ **35.** $7x^2 - x - 7$ **37.** $-2x^3 + 18x^2 - 18$ **39.** $2x^2 - 4x + 6$ **41.** $15y^2 - 27y + 30$

43. $x^3 + x^2 - 4x$ **45.** $-8x^5 - 10x^2$ **47.** $x^3 + 3x^2 - 2x - 4$ **49.** $x^2 + 6x + 8$ **51.** $2x^2 + 9x + 10$ **53.** $x^2 - 2x - 8$ **55.** $x^2 - 5x + 6$

57. $2x^2 - x - 6$ **59.** $-2x^2 + 11x - 12$ **61.** $2x^2 + 8x + 8$ **63.** $x^2 - xy - 2y^2$ **65.** $-6x^2 - 13xy - 6y^2$ **67.** $x^2 - 49$ **69.** $4x^2 - 9$

71. $x^2 + 8x + 16$ **73.** $x^2 - 8x + 16$ **75.** $9x^2 - 16$ **77.** $4x^2 - 12x + 9$ **79.** $x^2 - y^2$ **81.** $9x^2 - y^2$ **83.** $x^2 + 2xy + y^2$ **85.** $x^2 - 4xy + 4y^2$

87. $x^3 - 6x^2 + 12x - 8$ **89.** $8x^3 + 12x^2 + 6x + 1$ **91.** $4x^2 - 11x + 23$; remainder -45 **93.** $4x - 3$; remainder $x + 1$

95. $5x^2 - 13$; remainder $x + 27$ **97.** $2x^2$; remainder $-x^2 + x + 1$ **99.** $x^2 - 2x + \frac{1}{2}$; remainder $\frac{5}{2}x + \frac{1}{2}$ **101.** $-4x^2 - 3x - 3$; remainder -7

103. $x^2 - x - 1$; remainder $2x + 2$ **105.** $x^2 + ax + a^2$; remainder 0

R.5 Assess Your Understanding (page 57)

1. $3x(x-2)(x+2)$ **2.** prime **3.** c **4.** b **5.** d **6.** c **7.** T **8.** F **9.** $3(x+2)$ **11.** $a(x^2+1)$ **13.** $x(x^2+x+1)$ **15.** $2x(x-1)$
17. $3xy(x-2y+4)$ **19.** $(x+1)(x-1)$ **21.** $(2x+1)(2x-1)$ **23.** $(x+4)(x-4)$ **25.** $(5x+2)(5x-2)$ **27.** $(x+1)^2$ **29.** $(x+2)^2$
31. $(x-5)^2$ **33.** $(2x+1)^2$ **35.** $(4x+1)^2$ **37.** $(x-3)(x^2+3x+9)$ **39.** $(x+3)(x^2-3x+9)$ **41.** $(2x+3)(4x^2-6x+9)$
43. $(x+2)(x+3)$ **45.** $(x+6)(x+1)$ **47.** $(x+5)(x+2)$ **49.** $(x-8)(x-2)$ **51.** $(x-8)(x+1)$ **53.** $(x+8)(x-1)$
55. $(x+2)(2x+3)$ **57.** $(x-2)(2x+1)$ **59.** $(2x+3)(3x+2)$ **61.** $(3x+1)(x+1)$ **63.** $(z+1)(2z+3)$ **65.** $(x+2)(3x-4)$
67. $(x-2)(3x+4)$ **69.** $(x+4)(3x+2)$ **71.** $(x+4)(3x-2)$ **73.** 25; $(x+5)^2$ **75.** 9; $(y-3)^2$ **77.** $\frac{1}{16}$; $\left(x-\frac{1}{4}\right)^2$ **79.** $(x+6)(x-6)$
81. $2(1+2x)(1-2x)$ **83.** $(x+1)(x+10)$ **85.** $(x-7)(x-3)$ **87.** $4(x^2-2x+8)$ **89.** Prime **91.** $-(x-5)(x+3)$
93. $3(x+2)(x-6)$ **95.** $y^2(y+5)(y+6)$ **97.** $(2x+3)^2$ **99.** $2(3x+1)(x+1)$ **101.** $(x-3)(x+3)(x^2+9)$ **103.** $(x-1)^2(x^2+x+1)^2$
105. $x^5(x-1)(x+1)$ **107.** $(4x+3)^2$ **109.** $-(4x-5)(4x+1)$ **111.** $(2y-5)(2y-3)$ **113.** $-(3x-1)(3x+1)(x^2+1)$
115. $(x+3)(x-6)$ **117.** $(x+2)(x-3)$ **119.** $(3x-5)(9x^2-3x+7)$ **121.** $(x+5)(3x+11)$ **123.** $(x-1)(x+1)(x+2)$
125. $(x-1)(x+1)(x^2-x+1)$ **127.** $2(3x+4)(9x+13)$ **129.** $2x(3x+5)$ **131.** $5(x+3)(x-2)^2(x+1)$ **133.** $3(4x-3)(4x-1)$
135. $6(3x-5)(2x+1)^2(5x-4)$ **137.** The possibilities are $(x\pm 1)(x\pm 4)=x^2\pm 5x+4$ or $(x\pm 2)(x\pm 2)=x^2\pm 4x+4$, none of
which equals x^2+4.

R.6 Assess Your Understanding (page 61)

1. quotient; divisor; remainder **2.** $-3\overline{)2\,0\,-5\,1}$ **3.** d **4.** a **5.** T **6.** T **7.** x^2+x+4; remainder 12 **9.** $3x^2+11x+32$; remainder 99
11. $x^4-3x^3+5x^2-15x+46$; remainder -138 **13.** $4x^5+4x^4+x^3+x^2+2x+2$; remainder 7 **15.** $0.1x^2-0.11x+0.321$; remainder -0.3531
17. $x^4+x^3+x^2+x+1$; remainder 0 **19.** No **21.** Yes **23.** Yes **25.** No **27.** Yes **29.** -9

R.7 Assess Your Understanding (page 70)

1. lowest terms **2.** least common multiple **3.** d **4.** a **5.** T **6.** F **7.** $\dfrac{3}{x-3}$ **9.** $\dfrac{x}{3}$ **11.** $\dfrac{4x}{2x-1}$ **13.** $\dfrac{y+5}{2(y+1)}$ **15.** $\dfrac{x+5}{x-1}$ **17.** $-(x+7)$

19. $\dfrac{3}{5x(x-2)}$ **21.** $\dfrac{2x(x^2+4x+16)}{x+4}$ **23.** $\dfrac{8}{3x}$ **25.** $\dfrac{x-3}{x+7}$ **27.** $\dfrac{4x}{(x-2)(x-3)}$ **29.** $\dfrac{4}{5(x-1)}$ **31.** $-\dfrac{(x-4)^2}{4x}$ **33.** $\dfrac{(x+3)^2}{(x-3)^2}$ **35.** $\dfrac{(x-4)(x+3)}{(x-1)(2x+1)}$

37. $\dfrac{x+5}{2}$ **39.** $\dfrac{(x-2)(x+2)}{2x-3}$ **41.** $\dfrac{3x-2}{x-3}$ **43.** $\dfrac{x+9}{2x-1}$ **45.** $\dfrac{4-x}{x-2}$ **47.** $\dfrac{2(x+5)}{(x-1)(x+2)}$ **49.** $\dfrac{3x^2-2x-3}{(x+1)(x-1)}$ **51.** $\dfrac{-(11x+2)}{(x+2)(x-2)}$

53. $\dfrac{2(x^2-2)}{x(x-2)(x+2)}$ **55.** $(x-2)(x+2)(x+1)$ **57.** $x(x-1)(x+1)$ **59.** $x^3(2x-1)^2$ **61.** $x(x-1)^2(x+1)(x^2+x+1)$

63. $\dfrac{5x}{(x-6)(x-1)(x+4)}$ **65.** $\dfrac{2(2x^2+5x-2)}{(x-2)(x+2)(x+3)}$ **67.** $\dfrac{5x+1}{(x-1)^2(x+1)^2}$ **69.** $\dfrac{-x^2+3x+13}{(x-2)(x+1)(x+4)}$ **71.** $\dfrac{x^3-2x^2+4x+3}{x^2(x+1)(x-1)}$

73. $\dfrac{-1}{x(x+h)}$ **75.** $\dfrac{x+1}{x-1}$ **77.** $\dfrac{(x-1)(x+1)}{2x(2x+1)}$ **79.** $\dfrac{2(5x-1)}{(x-2)(x+1)^2}$ **81.** $\dfrac{-2x(x^2-2)}{(x+2)(x^2-x-3)}$ **83.** $\dfrac{-1}{x-1}$ **85.** $\dfrac{3x-1}{2x+1}$ **87.** $\dfrac{19}{(3x-5)^2}$

89. $\dfrac{(x+1)(x-1)}{(x^2+1)^2}$ **91.** $\dfrac{x(3x+2)}{(3x+1)^2}$ **93.** $-\dfrac{(x+3)(3x-1)}{(x^2+1)^2}$ **95.** $f=\dfrac{R_1\cdot R_2}{(n-1)(R_1+R_2)}$; $\dfrac{2}{15}$ m

R.8 Assess Your Understanding (page 78)

3. index **4.** cube root **5.** b **6.** d **7.** c **8.** c **9.** T **10.** F **11.** 3 **13.** -2 **15.** $2\sqrt{2}$ **17.** $-2x\sqrt[3]{x}$ **19.** x^3y^2 **21.** x^2y **23.** $6\sqrt{x}$ **25.** $3x^2y^3\sqrt[4]{2x}$
27. $6x\sqrt{x}$ **29.** $15\sqrt[3]{3}$ **31.** $12\sqrt{3}$ **33.** $7\sqrt{2}$ **35.** $\sqrt{2}$ **37.** $2\sqrt{3}$ **39.** $-\sqrt[3]{2}$ **41.** $x-2\sqrt{x}+1$ **43.** $(2x-1)\sqrt[3]{2x}$ **45.** $(2x-15)\sqrt{2x}$
47. $-(x+5y)\sqrt[3]{2xy}$ **49.** $\dfrac{\sqrt{2}}{2}$ **51.** $-\dfrac{\sqrt{15}}{5}$ **53.** $\dfrac{(5+\sqrt{2})\sqrt{3}}{23}$ **55.** $\dfrac{8\sqrt{5}-19}{41}$ **57.** $5\sqrt{2}+5$ **59.** $\dfrac{5\sqrt[3]{4}}{2}$ **61.** $\dfrac{2x+h-2\sqrt{x^2+xh}}{h}$ **63.** 4
65. -3 **67.** 64 **69.** $\dfrac{1}{27}$ **71.** $\dfrac{27\sqrt{2}}{32}$ **73.** $\dfrac{27\sqrt{2}}{32}$ **75.** $-\dfrac{1}{10}$ **77.** $\dfrac{25}{16}$ **79.** $x^{7/12}$ **81.** xy^2 **83.** $x^{2/3}y$ **85.** $\dfrac{8x^{5/4}}{y^{3/4}}$ **87.** $\dfrac{3x+2}{(1+x)^{1/2}}$ **89.** $\dfrac{x(3x^2+2)}{(x^2+1)^{1/2}}$
91. $\dfrac{22x+5}{10\sqrt{x}-5\sqrt{4x+3}}$ **93.** $\dfrac{2+x}{2(1+x)^{3/2}}$ **95.** $\dfrac{4-x}{(x+4)^{3/2}}$ **97.** $\dfrac{1}{x^2(x^2-1)^{1/2}}$ **99.** $\dfrac{1-3x^2}{2\sqrt{x}(1+x^2)^2}$ **101.** $\dfrac{1}{2}(5x+2)(x+1)^{1/2}$
103. $2x^{1/2}(3x-4)(x+1)$ **105.** $(x^2+4)^{1/3}(11x^2+12)$ **107.** $(3x+5)^{1/3}(2x+3)^{1/2}(17x+27)$ **109.** $\dfrac{3(x+2)}{2x^{1/2}}$ **111.** 1.41 **113.** 1.59
115. 4.89 **117.** 2.15 **119. (a)** 15,660.4 gal **(b)** 390.7 gal **121.** $2\sqrt{2}\pi\approx 8.89$ sec

CHAPTER 1 Equations and Inequalities

1.1 Assess Your Understanding (page 90)

4. F **5.** identity **6.** linear; first-degree **7.** F **8.** T **9.** b **10.** d **11.** $\{3\}$ **13.** $\{-5\}$ **15.** $\left\{\dfrac{3}{2}\right\}$ **17.** $\left\{\dfrac{5}{4}\right\}$ **19.** $\{-2\}$ **21.** $\{3\}$ **23.** $\{-1\}$ **25.** $\{-2\}$

27. $\{-18\}$ **29.** $\{-4\}$ **31.** $\left\{-\dfrac{3}{4}\right\}$ **33.** $\{-20\}$ **35.** $\{2\}$ **37.** $\{0.5\}$ **39.** $\left\{\dfrac{29}{10}\right\}$ **41.** $\{2\}$ **43.** $\{8\}$ **45.** $\{2\}$ **47.** $\{-1\}$ **49.** $\{3\}$ **51.** No solution

53. No solution **55.** $\{-6\}$ **57.** $\{34\}$ **59.** $\left\{-\dfrac{20}{39}\right\}$ **61.** $\{-1\}$ **63.** $\left\{-\dfrac{11}{6}\right\}$ **65.** $\{-6\}$ **67.** $\{5.91\}$ **69.** $\{0.41\}$ **71.** $x=\dfrac{b+c}{a}$ **73.** $x=\dfrac{abc}{a+b}$

75. $a=3$ **77.** $R=\dfrac{R_1R_2}{R_1+R_2}$ **79.** $R=\dfrac{mv^2}{F}$ **81.** $r=\dfrac{S-a}{S}$ **83.** \$11,500 will be invested in bonds and \$8500 in CDs.

85. The regular hourly rate is $10.50. **87.** Brooke needs a score of 85. **89.** The original price was $500,000; purchasing the model saves $75,000.
91. The theater paid $0.80 for the candy. **93.** There were 2187 adults. **95.** The length is 19 ft; the width is 11 ft.
97. breakfast: 675 cal; lunch: 550 cal; dinner: 800 cal **99.** Judy pays $10.80 and Tom, $7.20.

Historical Problems (page 100)

1. The area of each shaded square is 9, so the larger square will have area $85 + 4(9) = 121$. The area of the larger square is also given by the
expression $(x + 6)^2$ so $(x + 6)^2 = 121$. Taking the positive square root of each side, $x + 6 = 11$ or $x = 5$.
2. Let $z = -6$, so $z^2 + 12z - 85 = -121$. We get the equation $u^2 - 121 = 0$ or $u^2 = 121$. Thus $u = \pm 11$, so $x = \pm 11 - 6$. $x = -17$ or $x = 5$.

3.
$$\left(x + \frac{b}{2a}\right)^2 = \left(\frac{\sqrt{b^2 - 4ac}}{2a}\right)^2$$
$$\left(x + \frac{b}{2a}\right)^2 - \left(\frac{\sqrt{b^2 - 4ac}}{2a}\right)^2 = 0$$
$$\left(x + \frac{b}{2a} - \frac{\sqrt{b^2 - 4ac}}{2a}\right)\left(x + \frac{b}{2a} + \frac{\sqrt{b^2 - 4ac}}{2a}\right) = 0$$
$$\left(x + \frac{b - \sqrt{b^2 - 4ac}}{2a}\right)\left(x + \frac{b + \sqrt{b^2 - 4ac}}{2a}\right) = 0$$
$$x = \frac{-b + \sqrt{b^2 - 4ac}}{2a} \text{ or } x = \frac{-b - \sqrt{b^2 - 4ac}}{2a}$$

1.2 Assess Your Understanding (page 101)

6. discriminant; negative **7.** F **8.** F **9.** b **10.** d **11.** $\{0, 9\}$ **13.** $\{-5, 5\}$ **15.** $\{-3, 2\}$ **17.** $\left\{-\frac{1}{2}, 3\right\}$ **19.** $\{-4, 4\}$ **21.** $\{2, 6\}$ **23.** $\left\{\frac{3}{2}\right\}$
25. $\left\{-\frac{2}{3}, \frac{3}{2}\right\}$ **27.** $\left\{-\frac{2}{3}, \frac{3}{2}\right\}$ **29.** $\left\{-\frac{3}{4}, 2\right\}$ **31.** $\{-5, 5\}$ **33.** $\{-1, 3\}$ **35.** $\{-3, 0\}$ **37.** $\{-7, 3\}$ **39.** $\left\{-\frac{1}{4}, \frac{3}{4}\right\}$ **41.** $\left\{\frac{-1 - \sqrt{7}}{6}, \frac{-1 + \sqrt{7}}{6}\right\}$
43. $\{2 - \sqrt{2}, 2 + \sqrt{2}\}$ **45.** $\{2 - \sqrt{5}, 2 + \sqrt{5}\}$ **47.** $\left\{1, \frac{3}{2}\right\}$ **49.** No real solution **51.** $\left\{\frac{-1 - \sqrt{5}}{4}, \frac{-1 + \sqrt{5}}{4}\right\}$ **53.** $\left\{0, \frac{9}{4}\right\}$ **55.** $\left\{\frac{1}{3}\right\}$
57. $\left\{-\frac{2}{3}, 1\right\}$ **59.** $\left\{\frac{3 - \sqrt{29}}{10}, \frac{3 + \sqrt{29}}{10}\right\}$ **61.** $\left\{\frac{-2 - \sqrt{10}}{2}, \frac{-2 + \sqrt{10}}{2}\right\}$ **63.** $\left\{\frac{1 - \sqrt{33}}{8}, \frac{1 + \sqrt{33}}{8}\right\}$ **65.** $\left\{\frac{9 - \sqrt{73}}{2}, \frac{9 + \sqrt{73}}{2}\right\}$
67. $\{0.63, 3.47\}$ **69.** $\{-2.80, 1.07\}$ **71.** $\{-0.85, 1.17\}$ **73.** No real solution **75.** Repeated real solution **77.** Two unequal real solutions
79. $\{-\sqrt{5}, \sqrt{5}\}$ **81.** $\left\{\frac{1}{4}\right\}$ **83.** $\left\{-\frac{3}{5}, \frac{5}{2}\right\}$ **85.** $\left\{-\frac{1}{2}, \frac{2}{3}\right\}$ **87.** $\left\{\frac{-\sqrt{2} + 2}{2}, \frac{-\sqrt{2} - 2}{2}\right\}$ **89.** $\left\{\frac{-1 - \sqrt{17}}{2}, \frac{-1 + \sqrt{17}}{2}\right\}$ **91.** $\{5\}$
93. 2; 5 meters, 12 meters, 13 meters; 20 meters, 21 meters; 29 meters **95.** The dimensions are 11 ft by 13 ft. **97.** The dimensions are 5 m by 8 m.
99. The dimensions should be 4 ft by 4 ft. **101. a.** The ball strikes the ground after 6 sec. **b.** The ball passes the top of the building on its way down
after 5 sec. **103.** The dimensions should be 11.55 cm by 6.55 cm by 3 cm. **105.** The border will be 2.71 ft wide.
107. The border will be 2.56 ft wide. **109.** The screen of an iPad Air in 4:3 format has an area of 45.16 square inches; the screen of the Google Nexus in
16:10 format has an area of 44.94 square inches. The iPad Air has a larger screen. **111.** 1.1 ft **113.** 29 hr **115.** 37 consecutive integers must be added.
117. $\dfrac{-b + \sqrt{b^2 - 4ac}}{2a} + \dfrac{-b - \sqrt{b^2 - 4ac}}{2a} = \dfrac{-2b}{2a} = -\dfrac{b}{a}$ **119.** $k = \dfrac{1}{2}$ or $k = -\dfrac{1}{2}$
121. $ax^2 + bx + c = 0, x = \dfrac{-b \pm \sqrt{b^2 - 4ac}}{2a}$; $ax^2 - bx + c = 0, x = \dfrac{b \pm \sqrt{(-b)^2 - 4ac}}{2a} = \dfrac{b \pm \sqrt{b^2 - 4ac}}{2a} = -\dfrac{-b \pm \sqrt{b^2 - 4ac}}{2a}$ **123.** b

1.3 Assess Your Understanding (page 111)

4. real; imaginary; imaginary unit **5.** F **6.** T **7.** F **8.** b **9.** a **10.** c **11.** $8 + 5i$ **13.** $-7 + 6i$ **15.** $-6 - 11i$ **17.** $6 - 18i$ **19.** $6 + 4i$
21. $10 - 5i$ **23.** 37 **25.** $\dfrac{6}{5} + \dfrac{8}{5}i$ **27.** $1 - 2i$ **29.** $\dfrac{5}{2} - \dfrac{7}{2}i$ **31.** $-\dfrac{1}{2} + \dfrac{\sqrt{3}}{2}i$ **33.** $2i$ **35.** $-i$ **37.** i **39.** -6 **41.** $-10i$ **43.** $-2 + 2i$ **45.** 0 **47.** 0
49. $2i$ **51.** $5i$ **53.** $5i$ **55.** $\{-2i, 2i\}$ **57.** $\{-4, 4\}$ **59.** $\{3 - 2i, 3 + 2i\}$ **61.** $\{3 - i, 3 + i\}$ **63.** $\left\{\dfrac{1}{4} - \dfrac{1}{4}i, \dfrac{1}{4} + \dfrac{1}{4}i\right\}$ **65.** $\left\{\dfrac{1}{5} - \dfrac{2}{5}i, \dfrac{1}{5} + \dfrac{2}{5}i\right\}$
67. $\left\{-\dfrac{1}{2} - \dfrac{\sqrt{3}}{2}i, -\dfrac{1}{2} + \dfrac{\sqrt{3}}{2}i\right\}$ **69.** $\{2, -1 - \sqrt{3}i, -1 + \sqrt{3}i\}$ **71.** $\{-2, 2, -2i, 2i\}$ **73.** $\{-3i, -2i, 2i, 3i\}$
75. Two complex solutions that are conjugates of each other **77.** Two unequal real solutions **79.** A repeated real solution
81. $2 - 3i$ **83.** 6 **85.** 25 **87.** $2 + 3i$ ohms **89.** $z + \bar{z} = (a + bi) + (a - bi) = 2a$; $z - \bar{z} = (a + bi) - (a - bi) = 2bi$
91. $\bar{z} + \bar{w} = (a + bi) + (c + di) = (a + c) + (b + d)i = (a + c) - (b + d)i = (a - bi) + (c - di) = \bar{z} + \bar{w}$

1.4 Assess Your Understanding (page 117)

4. F **5.** quadratic in form **6.** T **7.** a **8.** c **9.** $\{1\}$ **11.** No real solution **13.** $\{-13\}$ **15.** $\{-1\}$ **17.** $\{0, 64\}$ **19.** $\{3\}$ **21.** $\{2\}$ **23.** $\left\{-\dfrac{8}{5}\right\}$
25. $\{8\}$ **27.** $\{2\}$ **29.** $\{-1, 3\}$ **31.** $\{1, 5\}$ **33.** $\{1\}$ **35.** $\{5\}$ **37.** $\{2\}$ **39.** $\{-4, 4\}$ **41.** $\{0, 3\}$ **43.** $\{-2, -1, 1, 2\}$ **45.** $\{-1, 1\}$ **47.** $\{-2, 1\}$ **49.** $\{-6, -5\}$
51. $\left\{-\dfrac{1}{3}\right\}$ **53.** $\left\{-\dfrac{3}{2}, 2\right\}$ **55.** $\left\{0, \dfrac{1}{16}\right\}$ **57.** $\{16\}$ **59.** $\{1\}$ **61.** $\left\{\left(\dfrac{9 - \sqrt{17}}{8}\right)^4, \left(\dfrac{9 + \sqrt{17}}{8}\right)^4\right\}$ **63.** $\{\sqrt{2}, \sqrt{3}\}$ **65.** $\{-4, 1\}$ **67.** $\left\{-2, -\dfrac{1}{2}\right\}$
69. $\left\{-\dfrac{3}{2}, \dfrac{1}{3}\right\}$ **71.** $\left\{-\dfrac{1}{8}, 27\right\}$ **73.** $\left\{-2, -\dfrac{4}{5}\right\}$ **75.** $\{-3, 0, 3\}$ **77.** $\left\{0, \dfrac{3}{4}\right\}$ **79.** $\{-5, 0, 4\}$ **81.** $\{-1, 1\}$ **83.** $\{-2, 2, 3\}$ **85.** $\left\{-2, \dfrac{1}{2}, 2\right\}$

56. **(a)** 6.5 in. by 6.5 in.; 12.5 in. by 12.5 in.   **(b)** $8\frac{2}{3}$ in. by $4\frac{1}{3}$ in.; $14\frac{2}{3}$ in. by $10\frac{1}{3}$ in.

57. Scott receives \$400,000, Alice receives \$300,000, and Tricia receives \$200,000.   **58.** It would take the older copier 180 min or 3 hr.

59. **(a)** No   **(b)** Todd   **(c)** $\frac{1}{4}$ m   **(d)** 5.26 m   **(e)** Yes   **60.** The freight train is 190.67 feet long.

Chapter Test *(page 147)*

1. $\left\{\frac{5}{2}\right\}$ **2.** $\{-2, 3\}$ **3.** $\{-2, 2\}$ **4.** $\left\{\frac{9}{2}\right\}$ **5.** $\{0, 3\}$ **6.** $\left\{-2, -\frac{2}{3}, 2\right\}$ **7.** No real solution   **8.** $\left[-\frac{2}{3}, \frac{16}{3}\right]$

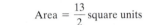

9. $\left(-4, \frac{4}{3}\right)$ **10.** $(-\infty, -1] \cup [6, \infty)$ **11.** $-\frac{3}{5} - \frac{1}{5}i$ **12.** $\left\{\frac{1}{2} - i, \frac{1}{2} + i\right\}$ **13.** Add $6\frac{2}{3}$ lb of \$8/lb coffee to get $26\frac{2}{3}$ lb of \$5/lb coffee.

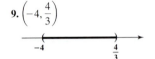

CHAPTER 2 Graphs

2.1 Assess Your Understanding *(page 154)*

7. *x*-coordinate or abscissa; *y*-coordinate or ordinate   **8.** quadrants   **9.** midpoint   **10.** F   **11.** F   **12.** T   **13.** b   **14.** a

15. **(a)** Quadrant II   **(b)** *x*-axis   **(c)** Quadrant III   **(d)** Quadrant I   **(e)** *y*-axis   **(f)** Quadrant IV

17. The points will be on a vertical line that is 2 units to the right of the *y*-axis.

19. $\sqrt{5}$ **21.** $\sqrt{10}$ **23.** $2\sqrt{17}$ **25.** $\sqrt{85}$ **27.** $\sqrt{53}$ **29.** $\sqrt{a^2 + b^2}$

31. $d(A, B) = \sqrt{13}$
$d(B, C) = \sqrt{13}$
$d(A, C) = \sqrt{26}$
$(\sqrt{13})^2 + (\sqrt{13})^2 = (\sqrt{26})^2$
Area $= \frac{13}{2}$ square units

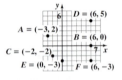

33. $d(A, B) = \sqrt{130}$
$d(B, C) = \sqrt{26}$
$d(A, C) = 2\sqrt{26}$
$(\sqrt{26})^2 + (2\sqrt{26})^2 = (\sqrt{130})^2$
Area $= 26$ square units

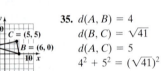

35. $d(A, B) = 4$
$d(B, C) = \sqrt{41}$
$d(A, C) = 5$
$4^2 + 5^2 = (\sqrt{41})^2$
Area $= 10$ square units

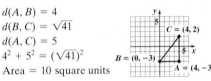

37. $(4, 0)$ **39.** $\left(\frac{3}{2}, 1\right)$ **41.** $(5, -1)$ **43.** $\left(\frac{a}{2}, \frac{b}{2}\right)$ **45.** $(5, 3)$ **47.** $(3, -13), (3, 11)$

49. $(4 + 3\sqrt{3}, 0); (4 - 3\sqrt{3}, 0)$ **51.** **(a)** $(-1, 1)$ **(b)** $(0, 13)$ **53.** $(1, 2)$ **55.** $\sqrt{17}; 2\sqrt{5}; \sqrt{29}$ **57.** $\left(\frac{s}{2}, \frac{s}{2}\right)$

59. $d(P_1, P_2) = 6; d(P_2, P_3) = 4; d(P_1, P_3) = 2\sqrt{13}$; right triangle   **61.** $d(P_1, P_2) = 2\sqrt{17}; d(P_2, P_3) = \sqrt{34}; d(P_1, P_3) = \sqrt{34}$; isosceles right triangle

63. $90\sqrt{2} \approx 127.28$ ft   **65.** **(a)** $(90, 0), (90, 90), (0, 90)$ **(b)** $5\sqrt{2161} \approx 232.43$ ft   **(c)** $30\sqrt{149} \approx 366.20$ ft

67. $d = 50t$ mi   **69.** **(a)** $(2.65, 1.6)$ **(b)** Approximately 1.285 units

71. \$21,142; a slight underestimate   **73.** $\left\{x \,\middle|\, x \neq \frac{5}{2}\right\}$ **74.** $\left\{-\frac{5}{3}, 4\right\}$ **75.** $13 - 11i$ **76.** $(-\infty, 2]$

2.2 Assess Your Understanding *(page 164)*

3. intercepts   **4.** $y = 0$ **5.** *y*-axis   **6.** 4   **7.** $(-3, 4)$ **8.** T   **9.** F   **10.** F   **11.** d   **12.** c   **13.** $(0, 0)$ is on the graph.   **15.** $(0, 3)$ is on the graph.

17. $(0, 2)$ and $(\sqrt{2}, \sqrt{2})$ are on the graph.

19. $(-2, 0), (0, 2)$ **21.** $(-4, 0), (0, 8)$ **23.** $(-1, 0), (1, 0), (0, -1)$ **25.** $(-2, 0), (2, 0), (0, 4)$ **27.** $(3, 0), (0, 2)$

29. $(-2, 0), (2, 0), (0, 9)$ **31.**   **33.**   **35.**   **37.**

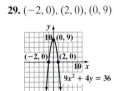
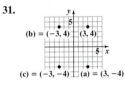
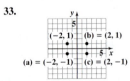
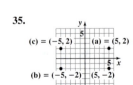
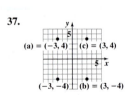

39.

41. (a) $(-1, 0), (1, 0)$
(b) Symmetric with respect to the x-axis, the y-axis, and the origin
47. (a) $(-2, 0), (0, 0), (2, 0)$
(b) Symmetric with respect to the origin

43. (a) $\left(-\frac{\pi}{2}, 0\right), (0, 1), \left(\frac{\pi}{2}, 0\right)$
(b) Symmetric with respect to the y-axis
49. (a) $(x, 0), -2 \le x \le 1$
(b) No symmetry

45. (a) $(0, 0)$
(b) Symmetric with respect to the x-axis
51. (a) No intercepts
(b) Symmetric with respect to the origin

53.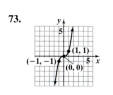

55.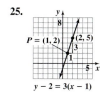

57. $(-4, 0), (0, -2), (0, 2)$; symmetric with respect to the x-axis   **59.** $(0, 0)$; symmetric with respect to the origin   **61.** $(0, 9), (3, 0), (-3, 0)$; symmetric with respect to the y-axis
63. $(-2, 0), (2, 0), (0, -3), (0, 3)$; symmetric with respect to the x-axis, y-axis, and origin
65. $(0, -27), (3, 0)$; no symmetry   **67.** $(0, -4), (4, 0), (-1, 0)$; no symmetry   **69.** $(0, 0)$; symmetric with respect to the origin   **71.** $(0, 0)$; symmetric with respect to the origin

73.

75.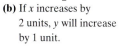

77. $a = -4$ or $a = 1$ **79.** $(-1, -2)$ **81.** 4   **83. (a)** $(0, 0), (2, 0), (0, 1), (0, -1)$
(b) x-axis symmetry   **85. (a)** $y = \sqrt{x^2}$ and $y = |x|$ have the same graph.   **(b)** $\sqrt{x^2} = |x|$
(c) $x \ge 0$ for $y = (\sqrt{x})^2$, while x can be any real number for $y = x$.   **(d)** $y \ge 0$ for $y = \sqrt{x^2}$
92. $\frac{1}{2}$ **93.** $3(x - 5)^2$ **94.** $14i$ **95.** $\{4 - 2\sqrt{3}, 4 + 2\sqrt{3}\}$

2.3 Assess Your Understanding *(page 178)*

1. undefined; 0   **2.** 3; 2   **3.** T   **4.** F   **5.** T   **6.** $m_1 = m_2$; y-intercepts; $m_1 m_2 = -1$ **7.** 2   **8.** $-\frac{1}{2}$ **9.** F   **10.** d   **11.** c   **12.** b

13. (a) Slope $= \frac{1}{2}$
(b) If x increases by 2 units, y will increase by 1 unit.

15. (a) Slope $= -\frac{1}{3}$
(b) If x increases by 3 units, y will decrease by 1 unit.

17. Slope $= -\frac{3}{2}$

19. Slope $= -\frac{1}{2}$

21. Slope $= 0$

23. Slope undefined

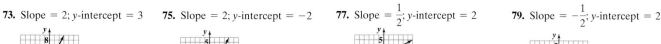

25.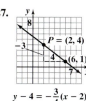

$y - 2 = 3(x - 1)$

27.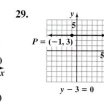

$y - 4 = -\frac{3}{4}(x - 2)$

29.

$y - 3 = 0$

31.

$x = 0$

33. $(2, 6); (3, 10); (4, 14)$
35. $(4, -7); (6, -10); (8, -13)$
37. $(-1, -5); (0, -7); (1, -9)$
39. $x - 2y = 0$ or $y = \frac{1}{2}x$

41. $x + y = 2$ or $y = -x + 2$ **43.** $2x - y = 3$ or $y = 2x - 3$ **45.** $x + 2y = 5$ or $y = -\frac{1}{2}x + \frac{5}{2}$ **47.** $3x - y = -9$ or $y = 3x + 9$
49. $2x + 3y = -1$ or $y = -\frac{2}{3}x - \frac{1}{3}$ **51.** $x - 2y = -5$ or $y = \frac{1}{2}x + \frac{5}{2}$ **53.** $3x + y = 3$ or $y = -3x + 3$ **55.** $x - 2y = 2$ or $y = \frac{1}{2}x - 1$
57. $x = 2$; no slope–intercept form   **59.** $y = 2$ **61.** $2x - y = -4$ or $y = 2x + 4$ **63.** $2x - y = 0$ or $y = 2x$ **65.** $x = 4$; no slope–intercept form
67. $2x + y = 0$ or $y = -2x$ **69.** $x - 2y = -3$ or $y = \frac{1}{2}x + \frac{3}{2}$ **71.** $y = 4$

73. Slope $= 2$; y-intercept $= 3$

75. Slope $= 2$; y-intercept $= -2$

77. Slope $= \frac{1}{2}$; y-intercept $= 2$

79. Slope $= -\frac{1}{2}$; y-intercept $= 2$

81. Slope $= \frac{2}{3}$; y-intercept $= -2$

83. Slope $= -1$; y-intercept $= 1$

85. Slope undefined; no y-intercept

87. Slope $= 0$; y-intercept $= 5$

89. Slope $= 1$; y-intercept $= 0$

91. Slope $= \frac{3}{2}$; y-intercept $= 0$

93. (a) x-intercept: 3; y-intercept: 2
(b)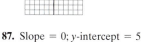

95. (a) x-intercept: -10; y-intercept: 8
(b)

79. (a)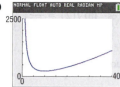

(b) 10 riding lawn mowers
(c) $239/mower

81. (a), (b)

(c) $5/gigabyte
(d) $6.25/gigabyte
(e) $7.50/gigabyte
(f) The average rate of change is increasing as the number of gigabytes increases.

83. (a) On average, the population is increasing at a rate of 0.036 g/h from 0 to 2.5 h. **(b)** On average, from 4.5 to 6 h, the population is increasing at a rate of 0.1 g/h. **(c)** The average rate of change is increasing over time.

85. (a) 1 **(b)** 0.5 **(c)** 0.1 **(d)** 0.01
(e) 0.001
(f)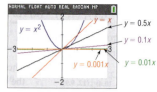

(g) They are getting closer to the tangent line at (0, 0).
(h) They are getting closer to 0.

87. (a) 2
(b) 2; 2; 2; 2
(c) $y = 2x + 5$
(d)

89. (a) $2x + h + 2$
(b) 4.5; 4.1; 4.01; 4
(c) $y = 4.01x - 1.01$
(d)

91. (a) $4x + 2h - 3$
(b) 2; 1.2; 1.02; 1
(c) $y = 1.02x - 1.02$
(d)

93. (a) $-\dfrac{1}{(x+h)x}$
(b) $-\dfrac{2}{3}; -\dfrac{10}{11}; -\dfrac{100}{101}; -1$
(c) $y = -\dfrac{100}{101}x + \dfrac{201}{101}$
(d)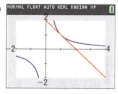

97. At most one **99.** Yes; the function $f(x) = 0$ is both even and odd. **101.** Not necessarily. It just means $f(5) > f(2)$.
103. (a) 7.01×10^{-6} **(b)** 2.305×10^9 **104.** $6\sqrt{15}$ **105.** $[-8, -3)$ **106.** 8.25 days

3.4 Assess Your Understanding *(page 244)*

4. $(-\infty, 0)$ **5.** piecewise-defined **6.** T **7.** F **8.** F **9.** b **10.** a **11.** C **13.** E **15.** B **17.** F

19.

21.

23.

25.

27. (a) 4 **(b)** 2 **(c)** 5 **29. (a)** -4 **(b)** -2 **(c)** 0 **(d)** 25

31. (a) All real numbers
(b) (0, 1)
(c)

(d) $\{y | y \neq 0\}$; $(-\infty, 0) \cup (0, \infty)$
(e) Discontinuous at $x = 0$

33. (a) All real numbers
(b) (0, 3)
(c)

(d) $\{y | y \geq 1\}$; $[1, \infty)$
(e) Continuous

35. (a) $\{x | x \geq -2\}$; $[-2, \infty)$
(b) (0, 3), (2, 0)
(c)

(d) $\{y | y < 4, y = 5\}$; $(-\infty, 4) \cup \{5\}$
(e) Discontinuous at $x = 1$

37. (a) All real numbers
(b) $(-1, 0)$, (0, 0)
(c)

(d) All real numbers
(e) Discontinuous at $x = 0$

39. (a) $\{x | x \geq -2, x \neq 0\}$; $[-2, 0) \cup (0, \infty)$
(b) No intercepts
(c)

(d) $\{y | y > 0\}$; $(0, \infty)$
(e) Discontinuous at $x = 0$

41. (a) All real numbers
(b) $(x, 0)$ for $0 \leq x < 1$
(c)

(d) Set of even integers
(e) Discontinuous at $\{x | x \text{ is an integer}\}$

43. $f(x) = \begin{cases} -x & \text{if } -1 \leq x \leq 0 \\ \frac{1}{2}x & \text{if } 0 < x \leq 2 \end{cases}$ (Other answers are possible.)

45. $f(x) = \begin{cases} -x & \text{if } x \leq 0 \\ -x + 2 & \text{if } 0 < x \leq 2 \end{cases}$ (Other answers are possible.)

47. (a) 2 **(b)** 3 **(c)** -4 **49. (a)** $34.99 **(b)** $64.99 **(c)** $184.99

51. (a) $44.46 **(b)** $126.03

(c) $C(x) = \begin{cases} 1.24816x + 19.50 & \text{if } 0 \leq x \leq 30 \\ 0.5757x + 39.6738 & \text{if } x > 30 \end{cases}$

(d)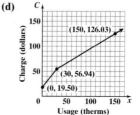

53. $f(x) = \begin{cases} 0.10x & \text{if} & 0 < x \le 9075 \\ 907.50 + 0.15(x - 9075) & \text{if} & 9075 < x \le 36,900 \\ 5081.25 + 0.25(x - 36,900) & \text{if} & 36,900 < x \le 89,350 \\ 18,193.75 + 0.28(x - 89,350) & \text{if} & 89,350 < x \le 186,350 \\ 45,353.75 + 0.33(x - 186,350) & \text{if} & 186,350 < x \le 405,100 \\ 117,541.25 + 0.35(x - 405,100) & \text{if} & 405,100 < x \le 406,750 \\ 118,188.75 + 0.396(x - 406,750) & \text{if} & x > 406,750 \end{cases}$

55. (a)
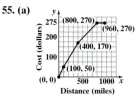
(b) $C(x) = 10 + 0.4x$ **(c)** $C(x) = 70 + 0.25x$

57. (a) $C(s) = \begin{cases} 9000 & \text{if} & s \le 659 \\ 7500 & \text{if} & 660 \le s \le 679 \\ 5250 & \text{if} & 680 \le s \le 699 \\ 3000 & \text{if} & 700 \le s \le 719 \\ 1500 & \text{if} & 720 \le s \le 739 \\ 750 & \text{if} & s \ge 740 \end{cases}$ **(b)** \$1500 **(c)** \$7500

59. (a) $10°C$ **(b)** $4°C$ **(c)** $-3°C$ **(d)** $-4°C$
  (e) The wind chill is equal to the air temperature.
  (f) At wind speed greater than 20 m/s, the wind
    chill factor depends only on the air temperature.

61. $C(x) = \begin{cases} 0.98 & \text{if} & 0 < x \le 1 \\ 1.19 & \text{if} & 1 < x \le 2 \\ 1.40 & \text{if} & 2 < x \le 3 \\ 1.61 & \text{if} & 3 < x \le 4 \\ 1.82 & \text{if} & 4 < x \le 5 \\ 2.03 & \text{if} & 5 < x \le 6 \\ 2.24 & \text{if} & 6 < x \le 7 \\ 2.45 & \text{if} & 7 < x \le 8 \\ 2.66 & \text{if} & 8 < x \le 9 \\ 2.87 & \text{if} & 9 < x \le 10 \\ 3.08 & \text{if} & 10 < x \le 11 \\ 3.29 & \text{if} & 11 < x \le 12 \\ 3.50 & \text{if} & 12 < x \le 13 \end{cases}$
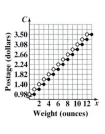

63. Each graph is that of $y = x^2$, but shifted horizontally. If $y = (x - k)^2$, $k > 0$, the shift is right k units; if $y = (x + k)^2$, $k > 0$, the shift is left k units. **65.** The graph of $y = -f(x)$ is the reflection about the x-axis of the graph of $y = f(x)$. **67.** Yes. The graph of $y = (x - 1)^3 + 2$ is the graph of $y = x^3$ shifted right 1 unit and up 2 units. **69.** They all have the same general shape. All three go through the points $(-1, -1)$, $(0, 0)$, and $(1, 1)$. As the exponent increases, the steepness of the curve increases (except near $x = 0$). **72.** $22 - 7i$ **73.** $(h, k) = (0, 3); r = 5$
74. $\{-8\}$ **75.** CD: \$22,000; Mutual fund: \$38,000

3.5 Assess Your Understanding (page 256)

1. horizontal; right **2.** y **3.** F **4.** T **5.** d **6.** a **7.** B **9.** H **11.** I **13.** L **15.** F **17.** G **19.** $y = (x - 4)^3$ **21.** $y = x^3 + 4$ **23.** $y = -x^3$
25. $y = 4x^3$ **27.** $y = -(\sqrt{-x} + 2)$ **29.** $y = -\sqrt{x + 3} + 2$ **31.** (c) **33.** (c) **35. (a)** -7 and 1 **(b)** -3 and 5 **(c)** -5 and 3 **(d)** -3 and 5
37. (a) $(-3, 3)$ **(b)** $(4, 10)$ **(c)** Decreasing on $(-1, 5)$ **(d)** Decreasing on $(-5, 1)$

39.
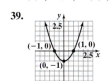
Domain: $(-\infty, \infty)$;
Range: $[-1, \infty)$

41.

Domain: $(-\infty, \infty)$;
Range: $(-\infty, \infty)$

43

Domain: $[-2, \infty)$;
Range: $[0, \infty)$

45.

Domain: $(-\infty, \infty)$;
Range: $(-\infty, \infty)$

47.

Domain: $[0, \infty)$;
Range: $[0, \infty)$

49.
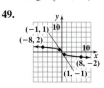
Domain: $(-\infty, \infty)$;
Range: $(-\infty, \infty)$

51.

Domain: $(-\infty, \infty)$;
Range: $[-3, \infty)$

53.

Domain: $[2, \infty)$;
Range: $[1, \infty)$

55.

Domain: $(-\infty, 0]$;
Range: $[-2, \infty)$

57.

Domain: $(-\infty, \infty)$;
Range: $(-\infty, \infty)$

59.

Domain: $(-\infty, \infty)$;
Range: $[0, \infty)$

61.
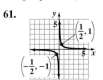
Domain: $(-\infty, 0) \cup (0, \infty)$
Range: $(-\infty, 0) \cup (0, \infty)$

63. (a) $F(x) = f(x) + 3$

(b) $G(x) = f(x + 2)$
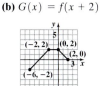

(c) $P(x) = -f(x)$

17. (a) Domain: $\{x | x \leq 4\}$ or $(-\infty, 4]$
Range: $\{y | y \leq 3\}$ or $(-\infty, 3]$
(b) Increasing on $(-\infty, -2)$ and $(2, 4)$; Decreasing on $(-2, 2)$
(c) Local maximum value is 1 and occurs at $x = -2$.
Local minimum value is -1 and occurs at $x = 2$.

(d) Absolute maximum: $f(4) = 3$
Absolute minimum: none
(e) No symmetry
(f) Neither
(g) x-intercepts: $-3, 0, 3$ y-intercept: 0

18. Odd **19.** Even **20.** Neither **21.** Odd

22.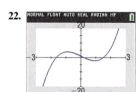

Local maximum value: 4.04 at $x = -0.91$
Local minimum value: -2.04 at $x = 0.91$
Increasing: $(-3, -0.91)$; $(0.91, 3)$
Decreasing: $(-0.91, 0.91)$

23.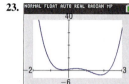

Local maximum value: 1.53 at $x = 0.41$
Local minima values: 0.54 at
$x = -0.34$ and -3.56 at $x = 1.80$
Increasing: $(-0.34, 0.41)$; $(1.80, 3)$
Decreasing: $(-2, -0.34)$; $(0.41, 1.80)$

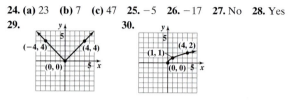

24. (a) 23 **(b)** 7 **(c)** 47 **25.** -5 **26.** -17 **27.** No **28.** Yes
29.
30.

31.

Intercepts: $(-4, 0)$, $(4, 0)$, $(0, -4)$
Domain: all real numbers
Range: $\{y | y \geq -4\}$ or $[-4, \infty)$

32.

Intercept: $(0, 0)$
Domain: all real numbers
Range: $\{y | y \leq 0\}$ or $(-\infty, 0]$

33.

Intercept: $(1, 0)$
Domain: $\{x | x \geq 1\}$ or $[1, \infty)$
Range: $\{y | y \geq 0\}$ or $[0, \infty)$

34.

Intercepts: $(0, 1)$, $(1, 0)$
Domain: $\{x | x \leq 1\}$ or $(-\infty, 1]$
Range: $\{y | y \geq 0\}$ or $[0, \infty)$

35.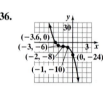

Intercept: $(0, 3)$
Domain: all real numbers
Range: $\{y | y \geq 2\}$ or $[2, \infty)$

36.

Intercepts: $(0, -24)$,
$(-2 - \sqrt[3]{4}, 0)$ or about
$(-3.6, 0)$
Domain: all real numbers
Range: all real numbers

37. (a) $\{x | x > -2\}$ or $(-2, \infty)$
(b) $(0, 0)$
(c)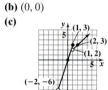

(d) $\{y | y > -6\}$ or $(-6, \infty)$
(e) Discontinuous at $x = 1$

38. (a) $\{x | x \geq -4\}$ or $[-4, \infty)$
(b) $(0, 1)$
(c)

(d) $\{y | -4 \leq y < 0$ or $y > 0\}$
or $[-4, 0) \cup (0, \infty)$
(e) Discontinuous at $x = 0$

39. $A = 11$

40. (a) $A(x) = 2x^2 + \dfrac{40}{x}$
(b) 42 ft^2 **(c)** 28 ft^2
(d)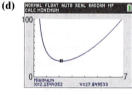

A is smallest when $x \approx 2.15$ ft.

41. (a) $A(x) = 10x - x^3$
(b) The largest area that can be enclosed by the
rectangle is approximately 12.17 square units.

Chapter Test *(page 270)*

1. (a) Function; domain: $\{2, 4, 6, 8\}$; range: $\{5, 6, 7, 8\}$ **(b)** Not a function **(c)** Not a function **(d)** Function; domain; all real numbers;
range: $\{y | y \geq 2\}$ **2.** Domain: $\left\{x | x \leq \dfrac{4}{5}\right\}$; $f(-1) = 3$ **3.** Domain: $\{x | x \neq -2\}$; $g(-1) = 1$ **4.** Domain: $\{x | x \neq -9, x \neq 4\}$; $h(-1) = \dfrac{1}{8}$
5. (a) Domain: $\{x | -5 \leq x \leq 5\}$; range: $\{y | -3 \leq y \leq 3\}$ **(b)** $(0, 2)$, $(-2, 0)$, and $(2, 0)$ **(c)** $f(1) = 3$ **(d)** $x = -5$ and $x = 3$
(e) $\{x | -5 \leq x < -2$ or $2 < x \leq 5\}$ or $[-5, -2) \cup (2, 5]$ **6.** Local maxima values: $f(-0.85) \approx -0.86$; $f(2.35) \approx 15.55$; local minimum value:
$f(0) = -2$; the function is increasing on the intervals $(-5, -0.85)$ and $(0, 2.35)$ and decreasing on the intervals $(-0.85, 0)$ and $(2.35, 5)$.
7. (a)   **(b)** $(0, -4)$, $(4, 0)$ **8.** 19 **9. (a)** $(f - g)(x) = 2x^2 - 3x + 3$ **10. (a)** **(b)**
(c) $g(-5) = -9$
(d) $g(2) = -2$
(b) $(f \cdot g)(x) = 6x^3 - 4x^2 + 3x - 2$
(c) $f(x + h) - f(x) = 4xh + 2h^2$

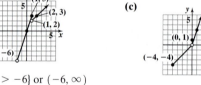

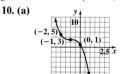

11. (a) 8.67% occurring in 1997 $(x \approx 5)$ **(b)** The model predicts that the interest rate will be -10.343%. This is not reasonable.

12. (a) $V(x) = \dfrac{x^2}{8} - \dfrac{5x}{4} + \dfrac{\pi x^2}{64}$ **(b)** 1297.61 ft^3

Cumulative Review *(page 271)*

1. $\{6\}$ **2.** $\left\{0, \dfrac{1}{3}\right\}$ **3.** $\{-1, 9\}$ **4.** $\left\{\dfrac{1}{3}, \dfrac{1}{2}\right\}$ **5.** $\left\{-\dfrac{7}{2}, \dfrac{1}{2}\right\}$ **6.** $\left\{\dfrac{1}{2}\right\}$

7. $\left\{x \,\middle|\, x < -\dfrac{4}{3}\right\}; \left(-\infty, -\dfrac{4}{3}\right)$ **8.** $\{x \,|\, 1 < x < 4\}; (1, 4)$ **9.** $\left\{x \,\middle|\, x \le -2 \text{ or } x \ge \dfrac{3}{2}\right\}; (-\infty, -2] \cup \left[\dfrac{3}{2}, \infty\right)$

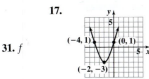

10. (a) distance: $\sqrt{29}$ **(b)** midpoint: $\left(\dfrac{1}{2}, -4\right)$ **(c)** slope: $-\dfrac{2}{5}$

11. 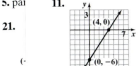 **12.** **13.** **14.** **15.** Intercepts: $(0, -3), (-2, 0), (2, 0)$; symmetry with respect to the y-axis

16. $y = \dfrac{1}{2}x + 5$

17. 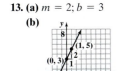 **18.** 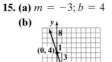 **19.**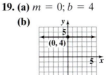

CHAPTER 4 Linear and Quadratic Functions

4.1 Assess Your Understanding *(page 280)*

7. slope; y-intercept **8.** positive **9.** T **10.** F **11.** a **12.** d

13. (a) $m = 2; b = 3$ **15. (a)** $m = -3; b = 4$ **17. (a)** $m = \dfrac{1}{4}; b = -3$ **19. (a)** $m = 0; b = 4$ **21.** Linear; -3 **23.** Nonlinear

(b) **(b)** **(b)** 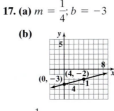 **(b)** 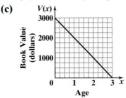 **25.** Nonlinear **27.** Linear; 0

(c) 2 **(d)** Increasing **(c)** -3 **(d)** Decreasing **(c)** 0 **(d)** Constant

(c) $\dfrac{1}{4}$ **(d)** Increasing

29. (a) $\dfrac{1}{4}$ **(b)** $\left\{x \,\middle|\, x > \dfrac{1}{4}\right\}$ or $\left(\dfrac{1}{4}, \infty\right)$ **31. (a)** 40 **(b)** 88 **(c)** -40 **(d)** $\{x \,|\, x > 40\}$ or $(40, \infty)$ **(e)** $\{x \,|\, x \le 88\}$ or $(-\infty, 88]$

(c) 1 **(d)** $\{x \,|\, x \le 1\}$ or $(-\infty, 1]$ **(f)** $\{x \,|\, -40 < x < 88\}$ or $(-40, 88)$ **33. (a)** -4 **(b)** $\{x \,|\, x < -4\}$ or $(-\infty, -4)$

(e) 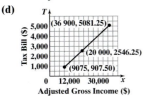 **35. (a)** -6 **(b)** $\{x \,|\, -6 \le x < 5\}$ or $[-6, 5)$ **37. (a)** \$59 **(b)** 180 mi **(c)** 300 mi

(d) $\{x \,|\, x \ge 0\}$ or $[0, \infty)$ **39. (a)** \$24; 600 T-shirts **(b)** $\$0 \le p < \$24$ **(c)** The price will increase.

41. (a) $\{x \,|\, 9075 \le x \le 36\,900\}$ or $[9075, 36\,900]$ **43. (a)** $x = 5000$ **45. (a)** $V(x) = -1000x + 3000$ **47. (a)** $C(x) = 90x + 1800$

(b) \$2546.25 **(b)** $x > 5000$ **(b)** $\{x \,|\, 0 \le x \le 3\}$ or $[0, 3]$ **(b)**

(c) The independent variable is adjusted gross income, x. The dependent variable is the tax bill, T.

(c)

(d)

(c) \$3060

(d) 22 bicycles

(d) \$1000

(e) After 1 year

(e) \$27,500

71. $a = 6, b = 0, c = 2$

73. (a), (c), (d)

(b) $\{-1, 3\}$

75. (a), (c), (d)

(b) $\{-1, 3\}$

77. (a), (c), (d)

(b) $\{-1, 2\}$

79. (a) $a = 1: f(x) = (x + 3)(x - 1) = x^2 + 2x - 3$
$a = 2: f(x) = 2(x + 3)(x - 1) = 2x^2 + 4x - 6$
$a = -2: f(x) = -2(x + 3)(x - 1) = -2x^2 - 4x + 6$
$a = 5: f(x) = 5(x + 3)(x - 1) = 5x^2 + 10x - 15$

(b) The value of a does not affect the x-intercepts, but it changes the y-intercept by a factor of a.
(c) The value of a does not affect the axis of symmetry. It is $x = -1$ for all values of a.
(d) The value of a does not affect the x-coordinate of the vertex. However, the y-coordinate of the vertex is multiplied by a.
(e) The mean of the x-intercepts is the x-coordinate of the vertex.

81. (a) $(-2, -25)$
(b) $-7, 3$
(c) $-4, 0; (-4, -21), (0, -21)$
(d)

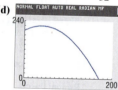

83. $(2, 2)$ **85.** $500; 1,000,000$ **87. (a)** 70,000 players   **(b)** $2500   **89. (a)** 187 or 188 watches; $7031.20
(b) $P(x) = -0.2x^2 + 43x - 1750$ **(c)** 107 or 108 watches; $561.20   **91. (a)** 171 ft   **(b)** 49 mph
(c) Reaction time   **93.** If x is even, then ax^2 and bx are even and $ax^2 + bx$ is even, which means that
$ax^2 + bx + c$ is odd. If x is odd, then ax^2 and bx are odd and $ax^2 + bx$ is even, which means that $ax^2 + bx + c$
is odd. In either case, $f(x)$ is odd.
95.

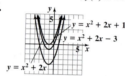

97. $b^2 - 4ac < 0$ **99.** No
101. Symmetric with respect to the x-axis, the y-axis, and the origin
102. $\{x | x \le 4\}$ or $(-\infty, 4]$
103. Center $(5, -2)$; radius $= 3$
104. $y = \sqrt{-x}$

4.4 Assess Your Understanding (page 307)

3. (a) $R(x) = -\frac{1}{6}x^2 + 100x$ **(b)** $\{x | 0 \le x \le 600\}$ **(c)** $13,333.33   **(d)** 300; $15,000   **(e)** $50

5. (a) $R(x) = -\frac{1}{5}x^2 + 20x$ **(b)** $255   **(c)** 50; $500   **(d)** $10   **(e)** Between $8 and $12

7. (a) $A(w) = -w^2 + 200w$ **(b)** A is largest when $w = 100$ yd.   **(c)** 10,000 yd^2 **9.** 2,000,000 m^2

11. (a) $\frac{625}{16} \approx 39$ ft   **(b)** $\frac{7025}{32} \approx 219.5$ ft   **(c)** About 170 ft

(d)

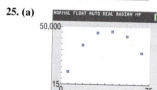

(f) When the height is 100 ft, the projectile is about 135.7 ft from the cliff.

13. 18.75 m   **15. (a)** 3 in. **(b)** Between 2 in. and 4 in.
17. $\frac{750}{\pi} \approx 238.73$ m by 375 m
19. $x = \frac{a}{2}$ **21.** $\frac{38}{3}$ **23.** $\frac{248}{3}$

25. (a)

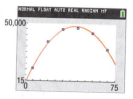

(b) $I(x) = -44.759x^2 + 4295.356x - 55,045.418$
(c) About 48.0 years of age
(d) Approximately $48,007

(e)

The data appear to follow a quadratic relation with $a < 0$.

27. (a)

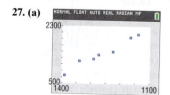

The data appear to be linearly related with positive slope.
(b) $R(x) = 1.229x + 917.385$ **(c)** $1993

29. (a)

The data appear to follow a quadratic relation with $a < 0$.
(b) $B(a) = -0.547a^2 + 31.190a - 342.218$ **(c)** 79.357

32. $15i$ **33.** 13   **34.** $(x + 6)^2 + y^2 = 7$ **35.** $\left\{ \dfrac{-4 - \sqrt{31}}{5}, \dfrac{-4 + \sqrt{31}}{5} \right\}$

4.5 Assess Your Understanding *(page 314)*

3. (a) $\{x|x < -2 \text{ or } x > 2\}$; $(-\infty, -2) \cup (2, \infty)$ **(b)** $\{x|-2 \le x \le 2\}$; $[-2, 2]$

5. (a) $\{x|-2 \le x \le 1\}$; $[-2, 1]$ **(b)** $\{x|x < -2 \text{ or } x > 1\}$; $(-\infty, -2) \cup (1, \infty)$

7. $\{x|-2 < x < 5\}$; $(-2, 5)$ **9.** $\{x|x < 0 \text{ or } x > 4\}$; $(-\infty, 0) \cup (4, \infty)$ **11.** $\{x|-3 < x < 3\}$; $(-3, 3)$

13. $\{x|x < -4 \text{ or } x > 3\}$; $(-\infty, -4) \cup (3, \infty)$ **15.** $\left\{x\left|-\frac{1}{2} < x < 3\right.\right\}$; $\left(-\frac{1}{2}, 3\right)$ **17.** No real solution   **19.** No real solution

21. $\left\{x\left|x < -\frac{2}{3} \text{ or } x > \frac{3}{2}\right.\right\}$; $\left(-\infty, -\frac{2}{3}\right) \cup \left(\frac{3}{2}, \infty\right)$ **23.** $\{x|x \le -4 \text{ or } x \ge 4\}$; $(-\infty, -4] \cup [4, \infty)$ **25. (a)** $\{-1, 1\}$ **(b)** $\{-1\}$ **(c)** $\{-1, 4\}$

(d) $\{x|x < -1 \text{ or } x > 1\}$; $(-\infty, -1) \cup (1, \infty)$ **(e)** $\{x|x \le -1\}$; $(-\infty, -1]$ **(f)** $\{x|x < -1 \text{ or } x > 4\}$; $(-\infty, -1) \cup (4, \infty)$

(g) $\{x|x \le -\sqrt{2} \text{ or } x \ge \sqrt{2}\}$; $(-\infty, -\sqrt{2}] \cup [\sqrt{2}, \infty)$

27. (a) $\{-1, 1\}$ **(b)** $\left\{-\frac{1}{4}\right\}$ **(c)** $\{-4, 0\}$ **(d)** $\{x|-1 < x < 1\}$; $(-1, 1)$ **(e)** $\left\{x\left|x \le -\frac{1}{4}\right.\right\}$; $\left(-\infty, -\frac{1}{4}\right]$ **(f)** $\{x|-4 < x < 0\}$; $(-4, 0)$ **(g)** $\{0\}$

29. (a) $\{-2, 2\}$ **(b)** $\{-2, 2\}$ **(c)** $\{-2, 2\}$ **(d)** $\{x|x < -2 \text{ or } x > 2\}$; $(-\infty, -2) \cup (2, \infty)$ **(e)** $\{x|x \le -2 \text{ or } x \ge 2\}$; $(-\infty, -2] \cup [2, \infty)$

(f) $\{x|x < -2 \text{ or } x > 2\}$; $(-\infty, -2) \cup (2, \infty)$ **(g)** $\{x|x \le -\sqrt{5} \text{ or } x \ge \sqrt{5}\}$; $(-\infty, -\sqrt{5}] \cup [\sqrt{5}, \infty)$

31. (a) $\{-1, 2\}$ **(b)** $\{-2, 1\}$ **(c)** $\{0\}$ **(d)** $\{x|x < -1 \text{ or } x > 2\}$; $(-\infty, -1) \cup (2, \infty)$ **(e)** $\{x|-2 \le x \le 1\}$; $[-2, 1]$

(f) $\{x|x < 0\}$; $(-\infty, 0)$ **(g)** $\left\{x\left|x \le \frac{1 - \sqrt{13}}{2} \text{ or } x \ge \frac{1 + \sqrt{13}}{2}\right.\right\}$; $\left(-\infty, \frac{1 - \sqrt{13}}{2}\right] \cup \left[\frac{1 + \sqrt{13}}{2}, \infty\right)$

33. (a) 5 sec   **(b)** The ball is more than 96 ft above the ground for time t between 2 and 3 sec, $2 < t < 3$.

35. (a) \$0, \$1000   **(b)** The revenue is more than \$800,000 for prices between \$276.39 and \$723.61, \$276.39 $< p <$ \$723.61.

37. (a) $\{c \mid 0.112 < c < 81.907\}$; $(0.112, 81.907)$ **(b)** It is possible to hit a target 75 km away if $c = 0.651$ or $c = 1.536$.

44. $\{x|x \le 5\}$ **45.** Odd   **46. (a)** $(9, 0), (0, -6)$ **(b)**   **47.** $19 - 26i$

Review Exercises *(page 316)*

1. (a) $m = 2$; $b = -5$ **(b)** 2   **2. (a)** $m = \frac{4}{5}$; $b = -6$ **(b)** $\frac{4}{5}$ **3. (a)** $m = 0$; $b = 4$ **(b)** 0   **4.** Linear; Slope: 5

(c)

5. Nonlinear

(d) Increasing
(d) Increasing
(d) Constant

6.   **7.**   **8.**   **9. (a)**

(b) Domain: $(-\infty, \infty)$
Range: $[2, \infty)$
(c) Decreasing: $(-\infty, 2)$
Increasing: $(2, \infty)$

10. (a)   **11. (a)**   **12. (a)**   **13. (a)**

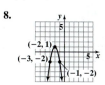

(b) Domain: $(-\infty, \infty)$
Range: $[-16, \infty)$
(c) Decreasing: $(-\infty, 0)$
Increasing: $(0, \infty)$

(b) Domain: $(-\infty, \infty)$
Range: $(-\infty, 1]$
(c) Increasing: $\left(-\infty, \frac{1}{2}\right)$
Decreasing: $\left(\frac{1}{2}, \infty\right)$

(b) Domain: $(-\infty, \infty)$
Range: $\left[\frac{1}{2}, \infty\right)$
(c) Decreasing: $\left(-\infty, -\frac{1}{3}\right)$
Increasing: $\left(-\frac{1}{3}, \infty\right)$

(b) Domain: $(-\infty, \infty)$
Range: $\left[-\frac{7}{3}, \infty\right)$
(c) Decreasing: $\left(-\infty, -\frac{2}{3}\right)$
Increasing: $\left(-\frac{2}{3}, \infty\right)$

14. Minimum value; 1   **15.** Maximum value; 12   **16.** Maximum value; 16   **17.** $\{x \mid -8 < x < 2\}$; $(-8, 2)$

18. $\left\{x \mid x \le -\dfrac{1}{3} \text{ or } x \ge 5\right\}$; $\left(-\infty, -\dfrac{1}{3}\right] \cup [5, \infty)$ **19.** $y = -3x^2 + 12x - 16$ **20.** $y = x^2 + 2x + 3$

21. (a) $S(x) = 0.01x + 25{,}000$ **(b)** $35{,}000$ **(c)** $7{,}500{,}000$ **(d)** $x > 12{,}500{,}000$

22. (a) $R(x) = -\dfrac{1}{10}x^2 + 150x$ **(b)** $14{,}000$ **(c)** 750; $56{,}250$ **(d)** 75 **23.** 50 ft by 50 ft   **24.** $4{,}166{,}666.7 \text{ m}^2$

25. The side with the semicircles should be $\dfrac{50}{\pi}$ ft; the other side should be 25 ft.   **26. (a)** 63 clubs   **(b)** $151.90   **27.** $A(x) = -x^2 + 10x$; 25 sq. units

28. 3.6 ft   **29. (a)**

(b) yes   **(c)** $y = 1.3902x + 1.1140$ **(d)** 37.95 mm

30. (a) Quadratic, $a < 0$ **(b)** About $26.5 thousand   **(c)** $6408 thousand   **(e)**

Chapter Test *(page 318)*

1. (a) Slope: -4; y-intercept: 3   **(b)** -4 **(c)** Decreasing   **(d)**

2. $\left(-\dfrac{4}{3}, 0\right)$, $(2, 0)$, $(0, -8)$

3. $\left(\dfrac{2 - \sqrt{6}}{2}, 0\right)$, $\left(\dfrac{2 + \sqrt{6}}{2}, 0\right)$, $(0, 1)$

4. $\{-1, 3\}$

5.

6. (a) Opens up   **(b)** $(2, -8)$ **(c)** $x = 2$ **(d)** x-intercepts: $\dfrac{6 - 2\sqrt{6}}{3}$, $\dfrac{6 + 2\sqrt{6}}{3}$; y-intercept: 4

(e)

7. Maximum value; 21   **8.** $\{x \mid x \le 4 \text{ or } x \ge 6\}$; $(-\infty, 4] \cup [6, \infty)$

9. (a) $C(m) = 0.15m + 129.50$ **(b)** $258.50   **(c)** 562 miles

Cumulative Review *(page 319)*

1. $5\sqrt{2}$; $\left(\dfrac{3}{2}, \dfrac{1}{2}\right)$ **2.** $(-2, -1)$ and $(2, 3)$ are on the graph.

3. $\left\{x \mid x \ge -\dfrac{3}{5}\right\}$; $\left[-\dfrac{3}{5}, \infty\right)$

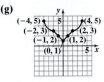

4. $y = -2x + 2$

5. $y = -\dfrac{1}{2}x + \dfrac{13}{2}$

6. $(x - 2)^2 + (y + 4)^2 = 25$

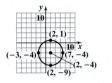

7. Yes   **8. (a)** -3 **(b)** $x^2 - 4x - 2$ **(c)** $x^2 + 4x + 1$ **(d)** $-x^2 + 4x - 1$ **(e)** $x^2 - 3$ **(f)** $2x + h - 4$ **9.** $\left\{z \mid z \ne \dfrac{7}{6}\right\}$

10. Yes   **11. (a)** No   **(b)** -1; $(-2, -1)$ is on the graph.   **(c)** -8; $(-8, 2)$ is on the graph.   **12.** Neither   **13.** Local maximum value is 5.30 and occurs at $x = -1.29$. Local minimum value is -3.30 and occurs at $x = 1.29$. Increasing: $(-4, -1.29)$ and $(1.29, 4)$; Decreasing: $(-1.29, 1.29)$

14. (a) -4 **(b)** $\{x \mid x > -4\}$ or $(-4, \infty)$

15. (a) Domain: $\{x \mid -4 \le x \le 4\}$; Range: $\{y \mid -1 \le y \le 3\}$ **(b)** $(-1, 0)$, $(0, -1)$, $(1, 0)$ **(c)** y-axis   **(d)** 1   **(e)** -4 and 4   **(f)** $\{x \mid -1 < x < 1\}$ **(g)** *[graph]*   **(h)** *[graph]*   **(i)** *[graph]*   **(j)** Even   **(k)** $(0, 4)$

CHAPTER 5 Polynomial and Rational Functions

5.1 Assess Your Understanding (*page 338*)

7. smooth; continuous **8.** touches **9.** $(-1, 1); (0, 0); (1, 1)$ **10.** r is a real zero of f; r is an x-intercept of the graph of f; $x - r$ is a factor of f.
11. turning points **12.** $y = 3x^4$ **13.** $\infty; -\infty$ **14.** As x increases in the positive direction, $f(x)$ decreases without bound. **15.** b **16.** d

17. Yes; degree 3; $f(x) = x^3 + 4x$; leading term: x^3; constant term: 0 **19.** Yes; degree 2; $g(x) = -\dfrac{1}{2}x^2 + \dfrac{1}{2}$; leading term: $-\dfrac{1}{2}x^2$; constant term: $\dfrac{1}{2}$

21. No; x is raised to the –1 power **23.** No; x is raised to the $\dfrac{3}{2}$ power **25.** Yes; degree 4; $F(x) = 5x^4 - \pi x^3 + \dfrac{1}{2}$; leading term: $5x^4$; constant term: $\dfrac{1}{2}$
27. Yes; degree 4; $G(x) = 2x^4 - 4x^3 + 4x^2 - 4x + 2$; leading term: $2x^4$; constant term: 2

29.

31.
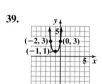

33.

35.

37.

39.

41.

43. $f(x) = x^3 - 3x^2 - x + 3$ for $a = 1$ **45.** $f(x) = x^3 - x^2 - 12x$ for $a = 1$
47. $f(x) = x^4 - 15x^2 + 10x + 24$ for $a = 1$ **49.** $f(x) = x^3 - 5x^2 + 3x + 9$ for $a = 1$
51. $f(x) = 2x^3 - 4x^2 - 22x + 24$ **53.** $f(x) = 16x^4 - 80x^3 + 32x^2 + 128x$
55. $f(x) = 5x^4 - 10x^2 + 5$ **57. (a)** 7, multiplicity 1; -3, multiplicity 2 **(b)** Graph touches the x-axis at -3 and crosses it at 7. **(c)** 2 **(d)** $y = 3x^3$

59. (a) 2, multiplicity 3 **(b)** Graph crosses the x-axis at 2. **(c)** 4 **(d)** $y = 4x^5$ **61. (a)** $-\dfrac{1}{2}$, multiplicity 2; -4, multiplicity 3 **(b)** Graph touches the x-axis at $-\dfrac{1}{2}$ and crosses at -4. **(c)** 4 **(d)** $y = -2x^5$ **63. (a)** 5, multiplicity 3; -4, multiplicity 2 **(b)** Graph touches the x-axis at -4 and crosses it at 5. **(c)** 4 **(d)** $y = x^5$ **65. (a)** No real zeros **(b)** Graph neither crosses nor touches the x-axis. **(c)** 5 **(d)** $y = 3x^6$ **67. (a)** 0, multiplicity 2; $-\sqrt{2}, \sqrt{2}$, multiplicity 1 **(b)** Graph touches the x-axis at 0 and crosses at $-\sqrt{2}$ and $\sqrt{2}$. **(c)** 3 **(d)** $y = -2x^4$ **69.** Could be; zeros: $-1, 1, 2$; Least degree is 3. **71.** Cannot be the graph of a polynomial; gap at $x = -1$ **73.** $f(x) = x(x - 1)(x - 2)$
75. $f(x) = -\dfrac{1}{2}(x + 1)(x - 1)^2(x - 2)$ **77.** $f(x) = 0.2(x + 4)(x + 1)^2(x - 3)$ **79.** $f(x) = -x(x + 3)^2(x - 3)^2$
81. Step 1: $y = x^3$
  Step 2: x-intercepts: 0, 3; y-intercept: 0
  Step 3: 0: multiplicity 2, touches; 3: multiplicity 1, crosses
  Step 4: At most 2 turning points
  Step 5: $f(-1) = -4; f(2) = -4; f(4) = 16$

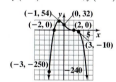

83. Step 1: $y = -x^3$
  Step 2: x-intercepts: -4, 1; y-intercept: 16
  Step 3: -4: multiplicity 2, touches; 1: multiplicity 1, crosses
  Step 4: At most 2 turning points
  Step 5: $f(-5) = 6; f(-2) = 12; f(2) = -36$

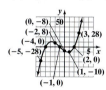

85. Step 1: $y = -2x^4$
  Step 2: x-intercepts: -2, 2; y-intercept: 32
  Step 3: -2: multiplicity 1, crosses; 2: multiplicity 3, crosses
  Step 4: At most 3 turning points
  Step 5: $f(-3) = -250; f(-1) = 54; f(3) = -10$

87. Step 1: $y = x^3$
  Step 2: x-intercepts: $-4, -1, 2$; y-intercept: -8
  Step 3: $-4, -1, 2$: multiplicity 1, crosses
  Step 4: At most 2 turning points
  Step 5: $f(-5) = -28; f(-2) = 8; f(1) = -10; f(3) = 28$

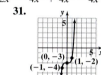

89. Step 1: $y = x^4$
  Step 2: x-intercepts: $-2, 0, 2$; y-intercept: 0
  Step 3: $-2, 2$: multiplicity 1, crosses; 0: multiplicity 2, touches
  Step 4: At most 3 turning points
  Step 5: $f(-3) = 45; f(-1) = -3$;
    $f(1) = -3; f(3) = 45$

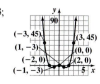

91. Step 1: $y = x^4$
  Step 2: x-intercepts: $-1, 2$; y-intercept: 4
  Step 3: $-1, 2$: multiplicity 2, touches
  Step 4: At most 3 turning points
  Step 5: $f(-2) = 16; f(1) = 4; f(3) = 16$

33. $\pm 1, \pm \dfrac{1}{3}$ **35.** $\pm 1, \pm 3$ **37.** $\pm 1, \pm 2, \pm \dfrac{1}{4}, \pm \dfrac{1}{2}$ **39.** $\pm 1, \pm 3, \pm 9, \pm \dfrac{1}{2}, \pm \dfrac{1}{3}, \pm \dfrac{1}{6}, \pm \dfrac{3}{2}, \pm \dfrac{9}{2}$

41. $\pm 1, \pm 2, \pm 3, \pm 4, \pm 6, \pm 12, \pm \dfrac{1}{2}, \pm \dfrac{3}{2}$ **43.** $\pm 1, \pm 2, \pm 4, \pm 5, \pm 10, \pm 20, \pm \dfrac{1}{2}, \pm \dfrac{5}{2}, \pm \dfrac{1}{3}, \pm \dfrac{2}{3}, \pm \dfrac{4}{3}, \pm \dfrac{5}{3}, \pm \dfrac{10}{3}, \pm \dfrac{20}{3}, \pm \dfrac{1}{6}, \pm \dfrac{5}{6}$

45. $-3, -1, 2; f(x) = (x + 3)(x + 1)(x - 2)$ **47.** $\dfrac{1}{2}; f(x) = 2\left(x - \dfrac{1}{2}\right)(x^2 + 1)$ **49.** $2, \sqrt{5}, -\sqrt{5}; f(x) = 2(x - 2)(x - \sqrt{5})(x + \sqrt{5})$

51. $-1, \dfrac{1}{2}, \sqrt{3}, -\sqrt{3}; f(x) = 2(x + 1)\left(x - \dfrac{1}{2}\right)(x - \sqrt{3})(x + \sqrt{3})$ **53.** $1,$ multiplicity 2; $-2, -1; f(x) = (x + 2)(x + 1)(x - 1)^2$

55. $-1, -\dfrac{1}{4}; f(x) = 4(x + 1)\left(x + \dfrac{1}{4}\right)(x^2 + 2)$ **57.** $\{-1, 2\}$ **59.** $\left\{\dfrac{2}{3}, -1 + \sqrt{2}, -1 - \sqrt{2}\right\}$ **61.** $\left\{\dfrac{1}{3}, \sqrt{5}, -\sqrt{5}\right\}$ **63.** $\{-3, -2\}$

65. $\left\{-\dfrac{1}{3}\right\}$ **67.** $\left\{\dfrac{1}{2}, 2, 5\right\}$ **69.** LB $= -2$; UB $= 2$ **71.** LB $= -1$; UB $= 1$ **73.** LB $= -2$; UB $= 2$ **75.** LB $= -1$; UB $= 1$

77. LB $= -2$; UB $= 3$ **79.** $f(0) = -1; f(1) = 10$ **81.** $f(-5) = -58; f(-4) = 2$ **83.** $f(1.4) = -0.17536; f(1.5) = 1.40625$

85. 0.21   **87.** -4.04 **89.** 1.15   **91.** 2.53

93.

95.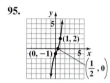

97.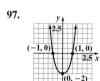

99.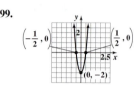

101.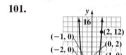

103.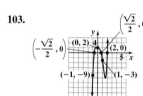

105. $-8, -4, -\dfrac{7}{3}$ **107.** $k = 5$ **109.** -7

111. If $f(x) = x^n - c^n$, then $f(c) = c^n - c^n = 0$, so $x - c$ is a factor of f.
113. 5   **115.** 7 in.   **117.** All the potential rational zeros are integers, so r either is an integer or is not a rational zero (and is therefore irrational).
119. 0.215

121. No; by the Rational Zeros Theorem, $\dfrac{1}{3}$ is not a potential rational zero.   **123.** No; by the Rational Zeros Theorem, $\dfrac{2}{3}$ is not a potential rational zero.

124. $y = \dfrac{2}{5}x - \dfrac{3}{5}$ **125.** $[3, 8)$ **126.** $(0, -2\sqrt{3}), (0, 2\sqrt{3}), (4, 0)$ **127.** $(-3, 2)$ and $(5, \infty)$

5.6 Assess Your Understanding (page 394)

3. one   **4.** $3 - 4i$ **5.** T   **6.** F   **7.** $4 + i$ **9.** $-i, 1 - i$ **11.** $-i, -2i$ **13.** $-i$ **15.** $2 - i, -3 + i$
17. $f(x) = x^4 - 14x^3 + 77x^2 - 200x + 208; a = 1$ **19.** $f(x) = x^5 - 4x^4 + 7x^3 - 8x^2 + 6x - 4; a = 1$

21. $f(x) = x^4 - 6x^3 + 10x^2 - 6x + 9; a = 1$ **23.** $-2i, 4$ **25.** $2i, -3, \dfrac{1}{2}$ **27.** $3 + 2i, -2, 5$ **29.** $4i, -\sqrt{11}, \sqrt{11}, -\dfrac{2}{3}$

31. $1, -\dfrac{1}{2} - \dfrac{\sqrt{3}}{2}i, -\dfrac{1}{2} + \dfrac{\sqrt{3}}{2}i; f(x) = (x - 1)\left(x + \dfrac{1}{2} + \dfrac{\sqrt{3}}{2}i\right)\left(x + \dfrac{1}{2} - \dfrac{\sqrt{3}}{2}i\right)$

33. $2, 3 - 2i, 3 + 2i; f(x) = (x - 2)(x - 3 + 2i)(x - 3 - 2i)$ **35.** $-i, i, -2i, 2i; f(x) = (x + i)(x - i)(x + 2i)(x - 2i)$

37. $-5i, 5i, -3, 1; f(x) = (x + 5i)(x - 5i)(x + 3)(x - 1)$ **39.** $-4, \dfrac{1}{3}, 2 - 3i, 2 + 3i; f(x) = 3(x + 4)\left(x - \dfrac{1}{3}\right)(x - 2 + 3i)(x - 2 - 3i)$

41. 130   **43. (a)** $f(x) = (x^2 - \sqrt{2}x + 1)(x^2 + \sqrt{2}x + 1)$ **(b)** $-\dfrac{\sqrt{2}}{2} - \dfrac{\sqrt{2}}{2}i, -\dfrac{\sqrt{2}}{2} + \dfrac{\sqrt{2}}{2}i, \dfrac{\sqrt{2}}{2} - \dfrac{\sqrt{2}}{2}i, \dfrac{\sqrt{2}}{2} + \dfrac{\sqrt{2}}{2}i$

45. Zeros that are complex numbers must occur in conjugate pairs; or a polynomial with real coefficients of odd degree must have at least one real zero.
47. If the remaining zero were a complex number, its conjugate would also be a zero, creating a polynomial of degree 5.

49.

50. $\{-22\}$
51. $6x^3 - 13x^2 - 13x + 20$
52. $A = 9\pi$ ft^2 (≈ 28.274 ft^2); $C = 6\pi$ ft (≈ 18.850 ft)

Review Exercises (page 397)

1. Polynomial of degree 5   **2.** Rational   **3.** Neither   **4.** Polynomial of degree 0

5.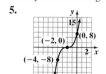

6.

7.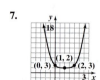

8. 1. $y = x^3$
2. x-intercepts: $-4, -2, 0$; y-intercept: 0
3. $-4, -2, 0$ (all multiplicity 1), crosses
4. 2
5. $f(-5) = -15; f(-3) = 3;$
$f(-1) = -3; f(1) = 15$

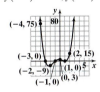

9. 1. $y = x^3$
2. x-intercepts: $-4, 2$; y-intercept: 16
3. -4, multiplicity 1, crosses; 2, multiplicity 2, touches
4. 2
5. $f(-5) = -49; f(-2) = 32; f(3) = 7$

10. 1. $y = -2x^3$
2. $f(x) = -2x^2(x-2)$
x-intercepts: 0, 2; y-intercept: 0
3. 0, multiplicity 2, touches; 2, multiplicity 1, crosses
4. 2
5. $f(-1) = 6; f(1) = 2; f(3) = -18$

11. 1. $y = x^4$
2. x-intercepts: $-3, -1, 1$; y-intercept: 3
3. $-3, -1$ (both multiplicity 1), crosses; 1, multiplicity 2, touches
4. 3
5. $f(-4) = 75; f(-2) = -9; f(2) = 15$

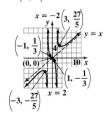

12. Domain: $\{x | x \neq -3, x \neq 3\}$; horizontal asymptote: $y = 0$;
vertical asymptotes: $x = -3, x = 3$
13. Domain: $\{x | x \neq 2\}$; oblique asymptote: $y = x + 2$;
vertical asymptote: $x = 2$
14. Domain: $\{x | x \neq -2\}$; horizontal asymptote: $y = 1$;
vertical asymptote: $x = -2$

15. 1. $R(x) = \dfrac{2(x-3)}{x}$; domain: $\{x | x \neq 0\}$ **2.** R is in lowest terms   **3.** no y-intercept; x-intercept: 3
4. R is in lowest terms; vertical asymptote: $x = 0$ **5.** Horizontal asymptote: $y = 2$; not intersected
6.

Interval	$(-\infty, 0)$	$(0, 3)$	$(3, \infty)$
Number Chosen	-2	1	4
Value of R	$R(-2) = 5$	$R(1) = -4$	$R(4) = \frac{1}{2}$
Location of Graph	Above x-axis	Below x-axis	Above x-axis
Point on Graph	$(-2, 5)$	$(1, -4)$	$\left(4, \frac{1}{2}\right)$

7.

16. 1. Domain: $\{x | x \neq 0, x \neq 2\}$ **2.** H is in lowest terms   **3.** no y-intercept; x-intercept: -2
4. H is in lowest terms; vertical asymptotes: $x = 0, x = 2$ **5.** Horizontal asymptote: $y = 0$; intersected at $(-2, 0)$
6.

Interval	$(-\infty, -2)$	$(-2, 0)$	$(0, 2)$	$(2, \infty)$
Number Chosen	-3	-1	1	3
Value of H	$H(-3) = -\frac{1}{15}$	$H(-1) = \frac{1}{3}$	$H(1) = -3$	$H(3) = \frac{5}{3}$
Location of Graph	Below x-axis	Above x-axis	Below x-axis	Above x-axis
Point on Graph	$\left(-3, -\frac{1}{15}\right)$	$\left(-1, \frac{1}{3}\right)$	$(1, -3)$	$\left(3, \frac{5}{3}\right)$

7.

17. 1. $R(x) = \dfrac{(x+3)(x-2)}{(x-3)(x+2)}$; domain: $\{x | x \neq -2, x \neq 3\}$ **2.** R is in lowest terms   **3.** y-intercept: 1; x-intercepts: $-3, 2$
4. R is in lowest terms; vertical asymptotes: $x = -2, x = 3$ **5.** Horizontal asymptote: $y = 1$; intersected at $(0, 1)$
6.

Interval	$(-\infty, -3)$	$(-3, -2)$	$(-2, 2)$	$(2, 3)$	$(3, \infty)$
Number Chosen	-4	$-\frac{5}{2}$	0	$\frac{5}{2}$	4
Value of R	$R(-4) = \frac{3}{7}$	$R\left(-\frac{5}{2}\right) = -\frac{9}{11}$	$R(0) = 1$	$R\left(\frac{5}{2}\right) = -\frac{11}{9}$	$R(4) = \frac{7}{3}$
Location of Graph	Above x-axis	Below x-axis	Above x-axis	Below x-axis	Above x-axis
Point on Graph	$\left(-4, \frac{3}{7}\right)$	$\left(-\frac{5}{2}, -\frac{9}{11}\right)$	$(0, 1)$	$\left(\frac{5}{2}, -\frac{11}{9}\right)$	$\left(4, \frac{7}{3}\right)$

7.

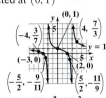

18. 1. $F(x) = \dfrac{x^3}{(x+2)(x-2)}$; domain: $\{x | x \neq -2, x \neq 2\}$ **2.** F is in lowest terms   **3.** y-intercept: 0; x-intercept: 0
4. F is in lowest terms; vertical asymptotes: $x = -2, x = 2$ **5.** Oblique asymptote: $y = x$; intersected at $(0, 0)$
6.

Interval	$(-\infty, -2)$	$(-2, 0)$	$(0, 2)$	$(2, \infty)$
Number Chosen	-3	-1	1	3
Value of F	$F(-3) = -\frac{27}{5}$	$F(-1) = \frac{1}{3}$	$F(1) = -\frac{1}{3}$	$F(3) = \frac{27}{5}$
Location of Graph	Below x-axis	Above x-axis	Below x-axis	Above x-axis
Point on Graph	$\left(-3, -\frac{27}{5}\right)$	$\left(-1, \frac{1}{3}\right)$	$\left(1, -\frac{1}{3}\right)$	$\left(3, \frac{27}{5}\right)$

7.

CHAPTER 6 Exponential and Logarithmic Functions

6.1 Assess Your Understanding (page 408)

4. composite function; $f(g(x))$ **5.** F **6.** c **7.** a **8.** F **9. (a)** -1 **(b)** -1 **(c)** 8 **(d)** 0 **(e)** 8 **(f)** -7 **11. (a)** 4 **(b)** 5 **(c)** -1 **(d)** -2

13. (a) 98 **(b)** 49 **(c)** 4 **(d)** 4 **15. (a)** 97 **(b)** $-\dfrac{163}{2}$ **(c)** 1 **(d)** $-\dfrac{3}{2}$ **17. (a)** $2\sqrt{2}$ **(b)** $2\sqrt{2}$ **(c)** 1 **(d)** 0 **19. (a)** $\dfrac{1}{17}$ **(b)** $\dfrac{1}{5}$ **(c)** 1 **(d)** $\dfrac{1}{2}$

21. (a) $\dfrac{3}{\sqrt[3]{4}+1}$ **(b)** 1 **(c)** $\dfrac{6}{5}$ **(d)** 0 **23. (a)** $(f \circ g)(x) = 6x + 3$; all real numbers **(b)** $(g \circ f)(x) = 6x + 9$; all real numbers

(c) $(f \circ f)(x) = 4x + 9$; all real numbers **(d)** $(g \circ g)(x) = 9x$; all real numbers **25. (a)** $(f \circ g)(x) = 3x^2 + 1$; all real numbers

(b) $(g \circ f)(x) = 9x^2 + 6x + 1$; all real numbers **(c)** $(f \circ f)(x) = 9x + 4$; all real numbers **(d)** $(g \circ g)(x) = x^4$; all real numbers

27. (a) $(f \circ g)(x) = x^4 + 8x^2 + 16$; all real numbers **(b)** $(g \circ f)(x) = x^4 + 4$; all real numbers **(c)** $(f \circ f)(x) = x^4$; all real numbers

(d) $(g \circ g)(x) = x^4 + 8x^2 + 20$; all real numbers **29. (a)** $(f \circ g)(x) = \dfrac{3x}{2-x}$; $\{x | x \neq 0, x \neq 2\}$ **(b)** $(g \circ f)(x) = \dfrac{2(x-1)}{3}$; $\{x | x \neq 1\}$

(c) $(f \circ f)(x) = \dfrac{3(x-1)}{4-x}$; $\{x | x \neq 1, x \neq 4\}$ **(d)** $(g \circ g)(x) = x$; $\{x | x \neq 0\}$ **31. (a)** $(f \circ g)(x) = \dfrac{4}{4+x}$; $\{x | x \neq -4, x \neq 0\}$

(b) $(g \circ f)(x) = \dfrac{-4(x-1)}{x}$; $\{x | x \neq 0, x \neq 1\}$ **(c)** $(f \circ f)(x) = x$; $\{x | x \neq 1\}$ **(d)** $(g \circ g)(x) = x$; $\{x | x \neq 0\}$

33. (a) $(f \circ g)(x) = \sqrt{2x+3}$; $\left\{x \Big| x \geq -\dfrac{3}{2}\right\}$ **(b)** $(g \circ f)(x) = 2\sqrt{x} + 3$; $\{x | x \geq 0\}$ **(c)** $(f \circ f)(x) = \sqrt[4]{x}$; $\{x | x \geq 0\}$

(d) $(g \circ g)(x) = 4x + 9$; all real numbers **35. (a)** $(f \circ g)(x) = x$; $\{x | x \geq 1\}$ **(b)** $(g \circ f)(x) = |x|$; all real numbers

(c) $(f \circ f)(x) = x^4 + 2x^2 + 2$; all real numbers **(d)** $(g \circ g)(x) = \sqrt{\sqrt{x-1}-1}$; $\{x | x \geq 2\}$ **37. (a)** $(f \circ g)(x) = -\dfrac{4x-17}{2x-1}$; $\left\{x \Big| x \neq 3; x \neq \dfrac{1}{2}\right\}$

(b) $(g \circ f)(x) = -\dfrac{3x-3}{2x+8}$; $\{x | x \neq -4; x \neq -1\}$ **(c)** $(f \circ f)(x) = -\dfrac{2x+5}{x-2}$; $\{x | x \neq -1; x \neq 2\}$ **(d)** $(g \circ g)(x) = -\dfrac{3x-4}{2x-11}$; $\left\{x \Big| x \neq \dfrac{11}{2}; x \neq 3\right\}$

39. $(f \circ g)(x) = f(g(x)) = f\left(\dfrac{1}{2}x\right) = 2\left(\dfrac{1}{2}x\right) = x$; $(g \circ f)(x) = g(f(x)) = g(2x) = \dfrac{1}{2}(2x) = x$

41. $(f \circ g)(x) = f(g(x)) = f(\sqrt[3]{x}) = (\sqrt[3]{x})^3 = x$; $(g \circ f)(x) = g(f(x)) = g(x^3) = \sqrt[3]{x^3} = x$

43. $(f \circ g)(x) = f(g(x)) = f\left(\dfrac{1}{2}(x+6)\right) = 2\left[\dfrac{1}{2}(x+6)\right] - 6 = x + 6 - 6 = x$; $(g \circ f)(x) = g(f(x)) = g(2x - 6) = \dfrac{1}{2}(2x - 6 + 6) = \dfrac{1}{2}(2x) = x$

45. $(f \circ g)(x) = f(g(x)) = f\left(\dfrac{1}{a}(x-b)\right) = a\left[\dfrac{1}{a}(x-b)\right] + b = x$; $(g \circ f)(x) = g(f(x)) = g(ax+b) = \dfrac{1}{a}(ax + b - b) = x$

47. $f(x) = x^4$; $g(x) = 2x + 3$ (Other answers are possible.) **49.** $f(x) = \sqrt{x}$; $g(x) = x^2 + 1$ (Other answers are possible.)

51. $f(x) = |x|$; $g(x) = 2x + 1$ (Other answers are possible.) **53.** $(f \circ g)(x) = 11$; $(g \circ f)(x) = 2$ **55.** $-3, 3$ **57. (a)** $(f \circ g)(x) = acx + ad + b$

(b) $(g \circ f)(x) = acx + bc + d$ **(c)** The domains of both $f \circ g$ and $g \circ f$ are all real numbers. **(d)** $f \circ g = g \circ f$ when $ad + b = bc + d$

59. $S(t) = \dfrac{16}{9}\pi t^6$ **61.** $C(t) = 15,000 + 800,000t - 40,000t^2$ **63.** $C(p) = \dfrac{2\sqrt{100-p}}{25} + 600, 0 \leq p \leq 100$ **65.** $V(r) = 2\pi r^3$

67. (a) $f(x) = 0.7235x$ **(b)** $g(x) = 141.119x$ **(c)** $g(f(x)) = g(0.7235x) = 102.0995965x$ **(d)** $102,099.5965$ yen **69. (a)** $f(p) = p - 200$

(b) $g(p) = 0.8p$ **(c)** $(f \circ g)(p) = 0.8p - 200$; $(g \circ f)(p) = 0.8p - 160$; The 20% discount followed by the $200 rebate is the better deal. **71.** 15

73. f is an odd function, so $f(-x) = -f(x)$. g is an even function, so $g(-x) = g(x)$. Then $(f \circ g)(-x) = f(g(-x)) = f(g(x)) = (f \circ g)(x)$. So $f \circ g$ is even.

Also, $(g \circ f)(-x) = g(f(-x)) = g(-f(x)) = g(f(x)) = (g \circ f)(x)$, so $g \circ f$ is even.

74. $(f + g)(x) = 4x + 3$; Domain: all real numbers

$(f - g)(x) = 2x + 13$; Domain: all real numbers

$(f \cdot g)(x) = 3x^2 - 7x - 40$; Domain: all real numbers

$\left(\dfrac{f}{g}\right)(x) = \dfrac{3x + 8}{x - 5}$; Domain: $\{x | x \neq 5\}$

75. $\dfrac{1}{4}, 4$

76.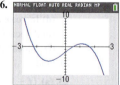

Local minimum: -5.08 at $x = -1.15$
Local maximum: 1.08 at $x = 1.15$
Decreasing: $(-3, -1.15); (1.15, 3)$
Increasing: $(-1.15, 1.15)$

77. Domain: $\{x | x \neq 3\}$
Vertical asymptote: $x = 3$
Oblique asymptote: $y = x + 9$

6.2 Assess Your Understanding (page 419)

5. $f(x_1) \neq f(x_2)$ **6.** one-to-one **7.** 3 **8.** $y = x$ **9.** $[4, \infty)$ **10.** T **11.** a **12.** d **13.** one-to-one **15.** not one-to-one

17. not one-to-one **19.** one-to-one **21.** one-to-one **23.** not one-to-one **25.** one-to-one

27.

Annual Rainfall (inches)		Location
49.7	→	Atlanta, Georgia
43.8	→	Boston, Massachusetts
4.2	→	Las Vegas, Nevada
61.9	→	Miami, Florida
12.8	→	Los Angeles, California

Domain: {49.7, 43.8, 4.2, 61.9, 12.8}
Range: {Atlanta, Boston, Las Vegas, Miami, Los Angeles}

29.

Monthly Cost of Life Insurance		Age
$10.59	→	30
$12.52	→	40
$15.94	→	45

Domain: {$10.59, $12.52, $15.94}
Range: {30, 40, 45}

31. $\{ (5, -3), (9, -2), (2, -1), (11, 0), (-5, 1) \}$
Domain: $\{ 5, 9, 2, 11, -5 \}$
Range: $\{ -3, -2, -1, 0, 1 \}$

33. $\{ (1, -2), (2, -3), (0, -10), (9, 1), (4, 2) \}$
Domain: $\{ 1, 2, 0, 9, 4 \}$
Range: $\{ -2, -3, -10, 1, 2 \}$

35. $f(g(x)) = f\left(\dfrac{1}{3}(x-4)\right) = 3\left[\dfrac{1}{3}(x-4)\right] + 4 = (x-4) + 4 = x$
$g(f(x)) = g(3x+4) = \dfrac{1}{3}[(3x+4) - 4] = \dfrac{1}{3}(3x) = x$

37. $f(g(x)) = f\left(\dfrac{x}{4} + 2\right) = 4\left[\dfrac{x}{4} + 2\right] - 8 = (x+8) - 8 = x$
$g(f(x)) = g(4x-8) = \dfrac{4x-8}{4} + 2 = (x-2) + 2 = x$

39. $f(g(x)) = f(\sqrt[3]{x+8}) = (\sqrt[3]{x+8})^3 - 8 = (x+8) - 8 = x$
$g(f(x)) = g(x^3 - 8) = \sqrt[3]{(x^3 - 8) + 8} = \sqrt[3]{x^3} = x$

41. $f(g(x)) = f\left(\dfrac{1}{x}\right) = \dfrac{1}{\left(\dfrac{1}{x}\right)} = x; x \neq 0, \ g(f(x)) = g\left(\dfrac{1}{x}\right) = \dfrac{1}{\left(\dfrac{1}{x}\right)} = x, \ x \neq 0$

43. $f(g(x)) = f\left(\dfrac{4x-3}{2-x}\right) = \dfrac{2\left(\dfrac{4x-3}{2-x}\right) + 3}{\dfrac{4x-3}{2-x} + 4}$
$= \dfrac{2(4x-3) + 3(2-x)}{4x-3 + 4(2-x)} = \dfrac{5x}{5} = x, \ x \neq 2$

$g(f(x)) = g\left(\dfrac{2x+3}{x+4}\right) = \dfrac{4\left(\dfrac{2x+3}{x+4}\right) - 3}{2 - \dfrac{2x+3}{x+4}}$
$= \dfrac{4(2x+3) - 3(x+4)}{2(x+4) - (2x+3)} = \dfrac{5x}{5} = x, \ x \neq -4$

45.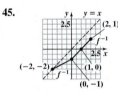

47.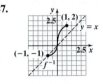

49.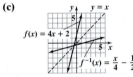

51. (a) $f^{-1}(x) = \dfrac{1}{3}x$

$f(f^{-1}(x)) = f\left(\dfrac{1}{3}x\right) = 3\left(\dfrac{1}{3}x\right) = x$

$f^{-1}(f(x)) = f^{-1}(3x) = \dfrac{1}{3}(3x) = x$

(b) Domain of f = Range of f^{-1} = All real numbers;
Range of f = Domain of f^{-1} = All real numbers

(c)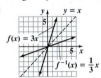

53. (a) $f^{-1}(x) = \dfrac{x}{4} - \dfrac{1}{2}$

$f(f^{-1}(x)) = f\left(\dfrac{x}{4} - \dfrac{1}{2}\right) = 4\left(\dfrac{x}{4} - \dfrac{1}{2}\right) + 2$
$= (x-2) + 2 = x$

$f^{-1}(f(x)) = f^{-1}(4x+2) = \dfrac{4x+2}{4} - \dfrac{1}{2}$
$= \left(x + \dfrac{1}{2}\right) - \dfrac{1}{2} = x$

(b) Domain of f = Range of f^{-1} = All real numbers;
Range of f = Domain of f^{-1} = All real numbers

(c)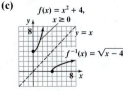

55. (a) $f^{-1}(x) = \sqrt[3]{x+1}$
$f(f^{-1}(x)) = f(\sqrt[3]{x+1})$
$= (\sqrt[3]{x+1})^3 - 1 = x$
$f^{-1}(f(x)) = f^{-1}(x^3 - 1)$
$= \sqrt[3]{(x^3 - 1) + 1} = x$

(b) Domain of f = Range of f^{-1} = All real numbers;
Range of f = Domain of f^{-1} = All real numbers

(c)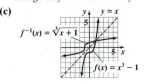

57. (a) $f^{-1}(x) = \sqrt{x-4}, x \geq 4$
$f(f^{-1}(x)) = f(\sqrt{x-4}) = (\sqrt{x-4})^2 + 4 = x$
$f^{-1}(f(x)) = f^{-1}(x^2 + 4) = \sqrt{(x^2 + 4) - 4} = \sqrt{x^2} = x, x \geq 0$

(b) Domain of f = Range of f^{-1} = $\{x | x \geq 0\}$;
Range of f = Domain of f^{-1} = $\{x | x \geq 4\}$

(c)

59. (a) $f^{-1}(x) = \dfrac{4}{x}$

$$f(f^{-1}(x)) = f\left(\dfrac{4}{x}\right) = \dfrac{4}{\left(\dfrac{4}{x}\right)} = x$$

$$f^{-1}(f(x)) = f^{-1}\left(\dfrac{4}{x}\right) = \dfrac{4}{\left(\dfrac{4}{x}\right)} = x$$

(b) Domain of f = Range of f^{-1} = $\{x \mid x \ne 0\}$;
Range of f = Domain of f^{-1} = $\{x \mid x \ne 0\}$

(c)

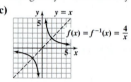

61. (a) $f^{-1}(x) = \dfrac{2x + 1}{x}$

$$f(f^{-1}(x)) = f\left(\dfrac{2x + 1}{x}\right) = \dfrac{1}{\dfrac{2x + 1}{x} - 2} = \dfrac{x}{(2x + 1) - 2x} = x$$

$$f^{-1}(f(x)) = f^{-1}\left(\dfrac{1}{x - 2}\right) = \dfrac{2\left(\dfrac{1}{x - 2}\right) + 1}{\dfrac{1}{x - 2}} = \dfrac{2 + (x - 2)}{1} = x$$

(b) Domain of f = Range of f^{-1} = $\{x \mid x \ne 2\}$;
Range of f = Domain of f^{-1} = $\{x \mid x \ne 0\}$

(c)

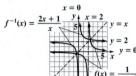

63. (a) $f^{-1}(x) = \dfrac{2 - 3x}{x}$

$$f(f^{-1}(x)) = f\left(\dfrac{2 - 3x}{x}\right) = \dfrac{2}{3 + \dfrac{2 - 3x}{x}} = \dfrac{2x}{3x + 2 - 3x} = \dfrac{2x}{2} = x$$

$$f^{-1}(f(x)) = f^{-1}\left(\dfrac{2}{3 + x}\right) = \dfrac{2 - 3\left(\dfrac{2}{3 + x}\right)}{\dfrac{2}{3 + x}} = \dfrac{2(3 + x) - 3 \cdot 2}{2} = \dfrac{2x}{2} = x$$

(b) Domain of f = Range of f^{-1} = $\{x \mid x \ne -3\}$; Range of f = Domain of f^{-1} = $\{x \mid x \ne 0\}$

65. (a) $f^{-1}(x) = \dfrac{-2x}{x - 3}$

$$f(f^{-1}(x)) = f\left(\dfrac{-2x}{x - 3}\right) = \dfrac{3\left(\dfrac{-2x}{x - 3}\right)}{\dfrac{-2x}{x - 3} + 2} = \dfrac{3(-2x)}{-2x + 2(x - 3)} = \dfrac{-6x}{-6} = x$$

$$f^{-1}(f(x)) = f^{-1}\left(\dfrac{3x}{x + 2}\right) = \dfrac{-2\left(\dfrac{3x}{x + 2}\right)}{\dfrac{3x}{x + 2} - 3} = \dfrac{-2(3x)}{3x - 3(x + 2)} = \dfrac{-6x}{-6} = x$$

(b) Domain of f = Range of f^{-1} = $\{x \mid x \ne -2\}$; Range of f = Domain of f^{-1} = $\{x \mid x \ne 3\}$

67. (a) $f^{-1}(x) = \dfrac{x}{3x - 2}$

$$f(f^{-1}(x)) = f\left(\dfrac{x}{3x - 2}\right) = \dfrac{2\left(\dfrac{x}{3x - 2}\right)}{3\left(\dfrac{x}{3x - 2}\right) - 1} = \dfrac{2x}{3x - (3x - 2)} = \dfrac{2x}{2} = x$$

$$f^{-1}(f(x)) = f^{-1}\left(\dfrac{2x}{3x - 1}\right) = \dfrac{\dfrac{2x}{3x - 1}}{3\left(\dfrac{2x}{3x - 1}\right) - 2} = \dfrac{2x}{6x - 2(3x - 1)} = \dfrac{2x}{2} = x$$

(b) Domain of f = Range of f^{-1} = $\left\{x \mid x \ne \dfrac{1}{3}\right\}$; Range of f = Domain of f^{-1} = $\left\{x \mid x \ne \dfrac{2}{3}\right\}$

69. (a) $f^{-1}(x) = \dfrac{3x + 4}{2x - 3}$

$$f(f^{-1}(x)) = f\left(\dfrac{3x + 4}{2x - 3}\right) = \dfrac{3\left(\dfrac{3x + 4}{2x - 3}\right) + 4}{2\left(\dfrac{3x + 4}{2x - 3}\right) - 3} = \dfrac{3(3x + 4) + 4(2x - 3)}{2(3x + 4) - 3(2x - 3)} = \dfrac{17x}{17} = x$$

$$f^{-1}(f(x)) = f^{-1}\left(\dfrac{3x + 4}{2x - 3}\right) = \dfrac{3\left(\dfrac{3x + 4}{2x - 3}\right) + 4}{2\left(\dfrac{3x + 4}{2x - 3}\right) - 3} = \dfrac{3(3x + 4) + 4(2x - 3)}{2(3x + 4) - 3(2x - 3)} = \dfrac{17x}{17} = x$$

(b) Domain of f = Range of f^{-1} = $\left\{x \mid x \ne \dfrac{3}{2}\right\}$; Range of f = Domain of f^{-1} = $\left\{x \mid x \ne \dfrac{3}{2}\right\}$

71. (a) $f^{-1}(x) = \dfrac{-2x + 3}{x - 2}$

$$f(f^{-1}(x)) = f\left(\dfrac{-2x + 3}{x - 2}\right) = \dfrac{2\left(\dfrac{-2x + 3}{x - 2}\right) + 3}{\dfrac{-2x + 3}{x - 2} + 2} = \dfrac{2(-2x + 3) + 3(x - 2)}{-2x + 3 + 2(x - 2)} = \dfrac{-x}{-1} = x$$

$$f^{-1}(f(x)) = f^{-1}\left(\dfrac{2x + 3}{x + 2}\right) = \dfrac{-2\left(\dfrac{2x + 3}{x + 2}\right) + 3}{\dfrac{2x + 3}{x + 2} - 2} = \dfrac{-2(2x + 3) + 3(x + 2)}{2x + 3 - 2(x + 2)} = \dfrac{-x}{-1} = x$$

(b) Domain of f = Range of f^{-1} = $\{x \mid x \neq -2\}$; Range of f = Domain of f^{-1} = $\{x \mid x \neq 2\}$

73. (a) $f^{-1}(x) = \dfrac{2}{\sqrt{1 - 2x}}$

$$f(f^{-1}(x)) = f\left(\dfrac{2}{\sqrt{1 - 2x}}\right) = \dfrac{\dfrac{4}{1 - 2x} - 4}{2 \cdot \dfrac{4}{1 - 2x}} = \dfrac{4 - 4(1 - 2x)}{2 \cdot 4} = \dfrac{8x}{8} = x$$

$$f^{-1}(f(x)) = f^{-1}\left(\dfrac{x^2 - 4}{2x^2}\right) = \dfrac{2}{\sqrt{1 - 2\left(\dfrac{x^2 - 4}{2x^2}\right)}} = \dfrac{2}{\sqrt{\dfrac{4}{x^2}}} = \sqrt{x^2} = x, \text{ since } x > 0$$

(b) Domain of f = Range of f^{-1} = $\{x \mid x > 0\}$; Range of f = Domain of f^{-1} = $\left\{x \mid x < \dfrac{1}{2}\right\}$

75. (a) 0 **(b)** 2 **(c)** 0 **(d)** 1 **77.** 7 **79.** Domain of f^{-1}: $[-2, \infty)$; range of f^{-1}: $[5, \infty)$ **81.** Domain of g^{-1}: $[0, \infty)$; range of g^{-1}: $(-\infty, 0]$

83. Increasing on the interval $(f(0), f(5))$ **85.** $f^{-1}(x) = \dfrac{1}{m}(x - b), m \neq 0$ **87.** Quadrant I

89. Possible answer: $f(x) = |x|, x \geq 0$, is one-to-one; $f^{-1}(x) = x, x \geq 0$

91. (a) $r(d) = \dfrac{d + 90.39}{6.97}$

(b) $r(d(r)) = \dfrac{6.97r - 90.39 + 90.39}{6.97} = \dfrac{6.97r}{6.97} = r$

$d(r(d)) = 6.97\left(\dfrac{d + 90.39}{6.97}\right) - 90.39 = d + 90.39 - 90.39 = d$

(c) 56 miles per hour

93. (a) 77.6 kg

(b) $h(W) = \dfrac{W - 50}{2.3} + 60 = \dfrac{W + 88}{2.3}$

(c) $h(W(h)) = \dfrac{50 + 2.3(h - 60) + 88}{2.3} = \dfrac{2.3h}{2.3} = h$

$W(h(W)) = 50 + 2.3\left(\dfrac{W + 88}{2.3} - 60\right)$
$= 50 + W + 88 - 138 = W$

(d) 73 inches

95. (a) $\{g \mid 36{,}900 \leq g \leq 89{,}350\}$
(b) $\{T \mid 5081.25 \leq T \leq 18{,}193.75\}$
(c) $g(T) = \dfrac{T - 5081.25}{0.25} + 36{,}900$

Domain: $\{T \mid 5081.25 \leq T \leq 18{,}193.75\}$
Range: $\{g \mid 36{,}900 \leq g \leq 89{,}350\}$

97. (a) t represents time, so $t \geq 0$.

(b) $t(H) = \sqrt{\dfrac{H - 100}{-4.9}} = \sqrt{\dfrac{100 - H}{4.9}}$

(c) 2.02 seconds

99. $f^{-1}(x) = \dfrac{-dx + b}{cx - a}; f = f^{-1}$ if $a = -d$ **103.** No

107. $6xh + 3h^2 - 7h$ **108.**

109. Zeros: $\dfrac{-5 - \sqrt{13}}{6}, \dfrac{-5 + \sqrt{13}}{6}$, x-intercepts: $\dfrac{-5 - \sqrt{13}}{6}, \dfrac{-5 + \sqrt{13}}{6}$

110. Domain: $\left\{x \mid x \neq -\dfrac{3}{2}, x \neq 2\right\}$; Vertical asymptote: $x = -\dfrac{3}{2}$;
Horizontal asymptote: $y = 3$

6.3 Assess Your Understanding *(page 434)*

6. Exponential function; growth factor; initial value **7.** a **8.** T **9.** T **10.** $\left(-1, \dfrac{1}{a}\right)$; $(0, 1)$; $(1, a)$ **11.** 4 **12.** F **13.** b **14.** c

15. (a) 8.815 **(b)** 8.821 **(c)** 8.824 **(d)** 8.825 **17. (a)** 21.217 **(b)** 22.217 **(c)** 22.440 **(d)** 22.459 **19.** 1.265 **21.** 0.347 **23.** 3.320 **25.** 149.952
27. Neither **29.** Exponential; $H(x) = 4^x$ **31.** Exponential; $f(x) = 3(2^x)$ **33.** Linear; $H(x) = 2x + 4$ **35.** B **37.** D **39.** A **41.** E

43.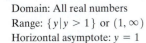

45.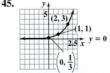

47.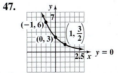

49.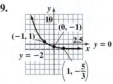

Domain: All real numbers
Range: $\{y \mid y > 1\}$ or $(1, \infty)$
Horizontal asymptote: $y = 1$

Domain: All real numbers
Range: $\{y \mid y > 0\}$ or $(0, \infty)$
Horizontal asymptote: $y = 0$

Domain: All real numbers
Range: $\{y \mid y > 0\}$ or $(0, \infty)$
Horizontal asymptote: $y = 0$

Domain: All real numbers
Range: $\{y \mid y > -2\}$ or $(-2, \infty)$
Horizontal asymptote: $y = -2$

51.

$\left(-1, \frac{33}{16}\right)$ $(2, 6)$ $(1, 3)$ $y = 2$

Domain: All real numbers
Range: $\{y \mid y > 2\}$ or $(2, \infty)$
Horizontal asymptote: $y = 2$

53.

$\left(-2, \frac{7}{3}\right)$ $(0, 3)$ $(2, 5)$ $y = 2$

Domain: All real numbers
Range: $\{y \mid y > 2\}$ or $(2, \infty)$
Horizontal asymptote: $y = 2$

55.

$(-2, e^2)$ $(-1, e)$ $\left(1, \frac{1}{e}\right)$ $(0, 1)$ $y = 0$

Domain: All real numbers
Range: $\{y \mid y > 0\}$ or $(0, \infty)$
Horizontal asymptote: $y = 0$

57.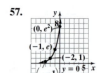

$(0, e^2)$ $(-1, e)$ $(-2, 1)$ $\left(-3, \frac{1}{e}\right)$ $y = 0$

Domain: All real numbers
Range: $\{y \mid y > 0\}$ or $(0, \infty)$
Horizontal asymptote: $y = 0$

59.

$y = 5$ $(0, 4)$

Domain: All real numbers
Range: $\{y \mid y < 5\}$ or $(-\infty, 5)$
Horizontal asymptote: $y = 5$

61.

$(0, 1)$ $y = 2$

Domain: All real numbers
Range: $\{y \mid y < 2\}$ or $(-\infty, 2)$
Horizontal asymptote: $y = 2$

63. $\{3\}$ **65.** $\{-4\}$ **67.** $\{2\}$ **69.** $\left\{\frac{3}{2}\right\}$ **71.** $\left\{-\sqrt{2}, 0, \sqrt{2}\right\}$

73. $\{6\}$ **75.** $\{-1, 7\}$ **77.** $\{-4, 2\}$ **79.** $\{-4\}$ **81.** $\{1, 2\}$ **83.** $\frac{1}{49}$

85. $\frac{1}{4}$ **87.** 5   **89.** $f(x) = 3^x$ **91.** $f(x) = -6^x$ **93.** $f(x) = 3^x + 2$

95. (a) 16; (4, 16)   **(b)** $-4; \left(-4, \frac{1}{16}\right)$ **97. (a)** $\frac{9}{4}; \left(-1, \frac{9}{4}\right)$ **(b)** 3; (3, 66)

99. (a) 60; $(-6, 60)$ **(b)** $-4; (-4, 12)$ **(c)** -2

101.

$(-2, e^2)$ $(2, e^2)$ $(-1, e)$ $(1, e)$ $(0, 1)$

Domain: $(-\infty, \infty)$
Range: $[1, \infty)$
Intercept: (0, 1)

103.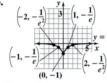

$\left(-2, -\frac{1}{e^2}\right)$ $\left(1, -\frac{1}{e}\right)$ $\left(-1, -\frac{1}{e}\right)$ $y = 0$ $\left(2, -\frac{1}{e^2}\right)$ $(0, -1)$

Domain: $(-\infty, \infty)$
Range: $[-1, 0)$
Intercept: $(0, -1)$

105. (a) 74%   **(b)** 47%   **(c)** Each pane allows only 97% of light to pass through.
107. (a) \$16,231   **(b)** \$8626   **(c)** As each year passes, the sedan is worth 90% of its value the previous year.   **109. (a)** 30%   **(b)** 9%   **(c)** Each year only 30% of the previous survivors survive again.   **111.** 3.35 mg; 0.45 mg   **113. (a)** 0.632   **(b)** 0.982   **(c)** 1

(d) 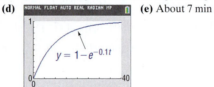 **(e)** About 7 min

$y = 1 - e^{-0.1t}$

115. (a) 0.0516   **(b)** 0.0888   **117. (a)** 70.95%   **(b)** 72.62%   **(c)** 100%
119. (a) 5.41 amp, 7.59 amp, 10.38 amp   **(b)** 12 amp
(d) 3.34 amp, 5.31 amp, 9.44 amp
(e) 24 amp
(c), (f)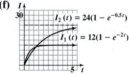

$I_2(t) = 24(1 - e^{-0.5t})$
$I_1(t) = 12(1 - e^{-2t})$

121. 36

123.

Final Denominator	Value of Expression	Compare Value to $e \approx$ 2.718281828
1 + 1	2.5	2.5 < e
2 + 2	2.8	2.8 > e
3 + 3	2.7	2.7 < e
4 + 4	2.721649485	2.721649485 > e
5 + 5	2.717770035	2.717770035 < e
6 + 6	2.718348855	2.718348855 > e

125. $f(A + B) = a^{A+B} = a^A \cdot a^B = f(A) \cdot f(B)$ **127.** $f(\alpha x) = a^{\alpha x} = (a^x)^\alpha = [f(x)]^\alpha$

129. (a) $f(-x) = \frac{1}{2}(e^{-x} + e^{-(-x)})$ **(b)**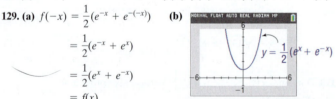

$= \frac{1}{2}(e^{-x} + e^x)$

$= \frac{1}{2}(e^x + e^{-x})$

$= f(x)$

$y = \frac{1}{2}(e^x + e^{-x})$

(c) $(\cosh x)^2 - (\sinh x)^2$

$= \left[\frac{1}{2}(e^x + e^{-x})\right]^2 - \left[\frac{1}{2}(e^x - e^{-x})\right]^2$

$= \frac{1}{4}[e^{2x} + 2 + e^{-2x} - e^{2x} + 2 - e^{-2x}]$

$= \frac{1}{4}(4) = 1$

131. 59 minutes   **135.** $a^{-x} = (a^{-1})^x = \left(\frac{1}{a}\right)^x$ **136.** $(-\infty, -5] \cup [-2, 2]$ **137.** $(2, \infty)$ **138.** $f(x) = -2x^2 + 12x - 13$

139. (a)

$(-3, 0)$ $(1, 0)$ $(-2, -3)$ $(0, -3)$ $(-1, -4)$ $x = -1$

(b) Domain: $(-\infty, \infty)$; Range: $[-4, \infty)$
(c) Decreasing: $(-\infty, -1)$; Increasing: $(-1, \infty)$

6.4 Assess Your Understanding *(page 448)*

4. $\{x \mid x > 0\}$ or $(0, \infty)$ **5.** $\left(\frac{1}{a}, -1\right)$, $(1, 0)$, $(a, 1)$ **6.** 1   **7.** F   **8.** T   **9.** a   **10.** c   **11.** $2 = \log_3 9$ **13.** $2 = \log_a 1.6$ **15.** $x = \log_2 7.2$ **17.** $x = \ln 8$

19. $2^3 = 8$ **21.** $a^6 = 3$ **23.** $3^x = 2$ **25.** $e^x = 4$ **27.** 0   **29.** 2   **31.** -4 **33.** $\frac{1}{2}$ **35.** 4   **37.** $\frac{1}{2}$ **39.** $\{x \mid x > 3\}$; $(3, \infty)$

41. All real numbers except 0; $\{x \mid x \neq 0\}$; $(-\infty, 0) \cup (0, \infty)$ **43.** $\{x \mid x > 10\}$; $(10, \infty)$ **45.** $\{x \mid x > -1\}$; $(-1, \infty)$
47. $\{x \mid x < -1 \text{ or } x > 0\}$; $(-\infty, -1) \cup (0, \infty)$ **49.** $\{x \mid x \geq 1\}$; $[1, \infty)$ **51.** 0.511   **53.** 30.099   **55.** 2.303   **57.** -53.991 **59.** $\sqrt{2}$

61.
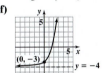

63.

65. B   **67.** D   **69.** A   **71.** E

73. (a) Domain: $(-4, \infty)$
(b)

(c) Range: $(-\infty, \infty)$
    Vertical asymptote: $x = -4$
(d) $f^{-1}(x) = e^x - 4$
(e) Domain of f^{-1}: $(-\infty, \infty)$
    Range of f^{-1}: $(-4, \infty)$
(f)

75. (a) Domain: $(0, \infty)$
(b)

(c) Range: $(-\infty, \infty)$
    Vertical asymptote: $x = 0$
(d) $f^{-1}(x) = e^{x-2}$
(e) Domain of f^{-1}: $(-\infty, \infty)$
    Range of f^{-1}: $(0, \infty)$
(f)

77. (a) Domain: $(0, \infty)$
(b)

(c) Range: $(-\infty, \infty)$
    Vertical asymptote: $x = 0$
(d) $f^{-1}(x) = \dfrac{1}{2}e^{x+3}$
(e) Domain of f^{-1}: $(-\infty, \infty)$
    Range of f^{-1}: $(0, \infty)$
(f)

79. (a) Domain: $(4, \infty)$
(b)

(c) Range: $(-\infty, \infty)$
    Vertical asymptote: $x = 4$
(d) $f^{-1}(x) = 10^{x-2} + 4$
(e) Domain of f^{-1}: $(-\infty, \infty)$
    Range of f^{-1}: $(4, \infty)$
(f)

81. (a) Domain: $(0, \infty)$
(b)

(c) Range: $(-\infty, \infty)$
    Vertical asymptote: $x = 0$
(d) $f^{-1}(x) = \dfrac{1}{2} \cdot 10^{2x}$
(e) Domain of f^{-1}: $(-\infty, \infty)$
    Range of f^{-1}: $(0, \infty)$
(f)

83. (a) Domain: $(-2, \infty)$
(b)
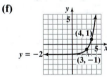
(c) Range: $(-\infty, \infty)$
    Vertical asymptote: $x = -2$
(d) $f^{-1}(x) = 3^{x-3} - 2$
(e) Domain of f^{-1}: $(-\infty, \infty)$
    Range of f^{-1}: $(-2, \infty)$
(f)
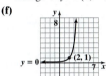

85. (a) Domain: $(-\infty, \infty)$
(b)

(c) Range: $(-3, \infty)$
    Horizontal asymptote: $y = -3$
(d) $f^{-1}(x) = \ln(x+3) - 2$
(e) Domain of f^{-1}: $(-3, \infty)$
    Range of f^{-1}: $(-\infty, \infty)$
(f)

87. (a) Domain: $(-\infty, \infty)$
(b)

(c) Range: $(4, \infty)$
    Horizontal asymptote: $y = 4$
(d) $f^{-1}(x) = 3\log_2(x-4)$
(e) Domain of f^{-1}: $(4, \infty)$
    Range of f^{-1}: $(-\infty, \infty)$
(f)

89. $\{9\}$ **91.** $\left\{\dfrac{7}{2}\right\}$ **93.** $\{2\}$ **95.** $\{5\}$ **97.** $\{3\}$

99. $\{2\}$ **101.** $\left\{\dfrac{\ln 10}{3}\right\}$ **103.** $\left\{\dfrac{\ln 8 - 5}{2}\right\}$

105. $\left\{-2\sqrt{2}, 2\sqrt{2}\right\}$ **107.** $\{-1\}$

109. $\left\{5\ln\dfrac{7}{5}\right\}$ **111.** $\left\{2 - \log\dfrac{5}{2}\right\}$

113. (a) $\left\{x \mid x > -\dfrac{1}{2}\right\}$; $\left(-\dfrac{1}{2}, \infty\right)$
    (b) 2; (40, 2)   **(c)** 121; (121, 3)   **(d)** 4

115.

Domain: $\{x \mid x \neq 0\}$
Range: $(-\infty, \infty)$
Intercepts: $(-1, 0), (1, 0)$

117.

Domain: $\{x \mid x > 0\}$
Range: $\{y \mid y \geq 0\}$
Intercept: $(1, 0)$

119. (a) 1 **(b)** 2 **(c)** 3
(d) It increases. **(e)** 0.000316
(f) 3.981×10^{-8}
121. (a) 5.97 km **(b)** 0.90 km
123. (a) 6.93 min **(b)** 16.09 min
125. $h \approx 2.29$, so the time between
injections is about 2 h, 17 min.

127. 0.2695 s
0.8959 s

129. 50 decibels (dB) **131.** 90 dB **133.** 8.1 **135. (a)** $k \approx 11.216$ **(b)** 6.73 **(c)** 0.41% **(d)** 0.14%

137. Because $y = \log_1 x$ means $1^y = 1 = x$, which cannot be true for $x \neq 1$ **139.** Zeros: $-3, -\dfrac{1}{2}, \dfrac{1}{2}, 3$; x-intercepts: $-3, -\dfrac{1}{2}, \dfrac{1}{2}, 3$

140. 12 **141.** $f(1) = -5; f(2) = 17$ **142.** $3 + i; f(x) = x^4 - 7x^3 + 14x^2 + 2x - 20; a = 1$

6.5 Assess Your Understanding (page 459)

1. 0 **2.** M **3.** r **4.** $\log_a M; \log_a N$ **5.** $\log_a M; \log_a N$ **6.** $r \log_a M$ **7.** 7 **8.** F **9.** F **10.** F **11.** b **12.** b **13.** 71 **15.** -4 **17.** 7 **19.** 1 **21.** 1

23. 3 **25.** $\dfrac{5}{4}$ **27.** 4 **29.** $a + b$ **31.** $b - a$ **33.** $3a$ **35.** $\dfrac{1}{5}(a + b)$ **37.** $2 + \log_5 x$ **39.** $3 \log_2 z$ **41.** $1 + \ln x$ **43.** $\ln x - x$ **45.** $2 \log_a u + 3 \log_a v$

47. $2 \ln x + \dfrac{1}{2} \ln (1 - x)$ **49.** $3 \log_2 x - \log_2 (x - 3)$ **51.** $\log x + \log (x + 2) - 2 \log (x + 3)$ **53.** $\dfrac{1}{3} \ln (x - 2) + \dfrac{1}{3} \ln (x + 1) - \dfrac{2}{3} \ln (x + 4)$

55. $\ln 5 + \ln x + \dfrac{1}{2} \ln (1 + 3x) - 3 \ln (x - 4)$ **57.** $\log_5 u^3 v^4$ **59.** $\log_3 \left(\dfrac{1}{x^{5/2}}\right)$ **61.** $\log_4 \left[\dfrac{x - 1}{(x + 1)^4}\right]$ **63.** $-2 \ln (x - 1)$ **65.** $\log_2 [x(3x - 2)^4]$

67. $\log_a \left(\dfrac{25x^6}{\sqrt{2x + 3}}\right)$ **69.** $\log_2 \left[\dfrac{(x + 1)^2}{(x + 3)(x - 1)}\right]$ **71.** 2.771 **73.** -3.880 **75.** 5.615 **77.** 0.874

79. $y = \dfrac{\log x}{\log 4}$

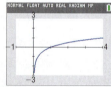

81. $y = \dfrac{\log (x + 2)}{\log 2}$

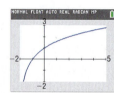

83. $y = \dfrac{\log (x + 1)}{\log (x - 1)}$

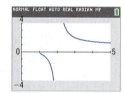

85. (a) $(f \circ g)(x) = x$; $\{x \mid x \text{ is any real number}\}$ or $(-\infty, \infty)$
(b) $(g \circ f)(x) = x$; $\{x \mid x > 0\}$ or $(0, \infty)$ **(c)** 5
(d) $(f \circ h)(x) = \ln x^2$; $\{x \mid x \neq 0\}$ or $(-\infty, 0) \cup (0, \infty)$ **(e)** 2

87. $y = Cx$ **89.** $y = Cx(x + 1)$ **91.** $y = Ce^{3x}$ **93.** $y = Ce^{-4x} + 3$

95. $y = \dfrac{\sqrt[3]{C}(2x + 1)^{1/6}}{(x + 4)^{1/9}}$ **97.** 3 **99.** 1

101. $\log_a (x + \sqrt{x^2 - 1}) + \log_a (x - \sqrt{x^2 - 1}) = \log_a [(x + \sqrt{x^2 - 1})(x - \sqrt{x^2 - 1})] = \log_a [x^2 - (x^2 - 1)] = \log_a 1 = 0$

103. $\ln (1 + e^{2x}) = \ln[e^{2x}(e^{-2x} + 1)] = \ln e^{2x} + \ln (e^{-2x} + 1) = 2x + \ln (1 + e^{-2x})$

105. $y = f(x) = \log_a x; a^y = x$ implies $a^y = \left(\dfrac{1}{a}\right)^{-y} = x$, so $-y = \log_{1/a} x = -f(x)$.

107. $f(x) = \log_a x; f\left(\dfrac{1}{x}\right) = \log_a \dfrac{1}{x} = \log_a 1 - \log_a x = -f(x)$

109. $\log_a \dfrac{M}{N} = \log_a (M \cdot N^{-1}) = \log_a M + \log_a N^{-1} = \log_a M - \log_a N$, since $a^{\log_a N^{-1}} = N^{-1}$ implies $a^{-\log_a N^{-1}} = N$; that is, $\log_a N = -\log_a N^{-1}$.

115. $\{-1.78, 1.29, 3.49\}$ **116.** A repeated real solution (double root) **117.** $-2, \dfrac{1}{5}, \dfrac{-5 - \sqrt{21}}{2}, \dfrac{-5 + \sqrt{21}}{2}$
118.

Domain: $\{x \mid x \leq 2\}$ or $(-\infty, 2]$
Range: $\{y \mid y \geq 0\}$ or $[0, \infty)$

6.6 Assess Your Understanding (page 465)

5. $\{16\}$ **7.** $\left\{\dfrac{16}{5}\right\}$ **9.** $\{6\}$ **11.** $\{16\}$ **13.** $\left\{\dfrac{1}{3}\right\}$ **15.** $\{3\}$ **17.** $\{5\}$ **19.** $\left\{\dfrac{21}{8}\right\}$ **21.** $\{-6\}$ **23.** $\{-2\}$ **25.** $\{-1 + \sqrt{1 + e^4}\} \approx \{6.456\}$

27. $\left\{\dfrac{-5 + 3\sqrt{5}}{2}\right\} \approx \{0.854\}$ **29.** $\{2\}$ **31.** $\left\{\dfrac{9}{2}\right\}$ **33.** $\{7\}$ **35.** $\{-2 + 4\sqrt{2}\}$ **37.** $\{-\sqrt{3}, \sqrt{3}\}$ **39.** $\left\{\dfrac{1}{3}, 729\right\}$ **41.** $\{8\}$

43. $\{\log_2 10\} = \left\{\dfrac{\ln 10}{\ln 2}\right\} \approx \{3.322\}$ **45.** $\{-\log_8 1.2\} = \left\{-\dfrac{\ln 1.2}{\ln 8}\right\} \approx \{-0.088\}$ **47.** $\left\{\dfrac{1}{3}\log_2 \dfrac{8}{5}\right\} = \left\{\dfrac{\ln \dfrac{8}{5}}{3 \ln 2}\right\} \approx \{0.226\}$

49. $\left\{\dfrac{\ln 3}{2 \ln 3 + \ln 4}\right\} \approx \{0.307\}$ **51.** $\left\{\dfrac{\ln 7}{\ln 0.6 + \ln 7}\right\} \approx \{1.356\}$ **53.** $\{0\}$ **55.** $\left\{\dfrac{\ln \pi}{1 + \ln \pi}\right\} \approx \{0.534\}$ **57.** $\left\{\dfrac{\ln 3}{\ln 2}\right\} \approx \{1.585\}$

59. $\{0\}$ **61.** $\left\{\log_4\left(-2 + \sqrt{7}\right)\right\} \approx \{-0.315\}$ **63.** $\{\log_5 4\} \approx \{0.861\}$ **65.** No real solution **67.** $\{\log_4 5\} \approx \{1.161\}$ **69.** $\{2.79\}$

71. $\{-0.57\}$ **73.** $\{-0.70\}$ **75.** $\{0.57\}$ **77.** $\{0.39, 1.00\}$ **79.** $\{1.32\}$ **81.** $\{1.31\}$ **83.** $\{1\}$ **85.** $\{16\}$ **87.** $\left\{-1, \dfrac{2}{3}\right\}$ **89.** $\{0\}$

91. $\left\{\ln\left(2 + \sqrt{5}\right)\right\} \approx \{1.444\}$ **93.** $\left\{e^{\frac{\ln 5 \cdot \ln 3}{\ln 15}}\right\} \approx \{1.921\}$ **95. (a)** $\{5\}; (5, 3)$ **(b)** $\{5\}; (5, 4)$ **(c)** $\{1\}$; yes, at $(1, 2)$ **(d)** $\{5\}$ **(e)** $\left\{-\dfrac{1}{11}\right\}$

97. (a), (b)
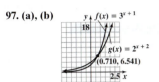
(c) $\{x | x > 0.710\}$ or $(0.710, \infty)$

99. (a), (b), (c)
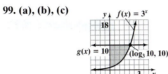

101. (a), (b), (c)
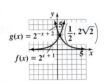

103. (a)
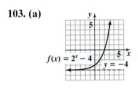

(b) 2 **(c)** $\{x | x < 2\}$ or $(-\infty, 2)$

105. (a) 2047 **(b)** 2059
107. (a) After 4.2 yr
(b) After 6.5 yr
(c) After 12.8 yr

110. $\left\{-3, \dfrac{1}{4}, 2\right\}$
111. one-to-one
112. $(f \circ g)(x) = \dfrac{x + 5}{-x + 11}; \{x | x \neq 3, x \neq 11\}$
113. $\{x | x \geq 1\}$, or $[1, \infty)$

6.7 Assess Your Understanding (page 474)

3. principal **4.** $I; Prt;$ simple interest **5.** 4 **6.** effective rate of interest **7.** $108.29 **9.** $609.50 **11.** $697.09 **13.** $1246.08 **15.** $88.72 **17.** $860.72

19. $554.09 **21.** $59.71 **23.** 5.095% **25.** 5.127% **27.** $6\dfrac{1}{4}$% compounded annually **29.** 9% compounded monthly **31.** 25.992% **33.** 24.573%

35. (a) About 8.69 yr **(b)** About 8.66 yr **37.** 6.823% **39.** 10.15 yr; 10.14 yr **41.** 15.27 yr or 15 yr, 3 mo **43.** $104,335 **45.** $12,910.62

47. About $30.17 per share or $3017 **49.** Not quite. Jim will have $1057.60. The second bank gives a better deal, since Jim will have $1060.62 after 1 yr.

51. Will has $11,632.73; Henry has $10,947.89. **53. (a)** $63,449 **(b)** $44,267 **55.** About $1019 billion; about $232 billion **57.** $940.90 **59.** 2.53%

61. 34.31 yr **63. (a)** $3686.45 **(b)** $3678.79 **65.** $6439.28

67. (a) 11.90 yr **(b)** 22.11 yr **(c)** $mP = P\left(1 + \dfrac{r}{n}\right)^{nt}$

$$m = \left(1 + \dfrac{r}{n}\right)^{nt}$$

$$\ln m = \ln\left(1 + \dfrac{r}{n}\right)^{nt} = nt \ln\left(1 + \dfrac{r}{n}\right)$$

$$t = \dfrac{\ln m}{n \ln\left(1 + \dfrac{r}{n}\right)}$$

69. (a) 1.59% **(b)** In 2029 or after 21 yr **71.** 22.7 yr **76.** $R = 0$; yes
77. $f^{-1}(x) = \dfrac{2x}{x - 1}$ **78.** $-2, 5; f(x) = (x + 2)^2(x - 5)(x^2 + 1)$ **79.** $\{6\}$

6.8 Assess Your Understanding (page 486)

1. (a) 500 insects **(b)** $0.02 = 2\%$ per day **(c)** About 611 insects **(d)** After about 23.5 days **(e)** After about 34.7 days
3. (a) $-0.0244 = -2.44\%$ per year **(b)** About 391.7 g **(c)** After about 9.1 yr **(d)** 28.4 yr **5. (a)** $N(t) = N_0 e^{kt}$ **(b)** 5832 **(c)** 3.9 days
7. (a) $N(t) = N_0 e^{kt}$ **(b)** 25,198 **9.** 9.797 g **11.** 9953 yr ago **13. (a)** 5:18 PM **(b)** About 14.3 min **(c)** The temperature of the pizza approaches 70°F.
15. 18.63°C; 25.1°C **17.** 1.7 ppm; 7.17 days, or 172 hr **19.** 0.26 M; 6.58 hr, or 395 min **21.** 26.6 days
23. (a) In 1984, 91.8% of households did not own a personal computer.

25. (a)

27. (a) 9.23×10^{-3}, or about 0
(b) 0.81, or about 1
(c) 5.01, or about 5
(d) 57.91°, 43.99°, 30.07°

(b)
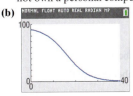

(c) 70.6% **(d)** During 2011

(b) 0.78, or 78%
(c) 50 people
(d) As n increases, the probability decreases.

29. (a) $P(t) = 50(3)^{t/20}$ **(b)** 661 **(c)** In 48 days **(d)** $P(t) = 50e^{0.055t}$ **30.** $f(x) = -\dfrac{3}{2}x + 7$ **31.** Neither **32.** $2 \ln x + \dfrac{1}{2}\ln y - \ln z$ **33.** $2\sqrt[3]{5}$

Chapter Test *(page 502)*

1. (a) $f \circ g = \dfrac{2x + 7}{2x + 3}$; domain: $\left\{ x \middle| x \neq -\dfrac{3}{2} \right\}$ **(b)** $(g \circ f)(-2) = 5$ **(c)** $(f \circ g)(-2) = -3$

2. (a) The function is not one-to-one. **(b)** The function is one-to-one.

3. $f^{-1}(x) = \dfrac{2 + 5x}{3x}$; domain of $f = \left\{ x \middle| x \neq \dfrac{5}{3} \right\}$, range of $f = \{ y | y \neq 0 \}$; domain of $f^{-1} = \{ x | x \neq 0 \}$; range of $f^{-1} = \left\{ y \middle| y \neq \dfrac{5}{3} \right\}$

4. The point $(-5, 3)$ must be on the graph of f^{-1}. **5.** $x = 5$ **6.** $b = 4$ **7.** $x = 625$ **8.** $e^3 + 2 \approx 22.086$ **9.** $\log 20 \approx 1.301$

10. $\log_3 21 = \dfrac{\ln 21}{\ln 3} \approx 2.771$ **11.** $\ln 133 \approx 4.890$

12. (a) Domain of f: $\{ x | -\infty < x < \infty \}$ or $(-\infty, \infty)$

(b)

(c) Range of f: $\{ y | y > -2 \}$ or $(-2, \infty)$;
Horizontal asymptote: $y = -2$

(d) $f^{-1}(x) = \log_4 (x + 2) - 1$

(e) Domain of f^{-1}: $\{ x | x > -2 \}$ or $(-2, \infty)$
Range of f^{-1}: $\{ y | -\infty < y < \infty \}$ or $(-\infty, \infty)$

(f)

13. (a) Domain of f: $\{ x | x > 2 \}$ or $(2, \infty)$

(b)

(c) Range of f: $\{ y | -\infty < y < \infty \}$ or $(-\infty, \infty)$;
vertical asymptote: $x = 2$

(d) $f^{-1}(x) = 5^{1-x} + 2$

(e) Domain of f^{-1}: $\{ x | -\infty < x < \infty \}$ or $(-\infty, \infty)$
Range of f^{-1}: $\{ y | y > 2 \}$ or $(2, \infty)$

(f)

14. $\{1\}$ **15.** $\{91\}$ **16.** $\{-\ln 2\} \approx \{-0.693\}$ **17.** $\left\{ \dfrac{1 - \sqrt{13}}{2}, \dfrac{1 + \sqrt{13}}{2} \right\} \approx \{-1.303, 2.303\}$ **18.** $\left\{ \dfrac{3 \ln 7}{1 - \ln 7} \right\} \approx \{-6.172\}$

19. $\{2\sqrt{6}\} \approx \{4.899\}$ **20.** $2 + 3 \log_2 x - \log_2(x - 6) - \log_2(x + 3)$ **21.** About 250.39 days **22. (a)** \$1033.82 **(b)** \$963.42 **(c)** 11.9 yr

23. (a) About 83 dB **(b)** The pain threshold will be exceeded if 31,623 people shout at the same time.

Cumulative Review *(page 502)*

1. Yes; no **2. (a)** 10 **(b)** $2x^2 + 3x + 1$ **(c)** $2x^2 + 4xh + 2h^2 - 3x - 3h + 1$ **3.** $\left(\dfrac{1}{2}, \dfrac{\sqrt{3}}{2} \right)$ is on the graph. **4.** $\{-26\}$

5.
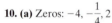

6. (a)
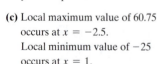

(b) $\{ x | -\infty < x < \infty \}$ **7.** $f(x) = 2(x - 4)^2 - 8 = 2x^2 - 16x + 24$

8.

9. $f(g(x)) = \dfrac{4}{(x - 3)^2} + 2$; domain: $\{ x | x \neq 3 \}$; 3

10. (a) Zeros: $-4, -\dfrac{1}{4}, 2$

(b) x-intercepts: $-4, -\dfrac{1}{4}, 2$;
y-intercept: -8

(c) Local maximum value of 60.75
occurs at $x = -2.5$.
Local minimum value of -25
occurs at $x = 1$.

(d)
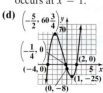

11. (a), (c)

Domain g = range g^{-1} = $(-\infty, \infty)$
Range g = domain g^{-1} = $(2, \infty)$

(b) $g^{-1}(x) = \log_3(x - 2)$

12. $\left\{ -\dfrac{3}{2} \right\}$ **13.** $\{2\}$

14. (a) $\{-1\}$ **(b)** $\{ x | x > -1 \}$ or $(-1, \infty)$

(c) $\{25\}$

15. (a)
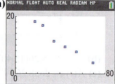

(b) Logarithmic; $y = 49.293 - 10.563 \ln x$
(c) Highest value of $|r|$

CHAPTER 7 Trigonometric Functions

7.1 Assess Your Understanding (page 514)

3. standard position **4.** central angle **5.** d **6.** $r\theta$; $\frac{1}{2}r^2\theta$ **7.** b **8.** $\frac{s}{t}$; $\frac{\theta}{t}$ **9.** T **10.** F

11.  **13.** 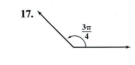 135° **15.** 450° **17.** $\frac{3\pi}{4}$ **19.** $-\frac{\pi}{6}$ **21.** 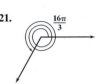 $\frac{16\pi}{3}$

23. $\frac{\pi}{6}$ **25.** $\frac{4\pi}{3}$ **27.** $-\frac{\pi}{3}$ **29.** π **31.** $-\frac{3\pi}{4}$ **33.** $-\frac{\pi}{2}$ **35.** 60° **37.** $-225°$ **39.** 90° **41.** 15° **43.** $-90°$ **45.** $-30°$ **47.** 0.30 **49.** -0.70 **51.** 2.18
53. 179.91° **55.** 114.59° **57.** 362.11° **59.** 40.17° **61.** 50.24° **63.** 9.15° **65.** 40°19′12″ **67.** 18°15′18″ **69.** 19°59′24″ **71.** 5 m **73.** 6 ft

75. 0.6 radian **77.** $\frac{\pi}{3} \approx 1.047$ in. **79.** 25 m^2 **81.** $2\sqrt{3} \approx 3.464$ ft **83.** 0.24 radian **85.** $\frac{\pi}{3} \approx 1.047$ in.^{2} **87.** $s = 2.094$ ft; $A = 2.094$ ft^2

89. $s = 14.661$ yd; $A = 87.965$ yd^2 **91.** $3\pi \approx 9.42$ in; $5\pi \approx 15.71$ in. **93.** $2\pi \approx 6.28$ m^2 **95.** $\frac{675\pi}{2} \approx 1060.29$ ft^2 **97.** $\frac{1075\pi}{3} \approx 1125.74$ in.^{2}

99. $\omega = \frac{1}{60}$ radian/s; $v = \frac{1}{12}$ cm/s **101.** ≈ 23.2 mph **103.** ≈ 120.6 km/h **105.** ≈ 452.5 rpm **107.** ≈ 359 mi **109.** ≈ 898 mi/h **111.** ≈ 2292 mi/h

113. $\frac{3}{4}$ rpm **115.** ≈ 2.86 mi/h **117.** ≈ 31.47 rpm **119.** 63π ft^2 **121.** ≈ 1037 mi/h **123.** Radius ≈ 3979 mi; circumference $\approx 25,000$ mi

125. $v_1 = r_1\omega_1$, $v_2 = r_2\omega_2$, and $v_1 = v_2$, so $r_1\omega_1 = r_2\omega_2$ and $\frac{r_1}{r_2} = \frac{\omega_2}{\omega_1}$. **134.** $x = -\frac{7}{3}$ **135.** $\left\{-3, \frac{1}{5}\right\}$ **136.** $y = -|x + 3| - 4$

137. HA: $y = 3$; VA: $x = 7$

7.2 Assess Your Understanding (page 527)

3. complementary **4.** cosine **5.** 62° **6.** c **7.** T **8.** a **9.** T **10.** F

11. $\sin\theta = \frac{5}{13}$; $\cos\theta = \frac{12}{13}$; $\tan\theta = \frac{5}{12}$; $\csc\theta = \frac{13}{5}$; $\sec\theta = \frac{13}{12}$; $\cot\theta = \frac{12}{5}$ **13.** $\sin\theta = \frac{2\sqrt{13}}{13}$; $\cos\theta = \frac{3\sqrt{13}}{13}$; $\tan\theta = \frac{2}{3}$; $\csc\theta = \frac{\sqrt{13}}{2}$; $\sec\theta = \frac{\sqrt{13}}{3}$;

$\cot\theta = \frac{3}{2}$ **15.** $\sin\theta = \frac{\sqrt{3}}{2}$; $\cos\theta = \frac{1}{2}$; $\tan\theta = \sqrt{3}$; $\csc\theta = \frac{2\sqrt{3}}{3}$; $\sec\theta = 2$; $\cot\theta = \frac{\sqrt{3}}{3}$ **17.** $\sin\theta = \frac{\sqrt{6}}{3}$; $\cos\theta = \frac{\sqrt{3}}{3}$; $\tan\theta = \sqrt{2}$; $\csc\theta = \frac{\sqrt{6}}{2}$;

$\sec\theta = \sqrt{3}$; $\cot\theta = \frac{\sqrt{2}}{2}$ **19.** $\sin\theta = \frac{\sqrt{5}}{5}$; $\cos\theta = \frac{2\sqrt{5}}{5}$; $\tan\theta = \frac{1}{2}$; $\csc\theta = \sqrt{5}$; $\sec\theta = \frac{\sqrt{5}}{2}$; $\cot\theta = 2$ **21.** $\tan\theta = \frac{\sqrt{3}}{3}$; $\csc\theta = 2$;

$\sec\theta = \frac{2\sqrt{3}}{3}$; $\cot\theta = \sqrt{3}$ **23.** $\tan\theta = \frac{2\sqrt{5}}{5}$; $\csc\theta = \frac{3}{2}$; $\sec\theta = \frac{3\sqrt{5}}{5}$; $\cot\theta = \frac{\sqrt{5}}{2}$ **25.** $\cos\theta = \frac{\sqrt{2}}{2}$; $\tan\theta = 1$; $\csc\theta = \sqrt{2}$; $\sec\theta = \sqrt{2}$; $\cot\theta = 1$

27. $\sin\theta = \frac{2\sqrt{2}}{3}$; $\tan\theta = 2\sqrt{2}$; $\csc\theta = \frac{3\sqrt{2}}{4}$; $\sec\theta = 3$; $\cot\theta = \frac{\sqrt{2}}{4}$ **29.** $\sin\theta = \frac{\sqrt{5}}{5}$; $\cos\theta = \frac{2\sqrt{5}}{5}$; $\csc\theta = \sqrt{5}$; $\sec\theta = \frac{\sqrt{5}}{2}$; $\cot\theta = 2$

31. $\sin\theta = \frac{2\sqrt{2}}{3}$; $\cos\theta = \frac{1}{3}$; $\tan\theta = 2\sqrt{2}$; $\csc\theta = \frac{3\sqrt{2}}{4}$; $\cot\theta = \frac{\sqrt{2}}{4}$ **33.** $\sin\theta = \frac{\sqrt{6}}{3}$; $\cos\theta = \frac{\sqrt{3}}{3}$; $\csc\theta = \frac{\sqrt{6}}{2}$; $\sec\theta = \sqrt{3}$; $\cot\theta = \frac{\sqrt{2}}{2}$

35. $\sin\theta = \frac{1}{2}$; $\cos\theta = \frac{\sqrt{3}}{2}$; $\tan\theta = \frac{\sqrt{3}}{3}$; $\sec\theta = \frac{2\sqrt{3}}{3}$; $\cot\theta = \sqrt{3}$ **37.** 1 **39.** 1 **41.** 0 **43.** 0 **45.** 1 **47.** 0 **49.** 0 **51.** 1 **53.** 1

55. (a) $\frac{1}{2}$ **(b)** $\frac{3}{4}$ **(c)** 2 **(d)** 2 **57. (a)** 17 **(b)** $\frac{1}{4}$ **(c)** 4 **(d)** $\frac{17}{16}$ **59. (a)** $\frac{1}{4}$ **(b)** 15 **(c)** 4 **(d)** $\frac{16}{15}$ **61. (a)** 0.78 **(b)** 0.79 **(c)** 1.27

(d) 1.28 **(e)** 1.61 **(f)** 0.78 **(g)** 0.62 **(h)** 1.27 **63.** 0.6 **65.** 20° **67.** $5\cos\theta$ **69.** $7\sec\theta$ **71.** $6\tan\theta$

73. (a) 10 min **(d)** Approximately 15.8 min **(f)** 70.5°; 177 ft; 9.7 min
  (b) 20 min **(e)** Approximately 10.4 min

  (c) $T(\theta) = 5\left(1 - \frac{1}{3\tan\theta} + \frac{1}{\sin\theta}\right)$

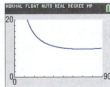

75. (a) $Z = 200\sqrt{13} \approx 721.1$ ohms **(b)** $\tan\phi = \frac{2}{3}$; $\sin\phi = \frac{2\sqrt{13}}{13}$; $\cos\phi = \frac{3\sqrt{13}}{13}$; $\cot\phi = \frac{3}{2}$; $\csc\phi = \frac{\sqrt{13}}{2}$; $\sec\phi = \frac{\sqrt{13}}{3}$

77. (a) $|OA| = |OC| = 1$; angle OAC = angle OCA; angle OAC + angle OAC + 180° $-\theta$ = 180°; angle $OAC = \frac{\theta}{2}$

  (b) $\sin\theta = \frac{|CD|}{|OC|} = |CD|$; $\cos\theta = \frac{|OD|}{|OC|} = |OD|$ **(c)** $\tan\frac{\theta}{2} = \frac{|CD|}{|AD|} = \frac{\sin\theta}{1 + |OD|} = \frac{\sin\theta}{1 + \cos\theta}$

79. $h = x \tan \theta$ and $h = (1 - x) \tan n\theta$; thus, $x \tan \theta = (1 - x) \tan n\theta$, so $x = \dfrac{\tan n\theta}{\tan \theta + \tan n\theta}$.

81. (a) Area $\triangle OAC = \dfrac{1}{2} |AC| \, |OC| = \dfrac{1}{2} \cdot \dfrac{|AC|}{1} \cdot \dfrac{|OC|}{1} = \dfrac{1}{2} \sin \alpha \cos \alpha$

(b) Area $\triangle OCB = \dfrac{1}{2} |BC| \, |OC| = \dfrac{1}{2} |OB|^2 \dfrac{|BC|}{|OB|} \cdot \dfrac{|OC|}{|OB|} = \dfrac{1}{2} |OB|^2 \sin \beta \cos \beta$

(c) Area $\triangle OAB = \dfrac{1}{2} |BD| \, |OA| = \dfrac{1}{2} |OB| \dfrac{|BD|}{|OB|} = \dfrac{1}{2} |OB| \sin(\alpha + \beta)$

(d) $\dfrac{\cos \alpha}{\cos \beta} = \dfrac{\dfrac{|OC|}{1}}{\dfrac{|OC|}{|OB|}} = |OB|$

(e) Area $\triangle OAB$ = area $\triangle OAC$ + area $\triangle OCB$

$\dfrac{1}{2} |OB| \sin (\alpha + \beta) = \dfrac{1}{2} \sin \alpha \cos \alpha + \dfrac{1}{2} |OB|^2 \sin \beta \cos \beta$

$\sin (\alpha + \beta) = \dfrac{\sin \alpha \cos \alpha + |OB|^2 \sin \beta \cos \beta}{|OB|}$

$\sin (\alpha + \beta) = \dfrac{\sin \alpha \, (|OB| \cos \beta) + |OB|^2 \sin \beta \left(\dfrac{\cos \alpha}{|OB|} \right)}{|OB|}$

$\sin (\alpha + \beta) = \sin \alpha \cos \beta + \cos \alpha \sin \beta$

83. $\sin \alpha = \tan \alpha \cos \alpha = \cos \beta \cos \alpha = \cos \beta \tan \beta = \sin \beta$;

$\sin^2 \alpha + \cos^2 \alpha = 1$

$\sin^2 \alpha + \tan^2 \beta = 1$

$\sin^2 \alpha + \dfrac{\sin^2 \beta}{\cos^2 \beta} = 1$

$\sin^2 \alpha + \dfrac{\sin^2 \alpha}{1 - \sin^2 \alpha} = 1$

$\sin^2 \alpha - \sin^4 \alpha + \sin^2 \alpha = 1 - \sin^2 \alpha$

$\sin^4 \alpha - 3 \sin^2 \alpha + 1 = 0$

$\sin^2 \alpha = \dfrac{3 \pm \sqrt{5}}{2}$

$\sin^2 \alpha = \dfrac{3 - \sqrt{5}}{2}$

$\sin \alpha = \sqrt{\dfrac{3 - \sqrt{5}}{2}}$

85. Since $a^2 + b^2 = c^2, a > 0, b > 0$, then $0 < a^2 < c^2$ or $0 < a < c$. Thus $0 < \dfrac{a}{c} < 1$ and $0 < \cos \theta < 1$ or $\dfrac{\cos \theta}{\cos \theta} < \dfrac{1}{\cos \theta}$; therefore, $\sec \theta > 1$.

89. $\left\{ x \middle| x > -\dfrac{2}{5} \right\}$ or $\left(-\dfrac{2}{5}, \infty \right)$ **90.** $4 - 3i, -5$ **91.** $R = 134$ **92.** 81π ft^2

7.3 Assess Your Understanding (page 538)

1. $\dfrac{3}{2}$ **2.** 0.84   **3.** T   **4.** F   **5.** b   **6.** c   **7.** $\sin 45° = \dfrac{\sqrt{2}}{2}$; $\cos 45° = \dfrac{\sqrt{2}}{2}$; $\tan 45° = 1$; $\csc 45° = \sqrt{2}$; $\sec 45° = \sqrt{2}$; $\cot 45° = 1$ **9.** $\dfrac{\sqrt{3}}{2}$ **11.** $\dfrac{1}{2}$ **13.** $\dfrac{3}{4}$

15. $\sqrt{3}$ **17.** $\dfrac{\sqrt{3}}{4}$ **19.** $\sqrt{2}$ **21.** 2   **23.** $\sqrt{2} + \dfrac{4\sqrt{3}}{3}$ **25.** $-\dfrac{8}{3}$ **27.** $\dfrac{1}{2}$ **29.** 0   **31.** 0.47   **33.** 0.38   **35.** 1.33   **37.** 0.31   **39.** 3.73   **41.** 1.04

43. 0.84   **45.** 0.02   **47.** 0.31   **49.** $\dfrac{1 + \sqrt{3}}{2}$ **51.** $\dfrac{1}{2}$ **53.** $\dfrac{\sqrt{3}}{2}$ **55.** $\dfrac{\sqrt{2}}{4}$ **57. (a)** $\dfrac{\sqrt{2}}{2}$; $\left(\dfrac{\pi}{4}, \dfrac{\sqrt{2}}{2} \right)$ **(b)** $\left(\dfrac{\sqrt{2}}{2}, \dfrac{\pi}{4} \right)$ **(c)** $\left(\dfrac{\pi}{4}, -2 \right)$ **59.** 4.59 in.; 6.55 in.

61. (a) 5.52 in. or 11.83 in.   **63.** $R \approx 310.56$ ft; $H \approx 77.64$ ft   **65.** $R \approx 19{,}541.95$ m; $H \approx 2278.14$ m   **67. (a)** 1.20 s   **(b)** 1.12 s   **(c)** 1.20 s

69. 1   **71.** $\dfrac{\sqrt{2}}{2}$ **73.** 70.02 ft   **75.** 985.91 ft   **77.** $50\dfrac{\sqrt{3}}{3} \approx 28.87$ m   **79.** 1978.09 ft   **81.** 60.27 ft   **83.** $190\sqrt{2} \approx 268.70$ ft   **85.** 3.83 mi

87. (a) $T(\theta) = 1 + \dfrac{2}{3 \sin \theta} - \dfrac{1}{4 \tan \theta}$

(b) 1.9 h; 0.57 h

(c) 1.69 h; 0.75 h

(d) 1.63 h; 0.86 h

(e) 1.67 h

(f) 2.75 h

(g)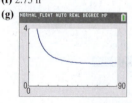

67.98°; 1.62 h; 0.9 h

89. The white ball should hit the top cushion 4.125 ft from the upper-left corner.

91. (a) $L(\theta) = \dfrac{15}{\cos \theta} + 5\sqrt{40 - 36 \tan \theta + 9 \tan^2 \theta}$

(b) $0° < \theta < 63.4°$

(c) $L(45°) = 15\sqrt{2} + 5\sqrt{13}$ ft ≈ 39.2 ft; 15 ft

(d)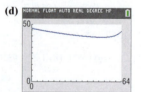

$\theta = 50.2°$

(e) $L_{\min} = 39.05$ ft; ≈ 18 ft from the 15-ft pole

93.

θ	0.5	0.4	0.2	0.1	0.01	0.001	0.0001	0.00001
$f(\theta) = \dfrac{\sin \theta}{\theta}$	0.9589	0.9735	0.9933	0.9983	1.0000	1.0000	1.0000	1.0000

$\dfrac{\sin \theta}{\theta}$ approaches 1 as $\theta \to 0$.

97. All real numbers   **98.** Vertex: $(3, 5)$; axis of symmetry: $x = 3$ **99.** $\{\ln 6 + 4\}$ **100.** 9

7.4 Assess Your Understanding *(page 551)*

1. tangent; cotangent **2.** coterminal **3.** 60° **4.** F **5.** T **6.** T **7.** 60° **8.** b **9.** $\dfrac{\pi}{2}, \dfrac{3\pi}{2}$ **10.** d **11.** $\sin\theta = \dfrac{4}{5}$; $\cos\theta = -\dfrac{3}{5}$;

$\tan\theta = -\dfrac{4}{3}$; $\csc\theta = \dfrac{5}{4}$; $\sec\theta = -\dfrac{5}{3}$; $\cot\theta = -\dfrac{3}{4}$ **13.** $\sin\theta = -\dfrac{3\sqrt{13}}{13}$; $\cos\theta = \dfrac{2\sqrt{13}}{13}$; $\tan\theta = -\dfrac{3}{2}$; $\csc\theta = -\dfrac{\sqrt{13}}{3}$; $\sec\theta = \dfrac{\sqrt{13}}{2}$; $\cot\theta = -\dfrac{2}{3}$

15. $\sin\theta = -\dfrac{\sqrt{2}}{2}$; $\cos\theta = -\dfrac{\sqrt{2}}{2}$; $\tan\theta = 1$; $\csc\theta = -\sqrt{2}$; $\sec\theta = -\sqrt{2}$; $\cot\theta = 1$

17. $\sin\theta = \dfrac{1}{2}$; $\cos\theta = \dfrac{\sqrt{3}}{2}$; $\tan\theta = \dfrac{\sqrt{3}}{3}$; $\csc\theta = 2$; $\sec\theta = \dfrac{2\sqrt{3}}{3}$; $\cot\theta = \sqrt{3}$ **19.** $\sin\theta = -\dfrac{\sqrt{2}}{2}$; $\cos\theta = \dfrac{\sqrt{2}}{2}$; $\tan\theta = -1$; $\csc\theta = -\sqrt{2}$;

$\sec\theta = \sqrt{2}$; $\cot\theta = -1$ **21.** $\dfrac{\sqrt{2}}{2}$ **23.** 1 **25.** 1 **27.** $\sqrt{3}$ **29.** $\dfrac{\sqrt{2}}{2}$ **31.** 0 **33.** II **35.** IV **37.** IV **39.** III **41.** 30° **43.** 60° **45.** 60°

47. $\dfrac{\pi}{4}$ **49.** $\dfrac{\pi}{3}$ **51.** 45° **53.** $\dfrac{\pi}{3}$ **55.** 80° **57.** $\dfrac{\pi}{4}$ **59.** $\dfrac{1}{2}$ **61.** $\dfrac{1}{2}$ **63.** $\dfrac{\sqrt{2}}{2}$ **65.** -2 **67.** $-\sqrt{3}$ **69.** $\dfrac{\sqrt{2}}{2}$ **71.** $-\dfrac{\sqrt{2}}{2}$ **73.** $-\dfrac{\sqrt{3}}{2}$ **75.** $-\sqrt{3}$

77. 0 **79.** 0 **81.** -1 **83.** $\cos\theta = -\dfrac{5}{13}$; $\tan\theta = -\dfrac{12}{5}$; $\csc\theta = \dfrac{13}{12}$; $\sec\theta = -\dfrac{13}{5}$; $\cot\theta = -\dfrac{5}{12}$

85. $\sin\theta = -\dfrac{3}{5}$; $\tan\theta = \dfrac{3}{4}$; $\csc\theta = -\dfrac{5}{3}$; $\sec\theta = -\dfrac{5}{4}$; $\cot\theta = \dfrac{4}{3}$ **87.** $\cos\theta = -\dfrac{12}{13}$; $\tan\theta = -\dfrac{5}{12}$; $\csc\theta = \dfrac{13}{5}$; $\sec\theta = -\dfrac{13}{12}$; $\cot\theta = -\dfrac{12}{5}$

89. $\sin\theta = -\dfrac{2\sqrt{2}}{3}$; $\tan\theta = 2\sqrt{2}$; $\csc\theta = -\dfrac{3\sqrt{2}}{4}$; $\sec\theta = -3$; $\cot\theta = \dfrac{\sqrt{2}}{4}$ **91.** $\cos\theta = -\dfrac{\sqrt{5}}{3}$; $\tan\theta = -\dfrac{2\sqrt{5}}{5}$; $\csc\theta = \dfrac{3}{2}$; $\sec\theta = -\dfrac{3\sqrt{5}}{5}$; $\cot\theta = -\dfrac{\sqrt{5}}{2}$

93. $\sin\theta = -\dfrac{\sqrt{3}}{2}$; $\cos\theta = \dfrac{1}{2}$; $\tan\theta = -\sqrt{3}$; $\csc\theta = -\dfrac{2\sqrt{3}}{3}$; $\cot\theta = -\dfrac{\sqrt{3}}{3}$ **95.** $\sin\theta = -\dfrac{3}{5}$; $\cos\theta = -\dfrac{4}{5}$; $\csc\theta = -\dfrac{5}{3}$; $\sec\theta = -\dfrac{5}{4}$; $\cot\theta = \dfrac{4}{3}$

97. $\sin\theta = \dfrac{\sqrt{10}}{10}$; $\cos\theta = -\dfrac{3\sqrt{10}}{10}$; $\csc\theta = \sqrt{10}$; $\sec\theta = -\dfrac{\sqrt{10}}{3}$; $\cot\theta = -3$ **99.** $\sin\theta = -\dfrac{1}{2}$; $\cos\theta = -\dfrac{\sqrt{3}}{2}$; $\tan\theta = \dfrac{\sqrt{3}}{3}$; $\sec\theta = -\dfrac{2\sqrt{3}}{3}$; $\cot\theta = \sqrt{3}$

101. 0 **103. (a)** $-\dfrac{\sqrt{2}}{2}$; $\left(315°, -\dfrac{\sqrt{2}}{2}\right)$ **(b)** $\sqrt{2}$; $(315°, \sqrt{2})$ **(c)** -1; $(315°, -1)$ **105. (a)** $-\dfrac{\sqrt{3}}{2}$; $\left(\dfrac{7\pi}{6}, -\dfrac{\sqrt{3}}{2}\right)$ **(b)** -2; $\left(\dfrac{7\pi}{6}, -2\right)$ **(c)** 1; $(-315°, 1)$

107. -0.2 **109.** 3 **111.** -5 **113.** 0 **115. (a)** Approximately 16.6 ft **(b)** 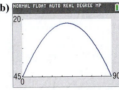 **(c)** 67.5°

119. $\{-10\}$ **120.** 2,476,000 units
121. $f(12) = 5$ **122.** $g(x) = x - 4$

7.5 Assess Your Understanding *(page 561)*

4. 2π; π **5.** b; a **6.** $\dfrac{b}{r}$; $\dfrac{a}{r}$ **7.** -0.2; 0.2 **8.** T **9.** b **10.** a **11.** $\sin t = -\dfrac{1}{2}$; $\cos t = \dfrac{\sqrt{3}}{2}$; $\tan t = -\dfrac{\sqrt{3}}{3}$; $\csc t = -2$; $\sec t = \dfrac{2\sqrt{3}}{3}$; $\cot t = -\sqrt{3}$

13. $\sin t = -\dfrac{\sqrt{2}}{2}$; $\cos t = -\dfrac{\sqrt{2}}{2}$; $\tan t = 1$; $\csc t = -\sqrt{2}$; $\sec t = -\sqrt{2}$; $\cot t = 1$

15. $\sin t = \dfrac{2}{3}$; $\cos t = \dfrac{\sqrt{5}}{3}$; $\tan t = \dfrac{2\sqrt{5}}{5}$; $\csc t = \dfrac{3}{2}$; $\sec t = \dfrac{3\sqrt{5}}{5}$; $\cot t = \dfrac{\sqrt{5}}{2}$ **17.** $\sin\theta = -\dfrac{4}{5}$; $\cos\theta = \dfrac{3}{5}$; $\tan\theta = -\dfrac{4}{3}$; $\csc\theta = -\dfrac{5}{4}$; $\sec\theta = \dfrac{5}{3}$; $\cot\theta = -\dfrac{3}{4}$

19. $\sin\theta = \dfrac{3\sqrt{13}}{13}$; $\cos\theta = -\dfrac{2\sqrt{13}}{13}$; $\tan\theta = -\dfrac{3}{2}$; $\csc\theta = \dfrac{\sqrt{13}}{3}$; $\sec\theta = -\dfrac{\sqrt{13}}{2}$; $\cot\theta = -\dfrac{2}{3}$

21. $\sin\theta = -\dfrac{\sqrt{2}}{2}$; $\cos\theta = -\dfrac{\sqrt{2}}{2}$; $\tan\theta = 1$; $\csc\theta = -\sqrt{2}$; $\sec\theta = -\sqrt{2}$; $\cot\theta = 1$ **23.** $\dfrac{\sqrt{2}}{2}$ **25.** 1 **27.** 1 **29.** $\sqrt{3}$ **31.** $\dfrac{\sqrt{2}}{2}$ **33.** 0 **35.** $\sqrt{2}$

37. $\dfrac{\sqrt{3}}{3}$ **39.** $-\dfrac{\sqrt{3}}{2}$ **41.** $-\dfrac{\sqrt{3}}{3}$ **43.** 2 **45.** -1 **47.** -1 **49.** $\dfrac{\sqrt{2}}{2}$ **51.** 0 **53.** $-\sqrt{2}$ **55.** $\dfrac{2\sqrt{3}}{3}$ **57.** -1 **59.** -2 **61.** $\dfrac{2-\sqrt{2}}{2}$ **63.** $(-\infty, \infty)$

65. At odd multiples of $\dfrac{\pi}{2}$ **67.** At odd multiples of $\dfrac{\pi}{2}$ **69.** $[-1, 1]$ **71.** $(-\infty, \infty)$ **73.** $(-\infty, -1] \cup [1, \infty)$ **75.** Odd; yes; origin

77. Odd; yes; origin **79.** Even; yes; y-axis **81.** 0.9 **83.** 9 **85. (a)** $-\dfrac{1}{3}$ **(b)** 1 **87. (a)** -2 **(b)** 6 **89. (a)** -4 **(b)** -12
91. (a) Using graph: $\sin 1 \approx 0.8$, $\cos 1 \approx 0.5$, $\tan 1 \approx 1.6$, $\csc 1 \approx 1.3$, $\sec 1 \approx 2$, $\cot 1 \approx 0.6$;
    Using calculator: $\sin 1 \approx 0.8$, $\cos 1 \approx 0.5$, $\tan 1 \approx 1.6$, $\csc 1 \approx 1.2$, $\sec 1 \approx 1.9$, $\cot 1 \approx 0.6$
  (b) Using graph: $\sin 5.1 \approx -0.9$, $\cos 5.1 \approx 0.4$, $\tan 5.1 \approx -2.3$, $\csc 5.1 \approx -1.1$, $\sec 5.1 \approx 2.5$, $\cot 5.1 \approx -0.4$;
    Using calculator: $\sin 5.1 \approx -0.9$, $\cos 5.1 \approx 0.4$, $\tan 5.1 \approx -2.4$, $\csc 5.1 \approx -1.1$, $\sec 5.1 \approx 2.6$, $\cot 5.1 \approx -0.4$
93. Suppose that there is a number $p, 0 < p < 2\pi$, for which $\sin(\theta + p) = \sin\theta$ for all θ. If $\theta = 0$, then $\sin(0 + p) = \sin p = \sin 0 = 0$,

so $p = \pi$. If $\theta = \dfrac{\pi}{2}$, then $\sin\left(\dfrac{\pi}{2} + p\right) = \sin\left(\dfrac{\pi}{2}\right)$. But $p = \pi$. Thus $\sin\left(\dfrac{3\pi}{2}\right) = -1 = \sin\left(\dfrac{\pi}{2}\right) = 1$. This is impossible. Therefore, the smallest

positive number p for which $\sin(\theta + p) = \sin\theta$ for all θ is $p = 2\pi$.

95. $\sec\theta = \dfrac{1}{\cos\theta}$; since $\cos\theta$ has period 2π, so does $\sec\theta$.

97. If $P = (a, b)$ is the point on the unit circle corresponding to θ, then $Q = (-a, -b)$ is the point on the unit circle corresponding to $\theta + \pi$.

Thus $\tan(\theta + \pi) = \dfrac{(-b)}{(-a)} = \dfrac{b}{a} = \tan\theta$. If there exists a number $p, 0 < p < \pi$, for which $\tan(\theta + p) = \tan\theta$ for all θ, then if $\theta = 0$,

$\tan(p) = \tan 0 = 0$. But this means that p is a multiple of π. Since no multiple of π exists in the interval $(0, \pi)$, this is impossible.
Therefore, the fundamental period of $f(\theta) = \tan\theta$ is π.

99. Let $P = (x, y)$ be the point on the unit circle that corresponds to t. Consider the equation $\tan t = \dfrac{y}{x} = a$. Then $y = ax$.

But $x^2 + y^2 = 1$, so $x^2 + a^2x^2 = 1$. Thus $x = \pm\dfrac{1}{\sqrt{1 + a^2}}$ and $y = \pm\dfrac{a}{\sqrt{1 + a^2}}$; that is, for any real number a, there is a point $P = (x, y)$ on the

unit circle for which $\tan t = a$. In other words, $-\infty < \tan t < \infty$, and the range of the tangent function is the set of all real numbers.

101. $m = \dfrac{\sin\theta - 0}{\cos\theta - 0} = \dfrac{\sin\theta}{\cos\theta} = \tan\theta$ **106.** $\dfrac{5 - \sqrt{10}}{3}, \dfrac{5 + \sqrt{10}}{3}$ **107.** $y = \dfrac{1}{4}x + \dfrac{31}{4}$ **108.** 6.12%

109. Vertical: $x = -3$; horizontal: $y = 5$

7.6 Assess Your Understanding (page 574)

3. $1; \dfrac{\pi}{2}$ **4.** $3; \pi$ **5.** $3; \dfrac{\pi}{3}$ **6.** T **7.** F **8.** T **9.** d **10.** d **11. (a)** 0 **(b)** $-\dfrac{\pi}{2} < x < \dfrac{\pi}{2}$ **(c)** 1 **(d)** $0, \pi, 2\pi$

(e) $f(x) = 1$ for $x = -\dfrac{3\pi}{2}, \dfrac{\pi}{2}; f(x) = -1$ for $x = -\dfrac{\pi}{2}, \dfrac{3\pi}{2}$ **(f)** $-\dfrac{5\pi}{6}, -\dfrac{\pi}{6}, \dfrac{7\pi}{6}, \dfrac{11\pi}{6}$ **(g)** $\{x \mid x = k\pi, k \text{ an integer}\}$ **13.** Amplitude = 2; period = 2π

15. Amplitude = 4; period = π **17.** Amplitude = 6; period = 2 **19.** Amplitude = $\dfrac{1}{2}$; period = $\dfrac{4\pi}{3}$ **21.** Amplitude = $\dfrac{5}{3}$; period = 3

23. F **25.** A **27.** H **29.** C **31.** J

33.

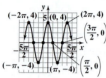

Domain: $(-\infty, \infty)$
Range: $[-4, 4]$

35.

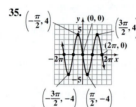

Domain: $(-\infty, \infty)$
Range: $[-4, 4]$

37.

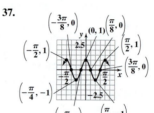

Domain: $(-\infty, \infty)$
Range: $[-1, 1]$

39.

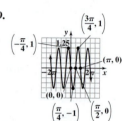

Domain: $(-\infty, \infty)$
Range: $[-1, 1]$

41.

Domain: $(-\infty, \infty)$
Range: $[-2, 2]$

43.

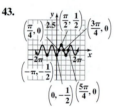

Domain: $(-\infty, \infty)$

Range: $\left[-\dfrac{1}{2}, \dfrac{1}{2}\right]$

45.

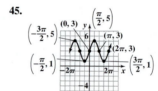

Domain: $(-\infty, \infty)$
Range: $[1, 5]$

47.

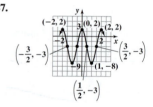

Domain: $(-\infty, \infty)$
Range: $[-8, 2]$

49.

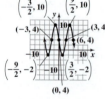

Domain: $(-\infty, \infty)$
Range: $[-2, 10]$

51.

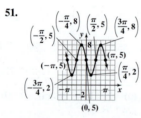

Domain: $(-\infty, \infty)$
Range: $[2, 8]$

53.

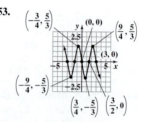

Domain: $(-\infty, \infty)$

Range: $\left[-\dfrac{5}{3}, \dfrac{5}{3}\right]$

55.

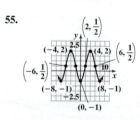

Domain: $(-\infty, \infty)$
Range: $[-1, 2]$

57. $y = \pm 3\sin(2x)$ **59.** $y = \pm 3\sin(\pi x)$ **61.** $y = 5\cos\left(\dfrac{\pi}{4}x\right)$ **63.** $y = -3\cos\left(\dfrac{1}{2}x\right)$ **65.** $y = \dfrac{3}{4}\sin(2\pi x)$ **67.** $y = -\sin\left(\dfrac{3}{2}x\right)$

69. $y = -\cos\left(\dfrac{4\pi}{3}x\right) + 1$ **71.** $y = 3\sin\left(\dfrac{\pi}{2}x\right)$ **73.** $y = -4\cos(3x)$ **75.** $\dfrac{2}{\pi}$ **77.** $\dfrac{\sqrt{2}}{\pi}$

79. $(f \circ g)(x) = \sin(4x)$ **81.** $(f \circ g)(x) = -2 \cos x$ **83.**   **85.** Period $= \dfrac{1}{30}$ s

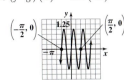

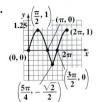

Amplitude $= 220$ amp

$(g \circ f)(x) = 4 \sin x$ $(g \circ f)(x) = \cos(-2x)$

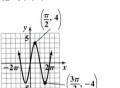

87. (a) Amplitude $= 220$ V   **(c)** $I(t) = 22 \sin(120\pi t)$

Period $= \dfrac{1}{60}$ s   **(d)** Amplitude $= 22$ amp

(b), (e)   Period $= \dfrac{1}{60}$ s

89. (a) $P(t) = \dfrac{[V_0 \sin(2\pi ft)]^2}{R} = \dfrac{V_0^2}{R}\sin^2(2\pi ft)$ **(b)** Since the graph of P has amplitude $\dfrac{V_0^2}{2R}$ and period $\dfrac{1}{2f}$ and is of the form $y = A\cos(\omega t) + B$,

then $A = -\dfrac{V_0^2}{2R}$ and $B = \dfrac{V_0^2}{2R}$. Since $\dfrac{1}{2f} = \dfrac{2\pi}{\omega}$, then $\omega = 4\pi f$. Therefore, $P(t) = -\dfrac{V_0^2}{2R}\cos(4\pi ft) + \dfrac{V_0^2}{2R} = \dfrac{V_0^2}{2R}[1 - \cos(4\pi ft)]$.

91. (a) Physical potential: $\omega = \dfrac{2\pi}{23}$; emotional potential: $\omega = \dfrac{\pi}{14}$; intellectual potential: $\omega = \dfrac{2\pi}{33}$

(b)

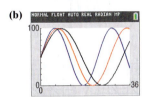

(c) No   **(d)** Physical potential peaks at 15 days after 20th birthday. Emotional potential is 50% at 17 days, with a maximum at 10 days and a minimum at 24 days. Intellectual potential starts fairly high, drops to a minimum at 13 days, and rises to a maximum at 29 days.   **93.**

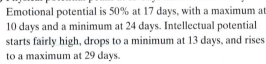

95. Answers may vary. $\left(-\dfrac{5\pi}{3}, \dfrac{1}{2}\right), \left(-\dfrac{\pi}{3}, \dfrac{1}{2}\right), \left(\dfrac{\pi}{3}, \dfrac{1}{2}\right), \left(\dfrac{5\pi}{3}, \dfrac{1}{2}\right)$ **97.** Answers may vary. $\left(-\dfrac{3\pi}{4}, 1\right), \left(\dfrac{\pi}{4}, 1\right), \left(\dfrac{5\pi}{4}, 1\right), \left(\dfrac{9\pi}{4}, 1\right)$

103. $2x + h - 5$ **104.** $(2, 5)$ **105.** $(0, 5), \left(-\dfrac{5}{3}, 0\right), \left(-\dfrac{7}{3}, 0\right)$ **106.** $\{-8\}$

7.7 Assess Your Understanding *(page 584)*

3. origin; odd multiples of $\dfrac{\pi}{2}$ **4.** y-axis; odd multiples of $\dfrac{\pi}{2}$ **5.** b   **6.** T   **7.** 0   **9.** 1

11. $\sec x = 1$ for $x = -2\pi, 0, 2\pi$; $\sec x = -1$ for $x = -\pi, \pi$ **13.** $-\dfrac{3\pi}{2}, -\dfrac{\pi}{2}, \dfrac{\pi}{2}, \dfrac{3\pi}{2}$ **15.** $-\dfrac{3\pi}{2}, -\dfrac{\pi}{2}, \dfrac{\pi}{2}, \dfrac{3\pi}{2}$

17.

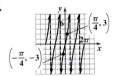

Domain: $\left\{ x \,\middle|\, x \neq \dfrac{k\pi}{2}, k \text{ is an odd integer} \right\}$
Range: $(-\infty, \infty)$

19.

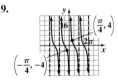

Domain: $\{x \,|\, x \neq k\pi, k \text{ is an integer}\}$
Range: $(-\infty, \infty)$

21.

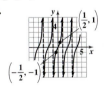

Domain: $\{x \,|\, x \text{ does not equal an odd integer}\}$
Range: $(-\infty, \infty)$

23.

Domain: $\{x \,|\, x \neq 4k\pi, k \text{ is an integer}\}$
Range: $(-\infty, \infty)$

25.

Domain: $\left\{ x \,\middle|\, x \neq \dfrac{k\pi}{2}, k \text{ is an odd integer} \right\}$
Range: $\{y \,|\, y \leq -2 \text{ or } y \geq 2\}$

27.

Domain: $\{x \,|\, x \neq k\pi, k \text{ is an integer}\}$
Range: $\{y \,|\, y \leq -3 \text{ or } y \geq 3\}$

8. $f^{-1}(x) = \frac{1}{3}(x + 2)$ **9.** -2

10.
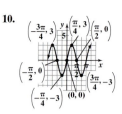

11. $3 - \frac{3\sqrt{3}}{2}$ **12.** $y = 2(3^x)$ **13.** $y = 3\cos\left(\frac{\pi}{6}x\right)$

14. (a) $f(x) = -3x - 3$;
$m = -3$; $(-1, 0)$, $(0, -3)$

(b) $f(x) = (x - 1)^2 - 6$; $(0, -5)$,
$(-\sqrt{6} + 1, 0)$, $(\sqrt{6} + 1, 0)$

(c) We have that $y = 3$ when $x = -2$ and $y = -6$ when $x = 1$. Both points satisfy $y = ae^x$. Therefore, for $(-2, 3)$ we have $3 = ae^{-2}$, which implies that $a = 3e^2$. But for $(1, -6)$ we have $-6 = ae^1$, which implies that $a = -6e^{-1}$. Therefore, there is no exponential function $y = ae^x$ that contains $(-2, 3)$ and $(1, -6)$.

15. (a) $f(x) = \frac{1}{6}(x + 2)(x - 3)(x - 5)$

(b) $R(x) = -\frac{(x + 2)(x - 3)(x - 5)}{3(x - 2)}$

CHAPTER 8 Analytic Trigonometry

8.1 Assess Your Understanding (page 616)

7. $x = \sin y$ **8.** $0 \le x \le \pi$ **9.** $-\infty < x < \infty$ **10.** F   **11.** T   **12.** T   **13.** d   **14.** a   **15.** 0   **17.** $-\frac{\pi}{2}$ **19.** 0   **21.** $\frac{\pi}{4}$ **23.** $\frac{\pi}{3}$ **25.** $\frac{5\pi}{6}$ **27.** 0. 10   **29.** 1.37

31. 0.51   **33.** -0.38 **35.** -0.12 **37.** 1.08   **39.** $\frac{4\pi}{5}$ **41.** $-\frac{3\pi}{8}$ **43.** $-\frac{\pi}{8}$ **45.** $-\frac{\pi}{5}$ **47.** $\frac{\pi}{4}$ **49.** Not defined   **51.** $\frac{1}{4}$ **53.** 4   **55.** Not defined   **57.** π

59. $f^{-1}(x) = \sin^{-1}\frac{x - 2}{5}$
Range of f = Domain of $f^{-1} = [-3, 7]$
Range of $f^{-1} = \left[-\frac{\pi}{2}, \frac{\pi}{2}\right]$

61. $f^{-1}(x) = \frac{1}{3}\cos^{-1}\left(-\frac{x}{2}\right)$
Range of f = Domain of $f^{-1} = [-2, 2]$
Range of $f^{-1} = \left[0, \frac{\pi}{3}\right]$

63. $f^{-1}(x) = -\tan^{-1}(x + 3) - 1$
Range of f = Domain of $f^{-1} = (-\infty, \infty)$
Range of $f^{-1} = \left(-1 - \frac{\pi}{2}, \frac{\pi}{2} - 1\right)$

65. $f^{-1}(x) = \frac{1}{2}\left[\sin^{-1}\left(\frac{x}{3}\right) - 1\right]$
Range of f = Domain of $f^{-1} = [-3, 3]$
Range of $f^{-1} = \left[-\frac{1}{2} - \frac{\pi}{4}, -\frac{1}{2} + \frac{\pi}{4}\right]$

67. $\left\{\frac{\sqrt{2}}{2}\right\}$ **69.** $\left\{-\frac{1}{4}\right\}$ **71.** $\{\sqrt{3}\}$ **73.** $\{-1\}$

75. (a) 13.92 h or 13 h, 55 min   **(b)** 12 h   **(c)** 13.85 h or 13 h, 51 min
77. (a) 13.3 h or 13 h, 18 min   **(b)** 12 h   **(c)** 13.26 h or 13 h, 15 min
79. (a) 12 h   **(b)** 12 h   **(c)** 12 h   **(d)** It is 12 h.   **81.** 3.35 min

83. (a) $\frac{\pi}{3}$ square units   **(b)** $\frac{5\pi}{12}$ square units   **85.** 4250 mi   **87.** $\left[-\frac{2}{3}, 2\right]$ **88.** The graph passes the horizontal-line test.

89. $f^{-1}(x) = \log_2(x - 1)$ **90.** $(2x + 1)^{-\frac{1}{2}}(x^2 + 3)^{-\frac{3}{2}}(-x^2 - x + 3)$

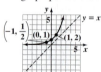

8.2 Assess Your Understanding (page 623)

4. $x = \sec y$; ≥ 1; 0; π **5.** cosine   **6.** F   **7.** T   **8.** T   **9.** $\frac{\sqrt{2}}{2}$ **11.** $-\frac{\sqrt{3}}{3}$ **13.** 2   **15.** $\sqrt{2}$ **17.** $-\frac{\sqrt{2}}{2}$ **19.** $\frac{2\sqrt{3}}{3}$ **21.** $\frac{3\pi}{4}$ **23.** $-\frac{\pi}{3}$ **25.** $\frac{\sqrt{2}}{4}$ **27.** $\frac{\sqrt{5}}{2}$

29. $-\frac{\sqrt{14}}{2}$ **31.** $-\frac{3\sqrt{10}}{10}$ **33.** $\sqrt{5}$ **35.** $-\frac{\pi}{4}$ **37.** $\frac{\pi}{6}$ **39.** $-\frac{\pi}{2}$ **41.** $\frac{\pi}{6}$ **43.** $\frac{2\pi}{3}$ **45.** 1.32   **47.** 0.46   **49.** -0.34 **51.** 2.72   **53.** -0.73 **55.** 2.55

57. $\frac{1}{\sqrt{1 + u^2}}$ **59.** $\frac{u}{\sqrt{1 - u^2}}$ **61.** $\frac{\sqrt{u^2 - 1}}{|u|}$ **63.** $\frac{\sqrt{u^2 - 1}}{|u|}$ **65.** $\frac{1}{u}$ **67.** $\frac{5}{13}$ **69.** $\frac{3\pi}{4}$ **71.** $-\frac{3}{4}$ **73.** $\frac{5}{13}$ **75.** $\frac{5\pi}{6}$ **77.** $-\sqrt{15}$

79. (a) $\theta = 31.89°$ **(b)** 54.64 ft in diameter   **(c)** 37.96 ft high   **81. (a)** $\theta = 22.3°$ **(b)** $v_0 = 2940.23$ ft/s

83. 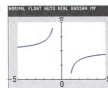 **87.** $-5i, 5i, -2, 2$ **88.** Neither **89.** $\dfrac{7\pi}{4}$ **90.** $\dfrac{5\pi}{2} \approx 7.85$ in.

8.3 Assess Your Understanding *(page 630)*

7. F **8.** T **9.** T **10.** F **11.** d **12.** a **13.** $\left\{\dfrac{7\pi}{6}, \dfrac{11\pi}{6}\right\}$ **15.** $\left\{\dfrac{7\pi}{6}, \dfrac{11\pi}{6}\right\}$ **17.** $\left\{\dfrac{3\pi}{4}, \dfrac{7\pi}{4}\right\}$ **19.** $\left\{\dfrac{2\pi}{3}, \dfrac{4\pi}{3}\right\}$ **21.** $\left\{\dfrac{3\pi}{4}, \dfrac{5\pi}{4}\right\}$ **23.** $\left\{\dfrac{\pi}{3}, \dfrac{2\pi}{3}, \dfrac{4\pi}{3}, \dfrac{5\pi}{3}\right\}$

25. $\left\{\dfrac{\pi}{4}, \dfrac{3\pi}{4}, \dfrac{5\pi}{4}, \dfrac{7\pi}{4}\right\}$ **27.** $\left\{\dfrac{\pi}{2}, \dfrac{7\pi}{6}, \dfrac{11\pi}{6}\right\}$ **29.** $\left\{\dfrac{\pi}{3}, \dfrac{2\pi}{3}, \dfrac{4\pi}{3}, \dfrac{5\pi}{3}\right\}$ **31.** $\left\{\dfrac{4\pi}{9}, \dfrac{8\pi}{9}, \dfrac{16\pi}{9}\right\}$ **33.** $\left\{\dfrac{3\pi}{4}, \dfrac{7\pi}{4}\right\}$ **35.** $\left\{\dfrac{11\pi}{6}\right\}$

37. $\left\{\theta \,\middle|\, \theta = \dfrac{\pi}{6} + 2k\pi, \theta = \dfrac{5\pi}{6} + 2k\pi\right\}; \dfrac{\pi}{6}, \dfrac{5\pi}{6}, \dfrac{13\pi}{6}, \dfrac{17\pi}{6}, \dfrac{25\pi}{6}, \dfrac{29\pi}{6}$ **39.** $\left\{\theta \,\middle|\, \theta = \dfrac{5\pi}{6} + k\pi\right\}; \dfrac{5\pi}{6}, \dfrac{11\pi}{6}, \dfrac{17\pi}{6}, \dfrac{23\pi}{6}, \dfrac{29\pi}{6}, \dfrac{35\pi}{6}$

41. $\left\{\theta \,\middle|\, \theta = \dfrac{\pi}{2} + 2k\pi, \theta = \dfrac{3\pi}{2} + 2k\pi\right\}; \dfrac{\pi}{2}, \dfrac{3\pi}{2}, \dfrac{5\pi}{2}, \dfrac{7\pi}{2}, \dfrac{9\pi}{2}, \dfrac{11\pi}{2}$ **43.** $\left\{\theta \,\middle|\, \theta = \dfrac{\pi}{3} + k\pi, \theta = \dfrac{2\pi}{3} + k\pi\right\}; \dfrac{\pi}{3}, \dfrac{2\pi}{3}, \dfrac{4\pi}{3}, \dfrac{5\pi}{3}, \dfrac{7\pi}{3}, \dfrac{8\pi}{3}$

45. $\left\{\theta \,\middle|\, \theta = \dfrac{8\pi}{3} + 4k\pi, \theta = \dfrac{10\pi}{3} + 4k\pi\right\}; \dfrac{8\pi}{3}, \dfrac{10\pi}{3}, \dfrac{20\pi}{3}, \dfrac{22\pi}{3}, \dfrac{32\pi}{3}, \dfrac{34\pi}{3}$ **47.** $\{0.41, 2.73\}$ **49.** $\{1.37, 4.51\}$ **51.** $\{2.69, 3.59\}$ **53.** $\{1.82, 4.46\}$

55. $\{2.08, 5.22\}$ **57.** $\{0.73, 2.41\}$ **59.** $\left\{\dfrac{\pi}{2}, \dfrac{2\pi}{3}, \dfrac{4\pi}{3}, \dfrac{3\pi}{2}\right\}$ **61.** $\left\{\dfrac{\pi}{2}, \dfrac{7\pi}{6}, \dfrac{11\pi}{6}\right\}$ **63.** $\left\{0, \dfrac{\pi}{4}, \dfrac{5\pi}{4}\right\}$ **65.** $\left\{\dfrac{\pi}{2}, \dfrac{2\pi}{3}, \dfrac{4\pi}{3}, \dfrac{3\pi}{2}\right\}$ **67.** $\{\pi\}$ **69.** $\left\{\dfrac{\pi}{4}, \dfrac{5\pi}{4}\right\}$

71. $\left\{0, \dfrac{\pi}{3}, \pi, \dfrac{5\pi}{3}\right\}$ **73.** $\left\{\dfrac{\pi}{6}, \dfrac{5\pi}{6}, \dfrac{3\pi}{2}\right\}$ **75.** $\left\{\dfrac{\pi}{2}\right\}$ **77.** $\{0\}$ **79.** $\left\{\dfrac{\pi}{3}, \dfrac{5\pi}{3}\right\}$ **81.** No real solution **83.** $-1.31, 1.98, 3.84$ **85.** 0.52

87. 1.26 **89.** $-1.02, 1.02$ **91.** $0, 2.15$ **93.** $0.76, 1.35$ **95.** $\dfrac{\pi}{3}, \dfrac{2\pi}{3}, \dfrac{4\pi}{3}, \dfrac{5\pi}{3}$

97. (a) $-2\pi, -\pi, 0, \pi, 2\pi, 3\pi, 4\pi$ **(b)**

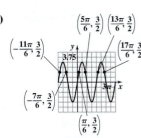

(c) $\left\{-\dfrac{11\pi}{6}, -\dfrac{7\pi}{6}, \dfrac{\pi}{6}, \dfrac{5\pi}{6}, \dfrac{13\pi}{6}, \dfrac{17\pi}{6}\right\}$

(d) $\left\{x \,\middle|\, -\dfrac{11\pi}{6} < x < -\dfrac{7\pi}{6} \text{ or } \dfrac{\pi}{6} < x < \dfrac{5\pi}{6} \text{ or } \dfrac{13\pi}{6} < x < \dfrac{17\pi}{6}\right\}$

99. (a) $\left\{x \,\middle|\, x = -\dfrac{\pi}{4} + k\pi, k \text{ is any integer}\right\}$ **(b)** $-\dfrac{\pi}{2} < x < -\dfrac{\pi}{4}$ or $\left(-\dfrac{\pi}{2}, -\dfrac{\pi}{4}\right)$

101. (a), (d) **(b)** $\left\{\dfrac{\pi}{12}, \dfrac{5\pi}{12}\right\}$

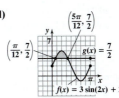

103. (a), (d)

(b) $\left\{\dfrac{2\pi}{3}, \dfrac{4\pi}{3}\right\}$

(c) $\left\{x \,\middle|\, \dfrac{\pi}{12} < x < \dfrac{5\pi}{12}\right\}$ or $\left(\dfrac{\pi}{12}, \dfrac{5\pi}{12}\right)$

(c) $\left\{x \,\middle|\, \dfrac{2\pi}{3} < x < \dfrac{4\pi}{3}\right\}$ or $\left(\dfrac{2\pi}{3}, \dfrac{4\pi}{3}\right)$

105. (a) 0 s, 0.43 s, 0.86 s **(b)** 0.21 s **(c)** $[0, 0.03] \cup [0.39, 0.43] \cup [0.86, 0.89]$ **107. (a)** 150 mi **(b)** $6.06, 8.44, 15.72, 18.11$ min **(c)** Before 6.06 min, between 8.44 and 15.72 min, and after 18.11 min **(d)** No **109.** $2.03, 4.91$

111. (a) $30°, 60°$ **(b)** 123.6 m **113.** $28.90°$ **115.** Yes; it varies from 1.25 to 1.34. **117.** 1.47

(c)

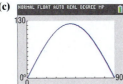

119. If θ is the original angle of incidence and ϕ is the angle of refraction, then $\dfrac{\sin\theta}{\sin\phi} = n_2$.

The angle of incidence of the emerging beam is also ϕ, and the index of refraction is $\dfrac{1}{n_2}$.

Thus, θ is the angle of refraction of the emerging beam.

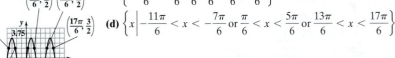

123. $x = \log_6 y$ **124.** $\dfrac{9 - \sqrt{17}}{4}, \dfrac{9 + \sqrt{17}}{4}$ **125.** $\tan\theta = -\dfrac{1}{3}; \csc\theta = -\sqrt{10}; \sec\theta = \dfrac{\sqrt{10}}{3}; \cot\theta = -3$ **126.** Amplitude: 2

Period: π

Phase shift: $\dfrac{\pi}{2}$

8.4 Assess Your Understanding (page 640)

3. identity; conditional   **4.** -1 **5.** 0   **6.** T   **7.** F   **8.** T   **9.** c   **10.** b   **11.** $\dfrac{1}{\cos\theta}$ **13.** $\dfrac{1 + \sin\theta}{\cos\theta}$ **15.** $\dfrac{1}{\sin\theta\cos\theta}$ **17.** 2   **19.** $\dfrac{3\sin\theta + 1}{\sin\theta + 1}$

21. $\csc\theta \cdot \cos\theta = \dfrac{1}{\sin\theta} \cdot \cos\theta = \dfrac{\cos\theta}{\sin\theta} = \cot\theta$ **23.** $1 + \tan^2(-\theta) = 1 + (-\tan\theta)^2 = 1 + \tan^2\theta = \sec^2\theta$

25. $\cos\theta(\tan\theta + \cot\theta) = \cos\theta\left(\dfrac{\sin\theta}{\cos\theta} + \dfrac{\cos\theta}{\sin\theta}\right) = \cos\theta\left(\dfrac{\sin^2\theta + \cos^2\theta}{\cos\theta\sin\theta}\right) = \cos\theta\left(\dfrac{1}{\cos\theta\sin\theta}\right) = \dfrac{1}{\sin\theta} = \csc\theta$

27. $\tan u\cot u - \cos^2 u = \tan u \cdot \dfrac{1}{\tan u} - \cos^2 u = 1 - \cos^2 u = \sin^2 u$ **29.** $(\sec\theta - 1)(\sec\theta + 1) = \sec^2\theta - 1 = \tan^2\theta$

31. $(\sec\theta + \tan\theta)(\sec\theta - \tan\theta) = \sec^2\theta - \tan^2\theta = 1$ **33.** $\cos^2\theta(1 + \tan^2\theta) = \cos^2\theta\sec^2\theta = \cos^2\theta \cdot \dfrac{1}{\cos^2\theta} = 1$

35. $(\sin\theta + \cos\theta)^2 + (\sin\theta - \cos\theta)^2 = \sin^2\theta + 2\sin\theta\cos\theta + \cos^2\theta + \sin^2\theta - 2\sin\theta\cos\theta + \cos^2\theta$

$$= \sin^2\theta + \cos^2\theta + \sin^2\theta + \cos^2\theta = 1 + 1 = 2$$

37. $\sec^4\theta - \sec^2\theta = \sec^2\theta(\sec^2\theta - 1) = (1 + \tan^2\theta)\tan^2\theta = \tan^4\theta + \tan^2\theta$

39. $\sec u - \tan u = \dfrac{1}{\cos u} - \dfrac{\sin u}{\cos u} = \dfrac{1 - \sin u}{\cos u} \cdot \dfrac{1 + \sin u}{1 + \sin u} = \dfrac{1 - \sin^2 u}{\cos u(1 + \sin u)} = \dfrac{\cos^2 u}{\cos u(1 + \sin u)} = \dfrac{\cos u}{1 + \sin u}$

41. $3\sin^2\theta + 4\cos^2\theta = 3\sin^2\theta + 3\cos^2\theta + \cos^2\theta = 3(\sin^2\theta + \cos^2\theta) + \cos^2\theta = 3 + \cos^2\theta$

43. $1 - \dfrac{\cos^2\theta}{1 + \sin\theta} = 1 - \dfrac{1 - \sin^2\theta}{1 + \sin\theta} = 1 - \dfrac{(1 + \sin\theta)(1 - \sin\theta)}{1 + \sin\theta} = 1 - (1 - \sin\theta) = \sin\theta$

45. $\dfrac{1 + \tan v}{1 - \tan v} = \dfrac{1 + \dfrac{1}{\cot v}}{1 - \dfrac{1}{\cot v}} = \dfrac{\dfrac{\cot v + 1}{\cot v}}{\dfrac{\cot v - 1}{\cot v}} = \dfrac{\cot v + 1}{\cot v - 1}$ **47.** $\dfrac{\sec\theta}{\csc\theta} + \dfrac{\sin\theta}{\cos\theta} = \dfrac{\dfrac{1}{\cos\theta}}{\dfrac{1}{\sin\theta}} + \tan\theta = \dfrac{\sin\theta}{\cos\theta} + \tan\theta = \tan\theta + \tan\theta = 2\tan\theta$

49. $\dfrac{1 + \sin\theta}{1 - \sin\theta} = \dfrac{1 + \dfrac{1}{\csc\theta}}{1 - \dfrac{1}{\csc\theta}} = \dfrac{\dfrac{\csc\theta + 1}{\csc\theta}}{\dfrac{\csc\theta - 1}{\csc\theta}} = \dfrac{\csc\theta + 1}{\csc\theta - 1}$

51. $\dfrac{1 - \sin v}{\cos v} + \dfrac{\cos v}{1 - \sin v} = \dfrac{(1 - \sin v)^2 + \cos^2 v}{\cos v(1 - \sin v)} = \dfrac{1 - 2\sin v + \sin^2 v + \cos^2 v}{\cos v(1 - \sin v)} = \dfrac{2 - 2\sin v}{\cos v(1 - \sin v)} = \dfrac{2(1 - \sin v)}{\cos v(1 - \sin v)} = \dfrac{2}{\cos v} = 2\sec v$

53. $\dfrac{\sin\theta}{\sin\theta - \cos\theta} = \dfrac{1}{\dfrac{\sin\theta - \cos\theta}{\sin\theta}} = \dfrac{1}{1 - \dfrac{\cos\theta}{\sin\theta}} = \dfrac{1}{1 - \cot\theta}$

55. $(\sec\theta - \tan\theta)^2 = \sec^2\theta - 2\sec\theta\tan\theta + \tan^2\theta = \dfrac{1}{\cos^2\theta} - \dfrac{2\sin\theta}{\cos^2\theta} + \dfrac{\sin^2\theta}{\cos^2\theta} = \dfrac{1 - 2\sin\theta + \sin^2\theta}{\cos^2\theta} = \dfrac{(1 - \sin\theta)^2}{1 - \sin^2\theta} = \dfrac{(1 - \sin\theta)^2}{(1 - \sin\theta)(1 + \sin\theta)}$

$$= \dfrac{1 - \sin\theta}{1 + \sin\theta}$$

57. $\dfrac{\cos\theta}{1 - \tan\theta} + \dfrac{\sin\theta}{1 - \cot\theta} = \dfrac{\cos\theta}{1 - \dfrac{\sin\theta}{\cos\theta}} + \dfrac{\sin\theta}{1 - \dfrac{\cos\theta}{\sin\theta}} = \dfrac{\cos\theta}{\dfrac{\cos\theta - \sin\theta}{\cos\theta}} + \dfrac{\sin\theta}{\dfrac{\sin\theta - \cos\theta}{\sin\theta}} = \dfrac{\cos^2\theta}{\cos\theta - \sin\theta} + \dfrac{\sin^2\theta}{\sin\theta - \cos\theta}$

$$= \dfrac{\cos^2\theta - \sin^2\theta}{\cos\theta - \sin\theta} = \dfrac{(\cos\theta - \sin\theta)(\cos\theta + \sin\theta)}{\cos\theta - \sin\theta} = \sin\theta + \cos\theta$$

59. $\tan\theta + \dfrac{\cos\theta}{1 + \sin\theta} = \dfrac{\sin\theta}{\cos\theta} + \dfrac{\cos\theta}{1 + \sin\theta} = \dfrac{\sin\theta(1 + \sin\theta) + \cos^2\theta}{\cos\theta(1 + \sin\theta)} = \dfrac{\sin\theta + \sin^2\theta + \cos^2\theta}{\cos\theta(1 + \sin\theta)} = \dfrac{\sin\theta + 1}{\cos\theta(1 + \sin\theta)} = \dfrac{1}{\cos\theta} = \sec\theta$

61. $\dfrac{\tan\theta + \sec\theta - 1}{\tan\theta - \sec\theta + 1} = \dfrac{\tan\theta + (\sec\theta - 1)}{\tan\theta - (\sec\theta - 1)} \cdot \dfrac{\tan\theta + (\sec\theta - 1)}{\tan\theta + (\sec\theta - 1)} = \dfrac{\tan^2\theta + 2\tan\theta(\sec\theta - 1) + \sec^2\theta - 2\sec\theta + 1}{\tan^2\theta - (\sec^2\theta - 2\sec\theta + 1)}$

$$= \dfrac{\sec^2\theta - 1 + 2\tan\theta(\sec\theta - 1) + \sec^2\theta - 2\sec\theta + 1}{\sec^2\theta - 1 - \sec^2\theta + 2\sec\theta - 1} = \dfrac{2\sec^2\theta - 2\sec\theta + 2\tan\theta(\sec\theta - 1)}{-2 + 2\sec\theta}$$

$$= \dfrac{2\sec\theta(\sec\theta - 1) + 2\tan\theta(\sec\theta - 1)}{2(\sec\theta - 1)} = \dfrac{2(\sec\theta - 1)(\sec\theta + \tan\theta)}{2(\sec\theta - 1)} = \tan\theta + \sec\theta$

63. $\dfrac{\tan\theta - \cot\theta}{\tan\theta + \cot\theta} = \dfrac{\dfrac{\sin\theta}{\cos\theta} - \dfrac{\cos\theta}{\sin\theta}}{\dfrac{\sin\theta}{\cos\theta} + \dfrac{\cos\theta}{\sin\theta}} = \dfrac{\dfrac{\sin^2\theta - \cos^2\theta}{\cos\theta\sin\theta}}{\dfrac{\sin^2\theta + \cos^2\theta}{\cos\theta\sin\theta}} = \dfrac{\sin^2\theta - \cos^2\theta}{1} = \sin^2\theta - \cos^2\theta$

65. $\dfrac{\tan u - \cot u}{\tan u + \cot u} + 1 = \dfrac{\dfrac{\sin u}{\cos u} - \dfrac{\cos u}{\sin u}}{\dfrac{\sin u}{\cos u} + \dfrac{\cos u}{\sin u}} + 1 = \dfrac{\dfrac{\sin^2 u - \cos^2 u}{\cos u\sin u}}{\dfrac{\sin^2 u + \cos^2 u}{\cos u\sin u}} + 1 = \sin^2 u - \cos^2 u + 1 = \sin^2 u + (1 - \cos^2 u) = 2\sin^2 u$

67. $\dfrac{\sec\theta + \tan\theta}{\cot\theta + \cos\theta} = \dfrac{\dfrac{1}{\cos\theta} + \dfrac{\sin\theta}{\cos\theta}}{\dfrac{\cos\theta}{\sin\theta} + \cos\theta} = \dfrac{\dfrac{1 + \sin\theta}{\cos\theta}}{\dfrac{\cos\theta + \cos\theta\sin\theta}{\sin\theta}} = \dfrac{1 + \sin\theta}{\cos\theta} \cdot \dfrac{\sin\theta}{\cos\theta(1 + \sin\theta)} = \dfrac{\sin\theta}{\cos\theta} \cdot \dfrac{1}{\cos\theta} = \tan\theta\sec\theta$

69. $\dfrac{1 - \tan^2\theta}{1 + \tan^2\theta} + 1 = \dfrac{1 - \tan^2\theta + 1 + \tan^2\theta}{1 + \tan^2\theta} = \dfrac{2}{1 + \tan^2\theta} = \dfrac{2}{\sec^2\theta} = 2\cos^2\theta$

71. $\dfrac{\sec\theta - \csc\theta}{\sec\theta\csc\theta} = \dfrac{\sec\theta}{\sec\theta\csc\theta} - \dfrac{\csc\theta}{\sec\theta\csc\theta} = \dfrac{1}{\csc\theta} - \dfrac{1}{\sec\theta} = \sin\theta - \cos\theta$

73. $\sec\theta - \cos\theta = \dfrac{1}{\cos\theta} - \cos\theta = \dfrac{1 - \cos^2\theta}{\cos\theta} = \dfrac{\sin^2\theta}{\cos\theta} = \sin\theta \cdot \dfrac{\sin\theta}{\cos\theta} = \sin\theta\tan\theta$

75. $\dfrac{1}{1 - \sin\theta} + \dfrac{1}{1 + \sin\theta} = \dfrac{1 + \sin\theta + 1 - \sin\theta}{(1 + \sin\theta)(1 - \sin\theta)} = \dfrac{2}{1 - \sin^2\theta} = \dfrac{2}{\cos^2\theta} = 2\sec^2\theta$

77. $\dfrac{\sec\theta}{1 - \sin\theta} = \dfrac{\sec\theta}{1 - \sin\theta} \cdot \dfrac{1 + \sin\theta}{1 + \sin\theta} = \dfrac{\sec\theta(1 + \sin\theta)}{1 - \sin^2\theta} = \dfrac{\sec\theta(1 + \sin\theta)}{\cos^2\theta} = \dfrac{1 + \sin\theta}{\cos^3\theta}$

79. $\dfrac{(\sec v - \tan v)^2 + 1}{\csc v(\sec v - \tan v)} = \dfrac{\sec^2 v - 2\sec v\tan v + \tan^2 v + 1}{\dfrac{1}{\sin v}\left(\dfrac{1}{\cos v} - \dfrac{\sin v}{\cos v}\right)} = \dfrac{2\sec^2 v - 2\sec v\tan v}{\dfrac{1}{\sin v}\left(\dfrac{1 - \sin v}{\cos v}\right)} = \dfrac{\dfrac{2}{\cos^2 v} - \dfrac{2\sin v}{\cos^2 v}}{\dfrac{1 - \sin v}{\sin v\cos v}} = \dfrac{2 - 2\sin v}{\cos^2 v} \cdot \dfrac{\sin v\cos v}{1 - \sin v}$

$= \dfrac{2(1 - \sin v)}{\cos v} \cdot \dfrac{\sin v}{1 - \sin v} = \dfrac{2\sin v}{\cos v} = 2\tan v$

81. $\dfrac{\sin\theta + \cos\theta}{\cos\theta} - \dfrac{\sin\theta - \cos\theta}{\sin\theta} = \dfrac{\sin\theta}{\cos\theta} + 1 - 1 + \dfrac{\cos\theta}{\sin\theta} = \dfrac{\sin^2\theta + \cos^2\theta}{\cos\theta\sin\theta} = \dfrac{1}{\cos\theta\sin\theta} = \sec\theta\csc\theta$

83. $\dfrac{\sin^3\theta + \cos^3\theta}{\sin\theta + \cos\theta} = \dfrac{(\sin\theta + \cos\theta)(\sin^2\theta - \sin\theta\cos\theta + \cos^2\theta)}{\sin\theta + \cos\theta} = \sin^2\theta + \cos^2\theta - \sin\theta\cos\theta = 1 - \sin\theta\cos\theta$

85. $\dfrac{\cos^2\theta - \sin^2\theta}{1 - \tan^2\theta} = \dfrac{\cos^2\theta - \sin^2\theta}{1 - \dfrac{\sin^2\theta}{\cos^2\theta}} = \dfrac{\cos^2\theta - \sin^2\theta}{\dfrac{\cos^2\theta - \sin^2\theta}{\cos^2\theta}} = \cos^2\theta$

87. $\dfrac{(2\cos^2\theta - 1)^2}{\cos^4\theta - \sin^4\theta} = \dfrac{[2\cos^2\theta - (\sin^2\theta + \cos^2\theta)]^2}{(\cos^2\theta - \sin^2\theta)(\cos^2\theta + \sin^2\theta)} = \dfrac{(\cos^2\theta - \sin^2\theta)^2}{\cos^2\theta - \sin^2\theta} = \cos^2\theta - \sin^2\theta = (1 - \sin^2\theta) - \sin^2\theta = 1 - 2\sin^2\theta$

89. $\dfrac{1 + \sin\theta + \cos\theta}{1 + \sin\theta - \cos\theta} = \dfrac{(1 + \sin\theta) + \cos\theta}{(1 + \sin\theta) - \cos\theta} \cdot \dfrac{(1 + \sin\theta) + \cos\theta}{(1 + \sin\theta) + \cos\theta} = \dfrac{1 + 2\sin\theta + \sin^2\theta + 2(1 + \sin\theta)\cos\theta + \cos^2\theta}{1 + 2\sin\theta + \sin^2\theta - \cos^2\theta}$

$= \dfrac{1 + 2\sin\theta + \sin^2\theta + 2(1 + \sin\theta)(\cos\theta) + (1 - \sin^2\theta)}{1 + 2\sin\theta + \sin^2\theta - (1 - \sin^2\theta)} = \dfrac{2 + 2\sin\theta + 2(1 + \sin\theta)(\cos\theta)}{2\sin\theta + 2\sin^2\theta}$

$= \dfrac{2(1 + \sin\theta) + 2(1 + \sin\theta)(\cos\theta)}{2\sin\theta(1 + \sin\theta)} = \dfrac{2(1 + \sin\theta)(1 + \cos\theta)}{2\sin\theta(1 + \sin\theta)} = \dfrac{1 + \cos\theta}{\sin\theta}$

91. $(a\sin\theta + b\cos\theta)^2 + (a\cos\theta - b\sin\theta)^2 = a^2\sin^2\theta + 2ab\sin\theta\cos\theta + b^2\cos^2\theta + a^2\cos^2\theta - 2ab\sin\theta\cos\theta + b^2\sin^2\theta$

$= a^2(\sin^2\theta + \cos^2\theta) + b^2(\cos^2\theta + \sin^2\theta) = a^2 + b^2$

93. $\dfrac{\tan\alpha + \tan\beta}{\cot\alpha + \cot\beta} = \dfrac{\tan\alpha + \tan\beta}{\dfrac{1}{\tan\alpha} + \dfrac{1}{\tan\beta}} = \dfrac{\tan\alpha + \tan\beta}{\dfrac{\tan\beta + \tan\alpha}{\tan\alpha\tan\beta}} = (\tan\alpha + \tan\beta) \cdot \dfrac{\tan\alpha\tan\beta}{\tan\alpha + \tan\beta} = \tan\alpha\tan\beta$

95. $(\sin\alpha + \cos\beta)^2 + (\cos\beta + \sin\alpha)(\cos\beta - \sin\alpha) = (\sin^2\alpha + 2\sin\alpha\cos\beta + \cos^2\beta) + (\cos^2\beta - \sin^2\alpha)$

$= 2\cos^2\beta + 2\sin\alpha\cos\beta = 2\cos\beta(\cos\beta + \sin\alpha) = 2\cos\beta(\sin\alpha + \cos\beta)$

97. $\ln|\sec\theta| = \ln|\cos\theta|^{-1} = -\ln|\cos\theta|$

99. $\ln|1 + \cos\theta| + \ln|1 - \cos\theta| = \ln(|1 + \cos\theta||1 - \cos\theta|) = \ln|1 - \cos^2\theta| = \ln|\sin^2\theta| = 2\ln|\sin\theta|$

101. $g(x) = \sec x - \cos x = \dfrac{1}{\cos x} - \cos x = \dfrac{1}{\cos x} - \dfrac{\cos^2 x}{\cos x} = \dfrac{1 - \cos^2 x}{\cos x} = \dfrac{\sin^2 x}{\cos x} = \sin x \cdot \dfrac{\sin x}{\cos x} = \sin x \cdot \tan x = f(x)$

103. $f(\theta) = \dfrac{1 - \sin\theta}{\cos\theta} - \dfrac{\cos\theta}{1 + \sin\theta} = \dfrac{1 - \sin\theta}{\cos\theta} \cdot \dfrac{1 + \sin\theta}{1 + \sin\theta} - \dfrac{\cos\theta}{1 + \sin\theta} \cdot \dfrac{\cos\theta}{\cos\theta} = \dfrac{1 - \sin^2\theta}{\cos\theta(1 + \sin\theta)} - \dfrac{\cos^2\theta}{\cos\theta(1 + \sin\theta)}$

$= \dfrac{\cos^2\theta}{\cos\theta(1 + \sin\theta)} - \dfrac{\cos^2\theta}{\cos\theta(1 + \sin\theta)} = 0 = g(\theta)$

105. $1200\sec\theta\,(2\sec^2\theta - 1) = 1200\,\dfrac{1}{\cos\theta}\left(\dfrac{2}{\cos^2\theta} - 1\right) = 1200\,\dfrac{1}{\cos\theta}\left(\dfrac{2}{\cos^2\theta} - \dfrac{\cos^2\theta}{\cos^2\theta}\right) = 1200\,\dfrac{1}{\cos\theta}\left(\dfrac{2 - \cos^2\theta}{\cos^2\theta}\right) = \dfrac{1200\,(1 + 1 - \cos^2\theta)}{\cos^3\theta}$

$= \dfrac{1200\,(1 + \sin^2\theta)}{\cos^3\theta}$

111. Maximum, 1250 **112.** $(f \circ g)(x) = \dfrac{x - 1}{x - 2}$ **113.** $\sin\theta = \dfrac{5}{13}$; $\cos\theta = -\dfrac{12}{13}$; $\tan\theta = -\dfrac{5}{12}$; $\csc\theta = \dfrac{13}{5}$; $\sec\theta = -\dfrac{13}{12}$; $\cot\theta = -\dfrac{12}{5}$ **114.** $-\dfrac{2}{\pi}$

8.5 Assess Your Understanding (page 652)

5. – **6.** – **7.** F **8.** F **9.** F **10.** T **11.** a **12.** d **13.** $-\dfrac{1}{4}\left(\sqrt{2} + \sqrt{6}\right)$ **15.** $2 - \sqrt{3}$ **17.** $\dfrac{1}{4}\left(\sqrt{6} + \sqrt{2}\right)$ **19.** $\dfrac{1}{4}\left(\sqrt{2} - \sqrt{6}\right)$

21. $-\dfrac{1}{4}\left(\sqrt{6} + \sqrt{2}\right)$ **23.** $\sqrt{6} - \sqrt{2}$ **25.** $\dfrac{1}{2}$ **27.** 0 **29.** 1 **31.** -1 **33.** $\dfrac{1}{2}$ **35. (a)** $\dfrac{2\sqrt{5}}{25}$ **(b)** $\dfrac{11\sqrt{5}}{25}$ **(c)** $\dfrac{2\sqrt{5}}{5}$ **(d)** 2

37. (a) $\dfrac{4 - 3\sqrt{3}}{10}$ **(b)** $\dfrac{-3 - 4\sqrt{3}}{10}$ **(c)** $\dfrac{4 + 3\sqrt{3}}{10}$ **(d)** $\dfrac{25\sqrt{3} + 48}{39}$ **39. (a)** $-\dfrac{5 + 12\sqrt{3}}{26}$ **(b)** $\dfrac{12 - 5\sqrt{3}}{26}$ **(c)** $\dfrac{-5 + 12\sqrt{3}}{26}$ **(d)** $\dfrac{-240 + 169\sqrt{3}}{69}$

41. (a) $-\dfrac{2\sqrt{2}}{3}$ **(b)** $\dfrac{-2\sqrt{2} + \sqrt{3}}{6}$ **(c)** $\dfrac{-2\sqrt{2} + \sqrt{3}}{6}$ **(d)** $\dfrac{9 - 4\sqrt{2}}{7}$ **43.** $\dfrac{1 - 2\sqrt{6}}{6}$ **45.** $\dfrac{\sqrt{3} - 2\sqrt{2}}{6}$ **47.** $\dfrac{8\sqrt{2} - 9\sqrt{3}}{5}$

49. $\sin\left(\dfrac{\pi}{2} + \theta\right) = \sin\dfrac{\pi}{2}\cos\theta + \cos\dfrac{\pi}{2}\sin\theta = 1 \cdot \cos\theta + 0 \cdot \sin\theta = \cos\theta$

51. $\sin(\pi - \theta) = \sin\pi\cos\theta - \cos\pi\sin\theta = 0 \cdot \cos\theta - (-1)\sin\theta = \sin\theta$

53. $\sin(\pi + \theta) = \sin\pi\cos\theta + \cos\pi\sin\theta = 0 \cdot \cos\theta + (-1)\sin\theta = -\sin\theta$

55. $\tan(\pi - \theta) = \dfrac{\tan\pi - \tan\theta}{1 + \tan\pi\tan\theta} = \dfrac{0 - \tan\theta}{1 + 0 \cdot \tan\theta} = -\tan\theta$

57. $\sin\left(\dfrac{3\pi}{2} + \theta\right) = \sin\dfrac{3\pi}{2}\cos\theta + \cos\dfrac{3\pi}{2}\sin\theta = (-1)\cos\theta + 0 \cdot \sin\theta = -\cos\theta$

59. $\sin(\alpha + \beta) + \sin(\alpha - \beta) = \sin\alpha\cos\beta + \cos\alpha\sin\beta + \sin\alpha\cos\beta - \cos\alpha\sin\beta = 2\sin\alpha\cos\beta$

61. $\dfrac{\sin(\alpha + \beta)}{\sin\alpha\cos\beta} = \dfrac{\sin\alpha\cos\beta + \cos\alpha\sin\beta}{\sin\alpha\cos\beta} = \dfrac{\sin\alpha\cos\beta}{\sin\alpha\cos\beta} + \dfrac{\cos\alpha\sin\beta}{\sin\alpha\cos\beta} = 1 + \cot\alpha\tan\beta$

63. $\dfrac{\cos(\alpha + \beta)}{\cos\alpha\cos\beta} = \dfrac{\cos\alpha\cos\beta - \sin\alpha\sin\beta}{\cos\alpha\cos\beta} = \dfrac{\cos\alpha\cos\beta}{\cos\alpha\cos\beta} - \dfrac{\sin\alpha\sin\beta}{\cos\alpha\cos\beta} = 1 - \tan\alpha\tan\beta$

65. $\dfrac{\sin(\alpha + \beta)}{\sin(\alpha - \beta)} = \dfrac{\sin\alpha\cos\beta + \cos\alpha\sin\beta}{\sin\alpha\cos\beta - \cos\alpha\sin\beta} = \dfrac{\dfrac{\sin\alpha\cos\beta + \cos\alpha\sin\beta}{\cos\alpha\cos\beta}}{\dfrac{\sin\alpha\cos\beta - \cos\alpha\sin\beta}{\cos\alpha\cos\beta}} = \dfrac{\dfrac{\sin\alpha\cos\beta}{\cos\alpha\cos\beta} + \dfrac{\cos\alpha\sin\beta}{\cos\alpha\cos\beta}}{\dfrac{\sin\alpha\cos\beta}{\cos\alpha\cos\beta} - \dfrac{\cos\alpha\sin\beta}{\cos\alpha\cos\beta}} = \dfrac{\tan\alpha + \tan\beta}{\tan\alpha - \tan\beta}$

67. $\cot(\alpha + \beta) = \dfrac{\cos(\alpha + \beta)}{\sin(\alpha + \beta)} = \dfrac{\cos\alpha\cos\beta - \sin\alpha\sin\beta}{\sin\alpha\cos\beta + \cos\alpha\sin\beta} = \dfrac{\dfrac{\cos\alpha\cos\beta - \sin\alpha\sin\beta}{\sin\alpha\sin\beta}}{\dfrac{\sin\alpha\cos\beta + \cos\alpha\sin\beta}{\sin\alpha\sin\beta}} = \dfrac{\dfrac{\cos\alpha\cos\beta}{\sin\alpha\sin\beta} - \dfrac{\sin\alpha\sin\beta}{\sin\alpha\sin\beta}}{\dfrac{\sin\alpha\cos\beta}{\sin\alpha\sin\beta} + \dfrac{\cos\alpha\sin\beta}{\sin\alpha\sin\beta}} = \dfrac{\cot\alpha\cot\beta - 1}{\cot\beta + \cot\alpha}$

69. $\sec(\alpha + \beta) = \dfrac{1}{\cos(\alpha + \beta)} = \dfrac{1}{\cos\alpha\cos\beta - \sin\alpha\sin\beta} = \dfrac{\dfrac{1}{\sin\alpha\sin\beta}}{\dfrac{\cos\alpha\cos\beta - \sin\alpha\sin\beta}{\sin\alpha\sin\beta}} = \dfrac{\dfrac{1}{\sin\alpha} \cdot \dfrac{1}{\sin\beta}}{\dfrac{\cos\alpha\cos\beta}{\sin\alpha\sin\beta} - \dfrac{\sin\alpha\sin\beta}{\sin\alpha\sin\beta}} = \dfrac{\csc\alpha\csc\beta}{\cot\alpha\cot\beta - 1}$

71. $\sin(\alpha - \beta)\sin(\alpha + \beta) = (\sin\alpha\cos\beta - \cos\alpha\sin\beta)(\sin\alpha\cos\beta + \cos\alpha\sin\beta) = \sin^2\alpha\cos^2\beta - \cos^2\alpha\sin^2\beta$

$= (\sin^2\alpha)(1 - \sin^2\beta) - (1 - \sin^2\alpha)(\sin^2\beta) = \sin^2\alpha - \sin^2\beta$

73. $\sin(\theta + k\pi) = \sin\theta\cos k\pi + \cos\theta\sin k\pi = (\sin\theta)(-1)^k + (\cos\theta)(0) = (-1)^k\sin\theta$, k any integer

75. $\dfrac{\sqrt{3}}{2}$ **77.** $-\dfrac{24}{25}$ **79.** $-\dfrac{33}{65}$ **81.** $\dfrac{63}{65}$ **83.** $\dfrac{48 + 25\sqrt{3}}{39}$ **85.** $\dfrac{4}{3}$ **87.** $u\sqrt{1 - v^2} - v\sqrt{1 - u^2}$: $-1 \le u \le 1$; $-1 \le v \le 1$

89. $\dfrac{u\sqrt{1 - v^2} - v}{\sqrt{1 + u^2}}$: $-\infty < u < \infty$; $-1 \le v \le 1$ **91.** $\dfrac{uv - \sqrt{1 - u^2}\sqrt{1 - v^2}}{v\sqrt{1 - u^2} + u\sqrt{1 - v^2}}$: $-1 \le u \le 1$; $-1 \le v \le 1$ **93.** $\left\{\dfrac{\pi}{2}, \dfrac{7\pi}{6}\right\}$ **95.** $\left\{\dfrac{\pi}{4}\right\}$ **97.** $\left\{\dfrac{11\pi}{6}\right\}$

99. Let $\alpha = \sin^{-1} v$ and $\beta = \cos^{-1} v$. Then $\sin\alpha = \cos\beta = v$, and since $\sin\alpha = \cos\left(\dfrac{\pi}{2} - \alpha\right)$, $\cos\left(\dfrac{\pi}{2} - \alpha\right) = \cos\beta$.

If $v \geq 0$, then $0 \leq \alpha \leq \dfrac{\pi}{2}$, so $\left(\dfrac{\pi}{2} - \alpha\right)$ and β both lie on $\left[0, \dfrac{\pi}{2}\right]$. If $v < 0$, then $-\dfrac{\pi}{2} \leq \alpha < 0$, so $\left(\dfrac{\pi}{2} - \alpha\right)$ and β both lie on $\left(\dfrac{\pi}{2}, \pi\right]$.

Either way, $\cos\left(\dfrac{\pi}{2} - \alpha\right) = \cos\beta$ implies $\dfrac{\pi}{2} - \alpha = \beta$, or $\alpha + \beta = \dfrac{\pi}{2}$.

101. Let $\alpha = \tan^{-1}\dfrac{1}{v}$ and $\beta = \tan^{-1}v$. Because $v \neq 0$, $\alpha, \beta \neq 0$. Then $\tan\alpha = \dfrac{1}{v} = \dfrac{1}{\tan\beta} = \cot\beta$, and since

$\tan\alpha = \cot\left(\dfrac{\pi}{2} - \alpha\right)$, $\cot\left(\dfrac{\pi}{2} - \alpha\right) = \cot\beta$. Because $v > 0, 0 < \alpha < \dfrac{\pi}{2}$, and so $\left(\dfrac{\pi}{2} - \alpha\right)$ and β both lie on $\left(0, \dfrac{\pi}{2}\right)$.

Then $\cot\left(\dfrac{\pi}{2} - \alpha\right) = \cot\beta$ implies $\dfrac{\pi}{2} - \alpha = \beta$, or $\alpha = \dfrac{\pi}{2} - \beta$.

103. $\sin(\sin^{-1}v + \cos^{-1}v) = \sin(\sin^{-1}v)\cos(\cos^{-1}v) + \cos(\sin^{-1}v)\sin(\cos^{-1}v) = (v)(v) + \sqrt{1 - v^2}\sqrt{1 - v^2} = v^2 + 1 - v^2 = 1$

105. $\dfrac{\sin(x + h) - \sin x}{h} = \dfrac{\sin x \cos h + \cos x \sin h - \sin x}{h} = \dfrac{\cos x \sin h - \sin x(1 - \cos h)}{h} = \cos x \cdot \dfrac{\sin h}{h} - \sin x \cdot \dfrac{1 - \cos h}{h}$

107. (a) $\tan(\tan^{-1}1 + \tan^{-1}2 + \tan^{-1}3) = \tan((\tan^{-1}1 + \tan^{-1}2) + \tan^{-1}3) = \dfrac{\tan(\tan^{-1}1 + \tan^{-1}2) + \tan(\tan^{-1}3)}{1 - \tan(\tan^{-1}1 + \tan^{-1}2)\tan(\tan^{-1}3)}$

$= \dfrac{\dfrac{\tan(\tan^{-1}1) + \tan(\tan^{-1}2)}{1 - \tan(\tan^{-1}1)\tan(\tan^{-1}2)} + 3}{1 - \dfrac{\tan(\tan^{-1}1) + \tan(\tan^{-1}2)}{1 - \tan(\tan^{-1}1)\tan(\tan^{-1}2)} \cdot 3} = \dfrac{\dfrac{1 + 2}{1 - 1\cdot 2} + 3}{1 - \dfrac{1 + 2}{1 - 1\cdot 2}\cdot 3} = \dfrac{\dfrac{3}{-1} + 3}{1 - \dfrac{3}{-1}\cdot 3} = \dfrac{-3 + 3}{1 + 9} = \dfrac{0}{10} = 0$

(b) From the definition of the inverse tangent function, $0 < \tan^{-1}1 < \dfrac{\pi}{2}, 0 < \tan^{-1}2 < \dfrac{\pi}{2}$, and $0 < \tan^{-1}3 < \dfrac{\pi}{2}$,

so $0 < \tan^{-1}1 + \tan^{-1}2 + \tan^{-1}3 < \dfrac{3\pi}{2}$.

On the interval $\left(0, \dfrac{3\pi}{2}\right)$, $\tan\theta = 0$ if and only if $\theta = \pi$. Therefore, from part (a), $\tan^{-1}1 + \tan^{-1}2 + \tan^{-1}3 = \pi$.

109. $\tan\theta = \tan(\theta_2 - \theta_1) = \dfrac{\tan\theta_2 - \tan\theta_1}{1 + \tan\theta_1\tan\theta_2} = \dfrac{m_2 - m_1}{1 + m_1 m_2}$

111. $2\cot(\alpha - \beta) = \dfrac{2}{\tan(\alpha - \beta)} = 2\left(\dfrac{1 + \tan\alpha\tan\beta}{\tan\alpha - \tan\beta}\right) = 2\left(\dfrac{1 + (x + 1)(x - 1)}{(x + 1) - (x - 1)}\right) = 2\left(\dfrac{1 + x^2 - 1}{x + 1 - x + 1}\right) = \dfrac{2x^2}{2} = x^2$

113. $\tan\dfrac{\pi}{2}$ is not defined; $\tan\left(\dfrac{\pi}{2} - \theta\right) = \dfrac{\sin\left(\dfrac{\pi}{2} - \theta\right)}{\cos\left(\dfrac{\pi}{2} - \theta\right)} = \dfrac{\cos\theta}{\sin\theta} = \cot\theta$.

114. $\left(-\dfrac{1}{3}, -\dfrac{5}{9}\right), (-5, 1)$ **115.** $510°$ **116.** $\dfrac{9\pi}{2}$ cm² ≈ 14.14 cm² **117.** $\sin\theta = -\dfrac{2\sqrt{5}}{5}$; $\cos\theta = \dfrac{\sqrt{5}}{5}$; $\csc\theta = -\dfrac{\sqrt{5}}{2}$; $\sec\theta = \sqrt{5}$; $\cot\theta = -\dfrac{1}{2}$

8.6 Assess Your Understanding *(page 662)*

1. $\sin^2\theta; 2\cos^2\theta; 2\sin^2\theta$ **2.** $1 - \cos\theta$ **3.** $\sin\theta$ **4.** T **5.** F **6.** F **7.** b **8.** c **9. (a)** $\dfrac{24}{25}$ **(b)** $\dfrac{7}{25}$ **(c)** $\dfrac{\sqrt{10}}{10}$ **(d)** $\dfrac{3\sqrt{10}}{10}$

11. (a) $\dfrac{24}{25}$ **(b)** $-\dfrac{7}{25}$ **(c)** $\dfrac{2\sqrt{5}}{5}$ **(d)** $-\dfrac{\sqrt{5}}{5}$ **13. (a)** $-\dfrac{2\sqrt{2}}{3}$ **(b)** $\dfrac{1}{3}$ **(c)** $\sqrt{\dfrac{3 + \sqrt{6}}{6}}$ **(d)** $\sqrt{\dfrac{3 - \sqrt{6}}{6}}$

15. (a) $\dfrac{4\sqrt{2}}{9}$ **(b)** $-\dfrac{7}{9}$ **(c)** $\dfrac{\sqrt{3}}{3}$ **(d)** $\dfrac{\sqrt{6}}{3}$ **17. (a)** $-\dfrac{4}{5}$ **(b)** $\dfrac{3}{5}$ **(c)** $\sqrt{\dfrac{5 + 2\sqrt{5}}{10}}$ **(d)** $\sqrt{\dfrac{5 - 2\sqrt{5}}{10}}$

19. (a) $-\dfrac{3}{5}$ **(b)** $-\dfrac{4}{5}$ **(c)** $\dfrac{1}{2}\sqrt{\dfrac{10 - \sqrt{10}}{5}}$ **(d)** $-\dfrac{1}{2}\sqrt{\dfrac{10 + \sqrt{10}}{5}}$ **21.** $\dfrac{\sqrt{2 - \sqrt{2}}}{2}$ **23.** $1 - \sqrt{2}$ **25.** $-\dfrac{\sqrt{2 + \sqrt{3}}}{2}$

27. $\dfrac{2}{\sqrt{2 + \sqrt{2}}} = (2 - \sqrt{2})\sqrt{2 + \sqrt{2}}$ **29.** $-\dfrac{\sqrt{2 - \sqrt{2}}}{2}$ **31.** $-\dfrac{4}{5}$ **33.** $\dfrac{\sqrt{10(5 - \sqrt{5})}}{10}$ **35.** $\dfrac{4}{3}$ **37.** $-\dfrac{7}{8}$ **39.** $\dfrac{\sqrt{10}}{4}$ **41.** $-\dfrac{\sqrt{15}}{3}$

43. $\sin^4\theta = (\sin^2\theta)^2 = \left(\dfrac{1 - \cos(2\theta)}{2}\right)^2 = \dfrac{1}{4}[1 - 2\cos(2\theta) + \cos^2(2\theta)] = \dfrac{1}{4} - \dfrac{1}{2}\cos(2\theta) + \dfrac{1}{4}\cos^2(2\theta)$

$= \dfrac{1}{4} - \dfrac{1}{2}\cos(2\theta) + \dfrac{1}{4}\left(\dfrac{1 + \cos(4\theta)}{2}\right) = \dfrac{1}{4} - \dfrac{1}{2}\cos(2\theta) + \dfrac{1}{8} + \dfrac{1}{8}\cos(4\theta) = \dfrac{3}{8} - \dfrac{1}{2}\cos(2\theta) + \dfrac{1}{8}\cos(4\theta)$

45. $\cos(3\theta) = 4\cos^3\theta - 3\cos\theta$ **47.** $\sin(5\theta) = 16\sin^5\theta - 20\sin^3\theta + 5\sin\theta$ **49.** $\cos^4\theta - \sin^4\theta = (\cos^2\theta + \sin^2\theta)(\cos^2\theta - \sin^2\theta) = \cos(2\theta)$

11. (a) $f(x) = (2x - 1)(x - 1)^2(x + 1)^2$;

$\dfrac{1}{2}$ multiplicity 1; 1 and -1 multiplicity 2

(b) $(0, -1)$; $\left(\dfrac{1}{2}, 0\right)$; $(-1, 0)$; $(1, 0)$ **(c)** $y = 2x^5$

(d)

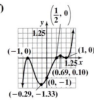

(e) Local minimum value -1.33 at $x = -0.29$,
Local minimum value 0 at $x = 1$
Local maximum value 0 at $x = -1$,
Local maximum value 0.10 at $x = 0.69$

(f)

```
                y    (1/2, 0)
              1.25
      (-1, 0)          (1, 0)
                     1.25 x
                     (0.69, 0.10)
                    (0, -1)
      (-0.29, -1.33)
```

(g) Increasing: $(-\infty, -1)$, $(-0.29, 0.69)$, $(1, \infty)$
Decreasing: $(-1, -0.29)$, $(0.69, 1)$

12. (a) $\left\{-1, -\dfrac{1}{2}\right\}$

(b) $\{-1, 1\}$

(c) $(-\infty, -1) \cup \left(-\dfrac{1}{2}, \infty\right)$

(d) $(-\infty, -1] \cup [1, \infty)$

CHAPTER 9   Applications of Trigonometric Functions

9.1 Assess Your Understanding *(page 679)*

5. b   **6.** direction; bearing   **7.** T   **8.** F   **9.** $a \approx 13.74$, $c \approx 14.62$, $A = 70°$ **11.** $b \approx 5.03$, $c \approx 7.83$, $A = 50°$ **13.** $a \approx 0.71$, $c \approx 4.06$, $B = 80°$
15. $b \approx 10.72$, $c \approx 11.83$, $B = 65°$ **17.** $b \approx 3.08$, $a \approx 8.46$, $A = 70°$ **19.** $c \approx 5.83$, $A \approx 59.0°$, $B \approx 31.0°$ **21.** $b \approx 4.58$, $A \approx 23.6°$, $B \approx 66.4°$
23. $23.6°$ and $66.4°$ **25.** 4.59 in.; 6.55 in.   **27.** 80.5°   **29. (a)** 111.96 ft/s or 76.3 mi/h   **(b)** 82.42 ft/s or 56.2 mi/h   **(c)** Under 18.8°   **31. (a)** 2.4898×10^{13} miles
(b) 0.000214°   **33.** S76.6°E   **35.** The embankment is 30.5 m high.   **37.** The buildings are 7984 ft apart.   **39.** 69.0°   **41.** 38.9°   **43.** 76.94 in.
46. Yes   **47.** $\dfrac{\sqrt{6} - \sqrt{2}}{4}$ or $\dfrac{\sqrt{2} - \sqrt{3}}{2}$ **48.** 0.236, 0.243, 0.248   **49.** $\left\{\dfrac{\pi}{2}, \dfrac{7\pi}{6}, \dfrac{11\pi}{6}\right\}$

9.2 Assess Your Understanding *(page 688)*

4. a   **5.** $\dfrac{\sin A}{a} = \dfrac{\sin B}{b} = \dfrac{\sin C}{c}$ **6.** F   **7.** F   **8.** ambiguous case   **9.** $a \approx 3.23$, $b \approx 3.55$, $A = 40°$ **11.** $a \approx 3.25$, $c \approx 4.23$, $B = 45°$
13. $C = 95°$, $c \approx 9.86$, $a \approx 6.36$ **15.** $A = 40°$, $a = 2$, $c \approx 3.06$ **17.** $C = 120°$, $b \approx 1.06$, $c \approx 2.69$ **19.** $A = 100°$, $a \approx 5.24$, $c \approx 0.92$
21. $B = 40°$, $a \approx 5.64$, $b \approx 3.86$ **23.** $C = 100°$, $a \approx 1.31$, $b \approx 1.31$ **25.** One triangle; $B \approx 30.7°$, $C \approx 99.3°$, $c \approx 3.86$ **27.** One triangle;
$C \approx 36.2°$, $A \approx 43.8°$, $a \approx 3.51$ **29.** No triangle   **31.** Two triangles; $C_1 \approx 30.9°$, $A_1 \approx 129.1°$, $a_1 \approx 9.07$ or $C_2 \approx 149.1°$, $A_2 \approx 10.9°$, $a_2 \approx 2.20$
33. No triangle   **35.** Two triangles; $A_1 \approx 57.7°$, $B_1 \approx 97.3°$, $b_1 \approx 2.35$ or $A_2 \approx 122.3°$, $B_2 \approx 32.7°$, $b_2 \approx 1.28$ **37.** 1490.48 ft   **39.** 335.16 ft
41. 153.42 ft; 136.59 ft   **43.** The tree is 39.4 ft high.   **45.** Adam receives 100.6 more frequent flyer miles.   **47. (a)** Station Able is about 143.33 mi from
the ship; Station Baker is about 135.58 mi from the ship.   **(b)** Approximately 41 min   **49.** 84.7°; 183.72 ft   **51.** 2.64 mi   **53.** 38.5 in.   **55.** 449.36 ft
57. 187,600,000 km or 101,440,000 km   **59.** The diameter is 252 ft.

61. $\dfrac{a - b}{c} = \dfrac{a}{c} - \dfrac{b}{c} = \dfrac{\sin A}{\sin C} - \dfrac{\sin B}{\sin C} = \dfrac{\sin A - \sin B}{\sin C} = \dfrac{2 \sin\left(\dfrac{A - B}{2}\right) \cos\left(\dfrac{A + B}{2}\right)}{2 \sin\dfrac{C}{2} \cos\dfrac{C}{2}} = \dfrac{\sin\left(\dfrac{A - B}{2}\right) \cos\left(\dfrac{\pi}{2} - \dfrac{C}{2}\right)}{\sin\dfrac{C}{2} \cos\dfrac{C}{2}} = \dfrac{\sin\left(\dfrac{A - B}{2}\right)}{\cos\dfrac{C}{2}}$

63. $\dfrac{a - b}{a + b} = \dfrac{\dfrac{a - b}{c}}{\dfrac{a + b}{c}} = \dfrac{\dfrac{\sin\left[\dfrac{1}{2}(A - B)\right]}{\cos\dfrac{C}{2}}}{\dfrac{\cos\left[\dfrac{1}{2}(A - B)\right]}{\sin\dfrac{C}{2}}} = \dfrac{\tan\left[\dfrac{1}{2}(A - B)\right]}{\cot\dfrac{C}{2}} = \dfrac{\tan\left[\dfrac{1}{2}(A - B)\right]}{\tan\left(\dfrac{\pi}{2} - \dfrac{C}{2}\right)} = \dfrac{\tan\left[\dfrac{1}{2}(A - B)\right]}{\tan\left[\dfrac{1}{2}(A + B)\right]}$

69. $\left\{-3, -\dfrac{4}{3}, 3\right\}$ **70.** $3\sqrt{5} \approx 6.71$ **71.** $-\dfrac{\sqrt{15}}{7}$ **72.**

```
                    y
       (-3π, 4)    5 (π, 4)
                          (2π, 0)
      (-4π, 0)            (4π, 0)
                          5π x
       (-2π, 0)
       (-π, -4)        (3π, -4)
                 (0, 0)
```

9.3 Assess Your Understanding *(page 695)*

3. Cosines   **4.** a   **5.** b   **6.** F   **7.** F   **8.** T   **9.** $b \approx 2.95$, $A \approx 28.7°$, $C \approx 106.3°$ **11.** $c \approx 3.75$, $A \approx 32.1°$, $B \approx 52.9°$
13. $A \approx 48.5°$, $B \approx 38.6°$, $C \approx 92.9°$ **15.** $A \approx 127.2°$, $B \approx 32.1°$, $C \approx 20.7°$ **17.** $c \approx 2.57$, $A \approx 48.6°$, $B \approx 91.4°$
19. $a \approx 2.99$, $B \approx 19.2°$, $C \approx 80.8°$ **21.** $b \approx 4.14$, $A \approx 43.0°$, $C \approx 27.0°$ **23.** $c \approx 1.69$, $A \approx 65.0°$, $B \approx 65.0°$
25. $A \approx 67.4°$, $B \approx 90°$, $C \approx 22.6°$ **27.** $A = 60°$, $B = 60°$, $C = 60°$ **29.** $A \approx 33.6°$, $B \approx 62.2°$, $C \approx 84.3°$
31. $A \approx 97.9°$, $B \approx 52.4°$, $C \approx 29.7°$ **33.** $A = 85°$, $a = 14.56$, $c = 14.12$ **35.** $A = 40.8°$, $B = 60.6°$, $C = 78.6°$ **37.** $A = 80°$, $b = 8.74$, $c = 13.80$
39. Two triangles: $B_1 = 35.4°$, $C_1 = 134.6°$, $c_1 = 12.29$; $B_2 = 144.6°$, $C_2 = 25.4°$, $c_2 = 7.40$ **41.** $B = 24.5°$, $C = 95.5°$, $a = 10.44$

43. 165 yd **45. (a)** 26.4° **(b)** 30.8 h **47. (a)** 63.7 ft **(b)** 66.8 ft **(c)** 92.8° **49. (a)** 492.6 ft **(b)** 269.3 ft
51. 342.33 ft **53.** The footings should be 7.65 ft apart.
55. Suppose $0 < \theta < \pi$. Then, by the Law of Cosines, $d^2 = r^2 + r^2 - 2r^2 \cos \theta = 4r^2 \left(\dfrac{1 - \cos \theta}{2} \right) \Rightarrow d = 2r\sqrt{\dfrac{1 - \cos \theta}{2}} = 2r \sin \dfrac{\theta}{2}$.

Since, for any angle in $(0, \pi)$, d is strictly less than the length of the arc subtended by θ, that is, $d < r\theta$, then $2r \sin \dfrac{\theta}{2} < r\theta$, or $2 \sin \dfrac{\theta}{2} < \theta$.
Since $\cos \dfrac{\theta}{2} < 1$, then, for $0 < \theta < \pi$, $\sin \theta = 2 \sin \dfrac{\theta}{2} \cos \dfrac{\theta}{2} < 2 \sin \dfrac{\theta}{2} < \theta$. If $\theta \geq \pi$, then, since $\sin \theta \leq 1$, $\sin \theta < \theta$. Thus $\sin \theta < \theta$ for all $\theta > 0$.

57. $\sin \dfrac{C}{2} = \sqrt{\dfrac{1 - \cos C}{2}} = \sqrt{\dfrac{1 - \dfrac{a^2 + b^2 - c^2}{2ab}}{2}} = \sqrt{\dfrac{2ab - a^2 - b^2 + c^2}{4ab}} = \sqrt{\dfrac{c^2 - (a - b)^2}{4ab}} = \sqrt{\dfrac{(c + a - b)(c + b - a)}{4ab}}$

$= \sqrt{\dfrac{(2s - 2b)(2s - 2a)}{4ab}} = \sqrt{\dfrac{(s - a)(s - b)}{ab}}$

64.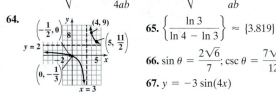

65. $\left\{ \dfrac{\ln 3}{\ln 4 - \ln 3} \right\} \approx \{3.819\}$

66. $\sin \theta = \dfrac{2\sqrt{6}}{7}$; $\csc \theta = \dfrac{7\sqrt{6}}{12}$; $\sec \theta = -\dfrac{7}{5}$; $\cot \theta = -\dfrac{5\sqrt{6}}{12}$

67. $y = -3 \sin(4x)$

9.4 Assess Your Understanding *(page 701)*

2. $\dfrac{1}{2} ab \sin C$ **3.** $\sqrt{s(s - a)(s - b)(s - c)}$; $\dfrac{1}{2}(a + b + c)$ **4.** T **5.** c **6.** c **7.** 2.83 **9.** 2.99 **11.** 14.98 **13.** 9.56 **15.** 3.86 **17.** 1.48 **19.** 2.82

21. 30 **23.** 1.73 **25.** 19.90 **27.** $K = \dfrac{1}{2} ab \sin C = \dfrac{1}{2} a \sin C \left(\dfrac{a \sin B}{\sin A} \right) = \dfrac{a^2 \sin B \sin C}{2 \sin A}$ **29.** 0.92 **31.** 2.27 **33.** 5.44 **35.** 9.03 sq ft **37.** \$5446.38

39. The area of home plate is about 216.5 in.² **41.** $K = \dfrac{1}{2} r^2 (\theta + \sin \theta)$ **43.** The ground area is 7517.4 ft².

45. (a) Area $\Delta OAC = \dfrac{1}{2} |OC||AC| = \dfrac{1}{2} \cdot \dfrac{|OC|}{1} \cdot \dfrac{|AC|}{1} = \dfrac{1}{2} \sin \alpha \cos \alpha$

(b) Area $\Delta OCB = \dfrac{1}{2} |BC||OC| = \dfrac{1}{2} |OB|^2 \dfrac{|BC|}{|OB|} \cdot \dfrac{|OC|}{|OB|} = \dfrac{1}{2} |OB|^2 \sin \beta \cos \beta$

(c) Area $\Delta OAB = \dfrac{1}{2} |BD||OA| = \dfrac{1}{2} |OB| \dfrac{|BD|}{|OB|} = \dfrac{1}{2} |OB| \sin(\alpha + \beta)$

(d) $\dfrac{\cos \alpha}{\cos \beta} = \dfrac{1}{\dfrac{|OC|}{|OB|}} = |OB|$

(e) Area $\Delta OAB =$ Area $\Delta OAC +$ Area ΔOCB

$\dfrac{1}{2} |OB| \sin(\alpha + \beta) = \dfrac{1}{2} \sin \alpha \cos \alpha + \dfrac{1}{2} |OB|^2 \sin \beta \cos \beta$

$\sin(\alpha + \beta) = \dfrac{1}{|OB|} \sin \alpha \cos \alpha + |OB| \sin \beta \cos \beta$

$\sin(\alpha + \beta) = \dfrac{\cos \beta}{\cos \alpha} \sin \alpha \cos \alpha + \dfrac{\cos \alpha}{\cos \beta} \sin \beta \cos \beta$

$\sin(\alpha + \beta) = \sin \alpha \cos \beta + \cos \alpha \sin \beta$

47. 31,145 ft² **49. (a)** The perimeter and area are both 36. **(b)** The perimeter and area are both 60.

51. $K = \dfrac{1}{2} ah = \dfrac{1}{2} ab \sin C \Rightarrow h = b \sin C = \dfrac{a \sin B \sin C}{\sin A}$

53. $\angle POQ = 180° - \left(\dfrac{A}{2} + \dfrac{B}{2} \right) = 180° - \dfrac{1}{2}(180° - C) = 90° + \dfrac{C}{2}$, and $\sin \left(90° + \dfrac{C}{2} \right) = \cos \left(-\dfrac{C}{2} \right) = \cos \dfrac{C}{2}$, since cosine is an even function.

Therefore, $r = \dfrac{c \sin \dfrac{A}{2} \sin \dfrac{B}{2}}{\sin \left(90° + \dfrac{C}{2} \right)} = \dfrac{c \sin \dfrac{A}{2} \sin \dfrac{B}{2}}{\cos \dfrac{C}{2}}$.

55. $\cot \dfrac{A}{2} + \cot \dfrac{B}{2} + \cot \dfrac{C}{2} = \dfrac{s - a}{r} + \dfrac{s - b}{r} + \dfrac{s - c}{r} = \dfrac{3s - (a + b + c)}{r} = \dfrac{3s - 2s}{r} = \dfrac{s}{r}$

60. Maximum value; 17 **61.** $(-\infty, -3) \cup [-1, 3)$

62. $\sin t = \dfrac{\sqrt{2}}{3}$, $\cos t = -\dfrac{\sqrt{7}}{3}$, $\tan t = -\dfrac{\sqrt{14}}{7}$, $\csc t = \dfrac{3\sqrt{2}}{2}$, $\sec t = -\dfrac{3\sqrt{7}}{7}$, $\cot t = -\dfrac{\sqrt{14}}{2}$

63. $\csc \theta - \sin \theta = \dfrac{1}{\sin \theta} - \sin \theta = \dfrac{1 - \sin^2 \theta}{\sin \theta} = \dfrac{\cos^2 \theta}{\sin \theta} = \cos \theta \cdot \dfrac{\cos \theta}{\sin \theta} = \cos \theta \cot \theta$

9.5 Assess Your Understanding *(page 711)*

2. Simple harmonic; amplitude **3.** Simple harmonic; damped **4.** T **5.** $d = -5 \cos(\pi t)$ **7.** $d = -6 \cos(2t)$ **9.** $d = -5 \sin(\pi t)$

11. $d = -6 \sin(2t)$ **13. (a)** Simple harmonic **(b)** 5 m **(c)** $\dfrac{2\pi}{3}$ sec **(d)** $\dfrac{3}{2\pi}$ oscillation/sec **15. (a)** Simple harmonic **(b)** 6 m **(c)** 2 sec

(d) $\dfrac{1}{2}$ oscillation/sec **17. (a)** Simple harmonic **(b)** 3 m **(c)** 4π sec **(d)** $\dfrac{1}{4\pi}$ oscillation/sec **19. (a)** Simple harmonic **(b)** 2 m

(c) 1 sec **(d)** 1 oscillation/sec

21.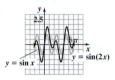

23.

25.

27.

29.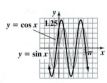

31.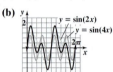

33. (a) $f(x) = \dfrac{1}{2}\left[\cos x - \cos(3x)\right]$

(b)

35. (a) $G(x) = \dfrac{1}{2}\left[\cos(6x) + \cos(2x)\right]$

(b)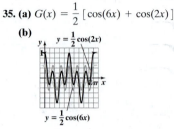

37. (a) $H(x) = \sin(4x) + \sin(2x)$

(b)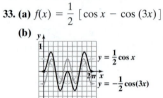

39. (a) $d = -10e^{-0.7t/50}\cos\left(\sqrt{\dfrac{4\pi^2}{25} - \dfrac{0.49}{2500}}\,t\right)$

(b)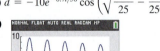

41. (a) $d = -18e^{-0.6t/60}\cos\left(\sqrt{\dfrac{\pi^2}{4} - \dfrac{0.36}{3600}}\,t\right)$

(b)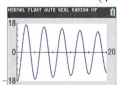

43. (a) $d = -5e^{-0.8t/20}\cos\left(\sqrt{\dfrac{4\pi^2}{9} - \dfrac{0.64}{400}}\,t\right)$

(b)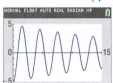

45. (a) The motion is damped. The bob has mass $m = 20$ kg with a damping factor of 0.7 kg/sec.
(b) 20 m leftward
(c)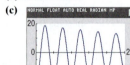
(d) 18.33 m leftward   **(e)** $d \to 0$

47. (a) The motion is damped. The bob has mass $m = 40$ kg with a damping factor of 0.6 kg/sec.
(b) 30 m leftward
(c)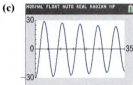
(d) 28.47 m leftward   **(e)** $d \to 0$

49. (a) The motion is damped. The bob has mass $m = 15$ kg with a damping factor of 0.9 kg/sec.
(b) 15 m leftward
(c)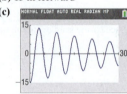
(d) 12.53 m leftward   **(e)** $d \to 0$

51. $\omega = 1040\pi$; $d = 0.80\cos(1040\pi t)$
53. $\omega = 880\pi$; $d = 0.01\sin(880\pi t)$
55. (a)

(b) At $t = 0$, $t = 2$; at $t = 1$, $t = 3$
(c) During the approximate intervals $0.35 < t < 0.67$, $1.29 < t < 1.75$, and $2.19 < t \le 3$

57.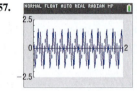

61.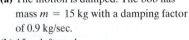

63. $y = \dfrac{1}{x}\sin x$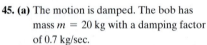

$y = \dfrac{1}{x^2}\sin x$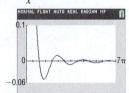

$y = \dfrac{1}{x^3}\sin x$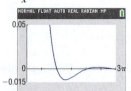

65. $f^{-1}(x) = \dfrac{4x-3}{x-1}$ **66.** $\log_7\left(\dfrac{xy^3}{x+y}\right)$ **67.** $\{4\}$ **68. (a)** $\dfrac{3\sqrt{10}}{10}$ **(b)** $\dfrac{\sqrt{10}}{10}$ **(c)** $\dfrac{1}{3}$

Review Exercises (page 714)

1. $A = 70°$, $b \approx 3.42$, $a \approx 9.40$ **2.** $a \approx 4.58$, $A \approx 66.4°$, $B \approx 23.6°$ **3.** $C = 100°$, $b \approx 0.65$, $c \approx 1.29$ **4.** $B \approx 56.8°$, $C \approx 23.2°$, $b \approx 4.25$
5. No triangle **6.** $b \approx 3.32$, $A \approx 62.8°$, $C \approx 17.2°$ **7.** $A \approx 36.2°$, $C \approx 63.8°$, $c \approx 4.55$ **8.** No triangle **9.** $A \approx 83.3°$, $B \approx 44.0°$, $C \approx 52.6°$
10. $c \approx 2.32$, $A \approx 16.1°$, $B \approx 123.9°$ **11.** $B \approx 36.2°$, $C \approx 63.8°$, $c \approx 4.55$ **12.** $A \approx 39.6°$, $B \approx 18.6°$, $C \approx 121.9°$ **13.** Two triangles:
$B_1 \approx 13.4°$, $C_1 \approx 156.6°$, $c_1 \approx 6.86$ or $B_2 \approx 166.6°$, $C_2 \approx 3.4°$, $c_2 \approx 1.02$ **14.** $b \approx 11.52$, $c \approx 10.13$, $C \approx 60°$
15. $a \approx 5.23$, $B \approx 46.0°$, $C \approx 64.0°$ **16.** 1.93 **17.** 18.79 **18.** 6 **19.** 3.80 **20.** 0.32 **21.** 12.7° **22.** 29.97 ft **23.** 6.22 mi

24. (a) 131.8 mi **(b)** 23.1° **(c)** 0.21 hr **25.** 8798.67 ft² **26.** $222,983.51 **27.** 1.92 in.² **28.** S4.0°E **29.** $d = -3 \cos\left(\dfrac{\pi}{2} t\right)$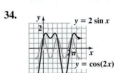

30. (a) simple harmonic **(b)** 6 ft **(c)** π s **(d)** $\dfrac{1}{\pi}$ oscillation/s **31. (a)** simple harmonic **(b)** 2 ft **(c)** 2 s **(d)** $\dfrac{1}{2}$ oscillation/s

32. (a) $d = -15 e^{-0.75t/80} \cos\left(\sqrt{\dfrac{4\pi^2}{25} - \dfrac{0.5625}{6400}} \, t\right)$ 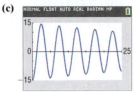 **33. (a)** The motion is damped. The bob has mass $m = 20$ kg with a damping factor of 0.6 kg/s.
(b) 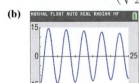 **(b)** 15 m leftward
(c)

34.

(d) 13.92 m leftward **(e)** $d \to 0$

Chapter Test (page 716)

1. 61.0° **2.** 1.3° **3.** $a = 15.88$, $B \approx 57.5°$, $C \approx 70.5°$ **4.** $b \approx 6.85$, $C = 117°$, $c \approx 16.30$ **5.** $A \approx 52.4°$, $B \approx 29.7°$, $C \approx 97.9°$
6. $b \approx 4.72$, $c \approx 1.67$, $B = 105°$ **7.** No triangle **8.** $c \approx 7.62$, $A \approx 80.5°$, $B \approx 29.5°$ **9.** 15.04 square units **10.** 19.81 square units
11. The area of the shaded region is 9.26 cm². **12.** 54.15 square units **13.** Madison will have to swim about 2.23 miles. **14.** 12.63 square units
15. The lengths of the sides are 15, 18, and 21. **16.** $d = 5(\sin 42°) \sin\left(\dfrac{\pi t}{3}\right)$ or $d \approx 3.346 \sin\left(\dfrac{\pi t}{3}\right)$

Cumulative Review (page 717)

1. $\left\{\dfrac{1}{3}, 1\right\}$ **2.** $(x+5)^2 + (y-1)^2 = 9$ **3.** $\{x \mid x \leq -1 \text{ or } x \geq 4\}$ **4.** **5.**

6. (a) $-\dfrac{2\sqrt{5}}{5}$ **(b)** $\dfrac{\sqrt{5}}{5}$ **(c)** $-\dfrac{4}{5}$ **(d)** $-\dfrac{3}{5}$ **(e)** $\sqrt{\dfrac{5-\sqrt{5}}{10}}$ **(f)** $-\sqrt{\dfrac{5+\sqrt{5}}{10}}$

7. (a) 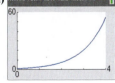 **(b)** 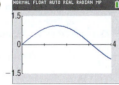 **(c)** 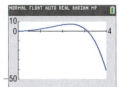 **(d)**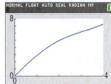

8. (a) **(b)** **(c)** **(d)** **(e)** **(f)**

(g)  **(h)** **(i)**

9. Two triangles: $A_1 \approx 59.0°$, $B_1 \approx 81.0°$, $b_1 \approx 23.05$ or $A_2 \approx 121.0°$, $B_2 \approx 19.0°$, $b_2 \approx 7.59$

10. $\left\{-2i, 2i, \dfrac{1}{3}, 1, 2\right\}$

11. $R(x) = \dfrac{(2x + 1)(x - 4)}{(x + 5)(x - 3)}$; domain: $\{x \mid x \neq -5, x \neq 3\}$

Intercepts: $\left(-\dfrac{1}{2}, 0\right), (4, 0), \left(0, \dfrac{4}{15}\right)$

No symmetry

Vertical asymptotes: $x = -5, x = 3$

Horizontal asymptote: $y = 2$

Intersects: $\left(\dfrac{26}{11}, 2\right)$

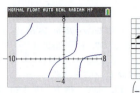

12. $\{2.26\}$ **13.** $\{1\}$ **14. (a)** $\left\{-\dfrac{5}{4}\right\}$ **(b)** $\{2\}$ **(c)** $\left\{\dfrac{-1 - 3\sqrt{13}}{2}, \dfrac{-1 + 3\sqrt{13}}{2}\right\}$ **(d)** $\left\{x \mid x > -\dfrac{5}{4}\right\}$ or $\left(-\dfrac{5}{4}, \infty\right)$

(e) $\{x \mid -8 \leq x \leq 3\}$ or $[-8, 3]$ **(f)**

(g)

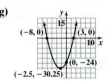

CHAPTER 10 Polar Coordinates; Vectors

10.1 Assess Your Understanding (page 727)

5. pole; polar axis **6.** $r \cos \theta$; $r \sin \theta$ **7.** b **8.** d **9.** T **10.** F **11.** A **13.** C **15.** B **17.** A

19.

21.

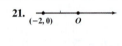

23.

25.

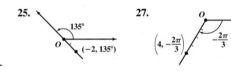

27.

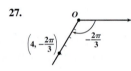

29.

31.

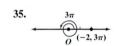

33.

(a) $\left(5, -\dfrac{4\pi}{3}\right)$ **(b)** $\left(-5, \dfrac{5\pi}{3}\right)$ **(c)** $\left(5, \dfrac{8\pi}{3}\right)$ **35.**

(a) $(2, -2\pi)$ **(b)** $(-2, \pi)$ **(c)** $(2, 2\pi)$

37.

(a) $\left(1, -\dfrac{3\pi}{2}\right)$

(b) $\left(-1, \dfrac{3\pi}{2}\right)$

(c) $\left(1, \dfrac{5\pi}{2}\right)$

39.

(a) $\left(3, -\dfrac{5\pi}{4}\right)$

(b) $\left(-3, \dfrac{7\pi}{4}\right)$

(c) $\left(3, \dfrac{11\pi}{4}\right)$

41. $(0, 3)$ **43.** $(-2, 0)$ **45.** $(-3\sqrt{3}, 3)$ **47.** $(\sqrt{2}, -\sqrt{2})$ **49.** $\left(-\dfrac{1}{2}, \dfrac{\sqrt{3}}{2}\right)$ **51.** $(2, 0)$

53. $(-2.57, 7.05)$ **55.** $(-4.98, -3.85)$ **57.** $(3, 0)$ **59.** $(1, \pi)$ **61.** $\left(\sqrt{2}, -\dfrac{\pi}{4}\right)$

63. $\left(2, \dfrac{\pi}{6}\right)$ **65.** $(2.47, -1.02)$ **67.** $(9.30, 0.47)$ **69.** $r^2 = \dfrac{3}{2}$ or $r = \dfrac{\sqrt{6}}{2}$

71. $r^2 \cos^2 \theta - 4r \sin \theta = 0$ **73.** $r^2 \sin 2\theta = 1$ **75.** $r \cos \theta = 4$

77. $x^2 + y^2 - x = 0$ or $\left(x - \dfrac{1}{2}\right)^2 + y^2 = \dfrac{1}{4}$ **79.** $(x^2 + y^2)^{3/2} - x = 0$

81. $x^2 + y^2 = 4$ **83.** $y^2 = 8(x + 2)$

85. (a) $(-10, 36)$ **(b)** $\left(2\sqrt{349}, 180° + \tan^{-1}\left(-\dfrac{18}{5}\right)\right) \approx (37.36, 105.5°)$ **(c)** $(-3, -35)$ **(d)** $\left(\sqrt{1234}, 180° + \tan^{-1}\left(\dfrac{35}{3}\right)\right) \approx (35.13, 265.1°)$

90. $\left\{\dfrac{19}{15}\right\}$ **91.** 2 or 0 positive real zeros; 1 negative real zero **92.** $\left(-\dfrac{5}{4}, \dfrac{9}{2}\right)$ **93.** $(0, -11)$

10.2 Assess Your Understanding *(page 741)*

7. polar equation   **8.** F   **9.** $-\theta$ **10.** $\pi - \theta$ **11.** T   **12.** $2n; n$ **13.** c   **14.** b

15. $x^2 + y^2 = 16$; circle, radius 4, center at pole

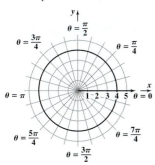

17. $y = \sqrt{3}\,x$; line through pole, making an angle of $\frac{\pi}{3}$ with polar axis

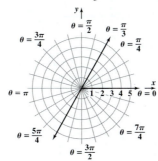

19. $y = 4$; horizontal line 4 units above the pole

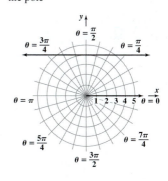

21. $x = -2$; vertical line 2 units to the left of the pole

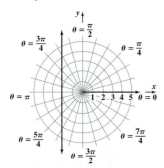

23. $(x - 1)^2 + y^2 = 1$; circle, radius 1, center $(1, 0)$ in rectangular coordinates

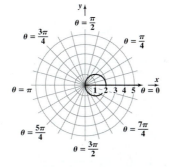

25. $x^2 + (y + 2)^2 = 4$; circle, radius 2, center at $(0, -2)$ in rectangular coordinates

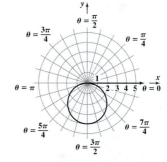

27. $(x - 2)^2 + y^2 = 4$, $x \neq 0$; circle, radius 2, center at $(2, 0)$ in rectangular coordinates, hole at $(0, 0)$

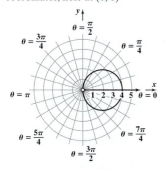

29. $x^2 + (y + 1)^2 = 1$, $x \neq 0$; circle, radius 1, center at $(0, -1)$ in rectangular coordinates, hole at $(0, 0)$

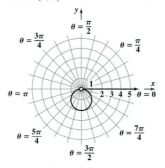

31. E **33.** F **35.** H **37.** D
39. Cardioid

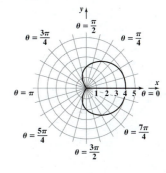

41. Cardioid

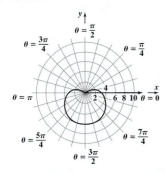

43. Limaçon without inner loop

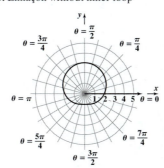

45. Limaçon without inner loop

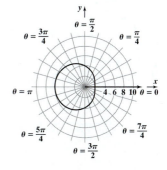

47. Limaçon with inner loop

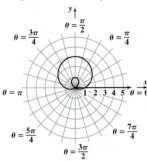

49. Limaçon with inner loop

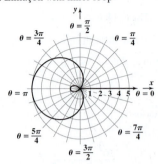

51. Rose

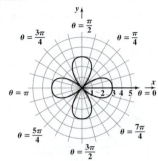

53. Rose

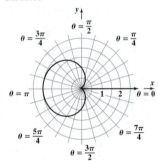

55. Lemniscate

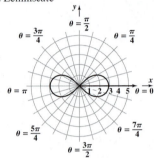

57. Spiral

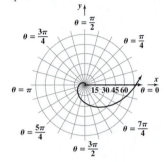

59. Cardioid

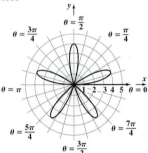

61. Limaçon with inner loop

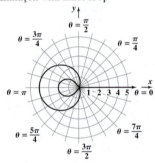

63.

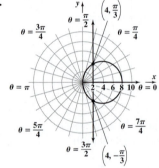

65.

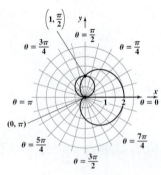

67.

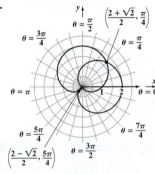

69. $r = 3 + 3\cos\theta$ **71.** $r = 4 + \sin\theta$
73.

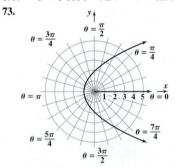

75.

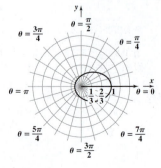

77.

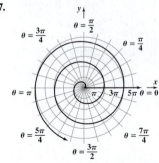

79.

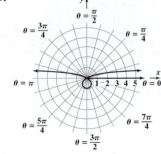

81.

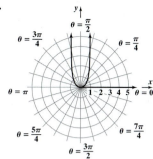

83. $r \sin \theta = a$
$y = a$

85.
$r = 2a \sin \theta$
$r^2 = 2ar \sin \theta$
$x^2 + y^2 = 2ay$
$x^2 + y^2 - 2ay = 0$
$x^2 + (y - a)^2 = a^2$
Circle, radius a, center at $(0, a)$
in rectangular coordinates

87.
$r = 2a \cos \theta$
$r^2 = 2ar \cos \theta$
$x^2 + y^2 = 2ax$
$x^2 - 2ax + y^2 = 0$
$(x - a)^2 + y^2 = a^2$
Circle, radius a, center at $(a, 0)$
in rectangular coordinates

89. (a) $r^2 = \cos \theta; r^2 = \cos(\pi - \theta)$
$r^2 = -\cos \theta$
Not equivalent; test fails.
$(-r)^2 = \cos(-\theta)$
$r^2 = \cos \theta$
New test works.

(b) $r^2 = \sin \theta; r^2 = \sin(\pi - \theta)$
$r^2 = \sin \theta$
Test works.
$(-r)^2 = \sin(-\theta)$
$r^2 = -\sin \theta$
Not equivalent; new test fails.

93. $\{x \mid 3 < x \leq 8\}$, or $(3, 8]$ **94.** $420°$ **95.** Amplitude $= 2$; period $= \dfrac{2\pi}{5}$ **96.** Horizontal asymptote: $y = 0$
Vertical asymptote: $x = 4$

Historical Problems (page 749)

1. (a) $1 + 4i, 1 + i$ **(b)** $-1, 2 + i$

10.3 Assess Your Understanding (page 750)

5. real; imaginary **6.** magnitude; modulus; argument **7.** $r_1 r_2$; $\theta_1 + \theta_2$; $\theta_1 + \theta_2$ **8.** r^n; $n\theta$; $n\theta$ **9.** three **10.** T **11.** c **12.** a

13.

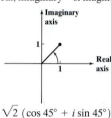

$\sqrt{2}(\cos 45° + i \sin 45°)$

15.

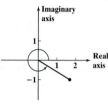

$2(\cos 330° + i \sin 330°)$

17.

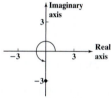

$3(\cos 270° + i \sin 270°)$

19.

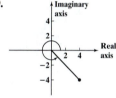

$4\sqrt{2}(\cos 315° + i \sin 315°)$

21.

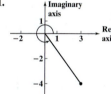

$5(\cos 306.9° + i \sin 306.9°)$

23.

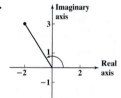

$\sqrt{13}(\cos 123.7° + i \sin 123.7°)$

25. $-1 + \sqrt{3}i$ **27.** $2\sqrt{2} - 2\sqrt{2}i$ **29.** $-3i$ **31.** $-0.035 + 0.197i$ **33.** $1.970 + 0.347i$

35. $zw = 8(\cos 60° + i \sin 60°); \dfrac{z}{w} = \dfrac{1}{2}(\cos 20° + i \sin 20°)$ **37.** $zw = 12(\cos 40° + i \sin 40°); \dfrac{z}{w} = \dfrac{3}{4}(\cos 220° + i \sin 220°)$

39. $zw = 4\left(\cos \dfrac{9\pi}{40} + i \sin \dfrac{9\pi}{40}\right); \dfrac{z}{w} = \cos \dfrac{\pi}{40} + i \sin \dfrac{\pi}{40}$ **41.** $zw = 4\sqrt{2}(\cos 15° + i \sin 15°); \dfrac{z}{w} = \sqrt{2}(\cos 75° + i \sin 75°)$

43. $-32 + 32\sqrt{3}i$ **45.** $32i$ **47.** $\dfrac{27}{2} + \dfrac{27\sqrt{3}}{2}i$ **49.** $-\dfrac{25\sqrt{2}}{2} + \dfrac{25\sqrt{2}}{2}i$ **51.** $-4 + 4i$ **53.** $-23 + 14.142i$

55. $\sqrt[6]{2}(\cos 15° + i \sin 15°), \sqrt[6]{2}(\cos 135° + i \sin 135°), \sqrt[6]{2}(\cos 255° + i \sin 255°)$

57. $\sqrt[4]{8}(\cos 75° + i \sin 75°), \sqrt[4]{8}(\cos 165° + i \sin 165°), \sqrt[4]{8}(\cos 255° + i \sin 255°), \sqrt[4]{8}(\cos 345° + i \sin 345°)$

59. $2(\cos 67.5° + i \sin 67.5°), 2(\cos 157.5° + i \sin 157.5°), 2(\cos 247.5° + i \sin 247.5°), 2(\cos 337.5° + i \sin 337.5°)$

61. $\cos 18° + i \sin 18°, \cos 90° + i \sin 90°, \cos 162° + i \sin 162°, \cos 234° + i \sin 234°, \cos 306° + i \sin 306°$

63. $1, i, -1, -i$

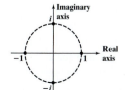

65. Look at formula (8). $|z_k| = \sqrt[n]{r}$ for all k.

67. Look at formula (8). The z_k are spaced apart by an angle of $\dfrac{2\pi}{n}$.

69. Assume the theorem is true for $n \geq 1$.

For $n = 0$:

$z^0 = r^0[\cos(0 \cdot \theta) + i\sin(0 \cdot \theta)]$

$1 = 1 \cdot [\cos(0) + i\sin(0)]$

$1 = 1 \cdot [1 + 0]$

$1 = 1$ True

For negative integers:

$z^{-n} = (z^n)^{-1} = (r^n[\cos(n\theta) + i\sin(n\theta)])^{-1}$ with $n \geq 1$

$= \dfrac{1}{r^n[\cos(n\theta) + i\sin(n\theta)]}$

$= \dfrac{1}{r^n[\cos(n\theta) + i\sin(n\theta)]} \cdot \dfrac{\cos(n\theta) - i\sin(n\theta)}{\cos(n\theta) - i\sin(n\theta)}$

$= \dfrac{\cos(n\theta) - i\sin(n\theta)}{r^n(\cos^2(n\theta) + \sin^2(n\theta))}$

$= \dfrac{\cos(n\theta) - i\sin(n\theta)}{r^n}$

$= r^{-n}[\cos(n\theta) - i\sin(n\theta)]$

$= r^{-n}[\cos(-n\theta) + i\sin(-n\theta)]$

Thus, De Moivre's Theorem is true for all integers.

71. ≈ 40.50 **72.** $\dfrac{4}{3}\pi$ **73.** $2y\sqrt[3]{3x^2y^2}$ **74.** Minimum: $f\left(\dfrac{6}{5}\right) = -\dfrac{16}{5}$

10.4 Assess Your Understanding (page 762)

1. vector   **2.** 0   **3.** unit   **4.** position   **5.** horizontal; vertical   **6.** resultant   **7.** T   **8.** F   **9.** a   **10.** b

11.   **13.**   **15.**   **17.**

19. T   **21.** F   **23.** F   **25.** T   **27.** 12   **29.** $\mathbf{v} = 3\mathbf{i} + 4\mathbf{j}$ **31.** $\mathbf{v} = 2\mathbf{i} + 4\mathbf{j}$ **33.** $\mathbf{v} = 8\mathbf{i} - \mathbf{j}$ **35.** $\mathbf{v} = -\mathbf{i} + \mathbf{j}$ **37.** 5   **39.** $\sqrt{2}$ **41.** $\sqrt{13}$ **43.** $-\mathbf{j}$

45. $\sqrt{89}$ **47.** $\sqrt{34} - \sqrt{13}$ **49.** $\mathbf{i}$ **51.** $\dfrac{3}{5}\mathbf{i} - \dfrac{4}{5}\mathbf{j}$ **53.** $\dfrac{\sqrt{2}}{2}\mathbf{i} - \dfrac{\sqrt{2}}{2}\mathbf{j}$ **55.** $\mathbf{v} = \dfrac{8\sqrt{5}}{5}\mathbf{i} + \dfrac{4\sqrt{5}}{5}\mathbf{j}$, or $\mathbf{v} = -\dfrac{8\sqrt{5}}{5}\mathbf{i} - \dfrac{4\sqrt{5}}{5}\mathbf{j}$

57. $\left\{-2 + \sqrt{21}, -2 - \sqrt{21}\right\}$ **59.** $\mathbf{v} = \dfrac{5}{2}\mathbf{i} + \dfrac{5\sqrt{3}}{2}\mathbf{j}$ **61.** $\mathbf{v} = -7\mathbf{i} + 7\sqrt{3}\mathbf{j}$ **63.** $\mathbf{v} = \dfrac{25\sqrt{3}}{2}\mathbf{i} - \dfrac{25}{2}\mathbf{j}$ **65.** $45°$ **67.** $150°$ **69.** $333.4°$ **71.** $258.7°$

73. $\mathbf{F} = 20\sqrt{3}\mathbf{i} + 20\mathbf{j}$ **75.** $\mathbf{F} = \left(20\sqrt{3} + 30\sqrt{2}\right)\mathbf{i} + \left(20 - 30\sqrt{2}\right)\mathbf{j}$

77. (a) $\mathbf{v}_a = 550\mathbf{j}$; $\mathbf{v}_w = 50\sqrt{2}\mathbf{i} + 50\sqrt{2}\mathbf{j}$

(b) $\mathbf{v}_g = 50\sqrt{2}\mathbf{i} + (550 + 50\sqrt{2})\mathbf{j}$

(c) $\|\mathbf{v}_g\| = 624.7$ mph; N6.5°E

97. (a) $(-1, 4)$

(b)

99.

79. $\mathbf{v} = (250\sqrt{2} - 30)\mathbf{i} + (250\sqrt{2} + 30\sqrt{3})\mathbf{j}$; 518.8 km/h; N38.6°E

81. Approximately 4031 lb   **83.** 8.6° left of direct heading across the river; 1.52 min

85. (a) N7.05°E   **(b)** 12 min   **87.** Tension in right cable: 1000 lb; tension in left cable: 845.2 lb

89. Tension in right part: 1088.4 lb; tension in left part: 1089.1 lb   **91.** $\mu = 0.36$

93. 13.68 lb   **95.** The truck must pull with a force of 4635.2 lb.

103. $\{29\}$ **104.** $-3x(x + 2)(x - 6)$ **105.** $\sqrt{3}$ **106.** Amplitude $= \dfrac{3}{2}$; period $= \dfrac{\pi}{3}$

Phase shift $= -\dfrac{\pi}{2}$

Historical Problem (page 771)

$(a\mathbf{i} + b\mathbf{j}) \cdot (c\mathbf{i} + d\mathbf{j}) = ac + bd$

Real part $[\overline{(a + bi)}(c + di)] = $ real part$[(a - bi)(c + di)] = $ real part$[ac + adi - bci - bdi^2] = ac + bd$

10.5 Assess Your Understanding (page 771)

2. dot product   **3.** orthogonal   **4.** parallel   **5.** T   **6.** F   **7.** d   **8.** b   **9. (a)** 0   **(b)** 90°   **(c)** orthogonal   **11. (a)** 0   **(b)** 90°   **(c)** orthogonal

13. (a) $\sqrt{3} - 1$ **(b)** 75°   **(c)** neither   **15. (a)** -50 **(b)** 180°   **(c)** parallel   **17. (a)** 0   **(b)** 90°   **(c)** orthogonal

19. $\dfrac{2}{3}$ **21.** $\mathbf{v}_1 = \dfrac{5}{2}\mathbf{i} - \dfrac{5}{2}\mathbf{j}$, $\mathbf{v}_2 = -\dfrac{1}{2}\mathbf{i} - \dfrac{1}{2}\mathbf{j}$ **23.** $\mathbf{v}_1 = -\dfrac{1}{5}\mathbf{i} - \dfrac{2}{5}\mathbf{j}$, $\mathbf{v}_2 = \dfrac{6}{5}\mathbf{i} - \dfrac{3}{5}\mathbf{j}$ **25.** $\mathbf{v}_1 = \dfrac{14}{5}\mathbf{i} + \dfrac{7}{5}\mathbf{j}$, $\mathbf{v}_2 = \dfrac{1}{5}\mathbf{i} - \dfrac{2}{5}\mathbf{j}$ **27.** Approximately 1.353

29. 9 ft-lb

31. (a) $\|\mathbf{I}\| \approx 0.022$; the intensity of the sun's rays is approximately 0.022 W/cm². $\|\mathbf{A}\| = 500$; the area of the solar panel is 500 cm².
(b) $W = 10$; ten watts of energy is collected.   **(c)** Vectors $\mathbf{I}$ and $\mathbf{A}$ should be parallel with the solar panels facing the sun.
33. Force required to keep the Sienna from rolling down the hill: 737.6 lb; force perpendicular to the hill: 5248.4 lb
35. Timmy must exert 85.5 lb.   **37.** 60°   **39.** Let $\mathbf{v} = a\mathbf{i} + b\mathbf{j}$. Then $\mathbf{0} \cdot \mathbf{v} = 0a + 0b = 0$.
41. $\mathbf{v} = \cos \alpha \mathbf{i} + \sin \alpha \mathbf{j}, 0 \le \alpha \le \pi$; $\mathbf{w} = \cos \beta \mathbf{i} + \sin \beta \mathbf{j}, 0 \le \beta \le \pi$. If θ is the angle between $\mathbf{v}$ and $\mathbf{w}$, then $\mathbf{v} \cdot \mathbf{w} = \cos \theta$,
since $\|\mathbf{v}\| = 1$ and $\|\mathbf{w}\| = 1$. Now $\theta = \alpha - \beta$ or $\theta = \beta - \alpha$. Since the cosine function is even,
$\mathbf{v} \cdot \mathbf{w} = \cos(\alpha - \beta)$. Also, $\mathbf{v} \cdot \mathbf{w} = \cos \alpha \cos \beta + \sin \alpha \sin \beta$. So $\cos(\alpha - \beta) = \cos \alpha \cos \beta + \sin \alpha \sin \beta$.
43. (a) If $\mathbf{u} = a_1\mathbf{i} + b_1\mathbf{j}$ and $\mathbf{v} = a_2\mathbf{i} + b_2\mathbf{j}$, then, since $\|\mathbf{u}\| = \|\mathbf{v}\|, a_1^2 + b_1^2 = \|\mathbf{u}\|^2 = \|\mathbf{v}\|^2 = a_2^2 + b_2^2$,
$(\mathbf{u} + \mathbf{v}) \cdot (\mathbf{u} - \mathbf{v}) = (a_1 + a_2)(a_1 - a_2) + (b_1 + b_2)(b_1 - b_2) = (a_1^2 + b_1^2) - (a_2^2 + b_2^2) = 0$.
(b) The legs of the angle can be made to correspond to vectors $\mathbf{u} + \mathbf{v}$ and $\mathbf{u} - \mathbf{v}$.
45. $(\|\mathbf{w}\|\mathbf{v} + \|\mathbf{v}\|\mathbf{w}) \cdot (\|\mathbf{w}\|\mathbf{v} - \|\mathbf{v}\|\mathbf{w}) = \|\mathbf{w}\|^2 \mathbf{v} \cdot \mathbf{v} - \|\mathbf{w}\|\|\mathbf{v}\|\mathbf{v} \cdot \mathbf{w} + \|\mathbf{v}\|\|\mathbf{w}\|\mathbf{w} \cdot \mathbf{v} - \|\mathbf{v}\|^2 \mathbf{w} \cdot \mathbf{w} = \|\mathbf{w}\|^2 \mathbf{v} \cdot \mathbf{v} - \|\mathbf{v}\|^2 \mathbf{w} \cdot \mathbf{w} = \|\mathbf{w}\|^2 \|\mathbf{v}\|^2 - \|\mathbf{v}\|^2 \|\mathbf{w}\|^2 = 0$
47. $\|\mathbf{u} + \mathbf{v}\|^2 - \|\mathbf{u} - \mathbf{v}\|^2 = (\mathbf{u} + \mathbf{v}) \cdot (\mathbf{u} + \mathbf{v}) - (\mathbf{u} - \mathbf{v}) \cdot (\mathbf{u} - \mathbf{v}) = (\mathbf{u} \cdot \mathbf{u} + \mathbf{u} \cdot \mathbf{v} + \mathbf{v} \cdot \mathbf{u} + \mathbf{v} \cdot \mathbf{v}) - (\mathbf{u} \cdot \mathbf{u} - \mathbf{u} \cdot \mathbf{v} - \mathbf{v} \cdot \mathbf{u} + \mathbf{v} \cdot \mathbf{v})$
$$= 2(\mathbf{u} \cdot \mathbf{v}) + 2(\mathbf{v} \cdot \mathbf{u}) = 4(\mathbf{u} \cdot \mathbf{v})$$
49. 12   **50.** $\dfrac{9}{2}$ **51.** $(1 - \sin^2\theta)(1 + \tan^2\theta) = (\cos^2\theta)(\sec^2\theta)$ **52.** $V(x) = x(19 - 2x)(13 - 2x)$, or $V(x) = 4x^3 - 64x^2 + 247x$
$$= \cos^2\theta \cdot \frac{1}{\cos^2\theta}$$
$$= 1$$

Review Exercises *(page 774)*

1. $\left(\dfrac{3\sqrt{3}}{2}, \dfrac{3}{2}\right)$

2. $(1, \sqrt{3})$

3. $(0, 3)$

4. $\left(3\sqrt{2}, \dfrac{3\pi}{4}\right), \left(-3\sqrt{2}, -\dfrac{\pi}{4}\right)$

5. $\left(2, -\dfrac{\pi}{2}\right), \left(-2, \dfrac{\pi}{2}\right)$

6. $(5, 0.93), (-5, 4.07)$

7. (a) $x^2 + (y - 1)^2 = 1$ **(b)** circle, radius 1, center $(0, 1)$ in rectangular coordinates

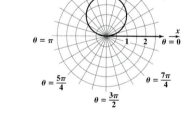

8. (a) $x^2 + y^2 = 25$ **(b)** circle, radius 5, center at pole

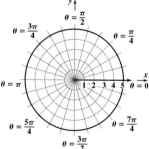

9. (a) $x - y = 0$ **(b)** line though pole, making an angle of $\dfrac{\pi}{4}$ with polar axis

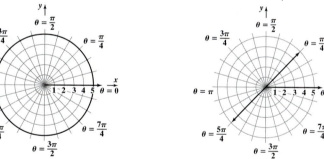

10. (a) $(x - 4)^2 + (y + 2)^2 = 25$ **(b)** circle, radius 5, center $(4, -2)$ in rectangular coordinates

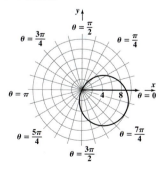

11. Circle; radius 2, center at $(2, 0)$ in rectangular coordinates; symmetric with respect to the polar axis

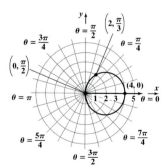

12. Cardioid; symmetric with respect to the line $\theta = \dfrac{\pi}{2}$

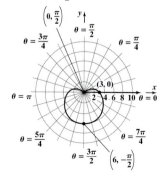

83. 35 million mi   **85.** $5\sqrt{5} - 4$

87. $Ax^2 + Cy^2 + Dx + Ey + F = 0$ $A \neq 0, C \neq 0$

$$Ax^2 + Dx + Cy^2 + Ey = -F$$

$$A\left(x^2 + \frac{D}{A}x\right) + C\left(y^2 + \frac{E}{Cy}\right) = -F$$

$$A\left(x + \frac{D}{2A}\right)^2 + C\left(y + \frac{E}{2C}\right)^2 = -F + \frac{D^2}{4A} + \frac{E^2}{4C}$$

(a) If $\dfrac{D^2}{4A} + \dfrac{E^2}{4C} - F$ is of the same sign as A (and C), this is the equation of an ellipse with center at $\left(-\dfrac{D}{2A}, -\dfrac{E}{2C}\right)$.

(b) If $\dfrac{D^2}{4A} + \dfrac{E^2}{4C} - F$, the graph is the single point $\left(-\dfrac{D}{2A}, -\dfrac{E}{2C}\right)$.

(c) If $\dfrac{D^2}{4A} + \dfrac{E^2}{4C} - F$ is of the sign opposite that of A (and C), the graph contains no points, because in this case, the left side has the sign opposite
that of the right side.

89. Zeros: $5 - 2\sqrt{3}, 5 + 2\sqrt{3}$; x-intercepts: $5 - 2\sqrt{3}, 5 + 2\sqrt{3}$ **90.** Domain: $\{x \mid x \neq 5\}$; Horizontal asymptote: $y = 2$; Vertical asymptote: $x = 5$
91. 617.1 ft-lb   **92.** $b \approx 10.94, c \approx 17.77, B = 38°$

11.4 Assess Your Understanding (page 809)

7. hyperbola   **8.** transverse axis   **9.** b   **10.** $(2, 4); (2, -2)$ **11.** $(2, 6); (2, -4)$ **12.** c   **13.** $2; 3; x$ **14.** $y = -\dfrac{4}{9}x; y = \dfrac{4}{9}x$ **15.** B   **17.** A

19. $x^2 - \dfrac{y^2}{8} = 1$

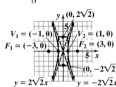

21. $\dfrac{y^2}{16} - \dfrac{x^2}{20} = 1$

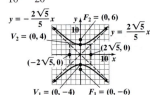

23. $\dfrac{x^2}{9} - \dfrac{y^2}{16} = 1$

25. $\dfrac{y^2}{36} - \dfrac{x^2}{9} = 1$

27. $\dfrac{x^2}{8} - \dfrac{y^2}{8} = 1$

29. $\dfrac{x^2}{25} - \dfrac{y^2}{9} = 1$

Center: $(0, 0)$
Transverse axis: x-axis
Vertices: $(-5, 0), (5, 0)$
Foci: $\left(-\sqrt{34}, 0\right), \left(\sqrt{34}, 0\right)$
Asymptotes: $y = \pm\dfrac{3}{5}x$

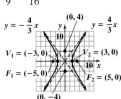

31. $\dfrac{x^2}{4} - \dfrac{y^2}{16} = 1$

Center: $(0, 0)$
Transverse axis: x-axis
Vertices: $(-2, 0), (2, 0)$
Foci: $\left(-2\sqrt{5}, 0\right), \left(2\sqrt{5}, 0\right)$
Asymptotes: $y = \pm 2x$

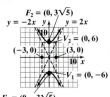

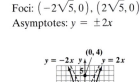

33. $\dfrac{y^2}{9} - x^2 = 1$

Center: $(0, 0)$
Transverse axis: y-axis
Vertices: $(0, -3), (0, 3)$
Foci: $\left(0, -\sqrt{10}\right), \left(0, \sqrt{10}\right)$
Asymptotes: $y = \pm 3x$

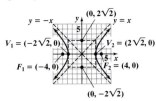

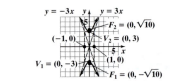

35. $\dfrac{y^2}{25} - \dfrac{x^2}{25} = 1$

Center: $(0, 0)$
Transverse axis: y-axis
Vertices: $(0, -5), (0, 5)$
Foci: $\left(0, -5\sqrt{2}\right), \left(0, 5\sqrt{2}\right)$
Asymptotes: $y = \pm x$

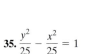

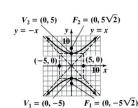

37. $x^2 - y^2 = 1$

39. $\dfrac{y^2}{36} - \dfrac{x^2}{9} = 1$

41. $\dfrac{(x-4)^2}{4} - \dfrac{(y+1)^2}{5} = 1$

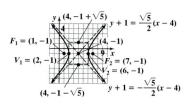

43. $\dfrac{(y+4)^2}{4} - \dfrac{(x+3)^2}{12} = 1$

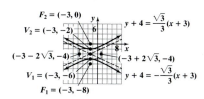

45. $(x-5)^2 - \dfrac{(y-7)^2}{3} = 1$

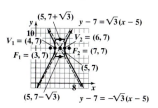

47. $\dfrac{(x-1)^2}{4} - \dfrac{(y+1)^2}{9} = 1$

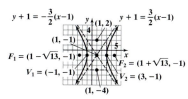

49. $\dfrac{(x-2)^2}{4} - \dfrac{(y+3)^2}{9} = 1$
Center: $(2, -3)$
Transverse axis: parallel to x-axis
Vertices: $(0, -3), (4, -3)$
Foci: $\left(2 - \sqrt{13}, -3\right), \left(2 + \sqrt{13}, -3\right)$
Asymptotes: $y + 3 = \pm\dfrac{3}{2}(x - 2)$

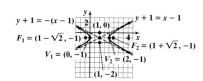

51. $\dfrac{(y-2)^2}{4} - (x+2)^2 = 1$
Center: $(-2, 2)$
Transverse axis: parallel to y-axis
Vertices: $(-2, 0), (-2, 4)$
Foci: $\left(-2, 2 - \sqrt{5}\right), \left(-2, 2 + \sqrt{5}\right)$
Asymptotes: $y - 2 = \pm 2(x + 2)$

53. $\dfrac{(x+1)^2}{4} - \dfrac{(y+2)^2}{4} = 1$
Center: $(-1, -2)$
Transverse axis: parallel to x-axis
Vertices: $(-3, -2), (1, -2)$
Foci: $\left(-1 - 2\sqrt{2}, -2\right), \left(-1 + 2\sqrt{2}, -2\right)$
Asymptotes: $y + 2 = \pm (x + 1)$

55. $(x-1)^2 - (y+1)^2 = 1$
Center: $(1, -1)$
Transverse axis: parallel to x-axis
Vertices: $(0, -1), (2, -1)$
Foci: $\left(1 - \sqrt{2}, -1\right), \left(1 + \sqrt{2}, -1\right)$
Asymptotes: $y + 1 = \pm (x - 1)$

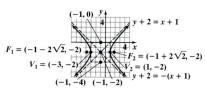

57. $\dfrac{(y-2)^2}{4} - (x+1)^2 = 1$
Center: $(-1, 2)$
Transverse axis: parallel to y-axis
Vertices: $(-1, 0), (-1, 4)$
Foci: $\left(-1, 2 - \sqrt{5}\right), \left(-1, 2 + \sqrt{5}\right)$
Asymptotes: $y - 2 = \pm 2(x + 1)$

59. $\dfrac{(x-3)^2}{4} - \dfrac{(y+2)^2}{16} = 1$
Center: $(3, -2)$
Transverse axis: parallel to x-axis
Vertices: $(1, -2), (5, -2)$
Foci: $\left(3 - 2\sqrt{5}, -2\right), \left(3 + 2\sqrt{5}, -2\right)$
Asymptotes: $y + 2 = \pm 2(x - 3)$

61. $\dfrac{(y-1)^2}{4} - (x+2)^2 = 1$
Center: $(-2, 1)$
Transverse axis: parallel to y-axis
Vertices: $(-2, -1), (-2, 3)$
Foci: $\left(-2, 1 - \sqrt{5}\right), \left(-2, 1 + \sqrt{5}\right)$
Asymptotes: $y - 1 = \pm 2(x + 2)$

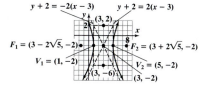

63.

65.

67. Center: $(3, 0)$
Transverse axis: parallel to x-axis
Vertices: $(1, 0), (5, 0)$
Foci: $(3 - \sqrt{29}, 0), (3 + \sqrt{29}, 0)$
Asymptotes: $y = \pm\dfrac{5}{2}(x - 3)$

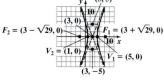

69. Vertex: $(0, 3)$; focus: $(0, 7)$;
directrix: $y = -1$

71. $\dfrac{(x-5)^2}{9} + \dfrac{y^2}{25} = 1$
Center: $(5, 0)$; vertices: $(5, 5), (5, -5)$;
foci: $(5, -4), (5, 4)$

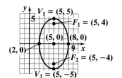

73. $(x-3)^2 = 8(y + 5)$
Vertex: $(3, -5)$; focus: $(3, -3)$;
directrix: $y = -7$

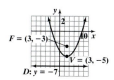

25. $y^2 + 2x - 1 = 0$ **27.** $16x^2 + 7y^2 + 48y - 64 = 0$ **29.** $3x^2 - y^2 + 12x + 9 = 0$ **31.** $4x^2 + 3y^2 - 16y - 64 = 0$

33. $9x^2 + 5y^2 - 24y - 36 = 0$ **35.** $3x^2 + 4y^2 - 12x - 36 = 0$ **37.** $r = \dfrac{1}{1 + \sin\theta}$ **39.** $r = \dfrac{12}{5 - 4\cos\theta}$ **41.** $r = \dfrac{12}{1 - 6\sin\theta}$

43. Use $d(D, P) = p - r\cos\theta$ in the derivation of equation (a) in Table 5.
45. Use $d(D, P) = p + r\sin\theta$ in the derivation of equation (a) in Table 5.

47. 27.81   **48.** Amplitude $= 4$; Period $= 10\pi$ **49.** $\left\{\dfrac{\pi}{3}, \pi, \dfrac{5\pi}{3}\right\}$ **50.** 26

11.7 Assess Your Understanding *(page 835)*

2. plane curve; parameter   **3.** ellipse   **4.** cycloid   **5.** F   **6.** T

7.
$x - 3y + 1 = 0$

9.
$y = \sqrt{x - 2}$

11.
$y = x - 8$

13.
$x = 3(y - 1)^2$

15.
$2y = 2 + x$

17.
$y = x^3$

19.
$\dfrac{x^2}{4} + \dfrac{y^2}{9} = 1$

21.
$\dfrac{x^2}{4} + \dfrac{y^2}{9} = 1$

23.
$x^2 - y^2 = 1$

25.
$x + y = 1$

27. $x = t$ or $x = \dfrac{t + 1}{4}$
$\quad y = 4t - 1 \qquad\quad y = t$

29. $x = t$ or $x = t^3$
$\quad y = t^2 + 1 \qquad y = t^6 + 1$

31. $x = t$ or $x = \sqrt[3]{t}$
$\quad y = t^3 \qquad\quad y = t$

33. $x = t$ or $x = t^3$
$\quad y = t^{2/3}, t \ge 0 \qquad y = t^2, t \ge 0$

35. $x = t + 2, y = t, 0 \le t \le 5$ **37.** $x = 3\cos t, y = 2\sin t, 0 \le t \le 2\pi$
39. $x = 2\cos(\pi t), y = -3\sin(\pi t), 0 \le t \le 2$
41. $x = 2\sin(2\pi t), y = 3\cos(2\pi t), 0 \le t \le 1$

43.

45.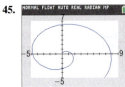

47.

49. (a) $x = 3$
$\quad y = -16t^2 + 50t + 6$
(b) 3.24 s
(c) 1.56 s; 45.06 ft
(d)

51. (a) Train: $x_1 = t^2, y_1 = 1$;
$\quad$ Bill: $x_2 = 5(t - 5), y_2 = 3$
(b) Bill won't catch the train.
(c)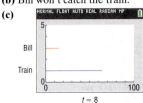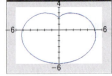

53. (a) $x = (145\cos 20°)t$
$\quad y = -16t^2 + (145\sin 20°)t + 5$
(b) 3.20 s
(c) 435.65 ft
(d) 1.55 s; 43.43 ft
(e)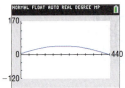

55. (a) $x = (40 \cos 45°)t$
$y = -4.9t^2 + (40 \sin 45°)t + 300$
(b) 11.23 s
(c) 317.52 m
(d) 2.89 s; 340.82 m
(e)

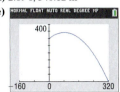

57. (a) Camry: $x = 40t - 5$, $y = 0$; Chevy Impala: $x = 0$, $y = 30t - 4$
(b) $d = \sqrt{(40t - 5)^2 + (30t - 4)^2}$
(c)

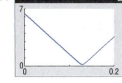

(d) 0.2 mi; 7.68 min
(e)

59. (a) $x = \dfrac{\sqrt{2}}{2} v_0 t$, $y = -16t^2 + \dfrac{\sqrt{2}}{2} v_0 t + 3$ **(b)** Maximum height is 139.1 ft. **(c)** The ball is 272.25 ft from home plate. **(d)** Yes, the ball will clear the wall by about 99.5 ft. **61.** The orientation is from (x_1, y_1) to (x_2, y_2).

65.

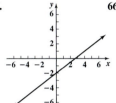

66.

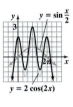

67. Approximately 2733 miles **68. (a)** Simple harmonic **(b)** 2 m **(c)** $\dfrac{\pi}{2}$ s **(d)** $\dfrac{2}{\pi}$ oscillations/s

Review Exercises *(page 839)*

1. Parabola; vertex $(0, 0)$, focus $(-4, 0)$, directrix $x = 4$ **2.** Hyperbola; center $(0, 0)$, vertices $(5, 0)$ and $(-5, 0)$, foci $(\sqrt{26}, 0)$ and $(-\sqrt{26}, 0)$, asymptotes $y = \dfrac{1}{5}x$ and $y = -\dfrac{1}{5}x$ **3.** Ellipse; center $(0, 0)$, vertices $(0, 5)$ and $(0, -5)$, foci $(0, 3)$ and $(0, -3)$

4. $x^2 = -4(y - 1)$: Parabola; vertex $(0, 1)$, focus $(0, 0)$, directrix $y = 2$ **5.** $\dfrac{x^2}{2} - \dfrac{y^2}{8} = 1$: Hyperbola; center $(0, 0)$, vertices $(\sqrt{2}, 0)$ and $(-\sqrt{2}, 0)$, foci $(\sqrt{10}, 0)$ and $(-\sqrt{10}, 0)$, asymptotes $y = 2x$ and $y = -2x$ **6.** $(x - 2)^2 = 2(y + 2)$: Parabola; vertex $(2, -2)$, focus $\left(2, -\dfrac{3}{2}\right)$, directrix $y = -\dfrac{5}{2}$

7. $\dfrac{(y - 2)^2}{4} - (x - 1)^2 = 1$: Hyperbola; center $(1, 2)$, vertices $(1, 4)$ and $(1, 0)$, foci $(1, 2 + \sqrt{5})$ and $(1, 2 - \sqrt{5})$, asymptotes $y - 2 = \pm 2(x - 1)$

8. $\dfrac{(x - 2)^2}{9} + \dfrac{(y - 1)^2}{4} = 1$: Ellipse; center $(2, 1)$, vertices $(5, 1)$ and $(-1, 1)$, foci $(2 + \sqrt{5}, 1)$ and $(2 - \sqrt{5}, 1)$

9. $(x - 2)^2 = -4(y + 1)$: Parabola; vertex $(2, -1)$, focus $(2, -2)$, directrix $y = 0$

10. $\dfrac{(x - 1)^2}{4} + \dfrac{(y + 1)^2}{9} = 1$: Ellipse; center $(1, -1)$, vertices $(1, 2)$ and $(1, -4)$, foci $(1, -1 + \sqrt{5})$ and $(1, -1 - \sqrt{5})$

11. $y^2 = -8x$

12. $\dfrac{y^2}{4} - \dfrac{x^2}{12} = 1$

13. $\dfrac{x^2}{16} + \dfrac{y^2}{7} = 1$

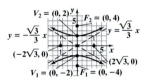

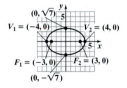

14. $(x - 2)^2 = -4(y + 3)$

15. $(x + 2)^2 - \dfrac{(y + 3)^2}{3} = 1$

16. $\dfrac{(x + 4)^2}{16} + \dfrac{(y - 5)^2}{25} = 1$

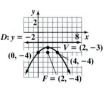

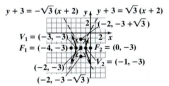

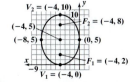

17. $\dfrac{(x+1)^2}{9} - \dfrac{(y-2)^2}{7} = 1$

$y - 2 = -\dfrac{\sqrt{7}}{3}(x+1)$ $(-1, 2+\sqrt{7})$
$V_1 = (-4, 2)$ $V_2 = (2, 2)$
$F_1 = (-5, 2)$ $F_2 = (3, 2)$
$(-1, 2)$
$y - 2 = \dfrac{\sqrt{7}}{3}(x+1)$ $(-1, 2-\sqrt{7})$

18. $\dfrac{(x-3)^2}{9} - \dfrac{(y-1)^2}{4} = 1$

$y - 1 = -\dfrac{2}{3}(x-3)$ $(3, 3)$ $(3, 1)$
$V_1 = (0, 1)$ $V_2 = (6, 1)$
$F_1 = (3 - \sqrt{13},\, 1)$ $F_2 = (3 + \sqrt{13},\, 1)$
$y - 1 = \dfrac{2}{3}(x-3)$ $(3, -1)$

19. Parabola   **20.** Ellipse
21. Parabola   **22.** Hyperbola
23. Ellipse

24. $x'^2 - \dfrac{y'^2}{9} = 1$

Hyperbola
Center at the origin
Transverse axis the x'-axis
Vertices at $(\pm 1, 0)$

$(-1, 0)$ $(1, 0)$

25. $\dfrac{x'^2}{2} + \dfrac{y'^2}{4} = 1$

Ellipse
Center at origin
Major axis the y'-axis
Vertices at $(0, \pm 2)$

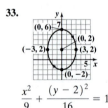

$(0, 2)$
$(0, -2)$

26. $y'^2 = -\dfrac{4\sqrt{13}}{13}x'$

Parabola
Vertex at the origin
Focus on the x'-axis at $\left(-\dfrac{\sqrt{13}}{13}, 0\right)$

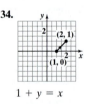

27. Parabola; directrix is perpendicular to the polar axis 4 units to the left of the pole; vertex is $(2, \pi)$.

Directrix
$\left(4, \dfrac{\pi}{2}\right)$
$(2, \pi)$ Polar
x axis
$\left(4, \dfrac{3\pi}{2}\right)$

28. Ellipse; directrix is parallel to the polar axis 6 units below the pole; vertices are $\left(6, \dfrac{\pi}{2}\right)$ and $\left(2, \dfrac{3\pi}{2}\right)$.

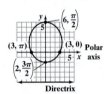

$\left(6, \dfrac{\pi}{2}\right)$
$(3, \pi)$ $(3, 0)$ Polar
x axis
$\left(2, \dfrac{3\pi}{2}\right)$
Directrix

29. Hyperbola; directrix is perpendicular to the polar axis 1 unit to the right of the pole; vertices are $\left(\dfrac{2}{3}, 0\right)$ and $(-2, \pi)$.

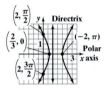

$\left(2, \dfrac{\pi}{2}\right)$ Directrix
$\left(\dfrac{2}{3}, 0\right)$ $(-2, \pi)$
Polar
x axis
$\left(2, \dfrac{3\pi}{2}\right)$

30. $y^2 - 8x - 16 = 0$ **31.** $3x^2 - y^2 - 8x + 4 = 0$

32.

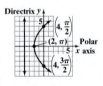

$(-2, 1)$
$(2, 0)$

$x + 4y = 2$

33.

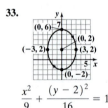

$(0, 6)$
$(-3, 2)$ $(0, 2)$ $(3, 2)$
$(0, -2)$

$\dfrac{x^2}{9} + \dfrac{(y-2)^2}{16} = 1$

34.

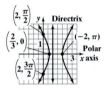

$(2, 1)$
$(1, 0)$

$1 + y = x$

35. $x = t,\ y = -2t + 4,\ -\infty < t < \infty$ **36.** $x = 4\cos\left(\dfrac{\pi}{2}t\right),\ y = 3\sin\left(\dfrac{\pi}{2}t\right),\ 0 \le t \le 4$ **37.** $\dfrac{x^2}{5} - \dfrac{y^2}{4} = 1$ **38.** The ellipse $\dfrac{x^2}{16} + \dfrac{y^2}{7} = 1$

$x = \dfrac{t-4}{-2},\ y = t,\ -\infty < t < \infty$

39. $\dfrac{1}{4}$ ft or 3 in.   **40.** 19.72 ft, 18.86 ft, 14.91 ft   **41.** 450 ft

42. (a) Train: $x_1 = \dfrac{3}{2}t^2,\ y_1 = 1$ **(c)**

Mary: $x_2 = 6(t-2),\ y_2 = 3$
(b) Mary won't catch the train.

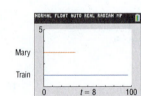

43. (a) $x = (80\cos 35°)t$ **(e)**
$y = -16t^2 + (80\sin 35°)t + 6$
(b) 2.9932 s
(c) 1.4339 s; 38.9 ft
(d) 196.15 ft

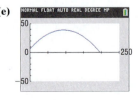

Chapter Test *(page 841)*

1. Hyperbola; center: $(-1, 0)$; vertices: $(-3, 0)$ and $(1, 0)$; foci: $\left(-1 - \sqrt{13}, 0\right)$ and $\left(-1 + \sqrt{13}, 0\right)$; asymptotes: $y = -\dfrac{3}{2}(x+1)$ and $y = \dfrac{3}{2}(x+1)$

2. Parabola; vertex: $\left(1, -\dfrac{1}{2}\right)$; focus: $\left(1, \dfrac{3}{2}\right)$; directrix: $y = -\dfrac{5}{2}$

3. Ellipse; center: $(-1, 1)$; foci: $\left(-1 - \sqrt{3}, 1\right)$ and $\left(-1 + \sqrt{3}, 1\right)$; vertices: $(-4, 1)$ and $(2, 1)$

4. $(x + 1)^2 = 6(y - 3)$

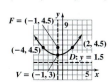

$F = (-1, 4.5)$
$(-4, 4.5)$ $(2, 4.5)$
$D: y = 1.5$
$V = (-1, 3)$

5. $\dfrac{x^2}{7} + \dfrac{y^2}{16} = 1$

$V_1 = (0, 4)$
$(-\sqrt{7}, 0)$ F_1 $(\sqrt{7}, 0)$
F_2
$V_2 = (0, -4)$

6. $\dfrac{(y - 2)^2}{4} - \dfrac{(x - 2)^2}{8} = 1$

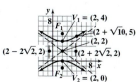

$V_1 = (2, 4)$
F_1 $(2 + \sqrt{10}, 5)$
$(2, 2)$
$(2 - 2\sqrt{2}, 2)$ $(2 + 2\sqrt{2}, 2)$
F_2
$V_2 = (2, 0)$

7. Hyperbola   **8.** Ellipse   **9.** Parabola

10. $x'^2 + 2y'^2 = 1$. This is the equation of an ellipse with center at $(0, 0)$ in the $x'y'$-plane. The vertices are at $(-1, 0)$ and $(1, 0)$ in the $x'y'$-plane.

11. Hyperbola; $(x + 2)^2 - \dfrac{y^2}{3} = 1$ **12.** $y = 1 - \sqrt{\dfrac{x + 2}{3}}$

$(-2, 1)$ $(1, 0)$
$45\ x$
$(25, -2)$
$(10, -1)$

13. The microphone should be located $\dfrac{2}{3}$ ft from the base of the reflector, along its axis of symmetry.

Cumulative Review *(page 841)*

1. $-6x + 5 - 3h$ **2.** $\left\{-5, -\dfrac{1}{3}, 2\right\}$ **3.** $\{x \,|\, -3 \le x \le 2\}$ or $[-3, 2]$ **4. (a)** Domain: $(-\infty, \infty)$; range: $(2, \infty)$

(b) $y = \log_3(x - 2)$; domain: $(2, \infty)$; range: $(-\infty, \infty)$ **5. (a)** $\{18\}$ **(b)** $(2, 18)$

6. (a) $y = 2x - 2$ **(b)** $(x - 2)^2 + y^2 = 4$ **(c)** $\dfrac{x^2}{9} + \dfrac{y^2}{4} = 1$ **(d)** $y = 2(x - 1)^2$ **(e)** $y^2 - \dfrac{x^2}{3} = 1$ **(f)** $y = 4^x$

7. $\theta = \dfrac{\pi}{12} \pm \pi k$, k is any integer; $\theta = \dfrac{5\pi}{12} \pm \pi k$, k is any integer   **8.** $\theta = \dfrac{\pi}{6}$

9. $r = 8 \sin \theta$

10. $\left\{ x \,\Big|\, x \ne \dfrac{3\pi}{4} \pm \pi k, k \text{ is an integer} \right\}$ **11.** $\{22.5°\}$ **12.** $y = \dfrac{x^2}{5} + 5$

CHAPTER 12 Systems of Equations and Inequalities

12.1 Assess Your Understanding *(page 854)*

3. inconsistent   **4.** consistent; independent   **5.** $(3, -2)$ **6.** consistent; dependent   **7.** b   **8.** a   **9.** $\begin{cases} 2(2) - (-1) = 5 \\ 5(2) + 2(-1) = 8 \end{cases}$

11. $\begin{cases} 3(2) - 4\left(\dfrac{1}{2}\right) = 4 \\ \dfrac{1}{2}(2) - 3\left(\dfrac{1}{2}\right) = -\dfrac{1}{2} \end{cases}$ **13.** $\begin{cases} 4 - 1 = 3 \\ \dfrac{1}{2}(4) + 1 = 3 \end{cases}$ **15.** $\begin{cases} 3(1) + 3(-1) + 2(2) = 4 \\ 1 - (-1) - 2 = 0 \\ 2(-1) - 3(2) = -8 \end{cases}$ **17.** $\begin{cases} 3(2) + 3(-2) + 2(2) = 4 \\ 2 - 3(-2) + 2 = 10 \\ 5(2) - 2(-2) - 3(2) = 8 \end{cases}$

19. $x = 6, y = 2; (6, 2)$

$x + y = 8$ $(6, 2)$
$x - y = 4$

21. $x = 3, y = -6; (3, -6)$

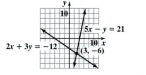

$5x - y = 21$
$2x + 3y = -12$ $(3, -6)$

23. $x = 8, y = -4; (8, -4)$

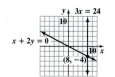

$3x = 24$
$x + 2y = 0$
$(8, -4)$

25. $x = \dfrac{1}{3}, y = -\dfrac{1}{6}; \left(\dfrac{1}{3}, -\dfrac{1}{6}\right)$

$5x + 4y = 1$ $\left(\dfrac{1}{3}, -\dfrac{1}{6}\right)$
$3x - 6y = 2$

27. Inconsistent

$2x + y = 1$
$4x + 2y = 3$

29. $x = \dfrac{3}{2}, y = 3; \left(\dfrac{3}{2}, 3\right)$

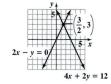

$\left(\dfrac{3}{2}, 3\right)$
$2x - y = 0$
$4x + 2y = 12$

31. $\{(x, y)\,|\,x = 4 - 2y, y \text{ is any real number}\}$, or $\left\{(x, y)\,\middle|\,y = \dfrac{4 - x}{2}, x \text{ is any real number}\right\}$ **33.** $x = 1, y = 1; (1, 1)$ **35.** $x = \dfrac{3}{2}, y = 1; \left(\dfrac{3}{2}, 1\right)$

37. $x = 4, y = 3; (4, 3)$ **39.** $x = \dfrac{4}{3}, y = \dfrac{1}{5}; \left(\dfrac{4}{3}, \dfrac{1}{5}\right)$ **41.** $x = \dfrac{1}{5}, y = \dfrac{1}{3}; \left(\dfrac{1}{5}, \dfrac{1}{3}\right)$ **43.** $x = 8, y = 2, z = 0; (8, 2, 0)$

45. $x = 2, y = -1, z = 1; (2, -1, 1)$ **47.** Inconsistent **49.** $\{(x, y, z)\,|\,x = 5z - 2, y = 4z - 3; z \text{ is any real number}\}$ **51.** Inconsistent

53. $x = 1, y = 3, z = -2; (1, 3, -2)$ **55.** $x = -3, y = \dfrac{1}{2}, z = 1; \left(-3, \dfrac{1}{2}, 1\right)$ **57.** Length 30 ft; width 15 ft

59. There were 23 commercial launches and 58 noncommercial launches in 2013. **61.** 22.5 lb **63.** Smartphone: \$325; tablet: \$640

65. Average wind speed 25 mph; average airspeed 175 mph **67.** 80 \$25 sets and 120 \$45 sets **69.** \$9.96

71. Mix 50 mg of first compound with 75 mg of second. **73.** $a = \dfrac{4}{3}, b = -\dfrac{5}{3}, c = 1$ **75.** $Y = 9000, r = 0.06$ **77.** $I_1 = \dfrac{10}{71}, I_2 = \dfrac{65}{71}, I_3 = \dfrac{55}{71}$

79. 100 orchestra, 210 main, and 190 balcony seats **81.** 1.5 chicken, 1 corn, 2 milk

83. If x = price of hamburgers, y = price of fries,

and z = price of colas, then $x = 2.75 - z, y = \dfrac{41}{60} + \dfrac{1}{3}z, \$0.60 \le z \le \$0.90$.

There is not sufficient information:

x	\$2.13	\$2.01	\$1.86
y	\$0.89	\$0.93	\$0.98
z	\$0.62	\$0.74	\$0.89

85. It will take Beth 30 hr, Bill 24 hr, and Edie 40 hr.

89.

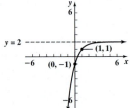

90. (a) $2(2x - 3)^3(x^3 + 5)(10x^3 - 9x^2 + 20)$ **(b)** $7(3x - 5)^{-1/2}(x + 3)^{-3/2}$

91. $\dfrac{\pi}{9}$ **92.** $2(\cos 150° + i\sin 150°)$

12.2 Assess Your Understanding (page 870)

1. matrix **2.** augmented **3.** third; fifth **4.** T **5.** b **6.** c **7.** $\begin{bmatrix} 1 & -5 & 5 \\ 4 & 3 & 6 \end{bmatrix}$ **9.** $\begin{bmatrix} 2 & 3 & 6 \\ 4 & -6 & -2 \end{bmatrix}$ **11.** $\begin{bmatrix} 0.01 & -0.03 & 0.06 \\ 0.13 & 0.10 & 0.20 \end{bmatrix}$

13. $\begin{bmatrix} 1 & -1 & 1 & 10 \\ 3 & 3 & 0 & 5 \\ 1 & 1 & 2 & 2 \end{bmatrix}$ **15.** $\begin{bmatrix} 1 & 1 & -1 & 2 \\ 3 & -2 & 0 & 2 \\ 5 & 3 & -1 & 1 \end{bmatrix}$ **17.** $\begin{bmatrix} 1 & -1 & -1 & 10 \\ 2 & 1 & 2 & -1 \\ -3 & 4 & 0 & 5 \\ 4 & -5 & 1 & 0 \end{bmatrix}$ **19.** $\begin{cases} x - 3y = -2 & (1) \\ 2x - 5y = 5 & (2) \end{cases}; \begin{bmatrix} 1 & -3 & -2 \\ 0 & 1 & 9 \end{bmatrix}$

21. $\begin{cases} x - 3y + 4z = 3 & (1) \\ 3x - 5y + 6z = 6 & (2) \\ -5x + 3y + 4z = 6 & (3) \end{cases}; \begin{bmatrix} 1 & -3 & 4 & 3 \\ 0 & 4 & -6 & -3 \\ 0 & -12 & 24 & 21 \end{bmatrix}$ **23.** $\begin{cases} x - 3y + 2z = -6 & (1) \\ 2x - 5y + 3z = -4 & (2) \\ -3x - 6y + 4z = 6 & (3) \end{cases}; \begin{bmatrix} 1 & -3 & 2 & -6 \\ 0 & 1 & -1 & 8 \\ 0 & -15 & 10 & -12 \end{bmatrix}$

25. $\begin{cases} 5x - 3y + z = -2 & (1) \\ 2x - 5y + 6z = -2 & (2) \\ -4x + y + 4z = 6 & (3) \end{cases}; \begin{bmatrix} 1 & 7 & -11 & 2 \\ 2 & -5 & 6 & -2 \\ 0 & -9 & 16 & 2 \end{bmatrix}$

27. $\begin{cases} x = 5 \\ y = -1 \end{cases}$
Consistent; $x = 5, y = -1$ or $(5, -1)$

29. $\begin{cases} x = 1 \\ y = 2 \\ 0 = 3 \end{cases}$
Inconsistent

31. $\begin{cases} x + 2z = -1 \\ y - 4z = -2 \\ 0 = 0 \end{cases}$
Consistent:
$\begin{cases} x = -1 - 2z \\ y = -2 + 4z \\ z \text{ is any real number} \end{cases}$ or
$\{(x, y, z)\,|\,x = -1 - 2z,$
$y = -2 + 4z, z \text{ is any real }$
number$\}$

33. $\begin{cases} x_1 = 1 \\ x_2 + x_4 = 2 \\ x_3 + 2x_4 = 3 \end{cases}$
Consistent:
$\begin{cases} x_1 = 1, x_2 = 2 - x_4 \\ x_3 = 3 - 2x_4 \\ x_4 \text{ is any real number} \end{cases}$ or
$\{(x_1, x_2, x_3, x_4)\,|\,x_1 = 1,$
$x_2 = 2 - x_4, x_3 = 3 - 2x_4,$
$x_4 \text{ is any real number}\}$

35. $\begin{cases} x_1 + 4x_4 = 2 \\ x_2 + x_3 + 3x_4 = 3 \\ 0 = 0 \end{cases}$
Consistent:
$\begin{cases} x_1 = 2 - 4x_4 \\ x_2 = 3 - x_3 - 3x_4 \\ x_3, x_4 \text{ are any real numbers} \end{cases}$ or
$\{(x_1, x_2, x_3, x_4)\,|\,x_1 = 2 - 4x_4,$
$x_2 = 3 - x_3 - 3x_4, x_3, x_4 \text{ are}$
any real numbers$\}$

37. $\begin{cases} x_1 + x_4 = -2 \\ x_2 + 2x_4 = 2 \\ x_3 - x_4 = 0 \\ 0 = 0 \end{cases}$
Consistent:
$\begin{cases} x_1 = -2 - x_4 \\ x_2 = 2 - 2x_4 \\ x_3 = x_4 \\ x_4 \text{ is any real number} \end{cases}$ or
$\{(x_1, x_2, x_3, x_4)\,|\,x_1 = -2 - x_4,$
$x_2 = 2 - 2x_4, x_3 = x_4, x_4 \text{ is any}$
real number$\}$

39. $x = 6, y = 2; (6, 2)$ **41.** $x = \dfrac{1}{2}, y = \dfrac{3}{4}; \left(\dfrac{1}{2}, \dfrac{3}{4}\right)$ **43.** $x = 4 - 2y, y$ is any real number; $\{ (x, y) \mid x = 4 - 2y, y$ is any real number $\}$

45. $x = \dfrac{3}{2}, y = 1; \left(\dfrac{3}{2}, 1\right)$ **47.** $x = \dfrac{4}{3}, y = \dfrac{1}{5}; \left(\dfrac{4}{3}, \dfrac{1}{5}\right)$ **49.** $x = 8, y = 2, z = 0; (8, 2, 0)$ **51.** $x = 2, y = -1, z = 1; (2, -1, 1)$ **53.** Inconsistent

55. $x = 5z - 2, y = 4z - 3$, where z is any real number; $\{ (x, y, z) \mid x = 5z - 2, y = 4z - 3, z$ is any real number $\}$ **57.** Inconsistent

59. $x = 1, y = 3, z = -2; (1, 3, -2)$ **61.** $x = -3, y = \dfrac{1}{2}, z = 1; \left(-3, \dfrac{1}{2}, 1\right)$ **63.** $x = \dfrac{1}{3}, y = \dfrac{2}{3}, z = 1; \left(\dfrac{1}{3}, \dfrac{2}{3}, 1\right)$

65. $x = 1, y = 2, z = 0, w = 1; (1, 2, 0, 1)$ **67.** $y = 0, z = 1 - x, x$ is any real number; $\{ (x, y, z) \mid y = 0, z = 1 - x, x$ is any real number $\}$

69. $x = 2, y = z - 3, z$ is any real number; $\{ (x, y, z) \mid x = 2, y = z - 3, z$ is any real number $\}$ **71.** $x = \dfrac{13}{9}, y = \dfrac{7}{18}, z = \dfrac{19}{18}; \left(\dfrac{13}{9}, \dfrac{7}{18}, \dfrac{19}{18}\right)$

73. $x = \dfrac{7}{5} - \dfrac{3}{5}z - \dfrac{2}{5}w, y = -\dfrac{8}{5} + \dfrac{7}{5}z + \dfrac{13}{5}w$, where z and w are any real numbers; $\left\{ (x, y, z, w) \,\middle|\, x = \dfrac{7}{5} - \dfrac{3}{5}z - \dfrac{2}{5}w, y = -\dfrac{8}{5} + \dfrac{7}{5}z + \dfrac{13}{5}w, \right.$

z and w are any real numbers $\Big\}$ **75.** $y = -2x^2 + x + 3$ **77.** $f(x) = 3x^3 - 4x^2 + 5$ **79.** 1.5 salmon steak, 2 baked eggs, 1 acorn squash

81. \$4000 in Treasury bills, \$4000 in Treasury bonds, \$2000 in corporate bonds   **83.** 8 Deltas, 5 Betas, 10 Sigmas   **85.** $I_1 = \dfrac{44}{23}, I_2 = 2, I_3 = \dfrac{16}{23}, I_4 = \dfrac{28}{23}$

87. (a)

Amount Invested At		
7%	**9%**	**11%**
0	10,000	10,000
1000	8000	11,000
2000	6000	12,000
3000	4000	13,000
4000	2000	14,000
5000	0	15,000

(b)

Amount Invested At		
7%	**9%**	**11%**
12,500	12,500	0
14,500	8500	2000
16,500	4500	4000
18,750	0	6250

(c) All the money invested at 7% provides \$2100, more than what is required.

89.

First Supplement	Second Supplement	Third Supplement
50 mg	75 mg	0 mg
36 mg	76 mg	8 mg
22 mg	77 mg	16 mg
8 mg	78 mg	24 mg

94. $\{x \mid -1 < x < 6\}$, or $(-1, 6)$
95.

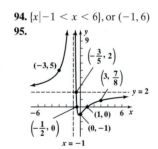

96. $\{x \mid x$ is any real nunber$\}$, or $(-\infty, \infty)$ **97.** 2.42

12.3 Assess Your Understanding (page 881)

1. $ad - bc$ **2.** $\begin{vmatrix} 5 & 3 \\ -3 & -4 \end{vmatrix}$ **3.** F   **4.** F   **5.** F   **6.** a   **7.** 22   **9.** -2 **11.** 10   **13.** -26 **15.** $x = 6, y = 2; (6, 2)$ **17.** $x = 3, y = 2; (3, 2)$

19. $x = 8, y = -4; (8, -4)$ **21.** $x = 4, y = -2; (4, -2)$ **23.** Not applicable   **25.** $x = \dfrac{1}{2}, y = \dfrac{3}{4}; \left(\dfrac{1}{2}, \dfrac{3}{4}\right)$ **27.** $x = \dfrac{1}{10}, y = \dfrac{2}{5}; \left(\dfrac{1}{10}, \dfrac{2}{5}\right)$

29. $x = \dfrac{3}{2}, y = 1; \left(\dfrac{3}{2}, 1\right)$ **31.** $x = \dfrac{4}{3}, y = \dfrac{1}{5}; \left(\dfrac{4}{3}, \dfrac{1}{5}\right)$ **33.** $x = 1, y = 3, z = -2; (1, 3, -2)$ **35.** $x = -3, y = \dfrac{1}{2}, z = 1; \left(-3, \dfrac{1}{2}, 1\right)$

37. Not applicable   **39.** $x = 0, y = 0, z = 0; (0, 0, 0)$ **41.** Not applicable   **43.** -4 **45.** 12   **47.** 8   **49.** 8   **51.** -5 **53.** $\dfrac{13}{11}$ **55.** 0 or -9

57. $(y_1 - y_2)x - (x_1 - x_2)y + (x_1y_2 - x_2y_1) = 0$

$$(y_1 - y_2)x + (x_2 - x_1)y = x_2y_1 - x_1y_2$$
$$(x_2 - x_1)y - (x_2 - x_1)y_1 = (y_2 - y_1)x + x_2y_1 - x_1y_2 - (x_2 - x_1)y_1$$
$$(x_2 - x_1)(y - y_1) = (y_2 - y_1)x - (y_2 - y_1)x_1$$
$$y - y_1 = \frac{y_2 - y_1}{x_2 - x_1}(x - x_1)$$

59. The triangle has an area of 5 square units.   **61.** 50.5 square units   **63.** $(x - 3)^2 + (y + 2)^2 = 25$

65. If $a = 0$, we have

$$by = s$$
$$cx + dy = t$$

Thus, $y = \dfrac{s}{b}$ and

$$x = \dfrac{t - dy}{c} = \dfrac{tb - ds}{bc}$$

Using Cramer's Rule, we get

$$x = \dfrac{sd - tb}{-bc} = \dfrac{tb - sd}{bc}$$

$$y = \dfrac{-sc}{-bc} = \dfrac{s}{b}$$

If $b = 0$, we have

$$ax = s$$
$$cx + dy = t$$

Since $D = ad \neq 0$, then $a \neq 0$ and $d \neq 0$.

Thus, $x = \dfrac{s}{a}$ and

$$y = \dfrac{t - cx}{d} = \dfrac{ta - cs}{ad}$$

Using Cramer's Rule, we get

$$x = \dfrac{sd}{ad} = \dfrac{s}{a}$$

$$y = \dfrac{ta - cs}{ad}$$

If $c = 0$, we have

$$ax + by = s$$
$$dy = t$$

Since $D = ad \neq 0$, then $a \neq 0$ and $d \neq 0$.

Thus, $y = \dfrac{t}{d}$ and

$$x = \dfrac{s - by}{a} = \dfrac{sd - bt}{ad}$$

Using Cramer's Rule, we get

$$x = \dfrac{sd - bt}{ad}$$

$$y = \dfrac{at}{ad} = \dfrac{t}{d}$$

If $d = 0$, we have

$$ax + by = s$$
$$cx = t$$

Since $D = -bc \neq 0$, then $b \neq 0$ and $c \neq 0$.

Thus, $x = \dfrac{t}{c}$ and

$$y = \dfrac{s - ax}{b} = \dfrac{sc - at}{bc}$$

Using Cramer's Rule, we get

$$x = \dfrac{-tb}{-bc} = \dfrac{t}{c}$$

$$y = \dfrac{at - sc}{-bc} = \dfrac{sc - at}{bc}$$

67.
$$\begin{vmatrix} a_{11} & a_{12} & a_{13} \\ ka_{21} & ka_{22} & ka_{23} \\ a_{31} & a_{32} & a_{33} \end{vmatrix} = -ka_{21}(a_{12}a_{33} - a_{32}a_{13}) + ka_{22}(a_{11}a_{33} - a_{31}a_{13}) - ka_{23}(a_{11}a_{32} - a_{31}a_{12})$$

$$= k[-a_{21}(a_{12}a_{33} - a_{32}a_{13}) + a_{22}(a_{11}a_{33} - a_{31}a_{13}) - a_{23}(a_{11}a_{32} - a_{31}a_{12})] = k\begin{vmatrix} a_{11} & a_{12} & a_{13} \\ a_{21} & a_{22} & a_{23} \\ a_{31} & a_{32} & a_{33} \end{vmatrix}$$

69.
$$\begin{vmatrix} a_{11} + ka_{21} & a_{12} + ka_{22} & a_{13} + ka_{23} \\ a_{21} & a_{22} & a_{23} \\ a_{31} & a_{32} & a_{33} \end{vmatrix} = (a_{11} + ka_{21})(a_{22}a_{33} - a_{32}a_{23}) - (a_{12} + ka_{22})(a_{21}a_{33} - a_{31}a_{23}) + (a_{13} + ka_{23})(a_{21}a_{32} - a_{31}a_{22})$$

$$= a_{11}a_{22}a_{33} - a_{11}a_{32}a_{23} + \cancel{ka_{21}a_{22}a_{33}} - \cancel{ka_{21}a_{32}a_{23}} - a_{12}a_{21}a_{33} + a_{12}a_{31}a_{23}$$
$$\quad - \cancel{ka_{22}a_{21}a_{33}} + \cancel{ka_{22}a_{31}a_{23}} + a_{13}a_{21}a_{32} - a_{13}a_{31}a_{22} + \cancel{ka_{23}a_{21}a_{32}} - \cancel{ka_{23}a_{31}a_{22}}$$
$$= a_{11}a_{22}a_{33} - a_{11}a_{32}a_{23} - a_{12}a_{21}a_{33} + a_{12}a_{31}a_{23} + a_{13}a_{21}a_{32} - a_{13}a_{31}a_{22}$$
$$= a_{11}(a_{22}a_{33} - a_{32}a_{23}) - a_{12}(a_{21}a_{33} - a_{31}a_{23}) + a_{13}(a_{21}a_{32} - a_{31}a_{22})$$
$$= \begin{vmatrix} a_{11} & a_{12} & a_{13} \\ a_{21} & a_{22} & a_{23} \\ a_{31} & a_{32} & a_{33} \end{vmatrix}$$

70. $\mathbf{v} = 9\mathbf{i} - 4\mathbf{j}$; $\sqrt{97}$ **71.** $\pm\dfrac{1}{2}, \pm\dfrac{5}{2}, \pm 1, \pm 2, \pm 5, \pm 10$

72.

73. 0

Historical Problems *(page 897)*

1. (a) $2 - 5i \longleftrightarrow \begin{bmatrix} 2 & -5 \\ 5 & 2 \end{bmatrix}$, $1 + 3i \longleftrightarrow \begin{bmatrix} 1 & 3 \\ -3 & 1 \end{bmatrix}$ **(b)** $\begin{bmatrix} 2 & -5 \\ 5 & 2 \end{bmatrix}\begin{bmatrix} 1 & 3 \\ -3 & 1 \end{bmatrix} = \begin{bmatrix} 17 & 1 \\ -1 & 17 \end{bmatrix}$ **(c)** $17 + i$ **(d)** $17 + i$

2. $\begin{bmatrix} a & b \\ -b & a \end{bmatrix}\begin{bmatrix} a & -b \\ b & a \end{bmatrix} = \begin{bmatrix} a^2+b^2 & 0 \\ 0 & b^2+a^2 \end{bmatrix}$; the product is a real number.

3. (a) $x = k(ar + bs) + l(cr + ds) = r(ka + lc) + s(kb + ld)$ **(b)** $A = \begin{bmatrix} ka + lc & kb + ld \\ ma + nc & mb + nd \end{bmatrix}$
$\qquad\quad y = m(ar + bs) + n(cr + ds) = r(ma + nc) + s(mb + nd)$

12.4 Assess Your Understanding *(page 897)*

1. square **2.** T **3.** F **4.** inverse **5.** T **6.** $A^{-1}B$ **7.** a **8.** d **9.** $\begin{bmatrix} 4 & 4 & -5 \\ -1 & 5 & 4 \end{bmatrix}$ **11.** $\begin{bmatrix} 0 & 12 & -20 \\ 4 & 8 & 24 \end{bmatrix}$ **13.** $\begin{bmatrix} -8 & 7 & -15 \\ 7 & 0 & 22 \end{bmatrix}$

15. $\begin{bmatrix} 28 & -9 \\ 4 & 23 \end{bmatrix}$ **17.** $\begin{bmatrix} 1 & 14 & -14 \\ 2 & 22 & -18 \\ 3 & 0 & 28 \end{bmatrix}$ **19.** $\begin{bmatrix} 15 & 21 & -16 \\ 22 & 34 & -22 \\ -11 & 7 & 22 \end{bmatrix}$ **21.** $\begin{bmatrix} 25 & -9 \\ 4 & 20 \end{bmatrix}$ **23.** $\begin{bmatrix} -13 & 7 & -12 \\ -18 & 10 & -14 \\ 17 & -7 & 34 \end{bmatrix}$ **25.** $\begin{bmatrix} -2 & 4 & 2 & 8 \\ 2 & 1 & 4 & 6 \end{bmatrix}$ **27.** $\begin{bmatrix} 5 & 14 \\ 9 & 16 \end{bmatrix}$

29. $\begin{bmatrix} 9 & 2 \\ 34 & 13 \\ 47 & 20 \end{bmatrix}$ **31.** $\begin{bmatrix} 1 & -1 \\ -1 & 2 \end{bmatrix}$ **33.** $\begin{bmatrix} 1 & -\dfrac{5}{2} \\ -1 & 3 \end{bmatrix}$ **35.** $\begin{bmatrix} 1 & -\dfrac{1}{a} \\ -1 & \dfrac{2}{a} \end{bmatrix}$ **37.** $\begin{bmatrix} 3 & -3 & 1 \\ -2 & 2 & -1 \\ -4 & 5 & -2 \end{bmatrix}$ **39.** $\begin{bmatrix} -\dfrac{5}{7} & \dfrac{1}{7} & \dfrac{3}{7} \\ \dfrac{9}{7} & \dfrac{1}{7} & -\dfrac{4}{7} \\ \dfrac{3}{7} & -\dfrac{2}{7} & \dfrac{1}{7} \end{bmatrix}$

41. $x = 3, y = 2; (3, 2)$ **43.** $x = -5, y = 10; (-5, 10)$ **45.** $x = 2, y = -1; (2, -1)$ **47.** $x = \frac{1}{2}, y = 2; \left(\frac{1}{2}, 2\right)$ **49.** $x = -2, y = 1; (-2, 1)$

51. $x = \frac{2}{a}, y = \frac{3}{a}; \left(\frac{2}{a}, \frac{3}{a}\right)$ **53.** $x = -2, y = 3, z = 5; (-2, 3, 5)$ **55.** $x = \frac{1}{2}, y = -\frac{1}{2}, z = 1; \left(\frac{1}{2}, -\frac{1}{2}, 1\right)$

57. $x = -\frac{34}{7}, y = \frac{85}{7}, z = \frac{12}{7}; \left(-\frac{34}{7}, \frac{85}{7}, \frac{12}{7}\right)$ **59.** $x = \frac{1}{3}, y = 1, z = \frac{2}{3}; \left(\frac{1}{3}, 1, \frac{2}{3}\right)$ **61.** $\begin{bmatrix} 4 & 2 & | & 1 & 0 \\ 2 & 1 & | & 0 & 1 \end{bmatrix} \rightarrow \begin{bmatrix} 1 & \frac{1}{2} & | & \frac{1}{4} & 0 \\ 2 & 1 & | & 0 & 1 \end{bmatrix} \rightarrow \begin{bmatrix} 1 & \frac{1}{2} & | & \frac{1}{4} & 0 \\ 0 & 0 & | & -\frac{1}{2} & 1 \end{bmatrix}$

63. $\begin{bmatrix} 15 & 3 & | & 1 & 0 \\ 10 & 2 & | & 0 & 1 \end{bmatrix} \rightarrow \begin{bmatrix} 1 & \frac{1}{5} & | & \frac{1}{15} & 0 \\ 10 & 2 & | & 0 & 1 \end{bmatrix} \rightarrow \begin{bmatrix} 1 & \frac{1}{5} & | & \frac{1}{15} & 0 \\ 0 & 0 & | & -\frac{2}{3} & 1 \end{bmatrix}$

65. $\begin{bmatrix} -3 & 1 & -1 & | & 1 & 0 & 0 \\ 1 & -4 & -7 & | & 0 & 1 & 0 \\ 1 & 2 & 5 & | & 0 & 0 & 1 \end{bmatrix} \rightarrow \begin{bmatrix} 1 & 2 & 5 & | & 0 & 0 & 1 \\ 1 & -4 & -7 & | & 0 & 1 & 0 \\ -3 & 1 & -1 & | & 1 & 0 & 0 \end{bmatrix} \rightarrow \begin{bmatrix} 1 & 2 & 5 & | & 0 & 0 & 1 \\ 0 & -6 & -12 & | & 0 & 1 & -1 \\ 0 & 7 & 14 & | & 1 & 0 & 3 \end{bmatrix}$

$\rightarrow \begin{bmatrix} 1 & 2 & 5 & | & 0 & 0 & 1 \\ 0 & 1 & 2 & | & 0 & -\frac{1}{6} & \frac{1}{6} \\ 0 & 1 & 2 & | & \frac{1}{7} & 0 & \frac{3}{7} \end{bmatrix} \rightarrow \begin{bmatrix} 1 & 2 & 5 & | & 0 & 0 & 1 \\ 0 & 1 & 2 & | & 0 & -\frac{1}{6} & \frac{1}{6} \\ 0 & 0 & 0 & | & \frac{1}{7} & \frac{1}{6} & \frac{11}{42} \end{bmatrix}$

67. $\begin{bmatrix} 0.01 & 0.05 & -0.01 \\ 0.01 & -0.02 & 0.01 \\ -0.02 & 0.01 & 0.03 \end{bmatrix}$ **69.** $\begin{bmatrix} 0.02 & -0.04 & -0.01 & 0.01 \\ -0.02 & 0.05 & 0.03 & -0.03 \\ 0.02 & 0.01 & -0.04 & 0.00 \\ -0.02 & 0.06 & 0.07 & 0.06 \end{bmatrix}$

71. $x = 4.57, y = -6.44, z = -24.07; (4.57, -6.44, -24.07)$ **73.** $x = -1.19, y = 2.46, z = 8.27; (-1.19, 2.46, 8.27)$

75. $x = -5, y = 7; (-5, 7)$ **77.** $x = -4, y = 2, z = \frac{5}{2}; \left(-4, 2, \frac{5}{2}\right)$

79. Inconsistent; $\varnothing$ **81.** $x = -\frac{1}{5}z + \frac{1}{5}, y = \frac{1}{5}z - \frac{6}{5}$, where z is any real number; $\left\{(x, y, z) \,\middle|\, x = -\frac{1}{5}z + \frac{1}{5}, y = \frac{1}{5}z - \frac{6}{5}, z \text{ is any real number}\right\}$

83. (a) $A = \begin{bmatrix} 6 & 9 \\ 3 & 12 \end{bmatrix}; B = \begin{bmatrix} 121.00 \\ 340.13 \end{bmatrix}$ **(b)** $AB = \begin{bmatrix} 3787.17 \\ 4444.56 \end{bmatrix}$; Nikki's total tuition is $3787.17, and Joe's total tuition is $4444.56.

85. (a) $\begin{bmatrix} 500 & 350 & 400 \\ 700 & 500 & 850 \end{bmatrix}; \begin{bmatrix} 500 & 700 \\ 350 & 500 \\ 400 & 850 \end{bmatrix}$ **(b)** $\begin{bmatrix} 15 \\ 8 \\ 3 \end{bmatrix}$ **(c)** $\begin{bmatrix} 11{,}500 \\ 17{,}050 \end{bmatrix}$ **(d)** $[0.10 \quad 0.05]$ **(e)** $2002.50

87. (a) $K^{-1} = \begin{bmatrix} 1 & 0 & -1 \\ -1 & 1 & 1 \\ 0 & -1 & 1 \end{bmatrix}$ **(b)** $M = \begin{bmatrix} 13 & 1 & 20 \\ 8 & 9 & 19 \\ 6 & 21 & 14 \end{bmatrix}$ **(c)** Math is fun.

89. If $D = ad - bc \neq 0$, then $a \neq 0$ and $d \neq 0$, or $b \neq 0$ and $c \neq 0$. Assuming the former,

$\begin{bmatrix} a & b & | & 1 & 0 \\ c & d & | & 0 & 1 \end{bmatrix} \rightarrow \begin{bmatrix} 1 & \frac{b}{a} & | & \frac{1}{a} & 0 \\ c & d & | & 0 & 1 \end{bmatrix} \rightarrow \begin{bmatrix} 1 & \frac{b}{a} & | & \frac{1}{a} & 0 \\ 0 & \frac{D}{a} & | & -\frac{c}{a} & 1 \end{bmatrix} \rightarrow \begin{bmatrix} 1 & \frac{b}{a} & | & \frac{1}{a} & 0 \\ 0 & 1 & | & -\frac{c}{D} & \frac{a}{D} \end{bmatrix} \rightarrow \begin{bmatrix} 1 & 0 & | & \frac{d}{D} & -\frac{b}{D} \\ 0 & 1 & | & -\frac{c}{D} & \frac{a}{D} \end{bmatrix}$

$R_1 = \frac{1}{a}r_1$ $R_2 = -cr_1 + r_2$ $R_2 = \frac{a}{D}r_2$ $R_1 = -\frac{b}{a}r_2 + r_1$

91. (a) $B_3 = A + A^2 + A^3 = \begin{bmatrix} 2 & 4 & 5 & 2 & 3 \\ 5 & 3 & 2 & 5 & 4 \\ 4 & 2 & 2 & 4 & 2 \\ 2 & 2 & 3 & 2 & 3 \\ 1 & 3 & 2 & 1 & 2 \end{bmatrix}$; Yes, all pages can reach every other page within 3 clicks.   **(b)** Page 3

93. (a) $(3 - 2\sqrt{3}, 3\sqrt{3} + 2)$ **(b)** $R^{-1} = \begin{bmatrix} \frac{1}{2} & \frac{\sqrt{3}}{2} & 0 \\ -\frac{\sqrt{3}}{2} & \frac{1}{2} & 0 \\ 0 & 0 & 1 \end{bmatrix}$; This is the rotation matrix needed to get the translated coordinates back to the original coordinates.

98. $x^6 - 4x^5 - 3x^4 + 18x^3$ **99.** $\mathbf{v} \cdot \mathbf{w} = -5; \theta = 180°$ **100.** $\{0, 3\}$ **101.** $\dfrac{\sqrt{u^2 - 1}}{|u|}$

12.5 Assess Your Understanding *(page 908)*

5. Proper   **7.** Improper; $1 + \dfrac{9}{x^2 - 4}$ **9.** Improper; $5x + \dfrac{22x - 1}{x^2 - 4}$ **11.** Improper; $1 + \dfrac{-2(x - 6)}{(x + 4)(x - 3)}$ **13.** $\dfrac{-4}{x} + \dfrac{4}{x - 1}$ **15.** $\dfrac{1}{x} + \dfrac{-x}{x^2 + 1}$

17. $\dfrac{-1}{x - 1} + \dfrac{2}{x - 2}$ **19.** $\dfrac{\frac{1}{4}}{x + 1} + \dfrac{\frac{3}{4}}{x - 1} + \dfrac{\frac{1}{2}}{(x - 1)^2}$ **21.** $\dfrac{\frac{1}{12}}{x - 2} + \dfrac{-\frac{1}{12}(x + 4)}{x^2 + 2x + 4}$ **23.** $\dfrac{\frac{1}{4}}{x - 1} + \dfrac{\frac{1}{4}}{(x - 1)^2} + \dfrac{-\frac{1}{4}}{x + 1} + \dfrac{\frac{1}{4}}{(x + 1)^2}$

25. $\dfrac{-5}{x + 2} + \dfrac{5}{x + 1} + \dfrac{-4}{(x + 1)^2}$ **27.** $\dfrac{\frac{1}{4}}{x} + \dfrac{1}{x^2} + \dfrac{-\frac{1}{4}(x + 4)}{x^2 + 4}$ **29.** $\dfrac{\frac{2}{3}}{x + 1} + \dfrac{\frac{1}{3}(x + 1)}{x^2 + 2x + 4}$ **31.** $\dfrac{\frac{2}{7}}{3x - 2} + \dfrac{\frac{1}{7}}{2x + 1}$ **33.** $\dfrac{\frac{3}{4}}{x + 3} + \dfrac{\frac{1}{4}}{x - 1}$

35. $\dfrac{1}{x^2 + 4} + \dfrac{2x - 1}{(x^2 + 4)^2}$ **37.** $\dfrac{-1}{x} + \dfrac{2}{x - 3} + \dfrac{-1}{x + 1}$ **39.** $\dfrac{4}{x - 2} + \dfrac{-3}{x - 1} + \dfrac{-1}{(x - 1)^2}$ **41.** $\dfrac{x}{(x^2 + 16)^2} + \dfrac{-16x}{(x^2 + 16)^3}$

43. $\dfrac{-\frac{8}{7}}{2x + 1} + \dfrac{\frac{4}{7}}{x - 3}$ **45.** $\dfrac{-\frac{2}{9}}{x} + \dfrac{-\frac{1}{3}}{x^2} + \dfrac{\frac{1}{6}}{x - 3} + \dfrac{\frac{1}{18}}{x + 3}$ **47.** $x - 2 + \dfrac{10x - 11}{x^2 + 3x - 4}; \dfrac{\frac{51}{5}}{x + 4} + \dfrac{-\frac{1}{5}}{x - 1}; x - 2 + \dfrac{\frac{51}{5}}{x + 4} - \dfrac{\frac{1}{5}}{x - 1}$

49. $x - \dfrac{x}{x^2 + 1}$ **51.** $x^2 - 4x + 7 + \dfrac{-11x - 32}{x^2 + 4x + 4}; \dfrac{-11}{x + 2} + \dfrac{-10}{(x + 2)^2}; x^2 - 4x + 7 - \dfrac{11}{x + 2} - \dfrac{10}{(x + 2)^2}$

53. $x + 1 + \dfrac{2x^3 + x^2 - x + 1}{x^4 - 2x^2 + 1}; \dfrac{1}{x + 1} + \dfrac{\frac{1}{4}}{(x + 1)^2} + \dfrac{1}{x - 1} + \dfrac{\frac{3}{4}}{(x - 1)^2}; x + 1 + \dfrac{1}{x + 1} + \dfrac{\frac{1}{4}}{(x + 1)^2} + \dfrac{1}{x - 1} + \dfrac{\frac{3}{4}}{(x - 1)^2}$

55. 3.85 years   **56.** -2 **57.** 1

58. $\left(\dfrac{\sqrt{2}}{2}, \dfrac{\sqrt{2}}{2}\right)$;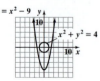

Historical Problem *(page 914)*

$x = 6$ units, $y = 8$ units

12.6 Assess Your Understanding *(page 914)*

5.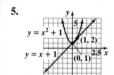

7.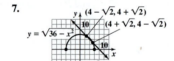

9.

11.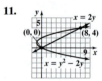

13.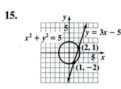

15.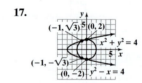

17.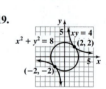

19.

21. No points of intersection

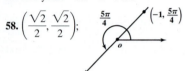

23.

25. $x = 1, y = 4; x = -1, y = -4; x = 2\sqrt{2}, y = \sqrt{2}; x = -2\sqrt{2}, y = -\sqrt{2}$ or $(1, 4), (-1, -4), (2\sqrt{2}, \sqrt{2}), (-2\sqrt{2}, -\sqrt{2})$

27. $x = 0, y = 1; x = -\dfrac{2}{3}, y = -\dfrac{1}{3}$ or $(0, 1), \left(-\dfrac{2}{3}, -\dfrac{1}{3}\right)$ **29.** $x = 0, y = -1; x = \dfrac{5}{2}, y = -\dfrac{7}{2}$ or $(0, -1), \left(\dfrac{5}{2}, -\dfrac{7}{2}\right)$

31. $x = 2, y = \dfrac{1}{3}; x = \dfrac{1}{2}, y = \dfrac{4}{3}$ or $\left(2, \dfrac{1}{3}\right), \left(\dfrac{1}{2}, \dfrac{4}{3}\right)$ **33.** $x = 3, y = 2; x = 3, y = -2; x = -3, y = 2; x = -3, y = -2$ or $(3, 2), (3, -2), (-3, 2),$

$(-3, -2)$ **35.** $x = \dfrac{1}{2}, y = \dfrac{3}{2}; x = \dfrac{1}{2}, y = -\dfrac{3}{2}; x = -\dfrac{1}{2}, y = \dfrac{3}{2}; x = -\dfrac{1}{2}, y = -\dfrac{3}{2}$ or $\left(\dfrac{1}{2}, \dfrac{3}{2}\right), \left(\dfrac{1}{2}, -\dfrac{3}{2}\right), \left(-\dfrac{1}{2}, \dfrac{3}{2}\right), \left(-\dfrac{1}{2}, -\dfrac{3}{2}\right)$

37. $x = \sqrt{2}, y = 2\sqrt{2}; x = -\sqrt{2}, y = -2\sqrt{2}$ or $(\sqrt{2}, 2\sqrt{2}), (-\sqrt{2}, -2\sqrt{2})$ **39.** No real solution exists.

41. $x = \dfrac{8}{3}, y = \dfrac{2\sqrt{10}}{3}; x = -\dfrac{8}{3}, y = \dfrac{2\sqrt{10}}{3}; x = \dfrac{8}{3}, y = -\dfrac{2\sqrt{10}}{3}; x = -\dfrac{8}{3}, y = -\dfrac{2\sqrt{10}}{3}$ or $\left(\dfrac{8}{3}, \dfrac{2\sqrt{10}}{3}\right), \left(-\dfrac{8}{3}, \dfrac{2\sqrt{10}}{3}\right), \left(\dfrac{8}{3}, -\dfrac{2\sqrt{10}}{3}\right), \left(-\dfrac{8}{3}, -\dfrac{2\sqrt{10}}{3}\right)$

43. $x = 1, y = \dfrac{1}{2}; x = -1, y = \dfrac{1}{2}; x = 1, y = -\dfrac{1}{2}; x = -1, y = -\dfrac{1}{2}$ or $\left(1, \dfrac{1}{2}\right), \left(-1, \dfrac{1}{2}\right), \left(1, -\dfrac{1}{2}\right), \left(-1, -\dfrac{1}{2}\right)$ **45.** No real solution exists.

47. $x = \sqrt{3}, y = \sqrt{3}; x = -\sqrt{3}, y = -\sqrt{3}; x = 2, y = 1; x = -2, y = -1$ or $(\sqrt{3}, \sqrt{3}), (-\sqrt{3}, -\sqrt{3}), (2, 1), (-2, -1)$
49. $x = 0, y = -2; x = 0, y = 1; x = 2, y = -1$ or $(0, -2), (0, 1), (2, -1)$ **51.** $x = 2, y = 8$ or $(2, 8)$ **53.** $x = 81, y = 3$ or $(81, 3)$

55.
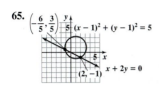

57. $x = 0.48, y = 0.62$ **59.** $x = -1.65, y = -0.89$
61. $x = 0.58, y = 1.86; x = 1.81, y = 1.05; x = 0.58, y = -1.86; x = 1.81, y = -1.05$
63. $x = 2.35, y = 0.85$

65.
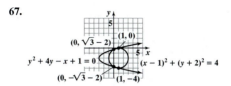

67.
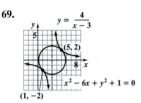

69.

71. 3 and 1; -1 and -3 **73.** 2 and 2; -2 and -2 **75.** $\dfrac{1}{2}$ and $\dfrac{1}{3}$ **77.** 5 **79.** 5 in. by 3 in. **81.** 2 cm and 4 cm **83.** tortoise: 7 m/hr, hare: $7\dfrac{1}{2}$ m/hr

85. 12 cm by 18 cm **87.** $x = 60$ ft; $y = 30$ ft **89.** $l = \dfrac{P + \sqrt{P^2 - 16A}}{4}; w = \dfrac{P - \sqrt{P^2 - 16A}}{4}$ **91.** $y = 4x - 4$ **93.** $y = 2x + 1$

95. $y = -\dfrac{1}{3}x + \dfrac{7}{3}$ **97.** $y = 2x - 3$ **99.** $r_1 = \dfrac{-b + \sqrt{b^2 - 4ac}}{2a}; r_2 = \dfrac{-b - \sqrt{b^2 - 4ac}}{2a}$ **101. (a)** 4.274 ft by 4.274 ft or 0.093 ft by 0.093 ft

102. $\left\{\dfrac{-3 - \sqrt{65}}{7}, \dfrac{-3 + \sqrt{65}}{7}\right\}$ **103.** $y = -\dfrac{2}{5}x - 3$ **104.** $\sin\theta = -\dfrac{7}{25}; \cos\theta = -\dfrac{24}{25}; \tan\theta = \dfrac{7}{24}; \csc\theta = -\dfrac{25}{7}; \sec\theta = -\dfrac{25}{24}$ **105.** $\approx 15.8°$

12.7 Assess Your Understanding *(page 923)*

7. dashes; solid **8.** half-planes **9.** F **10.** unbounded

11.

13.

15.
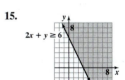

17.

19.
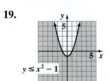

21.

23.
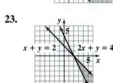

25.
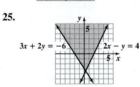

27.
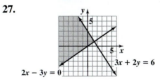

29.

31.
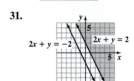

33. No solution
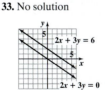

35.

37.
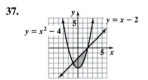

39.
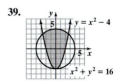

41.
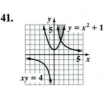

43. Bounded
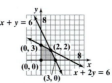

45. Unbounded

47. Bounded
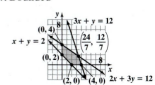

49. Bounded
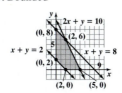

51. Bounded

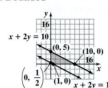

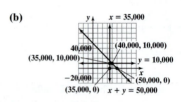

53. $\begin{cases} x \le 4 \\ x + y \le 6 \\ x \ge 0 \\ y \ge 0 \end{cases}$

55. $\begin{cases} x \le 20 \\ y \ge 15 \\ x + y \le 50 \\ x - y \le 0 \\ x \ge 0 \end{cases}$

57. (a) $\begin{cases} x + y \le 50{,}000 \\ x \ge 35{,}000 \\ y \le 10{,}000 \\ y \ge 0 \end{cases}$

(b)

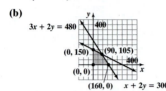

59. (a) $\begin{cases} x \ge 0 \\ y \ge 0 \\ x + 2y \le 300 \\ 3x + 2y \le 480 \end{cases}$

(b)

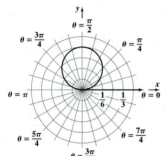

61. (a) $\begin{cases} 3x + 2y \le 160 \\ 2x + 3y \le 150 \\ x \ge 0 \\ y \ge 0 \end{cases}$

(b)

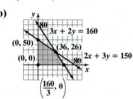

62. $\{-1 - 2i, -1 + 2i\}$

63. $x^2 + \left(y - \dfrac{1}{6}\right)^2 = \dfrac{1}{36}$;

circle with radius $\dfrac{1}{6}$ and center $\left(0, \dfrac{1}{6}\right)$;

64. $f(-1) = -5$;
$f(2) = 28$

65. $\left\{0, \dfrac{2\pi}{3}, \dfrac{4\pi}{3}\right\}$

12.8 Assess Your Understanding *(page 930)*

1. objective function   **2.** T   **3.** Maximum value is 11; minimum value is 3.   **5.** Maximum value is 65; minimum value is 4.
7. Maximum value is 67; minimum value is 20.   **9.** The maximum value of z is 12, and it occurs at the point $(6, 0)$.
11. The minimum value of z is 4, and it occurs at the point $(2, 0)$.   **13.** The maximum value of z is 20, and it occurs at the point $(0, 4)$.
15. The minimum value of z is 8, and it occurs at the point $(0, 2)$.   **17.** The maximum value of z is 50, and it occurs at the point $(10, 0)$.
19. Produce 8 downhill and 24 cross-country; $1760; $1920 which is the profit when producing 16 downhill and 16 cross-country.
21. Rent 15 rectangular tables and 16 round tables for a minimum cost of $1252.   **23. (a)** $10,000 in a junk bond and $10,000 in Treasury bills
(b) $12,000 in a junk bond and $8000 in Treasury bills   **25.** 100 lb of ground beef should be mixed with 50 lb of pork.
27. Manufacture 10 racing skates and 15 figure skates.   **29.** Order 2 metal samples and 4 plastic samples; $34
31. (a) Configure with 10 first class seats and 120 coach seats.   **(b)** Configure with 15 first class seats and 120 coach seats.

33. $\left\{-\dfrac{1}{32}, 1\right\}$

34.

Domain: $\{x \mid x \ne k\pi, k \text{ is an integer}\}$; range: $(-\infty, \infty)$
35. 89.1 years   **36.** $y = 3x + 7$

Review Exercises *(page 934)*

1. $x = 2, y = -1$ or $(2, -1)$ **2.** $x = 2, y = \dfrac{1}{2}$ or $\left(2, \dfrac{1}{2}\right)$ **3.** $x = 2, y = -1$ or $(2, -1)$ **4.** $x = \dfrac{11}{5}, y = -\dfrac{3}{5}$ or $\left(\dfrac{11}{5}, -\dfrac{3}{5}\right)$ **5.** Inconsistent

6. $x = 2, y = 3$ or $(2, 3)$ **7.** $y = -\dfrac{2}{5}x + 2$, where x is any real number, or $\left\{(x, y) \mid y = -\dfrac{2}{5}x + 2, x \text{ is any real number}\right\}$

8. $x = -1, y = 2, z = -3$ or $(-1, 2, -3)$

9. $x = \dfrac{7}{4}z + \dfrac{39}{4}, y = \dfrac{9}{8}z + \dfrac{69}{8}$, where z is any real number, or $\left\{(x, y, z) \mid x = \dfrac{7}{4}z + \dfrac{39}{4}, y = \dfrac{9}{8}z + \dfrac{69}{8}, z \text{ is any real number}\right\}$ **10.** Inconsistent

11. $\begin{cases} 3x + 2y = 8 \\ x + 4y = -1 \end{cases}$ **12.** $\begin{cases} x + 2y + 5z = -2 \\ 5x \quad\ -3z = 8 \\ 2x - y \quad\quad = 0 \end{cases}$ **13.** $\begin{bmatrix} 4 & -4 \\ 3 & 9 \\ 4 & 4 \end{bmatrix}$ **14.** $\begin{bmatrix} 6 & 0 \\ 12 & 24 \\ -6 & 12 \end{bmatrix}$ **15.** $\begin{bmatrix} 4 & -3 & 0 \\ 12 & -2 & -8 \\ -2 & 5 & -4 \end{bmatrix}$ **16.** $\begin{bmatrix} 9 & -31 \\ -6 & -3 \end{bmatrix}$

17. $\begin{bmatrix} \dfrac{1}{2} & -1 \\ -\dfrac{1}{6} & \dfrac{2}{3} \end{bmatrix}$ **18.** $\begin{bmatrix} -\dfrac{5}{7} & \dfrac{9}{7} & \dfrac{3}{7} \\ \dfrac{1}{7} & \dfrac{1}{7} & -\dfrac{2}{7} \\ \dfrac{3}{7} & -\dfrac{4}{7} & \dfrac{1}{7} \end{bmatrix}$ **19.** Singular **20.** $x = \dfrac{2}{5}, y = \dfrac{1}{10}$ or $\left(\dfrac{2}{5}, \dfrac{1}{10}\right)$ **21.** $x = 9, y = \dfrac{13}{3}, z = \dfrac{13}{3}$ or $\left(9, \dfrac{13}{3}, \dfrac{13}{3}\right)$

22. Inconsistent **23.** $x = -\dfrac{1}{2}, y = -\dfrac{2}{3}, z = -\dfrac{3}{4}$, or $\left(-\dfrac{1}{2}, -\dfrac{2}{3}, -\dfrac{3}{4}\right)$

24. $z = -1, x = y + 1$, where y is any real number, or $\{(x, y, z) \mid x = y + 1, z = -1, y \text{ is any real number}\}$
25. $x = 4, y = 2, z = 3, t = -2$ or $(4, 2, 3, -2)$ **26.** 5 **27.** 108 **28.** -100 **29.** $x = 2, y = -1$ or $(2, -1)$ **30.** $x = 2, y = 3$ or $(2, 3)$

31. $x = -1, y = 2, z = -3$ or $(-1, 2, -3)$ **32.** 16 **33.** -8 **34.** $\dfrac{-\dfrac{3}{2}}{x} + \dfrac{\dfrac{3}{2}}{x - 4}$ **35.** $\dfrac{-3}{x - 1} + \dfrac{3}{x} + \dfrac{4}{x^2}$ **36.** $\dfrac{-\dfrac{1}{10}}{x + 1} + \dfrac{\dfrac{1}{10}x + \dfrac{9}{10}}{x^2 + 9}$

37. $\dfrac{x}{x^2 + 4} + \dfrac{-4x}{(x^2 + 4)^2}$ **38.** $\dfrac{\dfrac{1}{2}}{x^2 + 1} + \dfrac{\dfrac{1}{4}}{x - 1} + \dfrac{-\dfrac{1}{4}}{x + 1}$ **39.** $x = -\dfrac{2}{5}, y = -\dfrac{11}{5}; x = -2, y = 1$ or $\left(-\dfrac{2}{5}, -\dfrac{11}{5}\right), (-2, 1)$

40. $x = 2\sqrt{2}, y = \sqrt{2}; x = -2\sqrt{2}, y = -\sqrt{2}$ or $(2\sqrt{2}, \sqrt{2}), (-2\sqrt{2}, -\sqrt{2})$
41. $x = 0, y = 0; x = -3, y = 3; x = 3, y = 3$ or $(0, 0), (-3, 3), (3, 3)$
42. $x = \sqrt{2}, y = -\sqrt{2}; x = -\sqrt{2}, y = \sqrt{2}; x = \dfrac{4}{3}\sqrt{2}, y = -\dfrac{2}{3}\sqrt{2}; x = -\dfrac{4}{3}\sqrt{2}, y = \dfrac{2}{3}\sqrt{2}$ or $(\sqrt{2}, -\sqrt{2}), (-\sqrt{2}, \sqrt{2}), \left(\dfrac{4}{3}\sqrt{2}, -\dfrac{2}{3}\sqrt{2}\right),$
$\left(-\dfrac{4}{3}\sqrt{2}, \dfrac{2}{3}\sqrt{2}\right)$ **43.** $x = 1, y = -1$ or $(1, -1)$

44.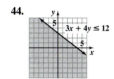

45.

46. Unbounded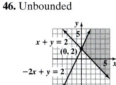

47. Bounded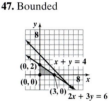

48. Bounded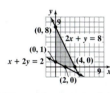

49.

50.

51. The maximum value is 32 when $x = 0$ and $y = 8$. **52.** The minimum value is 3 when $x = 1$ and $y = 0$. **53.** 10 **54.** A is any real number, $A \neq 10$.

55. $y = -\dfrac{1}{3}x^2 - \dfrac{2}{3}x + 1$ **56.** Mix 70 lb of \$6.00 coffee and 30 lb of \$9.00 coffee. **57.** Buy 1 small, 5 medium, and 2 large.

58. (a) $\begin{cases} x \geq 0 \\ y \geq 0 \\ 4x + 3y \leq 960 \\ 2x + 3y \leq 576 \end{cases}$

59. Speedboat: 36.67 km/hr; Aguarico River: 3.33 km/hr
60. Bruce: 4 hr; Bryce: 2 hr; Marty: 8 hr
61. Produce 35 gasoline engines and 15 diesel engines; the factory is producing an excess of 15 gasoline engines and 0 diesel engines.

(b)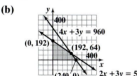

Chapter Test (page 936)

1. $x = 3, y = -1$ or $(3, -1)$ **2.** Inconsistent **3.** $x = -z + \dfrac{18}{7}, y = z - \dfrac{17}{7}$, where z is any real number, or

$\left\{(x, y, z) \mid x = -z + \dfrac{18}{7}, y = z - \dfrac{17}{7}, z \text{ is any real number}\right\}$ **4.** $x = \dfrac{1}{3}, y = -2, z = 0$ or $\left(\dfrac{1}{3}, -2, 0\right)$ **5.** $\begin{bmatrix} 4 & -5 & 1 & \bigm| & 0 \\ -2 & -1 & 0 & \bigm| & -25 \\ 1 & 5 & -5 & \bigm| & 10 \end{bmatrix}$

17. $r = \dfrac{3}{2}$; $t_1 = \dfrac{1}{2}$, $t_2 = \dfrac{3}{4}$, $t_3 = \dfrac{9}{8}$, $t_4 = \dfrac{27}{16}$ **19.** $a_5 = 162$; $a_n = 2 \cdot 3^{n-1}$ **21.** $a_5 = 5$; $a_n = 5 \cdot (-1)^{n-1}$ **23.** $a_5 = 0$; $a_n = 0$

25. $a_5 = 4\sqrt{2}$; $a_n = (\sqrt{2})^n$ **27.** $a_7 = \dfrac{1}{64}$ **29.** $a_9 = 1$ **31.** $a_8 = 0.00000004$ **33.** $a_n = 7 \cdot 2^{n-1}$ **35.** $a_n = -3 \cdot \left(-\dfrac{1}{3}\right)^{n-1} = \left(-\dfrac{1}{3}\right)^{n-2}$

37. $a_n = -(-3)^{n-1}$ **39.** $a_n = \dfrac{7}{15}(15)^{n-1} = 7 \cdot 15^{n-2}$ **41.** $-\dfrac{1}{4}(1 - 2^n)$ **43.** $2\left[1 - \left(\dfrac{2}{3}\right)^n\right]$ **45.** $1 - 2^n$

47.    **49.** 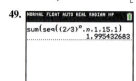   **51.**

53. Converges; $\dfrac{3}{2}$ **55.** Converges; 16   **57.** Converges; $\dfrac{8}{5}$ **59.** Diverges   **61.** Converges; $\dfrac{20}{3}$ **63.** Diverges   **65.** Converges; $\dfrac{18}{5}$ **67.** Converges; 6

69. Arithmetic; $d = 1$; 1375   **71.** Neither   **73.** Arithmetic; $d = -\dfrac{2}{3}$; -700 **75.** Neither   **77.** Geometric; $r = \dfrac{2}{3}$; $2\left[1 - \left(\dfrac{2}{3}\right)^{50}\right]$

79. Geometric; $r = -2$; $-\dfrac{1}{3}[1 - (-2)^{50}]$ **81.** Geometric; $r = 3^{1/2}$; $-\dfrac{\sqrt{3}}{2}(1 + \sqrt{3})(1 - 3^{25})$ **83.** -4 **85.** \$47,271.37

87. (a) 0.775 ft   **(b)** 8th   **(c)** 15.88 ft   **(d)** 20 ft   **89.** \$349,496.41   **91.** \$96,885.98   **93.** \$305.10   **95.** 1.845×10^{19} **97.** 10   **99.** \$72.67 per share
101. December 20, 2014; \$9999.92   **103.** Option A results in a higher salary in 5 years (\$50,499 versus \$49,522); option B results in a higher 5-year total
(\$225,484 versus \$233,602).   **105.** Option 2 results in the most: \$16,038,304; option 1 results in the least: \$14,700,000.
107. Yes. A constant sequence is both arithmetic and geometric. For example, $3, 3, 3, \ldots$ is an arithmetic sequence with $a_1 = 3$ and $d = 0$ and is a

geometric sequence with $a = 3$ and $r = 1$.   **111.** 2.121   **112.** $\dfrac{8}{17}\mathbf{i} - \dfrac{15}{17}\mathbf{j}$ **113.** $\dfrac{x^2}{4} - \dfrac{y^2}{12} = 1$ **114.** 54

13.4 Assess Your Understanding *(page 970)*

1. (I) $n = 1$: $2(1) = 2$ and $1(1 + 1) = 2$
  (II) If $2 + 4 + 6 + \cdots + 2k = k(k + 1)$, then $2 + 4 + 6 + \cdots + 2k + 2(k + 1) = (2 + 4 + 6 + \cdots + 2k) + 2(k + 1)$
  $= k(k + 1) + 2(k + 1) = k^2 + 3k + 2 = (k + 1)(k + 2) = (k + 1)[(k + 1) + 1]$.

3. (I) $n = 1$: $1 + 2 = 3$ and $\dfrac{1}{2}(1)(1 + 5) = \dfrac{1}{2}(6) = 3$

  (II) If $3 + 4 + 5 + \cdots + (k + 2) = \dfrac{1}{2}k(k + 5)$, then $3 + 4 + 5 + \cdots + (k + 2) + [(k + 1) + 2]$

  $= [3 + 4 + 5 + \cdots + (k + 2)] + (k + 3) = \dfrac{1}{2}k(k + 5) + k + 3 = \dfrac{1}{2}(k^2 + 7k + 6) = \dfrac{1}{2}(k + 1)(k + 6) = \dfrac{1}{2}(k + 1)[(k + 1) + 5]$.

5. (I) $n = 1$: $3(1) - 1 = 2$ and $\dfrac{1}{2}(1)[3(1) + 1] = \dfrac{1}{2}(4) = 2$

  (II) If $2 + 5 + 8 + \cdots + (3k - 1) = \dfrac{1}{2}k(3k + 1)$, then $2 + 5 + 8 + \cdots + (3k - 1) + [3(k + 1) - 1]$

  $= [2 + 5 + 8 + \cdots + (3k - 1)] + (3k + 2) = \dfrac{1}{2}k(3k + 1) + (3k + 2) = \dfrac{1}{2}(3k^2 + 7k + 4) = \dfrac{1}{2}(k + 1)(3k + 4)$

  $= \dfrac{1}{2}(k + 1)[3(k + 1) + 1]$.

7. (I) $n = 1$: $2^{1-1} = 1$ and $2^1 - 1 = 1$
  (II) If $1 + 2 + 2^2 + \cdots + 2^{k-1} = 2^k - 1$, then $1 + 2 + 2^2 + \cdots + 2^{k-1} + 2^{(k+1)-1} = (1 + 2 + 2^2 + \cdots + 2^{k-1}) + 2^k$
  $= 2^k - 1 + 2^k = 2(2^k) - 1 = 2^{k+1} - 1$.

9. (I) $n = 1$: $4^{1-1} = 1$ and $\dfrac{1}{3}(4^1 - 1) = \dfrac{1}{3}(3) = 1$

  (II) If $1 + 4 + 4^2 + \cdots + 4^{k-1} = \dfrac{1}{3}(4^k - 1)$, then $1 + 4 + 4^2 + \cdots + 4^{k-1} + 4^{(k+1)-1} = (1 + 4 + 4^2 + \cdots + 4^{k-1}) + 4^k$

  $= \dfrac{1}{3}(4^k - 1) + 4^k = \dfrac{1}{3}[4^k - 1 + 3(4^k)] = \dfrac{1}{3}[4(4^k) - 1] = \dfrac{1}{3}(4^{k+1} - 1)$.

11. (I) $n = 1$: $\dfrac{1}{1 \cdot 2} = \dfrac{1}{2}$ and $\dfrac{1}{1 + 1} = \dfrac{1}{2}$

  (II) If $\dfrac{1}{1 \cdot 2} + \dfrac{1}{2 \cdot 3} + \dfrac{1}{3 \cdot 4} + \cdots + \dfrac{1}{k(k + 1)} = \dfrac{k}{k + 1}$, then $\dfrac{1}{1 \cdot 2} + \dfrac{1}{2 \cdot 3} + \dfrac{1}{3 \cdot 4} + \cdots + \dfrac{1}{k(k + 1)} + \dfrac{1}{(k + 1)[(k + 1) + 1]}$

  $= \left[\dfrac{1}{1 \cdot 2} + \dfrac{1}{2 \cdot 3} + \dfrac{1}{3 \cdot 4} + \cdots + \dfrac{1}{k(k + 1)}\right] + \dfrac{1}{(k + 1)(k + 2)} = \dfrac{k}{k + 1} + \dfrac{1}{(k + 1)(k + 2)} = \dfrac{k(k + 2) + 1}{(k + 1)(k + 2)}$

  $= \dfrac{k^2 + 2k + 1}{(k + 1)(k + 2)} = \dfrac{(k + 1)^2}{(k + 1)(k + 2)} = \dfrac{k + 1}{k + 2} = \dfrac{k + 1}{(k + 1) + 1}$.

13. (I) $n = 1$: $1^2 = 1$ and $\frac{1}{6} \cdot 1 \cdot 2 \cdot 3 = 1$

  (II) If $1^2 + 2^2 + 3^2 + \cdots + k^2 = \frac{1}{6}k(k+1)(2k+1)$, then $1^2 + 2^2 + 3^2 + \cdots + k^2 + (k+1)^2$

    $= (1^2 + 2^2 + 3^2 + \cdots + k^2) + (k+1)^2 = \frac{1}{6}k(k+1)(2k+1) + (k+1)^2 = \frac{1}{6}(2k^3 + 9k^2 + 13k + 6)$

    $= \frac{1}{6}(k+1)(k+2)(2k+3) = \frac{1}{6}(k+1)[(k+1)+1][2(k+1)+1]$.

15. (I) $n = 1$: $5 - 1 = 4$ and $\frac{1}{2}(1)(9-1) = \frac{1}{2} \cdot 8 = 4$

  (II) If $4 + 3 + 2 + \cdots + (5-k) = \frac{1}{2}k(9-k)$, then $4 + 3 + 2 + \cdots + (5-k) + [5-(k+1)]$

    $= [4 + 3 + 2 + \cdots + (5-k)] + 4 - k = \frac{1}{2}k(9-k) + 4 - k = \frac{1}{2}(9k - k^2 + 8 - 2k) = \frac{1}{2}(-k^2 + 7k + 8)$

    $= \frac{1}{2}(k+1)(8-k) = \frac{1}{2}(k+1)[9-(k+1)]$.

17. (I) $n = 1$: $1 \cdot (1+1) = 2$ and $\frac{1}{3} \cdot 1 \cdot 2 \cdot 3 = 2$

  (II) If $1 \cdot 2 + 2 \cdot 3 + 3 \cdot 4 + \cdots + k(k+1) = \frac{1}{3}k(k+1)(k+2)$, then $1 \cdot 2 + 2 \cdot 3 + 3 \cdot 4 + \cdots + k(k+1)$

    $+ (k+1)[(k+1)+1] = [1 \cdot 2 + 2 \cdot 3 + 3 \cdot 4 + \cdots + k(k+1)] + (k+1)(k+2)$

    $= \frac{1}{3}k(k+1)(k+2) + \frac{1}{3} \cdot 3(k+1)(k+2) = \frac{1}{3}(k+1)(k+2)(k+3) = \frac{1}{3}(k+1)[(k+1)+1][(k+1)+2]$.

19. (I) $n = 1$: $1^2 + 1 = 2$, which is divisible by 2.
  (II) If $k^2 + k$ is divisible by 2, then $(k+1)^2 + (k+1) = k^2 + 2k + 1 + k + 1 = (k^2 + k) + 2k + 2$. Since $k^2 + k$ is divisible by 2 and $2k + 2$ is
    divisible by 2, $(k+1)^2 + (k+1)$ is divisible by 2.

21. (I) $n = 1$: $1^2 - 1 + 2 = 2$, which is divisible by 2.
  (II) If $k^2 - k + 2$ is divisible by 2, then $(k+1)^2 - (k+1) + 2 = k^2 + 2k + 1 - k - 1 + 2 = (k^2 - k + 2) + 2k$. Since $k^2 - k + 2$ is divisible
    by 2 and $2k$ is divisible by 2, $(k+1)^2 - (k+1) + 2$ is divisible by 2.

23. (I) $n = 1$: If $x > 1$, then $x^1 = x > 1$.
  (II) Assume, for an arbitrary natural number k, that if $x > 1$ then $x^k > 1$. Multiply both sides of the inequality $x^k > 1$ by x. If $x > 1$, then
    $x^{k+1} > x > 1$.

25. (I) $n = 1$: $a - b$ is a factor of $a^1 - b^1 = a - b$.
  (II) If $a - b$ is a factor of $a^k - b^k$, then $a^{k+1} - b^{k+1} = a(a^k - b^k) + b^k(a - b)$.
    Since $a - b$ is a factor of $a^k - b^k$ and $a - b$ is a factor of $a - b$, then $a - b$ is a factor of $a^{k+1} - b^{k+1}$.

27. (a) $n = 1$: $(1 + a)^1 = 1 + a \geq 1 + 1 \cdot a$
  (b) Assume that there is an integer k for which the inequality holds. So $(1 + a)^k \geq 1 + ka$. We need to show that $(1 + a)^{k+1} \geq 1 + (k+1)a$.
    $(1 + a)^{k+1} = (1 + a)^k(1 + a) \geq (1 + ka)(1 + a) = 1 + ka^2 + a + ka = 1 + (k+1)a + ka^2 \geq 1 + (k+1)a$.

29. If $2 + 4 + 6 + \cdots + 2k = k^2 + k + 2$, then $2 + 4 + 6 + \cdots + 2k + 2(k+1)$
  $= (2 + 4 + 6 + \cdots + 2k) + 2k + 2 = k^2 + k + 2 + 2k + 2 = k^2 + 3k + 4 = (k^2 + 2k + 1) + (k+1) + 2$
  $= (k+1)^2 + (k+1) + 2$.
  But $2 \cdot 1 = 2$ and $1^2 + 1 + 2 = 4$. The fact is that $2 + 4 + 6 + \cdots + 2n = n^2 + n$, not $n^2 + n + 2$ (Problem 1).

31. (I) $n = 1$: $[a + (1-1)d] = a$ and $1 \cdot a + d\dfrac{1 \cdot (1-1)}{2} = a$.

  (II) If $a + (a+d) + (a+2d) + \cdots + [a + (k-1)d] = ka + d\dfrac{k(k-1)}{2}$, then

    $a + (a+d) + (a+2d) + \cdots + [a + (k-1)d] + [a + ((k+1)-1)d] = ka + d\dfrac{k(k-1)}{2} + a + kd$

    $= (k+1)a + d\dfrac{k(k-1) + 2k}{2} = (k+1)a + d\dfrac{(k+1)(k)}{2} = (k+1)a + d\dfrac{(k+1)[(k+1)-1]}{2}$.

33. (I) $n = 3$: The sum of the angles of a triangle is $(3-2) \cdot 180° = 180°$.
  (II) Assume that for some $k \geq 3$, the sum of the angles of a convex polygon of k sides is $(k-2) \cdot 180°$. A convex polygon of $k+1$ sides consists of
    a convex polygon of k sides plus a triangle (see the illustration). The sum of the angles is
    $(k-2) \cdot 180° + 180° = (k-1) \cdot 180° = [(k+1) - 2] \cdot 180°$.

k sides
$k + 1$ sides

35. $\{251\}$ **36.** Left: 448.3 kg; right: 366.0 kg **37.** $x = \frac{1}{2}, y = -3; \left(\frac{1}{2}, -3\right)$ **38.** $\begin{bmatrix} 7 & -3 \\ -7 & 8 \end{bmatrix}$

13.5 Assess Your Understanding (page 976)

1. Pascal triangle **2.** $1; n$ **3.** F **4.** Binomial Theorem **5.** 10 **7.** 21 **9.** 50 **11.** 1 **13.** $\approx 1.8664 \times 10^{15}$ **15.** $\approx 1.4834 \times 10^{13}$
17. $x^5 + 5x^4 + 10x^3 + 10x^2 + 5x + 1$ **19.** $x^6 - 12x^5 + 60x^4 - 160x^3 + 240x^2 - 192x + 64$ **21.** $81x^4 + 108x^3 + 54x^2 + 12x + 1$
23. $x^{10} + 5x^8y^2 + 10x^6y^4 + 10x^4y^6 + 5x^2y^8 + y^{10}$ **25.** $x^3 + 6\sqrt{2}x^{5/2} + 30x^2 + 40\sqrt{2}x^{3/2} + 60x + 24\sqrt{2}x^{1/2} + 8$

27. $a^5x^5 + 5a^4bx^4y + 10a^3b^2x^3y^2 + 10a^2b^3x^2y^3 + 5ab^4xy^4 + b^5y^5$ **29.** $17{,}010$ **31.** $-101{,}376$ **33.** $41{,}472$ **35.** $2835x^3$
37. $314{,}928x^7$ **39.** 495 **41.** 3360 **43.** 1.00501

45. $\dbinom{n}{n-1} = \dfrac{n!}{(n-1)!\,[n-(n-1)]!} = \dfrac{n!}{(n-1)!\,1!} = \dfrac{n\cdot(n-1)!}{(n-1)!} = n;\ \dbinom{n}{n} = \dfrac{n!}{n!\,(n-n)!} = \dfrac{n!}{n!0!} = \dfrac{n!}{n!} = 1$

47. $2^n = (1+1)^n = \dbinom{n}{0}1^n + \dbinom{n}{1}(1)^{n-1}(1) + \cdots + \dbinom{n}{n}1^n = \dbinom{n}{0} + \dbinom{n}{1} + \cdots + \dbinom{n}{n}$ **49.** 1

51. $\left\{\dfrac{\ln 5}{\ln 6 - \ln 5}\right\} \approx \{8.827\}$ **52. (a)** 0 **(b)** $90°$ **(c)** Orthogonal **53.** $x = 1, y = 3, z = -2; (1, 3, -2)$
54. Bounded

Review Exercises *(page 978)*

1. $a_1 = -\dfrac{4}{3}, a_2 = \dfrac{5}{4}, a_3 = -\dfrac{6}{5}, a_4 = \dfrac{7}{6}, a_5 = -\dfrac{8}{7}$ **2.** $c_1 = 2, c_2 = 1, c_3 = \dfrac{8}{9}, c_4 = 1, c_5 = \dfrac{32}{25}$ **3.** $a_1 = 3, a_2 = 2, a_3 = \dfrac{4}{3}, a_4 = \dfrac{8}{9}, a_5 = \dfrac{16}{27}$

4. $a_1 = 2, a_2 = 0, a_3 = 2, a_4 = 0, a_5 = 2$ **5.** $6 + 10 + 14 + 18 = 48$ **6.** $\displaystyle\sum_{k=1}^{13}(-1)^{k+1}\dfrac{1}{k}$ **7.** Arithmetic; $d = 1; S_n = \dfrac{n}{2}(n + 11)$ **8.** Neither

9. Geometric; $r = 8; S_n = \dfrac{8}{7}(8^n - 1)$ **10.** Arithmetic; $d = 4; S_n = 2n(n-1)$ **11.** Geometric; $r = \dfrac{1}{2}; S_n = 6\left[1 - \left(\dfrac{1}{2}\right)^n\right]$ **12.** Neither

13. 9515 **14.** -1320 **15.** $\dfrac{1093}{2187} \approx 0.49977$ **16.** 682 **17.** 35 **18.** $\dfrac{1}{10^{10}}$ **19.** $9\sqrt{2}$ **20.** $\{a_n\} = \{5n - 4\}$ **21.** $\{a_n\} = \{n - 10\}$ **22.** Converges; $\dfrac{9}{2}$
23. Converges; $\dfrac{4}{3}$ **24.** Diverges **25.** Converges; 8

26. (I) $n = 1: 3\cdot 1 = 3$ and $\dfrac{3\cdot 1}{2}(1 + 1) = 3$

   (II) If $3 + 6 + 9 + \cdots + 3k = \dfrac{3k}{2}(k + 1)$, then $3 + 6 + 9 + \cdots + 3k + 3(k + 1) = (3 + 6 + 9 + \cdots + 3k) + (3k + 3)$

   $= \dfrac{3k}{2}(k + 1) + (3k + 3) = \dfrac{3k^2}{2} + \dfrac{3k}{2} + \dfrac{6k}{2} + \dfrac{6}{2} = \dfrac{3}{2}(k^2 + 3k + 2) = \dfrac{3}{2}(k + 1)(k + 2) = \dfrac{3(k + 1)}{2}[(k + 1) + 1].$

27. (I) $n = 1: 2\cdot 3^{1-1} = 2$ and $3^1 - 1 = 2$
   (II) If $2 + 6 + 18 + \cdots + 2\cdot 3^{k-1} = 3^k - 1$, then $2 + 6 + 18 + \cdots + 2\cdot 3^{k-1} + 2\cdot 3^{(k+1)-1} = (2 + 6 + 18 + \cdots + 2\cdot 3^{k-1}) + 2\cdot 3^k$
   $= 3^k - 1 + 2\cdot 3^k = 3\cdot 3^k - 1 = 3^{k+1} - 1.$

28. (I) $n = 1: (3\cdot 1 - 2)^2 = 1$ and $\dfrac{1}{2}\cdot 1\cdot[6(1)^2 - 3(1) - 1] = 1$

   (II) If $1^2 + 4^2 + 7^2 + \cdots + (3k - 2)^2 = \dfrac{1}{2}k(6k^2 - 3k - 1)$, then $1^2 + 4^2 + 7^2 + \cdots + (3k - 2)^2 + [3(k + 1) - 2]^2$

   $= [1^2 + 4^2 + 7^2 + \cdots + (3k - 2)^2] + (3k + 1)^2 = \dfrac{1}{2}k(6k^2 - 3k - 1) + (3k + 1)^2 = \dfrac{1}{2}(6k^3 - 3k^2 - k) + (9k^2 + 6k + 1)$

   $= \dfrac{1}{2}(6k^3 + 15k^2 + 11k + 2) = \dfrac{1}{2}(k + 1)(6k^2 + 9k + 2) = \dfrac{1}{2}(k + 1)[6(k + 1)^2 - 3(k + 1) - 1].$

29. 10 **30.** $x^5 + 10x^4 + 40x^3 + 80x^2 + 80x + 32$ **31.** $81x^4 - 432x^3 + 864x^2 - 768x + 256$ **32.** 144 **33.** 84

34. (a) 8 bricks **(b)** 1100 bricks **35.** 360 **36. (a)** $20\left(\dfrac{3}{4}\right)^3 = \dfrac{135}{16}$ ft **(b)** $20\left(\dfrac{3}{4}\right)^n$ ft **(c)** 13 times **(d)** 140 ft **37.** $151{,}873.77$ **38.** $58{,}492.93$

Chapter Test *(page 980)*

1. $0, \dfrac{3}{10}, \dfrac{8}{11}, \dfrac{5}{4}, \dfrac{24}{13}$ **2.** $4, 14, 44, 134, 404$ **3.** $2 - \dfrac{3}{4} + \dfrac{4}{9} = \dfrac{61}{36}$ **4.** $-\dfrac{1}{3} - \dfrac{14}{9} - \dfrac{73}{27} - \dfrac{308}{81} = -\dfrac{680}{81}$ **5.** $\displaystyle\sum_{k=1}^{10}(-1)^k\left(\dfrac{k + 1}{k + 4}\right)$ **6.** Neither

7. Geometric; $r = 4; S_n = \dfrac{2}{3}(1 - 4^n)$ **8.** Arithmetic: $d = -8; S_n = n(2 - 4n)$ **9.** Arithmetic; $d = -\dfrac{1}{2}; S_n = \dfrac{n}{4}(27 - n)$

10. Geometric; $r = \dfrac{2}{5}; S_n = \dfrac{125}{3}\left[1 - \left(\dfrac{2}{5}\right)^n\right]$ **11.** Neither **12.** Converges; $\dfrac{1024}{5}$ **13.** $243m^5 + 810m^4 + 1080m^3 + 720m^2 + 240m + 32$

14. First we show that the statement holds for $n = 1$. $\left(1 + \dfrac{1}{1}\right) = 1 + 1 = 2$. The equality is true for $n = 1$, so Condition I holds. Next we assume that

$\left(1 + \dfrac{1}{1}\right)\left(1 + \dfrac{1}{2}\right)\left(1 + \dfrac{1}{3}\right)\cdots\left(1 + \dfrac{1}{n}\right) = n + 1$ is true for some k, and we determine whether the formula then holds for $k + 1$. We assume that

$\left(1 + \dfrac{1}{1}\right)\left(1 + \dfrac{1}{2}\right)\left(1 + \dfrac{1}{3}\right)\cdots\left(1 + \dfrac{1}{k}\right) = k + 1$. Now we need to show that $\left(1 + \dfrac{1}{1}\right)\left(1 + \dfrac{1}{2}\right)\left(1 + \dfrac{1}{3}\right)\cdots\left(1 + \dfrac{1}{k}\right)\left(1 + \dfrac{1}{k + 1}\right)$

$= (k + 1) + 1 = k + 2$. We do this as follows:

$$\left(1 + \frac{1}{1}\right)\left(1 + \frac{1}{2}\right)\left(1 + \frac{1}{3}\right)\cdots\left(1 + \frac{1}{k}\right)\left(1 + \frac{1}{k+1}\right) = \left[\left(1 + \frac{1}{1}\right)\left(1 + \frac{1}{2}\right)\left(1 + \frac{1}{3}\right)\cdots\left(1 + \frac{1}{k}\right)\right]\left(1 + \frac{1}{k+1}\right)$$

$$= (k + 1)\left(1 + \frac{1}{k+1}\right)\text{(induction assumption)} = (k + 1)\cdot 1 + (k + 1)\cdot\frac{1}{k+1} = k + 1 + 1 = k + 2$$

Condition II also holds. Thus, the formula holds true for all natural numbers.
15. After 10 years, the Durango will be worth $6103.11.   **16.** The weightlifter will have lifted a total of 8000 pounds after 5 sets.

Cumulative Review *(page 980)*

1. $\{-3, 3, -3i, 3i\}$ **2. (a)**
(b) $\left\{\left(\sqrt{\dfrac{-1 + \sqrt{3601}}{18}}, \dfrac{-1 + \sqrt{3601}}{6}\right), \left(-\sqrt{\dfrac{-1 + \sqrt{3601}}{18}}, \dfrac{-1 + \sqrt{3601}}{6}\right)\right\}$
(c) The circle and the parabola intersect at
$\left(\sqrt{\dfrac{-1 + \sqrt{3601}}{18}}, \dfrac{-1 + \sqrt{3601}}{6}\right), \left(-\sqrt{\dfrac{-1 + \sqrt{3601}}{18}}, \dfrac{-1 + \sqrt{3601}}{6}\right)$.

3. $\left\{\ln\left(\dfrac{5}{2}\right)\right\}$ **4.** $y = 5x - 10$ **5.** $(x + 1)^2 + (y - 2)^2 = 25$ **6. (a)** 5   **(b)** 13   **(c)** $\dfrac{6x + 3}{2x - 1}$ **(d)** $\left\{x\,\middle|\,x \neq \dfrac{1}{2}\right\}$ **(e)** $\dfrac{7x - 2}{x - 2}$ **(f)** $\{x\,|\,x \neq 2\}$

(g) $g^{-1}(x) = \dfrac{1}{2}(x - 1)$; all reals   **(h)** $f^{-1}(x) = \dfrac{2x}{x - 3}$; $\{x\,|\,x \neq 3\}$ **7.** $\dfrac{x^2}{7} + \dfrac{y^2}{16} = 1$ **8.** $(x + 1)^2 = 4(y - 2)$

9. $r = 8\sin\theta$; $x^2 + (y - 4)^2 = 16$ **10.** $\left\{\dfrac{3\pi}{2}\right\}$ **11.** $\dfrac{2\pi}{3}$ **12. (a)** $-\dfrac{\sqrt{15}}{4}$ **(b)** $-\dfrac{\sqrt{15}}{15}$ **(c)** $-\dfrac{\sqrt{15}}{8}$ **(d)** $\dfrac{7}{8}$ **(e)** $\sqrt{\dfrac{1 + \dfrac{\sqrt{15}}{4}}{2}} = \dfrac{\sqrt{4 + \sqrt{15}}}{2\sqrt{2}}$

CHAPTER 14 Counting and Probability

14.1 Assess Your Understanding *(page 987)*

5. subset; $\subseteq$ **6.** finite   **7.** $n(A) + n(B) - n(A \cap B)$ **8.** T   **9.** $\varnothing$, $\{a\}$, $\{b\}$, $\{c\}$, $\{d\}$, $\{a, b\}$, $\{a, c\}$, $\{a, d\}$, $\{b, c\}$, $\{b, d\}$, $\{c, d\}$, $\{a, b, c\}$, $\{b, c, d\}$, $\{a, c, d\}$,
$\{a, b, d\}$, $\{a, b, c, d\}$ **11.** 25   **13.** 40   **15.** 25   **17.** 37   **19.** 18   **21.** 5   **23.** 15 different arrangements   **25.** 9000 numbers   **27.** 175; 125
29. (a) 15   **(b)** 15   **(c)** 15   **(d)** 25   **(e)** 40   **31. (a)** 14.0 million   **(b)** 79.3 million   **33.** 480 portfolios
36. 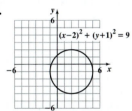   **37.** $A \approx 41.4°, B \approx 41.4°, C \approx 97.2°$ **38.** $2, 5, -2$ **39.** $\left\{\dfrac{1}{18}\right\}$

14.2 Assess Your Understanding *(page 994)*

3. permutation   **4.** combination   **5.** $\dfrac{n!}{(n - r)!}$ **6.** $\dfrac{n!}{(n - r)!\,r!}$ **7.** 30   **9.** 24   **11.** 1   **13.** 1680   **15.** 28   **17.** 35   **19.** 1   **21.** 10,400,600
23. $\{abc, abd, abe, acb, acd, ace, adb, adc, ade, aeb, aec, aed, bac, bad, bae, bca, bcd, bce, bda, bdc, bde, bea, bec, bed, cab, cad, cae, cba, cbd, cbe, cda,$
$cdb, cde, cea, ceb, ced, dab, dac, dae, dba, dbc, dbe, dca, dcb, dce, dea, deb, dec, eab, eac, ead, eba, ebc, ebd, eca, ecb, ecd, eda, edb, edc\}$; 60
25. $\{123, 124, 132, 134, 142, 143, 213, 214, 231, 234, 241, 243, 312, 314, 321, 324, 341, 342, 412, 413, 421, 423, 431, 432\}$; 24
27. $\{abc, abd, abe, acd, ace, ade, bcd, bce, bde, cde\}$; 10   **29.** $\{123, 124, 134, 234\}$; 4   **31.** 16   **33.** 8   **35.** 24   **37.** 60   **39.** 18,278   **41.** 35   **43.** 1024
45. 120   **47.** 132,860   **49.** 336   **51.** 90,720   **53. (a)** 63   **(b)** 35   **(c)** 1   **55.** 1.157×10^{76} **57.** 362,880   **59.** 660   **61.** 15
63. (a) 125,000; 117,600   **(b)** A better name for a *combination* lock would be a *permutation* lock because the order of the numbers matters.
67. 10 sq. ft   **68.** $(g \circ f)(x) = 4x^2 - 2x - 2$ **69.** $\sin 75° = \dfrac{\sqrt{2} + \sqrt{6}}{4}$; $\cos 15° = \dfrac{1}{2}\sqrt{2 + \sqrt{3}}$ or $\cos 15° = \dfrac{\sqrt{2} + \sqrt{6}}{4}$ **70.** $a_5 = 80$

Historical Problem *(page 1004)*

1. (a) $\{AAAA, AAAB, AABA, AABB, ABAA, ABAB, ABBA, ABBB, BAAA, BAAB, BABA, BABB, BBAA, BBAB, BBBA, BBBB\}$
(b) $P(A \text{ wins}) = \dfrac{C(4, 2) + C(4, 3) + C(4, 4)}{2^4} = \dfrac{6 + 4 + 1}{16} = \dfrac{11}{16}$; $P(B \text{ wins}) = \dfrac{C(4, 3) + C(4, 4)}{2^4} = \dfrac{4 + 1}{16} = \dfrac{5}{16}$

14.3 Assess Your Understanding *(page 1004)*

1. equally likely   **2.** complement   **3.** F   **4.** T   **5.** 0, 0.01, 0.35, 1   **7.** Probability model   **9.** Not a probability model
11. $S = \{HH, HT, TH, TT\}$; $P(HH) = \dfrac{1}{4}$, $P(HT) = \dfrac{1}{4}$, $P(TH) = \dfrac{1}{4}$, $P(TT) = \dfrac{1}{4}$

13. $S =$ {HH1, HH2, HH3, HH4, HH5, HH6, HT1, HT2, HT3, HT4, HT5, HT6, TH1, TH2, TH3, TH4, TH5, TH6, TT1, TT2, TT3, TT4, TT5, TT6}; each outcome has the probability of $\frac{1}{24}$.

15. $S =$ {HHH, HHT, HTH, HTT, THH, THT, TTH, TTT}; each outcome has the probability of $\frac{1}{8}$.

17. $S =$ {1 Yellow, 1 Red, 1 Green, 2 Yellow, 2 Red, 2 Green, 3 Yellow, 3 Red, 3 Green, 4 Yellow, 4 Red, 4 Green}; each outcome has the probability of $\frac{1}{12}$; thus, $P(2 \text{ Red}) + P(4 \text{ Red}) = \frac{1}{12} + \frac{1}{12} = \frac{1}{6}$.

19. $S =$ {1 Yellow Forward, 1 Yellow Backward, 1 Red Forward, 1 Red Backward, 1 Green Forward, 1 Green Backward, 2 Yellow Forward, 2 Yellow Backward, 2 Red Forward, 2 Red Backward, 2 Green Forward, 2 Green Backward, 3 Yellow Forward, 3 Yellow Backward, 3 Red Forward, 3 Red Backward, 3 Green Forward, 3 Green Backward, 4 Yellow Forward, 4 Yellow Backward, 4 Red Forward, 4 Red Backward, 4 Green Forward, 4 Green Backward}; each outcome has the probability of $\frac{1}{24}$; thus, $P(1 \text{ Red Backward}) + P(1 \text{ Green Backward}) = \frac{1}{24} + \frac{1}{24} = \frac{1}{12}$.

21. $S =$ {11 Red, 11 Yellow, 11 Green, 12 Red, 12 Yellow, 12 Green, 13 Red, 13 Yellow, 13 Green, 14 Red, 14 Yellow, 14 Green, 21 Red, 21 Yellow, 21 Green, 22 Red, 22 Yellow, 22 Green, 23 Red, 23 Yellow, 23 Green, 24 Red, 24 Yellow, 24 Green, 31 Red, 31 Yellow, 31 Green, 32 Red, 32 Yellow, 32 Green, 33 Red, 33 Yellow, 33 Green, 34 Red, 34 Yellow, 34 Green, 41 Red, 41 Yellow, 41 Green, 42 Red, 42 Yellow, 42 Green, 43 Red, 43 Yellow, 43 Green, 44 Red, 44 Yellow, 44 Green}; each outcome has the probability of $\frac{1}{48}$; thus, $E =$ {22 Red, 22 Green, 24 Red, 24 Green}; $P(E) = \frac{n(E)}{n(S)} = \frac{4}{48} = \frac{1}{12}$.

23. A, B, C, F   **25.** B   **27.** $P(H) = \frac{4}{5}$; $P(T) = \frac{1}{5}$ **29.** $P(1) = P(3) = P(5) = \frac{2}{9}$; $P(2) = P(4) = P(6) = \frac{1}{9}$ **31.** $\frac{3}{10}$ **33.** $\frac{1}{2}$ **35.** $\frac{1}{6}$ **37.** $\frac{1}{8}$

39. $\frac{1}{4}$ **41.** $\frac{1}{6}$ **43.** $\frac{1}{18}$ **45.** 0.55   **47.** 0.70   **49.** 0.30   **51.** 0.88   **53.** 0.63   **55.** 0.936   **57.** $\frac{17}{20}$ **59.** $\frac{11}{20}$ **61.** $\frac{1}{2}$ **63.** $\frac{17}{50}$ **65.** $\frac{12}{25}$

67. (a) 0.57   **(b)** 0.95   **(c)** 0.83   **(d)** 0.38   **(e)** 0.29   **(f)** 0.05   **(g)** 0.78   **(h)** 0.71

69. (a) $\frac{25}{33}$ **(b)** $\frac{25}{33}$ **71.** 0.167   **73.** $\frac{1}{175,223,510} \approx 0.00000000571$ **74.** 2; left; 3; down   **75.** $(-3, 3\sqrt{3})$ **76.** {22}   **77.** $(2, -3, -1)$

Review Exercises *(page 1008)*

1. $\varnothing$, {Dave}, {Joanne}, {Erica}, {Dave, Joanne}, {Dave, Erica}, {Joanne, Erica}, {Dave, Joanne, Erica}   **2.** 17   **3.** 24   **4.** 29   **5.** 34   **6.** 7
7. 45   **8.** 25   **9.** 7   **10.** 336   **11.** 56   **12.** 60   **13.** 128   **14.** 3024   **15.** 1680   **16.** 91   **17.** 1,600,000   **18.** 216,000
19. 256 (allowing numbers with initial zeros, such as 011)   **20.** 12,600   **21. (a)** 381,024   **(b)** 1260   **22. (a)** $8.634628387 \times 10^{45}$ **(b)** 0.6531   **(c)** 0.3469
23. (a) 0.074   **(b)** 0.926   **24.** $\frac{4}{9}$ **25.** 0.2; 0.26   **26. (a)** 0.68   **(b)** 0.58   **(c)** 0.32

Chapter Test *(page 1009)*

1. 22   **2.** 3   **3.** 8   **4.** 45   **5.** 5040   **6.** 151,200   **7.** 462   **8.** There are 54,264 ways to choose 6 different colors from the 21 available colors.
9. There are 840 distinct arrangements of the letters in the word REDEEMED.   **10.** There are 56 different exacta bets for an 8-horse race.
11. There are 155,480,000 possible license plates using the new format.   **12. (a)** 0.95   **(b)** 0.30   **13. (a)** 0.25   **(b)** 0.55   **14.** 0.19   **15.** 0.000033069
16. $P(\text{exactly 2 fours}) = \frac{625}{3888} \approx 0.1608$

Cumulative Review *(page 1010)*

1. $\left\{ \frac{1}{3} - \frac{\sqrt{2}}{3}i, \ \frac{1}{3} + \frac{\sqrt{2}}{3}i \right\}$ **2.**   **3.**   **4.** $\{x \mid 3.99 \le x \le 4.01\}$ or $[3.99, 4.01]$

5. $\left\{ -\frac{1}{2} + \frac{\sqrt{7}}{2}i, -\frac{1}{2} - \frac{\sqrt{7}}{2}i, -\frac{1}{5}, 3 \right\}$ **6.** 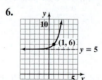   **7.** 2   **8.** $\left\{ \frac{8}{3} \right\}$ **9.** $x = 2, y = -5, z = 3$ **10.** 125; 700

Domain: all real numbers
Range: $\{y \mid y > 5\}$
Horizontal asymptote: $y = 5$

11.   **12.** $a \approx 6.09$, $B \approx 31.9°$, $C \approx 108.1°$; area ≈ 14.46 square units

APPENDIX Graphing Utilities

1 Exercises (page A2)

1. $(-1, 4)$; II   **3.** $(3, 1)$; I   **5.** $X\min = -6$, $X\max = 6$, $X\text{scl} = 2$, $Y\min = -4$, $Y\max = 4$, $Y\text{scl} = 2$
7. $X\min = -6$, $X\max = 6$, $X\text{scl} = 2$, $Y\min = -1$, $Y\max = 3$, $Y\text{scl} = 1$ **9.** $X\min = 3$, $X\max = 9$, $X\text{scl} = 1$, $Y\min = 2$, $Y\max = 10$, $Y\text{scl} = 2$
11. $X\min = -11$, $X\max = 5$, $X\text{scl} = 1$, $Y\min = -3$, $Y\max = 6$, $Y\text{scl} = 1$
13. $X\min = -30$, $X\max = 50$, $X\text{scl} = 10$, $Y\min = -90$, $Y\max = 50$, $Y\text{scl} = 10$
15. $X\min = -10$, $X\max = 110$, $X\text{scl} = 10$, $Y\min = -10$, $Y\max = 160$, $Y\text{scl} = 10$

2 Exercises (page A4)

1. (a)  **(b)** 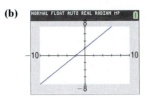 **3. (a)** 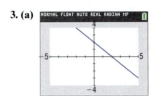 **(b)**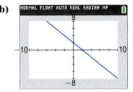

5. (a) 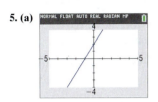 **(b)** 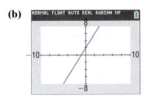 **7. (a)** 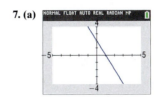 **(b)**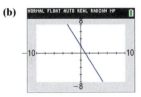

9. (a) 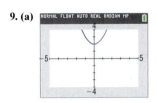 **(b)** 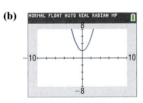 **11. (a)** 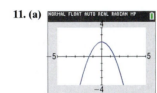 **(b)**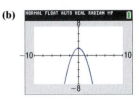

13. (a) 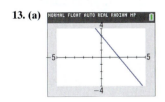 **(b)** 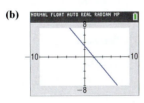 **15. (a)** 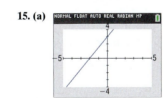 **(b)**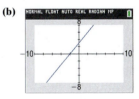

17.

X	Y₁
-5	-3
-4	-2
-3	-1
-2	0
-1	1
0	2
1	3
2	4
3	5
4	6
5	7

$Y_1 \equiv X+2$

19.

X	Y₁
-5	7
-4	6
-3	5
-2	4
-1	3
0	2
1	1
2	0
3	-1
4	-2
5	-3

$Y_1 \equiv -X+2$

21.

X	Y₁
-5	-8
-4	-6
-3	-4
-2	-2
-1	0
0	2
1	4
2	6
3	8
4	10
5	12

$Y_1 \equiv 2X+2$

23.

X	Y₁
-5	12
-4	10
-3	8
-2	6
-1	4
0	2
1	0
2	-2
3	-4
4	-6
5	-8

$Y_1 \equiv -2X+2$

25.

X	Y₁
-5	27
-4	18
-3	11
-2	6
-1	3
0	2
1	3
2	6
3	11
4	18
5	27

$Y_1 \equiv X^2+2$

27.

X	Y₁
-5	-23
-4	-14
-3	-7
-2	-2
-1	1
0	2
1	1
2	-2
3	-7
4	-14
5	-23

$Y_1 \equiv -X^2+2$

29.

X	Y₁
-5	10.5
-4	9
-3	7.5
-2	6
-1	4.5
0	3
1	1.5
2	0
3	-1.5
4	-3
5	-4.5

$Y_1 \equiv -(3/2)X+3$

31.

X	Y₁
-5	-4.5
-4	-3
-3	-1.5
-2	0
-1	1.5
0	3
1	4.5
2	6
3	7.5
4	9
5	10.5

$Y_1 \equiv (3/2)X+3$

3 Exercises (page A6)

1. -3.41 **3.** -1.71 **5.** -0.28 **7.** 3.00 **9.** 4.50 **11.** $1.00, 23.00$

5 Exercises (page A8)

1. Yes   **3.** Yes   **5.** No   **7.** Yes   **9.** Answers may vary. A possible answer is $Y\min = 0$, $Y\max = 10$, and $Y\text{scl} = 1$.

Credits

Photos

Front Matter Achieve Your Potential (Retain Your Knowledge), Mike Flippo/Shutterstock; Resources for Success, Sturti/E+/Getty Images

Chapter R Page 1, Andres Rodriguez/Fotolia; Page 31, Feraru Nicolae/Shutterstock.

Chapter 1 Pages 81 and 147, Rawpixel/Fotolia; Page 91, A.R. Monko/Shutterstock; Page 103(L), Stanca Sanda/Alamy; Page 103(R), Jeff Chiu/AP Images; Page 133, Nancy R. Cohen/Getty Images; Page 145, Hemera Technologies/Photos.com/Thinkstock.

Chapter 2 Pages 149 and 197, Maxx-Studio/Shutterstock; Page 160, Snapper/Fotolia; Page 166(L), Stockyimages/Fotolia; Page 166(R), Department of Energy; Page 180, Tetra Images/Alamy; Page 187, Tom Donoghue/Polaris/Newscom.

Chapter 3 Pages 198 and 271, Leigh Prather/Shutterstock/Asset Library; Page 212, NASA; Page 220, Exactostock/SuperStock; Page 259, Dreamstime LLC.

Chapter 4 Pages 273 and 320, Spaxiax/Shutterstock; Page 309, Geno EJ Sajko Photography/Shutterstock.

Chapter 5 Pages 321 and 400, John Foxx Collection/Imagestate/DK images; Page 367, Corbis.

Chapter 6 Pages 402 and 503, Rawpixel/Fotolia; Page 458, Hulton Archive/Handout/Getty Images; Page 467(L), Stockbyte/Getty Images; Page 467(R), Pascal Saez/Alamy; Page 472, Lacroix Serge/iStock/360/Getty Images; Page 487, Jupiterimages/Photos.com/360/Getty Images.

Chapter 7 Pages 505 and 604, NLSA; Page 517, Ryan McVay/Digital Vision/Getty Images; Page 540, Sergey Karpov/Shutterstock; Page 578, Srdjan Draskovic/Dreamstime

Chapter 8 Pages 606 and 674, Sebastian Kaulitzki/Fotolia.

Chapter 9 Pages 675 and 718, Jennifer Thermes/Photodisc/Getty Images; Page 703, Alexandre Fagundes De Fagundes/Dreamstime; Page 705, Anton Ignatenco/iStock/360/Getty Images.

Chapter 10 Pages 719 and 777, Aviator70/Fotolia; Page 741, Orth Wind Picture Archives/Alamy; Page 749, Science & Society Picture Library/Contributor/Getty Images; Page 761, Hulton Archive/Stringer/Getty Images.

Chapter 11 Pages 778 and 842, Marcel Clemens/Shutterstock; Page 796, Thomas Barrat/Shutterstock.

Chapter 12 Pages 843 and 937, Wavebreakmedia/Shutterstock; Page 897, Library of Congress Prints and Photographs Division [LC-USZ62-46864].

Chapter 13 Pages 939 and 981, Denis Cristo/Shutterstock; Pages 963 and 976, Pearson Education.

Chapter 14 Pages 982 and 1010, zentilia/Fotolia; Page 1003, Georgios Kollidas/Getty Images.

Text

TI 84 Plus C screenshots courtesy of Texas Instruments. Chapter 3, page 201: Diamond price. Used with permission of Diamonds.com. Copyright © Martin Rapaport. All rights reserved. Chapter 6, page 495: Cell Phone Towers by CTIA-The Wireless Association. Copyright © 2013 by CTIA-The Wireless Association. Used by permission of CTIA-The Wireless Association®.

Subject Index